MULTIVARIABLE CALCULUS SET FREE

Multivariable Calculus Set Free

Infinitesimals Ride Again

Charles Bryan Dawson

University Professor of Mathematics
Union University, USA

OXFORD

UNIVERSITY PRESS

Great Clarendon Street, Oxford, OX2 6DP,
United Kingdom

Oxford University Press is a department of the University of Oxford.
It furthers the University's objective of excellence in research, scholarship,
and education by publishing worldwide. Oxford is a registered trade mark of
Oxford University Press in the UK and in certain other countries.

Published in the United States of America by Oxford University Press
198 Madison Avenue, New York, NY 10016, United States of America

British Library Cataloguing in Publication Data
Data available

Library of Congress Control Number: 2026933564

ISBN 9780198984269
ISBN 9780198984252 (pbk.)

DOI: 10.1093/oso/9780198984269.001.0001

Printed and bound by
CPI Group (UK) Ltd., Croydon, CR0 4YY

The manufacturer's authorized representative in the EU for product safety is
Oxford University Press España S.A. of Parque Empresarial San Fernando de Henares,
Avenida de Castilla, 2 – 28830 Madrid (www.oup.es/en or product.safety@oup.com).
OUP España S.A. also acts as importer into Spain of products made by the manufacturer.

Contents

Preface

A sequel to *Calculus Set Free: Infinitesimals to the Rescue*, this text explores multi-variable calculus using the same infinitesimal framework, picking up where *Calculus Set Free* left off. The material contained herein corresponds to third-semester calculus at most universities in the United States. The organization is similar to that of other popular calculus texts.

What makes this textbook different is its use of infinitesimals, and hyperreals in general, instead of epsilon-delta style reasoning. Limit-based computations are simpler utilizing infinitesimals and approximation, making a wider variety of limits accessible. Definitions of double integrals, triple integrals, line integrals, and surface integrals, as well as related geometric and physical applications, are based on infinitely many subintervals (or subregions) of infinitesimal size, resulting more clearly in a sum, simplifying definitions. Local linearity is captured in an easily understood approximation formula, demystifying the use of straight objects instead of curved objects when developing integration formulas.

A brief historical introduction that may be of interest to all readers appears at the beginning of Section 0.1. Otherwise, readers familiar with *Calculus Set Free* may wish to skip the review chapter and jump right in to Chapter 1. For the benefit of readers who studied single-variable calculus from a non-infinitesimal perspective, a review chapter is included with full explanations of infinitesimals, the approximation relation, and evaluating functions at hyperreal numbers, with no prior knowledge of infinitesimal analysis required. The sections on approximation and evaluating functions at hyperreals collect all relevant material from throughout *Calculus Set Free*, to facilitate use as a reference. Then limits, continuity, differentiation, and integration are reviewed using these methods.

This text is written for learners of calculus of varying levels of ability and preparedness. There are, however, some topics, and some exercises, that explore the subject more deeply than many instructors (and learners) will wish to cover; these details are included for the sake of completeness, because of the lack of other resources.

Features of this textbook

Features of this textbook include:

- **A readable and student-friendly narrative.** The narrative is written in a storytelling style, with the goal of helping readers think through the development of concepts, think through solutions to examples, and more generally learn to think mathematically.

- **Reading exercises.** Reading exercises are meant to be worked when encountered during reading. The solution is placed in the margin one to three pages later.

- **Hundreds of diagrams.** Abundant use of figures aids the development of visual intuition and understanding.

- **Margin notes.** Margin notes are used to supplement understanding, including line-by-line commentary for many worked examples.

- **Examples.** Hundreds of examples with complete solutions are included. Some solutions are written compactly, demonstrating the level of detail expected of student work. Others include more details of how to think through the solution.

- **Thousands of exercises.** Exercises range from the simple and the routine to the challenging, offering flexibility to instructors and learners alike. Some sections include "rapid-response" exercises meant to help students distinguish between objects or algebraic forms; these exercises often work well in an interactive classroom setting. Some exercises are very similar to examples in the narrative. Other exercises require creative thinking or explore the ideas more deeply. Enough exercises are included to allow an instructor to have choices of which odd-numbered exercises to include in order to craft an appropriate homework set. Even-numbered exercises roughly correspond to odd-numbered exercises for additional practice. Although it is now common to find solutions to textbook exercises on the internet, I still follow the custom of only providing the answers to odd-numbered exercises in the textbook.

- **An extensive index.**

Among the differences from *Calculus Set Free* is that links to biographies on the MacTutor History of Mathematics website are not included. But readers can readily use the MacTutor resource for themselves.

Section dependencies

Some sections can be skipped, or possibly delayed, without much impact on understanding later material. Section 1.7 on quadric surfaces is occasionally alluded to, but can be skipped on a first reading. Although I consider extreme values and saddle points essential material, Section 3.9 is not significantly used later in the text. Sections 3.10 and 4.5 contain applications without later dependency. Sections 4.8–4.10 have occasional mention in Chapter 5 but can also be omitted. The order of Sections 5.8 and 5.9 can be swapped.

Although I have split coverage of vectors and of partial derivatives into two sections each based on my classroom preferences, some instructors may wish to cover Sections 1.2 and 1.3 together in one class period, and the same is true for

Sections 3.4 and 3.5. For the sake of completeness, Section 3.3 contains more details than many instructors will wish to cover; the exercise set is designed to allow for slimmer coverage, while still going beyond what most epsilon-delta textbooks include.

Section 5.6 can be fully covered as soon as after Section 4.6, and partially covered even earlier. Slight repositioning of a few other sections is also possible.

Finally, as is usual in calculus textbooks, numerous sections contain subsections upon which no later material depends. This allows instructors to cover favorite (sub)topics. However, coverage of every problem type demonstrated in the narrative may not always be practical.

I have often prayed that this textbook will be a blessing to all readers. May it always be so.

Charles Bryan Dawson
University Professor of Mathematics
Union University

September 2025

Acknowledgments

I am blessed with an incredibly supportive wife, Martha, who has shown her love and patience with me for nearly four decades. Her encouragement is inspiring. Our children Matthew, Patricia, and Tiffany, and daughter-in-law Alejandra, have been important parts of this journey as well. I wish that my father could have seen the culmination of these efforts, and I pray that my mother still will.

Colleagues at Union University have graciously given me their time and support as well as encouragement. Troy Riggs was the first to join me in using infinitesimal methods. George Moss has been extremely accommodating as my department chair. Matt Lunsford helped me see the importance of writing textbooks featuring infinitesimal methods. Chris Hail, in his unique way, keeps me grounded. All have been impacted in various ways by my pursuit, and they have my sincere gratitude. I am likewise grateful to Union's administration for granting a research leave in support of this project.

Nicholas Zoller (Southern Nazarene University) has classroom tested this book, providing helpful edits and additional perspective. Thanks also go to the many students who have provided encouragement and feedback.

No one exists in a vacuum, and people from the various communities of which I am a part have played a role in carrying me to this point, including many from the Mathematical Association of America, the Association of Christians in the Mathematical Sciences, Indigenous Mathematicians, and Grace Bible Fellowship.

Editors and others at Oxford University Press have been fantastic to work with, on both *Calculus Set Free* and on this book. May their investment in this project be rewarded!

Last, I thank my Creator, the Maker of Life, for assigning to me the task of bringing a simpler, more intuitive calculus to the world.

> Set me free and make me whole, and once again my heart will dance for joy.
> *—Psalm 51:12a FNV*

Charles Bryan Dawson
University Professor of Mathematics
Union University

September 2025

Single-Variable Calculus Review

0.1 Infinitesimals

What are infinitesimals? And how are they used in calculus?

0.1.1 Historical introduction

Infinitesimals can be thought of as infinitely small numbers. Their use in European mathematics goes back to a type of infinitesimal called an *indivisible*, which was used by the likes of Wallis, Cavalieri, and Gallileo to answer questions that we now consider to be part of calculus, such as determining areas. Leibniz and Newton, widely given credit for the invention of calculus due to their independent discovery of the Fundamental Theorem of Calculus, used infinitesimals, as did those who came after, such as the Bernoullis and Euler.

Using infinitesimals, however, was controversial from the beginning. In 1632, Jesuits banned the use of indivisibles in their sphere of influence worldwide. A century later, Berkeley's famous derision of infinitesimals as "ghosts of departed quantities" spurred further skepticism of calculus. Why all the fuss? Because infinitesimals were not considered to be proper mathematics; they appeared to be experimental in nature, not firmly rooted in a well-developed axiomatic system like geometry. For centuries, no one knew how to rigorously prove the existence of a number system expanding the real numbers to include these infinitely small numbers.

As a result, 19th-century mathematicians began exploring a workaround. Ideas such as the epsilon-delta definition of limit, the derivative as the limit of a difference quotient, and limits of Riemann sums all were developed to get around the use of infinitesimals and show that calculus can be placed on a firm mathematical foundation. Although less intuitive and more difficult to use, these methods eventually replaced infinitesimals in textbooks because of their status as legitimate, compared to the unknown, and therefore illegitimate, status of infinitesimals at the time. Although infinitesimals had been relegated to the sideline, they were not repudiated, just put on hold.

Invention or discovery? My take is that mathematical truths are discovered, but our expression of those truths (notation, terminology, conceptual understanding) is invented.

See *Infinitesimal: How a Dangerous Mathematical Theory Shaped the Modern World*, by Amir Alexander (Scientific American, 2014), for a fascinating account of this history.

Multivariable Calculus Set Free. Charles Bryan Dawson, Oxford University Press.
© C. Bryan Dawson (2026). DOI: 10.1093/oso/9780198984269.003.0001

Then came mid-20th-century mathematical logic. Using a powerful theorem of Jerzy Łoś, around 1960 Abraham Robinson finally achieved the goal, proving the existence of a number system called the *hyperreal numbers* that extends the real numbers to include infinitesimals, as well as infinitely large numbers. Finally, the mathematical community viewed the use of infinitesimals as legitimate! But change happens slowly, and the infinitesimal analysis of Robinson needed time to mature, which is why simpler, more intuitive infinitesimal calculations reminiscent of the subject's historical development are just now returning to the classroom.

Thankfully, to understand infinitesimals and their use in calculus, we do not need to understand how Robinson's proof works. We only need to understand what infinitesimals are, and how to calculate with them, which is a much easier task than using the workaround concepts and procedures in place for more than a century.

0.1.2 Infinitesimals

For a child first learning to count, numbers are 0, 1, 2, and so forth. But eventually there are operations that cannot be accomplished with only these numbers, so the number system expands.

Can't subtract 4 from 7? Call the answer a negative number: -3. Can't divide 5 by 3? Give it the name $\frac{5}{3}$ and introduce the rational numbers. Can't take the square root of -1? Call it i and expand our view of numbers to the complex numbers. Can't find a number that is positive yet smaller than every positive real number? Give it a name; call it an *infinitesimal*–a number that can be described as "infinitely small," and write it using the symbol ω. If we assume that arithmetic and algebra work as usual with such a number, what other numbers must exist? What does the resulting system of numbers, called the *hyperreal numbers*, look like? If we can get familiar with the numbers, learn how to add and multiply them and otherwise manipulate them, then we can get as comfortable with the hyperreals as with any other type of number. This is our task for the remainder of this section.

Arithmetic with infinitesimals is accomplished by treating ω the same as any other algebraic quantity (more on this idea later in this section).

Example 1 *Simplify* $7 + \omega - (4 + 3\omega)$.

Solution We distribute the negative through the parentheses and then collect like terms, just as we do if ω is a variable (think "x") instead:

$$7 + \omega - (4 + 3\omega) = 7 + \omega - 4 - 3\omega = 3 - 2\omega.$$

■

In other words, we already know how to do quite a bit of arithmetic involving infinitesimals. The next example should not be challenging either.

Robinson called the subject *nonstandard analysis*, which, like "imaginary numbers," often conveys an inaccurate idea of its status.

Readers who are familiar with the infinitesimal methods of *Calculus Set Free* may skip this review material. It is provided for the benefit of those who have not read *Calculus Set Free*.

Throughout this text, any reference to *Calculus Set Free* refers to *Calculus Set Free: Infinitesimals to the Rescue*, Oxford University Press, 2022.

ω is the lowercase Greek letter omega. It is not the letter "w" in the English alphabet and should not be pronounced "double-u."

Among the symbols used for the hyperreals are **H** and $^*\mathbf{R}$.

This review section, from before example 1 through reading exercise 5, is nearly identical to Section 1.1 of *Calculus Set Free*. Additional material from other sections is included after reading exercise 5.

Example 2 *Expand* $(3 + \omega)^2$.

Solution

$$(3 + \omega)^2 = 9 + 6\omega + \omega^2.$$

Don't forget to use $(a + b)^2 = a^2 + 2ab + b^2$.

■

Reading Exercise 1   Perform the arithmetic on infinitesimals: $5(7 + 4\omega)$.

Reading exercises are to be worked when encountered during reading. Answers to reading exercises appear in the margin, usually one to four pages after the exercise.

Example 2 brings up an interesting question: what is 6ω? How big is it? Knowing that an infinitesimal such as ω must have a decimal expansion starting with infinitely many zeros helps us answer the question. Computing 6ω, we have $6(0.00000\ldots) = 0.00000\ldots$, which also has to start with infinitely many zeros. It must therefore also be an infinitesimal, for it is infinitely small. The same is true for $38\,576.53\omega$, or even for -14ω, which should be negative because ω is positive, but still starts $-0.00000\ldots$. It seems that any nonzero real number times an infinitesimal is infinitesimal. But before writing and proving this theorem, we need a definition of just what an infinitesimal is.

Definition 1 INFINITESIMALS *A number h is* infinitesimal *if either (1) h > 0 and h < r for every positive real number r or (2) h < 0 and h > r for every negative real number r.*

The definition asserts that any number that is smaller than every positive real number but still positive is an infinitesimal, and that any number that is larger than every negative real number but still negative is also an infinitesimal. Notice that zero is not an infinitesimal.

Some sources include the number 0 as an infinitesimal, so care must be taken in interpreting statements found elsewhere.

Theorem 1 PRODUCT OF A REAL NUMBER AND AN INFINITESIMAL
    A nonzero real number times an infinitesimal is infinitesimal.

Proof. We want to show that for any nonzero real number k and any infinitesimal α that the number $k\alpha$ is infinitesimal. Let's show this for the case when both k and α are positive. Then, $k\alpha$ is also positive, and we need to show that $k\alpha < r$ for every positive real number r. Since $\frac{r}{k}$ is a positive real number and α is infinitesimal, by the definition of infinitesimal, $\alpha < \frac{r}{k}$. Multiply by k and we have $k\alpha < r$, as desired. The proof for the other cases is similar.   ■

This theorem is a fact to be remembered, similar to knowing that a negative times a negative is positive. Just as knowing such rules help us work with negative numbers, knowing facts such as theorem 1 helps us work with the hyperreal numbers.

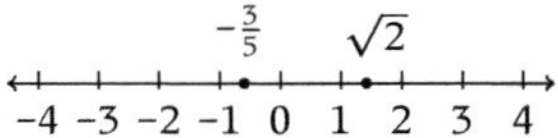

Figure 0.1 *Points on a number line*

Ans. to reading exercise 1:
$35 + 20\omega$

Over the years, as we learned different types of numbers, we learned how to place them on a number line, such as the one in Figure 0.1. So where should the infinitesimals be placed? They cannot go to the right of 0. Because they are less than every positive real number, they must go to the left of all the positive real numbers. Similarly, they must go to the right of all the negative real numbers because they are larger than all the negative real numbers. The only place remaining to put them on the real number line is at zero, but there are a lot of numbers to place there, such as ω, 12ω, and -14ω. The solution is to place them on their own number line, the ω number line.

In many video games there is a portal leading to another level. Going through that portal brings you to another world, another level of the game. Think of zero as a portal that takes you to a lower level of numbers, the ω level. It is pictured in Figure 0.2. The entire number line at the ω level, when infinitely shrunk, fits in the real number line at zero. And these are not the only two levels of hyperreal numbers.

0.1.3   More levels of infinitesimals

We now have a number-line picture of numbers of the form $k\omega$. But these are not the only infinitesimals. For instance, what about ω^2, a number we ran across in example 2? It is not pictured in Figure 0.2. Where does it lie?

We know that the infinitesimal ω is positive but smaller than any positive real number; we have $\omega < 0.003$ and $\omega < 0.000\,072$. Multiply each of these inequalities by the positive number ω and we have $\omega^2 < 0.003\omega$ and $\omega^2 < 0.000\,072\omega$. In fact, we must have $\omega^2 < 0.0000\ldots\omega$, so that ω^2 is infinitely smaller than ω. It is infinitesimal by comparison; it must be on a yet lower level. Just as zero on the real number line is a portal to the ω level, zero on the ω level is a portal to a yet lower level, the ω^2 level. And why stop there? The number ω^3 is on a yet lower level by the same reasoning, and we can continue ad infinitum. It's like many video games where you keep going to another level with seemingly no end in sight. A partial picture is in Figure 0.3.

And what about $\sqrt{\omega}$? We know that $\sqrt{\omega^2} = \omega$ is on a higher level than ω^2, and that $\sqrt{\omega^4} = \omega^2$ is on a higher level than ω^4. The square root of the infinitesimal is on

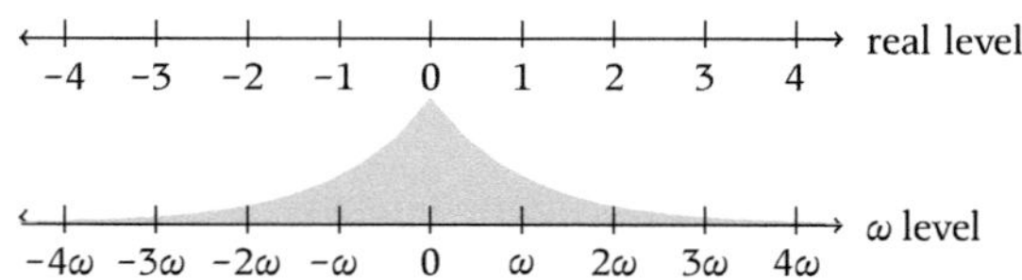

Figure 0.2 *Two levels of numbers*

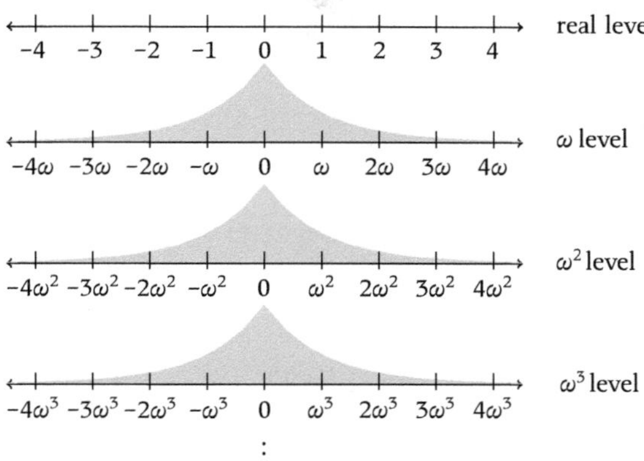

Figure 0.3 *Many levels of numbers*

a higher level than the number inside the square root. In the same way, $\sqrt{\omega}$ is on a higher level than ω, but it is still infinitesimal. There are levels in between the levels pictured in Figure 0.3; the higher the exponent on ω, the lower the level, and there are infinitely many exponents, so there are infinitely many levels—in fact, infinitely many levels between any two levels! For instance, The number $\omega^{5/2}$ is on a level in between ω^2 and ω^3. The number $\sqrt[4]{\omega}$ is on a higher level than $\sqrt[3]{\omega}$, which is on a higher level than $\sqrt{\omega}$, but all lie below the real level. A revised, but still incomplete, picture is in Figure 0.4.

Reading Exercise 2 Which number is on a lower level than the other? 3ω or $47\omega^3$?

0.1.4 Infinite numbers

You may recall that reciprocals of small numbers are large numbers. For instance,

$$\frac{1}{0.1} = 10, \quad \frac{1}{0.01} = 100, \quad \text{and} \quad \frac{1}{0.000\,000\,1} = 1\,000\,000.$$

The smaller the denominator, the larger the resulting number. Then what about

$$\frac{1}{\omega}?$$

The denominator is infinitely small, so the result must be infinitely large! This can be proved in the following manner. If we can show that $\frac{1}{\omega}$ is larger than any positive real number r, then it must be infinite. To this end, let r be any positive real number. Then $\frac{1}{r}$ is also a positive real number, and because ω is infinitesimal, we know that

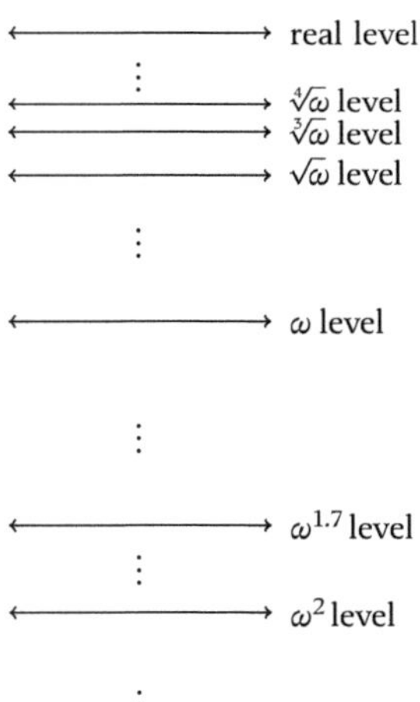

Figure 0.4 *Even more levels of numbers. The scales on the various number lines and the shading have been removed in this figure. The spacing between the levels is based on the value of the exponent. There are more levels between the ones pictured*

$\omega < \frac{1}{r}$. Taking reciprocals of numbers reverses the direction of the inequality (such as $\frac{1}{2} > \frac{1}{3}$); hence,

$$\frac{1}{\omega} > r,$$

as desired. The opposite is also true; the reciprocal of an infinite number is infinitesimal.

Theorem 2 INFINITESIMAL RECIPROCALS *The reciprocal of an infinitesimal is infinite. The reciprocal of an infinite number is infinitesimal.*

Because we will work with these infinite numbers often, it is convenient to write Ω in place of $\frac{1}{\omega}$. We do so throughout this text, and follow the convention that infinitesimals are represented by lowercase Greek letters whereas infinite numbers are represented by uppercase Greek letters–"little" letters for little numbers, "big" letters for big numbers.

Arithmetic with the infinite number Ω works just like arithmetic with ω.

Example 3 *Simplify $\Omega + 4 - (5\Omega)$.*

Solution We treat Ω like any other algebraic quantity and collect like terms:

$$\Omega + 4 - 5\Omega = 4 - 4\Omega.$$

$\blacksquare$

When working with both Ω and ω it is helpful to remember the reciprocal relationships.

Example 4 *Perform the arithmetic: $4\Omega^2 \cdot \omega$.*

Solution Remembering that $\omega = \frac{1}{\Omega}$, we write

$$4\Omega^2 \cdot \omega = 4\Omega^2 \cdot \frac{1}{\Omega} = 4\Omega.$$

$\blacksquare$

It is also sometimes helpful to think of cancellation of ω and Ω. Just as we would write $\cancel{7} \cdot \frac{1}{\cancel{7}} = 1$, we can write

$$\cancel{\Omega} \cdot \frac{1}{\cancel{\Omega}} = 1,$$

or

$$\cancel{\Omega} \cdot \cancel{\omega} = 1.$$

Ω is the uppercase Greek letter omega. When necessary, in order to distinguish Ω from ω when speaking, it is customary to refer to "capital omega" and "little omega."

Ans. to reading exercise 2:
   $47\omega^3$

Recall that $\frac{x^2}{x^1} = x^{2-1} = x^1 = x$ by the laws of exponents.

Reading Exercise 3   Simplify $6\Omega^5\omega^2$.

Now let's go back to the picture of different levels of numbers. Where does Ω fit? It certainly isn't infinitesimal, so it does not go on a level below the real numbers. And it is larger than any real number, so it cannot fit on the real number line, either. But it must go somewhere!

Recall that we think of ω as infinitely smaller than a real number. This is the same as saying that a real number is infinitely larger than an infinitesimal, and because the reals are on a higher level than ω, it seems that if one number is infinitely larger than another, then it goes on a higher level. In a similar way, infinite numbers, being infinitely larger than real numbers, belong on higher levels than the real numbers. The Ω level is above the real level, and the real numbers fit in the Ω level at 0Ω.

Now recall your playground days when someone said something like, "I'm infinity better than you," followed by the inevitable "Oh yeah? Well, I'm infinity times infinity better than you," or maybe even "I'm infinity to the infinity power better than you!" Even the intuition of children says that multiplying an infinite number by an infinite number gives a much larger number. This reasoning is reflected in the reality of higher levels of infinite numbers, such as the ones pictured in Figure 0.5. The larger the exponent on Ω, the higher the level. Just as with the levels of infinitesimals, there are infinitely many levels between each of the levels pictured, such as the $\sqrt{\Omega}$ level between the real level and the Ω level. The Ω^Ω level is

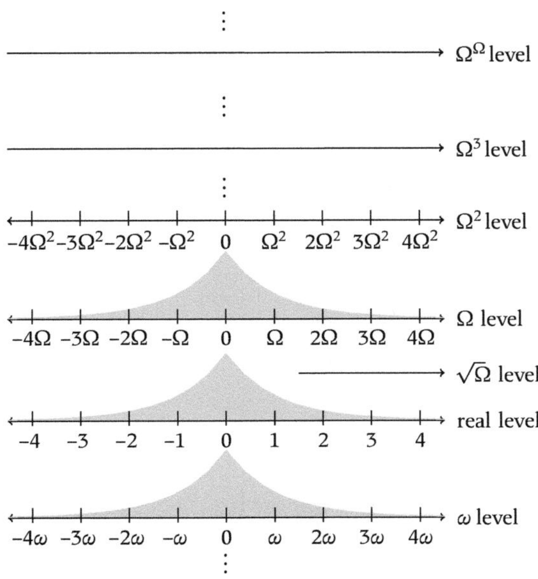

Figure 0.5 *A few of the infinitesimal and infinite levels of hyperreal numbers*

above all the real-number power of Ω levels, and there are even levels above that! It is a true video gamer's paradise of unending varieties of levels to explore. The collection of all of these numbers is called the *hyperreal numbers*. The hyperreals contain both infinitesimal and infinite numbers in addition to real numbers.

It seems as though the real number line is lost in this vast wonderland of numbers; the diagram may make it seem as though there is nothing special about the real numbers, but the real number level is the only one where you can multiply two numbers and stay on the same level; the real numbers are *closed under multiplication*. By contrast, if you multiply two infinitesimals, the result is much smaller; it is on a lower level than either of the numbers being multiplied. If you multiply two infinite numbers, the result is on an even higher level. The real numbers are special, after all!

Reading Exercise 4   Which number is larger, 5Ω or $2\Omega^3$?

We have seen what it means for two numbers to be on different levels, but we have not yet written the formal definition. To be on a lower level means to be infinitely smaller, or infinitesimal by comparison.

Definition 2 LOWER AND HIGHER LEVELS   *Let a and b be hyperreal numbers. Then (1) a is on a lower level than b if $\frac{a}{b}$ is infinitesimal and (2) a is on a higher level than b if $\frac{b}{a}$ is infinitesimal.*

Because the reciprocal of an infinitesimal is infinite, part (2) of definition 2 is the same as saying that a is on a higher level than b if $\frac{a}{b}$ is infinite.

Some hyperreal numbers have terms from more than one level, such as $5 + 2\omega$. On which level is that number? The 5 is on the real level, but the 2ω is on the ω level. This concept is made more precise in the next section, but the answer for now is that we choose the highest-level term; $5 + 2\omega$ is on the real number level.

Example 5   *On what level is $3\Omega^2 + \Omega - 8$?*

Solution   The highest-level term is $3\Omega^2$, so the number is on the Ω^2 level.   ■

0.1.5   Transfer principle

We asserted earlier that algebra in the hyperreal numbers works the same as it does for real numbers. This is part of what is called the *transfer principle*, which played a pivotal role in Robinson's proof of the existence of the hyperreal numbers and development of infinitesimal analysis.

Ans. to reading exercise 3:
$6\Omega^3$

- Multiply by a real number and the level stays the same.
- Multiply by an infinitesimal and the result is on a lower level.
- Multiply by an infinite number and the result is on a higher level.

For technical reasons, the formal definition of two numbers lying on the same level is not given until after we study approximation in the next section.

> ### TRANSFER PRINCIPLE (CALCULUS VERSION)
>
> Algebraic formulas are true in the real numbers if and only if they are true in the hyperreal numbers.

The full version of the transfer principle is much more complicated, and its technical details are beyond calculus. As long as a statement can be written using only certain types of symbols and quantifiers, then it is true in the reals if and only if it is true in the hyperreals. Although there are types of statements that cannot be written in a form for use with the transfer principle, these types of statements are easily avoided in calculus. Furthermore, all the algebraic statements transfer, so they can be used worry-free.

Applications of the transfer principle go as follows. First, we start with an algebraic statement that we know is true for the real numbers:

$$-1 \leq \sin x \leq 1 \quad \text{for every real number } x.$$

Notice that the statement, as written, is true for every real number. Then, by the transfer principle, the statement is also true for every hyperreal number:

$$-1 \leq \sin x \leq 1 \quad \text{for every hyperreal number } x.$$

Therefore, we can conclude, for instance, that

$$-1 \leq \sin(47\Omega - 500 + 2\omega^2) \leq 1.$$

We can put any hyperreal number we want inside sin and the result is still between -1 and 1.

The transfer principle is almost like a magic wand that we can wave over algebraic statements and say, "Be true in the hyperreals!" It is perhaps the most powerful tool used by *infinitesimal analysts*, mathematicians who study the hyperreal numbers and their applications to calculus and beyond.

Reading Exercise 5 True or false: $-1 \leq \cos \omega \leq 1$.

0.1.6   Logarithms of infinite numbers

An example of how the transfer principle is helpful is showing that $\ln \Omega$ is infinitely large. We know that $y = \ln x$ is an increasing function, that is, if $a > b$, then $\ln a > \ln b$. This statement is true for any real numbers a and b. By the transfer

Ans. to reading exercise 4:
$$2\Omega^3$$

principle, it is also true for hyperreal numbers! Because Ω is infinite, $\Omega > e^n$ for any real number n. Therefore

$$\ln \Omega > \ln e^n = n \ln e = n,$$

and $\ln \Omega$ is infinite because it is larger than any real number n.

The transfer principle also applies to laws of logarithms. Because $\ln a^n = n \ln a$ for any positive real number a and any real number n, the statement is true for any positive hyperreal number a and any hyperreal number n. Then

$$\ln \Omega^2 = 2 \ln \Omega,$$

and $\ln \Omega^2$ is on the same level as $\ln \Omega$.

So, where does $\ln \Omega$ fit among the levels of hyperreals? It can be shown that it fits above the real numbers but below roots of Ω. This level, along with other logarithmic levels and exponential levels, is shown in Figure 0.6.

There are many other levels above, below, and in between the levels shown in Figure 0.6, a fact that is true of any level diagram. The extent of the hyperreal numbers is truly wondrous.

It is not necessary to always be conscious of applying the transfer principle. All that is needed is to apply the usual algebraic rules to hyperreal numbers in the same manner as to real numbers.

Details are in *Calculus Set Free* Section 5.10.

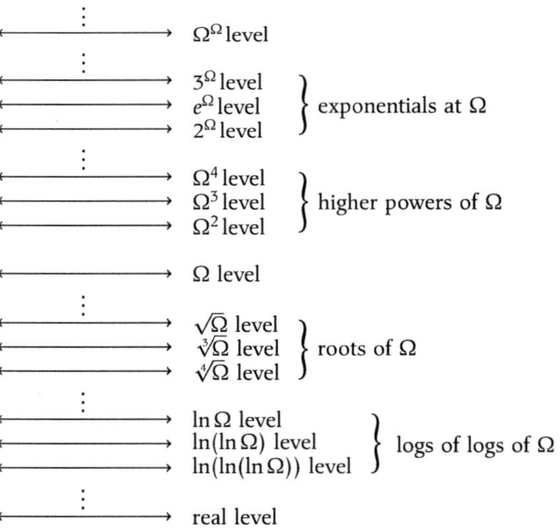

Figure 0.6 *Groups of levels of numbers are shown. Each wide arrow represents a number line at the stated level. There are more levels between, above, and below the ones pictured*

EXERCISES 0.1

1–12. Rapid response: is the given hyperreal number infinitesimal, infinite, or neither?

1. ω
2. Ω
3. 7ω
4. ω^4
5. 0
6. 5
7. $5 + \omega$
8. $\Omega^2 + 7$
9. $\dfrac{1}{\omega}$
10. $\dfrac{1}{\Omega}$
11. $\Omega^2\omega$
12. $\Omega\omega^2$

Rapid-response exercises are to be answered very quickly, and often involve object recognition. Students should not expect such exercises to appear on exams. Rapid-response exercises do not appear in every section.

13–32. Perform the arithmetic.

13. $2(7 - 4\omega)$
14. $3\omega - 5(\omega^2 - \omega)$
15. $\frac{1}{2}\omega - \frac{1}{4}(3\omega - \omega^2)$
16. $4\Omega(5 + 6\Omega)$
17. $(\Omega + 7) + 8(6 + \Omega)$
18. $(\omega + 3\omega^2) - \frac{1}{2}(4\omega - \omega^3)$
19. $(4 - \omega)^2$
20. $(2 - 3\omega)^2$
21. $(2 + \omega)^3$
22. $(\Omega + 1)^3$
23. $\dfrac{3 - \omega}{\omega}$
24. $(\omega + \Omega)^2$
25. $4\Omega(1 - \omega)$
26. $\dfrac{4\Omega}{\Omega + \Omega^2}$
27. $\dfrac{\Omega^3}{\omega}$
28. $\dfrac{\omega}{3\Omega}$
29. $\omega^3\Omega^5$
30. $\sqrt{\Omega^4}$
31. $(\Omega - \sqrt{2}\omega)^2$
32. $\omega(\Omega + 1)^2$

Ans. to reading exercise 5:
true

33–38. Which of the two given numbers is smaller?

33. $.000000000000000000000000000000000000000000000254$
    or $289475638239457856\sqrt{\omega}$
34. $\dfrac{\Omega}{10000000000000}$ or the number of dollars representing the national debt of the United States
35. $\Omega(3\omega^2)$ or ω

36. $12\omega^3$ or $3\omega^{12}$
37. $12\Omega^3$ or $3\Omega^{12}$
38. $5\omega^7$ or $7\Omega^5$

39–46. On which level is the number?

39. $\dfrac{5\omega^2}{2\sqrt{\omega}}$

40. $\Omega^7 \omega^4$

41. $\Omega^{12} \omega^3$

42. $\dfrac{3\Omega}{\sqrt{\omega}}$

43. $(\Omega + 5)^2 - \Omega^2$
44. $(\Omega - 2)^2 - 2\Omega^2$
45. $(\omega + 5)^2 - \omega^2$
46. $(4 + \omega) - 2(\omega + 2)$

47–52. True or false?

47. $-1 \le \cos(6\Omega - 127) \le 1$
48. $\sin^2 \omega + \cos^2 \omega = 1$
49. $\ln 3^\omega = \omega \ln 3$
50. $\ln \dfrac{4}{\omega} = \ln 4 - \ln \omega$
51. $(e^4)^\omega = e^{4\omega}$
52. $e^4 e^\omega = e^{4+\omega}$

53. Is the quantity $\ln 3^\omega$ infinitesimal, infinite, or neither?

54. Is the quantity $\ln \dfrac{4}{\omega}$ infinitesimal, infinite, or neither?

55. A rod has length 45 cm. It is divided into Ω pieces of equal length. What is the length of each piece?

56. A container of height 6 m is sliced horizontally into Ω pieces of equal height. What is the height of each piece?

57. A rod lies along the x-axis from $x = 3$ to $x = 7$. It is divided into Ω pieces of equal length. The x-axis is measured in centimeters. What is the length of each piece?

58. A container is positioned so that its bottom lies at $y = 1$ and its top lies at $y = 8$. It is sliced horizontally into Ω pieces of equal height. The y-axis is measured in meters. What is the height of each piece?

Although it is not physically possible to divide an actual object into infinitely many pieces, this type of abstract thinking is useful in calculus.

Through example 26, this review section is nearly identical to Section 1.2 of *Calculus Set Free*. Readers need familiarity with this material, which applies approximation to algebraic functions. This is followed by approximation with transcendental functions (trigonometric, logarithmic, exponential, and others), which can be used as a reference when needed. Students should check with their instructors to determine which material is essential.

0.2   Approximation

You are familiar with the idea of approximating numbers using a calculator:

Example 6 *Approximate $\sqrt{2}$ to four decimal places.*

Solution   Using a calculator, $\sqrt{2} \approx 1.4142$. ∎

You may be used to writing $\sqrt{2} = 1.4142$. However, technically speaking, this statement is false. The numbers $\sqrt{2}$ and 1.4142 are not exactly the same. But, for most practical purposes, they are the same, which is one reason why we often calculate with 1.4142 instead of $\sqrt{2}$.

We use decimal approximations because of their convenience. As long as we keep enough digits during intermediate calculations, our final answers are close enough. If we need a rod of length $10\sqrt{2}$ cm, we write the answer as 14.142 cm because we know how to measure the decimal expression much more easily than the square root expression, and if we are off in the fourth decimal place, it's just thousandths of a millimeter and, for most purposes, is totally inconsequential.

Remember approximations of π? You may have used $\pi = \frac{22}{7}$ or $\pi = 3.14$. Again, these statements as written are actually false, so we often write $\pi \approx \frac{22}{7}$ and $\pi \approx 3.14$ instead; but, for many applications, these approximations do just fine.

The difference between an approximation and the actual value it is meant to approximate is called the *error in the approximation*. For instance, the error in the approximation $\sqrt{2} \approx 1.4142$ can be calculated as

The error in the approximation is defined as approximation minus actual.

$$
\begin{array}{rcl}
1.4142 & = & 1.4142 \\
-\quad \sqrt{2} & = & 1.414213562\ldots \\
\hline
& & -0.000013562\ldots,
\end{array}
$$

whereas the error in $\pi \approx \frac{22}{7}$ is

$$
\begin{array}{rcl}
\frac{22}{7} & = & 3.14285714\ldots \\
-\quad \pi & = & 3.14159265\ldots \\
\hline
& & 0.00126449\ldots.
\end{array}
$$

Now consider approximating an expression such as $7 + \omega$. The infinitesimal ω is so much smaller than the real number 7 that $7 + \omega$ is, for all practical purposes, the same as 7. We therefore say that they are approximately equal.

Example 7 *Approximate $7 + \omega$.*

Solution   Because ω is infinitely smaller than 7, we write $7 + \omega \approx 7$.   ■

How close are we in this approximation? Let's calculate the error, recalling that the decimal expansion of an infinitesimal starts with infinitely many zeros:

$$
\begin{array}{rcl}
7 & = & 7.0000000\ldots \\
-\quad (7 + \omega) & = & 7.0000000\ldots \\
\hline
& & -0.0000000\ldots.
\end{array}
$$

This approximation is much closer than the approximations for $\sqrt{2}$ and π given earlier! The error in the approximation is $-\omega$, which is infinitesimal, whereas the earlier approximations had real-number errors.

A better way to compare approximations is to compare their *relative errors*, or their *percent relative errors*. The percent relative error is given by

$$\% \text{ relative error} = 100 \cdot \frac{\text{error}}{\text{actual value}}.$$

Example 8 *Determine the percent relative error in the approximation $\pi \approx \frac{22}{7}$.*

Solution   We have already calculated the error as $0.00126449\ldots$, so we have

$$\% \text{ relative error} = 100 \cdot \frac{0.00126449\ldots}{3.14159\ldots} = 0.04\%.$$

$\blacksquare$

Did the use of $=$ here bother you? We are so used to writing $=$ instead of $\approx$ in such instances that we often don't even catch that the two numbers are not equal, and that, instead, the given percent relative error has also been approximated! It is more accurate to write $\approx 0.04\%$.

Example 9 *Determine the percent relative error in the approximation $\sqrt{2} \approx 1.4142$.*

Solution   We have already calculated the error as $-0.000013562\ldots$, so we have

$$\% \text{ relative error} = 100 \cdot \frac{-0.000013562\ldots}{1.4142\ldots} = -0.00096\%.$$

$\blacksquare$

Not a bad approximation at all.

Example 10 *Determine the percent relative error in the approximation $7 + \omega \approx 7$.*

Solution   Calculating using decimals,

$$\% \text{ relative error} = 100 \cdot \frac{-\omega}{7 + \omega} = 100 \cdot \frac{0.00000\ldots}{7.00000\ldots} = 0.0000\ldots\%.$$

$\blacksquare$

The percent relative error is infinitesimal! The exact value of the percent relative error is $100 \cdot \frac{-\omega}{7+\omega}$. As long as what we "throw away" is infinitesimal by comparison to the original value, then the relative error is infinitesimal and the approximation is exceedingly good. Because terms on a lower level are infinitesimal compared to higher-level terms, we may throw away the lower-level terms.

Terms are separated by subtraction and addition. In the expression $39.6 + 2\omega - \omega^2$, there are three terms: 39.6, 2ω, and ω^2.

Example 11 *Approximate* $39.6 + 2\omega - \omega^2$.

Solution Because the terms 2ω and ω^2 are on lower levels than the term 39.6, we throw them away:

$$39.6 + 2\omega - \omega^2 \approx 39.6.$$

■

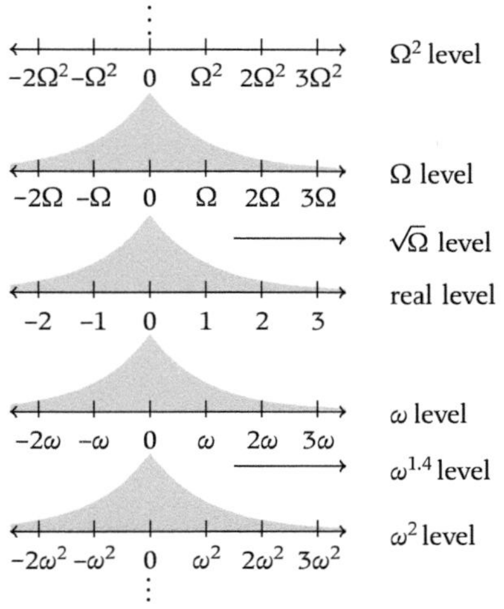

$-2\Omega^2$ $-\Omega^2$ 0 Ω^2 $2\Omega^2$ $3\Omega^2$
Ω^2 level
-2Ω $-\Omega$ 0 Ω 2Ω 3Ω
Ω level
$\sqrt{\Omega}$ level
-2 -1 0 1 2 3
real level
-2ω $-\omega$ 0 ω 2ω 3ω
ω level
$\omega^{1.4}$ level
$-2\omega^2$ $-\omega^2$ 0 ω^2 $2\omega^2$ $3\omega^2$
ω^2 level

Figure 0.7 *A few levels of hyperreals, as given in Section* 0.1

A picture of approximation may help. The figures of Section 0.1 show many levels of hyperreal numbers, such as those in Figure 0.7. Although one can find where any of the numbers 39.6 or 2ω or $-\omega^2$ fit in the vast array of number lines given in these pictures, where is $39.6 + 2\omega - \omega^2$? Using reasoning similar to that of Section 0.1, the number must fit in the real number line at 39.6, and this number acts as a portal to lower levels, containing all the numbers approximately equal to 39.6. In Figure 0.8, the location of $39.6 + 2\omega - \omega^2$ is marked with a dot on each level. The lower of the three dots marks the exact location of the number, whereas the higher of the three dots marks the location where it fits in the scheme of Figure 0.7. Determining the location of a point in the scheme of Figure 0.7, where we look for the number's nonzero appearance in that diagram, is one picture of approximation. Rising as high as possible in a diagram of the style of Figure 0.8 without rising to zero in a yet higher level is a second picture of approximation. When an expression is written in a form amenable to a Figure 0.8-style diagram, approximation can be accomplished by throwing away the lower-level terms.

Example 12 *Approximate* $27\omega + 3\omega^2$.

Solution The highest-level term is 27ω, so we throw away the lower-level term $3\omega^2$. Our approximation is

$$27\omega + 3\omega^2 \approx 27\omega.$$

■

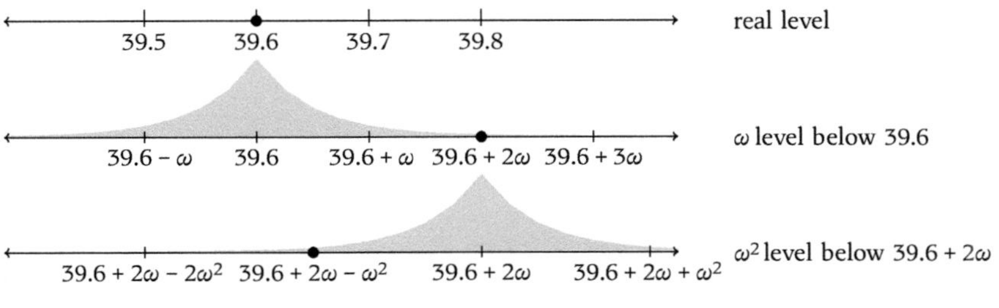

39.5 39.6 39.7 39.8
real level
$39.6 - \omega$ 39.6 $39.6 + \omega$ $39.6 + 2\omega$ $39.6 + 3\omega$
ω level below 39.6
$39.6 + 2\omega - 2\omega^2$ $39.6 + 2\omega - \omega^2$ $39.6 + 2\omega$ $39.6 + 2\omega + \omega^2$
ω^2 level below $39.6 + 2\omega$

Figure 0.8 *Levels of hyperreals below* 39.6*, with the location of* $39.6 + 2\omega - \omega^2$ *on each level marked by a dot*

Notice that we don't throw away all the infinitesimals, just the ones on the lower levels. The highest-level term remains.

Reading Exercise 6 Approximate $35 - 8.4\omega$.

Additional technical details and proofs appear in the web supplement to the article Dawson, C. Bryan, "Calculus Limits Unified and Simplified," *College Mathematics Journal* **50** No. 5 (November 2019), 331–342, doi: https://doi.org/10.1080/07468342.2019.1662706 (accessed May 10, 2021).

The phrase "$x \approx y$" can be read "x is approximately equal to y" or "x is approximately y" or "x approximates y."

Among other things, "approximation is an equivalence relation" means that if $x \approx y$, then $y \approx x$.

Based on the previous marginal note, this is the same as saying $y \approx y + x$.

0.2.1 Hyperreal approximations: a few details

Now that we have studied approximation, it is time to look at a few of the technical details before returning to more examples. Two hyperreal numbers are approximately equal if their relative difference is infinitesimal.

Definition 3 APPROXIMATION *Let x and y be hyperreal numbers. If $x \neq 0$, $x \approx y$ if $\frac{y-x}{x}$ is zero or infinitesimal. Also, $0 \approx 0$.*

By this definition, numbers that are equal are also approximately equal. Also, the only number that is approximately equal to zero is zero itself. This definition does not give approximation as an operation, but rather as an "equivalence relation" (think "=" rather than "+"). But theorem 3 shows how we can often use approximation as if it is an operation, by throwing away lower-level terms.

Theorem 3 SIMPLIFICATION THROUGH APPROXIMATION *If x is on a lower level than y, then $y + x \approx y$.*

Proof. Notice $\frac{y+x-y}{y} = \frac{x}{y}$ is infinitesimal by definition of lower level. Therefore, $y + x \approx y$ by definition of approximation. ∎

In Section 0.1, we looked at lower and higher levels, which were defined formally, but not the same level. Approximation allows us to state this definition now.

Definition 4 SAME LEVEL *Two hyperreral numbers x and y are on the same level if $\frac{x}{y} \approx r$ for some nonzero real number r.*

Both $3\omega + \omega^2$ and 12ω are on the ω level.

Therefore, $3\omega + \omega^2$ and 12ω are on the same level because $\frac{3\omega+\omega^2}{12\omega} \approx \frac{3\omega}{12\omega} = \frac{1}{4}$ is a nonzero real number.

0.2.2 Additional examples

When instructions are given to "approximate," use theorem 3 and throw away lower-level terms.

Example 13 *Approximate* $497 + 5\omega$.

Solution The number is on the real-number level, and by theorem 3, we can throw away the lower-level term 5ω:

$$497 + 5\omega \approx 497.$$

■

Throwing away lower-level terms works even if the number is on an infinite level.

Example 14 *Approximate* $3\Omega + 49\,241$.

Ans. to reading exercise 6:
35

Solution The number is on the Ω level, and by theorem 3, we can throw away the lower-level term $49\,241$:

$$3\Omega + 49\,241 \approx 3\Omega.$$

■

Notice that in example 14, we threw away a term that was not infinitesimal. As long as the term is on a lower level than the number as a whole, it can be thrown away. The term does not have to be infinitesimal; it just has to be infinitesimal by comparison.

This is not as foreign a concept as you may think. For instance, consider approximating the national debt of the United States. As of the time of this writing, the U.S. national debt is approximately \$18 trillion. It's not exactly \$18 trillion; in fact, that figure is off by hundreds of billions of dollars! But approximated to the nearest trillion, it's \$18 trillion. The error is very, very large, but the relative error is not that large. What is relatively small depends on context. We can even throw away an infinite amount when approximating, if the number is on an even higher level.

Example 15 *Approximate* $4\Omega^3 + 7\Omega$.

Solution The number is on the Ω^3 level, so we can throw away the lower-level term 7Ω:

$$4\Omega^3 + 7\Omega \approx 4\Omega^3.$$

■

Reading Exercise 7 Approximate $24 - 5\Omega$.

0.2.3 Approximation principle

Approximation requires us to know on which level the number is located. If we are wrong about the level of the number, we could make a mistake. For instance, suppose we wish to approximate the quantity

$$(\Omega + 7)^2 - \Omega^2.$$

If we say the number is on the Ω^2 level, throw away the lower-level terms Ω and 7, and approximate as

$$(\Omega + 7)^2 - \Omega^2 \approx -\Omega^2,$$

Be careful! This calculation is incorrect, demonstrating what not to do.

we are incorrect. The number is not on the Ω^2 level, but rather the Ω level, as a little arithmetic shows.

Example 16 *Approximate $(\Omega + 7)^2 - \Omega^2$.*

This is the correct version.

Solution Performing some arithmetic first to determine the level of the number,

$$(\Omega + 7)^2 - \Omega^2 = \Omega^2 + 14\Omega + 49 - \Omega^2 = 14\Omega + 49 \approx 14\Omega.$$

■

In contrast to examples 13–15, the preceding example contained an operation that could be performed to simplify the expression (squaring $\Omega + 7$) before determining the level of the number. This is what to look for. If there are such simplifications possible, then you may need to simplify before undertaking an approximation.

Example 17 *Approximate $(7 + \omega) - 4 - (3 + \omega^2)$.*

Solution We notice that some simplification can be performed, so we perform that arithmetic:

$$(7 + \omega) - 4 - (3 + \omega^2) = 7 + \omega - 4 - 3 - \omega^2 = \omega - \omega^2.$$

The number is then on the ω level and we may only throw away ω^2, arriving at the

This is the correct version.

answer

$$\approx \omega.$$

■

Suppose instead that we decided the number is on the real level and we throw away all the terms on lower levels than the reals. We then have

Be careful! This calculation is incorrect, demonstrating what not to do.

$$(7 + \omega) - 4 - (3 + \omega^2) \approx 7 - 4 - 3 = 0.$$

This is incorrect for multiple reasons. One is that when we approximate, we want the relative error to be infinitesimally small. But, if we approximate a number as zero, we throw away all of its value, and the percent relative error is 100%! Remember, the only number approximately equal to zero is zero itself. If we have a nonzero number, approximate, and get zero, we have made an error. We call this the *approximation principle*.

> ### APPROXIMATION PRINCIPLE (CALCULUS VERSION)
>
> If the result of an approximation is zero, then the approximation was performed incorrectly. Go back and perform exact arithmetic and possibly some other algebraic manipulation before determining the level of the number.

Example 18 *Approximate $(2 + \omega)^2 - 4$.*

Solution If we decide the number is on the real level and throw away ω, we have $(2 + \omega)^2 - 4 \approx 2^2 - 4 = 0$, which violates the approximation principle. Instead, let's try exact arithmetic first and see if anything cancels:

$$(2 + \omega)^2 - 4 = \cancel{4} + 4\omega + \omega^2 - \cancel{4} = 4\omega + \omega^2 \approx 4\omega.$$

It turns out that our number is on the ω level, not the real level. ∎

Reading Exercise 8 Approximate $(3 + \omega)^2 - 9$.

0.2.4 Approximations with fractions

Approximations with fractions are as simple as approximating the numerator and denominator separately and then dividing. You may already use this procedure in the real-number setting; for instance,

$$\frac{\sqrt{2} + 3}{5 - \sqrt{7}} \approx \frac{4.4142}{2.3542} \approx 1.8750.$$

Of course, we could enter the entire fraction into the calculator all at once, rather than calculate the numerator and denominator separately before dividing, in order to reduce roundoff error. This is good practice; but behind the scenes, the calculator is actually calculating the numerator and denominator separately before dividing, just to a greater precision than it shows on the screen. It is still an approximation of the type demonstrated.

Example 19 *Approximate $\dfrac{2 + \omega - \omega^2}{6 + \sqrt{\omega}}$.*

Actually, using the percent relative error formula gives a result of -100%, but it's the same idea; all the value of the number has been thrown away.

When we throw anything away and the result of our approximation is zero, we say that we have *violated the approximation principle*.

Ans. to reading exercise 7:
-5Ω

To enter the expression all at once in a calculator, be sure to use parentheses around numerator and denominator: $(\sqrt{}(2) + 3)/(5 - \sqrt{}(7))$.

Solution We approximate the numerator and denominator separately. The numerator is on the real level, so we throw away the infinitesimals, giving $2 + \omega - \omega^2 \approx 2$. The same is true for the denominator, and $6 + \sqrt{\omega} \approx 6$. Therefore,

$$\frac{2 + \omega - \omega^2}{6 + \sqrt{\omega}} \approx \frac{2}{6} = \frac{1}{3}.$$

■

The numerator and denominator do not need to be on the same level for the procedure to work; we still approximate the numerator and denominator separately.

Example 20 *Approximate* $\dfrac{\Omega^2 + 3}{2\omega + \omega^4}$.

Recall that the larger the exponent on Ω, the higher the level; the larger the exponent on ω, the lower the level.

Solution The numerator is on the Ω^2 level and the denominator is on the ω level. Throwing away the lower-level terms in each numerator and denominator yields

In step 2, the 2 remains in the denominator; $\frac{1}{2\omega} = \frac{1}{2}\Omega$, not 2Ω.

$$\frac{\Omega^2 + 3}{2\omega + \omega^4} \approx \frac{\Omega^2}{2\omega} = \frac{1}{2}\Omega^2 \cdot \Omega = \frac{1}{2}\Omega^3.$$

■

The second step of the calculation used the fact that $\frac{1}{\omega} = \Omega$. Why do this? If we stop at $\frac{\Omega^2}{2\omega}$, it is not clear where the number lies in the scheme of Figure 0.7, and we fall short of our goal. Approximating before using the reciprocal relationships is convenient, for if there is just one term in the numerator and denominator, the algebra of using the reciprocal relationships is much easier.

Ans. to reading exercise 8:
6ω

Reading Exercise 9 Approximate $\dfrac{6 + 7\omega^2}{2 + \sqrt{\omega}}$.

It is time to prove what we practiced.

Theorem 4 APPROXIMATION OF PRODUCTS AND FRACTIONS *Let $x \approx y$ and $x_1 \approx y_1$. Then,*

(a) $xx_1 \approx yy_1$, and
(b) $\dfrac{x}{x_1} \approx \dfrac{y}{y_1}$, provided $x_1, y_1 \neq 0$.

Proof. Suppose $x = 0$. Then $y = 0$ as well (since the only number approximately zero is zero), and $xx_1 = 0$, $yy_1 = 0$, and therefore $xx_1 \approx yy_1$. Similarly for part (b), $\frac{x}{x_1} = 0$, $\frac{y}{y_1} = 0$, and $\frac{x}{x_1} \approx \frac{y}{y_1}$.

Suppose now that none of x, x_1, y, y_1 are zero.

(a) Since $y \approx x$, y and x are on the same level and therefore $\frac{y}{x}$ is on the real level. Then,

$$\frac{yy_1 - xx_1}{xx_1} = \frac{yy_1 - yx_1 + yx_1 - xx_1}{xx_1} = \frac{y}{x} \cdot \frac{y_1 - x_1}{x_1} + \frac{y - x}{x},$$

which is real times (zero or infinitesimal) + (zero or infinitesimal), which then must be zero or infinitesimal. By definition of approximation, $xx_1 \approx yy_1$.

(b) First we wish to show that $\frac{1}{x_1} \approx \frac{1}{y_1}$. We have

$$\frac{\frac{1}{y_1} - \frac{1}{x_1}}{\frac{1}{x_1}} = x_1 \left(\frac{1}{y_1} - \frac{1}{x_1} \right) = \frac{x_1}{y_1} - 1 = \frac{x_1 - y_1}{y_1},$$

which is zero or infinitesimal since $x_1 \approx y_1$ (by definition of approximation). Therefore, $\frac{1}{x_1} \approx \frac{1}{y_1}$ by the definition of approximation. Last, by part (a), $x \cdot \frac{1}{x_1} \approx y \cdot \frac{1}{y_1}$.   ∎

Part (a) shows that approximating products can be accomplished by approximating each factor. It is also true that sums can be approximated by approximating each term separately, provided that we do not violate the approximation principle. This is more difficult to prove, so the proof is omitted.

Another useful fact: square roots can be approximated by approximating inside the square root.

Theorem 5 APPROXIMATION INSIDE SQUARE ROOTS *If $x \approx y$ and $y \geq 0$, then $\sqrt{x} \approx \sqrt{y}$.*

Example 21 *Approximate $\sqrt{9 + 4\omega} - \sqrt{4 - \omega^2}$.*

Solution Approximating inside the square root gives

$$\sqrt{9 + 4\omega} - \sqrt{4 - \omega^2} \approx \sqrt{9} - \sqrt{4} = 1.$$

■

An example that requires a little more thought is this: approximate $\frac{\sqrt{9+\omega}-3}{\omega}$. If we throw out the ω on top, then we will get $\frac{\sqrt{9}-3}{\omega} = \frac{0}{\omega} = 0$. But, with the approximation in the numerator being zero, we have made a mistake; the approximation principle has been violated. We must find a way of combining the real numbers in the numerator, or otherwise manipulating the expression, before we can proceed.

It is more accurate to say that sums can be approximated by approximating each term separately as long as there is no way to do so that violates the approximation principle.

See *Calculus Set Free* Section 2.6 for the proof.

Ans. to reading exercise 9:

3

Recall that $\sqrt{a+b} \neq \sqrt{a} + \sqrt{b}$, and therefore $\sqrt{9+\omega}$ is not the same as $\sqrt{9} + \sqrt{\omega}$.

We can't simplify $\sqrt{9+\omega}$ algebraically by itself, so we must try something different. In an algebra class, you learned how to rationalize denominators involving square roots by multiplying the numerator and the denominator by the *conjugate* of the expression involving the square root. The same strategy works here. If we multiply the numerator and the denominator by $\sqrt{9+\omega}+3$, then we can rid ourselves of the square root in the numerator and proceed from there. The full calculation is shown in the solution to example 22.

The conjugate of $\sqrt{9+\omega} - 3$ is $\sqrt{9+\omega}+3$, and vice versa. The conjugate changes $\sqrt{a} - b$ to $\sqrt{a}+b$ or vice versa, but it does not change what is inside the square root.

Example 22 *Approximate* $\dfrac{\sqrt{9+\omega}-3}{\omega}$.

Solution   If we approximate the numerator as $\sqrt{9+\omega}-3 \approx \sqrt{9}-3 = 0$, we are in violation of the approximation principle. Therefore, we multiply the numerator and the denominator by the conjugate of the square root expression:

It does not matter whether the square root expression is in the numerator or the denominator. We use the conjugate of the square root expression that causes the problem.

Recall that $(a-b)(a+b) = a^2 - b^2$, so $(\sqrt{9+\omega}-3)(\sqrt{9+\omega}+3) = (\sqrt{9+\omega})^2 - 3^2 = 9 + \omega - 9.$

$$\frac{\sqrt{9+\omega}-3}{\omega} = \frac{\sqrt{9+\omega}-3}{\omega} \cdot \frac{\sqrt{9+\omega}+3}{\sqrt{9+\omega}+3}$$

$$= \frac{9+\omega-9}{\omega(\sqrt{9+\omega}+3)}$$

$$= \frac{\cancel{\omega}}{\cancel{\omega}(\sqrt{9+\omega}+3)}$$

$$= \frac{1}{\sqrt{9+\omega}+3}$$

In line 2, it is convenient not to perform the multiplication in the denominator. With the denominator in factored form, it is easy to see when something from the numerator might cancel, leading to a simplified fraction.

Now the denominator may be approximated separately, and if we throw out ω, we do not get zero like we did before; we have a plus instead of a minus. We can continue without violating the approximation principle and get

$$= \frac{1}{\sqrt{9}+3} = \frac{1}{6}.$$

∎

Our first thought when approximating a square root should be to approximate inside the square root. But, if this leads to a violation of the approximation principle, try multiplying the numerator and the denominator by the conjugate of the square root expression, even if the expression is not originally a fraction.

0.2.5   Absolute value approximations

Recall that $|x| = \begin{cases} x & \text{if } x \geq 0 \\ -x & \text{if } x < 0 \end{cases}.$

Recall that the absolute value of a positive number is that number, whereas the absolute value of a negative number changes the sign of the number. Therefore, evaluating an absolute value requires us to know whether the number is positive or negative. Because ω is a positive infinitesimal,

$$|\omega| = \omega.$$

Because $-\omega$ is negative,

$$|-\omega| = -(-\omega) = \omega.$$

The sign changes to make the result positive. But what about more complicated hyperreals?

Consider $|\omega - \omega^2|$. To evaluate the absolute value, we need to know whether the number inside is positive or negative, and a first glance shows that ω is positive whereas $-\omega^2$ is negative. So which is it? Because $\omega - \omega^2 \approx \omega$, the number must be positive (it is approximately equal to a positive number). Therefore, absolute value does not change the number, and $|\omega - \omega^2| = \omega - \omega^2$.

Example 23 *Perform the arithmetic:* $|38 - 15\omega|$.

Solution Since $38 - 15\omega \approx 38$, the number inside is positive and

$$|38 - 15\omega| = 38 - 15\omega.$$

∎

If the number inside is negative, its sign is changed by taking the negative of what's inside: $|-5| = -(-5) = 5$.

Example 24 *Perform the arithmetic:* $|38 - 15\Omega|$.

Solution Since $38 - 15\Omega \approx -15\Omega$, the number is negative. Therefore

$$|38 - 15\Omega| = -(38 - 15\Omega) = -38 + 15\Omega.$$

∎

Approximations of absolute value quantities can be made by first approximating the quantity inside the absolute values, as long as the approximation principle is not violated.

Example 25 *Approximate* $\dfrac{|\Omega - 8\Omega^2|}{4 + \omega}$.

Solution Approximating inside the absolute values and approximating the numerator and denominator separately give

$$\frac{|\Omega - 8\Omega^2|}{4 + \omega} \approx \frac{|-8\Omega^2|}{4} = \frac{8\Omega^2}{4} = 2\Omega^2.$$

∎

Recall that $|a - b| \neq a + b$, so we cannot just change the subtraction to addition.

If two nonzero numbers are approximately equal, either both are positive or both are negative.

"Perform the arithmetic" should be interpreted as requiring exact arithmetic, not an approximation. But an approximation can be useful for determining how to proceed with the exact arithmetic.

Notice the difference in the instructions of Example 24 and Example 25.

Just as any other time, we must be careful to avoid violating the approximation principle.

Example 26 *Approximate* $\dfrac{|\omega - 5| - 5}{3 + \sqrt{\omega}}$.

Solution If we try to approximate inside the absolute value first,

$$\frac{|\omega - 5| - 5}{3 + \sqrt{\omega}} \approx \frac{|-5| - 5}{3} = \frac{0}{3},$$

we are in violation of the approximation principle. We must therefore perform exact arithmetic in the numerator, as in examples 23 and 24. Since $\omega - 5 \approx -5$, the quantity inside the absolute values is negative, and $|\omega - 5| = -(\omega - 5) = -\omega + 5$. Therefore,

$$\frac{|\omega - 5| - 5}{3 + \sqrt{\omega}} = \frac{-\omega + 5 - 5}{3 + \sqrt{\omega}} \approx \frac{-\omega}{3}.$$

∎

When approximating absolute value expressions, our first thought should be to try to approximate inside the absolute values. If this results in a violation of the approximation principle, we need to go back and perform exact arithmetic, evaluating the absolute value.

The remainder of this section includes information about approximation with various transcendental functions. Readers may skip this material for now, using it as a reference when needed. But the last paragraph of this section, immediately preceding the exercises, is not to be missed!

0.2.6 Trigonometric approximation formulas

An approximation that is useful in physics is that for small angles, $\sin \theta \approx \theta$. This is true for any infinitesimal (see *Calculus Set Free* Section 2.4 for a proof).

α is the lowercase Greek letter alpha. It should not be called by the same name as the letter "a" in the English alphabet.

USEFUL TRIGONOMETRIC APPROXIMATIONS

For any infinitesimal α,

$$\sin \alpha \approx \alpha$$
$$\cos \alpha \approx 1.$$

Example 27 *Approximate* $\dfrac{\sin 3\omega}{7\omega}$.

Solution Because 3ω is an infinitesimal, the approximation formula $\sin \alpha \approx \alpha$ applies, with $\alpha = 3\omega$. Using the formula, $\sin 3\omega \approx 3\omega$. Hence

$$\frac{\sin 3\omega}{7\omega} \approx \frac{3\omega}{7\omega} = \frac{3}{7}.$$

∎

Example 28 *Approximate* $\sec\left(5\omega - \omega^2\right).$

Solution An approximation formula for secant is not listed among the useful trigonometric approximations. However, by using the reciprocal identity for secant, $\sec\theta = \dfrac{1}{\cos\theta}$, we can rewrite the expression using cosine:

$$\sec\left(5\omega - \omega^2\right) = \frac{1}{\cos\left(5\omega - \omega^2\right)}.$$

Because $5\omega - \omega^2$ is infinitesimal, the approximation formula $\cos\alpha \approx 1$ applies, and

$$\sec\left(5\omega - \omega^2\right) = \frac{1}{\cos\left(5\omega - \omega^2\right)} \approx \frac{1}{1} = 1.$$

■

This strategy is helpful in a variety of contexts. Using the ratio and reciprocal identities $\tan x = \frac{\sin x}{\cos x}$, $\cot x = \frac{\cos x}{\sin x}$, $\sec x = \frac{1}{\cos x}$, and $\csc x = \frac{1}{\sin x}$, we can rewrite trig expressions using only sine and cosine, and then apply knowledge of these functions as appropriate.

The approximation principle applies in all situations, including when working with trig functions. If we try

$$\frac{-3\omega + \sin 3\omega}{\omega^2 + 2\omega^3} \approx \frac{-3\omega + 3\omega}{\omega^2} = 0,$$

we are in violation of the approximation principle. This difficulty can be overcome by use of a Maclaurin series approximation. Because

$$\sin x = x - \frac{1}{3!}x^3 + \frac{1}{5!}x^5 - \frac{1}{7!}x^7 + \cdots$$

for every real number x, by the transfer principle it is also true for every hyperreal number, including infinitesimals. Therefore, for any infinitesimal α,

$$\sin \alpha = \alpha - \frac{1}{3!}\alpha^3 + \frac{1}{5!}\alpha^5 - \frac{1}{7!}\alpha^7 + \cdots \approx \alpha - \frac{1}{3!}\alpha^3.$$

Be careful! This calculation is incorrect, demonstrating what not to do.

Example 29 *Approximate* $\dfrac{-3\omega + \sin 3\omega}{\omega^2 + 2\omega^3}.$

Solution Using the approximation formula $\sin\alpha \approx \alpha - \frac{1}{6}\alpha^3$, with $\alpha = 3\omega$, we have

$$\frac{-3\omega + \sin 3\omega}{\omega^2 + 2\omega^3} \approx \frac{-3\omega + \left(3\omega - \frac{1}{6}(3\omega)^3\right)}{\omega^2} = \frac{-\frac{27}{6}\omega^3}{\omega^2} = -\frac{9}{2}\omega.$$

■

Any number of terms of a Maclaurin series can be used for an approximation. See *Calculus Set Free* Section 10.10 for further examples.

This is the correct version.

0.2.7 Approximations inside functions

According to theorem 5, square roots can be approximated by approximating inside the square root. This is also true of logarithms, a proof of which can be found in *Calculus Set Free* Section 5.1.

In theorem 6, if $y = 1$, then $\ln y = \ln 1 = 0$ and the approximation principle is violated. In other words, we may approximate inside a logarithmic function as long as we do not violate the approximation principle.

Line 1 approximates inside the logarithm, and approximates the numerator and denominator of the fraction separately; line 2 uses the reciprocal relationship $\frac{1}{\omega} = \Omega$.

Be careful; $e^{\Omega+2} \not\approx e^{\Omega}$.

That is, a real-number multiple other than one; $e^2 \neq 1$.

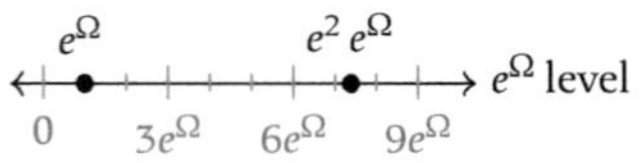

Figure 0.9 *The numbers e^{Ω} and $e^2 e^{\Omega}$ (which is about $7.389e^{\Omega}$) located on the e^{Ω} level*

Continuity is reviewed in Section 0.4. This theorem does not say anything about what happens when the quantity inside the continuous function is infinitesimal or infinite. See exercises 45 and 46.

Line 1 approximates inside the sine function, and approximates numerator and denominator separately; line 2 simplifies the quantity inside the sine function to a real number, which validates the use of theorem 7.

Theorem 6 APPROXIMATION INSIDE LOGARITHMS *If $x \approx y$, $y > 0$, and $y \neq 1$, then $\ln x \approx \ln y$.*

Example 30 *Approximate $\ln\left(\frac{3+5\omega}{\omega^2}\right)$.*

Solution Approximating inside the logarithm,

$$\ln\left(\frac{3+5\omega}{\omega^2}\right) \approx \ln\left(\frac{3}{\omega^2}\right)$$
$$= \ln\left(3\Omega^2\right),$$

which is positive infinite. ∎

Although it works for square roots and logarithms, it is not always the case that we can approximate inside a function. For instance, using laws of exponents,

$$e^{\Omega+2} = e^{\Omega}e^2,$$

which is a real number multiple of e^{Ω} and not approximately equal to e^{Ω}. Why is this true? Look again at Figures 0.7 and 0.8. Real-number multiples of a hyperreal occupy different positions on the same level (Figure 0.7). Numbers that are approximately equal occupy the same position on a number line of Figure 0.7, but can occupy different positions on levels below such a number, as in Figure 0.8. In this case, e^{Ω} and $e^2 e^{\Omega}$ occupy different positions on the e^{Ω} level (Figure 0.9).

Then, when can we approximate inside a function? Theorem 7 (see *Calculus Set Free* Section 1.7 for a proof) is a partial answer to the question.

Theorem 7 APPROXIMATING INSIDE A CONTINUOUS FUNCTION *If f is a continuous function, b is a real number in the domain of f, $k \approx b$, and $f(b) \neq 0$, then $f(k) \approx f(b)$.*

Theorem 7 says that we can approximate inside any continuous function as long as the inside approximation results in a real number (not an infinitesimal or infinite number) and we do not violate the approximation principle.

Example 31 *Approximate $\sin\left(\frac{4\omega+\omega^2}{\omega}\right)$.*

Solution Using theorem 7 to approximate inside the sine function, we have

$$\sin\left(\frac{4\omega+\omega^2}{\omega}\right) \approx \sin\left(\frac{4\omega}{\omega}\right)$$
$$= \sin 4.$$

Because the quantity inside the sine function is a real number, and because $\sin 4 \neq 0$, this use of theorem 7 is allowed. ∎

Look again at examples 27, 28, and 29. Because the expression inside the trig functions in these examples is infinitesimal and not a real number, theorem 7 does not apply. This is the reason that the "useful trigonometric approximations" are also needed. Similarly, theorem 7 does not apply to approximating an expression such as $\sin(\Omega + \pi)$, because the expression inside the trig function is infinite. The limited application of theorem 7 is why, when possible, separate theorems such as theorems 5 and 6 for specific functions are useful.

Admittedly, when dealing with nonalgebraic functions, some of the subtleties of approximation require caution. The approximation relation is actually quite sophisticated, embodying a great amount of analysis, with innumerable connections therein. This is the reason for the great power of approximation. But its basic concepts are simple enough to be easily understood and applied by those just beginning their study of calculus.

0.2.8 Accuracy

Because their relative errors are infinitesimal, the hyperreal approximations of this section are far more accurate than any decimal approximation you have encountered in the real-number system. In fact, a hyperreal approximation is a closer approximation than being off by one atom when counting the atoms of the universe!

A practical procedure is to tentatively try approximating inside the continuous function as if theorem 7 applies. Then if the expression inside the function turns out to be infinite or infinitesimal, or if the approximation principle is violated, abandon the calculation and try something else.

Because $\sin(x + \pi) = -\sin x$ for every real number x, by the transfer principle $\sin(\Omega + \pi) = -\sin \Omega$.

This amazing statement assumes that the current theories of a finite universe are correct and that such a count makes sense.

EXERCISES 0.2

1–4. Calculate the percent relative error for the given approximation.

1. $\sqrt{3} \approx 1.732$
2. $\sin(.05) \approx .05$
3. $2^{20} \approx 1\,000\,000$
4. $1.001^{1000} \approx 2.72$

5–10. Perform the arithmetic.

5. $|4 + \omega|$
6. $|4 - \Omega|$
7. $|-3\Omega + 5|$
8. $|\Omega - 4\Omega^2|$
9. $|\sqrt{\omega} - \omega|$
10. $|-3 - \omega^2 + 5|$

11–38. Approximate the given quantity.

11. $2(7 - 4\omega)$
12. $5\omega^2 - \omega^3$
13. $(\Omega + 7) + 8(6 + \Omega)$
14. $\sqrt{\Omega} + 2\Omega - \pi$
15. $(\omega + 1)^2 + 4$

Don't forget to make sure that your calculator is in radian mode before working exercise 2.

16. $\dfrac{3-\omega}{\omega}$

17. $\dfrac{4+3\omega+5\omega^2}{5-\omega}$

18. $\dfrac{\Omega^2-7}{4\Omega+3}$

19. $(\omega+\sqrt{\omega})^2-\omega$

20. $4+\dfrac{1}{5\omega}$

21. $(6+\omega)-2(\omega^2+3)$

22. $\dfrac{3+\omega-\omega^7}{12-\sqrt{\omega}}$

23. $\dfrac{(5+\omega)^2-5^2}{\omega}$

24. $\dfrac{\sqrt{5+\omega}-\sqrt{5}}{\omega}$

25. $\dfrac{|4-3\omega|-4}{2\omega}$

26. $\dfrac{|\omega^2-5\omega|}{2+\omega}$

27. $|3+2\omega|-|3-2\omega|$

28. $|2-\omega|+|5\omega-2|$

29. $\dfrac{\Omega^2+\sqrt{\Omega}-7\omega}{\sqrt{\Omega}+284}$

30. $\dfrac{\Omega-\Omega^2}{\omega-\omega^2}$

31. $\dfrac{\sqrt{49-\sqrt{\omega}-7}}{2\omega}$

32. $\dfrac{\sqrt{9\Omega^2-\Omega+1}}{4\Omega}$

33. $\dfrac{\sin 4\omega}{\omega+\sin(-5\omega)}$

34. $\dfrac{5\sin\omega^2}{\sin 3\omega}$

35. $\dfrac{\cos\omega^3}{\sin 4\omega}$

36. $\cot 2\omega$

37. $\csc\left(\dfrac{\sqrt{\omega}-\omega}{2}\right)$

38. $\dfrac{\omega+\tan\omega}{\omega-2\sin\omega}$

39. Draw a diagram in the style of Figure 0.8 showing the location of the number $13-\omega^2+\omega^3$.

40. Draw a diagram in the style of Figure 0.8 showing the location of the number $\Omega^2-3\Omega$.

41–44. Do as much approximation as can be justified by theorems or approximation formulas stated in this section.

41. (a) $\ln(\Omega+7)$; (b) $\sin(\Omega+7)$; (c) $e^{\Omega+7}$; (d) $\sqrt{\Omega+7}$

42. (a) $\ln(1+\omega)$; (b) $\sin(1+\omega)$; (c) $e^{1+\omega}$; (d) $\sqrt{1+\omega}$

43. (a) $\cosh(4-\omega)$;   (b) $\tan^{-1}(4-\omega)$;   (c) $\tan\left(\dfrac{\pi}{2}-\omega\right)$;
    (d) $\cos\left(\dfrac{\pi}{2}-\omega\right)$

44. (a) $\ln(\omega+\omega^2)$; (b) $\sin(\omega+\omega^2)$; (c) $e^{\omega+\omega^2}$; (d) $\sqrt{\omega+\omega^2}$

45. Theorem 7 does not apply to a continuous function at an infinitesimal. The following theorem does.

This theorem is related to the definition of continuity.

If f is continuous, $a\approx b$, b is infinitesimal, and $f(0)\neq 0$, then $f(a)\approx f(b)\approx f(0)$.

Use this theorem to approximate the quantity or state that this theorem does not apply: (a) $\cos(\omega^2+3\omega^4)$; (b) $\sin(\omega^2+3\omega^4)$; (c) $e^{\omega^2+3\omega^4}$

46. Theorem 7 does not apply to a continuous function at an infinite number. The following theorem does.

> If f is a continuous function with horizontal asymptote $y = k$ on the right, $k \neq 0$, $A \approx B$, and B is a positive infinite hyperreal, then $f(A) \approx f(B)$.
>
> If f is a continuous function with horizontal asymptote $y = k$ on the left, $k \neq 0$, $-A \approx -B$, and B is a positive infinite hyperreal, then $f(-A) \approx f(-B)$.

Use this theorem to approximate the quantity or state that this theorem does not apply: (a) $\tan^{-1}(-\Omega + 7)$; (b) $e^{-\Omega+7}$

This theorem appears in Section 3.6 of *Calculus Set Free.*

0.3 Hyperreals and Functions

To say that functions are important in calculus is an understatement. In algebra, we learned to evaluate functions at real-number inputs. In this section, we learn how to evaluate functions at hyperreal number inputs.

0.3.1 Evaluating functions at hyperreals

Let's begin with a review example.

Example 32 *For $f(x) = x^2 - 7x + 1$, find $f(1 + \sqrt{2})$.*

Solution We start by replacing the x in the function's output expression with the quantity $1 + \sqrt{2}$:

$$f(1 + \sqrt{2}) = (1 + \sqrt{2})^2 - 7(1 + \sqrt{2}) + 1.$$

We often then wish to simplify the expression. We can square the quantity in parentheses, distribute the -7 through the next parentheses, and collect like terms:

$$= 1 + 2\sqrt{2} + 2 - 7 - 7\sqrt{2} + 1 = -3 - 5\sqrt{2}.$$

∎

The idea of a function is sometimes represented as a machine on a production line. A real-number input rolls along a conveyor belt into the function machine, the function does its job, and a real-number output comes out the other side, as pictured in Figure 0.10. One of the main tricks in infinitesimal calculus is to extract

Through example 39, this review section is nearly identical to Section 1.3 of *Calculus Set Free.* Readers need familiarity with this material, which applies to algebraic functions. Additional material concerning transcendental functions follows, which can be used as a reference when needed. Students should check with their instructors to determine which material is essential.

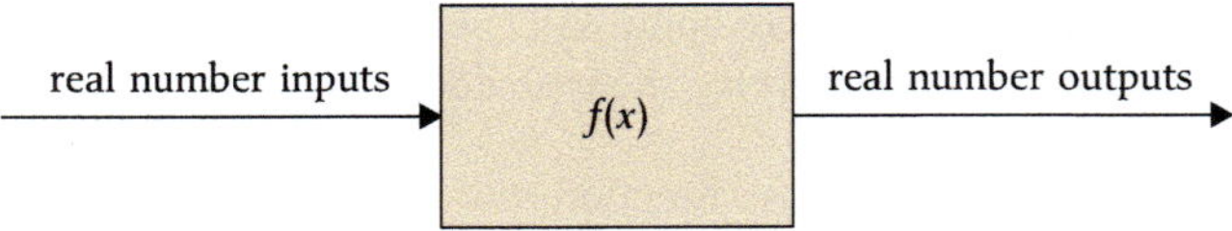

Figure 0.10 *Function as a machine*

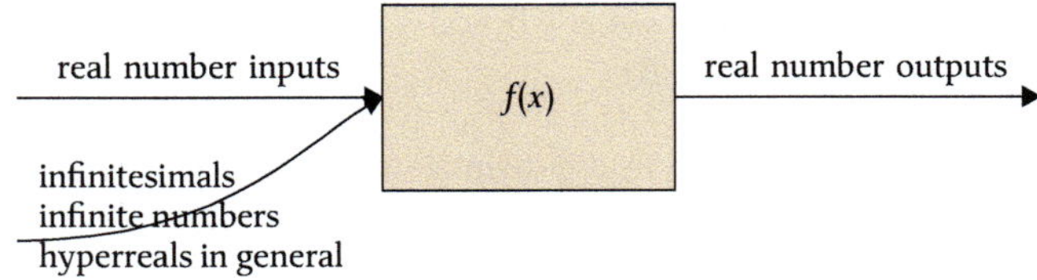

Figure 0.11 *Allowing hyperreal inputs in a real-number function*

information from a real-number function by using infinitesimals, or, more generally, by using hyperreal inputs instead of real inputs. It may seem like cheating, but any function designed to handle real-number inputs can also handle hyperreal inputs (more on this momentarily). However, we will still want a real-number output; we want our function machine to work something like Figure 0.11. Throwing hyperreal inputs into an algebraic function such as the one in example 32 is not difficult, because we know that algebra and arithmetic work the same with the hyperreals as it does with the reals. The only issue is to ensure a real-number output.

Example 33 *For $f(x) = x^2 - 7x + 1$, find $f(3 + \omega)$.*

Solution Just as in example 32, we evaluate the function at $3 + \omega$ by replacing the x in the function's output expression with the quantity $3 + \omega$:

$$f(3 + \omega) = (3 + \omega)^2 - 7(3 + \omega) + 1.$$

We can then simplify using the same algebraic steps as example 32, including collecting like terms:

$$= 9 + 6\omega + \omega^2 - 21 - 7\omega + 1 = -11 - \omega + \omega^2.$$

The result is not a real number, but a hyperreal number. We want a real number for our result. However, the answer is infinitely close to a real number, -11. We can get this result by approximating! The entire calculation, from beginning to end, therefore looks like this:

$$f(3 + \omega) = (3 + \omega)^2 - 7(3 + \omega) + 1$$
$$= 9 + 6\omega + \omega^2 - 21 - 7\omega + 1$$
$$= -11 - \omega + \omega^2$$
$$\approx -11.$$

■

This is not the whole story, but it is a good portion of the story. Our steps so far are to evaluate the expression, simplify as needed, and approximate.

Reading Exercise 10   For $f(x) = x^2$, find $f(1 - \omega)$.

Nonalgebraic functions such as the trigonometric functions can also be evaluated at hyperreal numbers. In fact, any function with a domain that includes an interval of real numbers can be evaluated at hyperreal numbers in that interval. For instance, if f is defined for real numbers x with $a \leq x \leq b$, then there is a hyperreal version of this function that is defined for hyperreals x with $a \leq x \leq b$. This is true for any type of interval. The proof is beyond the scope of this book.

Such intervals include the types $a < x < b$, $a < x \leq b$, $a \leq x < b$, $a \leq x \leq b$, $a < x$, $a \leq x$, $x < b$, $x \leq b$, and $x \in \mathbf{R}$.

0.3.2   The 3 Rs Principle

Evaluating a function at a hyperreal and then approximating does not always result in a real number. For instance, given $f(x) = x^2 - 4$ and trying $f(2 - \omega)$ yields

$$f(2 - \omega) = (2 - \omega)^2 - 4$$
$$= 4 - 4\omega + \omega^2 - 4$$
$$= -4\omega + \omega^2$$
$$\approx -4\omega.$$

The result is an infinitesimal instead of a real number. As stated before, we wish to have a real-number output from the function. But, an infinitesimal is infinitely close to zero, so our result should be zero. We say that we *render the real result* zero. Nor is an infinitesimal the only possibility; approximation can also result in an infinite value. In this case, we render the real result ∞ or $-\infty$, depending on whether the infinite number is positive or negative.

Recall that ∞ and $-\infty$ are not numbers, but rather descriptive symbols. For instance, in the interval $(3, \infty)$ the ∞ represents lack of an upper boundary. Hyperreal numbers such as 4Ω or $-27\Omega^3$ are infinite numbers, but ∞ and $-\infty$ are not specific hyperreal numbers. Therefore, we do not say "render the real number," but rather "render the real result," so that we are careful not to call ∞ or $-\infty$ a number.

Definition 5 3 Rs PRINCIPLE: RENDERING A REAL RESULT   *Let x be a hyperreal number.*

    (1) *If x is zero or infinitesimal, we say that x renders the real result* zero, *denoted $x \doteq 0$.*

The phrase "$x \doteq 0$" is read "*x* renders zero." Similarly, "$x \doteq \infty$" is read "*x* renders infinity." The use of the symbol $\doteq$ for rendering was suggested by and first used in the classroom by my colleague Troy Riggs.

Part (2) of the definition is included for technical reasons that may become clear later in this chapter. Parts (1) and (3) are the portions of the definition that are applied during calculations.

If the preliminary result after approximation is an infinitesimal, we finish the calculation by rendering the real result zero. This is part (1) of the 3 Rs principle.

Notice the use of the approximation symbol $\approx$ at the beginning of line 4 and the use of the rendering symbol $\doteq$ at the beginning of line 5. It is good practice to use the approximation symbol when, and only when, we throw something away in an approximation, and to use the rendering symbol when, and only when, we are using part (1) or part (3) of the 3 Rs principle.

Ans. to reading exercise 10:
1

(2) *If $x \approx r$ for a nonzero real number r, then we say that x renders the real result r, $x \doteq r$.*

(3) *If $x \approx A$ for an infinite hyperreal A, then we say that x renders the real result ∞ if A is positive or $-\infty$ if A is negative, denoted $x \doteq \infty$ or $x \doteq -\infty$, respectively.*

The previous calculation is then completed as follows:

Example 34 *For $f(x) = x^2 - 4$, find $f(2 - \omega)$.*

Solution We evaluate, simplify, approximate, and render a real result:

$$\begin{aligned}
f(2 - \omega) &= (2 - \omega)^2 - 4 && \text{(evaluate)} \\
&= 4 - 4\omega + \omega^2 - 4 && \text{(simplify)} \\
&= -4\omega + \omega^2 \\
&\approx -4\omega && \text{(approximate)} \\
&\doteq 0. && \text{(render a real result)}
\end{aligned}$$

■

This is called the *EAR procedure: evaluate* (and simplify), *approximate, render* a real result. The EAR procedure has many applications throughout calculus.

In example 34, when we reached $-4\omega + \omega^2$, we could skip the approximation step and render zero because we have an infinitesimal. It may also be tempting to skip other steps. However, it is generally unsafe to skip steps, because incorrect answers may result. *Using the EAR procedure in the correct order is the best practice every time.*

Reading Exercise 11 For $f(x) = x^2 - 1$, find $f(1 + \omega)$.

0.3.3 Another twist: arbitrary infinitesimals

α is the lowercase Greek letter alpha. It should not be called by the same name as the letter "a" in the English alphabet.

To say that α is an arbitrary infinitesimal is to say that α can be any infinitesimal. It can be ω. It can be $-14.3\omega^2$. It can be $-\omega + 3\omega^{3/2}$. It can be any other infinitesimal. An arbitrary infinitesimal α is a specific, but unidentified, infinitesimal. It can be positive; it can be negative. All we know is that it is an infinitesimal. Often, this is still enough to work a problem.

"Let α be an infinitesimal" is the same as saying, "Let α be an arbitrary infinitesimal."

Example 35 *Let $f(x) = \dfrac{x + 2}{x^2 - 4}$ and let α be an infinitesimal. Find (a) $f(1)$ and (b) $f(1 + \alpha)$.*

Solution Part (a) is algebra review: $f(1) = \dfrac{1+2}{1^2-4} = \dfrac{3}{-3} = -1.$

(b) Although we do not know much about α, all we need to know is that it is infinitesimal. Let's try the EAR procedure:

$$f(1+\alpha) = \frac{(1+\alpha)+2}{(1+\alpha)^2 - 4} = \frac{3+\alpha}{1+2\alpha+\alpha^2-4}$$

$$= \frac{3+\alpha}{-3+2\alpha+\alpha^2} \approx \frac{3}{-3} = -1.$$

In the approximation step, knowing that α is infinitesimal means that $3 + \alpha \approx 3$ and $-3 + 2\alpha + \alpha^2 \approx -3$. We do not need to know which specific infinitesimal is represented by α; this does not matter. All we need to know is that α is infinitesimal. ∎

If we compare parts (a) and (b) in the solution, we should notice similarities in the calculations.

Let's work a very similar problem without the extra discussion.

Example 36 *Let* $f(x) = \dfrac{x+2}{x^2-4}$ *and let* α *be an infinitesimal. Find (a)* $f(0)$ *and (b)* $f(\alpha)$.

Solution

(a) $f(0) = \dfrac{2}{-4} = -\dfrac{1}{2}.$

(b) $f(\alpha) = \dfrac{\alpha+2}{\alpha^2-4} \approx \dfrac{2}{-4} = -\dfrac{1}{2}.$

∎

In example 35, we first evaluated f at one and then at a hyperreal number infinitely close to one—namely, $1 + \alpha$. The answers were identical. The same thing happened in example 36; we first evaluated f at zero and then at a hyperreal number infinitely close to zero—namely, $0 + \alpha = \alpha$. Again, the results are identical. Does it always work this way? No. It does happen often, however, and we study this property in the next section.

Reading Exercise 12 Let $f(x) = \dfrac{x-3}{x^2-9}$ and let α be an infinitesimal. Find $f(\alpha)$.

0.3.4 Almost dividing by zero

Our next example continues to use the same function, but with a different outcome.

Example 37 *Let* $f(x) = \dfrac{x+2}{x^2-4}$ *and let* α *be an infinitesimal. Find (a)* $f(-2)$ *and (b)* $f(-2+\alpha)$.

If the preliminary result after approximation is a real number, the calculation is finished. According to the 3 Rs principle part (2), a real number renders itself, and nothing further needs to be done.

In this book, α is often used to represent an arbitrary infinitesimal.

Ans. to reading exercise 11:
0

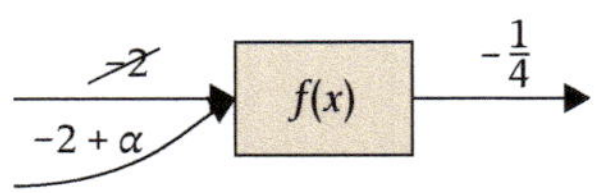

Figure 0.12 *Machine function diagram for example 37. Although* -2 *is not in the domain of* f, *using* $-2 + \alpha$ *instead gives a result*

The infinite hyperreal number A should be read "capital alpha." It may be the same symbol as the capital letter A used in the English language, but we are keeping with the heuristic of little Greek letters for infinitesimals and capital Greek letters for infinite numbers.

According to the 3 Rs Principle part (3), if the preliminary result after approximation is a positive infinite number, then we finish the calculation by rendering the real result ∞.

Solution

(a) $f(-2) = \dfrac{-2 + 2}{(-2)^2 - 4} = \dfrac{0}{0}$, which is undefined. In other words, -2 is not in the domain of the function f because we are then dividing by zero. But, if we scoot over an infinitesimal amount and evaluate the function at $-2 + \alpha$ instead (this is part (b); see Figure 0.12), then perhaps we can avoid division by zero.

(b) Using the EAR procedure, we have

$$f(-2 + \alpha) = \frac{(-2 + \alpha) + 2}{(-2 + \alpha)^2 - 4} = \frac{\alpha}{4 - 4\alpha + \alpha^2 - 4} = \frac{\alpha}{-4\alpha + \alpha^2}$$

$$\approx \frac{\alpha}{-4\alpha} = -\frac{1}{4}.$$

∎

We succeeded in evaluating the function at a number infinitely close to the hole in the domain; instead of dividing by zero in our calculation, we divided by a number that was not zero, but for all practical purposes is zero (the denominator in example 37(b) is an infinitesimal whereas the denominator in example 37(a) is zero). We may not have found a way to divide by zero, but we came awfully close!

Sometimes, instead of getting a real-number result when we almost divide by zero, we get an infinite result. The next example illustrates this. We need the notation $\frac{1}{\alpha} = A$, similar to $\frac{1}{\omega} = \Omega$. The difference is that although we know that Ω is positive (because ω is positive), we do not know the sign of A because α is an arbitrary infinitesimal and can be either positive or negative. If α is positive, so is A; but, if α is negative, A is negative as well.

Example 38 *Let* $f(x) = \dfrac{x + 2}{x^2 - 4}$ *and let* α *be an infinitesimal. Find (a)* $f(2)$, *(b)* $f(2 + \omega)$, *(c)* $f(2 - \omega)$, *and (d)* $f(2 + \alpha)$.

Solution

(a) $f(2) = \dfrac{2 + 2}{2^2 - 4} = \dfrac{4}{0}$, which is undefined because of division by zero. The number 2 is not in the domain of f.

(b) Using the EAR procedure, we have

$$f(2 + \omega) = \frac{(2 + \omega) + 2}{(2 + \omega)^2 - 4} = \frac{4 + \omega}{4 + 4\omega + \omega^2 - 4} \quad \text{(evaluate)}$$

$$\approx \frac{4}{4\omega} = \frac{1}{\omega} = \Omega \quad \text{(approximate)}$$

$$\doteq \infty. \quad \text{(render)}$$

(c) Similarly,

$$f(2 - \omega) = \frac{(2 - \omega) + 2}{(2 - \omega)^2 - 4} = \frac{4 - \omega}{4 - 4\omega + \omega^2 - 4} \quad \text{(evaluate)}$$

$$\approx \frac{4}{-4\omega} = -\frac{1}{\omega} = -\Omega \quad \text{(approximate)}$$

$$\doteq -\infty. \quad \text{(render)}$$

(d) Last,

$$f(2 + \alpha) = \frac{(2 + \alpha) + 2}{(2 + \alpha)^2 - 4} = \frac{4 + \alpha}{4 + 4\alpha + \alpha^2 - 4} \quad \text{(evaluate)}$$

$$\approx \frac{4}{4\alpha} = \frac{1}{\alpha} = A, \quad \text{(approximate)}$$

and we are stuck! We cannot render ∞ because A might be negative; we cannot render $-\infty$ because A might be positive. Remember that α is an arbitrary infinitesimal; it can be ω, as in part (b); it can be $-\omega$, as in part (c); or it can be something else. We know that the result is infinite; but, because we do not know its sign, we cannot go further.

■

Notice that division by zero, which is undefined, can produce multiple types of possible results nearby. In example 37, division by zero produced a real-number result nearby, whereas in example 38, division by zero produced an infinite result nearby. This is an important fact to remember: *division by zero does not always produce an infinite result.*

It is also true that in example 38 (b) and (c), the inputs $2 + \omega$ and $2 - \omega$ are infinitely close (their difference is 2ω, an infinitesimal), but the outputs are not identical; in fact, they are infinitely far apart! Therefore, infinitely close inputs do not always give the same results. More on this phenomenon in the next section.

Reading Exercise 13 Let $f(x) = \dfrac{x - 3}{x^2 - 9}$ and let α be an infinitesimal. Find $f(3 + \alpha)$.

0.3.5 Evaluating a function at an infinite number

If a function is defined on an infinite interval such as (a, ∞), then it can be evaluated at infinite hyperreal numbers. The EAR procedure still applies.

Example 39 *Let* $f(x) = \dfrac{x + 2}{x^2 - 4}$. *Find* $f(\Omega)$.

Ans. to reading exercise 12:
$$\frac{1}{3}$$

According to the 3 Rs principle part (3), if the preliminary result after approximation is a negative infinite number, then we finish the calculation by rendering the real result $-\infty$.

An infinite result with a sign that is unknown cannot be rendered ∞ or $-\infty$. We may either stop the computation, leaving the infinite hyperreal as written in example 38(d), or we may finish by writing the word *stuck* instead.

Division by zero does not always produce an infinite result.

Solution

$$f(\Omega) = \frac{\Omega + 2}{\Omega^2 - 4} \quad \text{(evaluate)}$$

$$\approx \frac{\Omega}{\Omega^2} = \frac{1}{\Omega} = \omega \quad \text{(approximate)}$$

$$\doteq 0. \quad \text{(render)}$$

∎

Evaluating a function at an infinite number can have any variety of results, finite or infinite, just as when evaluating at finite values.

Reading Exercise 14 Let $f(x) = \dfrac{x^2 - 3}{x^2 - 9}$. Find $f(\Omega)$.

0.3.6 Rendering facts for logarithmic and exponential functions

The following rendering facts about logarithms and exponentials of hyperreals can be remembered by their associations with the graphs of $y = \ln x$ and $y = e^x$. Proofs are in *Calculus Set Free* Sections 5.1 and 5.4.

RENDERING FACTS FOR LOGS AND EXPONENTIALS

For any positive infinitesimal ω and any positive infinite hyperreal Ω,

$$\ln(\omega) \doteq -\infty$$

$$\ln(\Omega) \doteq \infty$$

$$e^{\Omega} \doteq \infty$$

$$e^{-\Omega} \doteq 0.$$

Example 40 *Let $f(x) = \ln(x - 4)$. Find $f(4 + 3\omega)$.*

Solution Using the EAR procedure,

Line 3 uses the first rendering fact in the list, which applies for any positive infinitesimal.

$$f(4 + 3\omega) = \ln(4 + 3\omega - 4) \quad \text{(evaluate)}$$

$$= \ln(3\omega)$$

$$\doteq -\infty \quad \text{(render)}.$$

∎

Recall that approximation is allowed inside a logarithm, as long as the approximation principle is not violated.

Example 41 *Let $f(x) = \ln\left(2 + \frac{3}{x}\right)$. Find $f(\omega)$.*

Solution Using the EAR procedure,

$$
\begin{aligned}
f(\omega) &= \ln\left(2 + \frac{3}{\omega}\right) \quad \text{(evaluate)} \\
&= \ln(2 + 3\Omega) \\
&\approx \ln(3\Omega) \quad \text{(approximate)} \\
&\doteq \infty. \quad \text{(render)}
\end{aligned}
$$

■

Theorem 6 is very helpful for logarithmic functions, but there is no such theorem for exponential functions. As learned in the previous section, theorem 7 is nice when it applies, but otherwise caution must be taken when approximating in an exponent. Rendering, however, gives us a way of overcoming such difficulties.

Example 42 *Let $f(x) = e^{x/(4-x)}$ and let α be infinitesimal. Find (a) $f(1 + \alpha)$, (b) $f(4 + \omega)$, and (c) $f(4 - \omega)$.*

Solution

(a) Using the EAR procedure and tentatively using theorem 7 to approximate the exponent,

$$
\begin{aligned}
f(1 + \alpha) &= e^{\frac{1+\alpha}{4-(1+\alpha)}} \quad \text{(evaluate)} \\
&= e^{\frac{1+\alpha}{3-\alpha}} \\
&\approx e^{\frac{1}{3}}. \quad \text{(approximate)}
\end{aligned}
$$

Because the exponent is a real number, this use of theorem 7 is correct.

(b) Using the EAR procedure and tentatively using theorem 7 to approximate the exponent,

$$
\begin{aligned}
f(4 + \omega) &= e^{\frac{4+\omega}{4-(4+\omega)}} \\
&\approx e^{\frac{4}{-\omega}} = e^{-4\Omega}.
\end{aligned}
$$

Because the exponent is infinite, theorem 7 does not apply and this calculation, as written, must be abandoned. But, knowing that the exponent

Line 2 uses the reciprocal relationship $\frac{1}{\omega} = \Omega$; line 3 uses theorem 6; and line 4 uses the second rendering fact, which applies to any positive infinite hyperreal. Alternately, line 3 can be skipped; knowing that the expression inside the logarithm, $2 + 3\Omega$, is positive infinite is enough to render the real result ∞.

There is no need to render a real result because approximation results in a real number.

Be careful! The tentative calculation of part (b) must be abandoned because theorem 7 does not apply.

Although ignoring these subtleties, approximating in the exponent, and then rendering gives the correct result (and always will), it is an abuse of notation to write $\quad f(4 + \omega) = \cdots \approx e^{\frac{4}{-\omega}} = e^{-4\Omega} \doteq 0$, with the notational error located at the approximation step.

is negative infinite means that a rendering fact applies! We merely need to rewrite the calculation in a correct manner.

First we evaluate, but stop short of approximating in the exponent:

$$f(4 + \omega) = e^{\frac{4+\omega}{4-(4+\omega)}} = e^{\frac{4+\omega}{-\omega}}.$$

This is the correct version.

Next, we approximate the exponent separately:

$$\frac{4 + \omega}{-\omega} \approx \frac{4}{-\omega} = -4\Omega,$$

which is negative infinite. Therefore,

$$f(4 + \omega) = e^{\frac{4+\omega}{-\omega}} \doteq 0,$$

by the fourth rendering fact in the list, which applies for any negative infinite exponent.

(c) Suspecting that theorem 7 does not apply, we begin by evaluating but stop short of approximating in the exponent:

$$f(4 - \omega) = e^{\frac{4-\omega}{4-(4-\omega)}} = e^{\frac{4-\omega}{\omega}}.$$

Next, we approximate the exponent separately:

$$\frac{4 - \omega}{\omega} \approx \frac{4}{\omega} = 4\Omega,$$

which is positive infinite. Therefore,

$$f(4 - \omega) = e^{\frac{4-\omega}{\omega}} \doteq \infty,$$

by the third rendering fact in the list, which applies to any positive infinite exponent.

■

Rendering facts for other transcendental functions, such as $\tan^{-1} \Omega \doteq \frac{\pi}{2}$ and $\tan^{-1}(-\Omega) \doteq -\frac{\pi}{2}$, may also be remembered from their graphs.

Rendering a real result is the final piece of the puzzle for using infinitesimal methods for calculating limits. I suspect that you have already noticed this connection. The connection is made explicit in the next section.

Any time a function is to be evaluated at a hyperreal number, the EAR procedure is to be used. In other words, final answers should not be infinitesimals (they are rendered as zero), nor should they be positive infinite hyperreal numbers (rendered ∞) or negative infinite hyperreal numbers (rendered $-\infty$). Infinite numbers that cannot be rendered because the sign is unknown may be left as is or may be given the final answer *stuck*.

EXERCISES 0.3

1–5. Use $f(x) = x^3 - 1$ and find the given quantity, assuming that α is an arbitrary infinitesimal.

1. $f(1)$
2. $f(1 + \omega)$
3. $f(1 - \omega)$
4. $f(1 + \alpha)$

5. $f(1 + \sqrt{\omega - 14\omega^2 + 3\omega^{14}})$, without calculations, based on the answer to exercise 4.

6–14. Use $g(x) = \dfrac{x - 7}{x^2 - 9x + 14}$ and find the given quantity, assuming that α is an arbitrary infinitesimal.

6. $g(2)$
7. $g(2 + \omega)$
8. $g(2 - \omega)$
9. $g(2 + \alpha)$
10. $g(7)$

11. $g(7 + \alpha)$
12. $g(\Omega)$
13. $g(\Omega^2)$
14. $g(-\Omega)$

15–22. Use $h(x) = \dfrac{x^2 - 3x}{x^2 + 2x - 15}$ and find the given quantity, assuming that α is an arbitrary infinitesimal.

15. $h(3)$
16. $h(3 + \alpha)$
17. $h(-5)$
18. $h(-5 + \alpha)$

19. $h(-5 + \omega)$
20. $h(-5 - \omega)$
21. $h(\Omega)$
22. $h(-\Omega)$

23–26. Use $g(x) = \ln \dfrac{x + 4}{2x - 1}$ and find the given quantity.

23. $g(1 + \omega)$
24. $g(-4 - \omega)$

25. $g(\Omega)$
26. $g\left(\frac{1}{2} + \omega\right)$

27–32. Use $h(x) = \ln(x + 1)$ and find the given quantity, assuming that α is an arbitrary infinitesimal.

27. $h\left(\dfrac{3\Omega + 1}{\Omega - 1}\right)$
28. $h(1 + \alpha)$
29. $h(\Omega)$

30. $h(-1 + \omega)$
31. $h(-1 + 4\omega)$
32. $h(\Omega^2)$

33–38. Use $f(x) = e^{3 - 4x}$ and find the given quantity.

33. $f(\Omega)$
34. $f(-\Omega)$
35. $f(1 + \omega)$

36. $f(\omega)$
37. $f(-\sqrt{\Omega})$
38. $f(2\Omega + 15)$

39. For $f(x) = x^2$, find $\dfrac{f(1 + \alpha) - f(1)}{\alpha}$.

40. For $f(x) = 3x$, find $\dfrac{f(2 + \alpha) - f(2)}{\alpha}$.

41. For $f(x) = |x - 3|$, find $f(3 + \omega)$ and $f(3 - \omega)$.

42. For $f(x) = \dfrac{|4 - x|}{x - 4}$, find $f(4 + \omega)$ and $f(4 - \omega)$.

43. For $f(x) = \dfrac{x^6 - x^3 + 4}{x^3 - 465\sqrt{2}}$, find $f(\Omega)$ and $f(-\Omega)$.

44. For $f(x) = \dfrac{x^2 - x}{x^3 + \sqrt{x}}$, find $f(\Omega)$ and $f(-\Omega)$.

45. For $f(x) = \dfrac{x + 4}{\sqrt{x^2 - 7x + 1}}$, find $f(\Omega)$ and $f(-\Omega)$.

46. For $f(x) = \dfrac{\sqrt{x^2 + 6}}{2x + 1}$, find $f(\Omega)$ and $f(-\Omega)$.

47. For $f(x) = \sqrt{x - 7}$, find $f(11 - \omega)$.

48. For $f(x) = 2 + \sqrt{5x + 1}$, find $f(3 - 8\omega)$.

0.4 Limits and Continuity

Now that we have gained experience with the EAR procedure, we are ready to review limits and continuity, all from an infinitesimal perspective.

0.4.1 Limits: basic idea

A limit is often calculated when a function cannot be evaluated at a particular input, such as when division by zero is encountered. The result of a limit can tell us what is happening on the graph of the function, potentially indicating a removable discontinuity (hole in the graph), a jump discontinuity, or an infinite discontinuity (vertical asymptote on the graph), among other possibilities.

This section is a composite of portions of several sections of *Calculus Set Free*. As with Section 0.2, portions of this review section are included for reference purposes and may be skipped on a first reading. Students should check with their instructors to determine which material is essential.

To illustrate the idea of finding a limit using infinitesimals, consider the function $f(x) = \frac{x^2 + 2x - 15}{x - 3}$. This function is not defined for $x = 3$, because this input causes division by zero. As a result, there is no point on the graph of $y = f(x)$ for $x = 3$. However, if we evaluate the function not at $x = 3$ but at a hyperreal number infinitely close to $x = 3$, such as $x = 3 + \alpha$ where α is an arbitrary infinitesimal, then we can avoid division by zero:

Line 1 evaluates f at the input $x = 3 + \alpha$; line 2 simplifies; and line 3 approximates, arriving at a real-number output.

$$f(3 + \alpha) = \frac{(3 + \alpha)^2 + 2(3 + \alpha) - 15}{(3 + \alpha) - 3}$$

$$= \frac{9 + 6\alpha + \alpha^2 + 6 + 2\alpha - 15}{\alpha} = \frac{8\alpha + \alpha^2}{\alpha}$$

Instead of dividing by zero, line 2 features division by an infinitesimal, allowing the calculation to continue rather than be abandoned as undefined. This is a key insight into what makes calculus with infinitesimals work.

$$\approx \frac{8\alpha}{\alpha} = 8.$$

The calculation shows that if the input to our function is infinitely close to 3, then the output from our function is infinitely close to 8. We call 8 the *limit* of the function at $x = 3$.

Definition 6 LIMIT *Let f be a function and let b be a real number. If $f(b + \alpha)$ is defined and renders the same real result L for every infinitesimal α, then we write*

$$\lim_{x \to b} f(x) = L.$$

When L is a real number (i.e., not ∞ or $-\infty$), then we say that the limit exists.

Ways of reading $\lim_{x \to b} f(x) = L$ include "the limit of $f(x)$ as x approaches b is equal to L" and "the limit as x goes to b of $f(x)$ is L."

For the purposes of calculation we write

$$\lim_{x \to b} f(x) = f(b + \alpha),$$

with the understanding that α is an arbitrary infinitesimal (and therefore may be either positive or negative) and that we must render the same real result for every such α or the limit does not exist.

The earlier calculation can be rewritten as shown in example 43.

Infinitesimal limit definitions and calculations differ significantly in form from the epsilon-delta approach, but the two approaches are equivalent.

Example 43 *Find* $\lim_{x \to 3} \dfrac{x^2 + 2x - 15}{x - 3}$.

Solution   We evaluate the expression at $x = 3 + \alpha$ and use the EAR procedure:

$$\lim_{x \to 3} \frac{x^2 + 2x - 15}{x - 3} = \frac{(3 + \alpha)^2 + 2(3 + \alpha) - 15}{(3 + \alpha) - 3}$$

$$= \frac{9 + 6\alpha + \alpha^2 + 6 + 2\alpha - 15}{\alpha} = \frac{8\alpha + \alpha^2}{\alpha}$$

$$\approx \frac{8\alpha}{\alpha} = 8.$$

The limit exists since 8 is a real number. The value of the limit is 8.   ∎

Comparing the example to the definition, $b = 3, f(x) = \frac{x^2+2x-15}{x-3}$, and $L = 8$.

The graphical meaning of limit is pictured in Figure 0.13. If $\lim_{x \to b} f(x) = L$ exists, then the graph of f appears to contain the point (b, L). Whether the point (b, L) is actually on the graph is another matter; it might or might not be there. But, it appears to be there, because when the values of x are infinitely close to b, the values of $f(x)$ are infinitely close to L.

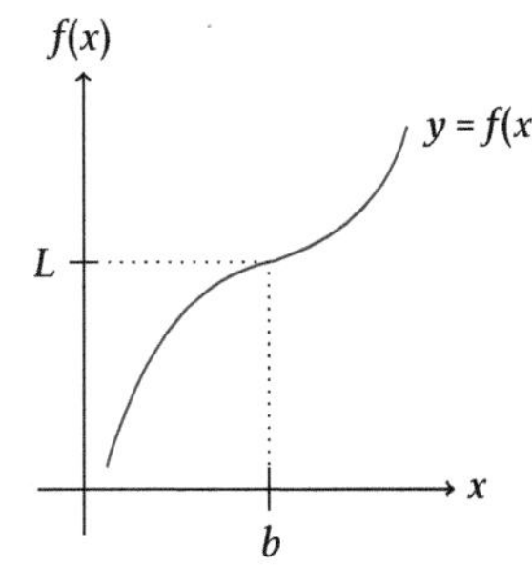

Figure 0.13 *A picture of* $\lim_{x \to b} f(x) = L$. *There may or may not actually be a point on the graph at (b, L)*

Reading Exercise 15   Find $\lim_{x \to 1} \dfrac{x^2 - 1}{x - 1}$.

0.4.2 Limits: holes in a graph (removable discontinuities)

Let's look at another example of calculating a limit and then interpret the answer graphically.

Example 44 *Find* $\lim\limits_{x\to 4} \dfrac{x-4}{x^2+x-20}$.

When compared to the definition of limit, $f(x) = \dfrac{x-4}{x^2+x-20}$ and $b=4$. We calculate $f(b+\alpha)=f(4+\alpha)$ for an arbitrary infinitesimal α and see if it renders the same real result every time. Because it renders $\frac{1}{9}$ for any such α, the limit exists.

Solution Evaluate the expression at $x = 4 + \alpha$:

$$\lim_{x\to 4} \frac{x-4}{x^2+x-20} = \frac{4+\alpha-4}{(4+\alpha)^2+(4+\alpha)-20}$$

$$= \frac{\alpha}{16+8\alpha+\alpha^2+4+\alpha-20} = \frac{\alpha}{9\alpha+\alpha^2}$$

$$\approx \frac{\alpha}{9\alpha} = \frac{1}{9}.$$

The limit is $\frac{1}{9}$. ∎

A graphical interpretation of the limit is that the point $(4, \frac{1}{9})$ appears to be on the graph of $f(x) = \dfrac{x-4}{x^2+x-20}$. But is this point really on the graph? Let's try evaluating the function at $x = 4$:

$$f(4) = \frac{4-4}{4^2+4-20} = \frac{0}{0} \quad \text{(undefined)}.$$

There is no point on the graph with $x = 4$, because the function is undefined there as a result of division by zero; $x = 4$ is not in the domain of f.

Putting all this information together, we see that there must be a "hole" in the graph of f at $(4, \frac{1}{9})$. There is no point there; but, on a graph produced by a calculator or computer, it appears there is. Because points have no width, the hole is too small to show up in the graph, so we must make it show up. We denote the hole with an unfilled circle (an "open circle") in the graph (Figure 0.14). The hole in the graph is also called a *removable discontinuity*.

To summarize, a removable discontinuity appears in a graph at (b, L) when $f(b)$ is undefined and $\lim_{x\to b} f(x) = L$. A rough sketch of the hole can be given even without knowledge of the rest of the function by drawing a small circle at the point (b, L) and drawing a very short snippet of curve on both sides of the hole, as in Figure 0.15. Once more is known about the function, the general direction of the curve near the hole can be fixed.

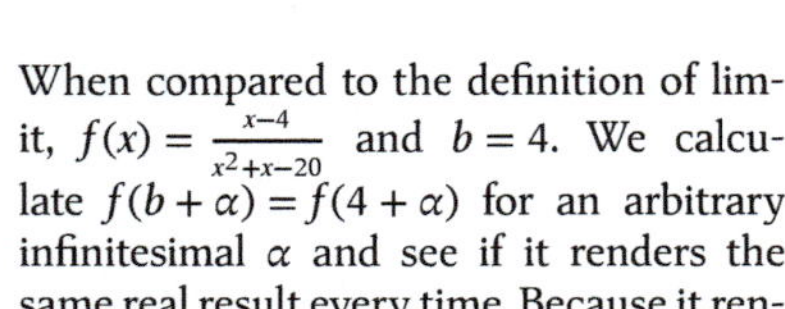

Figure 0.14 *The hole in the graph of* $f(x) = \dfrac{x-4}{x^2+x-20}$ *at* $x = 4$

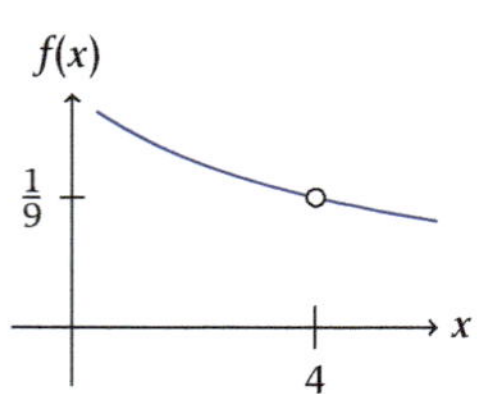

Figure 0.15 *Sketching the hole in the graph of* f *when* f *is undefined at* $x = b$ *and* $\lim_{x\to b} f(x) = L$

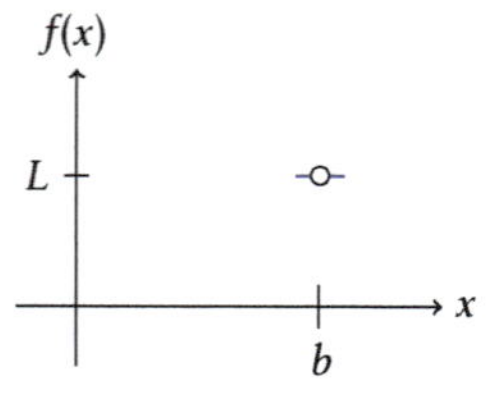

Reading Exercise 16 For $\lim\limits_{x\to 1} \dfrac{x^2-1}{x-1} = 2$, make a rough sketch of the "hole" in the graph corresponding to the limit.

0.4.3 Limit example with a square root

The next example involves the limit of a square root expression.

Example 45 *Find* $\displaystyle\lim_{x \to 2} \frac{\sqrt{x+7}-3}{x-2}$.

Solution We begin by evaluating the expression at $x = 2 + \alpha$:

$$\lim_{x \to 2} \frac{\sqrt{x+7}-3}{x-2} = \frac{\sqrt{2+\alpha+7}-3}{2+\alpha-2} = \frac{\sqrt{9+\alpha}-3}{\alpha}.$$

As in examples 43 and 44, the limit notation is dropped when the expression is evaluated at $x = b + \alpha$.

Our first thought when approximating a square root expression is to approximate under the square root, so let's try that:

$$\approx \frac{\sqrt{9}-3}{\alpha} = \frac{0}{\alpha}.$$

Ans. to reading exercise 15:
2

But, in doing so, we violate the approximation principle because the result of the approximation in the numerator is zero. We must therefore try something different, and as we learned before, multiplying the numerator and the denominator by the conjugate of the square root expression can help:

A more detailed algebraic explanation of the steps shown here is found in Section 0.2 example 22.

$$\frac{\sqrt{9+\alpha}-3}{\alpha} = \frac{\sqrt{9+\alpha}-3}{\alpha} \cdot \frac{\sqrt{9+\alpha}+3}{\sqrt{9+\alpha}+3}$$

$$= \frac{9+\alpha-9}{\alpha(\sqrt{9+\alpha}+3)}$$

$$= \frac{\alpha}{\alpha(\sqrt{9+\alpha}+3)} = \frac{1}{\sqrt{9+\alpha}+3}$$

$$\approx \frac{1}{\sqrt{9}+3} = \frac{1}{6}.$$

Notice in line 2 that we do not multiply the expressions in the denominator, leaving it as $\alpha(\sqrt{9+\alpha}+3)$. The purpose is to facilitate the cancellation of α from the numerator and the denominator in the next step.

The limit is $\frac{1}{6}$. ∎

Because the expression is undefined at $x = 2$ (division by zero), the graphical interpretation is that there is a removable discontinuity in the graph of $f(x) = \frac{\sqrt{x+7}-3}{x-2}$ at the point $(2, \frac{1}{6})$. See Figure 0.16.

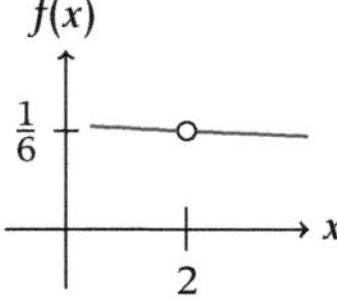

Figure 0.16 *The removable discontinuity in the graph of* $f(x) = \frac{\sqrt{x+7}-3}{x-2}$ *at* $x = 2$

0.4.4 Limit examples with trig

The approximation formulas $\sin \alpha \approx \alpha$ and $\cos \alpha \approx 1$, valid for any infinitesimal α, are sometimes useful in evaluating limits.

Example 46 *Find* $\displaystyle\lim_{x\to 0} \frac{\sin 3x}{7x}$.

Solution We evaluate the expression at $x = 0 + \alpha$, which is the same as $x = \alpha$:

$$\lim_{x\to 0} \frac{\sin 3x}{7x} = \frac{\sin 3\alpha}{7\alpha} \approx \frac{3\alpha}{7\alpha} = \frac{3}{7}.$$

The approximation step uses the sine approximation formula applied to the infinitesimal 3α, which is $\sin 3\alpha \approx 3\alpha$.

■

Example 47 *Find* $\displaystyle\lim_{x\to 0} x\sin \frac{1}{x}$.

Solution We evaluate the expression at $x = 0 + \alpha$, which is the same as $x = \alpha$:

$$\lim_{x\to 0} x\sin \frac{1}{x} = \alpha \sin \frac{1}{\alpha} = \alpha \sin A.$$

Ans. to reading exercise 16:

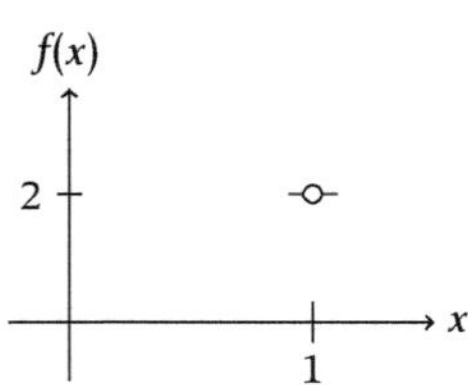

At this point, perhaps we are stuck, because we do not know how to evaluate trigonometric functions at infinite hyperreal numbers. However, the transfer principle does tell us something about the trig functions. Because $-1 \le \sin x \le 1$ for any real number x, the same is true for any hyperreal number x. Therefore, $-1 \le \sin A \le 1$, which means that $\sin A$ is on the real number level. Because α is infinitesimal, and a real number times an infinitesimal is infinitesimal, the expression $\alpha \sin A$ must also be infinitesimal and therefore renders the real result zero. The entire calculation can be written as follows:

$$\lim_{x\to 0} x\sin \frac{1}{x} = \alpha \sin \frac{1}{\alpha} = \alpha \cdot \underbrace{\sin A}_{\substack{\text{between} \\ \text{1 and } -1}} \doteq 0.$$
$$\underbrace{}_{\text{infinitesimal}}$$

The limit is 0.

■

Because the expression is undefined at $x = 0$ (division by zero), the graphical interpretation is that there is a removable discontinuity in the graph of $f(x) = x\sin \frac{1}{x}$ at the point $(0, 0)$.

0.4.5 One-sided limits

The definition of limit requires that we render the same real result for every infinitesimal α. In our previous examples, this has happened, but sometimes it does not happen.

Example 48 *Find* $\displaystyle\lim_{x\to 3} \frac{|x - 3|}{x - 3}$.

Solution   We evaluate the expression at $x = 3 + \alpha$:

$$\lim_{x \to 3} \frac{|x - 3|}{x - 3} = \frac{|3 + \alpha - 3|}{3 + \alpha - 3} = \frac{|\alpha|}{\alpha}.$$

At this point we are stuck! To evaluate $|\alpha|$, we need to know whether α is positive or negative, but because α is an arbitrary infinitesimal, it could be either. If $\alpha > 0$, we continue with

$$\frac{|\alpha|}{\alpha} = \frac{\alpha}{\alpha} = 1.$$

If $\alpha < 0$, we continue with

$$\frac{|\alpha|}{\alpha} = \frac{-\alpha}{\alpha} = -1.$$

For some values of α, the result is 1; for other values of α, the result is -1. Because we do not render the same real result for every infinitesimal α, the limit does not exist. ∎

A peek at the graph (Figure 0.17) reveals why the limit of example 48 does not exist. If x is close to 3 but larger than 3 (to the right of 3), then the values of the expression are close to 1. If x is close to 3 but smaller than 3 (to the left of 3), the values of the expression are close to -1.

Instead of a hole in the graph, we have a jump in the graph (a *jump discontinuity*). It appears there are limits from each side, but they are different. We call them *one-sided limits*. To distinguish these one-sided limits from the regular (two-sided) limit, we denote the limit from the right by $\lim_{x \to b^+} f(x)$ and we denote the limit from the left by $\lim_{x \to b^-} f(x)$.

How can these one-sided limits be calculated? If we want to see what happens infinitely close to $x = 3$, but only to the right of $x = 3$, this means we are adding a positive infinitesimal amount to 3. Instead of calculating $f(3 + \alpha)$ for an arbitrary infinitesimal, which checks both sides, we can calculate $f(3 + \omega)$ for an arbitrary positive infinitesimal ω, which only checks to the right of 3. Similarly, if we check $f(3 - \omega)$, still with ω positive, we are checking values just smaller than 3–that is, to the left of 3. The definition of a one-sided limit is like that of a limit, but with the modifications just discussed.

Definition 7 ONE-SIDED LIMIT   *Let f be a function and let b be a real number.*

    (a)   *If $f(b + \omega)$ is defined and renders the same real result L for every positive infinitesimal ω, then we write*

Recall the definition of $|x|$:
$$|x| = \begin{cases} x & \text{if } x \geq 0 \\ -x & \text{if } x < 0. \end{cases}$$

We may write $\lim_{x \to 3} \dfrac{|x - 3|}{x - 3}$ DNE or $\lim_{x \to 3} \dfrac{|x - 3|}{x - 3}$ d.n.e. as abbreviated ways to express that the limit does not exist. Also, some instructors may prefer the phrase "does not exist as a real number."

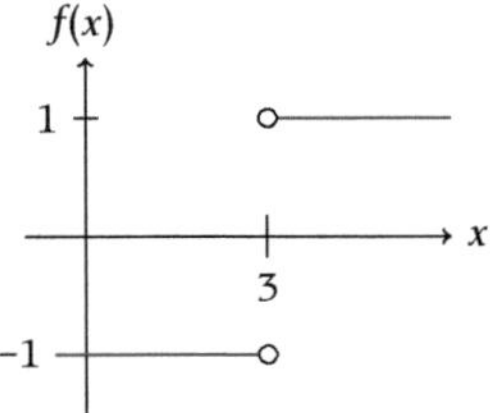

Figure 0.17 *The jump in the graph of $f(x) = \dfrac{|x-3|}{x-3}$ at $x = 3$*

$$\lim_{x \to b^+} f(x) = L.$$

When L is a real number (i.e., not ∞ or $-\infty$), then we say that the limit exists.

The phrase $\lim_{x \to b^+} f(x) = L$ can be read "the limit of $f(x)$ as x goes to b from the right is L," whereas $\lim_{x \to b^-} f(x) = L$ can be read "the limit of $f(x)$ as x goes to b from the left is L."

(b)   *If $f(b - \omega)$ is defined and renders the same real result L for every positive infinitesimal ω, then we write*

$$\lim_{x \to b^-} f(x) = L.$$

When L is a real number (i.e., not ∞ or $-\infty$), then we say that the limit exists.

For the purposes of calculation, we write

Notice that the notation is suggestive of the formula for calculation: for $x \to b^+$, we use $x = b + \omega$ whereas for $x \to b^-$, we use $x = b - \omega$.

$$\lim_{x \to b^+} f(x) = f(b + \omega)$$

$$\lim_{x \to b^-} f(x) = f(b - \omega)$$

with the understanding that ω is an arbitrary positive infinitesimal and that we must render the same real result for every such ω or the limit does not exist.

We can revisit the previous example to examine the calculation of one-sided limits.

Example 49 *Find $\lim\limits_{x \to 3^+} \dfrac{|x - 3|}{x - 3}$ and $\lim\limits_{x \to 3^-} \dfrac{|x - 3|}{x - 3}$.*

Solution   For the limit from the right, we evaluate at $x = 3 + \omega$:

$$\lim_{x \to 3^+} \frac{|x - 3|}{x - 3} = \frac{|3 + \omega - 3|}{3 + \omega - 3} = \frac{|\omega|}{\omega} = \frac{\omega}{\omega} = 1.$$

The limit from the right is 1. For this calculation we are not stuck, because we know ω is positive.

For the limit from the left, we evaluate at $x = 3 - \omega$:

Look again at Figure 0.17 and consider how these one-sided limits describe the jump discontinuity.

$$\lim_{x \to 3^-} \frac{|x - 3|}{x - 3} = \frac{|3 - \omega - 3|}{3 - \omega - 3} = \frac{|-\omega|}{-\omega} = \frac{\omega}{-\omega} = -1.$$

The limit from the left is -1. We are not stuck on this calculation either, because we know ω is positive.

∎

Note that the only way for the two-sided limit to exist is for both one-sided limits to exist and have the same value.

EFFECT OF ONE-SIDED LIMITS ON TWO-SIDED LIMITS

- If either $\lim_{x \to b^+} f(x)$ or $\lim_{x \to b^-} f(x)$ does not exist, then $\lim_{x \to b} f(x)$ does not exist.
- If $\lim_{x \to b^+} f(x) = L$ and $\lim_{x \to b^-} f(x) = M$ with $L \neq M$, then $\lim_{x \to b} f(x)$ DNE.
- If $\lim_{x \to b^+} f(x) = L$ and $\lim_{x \to b^-} f(x) = L$, then $\lim_{x \to b} f(x) = L$.

Reading Exercise 17 Determine $\lim_{x \to 2^-} \dfrac{|x - 2|}{x - 2}$.

To summarize, a jump discontinuity appears in a graph at $x = b$ when $\lim_{x \to b^+} f(x) = L$ and $\lim_{x \to b^-} f(x) = M$ each exist but have different values. This is true regardless of whether $f(b)$ exists. A rough sketch of the jump can be given even without knowing the rest of the function by drawing small circles at the points (b, L) and (b, M), and drawing a very short snippet of curve to the right of (b, L) and to the left of (b, M), as in Figure 0.18. When more is known about the function, the general direction of the curve near the jump can be fixed and the point $f(b)$ can be placed in the diagram if it exists.

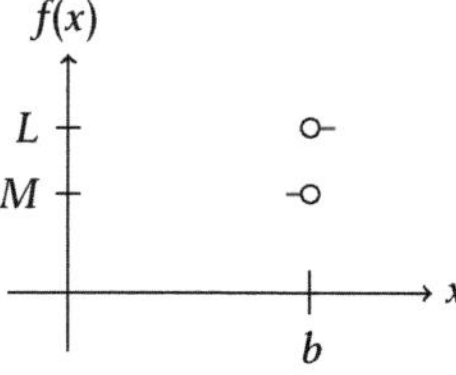

Figure 0.18 *Sketching the jump in the graph of f when $\lim_{x \to b^+} f(x) = L$ and $\lim_{x \to b^-} f(x) = M$. No point is shown at $x = b$, but such a point may be added if it exists*

0.4.6 Vertical asymptotes

One way for a limit not to exist is if it renders the real result ∞ or $-\infty$.

Example 50 *Determine* $\lim_{x \to 4} \dfrac{x + 3}{(x - 4)^2}$.

Solution This is not a one-sided limit (we have $x \to 4$, not $x \to 4^+$ or $x \to 4^-$). We therefore use $x = 4 + \alpha$:

$$\lim_{x \to 4} \frac{x + 3}{(x - 4)^2} = \frac{4 + \alpha + 3}{(4 + \alpha - 4)^2} = \frac{7 + \alpha}{\alpha^2}$$

$$\approx \frac{7}{\alpha^2} = 7A^2$$

$$\doteq \infty.$$

We are not stuck, because squared values cannot be negative; we know that $7A^2$ is positive. Because the limit is infinite, the limit does not exist. ■

Although writing $\lim_{x \to 4} \frac{x+3}{(x-4)^2}$ DNE is mathematically correct, writing $\lim_{x \to 4} \frac{x+3}{(x-4)^2} = \infty$ conveys more information; it says in what manner the limit does not exist. For this reason, $\lim_{x \to 4} \frac{x+3}{(x-4)^2} = \infty$ is the preferred expression of the answer.

What is the graphical interpretation of this limit? Because the limit's value is ∞, we know that near $x = 4$, the values of the expression must become infinite. Therefore, the graph must rise above a y-coordinate of 100, above 1 000, above 1 000 000,

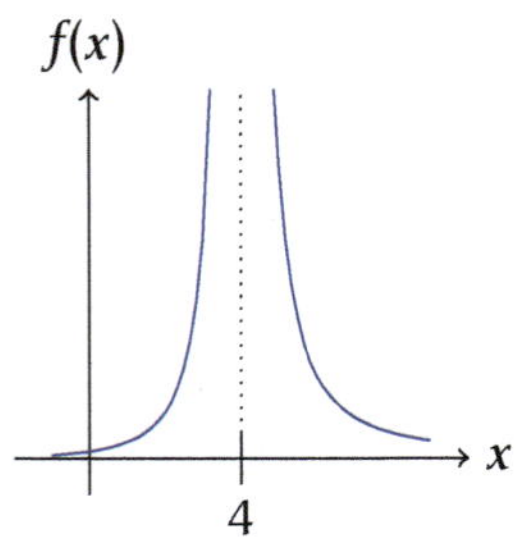

Figure 0.19 *The graph of* $f(x) = \frac{x+3}{(x-4)^2}$, *with the vertical asymptote at* $x = 4$ *marked by a dotted line*

Ans. to reading exercise 17:
-1

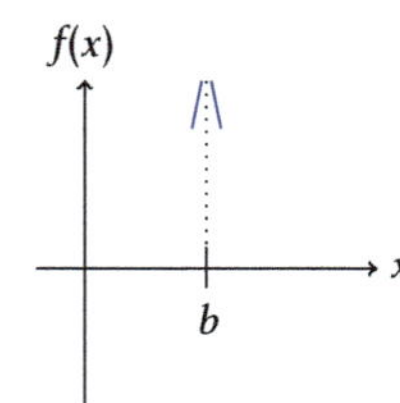

Figure 0.20 *Sketching the vertical asymptote in the graph of* f *when* $\lim_{x \to b} f(x) = \infty$

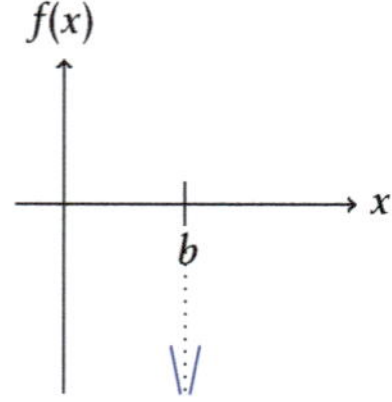

Figure 0.21 *Sketching the vertical asymptote in the graph of* f *when* $\lim_{x \to b} f(x) = -\infty$

Although we are stuck at $5A$, we do know that the result is infinite, so we may safely conclude there is a vertical asymptote on the graph of the function at $x = 3$.

above any real number! The function's values must therefore rise above the top of our picture. We express this idea by saying that the function has a *vertical asymptote* at $x = 4$. (We also say that the function has an *infinite discontinuity* at $x = 4$). We denote the vertical asymptote in a graph by a dotted line. The graph of this function is in Figure 0.19.

In example 44, division by zero resulted in a removable discontinuity. In example 50, division by zero resulted in an infinite discontinuity. Merely knowing that we are dividing by zero does not tell us what is happening on the graph; we can only tell by determining the limit.

DIVISION BY ZERO, TWO-SIDED LIMIT
___

- If the two-sided limit exists and is a real number, then there is a removable discontinuity.
- If the two-sided limit is ∞ or $-\infty$, then there is an infinite discontinuity.
- If the two-sided limit cannot be rendered, check the one-sided limits to see what is happening.

If $\lim_{x \to b} f(x) = \infty$, a rough sketch of the vertical asymptote can be made, even without knowing of the rest of the function, by drawing a dotted line at $x = b$ and drawing very short snippets of curve at the top of the picture near $x = b$, as in Figure 0.20. When more is known about the function, these snippets can be connected to the rest of the curve. A very similar sketch can be made if the limit is $-\infty$ instead; see Figure 0.21.

Reading Exercise 18 Determine $\lim\limits_{x \to 1} \dfrac{-3}{(x-1)^2}$.

Remember, if we are stuck on a two-sided limit, we can check the one-sided limits.

Example 51 *Determine* $\lim\limits_{x \to 3} \dfrac{5}{x-3}$. *Check one-sided limits if necessary.*

Solution The limit is two-sided, so we begin by using $x = 3 + \alpha$:

$$\lim_{x \to 3} \frac{5}{x-3} = \frac{5}{3+\alpha-3} = \frac{5}{\alpha} = 5A, \text{ DNE.}$$

Because A can be either positive or negative, we are stuck. Because the two-sided limit cannot be rendered, we check the one-sided limits instead. First, we check the limit from the right using $x = 3 + \omega$:

$$\lim_{x \to 3^+} \frac{5}{x-3} = \frac{5}{3+\omega-3} = \frac{5}{\omega} = 5\Omega \doteq \infty.$$

We are not stuck because we know that Ω is positive. Now let's try the limit from the left using $x = 3 - \omega$:

$$\lim_{x \to 3^-} \frac{5}{x-3} = \frac{5}{3-\omega-3} = \frac{5}{-\omega} = -5\Omega \doteq -\infty.$$

The one-sided limits exist and are different. This also means the two-sided limit does not exist. ■

In this example the limits were infinite, but from the right the limit was positive infinite and from the left the limit was negative infinite. This means the graph exits the top of the picture just to the right of $x = 3$ and the graph exits the bottom of the picture just to the left of $x = 3$. See Figure 0.22.

If $\lim_{x \to b^+} f(x) = \infty$ and $\lim_{x \to b^-} f(x) = -\infty$, a rough sketch of the vertical asymptote can be given even without knowledge of the rest of the function by drawing a dotted line at $x = b$, drawing a very short snippet of curve at the top of the picture just to the right of $x = b$, and drawing a very short snippet of curve at the bottom of the picture just to the left of $x = b$, as in Figure 0.23. Once more is known about the function, these snippets can be connected to the rest of the curve. A very similar sketch can be made if the ∞ and the $-\infty$ are swapped; see Figure 0.24.

In example 48, division by zero resulted in a jump discontinuity. In example 51, division by zero resulted in an infinite discontinuity. Even when the one-sided limits are different, there is more than one possible graphical outcome.

DIVISION BY ZERO, ONE-SIDED LIMITS

- If the one-sided limits exist and are different real numbers, then there is a jump discontinuity.
- If one of the one-sided limits is ∞ and the other is $-\infty$, then there is a vertical asymptote on the graph (an infinite discontinuity).
- If the one-sided limits are identical, then the two-sided limit can be rendered; check the box "division by zero, two-sided limit" for graphical interpretations.

Although the two-sided limit is infinite, we must write $\lim_{x \to 3} \frac{5}{x-3}$ DNE because we cannot render either ∞ or $-\infty$. We can render ∞ and $-\infty$ for the right- and left-hand limits, respectively, so we do so and write the results accordingly.

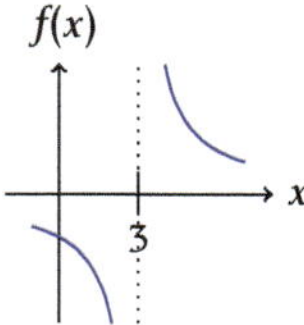

Figure 0.22 *The graph of* $f(x) = \frac{5}{x-3}$, *with a vertical asymptote at* $x = 3$ *marked by a dotted line*

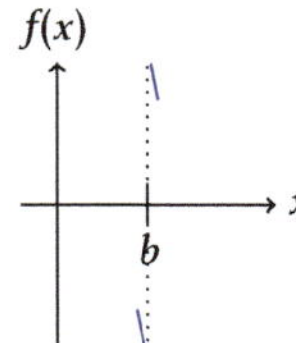

Figure 0.23 *Sketching the vertical asymptote in the graph of* f *when* $\lim_{x \to b^+} f(x) = \infty$ *and* $\lim_{x \to b^-} f(x) = -\infty$

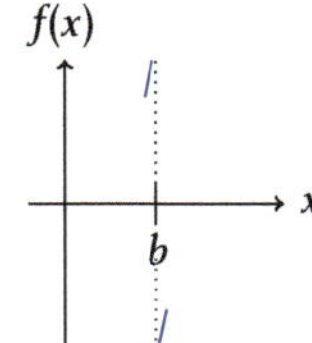

Figure 0.24 *Sketching the vertical asymptote in the graph of* f *when* $\lim_{x \to b^+} f(x) = -\infty$ *and* $\lim_{x \to b^-} f(x) = \infty$

0.4.7 Continuity

There is nothing in the definition of limit that requires us to find limits only at places where we are dividing by zero. The procedure is the same, regardless of whether it is at a place where we are dividing by zero.

Example 52 *Find* $\lim\limits_{x \to 2}(x^2 + 7x - 3)$.

Solution We evaluate the expression at $x = 2 + \alpha$:

$$\lim_{x \to 2}(x^2 + 7x - 3) = (2 + \alpha)^2 + 7(2 + \alpha) - 3$$

$$= 4 + 4\alpha + \alpha^2 + 14 + 7\alpha - 3 = 15 + 11\alpha + \alpha^2$$

$$\approx 15.$$

The limit is 15. ■

In this example, the function is defined at $x = 2$. In fact, $f(2) = 2^2 + 7(2) - 3 = 15$, and the value of the function matches the limit. In this case there is no hole in the graph (Figure 0.25); the point $(2, 15)$ is on the graph. The function is *continuous* at $x = 2$. A function is continuous at a real number in its domain if the function's two-sided limit at that number agrees with the function's value at that number.

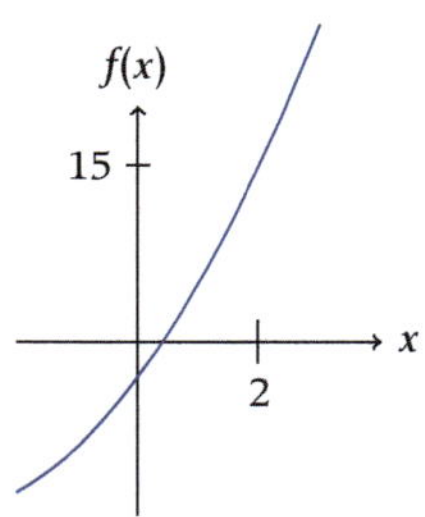

Figure 0.25 *The graph of $f(x) = x^2 + 7x - 3$, including the point with $x = 2$. There is no hole in the graph. The function is continuous at $x = 2$*

Definition 8 CONTINUITY (VERSION 1) *Let f be a function and let b be a real number in the domain of f. Then, f is continuous at $x = b$ if $\lim_{x \to b} f(x) = f(b)$.*

0.4.8 Alternate definition of continuity

The definition of continuity can be rephrased to give additional insight into the meaning of continuity. We know that $\lim_{x \to b} f(x) = L$ means that $f(b + \alpha)$ renders L for every infinitesimal α. The definition of continuity simply puts $f(b)$ in the place of L.

Except for the use of the rendering terminology and the use of the letter b, this is identical to Cauchy's definition of continuity two centuries ago.

Definition 9 CONTINUITY (VERSION 2) *Let f be a function and let b be a real number in the domain of f. Then, f is continuous at $x = b$ if $f(b + \alpha)$ renders the real result $f(b)$ for every infinitesimal α.*

This means that a small change in the value of x (from $x = b$ to $x = b + \alpha$) produces only a small change in the value of $y = f(x)$ (because $f(b + \alpha)$ renders $f(b)$ and therefore must differ from $f(b)$ by, at most, an infinitesimal amount). In other words, small changes in the x-coordinate on the graph produce small changes in the y-coordinate on the graph. This is what keeps the points close together and prevents the graph from jumping or heading off to infinity. Small changes in x producing small changes in y, rather than the ability to draw a graph in one piece, is the traditional intuitive concept of continuity.

0.4.9   Continuity of functions

Definitions 8 and 9 define continuity of a function at one number in its domain. Definition 10 applies the term to a function as a whole.

Definition 10 CONTINUOUS FUNCTION *A function f is* continuous *if it is continuous at $x = k$ for every real number k in its domain.*

As stated previously, every one of the common functions in calculus is continuous at every number in its domain.

Theorem 8 CONTINUITY OF VARIOUS FUNCTIONS *Polynomial, rational, root, trigonometric, logarithmic, and exponential functions, as well as combinations and compositions thereof, are continuous.*

Portions of Theorem 8 are proven separately in various locations in *Calculus Set Free*.

0.4.10   Using continuity to evaluate limits

One very useful consequence of knowing the continuity of certain types of functions lies in the evaluation of limits. Suppose that f is a continuous function. If k is in the domain of f, then f is continuous at $x = k$, and therefore, by definition of continuity, $\lim_{x \to k} f(x) = f(k)$. This gives us a simpler, alternate way of calculating limits:

EVALUATING LIMITS USING CONTINUITY

If f is a continuous function and k is a real number in the domain of f, then $\lim_{x \to k} f(x) = f(k)$; that is, we can evaluate $\lim_{x \to k} f(x)$ by finding $f(k)$.

If k is not in the domain of f, then continuity is of no help and we must use other methods to determine the limit.

Let's take a look at this by comparing two methods for finding $\lim_{x \to 3}(x^2 + 7)$:

using the definition of limit, we "plug in" $x = 3 + \alpha$:	using continuity, we "plug in" $x = 3$:
$\lim_{x \to 3}(x^2 + 7) = (3 + \alpha)^2 + 7$	$\lim_{x \to 3}(x^2 + 7) = 3^2 + 7$
$= 9 + 6\alpha + \alpha^2 + 7$	$= 9 + 7$
≈ 16	$= 16$

Because $f(x) = x^2 + 7$ is continuous at $x = 3$, the two answers must be the same; the limit $\lim_{x \to 3} f(x)$ (calculated on the left) must equal the value of the function $f(3)$ (calculated on the right). This means that when finding limits of continuous functions, we can try to evaluate the function first. If the function is defined, then we have found the limit. Otherwise, we must try other methods.

Continuous functions include any function type studied in calculus with the possible exception of piecewise-defined functions at an input across which the evaluation rules change.

Example 53 *Find* $\lim_{x \to 5} \dfrac{x^2 - 1}{x + 4}$.

Solution Rational functions are continuous, so we start by trying to evaluate using continuity:

$$\lim_{x \to 5} \frac{x^2 - 1}{x + 4} = \frac{5^2 - 1}{5 + 4} = \frac{24}{9}.$$

Because we evaluated the limit by substituting 5 for x, this method of calculating limits is sometimes called the *direct substitution property*.

Because the expression is defined at $x = 5$ (the answer is a real number, $\frac{24}{9}$), evaluating using continuity is successful and the limit is $\frac{24}{9}$. ∎

Reading Exercise 19 Use continuity to evaluate $\lim_{x \to 2} \dfrac{x + 1}{x^2 + 3}$.

Example 54 *Find* $\lim_{x \to 2} \dfrac{x^2 - 4}{x - 2}$.

Solution We start by trying to evaluate using continuity:

$$\lim_{x \to 2} \frac{x^2 - 4}{x - 2} = \frac{2^2 - 4}{2 - 2} = \frac{0}{0},$$

Any time we evaluate the function at the given value of x and the result is undefined (such as division by zero), we are not finished. We must use other methods to determine the limit.

which is undefined. Therefore, $x = 2$ is not in the domain of the expression and we cannot evaluate the limit using continuity; we must use other methods.

Returning to our previous methods, we use $x = 2 + \alpha$:

$$\lim_{x \to 2} \frac{x^2 - 4}{x - 2} = \frac{(2 + \alpha)^2 - 4}{2 + \alpha - 2} = \frac{4 + 4\alpha + \alpha^2 - 4}{\alpha}$$

This is the type of situation for which infinitesimal methods are designed. The function is undefined, but using hyperreals helps us because we can divide by an infinitesimal instead of dividing by zero.

$$\approx \frac{4\alpha}{\alpha} = 4.$$

The value of the limit is 4. ∎

From now on, our first thought when evaluating the limit of a continuous function should be to try to evaluate using continuity. When that fails, we use other methods.

0.4.11   Horizontal asymptotes and limits at infinity

Limits at infinity can also be determined using hyperreals, by evaluating the function at infinite numbers. For instance, if we evaluate $f(x) = 4 + \dfrac{1}{x}$ at a positive infinite hyperreal Ω, the result is

$$f(\Omega) = 4 + \frac{1}{\Omega} = 4 + \omega \approx 4.$$

If the input into f is infinite, then the output of the function is infinitely close to 4; in other words, as x gets "large," y gets "close" to 4. On the graph of f, this plays out as a horizontal asymptote on the right-hand side of the graph (Figure 0.26).

Similarly, evaluating f at a negative infinite hyperreal reveals the horizontal asymptote on the left side, also $y = 4$:

$$f(-\Omega) = 4 + \frac{1}{-\Omega} = 4 - \omega \approx 4.$$

Sometimes the calculation can reveal the direction from which the horizontal asymptote is approached; notice that before approximating, $f(\Omega)$ is a little larger than 4 whereas $f(-\Omega)$ is a little less than 4.

Definition 11 LIMITS AT INFINITY

(a) *If $f(\Omega)$ is defined and renders the same real result L for every positive infinite hyperreal Ω, then we write*

$$\lim_{x \to \infty} f(x) = L.$$

When L is a real number (i.e., not ∞ or $-\infty$), then we say that the limit exists and that f has a horizontal asymptote of $y = L$ (on the right).

(b) *If $f(-\Omega)$ is defined and renders the same real result L for every positive infinite hyperreal Ω, then we write*

$$\lim_{x \to -\infty} f(x) = L.$$

When L is a real number (i.e., not ∞ or $-\infty$), then we say that the limit exists and that f has a horizontal asymptote of $y = L$ (on the left).

For the purposes of calculation we write

$$\lim_{x \to \infty} f(x) = f(\Omega)$$

$$\lim_{x \to -\infty} f(x) = f(-\Omega),$$

with the understanding that Ω is an arbitrary positive infinite hyperreal and that we must render the same real result for every such Ω or the limit does not exist.

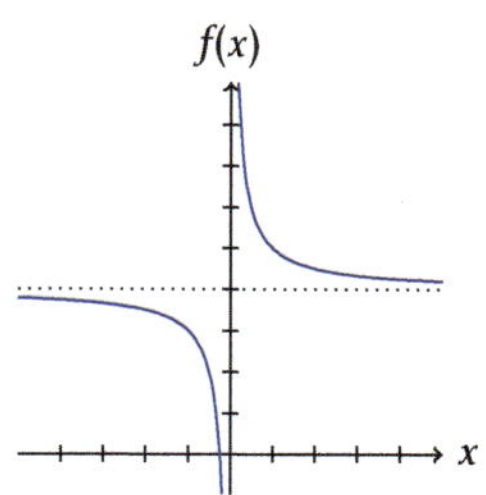

Figure 0.26 *The graph of $y = 4 + \frac{1}{x}$, with horizontal asymptote $y = 4$*

Notice that if Ω is positive, then $-\Omega$ is negative, so that in part (b) we are evaluating f at a negative infinite hyperreal.

Ways of reading the notation $\lim_{x \to \infty} f(x)$ include "the limit of f of x as x goes to infinity."

0.4.12   Examples of limits at infinity

The evaluate-approximate-render (EAR) procedure is used for this type of limit as well.

Example 55 *Determine* $\displaystyle\lim_{x\to\infty} \frac{2x + 8}{x^3 + 4x + 2}$.

Solution   We evaluate the expression at $x = \Omega$:

$$\lim_{x\to\infty} \frac{2x + 8}{x^3 + 4x + 2} = \frac{2\Omega + 8}{\Omega^3 + 4\Omega + 2}$$

$$\approx \frac{2\Omega}{\Omega^3} = \frac{2}{\Omega^2} = 2\omega^2$$

$$\doteq 0.$$

The limit is 0.   ■

The graph of the function with the limit that is taken in example 55 has a horizontal asymptote of $y = 0$ (on the right).

Example 56 *Evaluate* $\displaystyle\lim_{x\to-\infty} \frac{x^2 + 4}{3x^2 + 5x - 9}$.

Solution   We evaluate the expression at $x = -\Omega$:

$$\lim_{x\to-\infty} \frac{x^2 + 4}{3x^2 + 5x - 9} = \frac{(-\Omega)^2 + 4}{3(-\Omega)^2 + 5(-\Omega) - 9} = \frac{\Omega^2 + 4}{3\Omega^2 - 5\Omega - 9}$$

$$\approx \frac{\Omega^2}{3\Omega^2} = \frac{1}{3}.$$

The limit is $\frac{1}{3}$.   ■

The graph of the function with the limit that is taken in example 56 has a horizontal asymptote of $y = \frac{1}{3}$ (on the left).

Example 57 *Find the limit:* $\displaystyle\lim_{x\to\infty} \frac{5x^3 - 2x + 7}{x^2 + 7x - 1}$.

Solution   We evaluate the expression at $x = \Omega$:

$$\lim_{x\to\infty} \frac{5x^3 - 2x + 7}{x^2 + 7x - 1} = \frac{5\Omega^3 - 2\Omega + 7}{\Omega^2 + 7\Omega - 1}$$

$$\approx \frac{5\Omega^3}{\Omega^2} = 5\Omega$$

$$\doteq \infty.$$

The limit is ∞.   ■

Ans. to reading exercise 19:
$\frac{3}{7}$

Recall that the larger the power of Ω, the higher the level of the number.

Recall that an infinitesimal such as $2\omega^2$ renders the real result 0.

Since $2\omega^2 > 0$, the function approaches the horizontal asymptote $y = 0$ on the right side from above.

Failure to use parentheses with $-\Omega$ often results in sign errors.

In the limit calculation of example 56, it is not immediately apparent whether the exact value before approximation is a little larger or a little smaller than $\frac{1}{3}$, so we need additional reasoning to determine whether the function approaches the horizontal asymptote (on the left) from above or from below.

Recall that if the result of an approximation is positive infinite, then we render the real result ∞.

Because the limit in example 57 is ∞, the limit does not exist and there is no horizontal asymptote on the right side of the graph of the function.

Some algebra textbooks give (without proof) a method for determining the horizontal asymptotes of a rational function by considering the degrees of the polynomials in the numerator and denominator and, if necessary, the leading coefficients. If you have studied this method, then reconsider examples 55, 56, and 57 and see if you can determine why this method works.

Reading Exercise 20   Find $\lim\limits_{x\to\infty} \dfrac{4x^2 + 7x - 9}{2x^2 - 1}$.

Example 58 *Calculate* $\lim\limits_{x\to\infty}(5x^2 - 4x^3)$.

Solution   We evaluate the expression at $x = \Omega$:

$$\lim_{x\to\infty}(5x^2 - 4x^3) = 5\Omega^2 - 4\Omega^3$$

$$\approx -4\Omega^3$$

$$\doteq -\infty.$$

The limit is $-\infty$. ∎

The calculation shows there is no horizontal asymptote (on the right) on the graph of the polynomial $y = 5x^2 - 4x^3$. In fact, it is not difficult to see that the same is true for any polynomial.

Although there are no horizontal asymptotes for polynomials, limits at infinity still describe end behavior. For instance, $\lim_{x\to\infty}(5x^2 - 4x^3) = -\infty$ means that on the far right side of the graph, the polynomial's values plunge downward toward $-\infty$.

0.4.13   Examples of finding asymptotes

Determining whether a function has a horizontal asymptote (or two) is accomplished by checking the limits at infinity. Since $\lim_{x\to\infty} f(x)$ may be different from $\lim_{x\to-\infty} f(x)$, both limits should be checked.

Example 59 *Find all horizontal asymptotes on the graph of* $f(x) = \dfrac{3x - 7}{4x + \sqrt[3]{x^3 + 5x}}$.

Solution   We first check the right side by evaluating the limit as $x \to \infty$:

There do exist functions, such as $f(x) = e^x$, that possess a horizontal asymptote on one side only; that is, a function can have a horizontal asymptote on the left side even if it does not have one on the right side.

Recall that if the result of an approximation is negative infinite, then we render the real result $-\infty$.

For the degree n polynomial

$$a_n x^n + a_{n-1}x^{n-1} + \cdots + a_1 x + a_0,$$

finding limits at infinity gives

$$\lim_{x\to\infty}(a_n x^n + a_{n-1}x^{n-1} + \cdots)$$

$$= a_n\Omega^n + a_{n-1}\Omega^{n-1} + \cdots$$

$$\approx a_n\Omega^n,$$

which renders either ∞ or $-\infty$, depending on the value of a_n. Similarly,

$$\lim_{x\to-\infty}(a_n x^n + a_{n-1}x^{n-1} + \cdots) \approx a_n(-\Omega)^n,$$

which renders either ∞ or $-\infty$, depending on the values of n and a_n.

For a rational function, if there is a horizontal asymptote on one side (left or right), then the function has the same horizontal asymptote on the other side. (For readers who are students, check with your instructor to find out whether both limits must still be checked for rational functions.)

Because the function of example 59 involves a cube root, it is not a rational function.

Recall that we may approximate underneath (inside) roots, but this must take place entirely inside the root. In this case, the 5Ω inside the root can be discarded, but the 4Ω outside the root must be kept. After the root is taken, we approximate the (simplified) expression again if necessary. The approximation principle still applies.

Ans. to reading exercise 20:

2

Although $\sqrt{x^2} = |x|$, because Ω is positive we can ignore the absolute values:

$$\sqrt{4\Omega^2} = 2\sqrt{\Omega^2} = 2|\Omega| = 2\Omega.$$

Recall that

$$\sqrt{(-4)^2} \neq -4.$$

Similarly,

$$\sqrt{(-\Omega)^2} \neq -\Omega.$$

Just as $\sqrt{(-4)^2} = \sqrt{16} = 4$, we square first and then take the square root:

$$\sqrt{(-\Omega)^2} = \sqrt{\Omega^2} = \Omega.$$

$$\lim_{x\to\infty} \frac{3x-7}{4x+\sqrt[3]{x^3+5x}} = \frac{3\Omega-7}{4\Omega+\sqrt[3]{\Omega^3+5\Omega}}$$

$$\approx \frac{3\Omega}{4\Omega+\sqrt[3]{\Omega^3}} = \frac{3\Omega}{4\Omega+\Omega} = \frac{3\Omega}{5\Omega} = \frac{3}{5}.$$

The function has a horizontal asymptote on the right, $y = \frac{3}{5}$. We also need to check the left side to see if the result is different. We therefore evaluate the limit as $x \to -\infty$:

$$\lim_{x\to-\infty} \frac{3x-7}{4x+\sqrt[3]{x^3+5x}} = \frac{3(-\Omega)-7}{4(-\Omega)+\sqrt[3]{(-\Omega)^3+5(-\Omega)}}$$

$$= \frac{-3\Omega-7}{-4\Omega+\sqrt[3]{-\Omega^3-5\Omega}}$$

$$\approx \frac{-3\Omega}{-4\Omega+\sqrt[3]{-\Omega^3}} = \frac{-3\Omega}{-4\Omega-\Omega} = \frac{-3\Omega}{-5\Omega} = \frac{3}{5}.$$

The function has a horizontal asymptote of $y = \frac{3}{5}$ on the left side as well. ∎

Reading Exercise 21   Find any horizontal asymptotes on the graph of $y = \dfrac{4x-1}{x+3}$.

Example 60 *Find all horizontal asymptotes on the graph of* $y = \dfrac{\sqrt{4x^2+2}}{3x-1}$.

Solution   Determining horizontal asymptotes requires that we find the limits as $x \to \infty$ and as $x \to -\infty$. We start with the right-hand side, evaluating at $x = \Omega$:

$$\lim_{x\to\infty} \frac{\sqrt{4x^2+2}}{3x-1} = \frac{\sqrt{4\Omega^2+2}}{3\Omega-1}$$

$$\approx \frac{\sqrt{4\Omega^2}}{3\Omega} = \frac{2\Omega}{3\Omega} = \frac{2}{3}.$$

The function has the horizontal asymptote $y = \frac{2}{3}$ on the right. Next we check the left side, evaluating at $x = -\Omega$:

$$\lim_{x\to-\infty} \frac{\sqrt{4x^2+2}}{3x-1} = \frac{\sqrt{4(-\Omega)^2+2}}{3(-\Omega)-1} = \frac{\sqrt{4\Omega^2+2}}{-3\Omega-1}$$

$$\approx \frac{\sqrt{4\Omega^2}}{-3\Omega} = \frac{2\Omega}{-3\Omega} = -\frac{2}{3}.$$

The function has the horizontal asymptote $y = -\frac{2}{3}$ on the left. ∎

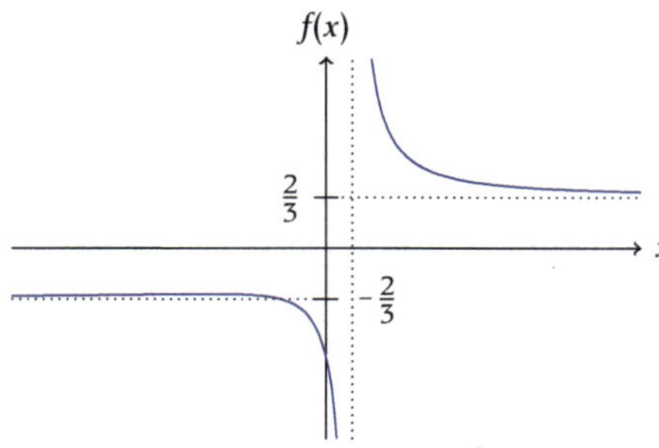

Figure 0.27 *The graph of the function in example 60, with dotted lines indicating the vertical and horizontal asymptotes. Notice that on the left side of the graph, the graph crosses the horizontal asymptote and approaches that asymptote from above, not from below*

When a function exhibits two different horizontal asymptotes, the dotted lines indicating the asymptotes should not be drawn all the way across the picture. The dotted line for the asymptote on the right is drawn on the right side only and the one for the asymptote on the left is drawn on the left side only, as in Figure 0.27.

The reason that a continuous function (continuous where defined) may not cross a vertical asymptote is that it is not defined at that value of x. The infinite discontinuity prevents us from connecting the pieces on either side of the asymptote, but there is no corresponding restriction on horizontal asymptotes. The y-coordinate of a horizontal asymptote does not arise from division by zero or any similar hole in the function's domain. Therefore, a function might cross a horizontal asymptote, as in Figure 0.27. Because the scale makes this crossing difficult to see, a close-up view of the crossing is given in Figure 0.28.

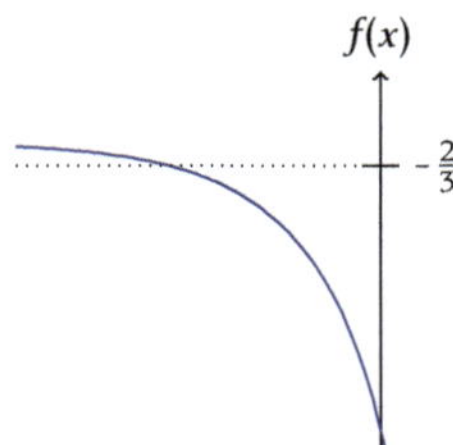

Figure 0.28 *A portion the graph of the function in example 60, showing the crossing of the horizontal asymptote. The curve then approaches the asymptote from above*

0.4.14   The approximation principle, continuous functions, and rendering

Sometimes, the use of rendering can help us overcome an instance of violation of the approximation principle.

Example 61 *Evaluate* $\displaystyle\lim_{x\to\infty} \ln\left(1 - \frac{1}{x}\right).$

Solution For any positive infinite hyperreal integer Ω,

Line 2 uses the reciprocal relationship $\frac{1}{\Omega} = \omega$; line 3 attempts to approximate inside the logarithm using theorem 6; line 4 shows that the approximation principle has been violated.

$$\lim_{x\to\infty} \ln\left(1 - \frac{1}{x}\right) = \ln\left(1 - \frac{1}{\Omega}\right)$$
$$= \ln(1 - \omega)$$
$$\approx \ln 1$$
$$= 0.$$

Be careful! This calculation is incorrect, demonstrating what not to do.

Because the approximation yields zero, we have violated the approximation principle. However, because no additional arithmetic is being performed afterward, the definition of continuity allows us to render a real result instead of using approximation:

Line 3 relies directly on version 2 of the definition of continuity (definition 9). Because $y = \ln x$ is continuous at $x = 1$, $\ln(1 + \alpha)$ must render the real result $\ln 1$ for any infinitesimal α.

$$\lim_{x\to\infty} \ln\left(1 - \frac{1}{x}\right) = \ln\left(1 - \frac{1}{\Omega}\right)$$
$$= \ln(1 - \omega)$$
$$\doteq \ln 1$$
$$= 0.$$

This is the correct version.

Although the result of the calculation is the same, the use of approximation is incorrect. The notation used in the second calculation is correct, and the limit is zero. ∎

Ans. to reading exercise 21:
$\quad y = 4$ (on both sides)

If the approximation principle is violated when approximating inside a continuous function, check to see if rendering can be used instead.

0.4.15   Level analysis

Sometimes, the use of rendering can be combined with knowledge of the relative position of levels of hyperreals to evaluate limits efficiently. For instance, if we know that a numerator is on a lower level than a denominator, the fraction must be infinitesimal and therefore renders zero.

See Section 0.1 Figure 0.6 for a level diagram.

Example 62 *Evaluate* $\displaystyle\lim_{x\to\infty} \frac{x}{e^x - 15}.$

Solution For any positive infinite hyperreal integer Ω,

$$\lim_{x\to\infty} \frac{x}{e^x - 15} = \frac{\Omega}{e^\Omega - 15}$$

$$\approx \frac{\Omega}{e^\Omega} \begin{array}{l} \leftarrow \text{ lower level} \\ \leftarrow \text{ higher level} \end{array}$$

$$\doteq 0.$$

L'Hospital's rule is also an option, but is less efficient.

The limit is zero. ∎

Similarly, if we know that a numerator is on a higher level than a denominator, the fraction must be infinite and therefore renders ∞ or $-\infty$, as appropriate.

Example 63 *Evaluate* $\displaystyle\lim_{x\to\infty} \frac{-\sqrt{x+1}}{\ln x^2}$.

Solution For any positive infinite hyperreal integer Ω,

$$\lim_{x\to\infty} \frac{-\sqrt{x+1}}{\ln x^2} = \frac{-\sqrt{\Omega+1}}{\ln \Omega^2}$$

L'Hospital's rule is also an option, but is less efficient.

$$\approx \frac{-\sqrt{\Omega}}{\ln \Omega^2}$$

Line 3 uses a law of logarithms to simplify, making the level of the number more apparent. This line may be skipped, with the level analysis taking place one line sooner.

$$= \frac{-\sqrt{\Omega}}{2\ln \Omega} \begin{array}{l} \leftarrow \text{ higher level } (\Omega^{\frac{1}{2}} \text{ level}) \\ \leftarrow \text{ lower level } (\ln \Omega \text{ level}) \end{array}$$

$$\doteq -\infty.$$

∎

0.4.16 Additional techniques

You may have noticed that many of the calculations involved in finding limits are practiced in Sections 0.2 and 0.3. Some of the techniques mentioned there, such as the use of Maclaurin series in an approximation and knowledge of rendering facts for logs and exponentials, can be useful for calculating limits.

In specific circumstances, the squeeze theorem or L'Hospital's rule can be useful as well. They are not reviewed here.

EXERCISES 0.4

1–4. Sketch the removable discontinuity (the hole in the graph) indicated by the limit.

1. $\lim_{x\to3} f(x) = -2$
2. $\lim_{x\to-4} f(x) = 3$
3. $\lim_{x\to5} f(x) = 7$
4. $\lim_{x\to-2} f(x) = 0$

The notation $\lim_{x\to3} f(x) = -2$ and the notation $\lim\limits_{x\to3} f(x) = -2$ mean the same thing.

5–8. Sketch the jump discontinuity indicated by the limits.

5. $\lim_{x\to 1^+} f(x) = 2$, $\lim_{x\to 1^-} f(x) = 4$
6. $\lim_{x\to 0^+} f(x) = -3$, $\lim_{x\to 0^-} f(x) = \sqrt{2}$
7. $\lim_{x\to -1^+} f(x) = -1$, $\lim_{x\to -1^-} f(x) = 1$
8. $\lim_{x\to 3^+} f(x) = 2$, $\lim_{x\to 3^-} f(x) = 2$

9–14. Sketch the vertical asymptote in the graph indicated by the limit(s).

9. $\lim_{x\to 1} f(x) = \infty$
10. $\lim_{x\to -1} f(x) = -\infty$
11. $\lim_{x\to -1^+} f(x) = \infty$, $\lim_{x\to -1^-} f(x) = -\infty$
12. $\lim_{x\to -2^+} f(x) = -\infty$, $\lim_{x\to -2^-} f(x) = \infty$
13. $\lim_{x\to 0^+} f(x) = -\infty$, $\lim_{x\to 0^-} f(x) = -\infty$
14. $\lim_{x\to 4^+} f(x) = 2$, $\lim_{x\to 4^-} f(x) = \infty$

15–22. Rapid response: state the equation of the horizontal asymptote, if any, indicated by the limit.

15. $\lim_{x\to\infty} \dfrac{x^2 + 3x - 9}{4x^2 - 7} = \dfrac{1}{4}$

16. $\lim_{x\to -\infty} \dfrac{x^2 - 8x + 6}{4x^3 - 7x^2 + x} = 0$

17. $\lim_{x\to -\infty} f(x) = 14$

18. $\lim_{x\to\infty} g(x) = -\infty$

19. $\lim_{x\to\infty} (x^2 - 7) = \infty$

20. $\lim_{x\to\infty} g(x) = \pi$

21. $\lim_{x\to 3} f(x) = 4$

22. $\lim_{x\to 2^+} g(x) = \infty$

"Calculate the limit" implies the use of the EAR procedure as demonstrated in the examples in this section.

23–68. Calculate the limit.

23. $\lim_{x\to 1} \dfrac{x^2 + 3x - 4}{2x - 2}$

24. $\lim_{x\to 2} \dfrac{4x - 8}{x^2 - 5x + 6}$

25. $\lim_{x\to 3^-} \dfrac{-4}{3 - x}$

26. $\lim_{x\to 4^-} \dfrac{x + 1}{x - 4}$

27. $\lim_{x\to\infty} \dfrac{5x^3 + 2x^2 + 4}{x^2 - 6x + 7}$

28. $\lim_{x\to\infty} \dfrac{3x^5 - 9x^2 + 1}{8x^5 - 7x^3 + x^2}$

29. $\lim_{x\to -2} \dfrac{x + 2}{\sqrt{x + 6} - 2}$

30. $\lim_{x\to 3} \dfrac{\sqrt{x - 2} - 1}{x - 3}$

31. $\lim_{x\to 2^+} \dfrac{x + 1}{2 - x}$

32. $\lim_{x\to 3} \dfrac{3x - 9}{x^2 - 2x - 3}$

33. $\lim_{x\to 9} \dfrac{\sqrt{x} + 2}{3x - 20}$

34. $\lim_{x\to 1} \dfrac{x^2 + 5}{x + 4}$

35. $\lim_{x\to 0} \dfrac{\sin 4x}{9x}$

36. $\lim_{x\to 0} \dfrac{\sin 3x}{6x - \sin 2x}$

37. $\lim_{x\to -7} \dfrac{3x + 21}{x^2 + 12x + 35}$

38. $\lim_{x\to 3^-} \dfrac{2 - x}{x^2 - 9}$

39. $\displaystyle\lim_{x\to 5^-}\frac{\sqrt{x+4}-3}{x-5}$

40. $\displaystyle\lim_{x\to 4^+}\frac{(x-4)^2}{\sqrt{2x+1}-3}$

41. $\displaystyle\lim_{x\to 0}\frac{\sin 4x+\cos 2x}{6x-1}$

42. $\displaystyle\lim_{x\to 0}\cos\left(\frac{\sin 4x^2}{5x}\right)$

43. $\displaystyle\lim_{x\to 3^-}\frac{4|2x-6|}{3-x}$

44. $\displaystyle\lim_{x\to 2^-}\frac{4|x-2|}{6-3x}$

45. $\displaystyle\lim_{x\to 0}\sin\left(\frac{x+\pi}{x^2-2}\right)$

46. $\displaystyle\lim_{x\to \pi/2}\frac{\sin x}{x}$

47. $\displaystyle\lim_{x\to 0}(4x^2-5x+12)$

48. $\displaystyle\lim_{x\to 1}(4x^2-5x+12)$

49. $\displaystyle\lim_{x\to -1^+} e^{x/(x+1)}$

50. $\displaystyle\lim_{x\to 5} e^{x+\ln(x-4)}$

51. $\displaystyle\lim_{x\to 0} e^{x/(x+1)}$

52. $\displaystyle\lim_{x\to \infty}\ln\left(\frac{x+1}{x+2}+e^x\right)$

53. $\displaystyle\lim_{x\to 2}\frac{\frac{1}{x}-\frac{1}{2}}{x-2}$

54. $\displaystyle\lim_{x\to 0^+}\frac{\frac{1}{x}-\frac{1}{2}}{x-2}$

55. $\displaystyle\lim_{x\to -\infty}\frac{4x^2-3x+12}{7x^2-11x}$

56. $\displaystyle\lim_{x\to -\infty}\frac{2x-114}{x^2+9x}$

57. $\displaystyle\lim_{x\to -\infty}\frac{6x\cos x}{7x^2+1}$

58. $\displaystyle\lim_{x\to \infty}\frac{2x\sin x}{3x+11}$

59. $\displaystyle\lim_{x\to 0} x\sin(2x^2+1)$

60. $\displaystyle\lim_{x\to 0} x^2\cos x$

61. $\displaystyle\lim_{x\to 0}\frac{1-x\sin x}{3+x}$

62. $\displaystyle\lim_{x\to 1^+}\frac{1-x\sin 3}{x-1}$

63. $\displaystyle\lim_{x\to 2^+}\ln(4x-8)$

64. $\displaystyle\lim_{x\to \infty}\ln\sqrt{x}$

65. $\displaystyle\lim_{x\to \infty}\ln(4x-8)$

66. $\displaystyle\lim_{x\to 5^+}\frac{\ln(x-5)}{x}$

67. $\displaystyle\lim_{x\to 4^+}\frac{\sqrt{x-4}}{\sqrt{2x+1}-3}$

68. $\displaystyle\lim_{x\to 2^+}\frac{\sqrt{8x+9}-5}{\sqrt{x-2}}$

69–74. Determine the limit. If the two-sided limit cannot be rendered, check the one-sided limits.

69. $\displaystyle\lim_{x\to 1}\frac{|x+2|}{(x-1)^2}$

70. $\displaystyle\lim_{x\to 3}\frac{-2}{(x-3)^2}$

71. $\displaystyle\lim_{x\to 0}\frac{4}{x}$

72. $\displaystyle\lim_{x\to 3}\frac{-2}{(x-3)^3}$

73. $\displaystyle\lim_{x\to 3}\frac{x^2-9}{|x-3|}$

74. $\displaystyle\lim_{x\to 1}\frac{|x^2-1|}{x}$

75–78. Find all horizontal asymptotes on the graph of the given function.　　It is implied that limits should be used.

75. $f(x)=\dfrac{4x^3-4x^2+1}{x^3-5}$

76. $y=\dfrac{5x^2+6x+1}{x-7}$

77. $y=\dfrac{3x+4}{\sqrt{x^2+8x+205}}$

78. $f(x)=\dfrac{5x-1}{\sqrt{x^2+12x+927}}$

79–86. Evaluate the limit.

79. $\lim\limits_{x\to\infty} \sinh\left(\dfrac{1}{x^2 + x}\right)$

80. $\lim\limits_{x\to\infty} \cos\left(\dfrac{\pi}{2} + \dfrac{1}{x}\right)$

81. $\lim\limits_{x\to\infty} \ln\left(\dfrac{3x^2 - 2x + 1}{3x^2 - 7}\right)$

82. $\lim\limits_{x\to-\infty} \sin\left(\dfrac{\pi x + 4}{x - 7}\right)$

83. $\lim\limits_{x\to\infty} \dfrac{2^x}{x^2}$

84. $\lim\limits_{x\to\infty} \dfrac{\ln x}{\sqrt[3]{x}}$

85. $\lim\limits_{x\to\infty} \dfrac{\ln(\ln x)}{\ln x^{47}}$

86. $\lim\limits_{x\to\infty} \dfrac{3^x + x^3}{e^x + 7}$

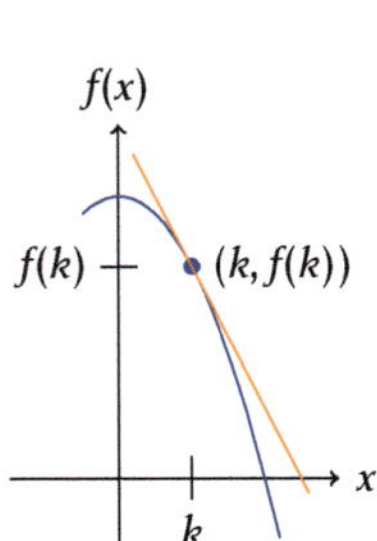

Figure 0.29 *A curve* $y = f(x)$ *(blue) with its tangent line at* $x = k$ *(orange)*

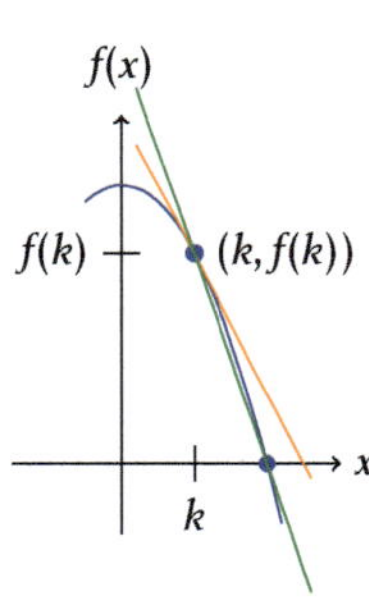

Figure 0.30 *A curve* $y = f(x)$ *with tangent line (orange) at* $x = k$ *and secant line (green) through two points on the curve. The slopes of the tangent line and secant line are not the same*

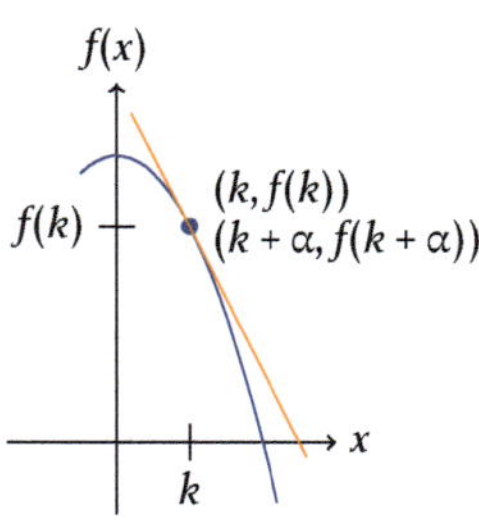

Figure 0.31 *The curve* $y = f(x)$ *with tangent line at* $x = k$*. The curve passes through the points* $(k, f(k))$ *and* $(k + \alpha, f(k + \alpha))$*, which are both located at the blue dot*

0.5 Derivatives

In this section, we review the derivative concept from an infinitesimal point of view, followed by local linearity and a review of derivative rules and formulas.

0.5.1 Slope of a tangent line

One interpretation of the derivative is that it represents the slope of the tangent line. To develop the derivative using infinitesimals, suppose we have a curve $y = f(x)$ and we wish to know the slope of the tangent line at the point $(k, f(k))$ (Figure 0.29). The slope formula,

$$m = \frac{y_2 - y_1}{x_2 - x_1},$$

needs two points on the line, (x_1, y_1) and (x_2, y_2). We only have one point, $(x_1, y_1) = (k, f(k))$.

What should we use for the second point? If we use another point on the curve that is a real-number distance away from our point of tangency, then the line through the two points (called a *secant line*) is not the same as the tangent line (Figure 0.30). And we can't use the same point $(k, f(k))$ for both (x_1, y_1) and (x_2, y_2), for then the slope formula results in division by zero.

We have successfully avoided division by zero in other instances by using infinitesimals, and that works here as well. For the second point we can choose a point that is infinitely close to the first one, namely $(k + \alpha, f(k + \alpha))$ (Figure 0.31). As long as the function is continuous at $x = k$, this second point is indistinguishable from the first, avoiding not only division by zero but also the problem of producing a different line.

Using the slope formula,

$$m = \frac{y_2 - y_1}{x_2 - x_1} = \frac{f(k + \alpha) - f(k)}{k + \alpha - k} = \frac{f(k + \alpha) - f(k)}{\alpha}.$$

As long as the formula renders the same real result for every infinitesimal α, then we have found the slope of the tangent line, which is the value of the derivative at $x = k$.

0.5.2   The definition of derivative

The derivative is defined in a similar manner as limit and continuity.

Definition 12 THE DERIVATIVE *Let f be a function and let k be a real number in the domain of f. If*

$$\frac{f(k + \alpha) - f(k)}{\alpha}$$

is defined and renders the same real result L for every infinitesimal α, then we write $f'(k) = L$ and call $f'(k)$ the derivative *of f at k. When L is a real number (not ∞ or $-\infty$), then we say that f is* differentiable *at k.*

For the purpose of calculations we write

$$f'(k) = \frac{f(k + \alpha) - f(k)}{\alpha},$$

with the understanding that α is an arbitrary infinitesimal that can be either positive or negative and that we must render the same real result for every α.

Example 64 *Let $f(x) = x^2$. Use the definition of derivative to find $f'(5)$.*

The phrase "use the definition of derivative" means that the definition must be used. Derivative rules are not to be used.

Solution   For any infinitesimal α,

$$f'(5) = \frac{f(5 + \alpha) - f(5)}{\alpha} = \frac{(5 + \alpha)^2 - 5^2}{\alpha}$$

$$= \frac{25 + 10\alpha + \alpha^2 - 25}{\alpha} = \frac{10\alpha + \alpha^2}{\alpha}$$

$$\approx \frac{10\alpha}{\alpha} = 10.$$

As per the definition, we write $f'(5) = 10$.   ∎

This concluding sentence is optional.

Reading Exercise 22   Let $f(x) = 5x + 1$. Use the definition of derivative to find $f'(3)$.

Two important theorems can be derived from the definition of derivative. First, note that f differentiable at $x = k$ means $\frac{f(k+\alpha)-f(k)}{\alpha}$ renders the real number $f'(k)$ for every infinitesimal α. As long as $f'(k) \neq 0$, this is the same as

$$\frac{f(k + \alpha) - f(k)}{\alpha} \approx f'(k).$$

If $f'(k) = 0$, then $\frac{f(k+\alpha)-f(k)}{\alpha}$ is infinitesimal. This means $f(k + \alpha) - f(k)$ is on a lower level than α, and $f(k + \alpha) \doteq f(k)$.

Rearranging to solve for $f(k + \alpha)$ gives

$$f(k + \alpha) - f(k) \approx \alpha f'(k)$$

$$f(k + \alpha) \approx \underbrace{f(k)}_{\text{real}} + \underbrace{\alpha f'(k)}_{\text{infinitesimal}} \approx f(k).$$

If $f(k) = 0$, then the quantity is infinitesimal and renders zero, which is the same as $f(k)$.

Then, for every infinitesimal α, the quantity $f(k + \alpha)$ renders the real result $f(k)$, so f is continuous at $x = k$.

Theorem 9 DIFFERENTIABILITY IMPLIES CONTINUITY *If f is differentiable at $x = k$, then f is continuous at $x = k$.*

Another fact from the previous calculation that is very useful is that $f(k + \alpha) \approx f(k) + \alpha f'(k)$. This forms the basis of the local linearity theorem.

Theorem 10 LOCAL LINEARITY THEOREM *Let f be differentiable at $x = k$ and let α be an infinitesimal.*
(1) If not both $f(k) = 0$ and $f'(k) = 0$, then $f(k + \alpha) \approx f(k) + \alpha f'(k)$.
(2) If $f(k) = 0 = f'(k)$, then $f(k + \alpha)$ is 0 or is on a lower level than α.

The local linearity theorem gets its name from the fact that at an infinitesimal scale, the graph of a differentiable function appears to be a straight line, indistinguishable from its tangent line. Because $f(k + \alpha) \approx f(k) + \alpha f'(k)$, the difference between the quantities $f(k + \alpha)$ and $f(k) + \alpha f'(k)$ must be on a lower level than α. This means that at the scale in which α is visible (Figure 0.32), at the input $x = k + \alpha$, the difference between the curve's y-coordinate $f(k + \alpha)$ and the tangent line's y-coordinate $f(k) + \alpha f'(k)$ is not visible. This is true for every point along the curve and tangent line at this scale. The curve and its tangent line are indistinguishable.

The slope of the line between the points $(k, f(k))$ and $(k + \alpha, f(k) + \alpha f'(k))$ is $\frac{\text{rise}}{\text{run}} = \frac{\alpha f'(k)}{\alpha} = f'(k)$, which is the slope of the tangent line at $x = k$.

Local linearity shows up even when the tangent line is horizontal. The key is keeping the scale on the two axes at the same infinitesimal level. When using a CAS (computer algebra system) to explore local linearity at a real-number scale, care must be taken to keep successive pictures at the same aspect ratio.

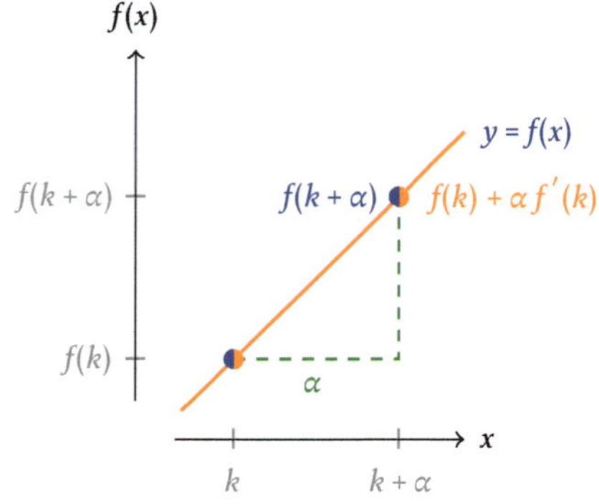

Figure 0.32 *The curve $y = f(x)$ (blue) and its tangent line at $x = k$ (orange), which are indistinguishable at an infinitesimal scale*

Local linearity is often used geometrically in infinitesimal calculus, straightening curved objects to allow for computation on infinitesimally sized pieces, which are then summed to calculate the whole and produce a convenient formula. Local linearity can also be used algebraically, to prove theorems such as derivative rules.

0.5.3 Derivative rules

The definition of derivative is not often used when calculating the derivative. Instead, we use derivative rules, which are reviewed next. Most proofs are omitted.

DERIVATIVE OF A CONSTANT FUNCTION

If $f(x) = c$, then $f'(x) = 0$.

POWER RULE

Let $f(x) = x^n$ for a real number n. Then $f'(x) = nx^{n-1}$.

For a positive integer n, the proof of the power rule can be written as a calculation involving the definition of derivative, algebra, and approximation. Let $f(x) = x^n$. Then

$$f'(x) = \frac{f(x + \alpha) - f(x)}{\alpha} = \frac{(x + \alpha)^n - x^n}{\alpha}$$

$$= \frac{x^n + nx^{n-1}\alpha + \text{terms on level } \alpha^2 \text{ or below} - x^n}{\alpha}$$

$$= \frac{nx^{n-1}\alpha + \text{terms on level } \alpha^2 \text{ or below}}{\alpha}$$

$$\approx \frac{nx^{n-1}\alpha}{\alpha} = nx^{n-1}.$$

CONSTANT MULTIPLE RULE

If f is differentiable at x and $g(x) = c \cdot f(x)$, then $g'(x) = c \cdot f'(x)$.

SUM RULE

If f and g are differentiable at x, then

$$(f + g)'(x) = f'(x) + g'(x).$$

Ans. to reading exercise 22:

$$f'(3) = \frac{f(3 + \alpha) - f(3)}{\alpha}$$

$$= \frac{5(3 + \alpha) + 1 - 16}{\alpha}$$

$$= \frac{5\alpha}{\alpha} = 5$$

Derivative formulas can be compactly written using Leibniz notation. They may be more easily remembered in this form. For convenience, Leibniz-notation versions of some formulas appear in the margin.

$$\frac{d}{dx}c = 0$$

$$\frac{d}{dx}x^n = nx^{n-1}$$

From the binomial theorem, $(a+b)^n = a^n + na^{n-1}b + \frac{n(n-1)}{2}a^{n-2}b^2 + \cdots + nab^{n-1} + b^n$. The powers of b in the formula increase by 1 each term. Here, $a = x$ and $b = \alpha$. Therefore the terms have increasing powers of α.

The proof for negative integer exponents uses the quotient rule. The proof for rational exponents uses implicit differentiation. The proof for irrational exponents requires general exponential functions.

$$\frac{d}{dx}c \cdot f(x) = c \cdot f'(x)$$

$$\frac{d}{dx}(f(x) + g(x)) = f'(x) + g'(x).$$

$\frac{d}{dx}(f(x) - g(x)) = f'(x) - g'(x).$

DIFFERENCE RULE

If f and g are differentiable at x, then

$$(f - g)'(x) = f'(x) - g'(x).$$

The derivative of any polynomial can be found using the preceding derivative rules. Consciously recalling which rules are applied at each step of differentiating a polynomial is not necessary, but general knowledge that these rules are being used is important for avoiding errors.

Reading Exercise 23   Find $f'(x)$ for $f(x) = x^2 + 2x - 7$.

PRODUCT RULE

If f and g are differentiable at x, then

$$\frac{d}{dx}\left(f(x) \cdot g(x)\right) = f(x)g'(x) + g(x)f'(x).$$

The proof of the product rule demonstrates the algebraic use of local linearity, in this case for replacing $f(x + \alpha)$ with $f(x) + \alpha f'(x)$ and replacing $g(x + \alpha)$ with $g(x) + \alpha g'(x)$. For any infinitesimal α,

Line 1 uses the definition of derivative; line 2 uses local linearity for both f and g, which is an approximation (notice the $\approx$); line 3 multiplies and line 4 simplifies; line 5 approximates, throwing away the α^2-level term in the numerator; and line 6 simplifies.

$$\frac{d}{dx}(f(x) \cdot g(x)) = \frac{f(x + \alpha)g(x + \alpha) - f(x)g(x)}{\alpha}$$

$$\approx \frac{(f(x) + \alpha f'(x))(g(x) + \alpha g'(x)) - f(x)g(x)}{\alpha}$$

$$= \frac{f(x)g(x) + \alpha f(x)g'(x) + \alpha f'(x)g(x) + \alpha^2 f'(x)g'(x) - f(x)g(x)}{\alpha}$$

$$= \frac{\alpha f(x)g'(x) + \alpha f'(x)g(x) + \alpha^2 f'(x)g'(x)}{\alpha}$$

$$\approx \frac{\alpha f(x)g'(x) + \alpha f'(x)g(x)}{\alpha}$$

$$= f(x)g'(x) + f'(x)g(x).$$

Handling the case when the approximation principle is violated requires attention to detail, but it is not difficult and does not add much insight.

The product rule is often remembered as

$$\text{first} \cdot \text{second}' + \text{second} \cdot \text{first}'.$$

Because multiplication and addition are both commutative, other orders for remembering the product rule are in common usage as well, such as $\text{first}' \cdot \text{second} + \text{first} \cdot \text{second}'$.

Example 65 *Differentiate $y = x^2 \sin x$.*

Solution Using the product rule,

$$y' = (x^2)(\cos x) + (\sin x)(2x)$$

$$\text{first} \cdot \text{second}' + \text{second} \cdot \text{first}'$$

$$= x^2 \cos x + 2x \sin x.$$

Trig derivative formulas are listed later in this section.

For clarity, it is traditional to write polynomials preceding trig functions when they are multiplied, to avoid the need for parentheses. Although $2x \sin x$ is clear, $\sin x 2x$ appears to incorrectly place the polynomial inside the sine function.

■

QUOTIENT RULE

If f and g are differentiable at x and $g(x) \neq 0$, then

$$\frac{d}{dx}\left(\frac{f(x)}{g(x)}\right) = \frac{g(x)f'(x) - f(x)g'(x)}{(g(x))^2}.$$

The quotient rule can also be proven in calculation format using local linearity. The quotient rule is often remembered as

$$\left(\frac{\text{top}}{\text{bot}}\right)' = \frac{\text{bot} \cdot \text{top}' - \text{top} \cdot \text{bot}'}{(\text{bot})^2}.$$

Another common expression of the quotient rule is "lo d hi minus hi d lo over lo lo."

Because subtraction is not commutative, the terms in the numerator cannot be swapped.

Example 66 *Find $f'(x)$ for $f(x) = \dfrac{4x^2 - 7x + 8}{x^5 + x^3}$.*

Solution The function is a quotient, so the quotient rule is required:

$$f'(x) = \frac{\overset{\text{bot}}{(x^5+x^3)} \cdot \overset{\text{top}'}{(8x-7)} - \overset{\text{top}}{(4x^2-7x+8)} \cdot \overset{\text{bot}'}{(5x^4+3x^2)}}{\underset{\text{bot}^2}{(x^5+x^3)^2}}.$$

Be careful: $f'(x) \neq \frac{8x-7}{5x^4+3x^2}$. Do not calculate the derivative of a quotient by taking the quotient of the derivatives.

■

If a second derivative is required, the form of the solution in example 66 is not convenient. Simplify first before finding another derivative.

Because we can use the quotient rule to find the derivatives of all rational functions, and the derivative is undefined precisely when the rational function is undefined, all rational functions are differentiable throughout their domains. Therefore, graphs of rational functions are smooth, without corners.

Ans. to reading exercise 23:
$$f'(x) = 2x + 2$$

0.5.4 Trig derivatives

In addition to derivative rules, there are derivative formulas.

<hr>

TRIG DERIVATIVES

$$\frac{d}{dx}\sin x = \cos x \qquad \frac{d}{dx}\csc x = -\csc x \cot x$$

$$\frac{d}{dx}\cos x = -\sin x \qquad \frac{d}{dx}\sec x = \sec x \tan x$$

$$\frac{d}{dx}\tan x = \sec^2 x \qquad \frac{d}{dx}\cot x = -\csc^2 x$$

<hr>

Tips for remembering the formulas: (1) the derivative formulas come in cofunction pairs; (2) the derivatives of functions beginning with "co" (cosine, cotangent, cosecant) all have a negative sign in the formula, whereas the others do not.

Derivative rules such as the product and quotient rules apply (where appropriate) to all functions, including trig functions. Sometimes more than one derivative rule must be used.

Example 67 *Find y' given $y = \dfrac{x^2 \tan x}{3x - 7}$.*

Solution The function is a quotient, but there is also a product in the numerator. Which rule comes first? The key to deciding is to determine which is the "outside" or "overall" structure; in this case, the function is a quotient with a product inside. We start by using the quotient rule.

We can rewrite the function as a product with a quotient inside:

$$y = \frac{x^2}{3x - 7} \cdot \tan x.$$

To find the derivative from this form, start with the product rule.

When finding top$'$, we are finding $\frac{d}{dx} x^2 \tan x$, which is the derivative of a product. Therefore, the product rule is needed.

$$y' = \frac{\overset{\text{bot}}{(3x-7)}\overset{\overbrace{\text{top}'}^{\text{first second}' + \text{second first}'}}{(x^2 \sec^2 x + (\tan x)(2x))} - \overset{\text{top}}{(x^2 \tan x)}\overset{\text{bot}'}{(3)}}{\underset{\text{bot}^2}{(3x-7)^2}}$$

$$= \frac{(3x - 7)(x^2 \sec^2 x + 2x \tan x) - 3x^2 \tan x}{(3x - 7)^2}.$$

∎

The next example requests calculation of a second derivative.

Example 68 *Find $\dfrac{d^2 y}{dx^2}$ for $y = \sec x$.*

Solution The first derivative is one of the trig derivative formulas:

$$y' = \sec x \tan x.$$

We need a second derivative. Because y' is a product, we use the product rule:

Recall that $\sec^2 x$ means the same as $(\sec x)^2$.

$$y'' = \overset{\text{first}}{\sec x}\,\overset{\text{second}'}{\sec^2 x} + \overset{\text{second}}{(\tan x)}\overset{\text{first}'}{(\sec x \tan x)} = \sec^3 x + \sec x \tan^2 x.$$

∎

Reading Exercise 24 Find $\dfrac{d}{dx}\dfrac{\sin x}{x + \cos x}$.

Next, we practice finding the equation of a tangent line.

Example 69 *Find the equation of the tangent line to the curve $y = \sin x$ at $x = 0$.*

Solution Recall that there are three steps in finding the equation of a tangent line.

❶ Calculate the slope of the tangent line. To find the slope we need the derivative:

$$y' = \cos x.$$

Then, we evaluate the derivative at the requested value of x:

$$\text{slope} = y'(0) = \cos 0 = 1.$$

❷ Determine the point of tangency. We have not been given the y-coordinate of the point of tangency, so we calculate it next:

$$y(0) = \sin 0 = 0.$$

The point of tangency is $(0, 0)$.

❸ Use the results of ❶ and ❷ to determine the equation of the line. We finish by using the point-slope equation of the line with the point $(0, 0)$ and slope 1:

$$y - 0 = 1(x - 0)$$

$$y = x.$$

The equation of the tangent line is $y = x$. 

The graph of $y = \sin x$ with its tangent line at $x = 0$ is in Figure 0.33. Because the derivative of $y = \sin x$ is $y' = \cos x$, and the values of $\cos x$ are always between 1 and -1, the slopes on the sine curve are always between 1 and -1. Do not make the mistake of drawing the sine curve with slopes that are too steep. See Figure 0.34.

0.5.5   Chain rule

Derivative formulas such as $\dfrac{d}{dx}\sin x = \cos x$ are usually stated for "just plain x." If we wish to calculate $\dfrac{d}{dx}\sin 6x$ instead, we need the chain rule.

CHAIN RULE

Suppose $y = f(u)$ and $u = g(x)$, and also suppose that g is differentiable at x and f is differentiable at $g(x)$. Then, $f \circ g(x)$ is differentiable at x and

$$\frac{d}{dx}f(g(x)) = f'(g(x))g'(x).$$

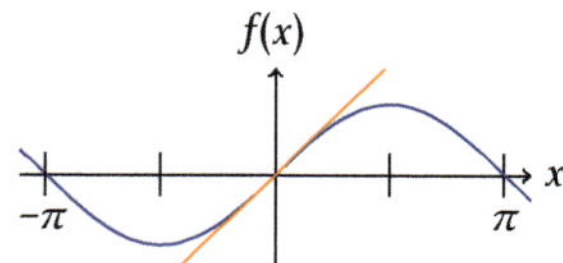

Figure 0.33 *The curve $y = \sin x$ (blue) with tangent line (orange) at $x = 0$*

Be careful–do not make the mistake of drawing $y = \sin x$ with slopes that are much too steep. If you are prone to this mistake, practice drawing the sine curve with appropriate slopes several times.

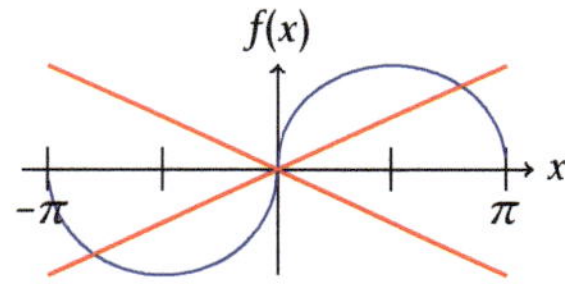

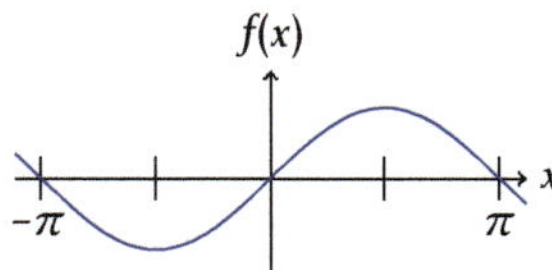

Figure 0.34 *(Top) incorrectly drawn graph of $y = \sin x$; (bottom) correctly drawn graph of $y = \sin x$*

In pure Leibniz notation, the chain rule formula is

$$\frac{dy}{dx} = \frac{dy}{du}\frac{du}{dx}.$$

The chain rule can be proved in the same manner as the product rule, with two applications of local linearity (first to $g(x + \alpha)$, then to $f(g(x) + \alpha g'(x))$, using $g(x)$ as the real number and $\alpha g'(x)$ as the infinitesimal).

When using a derivative formula designed for "just plain x," the chain rule says to multiply by the derivative of what's inside.

Example 70 *Find* $\frac{d}{dx} \sin(3x - 5)$.

Solution Using the chain rule with $f(x) = \sin x$ and $g(x) = 3x - 5$,

Line 1 rewrites $\sin(3x - 5)$ in terms of a composition of f and g for using the chain rule; line 2 is the chain rule formula; line 3 uses $f'(x) = \cos x$ and $g'(x) = 3$; line 4 rewrites for the sake of clarity.

$$\frac{d}{dx} \sin(3x - 5) = \frac{d}{dx} f(g(x))$$
$$= f'(g(x))g'(x)$$
$$= \cos(3x - 5) \cdot 3$$
$$= 3 \cos(3x - 5).$$

∎

It is customary to skip rewriting in terms of the functions f and g, simplifying the calculation of example 70 to

$$\frac{d}{dx} \sin(3x - 5) = \cos(3x - 5) \cdot 3 = 3 \cos(3x - 5).$$

Just remember to multiply by the derivative of what's inside.

Discriminating between a product (using the product rule) and a composition (using the chain rule) is sometimes tricky. The next examples illustrate how to tell the difference.

Ans. to reading exercise 24:

$$\frac{(x + \cos x) \cos x - (\sin x)(1 - \sin x)}{(x + \cos x)^2}$$
$$= \frac{x \cos x - \sin x + 1}{(x + \cos x)^2}$$

Example 71 *Find* $\frac{d}{dx} \sin(\cos x)$.

Solution Although $\cos x$ is a trig function of "just plain x," the sine function is not; it is $\sin(\cos x)$, not $\sin x$. We must use the chain rule for the derivative of sine:

Compare to example 70; $\cos x$ is in place of $3x - 5$.

$$\frac{d}{dx} \sin(\cos x) = \cos(\cos x) \cdot \frac{d}{dx} \cos x$$
$$= \cos(\cos x) \cdot (-\sin x).$$

Note that $\cos(\cos x)$ is not the same as $\cos x \cos x$. That is, $\cos(\cos x) \neq \cos^2 x$.

Writing the final answer as $(-\sin x)(\cos(\cos x))$ is perhaps more clear. ∎

Example 72 *Find $\frac{d}{dx}\sin x \cos x$.*

Solution This is not a composition of functions. Instead, it is a product, $\sin x \cdot \cos x$, and requires the product rule:

$$\frac{d}{dx}\sin x \cos x = \sin x \cdot (-\sin x) + \cos x \cdot \cos x = -\sin^2 x + \cos^2 x.$$

■

How can we tell the difference between the need for the chain rule in example 71 and the need for the product rule in example 72? The key is in recognizing a complete expression. The expression $\sin x$ is a complete trigonometric expression; sin is a name of a function and is not a complete expression. A complete trigonometric expression requires something inside the trig function, as in $\tan(x^2)$ or $\sec(3 + \tan x - x^4)$. The expression $\sin x \cos x$ has two complete trigonometric expressions, $\sin x$ and $\cos x$, which are multiplied. The expression $\sin \cos x$ cannot be the product of two expressions because sin by itself is not a complete expression; it needs something inside of it, so $\cos x$ (which is a complete expression) must be the "inside," or the argument, of sin. A summary is provided in Table 0.1.

Those who have cultivated the bad habit of writing sin when they mean $\sin x$ will be hard-pressed to tell the difference between a product and a composition. Using proper notation is important!

Table 0.1 *Discriminating between product and composition*

algebraic form:	product	composition
rule required:	product rule	chain rule
example:	$\underbrace{\sin x}_{\text{complete}} \underbrace{\cos x}_{\text{complete}}$	$\underbrace{\underbrace{\sin}_{\text{incomplete}} \cos x}_{\text{complete}}$
visual recognition:	x in more than one spot separated by (implied) multiplication	x in only one spot, no multiplication of complete expressions

0.5.6   Chain rule with product rule

Just as the product and quotient rules are sometimes needed in the same exercise, the chain rule can be required along with other rules as well.

Example 73 *Find y' for $y = 5x^2(11x - 8)^4$.*

Solution The expression is a product, $5x^2 \cdot (11x - 8)^4$. Therefore, the product rule is needed. Setting "first" as $5x^2$ and "second" as $(11x - 8)^4$, we have

$$y' = 5x^2(4(11x - 8)^3 \cdot 11) + (11x - 8)^4 10x.$$

$$\text{first} \cdot \text{second}' + \text{second} \cdot \text{first}'$$

$$= 220x^2(11x - 8)^3 + 10x(11x - 8)^4.$$

Because juxtaposition means multiplication, products are often harder to recognize visually than quotients, which present themselves as fractions. Looking intentionally for two expressions written next to one another can help.

In line 1, for the derivative of the second factor, $(11x - 8)^4$, the chain rule is required because it is of the form "not just plain x" to a power.

Further simplification of the answer is possible but is not necessarily advantageous. ∎

Reading Exercise 25 Find y' for $y = 3x^2(4x + 1)^4$.

0.5.7 Multiple-link chains

The word *chain* in "chain rule" evokes a picture of a chain with many links. The chain rule may have to be applied repeatedly when differentiating one expression.

Example 74 *Find y' for $y = \sin^2(5x^2)$.*

Tip: when differentiating a power of a trig function, the form $\tan^3 \theta$ is likely to conceal the correct order for use of the chain rule. The form $(\tan \theta)^3$ is helpful to indicate the correct order for use of the chain rule. Always rewrite in the helpful form.

Solution Recall that the expression $\sin^2 \theta$ is the same as $(\sin \theta)^2$. Rewriting, we have

$$y = (\sin(5x^2))^2.$$

Looking at the outermost part of the expression first, we see that it is of the form "not just plain x" to a power. The chain rule is needed:

$$y' = 2(\sin(5x^2))^1 \cdot \frac{d}{dx} \sin(5x^2).$$

When finding $\frac{d}{dx} \sin(5x^2)$, the chain rule is needed again; this is a trig function of "not just plain x":

If you have difficulty knowing where to stop in a multiple-link chain rule situation, the format demonstrated in example 74 can be helpful. Write one application of the chain rule at a time, keeping track of what still needs to be differentiated.

$$= 2 \sin(5x^2) \cdot \cos(5x^2) \cdot \frac{d}{dx} 5x^2$$

$$= 2 \sin(5x^2) \cdot \cos(5x^2) \cdot 10x$$

$$= 20x \sin(5x^2) \cos(5x^2).$$

∎

0.5.8 Derivatives of logarithmic and exponential functions

Notationally, "ln" and "log" are like "sin" or other trig function names. Recognizing the need for the product rule, the chain rule, and so on for logarithmic functions is the same as for trig functions.

DERIVATIVES OF LOGARITHMIC AND EXPONENTIAL FUNCTIONS

If $a > 0$,

$$\frac{d}{dx} \ln x = \frac{1}{x} \qquad \frac{d}{dx} \log_a x = \frac{1}{x \ln a}$$

$$\frac{d}{dx} e^x = e^x \qquad \frac{d}{dx} a^x = a^x \ln a$$

Example 75 *Find $f'(x)$ for $f(x) = 2\ln(4x - 7)$.*

Solution Because we have the natural log of "not just plain x," the chain rule is needed:

$$f'(x) = 2 \cdot \frac{d}{dx}\ln(4x - 7) = 2 \cdot \frac{1}{4x - 7} \cdot \frac{d}{dx}(4x - 7)$$

$$= 2 \cdot \frac{1}{4x - 7} \cdot 4 = \frac{8}{4x - 7}.$$

The constant multiple rule is also used here.

∎

Expanding an expression using the laws of logarithms can lead to simpler differentiation.

Example 76 *Differentiate $y = \ln\left(\dfrac{(x + 4)^5(x - 2)}{x^2}\right)$.*

Solution Rather than use the chain rule inside the product rule inside the quotient rule inside the chain rule, expand the expression using laws of logarithms first:

$$y = \ln\left(\frac{(x + 4)^5(x - 2)}{x^2}\right)$$

$$= \ln\left((x + 4)^5(x - 2)\right) - \ln x^2$$

$$= \ln(x + 4)^5 + \ln(x - 2) - \ln x^2$$

$$= 5\ln(x + 4) + \ln(x - 2) - 2\ln x.$$

Ans. to reading exercise 25:
$48x^2(4x + 1)^3 + 6x(4x + 1)^4$

Line 2 uses $\ln\frac{a}{x} = \ln a - \ln x$; line 3 uses $\ln(ax) = \ln a + \ln x$; and line 4 uses $\ln x^n = n\ln x$. Because we do not have a law for log of a sum or a difference, the expansion is finished after line 4.

Now the derivative is much simpler to compute:

$$y' = 5 \cdot \frac{1}{x + 4} \cdot 1 + \frac{1}{x - 2} \cdot 1 - 2 \cdot \frac{1}{x} = \frac{5}{x + 4} + \frac{1}{x - 2} - \frac{2}{x}.$$

Although the chain rule is still needed, the product and quotient rules are avoided.

∎

Try computing the derivative in example 76 without using the laws of logarithms first. Not only is the derivative much more complicated to compute, the resulting expression needs much simplification to arrive at the nice expression in the solution.

For an exponential function, if the exponent is not "just plain x," the chain rule is needed.

Example 77 *Find y' for $y = 2^{4x^2 + 3x - 8}$.*

Solution Using the chain rule and the formula $\frac{d}{dx} 2^x = 2^x \ln 2$,

$$\frac{d}{dx} 2^{4x^2+3x-8} = 2^{4x^2+3x-8} \ln 2 \cdot \frac{d}{dx}(4x^2 + 3x - 8)$$

$$= 2^{4x^2+3x-8} \ln 2 \cdot (8x + 3)$$

$$= 2^{4x^2+3x-8}(8x + 3) \ln 2.$$

Because the notation for logarithms is the same type as the notation for trig functions, the same admonition about clarity applies. Writing the logarithm at the end of the product is helpful.

■

Reading Exercise 26 Find $f'(x)$ for $f(x) = e^{5x}$.

0.5.9   Variable in base and exponent

The function $y = x^2$ is a polynomial; the variable x is in the base and the exponent is a constant. The power rule applies, and $y' = 2x$.

The function $y = 2^x$ is an exponential function; the variable is in the exponent and the base is a constant. The general exponential rule applies, and, $y' = 2^x \ln 2$.

What if the variable is in both the exponent and the base, such as $y = x^x$? This function is not a polynomial because the exponent is not a constant, and it is not an exponential function because the base is not a constant. Neither the power rule nor the general exponential rule applies. A different strategy is needed.

More than one strategy is effective. Perhaps the easiest is to rewrite the expression using the definition of exponential function, $a^x = e^{x \ln a}$.

The formula $a^x = e^{x \ln a}$ says that we may rewrite any exponential expression as e to the power "exponent natural log base."

Example 78 *Differentiate* $y = x^{3x}$.

Solution We first rewrite the function as $y = e^{3x \ln x}$. This is an exponential function, and we follow the usual procedure:

Line 1 applies the chain rule; line 2 calculates the derivative of the exponent using the product rule; and line 3 both simplifies and rewrites using the original notation.

$$\frac{d}{dx} e^{3x \ln x} = e^{3x \ln x} \cdot \frac{d}{dx}(3x \ln x)$$

$$= e^{3x \ln x}\left(3x \cdot \frac{1}{x} + 3 \ln x\right)$$

$$= x^{3x}(3 + 3 \ln x).$$

■

0.5.10   Other derivative formulas

Derivative formulas for other elementary functions are used in the same manner as for trig, logarithmic, and exponential functions. All the derivative rules apply, including the product, quotient, and chain rules.

INVERSE TRIG DERIVATIVES

$$\frac{d}{dx}\sin^{-1}x = \frac{1}{\sqrt{1-x^2}}$$

$$\frac{d}{dx}\cos^{-1}x = -\frac{1}{\sqrt{1-x^2}}$$

$$\frac{d}{dx}\tan^{-1}x = \frac{1}{x^2+1}$$

DERIVATIVES OF HYPERBOLIC FUNCTIONS

$$\frac{d}{dx}\cosh x = \sinh x \qquad\qquad \frac{d}{dx}\operatorname{sech} x = -\operatorname{sech} x \tanh x$$

$$\frac{d}{dx}\sinh x = \cosh x \qquad\qquad \frac{d}{dx}\operatorname{csch} x = -\operatorname{csch} x \coth x$$

$$\frac{d}{dx}\tanh x = \operatorname{sech}^2 x \qquad\qquad \frac{d}{dx}\coth x = -\operatorname{csch}^2 x$$

Derivatives of the hyperbolic functions have the same pattern as the derivatives of trig functions, with the exception of which formulas contain a negative.

The graphs of $y = \cosh x$ and $y = \sinh x$, as well as facts such as $\cosh 0 = 1$ and $\sinh 0 = 0$, should also be remembered.

Ans. to reading exercise 26:
$$f'(x) = 5e^{5x}$$

Reading Exercise 27 Find the derivative of $y = \cosh(4x)$.

The derivatives of inverse hyperbolic functions are rarely, if ever, used in this book, but are presented here for the sake of completeness.

Students should check with their instructors for expectations regarding knowledge of various derivative formulas.

DERIVATIVES OF INVERSE HYPERBOLIC FUNCTIONS

$$\frac{d}{dx}\sinh^{-1}x = \frac{1}{\sqrt{1+x^2}} \qquad\qquad \frac{d}{dx}\operatorname{csch}^{-1}x = \frac{-1}{|x|\sqrt{1+x^2}}$$

$$\frac{d}{dx}\cosh^{-1}x = \frac{1}{\sqrt{x^2-1}} \qquad\qquad \frac{d}{dx}\operatorname{sech}^{-1}x = \frac{-1}{x\sqrt{1-x^2}}$$

$$\frac{d}{dx}\tanh^{-1}x = \frac{1}{1-x^2},\quad |x|<1 \qquad\qquad \frac{d}{dx}\coth^{-1}x = \frac{1}{1-x^2},\quad |x|>1$$

EXERCISES 0.5

1–6. Use the definition of derivative to find $f'(x)$.

1. $f(x) = x^2$
2. $f(x) = 4x$
3. $f(x) = 3x$
4. $f(x) = \sqrt{x}$
5. $f(x) = x^2 + 3x$
6. $f(x) = 4x + \sqrt{x}$

Because there may be more than one way to work an exercise, your answer might not look like the answer in the back of the book, even if your answer is correct. If you suspect your answer is correct, try using a calculator or CAS to graph or simplify the quantity **your answer minus the book's answer** to see if the result is zero.

7–62. Differentiate the function.

7. $g(x) = \dfrac{x^2 - 7}{4x + 9}$

8. $s(t) = \dfrac{t^2 - 4}{t + 17}$

9. $f(t) = t^5 - 7t^3 + 4t$

10. $y = 15x^3 - 10x^2 + 14x$

11. $y = \dfrac{3x - 9}{x^3 + 5x}$

12. $f(x) = \dfrac{\sqrt{x}}{x^2}$

13. $y = \dfrac{5x^2 - 7}{x}$

14. $g(x) = \dfrac{x^3 + 4x^2 - 7x + 12}{x^3}$

15. $f(x) = \sqrt{x} \cdot x^2$

16. $y = \dfrac{\sqrt{t}}{t + 1}$

17. $f(x) = \dfrac{\frac{1}{x} + 4}{5x^2 + 3}$

18. $y = x\left(\dfrac{1}{x + 2} + \dfrac{2}{x + 5}\right)$

19. $y = \dfrac{\frac{2}{3}t^3 - \frac{1}{4}t}{22t - 1}$

20. $f(x) = \dfrac{x^4 + x^2}{3 - \frac{1}{x^2}}$

21. $y = x^2 \sec x$

22. $f(x) = x^4 \cos x$

23. $f(x) = \dfrac{4x^3 - 7}{\sec x}$

24. $f(t) = \dfrac{\cos t}{3t^2}$

25. $g(y) = \dfrac{6y}{\csc y}$

26. $y = (x^2 + 2x + 7)\sec x$

27. $z = \dfrac{x \sec x}{x^3 + 9x}$

28. $f(x) = \dfrac{4}{x^3 \sin x}$

29. $g(t) = \dfrac{(t^3 - 4t^2)\cos t}{t \sin t}$

30. $s = \dfrac{t^2 \tan t}{t^2 - 7}$

31. $y = \cos(4x - 3)$

32. $f(x) = \sqrt{5x - 7}$

33. $f(x) = (7x + 5)^4$

34. $y = \sin(x^2 + 1)$

35. $g(x) = \sqrt{\tan 5x}$

36. $g(y) = \dfrac{(y^3 - 7)^4}{2y + 9}$

37. $y = \dfrac{(2x + 1)^3}{x^3 - 5}$

38. $y = t^2 \sin(2t + 1)$

39. $f(t) = t^3 \cos(6t + 1)$

40. $g(x) = \tan((x^2 + \pi)^5)$

41. $g(x) = \sec((x^2 + 5)^3)$

42. $f(x) = (3 + \sin(x^2))^4$

43. $y = (\tan(x^2 + 1))^4$

44. $f(t) = \sin^3(1 - 2t)$

45. $f(x) = (5 + \cos(x^2))^7$

46. $k(x) = \dfrac{5x - 1}{(3x^2 + 4)^4}$

47. $y = \cos \tan \theta$

48. $f(x) = \sec \cot x$

49. $x = t^2 \sec(5t - 8)^4$

50. $f(x) = \sin^2((4x - 1)^{\frac{1}{2}})$

51. $y = \dfrac{x}{\ln x}$

52. $y = \ln x \cos x$

53. $f(x) = \ln\left(\dfrac{(x-1)^2}{3x+7}\right)$

54. $f(x) = \ln\left(\dfrac{x+3}{4x-1}\right)^3$

55. $y = x^2 \ln(5x^2)$

56. $g(z) = z^4 - \ln(6z^2)$

57. $f(x) = x^2 + e^4$

58. $f(t) = e^t \sin t$

59. $y = \dfrac{e^x}{4x + 3}$

60. $g(y) = e^{\sin 4y}$

61. $g(x) = 5e^{4x+3}$

62. $y = \ln(3 + e^{7x})$

63-68. Find the indicated derivative.

63. $y = -2x^2 - 5x + 1.3,\ y''$

64. $y = x^6 - 5x^4,\ y'''$

65. $y = \dfrac{x^3}{4},\ \dfrac{d^2y}{dx^2}$

66. $y = -x^{-4},\ \dfrac{d^2y}{dx^2}$

67. $f(t) = t^7,\ f^{(4)}(t)$

68. $g(x) = x^6 - 7x^3,\ g^{(10)}(x)$

69-76. Find the required derivative.

69. $y = \csc x,\ y''$

70. $y = x^2 \sin x,\ \dfrac{d^2y}{dx^2}$

71. $f(x) = \sin x,\ f^{(4)}(x)$

72. $g(t) = \cos t,\ g^{(4)}(t)$

73. $f(x) = \sin x,\ f^{(4000)}(x)$

74. $g(t) = \cos t,\ g^{(401)}(t)$

75. $\dfrac{d^2}{dx^2} \sin 5x$

76. $y = \sin x^2,\ y''$

77-84. Find the equation of the tangent line to the curve at the given value of the variable.

77. $y = \dfrac{x+1}{x-4}$ at $x = 0$

78. $f(t) = \dfrac{1}{t^3}$ at $t = 2$

79. $g(z) = 4 - z^2$ at $z = -5$

80. $y = (x^{12} - 4)(5 - 3x)$ at $x = 0$

81. $y = \cos x$ at $x = \dfrac{\pi}{2}$

82. $y = \sin x$ at $x = -\dfrac{\pi}{2}$

83. $f(x) = \ln(x^2)$ at $x = 1$

84. $y = e^{2x}$ at $x = \ln 5$

Ans. to reading exercise 27:
$$y' = 4\sinh(4x)$$

85–92. Differentiate.

85. $y = 3^{4x+1}$

86. $f(x) = 6^{2-7x}$

87. $g(x) = x^{7x+1}$

88. $y = x^{\cos x}$

89. $y = (\sin x)^{2x}$

90. $f(x) = (4x)^x$

91. $y = t^2 \cosh t$

92. $h(x) = \dfrac{\sinh t^2}{6t2^t}$

0.6 Definite Integrals: Summing Infinitesimal Amounts

This review is more important for conceptual understanding than for working specific exercises. Most of this review consists of portions of Sections 4.3, 4.4, and 9.1 of *Calculus Set Free*.

The idea of integration is to take small bits of something—area, length, volume, work, force, and so on—and sum them to obtain a formula for calculating the desired quantity, applicable even when the object has an unusual shape or the physical quantity varies. Before reviewing the process of producing such formulas, we review infinitesimal methods that allow us to calculate these sums directly, for the purpose of cultivating helpful intuition.

0.6.1 Summation notation

Can we write $1 + 2 + 3 + \cdots + 100$ without the ellipsis (the $\cdots$)? Yes, using *summation notation*:

Σ is the capital Greek letter "sigma."

$$1 + 2 + 3 + \cdots + 100 = \sum_{k=1}^{100} k.$$

With the notation $1 + 2 + 3 + \cdots + 100$, we are required to infer the pattern from the numbers given. With the notation $\sum_{k=1}^{100} k$, there is nothing to infer; the notation indicates precisely which terms are to be added.

The index variable k is incremented by one at a time, beginning at the starting value of 1 and continuing until the ending value 100. The expression following Σ indicates the terms to be added.

Example 79 *Rewrite without summation notation:* $\displaystyle\sum_{k=5}^{8} \sqrt{k}.$

We are only asked to rewrite the sum, not to calculate its value.

Solution The terms of the sum are the values of the expression $\sqrt{k}$. The values of k start with $k = 5$, end with $k = 8$, and use each integer in between; that is, we use $k = 5, 6, 7,$ and 8:

$$\sum_{k=5}^{8} \sqrt{k} = \sqrt{5} + \sqrt{6} + \sqrt{7} + \sqrt{8}.$$

∎

Example 80 *Evaluate* $\displaystyle\sum_{k=3}^{7} k^2$.

Solution The terms to add are the values of k^2, beginning with $k = 3$ and ending with $k = 7$:

To evaluate the sum means to add all the terms and state the resulting value.

$$\sum_{k=3}^{7} k^2 = 3^2 + 4^2 + 5^2 + 6^2 + 7^2 = 135.$$

∎

Example 81 *Evaluate* $\displaystyle\sum_{k=3}^{7} 2k^2$.

Solution The terms to add are the values of $2k^2$, beginning with $k = 3$ and ending with $k = 7$:

$$\sum_{k=3}^{7} 2k^2 = 2 \cdot 3^2 + 2 \cdot 4^2 + 2 \cdot 5^2 + 2 \cdot 6^2 + 2 \cdot 7^2.$$

Rather than just finish the calculation from here, notice that we can factor out the 2 from each term:

We calculated $3^2 + 4^2 + 5^2 + 6^2 + 7^2 = 135$ in example 80.

$$= 2(3^2 + 4^2 + 5^2 + 6^2 + 7^2) = 2 \cdot 135 = 270.$$

∎

In other words,

$$\sum_{k=3}^{7} 2k^2 = 2 \sum_{k=3}^{7} k^2.$$

This is one instance of a general rule: we can factor a constant multiple out from a sum.

SUMMATION CONSTANT MULTIPLE RULE

For any real number c,

$$\sum_{k=m}^{n} c \cdot a_k = c \sum_{k=m}^{n} a_k.$$

In the formula, the a_k is any function of k.

The summation constant multiple rule is really just the distributive property from algebra, rewritten for use with summation notation. There are sum and difference rules for summations as well. These rules are essentially versions of the associative and commutative properties from algebra.

The proof of the summation constant multiple rule is like the solution to example 81:

$$\sum_{k=m}^{n} c \cdot a_k = c \cdot a_m + c \cdot a_{m+1} + \cdots + c \cdot a_n =$$

$$c(a_m + a_{m+1} + \cdots + a_n) = c \sum_{k=m}^{n} a_k.$$

It is not technically necessary to state the difference rule, because it can be proved from the constant multiple and sum rules. This is true in other contexts, as well. If constant multiple and sum rules are present, the difference rule must also be valid.

SUMMATION SUM AND DIFFERENCE RULES

$$\sum_{k=m}^{n} (a_k + b_k) = \sum_{k=m}^{n} a_k + \sum_{k=m}^{n} b_k$$

$$\sum_{k=m}^{n} (a_k - b_k) = \sum_{k=m}^{n} a_k - \sum_{k=m}^{n} b_k.$$

Example 82 *Evaluate* $\displaystyle\sum_{k=1}^{4} (3k^2 - k)$.

Solution The terms to add are the values of $3k^2 - k$, beginning with $k = 1$ and ending with $k = 4$:

$$\sum_{k=1}^{4} (3k^2 - k) = (3 \cdot 1^2 - 1) + (3 \cdot 2^2 - 2) + (3 \cdot 3^2 - 3) + (3 \cdot 4^2 - 4)$$

$$= 2 + 10 + 24 + 44$$

$$= 80.$$

Alternately, we could have used the summation constant multiple rule and the summation difference rule:

Use of the summation rules is more important in a different form later in this section.

$$\sum_{k=1}^{4} (3k^2 - k) = 3 \sum_{k=1}^{4} k^2 - \sum_{k=1}^{4} k$$

$$= 3(1^2 + 2^2 + 3^2 + 4^2) - (1 + 2 + 3 + 4)$$

$$= 3(30) - 10 = 80.$$

∎

Reading Exercise 28 Evaluate $\displaystyle\sum_{k=2}^{5} 4k$.

0.6.2 Writing summation notation

What if we have a sum written with an ellipsis and we wish to write it using summation notation? The key is to look for something in each term that is changing.

Example 83 *Write* $\omega + 2\omega + 3\omega + 4\omega + \cdots + 48\omega$ *using summation notation.*

Solution   The terms being added are ω, 2ω, 3ω, and so on. Do you see something that is changing from term to term? The changing values, the coefficients on ω, are consecutive integers beginning with 1 and ending with 48, so we let these be the values of k. The terms are then of the form $k\omega$, which goes after Σ:

Remember that $1\omega = \omega$, so the pattern still works for that term.

$$\omega + 2\omega + 3\omega + 4\omega + \cdots + 48\omega = \sum_{k=1}^{48} k\omega.$$

∎

Example 84 *Write* $(5 \cdot 3\omega)^2 + (6 \cdot 3\omega)^2 + (7 \cdot 3\omega)^2 + (8 \cdot 3\omega)^2 + \cdots + (50 \cdot 3\omega)^2$ *using summation notation.*

Solution   What is changing from term to term? The 3ω is present in every term and is not changing. The exponent 2 is the same in every term. The changing values are consecutive integers from 5 to 50, so we represent these values by k. The terms being added are therefore $(k \cdot 3\omega)^2$, and the sum can be written as

$$\sum_{k=5}^{50} (k \cdot 3\omega)^2.$$

∎

Example 85 *Write* $(1 + \omega)^3 + (1 + 2\omega)^3 + (1 + 3\omega)^3 + (1 + 4\omega^3) + \cdots + (1 + 70\omega)^3$ *using summation notation.*

Solution   This time the $1+$, the ω, and the exponent 3 are not changing, but the coefficient on ω is changing. The terms are $(1 + k\omega)^3$ from $k = 1$ to $k = 70$, and the sum can be written as

There are techniques for handling nonconsecutive integers, such as when the changing numbers increase by two each time, and it is not difficult to do so. For our purposes in this section, however, we only need to work with consecutive integers.

$$\sum_{k=1}^{70} (1 + k\omega)^3.$$

∎

Reading Exercise 29   Write $(4 \cdot 2\omega)^3 + (5 \cdot 2\omega)^3 + (6 \cdot 2\omega)^3 + \cdots + (12 \cdot 2\omega)^3$ using summation notation.

0.6.3   Helpful summation formulas

You may know a clever trick for summing the first n integers. Other summation formulas are traditionally proved using mathematical induction.

Ans. to reading exercise 28:
56

The first formula is easily seen to be true:

$$\sum_{k=1}^{n} 1 = \underbrace{1 + 1 + \cdots + 1}_{n \text{ times}} = n.$$

The other formulas can be proven using mathematical induction, which is often taught in an introduction to proofs course.

In prior sections, we used Ω to represent a positive infinite hyperreal number, but because the previous summation formulas require an integer n, we must add the restriction that Ω is also an integer. Such numbers are sometimes called *hypernatural numbers*.

SUMMATION FORMULAS

For any positive integer n,

$$\sum_{k=1}^{n} 1 = n,$$

$$\sum_{k=1}^{n} k = \frac{n(n + 1)}{2},$$

$$\sum_{k=1}^{n} k^2 = \frac{n(n + 1)(2n + 1)}{6},$$

$$\text{and } \sum_{k=1}^{n} k^3 = \left(\frac{n(n + 1)}{2}\right)^2.$$

To extend the fun, what happens if we use a positive infinite hyperreal integer Ω for n? The transfer principle guarantees the formulas are still valid. Using approximation but stopping short of rendering, the summation formulas become

$$\sum_{k=1}^{\Omega} 1 = \Omega,$$

$$\sum_{k=1}^{\Omega} k = \frac{\Omega(\Omega + 1)}{2} = \frac{\Omega^2 + \Omega}{2} \approx \frac{\Omega^2}{2},$$

$$\sum_{k=1}^{\Omega} k^2 = \frac{\Omega(\Omega + 1)(2\Omega + 1)}{6} = \frac{2\Omega^3 + 3\Omega^2 + \Omega}{6} \approx \frac{2\Omega^3}{6} = \frac{\Omega^3}{3},$$

$$\text{and } \sum_{k=1}^{\Omega} k^3 = \left(\frac{\Omega(\Omega + 1)}{2}\right)^2 = \cdots \approx \frac{\Omega^4}{4}.$$

In each line of these formulas, compare the exponent on k in the summation on the left to the approximation on the right and you should notice a pattern. The pattern continues for higher exponents. For instance, the summation formula for seventh powers is

$$\sum_{k=1}^{n} k^7 = \frac{n^2(n + 1)^2(3n^4 + 6n^3 - n^2 - 4n + 2)}{24},$$

and because the transfer principle says the formula is valid for positive infinite hyperreal integers as well,

$$\sum_{k=1}^{\Omega} k^7 = \cdots \approx \frac{3\Omega^8}{24} = \frac{\Omega^8}{8}.$$

Notice how much these formulas resemble the antiderivative power rule! A proof that this always works can be found in *Calculus Set Free* Section 10.4.

SUM OF POWERS APPROXIMATION FORMULA

For any real number $m > -1$ and any positive infinite hyperreal integer Ω,

$$\sum_{k=1}^{\Omega} k^m \approx \frac{\Omega^{m+1}}{m+1}.$$

The sum of powers approximation formula is introduced and proved in the article Dawson, C. Bryan, "A New Extension of the Riemann Integral," *American Mathematical Monthly* **125** No. 2 (February 2018), 130–140, doi:https//doi.org/10.1080/00029890.2018.1401832.

Reading Exercise 30 Approximate $\displaystyle\sum_{k=1}^{\Omega} k^4$.

0.6.4 Omega sums

The summation constant multiple rule, summation sum rule, and summation difference rule are also valid for sums to Ω (by the transfer principle). Using these rules along with the sum of powers approximation formula helps us evaluate a variety of sums. Because these sums have ending value Ω, they are called Ω *sums* (*omega sums*).

Ans. to reading exercise 29:

$$\sum_{k=4}^{12} (k \cdot 2\omega)^3$$

The term *omega sum* was coined by my colleague Troy Riggs. Omega sums are a type of *hyperfinite sum*, so named because they exhibit many of the properties of finite sums, such as obeying the same sum rules.

OMEGA SUM CONSTANT MULTIPLE RULE

For any positive infinite hyperreal integer Ω and any hyperreal c,

$$\sum_{k=1}^{\Omega} c \cdot a_k = c \sum_{k=1}^{\Omega} a_k.$$

Example 86 *Evaluate* $\displaystyle\sum_{k=1}^{\Omega} 4\omega^2 k$.

Solution Using the omega sum constant multiple rule (first line) and then the sum of powers approximation formula (second line),

In an omega sum, it is implied that Ω is a positive infinite hyperreal integer.

The number $4\omega^2$ is a constant, so it can factor out of the sum. The number k is not constant; it is a number that is changing.

$$\sum_{k=1}^{\Omega} 4\omega^2 k = 4\omega^2 \sum_{k=1}^{\Omega} k$$

$$\approx 4\omega^2 \cdot \frac{\Omega^2}{2} = 2.$$

Recall that ω and Ω are reciprocals–that is, $\Omega = \frac{1}{\omega}$. Therefore, ω^2 and Ω^2 cancel, because $\omega^2 \cdot \Omega^2 = \omega^2 \cdot \frac{1}{\omega^2} = 1$.

■

Example 87 *Evaluate* $\displaystyle\sum_{k=1}^{\Omega} (3\omega k)^2 \cdot 7\omega.$

The *summand* is the expression that comes after Σ.

Solution   First, let's simplify the summand:

$$\sum_{k=1}^{\Omega} (3\omega k)^2 \cdot 7\omega = \sum_{k=1}^{\Omega} 9\omega^2 k^2 \cdot 7\omega = \sum_{k=1}^{\Omega} 63\omega^3 k^2.$$

Now, we can use the omega sum constant multiple rule followed by the sum of powers approximation formula:

The factors that do not involve k are constant and can be factored out. The factor k^2 is not constant, but rather changes with k.

$$\sum_{k=1}^{\Omega} 63\omega^3 k^2 = 63\omega^3 \sum_{k=1}^{\Omega} k^2$$

$$\approx 63\omega^3 \cdot \frac{\Omega^3}{3} = 21.$$

As in example 86, whenever the exponents on ω and Ω are the same, they cancel.

■

Explanation of names: The first rule is a sum rule for omega sums, so it is the "(omega sum) (sum rule)." The second is a difference rule for omega sums, the "(omega sum) (difference rule)."

OMEGA SUM SUM AND DIFFERENCE RULES

For any positive infinite hyperreal integer Ω,

$$\sum_{k=1}^{\Omega} (a_k + b_k) = \sum_{k=1}^{\Omega} a_k + \sum_{k=1}^{\Omega} b_k$$

$$\sum_{k=1}^{\Omega} (a_k - b_k) = \sum_{k=1}^{\Omega} a_k - \sum_{k=1}^{\Omega} b_k.$$

Ans. to reading exercise 30:
$$\frac{\Omega^5}{5}$$

The next example of an omega sum makes use of the sum rule.

Example 88 *Evaluate* $\displaystyle\sum_{k=1}^{\Omega} (2 + 5\omega k)^2 \cdot \omega.$

Solution   As in the previous example, we begin by manipulating the summand algebraically:

$$\sum_{k=1}^{\Omega} (2 + 5\omega k)^2 \cdot \omega = \sum_{k=1}^{\Omega} (4 + 20\omega k + 25\omega^2 k^2)\omega$$

$$= \sum_{k=1}^{\Omega} (4\omega + 20\omega^2 k + 25\omega^3 k^2).$$

To use the sum of powers approximation formula, we need expressions of the form $\sum_{k=1}^{\Omega} k^m$. Completing the operations of squaring and multiplying by ω are necessary steps toward accomplishing this goal.

Now, we can use the omega sum sum rule to break this into three separate omega sums:

$$= \sum_{k=1}^{\Omega} 4\omega + \sum_{k=1}^{\Omega} 20\omega^2 k + \sum_{k=1}^{\Omega} 25\omega^3 k^2.$$

We can then use the omega sum constant multiple rule on each of the omega sums, followed by the sum of powers approximation formula on each of the omega sums:

With practice, it is easy to skip some steps here and go straight from

$$= 4\omega \sum_{k=1}^{\Omega} 1 + 20\omega^2 \sum_{k=1}^{\Omega} k + 25\omega^3 \sum_{k=1}^{\Omega} k^2$$

$$\approx 4\omega \cdot \Omega + 20\omega^2 \cdot \frac{\Omega^2}{2} + 25\omega^3 \cdot \frac{\Omega^3}{3}$$

$$= 4 + 10 + \frac{25}{3} = \frac{67}{3}.$$

$$\sum_{k=1}^{\Omega} (4\omega + 20\omega^2 k + 25\omega^3 k^2)$$

to

$$\approx 4\omega \cdot \Omega + 20\omega^2 \cdot \frac{\Omega^2}{2} + 25\omega^3 \cdot \frac{\Omega^3}{3}.$$

∎

The omega sums in examples 86–88 are typical of the sums we encounter when solving area problems.

Reading Exercise 31   Evaluate $\displaystyle\sum_{k=1}^{\Omega} (2\omega k)^2 \cdot 3\omega$.

0.6.5   Areas

Let's turn to the idea of trying to find the area under the curve $y = x^2$ between $x = 1$ and $x = 5$. Estimating the area using $n = 8$ rectangles and right-hand endpoints, the width of each rectangle is $\Delta x = \frac{b-a}{n} = \frac{5-1}{8} = 0.5$.

Consider the locations of the right-hand endpoints of the subintervals in relation to the widths of the rectangles, as marked in Figure 0.35. The right-hand endpoints are $1, 2, 3, \ldots, 8$ rectangle widths to the right of $x = 1$. Because the width is 0.5, the endpoints can be calculated as $1 + 1 \cdot 0.5$, $1 + 2 \cdot 0.5$, $1 + 3 \cdot 0.5, \ldots$, $1 + 8 \cdot 0.5$. The heights of the rectangles must match the height of the curve

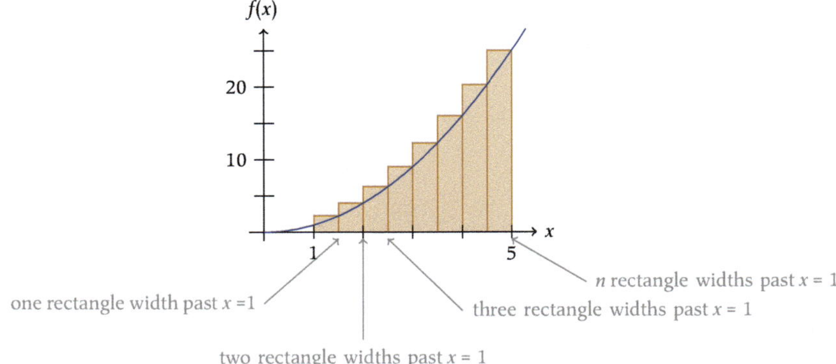

Figure 0.35 *Eight approximating rectangles using right-hand endpoints. The right-hand endpoints are $1, 2, 3, \ldots, 8$ rectangle widths past $x = 1$*

at the right-hand endpoints. The curve is $y = x^2$, so the heights of the curve are $(1 + 1 \cdot 0.5)^2$, $(1 + 2 \cdot 0.5)^2$, $(1 + 3 \cdot 0.5)^2, \ldots, (1 + 8 \cdot 0.5)^2$. The total area of the rectangles is found by multiplying heights by widths and adding the results. However, rather than calculating a numerical answer, let's write the result in summation notation. Noting the pattern in the heights of the rectangles, the total area is

$$\sum_{k=1}^{8} \underbrace{(1 + k \cdot 0.5)^2}_{\text{heights}} \cdot \underbrace{0.5}_{\text{widths}} \; .$$

Calculating the value of this sum gives us an overestimate of the area under the curve. Figure 0.35 makes it clear that the total area of the rectangles and the area under the curve are different. As one might surmise, using more rectangles makes the estimate better.

Now for the key idea that you may have already figured out: if using more rectangles is better, why not use infinitely many rectangles (Figure 0.36)? This ought to eliminate any "extra" area and give us the exact area under the curve!

Using $n = \Omega$ rectangles, where Ω is a positive infinite hyperreal integer, gives a width of

$$\Delta x = \frac{b - a}{n} = \frac{5 - 1}{\Omega} = 4\omega.$$

Because the approximating rectangles have infinitesimal width, there is not enough room for the straight tops of the rectangles to stray more than infinitesimally from the curve itself. Any real difference between the area of the rectangles and the area under the curve is therefore eliminated.

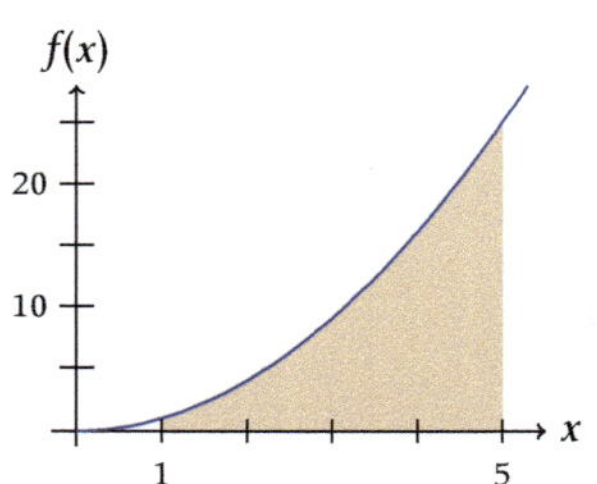

Figure 0.36 *Infinitely many (Ω) approximating rectangles using right-hand endpoints. The individual rectangles can only be seen when zoomed in to an infinitesimal scale. The total area of the rectangles appears to be equal to the area under the curve. Silly remark: drawing this takes a very, very long time*

The intuitive arguments given here rely on using a continuous curve.

The right-hand endpoints are $1, 2, 3, \dots, \Omega$ rectangle widths to the right of $x = 1$, or $1 + 1 \cdot 4\omega$, $1 + 2 \cdot 4\omega$, $1 + 3 \cdot 4\omega$, ..., $1 + \Omega \cdot 4\omega = 5$. The curve is $y = x^2$, so the heights of the rectangles are

$$(1 + 1 \cdot 4\omega)^2, (1 + 2 \cdot 4\omega)^2, (1 + 3 \cdot 4\omega)^2, \dots, (1 + \Omega \cdot 4\omega)^2.$$

The total area of the rectangles is found by multiplying heights by widths and adding the results. In summation notation, this is

$$\text{area under the curve} = \sum_{k=1}^{\Omega} \underbrace{(1 + k \cdot 4\omega)^2}_{\text{heights}} \cdot \underbrace{4\omega}_{\text{widths}}\ .$$

This is an omega sum! The area under the curve is therefore

$$\sum_{k=1}^{\Omega}(1 + k \cdot 4\omega)^2 \cdot 4\omega = \sum_{k=1}^{\Omega}(1 + 8\omega k + 16\omega^2 k^2)4\omega$$

$$= \sum_{k=1}^{\Omega}(4\omega + 32\omega^2 k + 64\omega^3 k^2)$$

$$= 4\omega \sum_{k=1}^{\Omega} 1 + 32\omega^2 \sum_{k=1}^{\Omega} k + 64\omega^3 \sum_{k=1}^{\Omega} k^2$$

$$\approx 4\omega\Omega + 32\omega^2\frac{\Omega^2}{2} + 64\omega^3\frac{\Omega^3}{3}$$

$$= 4 + 16 + \frac{64}{3} = \frac{124}{3}\,\text{units}^2.$$

Ans. to reading exercise 31:

4

We always do the indicated arithmetic on the "heights" of the rectangles (line 1) and multiply by the widths (line 2) before using the omega sum rules (line 3) and the sum of powers approximation formula (line 4).

This result ($\frac{124}{3}\,\text{units}^2$) is the exact area under the curve.

We didn't calculate the hyperreal number that represents the exact area of the approximating rectangles. Instead, we used approximation (specifically, the sum of powers approximation formula) to determine an appropriate real number that represents the exact area under the curve. If the two numbers differ, they differ by an infinitesimal amount.

0.6.6 Definition of the definite integral

The process of determining the area under a curve is typical of many geometric and physical applications. The same pattern is followed: we subdivide an object into Ω pieces of infinitesimal size, calculate the value we seek on an infinitesimal piece, and sum the results. This omega sum is then rewritten as a definite integral, providing an integration formula for calculating the quantity.

The next task, then, is to abstract the area procedure to develop the definition of definite integral.

The procedure for finding the area under the curve $y = f(x)$ over the interval $[a, b]$ begins with determining the width of the intervals. Using $n = \Omega$ rectangles, the width of each rectangle is

$$\Delta x = \frac{b-a}{n} = \frac{b-a}{\Omega} = (b-a)\omega.$$

The right-hand endpoints are $1, 2, 3, \dots, \Omega$ rectangle widths from $x = a$. Calling these endpoints x_1, x_2, x_3, ..., x_Ω, we have $x_1 = a + 1 \cdot \Delta x$, $x_2 = a + 2 \cdot \Delta x$, $x_3 = a + 3 \cdot \Delta x$, ..., $x_\Omega = a + \Omega \cdot \Delta x$. In other words,

$$x_k = a + k\Delta x.$$

The heights of the rectangles match the height of the curve $y = f(x)$ at the endpoints x_k, hence the heights are $f(x_k)$. We then form the omega sum. The area under the curve is found by multiplying heights by widths and adding the results:

$$\text{area under the curve} = \sum_{k=1}^{\Omega} \underbrace{f(x_k)}_{\text{heights}} \cdot \underbrace{\Delta x}_{\text{widths}} .$$

This is as far as we need to go to make our definition.

When defining the definite integral, most calculus textbooks use what is called the *Riemann integral*. The definite integral presented here is called the *omega integral* (or Ω integral) and is introduced in the same article as the sum of powers approximation formula.

Definition 13 DEFINITE INTEGRAL *Let f be a function defined on $[a, b]$. If $\sum_{k=1}^{\Omega} f(x_k)\Delta x$ renders the same real result L for every positive infinite hyperreal integer Ω (where $\Delta x = (b-a)\omega$ and $x_k = a + k\Delta x$ for $k = 1, 2, 3, \dots, \Omega$), then we write*

$$\int_a^b f(x)\, dx = L$$

and call $\int_a^b f(x)\, dx$ the definite integral of f from a to b.

When L is a real number (i.e., not ∞ or $-\infty$), then we say that f is integrable on $[a, b]$.

For the purposes of calculation (using the definition) we write

$$\int_a^b f(x)\, dx = \sum_{k=1}^{\Omega} f(x_k)\Delta x$$

with the understanding that Ω is an arbitrary positive infinite hyperreal integer and we must render the same real result for every such Ω or the integral does not exist.

Example 89 *Use the definition of definite integral to evaluate $\int_0^3 x^2\, dx$.*

Solution Comparing $\int_a^b f(x)\,dx$ (notation from the definition) to $\int_0^3 x^2\,dx$, we see that $a = 0$, $b = 3$, and $f(x) = x^2$. We then use the formulas to proceed:

$$\Delta x = (b - a)\omega = (3 - 0)\omega = 3\omega,$$

$$x_k = a + k\Delta x = 0 + k \cdot 3\omega = k \cdot 3\omega,$$

$$f(x_k) = (k \cdot 3\omega)^2.$$

The calculation here is much shorter than the corresponding limit of Riemann sums calculation used in the epsilon-delta calculus system.

The formula for the definite integral gives

$$\int_0^3 x^2\,dx = \sum_{k=1}^{\Omega} f(x_k)\,\Delta x = \sum_{k=1}^{\Omega} \underbrace{(k \cdot 3\omega)^2}_{f(x_k)} \cdot \underbrace{3\omega}_{\Delta x}$$

$$= \sum_{k=1}^{\Omega} k^2 \cdot 9\omega^2 \cdot 3\omega$$

$$= \sum_{k=1}^{\Omega} 27\omega^3 k^2$$

$$\approx 27\omega^3 \frac{\Omega^3}{3} = 9.$$

The value of the definite integral is 9.

Although we are no longer in the context of areas, the procedure is the same; only the notation is new. The first line is the same as multiplying heights ($f(x_k)$) by widths (Δx) and adding the results. The second line performs the indicated arithmetic on the heights, the third line multiplies by the widths, and the fourth line uses the sum rules and the sum of powers approximation formula.

Notice that "units^2" is not present in the final answer because we extracted the calculation from the context of area. Unless a context is given, no units are appropriate.

Reading Exercise 32 Use the definition of definite integral to evaluate $\displaystyle\int_0^4 x^2\,dx$.

0.6.7 Notation

In the definite integral notation $\int_a^b f(x)\,dx$, the function $f(x)$ is called the *integrand*, a is the *lower limit of integration*, b is the *upper limit of integration*, and x is the *variable of integration*. The symbol $\int$ is the *integral symbol*. The interval $[a, b]$ is also called the *interval of integration*.

The variable of integration is useful for telling the difference between the variable being used and other symbols representing constants. It also lets us know that a and b are values of the variable x, a fact that is important when making substitutions.

Because uses of the definite integral are recognized by the development of an omega sum, it is helpful to practice rewriting an omega sum as a definite integral.

The *limits of integration* are not limits as in Section 0.4. Here we are using *limit* synonymously with *edge* or *boundary*.

The integral symbol is an elongated S (for "sum") and is attributed to Leibniz.

Why the differential dx? The notation arose when differentials were dominant instead of the use of the derivative. One would use the differential $x^2\,dx$ rather than the function value x^2.

Example 90 *Express the sum as a definite integral on the given interval:*

$$\sum_{k=1}^{\Omega} \left(x_k^3 - \sqrt{x_k}\right) \Delta x \ \text{on}\ [2, 5].$$

Example 90 represents the most applicable skill from this review section.

Ans. to reading exercise 32:
$$\frac{64}{3}$$

Selected details:

$$\Delta x = (b - a)\omega = (4 - 0)\omega = 4\omega$$
$$x_k = a + k\Delta x = 0 + k \cdot 4\omega = k \cdot 4\omega$$
$$f(x_k) = (k \cdot 4\omega)^2$$

$$\int_0^4 x^2 \, dx = \sum_{k=1}^{\Omega} f(x_k)\Delta x$$
$$= \sum_{k=1}^{\Omega} (k \cdot 4\omega)^2 \cdot 4\omega = \cdots = \frac{64}{3}$$

The equivalent in physical applications is having a formula for dealing with a constant quantity, but needing to apply it to a variable quantity.

Solution   Compare the omega sum notation to the definite integral notation and the procedure is clear: we replace the summation symbol with the integral symbol, we insert the lower limit of integration 2 and the upper limit of integration 5, we replace each x_k with the variable x, and we replace the Δx with dx:

$$\int_2^5 \left(x^3 - \sqrt{x}\right) dx.$$

■

Reading Exercise 33   Express the sum as a definite integral on the given interval:

$$\sum_{k=1}^{\Omega} \left(x_k^2 + \sin x_k\right) \Delta x \text{ on } [1, 4].$$

0.6.8   Arc length: how long is a curved path?

To illustrate the development of a formula featuring a definite integral, we review the derivation of the arc length formula. This type of reasoning is repeated many times throughout this book.

Consider walking along the path in Figure 0.37 from the point $(a, f(a))$ to the point $(b, f(b))$. How long is the walk, not as the crow flies–which is the shortest path distance along a straight line and is found using the distance formula $d = \sqrt{(x_2 - x_1)^2 + (y_2 - y_1)^2}$–but as the length of the curve $y = f(x)$ between the two points?

We find ourselves in the position of wanting to deal with a curved object, and only having tools from geometry for dealing with straight objects. Our strategy is to break up the interval $[a, b]$ into smaller pieces and approximate using straight objects. In Figure 0.38, the interval is broken into five subintervals, with straight line segments between points on the curve. Adding the lengths of the five line segments, which can be computed using the distance formula, gives an approximation to the length of the curve.

It seems reasonable to assume that using $n = 10$ subintervals (10 line segments, as in Figure 0.39) gives a better approximation, and $n = 100$ is even better, so why not use $n = \Omega$ subintervals? Using infinitely many subintervals seems as though it

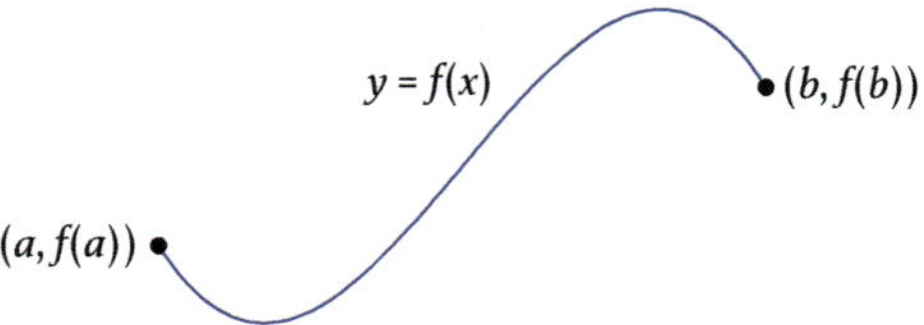

Figure 0.37   *A curve $y = f(x)$ from $x = a$ to $x = b$*

should give us the true length of the curve. But, there is one issue to account for. In Figure 0.38, it is apparent that every approximating line segment is shorter than the curved path it is meant to approximate. How do we know that infinitely many subintervals fixes this issue? The answer comes from considering local linearity.

The concept of local linearity reviewed in Section 0.5 states that, on an infinitesimal scale, a differentiable function and its tangent line are indistinguishable. When using Ω subintervals, the quantity $\Delta x = (b - a)\omega$ is infinitesimal; therefore, the curve $y = f(x)$ follows its tangent line, and the length of the portion of curve between the points $(x_{k-1}, f(x_{k-1}))$ and $(x_k, f(x_k))$ is the length c of the line segment between these points, as illustrated in Figure 0.40.

Many CASs graph functions by plotting a large number of points and "connecting the dots" with straight line segments. The computer-generated graphs in this book are produced in this manner. We are using the same basic idea to approximate the curve's length.

Ans. to reading exercise 33:
$$\int_1^4 (x^2 + \sin x)\, dx$$

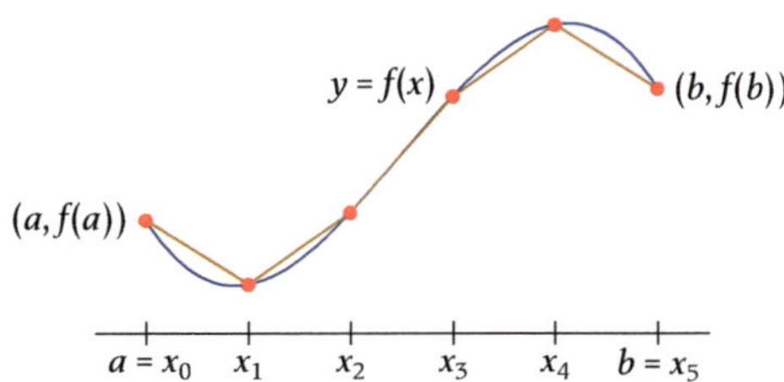

Figure 0.38 *A curve $y = f(x)$ (blue) from $x = a$ to $x = b$. The interval $[a, b]$ is partitioned into five subintervals. The length of the curve is approximated by the sum of the lengths of the five line segments (brown)*

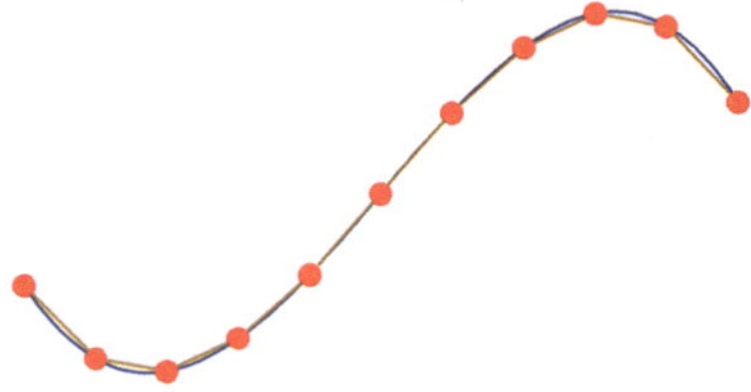

Figure 0.39 *A curve $y = f(x)$ (blue) from $x = a$ to $x = b$. The interval $[a, b]$ is partitioned into 10 subintervals. The length of the curve is approximated by the sum of the lengths of the 10 line segments (brown). These line segments follow the curve more closely than in Figure 0.38, so this approximation should be better*

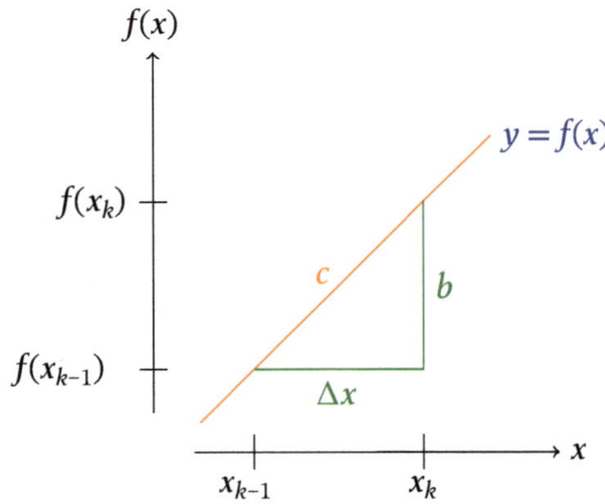

Figure 0.40 *Infinitesimally close to x_k, the curve $y = f(x)$ and its tangent line at $x = x_k$ (orange) are indistinguishable. The length of the portion of the curve between $x = x_{k-1}$ and $x = x_k$ is the length of this segment of its tangent line*

Using right-hand endpoints as usual, the slope of the tangent line in Figure 0.40 is

$$f'(x_k) = \frac{b}{\Delta x}.$$

In this context, instead of labeling the quantities in Figure 0.40 Δx, b, and c, they are traditionally labeled dx, dy, and ds. The Pythagorean relationship is then restated as

$$(ds)^2 = (dx)^2 + (dy)^2.$$

Then,

$$b = f'(x_k)\Delta x.$$

By the Pythagorean theorem,

$$(\Delta x)^2 + b^2 = c^2,$$

and we solve for the length c:

In line 1, the positive square root is used because the length c is positive, and b is replaced by $f'(x_k)\Delta x$ from an earlier equation. In line 2, $(\Delta x)^2$ is factored out. In line 3, the square root of $(\Delta x)^2$ is extracted.

$$c = \sqrt{(\Delta x)^2 + b^2} = \sqrt{(\Delta x)^2 + (f'(x_k)\Delta x)^2}$$

$$= \sqrt{(\Delta x)^2 + (f'(x_k))^2(\Delta x)^2} = \sqrt{(\Delta x)^2(1 + (f'(x_k))^2)}$$

$$= \sqrt{1 + (f'(x_k))^2} \cdot \Delta x.$$

Having found the length of the curve on one subinterval, we add the lengths over all the subintervals to obtain the total length of the curve:

Analysis of the derivation of the formula reveals that the unit of measure of the length of the curve is the same unit of measure as the variables x and y. The analysis is left to you to do as an exercise.

$$\text{length of curve} = \sum_{k=1}^{\Omega} \sqrt{1 + (f'(x_k))^2} \cdot \Delta x.$$

This is an omega sum! Following the procedure of example 90 for rewriting as a definite integral (replacing the summation symbol with the integral symbol, dropping the subscripts (that is, replacing x_k with x), replacing Δx by dx, and inserting the limits of integration), we have

$$\text{length of curve} = \int_a^b \sqrt{1 + (f'(x))^2}\, dx.$$

We add the condition that f' be continuous to guarantee integrability.

Technically speaking, this formula should be presented as the definition of arc length. In this book, we often present such formulas as formulas. The issue of whether they should be written as definitions is ignored.

ARC LENGTH

If f' is continuous on $[a, b]$, then the length of the curve $y = f(x)$ on $[a, b]$ is

$$L = \int_a^b \sqrt{1 + (f'(x))^2}\, dx.$$

The ubiquity of formulas developed in this manner indicates the need for the fundamental theorem of calculus, which allows us to avoid calculating omega sums directly. This flavor of integration is reviewed in the next section.

EXERCISES 0.6

1–4. Rewrite without summation notation.

The most important exercises in this section are 25–32.

1. $\displaystyle\sum_{k=2}^{5} \frac{1}{k}$

3. $\displaystyle\sum_{k=-2}^{1} k^2$

2. $\displaystyle\sum_{k=1}^{3} \sin 2k$

4. $\displaystyle\sum_{k=5}^{9} \frac{k+1}{k}$

5–8. Evaluate the sum.

5. $\displaystyle\sum_{k=2}^{5} (k-1)^2$

7. $\displaystyle\sum_{k=1}^{4} \sin\left(\frac{k\pi}{2}\right)$

6. $\displaystyle\sum_{k=1}^{4} (2k)^2$

8. $\displaystyle\sum_{k=2}^{4} \left(\frac{1}{k} - k\right)$

9–14. Write the expression using summation notation.

9. $5^4 + 6^4 + 7^4 + \cdots + 125^4$

10. $\dfrac{1}{2} + \dfrac{1}{3} + \dfrac{1}{4} + \cdots + \dfrac{1}{9}$

11. $\sqrt{\omega} + \sqrt{2\omega} + \sqrt{3\omega} + \cdots + \sqrt{14\omega}$

12. $\omega + \omega^2 + \omega^3 + \cdots + \omega^{27}$

13. $(4 + 5\omega)^3 + (4 + 6\omega)^3 + (4 + 7\omega)^3 + \cdots + (4 + 99\omega)^3$

14. $(6 \cdot 8\omega)^2 + (7 \cdot 8\omega)^2 + (8 \cdot 8\omega)^2 + \cdots + (22 \cdot 8\omega)^2$

15–18. Evaluate the omega sum.

15. $\displaystyle\sum_{k=1}^{\Omega} (2\omega k)^3 \cdot 5\omega$

17. $\displaystyle\sum_{k=1}^{\Omega} (2 + \omega k)^2 \cdot \omega$

16. $\displaystyle\sum_{k=1}^{\Omega} (6\omega k)^2 \cdot 2\omega$

18. $\displaystyle\sum_{k=1}^{\Omega} (1 + \omega k)^3 \cdot 5\omega$

19–24. Use the definition to evaluate the definite integral.

19. $\int_0^1 x^9 \, dx$

22. $\int_3^4 5x^2 \, dx$

20. $\int_0^7 x^2 \, dx$

23. $\int_0^3 (x^2 + 5) \, dx$

21. $\int_2^5 -x^2 \, dx$

24. $\int_{-2}^{-1} (x^2 - 4x + 11) \, dx$

25–32. Express the sum as a definite integral on the given interval.

25. $\displaystyle\sum_{k=1}^{\Omega}(x_k^5 - 4x_k + 1)\Delta x$ on $[3, 19]$

26. $\displaystyle\sum_{k=1}^{\Omega}\frac{x_k^2 + 1}{x_k^3}\Delta x$ on $[1, 9]$

27. $\displaystyle\sum_{k=1}^{\Omega}(\cos(x_k + \pi) + \sin(4x_k))\Delta x$ on $\left[\frac{\pi}{2}, \frac{3\pi}{2}\right]$

28. $\displaystyle\sum_{k=1}^{\Omega}19\sqrt[3]{1 + x_k}\,\Delta x$ on $[-1, 1]$

29. $\displaystyle\sum_{k=1}^{\Omega}4x_k\sqrt{1 - x_k^4}\,\Delta x$ on $[0, 1]$

30. $\displaystyle\sum_{k=1}^{\Omega}(9 - x_k + 12x_k \tan x_k)\Delta x$ on $[-1, 0]$

31. $\displaystyle\sum_{k=1}^{\Omega}\left((f(x_k))^2 + (g(x_k))^2\right)\Delta x$ on $[a, b]$

32. $\displaystyle\sum_{k=1}^{\Omega}\sqrt{1 + f'(x_k)^2}\,\Delta x$ on $[a, b]$

0.7 Integration

In this section, we review the most commonly used integration theorems, formulas, rules, and techniques from single-variable calculus. Most proofs and some explanations are omitted.

0.7.1 Antiderivatives

Many mathematical operations have inverse operations. Subtraction is the inverse of addition, division is the inverse of multiplication, and extracting roots is the inverse of exponentiation. In algebra, factoring is inverse to multiplying polynomials.

Issues often arise for inverse operations that were not present in the original operation. For instance, we can multiply any two numbers, but we cannot always divide any two numbers (division by zero is undefined). We can square any number, but if we consider only real numbers, then we cannot take the square root of negative numbers. Any pair of polynomials can be multiplied, but not every polynomial can be factored (for instance, $x^2 + 1$ cannot be factored using only real numbers).

Differentiation also has an inverse operation: *antidifferentiation*. Just as with other inverse operations, we encounter issues that are not present with differentiation. Although the derivative rules and formulas are sufficient for finding the derivative of any elementary function, the same is not true for antiderivatives.

Definition 14 ANTIDERIVATIVE *A function F is called an* antiderivative *of f on an interval I if $F'(x) = f(x)$ for all x in the interval I.*

Antiderivatives of the function $f(x) = 2x + 7$ include $F(x) = x^2 + 7x - 4$, $F(x) = x^2 + 7x + 3$, and $F(x) = x^2 + 7x + 428.72$. In fact, any function of the form

$$F(x) = x^2 + 7x + C,$$

where C is a real number, is an antiderivative of $f(x) = 2x + 7$. This amounts to infinitely many antiderivatives of the function $f(x) = 2x + 7$. But, have we found all of them? Could there be some other function, perhaps some function we have not yet studied, that also has the same derivative? Not according to a corollary of the mean value theorem. Intuitively, functions with the same derivative have the same slopes, and therefore the same shape, so the functions must be of the same type. We have found all the functions with derivative $2x + 7$.

Definition 15 MOST GENERAL ANTIDERIVATIVE *The* most general antiderivative *of f is the collection of all possible antiderivatives of f.*

Example 91 *Find the most general antiderivative of $f(x) = \cos x$.*

Solution To find the most general antiderivative, we start by finding an antiderivative. That is, we need a function with a derivative that is $f(x) = \cos x$. Recalling that

$$\frac{d}{dx} \sin x = \cos x,$$

we know that $F(x) = \sin x$ is an antiderivative of $f(x) = \cos x$. Other antiderivatives must differ from this one by a constant, so the most general antiderivative is the collection of functions

$$F(x) = \sin x + C.$$

∎

0.7.2 Indefinite integrals

An *indefinite integral* is another name for the most general antiderivative.

Notice the phrase *an antiderivative*, not *the antiderivative*. This is because there can be more than one antiderivative for the same function.

Definition 14 is for an antiderivative "on an interval *I*." When no interval is specified, we assume an appropriate interval is implied. For a polynomial such as $f(x) = 2x + 7$, the interval $(-\infty, \infty)$ can be used. For exercises of the form "find an antiderivative," an interval is often not stated explicitly in the exercise or in the solution.

The mean value theorem is the topic of Section 3.2 of *Calculus Set Free*.

Other antiderivatives include $F(x) = \sin x + 12$ and $F(x) = \sin x - \sqrt{77}$.

The expression "$+C$" takes care of both addition and subtraction of real numbers. This is because the value of C may be any real number. If C is a negative number, such as -11, then "$+C$" is the same as "$+(-11)$," which can be rewritten as "-11."

The notation $\int f(x)\,dx$ can be read "the integral of f of x with respect to x" or "the integral of f of $x\,dx$," where the latter letters are pronounced by their names individually, but with a pause before the dx.

Any derivative calculation can be reversed to give an antiderivative result, as illustrated here:

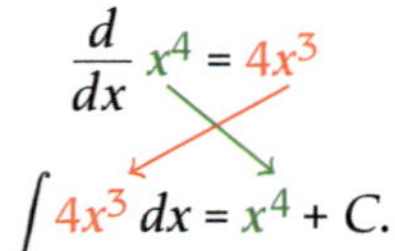

$$\frac{d}{dx}\,x^4 = 4x^3$$

$$\int 4x^3\,dx = x^4 + C.$$

Definition 16 INDEFINITE INTEGRAL *Suppose F is an antiderivative of f on the interval I. Then the* indefinite integral *of f on the interval I, written $\int f(x)\,dx$, is the most general antiderivative of f:*

$$\int f(x)\,dx = F(x) + C,$$

where C is a real number.

An indefinite integral is not one function but a collection of infinitely many functions, one for each real number C.

Derivative rules can be reversed to get antiderivative rules. The first few are reviewed next.

ANTIDERIVATIVE POWER RULE

For any real number $n \neq -1$,

$$\int x^n\,dx = \frac{x^{n+1}}{n+1} + C.$$

ANTIDERIVATIVE CONSTANT MULTIPLE RULE

For any real number k,

$$\int k \cdot f(x)\,dx = k \cdot \int f(x)\,dx,$$

assuming the latter integral exists.

The antiderivative sum and difference rules are also dependent on the existence of the integrals on the right side.

ANTIDERIVATIVE SUM AND DIFFERENCE RULES

$$\int (f(x) + g(x))\,dx = \int f(x)\,dx + \int g(x)\,dx$$

$$\int (f(x) - g(x))\,dx = \int f(x)\,dx - \int g(x)\,dx.$$

ANTIDERIVATIVE OF A CONSTANT

For any real number k,

$$\int k\,dx = kx + C.$$

These rules are sufficient to allow us to find antiderivatives for all polynomials (and some nonpolynomials). After a little practice, it is traditional to learn to calculate antiderivatives of polynomials in one step.

Example 92 *Calculate $\int (2x^5 + 9x^4 - \frac{1}{5}x^3 + 2x + 11.5)\,dx.$*

Solution Using the various antiderivative rules,

$$\int (2x^5 + 9x^4 - \frac{1}{5}x^3 + 2x + 11.5)\,dx$$

$$= 2 \cdot \frac{x^6}{6} + 9 \cdot \frac{x^5}{5} - \frac{1}{5} \cdot \frac{x^4}{4} + 2 \cdot \frac{x^2}{2} + 11.5x + C$$

$$= \frac{1}{3}x^6 + \frac{9}{5}x^5 - \frac{1}{20}x^4 + x^2 + 11.5x + C.$$

∎

Reading Exercise 34 Calculate $\int (5x^2 - 3x + 2)\,dx.$

0.7.3 Manipulating integrands

We have now reversed several derivative rules and given their antiderivative analogs, including the power rule, sum rule, difference rule, and constant multiple rule. Other derivative rules, such as the product rule, quotient rule, and chain rule, are more difficult to reverse. Although they have their analogs in antidifferentiation, they are not as straightforward to apply as recognizing a product or quotient and using the associated rule. For the moment, if we wish to find antiderivatives of products, quotients, or compositions, we must change the form of the expression into something we can antidifferentiate using the rules we have at our disposal. These ideas remain useful after the introduction of other techniques.

The reverse of the product rule is the integration technique known as integration by parts. The reverse of the chain rule is the integration technique known as substitution. These techniques are reviewed later in this section.

Example 93 *Evaluate $\int x(x^2 - 3x)\,dx.$*

Solution Because we have no product rule for antidifferentiation, we cannot evaluate the antiderivative in its current form. Instead, let's perform the multiplication in the integrand:

When manipulating the integrand (the expression inside the integral), the integral symbol at the beginning and the dx at the end do not change; only the expression inside changes to something equivalent.

$$\int x(x^2 - 3x)\, dx = \int (x^3 - 3x^2)\, dx.$$

Now the integrand is in a form we can handle using the available rules; there is no product in the integral on the right. Using the antiderivative difference rule, the antiderivative constant multiple rule, and the antiderivative power rule, we have

$$\int (x^3 - 3x^2)\, dx = \frac{x^4}{4} - 3 \cdot \frac{x^3}{3} + C = \frac{x^4}{4} - x^3 + C.$$

■

The expression $\frac{x^5 - 4x}{2x^3}$ has one term in its denominator–namely, $2x^3$. An expression of the form $\frac{2x^3}{x^5 - 4x}$ has two terms in its denominator, x^5 and $4x$ (recall that terms are separated by a $+$ or $-$). Performing the division in an expression with two terms in the denominator is more difficult, but sometimes necessary.

Details of the division:

$$\frac{x^5 - 4x}{2x^3} = \frac{x^5}{2x^3} - \frac{4x}{2x^3} = \frac{1}{2}x^{5-3} - 2x^{1-3}$$

$$= \frac{1}{2}x^2 - 2x^{-2}.$$

Be careful: $-2 + 1 = -1$;

$$\int x^{-2}\, dx \neq \frac{x^{-3}}{-3} + C.$$

Example 94 *Evaluate* $\displaystyle \int \frac{x^5 - 4x}{2x^3}\, dx.$

Solution Because we have no quotient rule for integration, we cannot evaluate the integral in its current form. We need to manipulate the expression inside the integral (the integrand) into a form for which we can use the available rules. Because there is only one term in the denominator, we may perform the division easily:

$$\int \frac{x^5 - 4x}{2x^3}\, dx = \int \left(\frac{1}{2}x^2 - 2x^{-2}\right) dx.$$

Now the integrand is in a form we can handle using the available rules; there is no quotient in the integral on the right. Using the antiderivative difference rule, the antiderivative constant multiple rule, and the antiderivative power rule, we have

$$\int \left(\frac{1}{2}x^2 - 2x^{-2}\right) dx = \frac{1}{2} \cdot \frac{x^3}{3} - 2 \cdot \frac{x^{-1}}{-1} + C = \frac{1}{6}x^3 + \frac{2}{x} + C.$$

■

Example 95 *Evaluate* $\int (2x^3 - 1)^2\, dx.$

Just as with differentiation, the antiderivative power rule is for "just plain x" to a power. The antiderivative power rule does not apply directly to $\int (2x^3 - 1)^2\, dx$ because we are squaring $2x^3 - 1$, not x.

Solution Because we have no chain rule for antidifferentiation, we cannot evaluate the integral in its current form. However, this issue can be fixed if we go ahead and square the integrand:

$$\int (2x^3 - 1)^2\, dx = \int (4x^6 - 4x^3 + 1)\, dx.$$

Now the integrand is in a form we can handle using the available rules; there is no composition in the integral on the right. Using the antiderivative difference rule, the antiderivative sum rule, the antiderivative constant multiple rule, the antiderivative power rule, and the rule for the antiderivative of a constant, we have

$$\int (4x^6 - 4x^3 + 1)\, dx = 4 \cdot \frac{x^7}{7} - 4 \cdot \frac{x^4}{4} + 1 \cdot x + C = \frac{4}{7}x^7 - x^4 + x + C.$$

Substitution is reviewed later in this section. Note that substitution fails for this integral. The only available solution method is the one described here.

THE RULES RULE

Knowing what "rules" do not exist is just as important as knowing which rules do exist for calculating antiderivatives.

In my experience observing integral calculations, more errors occur as a result of applying a nonexistent "rule" than failing to know a rule.

0.7.4   Trig antiderivatives

Reversing the derivative formulas for trigonometric functions yields six antiderivative formulas. Four additional trig antiderivatives are also included in the following list.

Ans. to reading exercise 34:
$$\frac{5x^3}{3} - \frac{3x^2}{2} + 2x + C$$

TRIG ANTIDERIVATIVES

$$\int \sin x \, dx = -\cos x + C \qquad \int \csc x \, dx = -\ln|\csc x + \cot x| + C$$

$$\int \cos x \, dx = \sin x + C \qquad \int \sec x = \ln|\sec x + \tan x| + C$$

$$\int \tan x \, dx = \ln|\sec x| + C \qquad \int \cot x \, dx = -\ln|\csc x| + C$$

$$\int \sec^2 x \, dx = \tan x + C \qquad \int \csc^2 x \, dx = -\cot x + C$$

$$\int \sec x \tan x \, dx = \sec x + C \qquad \int \csc x \cot x \, dx = -\csc x + C$$

As with any antiderivative formula, these formulas can be used in conjunction with the antiderivative rules to integrate a variety of functions.

Example 96 *Evaluate the indefinite integral:* $\int (4 \sin x + \sec^2 x)\, dx$.

Solution   Using the formulas for the antiderivatives of $\sin x$ and $\sec^2 x$, along with the antiderivative sum rule and the antiderivative constant multiple rule, gives

$$\int (4 \sin x + \sec^2 x)\, dx = 4(-\cos x) + \tan x + C = -4 \cos x + \tan x + C.$$

Be careful! Writing $4 - \cos x + \tan x + C$ is incorrect, because it turns the multiplication $4(-\cos x)$ into a subtraction: $4 - \cos x$.

■

Reading Exercise 35   Evaluate the indefinite integral: $\int 6 \sec x \tan x\, dx$.

0.7.5   Fundamental theorem of calculus, part II

Instead of evaluating definite integrals using the definition, it is easier to use the fundamental theorem of calculus part II (FTC II) when it applies.

Some say that theorem 11 is the most important computational discovery in history, for it was essential to the development of many aspects of modern science and engineering. Definite integrals aren't just for finding areas; they are useful in many other applications, some of which appear in later chapters of this text.

Theorem 11 FUNDAMENTAL THEOREM OF CALCULUS, PART II *If f is continuous on $[a, b]$ and H is any antiderivative of f on $[a, b]$, then*

$$\int_a^b f(t)\, dt = H(b) - H(a).$$

Example 97 *Evaluate* $\int_2^3 5x^3\, dx$.

The steps are ❶ find an antiderivative of the integrand and write it alongside the evaluation symbol, ❷ evaluate the antiderivative at the top number and subtract the antiderivative evaluated at the bottom number, and ❸ finish the arithmetic.

Solution   Since $y = 5x^3$ is continuous on $[2, 3]$, we use FTC II:

$$\int_2^3 5x^3\, dx = 5 \cdot \frac{x^4}{4}\bigg|_2^3 = 5 \cdot \frac{3^4}{4} - 5 \cdot \frac{2^4}{4} = \frac{325}{4}.$$

■

When rewriting an expression algebraically, all else stays the same. If we rewrite an integrand, the rest of the integral does not change. If we rewrite an antiderivative before evaluating, we carry along the evaluation bar.

Example 98 *Evaluate* $\int_0^6 \sqrt{x}\, dx$.

Solution   Because $y = \sqrt{x}$ is continuous on $[0, 6]$, we use FTC II:

$$\int_0^6 \sqrt{x}\,dx = \int_0^6 x^{1/2}\,dx$$

$$= \frac{x^{3/2}}{\frac{3}{2}}\Bigg|_0^6$$

$$= \frac{2}{3}x^{3/2}\Bigg|_0^6$$

$$= \frac{2}{3}\cdot 6^{3/2} - \frac{2}{3}\cdot 0^{3/2}$$

$$= \frac{2}{3}\cdot 6^{3/2} = \frac{2}{3}\cdot 6\sqrt{6} = 4\sqrt{6}.$$

Line 1 rewrites the integrand, leaving all else the same; line 2 finds an antiderivative, dropping the integral notation and using the evaluation bar; line 3 rewrites the antiderivative, leaving the evaluation bar unchanged; line 4 evaluates the expression at $x = 6$, at $x = 0$, and subtracts, dropping the evaluation bar; and line 5 completes the calculation.

Reading Exercise 36   Evaluate $\int_1^3 x^4\,dx$.

Examples 97 and 98 can be worked using the definition of definite integral (using omega sums), in the manner of Section 0.6 example 89; FTC II makes the solutions a little shorter. The real advantage of FTC II comes from the ability to calculate easily those definite integrals that are difficult or impossible to handle using the definition.

Example 99   *Evaluate $\int_0^{2\pi} \sin x\,dx$.*

Solution   Since $y = \sin x$ is continuous on $[0, 2\pi]$, we use FTC II:

$$\int_0^{2\pi} \sin x\,dx = -\cos x\big|_0^{2\pi} = -\cos 2\pi - (-\cos 0) = -1 - (-1) = 0.$$

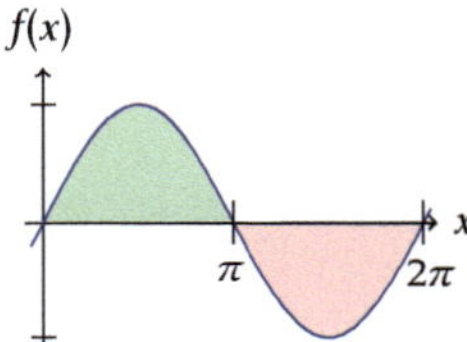

Figure 0.41 *The region between the curve $y = \sin x$ and the x-axis between $x = 0$ and $x = 2\pi$ (shaded). The definite integral takes the area of the green region (where the curve lies above the x-axis) and subtracts the area of the pink region (where the curve lies below the x-axis), resulting in the net area under the curve. As a result of symmetry, the net area is 0 units^2*

An interpretation of the solution to example 99 is that the net area under the curve $y = \sin x$ from $x = 0$ to $x = 2\pi$ is 0 units^2, as illustrated in Figure 0.41.

Be careful–FTC II can be used in the manner of examples 97–99 only when the integrand is continuous throughout the interval of integration.

Example 100   *Evaluate $\int_{-1}^{1} \frac{1}{t^2}\,dt$.*

Solution   Because $y = \frac{1}{t^2}$ is *not* continuous on $[-1, 1]$ (it is undefined at $t = 0$ for reason of division by zero), we cannot use FTC II in its current form. Instead, this is an *improper integral*, one that must be split into two pieces, one on each side of the discontinuity at $x = 0$.

The portion to the right of the discontinuity is evaluated by moving just to the right of $x = 0$, to $x = 0 + \omega$. Now the function is continuous throughout the interval $[\omega, 1]$, and FTC II applies:

Because FTC II requires that the integrand be continuous on the interval of integration, we must check for continuity every time. Failure to do so may result in a wrong answer:

$$\int_{-1}^{1} t^{-2}\,dt \neq \frac{t^{-1}}{-1}\Bigg|_{-1}^{1} = \cdots = -2.$$

Because the curve (the graph of the integrand) is above the x-axis, the definite integral cannot be negative.

The portion to the left of the discontinuity is evaluated by moving just to the left of $x = 0$, to $x = 0 - \omega$.

Line 1 finds an antiderivative; line 2 evaluates at $t = 1$, at $t = \omega$, and subtracts, then simplifies; and line 3 renders the positive infinite quantity.

$$\int_{\omega}^{1} \frac{1}{t^2}\, dt = -\frac{1}{t}\Big|_{\omega}^{1}$$

$$= -1 - -\frac{1}{\omega} = -1 + \Omega$$

$$\doteq \infty.$$

Be careful! Attempting to evaluate the two sides together can lead to a false conclusion, potentially canceling out infinite quantities (as for $\int_{-1}^{1} \frac{1}{t^3}\, dt$).

Because the integral on this side of the discontinuity is infinite, the integral does not exist, no matter the value on the other side of the discontinuity.

∎

0.7.6 Substitution: reversing the chain rule

The derivative calculation

$$\frac{d}{dx}\cos(x^2 + 3) = -\sin(x^2 + 3)\cdot 2x = -2x\sin(x^2 + 3)$$

can be reversed to give the antiderivative result

$$\int -2x\sin(x^2 + 3)\, dx = \cos(x^2 + 3) + C.$$

However, if we start from scratch given the integral

$$\int -2x\sin(x^2 + 3)\, dx,$$

how do we proceed to find the antiderivative if we do not already know it? The integrand contains both a product ($-2x$ times $\sin(x^2 + 3)$) and a composition (sine of "not just plain x"). We need a clear strategy for how to handle such integrands.

To this end, recall the chain rule formula:

$$\frac{d}{dx}f(g(x)) = f'(g(x))\cdot g'(x).$$

Suppose F is an antiderivative of f–that is, $F'(x) = f(x)$. We can rewrite the chain rule using F as

$$\frac{d}{dx}F(g(x)) = f(g(x))\cdot g'(x).$$

Reversing the derivative calculation yields the antiderivative result

$$\int f(g(x))\cdot g'(x)\, dx = F(g(x)) + C.$$

Next, we make the substitution

$$u = g(x)$$
$$du = g'(x)\,dx$$

into the antiderivative formula:

$$\int f(u)\,du = F(u) + C.$$

This integral makes sense given that F is an antiderivative of f. It also points the way to our procedure, for it gives us an easier integral to evaluate. The method of *substitution* takes an integral $\int f(g(x))g'(x)\,dx$ involving "not just plain x" (the expression $g(x)$), ❶ makes the previous substitution to obtain an integral $\int f(u)\,du$ with "just plain u," ❷ finds its (most general) antiderivative $F(u) + C$ using previously available antiderivative rules and formulas, and then ❸ reverses the substitution, giving the antiderivative we are seeking: $F(g(x)) + C$.

Example 101 *Evaluate $\int 3x^2(x^3 + 7)^4\,dx$.*

Solution Because we have "not just plain x" to a power, the antiderivative power rule does not apply and we need to try the method of substitution.
　　❶ ⓐ Let the "not just plain x" part be u. Here, this is the expression inside the parentheses:

$$u = x^3 + 7.$$

❶ ⓑ Calculate the differential du. For this substitution,

$$du = 3x^2\,dx.$$

❶ ⓒ Carry out the substitution. Tips: (1) begin by identifying the expression replaced by du; (2) never write an expression containing both the old and new variables. For this substitution,

$$\int 3x^2(x^3 + 7)^4\,dx$$

becomes

$$\int (u)^4\,du.$$

Notice that the resulting integral is much easier than the one we started with!

Recall that the derivative is $\frac{du}{dx} = g'(x)$, but the differential rearranges this expression to give $du = g'(x)\,dx$.

Ans. to reading exercise 36:
$\frac{242}{5} = 48.4$

We also do not have an antiderivative product rule, but the first barrier to tackle is always "not just plain x."

The "not just plain x" part is often in parentheses to a power, under a square root, inside a trig function, and so on, and is recognized in the same manner as recognizing when the chain rule is required for differentiation.

Be careful! Forgetting the dx when calculating du (or being lazy and skipping it) is the root cause of many errors.

A successful substitution switches completely from the old variable (in this case, x) to the new variable (in this case, u). The replacements must be carried out exactly, just like coding a message. If anything is missing, the replacement cannot be made; if there are portions that cannot be successfully replaced, the substitution fails.

❷ Integrate (find an antiderivative). This is completed in the new variable:

$$\int u^4 \, du = \frac{u^5}{5} + C.$$

We can always check our work by taking the derivative:

$$\frac{d}{dx} \frac{1}{5}(x^3 + 7)^5 = \frac{1}{5} \cdot 5(x^3 + 7)^4 \cdot 3x^2$$

$$= 3x^2(x^3 + 7)^4,$$

which is the original integrand, as required.

❸ Return to the original variable. Using the substitution formula, the antiderivative we seek is

$$= \frac{(x^3 + 7)^5}{5} + C.$$

∎

The antiderivative constant multiple rule allows us to adjust for a missing constant when making a substitution.

Example 102 *Calculate $\int s^3(s^4 + 7)^{12} \, ds$.*

Solution   Once again we have an integrand with "not just plain s" to a power, so we try substitution and let u be the "not just plain s" part:

$$u = s^4 + 7.$$

Next, we calculate the differential du:

Be careful! The expression for du must be calculated from u using the derivative. Do not merely look for whatever is "left over" in the integrand.

$$du = 4s^3 \, ds.$$

Now we have a problem. The substitution requires replacing $4s^3 \, ds$ with du, but the 4 is missing from the integrand in $\int s^3(s^4 + 7)^{12} \, ds$. However, we can always multiply any expression by one without changing its value, so we can multiply by $4 \cdot \frac{1}{4}$ to obtain

This is what is meant by "adjusting for the constant 4."

$$\int 4 \cdot \frac{1}{4} \cdot s^3(s^4 + 7)^{12} \, ds.$$

Line 1 makes the substitution, replacing $4s^3 \, ds$ with du, replacing $s^4 + 7$ with u, and carrying along the constant $\frac{1}{4}$ and other items that are not part of either u or du; line 2 finds an antiderivative with respect to the variable u; and line 3 replaces u by $s^4 + 7$, returning to the original variable.

Now everything we need for the substitution is present and we can move forward:

$$\int 4 \cdot \frac{1}{4} \cdot s^3(s^4 + 7)^{12} \, ds = \int \frac{1}{4}(u)^{12} \, du$$

$$= \frac{1}{4} \cdot \frac{u^{13}}{13} + C$$

$$= \frac{(s^4 + 7)^{13}}{52} + C.$$

∎

As long as we are off only by a constant factor in our *du*, then the trick of example 102 applies (multiply and divide by the missing constant factor) and the substitution works.

Reading Exercise 37 Evaluate $\int x^4 (x^5 + 1)^7 \, dx$.

0.7.7 Substitutions sometimes fail

Sometimes we can recognize that substitution appears to be required, yet substitution is not successful. For instance, consider

$$\int \sin x^2 \, dx.$$

The integrand is a trig function of "not just plain x," so we try substitution, letting u be the "not just plain x" part:

$$u = x^2$$

$$du = 2x \, dx.$$

As always–*always!*–*du* is calculated from *u*, without further reference to the integrand.

For this substitution to work, we need to replace $2x \, dx$ with du. The problem is that there is no $2x$ in the integrand. If what is missing is just a constant, such as in example 102, then we can adjust by multiplying and dividing by that constant. Unfortunately, $2x$ is not just a constant (in addition to the constant 2, the variable x is missing), causing the substitution to fail.

Why do we stop there and say the substitution fails? We can try to go forward using the same trick:

It is occasionally possible to move forward in this manner and complete a substitution, but not often.

$$\int \sin x^2 \, dx = \int 2x \cdot \frac{1}{2x} \cdot \sin x^2 \, dx.$$

We then can replace x^2 with u and replace $2x \, dx$ with du, leaving all else unchanged:

$$\int 2x \cdot \frac{1}{2x} \cdot \sin x^2 \, dx = \int \frac{1}{2x} \cdot \sin u \, du.$$

Caution: when using substitution, never write an integral expression containing both variables. Such an expression is not in a form that can be integrated.

This is an *incomplete substitution* because of the continued presence of the variable x, and the substitution still fails. Substitution requires us to switch completely to a function of the new variable u. In the development of the method of substitution, the switch is made to $\int f(u) \, du$, which no longer contains any reference to the variable x.

There are, of course, other techniques of integration. Even if other techniques are successful (they are not in this case), it is still the fact that substitution fails.

Details of the substitution:

$$\int x^2 \cos x^3 \, dx = \frac{1}{3} \int 3x^2 \cos x^3 \, dx$$

$$= \frac{1}{3} \int \cos u \, du.$$

The solution to the integral in part (b) involves a function you have not studied.

Ans. to reading exercise 37:
$$\frac{(x^5 + 1)^8}{40} + C$$

If substitution doesn't work, then what must we do to find the requested antiderivative? Unfortunately, there are times when an antiderivative involves a function that is not in our list of "elementary" functions. The solution to $\int \sin x^2 \, dx$ involves a function you have never studied (almost assuredly so).

To summarize, if the substitution formula for du contains a variable expression that is not present in the integrand, then the substitution fails.

Example 103 *Does the substitution succeed or fail? (a)* $\int x^2 \cos x^3 \, dx$, $u = x^3$; *(b)* $\int \sqrt{x^3 + 7x} \, dx$, $u = x^3 + 7x$.

Solution (a) Substitution is needed because the integrand contains a trig function of "not just plain x," and u is the "not just plain x" part, so the setup is correct. We still need to calculate du for the substitution:

$$du = 3x^2 \, dx.$$

We only need to adjust for a missing constant (3), so the substitution succeeds.

(b) Substitution is needed because the integrand contains a square root of "not just plain x," and u is the "not just plain x" part, so the setup is correct. Calculate du:

$$du = (3x^2 + 7) \, dx.$$

Because the variable expression $3x^2 + 7$ is not present in the integrand, the substitution fails. ∎

0.7.8 Substitution with definite integrals: method 1

If finding an antiderivative is relatively complicated, then for a definite integral it can help to ❶ write the antiderivative step as a separate problem, to be followed by ❷ application of the fundamental theorem. The next example illustrates the process, although with an uncomplicated antiderivative.

Example 104 *Evaluate* $\int_0^{\pi/4} \sin\left(2z - \frac{\pi}{2}\right) dz.$

Solution Recall that we should always check for continuity before attempting to use FTC II, and since $\sin(2z - \frac{\pi}{2})$ is continuous everywhere, we may proceed.

❶ We find an antiderivative first; that is, we calculate $\int \sin\left(2z - \frac{\pi}{2}\right) dz.$ Because we have sine of "not just plain z," substitution is called for:

$$u = 2z - \frac{\pi}{2}$$

$$du = 2 \, dz.$$

Then,

$$\int \sin\left(2z - \frac{\pi}{2}\right) dz = \frac{1}{2} \int 2\sin\left(2z - \frac{\pi}{2}\right) dz$$

$$= \frac{1}{2} \int \sin u \, du$$

$$= -\frac{1}{2} \cos u + C$$

$$= -\frac{1}{2} \cos\left(2z - \frac{\pi}{2}\right) + C.$$

❷ Next, we use FTC II to evaluate the definite integral using an antiderivative calculated in step ❶:

$$\int_0^{\pi/4} \sin\left(2z - \frac{\pi}{2}\right) dz = -\frac{1}{2} \cos\left(2z - \frac{\pi}{2}\right)\Big|_0^{\pi/4}$$

$$= -\frac{1}{2} \cos 0 - \left(-\frac{1}{2} \cos\left(-\frac{\pi}{2}\right)\right)$$

$$= \cdots = -\frac{1}{2}.$$

∎

Because method 1 consists solely of combining two previous procedures, it is not difficult to learn. Be sure to pay attention to correct notation.

Reading Exercise 38 Use method 1 to evaluate $\int_1^2 \sqrt{5x - 1}\, dx$.

0.7.9 Substitution with definite integrals: method 2

Although method 1 is well-suited for evaluating any definite integral for which substitution is needed, an alternate procedure is used more often because of its relative efficiency, as long as the antiderivative is not too complicated. The only drawback is that more care must be taken with its notation.

Method 2 combines portions of the two steps of example 104 as follows. We begin by recognizing that substitution is required and write

$$u = 2z - \frac{\pi}{2}$$

$$du = 2\, dz.$$

The promised efficiency is accomplished by switching the limits of integration from values of the original variable (z) to values of the substitution variable (u):

$$z = \frac{\pi}{4} : \quad u = 2 \cdot \frac{\pi}{4} - \frac{\pi}{2} = 0$$

$$z = 0 : \quad u = 2 \cdot 0 - \frac{\pi}{2} = -\frac{\pi}{2}.$$

Line 1 adjusts for the constant 2, using the constant multiple rule to place $\frac{1}{2}$ outside the integral; line 2 makes the substitution, replacing $2\,dz$ with du and replacing $2z - \frac{\pi}{2}$ with u; line 3 finds an antiderivative with respect to the variable u; and line 4 replaces u with $2z - \frac{\pi}{2}$ to return to the original variable. Although we only need one antiderivative to use FTC II, when we write an indefinite integral $\int f(x)\, dx$, the answer is the most general antiderivative and therefore $+C$ is required.

In line 1, any one antiderivative will do when applying FTC II; there is no need to use the $+C$ written in the answer to step ❶. Line 2 evaluates at $z = \frac{\pi}{4}$, at $z = 0$, and subtracts.

In method 1, these calculations occur during evaluation. See the evaluation bar and the expression that precedes it.

When we then carry out the substitution, we also switch the limits of integration:

$$\int_0^{\pi/4} \sin\left(2z - \frac{\pi}{2}\right) dz = \frac{1}{2} \int_0^{\pi/4} 2\sin\left(2z - \frac{\pi}{2}\right) dz$$

$$= \frac{1}{2} \int_{-\pi/2}^{0} \sin u \, du.$$

The lower limit stays the lower limit and the upper limit stays the upper limit when switching variables, regardless of which number is smaller or larger.

The calculation then continues in the new variable as its own definite integral without needing to switch back to the original variable, making the calculations shorter:

$$= -\frac{1}{2} \cos u \Big|_{-\pi/2}^{0} = -\frac{1}{2}\cos 0 - \left(-\frac{1}{2}\cos\left(-\frac{\pi}{2}\right)\right) = \cdots = -\frac{1}{2}.$$

Comparing these calculations to those of example 104 helps us see how all the necessary steps remain, but unnecessary steps are discarded. However, care must be taken always to remember to switch the limits of integration to values of the new variable.

Example 105 *Find* $\displaystyle\int_1^4 \frac{1}{2\sqrt{y}(1 + \sqrt{y})^2} \, dy.$

Details of continuity: $\sqrt{y}$ requires $y \geq 0$, and avoiding division by zero requires $y \neq 0$ (notice that the equation $1 + \sqrt{y} = 0$ has no solutions). The domain is $(0, \infty)$, which includes $[1, 4]$.

Solution We first note that the integrand is continuous on $[1, 4]$, therefore FTC II applies and we may proceed.

Because we have "not just plain y" to a power, substitution is needed:

$$u = 1 + \sqrt{y}$$

Details of the derivative:

$$\frac{d}{dy}(1 + \sqrt{y}) = \frac{d}{dy}(1 + y^{1/2})$$

$$= \frac{1}{2}y^{-1/2} = \frac{1}{2\sqrt{y}}.$$

$$du = \frac{1}{2\sqrt{y}} \, dy.$$

Using method 2, we also calculate the u values of the limits of integration:

Tip: Write the calculation for the upper limit on top and the lower limit on bottom to avoid confusion.

$$y = 4 : \quad u = 1 + \sqrt{4} = 3$$

$$y = 1 : \quad u = 1 + \sqrt{1} = 2.$$

In line 1, the substitution is made, including switching from values of the original variable y (the limits of integration 1 and 4) to values of the substitution variable u (the limits of integration 2 and 3). The new definite integral is then calculated easily. Details of the antiderivative:

We then perform the substitution (including the change of the limits of integration) and calculate the value of the definite integral using the new variable:

$$\int_1^4 \frac{1}{2\sqrt{y}(1 + \sqrt{y})^2} \, dy = \int_2^3 \frac{1}{u^2} \, du$$

$$\int \frac{1}{u^2} \, du = \int u^{-2} \, du$$

$$= \frac{u^{-1}}{-1} + C = -\frac{1}{u} + C.$$

$$= -\frac{1}{u}\Big|_2^3$$

$$= -\frac{1}{3} - \left(-\frac{1}{2}\right) = \frac{1}{6}.$$

∎

The most common mistake made when using method 2 is to fail to switch the limits of integration when making the substitution. If the original limits of integration are kept, then the resulting definite integral does not yield the correct answer, as illustrated using the integral from example 105:

$$\int_1^4 \frac{1}{u^2}\,du = -\frac{1}{u}\Big|_1^4 = -\frac{1}{4} - (-1) = \frac{3}{4} \neq \frac{1}{6}.$$

After this section, this text nearly always uses method 2 when evaluating definite integrals using substitution.

Reading Exercise 39 Use method 2 to evaluate $\int_0^4 2x\sqrt{x^2 + 9}\,dx$.

0.7.10 Antiderivatives of $\frac{1}{x}$ and e^x

The antiderivative power rule excludes the exponent -1 for reason of division by zero. Logarithms provide an antiderivative for this missing case.

ANTIDERIVATIVE OF $1/x$

$$\int \frac{1}{x}\,dx = \ln|x| + C$$

There is still no quotient rule for antidifferentiation, but this formula can help us find antiderivatives for some quotients.

Example 106 *Evaluate* $\displaystyle\int \frac{1}{2x - 3}\,dx$.

Solution We now have an antiderivative formula for $1/x$, but this integral is 1 over "not just plain x," so the formula does not apply directly. Perhaps substitution will help; as always, the "not just plain x" part is our u:

$$u = 2x - 3$$

$$du = 2\,dx.$$

Adjusting for the constant 2, we may make the substitution:

$$\int \frac{1}{2x - 3}\,dx = \frac{1}{2}\int \frac{1}{2x - 3}\cdot 2\,dx = \frac{1}{2}\int \frac{1}{u}\,du.$$

Although the correct answer can be obtained by keeping the original limits of integration but switching back to the original variable before evaluating the definite integral, the resulting work is still incorrect because it says that two unequal expressions are equal. Said colloquially, two wrongs don't make a right, they make two wrongs.

Ans. to reading exercise 38:

Using substitution with $u = 5x - 1$ and $du = 5\,dx$,

$$\int \sqrt{5x - 1}\,dx = \frac{1}{5}\int 5\sqrt{5x - 1}\,dx$$

$$= \frac{1}{5}\int u^{1/2}\,du = \frac{1}{5}\frac{u^{3/2}}{3/2} + C$$

$$= \frac{2}{15}(5x - 1)^{3/2} + C.$$

Then

$$\int_1^2 \sqrt{5x - 1}\,dx = \frac{2}{15}(5x - 1)^{3/2}\Big|_1^2$$

$$= \cdots = \frac{38}{15}.$$

The antiderivative of $1/x$ can be restated using u to make it clear how it can be used with substitution:

$$\int \frac{1}{u}\,du = \ln|u| + C.$$

The need for substitution here is harder to recognize than when "not just plain x" is inside trig functions or a square root. Practice is essential.

The solution can be checked by differentiation:

$$\frac{d}{dx}\left(\frac{1}{2}\ln|2x-3|+C\right)$$

$$=\frac{1}{2}\cdot\frac{1}{2x-3}\cdot 2+0$$

$$=\frac{1}{2x-3},$$

which is the original integrand, as required.

In some sources, $\int \frac{1}{u}\,du$ is occasionally written as $\int \frac{du}{u}$.

Practice looking for this pattern. For rational functions, a numerator with degree one less than the denominator might indicate using substitution.

We then finish by using the formula and returning to the original variable x:

$$=\frac{1}{2}\ln|u|+C$$

$$=\frac{1}{2}\ln|2x-3|+C.$$

∎

In the formula $\int \frac{1}{u}\,du$, the u is in the denominator whereas du is in the numerator. If you recognize that an integrand is a fraction where the derivative of the denominator is in the numerator (allowing for adjustment by a constant), then a substitution of $u =$ denominator is helpful.

Example 107 *Find* $\displaystyle\int \frac{2x}{x^2+1}\,dx.$

Solution Notice that the denominator is x^2+1 and the numerator is its derivative, $2x$. We therefore try substitution, letting u be the denominator:

$$u = x^2+1$$

$$du = 2x\,dx.$$

Making the substitution and continuing from there,

$$\int \frac{2x}{x^2+1}\,dx = \int \frac{1}{u}\,du = \ln|u|+C = \ln|x^2+1|+C.$$

∎

Tip: when in doubt, leave the absolute values. Both $\ln|x^2+1|+C$ and $\ln(x^2+1)+C$ are correct solutions to example 107.

Because the quantity x^2+1 is always positive, the absolute values can be dropped in the solution to example 107. In other words, the solution can be written as $\ln(x^2+1)+C$.

The natural exponential function is its own derivative. It is therefore its own antiderivative as well.

ANTIDERIVATIVE OF THE NATURAL EXPONENTIAL FUNCTION

$$\int e^x\,dx = e^x + C$$

Example 108 *Evaluate $\int e^{3x}\,dx$.*

Solution　Because the antiderivative formula requires "just plain x" and our exponent is $3x$, substitution is needed:

$$u = 3x$$

$$du = 3\,dx.$$

Then,

$$\int e^{3x}\,dx = \frac{1}{3}\int 3e^{3x}\,dx$$

$$= \frac{1}{3}\int e^{u}\,du$$

$$= \frac{1}{3}e^{u} + C$$

$$= \frac{1}{3}e^{3x} + C.$$

∎

Any time the exponent in an exponential function is anything other than just plain x, try substitution with u as the exponent.

Line 1 adjusts for the required constant 3, using the antiderivative constant multiple rule to move the $\frac{1}{3}$ out of the integral; line 2 makes the substitution, replacing $3\,dx$ with du and replacing $3x$ with u; line 3 finds an antiderivative; and line 4 replaces u with $3x$, returning to the original variable.

Integrals similar to that of example 108 come up often enough that it is convenient to memorize a formula for $\int e^{ax}\,dx$, and similar formulas for $\int \sin kx\,dx$ and $\int \cos kx\,dx$. Derivations are similar to the solutions to example 108 and 104.

Ans. to reading exercise 39:
$$\frac{196}{3} = 65 + \frac{1}{3}$$

Some details: Use $u = x^2 + 9$. Then,

$$\int_0^4 2x\sqrt{x^2 + 9}\,dx = \int_9^{25} \sqrt{u}\,du$$

$$= \left.\frac{u^{3/2}}{3/2}\right|_9^{25} = \dots$$

CONVENIENT ANTIDERIVATIVE FORMULAS

$$\int e^{ax}\,dx = \frac{e^{ax}}{a} + C$$

$$\int \sin kx\,dx = \frac{-\cos kx}{k} + C$$

$$\int \cos kx\,dx = \frac{\sin kx}{k} + C.$$

Reading Exercise 40　Evaluate $\int e^{5x}\,dx$.

0.7.11　Integration by parts

The product rule states that if f and g are differentiable at x, then

$$\frac{d}{dx}\left(f(x)\cdot g(x)\right) = f(x)g'(x) + g(x)f'(x).$$

Knowing that every derivative formula can be reversed to give an integral formula, we get

We rewrite

$$\frac{d}{dx}\text{left side} = \text{right side}$$

as

$$\int \text{right side } dx = \text{left side} + C.$$

Because the right side now contains an integral, the $+C$ can be considered part of the answer to this integral. It is not necessary to write "$+C$" at the end of the equation.

The integration-by-parts formula can be remembered as

$$\int fg'\, dx = fg - \int gf'\, dx.$$

The integration-by-parts formula can be written in differential format as

$$\int u\, dv = uv - \int v\, du.$$

When both substitution and parts are needed, it is convenient not to use u with integration by parts.

Any antiderivative of $g'(x)$ suffices. We do not need all of them, so there is no need to write "$+C$" at the end of $g(x)$.

$$\int \big(f(x)g'(x) + g(x)f'(x)\big)\, dx = f(x) \cdot g(x) + C.$$

Using the antiderivative sum rule on the left-hand side gives

$$\int f(x)g'(x)\, dx + \int g(x)f'(x)\, dx = f(x)g(x) + C.$$

This equation can be rearranged by subtracting $\int g(x)f'(x)\, dx$ from both sides:

$$\int f(x)g'(x)\, dx = f(x)g(x) - \int g(x)f'(x)\, dx.$$

What did this accomplish? If $\int f(x)g'(x)\, dx$ is a difficult integral but $\int g(x)f'(x)\, dx$ is an easier integral, then the method helps. We call this technique of integration *integration by parts*.

INTEGRATION BY PARTS

If the functions are integrable,

$$\int f(x)g'(x)\, dx = f(x)g(x) - \int g(x)f'(x)\, dx.$$

Example 109 *Use integration by parts to evaluate* $\int x \sin x\, dx.$

Solution The integration-by-parts formula applies to an integral of the form $\int f(x)g'(x)\, dx$. The integrand must be a product of two functions, f and g'. Our integrand is the product of x and $\sin x$. Try using

$$f(x) = x \qquad\qquad g'(x) = \sin x.$$

The right-hand side of the integration-by-parts formula uses $f'(x)$ and $g(x)$, so they must be calculated. We find $f'(x)$ by taking the derivative of $f(x)$, and we find $g(x)$ by finding an antiderivative of $g'(x)$. We write them below our f and g':

$$f(x) = x \qquad\qquad g'(x) = \sin x$$
$$f'(x) = 1 \qquad\qquad g(x) = -\cos x.$$

Applying the integration by parts formula,

$$\int f(x)g'(x)\,dx = f(x)g(x) - \int g(x)f'(x)\,dx$$

$$\int x\sin x\,dx = x(-\cos x) - \int (-\cos x)\cdot 1\,dx$$

$$= -x\cos x + \int \cos x\,dx$$

$$= -x\cos x + \sin x + C.$$

Line 1 is the integration-by-parts formula; line 2 applies this formula with the functions f, g', f', and g; line 3 simplifies algebraically; and line 4 completes the integration and gives the final answer, which includes $+C$.

∎

In example 109, we began with an integral that contains a product, but by using integration by parts, we reached an integral that does not contain a product, $\int \cos x\,dx$, which we can integrate easily. Although the right-hand side's integral is $\int g(x)f'(x)\,dx$, which is a product, because $f'(x) = 1$, the "product" melts away into a single factor!

Going from $f(x) = x$ to $f'(x) = 1$ makes for a simpler integral. Meanwhile, changing from $g'(x)$ to $g(x)$ merely changes from $\sin x$ to $-\cos x$, which is not more complicated. This is the key: let f be a factor that has a simpler derivative, and let g be the rest of the expression, which hopefully does not have a more complicated antiderivative.

Which types of functions have less-complicated derivatives? Some examples are polynomials, logarithmic functions, and inverse trig functions:

polynomial	$f(x) = x^2$	$f'(x) = 2x$
logarithmic	$f(x) = \ln x$	$f'(x) = \dfrac{1}{x}$
inverse trig	$f(x) = \tan^{-1} x$	$f'(x) = \dfrac{1}{1+x^2}$

For polynomials the degree gets smaller; for logarithmic and inverse trig, the derivative is algebraic. These functions are good candidates for f in integration by parts.

On the other hand, exponential, trig, and hyperbolic functions do not have less-complicated derivatives:

exponential	$f(x) = e^x$	$f'(x) = e^x$
trig	$f(x) = \sin x$	$f'(x) = \cos x$
hyperbolic	$f(x) = \cosh x$	$f'(x) = \sinh x$

The derivatives of these functions are the same type of function. These functions are not good candidates for f in integration by parts.

Ans. to reading exercise 40:
$$\tfrac{1}{5}e^{5x} + C$$

0.7.12   Other techniques

There are many other techniques of integration that are not reviewed here, including partial fractions and trigonometric substitution. These techniques are used less frequently than substitution and parts.

Other strategies for integration utilize identities such as these:

$$\sin^2 \theta = \frac{1}{2}(1 - \cos 2\theta)$$

$$\cos^2 \theta = \frac{1}{2}(1 + \cos 2\theta)$$

$$\cosh^2 t - \sinh^2 t = 1.$$

And speaking of hyperbolic functions, there are many antiderivative formulas not mentioned here that are the reverse of derivative formulas reviewed in Section 0.6. Even more are listed in tables of integrals. Use of a CAS is also an option for difficult antiderivative questions.

For definite integrals, use of a calculator, a CAS, or other apps can provide answers accurate enough for most purposes. These technologies are based on numerical techniques such as the midpoint rule or Simpson's rule (although usually more sophisticated), which are not reviewed here.

EXERCISES 0.7

You are not asked to evaluate the integral, just to determine whether FTC II applies.

1–10. Rapid response: does FTC II apply to the integral?

1. $\displaystyle\int_3^7 4x^2 \, dx$

2. $\displaystyle\int_6^{17} (x^8 + x^3) \, dx$

3. $\displaystyle\int_{-5}^8 \frac{1}{x^2} \, dx$

4. $\displaystyle\int_{-5}^{-2} \frac{1}{x^2} \, dx$

5. $\displaystyle\int_{-3\pi}^{3\pi} \cos x \, dx$

6. $\displaystyle\int_0^7 \frac{1}{x^4} \, dx$

7. $\displaystyle\int_{-3\pi}^{3\pi} \tan x \, dx$

8. $\displaystyle\int_4^0 \sqrt{x} \, dx$

9. $\displaystyle\int_{-4}^{-1} \frac{1}{x-3} \, dx$

10. $\displaystyle\int_0^9 \sec x \, dx$

11–20. Rapid response: is substitution needed?

11. $\int 3x^2 \cos(x^3 + 12)\, dx$

12. $\int \sqrt{4x - 9}\, dx$

13. $\int x^2 \cos x\, dx$

14. $\int (6x + 1)\sqrt{3x^2 + x}\, dx$

15. $\int \sec x \tan x\, dx$

16. $\int \frac{x+4}{x-1}\, dx$

17. $\int (4x - 7)^8\, dx$

18. $\int x^2 \sin 6\, dx$

19. $\int \frac{1}{\sqrt{6x-11}}\, dx$

20. $\int \tan \pi x\, dx$

21–32. Rapid response: what substitution might you try?

21. $\displaystyle \int \frac{2}{(5x - 7)^3}\, dx$

22. $\displaystyle \int x^2 \sin x^3\, dx$

23. $\displaystyle \int \frac{2}{(5x - 7)^2}\, dx$

24. $\displaystyle \int \frac{\sin \sqrt{x}}{\sqrt{x}}\, dx$

25. $\displaystyle \int \frac{2}{5x - 7}\, dx$

26. $\displaystyle \int \frac{\sin(\ln x)}{x}\, dx$

27. $\displaystyle \int 2x(x^2 + 4)^5\, dx$

28. $\displaystyle \int \frac{x^4}{x^5 + 3}\, dx$

29. $\displaystyle \int \cos x(\sin x)^5\, dx$

30. $\displaystyle \int \frac{1 + \sec x \tan x}{x + \sec x}\, dx$

31. $\displaystyle \int \frac{(\ln x)^5}{x}\, dx$

32. $\displaystyle \int \frac{2x - 3}{x^2 - 3x + 4}\, dx$

33–46. Manipulate the integrand into a form for which the available rules and formulas apply, and then finish evaluating the integral.

33. $\int (x - 4)(x + 1)\, dx$

34. $\int \frac{x^3 - 4x^2 + 1}{x^2}\, dx$

35. $\int \frac{x^2 + 5x}{x}\, dx$

36. $\int (x + 1)^3\, dx$

37. $\int (x - 1)^2\, dx$

38. $\int 3x^4(x^2 - 7)\, dx$

39. $\int \sqrt{(3x + 1)^4}\, dx$

40. $\int (3x - 5x^2(x + 6))\, dx$

41. $\int \tan x \cos x\, dx$

42. $\int \frac{1}{\cos^2 x}\, dx$

43. $\int \sec x \frac{\sin x}{\cos x}\, dx$

44. $\int \sqrt{x^4 + 2x^2 + 1}\, dx$

45. $\int \frac{x^2 + 3}{\sqrt{x}}\, dx$

46. $\int 5 \sin^2 x \csc x\, dx$

47–56. Does the substitution succeed or fail?

47. $\int \sin(7x - 92)\, dx,\ u = 7x - 92$

48. $\int \sin(7x^2 - 92)\, dx,\ u = 7x^2 - 92$

49. $\int \sqrt{6x^3 + 12}\, dx,\ u = 6x^3 + 12$

50. $\int \sqrt{6x + 12}\, dx,\ u = 6x + 12$

51. $\int \sec 5x \tan 5x \, dx$, $u = 5x$

52. $\int 5x \sec 5x \tan 5x \, dx$, $u = 5x$

53. $\int x^3 (x^2 - 17x + 1)^{14} \, dx$, $u = x^2 - 17x + 1$

54. $\int \cos x \sqrt{\sin x} \, dx$, $u = \sin x$

55. $\int 2x \sqrt{\sin x^2} \cos x^2 \, dx$, $u = \sin x^2$

56. $\int \sqrt{x} \cos(7x + 5) \, dx$, $u = 7x + 5$

57–96. Evaluate the integral, if possible. If not possible, state "not possible with available rules, formulas, and techniques."

57. $\int (x^3 - 4x^2 + 7x - 1) \, dx$

58. $\int (7x^2 - 11x + 9) \, dx$

59. $\int \cos(2x + 5) \, dx$

60. $\int -2 \sec x \tan x \, dx$

61. $\int (2x + 5 \cos x) \, dx$

62. $\int (3x + 7)^{10} \, dx$

63. $\int \dfrac{2}{x + 1} \, dx$

64. $\int 4\sqrt[3]{x} \, dx$

65. $\int 6 \csc^2 x \, dx$

66. $\int \sin(3x - 4) \, dx$

67. $\int \csc^2(6z - 15) \, dz$

68. $\int e^{6x} \, dx$

69. $\int 4e^{5x} \, dx$

70. $\int 2(x + 3)^2 \, dx$

71. $\int x^3 (4x - 9) \, dx$

72. $\int \dfrac{x}{\sqrt{x^2 + 5}} \, dx$

73. $\int \dfrac{3}{4x^2} \, dx$

74. $\int (x^2 + 4)^2 \, dx$

75. $\int \dfrac{2x}{x^2 + 4} \, dx$

76. $\int \dfrac{3}{4 - x} \, dx$

77. $\int \dfrac{x^3 - 7x^2 + 4}{x} \, dx$

78. $\int x^2 \sin x \, dx$

79. $\int (5x - 2)^2 \, dx$

80. $\int \sec x \, dx$

81. $\int x \cos x \, dx$

82. $\int \dfrac{x^3 - 1}{x^2} \, dx$

83. $\int x^2 \sin x^3 \, dx$

84. $\int \dfrac{x^2}{x^3 - 1} \, dx$

85. $\int \sec 7x \tan 7x \, dx$

86. $\int x \cos x^2 \, dx$

87. $\int \dfrac{x^3 - 7x}{4x} \, dx$

88. $\int \tan^4 x \sec^2 x \, dx$

89. $\int \dfrac{e^{\sqrt{x}}}{2\sqrt{x}} \, dx$

90. $\int x e^{x^2} \, dx$

91. $\int \sqrt{9 - x^5} \, dx$

92. $\int \dfrac{\sin(\ln x)}{x} \, dx$

93. $\int \left(\dfrac{1}{\sin^2 x} + \dfrac{1}{x^2} \right) dx$

94. $\int 3x e^x \, dx$

95. $\int x \ln x \, dx$

96. $\int \dfrac{\sqrt{z^3 - 7z}}{3z^2 - 7} \, dz$

97–132. Evaluate the integral, if possible. If not possible, say why.

97. $\displaystyle\int_{9}^{16} \frac{1}{\sqrt{x}}\, dx$

98. $\displaystyle\int_{2}^{10} \sqrt{2y+5}\, dy$

99. $\displaystyle\int_{-3}^{7} \frac{2}{x^3}\, dx$

100. $\displaystyle\int_{0}^{\sqrt{\pi}} 4x^2 \cos x\, dx$

101. $\displaystyle\int_{-4}^{-4} \frac{x^2-7}{\sqrt{x+19}}\, dx$

102. $\displaystyle\int_{0}^{\sqrt{\pi}} 4x \cos x^2\, dx$

103. $\displaystyle\int_{0}^{1} 2x(x^2+1)^7\, dx$

104. $\displaystyle\int_{1}^{2} \frac{1}{x-3}\, dx$

105. $\displaystyle\int_{0}^{\pi} \sin x\, dx$

106. $\displaystyle\int_{3}^{-3} \frac{4}{x^4}\, dx$

107. $\displaystyle\int_{0}^{1} e^{4x-1}\, dx$

108. $\displaystyle\int_{2}^{3} \frac{10}{2x-5}\, dx$

109. $\displaystyle\int_{0}^{\frac{\pi}{2}} \sin 3x\, dx$

110. $\displaystyle\int_{0}^{6} (1+x)^2\, dx$

111. $\displaystyle\int_{-\pi/4}^{0} \frac{\sec x \tan x}{\sec^5 x}\, dx$

112. $\displaystyle\int_{0}^{\frac{\pi}{4}} \cos 5x\, dx$

113. $\displaystyle\int_{2}^{5} \frac{1}{3x-2}\, dx$

114. $\displaystyle\int_{-1}^{5} \sqrt{x^2+2x+1}\, dx$

115. $\displaystyle\int_{1}^{4} (1-\sqrt{x})^2\, dx$

116. $\displaystyle\int_{0}^{\pi/3} (1-\cos x)^3 \sin x\, dx$

117. $\displaystyle\int_{2}^{3} x(x^2-4)^{2/3}\, dx$

118. $\displaystyle\int_{1}^{2} \frac{-x^2-6}{x^2}\, dx$

119. $\displaystyle\int_{0}^{\pi} x^2 \cos x\, dx$

120. $\displaystyle\int_{-2\pi}^{0} \frac{1}{2} \cos x\, dx$

121. $\displaystyle\int_{4}^{\infty} \frac{1}{x^2}\, dx$

122. $\displaystyle\int_{-1}^{0} \sqrt[4]{3x+5}\, dx$

123. $\displaystyle\int_{1}^{2} \frac{(\ln x)^2}{x}\, dx$

124. $\displaystyle\int_{0}^{\pi/3} \sec^2 x\, dx$

125. $\displaystyle\int_{-\sqrt{\pi}}^{\sqrt{\pi}} \sin x^2\, dx$

126. $\displaystyle\int_{1}^{-1} \frac{5x^2+10x-1}{5}\, dx$

127. $\displaystyle\int_{-\frac{1}{2}}^{0} \frac{1}{x^2+4x+4}\, dx$

128. $\displaystyle\int_{1}^{2} x^3\sqrt{x^4+x}\, dx$

129. $\displaystyle\int_{-1}^{1} (x+2x)^4\, dx$

130. $\displaystyle\int_{1}^{4} e^{\ln \sqrt{x}}\, dx$

131. $\displaystyle\int_{-4}^{-2} \sqrt{x^2}\, dx$

132. $\displaystyle\int_{0}^{1} \frac{2}{(x-4)^{-2}}\, dx$

133–136. Find the net area under the curve.

133. $y = \sin x + \cos x$ from $x = 0$ to $x = \pi$

134. $y = 1 - 2\sin x$ from $x = \dfrac{\pi}{2}$ to $x = \pi$

135. $y = 5\sqrt{5x + 4}$ from $x = 0$ to $x = 9$

136. $y = (2x - 1)^7$ from $x = 1$ to $x = 2$

Vectors and the Geometry of Space

1.1   A Three-Dimensional Coordinate System

In some cultures, it is customary to refer to four directions: North, south, east, and west. These roughly correspond to the two-dimensional coordinate system, aka the *Cartesian plane* or the xy-plane (Figure 1.1). The positive x-axis corresponds to the direction east, the negative x-axis to west, the positive y-axis to north, and the negative y-axis to south.

By contrast, in Cherokee culture, there are seven directions. In addition to the familiar north, south, east, and west, we add earth, sky, and self. The result is a three-dimensional coordinate system as in Figure 1.2, where the third axis is labeled the z-axis. Self corresponds to the origin, earth corresponds to "down" (the negative z-axis), and sky corresponds to "up" (the positive z-axis). Imagine yourself at the intersection of the three axes (the origin), and the other six directions should make sense.

There are, of course, major differences between Figures 1.1 and 1.2. We live in a three-dimensional world, and the image on the paper or the screen is two-dimensional. In Figure 1.1 we perceive the xy-plane as lying entirely within the

The directions north, south, east, and west are sometimes referred to as *cardinal directions*.

The human brain is accustomed to perceiving three dimensions in an image that is actually two-dimensional, through constant practice viewing pictures and videos of our three-dimensional world. There have been instances of mathematicians blind from birth who claim to have an advantage over sighted mathematicians in imagining four or more dimensions.

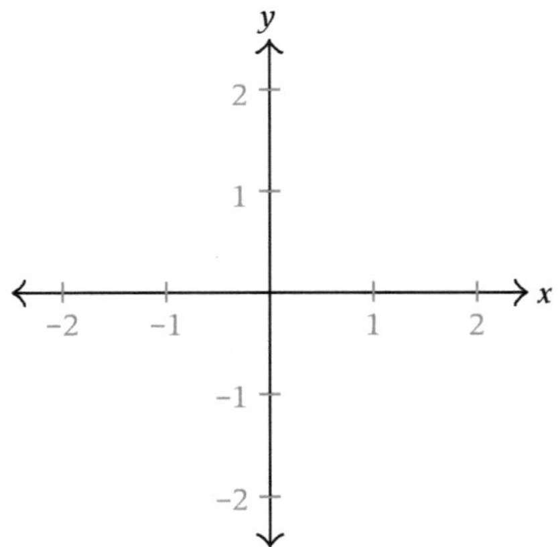

Figure 1.1 *The two-dimensional Cartesian plane*

Multivariable Calculus Set Free. Charles Bryan Dawson, Oxford University Press.
© C. Bryan Dawson (2026). DOI: 10.1093/oso/9780198984269.003.0002

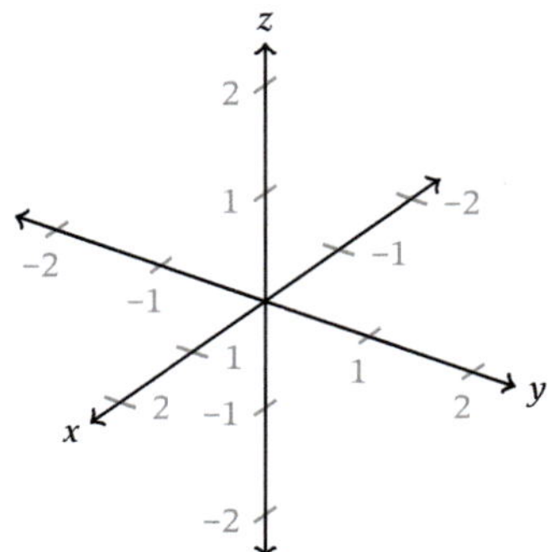

Figure 1.2 *The three-dimensional Cartesian coordinate system*

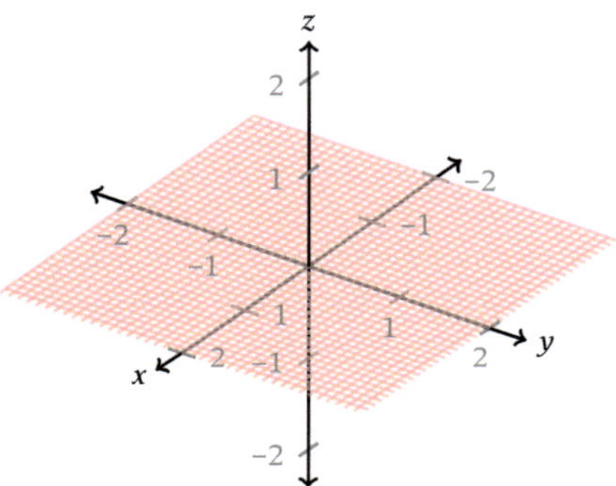

Figure 1.3 *The xy-plane indicated by cross-hatched shading*

page, with our eye lying outside the coordinate system. In Figure 1.2, the image is in two dimensions, but it is meant to represent something in three dimensions. Our eye, though, is not outside of the three-dimensional coordinate system, because that would require a fourth dimension, something we humans do not have at our disposal. These realities impact the way we produce and perceive three-dimensional diagrams.

Let's add to Figure 1.2 by shading the *xy*-plane, that is, the two-dimensional plane that includes the *x*- and *y*-axes (Figure 1.3). Think of it as horizontal, like a floor in a room, with the positive *z*-axis pointing upward and the negative *z*-axis pointing downward. In the figure, our eye appears to be above the *xy*-plane, so our viewpoint has a positive *z* coordinate. Although the *xy*-plane appears tilted in Figure 1.3, it is not difficult to imagine it being horizontal, due to our experience viewing pictures. Similarly, Figure 1.4 shows vertical shaded "walls" representing the *xz*-plane, which is the two-dimensional plane that includes the *x*- and *z*-axes, and the *yz*-plane, which includes the *y*- and *z*-axes. Don't be bothered by the

If the *xy*-plane is horizontal, the *z*-axis is vertical.

Remember that our eye is located inside the three-dimensional coordinate system, not outside of it.

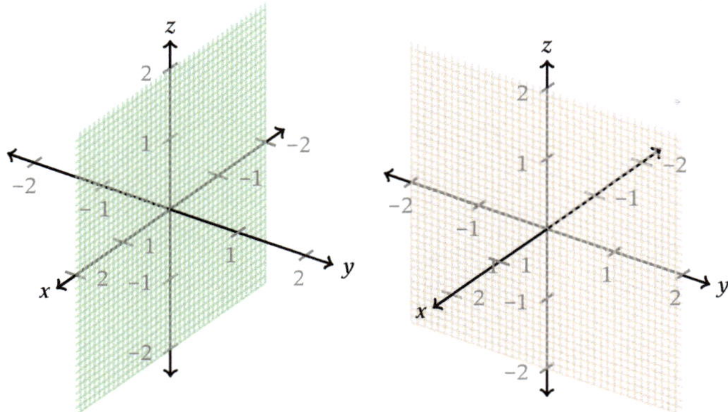

Figure 1.4 *The xz-plane (left) and yz-plane (right)*

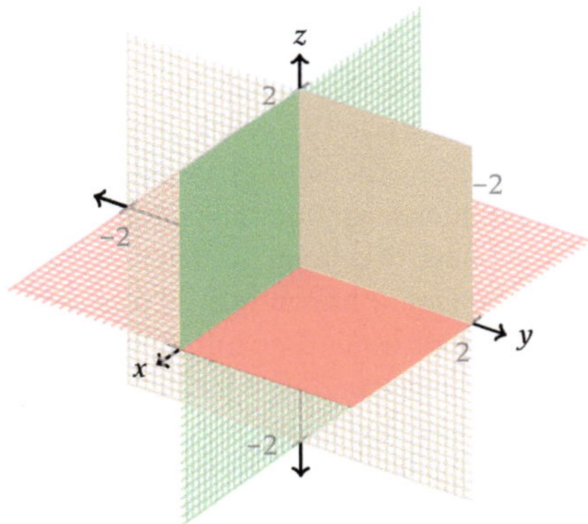

Figure 1.5 *The xy-plane, xz-plane, and yz-plane in one diagram. The solid portions indicate what would be seen from the chosen viewpoint*

impression that these walls extend below the floor; think of being in a second-floor apartment. Inspection of Figure 1.4 shows that our viewpoint has positive x- and y-coordinates, as well.

Now put all three planes in the same diagram (Figure 1.5). Think of yourself as being located in a second-floor apartment comprised of points with positive x-, y-, and z-coordinates, viewing your floor and two walls. The floor extends to other apartments, as do the walls. There are eight apartments, four on the first floor and four on the second floor. These eight regions are called *octants*, similar to quadrants in the two-dimensional coordinate system of Figure 1.1. Take a moment to locate, in Figure 1.5, the octant (apartment) consisting of positive x-coordinates, negative y-coordinates, and negative z-coordinates.

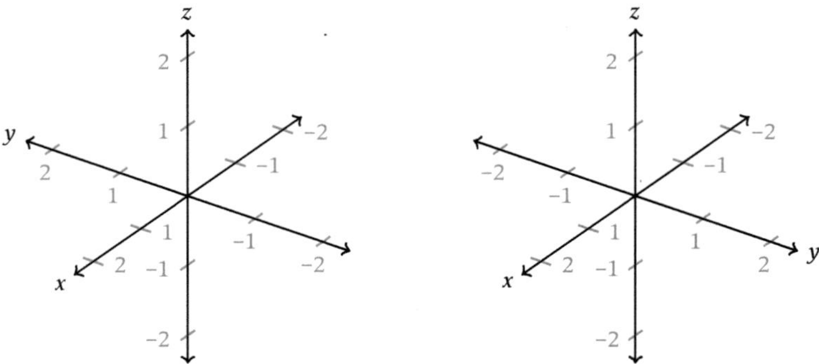

Figure 1.6 *A left-handed coordinate system (left) and a right-handed coordinate system (right)*

A four-dimensional being with a three-dimensional page does not have the same problem. A four-dimensional doctor treating a three-dimensional patient could see every point within the patient's body simultaneously, without the need for an MRI!

To read more about how beings in various dimensions would perceive objects of other dimensions, see the classic book *Flatland* by Edwin Abbott Abbott. Originally published in 1884, the book can be found in various editions today. The book also features satirical commentary on late 19th-century English society.

I realize that this is awkward when trying to match the figure.

Let's pause and look again at Figure 1.1. In a two-dimensional coordinate system, with our eye located outside the system, each location on the page corresponds to exactly one point in the coordinate system. But in Figure 1.5, with our eye inside the three-dimensional coordinate system and using only a two-dimensional page, we don't have that luxury. There are portions of the green wall that are hidden "behind" the (solid) brown wall of your apartment. Each location on the two-dimensional page corresponds to infinitely many different points in the three-dimensional coordinate system. But which points correspond to the same location on the page depend on the viewpoint. By moving your eye to a different viewpoint, you can see objects that might have been hidden from view in the original viewpoint; the correspondence between points and location on the page changes.

Here's yet another complication: there is more than one way that the three axes can relate to one another. Consider the differences between the coordinate systems in Figure 1.6, where the coordinate system of Figure 1.2 is on the right and a different choice is on the left. The system on the right is called a right-handed coordinate system, and the one on the left is left-handed. The reasoning for this nomenclature comes from positioning your hand to point in the direction of the positive x-axis with your thumb pointing up. Now allow your fingers to curl. Which direction are your fingers pointing? More to the point, in which side of Figure 1.6 are your fingers pointing in the direction of the positive y-axis? Using your right hand should match the diagram on the right, and your left hand should match the diagram on the left. With apologies to left-handed individuals, as indicated by Figures 1.2–1.5, we shall henceforth use the right-handed option.

1.1.1   Plotting points in three dimensions

Points in the two-dimensional Cartesian coordinate system have two coordinates, one for x and one for y. The three-dimensional Cartesian coordinate system adds a third coordinate, for z. The point $(2, 8.3, -9)$ has x-coordinate 2, y-coordinate 8.3, and z-coordinate -9.

To plot a point, we first need to decide on a viewpoint. It is customary to use a viewpoint in the first octant, where all coordinates are positive, as in Figures 1.2–1.5. Of course, other viewpoints are just as valid. But the steps shown in example 1 are designed for a first-octant viewpoint.

Example 1 *Plot the point $P = (1, 2, 3)$.*

Points are often named using capital letters, as you may have done when studying geometry.

Solution   ❶ Draw the positive coordinate axes (Figure 1.7, left). The axes should not look perpendicular on the page when viewed flat, but rather the angles should look larger than 90°; it is only in the three-dimensional image that we create in our minds that the three axes are perpendicular to one another.

❷ In the xy-plane, draw a rectangle with one corner at the origin and the opposite corner at $(1, 2, 0)$ (that is, using the x- and y-coordinates of the desired point; see Figure 1.7, right). Be sure to draw the sides of the rectangle parallel to the axes. Using dashed lines might help.

❸ Repeat step ❷, except that this rectangle is drawn 3 units higher (that is, in the plane $z = 3$ as indicated by the z-coordinate; anchor the rectangle at the point $(0, 0, 3)$), parallel to the xy-plane, as in Figure 1.8, left. Again, the sides of this rectangle should be drawn parallel to the x- and y-axes, as appropriate.

❹ Connect corresponding corners of the rectangles from steps ❷ and ❸. You should now have a rectangular box (called a *rectangular parallelepiped*). See Figure 1.8, right.

❺ The desired point is located at the corner of the rectangular box opposite from the origin; see Figure 1.9. ■

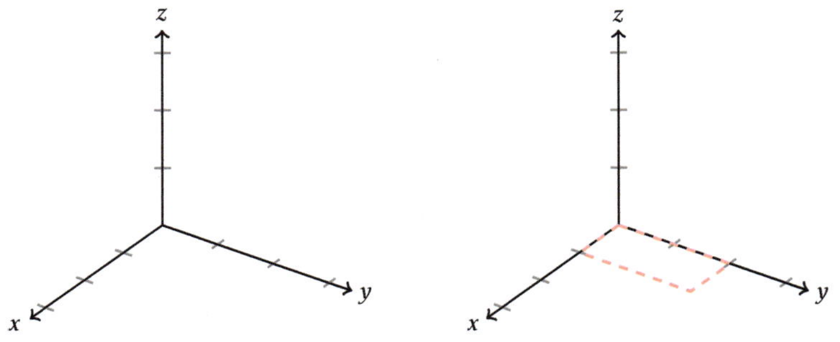

If a protractor is placed on the page to measure the angles between axes, the protractor should read greater than 90° if it is drawn correctly.

Figure 1.7 *Example 1 step 1 (left), step 2 (right)*

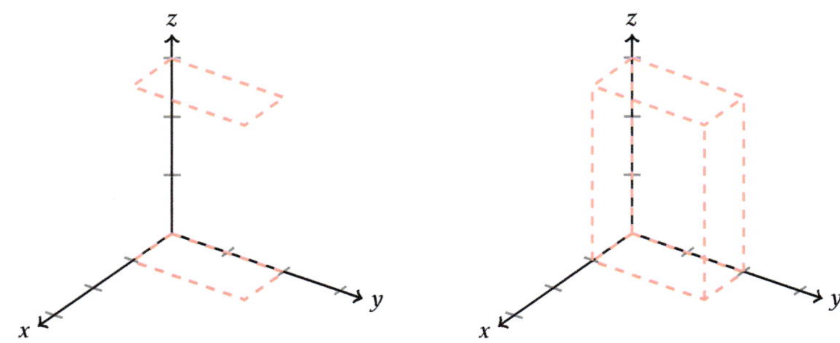

Figure 1.8 *Example 1 step 3 (left), step 4 (right)*

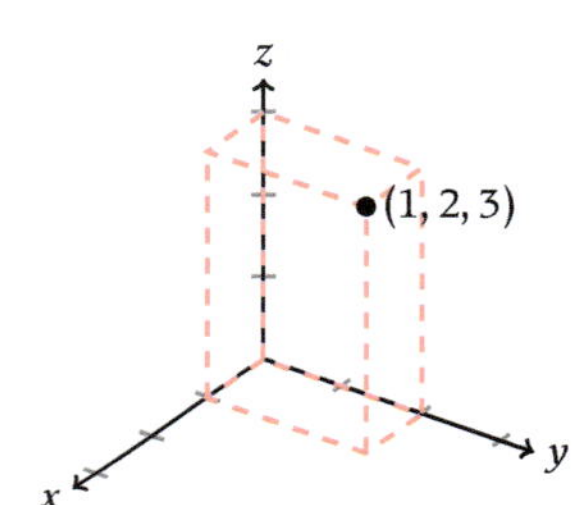

Figure 1.9 *Example 1 step 5*

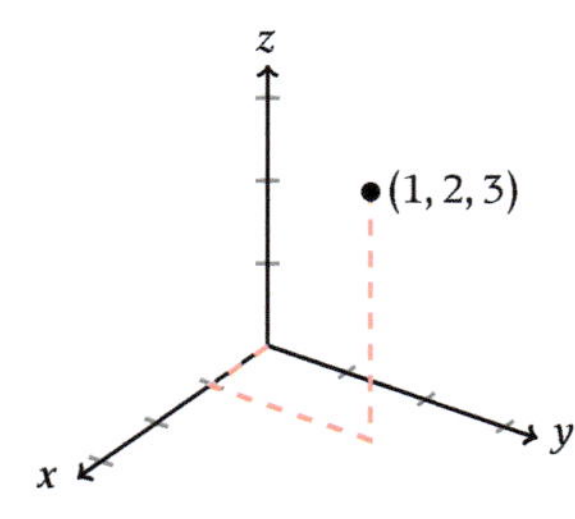

Figure 1.10 *Plotting the point* $(1, 2, 3)$. *Compare the location of the point, which has* z-*coordinate 3, to the* z-*axis. Similarly, compare the location of the point, which has* y-*coordinate 2, to the* y-*axis. Keeping these comparisons in mind can help avoid common errors. It also illustrates the usefulness of drawing the box, as in Figures 1.7–1.9*

Visualizing these possibilities in your mind can be an interesting exercise and can illustrate just how adept our minds are at creating three-dimensional images.

With practice, points can be located accurately without drawing the entire box, as illustrated in Figure 1.10. But practicing a few times in the manner shown in example 1 is helpful for avoiding common errors.

Next, let's look at the same point but without the dashed lines (Figure 1.11). Okay, we know it's supposed to be the point $(1, 2, 3)$. But suppose we didn't know that. Is the point, perhaps, in the yz-plane, at $(0, 1.1, 2)$? Or is it in the xz-plane at about $(-2, 0, 1)$? Or even in the xy-plane... there are many other possibilities as well.

Example 2 *Plot the point* $Q = (-1, 4, -2)$.

Solution ❶ Draw the positive coordinate axes. Next, because we have two negative coordinates, extend the x-axis and the z-axis as needed (Figure 1.12, left).

❷ In the xy-plane, draw a rectangle with one corner at the origin and the opposite corner at $(-1, 4, 0)$ (Figure 1.12, right).

For more details on these steps, follow the detailed explanation in example 1.

❸ Repeat step ❷, drawn 2 units lower (Figure 1.13, left).

❹ Connect corresponding corners of the rectangles from steps ❷ and ❸ to finish the rectangular box (Figure 1.13, right).

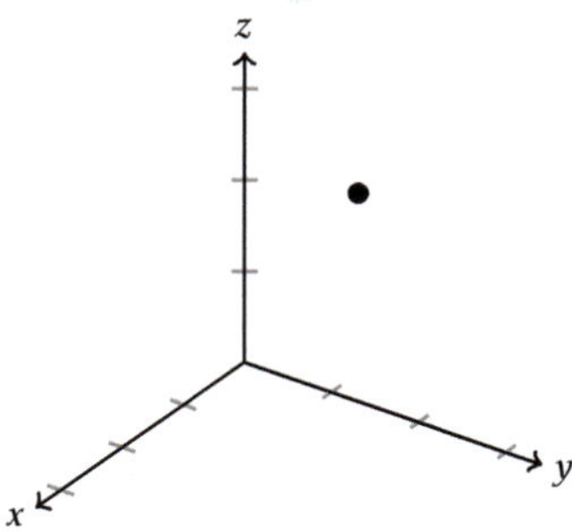

Figure 1.11 *The point from example 1. If we didn't already know the coordinates of the point, we would not be able to tell by only using this diagram which of infinitely many possibilities the point is meant to represent*

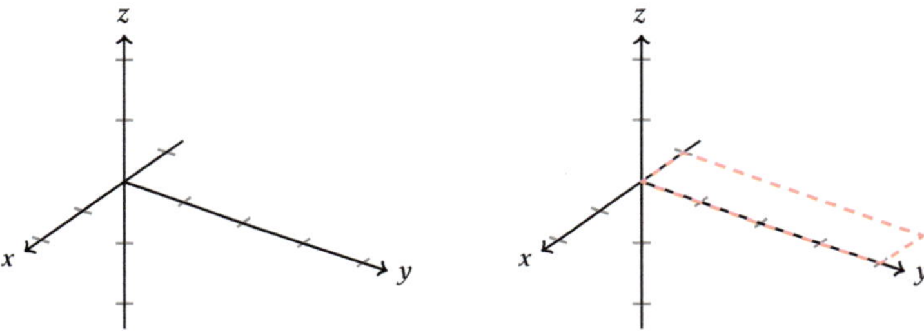

Figure 1.12 *Example 2 step 1 (left), step 2 (right)*

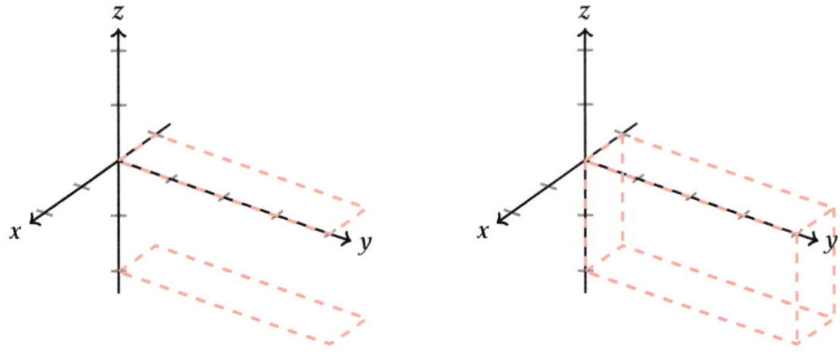

Figure 1.13 *Example 2 step 3 (left), step 4 (right)*

5 Place the point at the corner of the rectangular box opposite from the origin (Figure 1.14). ∎

Reading Exercise 1 Plot the point $(4, 2, 7)$.

You may have seen the two-dimensional Cartesian plane referred to as $\mathbf{R}^2$. Three-dimensional Cartesian space is referred to as $\mathbf{R}^3$.

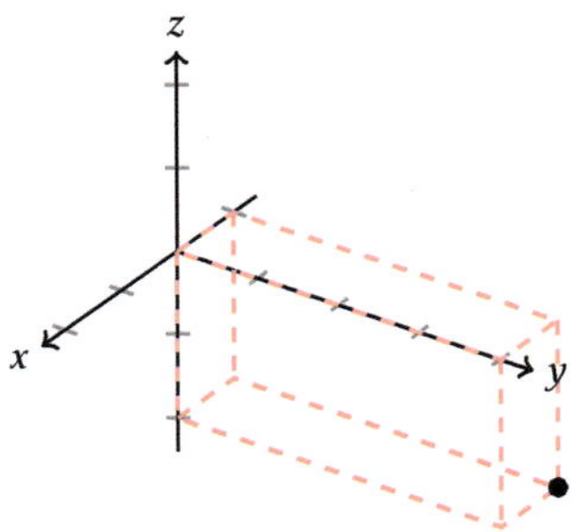

Figure 1.14 *Example 2 step 5. The dot is located at the point* $(-1, 4, -2)$

1.1.2   Equations and inequalities in three dimensions

In $\mathbf{R}^2$ (in two dimensions), the equation $x = 2$ represents a vertical line, as in Figure 1.15. Any point that satisfies the equation must have x-coordinate 2, but the y-coordinate, being unspecified by the equation, can be anything.

In $\mathbf{R}^3$ (in three dimensions), a point satisfying the equation $x = 2$ must still have x-coordinate 2, but now there are two coordinates that are unspecified, the y- and z-coordinates. The resulting set of points satisfying the equation is a plane.

Example 3 *Graph the equation $x = 2$.*

Solution   ❶ Draw the coordinate axes (Figure 1.16, left).

❷ From the point $(2, 0, 0)$ on the x-axis, draw arrows up and down parallel to the z-axis and arrows left and right parallel to the y-axis (Figure 1.16, right). Drawing these arrows of similar size to the previously drawn axes is appropriate.

❸ Finish framing the portion that will be shaded, drawing the frame lines parallel to the arrows of step ❷, as in Figure 1.17, left.

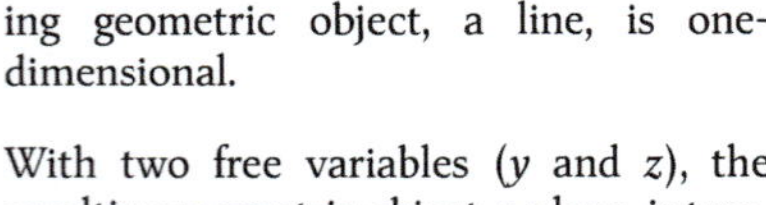

With one free variable (y), the resulting geometric object, a line, is one-dimensional.

With two free variables (y and z), the resulting geometric object, a plane, is two-dimensional.

Unless otherwise specified, equations and drawings in this section are assumed to be in $\mathbf{R}^3$.

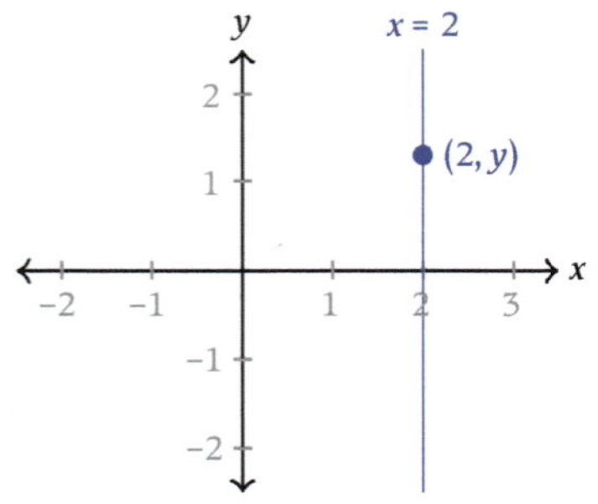

Figure 1.15 *The line $x = 2$ (blue) in the two-dimensional Cartesian plane. Any point on the line has the form $(2, y)$. The set of all points $(2, y)$ where y can be any real number forms the graph of the equation*

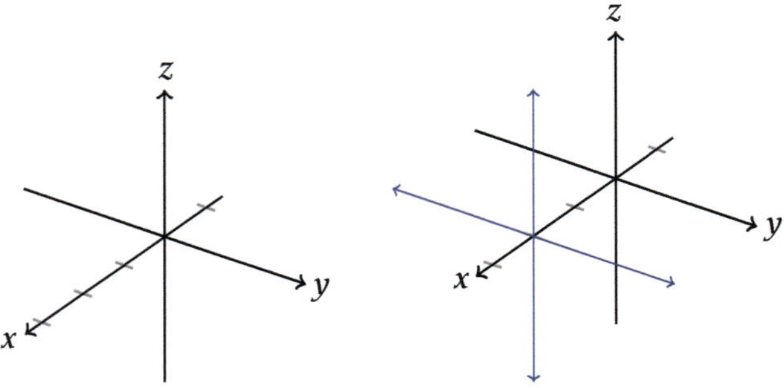

Figure 1.16 *Left: the coordinate axes for example 3. While there is nothing wrong with including tick marks for the y- and z-axes, they are not necessary for this particular graph. Right: step 2 for exercise 3*

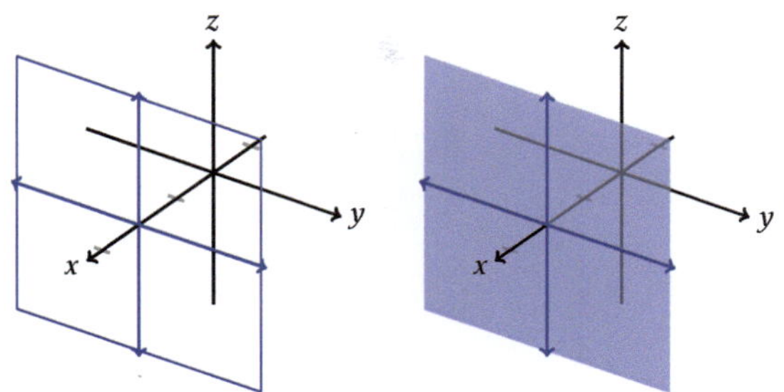

Figure 1.17 *Example 3, step 3 (left) and 4 (right). In three dimensions, the plane $x = 2$ (blue shaded) is parallel to the yz-plane, 2 units "in front" of the yz-plane. Any point in the plane has the form $(2, y, z)$. The set of all points $(2, y, z)$ where each of y and z can be any real number forms the graph of the equation*

4 Shade the portion of the plane inside the frame, resulting in the solution shown in Figure 1.17, right. ▮

Just as the portion of the line drawn in Figure 1.15 does not show the entire infinitely long line, the portion of the plane drawn in Figure 1.17 does not show the entire infinite extent of the plane. Using one's imagination is required!

The solution to example 3 illustrates the tricks that are sometimes necessary to help us visualize a three-dimensional coordinate system in a two-dimensional drawing. Such tricks are not necessary when graphing equations in $\mathbf{R}^2$, but they are essential for effective visualization in $\mathbf{R}^3$.

In example 3, the challenge was to draw a two-dimensional object in its appropriate location and orientation in three dimensions. Inequalities can be even more challenging, because the region described by the inequality may be three-dimensional. Drawing a three-dimensional object usually requires drawing the two-dimensional boundary of the object. Shading can then be added to indicate which "side" of the boundary is included in the object.

Example 4 *Graph the inequality $z < 0$.*

Solution **1** Draw the plane $z = 0$, using the techniques of example 3.
2 Shade "below" the plane. One method for accomplishing this is illustrated in Figure 1.18, which includes arrows to indicate that the region extends infinitely far downward. ▮

Does the solution shown here to example 4 seem inadequate to express what seems simpler in your mind's eye? Many variations are possible, and the reader is encouraged to experiment. If you wish to learn additional techniques for depicting three-dimensional objects on a two-dimensional page, consult one of the many

Even if you are "drawing challenged," as I am, it is my hope that some of these tricks can help you make three-dimensional drawings to the extent desired for calculus, and that more importantly you can learn to visualize graphs in $\mathbf{R}^3$ in your mind.

Think about an artist drawing an animal, which is a three-dimensional object. The artist doesn't draw the animal's innards; the artist draws the animal's fur or scales or other outward elements. The artist is drawing the surface only. You have previously studied volume, which is three-dimensional, and surface area, which is two-dimensional. The boundary, or surface, of the region to be depicted should be drawn first.

Ans. to reading exercise 1:

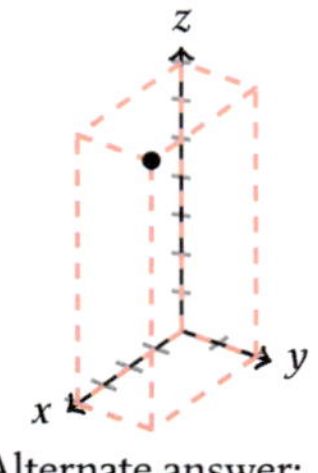

Alternate answer:

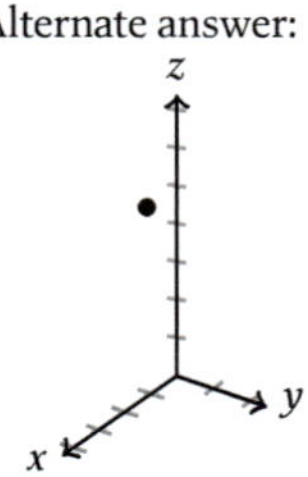

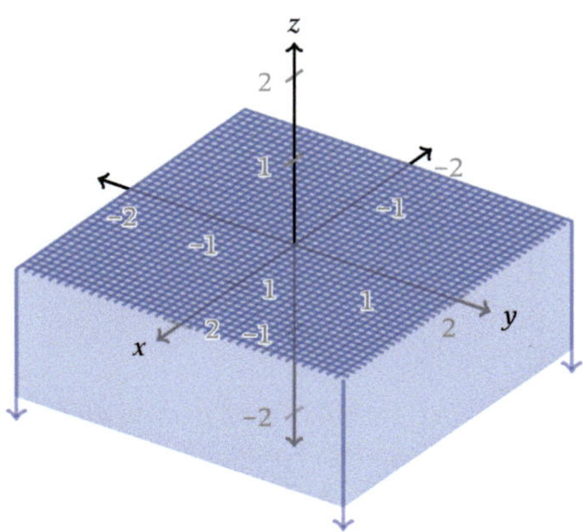

Figure 1.18 *The region satisfying the inequality $z < 0$ (shaded blue). The arrows indicate that the region continues below what's actually shaded. Because the plane continues indefinitely also, the shaded region extends in a like manner. Such pictures are meant as an aid to visualizing in one's mind the true extent of the region*

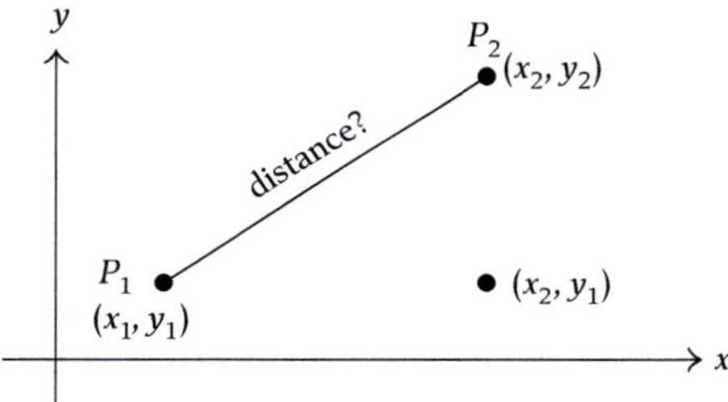

Figure 1.19 *Determining the distance between points P_1 and P_2*

available books by artists on three-dimensional drawing techniques. Technology, such as your favorite CAS (computer algebra system), can also help with visualization.

Reading Exercise 2   In $\mathbf{R}^3$, graph the equation $y = 2$.

1.1.3   Distance in three dimensions

Before presenting the distance formula in three dimensions, let's review the derivation of the distance formula in $\mathbf{R}^2$. Given two points $P_1 = (x_1, y_1)$ and $P_2 = (x_2, y_2)$, consider a third point (x_2, y_1) that aligns vertically with P_2 and horizontally with P_1, as pictured in Figure 1.19. Drawing the vertical and horizontal segments forms a right triangle, so the Pythagorean theorem applies. The length of the horizontal segment is the difference in the x-coordinates, $|x_2 - x_1|$. Likewise, the length of the vertical segment is $|y_2 - y_1|$; see Figure 1.20. Using $d(P_1, P_2)$ to represent the distance between P_1 and P_2, by the Pythagorean theorem,

$$d(P_1, P_2)^2 = |x_2 - x_1|^2 + |y_2 - y_1|^2$$
$$= (x_2 - x_1)^2 + (y_2 - y_1)^2.$$

To see that removing the absolute values in line 2 does not change the quantity, recall that, for instance, $(-4)^2 = 16$ and $|-4|^2 = 16$.

Therefore

$$d(P_1, P_2) = \sqrt{(x_2 - x_1)^2 + (y_2 - y_1)^2}.$$

To see how this extends to $\mathbf{R}^3$, we start with the simpler visualization of the distance between the points $(0, 0, 0)$ and (a, b, c), as shown in Figure 1.21. Notice that in the "floor," the xy-plane, there is a right triangle with legs of length a and b and hypotenuse of length k. Therefore, by the Pythagorean theorem,

$$a^2 + b^2 = k^2.$$

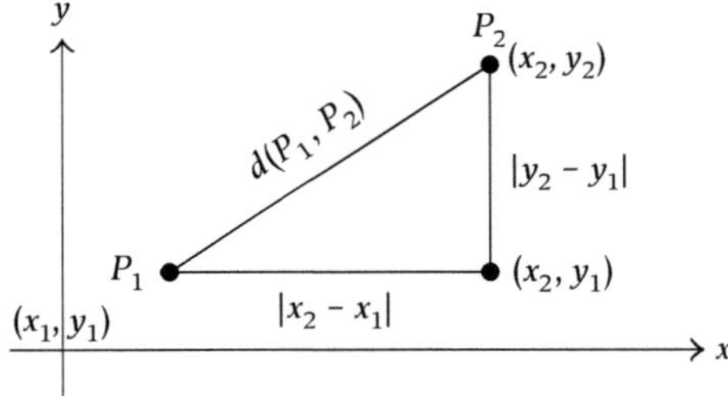

Figure 1.20 *Determining the distance between points P_1 and P_2*

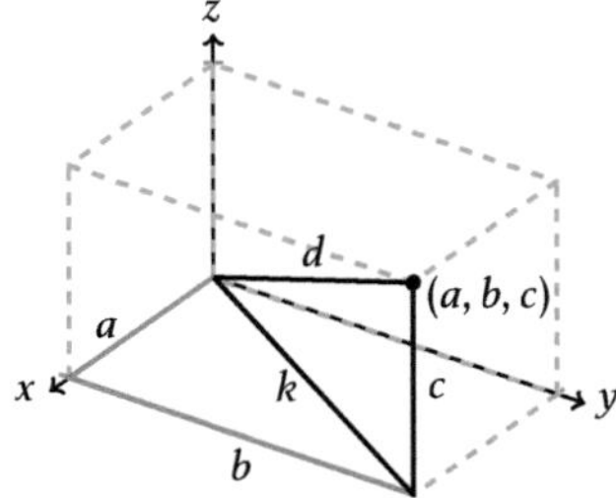

Figure 1.21 *Determining the distance between points $(0, 0, 0)$ and (a, b, c)*

Because the segment with length k is horizontal (in the floor) and the segment of length c is vertical (parallel to the z-axis), they are also perpendicular, and there is a right triangle with legs of lengths k and c and hypotenuse of length d; therefore

$$k^2 + c^2 = d^2.$$

Ans. to reading exercise 2:

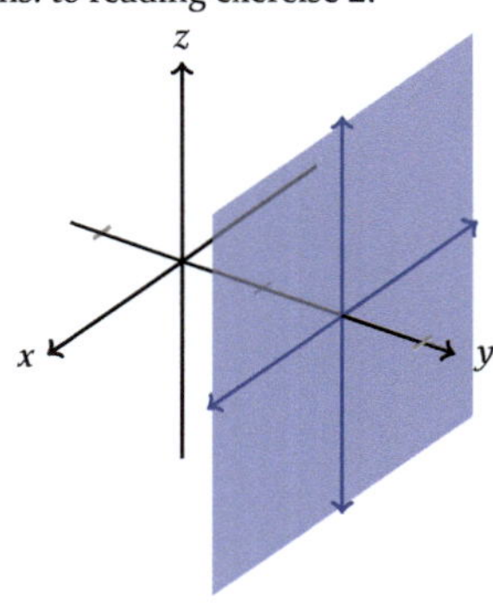

It is the distance d that we wish to find; by substitution,

$$k^2 + c^2 = d^2$$
$$a^2 + b^2 + c^2 = d^2.$$

Now consider two points (x_1, y_1, z_1) and (x_2, y_2, z_2), as in Figure 1.22. Comparing to Figure 1.21, we see that the geometry is identical, with $a = |x_2 - x_1|$, $b = |y_2 - y_1|$, and $c = |z_2 - z_1|$. Therefore

$$d^2 = |x_2 - x_1|^2 + |y_2 - y_1|^2 + |z_2 - z_1|^2,$$

and the distance between the points $P_1 = (x_1, y_1, z_1)$ and $P_2 = (x_2, y_2, z_2)$ is

$$d(P_1, P_2) = \sqrt{(x_2 - x_1)^2 + (y_2 - y_1)^2 + (z_2 - z_1)^2}.$$

Compare this formula for distance in $\mathbf{R}^3$ to the formula for distance in $\mathbf{R}^2$. For $\mathbf{R}^3$ we have added a term to deal with the difference in z-coordinates. Can you guess what the formula would be for distance between two points in $\mathbf{R}^4$? In $\mathbf{R}^{20}$? What about infinitely many dimensions–could the distance possibly be finite?

Definition 1 DISTANCE FORMULA FOR $\mathbf{R}^3$ *The* distance *between the points* $P_1 = (x_1, y_1, z_1)$ *and* $P_2 = (x_2, y_2, z_2)$ *is given by*

$$d(P_1, P_2) = \sqrt{(x_2 - x_1)^2 + (y_2 - y_1)^2 + (z_2 - z_1)^2}.$$

Example 5 *Determine the distance between the points* $(2, 0, 7)$ *and* $(1, 4, -3)$.

Solution Using the distance formula with $(x_1, y_1, z_1) = (2, 0, 7)$ and $(x_2, y_2, z_2) = (1, 4, -3)$, we have

$$d = \sqrt{(1 - 2)^2 + (4 - 0)^2 + (-3 - 7)^2}$$
$$= \sqrt{(-1)^2 + 4^2 + (-10)^2} = \sqrt{117}.$$

Alternate solution: using $(x_1, y_1, z_1) = (1, 4, -3)$ and $(x_2, y_2, z_2) = (2, 0, 7)$,

$$d = \sqrt{(2 - 1)^2 + (0 - 4)^2 + (7 - (-3))^2}$$
$$= \sqrt{1^2 + (-4)^2 + 10^2} = \sqrt{117}.$$

$\blacksquare$

Reading Exercise 3 Find the distance between the points $(4, 2, 7)$ and $(-3, 5, 1)$.

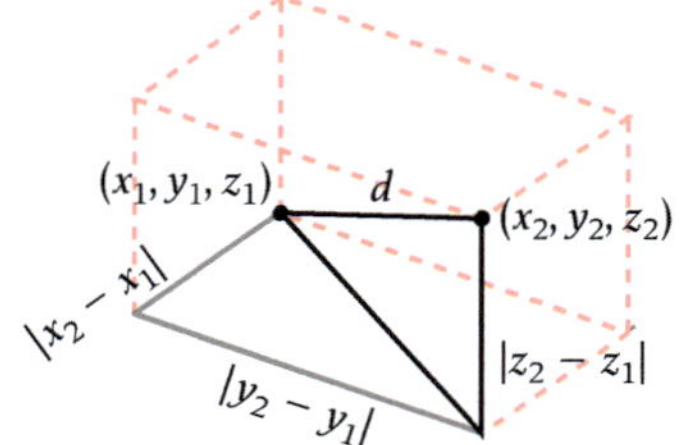

Figure 1.22 *Determining the distance between points* (x_1, y_1, z_1) *and* (x_2, y_2, z_2)

1.1.4   Equation of a sphere

Once again, let's return to two dimensions, this time to recall how the Cartesian equation of a circle is derived. If a circle is centered at the point (h, k) and has radius r, then the distance from any point (x, y) on the circle to the center (h, k) must be r (Figure 1.23). Using the distance formula,

$$r = \sqrt{(x - h)^2 + (y - k)^2},$$

and squaring both sides yields the familiar equation of a circle centered at (h, k) with radius r:

$$(x - h)^2 + (y - k)^2 = r^2.$$

Similarly, in $\mathbf{R}^3$ consider a sphere centered at the point (h, k, ℓ) with radius r. The distance from any point (x, y, z) on the sphere to the center (h, k, ℓ) must be r (Figure 1.24). Using the distance formula for $\mathbf{R}^3$,

$$r = \sqrt{(x - h)^2 + (y - k)^2 + (z - \ell)^2}.$$

Squaring both sides of the equation results in the following formula for the equation of a sphere centered at (h, k, ℓ) with radius r:

$$r^2 = (x - h)^2 + (y - k)^2 + (z - \ell)^2.$$

EQUATION OF A SPHERE

The equation of a sphere with center (h, k, ℓ) and radius r is

$$(x - h)^2 + (y - k)^2 + (z - \ell)^2 = r^2.$$

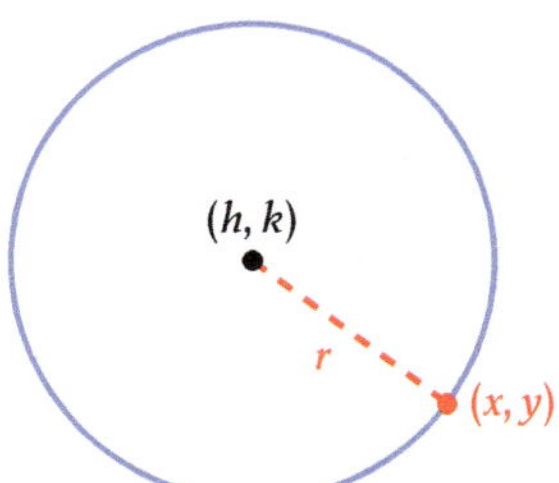

Figure 1.23 *A circle (blue) centered at (h, k) (black) with radius r, along with a point (x, y) (red) on the circle*

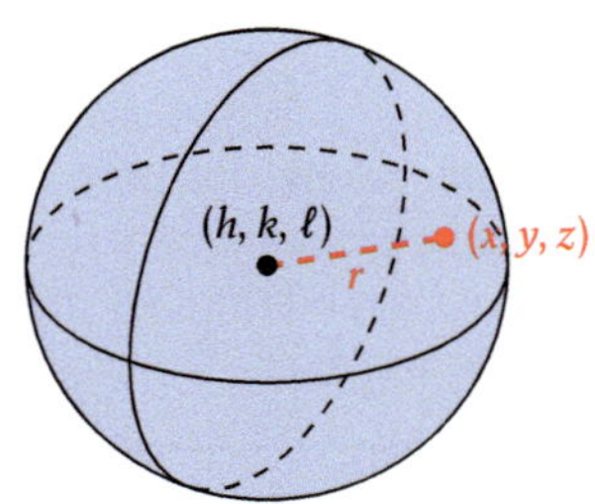

Figure 1.24 *A sphere (blue) centered at (h, k, ℓ) (black) with radius r, along with a point (x, y, z) (red) on the sphere. Compare to Figure 1.23. Three great circles are shown (black, solid in front or on the side and dashed on the back side of the sphere); a great circle is a circle on the surface of the sphere whose center is the center of the sphere. Adding one or more great circles adds depth to the picture and makes it easier to visualize as a sphere*

Example 6 *Find the equation of the sphere centered at the origin with radius 3.*

Solution In $\mathbf{R}^3$, the coordinates of the origin are $(0, 0, 0)$. Using the equation of a sphere with $(h, k, \ell) = (0, 0, 0)$ and $r = 3$, we have

$$(x - 0)^2 + (y - 0)^2 + (z - 0)^2 = 3^2$$
$$x^2 + y^2 + z^2 = 9.$$

Ans. to reading exercise 3:
$$\sqrt{94}$$

The equation of the sphere is $x^2 + y^2 + z^2 = 9$. ■

Reading Exercise 4 Find the equation of the sphere centered at $(1, 0, -3)$ with radius 5.

The next example is also analogous to exercises you have worked regarding equations of circles.

Example 7 *Find the center and radius of the sphere $x^2 + y^2 + z^2 - 2x + 4y - 11 = 0$.*

Solution If the equation had been given in the form $(x - h)^2 + (y - k)^2 + (z - \ell)^2 = r^2$, picking out the center (h, k, ℓ) and radius r would be straightforward. The task, therefore, is to manipulate the given equation to match this form. As with circles, completing the square does the trick.

The first step is to reorder the terms to group the x terms together, the y terms together, and the z terms together, leaving room for adding a constant:

$$x^2 + y^2 + z^2 - 2x + 4y - 11 = 0$$
$$x^2 - 2x \quad + y^2 + 4y \quad + z^2 \quad = 11.$$

Next, we complete the square on each variable. To complete the square on x, we take half the coefficient on the x term, $\frac{-2}{2} = -1$, and square the result, $(-1)^2 = 1$. We then add this amount to both sides of the equation:

$$x^2 - 2x + 1 + y^2 + 4y \quad + z^2 \quad = 11 + 1.$$

To complete the square on y, we take half the coefficient on the y term, $\frac{4}{2} = 2$, and square the result, $2^2 = 4$. We then add this amount to both sides of the equation:

$$x^2 - 2x + 1 + y^2 + 4y + 4 + z^2 \quad = 11 + 1 + 4.$$

There is no z term (so the coefficient is 0), and there is no need to complete the square for z. Next, we factor the perfect squares to finish rewriting in the desired form:

$$x^2 - 2x + 1 + y^2 + 4y + 4 + z^2 = 11 + 1 + 4$$

$$(x - 1)^2 + (y + 2)^2 + z^2 = 16$$

$$(x - 1)^2 + (y - (-2))^2 + (z - 0)^2 = 16.$$

The third line is meant to be helpful but is not traditionally written down.

The center of the circle is therefore $(1, -2, 0)$ and the radius is 4. ∎

Ans. to reading exercise 4:
$$(x - 1)^2 + y^2 + (z + 3)^2 = 25$$

EXERCISES 1.1

1–10. Plot the point.

All exercises in this section are assumed to be in $\mathbf{R}^3$ unless otherwise specified.

1. $(2, 1, 4)$	6. $(0, 3, -4)$
2. $(4, 1, 2)$	7. $(1, 0, -3)$
3. $(2, -1.5, 0)$	8. $(-2, 3, -1)$
4. $(-1, 4, 3)$	9. $(13, 22, -5)$
5. $(-1, 3, -2)$	10. $(-10, -44, -20)$

11–20. Graph the equation or inequality.

11. $x = -1$	16. $x = 4$
12. $y = 1$	17. $x > 0$
13. $y = -4$	18. $z > 2$
14. $z = -3$	19. $y < -2$
15. $z = 2$	20. $x < 1$

21–28. Find the distance between the points.

21. $(1, 4, 9)$ and $(-2, 0, 5)$
22. $(0, 4, 4)$ and $(3, 0, -8)$
23. $(3, 9, -2)$ and $(-3, 2, 4)$

24. $(4, 7, 0)$ and $(2, 8, 0)$
25. $(-5, 7, 1)$ and $(3, 6, 5)$
26. $(9, 4, -7)$ and $(11, 1, -1)$
27. $(11.4, 13.8, 19)$ and $(1, 17, 2)$
28. $(\pi, 3, 0)$ and $(0, 4, \sqrt{11})$

The *midpoint* of a segment lies on the segment; the distance from the midpoint to one endpoint is the same as the distance from the midpoint to the other endpoint. The method for finding the midpoint of a segment in $\mathbf{R}^3$ is analogous to the method for $\mathbf{R}^2$.

29–30. Find the midpoint of the segment joining the two points.

29. $(0, 4, 4)$ and $(3, 0, -8)$
30. $(3, 9, -2)$ and $(-3, 2, 4)$

31–40. Find the equation of the sphere with the given properties.

Here, the word *diameter* refers to a segment with endpoints on the sphere that passes through the center of the sphere.

31. center $(2, 9, -3)$, radius 11
32. center $(3, 0, 5)$, radius 6
33. center $(1, -2, 4)$, point on sphere $(2, 0, 1)$
34. center $(0, 4, 7)$, point on sphere $(8, 1, 0)$
35. endpoints of diameter $(1, 0, 5)$ and $(3, 4, -1)$
36. endpoints of diameter $(-10, 3, 5)$ and $(-10, 5, 7)$

A sphere is *tangent* to a plane if the intersection of the sphere and the plane consists of exactly one point.

37. center $(2.04, 1.67, 4)$, tangent to the xy-plane
38. center $(2, 4, 6)$, tangent to the yz-plane
39. tangent to the xz-plane at a point P; the other endpoint of a diameter containing P is $(2, 4, 9)$
40. radius 3, center on the positive z-axis, tangent to the xy-plane

41–48. Determine the center and radius of the sphere whose equation is given.

41. $(x - 3)^2 + (y + 2)^2 + (z - 11)^2 - 36 = 0$
42. $(x + 1)^2 + y^2 + (z - 8)^2 = 55$
43. $x^2 + y^2 + z^2 - 4x + 6y - 8z + 4 = 0$
44. $x^2 + y^2 + z^2 + x + y + z = 0$
45. $x^2 - 2y + y^2 + 4z + z^2 = 12$
46. $3x + 4y + 5z + x^2 + y^2 + z^2 - 3 = 7$
47. $4x^2 + 4y^2 + 4z^2 + 12x - 44z = 16$
48. $6x - x^2 + 10z - z^2 = 8y + y^2$

49. Use your favorite CAS to plot the point $(1, 2, 3)$. Make sure that the axes are marked. Spin the resulting picture to see the point from different viewpoints. How much does the ability to spin the picture help in visualizing the location of the point?
50. Plot the point $(1, 5, 3)$. Erase everything other than the axes and the point, as in Figure 1.11. Pretend that you do not know the coordinates of the point. (a) Visualize the point in your mind as being located in the yz-plane. What are the point's coordinates (approximately)? (b) Repeat part (a) for the xy-plane.

51. Using a viewpoint in the octant consisting of negative x-coordinates, negative y-coordinates, and positive z-coordinates, (a) draw the coordinate axes, ensuring that a right-handed coordinate system is used and (b) plot the point $(2, -3, 1)$.

52. Using a viewpoint in the octant consisting of negative x-coordinates, positive y-coordinates, and negative z-coordinates, (a) draw the coordinate axes, ensuring that a right-handed coordinate system is used and (b) plot the point $(1, 4, 2)$.

53. Graph the equation $x^2 + y^2 = 9$. Describe the shape.

54. A cylindrical can sits on the xy-plane. The height of the cylinder is 14 cm and the diameter of the cylinder is 8 cm. The center of the bottom of the can is located at the origin. Determine the equation(s) describing the lateral surface (side) of the can.

55. In $\mathbf{R}^2$, the equation of the x-axis is $y = 0$. What is the equation(s) of the x-axis in $\mathbf{R}^3$?

56. Sphere A has center $(1, 4, -2)$ and radius 5. Sphere B has center $(2, 0, 5)$ and is tangent to sphere A. Determine the equation of sphere B.

It might help to consult Figure 1.5.

1.2 Vectors, Part 1

At the moment that I am writing this, the weather app on my phone gives the wind as "SW 8 MPH." The description states both a *direction*, that the wind is from the southwest, and a *magnitude*, 8 miles per hour (denoted mph). Taken together, magnitude and direction comprise what is known as a *vector*.

Velocity, including wind velocity, is a vector. For instance, you may have studied the velocity of a particle moving on a number line. Such a particle (Figure 1.25) can move in one of two directions; a velocity of 3 cm/s means the particle is moving in the positive direction (increasing coordinates), and a velocity of -3 cm/s means the particle is moving in the negative direction (decreasing coordinates). In either case the speed is 3 cm/s.

Speed, having magnitude but not direction, is not a vector. Instead, we call speed a *scalar*. A scalar represents only a magnitude, such as speed or mass; it is a number. But a vector has both magnitude and direction, such as velocity or force.

Figure 1.25 *A particle moving along a number line with positive velocity (top) and negative velocity (bottom)*

A scalar can be a real number, a hyper-real number, a complex number, etc. as required by the context.

1.2.1 Geometric representation of a vector

A vector can be thought of as a directed line segment, such as presented in Figure 1.26. The magnitude of the vector is the length PQ, the distance between the points P and Q. The direction of the vector is from P toward Q, where P is the *initial point* and Q is the *terminal point*. Whether this diagram is in two dimensions,

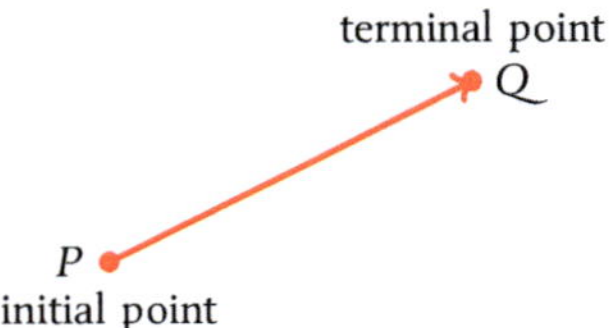

Figure 1.26 *A geometric representation of vector PQ. An arrow indicates the direction of the vector, and the length of the segment indicates the magnitude of the vector*

Notice the difference between the symbol used here for a vector and the symbol used in geometry for a line segment.

An alternative notation is $\overrightarrow{PQ}$, although this notation is usually associated with a ray in geometry.

three dimensions, or something else, is immaterial; the vector is still represented by an arrow from one point to another.

Various notations for vectors exist. The vector from P to Q (initial point P, terminal point Q) is represented as $\overrightarrow{PQ}$. We may also give the vector itself a name typeset in boldface, such as $\mathbf{v}$, setting $\mathbf{v} = \overrightarrow{PQ}$. Because boldface is difficult when writing by hand, a common practice is to write $\vec{v}$ in place of $\mathbf{v}$.

1.2.2   Algebraic representation of a vector

Consider a vector $\mathbf{v}$ whose initial point is $(2, 1)$ and whose terminal point is $(5, -1)$, as in Figure 1.27. The vector $\mathbf{v}$ can be described using an "ordered pair of travel." The vector moves 3 units to the right and 2 units down, a description which infers both magnitude and direction! To distinguish an ordered pair of travel from an ordered pair representing a point, we use angle brackets instead of parentheses:

A geometric representation is allowed to be placed in a coordinate system.

$$\mathbf{v} = \langle 3, -2 \rangle.$$

The numbers 3 and -2 are called *components* of the vector (not coordinates). Again, the vector $\langle 3, -2 \rangle$ and the point $(3, -2)$ are not only different, they are different types of objects. While the point $(3, -2)$ records a position, the vector $\langle 3, -2 \rangle$ records magnitude and direction, but does not record the position. There

It may be helpful, especially at first, to write the angle brackets with more exaggerated points, for instance $< 3, -2 >$, in order to make the difference between a vector and a point more easily recognizable.

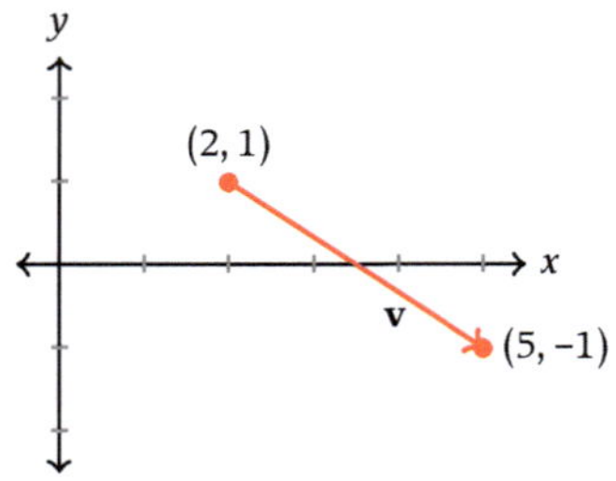

Figure 1.27 *The vector with initial point $(2, 1)$ and terminal point $(5, -1)$*

is nothing in $\langle 3, -2 \rangle$ that says that the initial point is located at $(2, 1)$ rather than somewhere else.

Example 8 *Determine the vector with initial point $A = (1, 0, 3)$ and terminal point $B = (-5, 1, 7)$.*

Solution Each component of the vector $\overrightarrow{AB}$ can be calculated by subtracting the coordinates of the points, using terminal − initial:

$$\overrightarrow{AB} = \langle -5 - 1,\ 1 - 0,\ 7 - 3 \rangle = \langle -6, 1, 4 \rangle.$$

As a check, notice that to move from point A to point B requires a move of 6 units in the negative x direction, 1 unit in the positive y direction, and 4 units in the positive z direction. ∎

A picture of the vector $\overrightarrow{AB}$ from example 8 is in Figure 1.28.

Reading Exercise 5 Determine the vector with initial point $A = (2, -3, -5)$ and terminal point $B = (1, 1, 4)$.

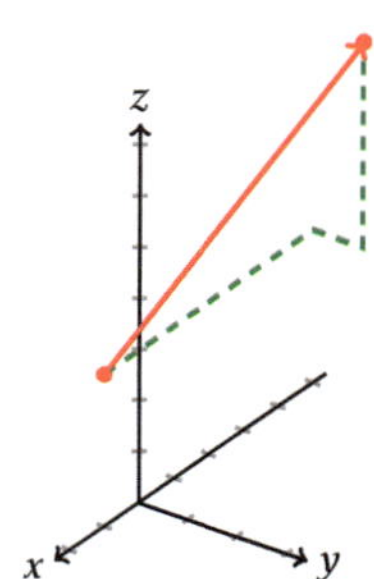

Figure 1.28 *The vector $\overrightarrow{AB}$ (red) from example 8. A path (green, dashed) moving from the initial point 6 units in the negative x direction, 1 unit in the positive y direction, and 4 units in the positive z direction is shown to create perspective*

1.2.3 Equality of vectors

Because a vector includes both magnitude and direction, in order for two vectors to be equal they must have both the same magnitude and the same direction. Vectors with the same magnitude but different directions (Figure 1.29, left) are not equal; vectors with the same direction but different magnitudes (Figure 1.29, middle) are not equal. Vectors with the same magnitude and the same direction (Figure 1.29, right) are equal, even though they are in different positions, because a vector does not include position; they are truly equal, not just congruent in a geometric sense.

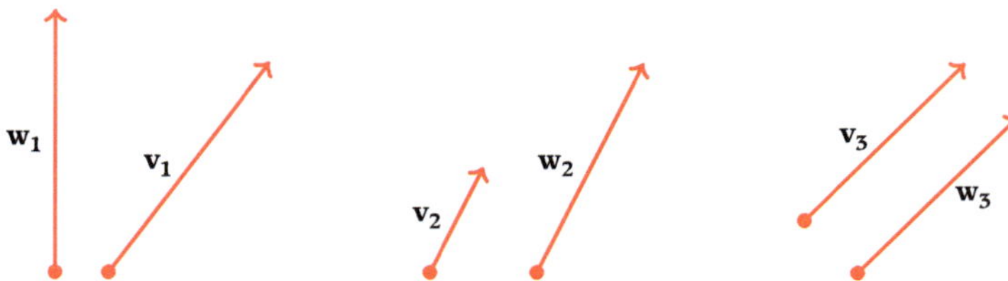

Figure 1.29 *Unequal vectors v_1 and w_1 (left); unequal vectors v_2 and w_2 (middle); and equal vectors v_3 and w_3 (right)*

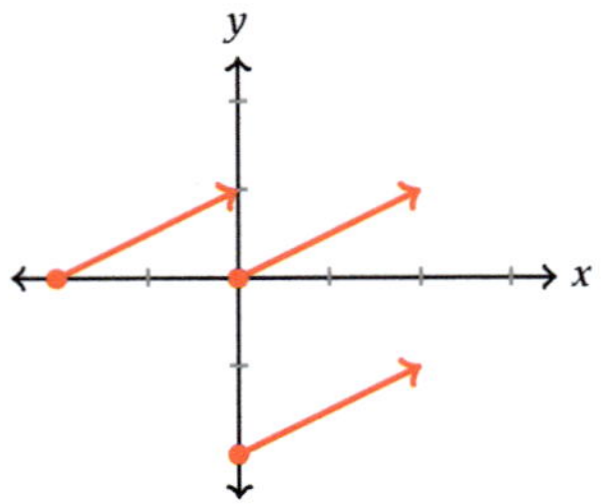

Figure 1.30 *Three copies of the vector $\langle 2, 1 \rangle$*

Three copies of the vector $\langle 2, 1 \rangle$ appear in Figure 1.30. The three copies all have the same direction and the same magnitude, so they are equal. They truly are copies of one another.

1.2.4   Zero vector

Suppose you are standing still, with velocity 0 cm/s. What direction are you going? Okay, that was a trick question–you are not moving in any particular direction. Likewise, there is one vector with no specific direction, the *zero vector*, written **0** (or, by hand, $\vec{0}$). In two dimensions the zero vector is

In the Cherokee seven directions, we can say that the zero vector has a direction, the direction "self."

$$\mathbf{0} = \langle 0, 0 \rangle,$$

and in three dimensions the zero vector is

Two-dimensional vectors have two components, and three-dimensional vectors have three components.

$$\mathbf{0} = \langle 0, 0, 0 \rangle.$$

1.2.5   Lengths of vectors

The vector $\mathbf{v} = \langle 3, -2 \rangle$ of Figure 1.27 has initial point $(2, 1)$ and terminal point $(5, -1)$. The distance between the initial and terminal points is the magnitude (length) of the vector. Using the distance formula, we get

$$d = \sqrt{(x_2 - x_1)^2 + (y_2 - y_1)^2}$$
$$= \sqrt{(5 - 2)^2 + ((-1) - 1)^2}$$
$$= \sqrt{(3)^2 + (-2)^2},$$

which is the same as

$$\sqrt{(x\text{-component})^2 + (y\text{-component})^2}.$$

Another name for the length or magnitude of a vector is the *norm* of the vector.

Definition 2 NORM OF A VECTOR, 2D *The norm of the vector* $\mathbf{a} = \langle a_1, a_2 \rangle$ *is*

$$\|\mathbf{a}\| = \sqrt{a_1^2 + a_2^2}.$$

For three dimensions, we include the z-component as well.

Definition 3 NORM OF A VECTOR, 3D *The norm of the vector* $\mathbf{a} = \langle a_1, a_2, a_3 \rangle$ *is*

$$\|\mathbf{a}\| = \sqrt{a_1^2 + a_2^2 + a_3^2}.$$

Example 9 *Find the length (magnitude, norm) of the vector* $\mathbf{v} = \langle 2, 1 \rangle$.

Solution Using definition 2,

$$\|\mathbf{v}\| = \sqrt{2^2 + 1^2} = \sqrt{5}.$$

The norm of the vector is $\sqrt{5}$.

Example 10 *Find the norm of the vector* $\mathbf{w} = \langle 2, 1, -2 \rangle$.

Solution Using definition 3,

$$\|\mathbf{w}\| = \sqrt{2^2 + 1^2 + (-2)^2} = \sqrt{9} = 3.$$

The norm of the vector is 3.

Reading Exercise 6 Find the length of the vector $\mathbf{w} = \langle 4, 4, 7 \rangle$.

1.2.6 Adding vectors

Suppose we wish to "add" the vectors $\mathbf{u} = \langle 2, 3 \rangle$ and $\mathbf{v} = \langle 1, -2 \rangle$. What might we mean by adding the two vectors?

When we define a new type of object (in this section, a vector), operations don't automatically come with it; they must be defined. And it helps if our definitions carry some meaning.

Ans. to reading exercise 5:
$\langle -1, 4, 9 \rangle$

From the definitions it is clear that $\|\mathbf{a}\| \geq 0$ for any vector $\mathbf{a}$, and the only vector of length 0 is the zero vector $\mathbf{0}$.

Notice that the norm of a vector is a scalar (a number), not a vector. This fact may be helpful in the exercises.

When exploring some new mathematical object, research mathematicians have the freedom to make their own definitions. But if they want others to have any interest at all in what they are doing, there must be some reasoning behind those definitions.

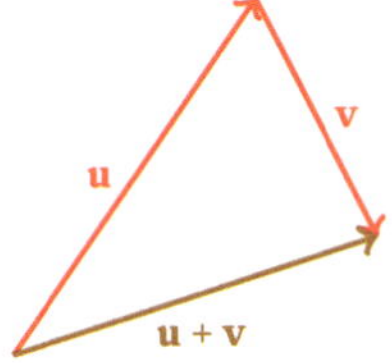

Figure 1.31 *Vector addition*

In elementary school, you may have been taught to add two numbers, such as 5 and −3, on a number line by starting at 0 (the origin), moving 5 units to the right (the positive direction), and then moving 3 units to the left (the negative direction) to arrive at the answer 2. This process helped build meaning to the concept of addition: $5 + (-3) = 2$.

In a similar manner, we may follow the vector **u** by moving 2 units in the positive x-direction and 3 units in the positive y-direction, and then follow the vector **v** by moving 1 unit in the positive x-direction and 2 units in the negative y-direction. In total, we would have moved $2 + 1 = 3$ units in the positive x-direction, and $3 + (-2) = 1$ unit in the positive y-direction. This suggests that we add two vectors by adding their components:

$$\mathbf{u} + \mathbf{v} = \langle 2, 3 \rangle + \langle 1, -2 \rangle = \langle 2 + 1, \, 3 + (-2) \rangle = \langle 3, 1 \rangle.$$

Geometrically, this is the same as following vector **u** and then **v** as in Figure 1.31. To determine $\mathbf{u} + \mathbf{v}$ we place the initial point of **v** at the terminal point of **u**. Then the vector $\mathbf{u} + \mathbf{v}$ is the overall result, with initial point the same as for **u** and terminal point the same as for **v**.

Recall that vectors do not record position, so we may position the vectors wherever we wish.

It is usually easier to use the algebraic representation in definitions.

Definition 4 VECTOR ADDITION *Given two vectors* $\mathbf{a} = \langle a_1, a_2 \rangle$ *and* $\mathbf{b} = \langle b_1, b_2 \rangle$, *their sum is*

$$\mathbf{a} + \mathbf{b} = \langle a_1 + b_1, \, a_2 + b_2 \rangle.$$

The formula for vector addition is sometimes called the *triangle law*; see Figure 1.31 for the reason why.

We will use a similar definition for three-dimensional vectors without formally stating it.

Example 11 *Determine* $\mathbf{a} + \mathbf{b}$ *given* $\mathbf{a} = \langle 2, -3, 10 \rangle$ *and* $\mathbf{b} = \langle 1, 0, -22 \rangle$.

Solution Following the definition,

$$\mathbf{a} + \mathbf{b} = \langle 2 + 1, \, -3 + 0, \, 10 + -22 \rangle = \langle 3, -3, -12 \rangle. \qquad \blacksquare$$

Because addition is commutative, $\mathbf{a} + \mathbf{b} = \mathbf{b} + \mathbf{a}$. Right?

The issue is that when we say "addition is commutative," that's for adding two real numbers. We have not yet shown that adding two vectors is commutative. Every time we define a new operation such as adding vectors, we must prove any properties that we wish to use. But thankfully the proof is not difficult.

Ans. to reading exercise 6:
$$9$$

Theorem 1 COMMUTATIVE PROPERTY OF VECTOR ADDITION *For any two vectors* $\mathbf{a}$ *and* $\mathbf{b}$,

$$\mathbf{a} + \mathbf{b} = \mathbf{b} + \mathbf{a}.$$

Proof. For two dimensions, let $\mathbf{a} = \langle a_1, a_2 \rangle$ and $\mathbf{b} = \langle b_1, b_2 \rangle$. By definition 4,

$$\mathbf{a} + \mathbf{b} = \langle a_1 + b_1, a_2 + b_2 \rangle.$$

But since addition of real numbers is commutative,

$$\langle a_1 + b_1, a_2 + b_2 \rangle = \langle b_1 + a_1, b_2 + a_2 \rangle.$$

The addition inside each component is addition of (real) numbers, which we already know is commutative; for any two numbers a_1 and b_1, $a_1 + b_1 = b_1 + a_1$.

Carefully interpreting definition 4 again, we see that

$$\langle b_1 + a_1, b_2 + a_2 \rangle = \mathbf{b} + \mathbf{a}.$$

Therefore

$$\mathbf{a} + \mathbf{b} = \mathbf{b} + \mathbf{a}.$$

In the definition of vector addition, the numbers on the left (b_1 and b_2) must be the components from the vector on the left ($\mathbf{b}$), and the numbers on the right (a_1 and a_2) must be the components from the vector on the right ($\mathbf{a}$).

The proof for three or more dimensions is analogous. ∎

The commutative property of vector addition is sometimes called the *parallelogram law*; see Figure 1.32 for the reason why.

Reading Exercise 7 Determine $\mathbf{a} + \mathbf{b}$ given $\mathbf{a} = \langle 1, 5 \rangle$ and $\mathbf{b} = \langle 4, -3 \rangle$.

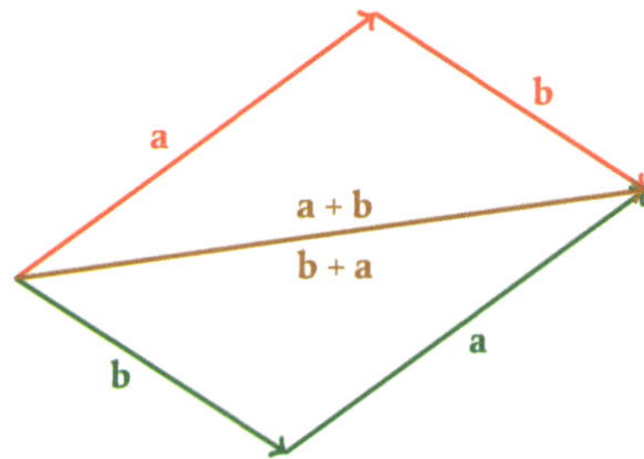

Figure 1.32 *Commutativity of vector addition*

1.2.7　Scalar multiplication

You may recall the idea of multiplication as repeated addition; for instance, $3 \cdot 5 = 5 + 5 + 5$. In the same manner, we could say that $3\mathbf{u} = \mathbf{u} + \mathbf{u} + \mathbf{u}$. For instance,

$$3\langle 2, 1 \rangle = \langle 2, 1 \rangle + \langle 2, 1 \rangle + \langle 2, 1 \rangle = \langle 6, 3 \rangle.$$

Geometrically, $3\mathbf{u}$ would have the same direction as $\mathbf{u}$ but 3 times the magnitude, as in Figure 1.33.

Because we are multiplying a scalar (a number) by a vector, we call this *scalar multiplication.*

For reasons that will eventually be apparent, we will not use the center dot $\cdot$ for scalar multiplication. We will always use juxtaposition instead.

Definition 5 SCALAR MULTIPLICATION *Let* $\mathbf{a} = \langle a_1, a_2 \rangle$ *be a vector and let* c *be a number (scalar). Then*

$$c\mathbf{a} = \langle ca_1, ca_2 \rangle.$$

As you would expect, the definition for three dimensions is analogous and will not be formally stated.

Using definition 5, the scalar does not have to be an integer. If the scalar c is positive, then the vector $c\mathbf{u}$ has the same direction as $\mathbf{u}$. If the scalar c is negative, then the vector $c\mathbf{u}$ has the opposite direction as $\mathbf{u}$.

An alternate instruction could be to "complete the vector arithmetic" or "complete the operation."

Example 12 *Complete the scalar multiplication: (a)* $2.5\langle -1, 7, 3 \rangle$*; (b)* $-2\langle 5, 3 \rangle$*.*

Solution　Using definition 5,

$$2.5\langle -1, 7, 3 \rangle = \langle -2.5, 17.5, 7.5 \rangle$$

and

$$-2\langle 5, 3 \rangle = \langle -10, -6 \rangle.$$

∎

Ans. to reading exercise 7:
$\langle 5, 2 \rangle$

A special case of scalar multiplication is multiplying by -1, which, just as with numbers, results in the negation (the *additive inverse*) of the vector. Given $\mathbf{w} = \langle 1, 3 \rangle$,

$$-1\mathbf{w} = \langle -1, -3 \rangle = -\mathbf{w}.$$

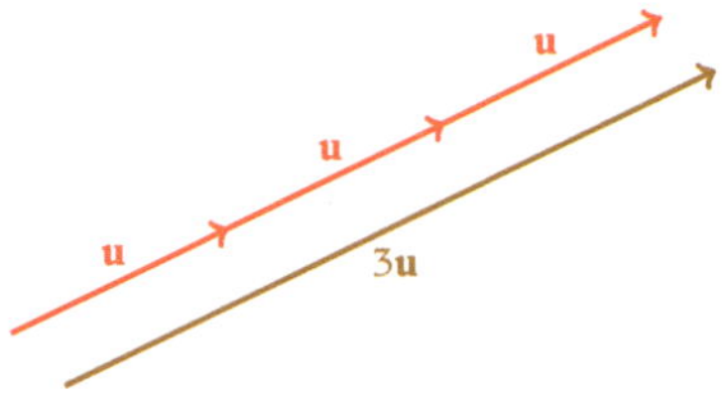

Figure 1.33 *Scalar multiplication of a vector, shown as the vector* 3**u**

Figure 1.34 *The vector* **w** *(red) and its negation, the vector* −**w** *(brown). The vector* −**w** *has the same magnitude but the opposite direction as the vector* **w**

The latter notation, −**w**, is to be interpreted as the scalar multiple −1**w**. See Figure 1.34 for a geometric interpretation.

Example 13 *Given* $\mathbf{v} = \langle 3, -6 \rangle$, *find* −**v**.

Solution We have

$$-\mathbf{v} = -1\mathbf{v} = -1\langle 3, -6 \rangle = \langle -3, 6 \rangle.$$

∎

Reading Exercise 8 Given $\mathbf{u} = \langle 1, 4, 3 \rangle$, find 2**u**.

Our introduction to vectors is not complete and will be continued in the next section.

EXERCISES 1.2

1–7. Rapid response: is it a vector?

1. velocity
2. mass
3. force
4. weight
5. length
6. volume
7. line segment

8–12. Rapid response: does it describe a vector?

8. a car traveling 50 mph eastbound on Highway 40
9. wind speed 20 mph
10. wind 30 mph from the south
11. a stationary ship (at anchor)
12. a balloon gaining altitude at 10 ft/s

13–22. For the given initial point P and terminal point Q, (a) draw the vector $\overrightarrow{PQ}$ and (b) determine the vector $\overrightarrow{PQ}$.

"Determine the vector $\overrightarrow{PQ}$" means to determine its algebraic representation, as in example 8.

13. $P = (1, 4)$, $Q = (2, 3)$
14. $P = (0, 4)$, $Q = (3, 0)$

15. $P = (2, 3), Q = (1, 4)$
16. $P = (3, 0), Q = (0, 4)$
17. $P = (-2, -2), Q = (1, 5)$
18. $P = (0, 0), Q = (1.5, 6)$
19. $P = (2, 5, 1), Q = (0, 2, 0)$
20. $P = (0, 2, 2), Q = (3, 0, 1)$
21. $P = (2, -4, 1), Q = (-3, 2, -2)$
22. $P = (-0.5, -1, 3), Q = (4, 1, -2)$

23–32. Find the norm of the vector.

23. $\mathbf{u} = \langle 5, 12 \rangle$
24. $\mathbf{v} = \langle 24, -7 \rangle$
25. $\mathbf{w} = \langle 8, -5 \rangle$
26. $\mathbf{a} = \langle 0, 7 \rangle$
27. $\mathbf{b} = \langle 4, 3, 0 \rangle$

28. $\mathbf{b} = \langle 3, 3, \sqrt{7} \rangle$
29. $\mathbf{a} = \langle 9, 0, 0 \rangle$
30. $\mathbf{v} = \langle 0, 0, -5 \rangle$
31. $\mathbf{w} = \langle -8, 1, 4 \rangle$
32. $\mathbf{u} = \langle 6, -6, 7 \rangle$

33. Find the length of the vector with initial point $(2, 7)$ and terminal point $(1, 10)$.
34. Find the length of the vector with initial point $(0, 1, 4)$ and terminal point $(3, -1, 5)$.
35. Draw three copies of the vector $\langle 1, -3 \rangle$.
36. Draw copies of the vector $\langle 0, 3 \rangle$ with initial points $(0, 0)$, $(1, 1)$, and $(-2, -1.5)$.

37–46. For the vectors in Figure 1.35, draw the specified vector.

37. $\mathbf{w} + \mathbf{a}$
38. $-\mathbf{u}$
39. $2\mathbf{b}$
40. $\frac{1}{2}\mathbf{w}$
41. $-\mathbf{a}$

42. $\mathbf{b} + \mathbf{u}$
43. $\mathbf{w} + 2\mathbf{a}$
44. $2\mathbf{a} + \mathbf{v}$
45. $0\mathbf{b}$
46. $\mathbf{b} + \mathbf{w} + \mathbf{a}$

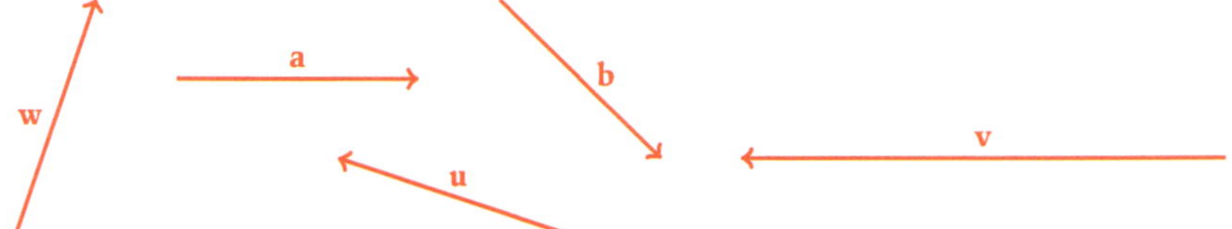

Figure 1.35 *The vectors for exercises 37–46*

47. Which vector in Figure 1.35 has the largest norm?
48. Which vectors in Figure 1.35 have opposite directions?
49. Describe the vector $\overrightarrow{AB} + \overrightarrow{BC}$ as one vector.
50. Describe the vector $\overrightarrow{AC} + -\overrightarrow{BC}$ as one vector.
51. Describe the vector $\overrightarrow{AB} + \overrightarrow{BC} + \overrightarrow{CA}$ as one vector.
52. Can the vectors $\langle 1, 0, 3 \rangle$, $\langle 0, 4, 2 \rangle$, and $\langle 1, 4, 5 \rangle$ form the sides of a triangle?

A diagram might be helpful for exercises 49–51.

53–68. Let $\mathbf{u} = \langle 2, -1, 4 \rangle$, $\mathbf{v} = \langle 7, -1.4, -5 \rangle$, $\mathbf{w} = \langle 8, 0, 12 \rangle$, and $\mathbf{a} = \langle 4, -2, -3 \rangle$. Find the requested item.

53. $\mathbf{u} + \mathbf{w}$
54. $\mathbf{w} + \mathbf{a}$
55. $-\mathbf{v}$
56. $\mathbf{a} + \mathbf{0}$
57. $\frac{1}{2}\mathbf{w}$
58. $4\mathbf{a}$
59. $2\mathbf{u} + \mathbf{v}$
60. $4\mathbf{w} + 10\mathbf{u}$

61. $\|\mathbf{u}\|$
62. $\|\mathbf{a}\|$
63. $\frac{1}{\|\mathbf{u}\|}\mathbf{u}$
64. $\frac{1}{\|\mathbf{a}\|}\mathbf{a}$
65. $\|2\mathbf{u}\|$
66. $\| - \mathbf{a}\|$
67. $-3\|\mathbf{u}\|$
68. $\left\|\frac{1}{\|\mathbf{a}\|}\mathbf{a}\right\|$

Ans. to reading exercise 8:
$2\mathbf{u} = \langle 2, 8, 6 \rangle$

1.3   Vectors, Part 2

In this section, we continue our introduction to vectors.

1.3.1   Vector subtraction

In arithmetic, subtraction is the same as addition of the negation:

$$5 - 2 = 5 + (-2).$$

In the previous section, we defined the negation of a vector (scalar multiplication using the scalar -1) as well as vector addition, so we are ready to define vector subtraction.

Definition 6 VECTOR SUBTRACTION *Given two vectors* $\mathbf{u} = \langle u_1, u_2 \rangle$ *and* $\mathbf{v} = \langle v_1, v_2 \rangle$,

$$\mathbf{u} - \mathbf{v} = \mathbf{u} + (-\mathbf{v}).$$

Alternately stated,

$$\mathbf{u} - \mathbf{v} = \langle u_1 - v_1, u_2 - v_2 \rangle.$$

A way to describe vector subtraction is that "vector subtraction is defined component-wise," meaning that the subtraction takes place in each component. Vector addition and negation are also said to be defined component-wise.

A less efficient solution follows the first portion of the definition:

$$\langle 2, 9, 4 \rangle - \langle 1, 3, -1 \rangle$$
$$= \langle 2, 9, 4 \rangle + (-\langle 1, 3, -1 \rangle)$$
$$= \langle 2, 9, 4 \rangle + \langle -1, -3, 1 \rangle$$
$$= \langle 2 + -1,\ 9 + -3,\ 4 + 1 \rangle$$
$$= \langle 1, 6, 5 \rangle.$$

A theorem from Euclidean geometry says that if a pair of opposite sides of a quadrilateral are congruent and parallel, then the quadrilateral is a parallelogram.

Example 14 *Determine the vector* $\langle 2, 9, 4 \rangle - \langle 1, 3, -1 \rangle$.

Solution   Using the latter part of definition 6, we subtract component-by-component:

$$\langle 2, 9, 4 \rangle - \langle 1, 3, -1 \rangle = \langle 2 - 1,\ 9 - 3,\ 4 - (-1) \rangle$$
$$= \langle 1, 6, 5 \rangle. \qquad \blacksquare$$

What does subtraction look like geometrically? Begin with two vectors **u** and **v**, as in Figure 1.36, left. Because $\mathbf{u} - \mathbf{v} = \mathbf{u} + (-\mathbf{v})$, we place vector **u** and add the vector $-\mathbf{v}$, placing the initial point of $-\mathbf{v}$ at the terminal point of **u**, as in Figure 1.36, right. The sum of these two vectors is the desired result, but it's somewhat unsatisfactory since the vector **v** is not in the diagram.

Exploring further, add the vector **v** in the diagram with the initial point of **v** at the initial point of **u**, as in Figure 1.37, left. Because **v** and $-\mathbf{v}$ have the same magnitude and are parallel, they form opposite sides of a parallelogram! Then the fourth side, dashed in Figure 1.37, left, must be congruent to and parallel to its opposite side, which is the vector $\mathbf{u} - \mathbf{v}$. In other words, the fourth side is a copy of the vector $\mathbf{u} - \mathbf{v}$, assuming we place the arrow the same direction (and not the opposite direction). The top portion of Figure 1.37, left is repeated in Figure 1.37, right, to reveal the final result: if we place the initial points of **u** and **v** together, then the vector $\mathbf{u} - \mathbf{v}$ has initial point at the terminal point of **v** and terminal point at the terminal point of **u**.

Reading Exercise 9   Calculate $\langle 1, 4, 0 \rangle - \langle 2, 2, 2 \rangle$.

Figure 1.36 *Subtraction of vectors, interpretation #1*

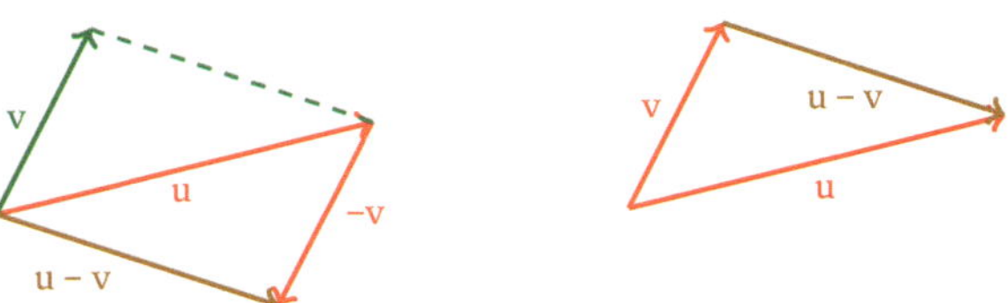

Figure 1.37 *Subtraction of vectors, interpretation #2*

1.3.2 Algebraic properties of vectors

You have studied algebraic properties of the real numbers, such as the commutative properties of addition and multiplication, the associative properties of addition and multiplication, the distributive property, and more. If you have studied matrix multiplication, you know that matrix multiplication is not commutative. So which algebraic properties hold for vector addition and scalar multiplication?

Theorem 2 PROPERTIES OF VECTORS *Let* **u**, **v**, *and* **w** *be vectors and let c and d be scalars. Then*

(1) $\mathbf{u} + \mathbf{v} = \mathbf{v} + \mathbf{u}$
(2) $\mathbf{u} + (\mathbf{v} + \mathbf{w}) = (\mathbf{u} + \mathbf{v}) + \mathbf{w}$
(3) $\mathbf{u} + \mathbf{0} = \mathbf{u}$
(4) $\mathbf{u} + (-\mathbf{u}) = \mathbf{0}$
(5) $c(\mathbf{u} + \mathbf{v}) = c\mathbf{u} + c\mathbf{v}$
(6) $(c + d)\mathbf{u} = c\mathbf{u} + d\mathbf{u}$
(7) $(cd)\mathbf{u} = c(d\mathbf{u})$
(8) $1\mathbf{u} = \mathbf{u}.$

Vector property (1), the commutative property of vector addition, was proven in Section 1.2. Other properties also have names, some of which are requested in exercises. We shall prove two more of these properties and leave the other proofs for exercises.

Proof. (5) We wish to show that $c(\mathbf{u} + \mathbf{v}) = c\mathbf{u} + c\mathbf{v}$. So that we can work with the components of the vectors and then use algebraic properties of real numbers, let $\mathbf{u} = \langle u_1, u_2 \rangle$ and $\mathbf{v} = \langle v_1, v_2 \rangle$.

One proof strategy is, for each side of the equation, to calculate the components of the vector; if the results are the same, the two expressions are equal.

Then the left-hand side is

$$c(\mathbf{u} + \mathbf{v}) = c(\langle u_1, u_2 \rangle + \langle v_1, v_2 \rangle)$$
$$= c\langle u_1 + v_1, u_2 + v_2 \rangle$$
$$= \langle c(u_1 + v_1), c(u_2 + v_2) \rangle$$
$$= \langle cu_1 + cv_1, cu_2 + cv_2 \rangle$$

Line 1 rewrites the vectors using their components; line 2 adds the vectors; line 3 completes the scalar multiplication; line 4 uses the distributive property of real numbers.

and the right-hand side is

$$cu + cv = c\langle u_1, u_2 \rangle + c\langle v_1, v_2 \rangle$$
$$= \langle cu_1, cu_2 \rangle + \langle cv_1, cv_2 \rangle$$
$$= \langle cu_1 + cv_1, cu_2 + cv_2 \rangle.$$

Line 1 rewrites the vectors using their components; line 2 performs the scalar multiplication; line 3 adds the vectors.

Because the expressions are identical, $c(\mathbf{u} + \mathbf{v}) = c\mathbf{u} + c\mathbf{v}$.

Notice that the structure of the proof of (7) is identical to the structure of the proof of (5).

(7) We wish to show that $(cd)\mathbf{u} = c(d\mathbf{u})$. So that we can work with the components of the vectors and then use algebraic properties of real numbers, let $\mathbf{u} = \langle u_1, u_2 \rangle$.

Recalling that the number cd is a scalar, the left-hand side is

Line 1 rewrites the vector using its components; line 2 completes the scalar multiplication.

$$(cd)\mathbf{u} = (cd)\langle u_1, u_2 \rangle$$
$$= \langle (cd)u_1, (cd)u_2 \rangle$$

and the right-hand side is

Line 1 rewrites the vector using its components; line 2 completes the scalar multiplication by d inside the parentheses; line 3 completes the scalar multiplication by c; line 4 uses the associative property of multiplication of real numbers.

$$c(d\mathbf{u}) = c\left(d\langle u_1, u_2 \rangle \right)$$
$$= c\langle du_1, du_2 \rangle$$
$$= \langle c(du_1), c(du_2) \rangle$$
$$= \langle (cd)u_1, (cd)u_2 \rangle.$$

Because the expressions are identical, $(cd)\mathbf{u} = c(d\mathbf{u})$. ∎

When trying to prove more of these properties, resist the temptation to combine algebraic steps; perform one step at a time, as shown in the proofs of properties (5) and (7). Combining steps risks missing the point of why the properties are true, and could result in an insufficient or faulty proof.

1.3.3 Vectors i, j, and k

A unit vector is a vector with magnitude 1, that is, 1 unit.

In two dimensions, two important *unit vectors* are

$$\mathbf{i} = \langle 1, 0 \rangle$$

and

$$\mathbf{j} = \langle 0, 1 \rangle.$$

If initial points are placed at the origin, these vectors lie along the positive x- and y-axes (Figure 1.38).

Consider the vector $\mathbf{u} = \langle 3, 5 \rangle$. Notice that

Ans. to reading exercise 9:
$\langle -1, 2, -2 \rangle$

$$\langle 3, 5 \rangle = \langle 3, 0 \rangle + \langle 0, 5 \rangle$$
$$= 3\langle 1, 0 \rangle + 5\langle 0, 1 \rangle$$
$$= 3\mathbf{i} + 5\mathbf{j}.$$

The vector names i and j are reserved for this purpose.

Then $\mathbf{u} = \langle 3, 5 \rangle = 3\mathbf{i} + 5\mathbf{j}$, and it is apparent how to quickly rewrite any vector in terms of the vectors i and j. Borrowing terms from the subject of linear algebra, the vectors i and j are called the *standard basis vectors*, and together they form a *basis* for the two-dimensional vector system.

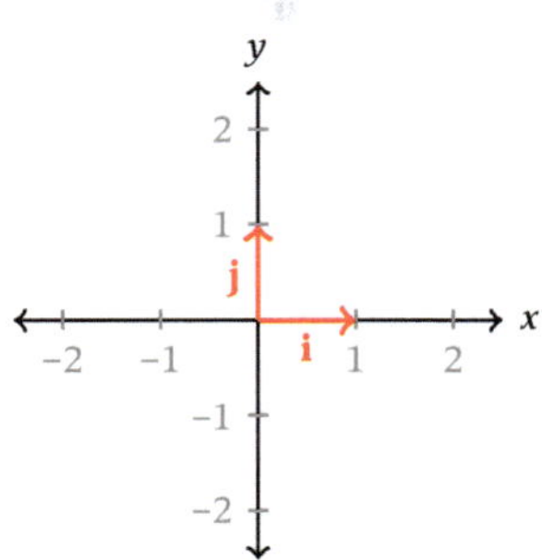

Figure 1.38 *The unit vectors* **i** *and* **j**, *in two dimensions*

Example 15 *Rewrite the vector* $\mathbf{w} = \langle 3.7, \cosh 4.92 \rangle$ *as a combination of the standard basis vectors.*

Solution We may quickly rewrite **w** as

$$\mathbf{w} = 3.7\mathbf{i} + \cosh 4.92\,\mathbf{j}.$$ ■

Notice that we do not use angle brackets or commas in the solution. Rather, the vector **w** is a *linear combination* of the vectors **i** and **j**.

Geometrically, when we rewrite a vector as the sum of its components times the standard basis vectors, we are picturing the vector as the sum of its horizontal and vertical *projections* (Figure 1.39).

By the word "times" here, we mean scalar multiplication.

In three dimensions (Figure 1.40), we need three standard basis vectors:

$$\mathbf{i} = \langle 1, 0, 0 \rangle,$$

$$\mathbf{j} = \langle 0, 1, 0 \rangle,$$

$$\text{and} \quad \mathbf{k} = \langle 0, 0, 1 \rangle.$$

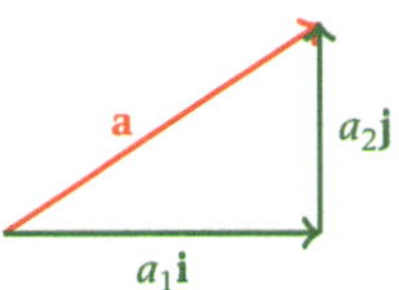

Figure 1.39 *The vector* $\mathbf{a} = \langle a_1, a_2 \rangle = a_1\mathbf{i} + a_2\mathbf{j}$ *(red) as the sum of its horizontal and vertical projections (green)*

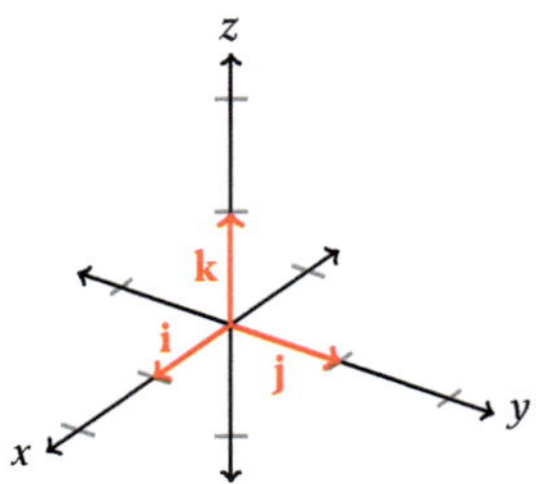

Figure 1.40 *The standard basis vectors* **i**, **j**, *and* **k** *in three dimensions, pictured with initial point at the origin*

Then any vector $\mathbf{a} = \langle a_1, a_2, a_3 \rangle$ can be rewritten as

$$\mathbf{a} = \langle a_1, a_2, a_3 \rangle = a_1\mathbf{i} + a_2\mathbf{j} + a_3\mathbf{k}.$$

Example 16 *Write $\langle 1, 0, -3 \rangle$ in terms of* **i**, **j**, *and* **k**.

Solution Rewriting,

$$\langle 1, 0, -3 \rangle = 1\mathbf{i} + 0\mathbf{j} + -3\mathbf{k}$$
$$= \mathbf{i} - 3\mathbf{k}. \qquad \blacksquare$$

Example 17 *Add the vectors* $3\mathbf{i} + 2\mathbf{j} + \mathbf{k}$ *and* $2\mathbf{i} - 5\mathbf{k}$.

An alternate solution is to rewrite using vector component notation:

$3\mathbf{i} + 2\mathbf{j} + \mathbf{k} + 2\mathbf{i} - 5\mathbf{k}$

$\qquad = \langle 3, 2, 1 \rangle + \langle 2, 0, -5 \rangle$

$\qquad = \langle 5, 2, -4 \rangle$

$\quad = 5\mathbf{i} + 2\mathbf{j} - 4\mathbf{k}.$

Solution Treating the vectors **i**, **j**, and **k** as algebraic quantities and collecting like terms,

$$3\mathbf{i} + 2\mathbf{j} + \mathbf{k} + 2\mathbf{i} - 5\mathbf{k} = 5\mathbf{i} + 2\mathbf{j} - 4\mathbf{k}. \qquad \blacksquare$$

Reading Exercise 10 Write $\langle 11, \pi, \sqrt{2} \rangle$ in terms of **i**, **j**, and **k**.

How do we know that we can treat **i**, **j**, and **k** as algebraic quantities and use standard algebraic manipulations on these vectors? The answer lies in the properties of vectors. Carefully adding the two quantities $3\mathbf{i} + 2\mathbf{j} + \mathbf{k}$ and $2\mathbf{i} - 5\mathbf{k}$ begins with

$$(3\mathbf{i} + 2\mathbf{j} + \mathbf{k}) + (2\mathbf{i} - 5\mathbf{k}).$$

Vector property (2), when generalized, allows us to group vectors in any manner; just as associativity of addition in the real numbers allows us to rewrite without the need for parentheses, we may write

$$3\mathbf{i} + 2\mathbf{j} + \mathbf{k} + 2\mathbf{i} - 5\mathbf{k}.$$

Next, we may use vector property (1), the commutative property, to re-order the terms:

$$3\mathbf{i} + 2\mathbf{i} + 2\mathbf{j} + \mathbf{k} - 5\mathbf{k}.$$

Property (8) is needed to get

$$3\mathbf{i} + 2\mathbf{i} + 2\mathbf{j} + 1\mathbf{k} - 5\mathbf{k}.$$

Then property (6) may be utilized, and we have

$$(3 + 2)\mathbf{i} + 2\mathbf{j} + (1 + -5)\mathbf{k}.$$

Finally, we use properties of real numbers inside the parentheses to obtain the final result

$$5\mathbf{i} + 2\mathbf{j} - 4\mathbf{k}.$$

By writing $5\mathbf{i} + 2\mathbf{j} - 4\mathbf{k}$ instead of $5\mathbf{i} + 2\mathbf{j} + -4\mathbf{k}$, we are also using the definition of vector subtraction.

The purpose of theorem 2, properties of vectors, is to allow us to work with vectors simply and quickly as in the solution to example 17. There is usually no need to consciously apply each property of vectors as a separate step.

1.3.4   Unit vectors

Let's state more formally the definition of a unit vector.

Definition 7 UNIT VECTOR *A* unit vector *is a vector of length* 1.

For example, the length of $\mathbf{i} = \langle 1, 0, 0 \rangle$ is

$$\|\mathbf{i}\| = \sqrt{1^2 + 0^2 + 0^2} = \sqrt{1} = 1.$$

As mentioned previously, $\mathbf{i}$ is a unit vector in the positive x-direction, $\mathbf{j}$ is a unit vector in the positive y-direction, and $\mathbf{k}$ is a unit vector in the positive z-direction. But, what if we want a unit vector in a different direction?

Ans. to reading exercise 10:
$$11\mathbf{i} + \pi\mathbf{j} + \sqrt{2}\mathbf{k}$$

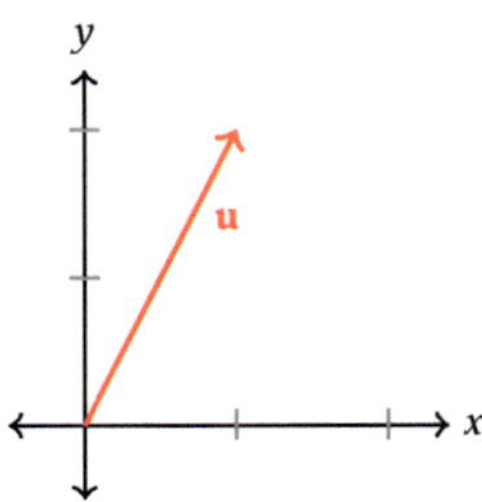

Figure 1.41 *A vector **u** that is not a unit vector*

Consider the vector $\mathbf{u} = \langle 1, 2 \rangle$ (Figure 1.41). The length of **u** is

$$\|\mathbf{u}\| = \sqrt{1^2 + 2^2} = \sqrt{5},$$

so **u** is not a unit vector. If we wish to have a unit vector in the same direction as **u**, we can't use **u** because it is too long. How can we make it shorter? Scalar multiplication, if the scalar is positive, keeps the same direction but changes the length. That's exactly what we want! Let's "divide" by the length of **u**, that is, multiply by the scalar $\dfrac{1}{\|\mathbf{u}\|} = \dfrac{1}{\sqrt{5}}$:

When we write a vector divided by a scalar, such as $\dfrac{\mathbf{u}}{\|\mathbf{u}\|}$, we mean the scalar multiplication $\dfrac{1}{\|\mathbf{u}\|}\mathbf{u}$.

$$\frac{\mathbf{u}}{\|\mathbf{u}\|} = \frac{1}{\|\mathbf{u}\|}\mathbf{u} = \frac{1}{\sqrt{5}}\langle 1, 2 \rangle = \left\langle \frac{1}{\sqrt{5}}, \frac{2}{\sqrt{5}} \right\rangle.$$

The length of this new vector is

$$\left\|\left\langle \frac{1}{\sqrt{5}}, \frac{2}{\sqrt{5}} \right\rangle\right\| = \sqrt{\left(\frac{1}{\sqrt{5}}\right)^2 + \left(\frac{2}{\sqrt{5}}\right)^2} = \sqrt{\frac{1}{5} + \frac{4}{5}} = \sqrt{1} = 1,$$

as desired. Figure 1.42 shows the unit vector $\dfrac{\mathbf{u}}{\|\mathbf{u}\|}$ with the same initial point as the vector **u**.

A proof of the following theorem, in any dimension we need, is analogous to the previous calculations.

Theorem 3 UNIT VECTORS *A unit vector in the same direction as **u** is $\dfrac{\mathbf{u}}{\|\mathbf{u}\|}$.*

Example 18 *Find a unit vector in the same direction as $\mathbf{w} = \langle 1, 4, -3 \rangle$.*

Solution   First, we determine the length (magnitude, norm) of the vector:

$$\|\mathbf{w}\| = \sqrt{1^2 + 4^2 + (-3)^2} = \sqrt{1 + 16 + 9} = \sqrt{26}.$$

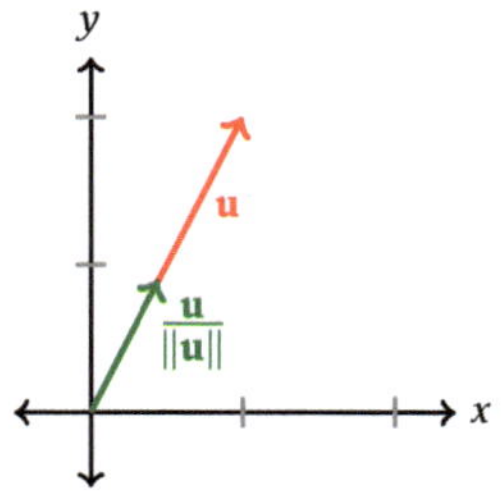

Figure 1.42 *A vector* **u** *with length greater than 1, along with a unit vector* $\frac{\mathbf{u}}{\|\mathbf{u}\|}$ *with the same initial point. Compare to Figure 1.41*

Then, we form the desired unit vector by dividing by the norm:

$$\frac{\mathbf{w}}{\|\mathbf{w}\|} = \frac{1}{\sqrt{26}}\langle 1, 4, -3\rangle = \left\langle \frac{1}{\sqrt{26}}, \frac{4}{\sqrt{26}}, \frac{-3}{\sqrt{26}}\right\rangle. \qquad \blacksquare$$

Reading Exercise 11   Find a unit vector in the same direction as $\mathbf{v} = \langle 2, 0, 9\rangle$.

EXERCISES 1.3

1–8. Let $\mathbf{u} = \langle 1, 7\rangle$, $\mathbf{v} = \langle -2, 4\rangle$, and $\mathbf{w} = \langle 8, 7\rangle$. Determine the stated vector.

1. $\mathbf{u} - \mathbf{v}$
2. $\mathbf{w} - \mathbf{u}$
3. $\mathbf{v} - \mathbf{u}$
4. $\mathbf{w} - \mathbf{v}$
5. $3\mathbf{u} - \mathbf{j}$
6. $\mathbf{i} - 4\mathbf{w}$
7. $2\mathbf{v} + 3\mathbf{u} - \mathbf{w}$
8. $4\mathbf{v} - 2\mathbf{w}$

9–14. Draw a diagram of the form given in Figure 1.37, right, to illustrate the subtraction of vectors.

9. $\mathbf{u} - \mathbf{v}$, where $\mathbf{u} = \langle 3, 6\rangle$, $\mathbf{v} = \langle 2, -1\rangle$
10. $\mathbf{u} - \mathbf{v}$, where $\mathbf{u} = \langle -1, 4\rangle$, $\mathbf{v} = \langle -3, 1\rangle$
11. $\mathbf{v} - \mathbf{u}$, where $\mathbf{u} = \langle 3, 6\rangle$, $\mathbf{v} = \langle 2, -1\rangle$
12. $\mathbf{v} - \mathbf{u}$, where $\mathbf{u} = \langle -1, 4\rangle$, $\mathbf{v} = \langle -3, 1\rangle$
13. $\mathbf{w} - \mathbf{a}$, where $\mathbf{w} = \langle -3, 0\rangle$, $\mathbf{a} = \langle -1, 5\rangle$
14. $\mathbf{i} - 2\mathbf{j}$

15–20. Write in terms of $\mathbf{i}$, $\mathbf{j}$, and $\mathbf{k}$.

15. $\langle 3, 9, -4 \rangle$
16. $\langle -3, 0, 5 \rangle$
17. $\langle 0, 0, 19 \rangle$
18. $\langle 0, -13, 0 \rangle$
19. $\langle 4, 0, 2 \rangle - \langle 2, 5, -1 \rangle$
20. $\langle 1, 1, 4 \rangle - \langle 2, 0, -3 \rangle$

21–26. Calculate the requested quantity.

21. $\|10\mathbf{k}\|$
22. $\|2\mathbf{i} + \mathbf{j}\|$
23. $\|4\mathbf{i} - 6\mathbf{j}\|$
24. $\|401\mathbf{k}\|$
25. $\|-\mathbf{i} + 5\mathbf{j} + \mathbf{k}\|$
26. $\|\mathbf{j}\|$

27–32. Is the vector a unit vector?

27. $\mathbf{k}$
28. $\langle 2, 0, 4 \rangle$
29. $\left\langle \frac{1}{2}, \frac{\sqrt{3}}{2} \right\rangle$
30. $\mathbf{i}$
31. $\langle 0.4, 0.2, 0.1 \rangle - \mathbf{i}$
32. $\frac{1}{2}\mathbf{i} - \frac{1}{2}\mathbf{j} + \frac{1}{\sqrt{2}}\mathbf{k}$

33–42. Find a unit vector in the same direction as the given vector.

33. $\langle 3, -1, 5 \rangle$
34. $\langle 1, 1, 0 \rangle$
35. $-16\mathbf{j}$
36. $\langle -1, -3, 8 \rangle$
37. $-4\mathbf{i} + 6\mathbf{j} - 12\mathbf{k}$
38. $3\mathbf{j} + 4\mathbf{k}$
39. $\left\langle \frac{3}{5}, \frac{7}{5} \right\rangle$
40. $\langle 2, 0, 0 \rangle + 3\mathbf{k}$
41. $\langle -10, 22 \rangle$
42. $\langle 0.8, 2.3 \rangle$

Notice in exercise 46 that $\mathbf{j}$ is not underneath the square root. This must be the case, for we have not defined the square root of a vector, nor will we.

43–46. Add the vectors.

43. $3\mathbf{i} + 4\mathbf{j}$ and $2\mathbf{i} + 6\mathbf{j} - 3\mathbf{k}$
44. $2.5\mathbf{i} + \mathbf{j}$ and $-3\mathbf{i} - 2\mathbf{j}$
45. $4\mathbf{j} + 5\mathbf{i}$ and $2\mathbf{i} - \mathbf{j}$
46. $5\mathbf{k} + \sqrt{7}\mathbf{j}$ and $3\mathbf{i} + 2\mathbf{j}$

47–52. Determine an appropriate name or descriptive phrase for the property of vectors.

47. property (2): $\mathbf{u} + (\mathbf{v} + \mathbf{w}) = (\mathbf{u} + \mathbf{v}) + \mathbf{w}$
48. property (3): $\mathbf{u} + \mathbf{0} = \mathbf{u}$
49. property (4): $\mathbf{u} + (-\mathbf{u}) = \mathbf{0}$
50. property (5): $c(\mathbf{u} + \mathbf{v}) = c\mathbf{u} + c\mathbf{v}$
51. property (6): $(c + d)\mathbf{u} = c\mathbf{u} + d\mathbf{u}$
52. property (7): $(cd)\mathbf{u} = c(d\mathbf{u})$

53–56. Prove the stated property of vectors.

    53. property (6): $(c + d)\mathbf{u} = c\mathbf{u} + d\mathbf{u}$
    54. property (4): $\mathbf{u} + (-\mathbf{u}) = \mathbf{0}$
    55. property (3): $\mathbf{u} + \mathbf{0} = \mathbf{u}$
    56. property (2): $\mathbf{u} + (\mathbf{v} + \mathbf{w}) = (\mathbf{u} + \mathbf{v}) + \mathbf{w}$

57. Suppose $\|\mathbf{u}\| = 5$ and $\|\mathbf{w}\| = 2$. Must $\|\mathbf{u} - \mathbf{w}\| = 3$?

58. In a drone programming contest, your drone is to launch from a specified point on the ground and fly through a hoop located on the top of a nearby building. The building is located 400 ft north and 200 ft east of the launch point and the center of the hoop is 50 ft higher elevation than the launch point. There is a steady wind out of the north (air flowing from north to south) that will push the drone toward the south at a rate of 7 ft/s. The drone may be set to fly at a rate that, in still air, would amount to either 15 ft/s, 20 ft/s, or 25 ft/s. Your task is to choose a speed and a still-air direction vector of the form $\langle\text{east}, \text{north}, \text{upward}\rangle$ that will make the drone fly through the hoop. The hoop is large enough that the drone can miss the center of the hoop by up to 2 ft and still pass through.

Ans. to reading exercise 11:
$$\frac{\mathbf{v}}{\|\mathbf{v}\|} = \left\langle \frac{2}{\sqrt{85}}, 0, \frac{9}{\sqrt{85}} \right\rangle$$

Hint: The vector $\langle 200, 400, 50 \rangle$ would not work because the wind would push it off course; you would end up flying south of the hoop.

1.4   Dot Product

Another useful tool is the *dot product* of two vectors. The dot product is found by multiplying corresponding components and adding the results.

Definition 8 DOT PRODUCT (2D) *Let* $\mathbf{a} = \langle a_1, a_2 \rangle$ *and* $\mathbf{b} = \langle b_1, b_2 \rangle$. *The dot product of* $\mathbf{a}$ *and* $\mathbf{b}$ *is*

$$\mathbf{a} \cdot \mathbf{b} = a_1 b_1 + a_2 b_2.$$

Because the result of the dot product is a scalar, the dot product is sometimes known as the *scalar product*. There is another type of product of two vectors to be discussed in Section 1.5, so we always use the dot notation when writing a dot product; careful use of notation is required to distinguish between the different operations.

It is important to keep in mind the type of object that is produced by an operation. For instance, scalar multiplication produces a vector; the dot product produces a scalar.

Example 19 *Compute the dot product* $\langle 3, 5 \rangle \cdot \langle -1, 7 \rangle$.

Solution We multiply the x-components of the two vectors, we multiply the y-components of the two vectors, and we add the results:

$$\langle 3, 5 \rangle \cdot \langle -1, 7 \rangle = 3(-1) + 5(7)$$
$$= -3 + 35 = 32.$$

The dot product of these two vectors is the number 32. ∎

The dot product in three dimensions is similar.

Definition 9 DOT PRODUCT (3D) *Let* $\mathbf{a} = \langle a_1, a_2, a_3 \rangle$ *and* $\mathbf{b} = \langle b_1, b_2, b_3 \rangle$. *The dot product of* $\mathbf{a}$ *and* $\mathbf{b}$ *is*

$$\mathbf{a} \cdot \mathbf{b} = a_1 b_1 + a_2 b_2 + a_3 b_3.$$

Example 20 *Compute* $\langle 1, 0, -3 \rangle \cdot \langle 10, \sqrt{17}, 4 \rangle$.

Solution Multiplying corresponding components and adding the results,

$$\langle 1, 0, -3 \rangle \cdot \langle 10, \sqrt{17}, 4 \rangle = 1 \cdot 10 + 0 \cdot \sqrt{17} + (-3)(4) = -2.$$

∎

If we are viewing the dot product similarly to multiplication, would we write $\mathbf{u} \cdot \mathbf{u} = \mathbf{u}^2$? No, but that's close to an important property. Consider the vector $\mathbf{u} = \langle 1, 4, -2 \rangle$. Take the dot product of $\mathbf{u}$ with itself:

$$\mathbf{u} \cdot \mathbf{u} = \langle 1, 4, -2 \rangle \cdot \langle 1, 4, -2 \rangle = 1^2 + 4^2 + (-2)^2 = 21.$$

But notice that

$$\|\mathbf{u}\| = \sqrt{1^2 + 4^2 + (-2)^2} = \sqrt{21}.$$

Squaring both sides gives

$$\|\mathbf{u}\|^2 = 1^2 + 4^2 + (-2)^2 = 21.$$

So we don't get $\mathbf{u}^2$, but instead we get $\|\mathbf{u}\|^2$. This is but one of the many properties of the dot product.

Theorem 4 PROPERTIES OF THE DOT PRODUCT *If* $\mathbf{u}$, $\mathbf{v}$, *and* $\mathbf{w}$ *are vectors and* a *is a scalar,*

(1) $\mathbf{u} \cdot \mathbf{v} = \mathbf{v} \cdot \mathbf{u}$
(2) $\mathbf{u} \cdot (\mathbf{v} + \mathbf{w}) = \mathbf{u} \cdot \mathbf{v} + \mathbf{u} \cdot \mathbf{w}$
(3) $(\mathbf{v} - \mathbf{w}) \cdot \mathbf{u} = \mathbf{v} \cdot \mathbf{u} - \mathbf{w} \cdot \mathbf{u}$
(4) $a(\mathbf{u} \cdot \mathbf{v}) = (a\mathbf{u}) \cdot \mathbf{v} = \mathbf{u} \cdot (a\mathbf{v})$
(5) $\mathbf{u} \cdot \mathbf{u} = \|\mathbf{u}\|^2.$

Property (3) is a combination of properties (1) and (2) and the definition of vector subtraction, so it is usually omitted from the list. But the form of property (3) is helpful in a derivation later in this section.

Proving properties of the dot product is similar to proving properties of vectors; see Section 1.3, theorem 2. Proofs are left as exercises, as are putting appropriate names to these properties.

Reading Exercise 12   Evaluate $\langle -1, 2, 5 \rangle \cdot \langle -3, 4, -3 \rangle$.

1.4.1   Angles between vectors

Given two vectors $\mathbf{v}$ and $\mathbf{w}$ placed at the same initial point, what is the angle between the vectors? That is, what is the angle θ in Figure 1.43? Including the vector $\mathbf{v} - \mathbf{w}$ creates a triangle whose side lengths are $\|\mathbf{v}\|$, $\|\mathbf{w}\|$, and $\|\mathbf{v} - \mathbf{w}\|$. Then by the law of cosines,

$$\|\mathbf{v} - \mathbf{w}\|^2 = \|\mathbf{v}\|^2 + \|\mathbf{w}\|^2 - 2\|\mathbf{v}\|\,\|\mathbf{w}\| \cos\theta.$$

Applying properties of the dot product (theorem 4),

$$
\begin{aligned}
\|\mathbf{v} - \mathbf{w}\|^2 &= (\mathbf{v} - \mathbf{w}) \cdot (\mathbf{v} - \mathbf{w}) \\
&= \mathbf{v} \cdot (\mathbf{v} - \mathbf{w}) - \mathbf{w} \cdot (\mathbf{v} - \mathbf{w}) \\
&= \mathbf{v} \cdot \mathbf{v} - \mathbf{v} \cdot \mathbf{w} - \mathbf{w} \cdot \mathbf{v} + \mathbf{w} \cdot \mathbf{w} \\
&= \|\mathbf{v}\|^2 - 2\mathbf{v} \cdot \mathbf{w} + \|\mathbf{w}\|^2.
\end{aligned}
$$

The two quantities equal to $\|\mathbf{v} - \mathbf{w}\|^2$ must be equal, hence

$$\|\mathbf{v}\|^2 + \|\mathbf{w}\|^2 - 2\|\mathbf{v}\|\,\|\mathbf{w}\| \cos\theta = \|\mathbf{v}\|^2 - 2\mathbf{v} \cdot \mathbf{w} + \|\mathbf{w}\|^2.$$

Simplifying by subtracting terms common to both sides leaves us with

$$-2\|\mathbf{v}\|\,\|\mathbf{w}\| \cos\theta = -2\mathbf{v} \cdot \mathbf{w},$$

and dividing both sides by -2 yields

$$\|\mathbf{v}\|\,\|\mathbf{w}\| \cos\theta = \mathbf{v} \cdot \mathbf{w}.$$

This latter formula and two of its rearrangements are all useful.

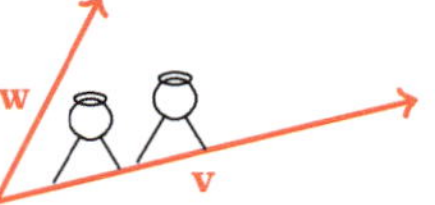

Not to be confused with "angels between vectors."

If you did not encounter the law of cosines in your study of trigonometry, it can be found in nearly any trigonometry textbook.

Line 1 applies property (5), line 2 applies property (3), line 3 applies property (2) combined with the definition of vector subtraction, and line 4 applies property (5) again, along with property (1) and collecting like terms.

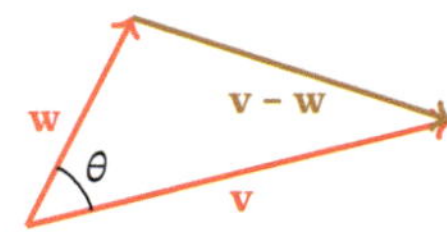

Figure 1.43 *Two vectors* v *and* w *(red) and the angle between them,* θ

The derivation was not specific to a particular dimension; the formulas work in both two and three (or more!) dimensions.

ANGLE BETWEEN VECTORS FORMULAS

Given two vectors $\mathbf{v}$ and $\mathbf{w}$ and the angle θ between them, we have

$$\|\mathbf{v}\|\,\|\mathbf{w}\|\cos\theta = \mathbf{v}\cdot\mathbf{w},$$

$$\cos\theta = \frac{\mathbf{v}\cdot\mathbf{w}}{\|\mathbf{v}\|\,\|\mathbf{w}\|}, \quad \text{and}$$

$$\theta = \cos^{-1}\left(\frac{\mathbf{v}\cdot\mathbf{w}}{\|\mathbf{v}\|\,\|\mathbf{w}\|}\right).$$

Consider the first of the formulas,

$$\mathbf{v}\cdot\mathbf{w} = \|\mathbf{v}\|\,\|\mathbf{w}\|\cos\theta.$$

Because $\|\mathbf{v}\| \geq 0$ and $\|\mathbf{w}\| \geq 0$, if $\mathbf{v}\cdot\mathbf{w}$ is positive then so is $\cos\theta$, and if $\mathbf{v}\cdot\mathbf{w}$ is negative then so is $\cos\theta$. Given that the angle must be between 0 and π (between $0°$ and $180°$; it is an angle of a triangle, see Figure 1.43), then

In degrees:

$$\text{if } \mathbf{v}\cdot\mathbf{w} > 0 \text{ then } 0 \leq \theta < \frac{\pi}{2}, \text{and}$$

$$\text{if } \mathbf{v}\cdot\mathbf{w} < 0 \text{ then } \frac{\pi}{2} < \theta \leq \pi.$$

if $\mathbf{v}\cdot\mathbf{w} > 0$ then $0° \leq \theta < 90°$, and

if $\mathbf{v}\cdot\mathbf{w} < 0$ then $90° < \theta \leq 180°$.

The latter of the three formulas can be used to determine the angle between two vectors.

Example 21 *Find the angle between the vectors* $\mathbf{v} = \langle 1, 2, 7\rangle$ *and* $\mathbf{w} = \langle 0, 3, -1\rangle$.

Ans. to reading exercise 12:
-4

Solution   First, we compute the norms of the vectors:

$$\|\mathbf{v}\| = \sqrt{1^2 + 2^2 + 7^2} = \sqrt{54},$$

$$\|\mathbf{w}\| = \sqrt{0^2 + 3^2 + (-1)^2} = \sqrt{10}.$$

We also need to compute the dot product:

$$\mathbf{v}\cdot\mathbf{w} = 1\cdot 0 + 2\cdot 3 + 7\cdot -1 = -1.$$

Because the dot product is negative, the angle must be between $\frac{\pi}{2}$ and π, that is, between $90°$ and $180°$.

Applying the angle between vectors formula,

$$\theta = \cos^{-1}\left(\frac{\mathbf{v}\cdot\mathbf{w}}{\|\mathbf{v}\|\,\|\mathbf{w}\|}\right) = \cos^{-1}\left(\frac{-1}{\sqrt{54}\sqrt{10}}\right) = 1.6138 \text{ rad} = 92.47°.$$

Using radians is standard in calculus, but using degrees is standard in many applications. Consequently, we shall often give the answer both ways.

The angle between the vectors is 1.6138 radians.   ∎

Reading Exercise 13   Find the angle between the vectors $\mathbf{v} = \langle -1, 2, 5\rangle$ and $\mathbf{w} = \langle -3, 4, -3\rangle$.

1.4.2   Orthogonal vectors

If $\mathbf{v} \cdot \mathbf{w} > 0$ then the angle between the vectors is acute, and if $\mathbf{v} \cdot \mathbf{w} < 0$ the angle between the vectors is obtuse. So what happens if $\mathbf{v} \cdot \mathbf{w} = 0$?

Definition 10 ORTHOGONAL VECTORS   *Two vectors* $\mathbf{v}$ *and* $\mathbf{w}$ *are* orthogonal *if*
$$\mathbf{v} \cdot \mathbf{w} = 0.$$

When are two vectors orthogonal? From the first of the angle between vectors formulas, $\mathbf{v} \cdot \mathbf{w} = \|\mathbf{v}\|\,\|\mathbf{w}\| \cos \theta$. If $\mathbf{v} \cdot \mathbf{w} = 0$, then

$$0 = \|\mathbf{v}\|\,\|\mathbf{w}\| \cos \theta.$$

This gives three possibilities:

$$\|\mathbf{v}\| = 0 \text{ or } \|\mathbf{w}\| = 0 \text{ or } \cos \theta = 0.$$

Therefore, either one of the vectors is the zero vector $\mathbf{0}$, or $\cos \theta = 0$. Because $0 \leq \theta \leq \pi$, the latter happens only when $\theta = \frac{\pi}{2}$, that is, when $\theta = 90°$. In other words, two nonzero vectors are orthogonal exactly when the angle between them is $90°$, which we associate with two straight objects being perpendicular. For this reason, we use the notation $\mathbf{v} \perp \mathbf{w}$ to signify that the two vectors are orthogonal.

So why not simply use the term "perpendicular" instead of the term "orthogonal?" One answer is that the term "orthogonal" includes the case when one of the vectors is $\mathbf{0}$, a case that is not considered when studying perpendicular objects in geometry.

Example 22   *Are* $\mathbf{v} = \langle 1, 7, 0 \rangle$ *and* $\mathbf{w} = \langle 0, 0, 4 \rangle$ *orthogonal?*

Solution   We merely need to check to see if $\mathbf{v} \cdot \mathbf{w} = 0$:

$$\mathbf{v} \cdot \mathbf{w} = 1 \cdot 0 + 7 \cdot 0 + 0 \cdot 4 = 0.$$

Yes, the two vectors are orthogonal.   ∎

Ans. to reading exercise 13:
1.6964 radians $= 97.19°$

If the vectors $\mathbf{v}$ and $\mathbf{w}$ of example 22 are placed with initial point at the origin as in Figure 1.44, then $\mathbf{v}$ lies in the horizontal "floor" and $\mathbf{w}$ is vertical, making it clear that the two vectors are orthogonal. But the procedure is much more helpful when it is not immediately clear whether the vectors are orthogonal.

Example 23   *Are* $\mathbf{v} = \langle 3, 2, 4 \rangle$ *and* $\mathbf{w} = \langle 5, \frac{1}{2}, -4 \rangle$ *orthogonal?*

Solution   Check the dot product:

$$\mathbf{v} \cdot \mathbf{w} = 3 \cdot 5 + 2 \cdot \frac{1}{2} + 4(-4) = 15 + 1 - 16 = 0.$$

Yes, the two vectors are orthogonal.   ∎

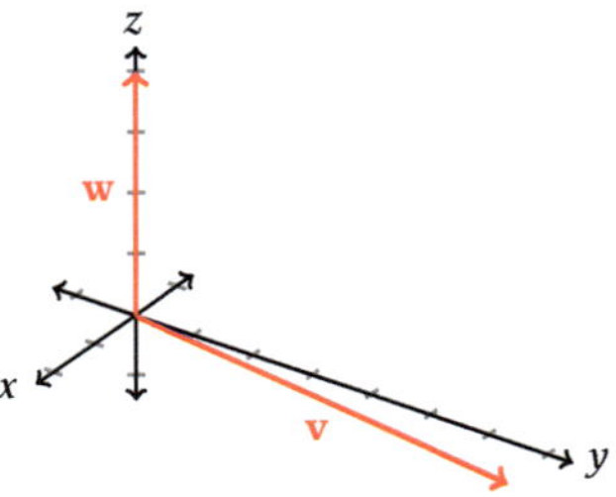

Figure 1.44 *The vectors* **v** *and* **w** *of example 22*

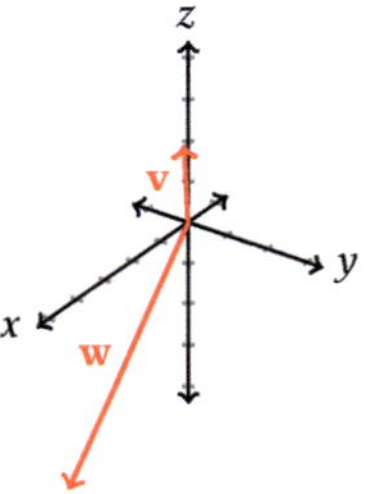

Figure 1.45 *The vectors* **v** *and* **w** *of example 23*

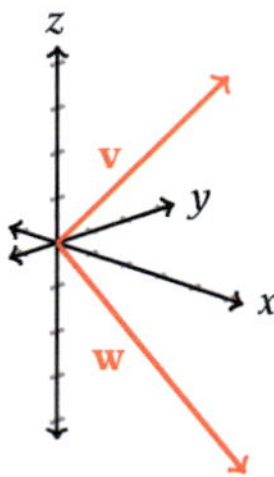

Figure 1.46 *The vectors* **v** *and* **w** *of example 23*

We know that the two vectors in example 23 are orthogonal, yet if we graph them with initial points at the origin using the same viewpoint we normally use (Figure 1.45), they may give the illusion of being far from perpendicular! Why? The issue is that in this viewpoint, we are looking nearly straight down the vector **v**, making it look much shorter than **w** and distorting our perception of its direction. In Figure 1.46 the same two vectors are graphed from a different viewpoint.

Now it looks much more believable that the vectors are orthogonal. Using technology, it is usually easy to grab and spin a three-dimensional graphic, and doing so is often helpful.

Of course, in the two-dimensional setting with our eye outside the graph, we do not have the same difficulties with visual illusions. But even then, it might be hard to tell visually if two vectors are truly orthogonal or just nearly orthogonal.

Think of rotating Figure 1.46 around the z-axis, moving the positive x-axis to our left, until it matches the position of Figure 1.45. In your mind, watch **v** as it swings around and you are looking almost directly down the vector, but from just a little below it instead.

Example 24 *Are* $\mathbf{v} = \langle 1, 4 \rangle$ *and* $\mathbf{w} = \langle 2, -1 \rangle$ *orthogonal?*

Solution   Check the dot product:

$$\mathbf{v} \cdot \mathbf{w} = 1(2) + 4(-1) = -2 \neq 0.$$

No, the two vectors are not orthogonal.   ∎

The zero vector does not have a specific direction. But it is orthogonal to every vector:

$$\mathbf{0} \cdot \mathbf{v} = \langle 0, 0, 0 \rangle \cdot \langle v_1, v_2, v_3 \rangle = 0(v_1) + 0(v_2) + 0(v_3) = 0.$$

This fact is sometimes listed among the properties of the dot product:

$$(6) \quad \mathbf{0} \cdot \mathbf{v} = 0.$$

In this equation, the bold zero on the left is the zero vector, and the zero on the right is the number zero.

Reading Exercise 14   Are $\mathbf{v} = \langle 1, 1, 7 \rangle$ and $\mathbf{w} = \langle 5, 2, -1 \rangle$ orthogonal?

1.4.3   Direction angles and direction cosines

The *direction angles* of a vector **v** are the angles between **v** and the positive x-, y-, and z-axes, as illustrated in Figure 1.47.

Definition 11 DIRECTION ANGLES AND DIRECTION COSINES *Let* $\mathbf{v} = \langle v_1, v_2, v_3 \rangle$. *The direction angle* α *is the angle between* **v** *and the positive x-axis. The direction angle* β *is the angle between* **v** *and the positive y-axis. The direction angle* γ *is the angle between* **v** *and the positive z-axis. The direction cosines are* $\cos \alpha$, $\cos \beta$, *and* $\cos \gamma$.

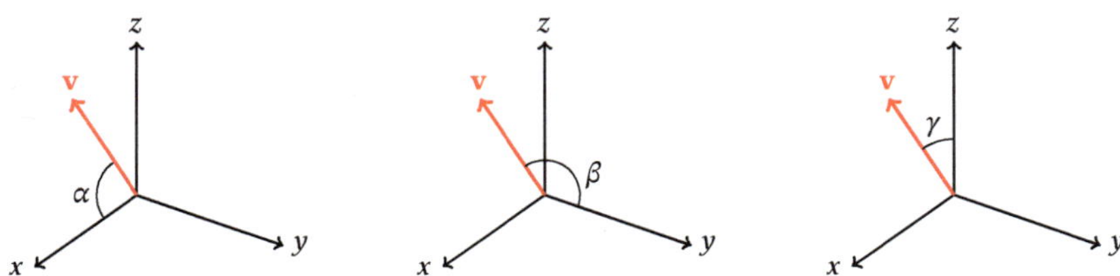

Figure 1.47 *A vector* **v** *and its direction angles* α, β, *and* γ

A vector in the direction of the positive x-axis is $\mathbf{i}$. Then α is the angle between the vectors $\mathbf{v}$ and $\mathbf{i}$, and the direction cosine can be found using the second angle between vectors formula:

$$
\begin{aligned}
\cos \alpha &= \frac{\mathbf{v} \cdot \mathbf{i}}{\|\mathbf{v}\| \, \|\mathbf{i}\|} \\[2mm]
&= \frac{\langle v_1, v_2, v_3 \rangle \cdot \langle 1, 0, 0 \rangle}{\|\mathbf{v}\|(1)} \\[2mm]
&= \frac{v_1}{\|\mathbf{v}\|}.
\end{aligned}
$$

Similarly, the other two direction cosines are

$$
\cos \beta = \frac{v_2}{\|\mathbf{v}\|} \quad \text{and}
$$

$$
\cos \gamma = \frac{v_3}{\|\mathbf{v}\|}.
$$

Once the direction cosines are known, the direction angles can be found using the inverse cosine.

DIRECTION ANGLE FORMULAS

For any vector $\mathbf{v} = \langle v_1, v_2, v_3 \rangle$, the direction angles are

$$
\alpha = \cos^{-1} \frac{v_1}{\|\mathbf{v}\|}, \quad \beta = \cos^{-1} \frac{v_2}{\|\mathbf{v}\|}, \quad \text{and} \quad \gamma = \cos^{-1} \frac{v_3}{\|\mathbf{v}\|}.
$$

Example 25 *Find the direction angles for* $\mathbf{u} = \langle 2, 7, 1 \rangle$, *in degrees.*

Solution First, we calculate the norm:

$$
\|\mathbf{u}\| = \sqrt{2^2 + 7^2 + 1^2} = \sqrt{54}.
$$

Using the direction angle formulas, we obtain

$$
\alpha = \cos^{-1} \frac{2}{\sqrt{54}} = 74.2°
$$

$$
\beta = \cos^{-1} \frac{7}{\sqrt{54}} = 17.7°
$$

$$
\gamma = \cos^{-1} \frac{1}{\sqrt{54}} = 82.2°. \qquad \blacksquare
$$

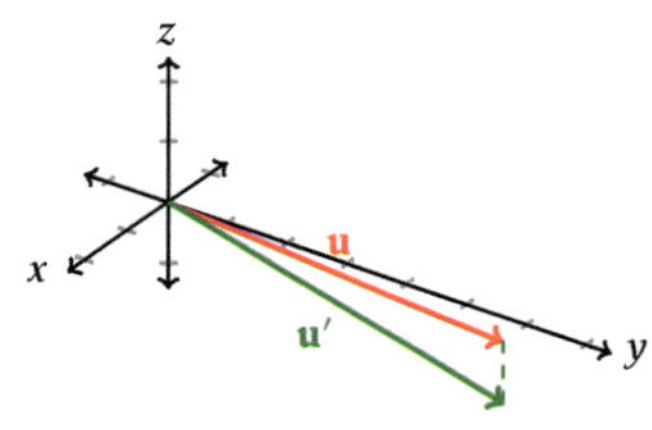

Figure 1.48 *The vector* $\mathbf{u}$ *(red) of example 25 and its shadow* $\mathbf{u}'$ *(green) in the xy-plane. Because* $\mathbf{u}'$ *is in the xy-plane, the angle between* $\mathbf{u}'$ *and the x-axis plus the angle between* $\mathbf{u}'$ *and the y-axis is 90°; those angles are approximately 74.05° and 15.95°, respectively. The vector* $\mathbf{u}$, *being out of the xy-plane, has slightly larger angles to the x- and y-axes, hence the sum of the angles is a little larger than 90°; in other words,* $\alpha + \beta > 90°$. *This is always true of any two direction angles for a vector with all nonzero coordinates*

Some of the calculations in example 25 should look a little familiar. In fact, if we were to find a unit vector in the direction of **u**, we would have

$$\frac{\mathbf{u}}{\|\mathbf{u}\|} = \left\langle \frac{2}{\sqrt{54}}, \frac{7}{\sqrt{54}}, \frac{1}{\sqrt{54}} \right\rangle.$$

In general, the direction cosines of a vector **u** are the components of the unit vector in the direction of **u**.

UNIT VECTORS AND DIRECTION COSINES

For any vector $\mathbf{u} = \langle u_1, u_2, u_3 \rangle$,

$$\frac{\mathbf{u}}{\|\mathbf{u}\|} = \langle \cos \alpha, \cos \beta, \cos \gamma \rangle.$$

1.4.4 Projections

We know that we can rewrite $\mathbf{w} = \langle 3, 4 \rangle$ as $3\mathbf{i} + 4\mathbf{j}$. Instead of thinking of 3 as the x-component and 4 as the y-component, we could think of 3 as the component in the **i** direction and 4 as the component in the **j** direction.

Consider parallel light rays, such as those produced by bouncing light off a paraboloid mirror. If all these rays are parallel to the x-axis and there is a wall at the y-axis (Figure 1.49), the shadow of vector $\mathbf{w} = \langle 3, 4 \rangle$ on the wall would be the vector $\langle 0, 4 \rangle = 4\mathbf{j}$. The vector $4\mathbf{j}$ is called the *vector projection of* **w** *onto* **j**, while the length of the vector projection, 4, is called the *scalar component of* **w** *in the* **j** *direction*.

The vector projection of $\mathbf{w} = 3\mathbf{i} + 4\mathbf{j}$ onto **i** would similarly be $3\mathbf{i}$. But what if we wanted to project in a different direction than parallel to an axis?

> Ans. to reading exercise 14:
> yes, since $\mathbf{v} \cdot \mathbf{w} = 0$

> In two dimensions, parallel to the x-axis is the same as perpendicular to the y-axis.

> It may seem odd to use the phrase "onto **j**" when the shadow is longer than **j** itself. When we say "onto **j**" we mean onto the line that contains **j**, placing the initial points of both vectors (**w** and **j** in this case) at the same point.

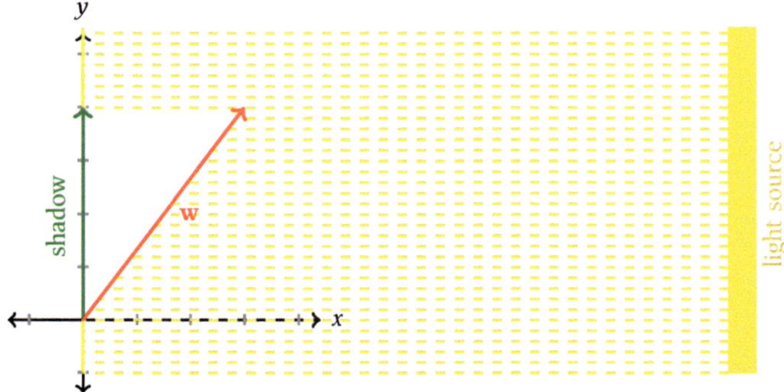

Figure 1.49 *The vector* **v** *(red) and its shadow (green)*

To project vector **v** onto the vector **u**, we place the initial points of the vectors at the same point (Figure 1.50, left). Then re-creating the idea of Figure 1.49, we place the light source parallel to vector **u** and consider light rays perpendicular to the line containing the vector **u**, which serves as the "wall" onto which the projection occurs; see Figure 1.50, right. Let d be the length of the shadow. Using trigonometry on the right triangle with hypotenuse **v** and with the shadow as one leg, we know that

$$\cos\theta = \frac{d}{\|\mathbf{v}\|}.$$

From the second angle between vectors formula, we also know that

$$\cos\theta = \frac{\mathbf{v}\cdot\mathbf{u}}{\|\mathbf{v}\|\,\|\mathbf{u}\|}.$$

By substitution,

$$\frac{d}{\|\mathbf{v}\|} = \frac{\mathbf{v}\cdot\mathbf{u}}{\|\mathbf{v}\|\,\|\mathbf{u}\|}.$$

Simplifying (multiplying both sides by $\|\mathbf{v}\|$) yields

$$d = \frac{\mathbf{v}\cdot\mathbf{u}}{\|\mathbf{u}\|}.$$

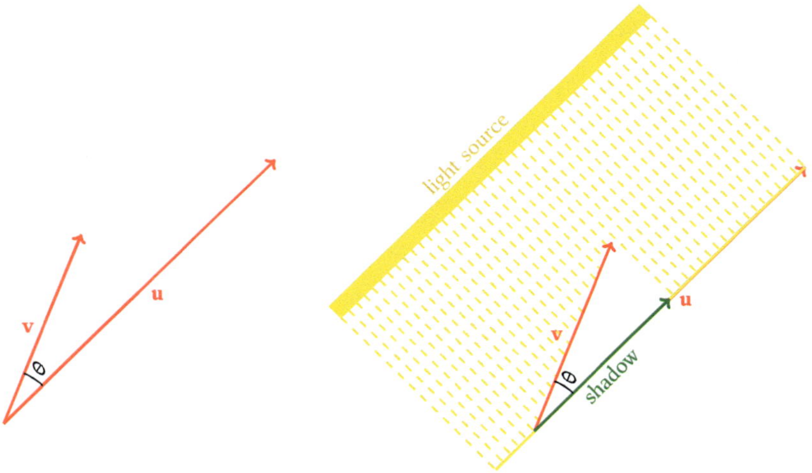

Figure 1.50 *Left: the vectors* **v** *and* **u**; *right: the vectors* **v** *and* **u**, *light rays perpendicular to the line containing the vector* **u**, *and the shadow of* **v** *along the line containing the vector* **u**. *Compare to Figure 1.49*

We call the length of the shadow the *scalar component of* **v** *in the direction of* **u**, which we denote $\text{comp}_{\mathbf{u}}\,\mathbf{v}$.

SCALAR COMPONENT OF **v** IN THE DIRECTION OF **u**

For vectors **v** and **u**, the *scalar component of* **v** *in the direction of* **u** is given by

$$\text{comp}_{\mathbf{u}}\,\mathbf{v} = \frac{\mathbf{v} \cdot \mathbf{u}}{\|\mathbf{u}\|}.$$

The quantity $\text{comp}_{\mathbf{u}}\,\mathbf{v}$ represents the length of the shadow when projecting **v** onto **u**.

An alternate presentation of the formula that might be easier to remember is to take the dot product of **v** with the unit vector in the direction of **u**:

$$\text{comp}_{\mathbf{u}}\,\mathbf{v} = \mathbf{v} \cdot \frac{\mathbf{u}}{\|\mathbf{u}\|}.$$

The scalar component gives us the length of the shadow. What if we wish to compute the vector of the shadow instead? We merely need to multiply the length by the unit vector in the direction of **u**; then it will have the correct direction (the direction of **u**) and the correct length (the length of the shadow times one). We would have

Here, "multiply" refers to scalar multiplication.

$$d\frac{\mathbf{u}}{\|\mathbf{u}\|} = \frac{\mathbf{v} \cdot \mathbf{u}}{\|\mathbf{u}\|}\frac{\mathbf{u}}{\|\mathbf{u}\|} = \left(\frac{\mathbf{v} \cdot \mathbf{u}}{\|\mathbf{u}\|^2}\right)\mathbf{u}.$$

We call this the *vector projection of* **v** *onto* **u**, denoted $\text{proj}_{\mathbf{u}}\mathbf{v}$.

VECTOR PROJECTION OF **v** ONTO **u**

For vectors **v** and **u**, the *vector projection of* **v** *onto* **u** is given by

$$\text{proj}_{\mathbf{u}}\mathbf{v} = \left(\frac{\mathbf{v} \cdot \mathbf{u}}{\|\mathbf{u}\|^2}\right)\mathbf{u}.$$

The dot product is a scalar, and the norm is a scalar. Therefore, the quantity in parentheses, $\frac{\mathbf{v} \cdot \mathbf{u}}{\|\mathbf{u}\|^2}$, is a scalar. So the vector $\text{proj}_{\mathbf{u}}\mathbf{v}$ is a scalar multiple of **u**. This should make sense; compare the shadow in Figure 1.50 to the vector **u**.

Example 26 *Find the scalar component and vector projection of* $\mathbf{v} = \langle 2, 0, 4\rangle$ *onto* $\mathbf{u} = \langle 1, -3, 9\rangle$.

Figures 1.49 and 1.50 show two dimensions only. For three dimensions, the light rays are in the plane containing the two vectors, perpendicular to the "onto" vector.

Solution First, we compute the dot product

$$\mathbf{v} \cdot \mathbf{u} = \langle 2, 0, 4\rangle \cdot \langle 1, -3, 9\rangle = 2 + 0 + 36 = 38$$

and the norm of **u**

$$\|\mathbf{u}\| = \sqrt{1^2 + (-3)^2 + 9^2} = \sqrt{91}.$$

Using the formula for the scalar component,

$$\text{comp}_{\mathbf{u}}\,\mathbf{v} = \frac{\mathbf{v} \cdot \mathbf{u}}{\|\mathbf{u}\|} = \frac{38}{\sqrt{91}}.$$

Using the formula for the vector projection,

$$\text{proj}_{\mathbf{u}}\mathbf{v} = \frac{\mathbf{v}\cdot\mathbf{u}}{\|\mathbf{u}\|^2}\mathbf{u} = \frac{38}{(\sqrt{91})^2}\langle 1, -3, 9\rangle = \left\langle \frac{38}{91}, \frac{-114}{91}, \frac{342}{91}\right\rangle.$$

■

Reading Exercise 15   Find the scalar component and vector projection of $\mathbf{v} = \langle 1, 1, 7\rangle$ onto $\mathbf{u} = \langle 2, 4, 1\rangle$.

Figures 1.49 and 1.50 show the case where the angle between the vectors $\mathbf{u}$ and $\mathbf{v}$ is acute. But what if the angle is obtuse? In this case, the shadow does not lie along the vector $\mathbf{u}$, but rather lies in the opposite direction; see Figure 1.51. Then $\mathbf{v}\cdot\mathbf{u} < 0$, which makes the scalar component negative as well; $\text{comp}_{\mathbf{u}}\,\mathbf{v} < 0$. Finally, the vector projection of $\mathbf{v}$ onto $\mathbf{u}$ has the opposite direction as the vector $\mathbf{u}$.

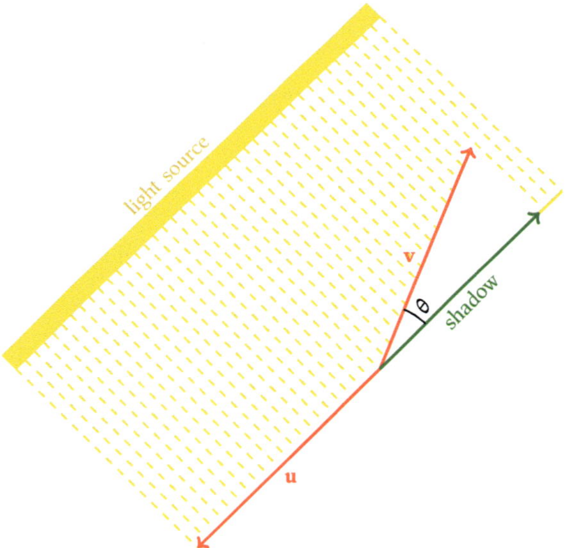

Figure 1.51 *The vectors* **v** *and* **u**, *light rays perpendicular to the line containing the vector* **u**, *and the shadow of* **v** *along the line containing the vector* **u**. *Compare to Figure 1.50, right*

EXERCISES 1.4

1–12. Rapid response: is the quantity a vector or a scalar?

1. $\mathbf{v} \cdot \mathbf{w}$

2. $\|\mathbf{u}\|$

3. $\mathbf{j}$

4. $\mathrm{comp}_{\mathbf{u}}\,\mathbf{v}$

5. $\mathrm{proj}_{\mathbf{u}}\mathbf{v}$

6. $\langle 1, 2, 3\rangle$

7. $\|\langle 1, 2, 3\rangle\|$

8. $\dfrac{\mathbf{u}}{\|\mathbf{v}\|}$

9. $\dfrac{\mathbf{v} \cdot \mathbf{u}}{\|\mathbf{u}\|}$

10. $\dfrac{\|\mathbf{u}\|}{\|\mathbf{v}\|}$

11. $\mathrm{comp}_{\mathbf{u}}\,\mathbf{v}\,\mathrm{proj}_{\mathbf{u}}\mathbf{v}$

12. $\mathrm{proj}_{\mathbf{u}}\mathbf{v} \cdot \mathrm{proj}_{\mathbf{u}}\mathbf{v}$

13–18. Rapid response: what's wrong with the expression?

13. $\langle 1, 2, 3\rangle \cdot \langle 4, 5\rangle$

14. $\langle 6, 7\rangle\langle 8, 9\rangle$

15. $10 \cdot \mathbf{v}$

16. $\langle 11, 12\rangle \cdot \mathbf{k}$

17. $\dfrac{\|\mathbf{w}\|}{\mathbf{w}}$

18. $(\mathbf{u} \cdot \mathbf{v}) \cdot \mathbf{w}$

19–28. Compute the dot product.

19. $\langle 2, 8\rangle \cdot \langle -4, 5\rangle$

20. $\langle 3, 0, 4\rangle \cdot \langle 0, 2, 5\rangle$

21. $\langle -3, 9, 10\rangle \cdot \langle 4, 1, 2\rangle$

22. $\langle 2, 5\rangle \cdot \langle -3, -7\rangle$

23. $\langle 1, 2, 3\rangle \cdot \mathbf{j}$

24. $\langle 0, 2, 5\rangle \cdot \langle 0, -3, -7\rangle$

25. $(\mathbf{i} + 3\mathbf{k}) \cdot (2\mathbf{j} + \mathbf{i})$

26. $(\mathbf{i} - 4\mathbf{j}) \cdot (2\mathbf{i} - \mathbf{j})$

27. $\mathbf{i} \cdot \mathbf{k}$

28. $\mathbf{j} \cdot (-\mathbf{k})$

29–38. Find the angle between the two vectors. State your answer in (a) radians and (b) degrees.

29. $\mathbf{u} = \langle 2, 6\rangle,\ -\mathbf{j}$

30. $\mathbf{w} = \langle 2, 1\rangle, \mathbf{u} = \langle -7, -3\rangle$

31. $\mathbf{a} = \langle 2, 5, 7\rangle, \mathbf{b} = \langle 1, 6, 4\rangle$

32. $\mathbf{a} = \langle 1, 5, 3\rangle, 2\mathbf{i} + 3\mathbf{j} + 8\mathbf{k}$

33. $\mathbf{v} = \langle 0, 5, -4\rangle, \mathbf{w} = \langle 1, 4, 0\rangle$

34. $\mathbf{w} = \langle 2, 3, 7\rangle, \mathbf{v} = \langle 4, 6, 14\rangle$

35. $\mathbf{v} = \langle 1, 4\rangle, \mathbf{u} = \langle -4, 1\rangle$

36. $(\mathbf{i} + \mathbf{j}), (\mathbf{j} - 3\mathbf{k})$

37. $4\mathbf{i} + 2\mathbf{j}, \mathbf{i} + 3\mathbf{k}$

38. $\mathbf{u} = \langle 4, 0, 4\rangle, \pi\mathbf{j}$

Ans. to reading exercise 15:
$$\mathrm{comp}_{\mathbf{u}}\,\mathbf{v} = \frac{13}{\sqrt{21}},\ \mathrm{proj}_{\mathbf{u}}\mathbf{v} = \left\langle \frac{26}{21}, \frac{52}{21}, \frac{13}{21}\right\rangle$$

39–46. Are the vectors orthogonal?

39. $\mathbf{i}, \mathbf{j}$
40. $\mathbf{u} = \langle 0, 1, 0 \rangle, \mathbf{v} = \langle 2.089, 0, 5.6 \rangle$
41. $\langle 3, 5, -1 \rangle, \langle 1, 0, 4 \rangle$
42. $\mathbf{i}, 2\mathbf{i}$
43. $\mathbf{a} = \langle 3142, 2.718 \rangle, \mathbf{b} = \langle 2.718, -3142 \rangle$
44. $\mathbf{w} = \langle 0.72, -0.46, 1.26 \rangle, \mathbf{v} = \langle 413.1, 88.9, -105.5 \rangle$
45. $\langle 0, 11, 24 \rangle, \mathbf{0}$
46. $\langle \sqrt{2}, 0, 3 \rangle, \left\langle \sqrt{2}, \pi, -\frac{2}{3} \right\rangle$

47–56. Find (a) the scalar component and (b) the vector projection of $\mathbf{v}$ onto $\mathbf{u}$.

47. $\mathbf{u} = \mathbf{i}, \mathbf{v} = \langle 12, -5 \rangle$
48. $\mathbf{u} = \langle 1, 4, 6 \rangle, \mathbf{v} = \langle 2, 7, 1 \rangle$
49. $\mathbf{u} = \langle 5, 0, 5 \rangle, \mathbf{v} = \langle 2, 1, 9 \rangle$
50. $\mathbf{u} = \langle 2, 5 \rangle, \mathbf{v} = \langle -3, 4 \rangle$
51. $\mathbf{u} = \langle 3, 0, 3 \rangle, \mathbf{v} = \langle 2, 1, 9 \rangle$
52. $\mathbf{u} = \mathbf{j}, \mathbf{v} = 2\mathbf{i} - 4\mathbf{j} + \mathbf{k}$
53. $\mathbf{u} = \langle 1, 19 \rangle, \mathbf{v} = \langle -3, 0 \rangle$
54. $\mathbf{u} = \langle 4, 9, 6 \rangle, \mathbf{v} = \langle 6, 2, -7 \rangle$
55. $\mathbf{u} = \langle 104, 18, -5 \rangle, \mathbf{v} = \langle 0, 0, 0 \rangle$
56. $\mathbf{u} = \langle \pi, 0, 0 \rangle, \mathbf{v} = \langle 72, 921, 14.8 \rangle$

57–62. Compute the direction angles, in degrees.

57. $\langle 2.5, 3, 0 \rangle$
58. $\langle 2, 9, 6 \rangle$
59. $\langle -3, -1, -5 \rangle$
60. $\langle 0, 5, 0 \rangle$
61. $\langle 2, 2, 2 \rangle$
62. $\langle 3, -3, 3 \rangle$

63–66. Prove the property of the dot product.

63. (1) $\mathbf{u} \cdot \mathbf{v} = \mathbf{v} \cdot \mathbf{u}$
64. (2) $\mathbf{u} \cdot (\mathbf{v} + \mathbf{w}) = \mathbf{u} \cdot \mathbf{v} + \mathbf{u} \cdot \mathbf{w}$
65. (3) $(\mathbf{v} - \mathbf{w}) \cdot \mathbf{u} = \mathbf{v} \cdot \mathbf{u} - \mathbf{w} \cdot \mathbf{u}$
66. (4) $a(\mathbf{u} \cdot \mathbf{v}) = (a\mathbf{u}) \cdot \mathbf{v} = \mathbf{u} \cdot (a\mathbf{v})$

67. (a) Write a definition for the dot product of four-dimensional vectors. (b) Compute $\langle 2, 5, -1, 7 \rangle \cdot \langle 3, 8, 2, -3 \rangle$.
68. Light rays in the direction of the vector $\langle -3, 0, 0 \rangle$ shine on a pole described as the vector $\langle 7, 7, \sqrt{2} \rangle$. A screen larger than you have ever seen before covers the yz-plane. The shadow of the pole on the screen can be described as what vector?

69. Vectors $\mathbf{v} = \langle 2, 1, 5 \rangle$ and $\mathbf{w} = \langle 100, 100, 100 \rangle$ share the same initial point. Light rays are in the plane containing the two vectors and perpendicular to the vector $\mathbf{w}$, and those light rays cause $\mathbf{v}$ to cast a shadow on $\mathbf{w}$. What percentage of $\mathbf{w}$ is covered by the shadow of $\mathbf{v}$?

70. On the same graph, draw vectors $\mathbf{a} = \langle 2, 1, -1 \rangle$, $\mathbf{b} = \langle 1, 5, 7 \rangle$, and $\mathrm{proj}_{\mathbf{b}}\, \mathbf{a}$.

1.5   Cross Product

Before introducing the cross product, we need to review determinants.

1.5.1   Determinants

A *determinant* is a quantity computed from the entries of a square matrix as defined below.

Readers that are familiar with determinants may skip this subsection. It is provided for the convenience of those who have never studied determinants or who may not remember how they are calculated.

Definition 12 2×2 **DETERMINANT**   *A 2×2 determinant, that is, the determinant of a 2×2 matrix, is computed as*

$$\begin{vmatrix} a & b \\ c & d \end{vmatrix} = ad - bc.$$

The name is pronounced, and can be written, as "two-by-two determinant."

The vertical bars on either side of the matrix form the symbol for the determinant, and their use is similar to the absolute value symbol or the symbol for the norm of a vector. Recall that brackets are used for the sides of a matrix; when we use bars instead, we are referring to the determinant.

Example 27 *Compute* $\begin{vmatrix} 4 & 2 \\ -1 & 5 \end{vmatrix}$.

Solution   Following the definition, we compute the product of the top left and bottom right entries, then subtract the product of the top right and bottom left entries:

$$\begin{vmatrix} 4 & 2 \\ -1 & 5 \end{vmatrix} = 4(5) - 2(-1) = 22.$$

∎

Care is needed to avoid making sign errors. Whenever there is a negative entry in the matrix, writing out the arithmetic can be a good idea.

Example 28 *Evaluate* $\begin{vmatrix} 2 & 3 \\ 7 & -4 \end{vmatrix}$.

Solution The definition of determinant yields

$$\begin{vmatrix} 2 & 3 \\ 7 & -4 \end{vmatrix} = 2(-4) - 3(7) = -8 - 21 = -29.$$

∎

It is sometimes helpful to think of performing the arithmetic by following a "swoop" around the matrix. Beginning at the upper left, we multiply the first pair, swoop around, and subtract the product of the second pair, as illustrated in the next example.

Example 29 *Find* $\begin{vmatrix} 1 & 7 \\ 9 & 4 \end{vmatrix}$.

Solution Using the definition of determinant,

$$\begin{vmatrix} 1 & 7 \\ 9 & 4 \end{vmatrix} = 4 - 63 = -59.$$

∎

Reading Exercise 16 Compute $\begin{vmatrix} 1 & 4 \\ 9 & 3 \end{vmatrix}$.

The determinant for a 3×3 matrix builds on the 2×2 determinant.

Definition 13 3×3 **DETERMINANT** *A 3×3 determinant, that is, the determinant of a 3×3 matrix, is computed as*

$$\begin{vmatrix} a_1 & a_2 & a_3 \\ b_1 & b_2 & b_3 \\ c_1 & c_2 & c_3 \end{vmatrix} = a_1 \begin{vmatrix} b_2 & b_3 \\ c_2 & c_3 \end{vmatrix} - a_2 \begin{vmatrix} b_1 & b_3 \\ c_1 & c_3 \end{vmatrix} + a_3 \begin{vmatrix} b_1 & b_2 \\ c_1 & c_2 \end{vmatrix}.$$

There are other, equivalent definitions of the 3×3 determinant. See a linear algebra textbook for additional information, including proofs of theorems and properties as well as motivation for these definitions.

Example 30 *Compute* $\begin{vmatrix} 1 & 4 & 9 \\ 2 & -3 & 7 \\ 4 & 5 & -1 \end{vmatrix}$.

Solution   Using the definition,

$$\begin{vmatrix} 1 & 4 & 9 \\ 2 & -3 & 7 \\ 4 & 5 & -1 \end{vmatrix} = 1 \begin{vmatrix} -3 & 7 \\ 5 & -1 \end{vmatrix} - 4 \begin{vmatrix} 2 & 7 \\ 4 & -1 \end{vmatrix} + 9 \begin{vmatrix} 2 & -3 \\ 4 & 5 \end{vmatrix}$$

$$= 1(3 - 35) - 4(-2 - 28) + 9(10 - (-12))$$

$$= -32 - 4(-30) + 9(22) = 286.$$

∎

Reading Exercise 17   Calculate the value of the determinant $\begin{vmatrix} 2 & 7 & 1 \\ 0 & 4 & -3 \\ -2 & 1 & -5 \end{vmatrix}$.

Familiarity with 3×3 determinants is helpful during the remainder of this section.

1.5.2   Cross product

The dot product of two vectors is a scalar. The *cross product* of two vectors is a vector; it is defined only in three dimensions.

Definition 14 CROSS PRODUCT *Given two vectors in three dimensions,* $\mathbf{v} = \langle v_1, v_2, v_3 \rangle$ *and* $\mathbf{w} = \langle w_1, w_2, w_3 \rangle$, *the* cross product *of* $\mathbf{v}$ *and* $\mathbf{w}$ *is*

$$\mathbf{v} \times \mathbf{w} = (v_2 w_3 - v_3 w_2)\mathbf{i} - (v_1 w_3 - v_3 w_1)\mathbf{j} + (v_1 w_2 - v_2 w_1)\mathbf{k}$$

$$= \langle v_2 w_3 - v_3 w_2, \ v_3 w_1 - v_1 w_3, \ v_1 w_2 - v_2 w_1 \rangle.$$

Several questions come to mind, such as "why do we care?" and "how am I going to remember that?." We care because it turns out that the cross product of two vectors $\mathbf{v}$ and $\mathbf{w}$ is orthogonal to both $\mathbf{v}$ and $\mathbf{w}$, a fact which we will explore shortly. And one way to remember the formula is by the determinant

$$\begin{vmatrix} \mathbf{i} & \mathbf{j} & \mathbf{k} \\ v_1 & v_2 & v_3 \\ w_1 & w_2 & w_3 \end{vmatrix}.$$

The fact that such a determinant hasn't yet been defined (this one has some vectors in place of numbers) doesn't really matter; the computational procedure of the 3×3 determinant produces the cross product.

Definition 13 gives one among many *cofactor expansion* methods for calculating a 3×3 determinant. Notice that a_1, a_2, and a_3 are all on one row. The 2×2 determinate that goes with each can be found by blocking out the remainder of the row and column, as visually indicated below.

$$\begin{vmatrix} a_1 & & \\ & b_2 & b_3 \\ & c_2 & c_3 \end{vmatrix}$$

$$\begin{vmatrix} & a_2 & \\ b_1 & & b_3 \\ c_1 & & c_3 \end{vmatrix}$$

$$\begin{vmatrix} & & a_3 \\ b_1 & b_2 & \\ c_1 & c_2 & \end{vmatrix}$$

The signs that go with each of a_1, a_2, and a_3 can be found in the *alternating signs matrix*:

$$\begin{vmatrix} + & - & + \\ - & + & - \\ + & - & + \end{vmatrix}.$$

Any row or column can be used in this manner to find the determinant. For a proof of this fact, consult a linear algebra textbook.

Try it! Compute the determinant and compare to definition 14.

Example 31 *Compute* $\langle 1, 4, 9\rangle \times \langle 2, 5, -1\rangle$.

Notice that the vector on the left is **v** in the definition, and the vector on the right is **w**. The order is important!

Solution   To follow the definition, we set $\mathbf{v} = \langle 1, 4, 9\rangle$ and $\mathbf{w} = \langle 2, 5, -1\rangle$, meaning that $v_1 = 1$, $v_2 = 4$, $v_3 = 9$, $w_1 = 2$, $w_2 = 5$, and $w_3 = -1$. Then

$$\langle 1, 4, 9\rangle \times \langle 2, 5, -1\rangle = \langle 4(-1) - 9(5), \, 9(2) - 1(-1), \, 1(5) - 4(2)\rangle$$

$$= \langle -49, 19, -3\rangle.$$

$\blacksquare$

Ans. to reading exercise 16:
  -33

An alternate solution is to follow the determinant method, which we shall write as if the determinant is actually defined.

Alternate solution   To follow the determinant method, we place the unit vectors **i**, **j**, and **k** in the first row, the vector on the left in the second row, and the vector on the right in the third row. Then

Ans. to reading exercise 17:
  16

$$\begin{vmatrix} \mathbf{i} & \mathbf{j} & \mathbf{k} \\ 1 & 4 & 9 \\ 2 & 5 & -1 \end{vmatrix} = \mathbf{i}\begin{vmatrix} 4 & 9 \\ 5 & -1 \end{vmatrix} - \mathbf{j}\begin{vmatrix} 1 & 9 \\ 2 & -1 \end{vmatrix} + \mathbf{k}\begin{vmatrix} 1 & 4 \\ 2 & 5 \end{vmatrix}$$

$$= \mathbf{i}(-4 - 45) - \mathbf{j}(-1 - 18) + \mathbf{k}(5 - 8)$$

$$= \langle -49, 19, -3\rangle.$$

$\blacksquare$

It was mentioned previously that the cross product is orthogonal to both of the vectors. Let's verify that fact for the vectors of example 31:

Recall that two vectors are orthogonal if their dot product is 0. The dot product is a scalar; the cross product is a vector.

$$\langle -49, 19, -3\rangle \cdot \langle 1, 4, 9\rangle = -49 + 76 - 27 = 0$$

$$\langle -49, 19, -3\rangle \cdot \langle 2, 5, -1\rangle = -98 + 95 + 3 = 0.$$

A more formal proof that this always works is next.

Reading Exercise 18   Compute the cross product $\langle 3, 1, 4\rangle \times \langle 2, 0, 6\rangle$.

1.5.3   Orthogonality of the cross product

Theorem 5 CROSS PRODUCT ORTHOGONALITY *For any two vectors* $\mathbf{v} = \langle v_1, v_2, v_3\rangle$ *and* $\mathbf{w} = \langle w_1, w_2, w_3\rangle$, $\mathbf{v} \times \mathbf{w}$ *is orthogonal to both* **v** *and* **w**.

Proof. Using the definition,

$$\mathbf{v} \cdot (\mathbf{v} \times \mathbf{w}) = \langle v_1, v_2, v_3 \rangle \cdot \langle v_2 w_3 - v_3 w_2, v_3 w_1 - v_1 w_3, v_1 w_2 - v_2 w_1 \rangle$$

$$= v_1(v_2 w_3 - v_3 w_2) + v_2(v_3 w_1 - v_1 w_3) + v_3(v_1 w_2 - v_2 w_1)$$

$$= v_1 v_2 w_3 - v_1 v_3 w_2 + v_2 v_3 w_1 - v_2 v_1 w_3 + v_3 v_1 w_2 - v_3 v_2 w_1$$

$$= 0.$$

Line 1 uses the definition of cross product. In line 3, notice that the first and fourth terms, $v_1 v_2 w_3$ and $-v_2 v_1 w_3$, nicely add to 0; the same is true for the second and fifth terms, and for the third and sixth terms.

Thus, v is orthogonal to $\mathbf{v} \times \mathbf{w}$. The proof that $\mathbf{w}$ is orthogonal to $\mathbf{v} \times \mathbf{w}$ is similar. ∎

Theorem 5 gives us a method for finding a vector that is mutually perpendicular to two other vectors.

Example 32 *Find a vector orthogonal to both $\langle 2, 0, 5 \rangle$ and $\langle 7, -1, -3 \rangle$.*

Solution According to theorem 5, all we need to do is find the cross product. Setting $\mathbf{v} = \langle 2, 0, 5 \rangle$ and $\mathbf{w} = \langle 7, -1, -3 \rangle$, we have

$$\mathbf{v} \times \mathbf{w} = \begin{vmatrix} \mathbf{i} & \mathbf{j} & \mathbf{k} \\ 2 & 0 & 5 \\ 7 & -1 & -3 \end{vmatrix} = \mathbf{i} \begin{vmatrix} 0 & 5 \\ -1 & -3 \end{vmatrix} - \mathbf{j} \begin{vmatrix} 2 & 5 \\ 7 & -3 \end{vmatrix} + \mathbf{k} \begin{vmatrix} 2 & 0 \\ 7 & -1 \end{vmatrix}$$

$$= \mathbf{i}(0 - (-5)) - \mathbf{j}(-6 - 35) + \mathbf{k}(-2 - 0)$$

$$= \langle 5, 41, -2 \rangle.$$

A vector orthogonal to both $\langle 2, 0, 5 \rangle$ and $\langle 7, -1, -3 \rangle$ is $\langle 5, 41, -2 \rangle$. ∎

Figure 1.52 shows the vectors of example 32 and their cross product. Consider the plane containing the two vectors $\mathbf{v}$ and $\mathbf{w}$. If you think of the plane as a wall, then the wall separates space into two pieces, one of which contains $\mathbf{v} \times \mathbf{w}$. Is there a way to know which side of the wall (which side of the plane) should contain

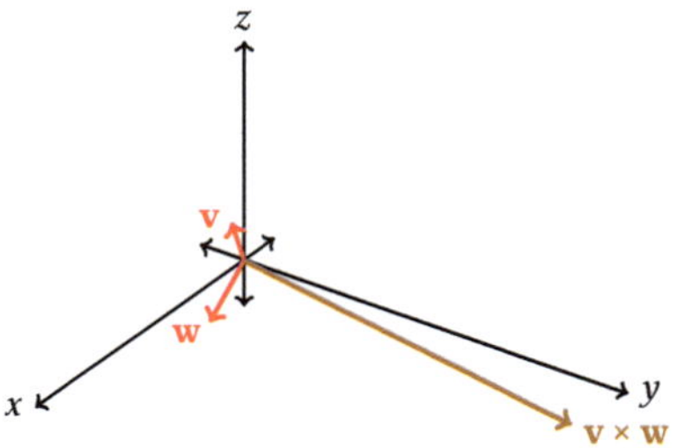

Figure 1.52 *The vectors* **v** *and* **w** *(red) of example 32, along with their cross product* **v** × **w** *(brown). The cross product* **v** × **w** *is orthogonal to both* **v** *and* **w**

the cross product before calculating it? Yes–the right-hand rule tells us! Point the fingers of your right hand along the first vector (**v** in this case) and let your fingers curl toward the second vector (**w** in this case). If necessary, re-position your hand so that this works. Let your thumb point outward; it then points to the correct side of the wall that contains the cross product. Try it on Figure 1.52. Then, try the procedure again with your left hand, and your thumb should point in the opposite direction.

Ans. to reading exercise 18:
$\langle 6, -10, -2 \rangle$

Example 33 *Find a vector orthogonal to both* **i** *and* **j**.

Solution　One such vector is **i** × **j**:

$$\langle 1, 0, 0 \rangle \times \langle 0, 1, 0 \rangle = \begin{vmatrix} \mathbf{i} & \mathbf{j} & \mathbf{k} \\ 1 & 0 & 0 \\ 0 & 1 & 0 \end{vmatrix}$$

Because the components of the two vectors in the cross product are needed for the computation, it helps to write **i** × **j** in terms of components. This also helps avoid confusion with the **i**, **j**, and **k** that always appear in the first row of the determinant.

$$= \mathbf{i} \begin{vmatrix} 0 & 0 \\ 1 & 0 \end{vmatrix} - \mathbf{j} \begin{vmatrix} 1 & 0 \\ 0 & 0 \end{vmatrix} + \mathbf{k} \begin{vmatrix} 1 & 0 \\ 0 & 1 \end{vmatrix}$$

$$= \mathbf{i}(0 - 0) - \mathbf{j}(0 - 0) + \mathbf{k}(1 - 0)$$

$$= \mathbf{k}.$$

The vector **k** is orthogonal to both **i** and **j**.　∎

That's not a surprise; we already knew that the standard basis vectors **i**, **j**, and **k** are orthogonal. But what happens if we switch the order and try **j** × **i** instead?

Alternate solution　One such vector is **j** × **i**:

$$\langle 0, 1, 0 \rangle \times \langle 1, 0, 0 \rangle = \begin{vmatrix} \mathbf{i} & \mathbf{j} & \mathbf{k} \\ 0 & 1 & 0 \\ 1 & 0 & 0 \end{vmatrix}$$

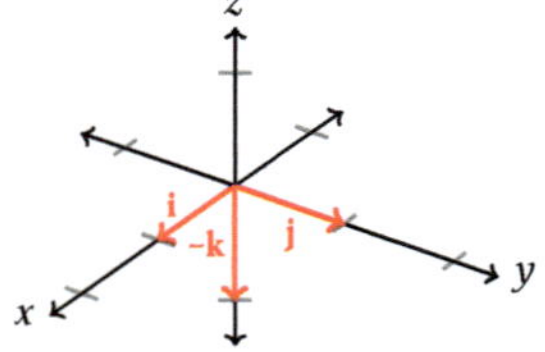

Figure 1.53 *The orthogonal vectors* **i**, **j**, *and* **j** × **i** = −**k** *from the alternate solution to example 33. As per the right-hand rule, the vector* **j** × **i** *points downward*

$$= \mathbf{i} \begin{vmatrix} 1 & 0 \\ 0 & 0 \end{vmatrix} - \mathbf{j} \begin{vmatrix} 0 & 0 \\ 1 & 0 \end{vmatrix} + \mathbf{k} \begin{vmatrix} 0 & 1 \\ 1 & 0 \end{vmatrix}$$

$$= \mathbf{i}(0 - 0) - \mathbf{j}(0 - 0) + \mathbf{k}(0 - 1)$$

$$= -\mathbf{k}.$$

The vector −**k** is orthogonal to both **i** and **j**.　∎

Notice that in the solution and alternate solution to example 33, reversing the order of the cross product changed the sign of the result, pointing it in the opposite direction. The results are not the same, and the cross product is not commutative.

In fact, the cross product is *anticommutative*—changing the order always reverses the direction. This is one of the properties of the cross product to be stated soon.

It can be helpful to know the cross products of pairs of standard basis vectors. Computation quickly yields the following facts:

$$\mathbf{i} \times \mathbf{j} = \mathbf{k} \qquad \mathbf{j} \times \mathbf{k} = \mathbf{i} \qquad \mathbf{k} \times \mathbf{i} = \mathbf{j}$$
$$\mathbf{j} \times \mathbf{i} = -\mathbf{k} \qquad \mathbf{k} \times \mathbf{j} = -\mathbf{i} \qquad \mathbf{i} \times \mathbf{k} = -\mathbf{j}.$$

A way to remember these facts is through the diagram in Figure 1.54, which features the vectors $\mathbf{i}$, $\mathbf{j}$, and $\mathbf{k}$ clockwise around a circle. To take the cross product of two vectors moving clockwise, such as $\mathbf{k} \times \mathbf{i}$, the answer is the next vector moving clockwise, $\mathbf{j}$. Moving counterclockwise, such as $\mathbf{j} \times \mathbf{i}$, do the same thing but insert a negative sign, $-\mathbf{k}$.

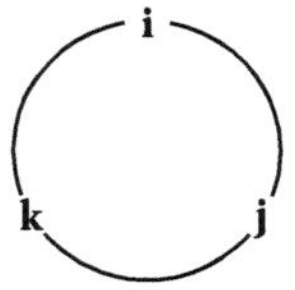

Reading Exercise 19   Find a vector orthogonal to both $\langle 1, 1, 1 \rangle$ and $\langle 2, 3, -1 \rangle$.

Figure 1.54 *Cross products of standard basis vectors; moving clockwise is positive, counterclockwise is negative*

1.5.4   Properties of the cross product

We have already mentioned that the cross product is not commutative, but rather anticommutative. The cross product is also not associative; in general, $(\mathbf{a} \times \mathbf{b}) \times \mathbf{c} \neq \mathbf{a} \times (\mathbf{b} \times \mathbf{c})$. Yet there remain some properties of the cross product that are useful.

This is a good example of why we must prove properties of an operation. Just because we want to manipulate them algebraically in particular ways, such as using the commutative or associative properties, doesn't mean we get to. Any such algebraic manipulation must be justified before being used.

Theorem 6 PROPERTIES OF THE CROSS PRODUCT *For any three-dimensional vectors* $\mathbf{u}$, $\mathbf{v}$, *and* $\mathbf{w}$ *and real number a,*

(1) $\mathbf{v} \times \mathbf{w} = -(\mathbf{w} \times \mathbf{v})$
(2) $(a\mathbf{v}) \times \mathbf{w} = \mathbf{v} \times (a\mathbf{w}) = a(\mathbf{v} \times \mathbf{w})$
(3) $\mathbf{u} \times (\mathbf{v} + \mathbf{w}) = (\mathbf{u} \times \mathbf{v}) + (\mathbf{u} \times \mathbf{w})$
(4) $\mathbf{u} \cdot (\mathbf{v} \times \mathbf{w}) = (\mathbf{u} \times \mathbf{v}) \cdot \mathbf{w}$
(5) $\mathbf{v} \times \mathbf{v} = \mathbf{0}.$

Proving these properties is similar to proving properties of vectors or properties of the dot product. We prove two of the properties and leave the others for exercises.

Proof.   (1) We wish to show that $\mathbf{v} \times \mathbf{w} = -(\mathbf{w} \times \mathbf{v})$. Let $\mathbf{v} = \langle v_1, v_2, v_3 \rangle$ and $\mathbf{w} = \langle w_1, w_2, w_3 \rangle$. Then, the left-hand side is

This is the anticommutative property.

$$\mathbf{v} \times \mathbf{w} = \langle v_2 w_3 - v_3 w_2, \ v_3 w_1 - v_1 w_3, \ v_1 w_2 - v_2 w_1 \rangle,$$

The definition of cross product is used.

Line 1 begins the determinant computation of the cross product, which is continued in lines 2 and 3; line 4 accomplishes the scalar multiplication by -1 and (subtly) uses the distributive property (multiplication distributes over addition); and line 5 uses the commutative properties of addition and multiplication.

A shorter derivation is to use the definition of cross product for the right-hand side as well, but doing so requires great care to avoid confusion and might not yield as much insight into why the anticommutative property is true.

and the right-hand side is

$$-(\mathbf{w} \times \mathbf{v}) = -\begin{vmatrix} \mathbf{i} & \mathbf{j} & \mathbf{k} \\ w_1 & w_2 & w_3 \\ v_1 & v_2 & v_3 \end{vmatrix}$$

$$= -\left(\mathbf{i}\begin{vmatrix} w_2 & w_3 \\ v_2 & v_3 \end{vmatrix} - \mathbf{j}\begin{vmatrix} w_1 & w_3 \\ v_1 & v_3 \end{vmatrix} + \mathbf{k}\begin{vmatrix} w_1 & w_2 \\ v_1 & v_2 \end{vmatrix} \right)$$

$$= -\langle w_2 v_3 - w_3 v_2,\ -(w_1 v_3 - w_3 v_1),\ w_1 v_2 - w_2 v_1 \rangle$$

$$= \langle -w_2 v_3 + w_3 v_2,\ w_1 v_3 - w_3 v_1,\ -w_1 v_2 + w_2 v_1 \rangle$$

$$= \langle v_2 w_3 - v_3 w_2,\ v_3 w_1 - v_1 w_3,\ v_1 w_2 - v_2 w_1 \rangle.$$

Because the expressions are identical, $\mathbf{v} \times \mathbf{w} = -(\mathbf{w} \times \mathbf{v})$.

(5) We wish to show that $\mathbf{v} \times \mathbf{v} = \mathbf{0}$. Let $\mathbf{v} = \langle v_1, v_2, v_3 \rangle$. Then

Line 1 begins the determinant computation of the cross product, which is continued in lines 2 and 3; line 4 completes the arithmetic, subtly using the commutative property of multiplication.

$$\mathbf{v} \times \mathbf{v} = \begin{vmatrix} \mathbf{i} & \mathbf{j} & \mathbf{k} \\ v_1 & v_2 & v_3 \\ v_1 & v_2 & v_3 \end{vmatrix}$$

$$= \mathbf{i}\begin{vmatrix} v_2 & v_3 \\ v_2 & v_3 \end{vmatrix} - \mathbf{j}\begin{vmatrix} v_1 & v_3 \\ v_1 & v_3 \end{vmatrix} + \mathbf{k}\begin{vmatrix} v_1 & v_2 \\ v_1 & v_2 \end{vmatrix}$$

$$= \mathbf{i}(v_2 v_3 - v_3 v_2) - \mathbf{j}(v_1 v_3 - v_3 v_1) + \mathbf{k}(v_1 v_2 - v_2 v_1)$$

$$= \mathbf{i}(0) - \mathbf{j}(0) + \mathbf{k}(0) = \mathbf{0}.$$

Therefore, $\mathbf{v} \times \mathbf{v} = \langle 0, 0, 0 \rangle = \mathbf{0}$. ■

Ans. to reading exercise 19:
$\langle -4, 3, 1 \rangle$; alternate answer $\langle 4, -3, -1 \rangle$

According to property (5), a vector cross product itself is the zero vector. Combine that with property (2), and we see that for any real number a, including a negative number, $(a\mathbf{v}) \times \mathbf{v} = a(\mathbf{v} \times \mathbf{v}) = a\mathbf{0} = \mathbf{0}$. In other words, if two vectors have the same or opposite direction, then their cross product is the zero vector. It turns out that the converse is also true; we shall delay the proof of this fact until later. For now we use the fact to motivate the following definition.

Vectors such as $3\mathbf{v}$ and $8.4\mathbf{v}$ have the same direction as $\mathbf{v}$, whereas $-2.78\mathbf{v}$ has the opposite direction.

This is similar to the idea of parallel lines in geometry, but in geometry parallel lines are often defined so that a line is not parallel to itself; for our purposes, a vector is considered parallel to itself. The difference is rooted in the concept of position. In geometry, lines have position. Vectors do not record position, only magnitude and direction.

Definition 15 PARALLEL VECTORS *Two vectors* $\mathbf{v}$ *and* $\mathbf{w}$ *are* parallel *if* $\mathbf{v} \times \mathbf{w} = \mathbf{0}$.

Two vectors are orthogonal if their dot product is zero; two vectors are parallel if their cross product is the zero vector.

Example 34 *Are* $\langle 1, 4, -2 \rangle$ *and* $\langle -3, -12, 6 \rangle$ *parallel?*

Solution   To determine if the vectors are parallel, we check their cross product:

$$\langle 1, 4, -2\rangle \times \langle -3, -12, 6\rangle = \begin{vmatrix} \mathbf{i} & \mathbf{j} & \mathbf{k} \\ 1 & 4 & -2 \\ -3 & -12 & 6 \end{vmatrix}$$

$$= \mathbf{i}\begin{vmatrix} 4 & -2 \\ -12 & 6 \end{vmatrix} - \mathbf{j}\begin{vmatrix} 1 & -2 \\ -3 & 6 \end{vmatrix} + \mathbf{k}\begin{vmatrix} 1 & 4 \\ -3 & -12 \end{vmatrix}$$

$$= \mathbf{i}(24 - 24) - \mathbf{j}(6 - 6) + \mathbf{k}(-12 - (-12)) = \mathbf{0}.$$

Yes, the two vectors are parallel.   ∎

You might have observed that one of the vectors is a negative scalar multiple of the other, and therefore the vectors have opposite directions. For now, however, the way we have defined parallel vectors dictates the use of the cross product for verification.

Reading Exercise 20   Are $\langle 1, 1, 4\rangle$ and $\langle 2, 1, 5\rangle$ parallel?

1.5.5   Magnitude of the cross product

So far, much of our discussion of the cross product has dealt with direction. The cross product is orthogonal to each of the vectors; we can use the cross product to see if two vectors are parallel. We have not yet investigated the magnitude of the cross product. The result is remarkably similar to the first angle between vectors formula. Whereas

$$\mathbf{v} \cdot \mathbf{w} = \|\mathbf{v}\|\ \|\mathbf{w}\| \cos\theta,$$

it turns out that

$$\|\mathbf{v} \times \mathbf{w}\| = \|\mathbf{v}\|\ \|\mathbf{w}\| \sin\theta.$$

To justify this formula, let $\mathbf{v} = \langle v_1, v_2, v_3\rangle$ and $\mathbf{w} = \langle w_1, w_2, w_3\rangle$. Let θ be the angle between the vectors. To make things slightly easier, we begin by showing that

$$\|\mathbf{v}\|^2\|\mathbf{w}\|^2 \sin^2\theta = \|\mathbf{v} \times \mathbf{w}\|^2.$$

Beginning with the left-hand side and using a Pythagorean identity,

$$\|\mathbf{v}\|^2\|\mathbf{w}\|^2 \sin^2\theta = \|\mathbf{v}\|^2\|\mathbf{w}\|^2(1 - \cos^2\theta)$$
$$= \|\mathbf{v}\|^2\|\mathbf{w}\|^2 - \|\mathbf{v}\|^2\|\mathbf{w}\|^2 \cos^2\theta.$$

Line 1 uses the identity $\sin^2\theta = 1 - \cos^2\theta$; line 2 uses the distributive property.

Now, we can use the first angle between vectors formula to rewrite as

$$= \|\mathbf{v}\|^2\|\mathbf{w}\|^2 - (\mathbf{v} \cdot \mathbf{w})^2.$$

The formula $\mathbf{v} \cdot \mathbf{w} = \|\mathbf{v}\|\ \|\mathbf{w}\| \cos\theta$ is used.

Next, we rewrite the expression using components, expand, and notice that some terms subtract out:

$$= (v_1^2 + v_2^2 + v_3^2)(w_1^2 + w_2^2 + w_3^2) - (v_1 w_1 + v_2 w_2 + v_3 w_3)^2$$

$$= v_1^2 w_1^2 + v_1^2 w_2^2 + v_1^2 w_3^2 + v_2^2 w_1^2 + v_2^2 w_2^2 + v_2^2 w_3^2 + v_3^2 w_1^2 + v_3^2 w_2^2 + v_3^2 w_3^2$$

$$\quad - (v_1^2 w_1^2 + 2v_1 w_1 v_2 w_2 + 2v_1 w_1 v_3 w_3 + v_2^2 w_2^2 + 2v_2 w_2 v_3 w_3 + v_3^2 w_3^2)$$

$$= v_1^2 w_2^2 + v_1^2 w_3^2 + v_2^2 w_1^2 + v_2^2 w_3^2 + v_3^2 w_1^2 + v_3^2 w_2^2$$

$$\quad - 2v_1 w_1 v_2 w_2 - 2v_1 w_1 v_3 w_3 - 2v_2 w_2 v_3 w_3.$$

Line 1 rewrites using the definitions of norm and dot product; lines 2 and 3 expand the expression by multiplying the two trinomial factors on the left (line 2) and squaring the trinomial on the right (line 3); lines 4 and 5 eliminate the terms that subtract out, which are denoted by color-coded underlines.

Next we reorder the terms, with the 2's and 3's first, followed by the 3's and 1's, and then the 1's and 2's:

$$= v_2^2 w_3^2 - 2v_2 w_3 v_3 w_2 + v_3^2 w_2^2$$

$$\quad + v_3^2 w_1^2 - 2v_3 w_1 v_1 w_3 + v_1^2 w_3^2$$

$$\quad + v_1^2 w_2^2 - 2v_1 w_2 v_2 w_1 + v_2^2 w_1^2.$$

The order of the four factors in each of the middle terms was changed, which uses the commutative property of multiplication. The order in which terms are written is meant to be helpful for recognizing the perfect squares, one in each of the three lines.

Notice that each of the three lines is now a perfect square; rewriting, we see that the expression is the square of the norm of a vector, which turns out to be the cross product:

$$= (v_2 w_3 - v_3 w_2)^2 + (v_3 w_1 - v_1 w_3)^2 + (v_1 w_2 - v_2 w_1)^2$$

$$= \|\langle v_2 w_3 - v_3 w_2, \, v_3 w_1 - v_1 w_3, \, v_1 w_2 - v_2 w_1 \rangle\|^2$$

$$= \|\mathbf{v} \times \mathbf{w}\|^2.$$

Line 1 factors the perfect squares; line 2 uses the definition of norm; and line 3 uses the definition of cross product. To verify the details of some of these steps, it may be helpful to start from the end and work backward.

Now that we have shown that $\|\mathbf{v}\|^2 \|\mathbf{w}\|^2 \sin^2 \theta = \|\mathbf{v} \times \mathbf{w}\|^2$, we may take square roots to obtain the promised formula,

Because $0 \le \theta \le \pi$, we know that $\sin \theta \ge 0$ and the expression on the right is nonnegative.

$$\|\mathbf{v} \times \mathbf{w}\| = \|\mathbf{v}\| \, \|\mathbf{w}\| \sin \theta.$$

MAGNITUDE OF THE CROSS PRODUCT

For any three-dimensional vectors $\mathbf{v}$ and $\mathbf{w}$,

$$\|\mathbf{v} \times \mathbf{w}\| = \|\mathbf{v}\| \, \|\mathbf{w}\| \sin \theta,$$

where θ is the angle between the vectors $\mathbf{v}$ and $\mathbf{w}$.

Recall again that $0 \le \theta \le \pi$ and that π radians is $180°$.

Finally, notice that because $\sin \theta = 0$ only when $\theta = 0$ or $\theta = \pi$, the magnitude of the cross product of nonzero vectors is 0 only when the vectors have the same or

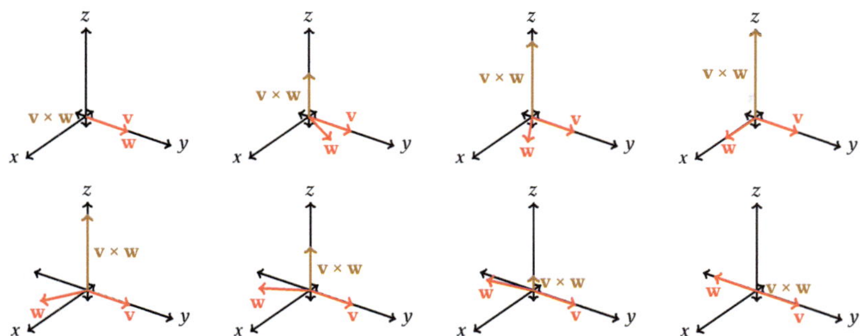

Figure 1.55 *Vectors* **v** *and* **w** *(red) in the xy-plane and their cross product* **v** × **w** *(brown), which is vertical, for various sizes of the angle θ between the vectors. The magnitude of the cross product is 0 when θ = 0 or θ = π, and reaches its maximum when* $\theta = \frac{\pi}{2}$

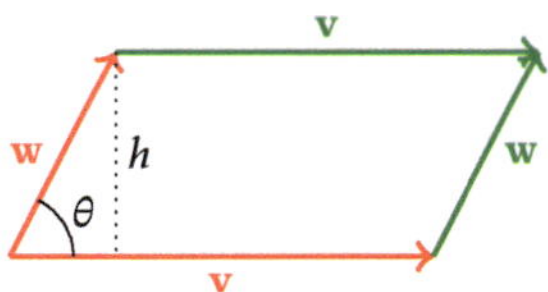

Figure 1.56 *Vectors* **v** *and* **w** *(both red and green) serving as the sides of a parallelogram, along with the height h of the parallelogram and the angle θ between the vectors. The area of the parallelogram is* $\|\mathbf{v} \times \mathbf{w}\|$

What would happen in Figure 1.56 if **v** and **w** have the same direction? What would be the area of the parallelogram? What would be the value of $\|\mathbf{v} \times \mathbf{w}\|$? Is the same true if **v** and **w** have opposite directions?

opposite directions, justifying definition 15. Figure 1.55 illustrates that as θ increases from 0 to $\frac{\pi}{2}$, the magnitude of the cross product increases, and as θ increases from $\frac{\pi}{2}$ to π, the magnitude of the cross product decreases and eventually reaches 0 again.

Another way to visualize the magnitude of the cross product **v** × **w** is as the area of a parallelogram with sides **v** and **w**. Consider the parallelogram in Figure 1.56, where **v** serves as the "base" and the "height" is represented by the dotted segment of length h. Notice that $\sin\theta = \dfrac{h}{\|\mathbf{w}\|}$, and thus $h = \|\mathbf{w}\|\sin\theta$. Also, the length of the base is $\|\mathbf{v}\|$. Recalling that the area of a parallelogram is base · height, we have

$$\text{area of parallelogram} = b \cdot h = \|\mathbf{v}\| \cdot \|\mathbf{w}\|\sin\theta = \|\mathbf{v} \times \mathbf{w}\|.$$

Similarly, the area of a triangle with sides **v** and **w** is $\frac{1}{2}bh = \frac{1}{2}\|\mathbf{v} \times \mathbf{w}\|$ (Figure 1.57).

Ans. to reading exercise 20:
no, because $\langle 1, 1, 4 \rangle \times \langle 2, 1, 5 \rangle = \langle 1, 3, -1 \rangle \neq \mathbf{0}$.

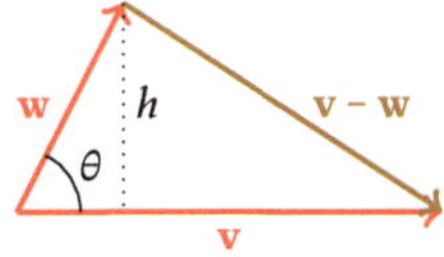

Figure 1.57 *Vectors* **v** *and* **w** *(red) serving as two sides of a triangle, along with the height h of the triangle and the angle θ between the vectors. The area of the triangle is* $\frac{1}{2}\|\mathbf{v} \times \mathbf{w}\|$

AREA FORMULAS RELATING TO THE CROSS PRODUCT

If $\mathbf{v}$ and $\mathbf{w}$ form the sides of a parallelogram, then

$$\text{area of parallelogram} = \|\mathbf{v} \times \mathbf{w}\|.$$

If $\mathbf{v}$ and $\mathbf{w}$ form two sides of a triangle, then

$$\text{area of triangle} = \frac{1}{2}\|\mathbf{v} \times \mathbf{w}\|.$$

Example 35 *Find the area of a triangle with sides given by the vectors* $\langle 1, 0, 7 \rangle$ *and* $\langle 2, -1, 4 \rangle$.

Solution Setting $\mathbf{v} = \langle 1, 0, 7 \rangle$ and $\mathbf{w} = \langle 2, -1, 4 \rangle$, we find the cross product

$$\mathbf{v} \times \mathbf{w} = \begin{vmatrix} \mathbf{i} & \mathbf{j} & \mathbf{k} \\ 1 & 0 & 7 \\ 2 & -1 & 4 \end{vmatrix}$$

$$= \mathbf{i}\begin{vmatrix} 0 & 7 \\ -1 & 4 \end{vmatrix} - \mathbf{j}\begin{vmatrix} 1 & 7 \\ 2 & 4 \end{vmatrix} + \mathbf{k}\begin{vmatrix} 1 & 0 \\ 2 & -1 \end{vmatrix}$$

$$= \mathbf{i}(0 - (-7)) - \mathbf{j}(4 - 14) + \mathbf{k}(-1 - 0)$$

$$= \langle 7, 10, -1 \rangle,$$

find its norm

$$\|\mathbf{v} \times \mathbf{w}\| = \sqrt{7^2 + 10^2 + (-1)^2} = \sqrt{150} = 5\sqrt{6},$$

and then use the formula:

$$\text{area of triangle} = \frac{1}{2}\|\mathbf{v} \times \mathbf{w}\| = \frac{5\sqrt{6}}{2} \text{ units}^2.$$

$\blacksquare$

1.5.6 Scalar triple product

Property (4) of the cross product states that $\mathbf{u} \cdot (\mathbf{v} \times \mathbf{w}) = (\mathbf{u} \times \mathbf{v}) \cdot \mathbf{w}$. The result of this product is a scalar (a number), and with three vectors involved, it is sometimes called the *scalar triple product*.

The scalar triple product is also sometimes called the *triple scalar product* or the *box product*.

Definition 16 SCALAR TRIPLE PRODUCT *The scalar triple product of vectors* $\mathbf{u}$, $\mathbf{v}$, *and* $\mathbf{w}$ *is*

$$\mathbf{u} \cdot (\mathbf{v} \times \mathbf{w}).$$

Proving the following formula is an exercise.

SHORTCUT FORMULA FOR THE SCALAR TRIPLE PRODUCT

A shortcut for calculating the scalar triple product of $\mathbf{u} = \langle u_1, u_2, u_3 \rangle$, $\mathbf{v} = \langle v_1, v_2, v_3 \rangle$, and $\mathbf{w} = \langle w_1, w_2, w_3 \rangle$ is

$$\mathbf{u} \cdot (\mathbf{v} \times \mathbf{w}) = \begin{vmatrix} u_1 & u_2 & u_3 \\ v_1 & v_2 & v_3 \\ w_1 & w_2 & w_3 \end{vmatrix}.$$

In contrast to the determinant formula for the cross product, this determinant has numbers for all its entries.

Example 36 *Calculate the scalar triple product of* $\mathbf{u} = \langle 2, 0, -9 \rangle$, $\mathbf{v} = \langle \frac{1}{2}, -3, 5 \rangle$, *and* $\mathbf{w} = \langle -1, 4, 1 \rangle$.

Solution Using the shortcut formula,

$$\mathbf{u} \cdot (\mathbf{v} \times \mathbf{w}) = \begin{vmatrix} 2 & 0 & -9 \\ \frac{1}{2} & -3 & 5 \\ -1 & 4 & 1 \end{vmatrix}$$

$$= 2 \begin{vmatrix} -3 & 5 \\ 4 & 1 \end{vmatrix} - 0 + (-9) \begin{vmatrix} \frac{1}{2} & -3 \\ -1 & 4 \end{vmatrix}$$

$$= 2(-3 - 20) - 0 + (-9)(2 - 3) = -37.$$

The value of the scalar triple product of $\mathbf{u}$, $\mathbf{v}$, and $\mathbf{w}$ is -37. ∎

In line 2, because $0 \begin{vmatrix} \frac{1}{2} & 5 \\ -1 & 1 \end{vmatrix} = 0$, whatever the entries are in that determinant does not matter. This illustrates the usefulness of the cofactor expansion by any row or column, as one can make a convenient choice that reduces the amount of computation necessary.

Re-ordering the vectors may cause a change in sign.

The scalar triple product has a geometric interpretation. Consider the parallelepiped determined by the vectors $\mathbf{u}$, $\mathbf{v}$, and $\mathbf{w}$, as in Figure 1.58. The volume of a parallelepiped is found by

A parallelepiped is a three-dimensional version of a parallelogram. All six faces of the parallelepiped are parallelograms. A cube is an example of a parallelepiped.

By the height h of the parallelepiped, we mean the length of an altitude, where an altitude extends from one vertex (corner) of the parallelepiped to the base of the parallelepiped such that the altitude is orthogonal to the plane containing the base; such orthogonality will be explored in the next section.

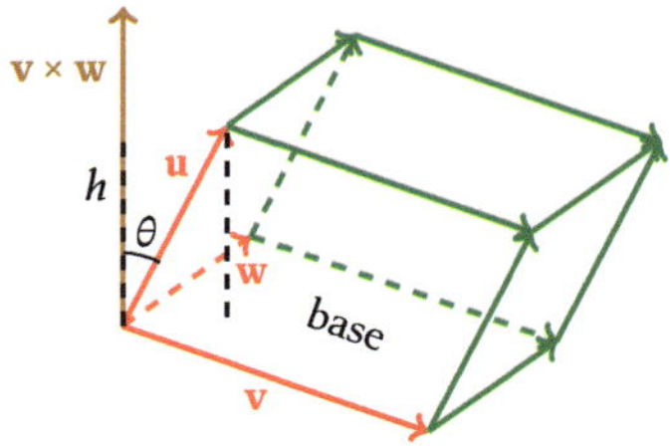

Figure 1.58 *Vectors* $\mathbf{u}$, $\mathbf{v}$, *and* $\mathbf{w}$ *(both red and green) serving as the sides of a parallelepiped, along with* $\mathbf{v} \times \mathbf{w}$ *(brown), the height h of the parallelepiped and the angle θ between* $\mathbf{v} \times \mathbf{w}$ *and* $\mathbf{u}$. *The volume of the parallelepiped is* $|\mathbf{u} \cdot (\mathbf{v} \times \mathbf{w})|$

$$\text{volume} = (\text{area of base}) \cdot \text{height}.$$

We use $|\cos\theta|$ because $\cos\theta$ is negative when $\frac{\pi}{2} < \theta \leq \pi$.

The area of the base, being the area of the parallelogram with sides $\mathbf{v}$ and $\mathbf{w}$, is $\|\mathbf{v} \times \mathbf{w}\|$. Let θ be the angle between $\mathbf{v} \times \mathbf{w}$ and $\mathbf{u}$. Then, $|\cos\theta| = \dfrac{h}{\|\mathbf{u}\|}$, and thus $h = \|\mathbf{u}\| \cdot |\cos\theta|$. Using the formula for the volume of a parallelepiped,

The dots here represent multiplication of real numbers. Line 2 uses the values determined for the area of the base and the height. Because $\|\mathbf{v} \times \mathbf{w}\|$ and $\|\mathbf{u}\|$ are non-negative, they can be moved inside the absolute values (line 3).

$$\text{volume} = \text{area of base} \cdot \text{height}$$

$$= \|\mathbf{v} \times \mathbf{w}\| \cdot (\|\mathbf{u}\| \cdot |\cos\theta|)$$

$$= \left| \|\mathbf{v} \times \mathbf{w}\| \cdot \|\mathbf{u}\| \cdot \cos\theta \right|.$$

Applying the first angle between vectors formula to the vectors $\mathbf{u}$ and $\mathbf{v} \times \mathbf{w}$, we have

$$\mathbf{u} \cdot (\mathbf{v} \times \mathbf{w}) = \|\mathbf{u}\|\, \|\mathbf{v} \times \mathbf{w}\| \cos\theta,$$

where θ is, as used in this derivation, the angle between $\mathbf{u}$ and $\mathbf{v} \times \mathbf{w}$. Substituting into the volume formula gives

$$\text{volume} = |\mathbf{u} \cdot (\mathbf{v} \times \mathbf{w})|,$$

and the absolute value of the scalar triple product is the volume of the parallelepiped.

VOLUME OF A PARALLELEPIPED

The volume of a parallelepiped with sides formed by the vectors $\mathbf{u}$, $\mathbf{v}$, and $\mathbf{w}$ is the absolute value of their scalar triple product:

$$\text{volume} = |\mathbf{u} \cdot (\mathbf{v} \times \mathbf{w})|.$$

Example 37 *Find the volume of the parallelepiped determined by the vectors* $\mathbf{u} = \langle 1, 1, -3 \rangle$, $\mathbf{v} = \langle 2, 5, 1 \rangle$, *and* $\mathbf{w} = \langle -1, 4, -2 \rangle$.

A fact from linear algebra is that if the rows of a determinant are written in a different order, then the absolute value of the determinant remains the same, although the sign might change. So it does not matter in which order we write the scalar triple product when finding the volume of a parallelepiped.

Solution   We begin by calculating the scalar triple product:

$$\mathbf{u} \cdot (\mathbf{v} \times \mathbf{w}) = \begin{vmatrix} 1 & 1 & -3 \\ 2 & 5 & 1 \\ -1 & 4 & -2 \end{vmatrix}$$

$$= 1 \begin{vmatrix} 5 & 1 \\ 4 & -2 \end{vmatrix} - 1 \begin{vmatrix} 2 & 1 \\ -1 & -2 \end{vmatrix} + (-3) \begin{vmatrix} 2 & 5 \\ -1 & 4 \end{vmatrix}$$

$$= 1(-10 - 4) - 1(-4 - (-1)) + (-3)(8 - (-5))$$

$$= -50.$$

Therefore, the volume of the parallelepiped is $|-50| = 50 \, \text{units}^3$. ∎

Reading Exercise 21 Find the volume of the parallelepiped determined by the vectors $\langle 2, 3, 0 \rangle$, $\langle 0, 4, 0 \rangle$, and $\langle 0, 0, -5 \rangle$.

Look again at Figure 1.58. What would it mean for the scalar triple product of three nonzero vectors to be 0? The volume of the parallelepiped would have to be zero, which means that the parallelepiped must be flat, that is, the height is zero and all three vectors are in the same plane.

We now have three similar geometric interpretations: if the dot product of two vectors is zero, the vectors are orthogonal; if the cross product of two vectors is zero, the vectors are parallel; and if the scalar triple product of three vectors is zero, the vectors are *coplanar*.

Three vectors are *coplanar* if they can be positioned so that they lie in the same plane.

Example 38 *Are the vectors* $\mathbf{u} = \langle 2, 1, 0 \rangle$, $\mathbf{v} = \langle 4, 2, 3 \rangle$, *and* $\mathbf{w} = \langle 6, 3, 9 \rangle$ *coplanar?*

Solution We check to see if the scalar triple product is zero:

$$
\mathbf{u} \cdot (\mathbf{v} \times \mathbf{w}) = \begin{vmatrix} 2 & 1 & 0 \\ 4 & 2 & 3 \\ 6 & 3 & 9 \end{vmatrix}
$$

$$
= 2 \begin{vmatrix} 2 & 3 \\ 3 & 9 \end{vmatrix} - 1 \begin{vmatrix} 4 & 3 \\ 6 & 9 \end{vmatrix} + 0
$$

$$
= 2(18 - 9) - 1(36 - 18) = 0.
$$

Because the scalar triple product is 0, the three vectors are coplanar. ∎

If the scalar triple product is not 0, then the three vectors are not coplanar.

1.5.7 Torque

Suppose you need to tighten a bolt, using a wrench, as in Figure 1.59. Assuming right-hand threads, you apply force to the wrench to make it turn clockwise. Then the torque τ is defined as a vector, the cross product of $\mathbf{r}$ and $\mathbf{F}$:

The Greek letter τ is pronounced "tau."

$$
\tau = \mathbf{r} \times \mathbf{F}.
$$

The right-hand rule can be used to determine the direction of the cross product; in Figure 1.59, the direction of τ is into the page (away from the viewer), and the bolt is being tightened. Of course, the bolt can be loosened by applying force as in Figure 1.60, where the direction of τ is out of the page (toward the viewer).

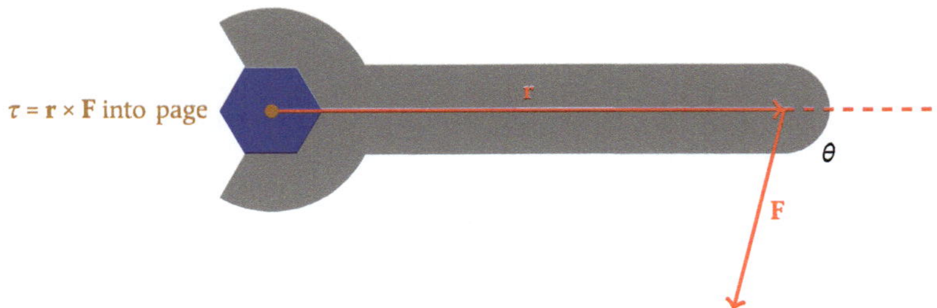

Figure 1.59 *A bolt head (blue-gray), a wrench (gray), a vector **r** with initial point at the center of the bolt head and terminal point at the center of the wrench handle where force is applied (red), a force vector **F** (red), and the angle θ between **r** and **F**. The torque vector τ = **r** × **F** goes into the page, which is tightening the bolt (assuming right-hand threads).*

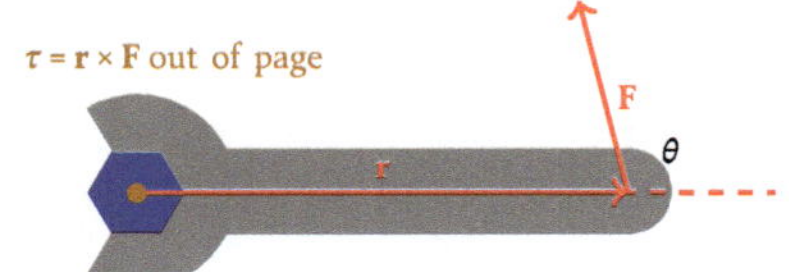

Figure 1.60 *A bolt head (blue-gray), a wrench (gray), a vector **r** with initial point at the center of the bolt head and terminal point at the center of the wrench handle where force is applied (red), a force vector **F** (red), and the angle θ between **r** and **F**. The torque vector τ = **r** × **F** goes out the page, which is loosening the bolt (assuming right-hand threads)*

If the force is applied in the middle of the handle of the wrench instead of the end of the handle, it is as if you are using a shorter wrench.

Recall that radians are dimensionless, so values of trig functions do not contribute to units.

Ans. to reading exercise 21:
   40 units^3

Applying the formula for the magnitude of the cross product, the magnitude of torque is

$$\|\tau\| = \|\mathbf{r} \times \mathbf{F}\| = \|\mathbf{r}\|\ \|\mathbf{F}\| \sin\theta.$$

The torque is greatest when $\sin\theta = 1$, which is at $\theta = \frac{\pi}{2}$ (when $\theta = 90°$); the force vectors in Figures 1.59 and 1.60 are less than optimal. The larger the value of $\|\mathbf{r}\|$ the larger the torque; in other words, longer wrenches give more torque. And the larger the magnitude of the force, $\|\mathbf{F}\|$, the larger the torque. Units for torque are the units of $\|\mathbf{r}\|$ (length of the wrench) times the units of $\|\mathbf{F}\|$ (force). An example of units in the SI system is Newton-meter (N·m), and in the English system is pound-foot (lb·ft).

Example 39 *Find the magnitude of the torque when tightening a bolt with a 1.5 ft long wrench and applying a force of 80 lb at a distance of 1.4 ft from the center of the bolt at an angle of 85°.*

Solution   Using the formula for the magnitude of torque,

$$\|\tau\| = \|\mathbf{r}\|\ \|\mathbf{F}\| \sin\theta = 1.4\,\text{ft} \cdot 80\,\text{lb} \cdot \sin 85° = 111.57\,\text{ft} \cdot \text{lb}.$$

∎

EXERCISES 1.5

1–4. Compute the determinant.

1. $\begin{vmatrix} 4 & 7 \\ -1 & -3 \end{vmatrix}$

2. $\begin{vmatrix} -2 & 8 \\ 1 & 5 \end{vmatrix}$

3. $\begin{vmatrix} 2 & 3 & -7 \\ 0 & -2 & 5 \\ -1 & 2 & 4 \end{vmatrix}$

4. $\begin{vmatrix} 4 & -2 & -5 \\ 6 & -1 & 10 \\ 0 & 5 & 11 \end{vmatrix}$

5–14. Determine the cross product.

5. $\langle 1, 2, 3 \rangle \times \langle 5, -7, 4 \rangle$

6. $\langle 1, 0, 5 \rangle \times \langle 2, 7, 7 \rangle$

7. $\langle 10, 20, 30 \rangle \times \langle 5, -7, 4 \rangle$

8. $\langle 2, 7, 7 \rangle \times \langle 1, 0, 5 \rangle$

9. $(\mathbf{i} - \mathbf{j}) \times (\mathbf{k} - 3\mathbf{i} + 2\mathbf{j})$

10. $\mathbf{i} \times (2\mathbf{i} - 4\mathbf{j} + \mathbf{k})$

11. $\left\langle 471, \sqrt{87}, \frac{13}{49} \right\rangle \times \left\langle 471, \sqrt{87}, \frac{13}{49} \right\rangle$

12. $\langle 9, 4, 12 \rangle \times (5\langle 9, 4, 12 \rangle)$

13. $\langle 2, -1, -4 \rangle \times (\langle 0, 4, 6 \rangle + \langle 3, -11, -2 \rangle)$

14. $\langle 0.5, 1.3, 2.87 \rangle \times \langle 0, 9.96, -1 \rangle$

15–18. Calculate the scalar triple product $\mathbf{u} \cdot (\mathbf{v} \times \mathbf{w})$.

15. $\mathbf{u} = \langle 2, 3, 1 \rangle, \mathbf{v} = \langle -1, -4, -5 \rangle,$ and $\mathbf{w} = \langle 6, 4, -3 \rangle$

16. $\mathbf{u} = \langle -1, 8, 2 \rangle, \mathbf{v} = \langle 3, -2, 4 \rangle,$ and $\mathbf{w} = \langle 5, 6, 7 \rangle$

17. $\mathbf{u} = \langle 5, 1, 0 \rangle, \mathbf{v} = \langle 0, 0, 6 \rangle,$ and $\mathbf{w} = \langle -4, 10, 100 \rangle$

18. $\mathbf{v} = \langle 1, 0, 4 \rangle, \mathbf{w} = \langle 2, 2, 1 \rangle,$ and $\mathbf{u} = \langle 0, -2, -3 \rangle$

19–24. Find a vector orthogonal to both of the given vectors.

19. $\mathbf{a} = \langle 2, -5, 10 \rangle, \mathbf{b} = \langle 4, 1, 3 \rangle$

20. $\mathbf{v} = \langle 10, -1, 7 \rangle, \mathbf{w} = \langle 3, 2, 8 \rangle$

21. $\langle \pi, 2, 9 \rangle$ and $\langle 4, 0, 0 \rangle$

22. $\mathbf{i}$ and $\mathbf{k}$

23. $\mathbf{v} = \mathbf{i} - 3\mathbf{j}, \mathbf{w} = 5\mathbf{i} + 2\mathbf{j}$

24. $\mathbf{v} = \langle 1, 0, 4 \rangle, \mathbf{w} = \langle -4, -1, -7 \rangle$

25–28. Are the two vectors parallel?

25. $\langle 8, 11, 15 \rangle, \langle -4, -5, -7 \rangle$

26. $\langle 2, 8, 12 \rangle, \langle -3, -12, -18 \rangle$

27. $\langle 4, 7, -2 \rangle, \langle 6, 10.5, -3 \rangle$

28. $\langle 2, 5, 4 \rangle, \langle 5, 13, 10 \rangle$

29–32. Are the three vectors coplanar?

29. $\langle 2, 8, 8 \rangle, \langle 3, -12, 9 \rangle,$ and $\langle 2, 0, 7 \rangle$

30. $\langle 1, 9, 7 \rangle, \langle 2, -11, 8 \rangle,$ and $\langle 1, 1, 7 \rangle$

31. $\langle -0.3, 2.7, -0.4\rangle$, $\langle 0.5, -1.5, 0.2\rangle$, and $\langle 0.2, 1.2, 0.1\rangle$
32. $\left\langle 2, \frac{1}{2}, \frac{1}{3}\right\rangle$, $\langle 4, 7, 1\rangle$, and $\langle 16, 10, 3\rangle$

33–36. Find the area of a parallelogram with sides given by the two vectors.

33. $\langle 0, 4, 9\rangle$ and $\langle 3, 0, -5\rangle$
34. $\langle 2, 0, 1\rangle$ and $\langle 1, 1, 0\rangle$
35. $\mathbf{i}$ and $2\mathbf{j}$
36. $\langle 9, 1, 3\rangle$ and $\langle -4, -4, 2\rangle$

37–40. Find the area of a triangle with two sides given by the two vectors.

37. $\langle 0, 4, 9\rangle$ and $\langle 3, 0, -5\rangle$
38. $\langle 2, 0, 1\rangle$ and $\langle 1, 1, 0\rangle$
39. $\langle 5, 5, 6\rangle$ and $\langle -2, 4, 8\rangle$
40. $\mathbf{k} - \mathbf{i}$ and $4\mathbf{j}$

41–44. Find the volume of the parallelepiped determined by the three vectors.

41. $\mathbf{u} = \langle 8, 1, 4\rangle$, $\mathbf{v} = \langle -1, -1, -1\rangle$, and $\mathbf{w} = \langle 3, 5, 4\rangle$
42. $\mathbf{v} = \langle 2, 1, 3\rangle$, $\mathbf{w} = \langle -4, 1, 1\rangle$, and $\mathbf{u} = \langle 5, 0, 2\rangle$
43. $\mathbf{w} = \langle 11, 19, 147\rangle$, $\mathbf{v} = \langle 0, 0, 3\rangle$, and $\mathbf{u} = \langle 2, 4, -77\rangle$
44. $2\mathbf{i}$, $3\mathbf{j}$, and $-5\mathbf{k}$

45. Find the magnitude of the torque when tightening a bolt on a piece of farm equipment with a very long wrench and applying a force of 150 lb at a distance of 2.4 ft from the center of the bolt at an angle of 105°.
46. A force of 240 N is applied to a wrench at a distance of 0.3 m from the center of the bolt, at an angle of 30°. Find the magnitude of the torque.
47. Ignoring proper safety measures, to tighten a bolt as tight as you can get it you stand on a wrench, applying a force of 780 Newtons at a distance of 0.4 meters from the center of the bolt, at an angle of 72°. What is the magnitude of the torque generated by this inadvisable procedure?
48. What is the maximum magnitude of torque that can be generated by a wrench that is 1.2 ft long and a person that can apply a force of 60 pounds?
49. Prove the cross product property (4) $\mathbf{u} \cdot (\mathbf{v} \times \mathbf{w}) = (\mathbf{u} \times \mathbf{v}) \cdot \mathbf{w}$.
50. Prove the cross product property (2) $(a\mathbf{v}) \times \mathbf{w} = \mathbf{v} \times (a\mathbf{w})$ $= a(\mathbf{v} \times \mathbf{w})$.
51. Give an example of vectors $\mathbf{a}$, $\mathbf{b}$, and $\mathbf{c}$ such that $(\mathbf{a} \times \mathbf{b}) \times \mathbf{c} \neq \mathbf{a} \times (\mathbf{b} \times \mathbf{c})$.

Do not try this, as injury could result.

52. Prove that for any two three-dimensional vectors $\mathbf{v}$ and $\mathbf{w}$, $\|\mathbf{v} \times \mathbf{w}\| = \|\mathbf{w} \times \mathbf{v}\|$.

53. Prove the scalar triple product shortcut formula.

54. The diagrams in Figure 1.55 illustrate the magnitude of the cross product for various angles between the vectors $\mathbf{v}$ and $\mathbf{w}$. Use variations of Figure 1.56 to accomplish the same purpose.

55. What must be true about the vectors $\mathbf{u}$, $\mathbf{v}$, and $\mathbf{w}$ in Figure 1.58 for the parallelepiped to be a cube?

56. How many copies of the vector $\mathbf{u}$ are in Figure 1.58?

57. An aging worker only has half the arm strength as compared to 40 years ago, but must still apply the same torque as before. What equipment does the worker need to accomplish the task?

58. Find a vector orthogonal to $\langle 1, 5, 8 \rangle$.

59. Suppose vectors $\mathbf{v} = \langle v_1, v_2 \rangle$ and $\mathbf{w} = \langle w_1, w_2 \rangle$ form the sides of a parallelogram in the xy-plane. By thinking of this parallelogram as being in three dimensions, derive a formula for the area of the parallelogram in terms of the components v_1, v_2, w_1, and w_2.

60. Prove that if $\mathbf{w} = k\mathbf{v}$, where k is a scalar, then $\mathbf{v}$ and $\mathbf{w}$ are parallel.

> Use of the definition of parallel is required for exercise 60.

1.6   Lines and Planes in Space

What information do we need to find the equation of a line in two dimensions? In your study of algebra, you learned the point-slope formula for the equation of a line; given a point (x_0, y_0) on the line and the slope m, the (Cartesian) equation of the line is

> There are other options, such as parametric equations or polar coordinates.

$$y - y_0 = m(x - x_0).$$

Any point (x, y) on the line satisfies the equation, and any point (x, y) that satisfies the equation lies on the line.

It is also clear that only knowing a point on the line is not enough; many lines pass through a given point (Figure 1.61, left). Similarly, only knowing the slope is

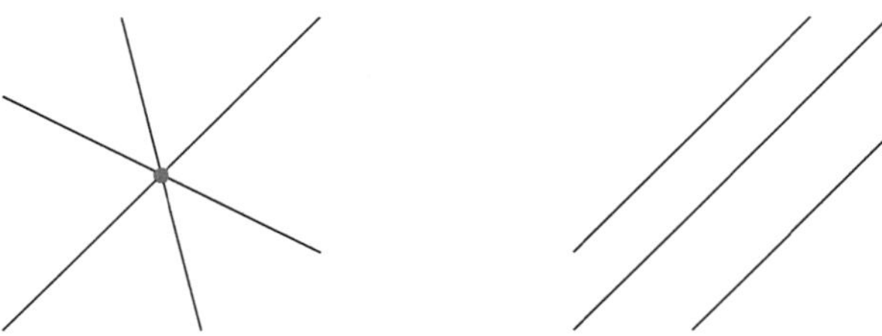

Figure 1.61 *(Left) a point (red) and three different lines through the point, each with a different slope; (right) three different lines with the same slope, passing through different points*

Another form of the equation of a line that you may be familiar with is the slope-intercept form; given the slope m and y-intercept b, the equation of the line is

$$y = mx + b.$$

But the y-intercept b is the point $(0, b)$, so we are still using both a point and the slope.

not enough; many different lines have the same slope (Figure 1.61, right). Both the slope and a point are required to determine a line.

What happens in three dimensions? Can we simply modify the point-slope formula? Is knowing a point and a direction of some kind enough?

1.6.1   Vectors and linear equations

Imagine standing in a park holding a rod of some kind, such as a straight stick, which can be used to represent a line in a three-dimensional world. In two dimensions, the slope, which is a number, is described as $\frac{\text{rise}}{\text{run}}$; think $\frac{\text{vertical displacement}}{\text{horizontal displacement}}$. If you turn around and point the stick a different direction while still holding it at the same angle to the ground, the rise and run of the stick do not change; only the direction is different. In two dimensions, the horizontal run is one-dimensional, along the x-axis. But in three dimensions the horizontal run is two-dimensional, along the xy-plane. So one number describing the slope is insufficient to capture the direction of a line in three dimensions. We need a different method to describe direction. Enter vectors, which can describe direction in any number of dimensions!

Let's try to describe a line in two dimensions using vectors. Start with a line as pictured in Figure 1.62, left. Place a vector **w** along the line as in Figure 1.62, middle, to describe the direction of the line. But because vectors do not record position, the same vector can be placed in a different location, and it no longer lies on the line, as in Figure 1.62, right. We are still lacking a way of describing a point on the line.

Points have position; vectors do not. So if a vector is to describe a position, we must designate some manner for this to happen. To that end, a *position vector* is a vector with initial point at the origin.

Now consider a point (x_0, y_0) on a line. The position vector $\langle x_0, y_0 \rangle$ would have initial point at $(0, 0)$ and terminal point at (x_0, y_0). Then following the direction of the line, we may add any scalar multiple of **w** and the terminal point of the position vector $\mathbf{v} + t\mathbf{w}$ lies on the line, for any scalar t, as in Figure 1.63, where the scalar is $t = 1.2$.

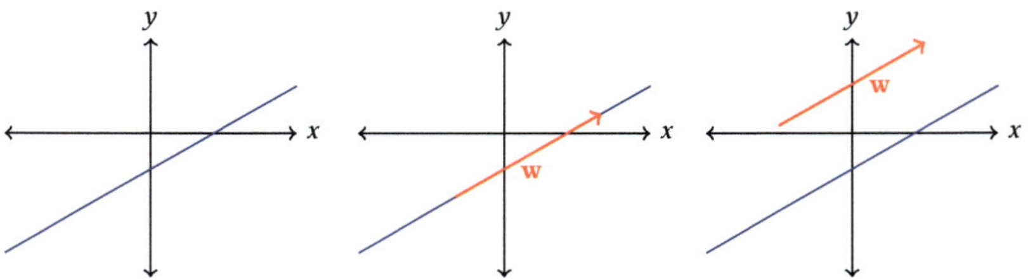

Figure 1.62 *(Left) a line (blue); (middle) a direction vector **w** is added (red); (right) the same vector positioned differently*

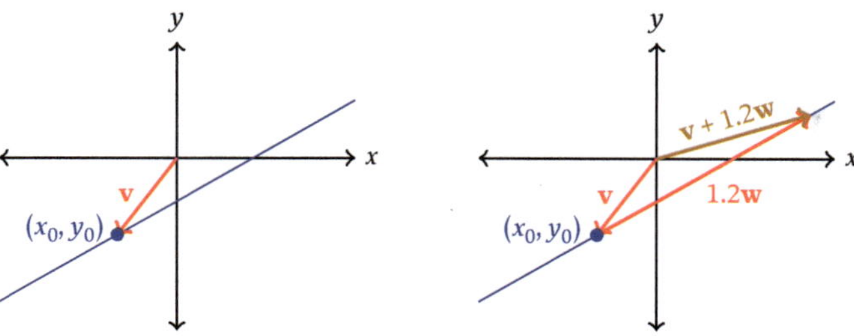

Figure 1.63 *(Left) a position vector (red) with terminal point (x_0, y_0) on the line (blue); (right) a scalar multiple of the direction vector is added, resulting in a position vector (brown) whose terminal point is on the line (blue)*

Because the terminal point of the position vector $\mathbf{v} + t\mathbf{w}$ lies on the line, we call this the *vector form of the equation of a line.*

VECTOR FORM OF THE EQUATION OF A LINE

A line with position vector $\mathbf{v}$ and direction vector $\mathbf{w}$ has equation

$$\mathbf{v} + t\mathbf{w},$$

where t is a scalar.

The terminal point of the position vector $\mathbf{v}$ must lie on the line. Then, the points on the line are the terminal points of every position vector of the form $\mathbf{v} + t\mathbf{w}$.

Imagine yourself with the stick in the park again. If you have a direction vector describing the direction of the stick, as well as a point (such as the point where you are holding the stick), then you have described that specific line. The vector form of the equation of a line works not only in two dimensions, but also in three (or more!).

Example 40 *Find a vector form of the equation of a line with direction vector $\mathbf{w} = \langle 2, 4, 6 \rangle$ containing the point $(1, -3, 5)$.*

Solution   Using the position vector $\mathbf{v} = \langle 1, -3, 5 \rangle$, a vector form of the equation of the line is

$$\begin{aligned}
\mathbf{v} + t\mathbf{w} &= \langle 1, -3, 5 \rangle + t\langle 2, 4, 6 \rangle \\
&= \langle 1, -3, 5 \rangle + \langle 2t, 4t, 6t \rangle \\
&= \langle 1 + 2t, -3 + 4t, 5 + 6t \rangle.
\end{aligned}$$

∎

Points on the line in example 40 can be found using any value of the scalar t. The x-coordinate of the point would be $x = 1 + 2t$, the y-coordinate would be $y = -3 + 4t$, and the z-coordinate would be $z = 5 + 6t$. Notice that these equations are a three-dimensional version of parametric equations, with parameter t. We now have a parametric form of the equation of the line.

PARAMETRIC FORM OF THE EQUATION OF A LINE, 3D

Parametric equations for the line containing the point (x_0, y_0, z_0) with direction vector $\mathbf{w} = \langle a, b, c \rangle$ are

$$x = x_0 + at, \quad y = y_0 + bt, \quad \text{and} \quad z = z_0 + ct.$$

Just as in two dimensions, any information that allows us to determine a point on the line and the direction of the line is sufficient for forming the equation of the line.

Example 41 *Find vector and parametric equations of the line through the points $P(1, 0, 4)$ and $Q(2, -3, 2)$.*

This notation means that P is the point $(1, 0, 4)$ and Q is the point $(2, -3, 2)$.

Solution   We may use either point for the position vector $\mathbf{v}$; we choose $\mathbf{v} = \langle 1, 0, 4 \rangle$. For a direction vector, convenient choices include $\overrightarrow{PQ}$ or $\overrightarrow{QP}$; we choose $\mathbf{w} = \overrightarrow{PQ} = \langle 2 - 1, -3 - 0, 2 - 4 \rangle = \langle 1, -3, -2 \rangle$. Then, a vector form of the equation is

Notice the phrase "a vector form" rather than "the vector form." That's because other choices could be made. Using Q instead of P to make a position vector results in

$$\mathbf{v} + t\mathbf{w} = \langle 1, 0, 4 \rangle + t\langle 1, -3, -2 \rangle$$
$$= \langle 1 + t, -3t, 4 - 2t \rangle.$$

$$\mathbf{v} + t\mathbf{w} = \langle 2 + t, -3 - 3t, 2 - 2t \rangle,$$

whereas using $\overrightarrow{QP}$ instead of $\overrightarrow{PQ}$ results in

Parametric equations may be read off a vector equation, but they may also be formed directly by writing the coordinates of the point first and adding t times the direction vector:

$$\mathbf{v} + t\mathbf{w} = \langle 1 - t, 3t, 4 + 2t \rangle.$$

$$x = 1 + t$$

There are infinitely many possible ways of writing the same line.

$$y = -3t$$
$$z = 4 - 2t.$$

■

Example 42 *Find parametric equations of the line through $P(2, 5, 0.1)$ parallel to the line $x = 2 + 5t$, $y = 3 - 4t$, $z = 8 - t$.*

Solution The direction of the line $\begin{cases} x = 2 + 5t \\ y = 3 - 4t \\ z = 8 - t \end{cases}$ can be read from the coefficients

on t in the parametric equations. Therefore, a direction vector is

$$\mathbf{w} = \langle 5, -4, -1 \rangle.$$

Using the position vector $\mathbf{v} = \langle 2, 5, 0.1 \rangle$ as well, we have the parametric equations

$$x = 2 + 5t$$
$$y = 5 - 4t$$
$$z = 0.1 - t.$$

$\blacksquare$

This is the inverse procedure from the second half of example 41.

Reading Exercise 22 Find parametric equations of the line through the points $(2, 2, -4)$ and $(5, 1, 0)$.

1.6.2 Distance from a point to a line in $\mathbb{R}^3$

Consider a line ℓ, a point Q on ℓ, and a point P not on ℓ, as in Figure 1.64, left. What is the distance from P to ℓ? The distance h would be measured along a segment from P to ℓ that is perpendicular to ℓ. Let $\mathbf{w}$ be a direction vector for the line ℓ, and consider the angle θ between the vectors $\mathbf{w}$ and $\overrightarrow{QP}$ (Figure 1.64, right). A right triangle is formed, and

$$\sin \theta = \frac{h}{\|\overrightarrow{QP}\|},$$

which can be rearranged to give

$$h = \|\overrightarrow{QP}\| \sin \theta.$$

Figure 1.64 *(Left) a point Q on a line ℓ (blue) and a point P not on ℓ; (right) the distance h from P to ℓ is measured along a segment (dashed) perpendicular to ℓ*

We also know from our study of the magnitude of the cross product that

$$\|\overrightarrow{QP} \times \mathbf{w}\| = \|\overrightarrow{QP}\| \, \|\mathbf{w}\| \sin\theta.$$

Dividing the latter by $\|\mathbf{w}\|$ yields $\dfrac{\|\overrightarrow{QP}\times\mathbf{w}\|}{\|\mathbf{w}\|} = \|\overrightarrow{QP}\| \sin\theta$, and then by substitution

$$h = \frac{\|\overrightarrow{QP} \times \mathbf{w}\|}{\|\mathbf{w}\|}.$$

DISTANCE FROM A POINT TO A LINE, 3D

If Q is a point on a line ℓ, P is not on ℓ, and $\mathbf{w}$ is a direction vector for the line ℓ, then the distance between P and ℓ is

$$h = \frac{\|\overrightarrow{QP} \times \mathbf{w}\|}{\|\mathbf{w}\|}.$$

When applying the formula, any point Q on the line ℓ may be used, as well as any direction vector $\mathbf{w}$, no matter which way $\mathbf{w}$ points along the line.

Example 43 *Find the distance from the point $P(-1, 3, 5)$ to the line $x = 4 - 7t, y = 2,$*
$z = 1 - t.$

Solution We have the point P; we need a point Q on the line. One possibility can be read from the parametric equations $\begin{cases} x = 4 - 7t \\ y = 2 \\ z = 1 - t \end{cases}$, $Q = (4, 2, 1)$. We need the direction vector $\mathbf{w}$, which can also be read from the parametric equations: $\mathbf{w} = \langle -7, 0, -1 \rangle$. Next, we need to compute the vector $\overrightarrow{QP}$:

$$\overrightarrow{QP} = \langle -1 - 4, \, 3 - 2, \, 5 - 1 \rangle = \langle -5, 1, 4 \rangle.$$

The formula requires the norm of the cross product, so we calculate the cross product

$$\overrightarrow{QP} \times \mathbf{w} = \begin{vmatrix} \mathbf{i} & \mathbf{j} & \mathbf{k} \\ -5 & 1 & 4 \\ -7 & 0 & -1 \end{vmatrix}$$

$$= \mathbf{i}\begin{vmatrix} 1 & 4 \\ 0 & -1 \end{vmatrix} - \mathbf{j}\begin{vmatrix} -5 & 4 \\ -7 & -1 \end{vmatrix} + \mathbf{k}\begin{vmatrix} -5 & 1 \\ -7 & 0 \end{vmatrix}$$

$$= \mathbf{i}(-1 - 0) - \mathbf{j}(5 - -28) + \mathbf{k}(0 - -7)$$

$$= \langle -1, -33, 7 \rangle.$$

It is visually simpler to pick out a point on the line and a direction vector when the parametric equations are stacked as illustrated here. The point is the set of constants and the direction vector is the set of coefficients on the parameter t.

The norm of the cross product is

$$\|\overrightarrow{QP} \times \mathbf{w}\| = \sqrt{(-1)^2 + (-33)^2 + 7^2} = \sqrt{1139}.$$

Lastly, the norm of the direction vector is

$$\|\mathbf{w}\| = \sqrt{(-7)^2 + 0^2 + (-1)^2} = \sqrt{50}.$$

Using the formula, the distance from P to ℓ is

$$h = \frac{\|\overrightarrow{QP} \times \mathbf{w}\|}{\|\mathbf{w}\|} = \frac{\sqrt{1139}}{\sqrt{50}} \approx 4.773 \text{ units.} \qquad \blacksquare$$

Ans. to reading exercise 22:
Possible answers include

$$\begin{cases} x = 2 + 3t \\ y = 2 - t \\ z = -4 + 4t \end{cases}, \qquad \begin{cases} x = 5 + 3t \\ y = 1 - t \\ z = 4t \end{cases},$$

$$\begin{cases} x = 2 - 3t \\ y = 2 + t \\ z = -4 - 4t \end{cases}, \text{ and } \begin{cases} x = 5 - 3t \\ y = 1 + t \\ z = -4t \end{cases}.$$

1.6.3 Vector and parametric equations of planes

When we specify the xy-plane in our three-dimensional coordinate system, we are using two direction vectors, not just one. Using only one gives the direction of a line; any plane requires two. In Figure 1.65, a plane is drawn, containing a point and two direction vectors. Just as we did for lines, we can specify a position vector for a point in the plane; in Figure 1.66 that point is P and the position vector is $\mathbf{v}$. Then the line containing point P with direction vector $\mathbf{w}$, pictured in Figure 1.66, left, has equation $\mathbf{v} + t\mathbf{w}$, whereas the line containing P with direction vector $\mathbf{u}$, pictured in Figure 1.66, right, has equation $\mathbf{v} + s\mathbf{u}$. But many points in the plane are not on those two lines. The parallelogram with sides formed by $\mathbf{w}$ and $\mathbf{u}$ would lie completely in the plane, and the corner opposite P would be found by following $\mathbf{v}$ and then $\mathbf{w} + \mathbf{u}$, that is, it would be at the terminal point of the position vector $\mathbf{v} + (\mathbf{w} + \mathbf{u}) = \mathbf{v} + 1\mathbf{w} + 1\mathbf{u}$, as in Figure 1.67, left. Notice that instead of $\mathbf{v} + t\mathbf{w}$ or $\mathbf{v} + s\mathbf{u}$, each of which describe a line, reaching that corner of the parallelogram requires an expression of the form $\mathbf{v} + t\mathbf{w} + s\mathbf{u}$. In fact, any point in the plane can be described as the terminal point of such a position vector, as illustrated

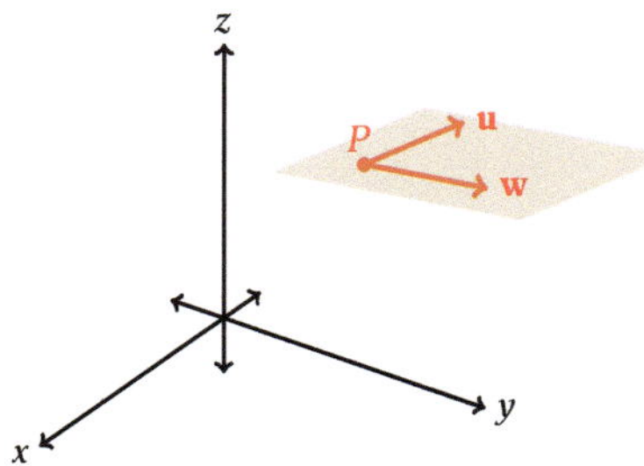

Figure 1.65 *A plane (brown), a point P in the plane (red), and two direction vectors $\mathbf{w}$ and $\mathbf{u}$ in the plane (red). Only a portion of the plane is shown; the plane is infinite in extent.*

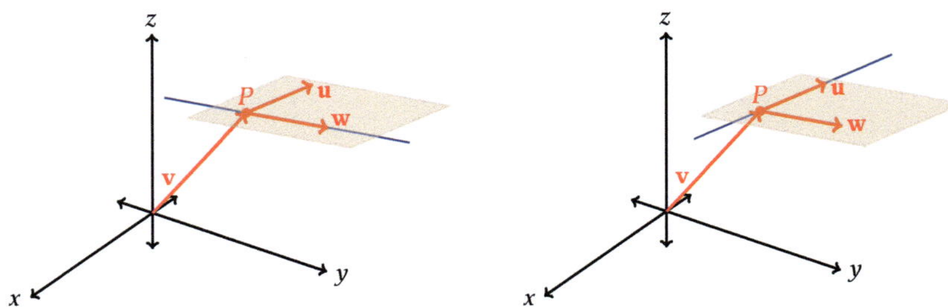

Figure 1.66 *(Left) the line* **v** + *t***w** *(blue); (right) the line* **v** + *s***u** *(blue)*

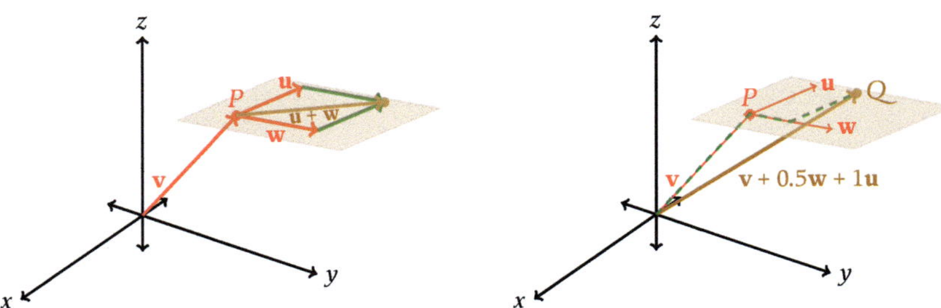

Figure 1.67 *(Left) the parallelogram with sides* **u** *and* **w** *lying in the plane; (right) point Q (brown) given by following the path of the position vector* **v** *to point P, then proceeding in the direction of* **w** *for half the length of* **w**, *and finally following a copy of* **u**, *arriving at point Q, which is at the terminal point of the position vector* **v** + 0.5**w** + 1**u** *(brown)*

in Figure 1.67, right, where $t = 0.5$ and $s = 1$. We now have the vector form of the equation of a plane.

VECTOR FORM OF THE EQUATION OF A PLANE

The terminal point of the position vector **v** must lie in the plane. Then, the points lying in the plane are the terminal points of every position vector of the form **v** + *t***w** + *s***u**.

A plane with position vector **v** and nonparallel direction vectors **w** and **u** has equation

$$\mathbf{v} + t\mathbf{w} + s\mathbf{u},$$

where *t* and *s* are scalars.

If the two direction vectors are parallel, then the result is a line, not a plane.

 The parametric form of the equation of a plane follows in the same manner as for lines.

PARAMETRIC FORM OF THE EQUATION OF A PLANE, 3D

Parametric equations for the plane containing the point (x_0, y_0, z_0) with nonparallel direction vectors $\mathbf{w} = \langle w_1, w_2, w_3 \rangle$ and $\mathbf{u} = \langle u_1, u_2, u_3 \rangle$ are

$$x = x_0 + w_1 t + u_1 s, \quad y = y_0 + w_2 t + u_2 s, \quad \text{and} \quad z = z_0 + w_3 t + u_3 s,$$

where the parameters t and s are scalars.

Example 44 *Find vector and parametric equations of the plane through $P(1, 2, 4)$, $Q(2, 5, 9)$, and $R(-1, 5, 0)$.*

Solution Just as in example 41, we may use any of the points in the plane for the position vector; we choose $\mathbf{v} = \langle 1, 2, 4 \rangle$. For the first direction vector we choose $\mathbf{w} = \overrightarrow{PQ} = \langle 1, 3, 5 \rangle$ and for the second direction vector we choose $\mathbf{u} = \overrightarrow{PR} = \langle -2, 3, -4 \rangle$. Then a vector form of the equation of the plane is

$$\mathbf{v} + t\mathbf{w} + s\mathbf{u} = \langle 1, 2, 4 \rangle + t\langle 1, 3, 5 \rangle + s\langle -2, 3, -4 \rangle$$

$$= \langle 1 + t - 2s, \ 2 + 3t + 3s, \ 4 + 5t - 4s \rangle.$$

Parametric equations may be read off the vector equation or formed directly in the manner of example 41:

$$x = 1 + t - 2s$$

$$y = 2 + 3t + 3s$$

$$z = 4 + 5t - 4s.$$

■

The vectors $\mathbf{v}$, $\mathbf{w}$, and $\mathbf{u}$ may be seen vertically in the parametric equation of the plane given in the solution to example 44.

Reading Exercise 23 Find parametric equations for the plane through the points $(2, 2, -4)$, $(5, 1, 0)$, and $(-3, 4, 1)$.

1.6.4 Cartesian equation of a plane

Take a piece of paper, draw a point P, and then draw five different vectors with initial point P. Finally, lay the paper on a horizontal surface and position your pencil vertically at point P. The vector represented by the pencil is orthogonal to every one of the vectors you drew on the paper, and every vector you could

---

Two points determine a line; three non-collinear points determine a plane. Take two fingertips as two points and put a piece of cardboard on those two fingertips; that's not enough to determine exactly how the cardboard will lie. But three fingertips that aren't all on one line are enough.

The point P corresponds to $t = 0$, $s = 0$. What values of the parameters t and s correspond to points Q and R?

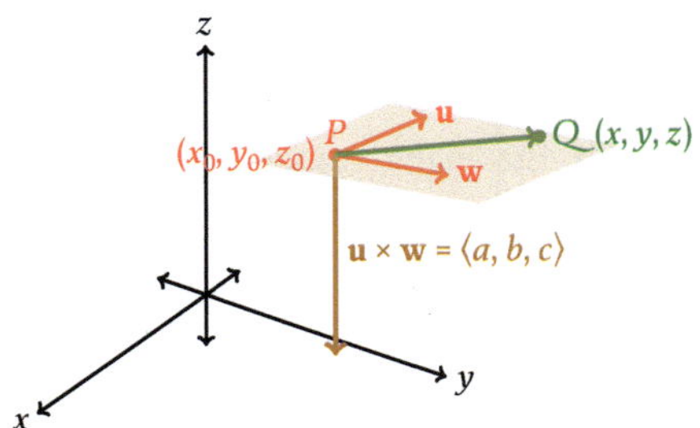

Figure 1.68 *The vector* $\mathbf{v} \times \mathbf{w}$ *(brown) is orthogonal to every vector in the plane, including* $\overrightarrow{PQ}$ *(green)*

have drawn on the paper. Consider Figure 1.68, where $P(x_0, y_0, z_0)$ is a point in the plane, $\mathbf{u}$ and $\mathbf{w}$ are vectors in the plane, and $\mathbf{u} \times \mathbf{w}$ is their cross product. Then, for any other point $Q(x, y, z)$ in the plane, $\mathbf{u} \times \mathbf{w}$ must be orthogonal to the vector $\overrightarrow{PQ}$, which is in the plane. If $\mathbf{u} \times \mathbf{w} = \langle a, b, c \rangle$, then the dot product of $\mathbf{u} \times \mathbf{w}$ with $\overrightarrow{PQ} = \langle x - x_0,\ y - y_0,\ z - z_0 \rangle$ must be zero:

$$\langle a, b, c \rangle \cdot \langle x - x_0,\ y - y_0,\ z - z_0 \rangle = 0.$$

Computing that dot product gives a Cartesian equation of the plane:

$$a(x - x_0) + b(y - y_0) + c(z - z_0) = 0.$$

We call a vector that is orthogonal to a plane, such as $\mathbf{u} \times \mathbf{w}$ here, a *normal vector*.

CARTESIAN EQUATION OF A PLANE

A Cartesian equation of a plane through the point (x_0, y_0, z_0) normal to $\langle a, b, c \rangle$ is

$$a(x - x_0) + b(y - y_0) + c(z - z_0) = 0.$$

This is also called the *scalar equation of the plane.*

Example 45 *Find a Cartesian equation of the plane containing the points* $P(1, 2, 4)$, $Q(2, 5, 9)$, *and* $R(-1, 5, 0)$.

These are the same points, and therefore the same plane, as in example 44.

Solution   We may choose any of the points to give us x_0, y_0, and z_0; we choose $P(1, 2, 4) = (x_0, y_0, z_0)$ so that $x_0 = 1$, $y_0 = 2$, and $z_0 = 4$. To obtain a normal vector, we find two vectors in the plane and take their cross product. Two vectors in the plane are $\overrightarrow{PQ} = \langle 1, 3, 5 \rangle$ and $\overrightarrow{PR} = \langle -2, 3, -4 \rangle$. Their cross product is

The two vectors used in this cross product must be nonparallel.

The vectors $\overrightarrow{PQ}$ and $\overrightarrow{PR}$ correspond to the vectors $\mathbf{u}$ and $\mathbf{w}$ of Figure 1.68.

$$\vec{PQ} \times \vec{PR} = \begin{vmatrix} \mathbf{i} & \mathbf{j} & \mathbf{k} \\ 1 & 3 & 5 \\ -2 & 3 & -4 \end{vmatrix} = \cdots = \langle -27, -6, 9 \rangle.$$

Then, with $a = -27$, $b = -6$, and $c = 9$, a Cartesian equation of the plane is

$$a(x - x_0) + b(y - y_0) + c(z - z_0) = 0$$

$$-27(x - 1) - 6(y - 2) + 9(z - 4) = 0.$$

It is easy to see that the point $P(1, 2, 4)$ satisfies the equation. You should verify that the points $Q(2, 5, 9)$ and $R(-1, 5, 0)$ also satisfy the equation.

The solution to example 45 may be expanded to

$$-27x + 27 - 6y + 12 + 9z - 36 = 0$$

$$-27x - 6y + 9z + 3 = 0.$$

The latter, which is in the form $Ax + By + Cz + D = 0$, is called the *general form of the equation of the plane*.

Reading Exercise 24   Find a Cartesian equation for the plane through the points $(2, 2, -4)$, $(5, 1, 0)$, and $(-3, 4, 1)$.

These are the same points, and therefore the same plane, as for reading exercise 23.

1.6.5   Intersection of a line and a plane

In three dimensions, if a line does not lie entirely within a plane, then either the line and the plane intersect in one point or no points. Given parametric equations of a line and a Cartesian (or general) equation of a plane, finding their intersection, if it exists, is a straightforward two-step process.

Example 46 *Determine the point of intersection of the line $x = 2 - t$, $y = 3t$, $z = 4 + t$ and the plane $4x + 2y - z + 4 = 0$.*

Solution   ❶ Substitute the line's parametric equation for each variable x, y, and z into the Cartesian equation of the plane and solve for the parameter t.

$$4x + 2y - z + 4 = 0$$

$$4(2 - t) + 2(3t) - (4 + t) + 4 = 0$$

$$8 - 4t + 6t - 4 - t + 4 = 0$$

$$t + 8 = 0$$

$$t = -8.$$

Ans. to reading exercise 23:
  Possible answers include

$$\begin{cases} x = 2 + 3t - 5s \\ y = 2 - t + 2s \\ z = -4 + 4t + 5s \end{cases}, \quad \begin{cases} x = 5 + 3t - 5s \\ y = 1 - t + 2s \\ z = 4t + 5s \end{cases},$$

$$\begin{cases} x = -3 + 3t - 5s \\ y = 4 - t + 2s \\ z = 1 + 4t + 5s \end{cases}, \quad \begin{cases} x = 5 - 3t - 8s \\ y = 1 + t + 3s \\ z = -4t + s \end{cases},$$

$$\text{and} \begin{cases} x = -3 + 5t + 8s \\ y = 4 - 2t - 3s \\ z = 1 - 5t - s \end{cases}.$$

Line 1 is the equation of the plane; line 2 substitutes the equations for x, y, and z, reducing the equation to one variable; line 3 expands; line 4 collects like terms; and line 5 finishes the solution.

❷ Put the calculated value of t into the parametric equations of the line to find the point of intersection.

$$x = 2 - (-8) = 10$$

$$y = 3(-8) = -24$$

$$z = 4 + (-8) = -4.$$

It is a good idea to verify that the point that has been calculated satisfies the equation of the plane.

The point of intersection is $(10, -24, -4)$. ■

If there is no solution to the equation in step ❶, then the line and the plane are parallel.

What if you are given parametric equations for both the line and the plane? First determine a Cartesian form of the equation of the plane, then proceed as in example 46.

Careful–if you are given equations such as

$$\begin{cases} x = 1 + t \\ y = 2 + 5t \\ z = 4 - t \end{cases}$$

and

$$\begin{cases} x = 2 + t + s \\ y = 5 - 3t - s \\ z = 1 + t - 2s \end{cases},$$

the values of the parameter t in the line and the parameter t in the plane might not be the same at a point of intersection.

1.6.6 Parallel planes

Place a pencil vertically on a horizontal surface such as a desk. The surface of the desk is a plane, and the vertical pencil represents a normal vector to that plane. Now take another flat surface, such as the cover of a hardback book or stiff notebook, and use another pencil to represent a normal vector to that plane. When are the two planes parallel? When the second plane is horizontal, which is also the only time that its normal vector is vertical. It is apparent that two planes are parallel only when normal vectors to the planes are parallel. We already know how to tell if two vectors are parallel, which is when their cross product is the zero vector. This gives us the outline of a procedure for deciding if two planes are parallel.

Example 47 *Are the planes* $4x - 2y + z + 5 = 0$ *and* $\begin{cases} x = 2 - t - 3s \\ y = 4 + t + s \\ z = 9 + 2s \end{cases}$ *parallel?*

Solution Strategy: find normal vectors to the two planes, and check to see if the normal vectors are parallel.

See the formula for the Cartesian equation of a plane for insight into why this works.

For a plane given in general or Cartesian form, finding a normal vector is easy; use the coefficients on the variables x, y, and z for the components of a normal vector. For $4x - 2y + z + 5 = 0$, a normal vector is

$$\mathbf{n_1} = \langle 4, -2, 1 \rangle.$$

To find a normal vector for a plane given by parametric equations, first we find two nonparallel vectors in the plane and then find their cross product. The two direction vectors (given by coefficients to the parameters t and s) are the most convenient vectors to use; to that end, for the plane given by

$$x = 2 - t - 3s$$
$$y = 4 + t + s \quad,$$
$$z = 9 + 2s$$

we set

$$\mathbf{w} = \langle -1, 1, 0 \rangle,$$
$$\mathbf{u} = \langle -3, 1, 2 \rangle.$$

The normal vector for this plane is found by the cross product

$$\mathbf{n}_2 = \mathbf{w} \times \mathbf{u} = \begin{vmatrix} \mathbf{i} & \mathbf{j} & \mathbf{k} \\ -1 & 1 & 0 \\ -3 & 1 & 2 \end{vmatrix} = \cdots = \langle 2, 2, 2 \rangle.$$

If you do not already recognize whether the two normal vectors are parallel, we finish by checking their cross product to see if it is the zero vector:

$$\mathbf{n}_1 \times \mathbf{n}_2 = \begin{vmatrix} \mathbf{i} & \mathbf{j} & \mathbf{k} \\ 4 & -2 & 1 \\ 2 & 2 & 2 \end{vmatrix} = \cdots = \langle -6, -6, 12 \rangle \neq \mathbf{0}.$$

Because the two normal vectors are not parallel, the two planes are not parallel. ∎

Reading Exercise 25 Are the planes $2x + y + 3z + 5 = 0$ and $4x + 2y + 6z + 1 = 0$ parallel?

1.6.7 Parallel, intersecting, skew

In two dimensions, two lines are either parallel, as in Figure 1.69, right, or intersecting, as in Figure 1.69, left.

Consider two airplanes, one flying from Atlanta to Los Angeles and one flying from Dallas/Fort Worth to Chicago. The two paths appear to intersect on a flat map (Figure 1.70), and if the planes are heading for what appears to be the intersection point at the exact same moment in time, we see impending disaster! Thankfully for the passengers, we live in a three-dimensional world and the air traffic controllers merely need to have the planes fly at different altitudes. Although the paths of the airplanes are not parallel, neither do they intersect. They are *skew lines*.

The strategy for determining if two lines are parallel, intersecting, or skew, is to check first for parallelism; then if they are not parallel, look for a point of intersection; if they are neither parallel nor intersecting, then they are skew.

Ans. to reading exercise 24:
Possible answers include

$$13(x - 5) + 35(y - 1) - 1(z - 0) = 0,$$
$$13(x - 2) + 35(y - 2) - 1(z + 4) = 0,$$
$$13(x + 3) + 35(y - 4) - 1(z - 1) = 0, \text{ and}$$
$$-13(x - 2) - 35(y - 2) + 1(z + 4) = 0,$$

all of which may be simplified to the general form $13x + 35y - z - 100 = 0$.

If the cross product is the zero vector, then the two vectors you chose are parallel; try again, choosing two nonparallel vectors in the plane.

If the two direction vectors formed from the coefficients of the parameters t and s are parallel, then the geometric object is not a plane, but rather a line in disguise.

If the two normal vectors are parallel, check to make sure that the two planes are not identical. One strategy is to choose a point from one plane and see if it lies in the other plane.

In three dimensions, when two planes are not parallel they will intersect in a line. This can be seen by taking two cards and placing them against one another. If you think that the intersection could be just one point, recall that the planes extend infinitely and follow the planes to where they intersect. Finding the equation of the intersection of two planes is explored in exercises 47 and 48.

Two lines are *skew* if they are nonparallel and nonintersecting.

Another possibility is that the lines are identical, that is, they are the same line; recall that in geometry a line is not considered parallel to itself. The time to check for this possibility is after determining that the direction vectors for the lines are parallel.

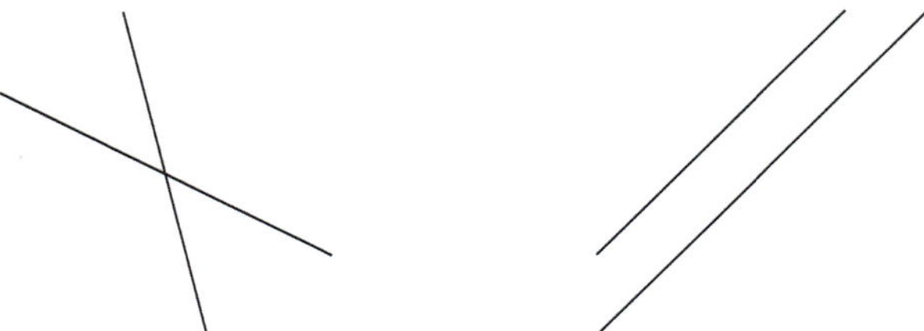

Figure 1.69 *(Left) intersecting lines; (right) parallel lines. In two dimensions, these are the only two possibilities.*

Although flight paths are not straight lines in three dimensions of the type we have studied in this chapter (such lines would pass deeply underground), they generally follow the shortest paths between the airports, which lie along "great circles," which in turn comprise the lines in *spherical geometry.*

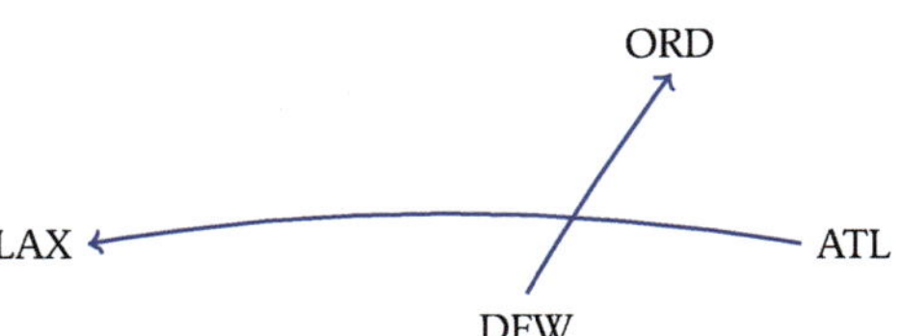

Figure 1.70 *Flight paths*

Example 48 *Are the two lines parallel, intersecting, or skew?* $\ell_1 = \begin{cases} x = 2 + t \\ y = -1 + 2t \\ z = 5 + 3t \end{cases}$,

$$\ell_2 = \begin{cases} x = 1 + s \\ y = 5s \\ z = 1 - s \end{cases}.$$

Solution   To check for parallel lines, we check to see if their direction vectors are parallel. Direction vectors may be picked out from the parametric equations; a direction vector for ℓ_1 is

The components of these direction vectors are the coefficients of the parameters t and s.

$$\mathbf{w_1} = \langle 1, 2, 3 \rangle,$$

and a direction vector for ℓ_2 is

$$\mathbf{w_2} = \langle 1, 5, -1 \rangle.$$

If necessary, take the cross product of the direction vectors to determine if they are parallel:

$$\mathbf{w_1} \times \mathbf{w_2} = \begin{vmatrix} \mathbf{i} & \mathbf{j} & \mathbf{k} \\ 1 & 2 & 3 \\ 1 & 5 & -1 \end{vmatrix} = \cdots = \langle -17, 4, 3 \rangle \neq \mathbf{0},$$

and the two lines are not parallel.

Next, we check to see if they are intersecting. Are there values of the parameters t and s that yield the same point lying on both lines? If so, the values of x must match, while simultaneously matching the values of both y and z. In other words, we would need three equations to be satisfied simultaneously:

$$2 + t = 1 + s$$

$$-1 + 2t = 5s$$

$$5 + 3t = 1 - s.$$

This system is overdetermined, with three equations but just two unknowns. A strategy for solving such a system, or determining that there is no solution, is to pick two of the equations, solve, and then check to see if the third equation is satisfied. To that end, let's pick the first two and solve the simultaneous linear equations

$$2 + t = 1 + s$$

$$-1 + 2t = 5s.$$

You may use your favorite method to solve these equations. Using substitution, we solve the second equation for s to get $s = -\frac{1}{5} + \frac{2}{5}t$ and then substitute into the first equation:

$$2 + t = 1 + \left(-\tfrac{1}{5} + \tfrac{2}{5}t\right)$$

$$10 + 5t = 5 - 1 + 2t$$

$$3t = -6$$

$$t = -2.$$

Substituting $t = -2$ into the second equation yields

$$-1 + 2(-2) = 5s$$

$$-1 = s.$$

Now the question is, when the x-coordinates match and the y-coordinates match, do the z-coordinates also match? Substituting $t = -2$ and $s = -1$ into the third equation yields

$$5 + 3(-2) = 1 - (-1)$$

$$-1 = 2.$$

If the two direction vectors do turn out to be parallel, check to see if the lines are parallel or if instead they are equal (identical). Although there are often more efficient methods, one way to accomplish this is to see if the lines intersect.

In what follows, line 1 is referred to as the first equation, line 2 the second equation, and line 3 the third equation.

Ans. to reading exercise 25:

Yes, because their normal vectors $\mathbf{n}_1 = \langle 2, 1, 3 \rangle$ and $\mathbf{n}_2 = \langle 4, 2, 6 \rangle$ are parallel (notice $\mathbf{n}_2 = 2\mathbf{n}_1$).

Among the methods for solving simultaneous linear equations are the addition method, the substitution method, Cramer's rule, and Gaussian elimination.

Line 1 substitutes $-\frac{1}{5} + \frac{2}{5}t$ for s in the first equation; line 2 multiplies both sides by 5 to clear the fractions; and line 3 collects like terms.

When $t = -2$ and $s = -1$ the points on the two lines are $(0, -5, -1)$ and $(0, -5, 2)$. The x- and y-coordinates match, but the z-coordinates are different. This is the same as the two airplanes of Figure 1.70 reaching the same x- and y-coordinates at the same time, but being at different altitudes and not colliding.

The z-coordinates do not match, so there is no solution to the system of equations, and the two lines do not intersect.

The lines are neither parallel nor intersecting, so we conclude that the two lines are skew. ∎

EXERCISES 1.6

1–10. Find (a) a vector equation and (b) parametric equations of the line.

1. line through the point $(2, 4, 7)$ with direction vector $\langle -3, 0, 9 \rangle$
2. line through the points $(3, 9, 7)$ and $(1, 8, 4)$
3. line through the points $(1, 3, 7)$ and the origin
4. line through the point $(1, 4, 4)$ parallel to the line $x = 2 + 3t$, $y = 2 - t, z = -4 + 4t$
5. line through the points $(8, 0, -3)$ and $(-2, 5, 5)$
6. line through the point $(0, 10, -3)$ orthogonal to the plane $x = 5 - 3t - 8s, y = 1 + t + 3s, z = -4t + s$
7. line through the point $(9, 1, -7)$ parallel to the line $x = 2 - 6t$, $y = 13 - 4t, z = 66 - 2t$
8. line through the origin orthogonal to $4x + 2y - 3z + 8 = 0$
9. line through the point $(3, 2, 1)$ orthogonal to the lines $x = 2t$, $y = 5t, z = 7t$ and $x = 3 + s, y = 2s, z = 9 - s$
10. line through $(2, 5, 3)$ parallel to both the xy-plane and the xz-plane

11–18. Find the designated equation of the plane.

11. plane through the points $(-2, 4, -1)$, $(0, 9, 9)$, and $(3, 5, 11)$; (a) vector equation, (b) Cartesian equation
12. plane through the points $(1, -5, 6)$, $(3, 8, 1)$, and $(1, 0, 6)$; (a) parametric equations, (b) Cartesian equation
13. plane with normal vector $\langle 2, 9, -6 \rangle$ through the point $(3, 0, 0)$; (a) Cartesian equation, (b) parametric equations
14. plane through $(1, 2, 3)$ parallel to $x = -1 + t - 2s, y = 4 - 4t + 5s, \; z = 8t + 12s$; (a) vector equation, (b) parametric equations
15. plane containing the line $x = 2 - 2t, y = 2 + 3t, z = 2 + 4t$ and the point $(0, 1, 5)$; parametric equations
16. plane through $(1, 2, 3)$ parallel to $4x - 3y + 2z + 7 = 0$; Cartesian equation

17. plane through the points $(1, 0, 5)$ and $(2, 7, 2)$ parallel to the line $x = 2 - t, y = 4 + 2t, z = -3t$; your choice of equation type
18. plane through the origin orthogonal to the line $x = 2 + 7t$, $y = 99, z = 3 - 5t$; your choice of equation type

19–24. Find the distance from the point to the line.

19. point $(0, 0, 3)$, line $x = 7t, y = 7 + t, z = 7 + 7t$
20. the origin, line $\mathbf{v} + t\mathbf{w} = \langle 1 - t, 3 + t, -2 + 6t \rangle$
21. point $(2, 8, 1)$, the z-axis
22. point $(2, -4, 9)$, line $x = 2 + t, y = 2 - t, z = 2$
23. point halfway between the origin and $(2, 8, 10)$, line $x = 3$, $y = 7, z = t$
24. point halfway between $(2, 4, 6)$ and the x-axis, line $x = t, y = t$, $z = t$

25–28. Determine the point of intersection of the line and the plane.

25. line $\langle 3 + t, 4 + t, 5 + 6t \rangle$, plane $4x - 3y + z = 7$
26. line $x = 2 + 3t, y = 2 - t, z = -4 + 4t$, plane $x - y + 2z = 1$
27. z-axis, plane $x = 2 + t + 3s, y = 6t - 3s, z = 92 + 4t + 10s$
28. line $\langle 3t, 5 - t, t - 7 \rangle$, plane $\langle 2 + t + s, -1 + 2t + 4s, 0 + s \rangle$

29–32. Are the planes parallel?

29. $3(x - 8) + 4(y - 1) - 2(z + 5) + 18 = 0$ and $-6x - 8y + 4z = 10$
30. $x + 2y + 3z = 4$ and $2x + 4y - z = 1$
31. $x + y + z = 1$ and $x = t, y = s, z = 2$
32. $3x + 8y + 2z = 5$ and $x = 5 - 8s, y = 1 + t + 3s, z = -4t$

33–38. Are the lines parallel, intersecting, or skew?

33. $x = 4t, \; y = 5 + 5t, \; z = 6 - t$ and $x = 12 + 12s, \; y = 2 + 15s$, $z = 7 - 3s$
34. $x = 2 + 3t, y = 2 - t, z = 5t$ and $x = 4 + s, y = s, z = 2 + s$
35. $x = 3 - t, y = 2t, z = 1 + 7t$ and $x = 2s, y = 1 + 7s, z = 3 - s$
36. $x = 7 + t, \; y = -3 - 2t, \; z = 6 + 3t$ and $x = 5 - 4s, \; y = 1 + s$, $z = -5s$
37. $x = 5 + 4t, \; y = 2 - t, \; z = 3 + 3t$ and $x = 1 + 2s, \; y = 8 - 3s$, $z = 5 - s$
38. $x = 5t, \; y = 7t, \; z = 3 + 6t$ and $x = 12 - 10s, \; y = 9 - 14s$, $z = -42 - 12s$

Hint: If **n** is a normal vector to a plane, then so is −**n**.

39–42. The *angle between two planes* is defined to be the angle between their normal vectors, with the caveat that the normal vectors must be chosen so that the angle between them is acute (less than 90°) or right. Find the angle between the two planes, in (a) radians and (b) degrees.

39. $z - 2 = 0$ and $2x + 3y + 3z + 351.27 = 0$
40. $x - y + 5z + 1 = 0$ and $3x - 5y + z = 0$
41. $x + 9y + 2z - 7 = 0$ and $-3x - y - 4z + 8 = 0$
42. $12x - 25y + z = 0$ and $8x - 13y - 2z + 18 = 0$

43. A margin note for example 41 lists two possibilities for a vector form of the equation of the line, but states that there are infinitely many possibilities. State at least three more.

44. Is the vector form of the equation of a line valid in four dimensions? If so, find an equation of the line through the points $(1, 2, 3, 4)$ and $(2, -5, 7, 11)$.

45. Is the vector form of the equation of a plane valid in four dimensions? If so, find an equation of the plane through the points $(1, 2, 3, 4), (0, -2, -1, 5)$, and $(9, 0, 3, 6)$.

46. This exercise is in four dimensions. (a) What would be the vector form of the equation of a three-dimensional *hyperplane*? (b) The vector form of the equation of a plane requires the two direction vectors to be nonparallel. What similar condition is required of the direction vectors for a three-dimensional hyperplane?

47. **Intersection of two planes, strategy** 1. Let plane A have the equation $3x - 2y + 4z - 10 = 0$ and let plane B have the equations $x = 1 + t + 3s, y = 2 - t$, and $z = 2 - 2t + s$. Find the equation of the line that forms the intersection of the two planes by performing the following steps.

   (a) Find parametric equations of a line in plane B (Hint: point and one direction vector).
   (b) Determine the intersection of the line from part (a) and plane A. This is one point in the intersection of the two planes.
   (c) Repeat parts (a) and (b) using a different direction vector. This is a second point in the intersection of the two planes.
   (d) Determine an equation of the line through the two points found in (b) and (c); this is the line of intersection of the two planes.

Variations and combinations of the two strategies outlined in exercises 47 and 48 are possible.

48. **Intersection of two planes, strategy** 2. Let plane A have the equation $x + 5y + 2z - 1 = 0$ and let plane B have the equation

$4x + 9y - 6z + 2 = 0$. Find the equation of the line that forms the intersection of the two planes by performing the following steps.

(a) Because a line lying in both planes is orthogonal to both normal vectors, determine normal vectors for the two planes and take their cross product. The result serves as a direction vector for the intersection of the two planes.

(b) Find a point in plane B by setting $y = 0$ and $z = 0$ and then solving for x to get a point of the form $(a, 0, 0)$.

(c) Find a point in plane B by setting $x = 0$ and $y = 0$ and then solving for z to get a point of the form $(0, 0, c)$.

(d) Determine an equation of the line through the points found in (b) and (c).

(e) Determine the intersection of the line found in (d) and plane A.

(f) Using the point found in (e) and the direction vector found in (a), write the equation of the line of intersection of the two planes.

49. **Distance from a point to a plane.** The distance d from a point $P(x_0, y_0, z_0)$ to a plane $ax + by + cz + k = 0$ is given by

$$d = \frac{|ax_0 + by_0 + cz_0 + k|}{\sqrt{a^2 + b^2 + c^2}}.$$

Derive this formula by forming a normal vector $\mathbf{n}$, choosing a point $Q(x_1, y_1, z_1)$ in the plane, forming the vector $\overrightarrow{QP}$, and observing that $d = |\mathrm{comp}_{\mathbf{n}}\, \mathbf{b}|$.

50. Using the formula from exercise 49, write a formula for the distance from the origin to the plane $ax + by + cz + k = 0$.

51-54. Using a formula from exercise 49 or 50, find the distance from the point to the plane.

51. origin and $3x - 5y + 10z = 4$
52. $(6, 7, -3)$ and $50x + 50y + 100z + 300 = 0$
53. $(1, 0, 4)$ and $5x + 7y + 2z + 6 = 0$
54. origin and $3x - 2y - 4z + 15 = 7$

55. Find the distance between the two parallel planes $3x + 2y - z + 5 = 0$ and $6x + 4y - 2z + 3 = 0$.

56. Find the closest distance between the skew lines in example 48. (Hint: sometimes exercises are in a particular order for a reason.)

1.7 Quadric Surfaces

In an algebraic sense, quadric surfaces are three-dimensional versions of conic sections. For that reason, we begin with some brief reminders.

1.7.1 Conic section review

The graph of $y = cx^2$ is a parabola with vertex at the origin. If the constant c is positive then the parabola opens upward, and if c is negative the parabola opens downward (Figure 1.71). The larger the absolute value of c, the narrower the parabola is (it rises more rapidly); the smaller the value of c, the wider the parabola is.

More details about conic sections, including focus, directrix, vertex, eccentricity, and more, can be found in Sections 8.5 and 8.6 of *Calculus Set Free*.

Swapping the variables x and y swaps the orientation of a conic section. Whereas the parabola $y = cx^2$ is oriented vertically (opening upward or downward), the parabola $x = cy^2$ is oriented horizontally, opening right if c is positive and opening left if c is negative (Figure 1.72).

The equation $\frac{x^2}{a^2} + \frac{y^2}{b^2} = 1$ is an ellipse centered at the origin (Figure 1.73); if $a = b$ then the ellipse is a circle. The values of a and b function similarly to a circle's radius, in the direction of the variable they are associated with.

The equation $\frac{x^2}{a^2} - \frac{y^2}{b^2} = 1$ is a hyperbola centered at the origin, opening left and right (Figure 1.74). Swapping the variables changes the orientation (Figure 1.75); the graph of $\frac{y^2}{b^2} - \frac{x^2}{a^2} = 1$ opens up and down.

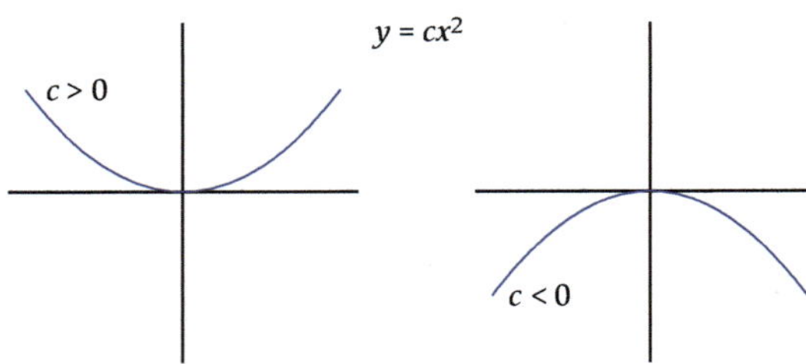

Figure 1.71 *Parabolas opening up (if $c > 0$) and down (if $c < 0$)*

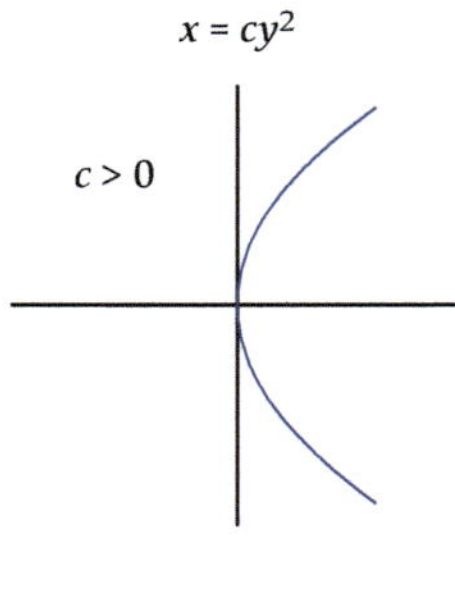

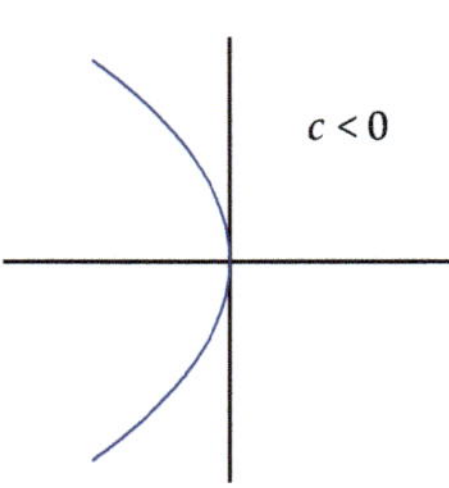

Figure 1.72 *Parabolas opening right (if $c > 0$) and left (if $c < 0$)*

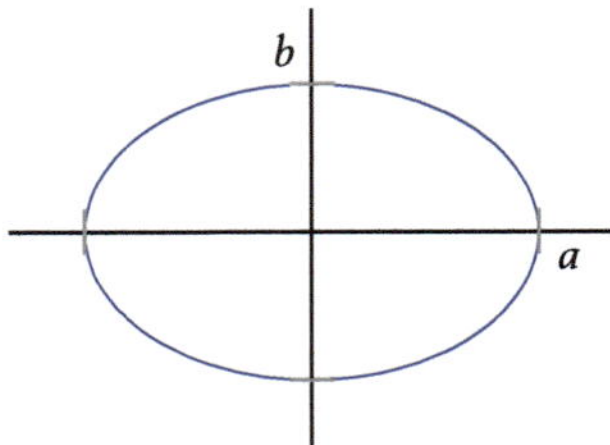

Figure 1.73 *An ellipse centered at the origin*

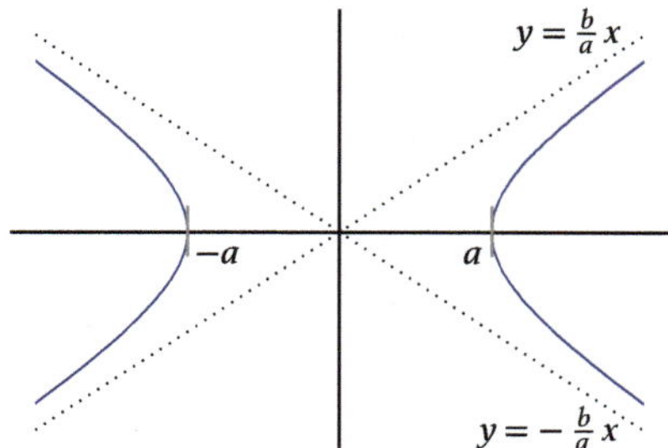

Figure 1.74 *A hyperbola $\frac{x^2}{a^2} - \frac{y^2}{b^2} = 1$, centered at the origin, opening horizontally*

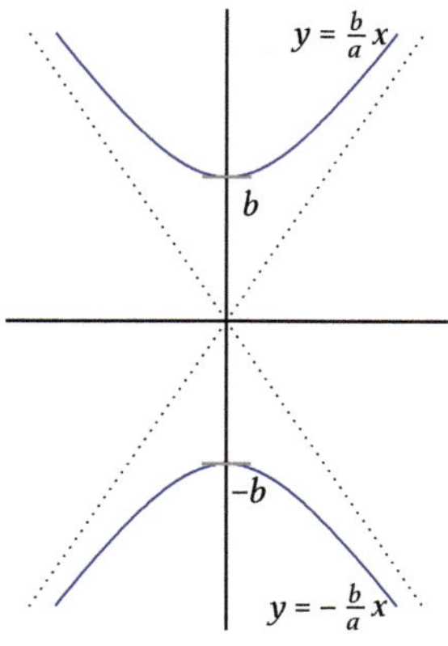

Figure 1.75 *A hyperbola $\frac{y^2}{b^2} - \frac{x^2}{a^2} = 1$, centered at the origin, opening vertically*

One type of "degenerate" conic section is of the form $\frac{x^2}{a^2} - \frac{y^2}{b^2} = 0$ (notice the 0 instead of 1), which can be simplified as

$$\frac{x^2}{a^2} - \frac{y^2}{b^2} = 0$$

$$\frac{x^2}{a^2} = \frac{y^2}{b^2}$$

$$y^2 = \frac{b^2}{a^2}x^2$$

$$|y| = \left|\frac{b}{a}x\right|$$

$$y = \pm\frac{b}{a}x,$$

Line 4 takes square roots of both sides; recall that $\sqrt{x^2} = |x|$. Line 5 simplifies to the two possibilities given by the absolute value expressions, yielding two intersecting lines.

which represents two lines through the origin, one with slope $\frac{b}{a}$ and one with slope $-\frac{b}{a}$ (Figure 1.76).

Quadratic expressions in the variables x and y can be manipulated into the form of a conic section. For instance, the equation $4x^2 + 16x + y^2 - 6y = 11$ can be placed in the form of a shifted ellipse through the use of completing the square:

A quadratic expression is an expression of degree 2.

$$4x^2 + 16x + y^2 - 6y = 11$$

$$4(x^2 + 4x \quad) + y^2 - 6y \quad = 11$$

$$4(x^2 + 4x + 4) + y^2 - 6y + 9 = 11 + 4(4) + 9$$

$$4(x + 2)^2 + (y - 3)^2 = 36$$

$$\frac{(x + 2)^2}{9} + \frac{(y - 3)^2}{36} = 1.$$

Line 1 is the original expression; line 2 factors out the coefficient on the x^2 term from all of the x terms, and leaves room for the constant to be added; line 3 completes the square for both variables (half the coefficient on the x term is $\frac{4}{2} = 2$, squaring gives 4, which is added on the left side inside the parentheses, and the total effect of $4(4)$ on the left side is added to the right side; half the coefficient on the y term is $\frac{-6}{2} = -3$, squaring gives 9, which is added to both sides); line 4 factors the perfect squares; line 5 divides both sides by 36 to obtain the desired form of an ellipse.

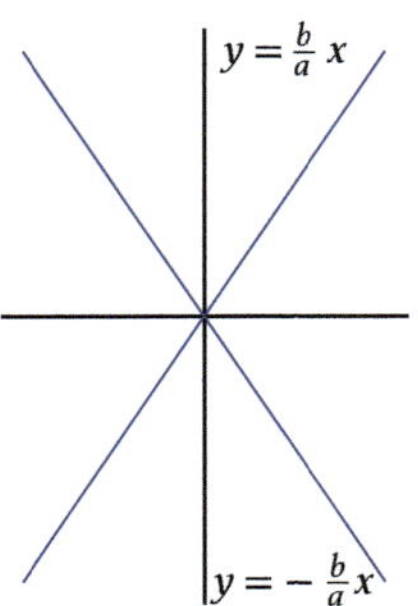

Figure 1.76 *The degenerate conic $\frac{x^2}{a^2} - \frac{y^2}{b^2} = 0$, which yields a pair of intersecting lines*

For a more thorough review of shifting, stretching, and reflecting graphs, see *Calculus Set Free* Section 0.5.

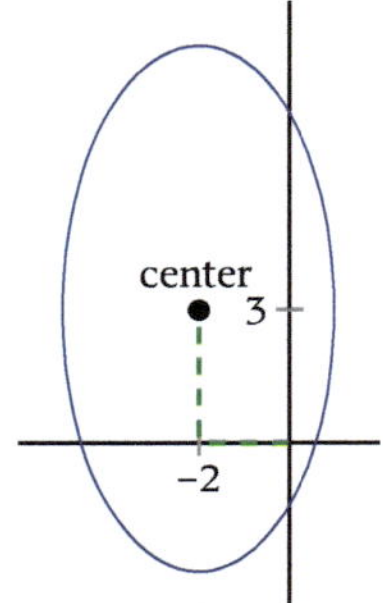

Figure 1.77 *The ellipse $\frac{(x+2)^2}{9} + \frac{(y-3)^2}{36} = 1$*

The effect is an ellipse that is shifted horizontally by 2 units to the left and vertically by 3 units upward, as shown in Figure 1.77.

We will not discuss rotation of axes in this text.

Add in the mixed term Bxy, which requires a rotation of axes, and all quadratic expressions of the form

$$Ax^2 + Bxy + Cy^2 + Dx + Ey + F = 0,$$

where not all of A, B, and C are zero, are conic sections. As long as $B = 0$, completing the square can rewrite any of these expressions into a form recognizable as a conic section.

Example 49 *Which conic section, if any, does the equation represent?*

(a) $7x^2 + 5y^2 = 43$
(b) $7x^2 - 5y^2 = 43$
(c) $7x - 5y^2 = 43$

(d) $7x - 5y^2 = 0$
(e) $7x^2 - 5y^2 = 0$

Solution

(a) For $7x^2 + 5y^2 = 43$, we see a quadratic expression in two variables (x and y, at least one is squared, no term of degree higher than 2, polynomial terms only), so it is a conic section. There are squared terms for both variables, so it is not a parabola. The coefficients on the squared terms have the same sign, so it is not a hyperbola or a pair of intersecting lines; it is an ellipse. Because those coefficients are not equal, the ellipse is not a circle.

> When checking squared terms for the same sign (positive or negative), the terms must be on the same side of the equation.

    Note: it is not necessary to place the ellipse in the form $\frac{x^2}{a^2} + \frac{y^2}{b^2} = 1$ to recognize the equation as an ellipse. If we wish to do so, we may divide both sides by 43 and then rewrite the coefficients:

$$7x^2 + 5y^2 = 43$$

$$\frac{7x^2}{43} + \frac{5y^2}{43} = 1$$

$$\frac{x^2}{\frac{43}{7}} + \frac{y^2}{\frac{43}{5}} = 1.$$

As long as there is a constant on the opposite side of the equation from the second-degree terms with the same sign as the coefficients of the squared terms, the form $\frac{x^2}{a^2} + \frac{y^2}{b^2} = 1$ can be reached. We will not usually bother to do so.

> The equation $-7x^2 - 5y^2 = -43$ is the same ellipse.

(b) For $7x^2 - 5y^2 = 43$, we see a quadratic expression in two variables with squared terms for both variables (not a parabola) and coefficients on those squared terms of opposite signs (not an ellipse). A nonzero constant term is present (43), so it is not intersecting lines; it is a hyperbola. The constant on the right-hand side of the equation is positive and so is the coefficient of the term $7x^2$, so the hyperbola opens left and right.

> The hyperbola $7x^2 - 5y^2 = -43$ opens upward and downward.

(c) For $7x - 5y^2 = 43$, we see a quadratic expression in two variables with a squared term for only one variable; it is a parabola. But there is a nonzero constant term as well, meaning the parabola is not of the form $y = cx^2$ or $x = cy^2$, and we conclude that shifting is present, so the vertex is not at the origin. Moving the squared term to the opposite side from the single-degree variable, $7x = 5y^2 + 43$, we see that the coefficient on x is positive and so is the coefficient on y^2, meaning $c > 0$ and the parabola opens to the right.

> This is an expression in two variables x and y. It is a quadratic expression because of the term $-5y^2$.

(d) For $7x - 5y^2 = 0$, we again see a quadratic expression in two variables with a squared term for only one variable; it is a parabola. This time there is no nonzero constant term, so the vertex of the parabola is at the origin. This parabola opens to the right.

(e) For $7x^2 - 5y^2 = 0$ we see a quadratic expression in two variables with squared terms for both variables (not a parabola) and coefficients on those squared terms of opposite signs (not an ellipse). But there is no nonzero constant term, so it is not a hyperbola; it is a pair of intersecting lines. ∎

Even when the constants are not specified, the same criteria apply.

Example 50 *Which conic section, if any, does the equation represent? Assume that every symbol other than the variables x and y represents a constant.*

(a) $\dfrac{x^2}{a^2} + \dfrac{y^2}{b^2} + \dfrac{k}{c} = 0$

(b) $\dfrac{x^2}{a^2} - \dfrac{k^2}{c^2} - \dfrac{y^2}{b^2} = 0$

Solution

(a) For $\dfrac{x^2}{a^2} + \dfrac{y^2}{b^2} + \dfrac{k}{c} = 0$, we see a quadratic expression in two variables (x and y) with squared terms for both variables (not a parabola) and coefficients on those squared terms of the same sign, so it is an ellipse.

Note that both a^2 and b^2 are positive.

(b) For $\dfrac{x^2}{a^2} - \dfrac{k^2}{c^2} - \dfrac{y^2}{b^2} = 0$, we see a quadratic expression in two variables with squared terms for both variables (not a parabola) and coefficients on those squared terms of opposite signs (not an ellipse). We also see a constant term, $-\dfrac{k^2}{c^2}$, so it is not intersecting lines; it is a hyperbola. Moving the constant term to the right-hand side yields a positive constant, which matches the sign of the coefficient on x^2, meaning that the hyperbola opens left and right. ∎

In part (a), moving the constant $\dfrac{k}{c}$ to the opposite side from the squared terms, we see that k and c must have opposite signs so that $-\dfrac{k}{c}$ is positive. There are no solutions to the equation if k and c have the same sign.

Reading Exercise 26 (a) Identify the conic section $\dfrac{x^2}{16} + \dfrac{y^2}{9} = 1$ and (b) sketch its graph.

1.7.2   Graphing conic sections in planes in $\mathbf{R}^3$

Graphing in three dimensions requires coordinates for three variables, whereas conic sections have two variables. But if we specify one of the three variables as having a constant value, such as $z = 2$, then we have specified a plane; an equation in the remaining two variables may be a conic section in that plane.

Points on the graph must satisfy both equations.

Example 51 *Graph the ellipse $x = 0$, $\dfrac{y^2}{9} + \dfrac{z^2}{4} = 1$.*

Solution   The plane $x = 0$ is the yz-plane. The ellipse is centered at the origin, containing the points $(0, \pm 3, 0)$ and $(0, 0, \pm 2)$. The resulting graph is in Figure 1.78. ∎

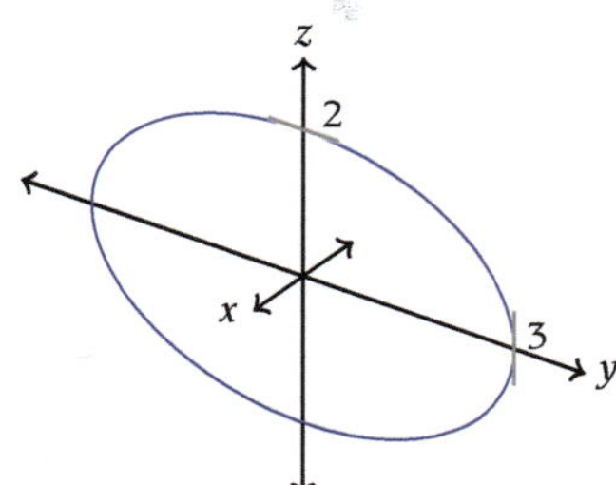

Figure 1.78 *The ellipse* $x = 0$, $\dfrac{y^2}{9} + \dfrac{z^2}{4} = 1$

The shadow is the hyperbola $z = 0$, $\dfrac{x^2}{9} - y^2 = 1$, which lies in the xy-plane. The hyperbola of example 52 is shifted 2 units in the positive z-direction from the shadow. The shadow is shown for perspective only and is not part of the graph requested in example 52.

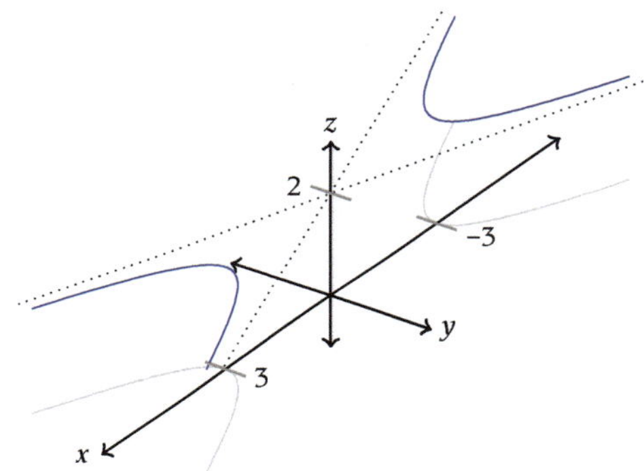

Figure 1.79 *The hyperbola* $z = 2$, $\dfrac{x^2}{9} - y^2 = 1$ *(blue), along with its asymptotes (dotted) and its shadow in the xy-plane (gray)*

Example 52 *Graph the conic section* $z = 2$, $\dfrac{x^2}{9} - y^2 = 1$.

Solution   The graph lies in the plane $z = 2$, which is parallel to the xy-plane. The second equation is in the standard form of a hyperbola opening in the x-direction, with $a = 3$ and $b = 1$. The equations of the asymptotes in the plane $z = 2$ are $y = \pm\dfrac{1}{3}x$. The vertices are at $(\pm 3, 0, 2)$. The resulting graph is in Figure 1.79.   ∎

Reading Exercise 27   Identify and graph the conic section $y = -1$, $x = z^2$.

1.7.3   Quadric surfaces introduction: ellipsoids

Describing a conic section in $\mathbf{R}^3$ requires two equations, as in examples 51 and 52. The graph of a single quadratic equation in $\mathbf{R}^3$ is not a conic section. But cross sections can be. The graph of a three-variable equation whose cross sections are conic sections is a type of *quadric surface*.

Definition 17 QUADRIC SURFACE *A quadric surface is the graph in* $\mathbf{R}^3$ *of an equation of the form*

$$Ax^2 + By^2 + Cz^2 + Dxy + Exz + Fyz + Gx + Hy + Iz + J = 0,$$

where at least one of A, B, C, D, E, and F is nonzero.

Equations of quadric surfaces whose cross sections are conic sections can be categorized into a few relatively simple forms. The terms Gx, Hy, and Iz, if necessary, can be dealt with by completing the square so that shifting (horizontally in the x- and y-directions and vertically in the z-direction) can be recognized. The terms Dxy, Exz, and Fyz, when present, require rotation of axes and will not be discussed in this text.

The first type of quadric surface to discuss is an ellipsoid.

ELLIPSOID

An *ellipsoid* has the form

$$\frac{x^2}{a^2} + \frac{y^2}{b^2} + \frac{z^2}{c^2} = 1,$$

where a, b, and c are constants. If $a = b = c$, then the ellipsoid is also called a *sphere*.

The name "ellipsoid" is reminiscent of "ellipse," as is the equation; in fact, the equation can be remembered by inserting a z term into the equation of an ellipse. The look is similar, as well, appearing to be a three-dimensional version of an ellipse (Figure 1.80). Or, if you like analogies, an ellipsoid is to a sphere as an ellipse is to a circle.

Slice a spherical fruit into two pieces and you will see a cross section that resembles a circular disk. The outline of that disk, around the rind of the fruit, is a circle. That's the idea of a *trace*. The path of the knife cutting the fruit is a plane. If we take a plane and intersect it with a quadric surface, the result is often a conic section. By taking traces of the form $x = k$, $y = k$, and $z = k$, where k is a constant, we can gain an understanding of the shape of a quadric surface. The traces are also helpful in graphing and visualizing quadric surfaces, as demonstrated in Figure 1.80. Let's investigate the ellipsoid $\frac{x^2}{a^2} + \frac{y^2}{b^2} + \frac{z^2}{c^2} = 1$ by taking cross sections and identifying the equations of the resulting traces.

Consider the points on the ellipsoid with y-coordinate k, where k is a constant. These points are on the ellipsoid and also in the plane $y = k$, which is a vertical

Ans. to reading exercise 26:

(a) ellipse;

(b)

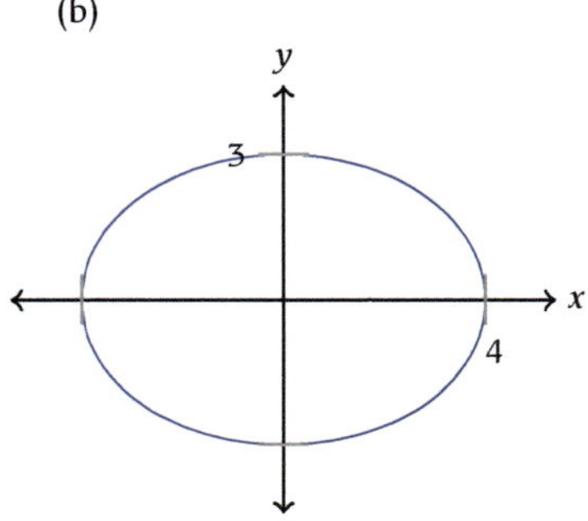

The shape of an ellipsoid may remind you of a watermelon.

Ans. to reading exercise 27: parabola (opening in the positive x-direction);

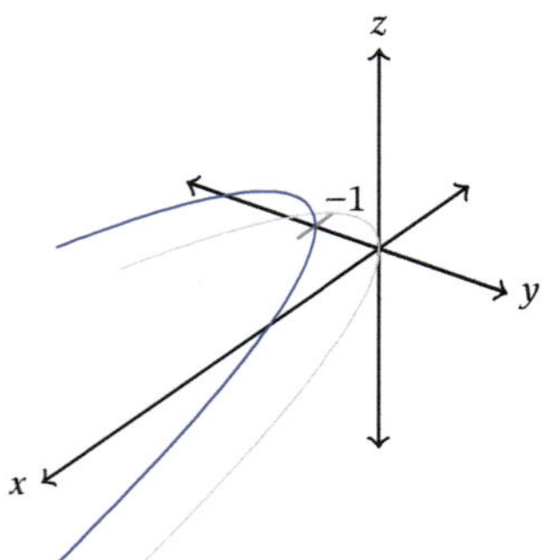

Note: drawing the shadow (gray) in the xz-plane is not required. It is shown here for added perspective.

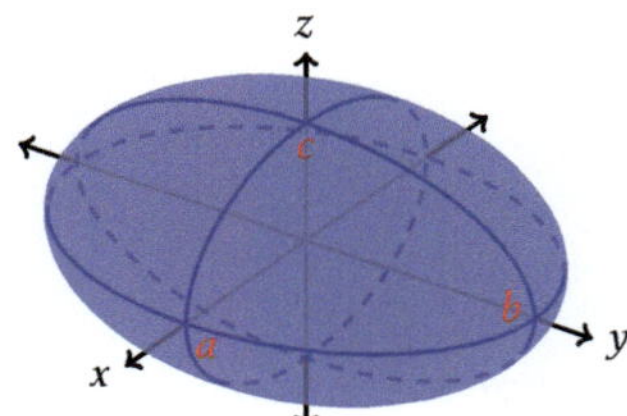

Figure 1.80 *The ellipsoid* $\frac{x^2}{a^2} + \frac{y^2}{b^2} + \frac{z^2}{c^2} = 1$, *with traces marked in the xy-plane, the xz-plane, and the yz-plane*

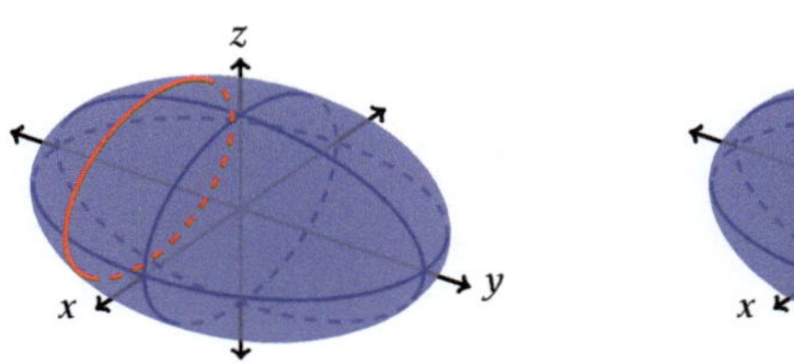

Figure 1.81 *The ellipsoid* $\frac{x^2}{9} + \frac{y^2}{16} + \frac{z^2}{4} = 1$, *with (left) the trace* $y = -2$ *and (right) the traces* $y = 0$, $y = 1$, $y = 2$, $y = 3$, *and* $y = 3.8$. *The trace* $y = 4$ *would be just a single point. All traces named here are red*

plane parallel to the xz-plane; they comprise a vertical slice, or cross section, of the ellipsoid. The equation of this trace can be found by replacing y by k in the ellipsoid's equation to obtain

$$\frac{x^2}{a^2} + \frac{k^2}{b^2} + \frac{z^2}{c^2} = 1.$$

Classifying the conic section as in example 50, we see a quadratic equation in two variables (x and z) with squared terms for both variables (not a parabola) and coefficients on those squared terms of the same sign, so it is an ellipse, as in Figure 1.81, left.

Sometimes analyzing the effect of various values of the constant k can be helpful. The previous equation can be further rearranged to

$$\frac{x^2}{a^2(1 - \frac{k^2}{b^2})} + \frac{z^2}{c^2(1 - \frac{k^2}{b^2})} = 1.$$

The larger the absolute value of k, the smaller the values of $a^2(1 - \frac{k^2}{b^2})$ and $c^2(1 - \frac{k^2}{b^2})$, meaning that as k gets further from the center the ellipses shrink. Figure 1.81, right, illustrates these shrinking ellipses.

The trace $y = -2$ on the ellipsoid $\frac{x^2}{9} + \frac{y^2}{16} + \frac{z^2}{4} = 1$ would have equation $\frac{x^2}{9} + \frac{(-2)^2}{16} + \frac{z^2}{4} = 1$, which could be rearranged as

$$\frac{x^2}{9\left(1 - \frac{4}{16}\right)} + \frac{z^2}{4\left(1 - \frac{4}{16}\right)} = 1,$$

which in turn is the same as

$$\frac{x^2}{\frac{27}{4}} + \frac{z^2}{3} = 1.$$

Details: move the constant term to the right-hand side,

$$\frac{x^2}{a^2} + \frac{z^2}{c^2} = 1 - \frac{k^2}{b^2},$$

and then divide by that constant to obtain

$$\frac{x^2}{a^2(1 - \frac{k^2}{b^2})} + \frac{z^2}{c^2(1 - \frac{k^2}{b^2})} = 1.$$

In a similar way, we may take a trace of the form $z = k$, which is a horizontal slice, and still obtain an ellipse;

$$\frac{x^2}{a^2} + \frac{y^2}{b^2} + \frac{k^2}{c^2} = 1$$

is once again a quadratic equation in two variables (x and y), with squared terms for both variables and coefficients on those squared terms of the same sign. A horizontal trace of the form $z = k$ is shown in Figure 1.82. The ellipses also get smaller as k increases in absolute value.

Finally, a vertical slice of the form $x = k$ would be the ellipse

$$\frac{k^2}{a^2} + \frac{y^2}{b^2} + \frac{z^2}{c^2} = 1.$$

Two vertical slices of the form $x = k$ are shown in Figure 1.83.

Notice that in all these figures, traces in the coordinate planes are shown in blue to give perspective; other traces are parallel to one of these blue traces. Portions of traces that would be visible on the front of the ellipsoid in our given viewpoint are solid, and the portions that would be hidden on the back of the ellipsoid are dashed. This convention will continue in other figures.

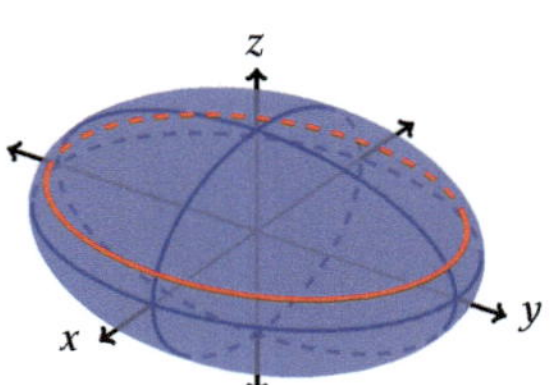

Figure 1.82 *The ellipsoid $\frac{x^2}{9} + \frac{y^2}{16} + \frac{z^2}{4} = 1$, with the trace $z = 0.5$ (red)*

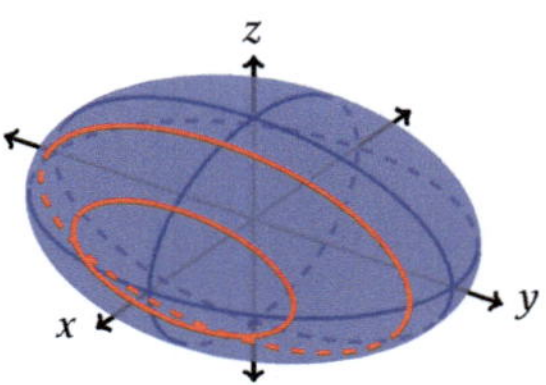

Figure 1.83 *The ellipsoid $\frac{x^2}{9} + \frac{y^2}{16} + \frac{z^2}{4} = 1$, with the traces $x = 0.8$ and $x = 2$ (red)*

As with conic sections, swapping the variables changes the orientation but not the shape; $\frac{x^2}{a^2} + \frac{z^2}{c^2} = \frac{y^2}{b^2}$ and $\frac{y^2}{b^2} + \frac{z^2}{c^2} = \frac{x^2}{a^2}$ are also elliptic cones.

1.7.4   Elliptic cone

Our second quadric surface also has some elliptic cross sections. As we did for the ellipsoid, we shall investigate these cross sections to understand the shape.

> **ELLIPTIC CONE**
>
> An *elliptic cone* has the form
>
> $$\frac{x^2}{a^2} + \frac{y^2}{b^2} = \frac{z^2}{c^2},$$
>
> where a, b, and c are constants.

The cross sections $z = k$, which lie in horizontal planes, have equation

$$\frac{x^2}{a^2} + \frac{y^2}{b^2} = \frac{k^2}{c^2},$$

which is an ellipse (or circle if $a = b$). The larger the absolute value of k, the larger the ellipse.

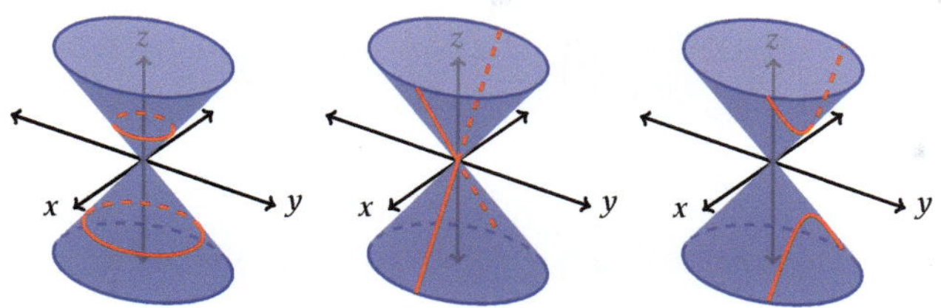

Figure 1.84 *The elliptic cone $\frac{x^2}{9} + \frac{y^2}{16} = \frac{z^2}{25}$, with (left) the traces $z = 1$ and $z = -2$, both ellipses; (middle) the trace $y = 0$, consisting of intersecting lines; and (right) the trace $y = 1$, a hyperbola. All traces named here are red. Only the portion of the cone lying between the planes $z = -3$ and $z = 3$ is shown*

Although the horizontal cross sections, which are ellipses, can be shown in their entirety, the vertical cross sections consisting of either intersecting lines or hyperbolas cannot be shown in their entirety. As with any depiction of a line or a hyperbola, you must use your imagination to see the remainder of the graph.

The cross sections $x = k$, which lie in vertical planes parallel to the yz-plane, have equation

$$\frac{k^2}{a^2} + \frac{y^2}{b^2} = \frac{z^2}{c^2},$$

which can be rearranged to

$$\frac{y^2}{b^2} - \frac{z^2}{c^2} = -\frac{k^2}{a^2}.$$

If $k \neq 0$, this is a hyperbola opening in the z-direction. If $k = 0$ then $\frac{y^2}{b^2} = \frac{z^2}{c^2}$, which is a pair of intersecting lines. It is the intersecting lines that regulate the size of the horizontal elliptic cross sections to the shape of a cone.

Similarly, the cross sections $y = k$, which lie in vertical planes parallel to the xz-plane, are hyperbolas if $k \neq 0$ and intersecting lines if $k = 0$.

An elliptic cone, along with horizontal and vertical cross sections, is shown in Figure 1.84. The elliptic cone is infinite in extent, so only a portion of the cone is actually shown.

1.7.5 Paraboloids

There are two types of paraboloids. One is the *elliptic paraboloid*.

ELLIPTIC PARABOLOID

An *elliptic paraboloid* has the form

$$\frac{x^2}{a^2} + \frac{y^2}{b^2} = \frac{z}{c},$$

where a, b, and c are constants.

The variables can be swapped; $\frac{x^2}{a^2} + \frac{z^2}{c^2} = \frac{y}{b}$ and $\frac{y^2}{b^2} + \frac{z^2}{c^2} = \frac{x}{a}$ are also elliptic paraboloids.

Although the horizontal cross sections, which are ellipses, can be shown in their entirety, the vertical cross sections, which are parabolas, cannot be shown in their entirety. As with any depiction of a parabola, you must use your imagination to see the remainder of the graph.

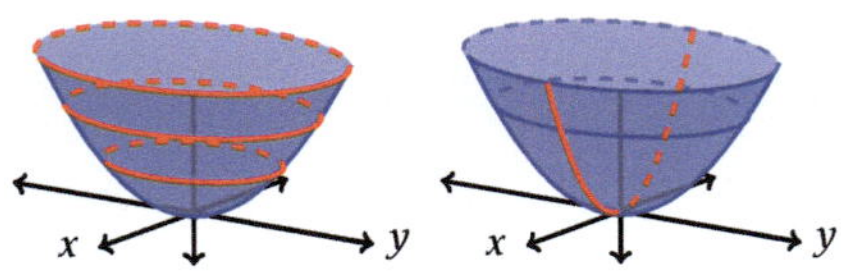

Figure 1.85 *The elliptic paraboloid $\frac{x^2}{9} + \frac{y^2}{16} = \frac{z}{5}$, with (left) the traces $z = 1$, $z = 2$, and $z = 3$, each of which is an ellipse; (right) the trace $y = 0$, a parabola. All traces named here are red. Only the portion of the paraboloid lying below the plane $z = 3$ is shown*

The vertical cross section $x = k$ yields

$$\frac{k^2}{a^2} + \frac{y^2}{b^2} = \frac{z}{c},$$

which is a parabola opening in the positive z-direction if $c > 0$ or the negative z-direction if $c < 0$. If $k = 0$ then the vertex is on the xy-plane (at the origin); otherwise the vertex is shifted vertically from the xy-plane. The vertical cross section $y = k$ is similarly a parabola.

Note that $c = 0$ is not a possibility because of division by zero. The same is true of a, b, and c for all of these quadric surfaces.

The horizontal cross section $z = k$ has equation

$$\frac{x^2}{a^2} + \frac{y^2}{b^2} = \frac{k}{c},$$

which is an ellipse assuming k and c have the same sign.

An elliptic paraboloid, along with horizontal and vertical cross sections, is shown in Figure 1.85.

The second type of paraboloid is the *hyperbolic paraboloid*.

HYPERBOLIC PARABOLOID

A *hyperbolic paraboloid* has the form

$$\frac{y^2}{b^2} - \frac{x^2}{a^2} = \frac{z}{c},$$

where a, b, and c are constants.

The variables can be swapped; $\frac{z^2}{c^2} - \frac{y^2}{b^2} = \frac{x}{a}$ and $\frac{x^2}{a^2} - \frac{z^2}{c^2} = \frac{y}{b}$ are also hyperbolic paraboloids.

The vertical cross section $x = k$ has equation

$$\frac{y^2}{b^2} - \frac{k^2}{a^2} = \frac{z}{c},$$

which is a parabola opening in the positive z-direction (upward) if $c > 0$ and the negative z-direction (downward) if $c < 0$. If $k = 0$ then the vertex is on the xy-plane (at the origin); otherwise the vertex is shifted vertically from the xy-plane.

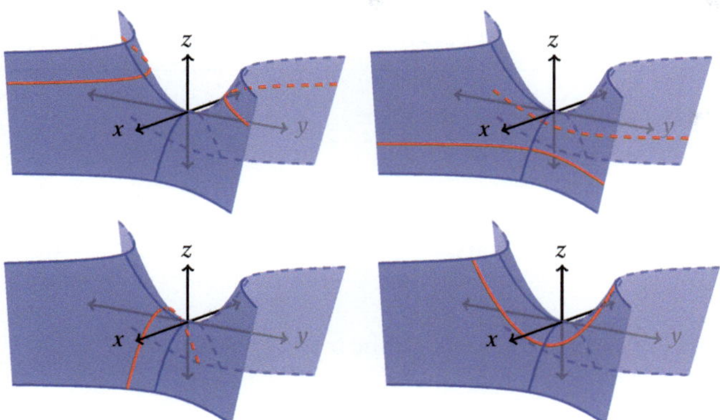

Figure 1.86 *The hyperbolic paraboloid $\frac{y^2}{16} - \frac{x^2}{9} = \frac{z}{5}$, with (top left) the trace $z = 1$, a hyperbola opening in the y-direction; (top right) the trace $z = -1$, a hyperbola opening in the x-direction; (bottom left) the trace $y = -1$, a parabola opening downward; and (bottom right) the trace $x = 1$, a parabola opening upward. All traces named here are red. Only a portion of the paraboloid is shown*

The vertical cross section $y = k$ has equation

$$\frac{k^2}{b^2} - \frac{x^2}{a^2} = \frac{z}{c},$$

which is a parabola opening in the negative z-direction (downward) if $c > 0$ and the positive z-direction (upward) if $c < 0$; notice that these parabolas open the opposite direction from the $x = k$ cross sections. If $k = 0$ then the vertex is on the xy-plane (at the origin); otherwise the vertex is shifted vertically from the xy-plane.

The horizontal cross section $z = k$ yields

$$\frac{y^2}{b^2} - \frac{x^2}{a^2} = \frac{k}{c},$$

which is a hyperbola opening in the y-direction if k and c have the same sign, and opening in the x-direction if k and c have opposite signs. See Figure 1.86.

1.7.6 Hyperboloids

There are two types of hyperboloids, those with one sheet and those with two.

The variables can be swapped; $\frac{x^2}{a^2} - \frac{y^2}{b^2} + \frac{z^2}{c^2} = 1$ and $-\frac{x^2}{a^2} + \frac{y^2}{b^2} + \frac{z^2}{c^2} = 1$ are also hyperboloids of one sheet.

HYPERBOLOID OF ONE SHEET

A *hyperboloid of one sheet* has the form

$$\frac{x^2}{a^2} + \frac{y^2}{b^2} - \frac{z^2}{c^2} = 1,$$

where a, b, and c are constants.

The vertical cross section $x = k$ can be rearranged to

$$\frac{y^2}{b^2} - \frac{z^2}{c^2} = 1 - \frac{k^2}{a^2},$$

which is a hyperbola opening in the z-direction if $k > a$ and a hyperbola opening in the y-direction if $k < a$. If $k = a$, the equation reduces to $\frac{y^2}{b^2} = \frac{z^2}{c^2}$ resulting in a pair of intersecting lines. The analysis for the vertical cross section $y = k$ is similar.

The horizontal cross section $z = k$ can be rearranged to

$$\frac{x^2}{a^2} + \frac{y^2}{b^2} = 1 + \frac{k^2}{c^2},$$

which is an ellipse. See Figure 1.87.

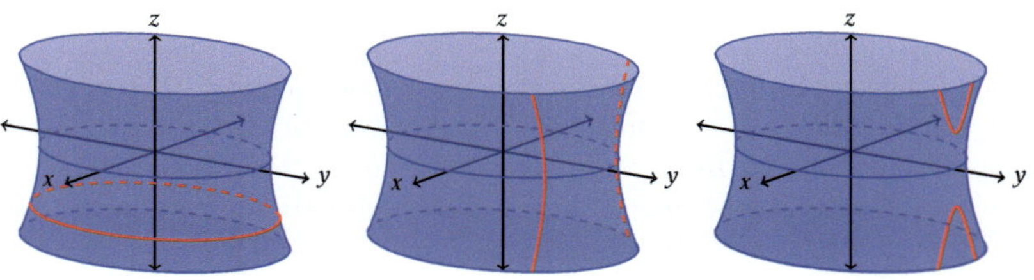

Figure 1.87 *The hyperboloid of one sheet $\frac{x^2}{9} + \frac{y^2}{16} - \frac{z^2}{25} = 1$, with (left) the trace $z = -2$, an ellipse; (middle) the trace $y = 3$, a hyperbola opening in the x-direction; and (right) the trace $y = 4.1$, a hyperbola opening in the z-direction. All traces named here are red. Only a portion of the hyperboloid is shown*

HYPERBOLOID OF TWO SHEETS

A *hyperboloid of two sheets* has the form

$$\frac{z^2}{c^2} - \frac{x^2}{a^2} - \frac{y^2}{b^2} = 1,$$

where a, b, and c are constants.

The variables can be swapped; $\frac{y^2}{b^2} - \frac{z^2}{c^2} - \frac{x^2}{a^2} = 1$ and $\frac{x^2}{a^2} - \frac{y^2}{b^2} - \frac{z^2}{c^2} = 1$ are also hyperboloids of two sheets.

The vertical cross section $x = k$ can be rearranged to

$$\frac{z^2}{c^2} - \frac{y^2}{b^2} = 1 + \frac{k^2}{a^2},$$

which is a hyperbola opening in the z-direction. The vertical cross section $y = k$ is also a hyperbola opening in the z-direction.

The horizontal cross section $z = k$ can be rearranged to

$$\frac{k^2}{c^2} - 1 = \frac{x^2}{a^2} + \frac{y^2}{b^2},$$

which is an ellipse as long as $k > c$. See Figure 1.88.

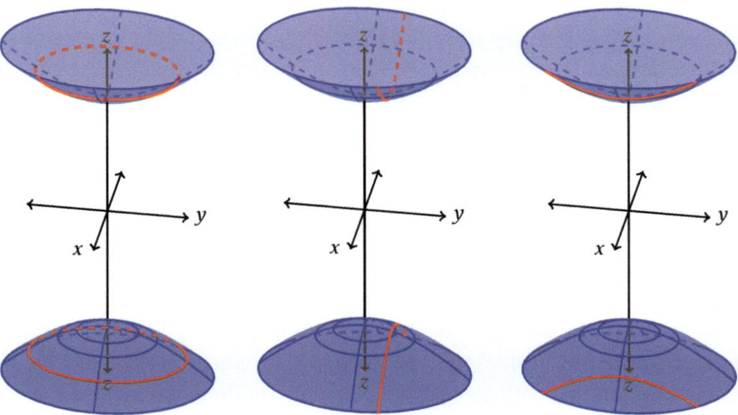

Figure 1.88 *The hyperboloid of two sheets $\frac{z^2}{25} - \frac{x^2}{9} - \frac{y^2}{16} = 1$, with (left) the traces $z = 6$ and $z = -6.2$, ellipses; (middle) the trace $y = 1$, a hyperbola opening in the z-direction; and (right) the trace $x = 2$, a hyperbola opening in the z-direction. All traces named here are red. Only a portion of the hyperboloid is shown*

1.7.7   Classifying quadric surfaces

Classifying quadric surfaces is similar to classifying conic sections, as in examples 49 and 50. Knowing which variables are squared and which are not, the signs of the coefficients on the squared terms, and the presence of a constant are enough to make the determination.

Notice that among the six standard-form equations of quadric surfaces, with one exception all squared terms are on one side of the equation (the left side), and any linear term for which there is no squared term is on the other side (the right side). The exception is for three squared terms with no constant term.

Example 53 *Identify the quadric surface.*

(a) $\dfrac{x^2}{9} + \dfrac{y^2}{7} = \dfrac{z^2}{4}$

(b) $x^2 + 3y^2 - 4z^2 - 5 = 0$

(c) $x + y^2 - z^2 = 0$

Solution

(a) For $\dfrac{x^2}{9} + \dfrac{y^2}{7} = \dfrac{z^2}{4}$, we see a quadratic equation in three variables $(x, y,$ and $z)$, with squared terms for all three variables (not a paraboloid) and no constant term (not a hyperboloid or ellipsoid), so it must be an elliptic cone. The equation is, in fact, in the standard form of an elliptic cone.

Comparing to the standard equation of an elliptic cone, we have $a^2 = 9$, $b^2 = 7$, and $c^2 = 4$.

(b) For $x^2 + 3y^2 - 4z^2 - 5 = 0$, we see a quadratic equation in three variables $(x, y,$ and $z)$, with squared terms for all three variables (not a paraboloid), and a constant term (not an elliptic cone). With all three squared terms on one side of the equation, the signs of the squared terms are not identical (not an ellipsoid). Therefore, the equation represents a hyperboloid, but we need to determine whether the hyperboloid has one or two sheets. To tell them apart, we need the squared terms on one side of the equation and the constant on the other side. Rearranging, we have

$$x^2 + 3y^2 - 4z^2 = 5.$$

Comparing to the standard form of a hyperboloid of one sheet after rearranging to $\dfrac{x^2}{5} + \dfrac{y^2}{5/3} - \dfrac{z^2}{5/4} = 1$, we have $a^2 = 5$, $b^2 = \dfrac{5}{3}$, and $c^2 = \dfrac{5}{4}$.

The constant on the right-hand side is positive (if it isn't, multiply both sides of the equation by -1.) Two of the squared terms' coefficients are positive and one is negative, so we have a hyperboloid of one sheet. The axis of the hyperboloid, which lies "inside" the hyperboloid's one sheet, is the z-axis because that is the variable with a negative coefficient on the squared term.

Before checking the number of positive and negative coefficients, make sure that the constant term is on one side of the equation by itself and that the constant is positive.

(c) For $x + y^2 - z^2 = 0$, we see a quadratic equation in three variables $(x, y,$ and $z)$, with squared terms for two of the three variables, so it is a paraboloid. To tell the difference between an elliptic and a hyperbolic paraboloid, notice that with the squared terms on one side of the equation, the coefficients of the squared terms of an elliptic paraboloid have the same sign, whereas for

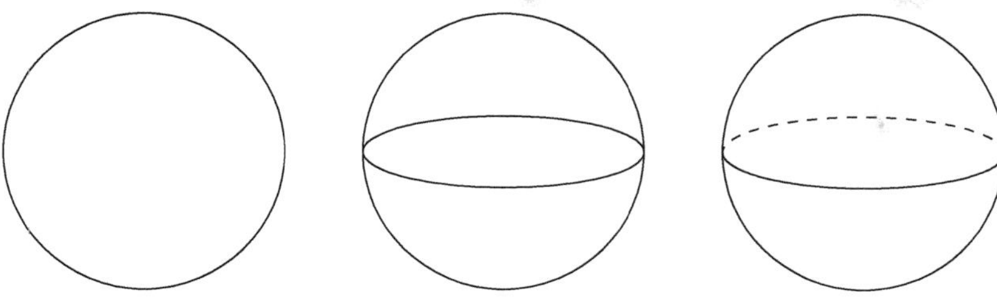

Figure 1.89 *The outline of a sphere (left); an equator is included to add depth (middle); dashing is used to improve perception (right)*

a hyperbolic paraboloid they have different signs. The coefficients on the squared terms here are 1 and -1, so this is a hyperbolic paraboloid. ∎

Reading Exercise 28 Identify the quadric surface $\frac{x^2}{9} + \frac{y^2}{4} = 1 - \frac{z^2}{16}$.

1.7.8 Sketching quadric surfaces

It's possible that you have sketched a sphere, which is an ellipsoid with $a = b = c$. A common method for drawing a sphere is to start by drawing a circle (Figure 1.89, left), followed by an "equator" of the sphere (Figure 1.89, middle). While that looks fine, drawing the portion in the back of the sphere dashed can help improve the picture (Figure 1.89, right). Sketching a quadric surface uses these same techniques.

The first step in sketching a quadric surface is to identify the surface. Then consulting Figures 1.80–1.88, as appropriate, determine cross sections and sketch traces to help picture the surface. Sketching the traces is the same skill demonstrated in examples 51 and 52. Finally, if desired, add shading. We will not add shading to the sketches produced in examples 54 and 55.

Example 54 *Sketch the quadric surface $z = \frac{y^2}{4} + \frac{x^2}{9}$.*

Solution The equation $z = \frac{y^2}{4} + \frac{x^2}{9}$ is a quadratic equation in three variables, with squared terms in just two of the variables; it is a paraboloid. The squared terms are on the same side of the equation and their signs match, so it is an elliptic paraboloid. Those signs match the sign of the linear term z (all are positive), so the paraboloid opens upward.

Consulting Figure 1.85, perhaps one or two horizontal traces and two vertical traces would be sufficient. For a horizontal trace, choose a value of z; it doesn't

These same techniques were used in the production of Figures 1.80–1.88. Remove the traces from Figures 1.80–1.83 and the shaded area looks like the interior of an ellipse. In fact, that's what the shading actually is and how it was produced–a shaded ellipse interior in two dimensions. The axes and traces are what add the perception of depth. Remove the axes, traces, and color differences from Figure 1.86 and it would be difficult to interpret if you didn't already know what it was supposed to look like.

Comparing to the standard form of a hyperbolic paraboloid after rearranging to $x = z^2 - y^2$, we have $a = 1$, $b^2 = 1$, and $c^2 = 1$.

The circle in Figure 1.89, left, does not give any sense of depth, of a third dimension. In general, by itself, the profile of a quadric surface is insufficient to give the impression of a three-dimensional object.

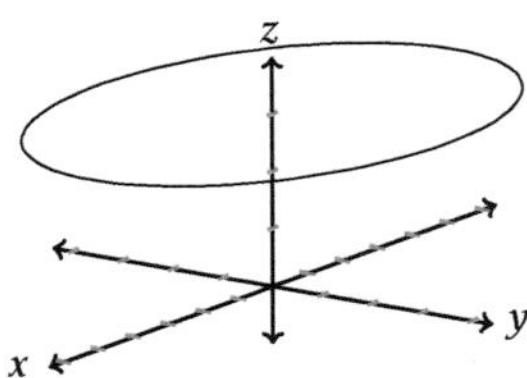

Figure 1.90 *The trace $z = 3$, $3 = \frac{y^2}{4} + \frac{x^2}{9}$*

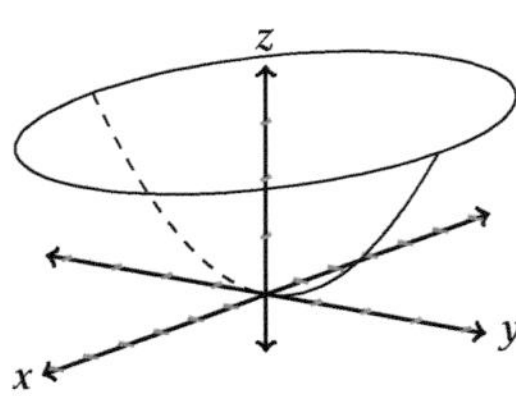

Figure 1.91 *The trace $x = 0$, $z = \frac{y^2}{4}$ is added*

Unsatisfied? If desired, shading can be added to make the paraboloid look more like Figure 1.85. If adding shading, however, notice that in this viewpoint, the bottom of the shading should appear to be "below" the vertex of the paraboloid (the point $(0, 0, 0)$), just as the trace $y = 0$ in the xz-plane looks like it dips below the y-axis, even though it does not. Because the paraboloid is tipped toward the viewer, the vertex is hidden from view. It is issues like these that make accurate sketches by hand challenging. It is for these reasons that sketching some traces, as demonstrated in example 54, and then using one's imagination is all that can be expected by hand. A CAS can be used if greater accuracy is needed.

much matter which one. We'll use $z = 3$. That trace is

$$3 = \frac{y^2}{4} + \frac{x^2}{9},$$

which rearranges to

$$\frac{y^2}{12} + \frac{x^2}{27} = 1,$$

which is an ellipse with $a = \sqrt{27}$, $b = \sqrt{12}$. The axes and the trace we have just calculated are sketched in Figure 1.90.

Next, we calculate the trace $x = 0$, which is in the yz-plane:

$$z = \frac{y^2}{4} + 0,$$

which is a parabola opening upward (the positive z-direction). For sketching this trace, just draw the parabola in the yz-plane through the origin and up to the previously sketched elliptical trace; see Figure 1.91. Draw what appears to be on the front side solid and on the back side dashed.

Finally, we calculate the trace $y = 0$, which is in the xz-plane:

$$z = 0 + \frac{x^2}{9},$$

which is a parabola opening upward. Again, we draw the parabola through the origin and up to the previously sketched elliptical trace, drawing what appears to be on the front side solid and on the back side dashed. The final result is in Figure 1.92.

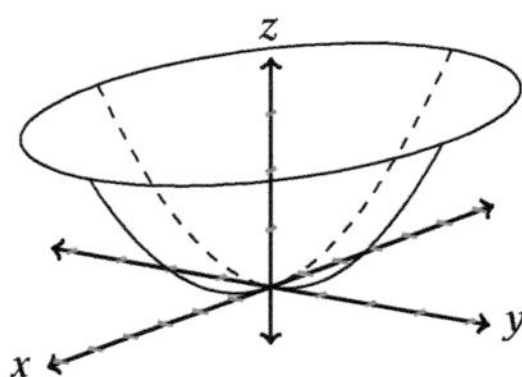

Figure 1.92 *Horizontal and vertical traces comprising the sketch of the elliptical paraboloid $z = \frac{y^2}{4} + \frac{x^2}{9}$*

∎

Example 55 *Sketch the quadric surface $z = \frac{y^2}{4} - \frac{x^2}{9}$.*

Solution The equation $z = \frac{y^2}{4} - \frac{x^2}{9}$ is a quadratic equation in three variables, with squared terms in just two of the variables; it is a paraboloid. The squared terms are on the same side of the equation, and their signs are different, so it is a hyperbolic paraboloid.

My advice for sketching a hyperbolic paraboloid by hand is based on the fact that vertical cross sections are either parabolas opening upward or parabolas opening downward. First, we draw either the trace $x = 0$ or $y = 0$, whichever is a parabola opening upward. Then, we draw a few traces using the other variable, traces of the form $x = k$ or $y = k$ as appropriate, that are parabolas opening downward with vertices on the previously sketched upward parabolic trace.

For this example, the trace $x = 0$ yields a parabola opening upward:

$$z = \frac{y^2}{4} - 0.$$

This parabolic trace, along with the coordinate axes, is sketched in Figure 1.93.

Traces of the form $y = k$ have equation

$$z = -\frac{x^2}{9} + \frac{k^2}{4},$$

which are parabolas opening downward, with vertex when $x = 0$. The traces $y = -4, y = -3, ..., y = 4$ are included in Figure 1.94. We have drawn these traces solid when $x > 0$ and dashed when $x < 0$, but after drawing the traces it is apparent that portions of dashed traces would be visible and portions of solid traces would not. This is common for a hyperbolic paraboloid, and the dashing can be corrected if desired. However, even as pictured in Figure 1.94, the quadric surface can be visualized reasonably effectively. Use a CAS if greater accuracy is required.

∎

1.7.9 Cylinders

Look again at definition 17. The graph in $\mathbf{R}^3$ of the quadratic equation $x^2 + y^2 = 9$ meets the definition, even though the variable z does not appear in the equation.

Compare the equations in examples 54 and 55. Only a sign change has been made.

Ans. to reading exercise 28:
   ellipsoid (Hint: rewrite the equation as $\frac{x^2}{9} + \frac{y^2}{4} + \frac{z^2}{16} = 1$ before classifying.)

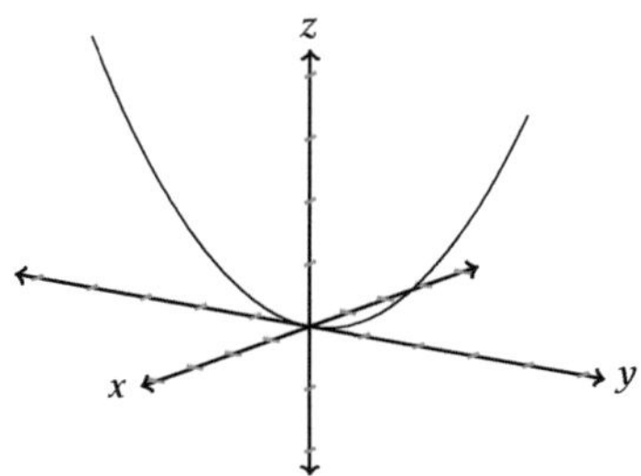

Figure 1.93 *The trace* $x = 0$, $z = \frac{y^2}{4}$

Notice that for a hyperbolic paraboloid we did not include horizontal traces. They can be added if desired, but are relatively difficult. The lack of emphasizing horizontal traces is what makes Figure 1.94 appear different from Figure 1.86, even though the shapes are the same.

One easier addition to make is to add two more traces that are parabolas opening upward, at the edges of the drawn parabolic traces that open downward.

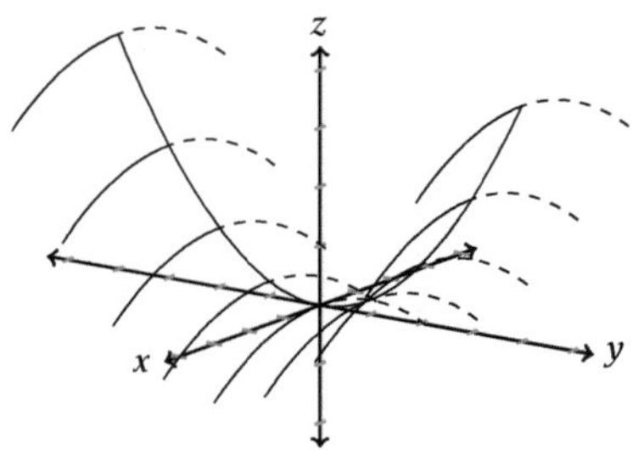

Figure 1.94 *Vertical traces comprising the sketch of the hyperbolic paraboloid* $z = \frac{y^2}{4} - \frac{x^2}{9}$

The definition requires that at least one of A, B, C, D, E, and F is nonzero. For $x^2 + y^2 - 9 = 0$, $A = 1$ and $B = 1$. It does not matter than $C = D = E = F = 0$.

While we recognize that in two dimensions the graph of $x^2 + y^2 = 9$ is a circle centered at the origin with radius 3, in three dimensions the set of points (x, y, z) that satisfy the equation includes points with any z-coordinate as long as the x- and y-coordinates satisfy $x^2 + y^2 = 9$. The result is an infinite-height circular cylinder, as partially pictured in Figure 1.95, left, along with a horizontal cross section.

What would a vertical cross section of the cylinder look like? Either one or two vertical lines. The vertical cross section $y = 1.3077$ is pictured in Figure 1.95, right.

The circular cylinder of Figure 1.95 can be pictured as being composed of horizontal circles translated vertically, as if the red circle is sliding up and down the cylinder. But it can also be considered to be composed of vertical lines sliding around a circle. It is the latter idea that is key to more general cylinders. A *cylinder* is composed of parallel lines passing through a plane curve. That curve may be a circle, as in Figure 1.95, or it could be a parabola, an ellipse, or something else.

The plane curve through which the lines pass does not need to be in a plane parallel to any of the xy- or xz- or yz-planes.

Example 56 *In $\mathbf{R}^3$, sketch the graph of the cylinder $y = e^x$.*

This curve is not a quadric surface; it does not satisfy definition 17.

Solution Start by drawing the coordinate axes and graphing the curve $y = e^x$ in the plane $z = 0$, as in Figure 1.96, left. Then draw line segments of the same length parallel to the z-axis (the missing variable is z) along the curve, as in Figure 1.96, right. $\blacksquare$

The graph of a quadric surface for which only two variables are present in the equation is a cylinder. So is the graph of a quadric surface whose equation in three variables contains only one squared term, such as $z^2 + 3x - 2y = 4$. Some of these ideas are explored in the exercises.

For the equation $z^2 + 3x - 2y = 4$, consider cross sections of the form $z = k$. These cross sections can be rearranged to $2y = 3x - 4 + k^2$, which is a line with "slope" $\frac{3}{2}$ in the horizontal plane $z = k$. All such lines are parallel, forming a cylinder.

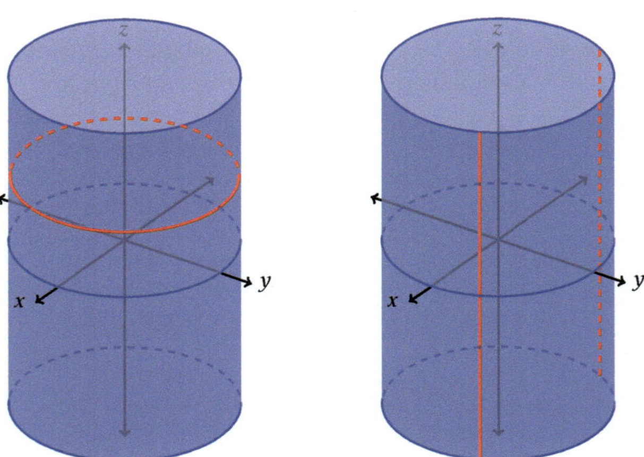

Figure 1.95 *The graph in $\mathbf{R}^3$ of $x^2 + y^2 = 9$, along with (left) the trace $z = 2$ and (right) the trace $y = 1.3077$. Only a portion of the cylinder is shown*

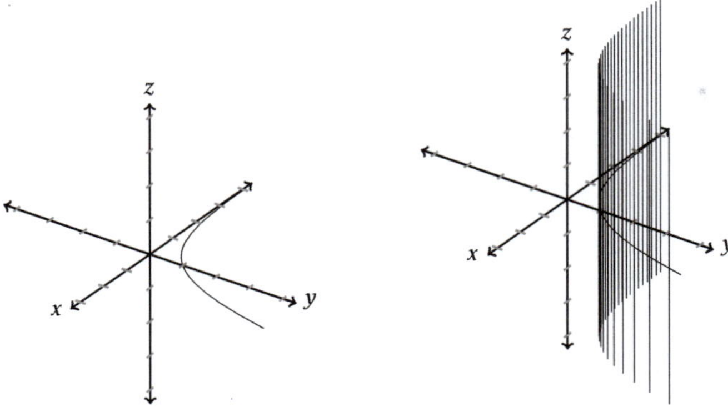

Figure 1.96 *(Left) the graph of the curve $z = 0$, $y = e^x$; (right) the graph of the cylinder $y = e^x$*

EXERCISES 1.7

1–6. Rapid response: the graph of a quadric surface has been produced by a CAS. Identify the quadric surface.

These graphs were produced using Mathematica.

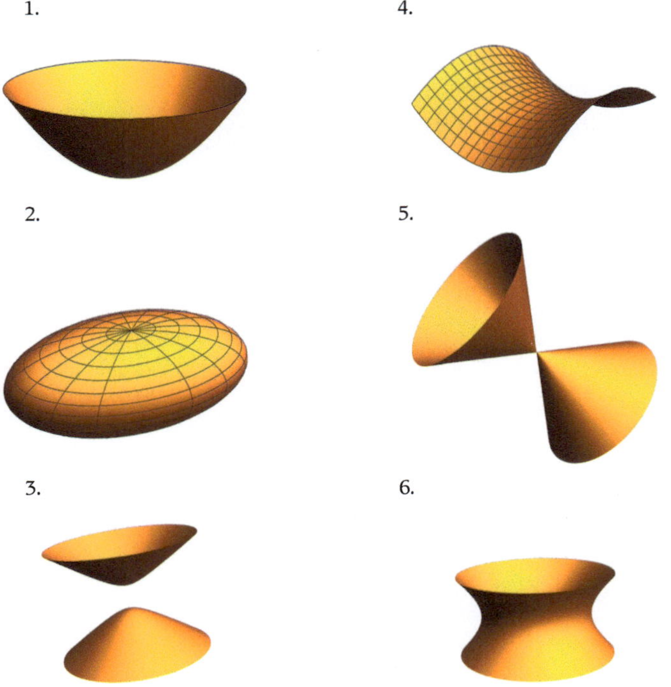

1.

2.

3.

4.

5.

6.

7–12. Rapid response: identify the quadric surface by its equation.

7. $\dfrac{x^2}{9} + y^2 - \dfrac{z^2}{25} = 0$

8. $\dfrac{x^2}{9} + y^2 - \dfrac{z^2}{25} = 5$

9. $\dfrac{x^2}{9} + y^2 - \dfrac{z^2}{25} = -5$

10. $\dfrac{x^2}{9} + y^2 + \dfrac{z^2}{25} = 5$

11. $\dfrac{x^2}{9} + y^2 + \dfrac{z^2}{25} = -5$

12. $\dfrac{x}{9} + y^2 - \dfrac{z^2}{25} = -5$

13–20. Graph the conic section in $\mathbf{R}^3$.

13. $z = 0, x^2 + y^2 = 16$

14. $z = 0, y = \dfrac{x^2}{2}$

15. $y = 0, z = 2x^2$

16. $x = 0, \dfrac{y^2}{9} + \dfrac{z^2}{4} = 1$

17. $z = -1, \dfrac{x^2}{4} - \dfrac{y^2}{9} = 1$

18. $z = 3, \dfrac{x^2}{4} - \dfrac{y^2}{9} = -1$

19. $x = 2, \dfrac{y^2}{9} + \dfrac{z^2}{4} = 16$

20. $y = -1, (x - 1)^2 + (z - 2)^2 = 4$

21–34. (a) Identify the quadric surface and (b) sketch its graph.

21. $\dfrac{x^2}{25} + \dfrac{y^2}{36} + \dfrac{z^2}{9} = 1$

22. $z = x^2 + y^2$

23. $\dfrac{x^2}{25} + \dfrac{y^2}{36} = \dfrac{z^2}{9}$

24. $\dfrac{z^2}{9} + \dfrac{y^2}{4} + \dfrac{x^2}{16} = 1$

25. $z = \dfrac{y^2}{9} - x^2$

26. $\dfrac{z^2}{25} + \dfrac{y^2}{36} = \dfrac{x^2}{9}$

27. $9x = \dfrac{y^2}{4} + \dfrac{z^2}{4}$

28. $z = \dfrac{x^2}{25} - \dfrac{y^2}{4}$

29. $\dfrac{y^2}{4} - \dfrac{z^2}{4} - \dfrac{x^2}{9} = 1$

30. $\dfrac{x^2}{25} - \dfrac{y^2}{9} + z^2 = 1$

31. $\dfrac{x^2}{8} + \dfrac{y^2}{5} - \dfrac{z^2}{11} = 1$

32. $\dfrac{x^2}{8} + \dfrac{y^2}{5} - \dfrac{z^2}{11} = -1$

33. $(x + 2)^2 + (y - 1)^2 + (z + 2)^2 = 16$

34. $z = (x - 1)^2 + (y - 3)^2 - 2$

35–42. (a) Use a CAS to sketch the graph of the quadric surface and (b) identify the quadric surface from the graph.

35. $z - xy = 5$
36. $(x + y)^2 - 3x + z^2 = 9$
37. $xy + 3xz - 7yz = 1$
38. $2x^2 + 3y^2 - 4z^2 + 5xy + 6x + 7y + (1/8)z = 10$
39. $x + y^2 + z^2 - yz = 9$
40. $(x + 2y)^2 = 4z^2$
41. $(x + 2y)^2 + (x + z + 1)^2 = 9z^2$
42. $x^2 + y^2 + z^2 + xy + xz + yz + x + y + z = 1$

Each of these quadric surfaces involve rotation of axes, which was not discussed in the narrative. However, a CAS can perform the necessary calculations seamlessly.

Desmos, which is available for free, includes a 3D-grapher that can produce any of these quadric surfaces, even though it is technically not a CAS.

43–46. (a) Sketch the cylinder. (b) Is this a quadric surface?

43. $\dfrac{x^2}{4} + y^2 = 1$
44. $z = y^2$
45. $z = \sqrt{y}$
46. $z = \ln(-x)$

47. Complete the square to find the center of the ellipsoid: $x^2 + 3y^2 + z^2 + 2x - 12y + 8z + 1 = 0$.

48. Complete the square to find the vertex of the paraboloid: $2x^2 + y^2 + 6x - 10y - 5z + 1 = 0$.

49. Investigate the cross sections of a Pringles potato crisp. Do these cross sections indicate that a Pringles crisp might be in the shape of a quadric surface? If so, which one?

Pringles are made in the city in which the author lives.

50. Investigate the cross sections of a flashlight reflector. Do these cross sections indicate that a flashlight reflector might be in the shape of a quadric surface? If so, which one?

A flashlight reflector is the mirrored surface that reflects light coming from the light source into a beam of light.

51. For the hyperboloid of one sheet $\dfrac{x^2}{a^2} + \dfrac{y^2}{b^2} - \dfrac{z^2}{c^2} = 1$, explain why the cross sections of form $y = k$ are hyperbolas opening in different directions if $k > b$ or $k < b$.

52. For the elliptic paraboloid $\dfrac{x^2}{a^2} + \dfrac{y^2}{b^2} = \dfrac{z}{c}$, show that the vertical cross section $y = k$ is a parabola. Under what conditions does the parabola open upward (in the positive z-direction)?

53. Why was there not a margin note next to the box containing the standard form of an ellipsoid to explain that swapping the variables still yields an ellipsoid?

54. (a) Consulting Figures 1.80–1.88, which of these quadric surfaces have a horizontal cross section consisting of exactly one point? (b) For each answer in part (a), under what conditions does this happen?

55. Sketch the graph of an ellipsoid in which one of a, b, or c is infinitesimal. What would such a surface be (approximately, or more accurately "render to")?

"What would such a surface be" means to use a different geometric description of the surface.

By "infinite" we mean an infinite hyperreal number.

56. Sketch the graph of an ellipsoid in which one of a, b, or c is infinite. What would such a surface be (approximately)?

57. Sketch the graph of a hyperboloid of one sheet $\dfrac{x^2}{a^2} + \dfrac{y^2}{b^2} - \dfrac{z^2}{c^2} = 1$ for which c is infinite. What would such a surface be (approximately)?

58. Sketch the graph of an elliptic cone $\dfrac{x^2}{a^2} + \dfrac{y^2}{b^2} = \dfrac{z^2}{c^2}$ for which c is infinitesimal. What would such a surface be (approximately, or more accurately "render to")?

59. Explore the graph of the quadric surface $y = x + z^2$ using the following steps.

    (a) Determine the trace $z = 0$ and describe it geometrically.

    (b) Determine the traces $z = \pm 1$ and $z = \pm 2$ and describe them geometrically.

    (c) Can the results of (a) and (b) be considered to be parallel? If so, is this quadric surface a cylinder?

    (d) Sketch the traces from parts (a) and (b) to visualize the surface.

60. Explore the graph of the quadric surface $x^2 + 2y + 5z = 0$ using the following steps.

    (a) Determine the trace $x = 0$ and describe it geometrically.

    (b) Determine the traces $x = \pm 1$ and $x = \pm 2$ and describe them geometrically.

    (c) Can the results of (a) and (b) be considered to be parallel? If so, is this quadric surface a cylinder?

    (d) Sketch the traces from parts (a) and (b) to visualize the surface.

61–62. Just as lines and planes have parametric forms of their equations, so do quadric surfaces. Look up parametric equations for the appropriate quadric surface and use a CAS to graph it.

61. $z = \dfrac{x^2}{16} + \dfrac{y^2}{25}$

62. $\dfrac{x^2}{7} + \dfrac{y^2}{4} + \dfrac{z^2}{9} = 1$

Curves in Three Dimensions

2.1 Vector-Valued Functions

Now that we have studied vectors and the geometry of space, the stage is set for three-dimensional curves. First, we briefly review functions and their graphical representations in two dimensions.

2.1.1 Review: functions

A function has the form

$$f(\text{input}) = \text{output},$$

where no more than one output is allowed per input. Typically, we have used numerical inputs and outputs, such as depicted in the "function machine" diagram in Figures 2.1 and 2.2. Using a generic input x, its associated output is denoted as $f(x)$, where f is the name of the function. An algebraic representation of a function is a rule for calculating the output, such as $f(x) = x^2$. The set of all possible inputs is called the *domain* of the function, and the set of all possible outputs is called the *range*. The outputs of a function are also called the *values* of the function, and a function whose outputs are real numbers is called a *real-valued function*.

Graphically in two dimensions, this restriction on multiple outputs is detected by the *vertical line test*.

A numerical input can be either a real number or a hyperreal number. The output can be rendered as a real result if needed.

A function can be depicted as a curve in two dimensions with the inputs measured along the horizontal x-axis and the outputs along the vertical y-axis, as in Figure 2.3.

You may have also studied parametrically defined curves, where the x- and y-coordinates of a point on the curve are described by two separate functions, $x(t)$ and $y(t)$, with the *parameter t* often representing time. In a parametric graph, the vertical line test does not apply and a curve can cross itself, as in Figure 2.4. A parametrically defined curve can wander anywhere in the plane, like an insect

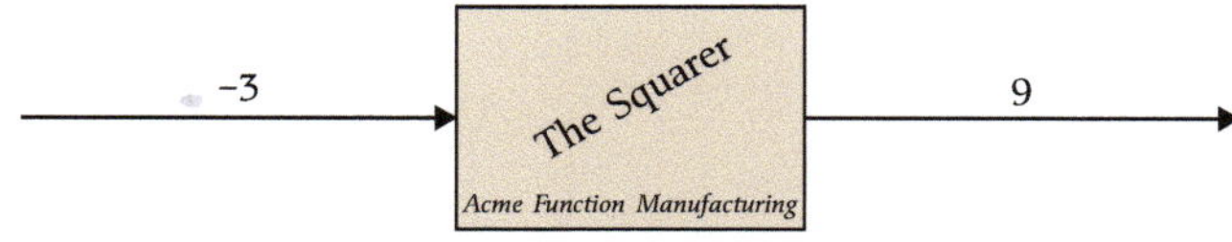

Figure 2.1 *The function $f(x) = x^2$, aka "The Squarer," shown with input -3 and output 9*

Multivariable Calculus Set Free. Charles Bryan Dawson, Oxford University Press.
© C. Bryan Dawson (2026). DOI: 10.1093/oso/9780198984269.003.0003

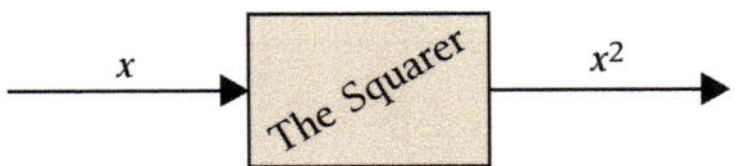

Figure 2.2 *"The Squarer" shown with numerical input x and numerical output x^2*

In a vector-valued function the output can be a vector of any dimension. We shall concentrate on three-dimensional vectors in this chapter.

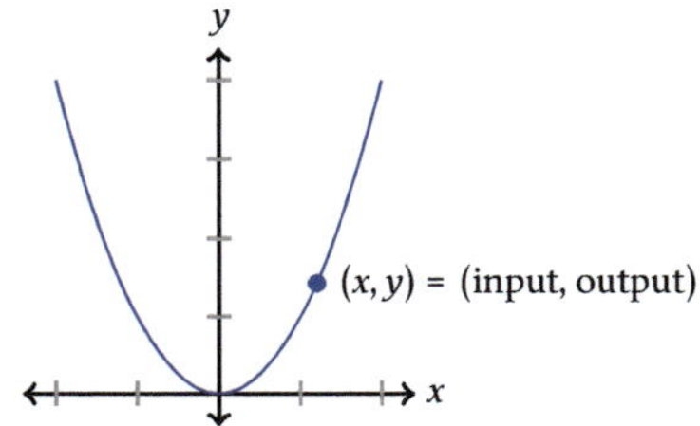

Figure 2.3 *The function $f(x) = x^2$ depicted graphically in two dimensions*

crawling on a floor. But what if the insect can fly, moving in three dimensions? Is it as simple as stating a third coordinate function? Yes–but with vectors at our disposal, we choose a slightly different method.

2.1.2 Vector-valued functions

The idea of a *vector-valued function* is that instead of spitting out a real number, the function outputs a vector, as depicted in the function machine diagram of Figure 2.5. Algebraic representations of this function include

$$\mathbf{r}(t) = \langle t^2,\, t + 1,\, e^t \rangle = t^2\mathbf{i} + (t + 1)\mathbf{j} + e^t\mathbf{k} = f(t)\mathbf{i} + g(t)\mathbf{j} + h(t)\mathbf{k},$$

where the real-valued functions $f(t) = t^2$, $g(t) = t + 1$, and $h(t) = e^t$ are called the *component functions* of $\mathbf{r}$.

With one input dimension and three output dimensions, a graph of the type depicted in Figure 2.3 would require a total of four dimensions, which presents a problem in our three-dimensional world. Therefore, a parametric graph of the type depicted in Figure 2.4 seems more practical. An example of a three-dimensional parametric curve (not the same function $\mathbf{r}$ as in Figure 2.5) is pictured in Figure 2.6. The points on the curve can be interpreted as the position at time t. Of course, all the difficulties of interpreting the three-dimensional location of points in a two-dimensional diagram apply.

2.1.3 Domain of a vector-valued function

When asked to find the domain of a real-valued function $y = f(x)$, the task is to determine which inputs make the output defined. For instance, the domain of $f(x) = \frac{x}{x+2}$ is all real numbers except $x = -2$, and the domain of $f(x) = \sqrt{x - 2}$ is $x \geq 2$.

The domain of a vector-valued function $\mathbf{r}(t) = \langle f(t), g(t), h(t) \rangle$ is the intersection of the domains of its component functions f, g, and h; in order for $\mathbf{r}$ to be defined at a given value of t, all its component functions must be defined at t. The same principles apply. For instance, we exclude values of t that cause division by zero, the quantity underneath a square root must be nonnegative, and the quantity inside a logarithm must be positive.

The *time stamps* indicate the value of the parameter t corresponding to the point.

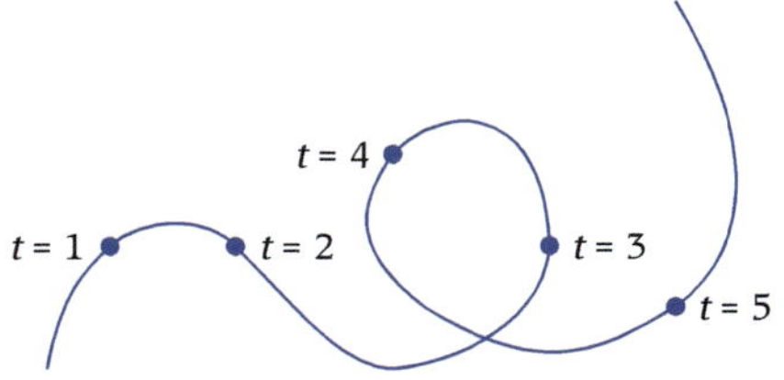

Figure 2.4 *A parametrically defined curve, along with a few time stamps*

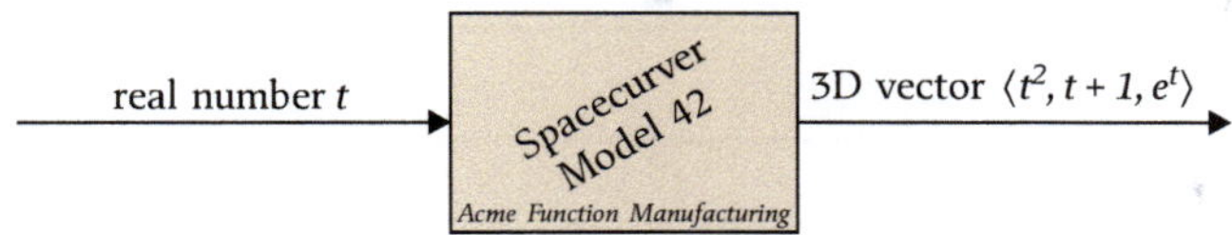

Figure 2.5 *A vector-valued function*

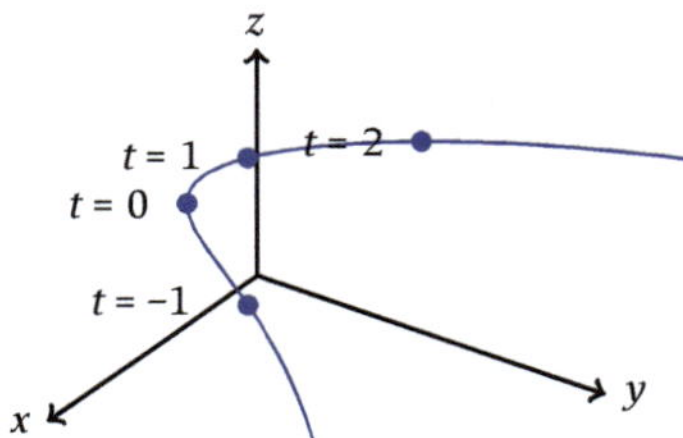

Figure 2.6 *The graph of a vector-valued function, along with a few time stamps*

Example 1 *Find the domain of* $\mathbf{r}(t) = \langle \sqrt{t+1}, \frac{1}{t-1}, \ln(4-t) \rangle$.

Solution We begin by finding the domains of the component functions. For the domain of $f(t) = \sqrt{t+1}$, we recognize that the quantity under the square root must be nonnegative:

$$t + 1 \geq 0$$

$$t \geq -1.$$

For the domain of $g(t) = \frac{1}{t-1}$, to avoid division by zero the domain is $t \neq 1$.

For the domain of $h(t) = \ln(4-t)$, the quantity inside the logarithm must be positive:

$$4 - t > 0$$

$$-t > -4$$

$$t < 4.$$

To find the intersection of these domains, it can be helpful to graph the domains of f, g, and h on number lines that are aligned vertically, as in Figure 2.7. Then, finding the intersection is as simple as looking vertically for numbers in all three domains. Excluding values less than -1, excluding 1, and excluding values greater than or equal to 4 gives our solution. The domain, expressed in interval notation, is $[-1, 1) \cup (1, 4)$. ∎

Reading Exercise 1 Find the domain of $\mathbf{r}(t) = \langle \frac{1}{t^2}, \ln(t+9), \frac{3}{t-2} \rangle$.

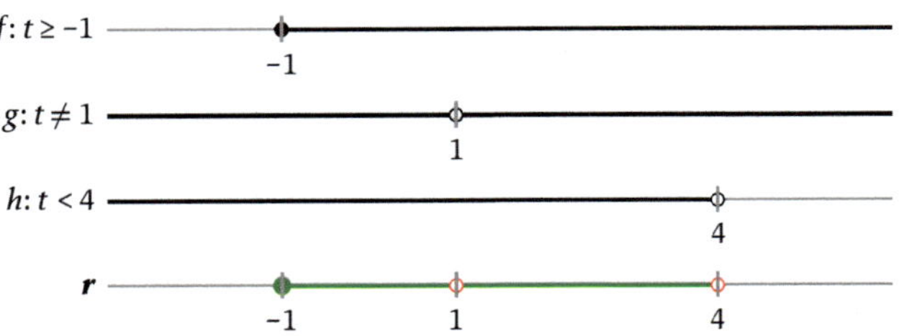

Figure 2.7 *Picturing the domain (green) of* $\mathbf{r}(t) = \langle \sqrt{t+1}, \frac{1}{t-1}, \ln(4-t) \rangle$

2.1.4   Graphing vector-valued functions

We shall investigate three strategies for graphing curves in three dimensions, also known as *spacecurves*.

Strategy 1: If the shape of a curve is known, plot enough points to determine its precise location and sketch the curve through those points.

Example 2 *Graph the curve* $\mathbf{r}(t) = \langle 1+t, 2+4t, -1+2t \rangle$.

Solution Notice that $\mathbf{r}(t)$ can be rewritten as

$$\mathbf{r}(t) = \langle 1, 2, -1 \rangle + t\langle 1, 4, 2 \rangle,$$

which is the vector form of the equation of a line containing the point $(1, 2, -1)$ having direction vector $\langle 1, 4, 2 \rangle$. Knowing that the graph is a line, and recalling from geometry that two points determine a line, we only need to plot two points and draw the line through them. Choosing two values of the parameter t, we calculate the coordinates of the corresponding points:

The point corresponding to $t = 0$ was already known.

t	x	y	z
0	1	2	−1
1	2	6	1

The two points $(1, 2, -1)$ and $(2, 6, 1)$, as well as the line through them, are plotted in Figure 2.8. ∎

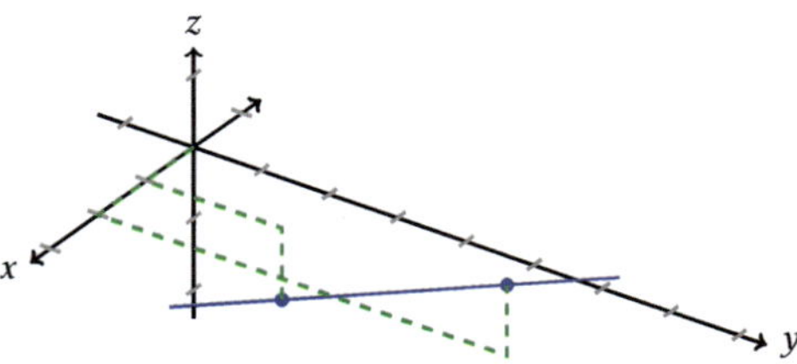

Figure 2.8 *The plotted points* $(1, 2, -1)$ *and* $(2, 6, 1)$ *on the line* $\mathbf{r}(t) = \langle 1+t, 2+4t, -1+2t \rangle$ *(blue). Paths to the plotted points (green) can aid in plotting the points as well as add perspective to the drawing*

Strategy 2: Graph the projections of the curve onto each of the xy-, xz-, and yz-planes in order to visualize the curve.

Example 3 *Graph* $\mathbf{r}(t) = \langle \cos t, \sin t, \sin t \rangle$.

Solution The projection of the curve onto the xy-plane is the "view from above." We plot the curve $x = \cos t$, $y = \sin t$. You might recognize this as a parameterization of the unit circle, which can be quickly drawn; see Figure 2.9.

The projection of the curve onto the xz-plane is the view looking straight down the y-axis toward the origin. We take the x- and z-component functions and graph the two-dimensional parametric curve $x = \cos t$, $z = \sin t$, ignoring the y-coordinate. Once again, we recognize a parameterization of the unit circle. See Figure 2.10, where the projection is on the left, compared to the view from down the y-axis on the right.

Finally, we consider the projection of the curve onto the yz-plane, which is looking straight down the x-axis toward the origin. We take the y- and z-component functions and graph the two-dimensional parametric curve $y = \sin t$, $z = \sin t$. This time, the y- and z-coordinates are equal, and the graph is along the line $z = y$. However, because $-1 \le \sin t \le 1$ for every value of the parameter t, it is only the portion of the line with y- and z-coordinates between -1 and 1 that is graphed; see Figure 2.11. ∎

The projection of $\mathbf{r}(t) = \langle f(t), g(t), h(t) \rangle$ onto the xy-plane is the graph of the parametric equation $x = f(t)$, $y = g(t)$. The z-coordinate is ignored. Think of looking at a satellite photo where you cannot determine the height (z-coordinate) of anything, but only the ground location of items, that is, their position on the earth's surface.

Ans. to reading exercise 1:
$$(-9, 0) \cup (0, 2) \cup (2, \infty)$$

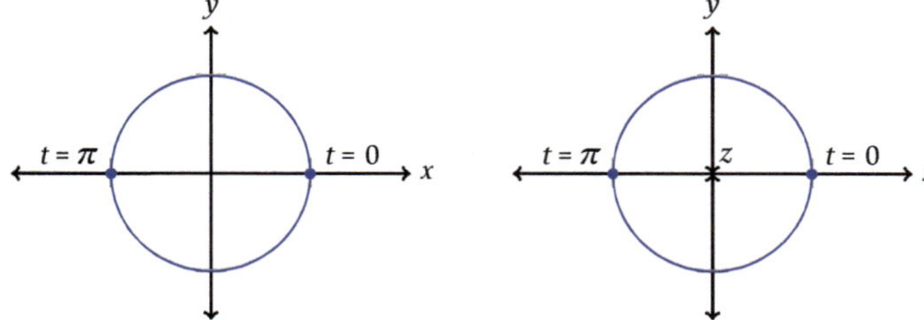

Figure 2.9 *(Left) the curve* $x = \cos t$, $y = \sin t$ *in the xy-plane; (right) the curve* $\mathbf{r}(t) = \langle \cos t, \sin t, \sin t \rangle$, *viewed from above*

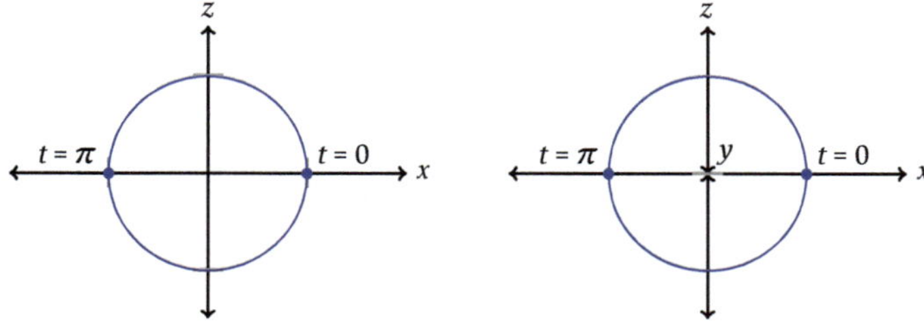

Figure 2.10 *(Left) the curve* $x = \cos t$, $z = \sin t$ *in the xz-plane; (right) the curve* $\mathbf{r}(t) = \langle \cos t, \sin t, \sin t \rangle$, *viewed straight down the y-axis*

The graph in the right-hand side of Figures 2.9, 2.10, and 2.11 was produced once, using $\mathbf{r}(t) = \langle \cos t, \sin t, \sin t \rangle$. The only difference between the three figures is the viewpoint. The same is true of Figure 2.12.

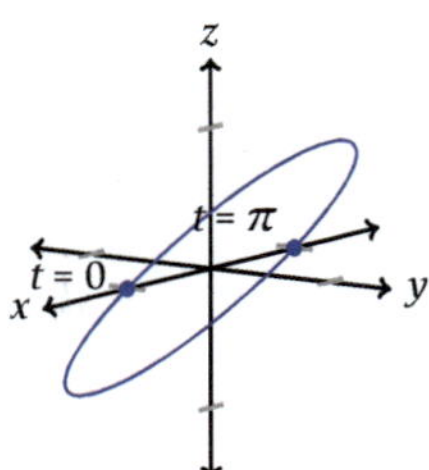

Figure 2.12 *The curve* $\mathbf{r}(t) = \langle \cos t, \sin t, \sin t \rangle$, *viewed from a point in the first octant*

In a t-joint where one pipe stops and does not continue across the other pipe, only half the ellipse will be visible. You may be able to see such a joint under a sink or in an attic or crawl space. Due to design differences, not all t-joints will show the actual intersection of the cylinders.

Desmos's free 3D-grapher can also produce graphs of spacecurves. Tip: use parentheses in place of angle brackets.

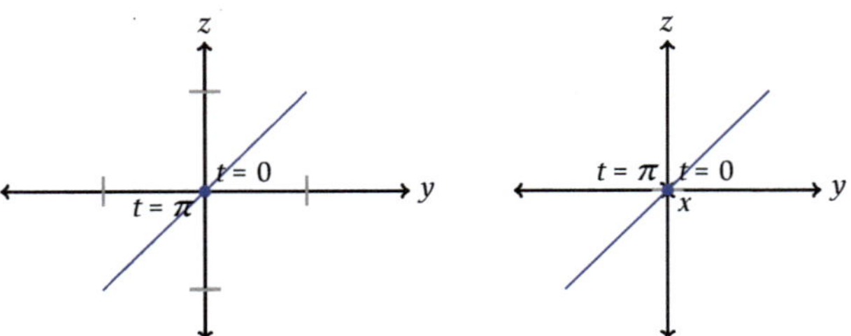

Figure 2.11 *(Left) the curve* $y = \sin t$, $z = \sin t$ *in the yz-plane; (right) the curve* $\mathbf{r}(t) = \langle \cos t, \sin t, \sin t \rangle$, *viewed straight down the x-axis*

Have you figured out what the graph of the curve in example 3 looks like? If not, perhaps looking at the curve from yet another viewpoint can help; see Figure 2.12. The curve is an ellipse.

Look again at Figure 2.9; the curve lies on the cylinder $x^2 + y^2 = 1$. And from Figure 2.10, the curve lies on the cylinder $x^2 + z^2 = 1$. Think of two cylindrical pipes meeting at a 90° angle, or go to your local hardware store and look at such a junction. In order to lie on both cylinders, the curve goes across the joint diagonally and is an ellipse, not a circle.

Perhaps example 3 has convinced you that our next strategy is preferable. Strategy 3: use technology.

2.1.5 Using technology to explore multiple viewpoints of a spacecurve

Many computer algebra systems (CAS) will draw a box around the graph of a spacecurve rather than draw coordinate axes. For instance, the curve of example 3 might be presented as in Figure 2.13. The CAS then often allows the user to spin

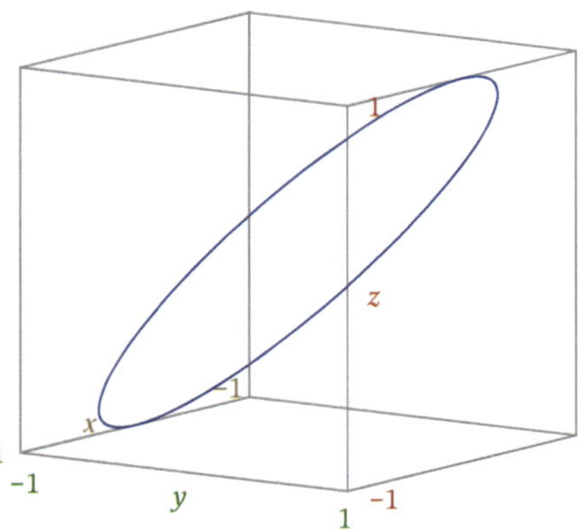

Figure 2.13 *The curve* $\mathbf{r}(t) = \langle \cos t, \sin t, \sin t \rangle$

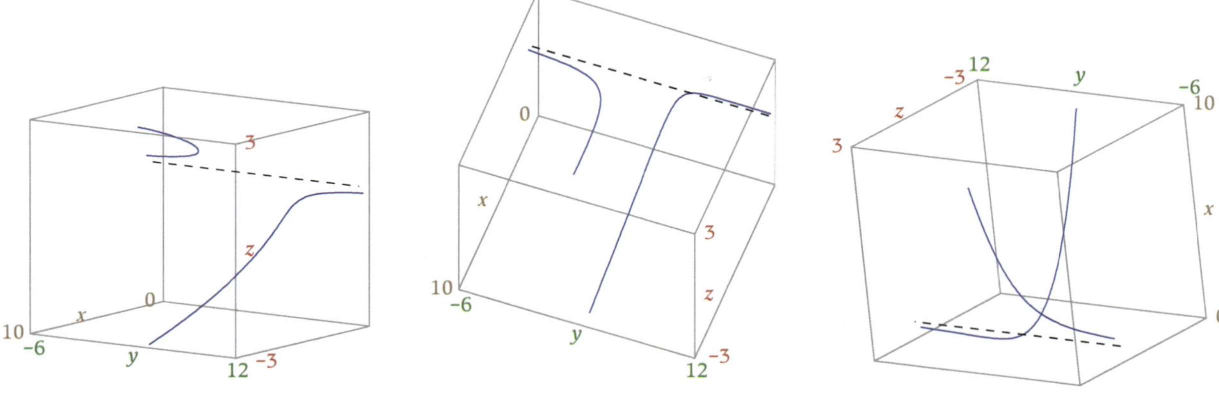

Figure 2.14 *Three viewpoints of the curve* $\mathbf{r}(t) = \left\langle t^2 - \frac{t}{t^2+4}, \frac{3t-5}{t-1}, t \right\rangle$ *(blue) along with the asymptote* $x = \frac{4}{5}, z = 1$ *(dashed)*

the box to see the curve from a different viewpoint, using only the mouse. To mimic this ability, we shall sometimes show multiple viewpoints in order to help the reader understand the curve.

Example 4 *Use technology to produce a picture of the curve* $\mathbf{r}(t) = \left\langle t^2 - \frac{t}{t^2+4}, \frac{3t-5}{t-1}, t \right\rangle.$

Solution The curve $\mathbf{r}(t) = \left\langle t^2 - \frac{t}{t^2+4}, \frac{3t-5}{t-1}, t \right\rangle$ for $t = -3$ to $t = 3$ (actually $t = -3$ to $t = \frac{7}{9}$ and $t = \frac{11}{9}$ to $t = 3$) is graphed in Figure 2.14, with three different viewpoints shown. ∎

In addition to the blue curve shown in Figure 2.14, an asymptote has been included. Calculating asymptotes will be discussed later in this section. For now, notice that $t = 1$ causes division by zero, which in two dimensions indicates the possibility of an asymptote. It is for this reason that the graph shown in Figure 2.14 does not show points corresponding to any values of t between $\frac{7}{9}$ and $\frac{11}{9}$, close to $t = 1$. Many CAS will handle this issue automatically, allowing the user to ask for the graph from $t = -3$ to $t = 3$, not showing the curve for values of t close to 1. Notice also that in three dimensions, there is a much greater variety of ways that a curve can approach an asymptote than there are in two dimensions.

Figure 2.14 also appears to indicate that points on the curve have larger x-coordinates as the curve gets further from the asymptote. The same is true for z-coordinates, although that may not be as readily apparent from these graphs. Are there (linear) asymptotes as $t \to \infty$ or $t \to -\infty$ as well? Such a question is not always easy to answer by exploring the graph. Using limits (later in this section) can be more illuminating.

Looking for features of a curve is both more interesting and more complicated in three dimensions than in two. The purpose of graphing a curve using multiple viewpoints, or spinning the picture as is usually possible using a CAS, is to help visualize features of the curve. Using various values of the parameter t to view different portions of the curve can also be helpful. The goal is understanding, and exploring the curve visually is a valuable tool.

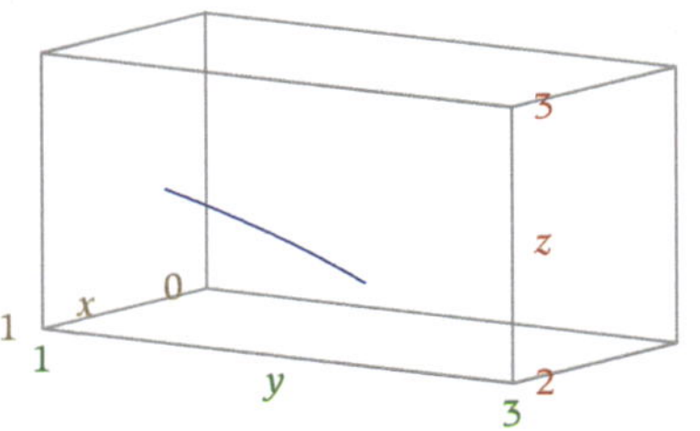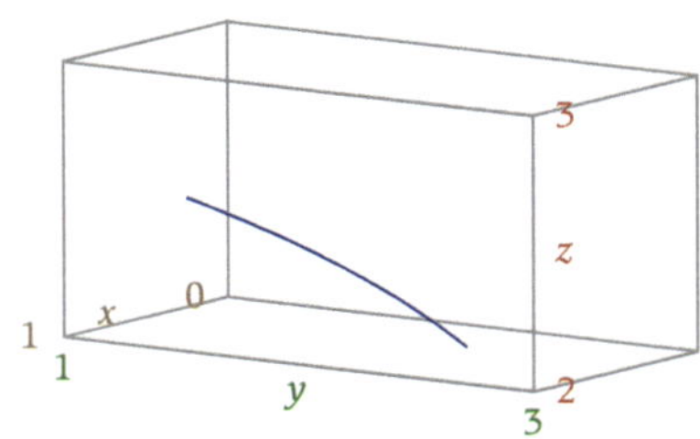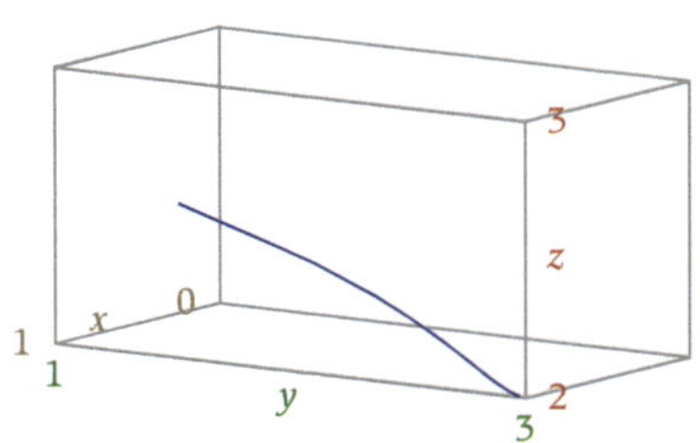

Figure 2.15 *The curve* $\mathbf{r}(t) = \left\langle \frac{t^2-t}{t^2+4}, \frac{3t-5}{t-1}, \frac{\sqrt{4t^2+7}}{t} \right\rangle$ *graphed (left) from* $t = 2$ *to* $t = 3$, *(middle) from* $t = 2$ *to* $t = 5$, *and (right) from* $t = 2$ *to* $t = 50$

Example 5 *Use technology to produce a picture of the curve* $\mathbf{r}(t) = \left\langle \frac{t^2-t}{t^2+4}, \frac{3t-5}{t-1}, \frac{\sqrt{4t^2+7}}{t} \right\rangle$.

Although the boxes in Figure 2.15 all have the same dimension, a CAS might use different dimensions for these three pictures unless the box is specified by the user.

As previously stated, calculating asymptotes is discussed later in this section.

Solution This curve features division by zero when $t = 0$ and when $t = 1$, so we first explore the curve for $t > 2$. Each of the three graphs in Figure 2.15 begins at $t = 2$, which is at the left side of each box. The curve is graphed to increasingly larger values of t using the same viewpoint, allowing us to see how we traverse the curve without the presence of time stamps in the picture. We get the sense of a curve that is "slowing down" and perhaps approaching one point. Checking the limits of each component function as $t \to \infty$, it does appear that instead of approaching an asymptote or moving off to infinitely large coordinates as t increases without bound, the curve stalls out at the point $(1, 3, 2)$. The same phenomenon happens as $t \to -\infty$, with the curve approaching the point $(1, 3, -2)$, pictured in Figure 2.16.

The remaining question is what happens near $t = 0$ and $t = 1$, which cause division by zero. The curve between $t = -2$ and $t = 3$ is in Figure 2.17. There are three portions to the curve: $t < 0$, the lowest piece in Figure 2.17; $0 < t < 1$, the upper right piece; and $t > 1$, the middle left piece. The asymptote in the z-direction, which occurs when $t = 0$, is the line $x = 0, y = 5$. The asymptote in the y-direction, which occurs when $t = 1$, is the line $x = 0, z = \sqrt{11}$. Combining Figures 2.15, 2.16, and 2.17, albeit from a different viewpoint, is accomplished in Figure 2.18, where the two stall points and the two asymptotes can be seen. ∎

We briefly explore a portion of one more curve graphically before turning our attention to calculating asymptotes.

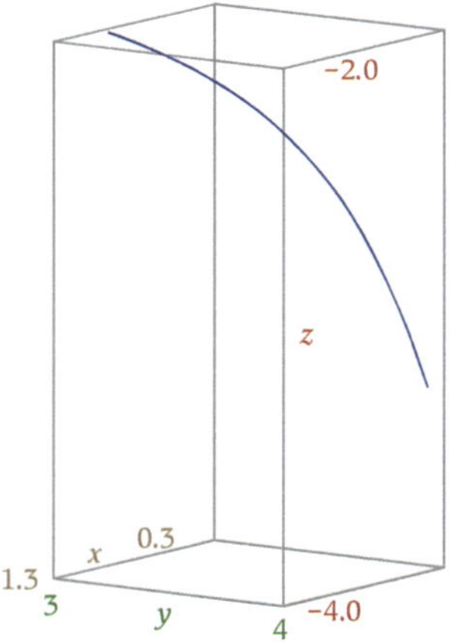

Figure 2.16 *The curve* $\mathbf{r}(t) = \left\langle \frac{t^2-t}{t^2+4}, \frac{3t-5}{t-1}, \frac{\sqrt{4t^2+7}}{t} \right\rangle$ *graphed from* $t = -1$ *(bottom right of curve portion) to* $t = -50$ *(top left of curve portion)*

Example 6 *Use technology to produce a picture of the curve* $\mathbf{r}(t) = \left\langle 4 + \frac{\sin t}{t}, 1 + \frac{\cos t}{t}, \frac{t}{10} \right\rangle$.

Solution The portion of the curve where $t > 3.84$ is in Figure 2.19, where we see what appears to be a spiral around an asymptote. This time the asymptote is

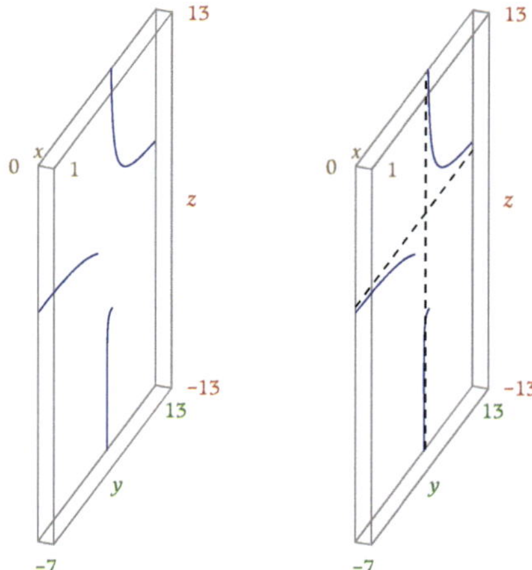

Figure 2.17 *The curve* $\mathbf{r}(t) = \left\langle \dfrac{t^2-t}{t^2+4}, \dfrac{3t-5}{t-1}, \dfrac{\sqrt{4t^2+7}}{t} \right\rangle$ *graphed from* $t = -2$ *to* $t = -0.2$, *from* $t = 0.2$ *to* $t = 0.8$, *and from* $t = 1.2$ *to* $t = 3$, *without asymptotes (left) and with asymptotes (right)*

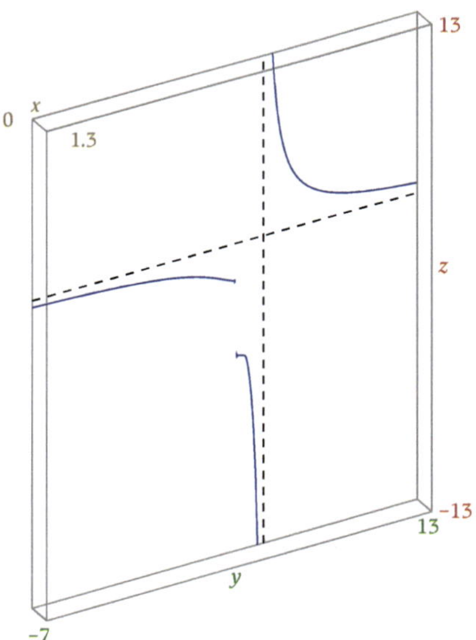

Figure 2.18 *The curve* $\mathbf{r}(t) = \left\langle \dfrac{t^2-t}{t^2+4}, \dfrac{3t-5}{t-1}, \dfrac{\sqrt{4t^2+7}}{t} \right\rangle$ *graphed from* $t = -50$ *to* $t = -0.2$, *from* $t = 0.2$ *to* $t = 0.8$, *and from* $t = 1.2$ *to* $t = 50$ *along with asymptotes* $x = 0, y = 5$ *and* $x = 0, z = \sqrt{11}$

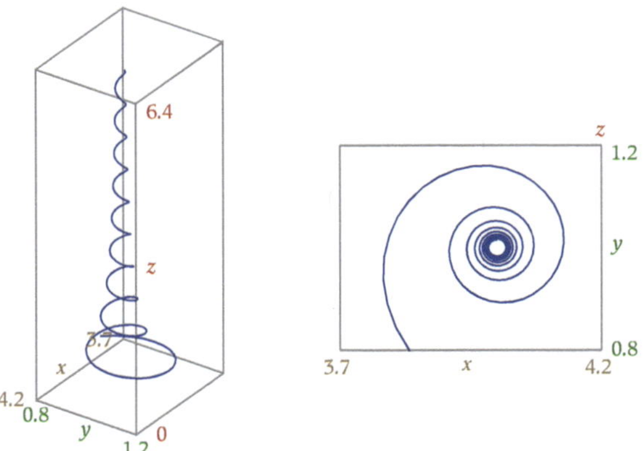

Figure 2.19 *The curve* $\mathbf{r}(t) = \left\langle 4 + \frac{\sin t}{t}, 1 + \frac{\cos t}{t}, \frac{t}{10} \right\rangle$ *graphed from approximately* $t = 3.84$ *to* $t = 62.8$, *(left) viewed from a point in the first octant and (right) viewed from above*

Why 3.84? This value was chosen for aesthetic reasons. Any value of t sufficiently far away from $t = 0$, which causes division by zero, works just as well.

produced as $t \to \infty$, not by division by zero. The equation of the asymptote, the line $x = 4$, $y = 1$, is relatively easily seen in the view from above, looking straight down the asymptote (Figure 2.19, right).

The portion of the graph where $t < -3.84$ is similar. Division by zero does occur when $t = 0$, and the portion of the graph near $t = 0$, which gives a different asymptote, is left to the exploratory pleasure of the reader. $\blacksquare$

2.1.6 Limits and asymptotes for vector-valued functions

Consider $\mathbf{r}(t) = \langle \frac{1}{t^2}, 2t, t + 1 \rangle$. The component functions are $f(t) = \frac{1}{t^2}$, $g(t) = 2t$, and $h(t) = t + 1$. The function is undefined at $t = 0$. So, what happens near $t = 0$? As with limits you have studied previously, the idea is to scoot over just a smidgen, an infinitesimal amount; we find the value of $\mathbf{r}(0 + \alpha)$, where α is an arbitrary infinitesimal. Calculating in each component, we would have

In line 1, we evaluate $\mathbf{r}$ by evaluating the component functions; line 2 carries out that evaluation; line 3 uses the fact that the infinitesimal α and the infinite number A are reciprocals of one another; line 4 approximates in the third component (the first and second component have only one term each, so there is nothing to throw away); and line 5 renders a real result, recalling that A^2 is positive infinite and therefore renders ∞, whereas 2α is infinitesimal and therefore renders 0. Note that it is acceptable to skip line 4 and go directly to line 5.

$$\mathbf{r}(0 + \alpha) = \langle f(0 + \alpha), g(0 + \alpha), h(0 + \alpha) \rangle$$

$$= \left\langle \frac{1}{\alpha^2}, 2\alpha, \alpha + 1 \right\rangle$$

$$= \langle A^2, 2\alpha, \alpha + 1 \rangle$$

$$\approx \langle A^2, 2\alpha, 1 \rangle$$

$$\doteq \langle \infty, 0, 1 \rangle.$$

Thus, when t is infinitely close to 0, the y-component of $\mathbf{r}$ is infinitely close to 0 and the z-component is infinitely close to 1, but the x-component is infinite.

Another way of stating this is that the graph is infinitely close to the line $y = 0$, $z = 1$, but with x infinite; the line $y = 0$, $z = 1$ is an asymptote on the function's graph.

In general, when the limit results in one component that is infinite and two components that are real numbers, the result is an asymptote in the direction of the infinite component with equation given by the other two variables equal to their real-number components.

Let's turn our attention to formally establishing the results of this discussion, beginning with a definition of limit.

Definition 1 LIMIT OF A VECTOR-VALUED FUNCTION *The* limit *of a vector-valued function* $\mathbf{r}(t) = \langle f(t), g(t), h(t) \rangle$ *as t approaches k, where k is a real number, the symbol ∞, or the symbol $-\infty$, is given by*

$$\lim_{t \to k} \mathbf{r}(t) = \left\langle \lim_{t \to k} f(t), \ \lim_{t \to k} g(t), \ \lim_{t \to k} h(t) \right\rangle.$$

One-sided limits are similarly calculated component-by-component.

For the purposes of calculation, if k is a real number,

$$\lim_{t \to k} \mathbf{r}(t) = \mathbf{r}(k + \alpha),$$

where α is an arbitrary infinitesimal and we must render the same real result for any α.

For one-sided limits,

The EAR (evaluate, approximate, render) procedure applies.

$$\lim_{t \to k^+} \mathbf{r}(t) = \mathbf{r}(k + \omega)$$

$$\lim_{t \to k^-} \mathbf{r}(t) = \mathbf{r}(k - \omega),$$

where ω is an arbitrary positive infinitesimal and we must render the same real result for any ω.

For limits at infinity,

$$\lim_{t \to \infty} \mathbf{r}(t) = \mathbf{r}(\Omega)$$

$$\lim_{t \to -\infty} \mathbf{r}(t) = \mathbf{r}(-\Omega),$$

where Ω is positive infinite and we must render the same real result for any Ω.

Notice that, in all cases, the result is not a single number (or infinite symbol) but rather a vector. Because we are essentially evaluating several limits of real-valued functions, one for each component, all techniques for evaluating function limits are allowed, including evaluating using continuity. Borrowing computer science terminology, each component's limit is computed separately, but in parallel.

Example 7 *Find* $\lim\limits_{t\to3^+}\left\langle 4,\ \dfrac{\sqrt{t+13}}{t+2},\ \dfrac{1}{3-t}\right\rangle.$

Solution For any positive infinitesimal ω,

$$\lim_{t\to3^+}\left\langle 4,\ \frac{\sqrt{t+13}}{t+2},\ \frac{1}{3-t}\right\rangle = \left\langle 4,\ \frac{\sqrt{16}}{5},\ \frac{1}{3-(3+\omega)}\right\rangle$$

$$= \left\langle 4,\ \frac{4}{5},\ \frac{1}{-\omega}\right\rangle$$

$$= \left\langle 4,\ \frac{4}{5},\ -\Omega\right\rangle$$

$$\doteq \left\langle 4,\ \frac{4}{5},\ -\infty\right\rangle. \qquad\blacksquare$$

In line 1, limits of the first two components are computed using continuity. Because using continuity fails for the third component for reason of division by zero, we use $t = 3 + \omega$ (one-sided limit). Line 2 simplifies, line 3 uses the reciprocal relationship, and line 4 renders a real result.

Reading Exercise 2 Find $\lim\limits_{t\to4}\left\langle \dfrac{t+1}{t+5},\ \dfrac{3}{(t-4)^2},\ \sin\pi t\right\rangle.$

Because two components of the limit in example 7 are real numbers and one is infinite, there is an asymptote on the graph of the vector-valued function, in the z-direction (because that is the infinite component) with equation $x = 4, y = \dfrac{4}{5}.$

More specifically, the asymptote is in the negative z-direction because the limit is $-\infty.$

ASYMPTOTES PARALLEL TO A COORDINATE AXIS

Let $\mathbf{r}$ be a vector-valued function and a, b, and c be real numbers.

(1) If $\lim\limits_{t\to k}\mathbf{r}(t) = \langle a, b, \pm\infty\rangle$, then $\mathbf{r}$ has an asymptote in the z-direction with equation $x = a, y = b.$

(2) If $\lim\limits_{t\to k}\mathbf{r}(t) = \langle a, \pm\infty, c\rangle$, then $\mathbf{r}$ has an asymptote in the y-direction with equation $x = a, z = c.$

(3) If $\lim\limits_{t\to k}\mathbf{r}(t) = \langle \pm\infty, b, c\rangle$, then $\mathbf{r}$ has an asymptote in the x-direction with equation $y = b, z = c.$

The same is true for one-sided limits and limits at infinity.

By $\lim\limits_{t\to k}\mathbf{r}(t) = \langle a, b, \pm\infty\rangle$ we mean either $\lim\limits_{t\to k}\mathbf{r}(t) = \langle a, b, \infty\rangle$ or $\lim\limits_{t\to k}\mathbf{r}(t) = \langle a, b, -\infty\rangle.$

Recall that the elementary functions include polynomial, root, power, trigonometric, exponential, logarithmic, hyperbolic, inverse trigonometric, and inverse hyperbolic functions, along with any algebraic combinations or compositions thereof.

From our experience with real-valued function limits, we know that for continuous functions such as the elementary functions, the only places to look for infinite limits are where there is division by zero (including in trig functions such as tangent), the edge of a domain (such as for a logarithmic function), or limits at infinity.

Example 8 *For the curve* $\mathbf{r}(t) = \left\langle t^2 - \dfrac{t}{t^2+4}, \dfrac{3t-5}{t-1}, t \right\rangle$, *determine any asymptotes in the* *x-, y-, or z-directions.*

Solution Division by zero occurs only at $t = 1$ (notice that $t^2 + 4 = 0$ does not have any real solutions). Checking the limit there,

This is the same function as in example 4.

$$\lim_{t\to 1} \left\langle t^2 - \frac{t}{t^2+4}, \frac{3t-5}{t-1}, t \right\rangle = \left\langle 1 - \frac{1}{5}, \frac{3(1+\alpha)-5}{(1+\alpha)-1}, 1 \right\rangle$$

$$= \left\langle \frac{4}{5}, \frac{3+3\alpha-5}{\alpha}, 1 \right\rangle$$

$$\approx \left\langle \frac{4}{5}, \frac{-2}{\alpha}, 1 \right\rangle$$

$$= \left\langle \frac{4}{5}, -2A, 1 \right\rangle.$$

Line 1 computes the limits in the x- and z-components using continuity; for the y-component, we use $t = 1 + \alpha$ (two-sided limit). Line 2 simplifies; line 3 approximates; and line 4 uses the reciprocal relationship between α and A.

Although we are stuck and cannot render either ∞ or $-\infty$ because we do not know the sign of A, we know that the limit is infinite and that is all that is necessary to determine asymptotes. Because the y-component of the limit is infinite and the other two components are real numbers, there is an asymptote in the y-direction with equation $x = \frac{4}{5}, z = 1$. The asymptote can be viewed in Figure 2.14.

We also need to check limits at infinity;

If desired, one-sided limits can be used to determine whether the curve approaches the asymptote in the positive y-direction or the negative y-direction, depending on whether t is less than 1 or greater than 1.

$$\lim_{t\to\infty} \left\langle t^2 - \frac{t}{t^2+4}, \frac{3t-5}{t-1}, t \right\rangle = \left\langle \Omega^2 - \frac{\Omega}{\Omega^2+4}, \frac{3\Omega-5}{\Omega-1}, \Omega \right\rangle$$

$$\approx \left\langle \Omega^2 - \frac{\Omega}{\Omega^2}, \frac{3\Omega}{\Omega}, \Omega \right\rangle$$

$$= \left\langle \Omega^2 - \omega, 3, \Omega \right\rangle$$

$$\approx \left\langle \Omega^2, 3, \Omega \right\rangle$$

$$\doteq \left\langle \infty, 3, \infty \right\rangle.$$

Line 1 uses $t = \Omega$ for an arbitrary positive infinite hyperreal Ω; line 2 approximates (the approximation is separate in each component, and numerator and denominator are approximated separately); line 3 simplifies and then uses the reciprocal relationship ($\frac{\Omega}{\Omega^2} = \frac{1}{\Omega} = \omega$); line 4 approximates again; and line 5 renders a real result.

Because two components are infinite, the limit does not indicate an asymptote in either the x-, y-, or z-direction.

The limit as $t \to -\infty$ is similar and does not indicate an asymptote in either the x-, y-, or z-direction.

We conclude that there is one asymptote parallel to a coordinate axis, in the y-direction with equation $x = \frac{4}{5}, z = 1$. ∎

Although no axis-parallel asymptote is indicated, the limit does describe a certain "end behavior" that can be seen in Figure 2.14. We shall not study all such implications.

Ans. to reading exercise 2:
$$\left\langle \frac{5}{9}, \infty, 0 \right\rangle$$

Example 9 *For the curve* $\mathbf{r}(t) = \left\langle \dfrac{t^2-t}{t^2+4}, \dfrac{3t-5}{t-1}, \dfrac{\sqrt{4t^2+7}}{t} \right\rangle$, *determine any asymptotes in* *the x-, y-, or z-directions.*

This is the same function as in example 5.

In line 1, the limits in the x- and y-components are evaluated using continuity; the two-sided limit in the z-component uses $t = 0 + \alpha = \alpha$. Line 2 approximates under the square root (this works as long as the approximation principle is not violated); line 3 uses the reciprocal relationship. Because we do not know whether A is positive or negative, we cannot render either ∞ or $-\infty$.

Solution   Division by zero occurs at $t = 0$ and $t = 1$. We first check the limit at $t = 0$:

$$\lim_{t \to 0}\left\langle \frac{t^2 - t}{t^2 + 4}, \frac{3t - 5}{t - 1}, \frac{\sqrt{4t^2 + 7}}{t}\right\rangle = \left\langle \frac{0}{4}, \frac{-5}{-1}, \frac{\sqrt{4\alpha^2 + 7}}{\alpha}\right\rangle$$

$$\approx \left\langle 0, 5, \frac{\sqrt{7}}{\alpha}\right\rangle$$

$$= \langle 0, 5, \sqrt{7}A\rangle,$$

and although we are stuck in the z-component, we know that the component is infinite, and there is an asymptote in the z-direction with equation $x = 0, y = 5$.

Next, we check the limit at $t = 1$, noting that the y-component is the same as in example 8:

See example 8 for details on calculating the limit of the y-component; the x- and z- component limits are evaluated using continuity.

$$\lim_{t \to 1}\left\langle \frac{t^2 - t}{t^2 + 4}, \frac{3t - 5}{t - 1}, \frac{\sqrt{4t^2 + 7}}{t}\right\rangle = \cdots = \langle 0, -2A, \sqrt{11}\rangle.$$

There is an asymptote in the y-direction with equation $x = 0, y = \sqrt{11}$.

We also check limits at infinity:

Line 1 evaluates the expression at $t = \Omega$, line 2 approximates numerators and denominators separately, and approximates underneath the square root; line 3 simplifies, including extracting the square root noting that Ω is a positive number.

$$\lim_{t \to \infty}\left\langle \frac{t^2 - t}{t^2 + 4}, \frac{3t - 5}{t - 1}, \frac{\sqrt{4t^2 + 7}}{t}\right\rangle = \left\langle \frac{\Omega^2 - \Omega}{\Omega^2 + 4}, \frac{3\Omega - 5}{\Omega - 1}, \frac{\sqrt{4\Omega^2 + 7}}{\Omega}\right\rangle$$

$$\approx \left\langle \frac{\Omega^2}{\Omega^2}, \frac{3\Omega}{\Omega}, \frac{\sqrt{4\Omega^2}}{\Omega}\right\rangle$$

$$= \left\langle 1, 3, \frac{2\Omega}{\Omega}\right\rangle = \langle 1, 3, 2\rangle,$$

and

Line 2 evaluates the expression at $t = -\Omega$, carefully using parentheses to avoid sign errors; line 3 simplifies recalling that the square of a negative number is positive; line 4 approximates numerators and denominators separately, and approximates underneath the square root; line 5 simplifies, including extracting the square root noting that Ω is a positive number.

$$\lim_{t \to -\infty}\left\langle \frac{t^2 - t}{t^2 + 4}, \frac{3t - 5}{t - 1}, \frac{\sqrt{4t^2 + 7}}{t}\right\rangle$$

$$= \left\langle \frac{(-\Omega)^2 - (-\Omega)}{(-\Omega)^2 + 4}, \frac{3(-\Omega) - 5}{-\Omega - 1}, \frac{\sqrt{4(-\Omega)^2 + 7}}{-\Omega}\right\rangle$$

$$= \left\langle \frac{\Omega^2 + \Omega}{\Omega^2 + 4}, \frac{-3\Omega - 5}{-\Omega - 1}, \frac{\sqrt{4\Omega^2 + 7}}{-\Omega}\right\rangle$$

$$\approx \left\langle \frac{\Omega^2}{\Omega^2}, \frac{-3\Omega}{-\Omega}, \frac{\sqrt{4\Omega^2}}{-\Omega}\right\rangle$$

$$= \left\langle 1, 3, \frac{2\Omega}{-\Omega}\right\rangle = \langle 1, 3, -2\rangle.$$

Neither of these limits has an infinite component, so these limits do not indicate asymptotes in the x-, y-, or z-directions. Instead, the graph approaches each of the points indicated by the limits, seemingly stalling as illustrated in Figures 2.15 and 2.16.

We conclude that there are two asymptotes and two stall points, all of which can be seen in Figure 2.18. ∎

STALL POINTS

Let $\mathbf{r}$ be a vector-valued function and a, b, and c be real numbers. If $\lim_{t \to \infty} \mathbf{r}(t) = \langle a, b, c \rangle$ or $\lim_{t \to -\infty} \mathbf{r}(t) = \langle a, b, c \rangle$, then $\mathbf{r}$ has *stall point* (a, b, c).

Next up is the spiral of example 6.

Example 10 *For the curve* $\mathbf{r}(t) = \left\langle 4 + \frac{\sin t}{t}, 1 + \frac{\cos t}{t}, \frac{t}{10} \right\rangle$, *determine any asymptotes in the* x-, y-, *or* z-*directions.*

Solution Division by zero occurs at $t = 0$; the sine and cosine functions are defined for all real numbers. Checking the limit, we have

$$\lim_{t \to 0} \left\langle 4 + \frac{\sin t}{t}, 1 + \frac{\cos t}{t}, \frac{t}{10} \right\rangle = \left\langle 4 + \frac{\sin \alpha}{\alpha}, 1 + \frac{\cos \alpha}{\alpha}, \frac{0}{10} \right\rangle$$

$$\approx \left\langle 4 + \frac{\alpha}{\alpha}, 1 + \frac{1}{\alpha}, 0 \right\rangle$$

$$= \langle 4 + 1, 1 + A, 0 \rangle$$

$$\approx \langle 5, A, 0 \rangle,$$

If we used tangent, cotangent, secant, or cosecant we would need to check places where they are undefined.

In line 1, the z-component is evaluated using continuity, but for the x- and y-components we use $t = 0 + \alpha = \alpha$. In line 2, we recall that the sine of an infinitesimal is approximately that same infinitesimal, but $\cos \alpha \approx 1$. Line 3 simplifies and uses the reciprocal relationship. Line 4 simplifies and approximates; note that the approximation symbol is used.

and there is an asymptote in the y-direction with equation $x = 5$, $z = 0$. This asymptote is not pictured in Figure 2.19, where only values of t larger than 3.84 are shown.

We also check limits at infinity:

$$\lim_{t \to \infty} \left\langle 4 + \frac{\sin t}{t}, 1 + \frac{\cos t}{t}, \frac{t}{10} \right\rangle = \left\langle 4 + \frac{\sin \Omega}{\Omega}, 1 + \frac{\cos \Omega}{\Omega}, \frac{\Omega}{10} \right\rangle.$$

When checking $t \to 0$, note that even though division by zero occurs in two components, only one component has infinite limit, so there is an asymptote. Beware of false assumptions!

Recalling that $-1 \le \sin \Omega \le 1$ and $-1 \le \cos \Omega \le 1$, we reason that $\frac{\sin \Omega}{\Omega} = \omega \sin \Omega$ is an infinitesimal times a number between 1 and -1 and is therefore infinitesimal. Continuing the limit calculation, we have

This is due to the transfer principle; because $-1 \le \sin x \le 1$ for any real number x, the same is true for any hyperreal number x.

$$\approx \left\langle 4, 1, \frac{1}{10}\Omega \right\rangle \doteq \langle 4, 1, \infty \rangle,$$

and there is an asymptote in the z-direction with equation $x = 4, y = 1$, which can be seen in Figure 2.19, where the curve spirals around the asymptote getting closer and closer but never reaching the line.

The limit as $t \to -\infty$ is similar, rendering $\langle 4, 1, -\infty \rangle$, meaning that the spiral approaches the same asymptote in the opposite direction. ∎

Reading Exercise 3   Does the limit indicate an asymptote? If so, in what direction and what is the equation of the asymptote?

Note: this is the same limit as reading exercise 2.

$$\lim_{t \to 4} \left\langle \frac{t+1}{t+5}, \ \frac{3}{(t-4)^2}, \ \sin \pi t \right\rangle$$

We end this section with one more example.

Example 11 *Find any asymptotes in the x-, y-, or z-directions on the graph of*
$$\mathbf{r}(t) = \left\langle \frac{3}{t(t-2)}, \ \frac{\sin 3t}{t}, \ \frac{3t}{t-2} \right\rangle.$$

Solution   Division by zero occurs at $t = 0$ and $t = 2$. We check $t = 0$ first:

Line 1 evaluates the limit in the z-component using continuity and evaluates the other two components at $t = 0 + \alpha = \alpha$; line 2 approximates, recalling that the sine of an infinitesimal is approximately that same infinitesimal; line 3 simplifies and uses the reciprocal relationship.

$$\lim_{t \to 0} \left\langle \frac{3}{t(t-2)}, \ \frac{\sin 3t}{t}, \ \frac{3t}{t-2} \right\rangle = \left\langle \frac{3}{\alpha^2 - 2\alpha}, \ \frac{\sin 3\alpha}{\alpha}, \ \frac{0}{-2} \right\rangle$$
$$\approx \left\langle \frac{3}{-2\alpha}, \ \frac{3\alpha}{\alpha}, \ 0 \right\rangle$$
$$= \left\langle -\tfrac{3}{2}A, 3, 0 \right\rangle.$$

There is an asymptote in the x-direction with equation $y = 3, z = 0$.

Next, we check $t = 2$:

Line 1 evaluates the limit in the y-component using continuity and evaluates the other two components at $t = 2 + \alpha$; line 2 approximates and simplifies; line 3 uses the reciprocal relationship.

$$\lim_{t \to 2} \left\langle \frac{3}{t(t-2)}, \ \frac{\sin 3t}{t}, \ \frac{3t}{t-2} \right\rangle = \left\langle \frac{3}{(2+\alpha)(2+\alpha-2)}, \ \frac{\sin 6}{2}, \ \frac{3(2+\alpha)}{2+\alpha-2} \right\rangle$$
$$\approx \left\langle \frac{3}{2\alpha}, \ \frac{\sin 6}{2}, \ \frac{6}{\alpha} \right\rangle$$
$$= \left\langle 1.5A, \ \frac{\sin 6}{2}, 6A \right\rangle,$$

and with two infinite components the limit does not indicate an asymptote parallel to a coordinate axis.

Finally, we check limits at infinity:

$$\lim_{t \to \infty} \left\langle \frac{3}{t(t-2)}, \frac{\sin 3t}{t}, \frac{3t}{t-2} \right\rangle = \left\langle \frac{3}{\Omega^2 - 2\Omega}, \frac{\sin(3\Omega)}{\Omega}, \frac{3\Omega}{\Omega-2} \right\rangle$$

$$\approx \left\langle \frac{3}{\Omega^2}, \frac{\sin(3\Omega)}{\Omega}, \frac{3\Omega}{\Omega} \right\rangle$$

$$= \langle 3\omega^2, \omega \sin(3\Omega), 3 \rangle$$

$$\doteq \langle 0, 0, 3 \rangle,$$

which does not indicate an asymptote, but rather a stall point. The limit as $t \to -\infty$ is the same stall point.

In summary, there is one asymptote parallel to a coordinate axis, in the x-direction with equation $y = 3$, $z = 0$. We also identified the stall point $(0, 0, 3)$. ∎

Look again at the result of the limit in example 11 as $t \to 2$, in which we are stuck at $\left\langle 1.5A, \frac{\sin 6}{2}, 6A \right\rangle$. This can be rewritten as

$$\left\langle 0, \frac{\sin 6}{2}, 0 \right\rangle + A\langle 1.5, 0, 6 \rangle.$$

This is the same as the line $\left\langle 0, \frac{\sin 6}{2}, 0 \right\rangle + t\langle 1.5, 0, 6 \rangle$, which is the equation of a line, evaluated at $t = A$. The line therefore serves as a *slant asymptote* on the graph of the curve.

EXERCISES 2.1

1–4.　Rapid response: is it a vector valued function?

1.　$f(t) = t^2$
2.　$\mathbf{r}(t) = \langle t, 3t, 5t \rangle$
3.　$\mathbf{r}(t) = \langle t, 3t \rangle$
4.　$\frac{x^2}{4} + \frac{y^2}{9} + z = 1$

5–10.　Find the domain of the vector-valued function.

5.　$\mathbf{r}(t) = \left\langle \sin^{-1} t, \frac{1}{t-2}, \ln t \right\rangle$
6.　$\mathbf{r}(t) = \left\langle \sqrt{t+1}, \ln(1-t), \csc t \right\rangle$
7.　$\mathbf{r}(t) = \left\langle \frac{2}{3t+1}, \frac{t^2+3t}{t^2+19}, \sqrt{3-t} \right\rangle$
8.　$\mathbf{r}(t) = \left\langle \frac{1}{t^2+4}, \frac{1}{t+5}, e^t \right\rangle$
9.　$\mathbf{r}(t) = \left\langle \sqrt{t^2+1}, e^{-2t+7}, \sin t \cos t^2 \right\rangle$
10.　$\mathbf{r}(t) = \left\langle \frac{1}{t^2+2t-3}, \cot t, \cos^{-1} t \right\rangle$

Line 1 evaluates using $t = \Omega$; line 2 approximates (notice that 3Ω is not infinitesimal, but rather infinite); line 3 uses the reciprocal relationship; and line 4 renders a real result, reasoning that $\omega \sin(3\Omega)$ is infinitesimal.

Ans. to reading exercise 3:
yes; the y-direction, equation $x = \frac{5}{9}$, $z = 0$

11–16. Find the limit.

11. $\lim\limits_{t\to 3^-} \left\langle \frac{t}{t-3},\ \sinh(t-3),\ t+3 \right\rangle$

12. $\lim\limits_{t\to -\infty} \left\langle \frac{t}{3t-1},\ \frac{\sqrt{t^2+1}}{3t-1},\ \frac{1}{3t-1} \right\rangle$

13. $\lim\limits_{t\to\infty} \left\langle \cos\frac{1}{t},\ \frac{\sqrt{t+7}}{t-5200},\ 2^t \right\rangle$

14. $\lim\limits_{t\to 5^+} \left\langle \sin\pi t,\ \frac{3}{5-t},\ \frac{4t-20}{t^2-6t+5} \right\rangle$

15. $\lim\limits_{t\to 4} \left\langle \frac{t+7}{t-1},\ \frac{4+\sqrt{t}}{(t-4)^2},\ \frac{3}{|4-t|} \right\rangle$

16. $\lim\limits_{t\to\infty} \left\langle 3^{2-t},\ 0.78^t,\ \frac{t^2}{e^t} \right\rangle$

17–26. (a) Find all asymptotes in the x-, y-, and z-directions on the graph of the given function, and (b) find all stall points on the graph.

17. $\mathbf{r}(t) = \langle t^2,\ e^t,\ t^{-2} \rangle$

18. $\mathbf{r}(t) = \left\langle \frac{1}{t},\ 2-t,\ \frac{1}{2-t} \right\rangle$

19. $\mathbf{r}(t) = \left\langle 1+\frac{1}{t},\ 2+\frac{1}{t+1},\ \frac{t^2}{t^2+1} \right\rangle$

20. $\mathbf{r}(t) = \langle e^t,\ \cosh t,\ 2\tan^{-1} t \rangle$

21. $\mathbf{r}(t) = \left\langle \frac{t^2-4t+3}{t^2-5t+4},\ \frac{1}{t^2+1},\ \sin t \right\rangle$

22. $\mathbf{r}(t) = \left\langle \frac{\sqrt{t^2+1}}{2t},\ \frac{4t}{5t^2+1},\ \frac{3}{1+e^t} \right\rangle$

23. $\mathbf{r}(t) = \left\langle \sqrt{t},\ \frac{1}{t},\ e^{-t} \right\rangle$

24. $\mathbf{r}(t) = \left\langle \ln t^2,\ \frac{t^2+7}{t^2-4},\ \frac{\sin t}{t} \right\rangle$

25. $\mathbf{r}(t) = \left\langle \ln(t^2-4),\ \cos\left((2+t)\frac{\pi}{4}\right),\ 11t+87 \right\rangle$

26. $\mathbf{r}(t) = \left\langle \frac{1}{t+2},\ \sqrt{t^2-9},\ \sin\frac{1}{t} \right\rangle$

For exercises 27–34, "Plot points" means to use strategy 1; "projections" means to use strategy 2 (all three projections are required); and "technology" means to use a CAS, considering multiple viewpoints as appropriate.

27. Graph the curve $\mathbf{r}(t) = \langle \sin 3t,\ \cos 3t,\ 0.1t \rangle$ by (a) plotting points, (b) graphing the projections, and (c) using technology. (d) Describe the curve.

28. (a) Describe the curve $\mathbf{r}(t) = \langle t,\ t^2,\ 3 \rangle$. (b) Graph the curve by plotting points.

29. Graph the curve $\mathbf{r}(t) = \langle \sin 2t,\ \cos t,\ \sin t \rangle$ by (a) graphing the projections and (b) using technology. (c) Describe the curve.

30. Graph the curve $\mathbf{r}(t) = \left\langle 3\sin t,\ 5\cos t,\ \frac{t^2}{5} \right\rangle$ for $-\pi \le t \le \pi$ by (a) graphing the projections and (b) using technology. (c) Describe the curve.

31. (a) Graph the curve $\mathbf{r}(t) = \langle (5+2\sin(25t))\cos t,\ (5+2\sin(25t))\sin t,\ \cos(25t) \rangle$ using technology. (b) Describe the curve.

32. (a) Graph the curve $\mathbf{r}(t) = \langle 100t,\ t^2\sin 3t,\ t^2\cos 3t \rangle$ using technology. (b) Describe the curve.

This is the same curve as in exercise 20.

33. (a) Graph the curve $\mathbf{r}(t) = \langle e^t,\ \cosh t,\ 2\tan^{-1} t \rangle$ for $-3 \le t \le 4$ using technology. (b) Describe the curve.

34. (a) Graph the curve $r(t) = \left\langle \frac{t}{3t-1}, \frac{\sqrt{t^2+1}}{3t-1}, \frac{1}{3t-1} \right\rangle$ using technology. This is the same curve as in exercise 12.

    (b) Describe the curve.

35–38. For the given vector-valued function and limit, does the limit indicate the presence of a slant asymptote? If so, state the vector equation of the asymptote.

35. $r(t) = \left\langle \frac{9}{t}, \frac{2}{t^2+t}, t \right\rangle$, $\lim\limits_{t \to 0} r(t)$

36. $r(t) = \left\langle 4t, \frac{2t^2}{3t+22}, 7 + \frac{1}{t} \right\rangle$, $\lim\limits_{t \to \infty} r(t)$

37. $r(t) = \left\langle e^t, \cosh t, \frac{5t}{t-7} \right\rangle$, $\lim\limits_{t \to \infty} r(t)$

38. $r(t) = \left\langle \frac{t+1}{4t-5}, 3t^2 - 7, \frac{t^3-4t}{t+1} \right\rangle$, $\lim\limits_{t \to -\infty} r(t)$

39. In example 8, the limit as $t \to \infty$ rendered $\langle \infty, 3, \infty \rangle$. Does this limit indicate a slant asymptote? To decide, see if you can rewrite the next-to-last step of the limit, $\langle \Omega^2, 3, \Omega \rangle$, in the form $\langle x_0, y_0, z_0 \rangle + t\langle a, b, c \rangle$, where the components of these vectors are real numbers and t is infinite.

40. Let $r(t) = \left\langle \frac{t+1}{t+5}, \frac{3}{(t-4)^2}, \sin \pi t \right\rangle$ (the same function as in reading exercise 2).

    (a) Use technology to graph the function for $5 \le t \le 10$, then for $5 \le t \le 15$, and then for $5 \le t \le 25$. What appears to be happening as $t \to \infty$?

    (b) Find $\lim\limits_{t \to \infty} r(t)$. Does this limit indicate an asymptote of some kind?

    (c) How does the limit describe what you see in the graph?

41. Suppose when calculating $\lim\limits_{t \to k} r(t)$ we arrive at $\langle 5A^3, 12, 95 \rangle$, which indicates an asymptote in the x-direction.

    (a) Following the style of a slant asymptote, rewrite $\langle 5A^3, 12, 95 \rangle$ in the form $\langle x_0, y_0, z_0 \rangle + t\langle a, b, c \rangle$, where t is infinite.

    (b) What is the direction vector from part (a)? Does this match the previously-noted asymptote in the x-direction?

42. (a) Find a vector function that represents the intersection curve of the plane $x + z = 3$ with the hyperbolic paraboloid $\frac{x^2}{4} - \frac{y}{9} - \frac{z^2}{4} = 0$. Hint: Let $x = t$, figure out y and z.

    (b) Describe the intersection curve.

2.2   Calculus of Vector-Valued Functions

We have studied limits of vector-valued functions. What's next? Continuity, derivatives, and integrals, of course!

2.2.1   Continuity of vector-valued functions

For real-valued functions, the idea of continuity is that a small (infinitesimal) change in the input produces a small (infinitesimal) change in the output. A consequence of this idea is that there are no "breaks" in the graph of the function. We also know that the elementary functions are continuous where they are defined, so that finding where an elementary function is continuous is identical to finding the function's domain.

Piecewise-defined functions can also be discontinuous where the function rule changes.

The idea of continuity for a vector-valued function is the same: an infinitesimal change in the input parameter t produces an infinitesimal change in the output vector $\mathbf{r}(t)$. And what would an infinitesimal change in the output vector mean? That the magnitude of the change, its norm, is infinitesimal.

Let $\Delta f(t)$, $\Delta g(t)$, and $\Delta h(t)$ be the changes in the component functions. Then the magnitude of the change in the output vector would be

$$\|\mathbf{r}(t + \alpha) - \mathbf{r}(t)\| = \sqrt{(\Delta f(t))^2 + (\Delta g(t))^2 + (\Delta h(t))^2},$$

Suppose the highest-level term under the square root is $(\Delta g(t))^2$. Then $\sqrt{(\Delta f(t))^2 + (\Delta g(t))^2 + (\Delta h(t))^2} \approx \sqrt{(\Delta g(t))^2} = |\Delta g(t)|$, and the norm of the change in the output vector is on the same level as the highest level of the changes in the component functions.

which is infinitesimal exactly when each of $\Delta f(t)$, $\Delta g(t)$, and $\Delta h(t)$ are infinitesimal. In other words, a vector-valued function is continuous as long as each of its component functions is continuous.

As long as the component functions are elementary functions, determining where a vector-valued function is continuous is identical to finding its domain.

Example 12 *Where is* $\mathbf{r}(t) = \left\langle \cos^{-1} t, \sqrt{t}, \dfrac{3+t}{2t-1} \right\rangle$ *continuous?*

Solution   The x-component function $f(t) = \cos^{-1}(t)$ has domain $[-1, 1]$, the y-component function $g(t) = \sqrt{t}$ has domain $[0, \infty)$, and the domain of the z-component function is found by avoiding division by zero, resulting in $t \neq \dfrac{1}{2}$.

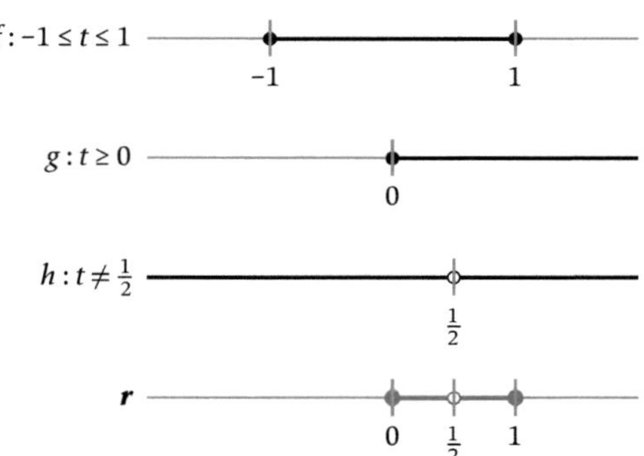

Figure 2.20 *The domain of* $\mathbf{r}(t) = \left\langle \cos^{-1} t, \sqrt{t}, \dfrac{3+t}{2t-1} \right\rangle$

The domain of **r**, which is pictured in Figure 2.20, is $\left[0, \frac{1}{2}\right) \cup \left(\frac{1}{2}, 1\right]$. The component functions are all elementary functions, therefore **r** is continuous on $\left[0, \frac{1}{2}\right) \cup \left(\frac{1}{2}, 1\right]$. ∎

2.2.2 Derivatives of vector-valued functions

Suppose a particle is moving along a path in three dimensions according to the vector-valued position function $\mathbf{r}(t) = \langle f(t), g(t), h(t) \rangle$. What is the instantaneous velocity of the particle at time t_0?

Recalling that average velocity is $\frac{\text{change in position}}{\text{change in time}}$, to find instantaneous velocity at time $t = t_0$ we compute the average velocity over an infinitesimal time period; for an arbitrary infinitesimal α, the time period is the interval between t_0 and $t_0 + \alpha$. Then

$$
\begin{aligned}
\text{velocity} &= \frac{\mathbf{r}(t_0 + \alpha) - \mathbf{r}(t_0)}{t_0 + \alpha - t_0} \\[2mm]
&= \frac{1}{\alpha} \left(\langle f(t_0 + \alpha), g(t_0 + \alpha), h(t_0 + \alpha) \rangle - \langle f(t_0), g(t_0), h(t_0) \rangle \right) \\[2mm]
&= \left\langle \frac{f(t_0 + \alpha) - f(t_0)}{\alpha}, \frac{g(t_0 + \alpha) - g(t_0)}{\alpha}, \frac{h(t_0 + \alpha) - h(t_0)}{\alpha} \right\rangle \\[2mm]
&= \langle f'(t_0), g'(t_0), h'(t_0) \rangle .
\end{aligned}
$$

Line 1 uses the formula for average velocity. The change in position is the ending position minus the beginning position, $\mathbf{r}(t_0 + \alpha) - \mathbf{r}(t_0)$, and the change in time is the ending time minus the beginning time, $(t_0 + \alpha) - t_0$. Line 2 moves the scalar in front and uses the formula $\mathbf{r}(t) = \langle f(t), g(t), h(t) \rangle$; line 3 performs the vector subtraction and scalar multiplication component-by-component; and line 4 recognizes the formula for the derivative in each component.

The velocity vector function is denoted $\mathbf{r}'(t)$ and called the derivative, just as in first-semester calculus. The velocity at time $t = t_0$ is therefore written $\mathbf{r}'(t_0)$. As illustrated by the derivation, because vector addition and subtraction and scalar multiplication are computed component-by-component, so is the derivative.

Definition 2 DERIVATIVE OF A VECTOR-VALUED FUNCTION *If* $\mathbf{r}(t) = \langle f(t), g(t), h(t) \rangle$, *then the derivative* $\mathbf{r}'(t)$ *is given by*

$$
\mathbf{r}'(t) = \langle f'(t), g'(t), h'(t) \rangle.
$$

Example 13 *Find* $\mathbf{r}'(t)$ *for* $\mathbf{r}(t) = \langle t^2, \cos t, e^{2t} \rangle$.

Solution We compute the derivative component-by-component:

$$
\mathbf{r}'(t) = \langle 2t, -\sin t, e^{2t} \cdot 2 \rangle. \quad ∎
$$

Because the derivative in each component is the derivative of a real-valued function, all of the derivative formulas and rules from your prior study of calculus still apply.

Example 14 *A particle moves according to the position function* $\mathbf{r}(t) = \left\langle \sqrt{t}, \frac{1}{t}, \frac{t}{t^2 + 1} \right\rangle$. *Find (a) the particle's velocity vector at time* $t = 1$ *and (b) the particle's speed at time* $t = 1$. *Units are meters and seconds.*

Solution

(a)   First, we compute the velocity vector function:

The velocity vector is denoted $\mathbf{v}(t)$. Line 1 uses the power rule for $t^{1/2}$ and t^{-1}, and the quotient rule for the z-component; line 2 simplifies.

$$\mathbf{v}(t) = \mathbf{r}'(t) = \left\langle \frac{1}{2}t^{-1/2},\ -1 \cdot t^{-2},\ \frac{(t^2+1) \cdot 1 - t(2t)}{(t^2+1)^2} \right\rangle$$

$$= \left\langle \frac{1}{2\sqrt{t}},\ \frac{-1}{t^2},\ \frac{1-t^2}{(t^2+1)^2} \right\rangle.$$

Then, we evaluate the velocity vector function at time $t = 1$:

When evaluating the velocity vector at a specified value of time t, such as $\mathbf{v}(1)$, the components are constants.

$$\mathbf{v}(1) = \mathbf{r}'(1) = \langle \tfrac{1}{2},\ -1,\ 0 \rangle.$$

(b)   Speed is the magnitude of velocity, so the speed at time $t = 1$ is

The velocity is a vector, the speed is a number.

$$\|\mathbf{v}(1)\| = \sqrt{\left(\tfrac{1}{2}\right)^2 + (-1)^2 + 0^2} = \sqrt{\tfrac{5}{4}} \approx 1.118\,\text{m/s}. \qquad \blacksquare$$

Reading Exercise 4   Find $\mathbf{r}'(t)$ for $\mathbf{r}(t) = \langle \sqrt{t}, e^t, 4 \rangle$.

2.2.3   Tangents to spacecurves

For real-valued functions, the derivative gives the direction (slope) of the curve. Then, for vector-valued functions, the x-component of the derivative, which is a real-valued function, gives the x-direction of the curve, the y-component of the derivative gives the y-direction of the curve, and the z-component of the derivative gives the z-direction of the curve. Taken together, the derivative of a vector-valued function gives the three-dimensional direction of the curve; it is tangent to the curve, as illustrated in Figure 2.21. For this reason, the derivative $\mathbf{r}'(t_0)$ is not only called the velocity vector at time $t = t_0$, it is also called the *tangent vector* at $t = t_0$.

The projections of the curve and tangent vector of Figure 2.21 onto the coordinate planes are shown in Figure 2.22. Although the tangent vector is not the same as the tangent line, the familiar look of tangency is still present.

The equation of the tangent line would be, of course, the equation of a line. The vector form of the equation of a line requires a point and a direction vector, and because the tangent vector points in the correct direction, it can serve as the direction vector.

Example 15 *Find an equation of the tangent line to the curve* $\mathbf{r}(t) = \langle t,\ e^t,\ \sin t \rangle$ *at* $t = 0$.

Solution   The point of tangency is

$$\mathbf{r}(0) = \langle 0,\ e^0,\ \sin 0 \rangle = \langle 0, 1, 0 \rangle.$$

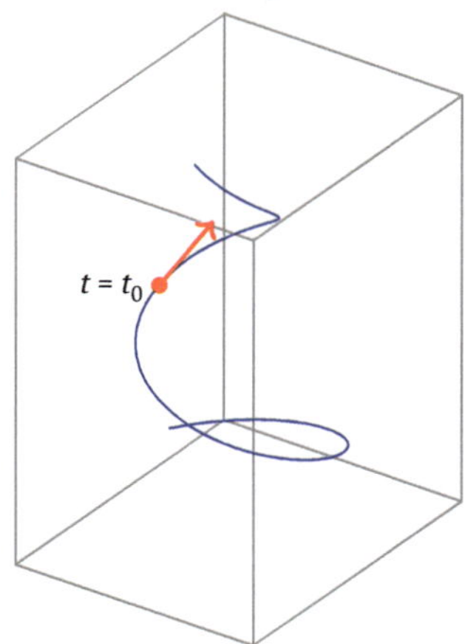

Figure 2.21 *A vector-valued function* **r**(t) *(blue) along with a tangent vector* **r**'(t_0) *(red)*

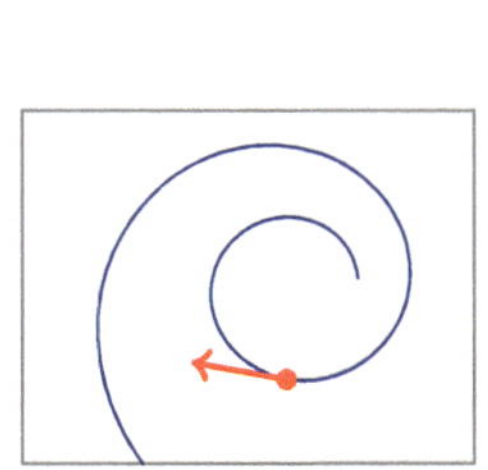 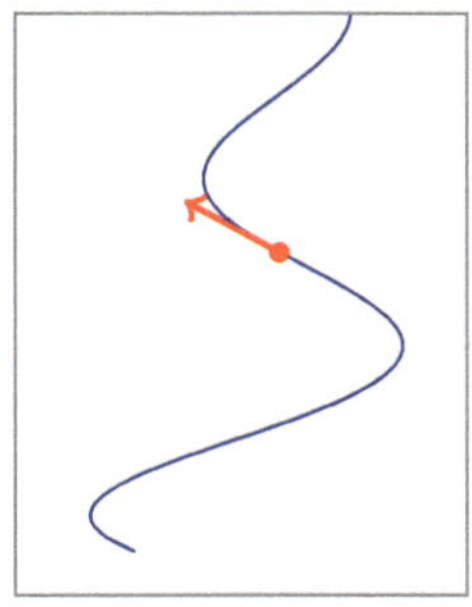 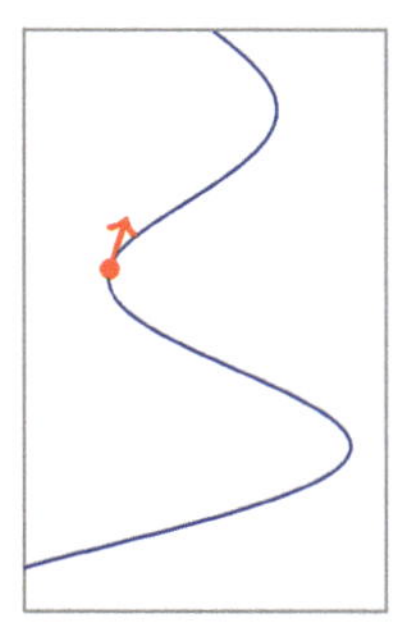

Figure 2.22 *A vector-valued function* **r**(t) *(blue) along with a tangent vector* **r**'(t_0) *(red), projected onto (left) the xy-plane, (middle) the xz-plane, and (right) the yz-plane*

Recall that for two-dimensional parametric equations, the slope of the curve $\frac{dy}{dx}$ is given by $\frac{dy}{dx} = \frac{\frac{dy}{dt}}{\frac{dx}{dt}}$, which is the apparent slope of the curve in the projection onto the xy-plane. For the projection onto the xz-plane where x is horizontal and z is vertical, the apparent slope is $\frac{dz}{dx} = \frac{\frac{dz}{dt}}{\frac{dx}{dt}}$, and for the projection onto the yz-plane where y is horizontal and z is vertical, the apparent slope is $\frac{dz}{dy} = \frac{\frac{dz}{dt}}{\frac{dy}{dt}}$.

To find the direction vector, we first calculate the derivative

$$\mathbf{r}'(t) = \langle 1, e^t, \cos t \rangle$$

and then evaluate the derivative at $t = 0$:

$$\mathbf{r}'(0) = \langle 1, e^0, \cos 0 \rangle = \langle 1, 1, 1 \rangle.$$

Then the vector form of the equation of the tangent line is

$$\langle 0, 1, 0 \rangle + t \langle 1, 1, 1 \rangle.$$

■ This solution can be rewritten in parametric form as $x = t, y = 1 + t, z = t$.

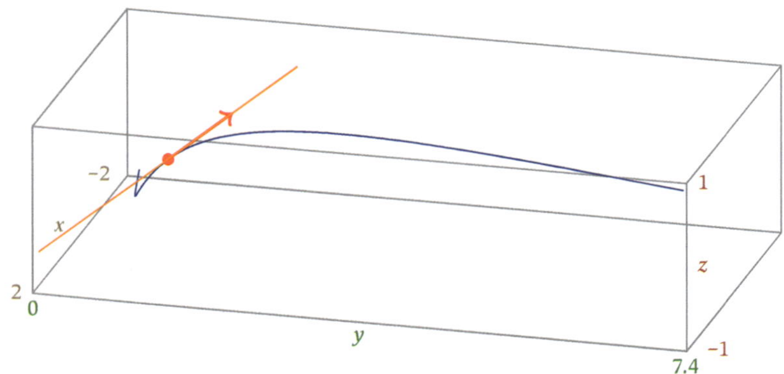

Figure 2.23 *The graph of the vector-valued function* $\mathbf{r}(r) = \langle t,\, e^t,\, \sin t \rangle$ *(blue) along with the tangent vector* $\mathbf{r}'(0)$ *(red) and the tangent line at* $t = 0$ *(orange)*

In example 15, the tangent vector is $\langle 1, 1, 1 \rangle$; the tangent line has equation $\langle 0, 1, 0 \rangle + t\langle 1, 1, 1 \rangle$. The curve from example 15, along with its tangent vector and tangent line at $t = 0$, is pictured in Figure 2.23.

Example 16 *Find an equation of the tangent line to the curve* $\mathbf{r}(t) = \langle t,\, e^t,\, \sin t \rangle$ *at the point* $(\pi, e^\pi, 0)$.

Solution This is the same curve as in example 15, but instead of being given the value of t at which we want the tangent, we are given the point on the curve at which we want the tangent. We have the point, so all we need now is the direction vector, which is given by the tangent vector. But to find the tangent vector we need to know the value of t. To determine t, we need the value of t at which

$$\langle t,\, e^t,\, \sin t \rangle = \langle \pi,\, e^\pi,\, 0 \rangle.$$

In this case, it is nearly immediately apparent that $t = \pi$. If it is not apparent, set the three component functions equal to the corresponding coordinates of the point and solve for t.

We therefore want $\mathbf{r}'(\pi)$:

$$\mathbf{r}'(\pi) = \langle 1,\, e^\pi,\, \cos \pi \rangle = \langle 1, e^\pi, -1 \rangle.$$

The vector equation of the tangent line is

$$\langle \pi,\, e^\pi,\, 0 \rangle + t\langle 1, e^\pi, -1 \rangle. \qquad\blacksquare$$

Ans. to reading exercise 4:
$$\mathbf{r}'(t) = \left\langle \frac{1}{2\sqrt{t}},\, e^t,\, 0 \right\rangle$$

In order for a value of t to work, it must give all three coordinates correctly. In this case, we check $\mathbf{r}(\pi) = \langle \pi, e^\pi, \sin \pi \rangle = \langle \pi, e^\pi, 0 \rangle$.

It is possible for there to be more than one value of t giving the same point, if the curve crosses itself. In that case, there could be more than one tangent line at the point, one for each value of t that gives the point.

Reading Exercise 5 Find an equation of the tangent line to the curve $\mathbf{r}(t) = \langle t,\, t^2,\, t^3 \rangle$ at $t = 1$.

2.2.4   Unit tangent vector

The *unit tangent vector*, which is a unit vector in the same direction as the tangent vector, can also be helpful.

UNIT TANGENT VECTOR

Let $\mathbf{r}(t)$ be a vector-valued function. Then, the *unit tangent vector function* is given by

$$\mathbf{T}(t) = \frac{\mathbf{r}'(t)}{\|\mathbf{r}'(t)\|}.$$

The unit tangent vector is the tangent vector divided by its norm.

Example 17 *Find* $\mathbf{T}(t)$ *for* $\mathbf{r}(t) = \langle \sin t,\ \cos t,\ 3t \rangle$.

Solution   In order to use the formula, we first need to calculate $\mathbf{r}'(t)$ and $\|\mathbf{r}'(t)\|$. To that end,

$$\mathbf{r}'(t) = \langle \cos t,\ -\sin t,\ 3 \rangle.$$

Then

$$\|\mathbf{r}'(t)\| = \sqrt{(\cos t)^2 + (-\sin t)^2 + 3^2} = \sqrt{1 + 9} = \sqrt{10}.$$

The identity $\sin^2 \theta + \cos^2 \theta = 1$ is used.

Using the formula,

$$\mathbf{T}(t) = \frac{\mathbf{r}'(t)}{\|\mathbf{r}'(t)\|} = \frac{\langle \cos t,\ -\sin t,\ 3 \rangle}{\sqrt{10}} = \left\langle \frac{\cos t}{\sqrt{10}},\ \frac{-\sin t}{\sqrt{10}},\ \frac{3}{\sqrt{10}} \right\rangle. \qquad \blacksquare$$

2.2.5   Derivative rules

Because derivatives are calculated component-by-component, the usual derivative rules such as the product rule, quotient rule, and chain rule apply in each component. But, we can also combine the derivative with other operations such as the dot product and the cross product.

DOT PRODUCT RULE

Suppose $\mathbf{u}$ and $\mathbf{v}$ are differentiable vector-valued functions. Then

$$\frac{d}{dt}\,(\mathbf{u}(t) \cdot \mathbf{v}(t)) = \mathbf{u}'(t) \cdot \mathbf{v}(t) + \mathbf{u}(t) \cdot \mathbf{v}'(t).$$

Where $\mathbf{r}'(t_0)$ exists, we say $\mathbf{r}$ is *differentiable* at t_0.

Because the dot product of two vectors is a real number, the function $\mathbf{u}(t) \cdot \mathbf{v}(t)$ is a real-valued function. This rule looks just like the product rule, and its proof is an exercise.

Example 18 *Find* $\frac{d}{dt}\left(\langle t, t^2, t^3 \rangle \cdot \langle e^t, \sin t, \cosh t \rangle\right)$.

Solution One solution method is to first find the dot product of the two vector-valued functions, and then find the derivative of the resulting real-valued function. The dot product is

$$\langle t, t^2, t^3 \rangle \cdot \langle e^t, \sin t, \cosh t \rangle = te^t + t^2 \sin t + t^3 \cosh t.$$

Then

Line 2 uses the product rule (for real-valued functions) for each of the three terms in the expression.

$$\frac{d}{dt}\left(te^t + t^2 \sin t + t^3 \cosh t\right)$$
$$= 1 \cdot e^t + t \cdot e^t + 2t \sin t + t^2 \cos t + 3t^2 \cosh t + t^3 \sinh t.$$

This solution uses the product rule, but not the dot product rule. ∎

Alternate solution Alternately, we can use the dot product rule:

Line 2 uses the dot product rule; line 3 computes the two dot products.

$$\frac{d}{dt}\left(\langle t, t^2, t^3 \rangle \cdot \langle e^t, \sin t, \cosh t \rangle\right)$$
$$= \langle 1, 2t, 3t^2 \rangle \cdot \langle e^t, \sin t, \cosh t \rangle + \langle t, t^2, t^3 \rangle \cdot \langle e^t, \cos t, \sinh t \rangle$$
$$= e^t + 2t \sin t + 3t^2 \cosh t + te^t + t^2 \cos t + t^3 \sinh t.$$

Ans. to reading exercise 5:
$\langle 1, 1, 1 \rangle + t\langle 1, 2, 3 \rangle$

Other than order of terms, this alternate solution is identical to the original solution. ∎

Is one method more advantageous than the other? We leave it to the reader to decide.

The cross product rule is similar. But notice that because the cross product is not commutative (it's anticommutative instead), the order of the cross products cannot be switched.

Because the result of the cross product is a vector, $\mathbf{u}(t) \times \mathbf{v}(t)$ is vector-valued, and its derivative is vector-valued.

CROSS PRODUCT RULE

Suppose $\mathbf{u}$ and $\mathbf{v}$ are differentiable vector-valued functions. Then

$$\frac{d}{dt}\left(\mathbf{u}(t) \times \mathbf{v}(t)\right) = \mathbf{u}'(t) \times \mathbf{v}(t) + \mathbf{u}(t) \times \mathbf{v}'(t).$$

We use the same functions from example 18 to illustrate the cross product rule. The original solution will be to compute the cross product first and then compute the derivative, and the alternate solution will use the cross product rule.

Example 19 *Find* $\frac{d}{dt}\left(\langle t,\, t^2,\, t^3\rangle \times \langle e^t,\, \sin t,\, \cosh t\rangle\right).$

Solution We compute the cross product first:

$$
\begin{vmatrix} \mathbf{i} & \mathbf{j} & \mathbf{k} \\ t & t^2 & t^3 \\ e^t & \sin t & \cosh t \end{vmatrix} = \mathbf{i}\left(t^2\cosh t - t^3\sin t\right)
$$

$$
-\,\mathbf{j}\left(t\cosh t - t^3 e^t\right) + \mathbf{k}\left(t\sin t - t^2 e^t\right).
$$

Then

$$
\frac{d}{dt}\left(\mathbf{i}\left(t^2\cosh t - t^3\sin t\right) - \mathbf{j}\left(t\cosh t - t^3 e^t\right) + \mathbf{k}\left(t\sin t - t^2 e^t\right)\right)
$$

$$
= \mathbf{i}\left(2t\cosh t + t^2\sinh t - (3t^2\sin t + t^3\cos t)\right)
$$

$$
-\,\mathbf{j}\left(\cosh t + t\sinh t - (3t^2 e^t + t^3 e^t)\right)
$$

$$
+\,\mathbf{k}\left(\sin t + t\cos t - (2te^t + t^2 e^t)\right).
$$

This solution uses the product rule, but not the cross product rule. ∎

Alternate solution Alternately, we can use the cross product rule:

$$
\frac{d}{dt}\left(\langle t,\, t^2,\, t^3\rangle \times \langle e^t,\, \sin t,\, \cosh t\rangle\right)
$$

$$
= \langle 1,\, 2t,\, 3t^2\rangle \times \langle e^t,\, \sin t,\, \cosh t\rangle + \langle t,\, t^2,\, t^3\rangle \times \langle e^t,\, \cos t,\, \sinh t\rangle.
$$

The cross product on the left is

$$
\begin{vmatrix} \mathbf{i} & \mathbf{j} & \mathbf{k} \\ 1 & 2t & 3t^2 \\ e^t & \sin t & \cosh t \end{vmatrix} = \mathbf{i}\left(2t\cosh t - 3t^2\sin t\right)
$$

$$
-\,\mathbf{j}\left(\cosh t - 3t^2 e^t\right) + \mathbf{k}\left(\sin t - 2te^t\right),
$$

and the cross product on the right is

$$
\begin{vmatrix} \mathbf{i} & \mathbf{j} & \mathbf{k} \\ t & t^2 & t^3 \\ e^t & \cos t & \sinh t \end{vmatrix} = \mathbf{i}\left(t^2\sinh t - t^3\cos t\right)
$$

$$
-\,\mathbf{j}\left(t\sinh t - t^3 e^t\right) + \mathbf{k}\left(t\cos t - t^2 e^t\right).
$$

The sum of the two cross products is

$$\mathbf{i}\left(2t\cosh t - 3t^2\sin t + t^2\sinh t - t^3\cos t\right)$$
$$-\mathbf{j}\left(\cosh t - 3t^2 e^t + t\sinh t - t^3 e^t\right)$$
$$+\mathbf{k}\left(\sin t - 2te^t + t\cos t - t^2 e^t\right).$$

Except for order of terms, the alternate solution is identical to the original solution. ∎

Several other derivative rules could be stated, such as the sum rule, the constant multiple rule, and the chain rule. Some of these are explored in the exercises.

2.2.6   Integrals of vector valued functions

Derivatives of vector-valued functions are computed component-by-component. Therefore, antiderivatives (indefinite integrals) of vector-valued functions are also computed component-by-component.

Definition 3 INDEFINITE INTEGRAL OF A VECTOR-VALUED FUNCTION *If* $\mathbf{r}(t) = \langle f(t), g(t), h(t)\rangle$, *then the* antiderivative (indefinite integral) $\int \mathbf{r}(t)\,dt$ *is given by*

$$\int \mathbf{r}(t)\,dt = \left\langle \int f(t)\,dt,\ \int g(t)\,dt,\ \int h(t)\,dt \right\rangle.$$

Example 20 *Calculate* $\int \left\langle \dfrac{1}{t},\ e^t,\ \dfrac{1}{1+t^2} \right\rangle dt.$

Solution   Calculating component-by-component,

The expression $\langle \ln|t| + C, e^t + C, \tan^{-1} t + C\rangle$ would be incorrect because this uses the same constant C for all three components; the constants in the components need not be equal.

$$\int \left\langle \frac{1}{t},\ e^t,\ \frac{1}{1+t^2} \right\rangle dt = \left\langle \ln|t| + C_1,\ e^t + C_2,\ \tan^{-1} t + C_3 \right\rangle$$
$$= \langle \ln|t|,\ e^t,\ \tan^{-1} t\rangle + \langle C_1,\ C_2,\ C_3\rangle$$
$$= \langle \ln|t|,\ e^t,\ \tan^{-1} t\rangle + \mathbf{C}.$$

Each of these above expressions can be used as the solution; note that the latter uses a constant vector $\mathbf{C}$. ∎

For real-valued functions, the definite integral is defined through the use of omega sums:

$$\int_a^b f(x)\,dx = \sum_{k=1}^{\Omega} f(x_k)\,\Delta x,$$

where the integral is from $x = a$ to $x = b$, $\Delta x = \frac{b-a}{\Omega} = (b-a)\omega$, and $x_k = a + k\Delta x$ for $k = 1, 2, ..., \Omega$, and we must render the same real result for every positive infinite hyperreal integer Ω.

We can do the same for a vector-valued function $\mathbf{r}$. Integrating from $t = a$ to $t = b$, splitting into Ω subintervals (with Ω a positive infinite integer), and using right-hand endpoints results in $\Delta t = (b-a)\omega$, $t_k = a + k\Delta t$, and the sum $\sum_{k=1}^{\Omega} \mathbf{r}(t_k)\,\Delta t$. If $\mathbf{r}(t) = \langle f(t),\, g(t),\, h(t) \rangle$, then

$$\int_a^b \mathbf{r}(t)\,dt = \sum_{k=1}^{\Omega} \mathbf{r}(t_k)\,\Delta t$$

$$= \sum_{k=1}^{\Omega} \langle f(t_k),\, g(t_k),\, h(t_k) \rangle\,\Delta t$$

$$= \sum_{k=1}^{\Omega} \langle f(t_k)\Delta t,\, g(t_k)\Delta t,\, h(t_k)\Delta t \rangle$$

$$= \left\langle \sum_{k=1}^{\Omega} f(t_k)\,\Delta t,\, \sum_{k=1}^{\Omega} g(t_k)\,\Delta t,\, \sum_{k=1}^{\Omega} h(t_k)\,\Delta t \right\rangle$$

$$= \left\langle \int_a^b f(t)\,dt,\, \int_a^b g(t)\,dt,\, \int_a^b h(t)\,dt \right\rangle,$$

Line 1 uses definite integral notation for the omega sum; line 2 uses $\mathbf{r}(t) = \langle f(t),\, g(t),\, h(t) \rangle$; line 3 performs the scalar multiplication (note that Δt is a scalar); line 4 performs the addition component-by-component, making use of the transfer principle; line 5 rewrites the omega sums using definite integral notation.

and definite integrals are also calculated component-by-component.

Definition 4 DEFINITE INTEGRAL OF A VECTOR-VALUED FUNCTION *If* $\mathbf{r}(t) = \langle f(t), g(t), h(t) \rangle$, *then the definite integral* $\int_a^b \mathbf{r}(t)\,dt$ *is given by*

$$\int_a^b \mathbf{r}(t)\,dt = \left\langle \int_a^b f(t)\,dt,\, \int_a^b g(t)\,dt,\, \int_a^b h(t)\,dt \right\rangle.$$

Example 21 *Find* $\displaystyle\int_0^1 \langle t, t^2, t^3 \rangle\,dt.$

Solution Computing the definite integral component-by-component yields

$$\int_0^1 \langle t, t^2, t^3 \rangle\,dt = \left\langle \int_0^1 t\,dt,\, \int_0^1 t^2\,dt,\, \int_0^1 t^3\,dt \right\rangle$$

$$= \left\langle \frac{t^2}{2}\bigg|_0^1,\, \frac{t^3}{3}\bigg|_0^1,\, \frac{t^4}{4}\bigg|_0^1 \right\rangle$$

$$= \left\langle \frac{1}{2}, \frac{1}{3}, \frac{1}{4} \right\rangle.$$

Line 1 rewrites the integral of the vector as the vector of integrals; line 2 uses the fundamental theorem of calculus in each component; and line 3 finishes the evaluation of the integrals.

Notice that the solution is a vector.   ∎

2.2.7 Vector-valued IVP

Because integration is accomplished component-by-component, so are initial value problems (IVPs).

Example 22 *Find* $\mathbf{r}(t)$ *if* $\mathbf{r}'(t) = \langle t,\ t^2,\ t^3 \rangle$ *and* $\mathbf{r}(1) = \langle 0, 1, -1 \rangle$.

Notice that the initial condition uses a vector.

Solution First, we find the indefinite integral, using three different constants, one for each component:

$$\mathbf{r}(t) = \left\langle \frac{t^2}{2} + C_1,\ \frac{t^3}{3} + C_2,\ \frac{t^4}{4} + C_3 \right\rangle.$$

We use the result to calculate $\mathbf{r}(1)$ and equate to the vector from the initial condition:

$$\mathbf{r}(1) = \left\langle \frac{1}{2} + C_1,\ \frac{1}{3} + C_2,\ \frac{1}{4} + C_3 \right\rangle = \langle 0, 1, -1 \rangle.$$

We now have three equations to solve, one in each component; first is $\frac{1}{2} + C_1 = 0$, with solution $C_1 = -\frac{1}{2}$; second is $\frac{1}{3} + C_2 = 1$, with solution $C_2 = \frac{2}{3}$; and third is $\frac{1}{4} + C_3 = -1$ with solution $C_3 = -\frac{5}{4}$.

We finish by replacing the constants with their values:

$$\mathbf{r}(t) = \left\langle \frac{t^2}{2} - \frac{1}{2},\ \frac{t^3}{3} + \frac{2}{3},\ \frac{t^4}{4} - \frac{5}{4} \right\rangle. \qquad \blacksquare$$

Reading Exercise 6 Evaluate $\displaystyle\int_0^1 \langle 2t,\ 4t,\ 6t \rangle\, dt$.

If you feel as though much of what we are doing is a review of calculus, three exercises at a time, you've caught on pretty well.

EXERCISES 2.2

1–4. Where is the function continuous?

1. $\mathbf{r}(t) = \left\langle \cos\left(\frac{1}{t+\pi}\right),\ \ln(2 - t),\ \sqrt{t^2 + 4} \right\rangle$

2. $\mathbf{r}(t) = \left\langle t^2,\ 2^{\sqrt{t}},\ \tan\left(\frac{1}{1+t^2}\right) \right\rangle$

3. $\mathbf{r}(t) = \left\langle t^3 - 1,\ \sqrt{t^2 - 4},\ \cos^{-1}\frac{t}{3} \right\rangle$

4. $\mathbf{r}(t) = \left\langle \frac{t}{t+4},\ \ln(2t + 6),\ \frac{t-7}{t^3 - 9t} \right\rangle$

5–8. Find $\mathbf{r}'(t)$.

5. $\mathbf{r}(t) = \left\langle \sec(4t + 1),\ \frac{3t-7}{t^2+6},\ 2^{1-t} \right\rangle$

6. $\mathbf{r}(t) = \left\langle \frac{1}{t},\ \sin 2t,\ e^{3t} \right\rangle$

7. $\mathbf{r}(t) = \mathbf{i} \sinh 3t + \mathbf{j} \sin^{-1} t - \mathbf{k} \ln(3t - 1)$

8. $\mathbf{r}(t) = (3t + 1)^4 \mathbf{i} - te^t \mathbf{j} + 5t\mathbf{k}$

9–12. Calculate the tangent vector to the curve at the indicated value of t.

9. $\mathbf{r}(t) = \langle 3t + 1,\ t^2,\ t^{-2} \rangle,\ t = 2$

10. $\mathbf{r}(t) = \left\langle 8t^2,\ 47.3,\ \frac{17}{t} \right\rangle,\ t = -1$

11. $\mathbf{r}(t) = \langle \sin t,\ \cos 2t,\ \tan 3t \rangle,\ t = \pi$

12. $\mathbf{r}(t) = \langle 3^t,\ 5\ln t,\ \sqrt{t} \rangle,\ t = 1$

13–16. Find an equation of the tangent line to the curve at the indicated value of t.

13. $\mathbf{r}(t) = \langle t + 1,\ 3t,\ t^3 \rangle,\ t = 5$

14. $\mathbf{r}(t) = \langle \cos 3t,\ \sin 3t,\ 3\cos t \rangle,\ t = 0$

15. $\mathbf{r}(t) = \langle e^{2t},\ e^{5t},\ t^2 + 7t \rangle,\ t = 0$

16. $\mathbf{r}(t) = \left\langle \frac{t}{3t+4},\ \frac{1}{t^2+5},\ \frac{2t+5}{t-1} \right\rangle,\ t = 2$

17. Find parametric equations of the tangent line to the curve $\mathbf{r}(t) = t^3\mathbf{j} + 3^t\mathbf{k}$ for $t = -1$.

18. Find parametric equations of the tangent line to the curve $\mathbf{r}(t) = 2t\mathbf{i} - \frac{3t}{t+7}\mathbf{k}$ for $t = 1$.

19. A particle moves according to the position function $\mathbf{r}(t) = \langle \sqrt{t},\ t^2 + 1,\ 5t^3 \rangle$. Find the speed of the vector at time $t = 1$. Units are meters and seconds.

20. A particle moves according to the position function $\mathbf{r}(t) = \langle 2\sin t,\ 5\cos t,\ \sin 3t \rangle$. Find the velocity vector at time $t = 0$.

21–24. (a) Find the unit tangent vector function $\mathbf{T}(t)$ for the curve; (b) find the unit tangent vector to the curve when $t = 0$.

21. $\mathbf{r}(t) = \langle t^2 + 3t,\ 4t + 1,\ 6t \rangle$

22. $\mathbf{r}(t) = \langle 3t + 1,\ \sin 5t,\ \cos 5t \rangle$

23. $\mathbf{r}(t) = \left\langle \frac{4t^{3/2}}{3},\ \frac{t^2}{2},\ 2t - 7 \right\rangle$

24. $\mathbf{r}(t) = \langle e^t,\ e^{-t},\ \sqrt{2}t \rangle$

Ans. to reading exercise 6:
$\langle 1, 2, 3 \rangle$

25. Find the unit tangent vector to the curve $\mathbf{r}(t) = \langle t^2,\ 4t,\ 3t + 5 \rangle$ at the point $(1, -4, 2)$.

26. Find the unit tangent vector to the curve $\mathbf{r}(t) = \langle t,\ t^3,\ t^5 \rangle$ at the point $(2, 8, 32)$.

27. Find $\frac{d}{dt}\left(\langle 2,\ 5t,\ 10t^2 \rangle \cdot \langle \ln t,\ 2^t,\ \sec t \rangle \right)$.

28. Find $\frac{d}{dt}\left(\langle \sin t,\ \sin 2t,\ \sin 3t \rangle \cdot \langle t^4,\ \sqrt{t},\ e^{2t} \rangle \right)$.

29. Find $\frac{d}{dt}\left(\langle 2,\ 5t,\ 10t^2 \rangle \times \langle \ln t,\ 2^t,\ \sec t \rangle \right)$.

30. Find $\frac{d}{dt}\left(\langle \sin t,\ \sin 2t,\ \sin 3t \rangle \times \langle t^4,\ \sqrt{t},\ e^{2t} \rangle \right)$.

31–38. Evaluate the integral.

31. $\displaystyle\int \langle \sin t,\ \sin 2t,\ \cos 3t \rangle\ dt$

32. $\displaystyle \int \left\langle \frac{1}{t}, \frac{1}{t^2}, \frac{1}{t^3} \right\rangle \, dt$

33. $\displaystyle \int \left\langle \frac{1}{1+t^2}, \frac{1}{\sqrt{1-t^2}}, \frac{t}{1+t^2} \right\rangle \, dt$

34. $\displaystyle \int \left\langle t\sin 1, \, t\sin t, \, t\sin t^2 \right\rangle \, dt$

35. $\displaystyle \int_{-1}^{5} \left\langle t^2, \, 4t-1, \, 4 \right\rangle \, dt$

36. $\displaystyle \int_{4}^{1} \left\langle t^3, \, 3^t, \, 3^3 \right\rangle \, dt$

37. $\displaystyle \int_{0}^{2} \left\langle t\sqrt{t}, \, te^t, \, t \right\rangle \, dt$

38. $\displaystyle \int_{0}^{1} \left\langle \frac{1}{t-2}, \, \frac{1}{3t+4}, \, e^{2t} \right\rangle \, dt$

39–42. Solve the IVP.

39. $\mathbf{r}'(t) = \left\langle 3t+1, \, 6t^2+5t, \, \sqrt{t} \right\rangle, \mathbf{r}(1) = \langle 2,5,11 \rangle$

40. $\mathbf{r}'(t) = \left\langle 2t\sqrt{t^2+1}, \, 2t, \, t^2+1 \right\rangle, \mathbf{r}(0) = \langle 1,3,5 \rangle$

41. $\mathbf{r}'(t) = \langle \cos t, \, \sin t, \, \tan t \rangle, \mathbf{r}(\pi) = \langle 1,1,4 \rangle$

42. $\mathbf{r}'(t) = \left\langle 2t, \, \sec^2 t, \, \sqrt{t+2} \right\rangle, \mathbf{r}(1) = \langle 3,7,4 \rangle$

43. Find $\mathbf{r}(t)$ given $\mathbf{r}''(t) = \langle 0,0,-32 \rangle$, $\mathbf{r}'(0) = \langle 2,1,100 \rangle$, and $\mathbf{r}(0) = \langle 14,72,3 \rangle$.

44. A particle moves in three dimensions such that its acceleration vector at time t is $\langle 2,5,4t \rangle$, its initial velocity (at time $t=0$) is $\langle 12,5,1 \rangle$, and its initial position is $\langle -40,13,29 \rangle$. Determine the position vector function.

45. Prove the dot product rule using the following steps. Let $\mathbf{u}(t) = \langle u_1(t), u_2(t), u_3(t) \rangle$ and $\mathbf{v}(t) = \langle v_1(t), v_2(t), v_3(t) \rangle$.

    (a) Use the method of the original solution to example 18 to calculate an expression for $\frac{d}{dt}(\mathbf{u}(t) \cdot \mathbf{v}(t))$. Note that this can be done using existing methods and does not require use of the dot product rule.

    (b) Use the method of the alternate solution to example 18 to calculate an expression for $\mathbf{u}'(t) \cdot \mathbf{v}(t) + \mathbf{u}(t) \cdot \mathbf{v}'(t)$.

    (c) Note that the results of (a) and (b) are equal, and conclude the truth of the dot product rule.

46. Prove the cross product rule using the following steps. Let $\mathbf{u}(t) = \langle u_1(t), u_2(t), u_3(t) \rangle$ and $\mathbf{v}(t) = \langle v_1(t), v_2(t), v_3(t) \rangle$.

    (a) Use the method of the original solution to example 19 to calculate an expression for $\frac{d}{dt}(\mathbf{u}(t) \times \mathbf{v}(t))$. Note that this can be done using existing methods and does not require use of the cross product rule.

   (b) Use the method of the alternate solution to example 19 to calculate an expression for $\mathbf{u}'(t) \times \mathbf{v}(t) + \mathbf{u}(t) \times \mathbf{v}'(t)$.

   (c) Note that the results of (a) and (b) are equal, and conclude the truth of the cross product rule.

47. State and prove a sum rule for derivatives of vector-valued functions.

48. State and prove a constant multiple rule for derivatives of vector-valued functions.

49. Derive a product rule for $\frac{d}{dt}(f(t)\mathbf{u}(t))$, where $\mathbf{u}$ is a vector-valued function and f is a real-valued function.

50. Derive a chain rule for $\frac{d}{dt}(\mathbf{u}(f(t)))$, where $\mathbf{u}$ is a vector-valued function and f is a real-valued function.

51. Definition 2 defines the derivative of a vector-valued function component-by-component. Rewrite the definition to use infinitesimals in the style of Chapter 0 definition 12.

2.3 Arc Length and Curvature

The tools developed in the previous section can be built upon to help us answer geometric questions about curves.

2.3.1 Arc length

Consider the spacecurve shown in its entirety in Figure 2.24. How long is it? That is, if the curve is a wire and we straighten it out and measure it, what would be its length?

You may have studied arc length in two dimensions. If $\frac{dx}{dt}$ and $\frac{dy}{dt}$ are both continuous on $[c, d]$, then the length of the parametric curve $x = f(t), y = g(t)$ from $t = c$ to $t = d$ is $L = \int_c^d \sqrt{\left(\frac{dx}{dt}\right)^2 + \left(\frac{dy}{dt}\right)^2}\, dt$, provided the curve is traversed only once. But, if we write such a curve in the language of vector-valued functions by setting $\mathbf{r}(t) = \langle f(t), g(t)\rangle$, then $\mathbf{r}'(t) = \langle f'(t), g'(t)\rangle$ and $\|\mathbf{r}'(t)\| = \sqrt{\left(f'(t)\right)^2 + \left(g'(t)\right)^2}$, and the arc length formula can be rewritten as

$$L = \int_c^d \sqrt{\left(\frac{dx}{dt}\right)^2 + \left(\frac{dy}{dt}\right)^2}\, dt$$

$$= \int_c^d \sqrt{\left(f'(t)\right)^2 + \left(g'(t)\right)^2}\, dt$$

$$= \int_c^d \|\mathbf{r}'(t)\|\, dt.$$

See Section 9.4 of *Calculus Set Free*.

Because $x = f(t)$ and $y = g(t)$, $\frac{dx}{dt} = f'(t)$ and $\frac{dy}{dt} = g'(t)$; these substitutions are made in line 2.

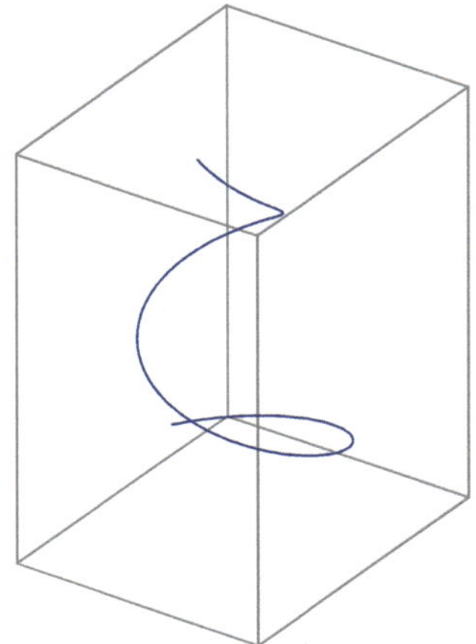

Figure 2.24 *A spacecurve*

This latter expression, which is in a form that does not specify the number of dimensions, remains true in any dimension, including three dimensions.

So, why does this formula work? If $\mathbf{r}(t)$ represents the position of a particle, then $\mathbf{r}'(t)$ is its velocity and $\|\mathbf{r}(t)\|$ is its speed. Then, the integral is adding small bits of $\|\mathbf{r}'(t_k)\|\Delta t$, which would be speed $\cdot$ time, which is distance. Add up all the small bits of distance traveled by the particle, and the result is the length of the curve.

ARC LENGTH FOR A VECTOR-VALUED FUNCTION

If the vector-valued function $\mathbf{r}$ is continuous on the interval $[c, d]$, then the length of the curve from $t = c$ to $t = d$ is

$$L = \int_c^d \|\mathbf{r}'(t)\| \, dt,$$

provided the curve is traversed only once.

Example 23 *Find the length of the curve* $\mathbf{r}(t) = \left\langle \frac{1}{2}t^2 - 3, \frac{(24t)^{3/2}}{36}, 12t + 7 \right\rangle$ *from* $t = 0$ *to* $t = 5$.

The expression $(24t)^{3/2}$ is the same as $\sqrt{(24t)^3}$, which has domain $t \geq 0$.

A reality check is to compute the points at each end of the curve and compute the distance between them; $t = 0$ yields the point $(-3, 0, 7)$ and $t = 5$ gives approximately $(9.5, 36.515, 67)$. The distance between these two points is 71.34 units. The length of the curve isn't much more than this, so the curve must be relatively straight, a fact that is verified by graphing the curve (Figure 2.25).

Solution The component functions are continuous for $t \geq 0$, so the hypotheses are satisfied and we may use the formula. First we compute $\mathbf{r}'(t)$:

$$\mathbf{r}'(t) = \left\langle \frac{1}{2}(2t), \frac{1}{36} \cdot \frac{3}{2}(24t)^{1/2} \cdot 24, 12 \right\rangle = \langle t, \sqrt{24t}, 12 \rangle.$$

Next, we compute $\|\mathbf{r}'(t)\|$:

$$\|\mathbf{r}'(t)\| = \sqrt{t^2 + 24t + 144} = \sqrt{(t + 12)^2} = |t + 12| = t + 12,$$

noting that $t \geq 0$.

Finally, we use the formula to compute the length of the curve:

$$L = \int_0^5 \|\mathbf{r}'(t)\|\, dt = \int_0^5 (t + 12)\, dt = \left(\frac{t^2}{2} + 12t \right)\Bigg|_0^5 = \cdots = 72.5 \text{ units.} \quad \blacksquare$$

Example 23 illustrates the special circumstances often needed to calculate arc length without use of numerical integration. More commonly, technology of some type is needed.

Example 24 *(a) Set up the integral for calculating the length of the curve* $\mathbf{r}(t) = \langle t, t^2, t^3 \rangle$ *from* $t = 1$ *to* $t = 2$. *(b) Use a calculator to determine the value of the integral.*

Solution

(a) The component functions are continuous, so we proceed by calculating $\mathbf{r}'(t)$,

$$\mathbf{r}'(t) = \langle 1, 2t, 3t^2 \rangle,$$

and its norm

$$\|\mathbf{r}'(t)\| = \sqrt{1 + 4t^2 + 9t^4}.$$

Using the formula, the arc length is

$$L = \int_1^2 \sqrt{1 + 4t^2 + 9t^4}\, dt.$$

(b) Using a numerical integration command on a calculator to evaluate the integral gives a length of 7.7075 units. $\blacksquare$

Reading Exercise 7   Find the length of the curve $\mathbf{r}(t) = \langle \cos t, \sin t, t \rangle$ from $t = 0$ to $t = 2\pi$.

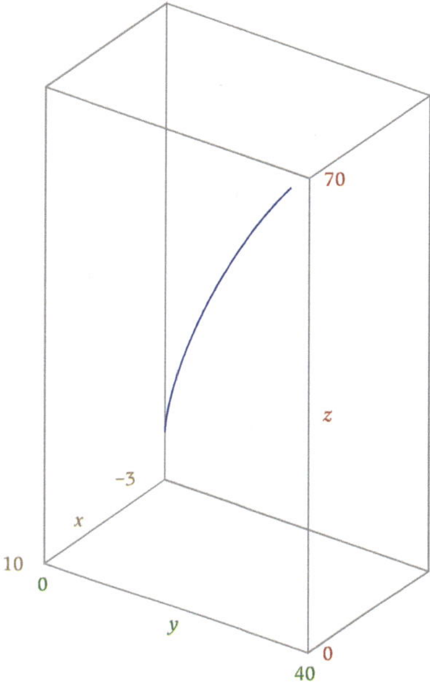

Figure 2.25 *The curve of example 23*

The indefinite integral of the integrand is non-elementary, that is, it involves functions that are not in the list of elementary functions.

In place of numerical integration on a calculator, a numerical integration method such as the trapezoid rule or Simpson's rule can be used.

2.3.2 Curvature

How curved is a curve? Consider the curves in Figures 2.24 and 2.25. At each point along the curve, a particle following the path of the curve is changing directions. Our task is to measure this curvature. Because the direction of a curve is one facet of velocity, perhaps measuring how fast velocity is changing will help. But we must be careful.

Suppose you are traveling in a car and come upon a curve in the road. The curvature of the road is a property of the road, not you or your car. Your car does not affect how curved the road is. Yet your experience of the curve does depend on you and your car! If you travel 50 km/h, you will not feel the curve as much as if you travel 150 km/h. The reason is that your direction of travel is changing more quickly with respect to time if you travel the higher speed, because you round the curve in a shorter amount of time. So neither $\mathbf{r}'(t)$ (velocity) nor $\mathbf{r}''(t)$ (change in velocity with respect to time), by themselves, measure how curved the road is. The curvature of the road depends on its geometry, without reference to time.

The root of the issue is that velocity is not just direction, but also magnitude. We need to consider only direction, taking speed (the magnitude of velocity) out of the equation. One way to accomplish this is to use the unit tangent vector $\mathbf{T}(t) = \dfrac{\mathbf{r}'(t)}{\|\mathbf{r}'(t)\|}$. Its length is always 1; no matter which parameterization is chosen, it eliminates the speed. The unit tangent is a property of the curve (the road, not the car).

Curvature is the change in direction, so perhaps we should try the derivative of direction, $\mathbf{T}'(t)$. Unfortunately, this derivative is the change with respect to time, so it also incorporates how fast we are traveling. But we can divide this by the speed as well! Doing so finally arrives at a description of the curvature of the road, regardless of how fast we travel on it.

Definition 5 CURVATURE *The curvature of a curve parameterized by $\mathbf{r}(t)$ is*

$$\kappa(t) = \frac{\|\mathbf{T}'(t)\|}{\|\mathbf{r}'(t)\|}.$$

An advantage of vector-valued function notation is that the formula does not need to specify the number of dimensions; it is valid in all dimensions. Our first example is in two dimensions.

Example 25 *Find the curvature of the curve given by $\mathbf{r}(t) = \langle 3\cos 10t, 3\sin 10t\rangle$.*

Solution To use the formula $\kappa(t) = \dfrac{\|\mathbf{T}'(t)\|}{\|\mathbf{r}'(t)\|}$, we need to calculate **❶** $\mathbf{r}'(t)$, **❷** $\|\mathbf{r}'(t)\|$, **❸** $\mathbf{T}(t)$, **❹** $\mathbf{T}'(t)$, **❺** $\|\mathbf{T}'(t)\|$, and finally **❻** $\kappa(t) = \dfrac{\|\mathbf{T}'(t)\|}{\|\mathbf{r}'(t)\|}$.

❶ First we calculate the velocity vector function:

$$\mathbf{r}'(t) = \langle -3\sin 10t \cdot 10,\ 3\cos 10t \cdot 10\rangle = \langle -30\sin 10t,\ 30\cos 10t\rangle.$$

Notice that 3 is the radius of the circle, which affects curvature, but the 10 represents how quickly the circle is being traversed and does not affect curvature.

❷ Next, we calculate the speed function:

$$\|\mathbf{r}'(t)\| = \sqrt{(-30\sin 10t)^2 + (30\cos 10t)^2}$$

$$= \sqrt{900\sin^2 10t + 900\cos^2 10t} = \sqrt{900(\sin^2 10t + \cos^2 10t)}$$

$$= \sqrt{900\cdot 1} = 30.$$

Professional driver, closed course.

Consider, in two dimensions, the vector-valued functions $\mathbf{r}_1(t) = \langle\cos t,\ \sin t\rangle$ and $\mathbf{r}_2(t) = \langle\cos 10t,\ \sin 10t\rangle$, both of which describe the unit circle. A particle traveling from $t = 0$ to $t = 2\pi$ according to $\mathbf{r}_1$ traverses the circle once, whereas a particle traveling according to $\mathbf{r}_2$ traverses the unit circle 10 times between $t = 0$ and $t = 2\pi$. In other words, $\mathbf{r}$ describes the car, not the road.

One caveat: there are still two possibilities depending on the specific parameterization, the unit tangent or its negation. Using either one is equivalent.

The symbol κ is the lowercase Greek letter "kappa." For our purposes, the expression $\kappa(t)$ can be read "curvature at t."

This curve is a circle of radius 3. Circles have uniform curvature, so the curvature should be identical for all values of t.

The chain rule is used when calculating the derivative.

These comments are meant to help the reader understand how the formula works, but they are not part of the solution itself.

The speed includes both the radius of the circle, 3, and the 10 that affects how often we traverse the circle.

❸ Then, we calculate the unit tangent vector function:

$$\mathbf{T}(t) = \frac{\mathbf{r}'(t)}{\|\mathbf{r}'(t)\|} = \frac{\langle -30 \sin 10t, 30 \cos 10t \rangle}{30} = \langle -\sin 10t, \cos 10t \rangle.$$

Now, both the radius of the circle and the speed are gone; the vector has magnitude 1 and describes the direction of the curve at t.

❹ Next up is the derivative of the unit tangent vector function:

$$\mathbf{T}'(t) = \langle -\cos 10t \cdot 10, -\sin 10t \cdot 10 \rangle = \langle -10 \cos 10t, -10 \sin 10t \rangle.$$

❺ We calculate the norm:

$$\|\mathbf{T}'(t)\| = \sqrt{(-10 \cos 10t)^2 + (-10 \sin 10t)^2}$$

$$= \sqrt{100 \cos^2 10t + 100 \sin^2 10t}$$

$$= \sqrt{100 \cdot 1} = 10.$$

The 10 represents how quickly the circle is being traversed, so it still describes the car and not just the road.

❻ Finally, we use the formula for curvature at t:

$$\kappa = \frac{\|\mathbf{T}'(t)\|}{\|\mathbf{r}'(t)\|} = \frac{10}{30} = \frac{1}{3}.$$

The numerator is how quickly we traverse the curve; the denominator is the product of how quickly we traverse the curve and the radius of the circle. The 10 in numerator and denominator cancel, leaving us with the radius of the circle in the denominator, which represents the curvature of the circle. ■

Following the calculations in example 25, if we replace the circle's radius of 3 with a generic radius of a, we see that a circle of radius a has curvature $\frac{1}{a}$. So why didn't we just define curvature as the reciprocal of the radius? Because most curves aren't circles.

The curvature of a circle of radius 50 is $\frac{1}{50} = 0.02$, whereas the curvature of a circle of radius 2 is $\frac{1}{2} = 0.5$. Why have a larger value of curvature for a smaller radius? Consider the two circles in Figure 2.26. If you were to take a stiff wire and bend it to make the circles, which circle would you make first? The larger one. You would have to continue bending the wire more to make a tighter curve to accomplish the smaller circle.

Reading Exercise 8 Find the curvature of the curve given by $\mathbf{r}(t) = \langle \cos t, 7, \sin t \rangle$.

Ans. to reading exercise 7:
$2\pi\sqrt{2}$ units (mid-exercise result: $\|\mathbf{r}'(t)\| = \sqrt{2}$)

The result is a constant because the curvature of a circle is constant. In general, the curvature function contains the variable t.

CURVATURE OF A CIRCLE

The curvature of a circle of radius a is $\frac{1}{a}$.

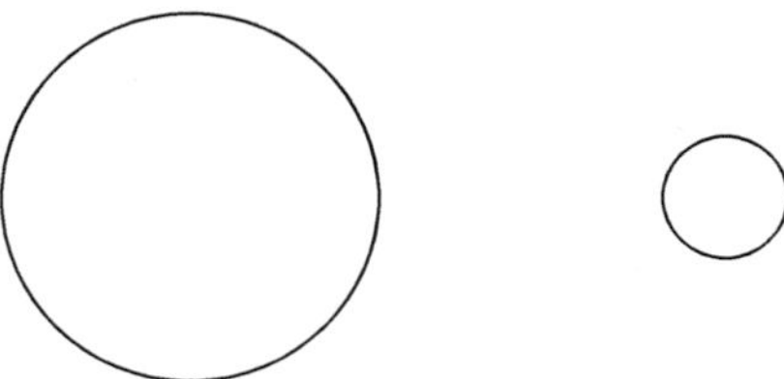

Figure 2.26 *Two circles. The circle with larger radius (left) has smaller curvature than the circle with smaller radius (right)*

In three dimensions, calculations using the definition of curvature can quickly become unwieldy. An alternate formula, whose proof is presented at the end of this section, is often easier to use.

ALTERNATE CURVATURE FORMULA

The curvature of a three-dimensional curve parameterized by $\mathbf{r}(t)$ is

$$\kappa(t) = \frac{\|\mathbf{r}'(t) \times \mathbf{r}''(t)\|}{\|\mathbf{r}'(t)\|^3}.$$

Example 26 *Find the curvature of the curve given by* $\mathbf{r}(t) = \langle t, t^2, t^3 \rangle$ *at the point* $(1, 1, 1)$.

Solution This time we use the alternate formula, for which we need to calculate ❶ $\mathbf{r}'(t)$, ❷ $\|\mathbf{r}'(t)\|$, ❸ $\mathbf{r}''(t)$, ❹ $\mathbf{r}'(t) \times \mathbf{r}''(t)$, ❺ $\|\mathbf{r}'(t) \times \mathbf{r}''(t)\|$, and finally ❻ $\kappa(t) = \frac{\|\mathbf{r}'(t) \times \mathbf{r}''(t)\|}{\|\mathbf{r}'(t)\|^3}$. Only the first two steps are the same as for using the definition.

❶ First, we calculate the velocity vector function:

$$\mathbf{r}'(t) = \langle 1,\ 2t,\ 3t^2 \rangle.$$

❷ Next, we calculate the speed function:

$$\|\mathbf{r}'(t)\| = \sqrt{1 + 4t^2 + 9t^4}.$$

❸ Then, we calculate the second derivative:

$$\mathbf{r}''(t) = \langle 0,\ 2,\ 6t \rangle.$$

4 Next up is the cross product:

$$\mathbf{r}'(t) \times \mathbf{r}''(t) = \begin{vmatrix} \mathbf{i} & \mathbf{j} & \mathbf{k} \\ 1 & 2t & 3t^2 \\ 0 & 2 & 6t \end{vmatrix}$$

$$= \langle 12t^2 - 6t^2, \ -(6t - 0), \ 2 - 0 \rangle = \langle 6t^2, \ -6t, \ 2 \rangle.$$

5 We calculate the norm:

$$\|\mathbf{r}'(t) \times \mathbf{r}''(t)\| = \sqrt{36t^4 + 36t^2 + 4}.$$

Notice that when the curve is not a circle, the curvature changes along the curve.

6 Finally, we use the alternate formula for curvature:

$$\kappa(t) = \frac{\|\mathbf{r}'(t) \times \mathbf{r}''(t)\|}{\|\mathbf{r}'(t)\|^3} = \frac{\sqrt{36t^4 + 36t^2 + 4}}{\left(\sqrt{1 + 4t^2 + 9t^4}\right)^3}.$$

We have calculated the curvature function. In this example, we wish to calculate the curvature at the point $(1, 1, 1)$, which occurs when $t = 1$. This curvature is

$$\kappa(1) = \frac{\sqrt{36 + 36 + 4}}{(1 + 4 + 9)^{3/2}} = \frac{\sqrt{76}}{14\sqrt{14}} \approx 0.1664,$$

which is close to $\frac{1}{6}$. We conclude that at the point $(1, 1, 1)$, the curvature is about the same as a circle of radius 6. ∎

Suppose for the curve in example 26 that we try using the definition to determine curvature. The first two steps would be the same as the featured solution. Then, step **3** would be to find the unit tangent vector,

$$\mathbf{T}(t) = \frac{\mathbf{r}'(t)}{\|\mathbf{r}(t)\|} = \frac{\langle 1, \ 2t, \ 3t^2 \rangle}{\sqrt{1 + 4t^2 + 9t^4}}$$

$$= \left\langle \frac{1}{\sqrt{1 + 4t^2 + 9t^4}}, \ \frac{2t}{\sqrt{1 + 4t^2 + 9t^4}}, \ \frac{3t^2}{\sqrt{1 + 4t^2 + 9t^4}} \right\rangle.$$

Then, step **4** would be to find $\mathbf{T}'(t)$, which would involve both the quotient and chain rules, and then we would need to find the norm of that messy expression, which is even worse! The alternate formula is often much simpler to use.

Ans. to reading exercise 8:
1 (mid-exercise result: $\mathbf{T}(t) = \langle -\sin t, 0, \cos t \rangle$)

2.3.3 Normal and binormal vectors

Consider a point on a curve. We can think of a tangent line as a line through that point on the curve matching the direction of the curve. We choose a line to be tangent because it has a constant direction.

At the same point, what if we wish to match the curvature of the curve in addition to the direction? A circle has constant curvature in the same manner that a line has constant direction, so perhaps we should use a circle to match the curvature at the point. We call such a circle an *osculating circle*. A curve with two instances of a tangent line and an osculating circle at a point on the curve is shown in Figure 2.27.

The osculating circle matches the direction of the curve, so it must also be tangent to the tangent line. You may recall from geometry that a line tangent to a circle is perpendicular (orthogonal) to a radius of the circle.

Our next task is to calculate a vector lying along that radius.

Think about a satellite in a circular orbit around the earth. It is "falling" toward the center of the earth under the influence of gravity, changing the direction of the satellite's path from its tangent direction. Its direction of change is toward the center of the earth, that is, toward the center of the osculating circle. Here, $\mathbf{T}'$ is the change in the unit tangent $\mathbf{T}$, so perhaps that vector points toward the osculating circle's center. But is it orthogonal to $\mathbf{T}$?

Recall that for any vector $\mathbf{v}$, $\mathbf{v} \cdot \mathbf{v} = \|\mathbf{v}\|^2$. Thus

$$\mathbf{T}(t) \cdot \mathbf{T}(t) = \|\mathbf{T}(t)\|^2 = 1^2 = 1.$$

Thus

$$\frac{d}{dt}\left(\mathbf{T}(t) \cdot \mathbf{T}(t)\right) = \frac{d}{dt}1 = 0.$$

But by the dot product rule,

$$\frac{d}{dt}\left(\mathbf{T}(t) \cdot \mathbf{T}(t)\right) = \mathbf{T}'(t) \cdot \mathbf{T}(t) + \mathbf{T}(t) \cdot \mathbf{T}'(t) = 2\mathbf{T}(t) \cdot \mathbf{T}'(t).$$

That is, the tangent line is perpendicular to the line $\overleftrightarrow{AO}$, where A is the point of tangency and O is the center of the circle.

In general, orbits are elliptical.

Just as a line is its own tangent line, a circle is its own osculating circle.

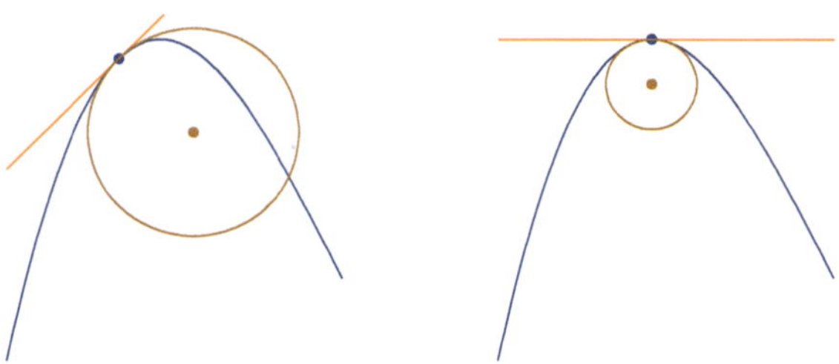

Figure 2.27 *A curve (blue), and at two points on the curve (blue) a tangent line (orange) and an osculating circle (brown). The tangent line is also tangent to the osculating circle*

Equating the two results,

$$\mathbf{T}(t) \cdot \mathbf{T}'(t) = 0,$$

and the unit tangent vector $\mathbf{T(t)}$ is orthogonal to $\mathbf{T}'(t)$, which could be called a *normal vector*. To isolate the direction, we divide by the norm to make it a *unit normal vector*.

Definition 6 UNIT NORMAL VECTOR AND OSCULATING PLANE *For a curve parameterized by* $\mathbf{r}(t)$ *with unit tangent vector function* $\mathbf{T}(t)$, *the* unit normal vector function $\mathbf{N}(t)$ *is given by*

$$\mathbf{N}(t) = \frac{\mathbf{T}'(t)}{\|\mathbf{T}'(t)\|}.$$

The unit normal vector $\mathbf{N}(t)$ is also known as the *principal unit normal vector.*

In three (or more) dimensions, the plane containing the point of tangency, the unit tangent vector, and the unit normal vector is called the osculating plane.

In three dimensions, we can also compute a *binormal vector*, which can help us determine a plane orthogonal to the tangent line.

Definition 7 BINORMAL VECTOR AND NORMAL PLANE *For a curve in* $\mathbf{R}^3$ *parameterized by* $\mathbf{r}(t)$ *with unit tangent vector function* $\mathbf{T}(t)$ *and unit normal vector function* $\mathbf{N}(t)$, *the vector function*

$$\mathbf{B}(t) = \mathbf{T}(t) \times \mathbf{N}(t)$$

is called the binormal vector function. *The plane containing the point of tangency, the unit normal vector, and the binormal vector is called the* normal plane.

The unit tangent, unit normal, and binormal vectors are illustrated in Figure 2.28. The osculating circle lies in the osculating plane. The line containing the unit normal vector also contains the center of the osculating circle. The normal plane is orthogonal to the tangent line, and it also contains the center of the osculating circle.

Example 27 *For* $\mathbf{r}(t) = \langle 3 \cos t,\ 4 \cos t,\ 5 \sin t \rangle$, *calculate (a)* $\mathbf{T}(t)$, *(b)* $\mathbf{N}(t)$, *(c)* $\mathbf{B}(t)$, *and find parametric equations for (d) the osculating plane at* $t = 0$ *and (e) the normal plane at* $t = 0$.

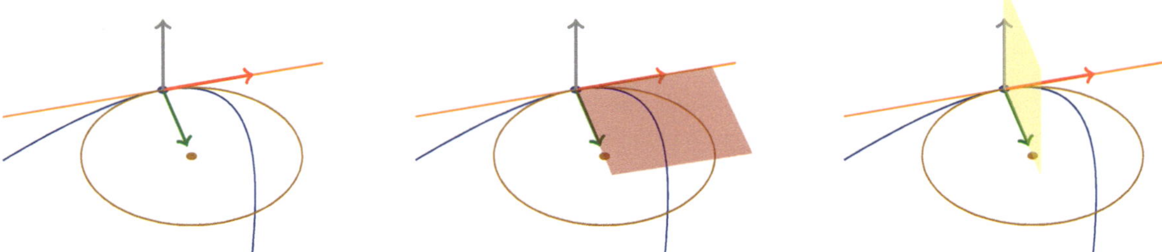

Figure 2.28 *(Left) a curve (blue), a tangent line (orange), an osculating circle (brown), $\mathbf{T}(t)$ (red), $\mathbf{N}(t)$ (green), and $\mathbf{B}(t)$ (gray); (middle) a portion of the osculating plane is shaded; (right) a portion of the normal plane is shaded*

Solution

(a) First, we need

$$\mathbf{r}'(t) = \langle -3\sin t,\ -4\sin t,\ 5\cos t \rangle$$

and

$$\|\mathbf{r}'(t)\| = \sqrt{9\sin^2 t + 16\sin^2 t + 25\cos^2 t} = \sqrt{25(\sin^2 t + \cos^2 t)} = 5.$$

Then

$$\mathbf{T}(t) = \frac{\mathbf{r}'(t)}{\|\mathbf{r}'(t)\|} = \frac{\langle -3\sin t,\ -4\sin t,\ 5\cos t \rangle}{5}$$
$$= \left\langle -\frac{3}{5}\sin t,\ -\frac{4}{5}\sin t,\ \cos t \right\rangle.$$

(b) First, we need

$$\mathbf{T}'(t) = \left\langle -\frac{3}{5}\cos t,\ -\frac{4}{5}\cos t,\ -\sin t \right\rangle$$

and

$$\|\mathbf{T}'(t)\| = \sqrt{\frac{9}{25}\cos^2 t + \frac{16}{25}\cos^2 t + \sin^2 t} = \sqrt{1} = 1.$$

Then

$$\mathbf{N}(t) = \frac{\mathbf{T}'(t)}{\|\mathbf{T}'(t)\|} = \frac{\left\langle -\frac{3}{5}\cos t,\ -\frac{4}{5}\cos t,\ -\sin t \right\rangle}{1}$$
$$= \left\langle -\frac{3}{5}\cos t,\ -\frac{4}{5}\cos t,\ -\sin t \right\rangle.$$

(c) We take the cross product:

$$\mathbf{B}(t) = \mathbf{T}(t) \times \mathbf{N}(t) = \begin{vmatrix} \mathbf{i} & \mathbf{j} & \mathbf{k} \\ -\dfrac{3}{5}\sin t & -\dfrac{4}{5}\sin t & \cos t \\ -\dfrac{3}{5}\cos t & -\dfrac{4}{5}\cos t & -\sin t \end{vmatrix}$$

$$= \mathbf{i}\left(\tfrac{4}{5}\sin^2 t + \tfrac{4}{5}\cos^2 t\right) - \mathbf{j}\left(\tfrac{3}{5}\sin^2 t + \tfrac{3}{5}\cos^2 t\right)$$

$$+ \mathbf{k}\left(\tfrac{12}{25}\sin t \cos t - \tfrac{12}{25}\sin t \cos t\right)$$

$$= \left\langle \tfrac{4}{5}, -\tfrac{3}{5}, 0 \right\rangle.$$

(d) The point on the curve corresponding to $t = 0$ is calculated using

$$\mathbf{r}(0) = \langle 3\cos 0,\ 4\cos 0,\ 5\sin 0 \rangle = \langle 3, 4, 0 \rangle.$$

The unit tangent and unit normal vectors are in the osculating plane, so two direction vectors are

$$\mathbf{T}(0) = \left\langle -\tfrac{3}{5}\sin 0,\ -\tfrac{4}{5}\sin 0,\ \cos 0 \right\rangle = \langle 0, 0, 1 \rangle$$

and

$$\mathbf{N}(0) = \left\langle -\tfrac{3}{5}\cos 0,\ -\tfrac{4}{5}\cos 0,\ -\sin 0 \right\rangle = \langle -\tfrac{3}{5}, -\tfrac{4}{5}, 0 \rangle.$$

Parametric equations for the osculating plane are therefore

$$x = 3 - \tfrac{3}{5}s, \quad y = 4 - \tfrac{4}{5}s, \quad z = t.$$

A vector form of the equation of the osculating plane would be $\langle 3, 4, 0 \rangle + t\langle 0, 0, 1 \rangle + s\langle -\tfrac{3}{5}, -\tfrac{4}{5}, 0 \rangle$.

(e) The unit normal and binormal vectors are in the normal plane. Using the point $(3, 4, 0)$ and direction vectors $\mathbf{N}(0) = \left\langle -\tfrac{3}{5}, -\tfrac{4}{5}, 0 \right\rangle$ and

$$\mathbf{B}(0) = \left\langle \tfrac{4}{5}, -\tfrac{3}{5}, 0 \right\rangle,$$

parametric equations for the normal plane are

$$x = 3 - \tfrac{3}{5}t + \tfrac{4}{5}s, \quad y = 4 - \tfrac{4}{5}t - \tfrac{3}{5}s, \quad z = 0. \qquad \blacksquare$$

2.3.4 Proof of the alternate curvature formula

Before beginning the proof, we first note that for a vector-valued function $\mathbf{r}$ we can define an arc length function s by

$$s(b) = \int_a^b \|\mathbf{r}'(t)\|\, dt,$$

with $s(b)$ denoting the length of the curve $\mathbf{r}$ from $t = a$ to $t = b$. Applying part I of the Fundamental Theorem of Calculus,

$$\frac{ds}{db} = \|\mathbf{r}'(b)\|.$$

Rewriting using our more commonly used variable t, $\|\mathbf{r}'(t)\| = \frac{ds}{dt}$. We wish to prove that in three dimensions,

$$\kappa(t) = \frac{\|\mathbf{T}'(t)\|}{\|\mathbf{r}'(t)\|} = \frac{\|\mathbf{r}'(t) \times \mathbf{r}''(t)\|}{\|\mathbf{r}'(t)\|^3}.$$

Recall that $\mathbf{T}(t) = \frac{\mathbf{r}'(t)}{\|\mathbf{r}'(t)\|}$. Rearranging,

$$\mathbf{r}'(t) = \|\mathbf{r}'(t)\|\mathbf{T}(t) = \frac{ds}{dt}\mathbf{T}(t).$$

Differentiating using the product rule,

Because $\frac{ds}{dt} = \|\mathbf{r}'(t)\|$ is real-valued and $\mathbf{T}(t)$ is vector-valued, this is the product rule developed in exercise 49 of Section 2.2.

$$\mathbf{r}''(t) = \frac{d^2 s}{dt^2}\mathbf{T}(t) + \frac{ds}{dt}\mathbf{T}'(t).$$

Then

Line 1 uses the two previous displayed formulas; line 2 uses cross product property (3); line 3 uses cross product property (2); and line 4 uses cross product property (5).

$$\mathbf{r}'(t) \times \mathbf{r}''(t) = \left(\frac{ds}{dt}\mathbf{T}(t)\right) \times \left(\frac{d^2 s}{dt^2}\mathbf{T}(t) + \frac{ds}{dt}\mathbf{T}'(t)\right)$$

$$= \frac{ds}{dt}\mathbf{T}(t) \times \frac{d^2 s}{dt^2}\mathbf{T}(t) + \frac{ds}{dt}\mathbf{T}(t) \times \frac{ds}{dt}\mathbf{T}'(t)$$

$$= \frac{ds}{dt}\frac{d^2 s}{dt^2}\left(\mathbf{T}(t) \times \mathbf{T}(t)\right) + \left(\frac{ds}{dt}\right)^2 \left(\mathbf{T}(t) \times \mathbf{T}'(t)\right)$$

$$= 0 + \left(\frac{ds}{dt}\right)^2 \left(\mathbf{T}(t) \times \mathbf{T}'(t)\right).$$

Then

$$\|\mathbf{r}'(t) \times \mathbf{r}''(t)\| = \left(\frac{ds}{dt}\right)^2 \|\mathbf{T}(t) \times \mathbf{T}'(t)\|$$

$$= \left(\frac{ds}{dt}\right)^2 \|\mathbf{T}(t)\| \, \|\mathbf{T}'(t)\| \sin\theta,$$

where θ is the angle between the vectors $\mathbf{T}(t)$ and $\mathbf{T}'(t)$. But we already know that these vectors are orthogonal, hence $\sin\theta = \sin\frac{\pi}{2} = 1$. We also know that the norm of the unit normal vector is 1. Thus, our expression becomes

$$= \left(\frac{ds}{dt}\right)^2 \cdot 1 \cdot \|\mathbf{T}'(t)\| \cdot 1 = \left(\frac{ds}{dt}\right)^2 \|\mathbf{T}'(t)\|.$$

Rearranging,

$$\|\mathbf{T}'(t)\| = \frac{1}{\left(\frac{ds}{dt}\right)^2}\|\mathbf{r}'(t) \times \mathbf{r}''(t)\| = \frac{\|\mathbf{r}'(t) \times \mathbf{r}''(t)\|}{\|\mathbf{r}'(t)\|^2}.$$

Then

$$\kappa(t) = \frac{\|\mathbf{T}'(t)\|}{\|\mathbf{r}'(t)\|} = \frac{1}{\|\mathbf{r}'(t)\|}\frac{\|\mathbf{r}'(t) \times \mathbf{r}''(t)\|}{\|\mathbf{r}'(t)\|^2} = \frac{\|\mathbf{r}'(t) \times \mathbf{r}''(t)\|}{\|\mathbf{r}'(t)\|^3}.$$

EXERCISES 2.3

1–6. Find the length of the curve as indicated.

   1. $\mathbf{r}(t) = \langle \sin 3t, \cos 3t, 0.1t \rangle$, from $t = -2$ to $t = 5$

   2. $\mathbf{r}(t) = \langle 5\cos t, 3\sin t, 4\sin t \rangle$, from $t = 0$ to $t = 2\pi$.

   3. $\mathbf{r}(t) = \langle 1 + t^2, 1 + 2t^2, 1 + 3t^2 \rangle$, from $t = -3$ to $t = -1$

   4. $\mathbf{r}(t) = \langle \sin t, \cos t, \cosh t \rangle$, from $t = 0$ to $t = \pi$

   5. $\mathbf{r}(t) = \left\langle 4 + \frac{\sin t}{t}, 1 + \frac{\cos t}{t}, \frac{t}{2} \right\rangle$, $1 \leq t \leq 3$

   6. $\mathbf{r}(t) = \left\langle t^{3/2}, \sqrt{t}, \sqrt{\frac{3}{2}}\,t \right\rangle$, $1 \leq t \leq 4$

7–10. (a) Set up the integral for finding the length of the indicated portion of the curve. (b) Use a calculator to determine the value of the integral.

   7. $\mathbf{r}(t) = \langle \sin 2t, \cos t, \sin t \rangle$, from $t = 0$ to $t = \pi$

   8. $\mathbf{r}(t) = \langle 3\sin t, 5\cos t, \frac{t^2}{5} \rangle$, $-\pi \leq t \leq \pi$

   9. $\mathbf{r}(t) = \langle e^t, \cosh t, 2\tan^{-1} t \rangle$, $-3 \leq t \leq 4$

  10. $\mathbf{r}(t) = \langle 100t, t^2 \sin 3t, t^2 \cos 3t \rangle$, $t = -10\pi$ to $t = 10\pi$

11. Find the length of the curve $\mathbf{r}(t) = \langle 1, t, \frac{1}{2}t^2 \rangle$, $0 \leq t \leq \sqrt{3}$ (a) using a table of integrals; (b) using a calculator; (c) using Simpson's rule with $n = 10$ to estimate the value of the integral to four decimal places.

12. Find the length of the curve $\mathbf{r}(t) = \langle 2t, t^2, \ln t \rangle$, $1 \leq t \leq 2$ (a) by hand, exactly; (b) using a calculator to evaluate the integral; (c) using Simpson's rule with $n = 10$ to estimate the value of the integral to four decimal places.

13–16. Find the curvature of the curve at the indicated point or value of t. Use the definition of curvature.

  13. $\mathbf{r}(t) = \langle \sin 3t, \cos 3t, 0.1t \rangle$, at $t = 1.37$

  14. $\mathbf{r}(t) = \langle 5\cos t, 3\sin t, 4\sin t \rangle$, at $t = \pi$

  15. $\mathbf{r}(t) = \langle 1 + t^2, 1 + 2t^2, 1 + 3t^2 \rangle$, at $t = 4$

  16. $\mathbf{r}(t) = \langle 2 + t, 4 - 5t, 7 - 2t \rangle$, anywhere

17–20. Find the curvature of the curve at the indicated point or value of t.

It is implied that these exercises are to be done by hand, exactly; do not use numerical integration.

The curve in exercise 5 is slightly modified from Section 2.1 example 6. Hint: Square root of a perfect square.

Many of these curves are graphing exercises in Section 2.1.

17. $\mathbf{r}(t) = \langle 1, t, \frac{1}{2}t^2 \rangle$, at $t = \sqrt{3}$

18. $\mathbf{r}(t) = \langle 2t, 5t, 7t^2 \rangle$, at the point $(-2, -5, 7)$

19. $\mathbf{r}(t) = \left\langle t^{3/2}, \sqrt{t}, \sqrt{\frac{3}{2}}\,t \right\rangle$, at the point $\left(1, 1, \sqrt{\frac{3}{2}}\right)$

20. $\mathbf{r}(t) = \langle 2t, t^2, \ln t \rangle$, at $t = 1$

21–24. For the given curve, find (a) $\mathbf{T}(t)$, (b) $\mathbf{N}(t)$, (c) $\mathbf{B}(t)$, (d) parametric equations for the osculating plane at the indicated value of t, and (e) parametric equations for the normal plane at the indicated value of t.

21. $\mathbf{r}(t) = \langle 1, t, \frac{1}{2}t^2 \rangle$, at $t = 1$

22. $\mathbf{r}(t) = \langle \sin 3t, \cos 3t, 0.1t \rangle$, at $t = \frac{\pi}{3}$

23. $\mathbf{r}(t) = \langle e^t \sin t, 3, e^t \cos t \rangle$, at $t = 0$

24. $\mathbf{r}(t) = \langle \cos t^2, \sin t^2, 5 \rangle$ for $t \geq 0$, at $t = \sqrt{\pi}$

25. (a) What is the curvature of a straight line? (b) What is the curvature of a circle of radius Ω, where Ω is positive infinite? (c) Explain the connection between (a) and (b).

26. Using the following steps, determine the curvature of a plane curve $y = f(x)$ at an inflection point $(t_0, f(t_0))$ for which $f''(t_0) = 0$.

The other possibility for an inflection point is that $f'(t_0)$ exists but $f''(t_0)$ does not exist; under this assumption, the curvature also does not exist.

Do not use a specific curve; work this exercise using $f(t), f'(t)$, etc.

(a) Parameterize the curve as $\mathbf{r}(t) = \langle t, f(t) \rangle$. Find $\mathbf{r}'(t)$.
(b) Determine $\|\mathbf{r}'(t)\|$.
(c) Determine $\mathbf{T}(t)$.
(d) Find $\mathbf{T}'(t)$.
(e) Using $f''(t_0) = 0$, determine $\mathbf{T}'(t_0)$.
(f) Finish by calculating $\kappa(t_0)$.

27. In a plane curve $y = f(x)$, at a point $(t_0, f(t_0))$ for which the curve is concave up, the unit normal vector should be pointing at least somewhat upward. Show that the y-component of $\mathbf{N}(t_0)$ is positive using the following steps.

Do not use a specific curve; work this exercise using $f(t), f'(t)$, etc. The first few steps are identical to the previous exercise.

(a) Parameterize the curve as $\mathbf{r}(t) = \langle t, f(t) \rangle$. Find $\mathbf{r}'(t)$.
(b) Determine $\|\mathbf{r}'(t)\|$.
(c) Determine the y-component of $\mathbf{T}(t)$.
(d) Find the y-component of $\mathbf{T}'(t)$.
(e) Simplify the y-component of $\mathbf{T}'(t)$ by multiplying numerator and denominator by $\sqrt{1 + (f'(t))^2}$. Then expand the numerator and simplify.
(f) Using $f''(t_0) > 0$ (because the curve is concave up there), conclude that the y-component of $\mathbf{T}'(t_0)$ is positive, and therefore so is the y-component of $\mathbf{N}(t_0)$.

28. Show that $\|\mathbf{B}(t)\| = 1$.

29. (a) Use the alternate curvature formula to find the curvature of $\mathbf{r}(t) = \langle \sin 3t, \cos 3t, 0.1t \rangle$ at $t = 1.37$.

(b) Compare to the work for exercise 13, where the definition of curvature is used. Which method is easier this time?

30. Let $\mathbf{r}(t) = \langle 1 + t^2,\, 1 + 2t^2,\, 1 + 3t^2 \rangle$.

    (a) Calculate $\|\mathbf{r}'(t)\|$ and $\|\mathbf{r}'(0)\|$.

    (b) Using either formula for curvature, what can be said about $\kappa(0)$? At what point on the curve does this refer?

    (c) When $t \neq 0$, the curvature is 0. What type of curve has curvature 0?

    (d) Write the relationship between the x- and y-components, and the relationship between the x- and z-components.

    (e) Re-parameterize the curve using the results from part (d). Does the result match the answer to part (c)?

    (f) Using the re-parameterization, find the curvature of this curve at the point from part (b). Is this the same conclusion about curvature as part (b)?

    (g) Graph the curve(s) given by the two different parameterizations. Are they really the same curve? How does this explain the answers to parts (b) and (f)?

2.4   Motion in Three Dimensions

We have already alluded to the ideas of velocity and speed for a vector-valued position function. In this section, we further our study of these ideas.

2.4.1   Vector-valued position, velocity, and acceleration

Let $\mathbf{r}(t)$ be vector-valued and represent the position of a particle in two, three, or even more dimensions at time t. Then velocity, the rate of change of position with respect to time, is given by the vector-valued function $\mathbf{v}(t) = \mathbf{r}'(t)$. The speed of the particle at time t is the magnitude of the velocity, $\|\mathbf{v}(t)\| = \|\mathbf{r}'(t)\|$, which is real-valued (not a vector). Acceleration is the rate of change of velocity, and just as with particle motion on a number line, acceleration at time t is $\mathbf{a}(t) = \mathbf{v}'(t) = \mathbf{r}''(t)$.

Example 28 *A particle moves through space according to the position function* $\mathbf{r}(t) = \left\langle 2t,\, \frac{t^2}{2},\, \frac{4}{3}t^{3/2} \right\rangle$. *Units are meters and seconds.*

    (a) *Find the particle's position, velocity, speed, and acceleration at time $t = 1$.*

    (b) *Does the particle ever stop?*

Solution (a) The position at time $t = 1$ is

$$\mathbf{r}(1) = \left\langle 2, \frac{1}{2}, \frac{4}{3} \right\rangle.$$

The velocity vector function is

$$\mathbf{v}(t) = \mathbf{r}'(t) = \left\langle 2, \frac{1}{2}(2t), \frac{4}{3}\left(\frac{3}{2}t^{1/2}\right) \right\rangle = \langle 2, t, 2\sqrt{t} \rangle.$$

Then, the velocity at time $t = 1$ is

$$\mathbf{v}(1) = \langle 2, 1, 2 \rangle,$$

which is a vector.

The speed at time $t = 1$ is

$$\|\mathbf{v}(1)\| = \sqrt{2^2 + 1^2 + 2^2} = \sqrt{9} = 3 \,\text{m/s},$$

Alternately, the speed function is $\|\mathbf{v}(t)\| = \sqrt{4 + t^2 + 4t}$, and then $\|\mathbf{v}(1)\| = \sqrt{4 + 1 + 4} = 3 \,\text{m/s}$.

which is a scalar.

The acceleration vector function is

$$\mathbf{a}(t) = \mathbf{v}'(t) = \left\langle 0, 1, \frac{1}{\sqrt{t}} \right\rangle.$$

The acceleration at time $t = 1$ is

$$\mathbf{a}(1) = \langle 0, 1, 1 \rangle.$$

(b) Method 1: The particle stops at time $t = k$ if the velocity vector is the zero vector, that is, if $\mathbf{v}(k) = \mathbf{0} = \langle 0, 0, 0 \rangle$. But $\mathbf{v}(t) = \langle 2, t, 2\sqrt{t} \rangle$, which has x-component $2 \neq 0$. The particle never stops.

Method 2: The particle stops when the speed is zero. The speed function is

$$\|\mathbf{v}(t)\| = \sqrt{2^2 + t^2 + (2\sqrt{t})^2} = \sqrt{4 + t^2 + 4t}.$$

Setting this expression equal to zero and solving for t, we have

$$\sqrt{4 + t^2 + 4t} = 0$$

$$4 + t^2 + 4t = 0$$

$$(t + 2)^2 = 0$$

$$t = -2.$$

When solving for t, always check the domain of the position function $\mathbf{r}$ to determine whether the solutions are valid.

But checking the domain of $\mathbf{r}$, the z-component $\frac{4}{3}\sqrt{t^3}$ requires $t \geq 0$, so the value $t = -2$ is not in the domain of the function. The particle never stops. ∎

Reading Exercise 9 For the position function $\mathbf{r}(t) = \langle t,\ \ln t,\ t^2 \rangle$, find the velocity, speed, and acceleration functions.

2.4.2 Tangential and normal components of acceleration

At any point on a spacecurve, the acceleration vector always lies in the osculating plane, as illustrated in Figure 2.29. Because $\mathbf{a}$, $\mathbf{T}$, and $\mathbf{N}$ all lie in a plane, it makes sense to look at the projections of $\mathbf{a}$ onto $\mathbf{T}$ and $\mathbf{N}$. The scalar components of those projections have special names; $\mathrm{comp}_{\mathbf{T}}\,\mathbf{a}$ is called the *tangential component of acceleration* and $\mathrm{comp}_{\mathbf{N}}\,\mathbf{a}$ is called the *normal component of acceleration*. The vector projections are illustrated in Figure 2.30; the scalar components are the lengths of the vector projections.

The osculating plane contains the tangent and normal lines. The binormal vector is orthogonal to the osculating plane.

Because the binormal vector is orthogonal to the osculating plane, $\mathrm{comp}_{\mathbf{B}}\,\mathbf{a} = 0$.

The tangential and normal components of acceleration can help us understand how the curve is changing at the point of tangency. If the tangential component is much larger than the normal component, such as illustrated in Figure 2.31, then

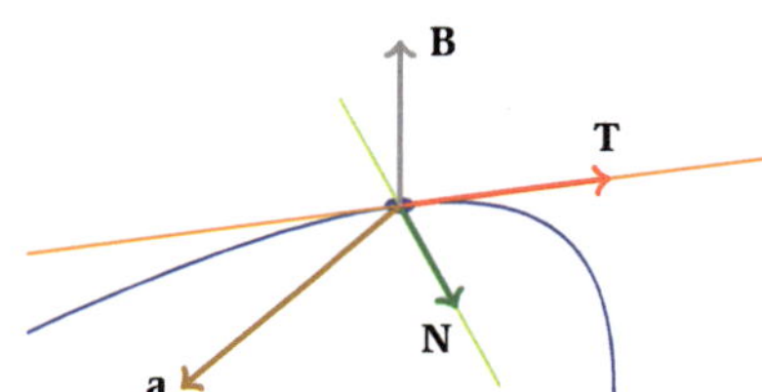

Figure 2.29 *A curve* $\mathbf{r}(t)$ *(blue), a point* $\mathbf{r}(t_0)$ *on the curve (blue), its tangent line (orange) and normal line (light green), unit tangent vector* $\mathbf{T}(t_0)$, *unit normal vector* $\mathbf{N}(t_0)$, *binormal vector* $\mathbf{B}(t_0)$, *and the acceleration vector* $\mathbf{a}(t_0)$. *The vectors* $\mathbf{a}$, $\mathbf{T}$, *and* $\mathbf{N}$ *lie in the osculating plane*

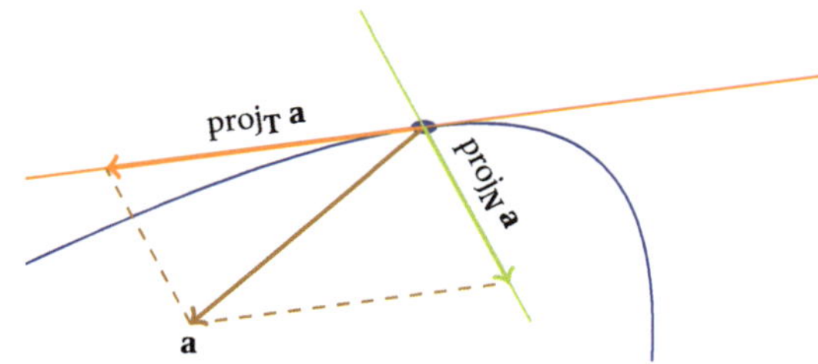

Figure 2.30 *A curve* $\mathbf{r}(t)$ *(blue), a point* $\mathbf{r}(t_0)$ *on the curve (blue), its tangent line (orange) and normal line (light green), the acceleration vector* $\mathbf{a}(t_0)$ *(brown), and the vector projections* $\mathrm{proj}_{\mathbf{T}}\mathbf{a}$ *(orange vector) and* $\mathrm{proj}_{\mathbf{N}}\mathbf{a}$ *(light green vector). Compare to Figure 2.29*

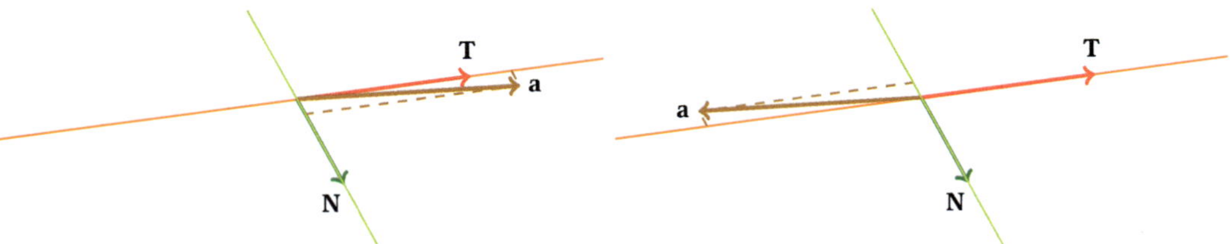

Figure 2.31 *A tangent line (orange), normal line (light green), unit tangent vector* **T** *(red), unit normal vector* **N** *(green), and acceleration vector* **a** *(brown). Here, the tangential component of acceleration is much larger than the normal component, so that a particle traveling the curve (not shown) is mostly speeding up (left) or slowing down (right)*

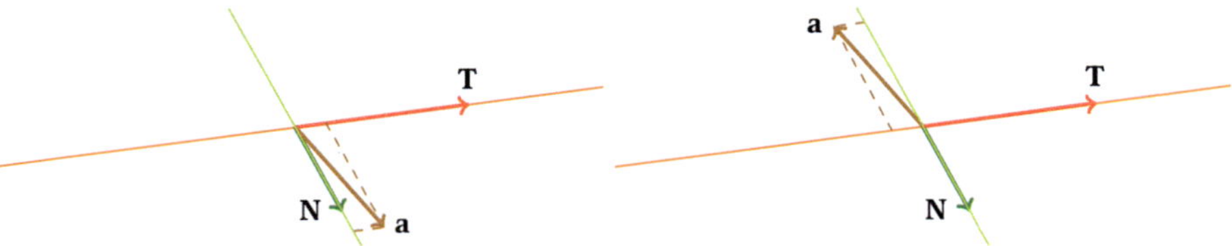

Figure 2.32 *A tangent line (orange), normal line (light green), unit tangent vector* **T** *(red), unit normal vector* **N** *(green), and acceleration vector* **a** *(brown). Here, the normal component of acceleration is much larger than the tangential component, so that a particle traveling the curve (not shown) is mostly changing direction without much change in forward speed*

there is little change in the direction of a particle traveling the curve; most of the change is the particle speeding up or slowing down.

On the other hand, if the normal component is much larger than the tangential component, then the greatest change is the direction of the particle, with little change of the forward speed of the particle; see Figure 2.32.

Using the formulas for the scalar component and the unit tangent, we have

Recall that $\text{comp}_{\mathbf{u}}\mathbf{v} = \frac{\mathbf{v}\cdot\mathbf{u}}{\|\mathbf{u}\|}$, $\|\mathbf{T}\| = 1$, and $\mathbf{T}(t) = \frac{\mathbf{r}'(t)}{\|\mathbf{r}'(t)\|}$.

$$\text{comp}_{\mathbf{T}}\,\mathbf{a} = \frac{\mathbf{a}\cdot\mathbf{T}}{\|\mathbf{T}\|} = \mathbf{a}\cdot\mathbf{T} = \mathbf{r}''(t)\cdot\frac{\mathbf{r}'(t)}{\|\mathbf{r}'(t)\|} = \frac{\mathbf{r}''(t)\cdot\mathbf{r}'(t)}{\|\mathbf{r}'(t)\|}.$$

Similarly,

$$\text{comp}_{\mathbf{N}}\,\mathbf{a} = \frac{\mathbf{a}\cdot\mathbf{N}}{\|\mathbf{N}\|} = \mathbf{a}\cdot\mathbf{N}.$$

Because **N** is sometimes difficult to compute, an alternate formula can be helpful:

Compare to the formula for $\text{comp}_{\mathbf{T}}\,\mathbf{a}$. Compare also to the formula for curvature.

$$\text{comp}_{\mathbf{N}}\,\mathbf{a} = \frac{\|\mathbf{r}'(t)\times\mathbf{r}''(t)\|}{\|\mathbf{r}'(t)\|}.$$

The derivation of this formula is explored in the exercises.

Example 29 *Find the tangential and normal components of acceleration for* $\mathbf{r}(t) = \langle t, t^2, t^3 \rangle$.

Solution   First, we calculate $\mathbf{r}'(t)$, $\|\mathbf{r}'(t)\|$, and $\mathbf{r}''(t)$:

$$\mathbf{r}'(t) = \langle 1, 2t, 3t^2 \rangle,$$

$$\|\mathbf{r}'(t)\| = \sqrt{1 + 4t^2 + 9t^4},$$

$$\mathbf{r}''(t) = \langle 0, 2, 6t \rangle.$$

For calculating the tangential component we also need a dot product:

$$\mathbf{r}'(t) \cdot \mathbf{r}''(t) = 0 + 4t + 18t^3.$$

Using the formula for the tangential component gives

$$\mathrm{comp}_T\, \mathbf{a} = \frac{\mathbf{r}'(t) \cdot \mathbf{r}''(t)}{\|\mathbf{r}'(t)\|} = \frac{4t + 18t^3}{\sqrt{1 + 4t^2 + 9t^4}}.$$

For calculating the normal component we need a cross product:

$$\mathbf{r}'(t) \times \mathbf{r}''(t) = \begin{vmatrix} \mathbf{i} & \mathbf{j} & \mathbf{k} \\ 1 & 2t & 3t^2 \\ 0 & 2 & 6t \end{vmatrix}$$

$$= \langle 12t^2 - 6t^2, -(6t - 0), 2 - 0 \rangle$$

$$= \langle 6t^2, -6t, 2 \rangle.$$

Then

$$\|\mathbf{r}'(t) \times \mathbf{r}''(t)\| = \sqrt{36t^4 + 36t^2 + 4}.$$

Using the alternate formula for the normal component gives

$$\mathrm{comp}_N\, \mathbf{a} = \frac{\|\mathbf{r}'(t) \times \mathbf{r}''(t)\|}{\|\mathbf{r}'(t)\|} = \frac{\sqrt{36t^4 + 36t^2 + 4}}{\sqrt{1 + 4t^2 + 9t^4}}. \qquad \blacksquare$$

Reading Exercise 10   Find the tangential component of acceleration for $\mathbf{r}(t) = \langle 2t, 4t, t^2 \rangle$.

2.4.3   Ideal projectile motion

Suppose that in two dimensions we wish to launch a projectile from the origin in the positive x-direction over horizontal ground, as illustrated in Figure 2.33. In *ideal*

Ans. to reading exercise 9:
$$\mathbf{v}(t) = \left\langle 1, \tfrac{1}{t}, 2t \right\rangle,$$
$$\|\mathbf{v}(t)\| = \sqrt{1 + \tfrac{1}{t^2} + 4t^2},\ \mathbf{a}(t) = \left\langle 0, -\tfrac{1}{t^2}, 2 \right\rangle$$

Recall that the dot product is commutative.

Notice that the difference in the formulas for the tangential and normal components is that the tangential component uses the dot product whereas the normal component uses the cross product. This is another reason to use the alternate formula for the normal component.

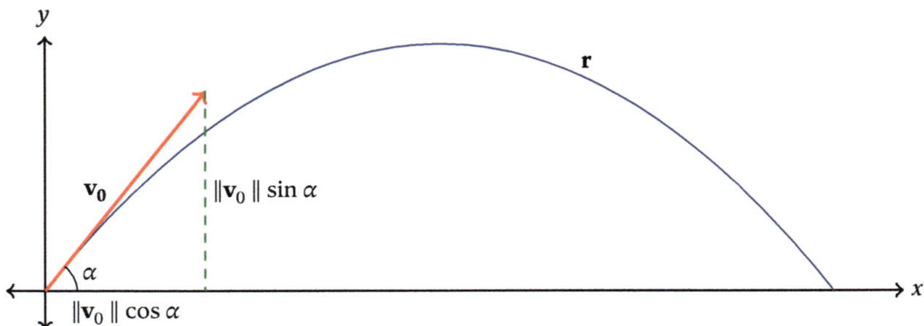

Figure 2.33 *A projectile path* $\mathbf{r}$ *(blue) and its initial velocity vector* $\mathbf{v_0}$ *(red), with launch angle* α

Each of these factors affect the flight of a real-life projectile, but not all must be considered in a particular application if the relative size of the effect is small enough.

projectile motion we assume that the only force acting on the object after launch is gravity; no aerodynamic drag (badminton shuttlecock), no lift (spinning golf ball), ignoring the turning of the earth on its axis, wind, and anything else. And there is no additional propulsion (think arrow, not rocket).

Let $\mathbf{v_0}$ be the initial velocity vector (velocity at time $t = 0$); this provides both the direction of the projectile at launch and the launch speed (initial speed), $\|\mathbf{v_0}\|$. The horizontal component of $\mathbf{v_0}$ is $\|\mathbf{v_0}\|\cos\alpha$, and the length of the vertical component of $\mathbf{v_0}$ is $\|\mathbf{v_0}\|\sin\alpha$. Therefore

The hypotenuse of the right triangle has length $\|\mathbf{v_0}\|$; if y_0 is the length of the vertical segment, then $\sin\alpha = \frac{y_0}{\|\mathbf{v_0}\|}$. Rearranging yields $y_0 = \|\mathbf{v_0}\|\sin\alpha$.

$$\mathbf{v}(0) = \mathbf{v_0} = \langle\|\mathbf{v_0}\|\cos\alpha,\ \|\mathbf{v_0}\|\sin\alpha\rangle.$$

Let $\mathbf{a}(t)$ be the acceleration due to gravity. This acceleration is directed toward the center of the earth, or the direction of $-\mathbf{j}$. Then, in the English system $\mathbf{a}(t) = -32\,\text{ft/s}^2\,\mathbf{j}$, and in the SI system $\mathbf{a}(t) = -9.8\,\text{m/s}^2\,\mathbf{j}$. Let $g = 32$ or $g = 9.8$ as appropriate; then

The IVP is $\mathbf{a}(t) = \langle 0, -g\rangle$, $\mathbf{v}(0) = \langle\|\mathbf{v_0}\|\cos\alpha, \|\mathbf{v_0}\|\sin\alpha\rangle$.

$$\mathbf{a}(t) = \langle 0, -g\rangle.$$

Solving the IVP, we take an antiderivative and get $\mathbf{v}(t) = \langle C_1, -gt + C_2\rangle$, followed by $\mathbf{v}(0) = \langle C_1, C_2\rangle$. Therefore, $C_1 = \|\mathbf{v_0}\|\cos\alpha$, $C_2 = \|\mathbf{v_0}\|\sin\alpha$, and

$$\mathbf{v}(t) = \langle\|\mathbf{v_0}\|\cos\alpha,\ -gt + \|\mathbf{v_0}\|\sin\alpha\rangle.$$

One more IVP to solve to reach the goal of having an equation for $\mathbf{r}$; once again finding an antiderivative gives $\mathbf{r}(t) = \langle(\|\mathbf{v_0}\|\cos\alpha)t + C_3,\ -\frac{g}{2}t^2 + (\|\mathbf{v_0}\|\sin\alpha)t + C_4\rangle$, and $\mathbf{r}(0) = \langle C_3, C_4\rangle$. But with the initial position of the projectile at the origin, $\mathbf{r}(0) = \langle 0, 0\rangle$, and thus $C_3 = 0$ and $C_4 = 0$. We therefore arrive at the following equation for ideal projectile motion.

IDEAL PROJECTILE MOTION

The position function of a projectile launched from the origin in the positive x-direction with initial velocity $\mathbf{v_0}$ and launch angle α is

$$\mathbf{r}(t) = \left\langle (\|\mathbf{v_0}\| \cos \alpha)t, \; -\tfrac{g}{2}t^2 + (\|\mathbf{v_0}\| \sin \alpha)t \right\rangle.$$

The x-component of $\mathbf{r}$ represents the distance downrange from the launch site, and the y-component represents the height of the projectile over the horizontal ground.

See Figure 2.33 for a visual explanation of the roles of the components.

You may have previously worked with vertical projectile motion. Some of the same principles apply; for instance, when the projectile impacts the ground its height above the ground is 0 ($y = 0$) and the maximum height is achieved when the vertical component of velocity is 0.

Ans. to reading exercise 10:
$$\operatorname{comp}_{\mathbf{T}} \mathbf{a} = \frac{4t}{\sqrt{20+4t^2}} = \frac{2t}{\sqrt{5+t^2}}$$

Example 30 *A projectile is launched with an initial speed (muzzle speed) of 250 ft/s and a launch angle of 40°.*

   (a) *How far from the launch position will the projectile land?*

   (b) *A 100-foot-tall tree stands 600 yards from the launch position, directly between the launch position and the target. Will the projectile clear the tree?*

It is common for the muzzle speed to be referred to as "muzzle velocity," but in our terminology muzzle velocity would also need to have a direction associated with it, so we use the term "muzzle speed" to avoid confusion.

Solution The information appears to describe ideal projectile motion. We are given that $\|\mathbf{v_0}\| = 250$ ft/s and $\alpha = 40°$. With units given in feet and seconds, we use $g = 32$. Then

Unless informed otherwise, assume that the ground is horizontal (no elevation change).

$$\mathbf{r}(t) = \left\langle (\|\mathbf{v_0}\| \cos \alpha)t, \; -\tfrac{g}{2}t^2 + (\|\mathbf{v_0}\| \sin \alpha)t \right\rangle$$

$$= \left\langle (250 \cos 40°)t, \; -16t^2 + (250 \sin 40°)t \right\rangle$$

$$= \left\langle 191.51t, \; -16t^2 + 160.70t \right\rangle.$$

(a) The projectile lands when $y = 0$, so we set the y-component equal to zero and solve:

$$-16t^2 + 160.70t = 0$$

$$t(-16t + 160.70) = 0$$

$$t = 0, \quad t = \frac{160.70}{16} = 10.044 \text{ s}.$$

One solution, $t = 0$, is when the projectile is launched. The other solution, $t = 10.044$ s, is when the projectile lands. The question asks how far from the launch position will the projectile land, that is how far downrange is the projectile when it lands. The x-component at $t = 10.044$ s is desired:

$$x = 191.51t = 191.51(10.044) = 1924 \text{ ft.}$$

Some roundoff error is present. Using more significant digits in the intermediate calculations results in a distance that rounds to 1923 ft.

1 yard $= 3$ feet.

(b) To determine whether the projectile clears the tree, we need to know the height of the projectile (y-component) when the x-component is 1800. But we know y in terms of t, not x. So we must determine the value of t corresponding to $x = 1800$:

$$1800 = 191.51t$$

$$t = \frac{1800}{191.51} = 9.399 \text{ s.}$$

An alternate solution method would be to determine when the y-component is 100 ft, which results in two values of t. Then determine the x-component at these two times. Between these two values of x, the projectile is at least 100 ft above the ground. If $x = 1800$ is included, then the projectile clears the tree. Otherwise, the projectile does not clear the tree.

When $t = 9.399$ s, the y-component is

$$y = -16(9.399)^2 + 160.70(9.399) = 96.96 \text{ ft.}$$

No! It's close, but with this launch position, muzzle speed, and launch angle, the projectile will not clear the 100-foot tree.   ∎

Success in projectile motion questions lies in determining the equation of motion and then reasoning correctly (and perhaps creatively) about the requested quantities.

Reading Exercise 11 Find $\mathbf{r}(t)$ for a projectile launched with initial speed 100 ft/s at a launch angle of 20°.

2.4.4   Ideal projectile motion in three dimensions

Now suppose that we are in three dimensions, but still launching our projectile over horizontal ground (now the xy-plane) with initial speed $\|\mathbf{v_0}\|$ and launch angle (from the horizontal) α. The same analysis shows that the distance downrange from the launch site is $(\|\mathbf{v_0}\| \cos \alpha)t$ and the height above the ground, which is now the z-component, is $-\frac{g}{2}t^2 + (\|\mathbf{v_0}\| \sin \alpha)t$. But instead of being restrained to launching the projectile in the positive x-direction, we can swivel the launcher to propel the projectile toward the north, the south-southwest, or any other desired direction over the xy-plane.

Suppose that the horizontal direction toward which we launch the projectile is represented by a unit vector $\langle x_0, y_0 \rangle$. Then, the distance downrange is the multiplier for that vector to determine the x- and y-components of the projectile's position. The equations of motion would then be as follows.

IDEAL PROJECTILE MOTION, THREE DIMENSIONS

The position function of a projectile launched from the origin toward the horizontal direction of the unit vector $\langle x_0, y_0 \rangle$, with initial speed $\|\mathbf{v_0}\|$ and launch angle α, is

$$\mathbf{r}(t) = \left\langle (x_0\|\mathbf{v_0}\| \cos\alpha)t, \ (y_0\|\mathbf{v_0}\| \cos\alpha)t, \ -\tfrac{g}{2}t^2 + (\|\mathbf{v_0}\| \sin\alpha)t \right\rangle.$$

The distance downrange from the launch site is $(\|\mathbf{v_0}\| \cos\alpha)t$, and the z-component represents the height of the projectile over the horizontal ground.

Example 31 *A projectile is launched with azimuth 105°, initial speed 200 m/s, and launch angle 32°.*

   (a) *Find the position of the projectile when it is 1 km downrange.*
   (b) *Determine the maximum height of the projectile.*

Solution We are given $\|\mathbf{v_0}\| = 200$ m/s and $\alpha = 32°$. Because units are meters and seconds, we use $g = 9.8$. We also need x_0 and y_0, which can be determined from the azimuth, as in Figure 2.34. The azimuth must be converted into the angle in standard position, and then the point on the unit circle can be used (because $\langle x_0, y_0 \rangle$ is a unit vector). Hence, $x_0 = \cos(-15°)$ and $y_0 = \sin(-15°)$. Using the ideal projectile motion in three dimensions formula,

The *azimuth* is the angle measured clockwise from due north. East has azimuth 90°, south has azimuth 180°, and west has azimuth 270°.

$$\mathbf{r}(t) = \left\langle (x_0\|\mathbf{v_0}\| \cos\alpha)t, \ (y_0\|\mathbf{v_0}\| \cos\alpha)t, \ -\tfrac{g}{2}t^2 + (\|\mathbf{v_0}\| \sin\alpha)t \right\rangle$$

$$= \langle (\cos(-15°) \cdot 200 \cos 32°)t, \ (\sin(-15°) \cdot 200 \cos 32°)t,$$

$$-4.9t^2 + (200 \sin 32°)t \rangle$$

$$= \langle 163.83t, \ -43.898t, \ -4.9t^2 + 105.98t \rangle.$$

(a) We can easily find position at time t, but we are asked to find the position when the projectile is 1 km = 1000 m downrange. The distance downrange is

$$d = (\|\mathbf{v_0}\| \cos\alpha)t = (200 \cos 32°)t = 169.61t.$$

Ans. to reading exercise 11:
$$\mathbf{r}(t) = \langle 93.97t, \ -16t^2 + 34.20t \rangle$$

The time when that occurs can be found by setting $d = 1000$ and solving for t:

$$169.61t = 1000$$

$$t = \frac{1000}{169.61} = 5.8959 \text{ s.}$$

The position of the projectile when it is 1 km downrange is therefore

$$\mathbf{r}(5.8959) = \langle 965.93, \ -258.82, \ 454.52 \rangle.$$

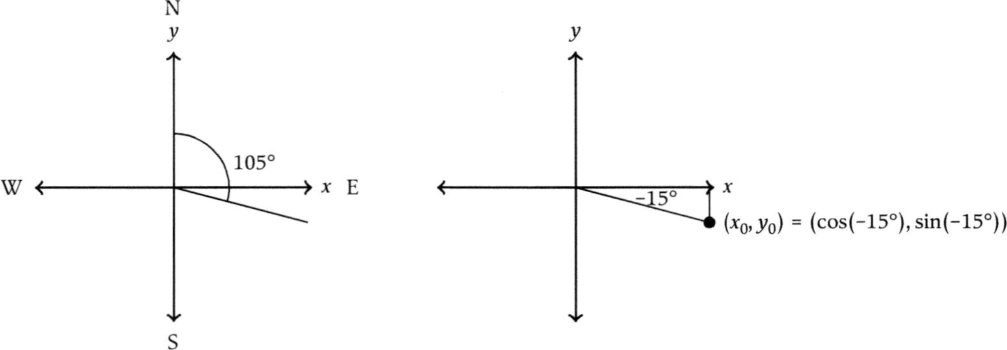

Figure 2.34 *(Left) azimuth 105°; (right) corresponding angle in standard position −15°, with (x_0, y_0) on the unit circle*

(b) The maximum height occurs when the z-component of velocity is zero (which is when the projectile stops rising and begins falling). The velocity vector function is

$$\mathbf{v}(t) = \langle 163.83, -43.898, -9.8t + 105.98 \rangle.$$

We set the z-component equal to zero and solve for t:

$$-9.8t + 105.98 = 0$$

$$t = \frac{105.98}{9.8} = 10.814 \,\text{s}.$$

The height of the projectile at $t = 10.814$ s is the z-component of $\mathbf{r}(10.814)$:

$$-4.9(10.814)^2 + 105.98(10.814) = 573.05 \,\text{m}.$$

The maximum height of the projectile is 573.05 m. ■

Note: for launching a long projectile such as an arrow, $\mathbf{r}$ gives the position of the tip of the arrowhead.

EXERCISES 2.4

1. A particle moves through space according to the position function
$\mathbf{r}(t) = \left\langle \frac{t^2}{2} - t + 3, \ \frac{t^3}{3} - t, \ \cos \pi t \right\rangle.$
    (a) Find the velocity, speed, and acceleration functions.
    (b) Find the velocity, speed, and acceleration at time $t = 0$.
    (c) Does the particle ever stop? If so, at what point does it stop? What is its acceleration at that time?

2. A particle moves through space according to the position function
$\mathbf{r}(t) = \left\langle 2t^2,\ t^3,\ \frac{8t}{3} \right\rangle$.

   (a) Find the velocity, speed, and acceleration functions.
   (b) Find the velocity, speed, and acceleration at time $t = 2$.
   (c) Does the particle ever stop? If so, at what point does it stop? What is its acceleration at that time?

3. A particle moves through space according to the position function
$\mathbf{r}(t) = \left\langle 3t,\ \frac{1}{t},\ \ln t \right\rangle$.

   (a) Find the particle's position, velocity, speed, and acceleration at time $t = 1$.
   (b) Does the particle ever stop? If so, when and where?
   (c) For what values of t is the particle traveling below the xy-plane?

4. A particle moves through space according to the position function
$\mathbf{r}(t) = \left\langle -\cos t,\ t^2,\ \cos 2t \right\rangle$.

   (a) Find the velocity, speed, and acceleration functions.
   (b) Find the velocity, speed, and acceleration at time $t = 2$.
   (c) Does the particle ever stop? If so, at what point does it stop? What is its acceleration at that time?

5–8. Find the tangential and normal components of acceleration for the given curve.

5. $\mathbf{r}(t) = \langle \sin 3t,\ \cos 3t,\ 0.1t \rangle$
6. $\mathbf{r}(t) = \langle 5\cos t,\ 3\sin t,\ 4\sin t \rangle$
7. $\mathbf{r}(t) = \langle 1 + t^2,\ 1 + 2t^2,\ 1 + 3t^2 \rangle$
8. $\mathbf{r}(t) = \langle 2 + t,\ 4 - 5t,\ 7 - 2t \rangle$

These are the same curves used in curvature exercises in Section 2.3.

9–12. (a) Find the tangential and normal components of acceleration for the given curve. (b) Find the tangential and normal components of acceleration at the indicated time or point.

9. $\mathbf{r}(t) = \langle 1,\ t,\ \frac{1}{2}t^2 \rangle,\ t = 1$
10. $\mathbf{r}(t) = \langle 2t,\ 5t,\ 7t^2 \rangle,\ t = 0$
11. $\mathbf{r}(t) = \langle t^{3/2},\ \sqrt{t},\ \sqrt{\frac{3}{2}}\,t \rangle,\ \text{point} \left(8, 2, 4\sqrt{\frac{3}{2}} \right)$
12. $\mathbf{r}(t) = \langle 2t,\ t^2,\ \ln t \rangle,\ \text{point}\ (2, 1, 0)$

These are the same curves used in curvature exercises in Section 2.3.

13. An arrow is launched with an initial speed of 60 m/s and a launch angle of 20°.

   (a) How far from the launch position will it land?
   (b) What is its maximum height?
   (c) The launch angle is changed to 86° in hopes that it will not go across the street into a neighbor's backyard. How far from the launch position will it land?

DO NOT TRY THIS AT HOME. My dad did–and it landed across the street in a fenced backyard. Fortunately, no one was harmed. It could have been tragic. Remember, safety first!

14. A projectile is launched with an initial speed of 110 m/s and a launch angle of 36°.

    (a) How far from the launch position will it land?

    (b) What is its maximum height?

15. An arrow is shot with an initial speed of 300 ft/s with a launch angle of 30°, from a point 5 ft above the ground.

    (a) How far from the launch position will it land?

    (b) Will the arrow clear a 30-ft-tall tree 100 ft from the launch point?

    (c) Instead of shooting the arrow over level ground, the arrow is shot from the top of a 40 ft cliff, over a flat plain below. How far from the launch position will it land?

16. A projectile is launched from a point 10 ft above the ground with an initial speed of 70 ft/s and a launch angle of 50°.

    (a) How far from the launch position will it land?

    (b) Your launch point is hidden 20 ft behind a 30 ft wall. Will your projectile hit the wall?

17. An arrow is shot with an initial speed of 80 m/s toward a target level with the release point at a distance of 150 m. What launch angle(s) could be used to hit the target?

18. A projectile is to be launched with an initial speed of 180 m/s, with an intended target 2 km away. What launch angle(s) will allow us to hit the target?

19. A projectile is launched with azimuth 245°, initial speed 600 m/s, and launch angle 40°.

    (a) Find the position of the projectile when it is 1 km downrange.

    (b) Determine the maximum height of the projectile.

20. A projectile is launched with azimuth 82°, initial speed 200 ft/s, and launch angle 20°.

    (a) Find the position of the projectile when it is 500 ft downrange.

    (b) Determine the maximum height of the projectile.

21. A projectile is launched toward the northeast with initial speed 100 m/s, and launch angle 35°.

    (a) At what coordinates will the projectile land?

    (b) One second before landing, how high is the projectile?

22. A projectile is launched with bearing N30°W with initial speed 680 m/s, and launch angle 35°.

    (a) At what coordinates will the projectile land?

    (b) A north-south road lies 2 km west of the launch point. When the projectile crosses above the road, how high is it?

23. In the game of Chunkey, a stone similar to a round wheel is rolled and two contestants throw Chunkey sticks (spears) to see who can

get closest to the stone. The catch is that the Chunkey sticks must be thrown before the stone stops rolling. The closest to the stone gets one point. The game is won by the first contestant to 11 points. Assume that the playing ground is horizontal.

(a) Suppose you judge that the stone will stop 50 ft downrange. Your favorite launch angle is 30°, from a point 6 ft above the ground. What initial speed should you use?

(b) Suppose you judge that the stone will stop 60 ft downrange. You favorite initial speed is 45 ft/s, again from a point 6 ft above the ground. What launch angle(s) should you use?

(c) In part (b) you can choose from two angles. Suppose that you are consistently able to throw within 1° of your intended launch angle. Keep the initial speed constant. For the first launch angle, add and subtract 1° and see how far from the intended target each throw will land. Repeat for the second launch angle. Is there a difference? If so, which launch angle is the better choice?

24. An enemy fort in the shape of a square is located on the flat top of a 100-ft-tall hill. The fort's walls run north/south and east/west and are 600 ft long, as measured on the outside. Its walls are 15 ft tall and 30 ft wide, made of stone. You are located 2000 ft west and 300 ft south of the southwest corner of the fort, in the flat plain below the hill. Your spies, whom you trust, inform you that the fort's ammunition storage is located in a 10 ft by 10 ft square pit dug out of the ground with a ground-level wooden roof, against the walls at the southwest corner of the fort. The fort's headquarters are in a square 20 ft by 20 ft building, 12 ft high, located against the north wall, halfway along that wall. You can fire projectiles with a 300 ft/s muzzle speed. In what directions (give a unit vector $\langle x_0, y_0 \rangle$ for the direction), and at what launch angles, should you fire projectiles in order to hit those two targets?

25. For ideal projectile motion in three dimensions, write the initial velocity vector $\mathbf{v}_0$.

26. Prove the normal component alternate formula using the following steps.

(a) Write an expression for $\mathbf{a} \cdot \mathbf{N}$ using the definition of $\mathbf{N}$. Your expression should use $\mathbf{r}''(t)$, $\mathbf{T}'(t)$, and a norm.

(b) In the derivation of the alternate curvature formula in Section 2.3, there is an expression equal to $\mathbf{r}''(t)$. Apply that to your answer to part (a).

(c) Expand the expression in (b) and use the fact that $\mathbf{T}$ and $\mathbf{T}'$ are orthogonal to simplify. Also recall that for any vector $\mathbf{w}$, $\mathbf{w} \cdot \mathbf{w} = \|\mathbf{w}\|^2$.

A blowgun contest is held annually at the Cherokee National Holiday, with blowguns made from river cane and naturally made darts as well.

(d) In the derivation of the alternate curvature formula in Section 2.3, there is an expression equal to $\|\mathbf{T}'(t)\|$. Apply that to your answer to part (c).

(e) If you haven't already done so, use the fact that $\frac{ds}{dt} = \|\mathbf{r}'(t)\|$ to finish the derivation.

27. A 5-ft-long blowgun is used to shoot a dart with initial velocity 200 ft/s. A target is 80 ft from the end of the blowgun.

(a) With a launch angle of $0°$, how far will the dart drop by the time it reaches the target?

(b) The shooter's eyes are $\frac{1}{4}$ ft above the mouth. If the shooter looks on a straight line across the tip of the blowgun, will the shooter be looking at, above, or below the landing spot of the dart at the target?

(c) What distance to the target would make the answer to part (b) "at" rather than above or below?

Partial Derivatives

3

3.1 Functions of Two or More Variables

In Chapter 2, we studied vector-valued functions, in which we had one-dimensional input and multi-dimensional output. In Chapter 3, we reverse this and study functions with multi-dimensional input and one-dimensional output.

3.1.1 Multivariable functions

You are familiar with functions of one variable (Figure 3.1). But for a machine on a production line in a factory, there may be more than one input; two or more pieces may be assembled, or ingredients combined. In fact, you have used such a multivariable process mathematically before. When you are given the lengths of the two legs of a right triangle and are asked to determine the length of the hypotenuse, you essentially evaluated a function of two variables (Figure 3.2). For a function of one variable, $g(x) = x^2$, we place the input variable in parentheses and write the output expression in terms of the input variable. We then evaluate the function by choosing a value of the input variable and substituting that value into the output expression: $g(3) = 3^2 = 9$. For functions of two variables, we place both variables in parentheses and write the output expression using both input variables:

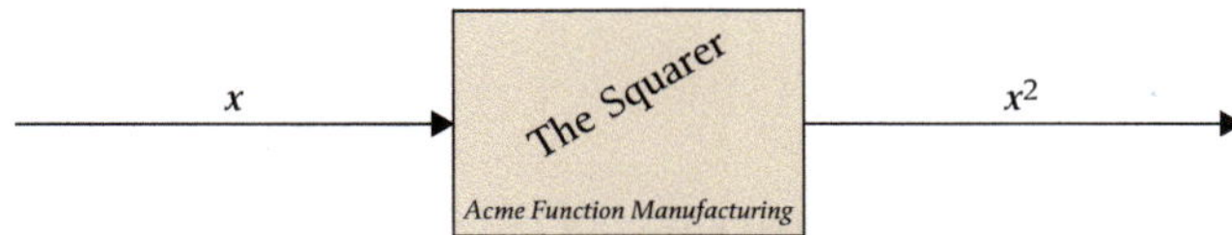

Figure 3.1 *A function of one variable (one input, one output)*

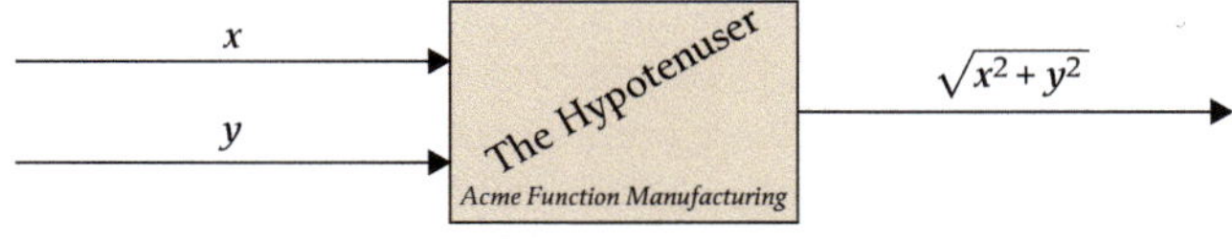

Figure 3.2 *A function of two variables (two inputs, one output)*

Multivariable Calculus Set Free. Charles Bryan Dawson, Oxford University Press.
© C. Bryan Dawson (2026). DOI: 10.1093/oso/9780198984269.003.0004

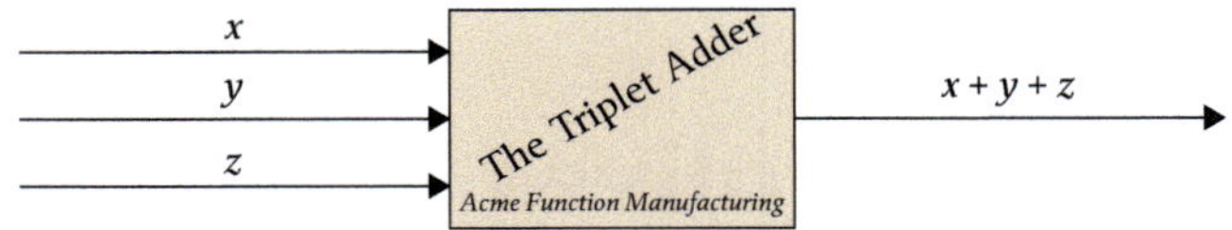

Figure 3.3 *A function of three variables (three inputs, one output)*

$$f(x, y) = \sqrt{x^2 + y^2}.$$

The function is evaluated by choosing values of the two input variables and substituting those values into the output expression:

$$f(3, 4) = \sqrt{3^2 + 4^2} = 5.$$

The input is an ordered pair; in this instance, the number 3 is a value of the variable x and the number 4 is a value of the variable y.

The same ideas apply to a function of three variables (Figure 3.3).

For

$$f(x, y, z) = x + y + z,$$

we have

$$f(3, -1, 4) = 3 + (-1) + 4 = 6.$$

Of course, we need not stop there; functions of four or more variables work in the same manner.

Collectively, functions of two or more variables are called *multivariable functions, multivariate functions,* or *functions of several variables.*

Reading Exercise 1 Find $f(2, -1, 0)$ for $f(x, y, z) = x^2 + y \cos z$.

3.1.2 Domains of functions of two variables

When asked for the domain of a function of one variable, we find the set of all values of the input variable for which the output expression is defined. For instance, to find the domain of $f(x) = \ln(x + 1)$, we note that the natural logarithm function is only defined for positive values; therefore, we need $x + 1 > 0$, resulting in a domain of $x > -1$. We may write the domain in interval notation as $(-1, \infty)$. We may also picture the domain on a number line (Figure 3.4).

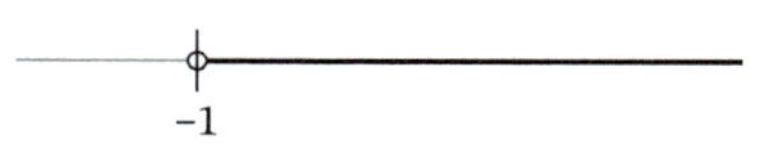

Figure 3.4 *The domain of* $f(x) =$ $\ln(x + 1)$

The same ideas apply to the domain of a function of two variables. For instance, consider the function

$$f(x, y) = \sqrt{x - y}.$$

Is $x = 1$ in the domain of f? While

$$f(1, 0) = \sqrt{1 - 0} = 1,$$

we also have

$$f(1, 2) = \sqrt{1 - 2} = \sqrt{-1},$$

which is undefined. So, is $x = 1$ in the domain of f? It depends on what y is!

Instead of considering the two variables x and y independently, they must be considered simultaneously. The question becomes which input pairs (x, y) are in the domain of f. The input pair $(1, 0)$ is in the domain, and the input pair $(1, 2)$ is not in the domain.

But notice that it still all comes down to the square root; the output is defined precisely when the expression under the square root is nonnegative. So our domain is

$$x - y \geq 0$$
$$x \geq y.$$

The input pair $(1, 0)$ satisfies $x \geq y$, but the input pair $(1, 2)$ does not.

Alternately, the domain can be written as $y \leq x$. Notice that the domain is not an interval, nor can it be; the domain is a collection of ordered pairs, not a collection of real numbers. We can picture the domain, but not on a one-dimensional number line; with two inputs, we need a two-dimensional plane. The domain of f is pictured in Figure 3.5.

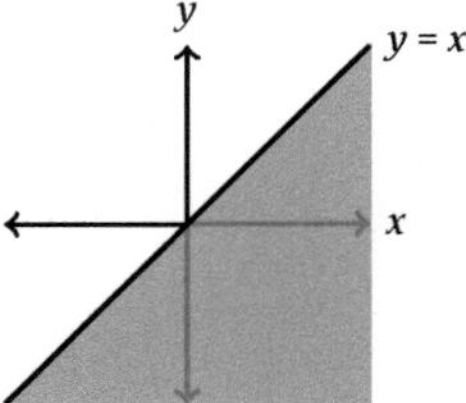

Figure 3.5 *A portion of the domain of $f(x, y) = \sqrt{x - y}$. The shaded region represents the domain, which consists of all points on or below (alternately, to the right of) the line $y = x$*

Writing the domain as $y \leq x$ emphasizes that the y-coordinates must be at or below (less than or equal to) the x-coordinate, meaning we shade below the line $y = x$. Writing the domain as $x \geq y$ emphasizes that the x-coordinate must be at or to the right of (greater than or equal to) the y-coordinate, meaning that we shade to the right of the line $y = x$. In either case, the line $y = x$ is included, so we draw it solid (as opposed to dashed).

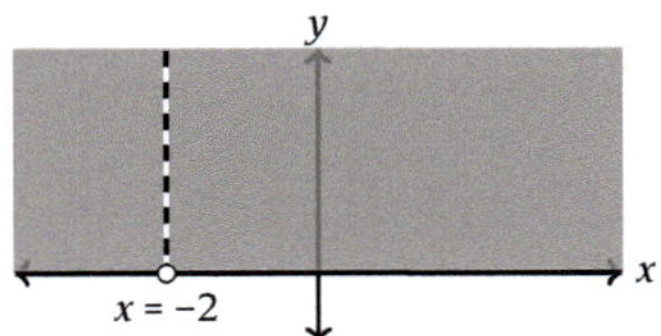

Figure 3.6 *A portion of the domain of $f(x,y) = \frac{\sqrt{y}}{x+2}$. The solid line indicates that the points on the line $y = 0$ are included (except the point $(-2, 0)$); the dashed line indicates that the points on the line $x = -2$ are not included*

Example 1 *Find and sketch the domain of $f(x,y) = \frac{\sqrt{y}}{x+2}$.*

Solution We cannot divide by zero, so we need $x \neq -2$. We also need the expression under the square root to be nonnegative: $y \geq 0$. The domain is therefore

$$y \geq 0 \text{ and } x \neq -2.$$

By "$y \geq 0$ and $x \neq -2$" we mean that both conditions must be true at the same time.

The domain is sketched in Figure 3.6. ∎

Ans. to reading exercise 1:
3

Reading Exercise 2 Find and sketch the domain of $f(x,y) = \ln(x - y)$.

3.1.3 Graphing functions of two variables

When we graph functions of one variable, we graph in two dimensions:

one input variable
one output variable
------
two dimensions needed.

Inputs are placed on the x-axis, outputs on the y-axis. For $f(x) = x^2$, the result is in Figure 3.7.

For functions of two variables such as $f(x,y) = 3x - 2y + 12$, we need three dimensions:

two input variables
one output variable
------
three dimensions needed.

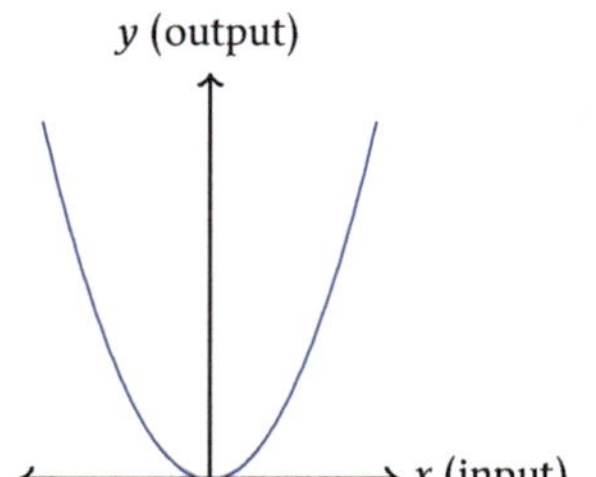

Figure 3.7 *The graph of $f(x) = x^2$, with one input dimension (x-axis) and one output dimension (y-axis)*

Instead of an input axis, we have an input plane, the xy-plane. The output is one-dimensional, and we will use the z-axis as the output axis (Figure 3.8).

Just as with a function of one variable, an option for graphing functions of two variables is to plot points. To do so we choose values of the two input variables and calculate the value of the output variable. For $f(x, y) = 3x - 2y + 12$, we set $z = 3x - 2y + 12$. A few points for plotting are below:

This is analogous to setting $y = x^2$ for graphing the function $f(x) = x^2$.

x	y	z	point
0	0	12	$(0, 0, 12)$
1	0	15	$(1, 0, 15)$
0	1	10	$(0, 1, 10)$
1	1	13	$(1, 1, 13)$
2	0	18	$(2, 0, 18)$
2	1	16	$(2, 1, 16)$

The values of x and y are chosen. The value of z is calculated using $z = 3x - 2y + 12$.

These points are plotted in Figure 3.9. Because the terms involving the variables x, y, and z are all linear, the graph of $z = 3x - 2y + 12$ is a plane. One way to finish the graph is to shade the portion of the plane bordered by the plotted points; the plane is infinite in extent, so we cannot draw the entire plane anyway. See Figure 3.10.

If you have been following along by plotting points yourself and drawing these figures by hand, you may have noticed how difficult it can be to produce an accurate three-dimensional graph. Not that it's impossible, but a lot of care must be taken to plot the points accurately. An alternate strategy that works for a plane and is much easier to draw, especially once you learn how to interpret the drawing, is to plot the x-, y-, and z-intercepts, connect them with straight lines, and shade the resulting triangular portion of the plane.

Ans. to reading exercise 2:
$$y < x$$

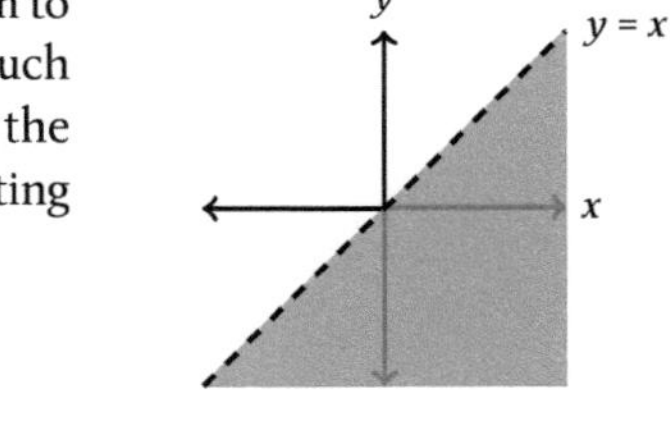

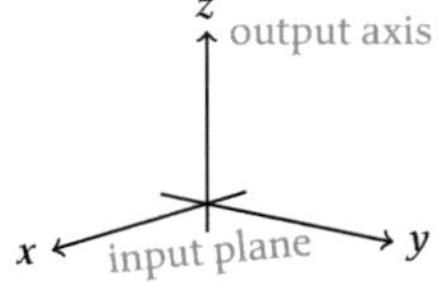

Figure 3.8 *The input plane (xy-plane) and output axis (z-axis) for graphing a function of two variables*

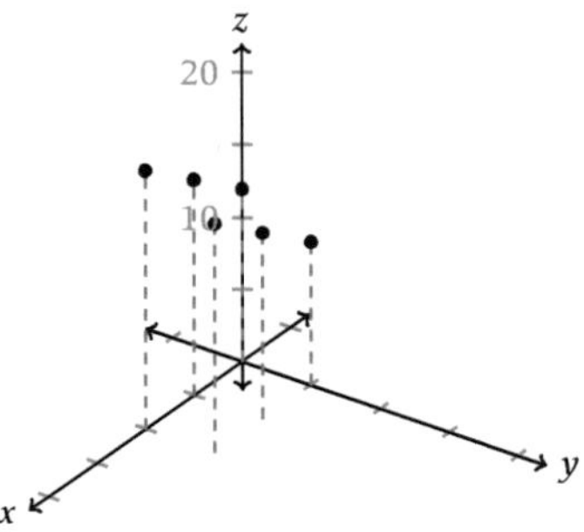

Figure 3.9 *Points on the graph of $f(x, y) = 3x - 2y + 12$*

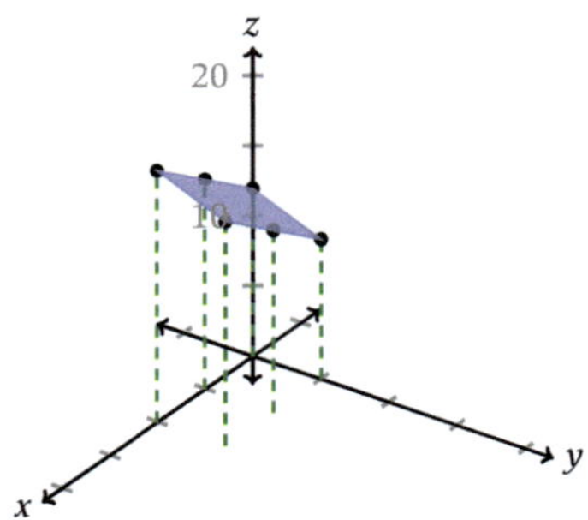

Figure 3.10 *The plane f(x, y) = 3x − 2y + 12*

The z-intercept is the point $(0, 0, 12)$, which we found when plotting points; it comes from evaluating $f(0, 0)$. The x- and y-intercepts may be found by rearranging the equation into the form $x =$ and the form $y =$, then setting the other variables to be 0.

For the x-intercept, we isolate x:

$$z = 3x - 2y + 12$$

$$z + 2y - 12 = 3x$$

$$\frac{1}{3}z + \frac{2}{3}y - 4 = x.$$

The x-intercept is therefore $(-4, 0, 0)$.

For the y-intercept we isolate y:

$$z = 3x - 2y + 12$$

$$2y = 3x - z + 12$$

$$y = \frac{3}{2}x - \frac{1}{2}z + 6.$$

The y-intercept is therefore $(0, 6, 0)$.

The three intercepts are plotted in Figure 3.11. Line segments between these points are drawn, and the region enclosed by these segments is shaded. The result all lies in one octant, one "apartment" of the complex. The z-intercept is a point in the crease where two walls meet, whereas the x- and y- intercepts are along the adjacent floorboards (or the ceiling-wall molding for a first-floor apartment). See Figure 3.11.

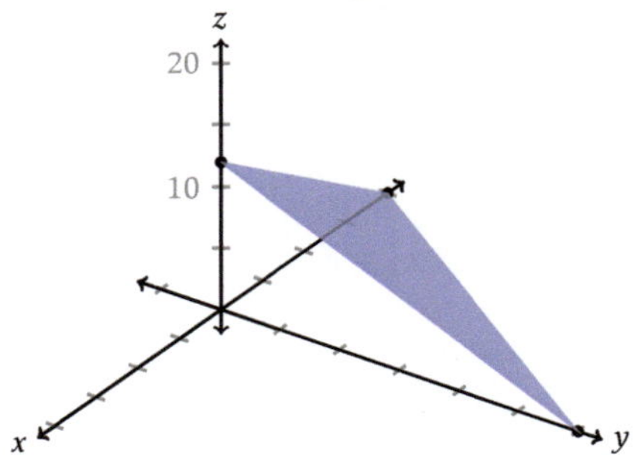

Using the apartment complex analogy, the three line segments bordering the shaded triangle in Figure 3.11 lie along the walls and floor of an apartment. The shaded portion of the plane is the portion of the plane that would be visible from inside that apartment.

Figure 3.11 *Another view of the plane* $f(x,y) = 3x - 2y + 12$. *Compare to Figure 3.10.*

Example 2 *Graph the function* $f(x,y) = x + y - 3$.

Solution Because the graph of $z = x + y - 3$ is a plane, we use the method of plotting the intercepts. The z-intercept is $(0, 0, -3)$. For the x-intercept we isolate the variable x:

$$z = x + y - 3$$
$$z - y + 3 = x.$$

The x-intercept is $(3, 0, 0)$.

Next, we isolate y:

$$z = x + y - 3$$
$$z - x + 3 = y.$$

The y-intercept is $(0, 3, 0)$.

The three intercepts are plotted and the resulting triangle is shaded in Figure 3.12.

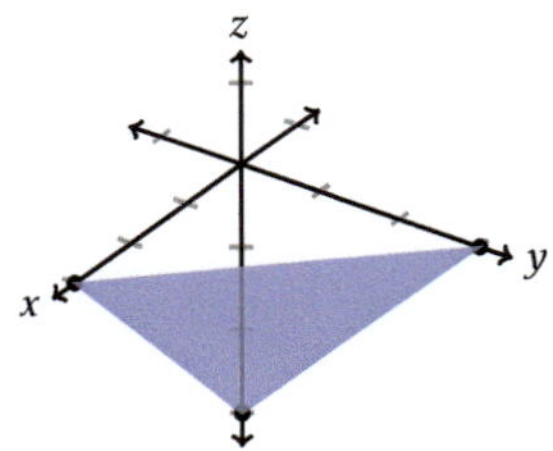

Figure 3.12 *A view of the plane* $f(x,y) = x + y - 3$

∎

Even when you know what the diagram is supposed to look like, it can sometimes be a challenge to adjust your mind's eye to see what you are supposed to see in a three-dimensional diagram such as Figure 3.12.

3.1.4 Analyzing 3D graphs produced by technology

A plane is a linear function of two variables, whereas a line is a linear function of one variable. Compare the difficulty of drawing a line in two dimensions and a plane in three dimensions. Similarly, compare the difficulty of drawing a parabola in two dimensions and an elliptic paraboloid in three dimensions. The third dimension truly adds a layer of sophistication and difficulty that can make anything beyond the simplest functions difficult for most of us to graph well without a great deal of practice and training. For this reason, using technology to produce graphs of functions of two variables is common. But, if we rely on the technology, we must be able to analyze and interpret what we see in such graphs.

Think about the concepts that have been useful in understanding two-dimensional graphs: domain, asymptotes, discontinuities, extrema, and more. We will carefully explore many of these concepts for functions of two variables during the course of this chapter. For now, let's begin the exploration graphically and see what we can learn. Such graphs are often called *surfaces*.

The function $f(x, y) = x^2 - y^2$ is graphed in Figure 3.13. This is the same as $z = x^2 - y^2$, which is a hyperbolic paraboloid.

When studying the quadric sections we learned to consider vertical and horizontal slices to understand the graph. This is helpful for surfaces in general. If we let $x = k$, the equation becomes $z = k^2 - y^2$, which is a parabola opening downward; remember to consider k as a constant and only consider x, y, and z as variables. At an end of the box surrounding the surface, the cross section $x = 3$ appears to be visible, showing the parabola $z = 9 - y^2$. Similarly, at another end of the box, the

It is common, although not necessarily universal, for a computer algebra system (CAS) to surround a three-dimensional graph with a box and mark scales for the three variables on edges of the box.

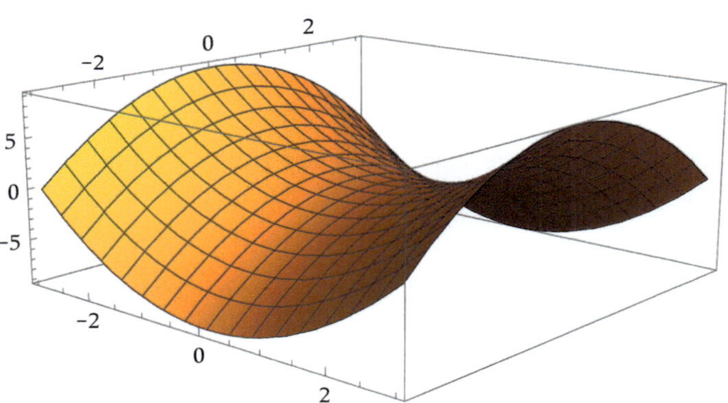

Figure 3.13 *The graph of $f(x, y) = x^2 - y^2$*

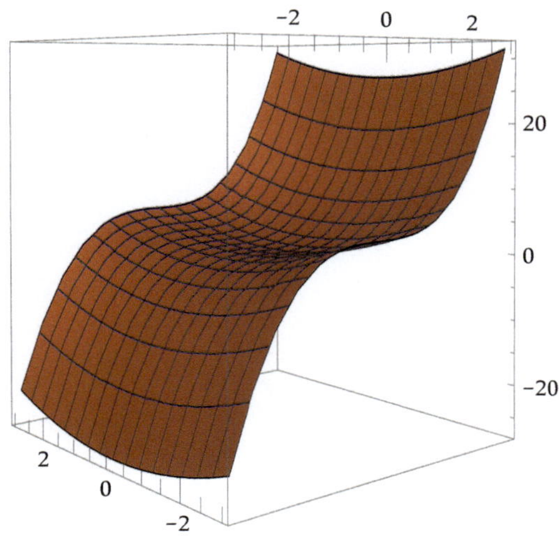

Figure 3.14 *The graph of f(x, y) = x³ + 0.5y², a "funky chair"*

Figures 3.13–3.15, 3.17, and 3.19–3.22 were produced using Mathematica. Desmos and CalcPlot3D can also produce graphs of these functions, as can many other apps.

slice $y = -3$ is visible, showing the parabola $z = x^2 - 9$, which opens upward. At the center of the box, where the slices $x = 0$ and $y = 0$ intersect, the point $(0, 0, 0)$ is a *saddle point*, which will be explored later in this chapter.

Because polynomials are defined everywhere, the domain of the (hyperbolic paraboloid) function is $\mathbf{R}^2$. The same is true of our next function, $f(x, y) = x^3 + 0.5y^2$, which is graphed in Figure 3.14.

If x is a constant, $z = C + 0.5y^2$, which is a parabola opening upward; such parabolas can be seen at the ends of the box representing the top and bottom of the funky chair. If y is a constant, $z = x^3 + C$, which is a cubic; cubics can be seen at the ends of the box representing the sides of the funky chair. Notice that in the seat of the chair there appears to be a minimum, a place where water would gather if spilled. We will explore extrema, including maxima and minima, later in this chapter.

Here, C represents a constant. For instance, if $x = 2$, then $C = 8$. The specific value of the constant is not always important for the analysis.

The "ends" of the box are more properly called *faces* of the rectangular prism.

Next is $f(x, y) = \ln(x + y)$, whose domain is $y > -x$ (Figure 3.15). Recall that the graph of $g(x) = \ln x$ (Figure 3.16) has a vertical asymptote at the edge of its domain, $x = 0$; the asymptote is a vertical line. But in the function $f(x, y) = \ln(x + y)$, the edge of the domain is the line $x + y = 0$ (same as $y = -x$), and the vertical asymptote is a plane!

Vertical asymptotes can be lines or planes. They can have other shapes, as well. The graph of $f(x, y) = \sec(x^2 + y^2)$ (Figure 3.17) has circular vertical asymptotes! The asymptotes of $g(\theta) = \sec\theta$ (Figure 3.18) occur at $\theta = \pm\frac{\pi}{2}$, $\theta = \pm\frac{3\pi}{2}$, ... The innermost asymptote in Figure 3.17 has equation $x^2 + y^2 = \frac{\pi}{2}$, and the next ring out has equation $x^2 + y^2 = \frac{3\pi}{2}$. Because the CAS graphed the function on a square

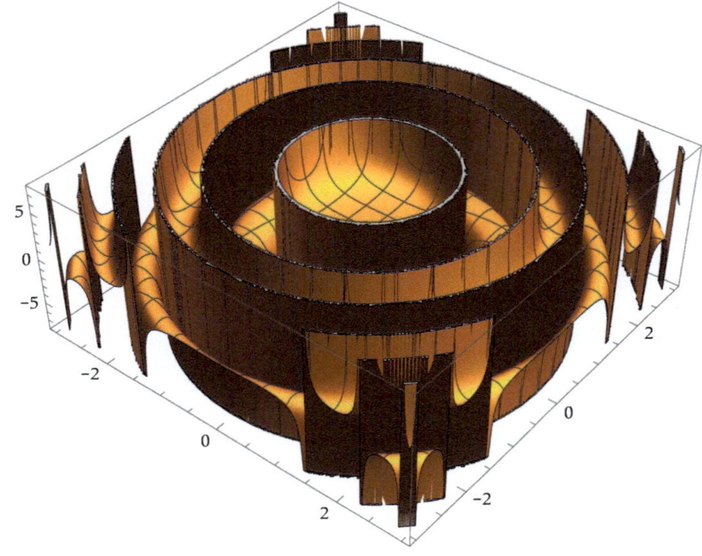

Figure 3.15 *The graph of $f(x, y) = \ln(x + y)$, a waterfall*

The waterfall in Figure 3.15 must be quite spectacular, because the water drops infinitely far along the vertical asymptote!

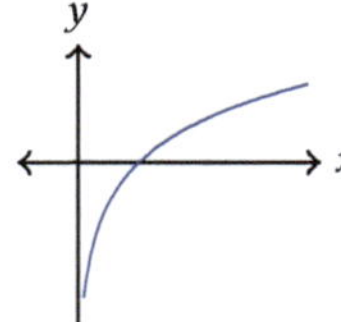

Figure 3.16 *The graph of $y = \ln x$, with vertical asymptote $x = 0$*

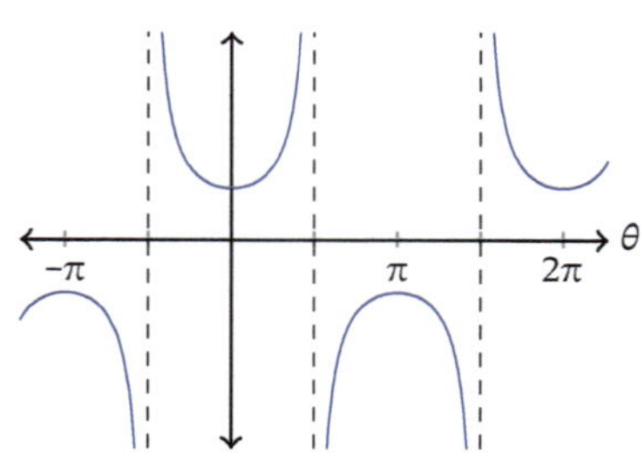

Figure 3.18 *The graph of $y = \sec\theta$*

Figure 3.17 *The graph of $f(x, y) = \sec(x^2 + y^2)$*

domain rather than circular, the corners contain pieces of asymptotes that are still circular but extend beyond the sides of the box.

The function $f(x, y) = 0.1\sqrt{x^2 + y}$ (Figure 3.19) has domain $y \geq -x^2$. The graph of $y = -x^2$ is a parabola, so the edge of the domain of f is a parabola. But this time there is no asymptote, just as there is no asymptote on the graph of $g(x) = \sqrt{x}$.

The crevasse in Figure 3.20 looks dangerous for travelers, and appears to be a vertical asymptote. The graph of $f(x, y) = \ln|x^3 + y^2|$ does indeed have a vertical

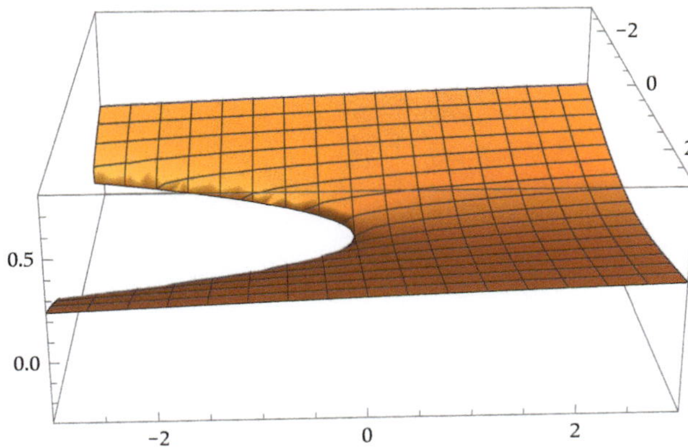

Figure 3.19 *The graph of $f(x, y) = 0.1\sqrt{x^2 + y}$, a shark bite*

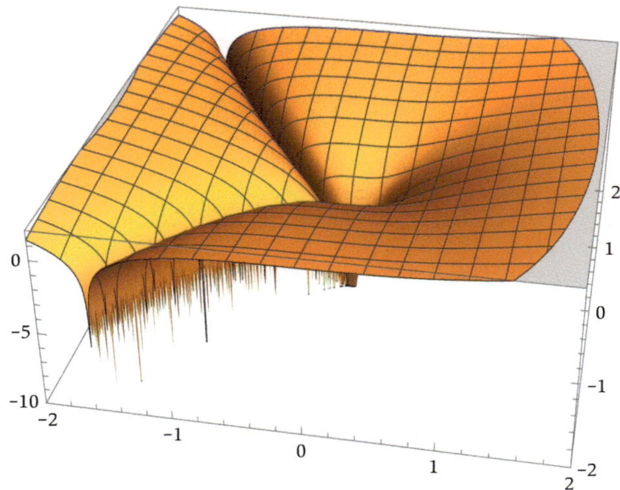

Figure 3.20 *The graph of $f(x, y) = \ln|x^3 + y^2|$, a scary crevasse*

asymptote with equation $x^3 + y^2 = 0$. In this manner, any two-dimensional curve can serve as the base for an asymptote on the graph of a function of two variables. The gray portion of Figure 3.20 is not part of the graph of the function, but rather a byproduct of the graphing routine used by the CAS. In this case, the gray indicates that the graph rises outside the box. Watch for such indicators when using a CAS.

More precisely, the asymptote is the cylinder (see Section 1.7) in $\mathbf{R}^3$ with equation $x^3 + y^2 = 0$.

The domain of $f(x, y) = 2\sin x \sin y$ (Figure 3.21) is $\mathbf{R}^2$, because the sine function is defined for all real numbers. It is also apparent that $-2 \le z \le 2$ throughout the surface. All the peaks have z-coordinate 2, and all the valley bottoms have z-coordinate -2. Slices of the form $x = k$ have equation of the form $z = C\sin y$, and

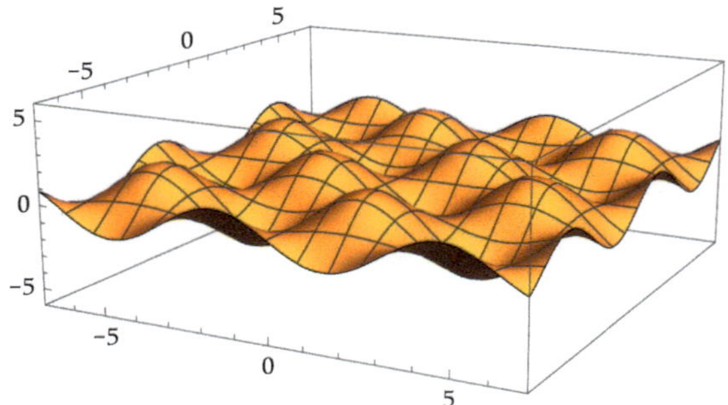

Figure 3.21 *The graph of $f(x, y) = 2 \sin x \sin y$, an egg crate mattress*

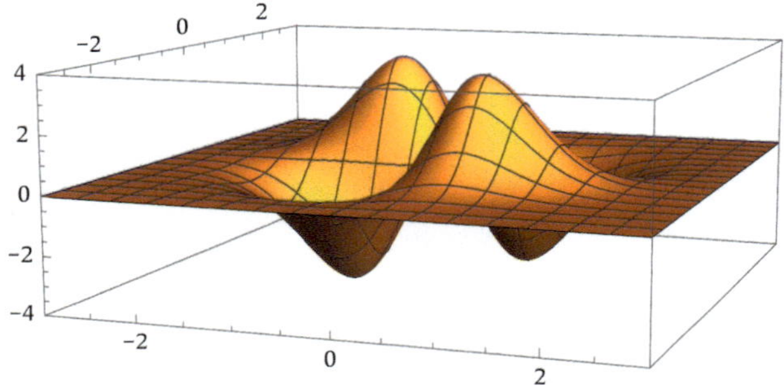

Figure 3.22 *The graph of $f(x, y) = 20xye^{-x^2-y^2}$, peaks and pits*

slices of the form $y = k$ have equation of the form $z = C \sin x$. The sine waves can be seen at the ends of the box. Because the sine function is periodic, the function f is also periodic, in any direction! There are also flat cross sections; for instance, the slice $x = \pi$ yields $z = 2 \sin \pi \sin y = 2 \cdot 0 \cdot \sin y = 0$. It may be difficult to see such a slice in the current view, but rotating the viewing box can be helpful in this regard.

The function $f(x, y) = 20xye^{-x^2-y^2}$ (Figure 3.22) features two peaks and two pits, in the midst of what appears to be a relatively flat plain. The plain appears similar to a horizontal asymptote. But instead of only two infinite directions ($x \to \infty$ and $x \to -\infty$) in the case of functions of one variable, we can wander away from the center in infinitely many directions toward an infinitely distant horizon. In this case, any such direction renders the same result, but that will not always be the case.

Think about standing in a flat desert landscape. In how many different directions could you walk?

The graphical analysis techniques demonstrated here are helpful beginning points for exploring the graph of a function of two variables, and additional graphical techniques are discussed in the next section. But much more is available through the application of calculus, which will be the focus of the majority of Chapter 3.

EXERCISES 3.1

1. Find $f(1, 0, -7)$ for $f(x, y, z) = \sqrt{15x^2 + 6y^2 + z^2}$.
2. Find $f(-1, 1)$ for $f(x, y) = x^2 + x \ln y$.
3. Find $f(2, 4)$ for $f(x, y) = \ln x + \ln \frac{x}{y}$.
4. Find $f(\pi, -\pi, 22\pi)$ for $f(x, y, z) = e^{x+y} \sin z$.
5. (a) Find $f(-2, 3, 5, 1)$ for $f(x_1, y_1, x_2, y_2) = \frac{y_2 - y_1}{x_2 - x_1}$.
   (b) What is this function called?
6. (a) Find $f(1, 3, -10)$ and $g(1, 3, -10)$ for $f(a, b, c) = \frac{-b + \sqrt{b^2 - 4ac}}{2a}$ and
   $$g(a, b, c) = \frac{-b - \sqrt{b^2 - 4ac}}{2a}.$$
   (b) What are these functions called?
   (c) Why write it as two separate functions?
7. (a) Find $f(-2, 3, 5, 1)$ for $f(x_1, y_1, x_2, y_2) = \sqrt{(x_2 - x_1)^2 + (y_2 - y_1)^2}$.
   (b) What is this function called?
8. (a) Find $f(2, -7, 3, 12)$ for $f(a, b, c, d) = ad - bc$.
   (b) Assuming you have taken the appropriate course, what is this function called?

9–16. (a) Find the domain of the function. (b) If the domain is not $\mathbf{R}^2$, sketch the domain of the function.

9. $f(x, y) = \dfrac{x + 2y}{3x - y}$
10. $f(x, y) = \sqrt{1 - x^2} + y$
11. $f(x, y) = \sqrt{x^2 + y^2 - 4}$
12. $f(x, y) = x^2 + \sin xy - e^{3x + 2y}$
13. $f(x, y) = x^5 - 3x^2 y^3 + 7x^4 y - \dfrac{1}{\sqrt{18}} y^2$
14. $f(x, y) = \dfrac{4}{9 - x^2 - y^2}$
15. $f(x, y) = \sqrt{x + 3} + \dfrac{2}{y + 3}$
16. $f(x, y) = \ln\left(\frac{1}{4} x^2 + y^2 - 4\right)$

17–22. Graph the function.

17. $f(x, y) = 2x - 4y + 8$
18. $f(x, y) = 20 - 5x - 4y$
19. $f(x, y) = 3x + 6y - 24$
20. $f(x, y) = 3x - y + 6$
21. $f(x, y) = 4$
22. $f(x, y) = 3x$

23. Match the graph to the function.

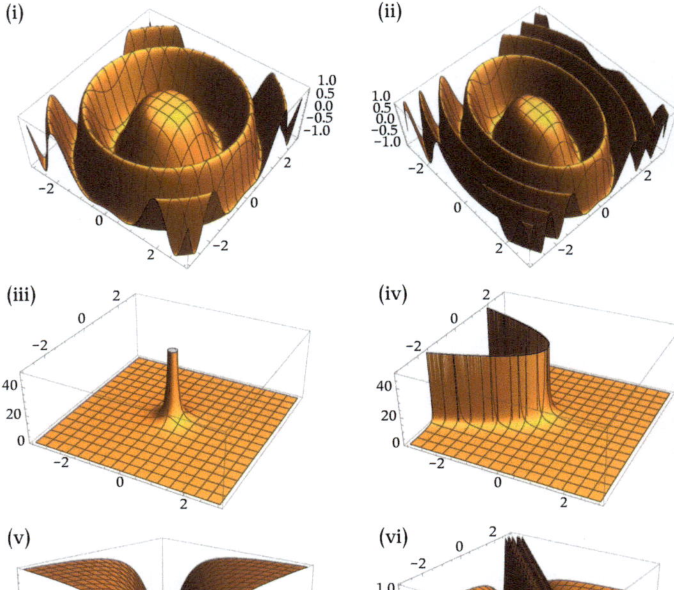

(a) $f(x,y) = \dfrac{1}{x^2+y^2}$

(b) $f(x,y) = e^{-1/(x+y)^2}$

(c) $f(x,y) = \dfrac{1}{x+y^2}$

(d) $f(x,y) = \sin\left(\dfrac{-1}{(x+y)^2}\right)$

(e) $f(x,y) = \cos\left(x^2 + 2y^2\right)$

(f) $f(x,y) = \cos\left(x^2 + y^2\right)$

24. Match the graph to the function.

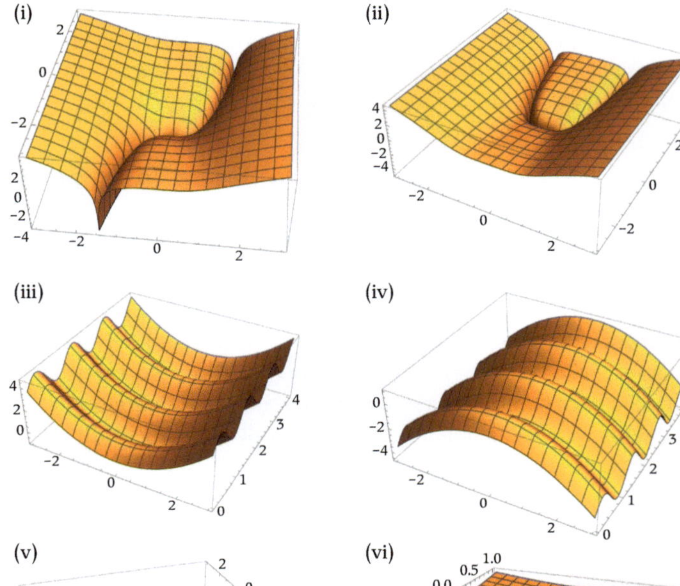

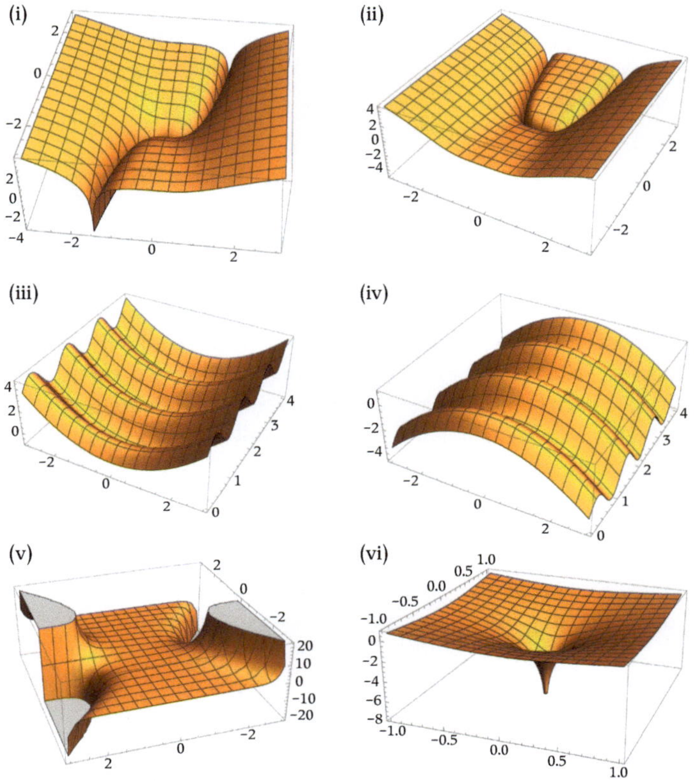

(a) $f(x,y) = 0.4x^2 + \sin 5y$
(b) $f(x,y) = \ln|y - x^3|$
(c) $f(x,y) = -0.4x^2 + \sin 5y$
(d) $f(x,y) = \ln(x^2 + y^2)$
(e) $f(x,y) = \ln|y - x^4|$
(f) $f(x,y) = xye^{y^2 - x^2}$

25. Which of the functions from exercise 23 have a vertical asymptote?
26. Which of the functions from exercise 24 have domain $\mathbb{R}^2$?

27–32. Use a CAS to graph the function.

27. $f(x,y) = xy + \dfrac{5x^2}{y^2 + 1}$

28. $f(x,y) = \sqrt{\sin(xy)}$

29. $f(x,y) = xy + \dfrac{5x^2}{y^2 - 1}$

30. $f(x,y) = \ln(x^2 - 3y + y^2 - 7)$

31. $f(x,y) = e^{-x^2} + e^{-3y^2}$

32. $f(x,y) = \dfrac{x}{x^3 - y} + \dfrac{y}{x^2 - 1}$

Be sure to choose an appropriate viewing box (more than one if necessary) to show features of the graph. Rotate the box to view from various viewpoints. It might be necessary to specify all three dimensions of the box, and not just two. It might also be necessary to tell the CAS to use more sampling points, that is, to plot more points, to get a smoother-looking graph.

3.2   Contour Plots

A topographical map, which is a map overlaid with elevation information, can be considered a version of the graph of a surface. In general, we call such graphs *contour plots*.

3.2.1   Topographical maps and the concept of contour plot

An example of a topographical map is in Figure 3.23. Toward the bottom of the map is a brown curve with the number 4600 written on it, indicating that all points along that curve have elevation 4600 feet above mean sea level (ASL). Move five brown curves (called *contours*) north on the map, and you will see a contour marked 4400 ft ASL. Each contour on the map represents 40 ft difference from the neighboring contours. Traveling north is going downhill; traveling south is going uphill. If you could walk the path of a contour, you would be walking a flat path, one that does not change elevation.

Move to the west (Figure 3.24), and you will see the contours get much closer together. This is the base of a mountain, where the elevation changes rapidly. Still, if you could walk along a contour your path would be flat; but such a path is more difficult in practice along a mountainside.

The author once lived at a location on this map.

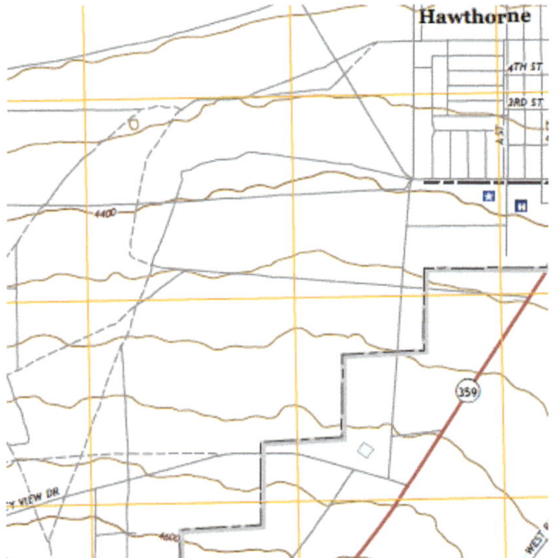

Figure 3.23 *Selection from USGS product number 474264, courtesy of the U.S. Geological Survey*

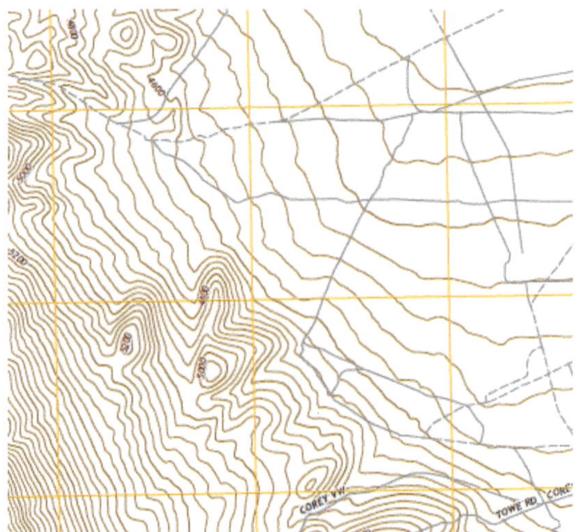

Figure 3.24 *Selection from USGS product number 474264, courtesy of the U.S. Geological Survey*

Now think of the map as representing the *xy*-plane, and the contours as indicating the value of the elevation function. The elevation function has two-dimensional input (location on the map) and one-dimensional output (elevation), and is therefore an example of a function of two variables. But the topographical maps in Figures 3.23 and 3.24 are not the same type of graph as we encountered in Section 3.1. Instead of a three-dimensional graph with *z*-coordinate indicating the value of the function, we have a two-dimensional graph with contours indicating values of the function.

From the contours, it is possible to construct a three-dimensional mental image of the terrain. Where is the mountainside steepest? Where are the places that appear to be local extrema? But if all we are after is a three-dimensional graph, why not produce one as we did in Section 3.1?

Contours can give information that is difficult to perceive directly from a three-dimensional image, even in the absence of such visual barriers as thick woods or city infrastructure. Additionally, there are some applications that lend themselves more naturally to the idea of a contour than a graph expressing values by *z*-coordinates, such as temperatures (Figure 3.25). Contours on a temperature map are called *isotherms*, meaning same temperature. In a similar manner, the contours on a map of barometric pressure are called *isobars*. Because each contour represents a constant value of the function, a contour is also called a *level curve*.

Ignore the curvature of the earth. If a topographical map is for a small enough region, the effect of the earth's curvature (distortion on a two-dimensional flat map) is negligible.

One example that you may have encountered is driving along a winding, up-and-down mountain road and having a dispute as to whether you are going uphill or downhill. Or walking along visually flat ground that is gently sloping. A three-dimensional image can be deceiving.

Try to imagine a three-dimensional graph where z-coordinates represent temperature. A contour plot is much simpler and more easily conveys the information.

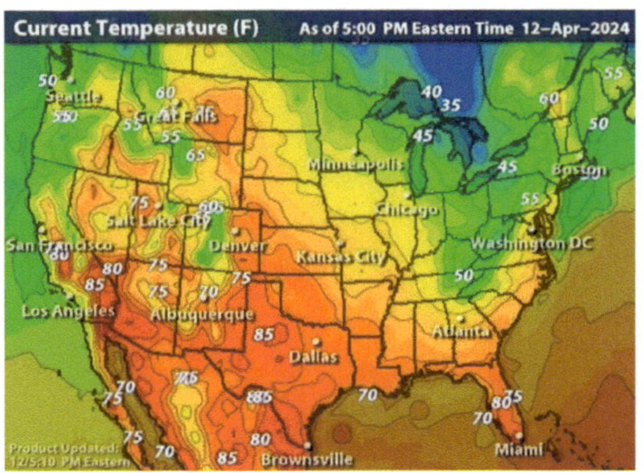

Figure 3.25 *Weather Underground temperature map, with 5° contours*

3.2.2   Drawing a contour plot

To draw a contour plot for a function of two variables, we first find equations of various contours and then graph those equations, with appropriate labeling. Choosing contours is an inexact science, but one good practice is to have the contours equally incremented, which is essential for the type of interpretations previously discussed. In Figures 3.23 and 3.24, the elevation contours are given in 40 ft increments, and in Figure 3.25, the isotherms (contours) are given in 5° increments.

Example 3 *Draw a contour plot for $f(x,y) = x^2 + y$.*

Solution   For lack of a better place to start, let's determine the contour $z = 0$, that is, $f(x, y) = 0$. We would have

$$0 = x^2 + y,$$

which can be rearranged to

$$y = -x^2.$$

If you think of f as an elevation function, the curve $y = -x^2$, as well as each of the other contours, would be a path of constant elevation.

Then for all points on the curve $y = -x^2$, the value of the function $f(x, y) = x^2 + y$ is 0. This is one contour for our contour plot, which can be easily graphed (Figure 3.26).

   If we equally increment contours by 1, we could also determine the contours $z = 1$, $z = 2$, $z = -1$, and $z = -2$. For $z = 1$, we would have $1 = x^2 + y$, which can be rearranged to $y = 1 - x^2$, and for $z = 2$, we would have $2 = x^2 + y$, which can be rearranged to $y = 2 - x^2$. Similarly, the contour $z = -1$ has equation $y = -1 - x^2$, and the contour $z = -2$ has equation $y = -2 - x^2$.

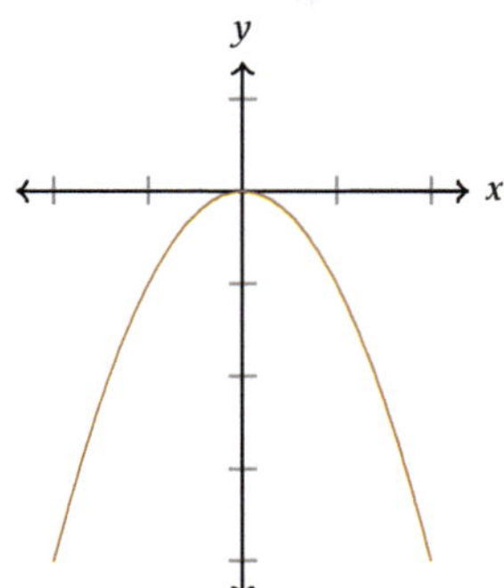

Figure 3.26 *The contour $z = 0$ for $f(x, y) = x^2 + y$*

Each of these contours are parabolas opening downward. Plotting points can help with accuracy:

x	contour $z = -2$ $y = -2 - x^2$	contour $z = -1$ $y = -1 - x^2$	contour $z = 1$ $y = 1 - x^2$	contour $z = 2$ $y = 2 - x^2$
-2	-6	-5	-3	-2
-1	-3	-2	0	1
0	-2	-1	1	2
1	-3	-2	0	1
2	-6	-5	-3	-2

Contours can be plotted and drawn one at a time; for instance, the contour $z = -1$ is added in Figure 3.27. A finished contour plot is in Figure 3.28. Notice the labeling of contours, similar to the labeling in Figures 3.23–3.25. ■

Notice that in Figure 3.28 the contours appear to get closer together (the surface gets steeper) as we move further from the y-axis. Yet we know from the equations of the contours, or from the chart of points to plot, that the y-coordinates of the contours are equally spaced! The vertical segments between contours in Figure 3.29 all have the same length. Using elevation as the interpretation, if we have a group of people on the surface all moving due north, they would all be going uphill at the same rate. But due north is not always the closest distance between contours; see the non-horizontal segments in Figure 3.29. Constant spacing between contours (not present in this contour plot) and constant vertical spacing between contours (present in this contour plot) are not the same thing; a common error is to conflate the two, drawing these contours with constant spacing. Carefully plotting points for each contour can help avoid this error.

Reading Exercise 3 For $f(x, y) = 3xy$, what is the equation of the contour $z = 2$?

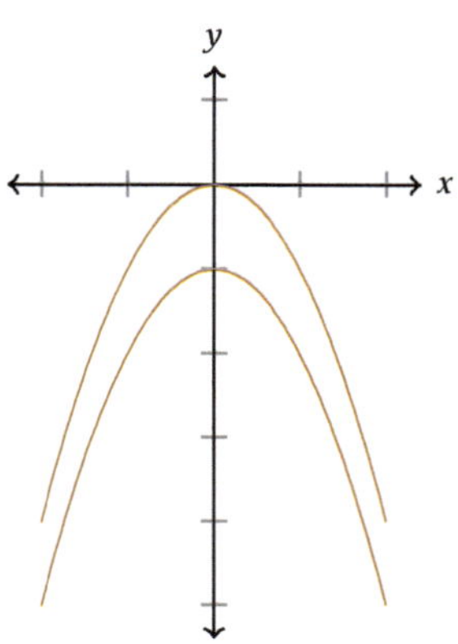

Figure 3.27 *The contour $z = -1$ is added*

Points on the contour $z = -2$ include $(-2, -6)$, $(-1, -3)$, $(0, -2)$, $(1, -3)$, and $(2, -6)$; similarly, points on the contour $z = -1$ include $(-2, -5)$, $(-1, -2)$, $(0, -1)$, $(1, -2)$, and $(2, -5)$. Interpret the columns for other contours similarly.

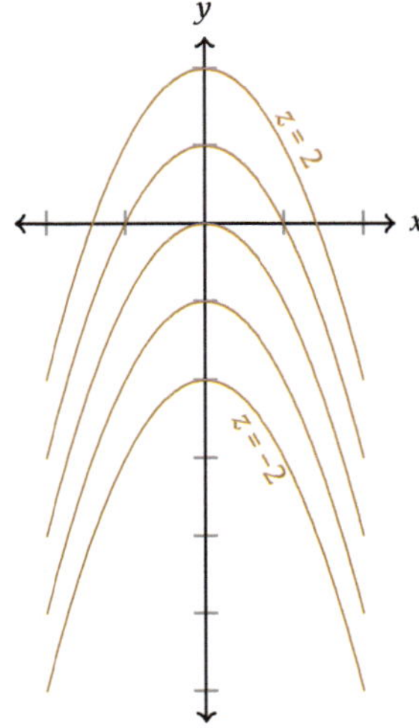

Figure 3.28 *A contour plot for $f(x, y) = x^2 + y$*

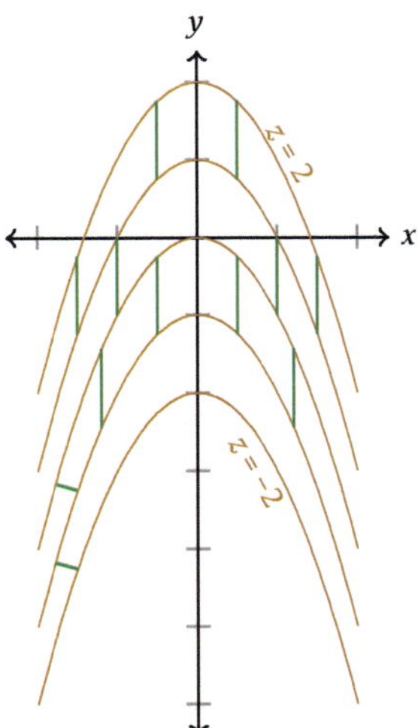

Figure 3.29 *Is the distance between two contours measured along the longer, vertical green segments or along shorter, non-vertical segments?*

3.2.3 Contour plots using technology

The same functions of two variables that are graphed using technology in Section 3.1 can also be explored using contour plots. Many of these contour plots are reasonable to draw by hand, but we will explore them here using technology.

Consider $f(x,y) = x^2 - y^2$, whose graph (Figure 3.30) is the hyperbolic paraboloid $z = x^2 - y^2$. A contour plot produced by a CAS is in Figure 3.31. Recall from Section 1.7 that the horizontal slices of a hyperbolic paraboloid with this orientation are hyperbolas; for instance, the slice $z = 4$ yields the hyperbola $4 = x^2 - y^2$, which is one of the contours in Figure 3.31. In fact, for any surface, horizontal slices are contours.

In addition to the contours themselves, and in place of labeling the contours, we see color-coding. A legend accompanies the contour plot that enables us to determine the location of contours incremented by 2 units, such as $z = 8$, $z = 6$, and $z = 4$, each contour forming the border between two shades. Compare the color coding to that of the temperature map in Figure 3.25; "warmer" colors are for larger values of the function and "cooler" colors are for lesser values of the function. The colors allow us to more easily create a mental image of the surface, as we imagine the surface dropping to the top and bottom of the picture with the cooler colors and rising to the left and right of the picture with the warmer colors.

Notice the crossing lines forming the contour $z = 0$. In any similarly oriented hyperbolic paraboloid, there is a level curve (contour) that is formed by two straight lines intersecting at the saddle point.

The graph of $f(x,y) = 20xye^{-x^2-y^2}$, peaks and pits (Figure 3.32), exhibits what appear to be two maximum values and two minimum values. The contour plot

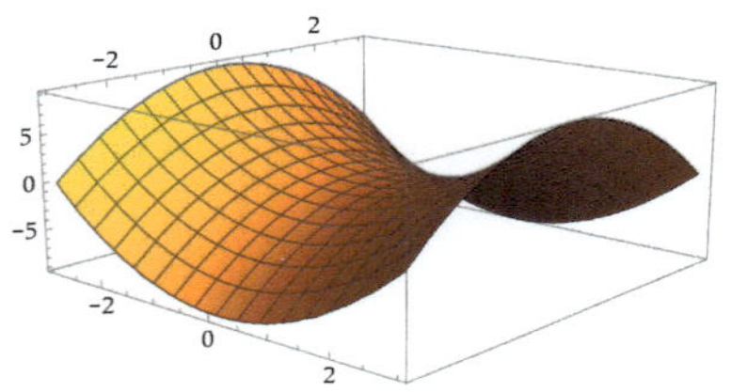

Figure 3.30 *The graph of* $f(x,y) = x^2 - y^2$

Using the 2D-grapher, Desmos can plot individual contours. Graphing several contours on the same graph, similar to Figure 3.28, can produce a contour plot.

Because color-coding improves rapid visual interpretation, it is common for technology to include colors in a contour plot.

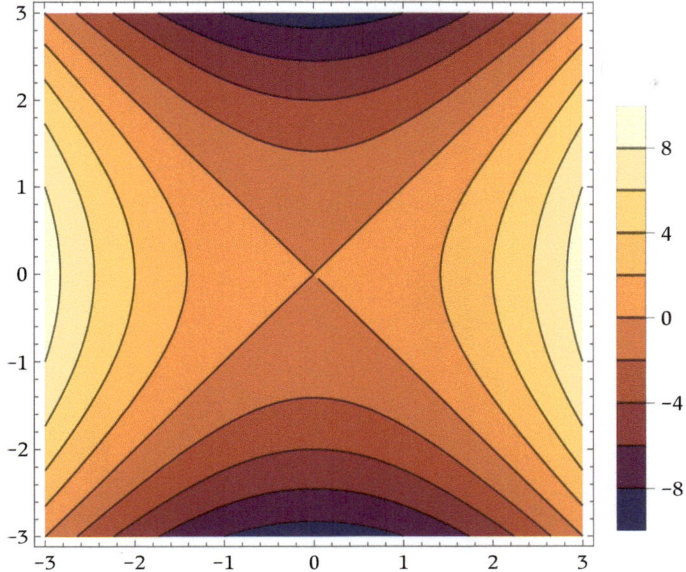

Figure 3.31 *A contour plot for* $f(x,y) = x^2 - y^2$, *with legend, produced by Mathematica*

Because the contours are hyperbolas or intersecting straight lines, the contour plot is easier to draw by hand than the surface itself. This is a common occurrence and one of the reasons for the development of contour plots.

(Figure 3.33) has concentric contours in four places. Two of these have contours with warmer colors, meaning as we pass from one contour to the next the elevation increases, indicating a peak (maximum value). Each peak lies in the interior of the innermost of the concentric contours. Similarly, each of the two pits (minimum values) lies within the innermost concentric contour with the coolest color. Look in Figure 3.24 at the mountainside and you should be able to identify a handful of locations in the topographical map of local extrema, although it may be difficult to determine whether they represent peaks or pits. Local extrema can also be identified in Figure 3.25, where the color coding makes it easier to distinguish between minima and maxima. For instance, there are local maxima in west Texas and a local minimum in northeastern Utah.

The funky chair $f(x,y) = x^3 + 0.5y^2$ (Figure 3.34) appears to have a minimum point in the seat. However, the contour plot (Figure 3.35) shows that the contour $z = 0$ appears to have a cusp, and it does not have the look of a minimum value like Figure 3.33. Because only selected contours are graphed, a contour plot might miss a subtle minimum or maximum. Is there really a minimum point in the funky chair? Later in this chapter we explore how to find all local extrema using calculus procedures.

The waterfall $f(x,y) = \ln(x+y)$ (Figure 3.36) has domain $y > -x$, which causes the contour plot, which is shown on a rectangular domain, to have a blank (white) triangular region (Figure 3.37). This surface has a vertical asymptote with

Ans. to reading exercise 3:
$$y = \frac{2}{3x}$$

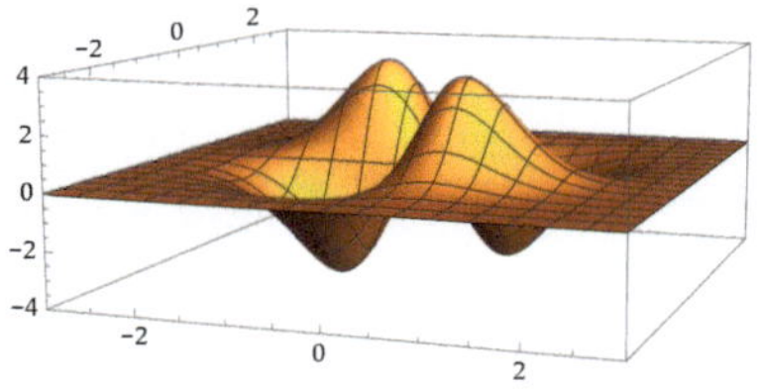

Figure 3.32 *The graph of* $f(x,y) = 20xye^{-x^2-y^2}$

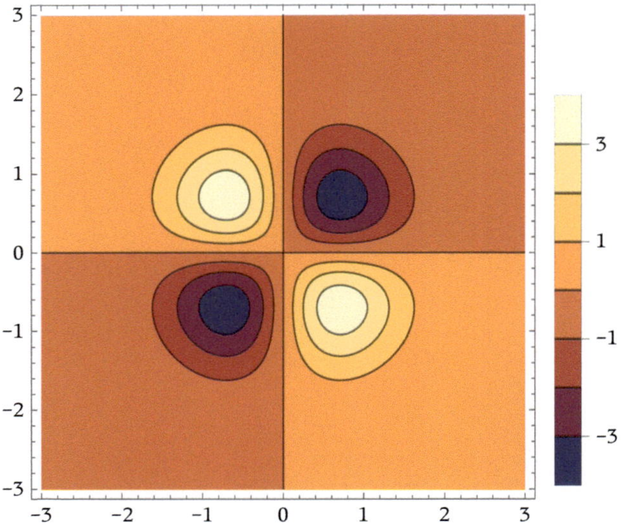

Figure 3.33 *A contour plot for $f(x, y) = 20xye^{-x^2-y^2}$*

The contour plots and three-dimensional plots in Figures 3.30–3.43 were produced using Mathematica.

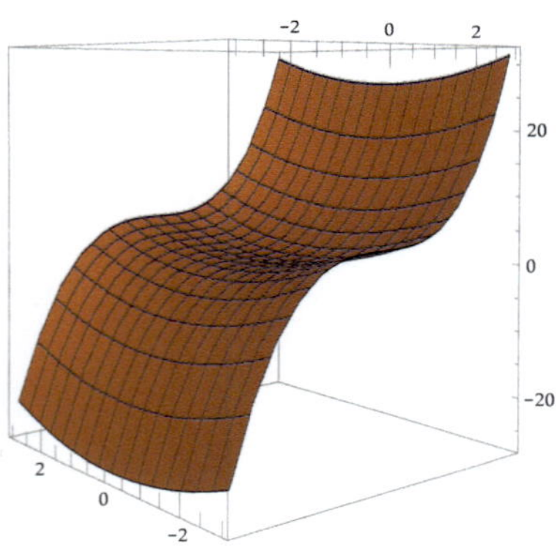

Figure 3.34 *The graph of $f(x, y) = x^3 + 0.5y^2$*

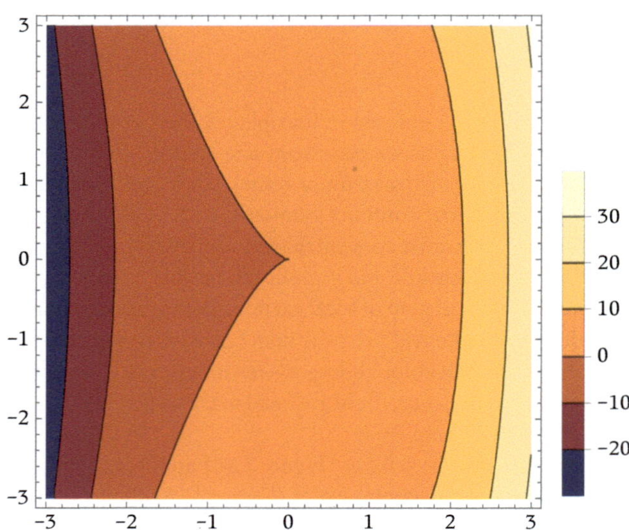

Figure 3.35 *A contour plot for $f(x, y) = x^3 + 0.5y^2$*

equation $y = -x$ (a plane) at the edge of the domain. In Figure 3.36 we can see the water falling down that vertical asymptote. In Figure 3.37 we see the contours getting closer and closer together as that vertical asymptote is approached; all the remaining contours incremented by 0.5 lie between the ones pictured and the asymptote. Accumulation of infinitely many equally spaced contours indicates a vertical asymptote, but because we cannot actually see all infinitely many contours such a picture indicates a possibility to be explored.

The circular vertical asymptotes of $f(x, y) = \sec(x^2 + y^2)$ (Figure 3.38) are easy to locate in the picture once we know they exist. Similarly, the contour plot (Figure 3.39) is challenging to interpret around those asymptotes. The squiggly curve in the white region is not a contour; all contours for this function are circular. The thick black circles appear to have white specks in them. The asymptotes are at circles not in the domain, but the picture shows thickness to those regions. The picture merely indicates that something is going on in or near those regions and should be explored. One technique for improving such a contour plot is to increase the sampling precision, which is an option present in some CAS, but this can dramatically increase the time required for computation.

The shark bite $f(x, y) = 0.1\sqrt{x^2 + y}$ (Figure 3.40) has domain $y \geq -x^2$, and at the edge of the domain there is no asymptote, in contrast with the waterfall. However, in the contour plot (Figure 3.41), the contours do get closer together as the edge of the domain is reached, just as with the waterfall. The surface appears vertical at the edge of the domain. The effect comes from the square root function, which does not have a vertical asymptote yet has infinite slope at the edge of its domain. Knowledge of the behavior of single-variable functions can help us analyze functions of two variables.

More precisely, contours for this function are collections of circles.

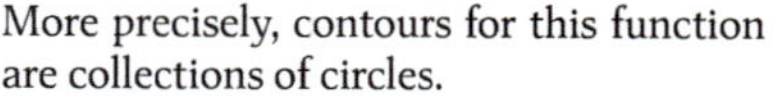
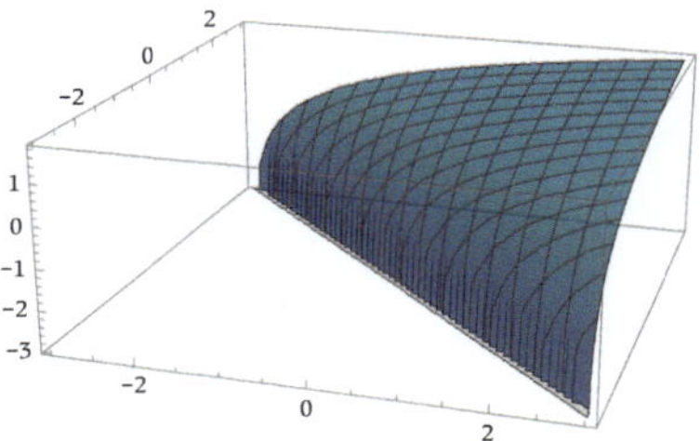

Figure 3.36 *The graph of* $f(x, y) = \ln(x + y)$

Technically, the white region consists not only of points outside the domain of the function but also points whose function value is outside of the values given in the legend.

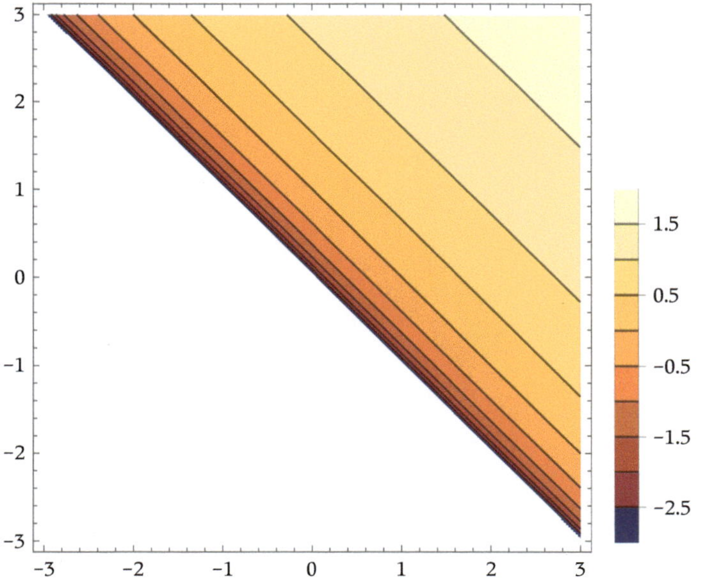

Figure 3.37 *A contour plot for* $f(x, y) = \ln(x + y)$. *The white-colored region is (mostly) not in the domain of* f

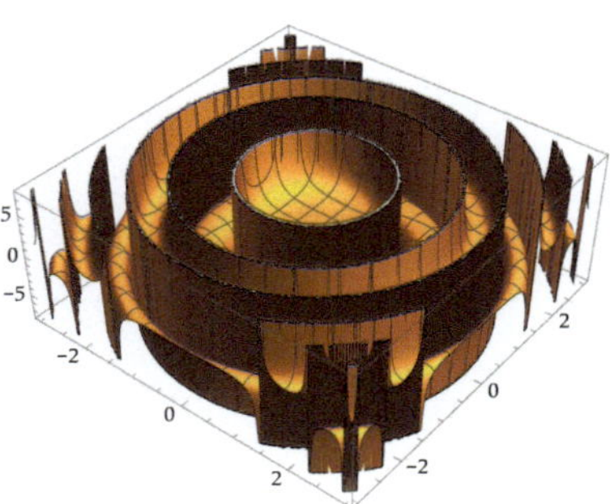

Figure 3.38 *The graph of* $f(x, y) = \sec(x^2 + y^2)$

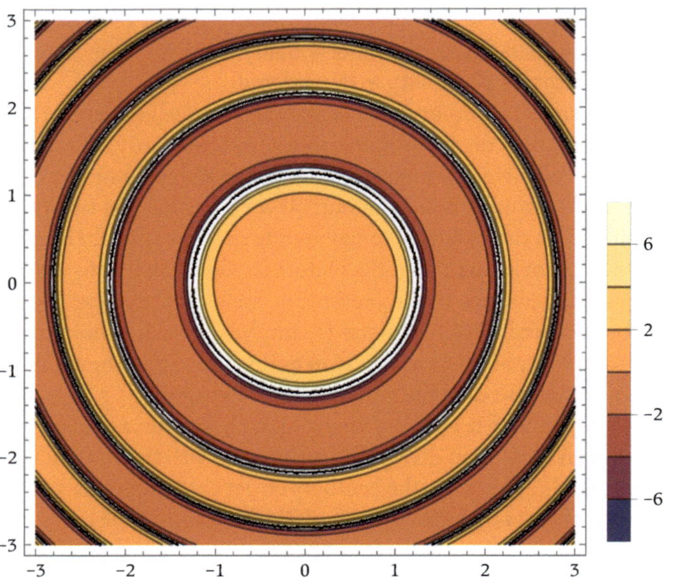

Figure 3.39 *A contour plot for $f(x, y) = \sec(x^2 + y^2)$. The white-colored regions and odd-looking noise contained in them have function values that are outside the range of values given in the legend*

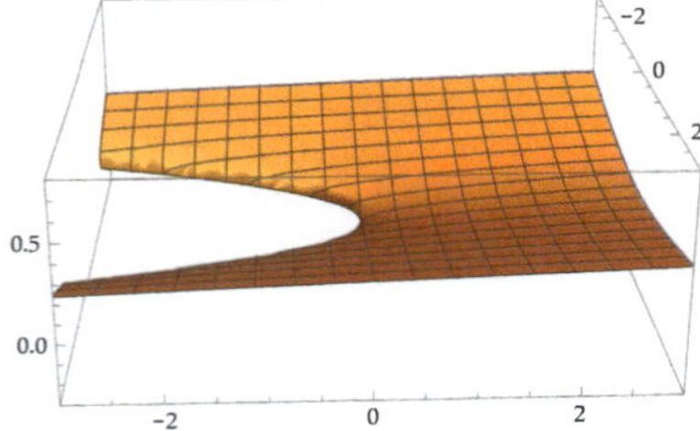

Figure 3.40 *The graph of $f(x, y) = 0.1\sqrt{x^2 + y}$*

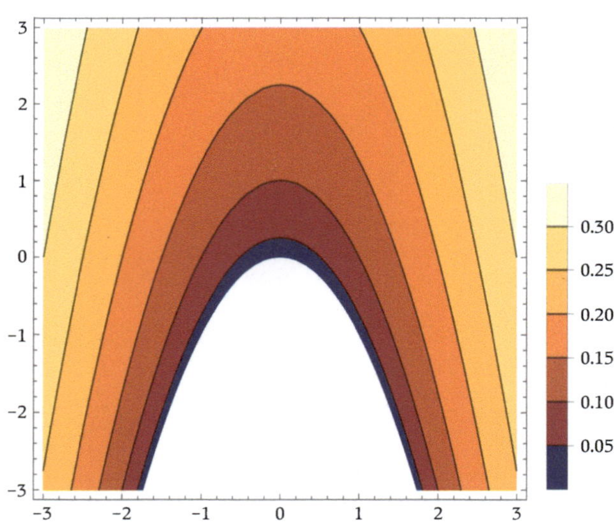

Figure 3.41 *A contour plot for $f(x, y) = 0.1\sqrt{x^2 + y}$. The white-colored region lies outside the domain of the function. Compare to Figure 3.37 at the edge of the domain*

The scary crevasse $f(x, y) = \ln |x^3 + y^2|$ (Figure 3.42) has vertical asymptote $x^3 + y^2 = 0$, a curve with a cusp. The contour plot (Figure 3.43) shows contours close together near portions of the asymptote like Figure 3.37, but then exhibits a large white region that is much thicker than a curve, and the contours are not so close together. The look is similar to that of Figure 3.41, which has no vertical asymptote and instead has a two-dimensional region outside the domain. But in this case the majority of the white region is comprised of points whose values lie outside the range of values in the legend. The same look in two contour plots may actually have two different interpretations. This same dilemma is present in any type of graphical representation, and contour plots are no exception. The exact algebraic, rather than graphical, methods of calculus can help us make precise what may still be vague from graphical analysis, as well as give us confidence that we have found what we are looking for and that we have not missed anything. The remaining sections of this chapter are devoted to that task.

By two-dimensional region, we mean a region that possesses area. A curve such as $x^3 + y^2 = 0$ has zero area.

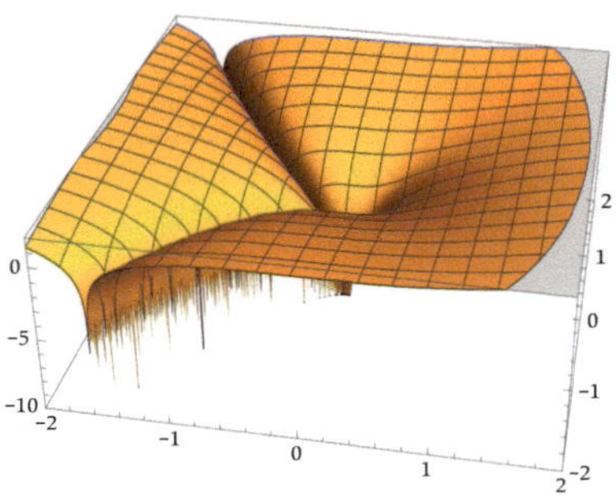

Figure 3.42 *The graph of* $f(x, y) = \ln |x^3 + y^2|$

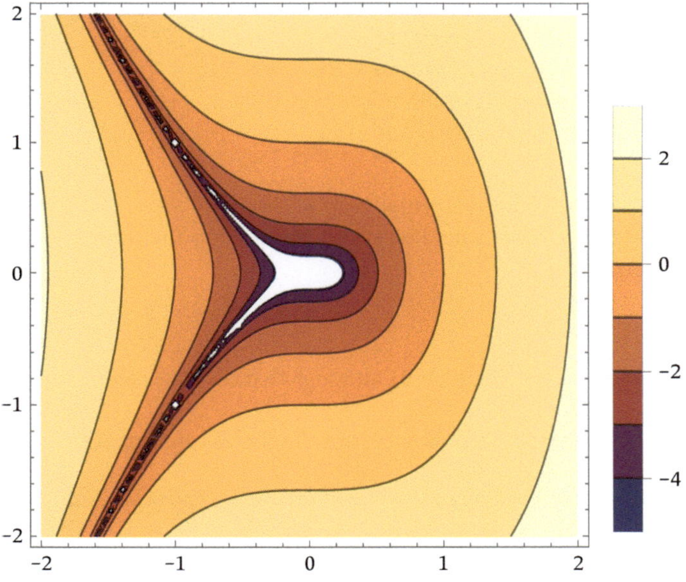

Figure 3.43 *A contour plot for $f(x, y) = \ln |x^3 + y^2|$. The white-colored region has points with function values that are outside the range of values given in the legend and also contains points not in the domain*

3.2.4 Level surfaces

Consider a function of three variables, such as the triplet adder $f(x,y,z) = x + y + z$ of Section 3.1. With three input variables (x, y, and z) and one output variable, we would need a total of four dimensions to produce an input-output graph. We humans live in a universe with three apparent dimensions, so a four-dimensional graph is very difficult to visualize.

By contrast, a contour plot of a function of two variables uses level curves and therefore utilizes just two dimensions. A contour plot for a function of three variables would only need three dimensions, making use of *level surfaces.*

Example 4 *Describe the level surfaces (contours) of the function* $f(x,y,z) = x^2 + y^2 + z^2.$

Solution The contour $f(x,y,z) = 1$ has equation

$$1 = x^2 + y^2 + z^2,$$

which is a sphere of radius 1 centered at the origin.

Similarly, the contour $f(x,y,z) = 2$ has equation

$$2 = x^2 + y^2 + z^2,$$

which is a sphere of radius $\sqrt{2}$ centered at the origin.

The only contour that would be slightly different is the contour $f(x,y,z) = 0$, which has equation

$$0 = x^2 + y^2 + z^2,$$

which consists of one point, $(0, 0, 0)$, although that could be considered a sphere of radius 0.

We conclude that the level surfaces are all spheres centered at the origin. ∎

Reading Exercise 4 For $f(x,y,z) = x^2 + y^2 - z$, what type of surface is the contour $f(x,y,z) = 0$?

Before concluding that level surfaces for a function of three variables are irrelevant to the real world, consider the situation of a cabin with an interior heat source on a cold day. The temperature of the cabin would be warmer at the source of the heat and colder near the exterior walls. Imagine now the isotherms (temperature contours). Likely they are not spheres, except possibly around the heat source, but they would be some type of three-dimensional surface.

EXERCISES 3.2

1. Find an equation of the given contour for the function $f(x,y) = \sqrt[3]{x^2 - y^2}$:
   (a) $f(x,y) = 1$; (b) $f(x,y) = -1$; (c) $z = 0$.
2. Find an equation of the given contour for the function $f(x,y) = y\sec x$:
   (a) $f(x,y) = 0$; (b) $f(x,y) = 1$; (c) $z = 2$; (d) $z = -\pi$.
3. Find an equation of the given contour for the function $f(x,y) = xe^{-y}$ and
   describe the contour: (a) $z = 0$; (b) $z = 1$; (c) $z = 2$; (d) $z = -1$.
4. Find an equation of the given contour for the function $f(x,y) = \dfrac{x^2}{4} - \dfrac{y^2}{9}$
   and describe the contour: (a) $z = 1$; (b) $z = 4$; (c) $z = -1$; (d) $z = 0$.
5. Find equations of the contours $f(x,y,z) = 4$ and $f(x,y,z) = -9$ for the
   function $f(x,y,z) = x^2 + y^2 - z^2$, and describe each contour.
6. Find an equation of the contour $f(x,y,z) = 1$ for the function
   $f(x,y,z) = \dfrac{x^2}{4} + \dfrac{y^2}{9} + \dfrac{z^2}{16}$ and describe the contour.

7–12. Draw a contour plot for the function.

   7. $f(x,y) = 1 - xy$
   8. $f(x,y) = x^2 + y^2$
   9. $f(x,y) = \dfrac{y}{\sqrt{x}}$
   10. $f(x,y) = \dfrac{y}{x^2}$
   11. $f(x,y) = \log_x y$
   12. $f(x,y) = e^{y/x}$

13. Match the graph to the contour plot.

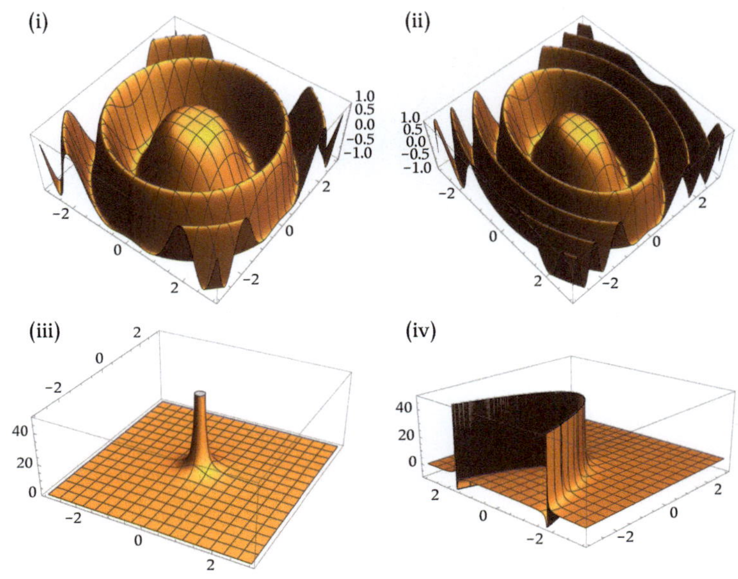

These same graphs are matched to functions in Section 3.1 exercise 23.

Ans. to reading exercise 4:
   an elliptic paraboloid (alternate answer: a paraboloid)

As in the narrative, the warmer colors on contour plots are for greater function values, and the cooler colors are for lesser values.

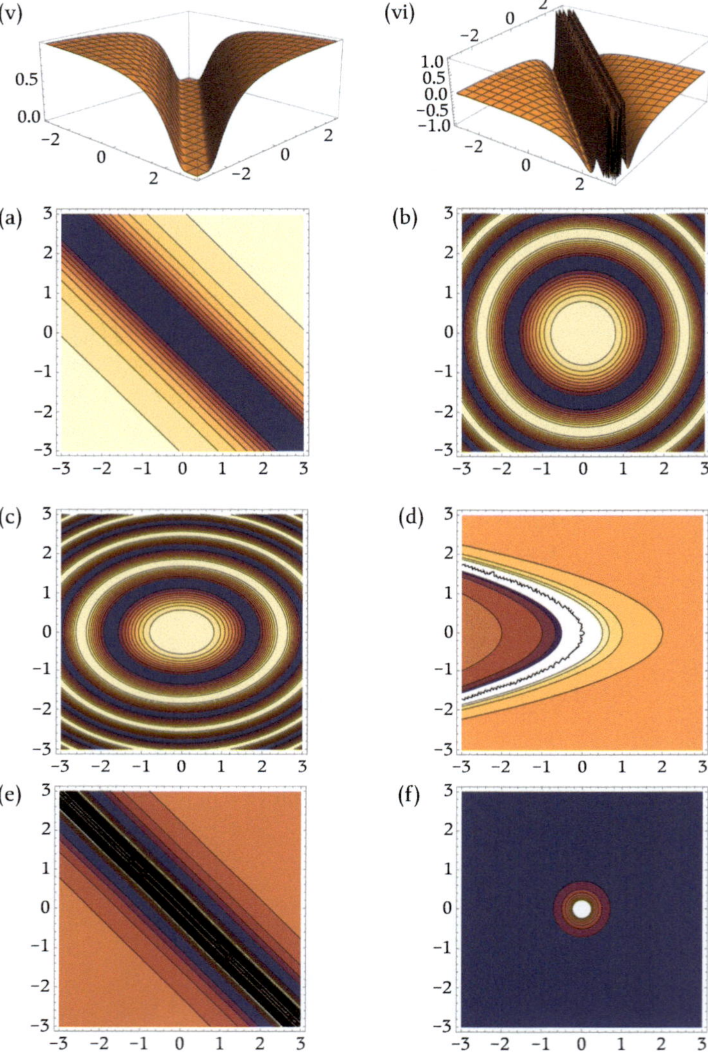

14. Match the graph to the contour plot.

These same graphs are matched to functions in Section 3.1 exercise 24.

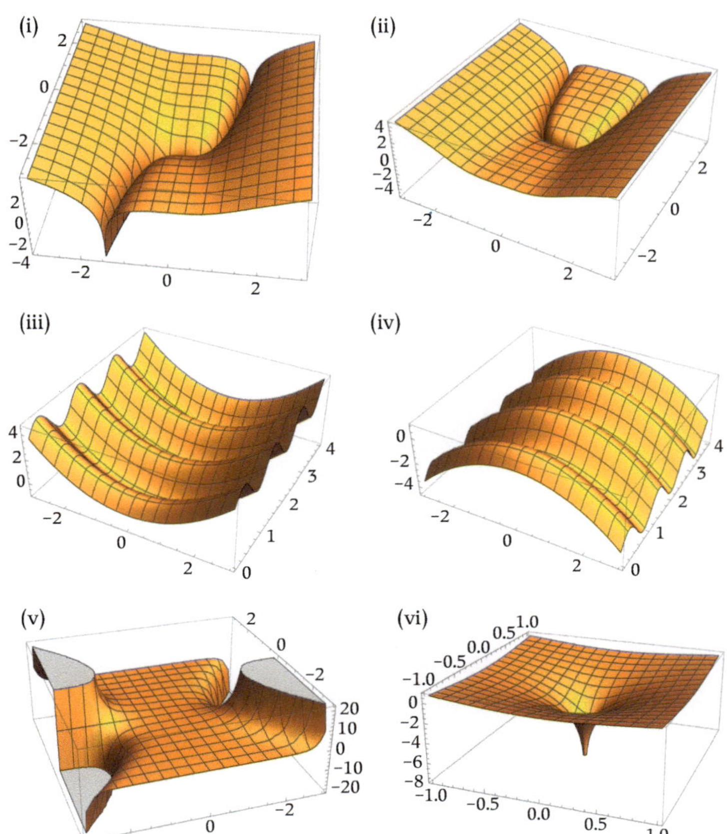

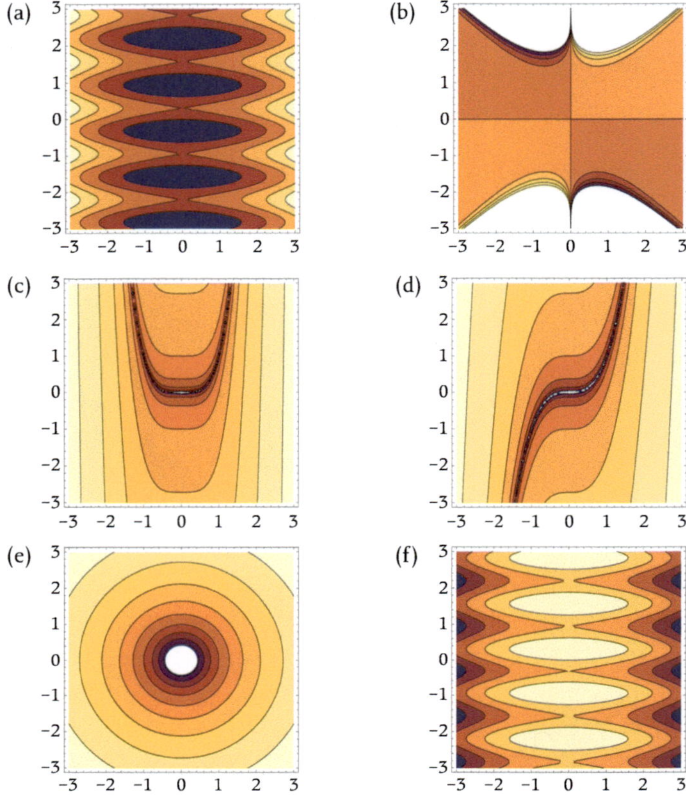

These same functions are to be graphed in Section 3.1 exercises 27–32.

15–20. Use a CAS to produce a contour plot of the function.

15. $f(x,y) = xy + \dfrac{5x^2}{y^2+1}$

16. $f(x,y) = \sqrt{\sin(xy)}$

17. $f(x,y) = xy + \dfrac{5x^2}{y^2-1}$

18. $f(x,y) = \ln(x^2 - 3y + y^2 - 7)$

19. $f(x,y) = e^{-x^2} + e^{-3y^2}$

20. $f(x,y) = \dfrac{x}{x^3-y} + \dfrac{y}{x^2-1}$

3.3   Limits of Multivariable Functions

As noted previously, graphical representations of functions are helpful but do not always give us the complete story. Just as with single-variable functions, limits can help us explore functions more precisely.

3.3.1   Two-variable limits: basic concepts

Suppose we wish to explore the function

$$f(x,y) = \frac{x}{x+y^2}.$$

We may begin by finding the domain of the function. In this case, we only need to avoid division by zero, so the domain is $x + y^2 \neq 0$, or $x \neq -y^2$. The domain is graphed in Figure 3.44.

The function is graphed in Figure 3.45 (a CAS was used). The graph indicates that function values may be approaching ∞ when $x < -y^2$ and approaching $-\infty$ when $x > -y^2$, although it isn't at all clear what's happening at $(x,y) = (0,0)$. Perhaps limits can help.

Reminder: when we encounter division by zero (at $x = a$) with one-variable functions, there are several possibilities. For instance, if $\lim\limits_{x \to a} f(x) = L$ with L a real number, there is a removable discontinuity (Figure 3.46, left).

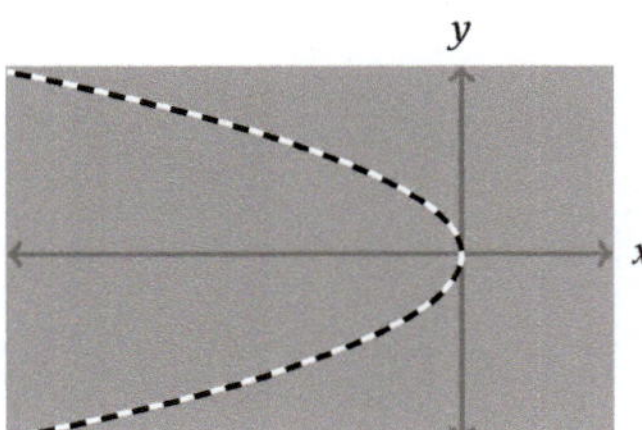

Figure 3.44 *The domain of $f(x,y) = \frac{x}{x+y^2}$. The dashed curve $x = -y^2$ is excluded from the domain*

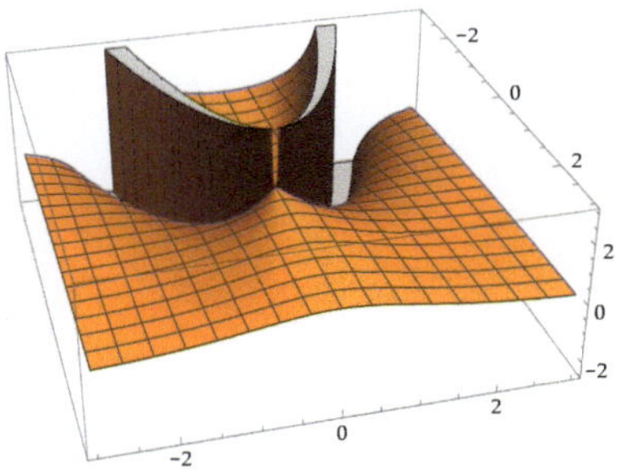

Figure 3.45 *The graph of $f(x,y) = \frac{x}{x+y^2}$*

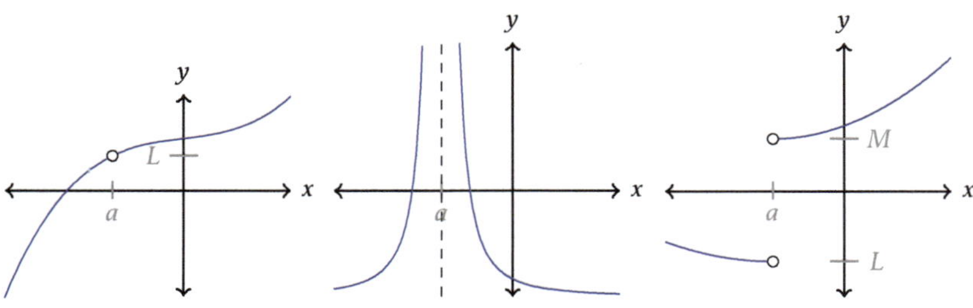

Figure 3.46 *(Left) removable discontinuity; (middle) infinite discontinuity; and (right) jump discontinuity*

There are other types of discontinuities of single-variable functions not represented in Figure 3.46.

The all-path limit for a two-variable function is analogous to a two-sided limit for a one-variable function.

The path $y = 0$ catches both an approach from the west and an approach from the east. If we want just the approach from the west, we could use $y = 0, x < 0$.

If $\lim_{x \to a} f(x) = \infty$ (Figure 3.46, middle), there is an infinite discontinuity. If $\lim_{x \to a^-} f(x) = L$, $\lim_{x \to a^+} f(x) = M$, and $L \neq M$, there is a jump discontinuity (Figure 3.46, right). We know that for the two-sided limit to exist, the one-sided limits must exist and be equal. Importantly, for one-variable functions there are only two ways for x to approach a number a, from the left and from the right.

Back to the function $f(x, y) = \frac{x}{x+y^2}$. Suppose we wish to use a limit to find out what's happening at $(x, y) = (0, 0)$. The problem is that there are more than just two ways to approach the point, because the domain is two-dimensional. Think of $(0, 0)$ as a point on a map. We could approach from the west (Figure 3.47, top left), from the north (Figure 3.47, top middle), or even from the southeast (Figure 3.47, top right). And there is no rule saying we must approach only on a straight line! We can dance along a curved path toward the point, such as along the path $y = \sqrt{x}$ (Figure 3.47, bottom left), or the path $y = x \sin(-5x)$ (Figure 3.47, bottom right). There are infinitely many ways to approach the point, and all of them must agree for the all-path limit to exist!

Infinitely many possible paths—let's get started. Suppose we approach from the west, as in Figure 3.47, top left. The y-coordinates on this path are all 0, so try using $y = 0$ as our path. Then

$$\lim_{(x,y)\to(0,0),\, y=0} \frac{x}{x + y^2} = \lim_{(x,y)\to(0,0)} \frac{x}{x + 0} = \lim_{(x,y)\to(0,0)} 1 = 1.$$

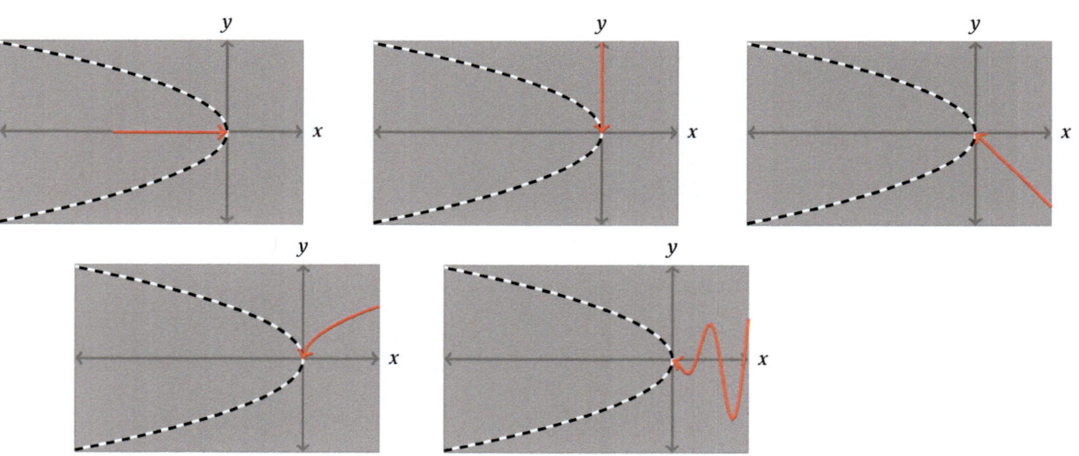

Figure 3.47 *Five of the infinitely many ways to approach the point* $(0, 0)$ *in the domain of the function* $f(x, y) = \frac{x}{x+y^2}$

Next, try the path $y = \sqrt{x}$ as in Figure 3.47, bottom left. We can rearrange the equation of the path to $y^2 = x$:

$$\lim_{(x,y)\to(0,0),\, x=y^2} \frac{x}{x + y^2} = \lim_{(x,y)\to(0,0)} \frac{y^2}{y^2 + y^2} = \lim_{(x,y)\to(0,0)} \frac{y^2}{2y^2} = \frac{1}{2}.$$

The limits on these two paths are different, so the all-path limit does not exist;

$$\lim_{(x,y)\to(0,0)} \frac{x}{x + y^2} \text{ DNE.}$$

By now you may have spotted a fundamental difficulty in this approach. If we need to have the limits on infinitely many different paths all agree in order to conclude that a limit exists, we can't check them all individually. We need a different approach.

Reading Exercise 5   If the limit along the path $x = 0$, the limit along the path $y = 0$, and the limit along the path $y = x$ all agree, must the limit exist?

3.3.2   Definition of limit

For single-variable functions, we find two-sided limits by scooting over just an infinitesimal amount, checking $f(k + \alpha)$ for an arbitrary infinitesimal α. Because the infinitesimal α can be either positive or negative, both sides of k are checked simultaneously. The idea is the same for a two-variable function, but we need to be able to scoot over an infinitesimal amount in any direction, along any path. To accomplish this we use two arbitrary infinitesimals, α and β, one for each variable. By allowing α and β to vary independently, including not just sign but also being on different levels, we can check all possible paths. How this is accomplished will be explained later in this section. First, we present the definition, in the same style as the single-variable definition.

Definition 1 LIMIT, TWO-VARIABLE FUNCTION   *Let $f(x, y)$ be a function of two variables and let a and b be real numbers. If $f(a + \alpha, b + \beta), f(a + \alpha, b),$ and $f(a, b + \beta)$ all render the same real result L for every choice of infinitesimals α and β for which the expressions are defined, then we write*

$$\lim_{(x,y)\to(a,b)} f(x,y) = L.$$

When L is a real number (i.e., not ∞ or $-\infty$), then we say that the limit exists.

For the purposes of calculation, we write

$$\lim_{(x,y)\to(a,b)} f(x,y) = f(a + \alpha, b + \beta),$$

We have not yet defined limits of two-variable functions. Think of these as experimental calculations to get an idea of how things might work.

The quantity $f(a + \alpha, b)$ checks the function values by scooting over an infinitesimal amount in the positive or negative x-direction. Likewise, $f(a, b + \beta)$ checks in the y-direction, and $f(a + \alpha, b + \beta)$ checks all the other directions/paths.

The reason for stipulating that the expressions be defined is that for some functions there are paths to the point (a, b) that are excluded from the domain in their entirety, as can be inferred from Figure 3.47.

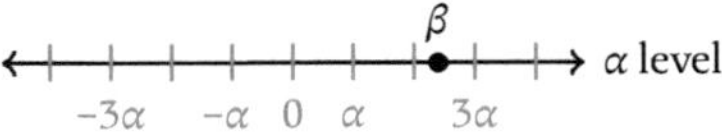

$$\beta$$

α level

$-3\alpha \quad -\alpha \quad 0 \quad \alpha \quad 3\alpha$

Figure 3.48 *β on the same level as α*

If β is on the same level as α, then β lies on the number line consisting of real-number multiples of α, as in Figure 3.48, and $\beta \approx k\alpha$ for some real number k. (Using $\beta = k\alpha$ is usually sufficient, but see later side notes.)

If it is more convenient, we could use $\alpha = k\beta$.

Notice that the idea of evaluating using continuity, which will be discussed later in this section, would not work for this limit; if $x = 0$ and $y = 0$, we would be dividing by zero.

Recall that an infinitesimal renders the real result 0 regardless of whether it is positive or negative, a positive infinite number renders the real result ∞, and a negative infinite number renders the real result $-\infty$.

As long as k is a real number, then so is $\frac{k^2-1}{k+1}$, and the quantity $\frac{k^2-1}{k+1}\beta$ is infinitesimal and renders 0. The only exception is when $\frac{k^2-1}{k+1}$ is undefined, that is, $k = -1$. Technically, we still need to check $\alpha = -1 \cdot \beta + \gamma$, where $\gamma \ll \beta$; this is still part of α and β on the same level (part of $\alpha \approx -1 \cdot \beta$), and in contrast to $\alpha = -\beta$ still allows the expression to be defined, hugging infinitesimally close to the domain exclusion. Although we will usually ignore this detail for limit calculations, it is possible for this final case to render a different result and cause the limit to not exist; see exercise 59.

with the understanding that the expression must render the same real result for any infinitesimals α and β, no matter their signs or what levels α and β are on; otherwise, the limit does not exist. The quantities $f(a + \alpha, b)$ and $f(a, b + \beta)$ must also render the same real result or the limit does not exist.

3.3.3   Two-variable function limit examples

The EAR procedure (evaluate, approximate, render) applies to two-variable limit calculations in the same manner as for single-variable limits, with one additional complication. There are times when an expression involves both α and β and we need to know how their levels compare. For instance, if we are approximating $\alpha + \beta$, there are three possibilities. Either α is on a lower level than β, written $\alpha \ll \beta$, in which case $\alpha + \beta \approx \beta$; α is on a higher level than β, written $\alpha \gg \beta$, in which case $\alpha + \beta \approx \alpha$; or α and β are on the same level (Figure 3.48), in which case we use $\beta = k\alpha$ for some real number k.

Example 5 *Evaluate* $\displaystyle\lim_{(x,y)\to(0,0)} \frac{x^2 - y^2}{x + y}$.

Solution   Using the definition of limit, we let α and β be arbitrary infinitesimals and use $x = 0 + \alpha = \alpha$, $y = 0 + \beta = \beta$:

$$\lim_{(x,y)\to(0,0)} \frac{x^2 - y^2}{x + y} = \frac{\alpha^2 - \beta^2}{\alpha + \beta}.$$

Before approximating, we need to know how the levels of α and β compare. There are three possibilities.

❶ Suppose $\alpha \gg \beta$. Then $\alpha^2 \gg \beta^2$ as well. Continuing the calculation,

$$\frac{\alpha^2 - \beta^2}{\alpha + \beta} \approx \frac{\alpha^2}{\alpha} = \alpha \doteq 0.$$

❷ Suppose $\alpha \ll \beta$. Then $\alpha^2 \ll \beta^2$ as well. Continuing the calculation,

$$\frac{\alpha^2 - \beta^2}{\alpha + \beta} \approx \frac{-\beta^2}{\beta} = -\beta \doteq 0.$$

❸ Suppose $\alpha = k\beta$, that is, α and β are on the same level. Continuing the calculation,

$$\frac{\alpha^2 - \beta^2}{\alpha + \beta} = \frac{k^2\beta^2 - \beta^2}{k\beta + \beta} = \frac{(k^2 - 1)\beta^2}{(k + 1)\beta} = \frac{k^2 - 1}{k + 1}\beta \doteq 0.$$

Following the definition, there are still two more quantities to check (but see the discussion following this solution). The first, $f(a + \alpha, b)$, uses $x = 0 + \alpha$ and $y = 0$, in which case we have

$$\frac{x^2 - y^2}{x + y} = \frac{\alpha^2 - 0}{\alpha + 0} = \alpha \doteq 0.$$

Compare to the calculation for the case $\alpha \gg \beta$.

The second, $f(a, b + \beta)$, uses $x = 0$ and $y = 0 + \beta$, yielding

$$\frac{x^2 - y^2}{x + y} = \frac{0 - \beta^2}{0 + \beta} = -\beta \doteq 0.$$

Compare to the case $\alpha \ll \beta$.

Because we render the same real result for all of these quantities for any choice of α and β, we conclude that the all-path limit exists;

$$\lim_{(x,y)\to(0,0)} \frac{x^2 - y^2}{x + y} = 0.$$

■

Notice that checking $f(a + \alpha, b)$ and $f(a, b + \beta)$ seems to repeat previous calculations in a slightly different manner. This is often true and under the usual circumstances it is customary to omit checking these two quantities.

LIMIT CALCULATIONS FOR ELEMENTARY FUNCTIONS

When calculating $\lim_{(x,y)\to(a,b)} f(x, y)$ for an elementary function f, it is customary to check only $f(a + \alpha, b + \beta)$. For a piecewise-defined or other nonelementary function, $f(a + \alpha, b)$ and $f(a, b + \beta)$ must also be checked.

Recall that elementary functions are comprised of polynomial, rational, root, trigonometric, hyperbolic, logarithmic, exponential, inverse trigonometric, and inverse hyperbolic functions, along with any algebraic combination thereof, or composition thereof.

Consider again Figure 3.46, left. For a discontinuity of a single-variable function, if the two-sided limit is a real number, then there is a removable discontinuity in the graph of the function. In example 5, the limit is a real number, and it appears that we have a removable discontinuity; see Figure 3.49. The graph indicates that instead of being at just one point, the removable discontinuity is along an entire line, $x + y = 0$. The limit we calculated, however, only shows what happens at the input point $(0, 0)$, not on the entire line $y = -x$. To explore other points on the line, which are of the form $(a, -a)$ for a real number a, we must calculate the limit $\lim_{(x,y)\to(a,-a)} \frac{x^2 - y^2}{x + y}$.

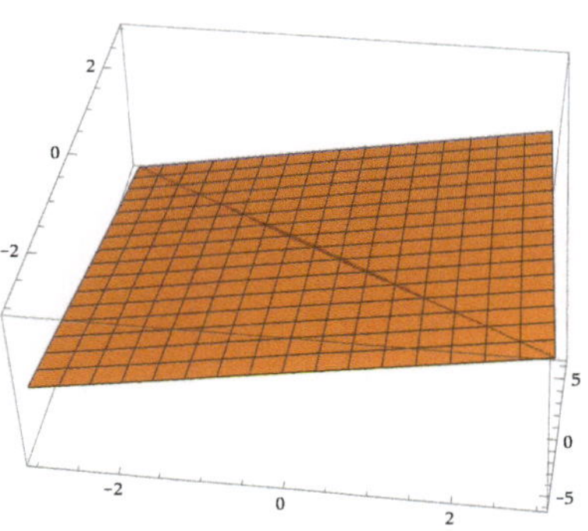

Figure 3.49 *The graph of* $f(x, y) = \frac{x^2 - y^2}{x + y}$, *produced by a CAS. The removable discontinuity line* $y = -x$ *is visible in the graph*

Example 6 *Verify the removable discontinuities along the line $y = -x$ on the graph of $f(x, y) = \dfrac{x^2 - y^2}{x + y}$ by calculating $\lim_{(x,y)\to(a,-a)} \dfrac{x^2 - y^2}{x + y}$ where a is a real number.*

Line 1 uses the definition of limit; line 2 expands the binomials in the numerator and simplifies the denominator; line 3 simplifies; line 4 approximates by throwing away terms that must be lower level terms no matter how the levels of α and β compare.

If $a = 0$, line 4 would violate the approximation principle, because we throw away terms and the result of the approximation is 0. The case $a = 0$ is taken care of in example 5.

Ans. to reading exercise 5:
  No, there are infinitely many other paths that have not yet been checked.

Solution Using the definition of limit, we let $x = a + \alpha$, $y = -a + \beta$:

$$\lim_{(x,y)\to(a,-a)} \frac{x^2 - y^2}{x + y} = \frac{(a + \alpha)^2 - (-a + \beta)^2}{a + \alpha + (-a + \beta)}$$

$$= \frac{a^2 + 2a\alpha + \alpha^2 - (a^2 - 2a\beta + \beta^2)}{\alpha + \beta}$$

$$= \frac{2a\alpha + \alpha^2 + 2a\beta - \beta^2}{\alpha + \beta}$$

$$\approx \frac{2a\alpha + 2a\beta}{\alpha + \beta} = \frac{2a(\alpha + \beta)}{\alpha + \beta} = 2a.$$

The value of the all-path limit is the real number $2a$, and at every point $(a, -a)$ on the line $y = -x$, the discontinuity is removable. ∎

Notice that in example 6, as long as $a \neq 0$, the use of three separate cases is not necessary.

Reading Exercise 6 Evaluate $\displaystyle\lim_{(x,y)\to(0,0)} \frac{y(x + 1)}{y^2 - y}$.

Example 7 *Find* $\displaystyle\lim_{(x,y)\to(2,4)} \frac{y}{x - 2}$.

Solution Using the definition of limit, we let α and β be arbitrary infinitesimals and use $x = 2 + \alpha$, $y = 4 + \beta$:

$$\lim_{(x,y)\to(2,4)} \frac{y}{x - 2} = \frac{4 + \beta}{2 + \alpha - 2} \approx \frac{4}{\alpha} = 4A,$$

which is infinite. Because α is arbitrary and can be either positive or negative, we cannot render either ∞ or $-\infty$ for the all-path limit. ∎

Although we cannot render either ∞ or $-\infty$ for the limit in example 7, we can conclude that the limit is infinite and there is a vertical asymptote on the graph of the function. Note that this particular limit only shows what happens near the input point $(2, 4)$, although we can surmise more from the graph of the function (Figure 3.50). And by working with a generic y-coordinate instead of $y = 4$, we can more fully explore the asymptote.

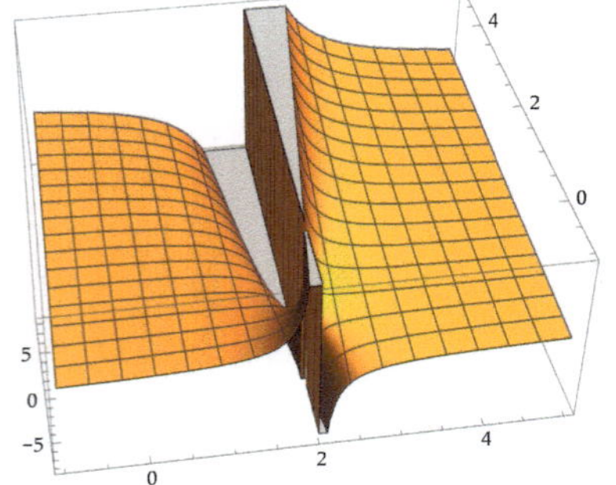

Figure 3.50 *The graph of* $f(x, y) = \frac{y}{x-2}$, *produced by a CAS. See examples 7 and 8*

Example 8 *Verify and explore the vertical asymptote $x = 2$ (a plane) on the graph of $f(x, y) = \dfrac{y}{x - 2}$ using the following steps. (a) For $y = b$ with $b > 0$, calculate $\displaystyle\lim_{(x,y)\to(2,b)} \dfrac{y}{x - 2}$. If we approach $(2, b)$ from the left, what is the limit? If we*

*approach $(2, b)$ from the right, what is the limit? (b) Repeat for $y = b$, $b < 0$.
(c) What happens if $b = 0$?*

Solution

(a) With $y = b$ and $b > 0$, using the definition of limit, we let α and β be arbitrary infinitesimals and use $x = 2 + \alpha$, $y = b + \beta$:

$$\lim_{(x,y)\to(2,b)} \frac{y}{x-2} = \frac{b+\beta}{2+\alpha-2} \approx \frac{b}{\alpha} = bA,$$

which is infinite, indicating a vertical asymptote. From the left of the point $(2, b)$, α is negative. Because b is positive, bA is negative, and the limit is $-\infty$. From the right of the point $(2, b)$, α is positive, and the limit is ∞. These match what appears on the graph in Figure 3.50.

(b) The limit calculation is identical. With $b < 0$, the signs of the limits are reversed; from the left of $(2, b)$, the limit is ∞, and from the right of $(2, b)$, the limit is $-\infty$. Again, these match what appears on the graph in Figure 3.50.

(c) If $b = 0$, then

$$\lim_{(x,y)\to(2,0)} \frac{y}{x-2} = \frac{0+\beta}{2+\alpha-2} = \frac{\beta}{\alpha}.$$

If $\beta \gg \alpha$, the limit is infinite; if $\beta \ll \alpha$, the limit renders 0; if $\beta = k\alpha$, the limit is k. Anything can be achieved by choosing an appropriate path!

We conclude that the graph of $z = \frac{y}{x-2}$ has vertical asymptote $x = 2$ for $y \neq 0$. ∎

It is possible to use notation analogous to one-sided limits, but we shall decline to do so.

This type of asymptote can also be described as a *wall asymptote*.

Ans. to reading exercise 6:
-1

Reading Exercise 7 Evaluate $\displaystyle\lim_{(x,y)\to(1,3)} \frac{y}{x-1}$.

Sometimes it is not α and β that need to be compared in an approximation, but rather α and β^2 or some other comparison.

Example 9 *Calculate* $\displaystyle\lim_{(x,y)\to(0,0)} \frac{x^2 y}{x^4 + y^2}$.

Solution Using the definition of limit, we let α and β be arbitrary infinitesimals and use $x = 0 + \alpha = \alpha$, $y = 0 + \beta = \beta$:

$$\lim_{(x,y)\to(0,0)} \frac{x^2 y}{x^4 + y^2} = \frac{\alpha^2 \beta}{\alpha^4 + \beta^2}.$$

Taking square roots works. Alternately, if $\alpha^2 \gg \beta$, then $\alpha^2\alpha^2 \gg \alpha^2\beta \gg \beta\beta = \beta^2$.

To approximate the denominator, we need to know how the levels of α^4 and β^2 compare.

❶ Suppose $\alpha^4 \gg \beta^2$, which is the same as $\alpha^2 \gg \beta$. We shall defer this calculation until later, if necessary.

❷ Suppose $\alpha^2 \ll \beta$. We shall also defer this calculation until later, if necessary.

❸ Suppose α^2 and β are on the same level, that is, suppose $\beta = k\alpha^2$ for some real number k. Continuing the calculation,

$$\frac{\alpha^2\beta}{\alpha^4 + \beta^2} = \frac{\alpha^2 k\alpha^2}{\alpha^4 + (k\alpha^2)^2} = \frac{k\alpha^4}{\alpha^4 + k^2\alpha^4} = \frac{k\alpha^4}{\alpha^4(1 + k^2)} = \frac{k}{1 + k^2},$$

which varies according to the value of k. (For instance, if $k = 1$, the expression is $\frac{1}{1+1^2} = \frac{1}{2}$, and if $k = 2$, the expression is $\frac{2}{1+2^2} = \frac{2}{5}$.) Because we do not render the same real result for all possible infinitesimals α and β, the all-path limit does not exist, and there is no need to compute the cases ❶ and ❷. ∎

Notice that in example 9 there are not just two different limits as we approach along different paths, but infinitely many! Rather than a jump discontinuity as illustrated in Figure 3.46, right, the z-coordinates we can approach vary from $-\frac{1}{2}$ to $\frac{1}{2}$; see Figure 3.51. Three dimensions allow for many interesting possibilities at a discontinuity. A jump discontinuity with two (rather than infinitely many) different limits is still possible; at least one is presented in the exercises.

Notice the strategy in example 9. Very often, if a limit does not exist, it is because the same-level case results in more than one possible value. Computing the same-level case first is a good strategy to save work. If the limit does exist, all three cases need to be calculated anyway, and no effort is lost computing the same-level case first.

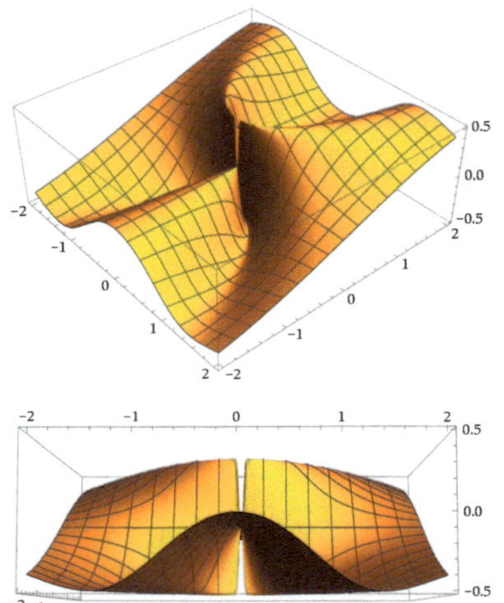

Figure 3.51 *Two views of the graph of* $f(x,y) = \frac{x^2 y}{x^4+y^2}$, *produced by a CAS. At the input* $(0, 0)$ *a vertical slit can be seen in the surface. The limit calculations in example 9 show that infinitely close to the slit there are points on the surface with any z-coordinate from* $-\frac{1}{2}$ *to* $\frac{1}{2}$

3.3.4 Arbitrary infinitesimals check all paths

Let's return to the earlier discussion of the function $f(x, y) = \frac{x}{x+y^2}$, whose graph is in Figure 3.45. When exploring $\lim\limits_{(x,y)\to(0,0)} \dfrac{x}{x + y^2}$, we looked at calculating the limit using different paths as in Figure 3.47. How does using the infinitesimal definition of limit check all of those paths?

To see how, first we calculate the limit using the definition. We let α and β be arbitrary infinitesimals and use $x = 0 + \alpha = \alpha, y = 0 + \beta = \beta$:

$$\lim_{(x,y)\to(0,0)} \frac{x}{x + y^2} = \frac{\alpha}{\alpha + \beta^2}.$$

❶ Suppose $\alpha \ll \beta^2$. Continuing the calculation,

$$\frac{\alpha}{\alpha + \beta^2} \approx \frac{\alpha}{\beta^2} \doteq 0.$$

Ans. to reading exercise 7:

DNE; the limit is infinite (the calculation results in $3A$)

Because α is on a lower level than β^2, the quantity $\frac{\alpha}{\beta^2}$ is infinitesimal.

2 Suppose $\alpha \gg \beta^2$. Continuing the calculation,

$$\frac{\alpha}{\alpha + \beta^2} \approx \frac{\alpha}{\alpha} = 1.$$

3 Suppose α and β^2 are on the same level, that is, $\alpha = k\beta^2$. Continuing the calculation,

$$\frac{\alpha}{\alpha + \beta^2} = \frac{k\beta^2}{k\beta^2 + \beta^2} = \frac{k\beta^2}{(k+1)\beta^2} = \frac{k}{k+1}.$$

Every real number is a possibility according to **1**, **2**, and **3**, and the limit does not exist.

Do you see where we checked the path $x = y^2$? With $\alpha = \beta^2$, which is $k = 1$ in **3**. Because β is arbitrary, every possible infinitesimal pair of coordinates (β^2, β) on the path $x = y^2$ leads to the approximation $\frac{1}{1+1} = \frac{1}{2}$, and the limit along that path is $\frac{1}{2}$.

What if we want the path from the southeast, which is $y = -x$ with $x > 0$? This path requires us to check $\beta = -\alpha$, with $\alpha > 0$. But because $\beta \gg \beta^2$, we have $\alpha \gg \beta^2$, and therefore all the points $(\alpha, -\alpha)$ are checked in **2**. The limit along the path from the southeast is 1.

The path $y = x\sin(-5x)$ is also checked in **2**. If $\beta = \alpha\sin(-5\alpha)$, then $\beta \approx \alpha(-5\alpha) = -5\alpha^2 \ll \alpha$, and $\beta^2 \ll \beta \ll \alpha$. There is no need to check each such path explicitly; they are all checked implicitly somewhere along the way!

The paths $x = 0$ and $y = 0$ correspond to checking $f(0, 0 + \beta)$ and $f(0 + \alpha, 0)$, respectively. As is customary for an elementary function, we skipped calculating these quantities. But in a manner similar to the solution to example 5, the calculations are subsumed in cases that are checked. Compare the earlier calculation along the path $y = 0$ to the calculation in **2**. In each, the quantity representing y is discarded. Similarly, the path $x = 0$ is implicitly checked in **1**.

Not only is it unnecessary to determine which cases check particular paths, the paths that cause issues are automatically found during the process of comparing levels for the purpose of completing approximations. The efficiency of the infinitesimal method is quite nice.

3.3.5   Horizon limits

In the graph of $f(x, y) = 20xye^{-x^2 - y^2}$ (Figure 3.52), it appears that as we wander away from the peaks and pits the surface flattens out toward a z-coordinate of 0. This is an example of a *horizon limit*, which checks all paths to an infinite distance away from the origin.

Definition 2 HORIZON LIMIT *Let $f(x, y)$ be a function of two variables. If $f(A, B)$, $f(A, k)$, and $f(k, B)$ all render the same real result L for every choice of infinite*

At this point we already know that the limit does not exist, because not every choice of infinitesimals renders the same real result.

The case $k = -1$ would cause division by zero in this result. One last case, $\alpha = -1 \cdot \beta^2 + \gamma$ with $\gamma \ll \beta^2$, remains and catches paths that are infinitely close to the domain exclusion.

In example 5, the parabolic path $x = y^2$, which uses $\alpha = \beta^2$, is implicitly checked in the case $\alpha \ll \beta$.

In example 5, the path $y = x\sin(-5x)$ is implicitly checked in the case $\beta \ll \alpha$.

A more concise argument is as follows. In general, any path approaching (a, b), even ones that spiral around the point, must eventually get infinitely close to (a, b). The points on the path that are infinitely close to (a, b) must be within an infinitesimal distance, and therefore must be of the form $(a + \alpha, b + \beta)$ or $(a + \alpha, b)$ or $(a, b + \beta)$, for some infinitesimals α and β. If all of these expressions render the same real result for all infinitesimals α and β, then all the points on the path that are infinitely close to (a, b) must render the same real result.

The symbols A and B are the capital Greek letters "alpha" and "beta." As usual, we will use the infinitesimals α and β, respectively, for their reciprocals.

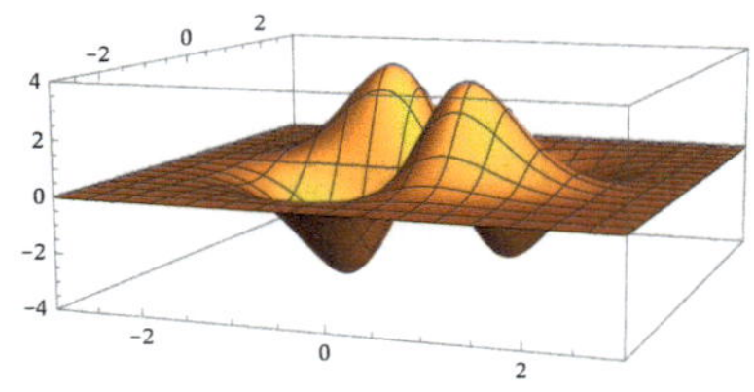

Figure 3.52 *The graph of* $f(x,y) = 20xye^{-x^2-y^2}$

The symbol ∞ in the notation $\lim\limits_{(x,y)\to\infty}$ signifies that the distance from the origin becomes infinite, that is, $\sqrt{x^2+y^2} \to \infty$. We choose not to write $(x,y) \to (\infty,\infty)$, because it is possible to be infinitely far from the origin with only one coordinate infinite, not both.

If $A \gg B$, the numerator of $f(A,B)$ is on a lower level than A^2, while the denominator is on a higher level than e^{A^2}. Recall that $e^\Omega \gg \Omega$ for any positive infinite hyperreal Ω, including A^2. The case $B \gg A$ and the same-level case are similar.

The numerator of $f(A,K)$ is on the A level. The denominator is on the e^{A^2} level, which is a higher level. Therefore, the fraction is an infinitesimal.

Be ye not dismayed; the level analysis used in this example is more sophisticated than most exercises.

hyperreal numbers A and B and real number k, then we write

$$\lim_{(x,y)\to\infty} f(x,y) = L.$$

When L is a real number (i.e., not ∞ or $-\infty$), then we say that the limit exists.

To move infinitely far from the origin does not require both the x- and y-coordinates to be infinite; as long as at least one coordinate is infinite, we are infinitely far out. The first example verifies what we see in the peaks and pits graph.

Example 10 *Calculate* $\lim\limits_{(x,y)\to\infty} 20xye^{-x^2-y^2}$.

Solution To use the definition of horizon limit we must check three quantities. First,

$$f(A,B) = 20ABe^{-A^2-B^2} = \frac{20AB}{e^{A^2+B^2}} \doteq 0,$$

recognizing that because the denominator is on a higher level than the numerator the fraction is an infinitesimal. Next,

$$f(A,k) = 20Ake^{-A^2-k^2} = \frac{20kA}{e^{A^2+k^2}} \doteq 0$$

and

$$f(k,B) = 20kBe^{-k^2-B^2} = \frac{20kB}{e^{k^2+B^2}} \doteq 0.$$

Because all of these quantities render 0 for any choice of A, B, and k,

$$\lim_{(x,y)\to\infty} 20xye^{-x^2-y^2} = 0. \qquad \blacksquare$$

In the egg crate mattress $f(x,y) = 2\sin x \sin y$ (Figure 3.53), the z-coordinates are not becoming infinite, but neither do they appear to settle down toward a particular limit like the peaks and pits graph.

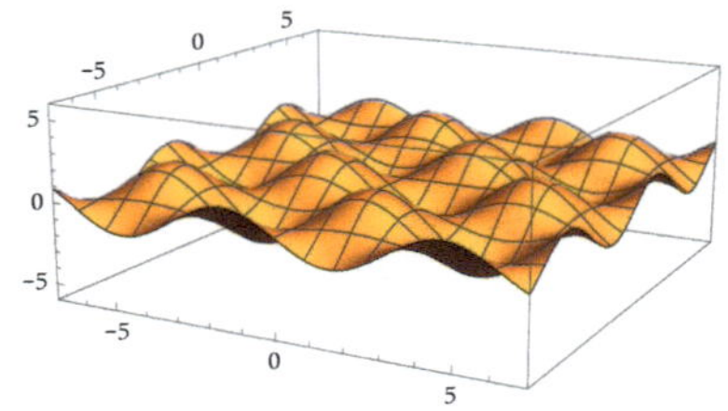

Figure 3.53 *The graph of* $f(x,y) = 2\sin x \sin y$

Example 11 *Evaluate* $\lim\limits_{(x,y)\to\infty} 2\sin x \sin y$.

Solution To use the definition of horizon limit we first check $f(A,B)$:

$$f(A,B) = 2\sin A \sin B.$$

Because $\sin A$ and $\sin B$ each vary from -1 to 1 (a consequence of the transfer principle), the value of $f(A,B)$ varies from -2 to 2. Because we do not render the same real result for every choice of A, B, and k, the limit does not exist;

$$\lim_{(x,y)\to\infty} 2\sin x \sin y \ \text{DNE}. \qquad \blacksquare$$

The limit in example 11 tells us what we see in Figure 3.53, that the z-coordinates vary between -2 and 2 forever outward.

The horizon limit can also be infinite, as evidenced by the elliptic paraboloid in Figure 3.54.

Example 12 *Find* $\lim\limits_{(x,y)\to\infty} (x^2 + 4y^2)$.

Solution To use the definition of horizon limit we first check $f(A, B)$:

$$f(A, B) = A^2 + 4B^2 \doteq \infty.$$

Similarly,

$$f(A, k) = A^2 + 4k^2 \approx A^2 \doteq \infty$$

and

$$f(k, B) = k^2 + 4B^2 \approx 4B^2 \doteq \infty.$$

Therefore,

$$\lim_{(x,y)\to\infty} (x^2 + 4y^2) = \infty.$$

∎

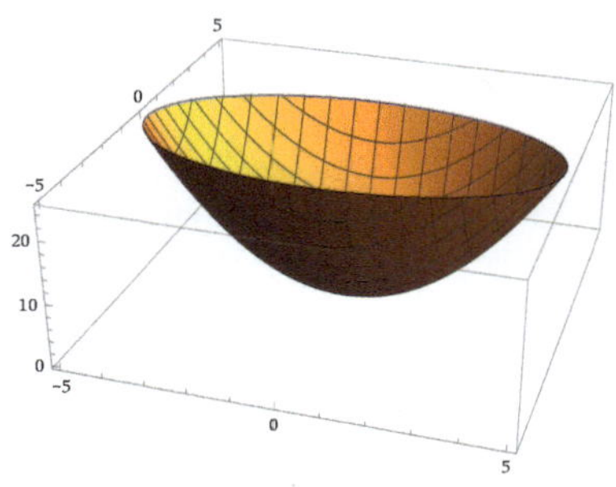

Figure 3.54 *The graph of the elliptic paraboloid* $f(x, y) = x^2 + 4y^2$

Because $A^2 > 0$ and $B^2 > 0$ no matter the signs of A and B, we know that the quantity is positive infinite.

The limit does not exist (it is not a real number), but writing ∞ for the value of the limit indicates the manner in which it does not exist.

In a hyperbolic paraboloid (Figure 3.55), the z-coordinates are becoming positive infinite in some directions but negative infinite in others; in such a case the limit cannot be rendered as either ∞ or $-\infty$.

Example 13 *Evaluate* $\lim\limits_{(x,y)\to\infty} (x^2 - y^2 + 3)$.

Solution To use the definition of horizon limit we first check $f(A, B)$:

$$f(A, B) = A^2 - B^2 + 3.$$

In order to approximate and/or render, there are three cases to consider:
❶ Suppose $A \gg B$. Continuing the calculation,

$$A^2 - B^2 + 3 \approx A^2 \doteq \infty.$$

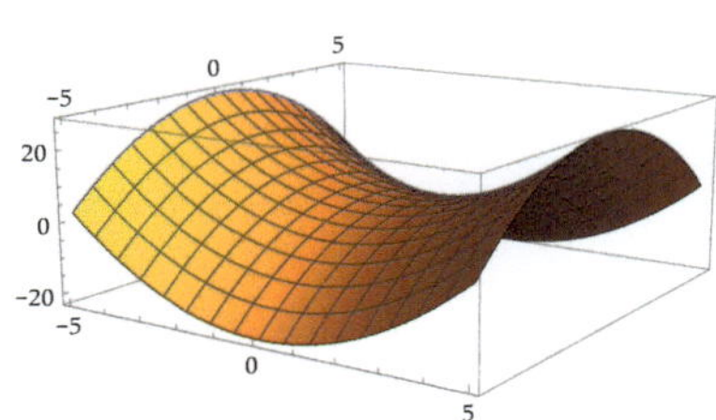

Figure 3.55 *The graph of the hyperbolic paraboloid* $f(x, y) = x^2 - y^2 + 3$

② Suppose $B \gg A$. Continuing the calculation,

$$A^2 - B^2 + 3 \approx -B^2 \doteq -\infty.$$

Having rendered two different results we may stop here;

$$\lim_{(x,y)\to\infty} (x^2 - y^2 + 3) \text{ DNE.} \qquad \blacksquare$$

Case **③** is $A = cB$ for a real number c. If $c = 1$, we have $A = B$ and $A^2 - B^2 + 3 = 3$, showing that some paths on the hyperbolic paraboloid do not lead to an infinite z-coordinate.

The reason for checking $f(A,k)$ and $f(k,B)$ is illustrated by the cylinder $f(x,y) = y^2$ (Figure 3.56). On some paths, we can go infinitely far without the z-coordinate becoming infinite.

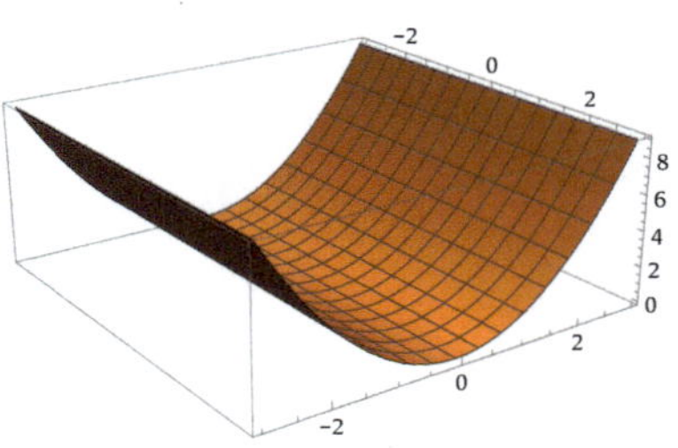

Figure 3.56 *The graph of the cylinder* $f(x,y) = y^2$

Example 14 *Calculate* $\displaystyle\lim_{(x,y)\to\infty} y^2$.

Solution To use the definition of horizon limit we first check $f(A,B)$:

$$f(A,B) = B^2 \doteq \infty.$$

Next, we check $f(A,k)$ and $f(k,B)$:

$$f(k,B) = B^2 \doteq \infty$$

and

$$f(A,k) = k^2,$$

which is a real number. Because not all infinite paths render the same real result, the all-path horizon limit does not exist;

$$\lim_{(x,y)\to\infty} y^2 \text{ DNE.} \qquad \blacksquare$$

Again, only one coordinate need be infinite to be infinitely far out. On the cylinder of example 14 and Figure 3.56, we can go infinitely far along the path $y = 3$, allowing the x-coordinate to become infinite, and stay on the level curve $z = 9$ the entire time. Checking only $f(A,B)$ forces both coordinates to be infinite simultaneously, ignoring infinitely long paths for which one coordinate remains finite.

Reading Exercise 8 Find $\displaystyle\lim_{(x,y)\to\infty} \frac{1}{x^2 + y^2}$.

3.3.6   Continuity

The definition of continuity for functions of two variables is similar to the definition for single-variable functions.

Definition 3 CONTINUITY FOR FUNCTIONS OF TWO VARIABLES *Let f be a function of two variables and let (a, b) be in the domain of f. Then f is continuous at (a, b) if $\lim\limits_{(x,y)\to(a,b)} f(x, y) = f(a, b)$.*

An alternative definition of continuity is that f is continuous at (a, b) if (a, b) is in the domain of f and $f(a + \alpha, b + \beta)$, $f(a + \alpha, b)$, and $f(a, b + \beta)$ all render the real result $f(a, b)$ for every choice of infinitesimals α and β for which the quantities are defined.

Intuitively, continuity means that in the graph of f there are no asymptotes, jumps, holes, tears, and so on at the input (a, b).

Continuity can be verified by calculating the value of the function and the value of the limit and verifying that the two quantities are equal.

Example 15 *Verify that $f(x, y) = xy + \frac{3y}{x+1}$ is continuous at the input $(3, 5)$.*

Solution First, we calculate $f(3, 5)$:

$$f(3, 5) = 3 \cdot 5 + \frac{3(5)}{3 + 1} = 15 + \frac{15}{4} = \frac{75}{4}.$$

Next, we calculate $\lim_{(x,y)\to(3,5)} f(x, y)$ using the definition of limit. Letting α and β be arbitrary infinitesimals, we use $x = 3 + \alpha$ and $y = 5 + \beta$:

Because we are trying to verify continuity, we cannot evaluate the limit using continuity; that would be saying that the function is continuous because it's continuous, a logical flaw. The definition of limit must be used to calculate the limit in this type of exercise.

$$\lim_{(x,y)\to(3,5)} \left(xy + \frac{3y}{x+1} \right) = (3 + \alpha)(5 + \beta) + \frac{3(5 + \beta)}{3 + \alpha + 1}$$

$$= 15 + 3\beta + 5\alpha + \alpha\beta + \frac{15 + 3\beta}{4 + \alpha}$$

$$\approx 15 + \frac{15}{4} = \frac{75}{4}.$$

Because the two quantities are the same, that is, $f(3, 5) = \lim\limits_{(x,y)\to(3,5)} f(x, y)$, the function f is continuous at the input $(3, 5)$. ∎

Example 16 *Show that the linear function $f(x, y) = 3x - 4y + 5$ is continuous at every input (a, b).*

The graph of a linear function of two variables is a plane, which is continuous everywhere. This example asks us to verify that continuity.

Solution First, we calculate $f(a, b)$:

$$f(a, b) = 3a - 4b + 5.$$

Compare this solution with the solution to example 15.

Next, we calculate $\lim_{(x,y)\to(a,b)} f(x,y)$ using the definition of limit. Letting α and β be arbitrary infinitesimals, we use $x = a + \alpha$ and $y = b + \beta$:

$$\lim_{(x,y)\to(a,b)} (3x - 4y + 5) = 3(a + \alpha) - 4(b + \beta) + 5$$

$$= 3a + 3\alpha - 4b - 4\beta + 5$$

$$\approx 3a - 4b + 5.$$

Because $f(a,b) = \lim_{(x,y)\to(a,b)} f(x,y)$, the function f is continuous at the input (a,b). ∎

For the instances in which $3a - 4b + 5 = 0$, the approximation in line 3 is incorrect because it would violate the approximation principle. But in that case, the quantity in line 2 is $3\alpha - 4\beta \doteq 0$, which then matches $f(a,b)$, as required. We will usually not include such details.

Showing that any linear function $f(x,y) = cx + dy + k$, where c, d, and k are real numbers, is continuous at any input (a,b) follows the pattern of example 16. In a similar fashion, many other types of functions can be shown to be continuous throughout their domains. By also proving theorems about combinations of continuous functions and compositions of continuous functions, the following theorem can be obtained.

Theorem 1 CONTINUITY OF ELEMENTARY FUNCTIONS *If $f(x,y)$ is a function of two variables utilizing only elementary functions, then f is continuous where defined.*

Once we know that a function f is continuous at (a,b), the value of $\lim_{(x,y)\to(a,b)} f(x,y)$ must be the same as $f(a,b)$, allowing us to evaluate the limit of an elementary function by calculating the function value, as long as the result is defined. We call this method of evaluating a limit *evaluating using continuity*.

Evaluating using continuity is also known as the *direct substitution method*.

Example 17 *Evaluate* $\displaystyle \lim_{(x,y)\to(2,3)} \frac{x^2 + 2}{xy}$.

All functions of two variables encountered so far in this chapter are elementary functions.

Solution This function in this limit is a rational function of two variables (the numerator and denominator are both polynomials in the variables x and y), hence it is continuous where defined. We may try evaluating the limit using continuity:

$$\lim_{(x,y)\to(2,3)} \frac{x^2 + 2}{xy} = \frac{2^2 + 2}{2 \cdot 3} = \frac{6}{6} = 1.$$

∎

Ans. to reading exercise 8:
  0 (each of $f(A,B), f(A,k),$ and $f(k,B)$ is infinitesimal)

Because the limit calculation in example 17 does not result in division by zero or something else undefined, evaluating using continuity succeeds. Notice that for the limits in examples 5–9, evaluating using continuity fails because of division by zero; for those examples, the definition of limit must be used.

Reading Exercise 9 Evaluate $\displaystyle\lim_{(x,y)\to(1,4)} \frac{x+1}{y-1}$.

As long as our functions are elementary, theorem 18 guarantees that questions asking "where is f continuous" are identical to questions instructing us to "find the domain of f."

Example 18 *Where is* $f(x,y) = \frac{x\sin y}{x+y}$ *continuous?*

Solution Because the sine function is defined for all real numbers, we need only be concerned about where we are dividing by zero. The function is continuous as long as $x + y \neq 0$, that is, everywhere except $y = -x$. ∎

Be careful; although sine and cosine are defined for all real numbers, tangent, cotangent, secant, and cosecant are not.

EXERCISES 3.3

1-4. Rapid response:

    1. If the limit along the path $y = 0$, the limit along the path $x = 0$, the limit along the path $y = x$, the limit along the path $y = x^2$, and the limit along the path $y = -x^2$ all agree, must the limit exist?

    2. If the limits along all straight paths exist, must the limit exist?

    3. If the limits along all paths exist, must the limit exist?

    4. If the limits along all paths exist and are equal, must the limit exist?

5-20. Evaluate the limit.

None of these limit exercises involve the "last case" mentioned in the side notes.

5. $\displaystyle\lim_{(x,y)\to(2,1)} \frac{x^2 + y - 5}{y^2 - 1}$

6. $\displaystyle\lim_{(x,y)\to(0,0)} \frac{x + y}{x - y}$

7. $\displaystyle\lim_{(x,y)\to(3,2)} \frac{2x - xy}{y^2 - 4}$

8. $\displaystyle\lim_{(x,y)\to(-1,1)} \frac{xy + y}{x^2 + 4x + 3}$

9. $\displaystyle\lim_{(x,y)\to(2,0)} \left(x^2 - \frac{y^2 + 2}{y + 1} \right)$

10. $\displaystyle\lim_{(x,y)\to(4,-2)} \frac{x}{y + 2}$

11. $\displaystyle\lim_{(x,y)\to(0,0)} \frac{3}{x^2 + y^2}$

12. $\displaystyle\lim_{(x,y)\to(1,6)} \frac{x^2 y + 3x - y}{y^2 - 3x + 5}$

13. $\displaystyle\lim_{(x,y)\to(-1,2)} \frac{x^2 + 3y - 7}{y^2 - 2x - 6}$

14. $\displaystyle\lim_{(x,y)\to(0,3)} \frac{\sqrt{x^2 + 1} - 1}{xy}$

15. $\displaystyle\lim_{(x,y)\to(0,0)} \frac{x^3 y}{\sqrt{x^4 + y^4}}$

16. $\displaystyle\lim_{(x,y)\to(0,0)} \frac{xy}{\sqrt{x^2 + y^2}}$

17. $\displaystyle\lim_{(x,y)\to(0,0)} \ln(x^2 + y^2)$

18. $\displaystyle\lim_{(x,y)\to(0,-1)} e^{y/x^2}$

19. $\displaystyle\lim_{(x,y)\to(1,0)} \tan^{-1} \frac{x^2}{y^2}$

20. $\displaystyle\lim_{(x,y)\to(1,1)} \cos \frac{(2x - 2y)\pi}{x^2 - y^2}$

21-28. Verify that f is continuous at the given input.

    21. $f(x,y) = \sqrt{x^2 - y}$, at $(3,5)$

Using the method of example 15 is required for these exercises.

22. $f(x,y) = xy - 11$, at $(2,-5)$
23. $f(x,y) = x + 3y + x^2 + 1$, at $(1,-1)$
24. $f(x,y) = x^5 - \frac{3y}{x+1}$, at $(0,4)$
25. $f(x,y) = \sin(2x - y)$, at $(2,4)$
26. $f(x,y) = \frac{\sqrt[3]{y-1}}{x}$, at $(4,9)$
27. $f(x,y) = \frac{x+2}{y-5}$, at $(5,-2)$
28. $f(x,y) = xy^2 - 7y + 1$, at $(0,1)$

29–34. State the set of points for which the function is continuous.

29. $f(x,y) = 3\sqrt{x} + \frac{4}{y-1}$

30. $f(x,y) = \frac{\ln(x-3)}{y}$

31. $f(x,y) = \frac{x-7+y}{xy}$

32. $f(x,y) = \sqrt{3 - \sin xy}$

33. $f(x,y) = \ln\left(\frac{x}{y}\right)$

34. $f(x,y) = \frac{x+3y}{x^2+y^2}$

35–42. Evaluate the horizon limit.

35. $\displaystyle\lim_{(x,y)\to\infty} \ln\left(x^2 + y^2\right)$

36. $\displaystyle\lim_{(x,y)\to\infty} \left(x^2 - 2x + y^4 + 3y^2 - 2y + 7\right)$

37. $\displaystyle\lim_{(x,y)\to\infty} \left(e^{-x^2} + e^{-3y^2}\right)$

38. $\displaystyle\lim_{(x,y)\to\infty} \frac{\sin y}{1 + x^2}$

39. $\displaystyle\lim_{(x,y)\to\infty} \left(x^3 - 2x^2 + y^4 - 3y^2 + 4x + 2\right)$

40. $\displaystyle\lim_{(x,y)\to\infty} e^{12/(x^4+y^2)}$

41. $\displaystyle\lim_{(x,y)\to\infty} \cos\left(\frac{\pi}{x^2 + y^2}\right)$

42. $\displaystyle\lim_{(x,y)\to\infty} \frac{x^3 + 5}{x^4 + 2x^2 + 17}$

43. The function $f(x,y) = \frac{\sin x}{(1+y^2)\sqrt{x^2}}$ is undefined when $x = 0$. Use the limit $\displaystyle\lim_{(x,y)\to(0,b)} \frac{\sin x}{(1 + y^2)\sqrt{x^2}}$, where b is a real number, to determine what is happening on the graph and describe the type of discontinuity.

44. For the waterfall $f(x,y) = \ln(x + y)$, evaluate $\displaystyle\lim_{(x,y)\to(a,-a)} \ln(x + y)$, where a is a real number, and describe what is happening at the edge of the domain of the function.

45. For the function in example 5 and graphed in Figure 3.49, show that $y = -x$ is a *removable discontinuity line* by evaluating $\displaystyle\lim_{(x,y)\to(a,-a)} \frac{x^2 - y^2}{x + y}$, where a is a real number.

46. For the function $f(x,y) = \frac{\sin(x^2+y^2-4)}{x^2+y^2-4}$, show that $x^2 + y^2 = 4$ is a *removable discontinuity circle* by evaluating $\lim\limits_{(x,y)\to(a,b)} \dfrac{\sin(x^2+y^2-4)}{x^2+y^2-4}$, where a and b are real numbers such that $a^2 + b^2 = 4$.

47. Show that the function $f(x,y) = \frac{\sin(y-e^x)}{y-e^x}$ has *removable discontinuity curve* $y = e^x$. (Hints: use $(x,y) \to (a, e^a)$. Because $y = e^x$ is continuous, if α is infinitesimal, then $e^{a+\alpha} = e^a + \gamma$ for some infinitesimal γ.) Conclude that any continuous curve $y = f(x)$ can serve as the shape of a removable discontinuity curve on the graph of a function of two variables.

48. Find an example of a function of two variables with a removable discontinuity at every point along the cubic $y = x^3$.

49. Show that the elliptic hyperboloid $f(x,y) = x^2 + y^2$ is continuous at every input (a, b).

50. Show that the hyperbolic paraboloid $f(x,y) = 2x^2 - y^2$ is continuous at every input (a, b).

51. Show that the linear function $f(x,y) = cx + dy + k$, where c, d, and k are real numbers, is continuous at every input (a, b).

52. Show that the elliptic paraboloid $f(x,y) = cx^2 + dy^2 + k$, where c, d, and k are real numbers, is continuous at every input (a, b).

53. In example 9, the limit calculation gives rise to values of the form $\frac{k}{1+k^2}$ where k is a real number. Find the maximum and minimum values of the single-variable function $f(x) = \frac{x}{1+x^2}$ and compare to Figure 3.51.

54. In the discussion of arbitrary infinitesimals checking all paths, the calculation of $\lim\limits_{(x,y)\to(0,0)} \dfrac{x}{x+y^2}$ shows that any real number can be obtained as a limit along a path, without showing that ∞ or $-\infty$ can be obtained. Does this make intuitive sense? Check the remaining case mentioned in the side note, which checks paths infinitely close to the domain exclusion, to see if you can obtain infinite limits along such paths.

55. Complete example 5 by checking the final case mentioned in the side note, $\alpha = -\beta + \gamma$ where $\gamma \ll \beta$.

56. In exercise 13, when checking the case $\alpha = k\beta$, the expression $\frac{-2k+3}{-2k+4}$ is obtained. (a) What value of k causes division by zero? (b) What final case should be checked? (c) Check that final case. Are these path limits infinite?

57. Write a definition of limit of a three-variable function, that is, define $\lim\limits_{(x,y,z)\to(a,b,c)} f(x,y,z)$.

58. Suppose we are evaluating a limit of a function of three variables and have an expression with $\alpha + \beta + \gamma$ in the denominator. What are the different cases that must be considered?

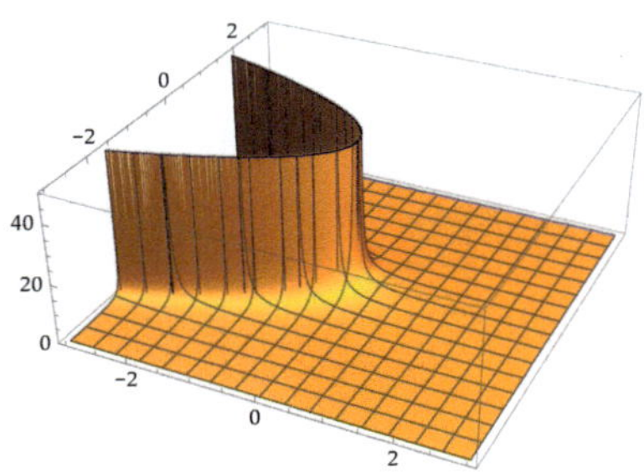

Figure 3.57 *The graph of $f(x,y) = \dfrac{1}{x+y^2}$*

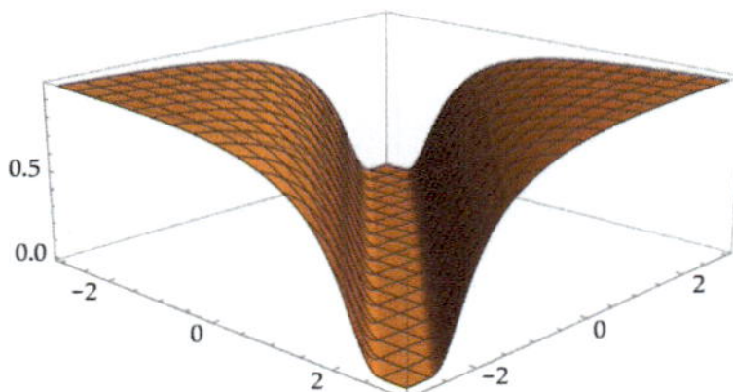

Figure 3.58 *The graph of $f(x,y) = e^{-1/(x+y)^2}$*

Exercises 65–67 were suggested by Wesley Perkins of Lyon College.

59. Investigate $\displaystyle\lim_{(x,y)\to(0,0)} \frac{xy}{y+\sin x}$. (a) Use the three cases $\beta \ll \sin\alpha$, $\beta \gg \sin\alpha$, and $\beta = k\sin\alpha$, all of which should render the same real result. (b) In the case $\beta = k\sin\alpha$, which value of k results in division by zero? What case should still be checked? (c) Check the final case. After continuing the calculation, even more subcases arise. If $\gamma \gg \alpha^2$, what is the result? If $\gamma \ll \alpha^2$, what is the result? What can we conclude?

60. Revise the investigation of $\displaystyle\lim_{(x,y)\to(0,0)} \frac{xy}{y+\sin x}$, this time using the approximation $\sin\alpha \approx \alpha$ before choosing the three cases. (a) Use the three cases $\alpha \ll \beta$, $\beta \ll \alpha$, and $\alpha = k\beta$. (b) For the "last case" $\alpha = -\beta$, back up a step to the expression $\dfrac{\alpha\beta}{\beta+\sin\alpha}$ to make the substitution $\alpha = -\beta$. Use two terms of the Maclaurin series for sine to make the approximation, and then complete the calculation. (c) What path (state the equation using variables x and y) gives the infinite result of part (b)?

61. Investigate the horizon limit $\displaystyle\lim_{(x,y)\to\infty} \frac{1}{x+y^2}$ (Figure 3.57). A last case is needed. (a) Calculate using the case $A = -B^2 + c$ for a real number c. (b) Calculate using the case $A = -B^2 + \gamma$ for an infinitesimal γ. (c) What can you conclude about this horizon limit? (d) Describe how the answers to (a) and (b) explain features of the graph.

62. Investigate the horizon limit $\displaystyle\lim_{(x,y)\to\infty} e^{-1/(x+y)^2}$ (Figure 3.58). A last case is needed. (a) Calculate using the case $A = -B + c$ for a real number c. (b) Calculate using the case $A = -B + \gamma$ for an infinitesimal γ. (c) What can you conclude about this horizon limit? (d) Describe how the answers to (a) and (b) explain features of the graph.

63. Let $f(x,y) = \begin{cases} 0, & x \neq 0 \\ 1, & x = 0 \end{cases}$. Investigate $\displaystyle\lim_{(x,y)\to(0,0)} f(x,y)$.

64. Factoring the numerator and denominator of an expression and simplifying by cancelling identical factors from numerator and denominator can help simplify a limit calculation. Use this technique for the limits in (a) exercise 8, (b) example 5, and (c) reading exercise 6. (d) These cancellations appear to be of the form $\dfrac{0}{0}$; how is this justified?

65. One option for finding a limit as $(x,y) \to (0,0)$ is to convert the Cartesian equation to polar coordinates and find the limit as $r \to 0$. (a) Find $\displaystyle\lim_{(x,y)\to(0,0)} \frac{\sin(x^2+y^2)}{x^2+y^2}$ by converting to polar coordinates; (b) find the same limit using the methods of this section.

66. (a) Find $\displaystyle\lim_{(x,y)\to(0,0)} \frac{\sqrt{x^2 + y^2 + 1} - 1}{x^2 + y^2}$ by converting to polar coordinates;

    (b) find the same limit using the methods of this section.

67. (a) Find $\displaystyle\lim_{(x,y)\to(0,0)} \frac{x^3 y \sin y}{x^4 + y^4}$ by converting to polar coordinates; (b) find the same limit using the methods of this section.

3.4   Partial Derivatives, Part 1

Imagine yourself standing on the side of a hill. What is the slope where you are standing? Doesn't it depend on what direction you are looking? Think of the hill as a surface, the graph of a function of two variables, such as in Figure 3.59. You are standing at the red dot, the point $(a, b, f(a, b))$. If you walk along the surface in the y-direction, your elevation on the hill (z-coordinate on the surface) changes more rapidly than if you walk in the x-direction. And there is no rule saying that we must only travel in the x- or y-directions, just as there is no rule that we may only travel in east-west or north-south directions when standing on a real-life hill. But these cardinal directions are a good place to begin our study of slope.

The particular viewpoint chosen can affect our perception of uphill and downhill on a surface.

3.4.1   Partial derivatives

Consider walking in the x-direction along the surface. What slope do we experience at the point $(a, b, f(a, b))$? It should be the slope of the tangent line in the x-direction, as viewed in Figure 3.60 (or the negative of the slope, depending on which way you are traveling the path). What affects the slope along that path? It's only points along that path, not to either side. In other words, the slice $y = b$ yields

The input (a, b), with x-coordinate a and y-coordinate b, produces the output $f(a, b)$, which serves as the z-coordinate of the point. From (a, b) in the xy-plane, we move $f(a, b)$ units in the z-direction to find the point on the surface. The three-dimensional coordinates of the point are $(a, b, f(a, b))$.

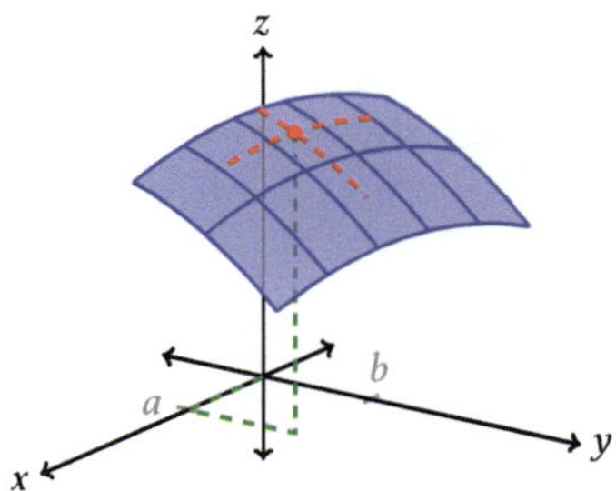

Figure 3.59 *A surface $z = f(x, y)$ (blue), along with a point on the surface $(a, b, f(a, b))$ (red). Which path we walk (red, dashed) determines the slope we experience*

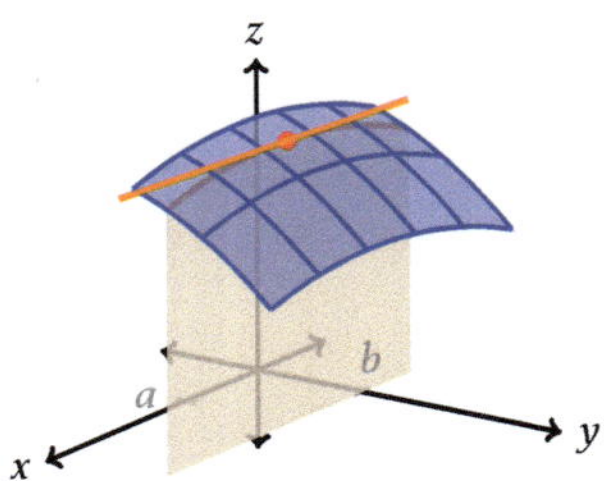

Figure 3.60 *A surface $z = f(x, y)$ (blue), along with a point on the surface $(a, b, f(a, b))$ (red), the slice $y = b$ (tan), and the path (brown) and tangent line (orange) in the x-direction*

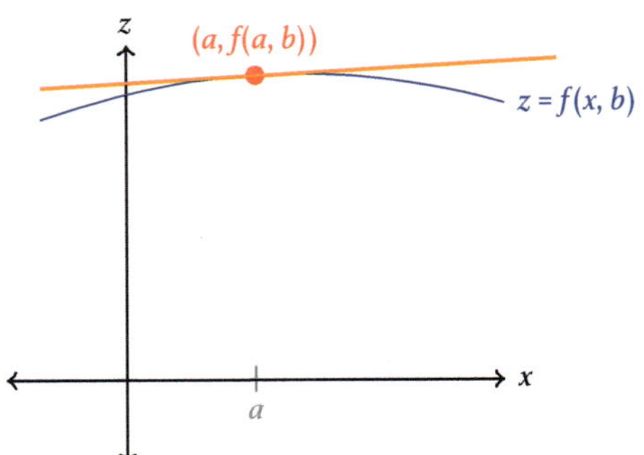

Figure 3.61 *The slice $y = b$ extracted from Figure 3.60. Everywhere in this slice, the y-coordinate of the input to the function f is b. The blue curve here corresponds to the brown curve in Figure 3.60*

a curve for which we can find the slope in the same manner as for a single-variable function! The slice $y = b$ is extracted in Figure 3.61, where the z-coordinates on the curve come from evaluating $z = f(x, y)$ with y-coordinate always equal to b. Following the procedure for the single-variable case, we then find the slope of the tangent line by using the slope formula, which requires two points. The first point is $(a, f(a, b))$. The second point is found by scooting over an infinitesimal amount α in the x-direction, staying in the slice $y = b$, to the point $(a + \alpha, f(a + \alpha, b))$:

We use the slope formula $m = \frac{y_2 - y_1}{x_2 - x_1}$ with $(x_1, y_1) = (a, f(a, b))$ and $(x_2, y_2) = (a + \alpha, f(a + \alpha, b))$.

$$\text{slope in } x\text{-direction} = \frac{f(a + \alpha, b) - f(a, b)}{a + \alpha - a} = \frac{f(a + \alpha, b) - f(a, b)}{\alpha}.$$

This formula is used in the same manner as the definition of the derivative, only this time it is for the *partial derivative $f_x(a, b)$*.

For the slope in the y-direction, we use the slice $x = a$ (Figure 3.62). The slice $x = a$ is extracted in Figure 3.63, where the z-coordinates on the curve come from evaluating $z = f(x, y)$ with x-coordinate always equal to a. Our first point to use is $(b, f(a, b))$, and for the second point we stay in the slice $x = a$ and scoot over an infinitesimal amount β in the y-direction to $(b + \beta, f(a, b + \beta))$. Using the slope formula,

$$\text{slope in } y\text{-direction} = \frac{f(a, b + \beta) - f(a, b)}{b + \beta - b} = \frac{f(a, b + \beta) - f(a, b)}{\beta},$$

yielding the partial derivative $f_y(a, b)$.

For the definition, instead of using a specific point (a, b), we use variables x and y to find the partial derivative functions.

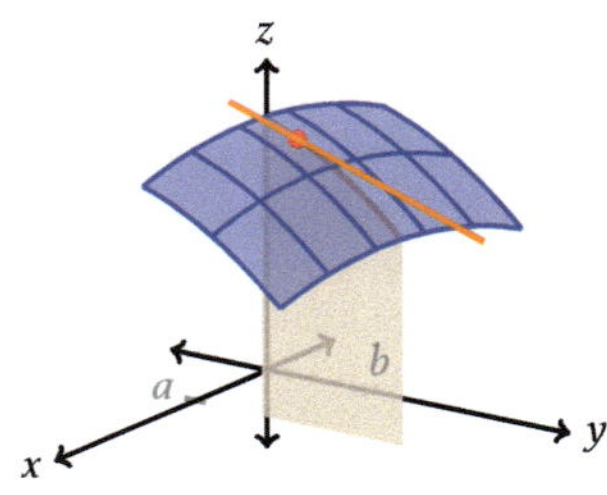

Figure 3.62 *A surface $z = f(x, y)$ (blue), along with a point on the surface $(a, b, f(a, b))$ (red), the slice $x = a$ (tan), and the path (brown) and tangent line (orange) in the y-direction*

Definition 4 PARTIAL DERIVATIVES *Let $z = f(x, y)$ be a function of two variables.*

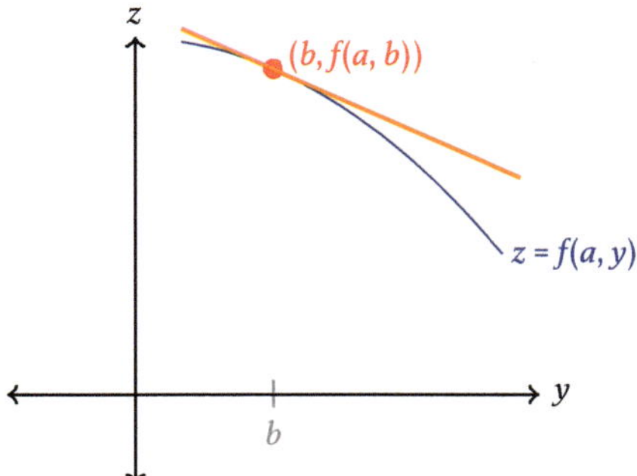

Figure 3.63 *The slice x = a extracted from Figure 3.62. Everywhere in this slice, the x-coordinate of the input to the function f is a. The blue curve here corresponds to the brown curve in Figure 3.62*

The partial derivative of *f* with respect to *x is given by*

$$f_x(x, y) = \frac{f(x + \alpha, y) - f(x, y)}{\alpha},$$

provided we render the same real result for every infinitesimal α. Likewise, the partial derivative of *f with respect to y is given by*

$$f_y(x, y) = \frac{f(x, y + \beta) - f(x, y)}{\beta},$$

provided we render the same real result for every infinitesimal β.

These formulas are the same as for the slopes in the *x*- and *y*-directions, with *a* and *b* replaced by the variables *x* and *y*, respectively.

Leibniz notation uses the partial derivative symbol ∂ instead of the derivative symbol *d*:

$$f_x(x, y) = \frac{\partial f}{\partial x} = \frac{\partial}{\partial x} f(x, y) = \frac{\partial z}{\partial x} = \frac{\partial}{\partial x} z,$$

$$f_y(x, y) = \frac{\partial f}{\partial y} = \frac{\partial}{\partial y} f(x, y) = \frac{\partial z}{\partial y} = \frac{\partial}{\partial y} z.$$

Other notations used for $f_x(x, y)$ include operator notation $D_x f$ and coordinate number instead of variable, $f_1(x, y)$ or $D_1 f$. For $f_y(x, y)$ the corresponding notations are $D_y f, f_2(x, y)$, and $D_2 f$.

The notation $\frac{\partial f}{\partial x}$ is read "partial derivative of *f* with respect to *x*."

3.4.2   Using the definition of partial derivative

The definition of partial derivative is used in the same manner as the definition of derivative.

Example 19 *Use the definition of partial derivative to find $f_x(x,y)$ for $f(x,y) = x^2y + 3x^2$.*

Solution Let α be an arbitrary infinitesimal. Then

$$
\begin{aligned}
f_x(x,y) &= \frac{f(x+\alpha, y) - f(x,y)}{\alpha} \\[2mm]
&= \frac{(x+\alpha)^2 y + 3(x+\alpha)^2 - (x^2y + 3x^2)}{\alpha} \\[2mm]
&= \frac{(x^2 + 2x\alpha + \alpha^2)y + 3(x^2 + 2x\alpha + \alpha^2) - x^2y - 3x^2}{\alpha} \\[2mm]
&= \frac{x^2y + 2x\alpha y + \alpha^2 y + 3x^2 + 6x\alpha + 3\alpha^2 - x^2y - 3x^2}{\alpha} \\[2mm]
&= \frac{2x\alpha y + \alpha^2 y + 6x\alpha + 3\alpha^2}{\alpha} \\[2mm]
&\approx \frac{2xy\alpha + 6x\alpha}{\alpha} = \frac{\alpha(2xy + 6x)}{\alpha} = 2xy + 6x.
\end{aligned}
$$

Line 1 is the definition; line 2 evaluates the function f; lines 3 and 4 expand the expressions; line 5 simplifies; line 6 approximates by throwing away the lower-level terms involving α^2, then continues by factoring α from the numerator to cancel with the α in the denominator.

Notice that in line 5, all that remains in the numerator are infinitesimal terms; the real-number-level terms subtract out from the previous line. Knowing that this should happen (for elementary functions) can help catch algebraic errors.

Notice in the calculations for example 19 that the constant 3 and the variable y are treated identically! When finding $f_x(x,y)$, we are staying in a slice with a constant y-coordinate; in Figures 3.60 and 3.61 that y-coordinate is b. We can therefore treat the variable y as if it is a constant. Because the definition of f_x is otherwise identical to the definition of derivative, all of the derivative rules can be derived in the same manner as for the derivative. We may use all of the usual derivative rules, including the product, quotient, and chain rules, as well as derivative formulas such as $\frac{d}{dx}\ln x = \frac{1}{x}$ and $\frac{d}{dx}\sin x = \cos x$, when calculating f_x, treating y as a constant.

Example 20 *Find $f_x(x,y)$ for $f(x,y) = x^2y + 3x^2$.*

Solution This time we are not instructed to use the definition of partial derivative, so we may use the derivative rules and formulas, treating y as a constant. Then

$$f_x(x,y) = 2x \cdot y + 3 \cdot 2x = 2xy + 6x.$$

When finding $f_y(x,y)$, we are staying in a slice with constant x-coordinate; in Figures 3.62 and 3.63, that x-coordinate is a.

To find $f_y(x,y)$, we treat x as a constant.

Example 21 *Find $f_y(x,y)$ for $f(x,y) = x^2y + 3x^2$.*

Solution This time we treat x as a constant. Then

$$f_y(x,y) = x^2 + 0 = x^2.$$

Because we are treating x as a constant, $3x^2$ is a constant and its derivative is 0. Also, just as $\frac{d}{dy}5y = 5$, the partial derivative of $x^2 \cdot y$ with respect to y is x^2.

3.4.3   Partial derivative examples

Example 22 *Find $f_x(x,y)$ and $f_y(x,y)$ for $f(x,y) = 5x - 3y + 7 + 4x^2y^3$.*

Solution To find the partial derivative $f_x(x,y)$, we treat y as a constant and use derivative rules and formulas:

$$f_x(x,y) = 5 - 0 + 0 + 4y^3 \cdot 2x = 5 + 8xy^3.$$

Treating y as a constant, the derivative of $4x^2y^3 = 4y^3x^2$ is the constant $4y^3$ times $2x$.

To find the partial derivative $f_y(x,y)$, we treat x as a constant and use derivative rules and formulas:

$$f_y(x,y) = 0 - 3 + 0 + 4x^2 \cdot 3y^2 = -3 + 12x^2y^2.$$ ∎

Treating x as a constant, the derivative of $4x^2y^3$ is the constant $4x^2$ times $3y^2$.

Notice from example 22 that each partial derivative can contain both variables, and the two partial derivatives are not necessarily equal.

Example 23 *Find the partial derivatives of $f(x,y) = x^2 e^{5x+3y}$.*

Solution We have two partial derivatives to find, $f_x(x,y)$ and $f_y(x,y)$.
To find $f_x(x,y)$, we treat y as a constant. We need the product rule because the variable x is in both factors of $x^2 \cdot e^{5x+3y}$:

$$f_x(x,y) = x^2 \cdot e^{5x+3y} \cdot 5 + e^{5x+3y} \cdot 2x$$

$$= (5x^2 + 2x)e^{5x+3y}.$$

The form used for the product rule is first $\cdot$ second$'$ + second $\cdot$ first$'$. When finding the derivative of e^{5x+3y}, the chain rule is needed because the exponent is not "just plain x".

To find $f_y(x,y)$ we treat x as a constant. We do not need the product rule, because the variable y is in only one factor of $x^2 \cdot e^{5x+3y}$; we use the constant multiple rule instead:

$$f_y(x,y) = x^2 \cdot e^{5x+3y} \cdot 3 = 3x^2 e^{5x+3y}.$$ ∎

When finding the derivative of e^{5x+3y}, the chain rule is needed because the exponent is not "just plain y."

Reading Exercise 10 Find the partial derivatives of $f(x,y) = x^2 \sin y$.

Example 24 *Find $\frac{\partial z}{\partial x}$ and $\frac{\partial z}{\partial y}$ for $z = \cos(x^2 y)$.*

Solution To find $\frac{\partial z}{\partial x}$ we treat y as a constant. The chain rule is needed because $x^2 y$ is not "just plain x":

$$\frac{\partial z}{\partial x} = -\sin(x^2 y) \cdot y \cdot 2x = -2xy \sin(x^2 y).$$

Treating y as a constant, the derivative of $x^2 y = y \cdot x^2$ is found using the constant multiple rule, giving $y \cdot 2x$.

To find $\frac{\partial z}{\partial y}$ we treat x as a constant. The chain rule is needed because $x^2 y$ is not "just plain y":

$$\frac{\partial z}{\partial y} = -\sin(x^2 y) \cdot x^2 = -x^2 \sin(x^2 y). \qquad \blacksquare$$

Treating x as a constant, the derivative of $x^2 \cdot y$ is found using the constant multiple rule, giving $x^2 \cdot 1$.

The solution to example 23 illustrates the fact that different rules can be needed for f_x and f_y. The next example illustrates that we may also need different formulas for calculating the two partial derivatives.

Example 25 *Find the partial derivatives of $f(x, y) = x^y$.*

Solution To find $\frac{\partial f}{\partial x}$ we treat y as a constant. Then x^y has the same form as x^4, a power function. We therefore use the power rule:

$$\frac{\partial f}{\partial x} = y \cdot x^{y-1}.$$

To find $\frac{\partial f}{\partial y}$ we treat x as a constant. Then x^y has the same form as 4^y, a general exponential function. We therefore use the formula for a general exponential function:

$$\frac{\partial f}{\partial y} = x^y \cdot \ln x. \qquad \blacksquare$$

Recall that $\frac{d}{dy} a^y = a^y \ln a$.

As illustrated in the solution to example 25, it is sometimes helpful as preparatory scratch work to rewrite a function's expression (such as x^y) by replacing x or y with a constant (such as 4) in order to determine the correct formulas or rules to use.

Partial derivatives will be used extensively throughout the remainder of this chapter and beyond. More about partial derivatives will be explored in the next section, after the opportunity for some basic practice.

EXERCISES 3.4

1–6. Rapid response: is it a derivative or a partial derivative?

1. $\dfrac{dz}{dx}$

2. $\dfrac{\partial z}{\partial x}$

3. $s'(t)$

4. $s_t(t, u)$

5. $\dfrac{\partial}{\partial x} f$

6. $\dfrac{df}{dx}$

7–12. Rapid response: Which variable can be treated as a constant?

7. $f_x(x, y) = x^2 + y^2$

8. $\dfrac{\partial}{\partial y}\left(x^2 + y^2\right)$

9. $s_t(t, u)$

10. $\dfrac{\partial z}{\partial s}$ for $z = \left(r^2 + s^2\right)$

11. $\dfrac{\partial}{\partial y}\left(y^2 + z^2\right)$

12. $x_g(f, g)$

13–18. Rapid response: find $f_y(x, y)$.

Ans. to reading exercise 10:
$f_x(x, y) = 2x \sin y,\ f_y(x, y) = x^2 \cos y$

13. $f(x, y) = 4y + \sqrt{3x^2 + 4x - 7}\left(e^{(x^5)} \sin(x^3 - 6)\right)$

14. $f(x, y) = \dfrac{x^3 e^{\sqrt{x^2 - 1/x}}}{\cosh(\sin^3 x^4)} + 7y^3$

15. $f(x, y) = 4y\sqrt{3x^2 + 4x - 7}\left(e^{(x^5)} \sin(x^3 - 6)\right)$

16. $f(x, y) = e^{\sqrt{x^2 - 1/x}} \cdot 7y^3$

17. $f(x, y) = \dfrac{4y}{\sqrt{3x^2 + 4x - 7}\left(e^{(x^5)} \sin(x^3 - 6)\right)}$

18. $f(x, y) = \dfrac{x^3 e^{\sqrt{x^2 - 1/x}}}{\cosh\left(\sin^3 x^4\right)}$

19–32. Find $f_x(x, y)$ and $f_y(x, y)$.

19. $f(x, y) = x^2 + y^3 - \sin x + 2^y$

20. $f(x, y) = 4x - 7y^5 + \cosh x - \ln y$

21. $f(x, y) = \sqrt{y^3 + y} + \cos(7x^2)$

22. $f(x, y) = \sec(5x + \pi) - e^{y^2 + 2}$

23. $f(x, y) = 4 + \dfrac{x^2 + 2}{3x - 1} + y \sinh y$

24. $f(x, y) = x\sqrt{x^2 - 7} + \dfrac{\tan^{-1} y}{y^2}$

25. $f(x, y) = x^2 + 3xy - 7y^2 + 2x - 3y + 8$

26. $f(x, y) = x^3 + 4x^2y^2 - y^5x + 12x - 9y + 11$

27. $f(x, y) = x^2y - 7xy^3 + 4x \sin y$

28. $f(x, y) = x\sqrt{y} + y^3 e^x$

29. $f(x, y) = y^2 \sin(xy)$

30. $f(x, y) = \dfrac{x^2 - y}{x + y^2}$

31. $f(x, y) = \dfrac{\ln(x + y)}{x}$

32. $f(x, y) = (x + y) \sin x^2$

33–46. Find the partial derivatives of the function.

33. $z = x^2 \tan y^2$

34. $z = \sqrt{x^3 - y^2}$

35. $z = \tan^{-1}\left(x^2 y^4\right)$

36. $z = \cos^2(7x - 3y)$

37. $z = (\sin t)^k$

38. $z = t \cosh(x^2 - 7)$

39. $f(u, v) = \sqrt{u^2 + v^2}$
40. $f(x, y) = 7\pi x e^{xy^2}$
41. $z = (x + 3y)(x^2 - 7)(y^2 + 1)$
42. $z = 3x^y$
43. $g(x, y) = \dfrac{x}{y + \frac{3x}{y^2+2}}$
44. $f(r, s) = r^2 s^3 \ln r^4$
45. $s = \sec(ut - 1)\tan(u^2 + t)$
46. $x(f, g) = \sin \ln(f^2 g + 1)$

Use definition 4. Do not use the derivative rules and formulas.

47–54. Use the definition to find the requested partial derivative.

47. $\dfrac{\partial z}{\partial x}$ for $z = x^2 + 3x - y^2 + 5y$
48. $\dfrac{\partial z}{\partial y}$ for $z = x^2 + 3x - y^2 + 5y$
49. $f_x(x, y)$ for $f(x, y) = \dfrac{x}{y}$
50. $f_x(x, y)$ for $f(x, y) = x^2 + \sqrt{y}$
51. $f_y(x, y)$ for $f(x, y) = x^2 + \sqrt{y}$
52. $f_y(x, y)$ for $f(x, y) = \dfrac{x}{y}$
53. $f_y(x, y)$ for $f(x, y) = \sin xy$ (Hint: use sine of a sum formula, approximation formulas)
54. $f_x(x, y)$ for $f(x, y) = e^{xy}$ (Hint: use series approximation to rewrite $e^{\alpha y}$)

3.5 Partial Derivatives, Part 2

In this section, we continue our study of partial derivatives.

3.5.1 Three-variable partial derivatives

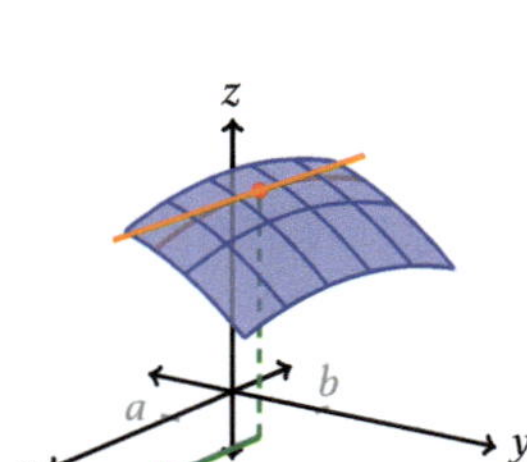

Figure 3.64 *A surface $z = f(x, y)$ (blue), along with a point on the surface $(a, b, f(a, b))$ (red) and the tangent line at that point in the x-direction (orange). The green vector lies in the domain of f (the xy-plane), indicating movement in the positive x-direction*

The graph of a function f of two variables appears in Figure 3.64. Two variables, x and y, are input (domain) variables, and one variable, z, is the output variable. The xy-plane comprises the domain of the function.

The input (a, b) lies in the xy-plane at the initial point of the green vector in Figure 3.64; the point $(a, b, f(a, b))$ on the surface corresponding to that input is pictured, as is the tangent line in the x-direction. If we walk along the surface in the positive x-direction, the corresponding inputs lie along the green vector. These inputs all have the same y-coordinate, b. As a result, when finding the partial derivative of f with respect to x, we can consider y to be constant; only x is changing.

Now consider a function g of three variables x, y, and z. The domain consists of three dimensions, and the output requires a fourth dimension. Figure 3.65 shows only the input (domain) space, along with the input (a, b, c). We can still picture moving in the x-direction from our input point (a, b, c), analogous to the green

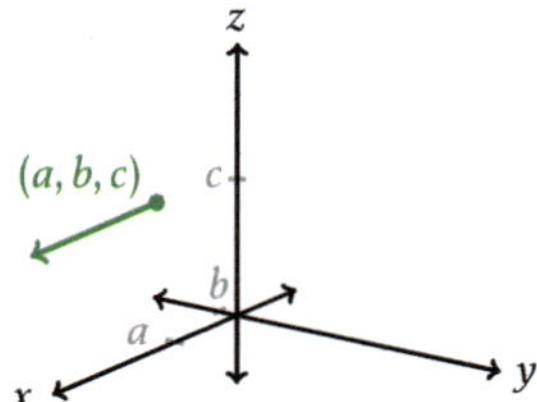

Figure 3.65 *The domain space of a function $t = g(x, y, z)$ of three variables, along with an input (a, b, c) and a vector pointing in the x-direction*

vector in the input plane in Figure 3.64. Still, only the x-coordinate of the input is changing. But both the y- and z-coordinates of the input in Figure 3.65 are constant; the y-coordinate is always b, and the z-coordinate is always c. As a result, when finding the partial derivative of g with respect to x, we can consider both y and z to be constant.

Definition 5 PARTIAL DERIVATIVES, FUNCTION OF THREE VARIABLES
Let $t = g(x, y, z)$ be a function of three variables.
The partial derivative of g with respect to x is given by

$$g_x(x, y, z) = \frac{g(x + \alpha, y, z) - g(x, y, z)}{\alpha},$$

provided we render the same real result for every infinitesimal α.
The partial derivative of g with respect to y is given by

$$g_y(x, y, z) = \frac{g(x, y + \beta, z) - g(x, y, z)}{\beta},$$

provided we render the same real result for every infinitesimal β.
The partial derivative of g with respect to z is given by

$$g_z(x, y, z) = \frac{g(x, y, z + \gamma) - g(x, y, z)}{\gamma},$$

provided we render the same real result for every infinitesimal γ.

As you might surmise, the definition of partial derivative for a function of four variables is analogous, with four partial derivatives, one for each variable, and still only one variable changing at a time, while all the other variables are held constant.

There are three partial derivatives to calculate. For each, we use the usual derivative rules and formulas for the variable of interest, treating the other two variables as constants.

Example 26 *Find the partial derivatives of $f(x, y, z) = x^2 y + \cos z$.*

Solution To find $f_x(x, y, z)$, we treat y and z as constants:

$$f_x(x, y, z) = y \cdot 2x + 0 = 2xy.$$

Just as we would for the two-variable case, the partial derivative of $x^2 y = y \cdot x^2$ with respect to x is found using the constant multiple rule, resulting in $y \cdot 2x$.

Just as we would for the two-variable case, the partial derivative of $x^2 \cdot y$ with respect to y is found using the constant multiple rule, resulting in $x^2 \cdot 1$.

To find $f_y(x, y, z)$, we treat x and z as constants:

$$f_y(x, y, z) = x^2 \cdot 1 + 0 = x^2.$$

Finally, to find $f_z(x, y, z)$, we treat x and y as constants:

The term $x^2 y$ is a constant when finding the partial derivative with respect to z.

$$f_z(x, y, z) = 0 - \sin z = -\sin z.$$ ∎

Reading Exercise 11   Find the partial derivatives of $f(x, y, z) = 4x^2 + yz^3$.

3.5.2   Slopes in the x- and y-directions

See the discussion before definition 4 in Section 3.4, and compare to the definition.

Suppose that we wish to walk on the surface $z = f(x, y)$ from the point $(a, b, f(a, b))$ in the positive x-direction, as in Figure 3.66, left. The slope is the value of $f_x(a, b)$, the partial derivative of f with respect to x evaluated at the input (a, b).

If you are walking uphill on a trail and turn around to go the opposite direction along the trail, you begin walking downhill. The slopes are not the same! The value of the downhill slope is the negative of the value of the uphill slope. For this reason, the slope when walking from $(a, b, f(a, b))$ in the negative x-direction is $-f_x(a, b)$.

Example 27   *Let $f(x, y) = x^2 y - y^3$. Find the slope at the point $(1, 2, -6)$ (a) in the positive x-direction and (b) in the negative x-direction.*

Solution   To find the slope in either the positive or negative x-direction, we need to find the partial derivative $f_x(x, y)$:

$$f_x(x, y) = 2x \cdot y - 0 = 2xy.$$

Points on the surface are of the form $(a, b, f(a, b))$; for $(1, 2, -6)$, $a = 1$, $b = 2$, and $f(a, b) = f(1, 2) = 1^2(2) - 2^3 = -6$.

The input at which we wish to know the slope is $(1, 2)$.

(a)   The slope in the positive x-direction is $f_x(1, 2) = 2(1)(2) = 4$.

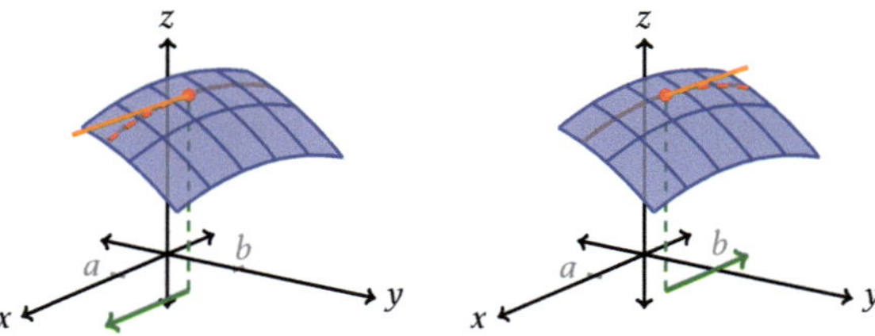

Figure 3.66   *A point $(a, b, f(a, b))$ (red) on a surface $z = f(x, y)$ (blue), (left) walking in the positive x-direction and (right) walking in the negative x-direction*

(b) The slope in the negative x-direction is $-f_x(1, 2) = -4$. ∎

The sign of the slope tells us whether our elevation is increasing or decreasing. At the point $(1, 2, -6)$ on the surface of example 27, if we set out walking in the positive x-direction we are traveling uphill, and if we set out walking in the negative x-direction we are traveling downhill.

Slopes in the y-direction are found in the same manner; the slope at the input (a, b) in the positive y-direction is $f_y(a, b)$, and the slope in the negative y-direction is $-f_y(a, b)$.

Example 28 *Let $f(x, y) = x^2 y - y^3$. Find the slope at the point $(1, 2, -6)$ (a) in the positive y-direction and (b) in the negative y-direction.*

Solution To find the slope in either the positive or negative y-direction, we need to find the partial derivative $f_y(x, y)$:

$$f_y(x, y) = x^2 - 3y^2.$$

(a) The slope in the positive y-direction is $f_y(1, 2) = 1^2 - 3(2^2) = -11$ (downhill).

(b) The slope in the negative y-direction is $-f_y(1, 2) = -(-11) = 11$ (uphill).

∎

Reading Exercise 12 Find the slope in the positive x-direction at the point $(0, 0, 1)$ on the surface $f(x, y) = \sin x + \cos y$.

Ans. to reading exercise 11:
$f_x(x, y, z) = 8x, f_y(x, y, z) = z^3,$
$f_z(x, y, z) = 3yz^2$

3.5.3 Implicit partial differentiation

Implicit partial differentiation works in the same manner as implicit differentiation for single-variable functions, and for the same reasons; it is an application of the chain rule. Consider, for instance, finding $\frac{\partial}{\partial x}\left(x^3 + 3xy^2\right)^8$. Using the chain rule, we would have

$$\frac{\partial}{\partial x}\left(x^3 + 3xy^2\right)^8 = 8\left(x^3 + 3xy^2\right)^7 \cdot \frac{\partial}{\partial x}\left(x^3 + 3xy^2\right)$$

$$= 8\left(x^3 + 3xy^2\right)^7 \cdot \left(3x^2 + 3y^2\right).$$

If $z = x^3 + 3xy^2$, then the preceding derivation can be rewritten as

$$\frac{\partial}{\partial x} z^8 = 8z^7 \cdot \frac{\partial}{\partial x} z$$

$$= 8z^7 \cdot \frac{\partial z}{\partial x}.$$

In the one case in which $z = x$ so that z actually is "just plain x," the partial derivative is $\frac{\partial z}{\partial x} = 1$ and the procedure still works.

Because z consists of x-stuff and y-stuff (more formally, z is a function of x and y), it is not "just plain x" and the chain rule applies when finding the partial derivative with respect to x.

Therefore, when z is given implicitly as a function of x and y, to find $\frac{\partial z}{\partial x}$ we ❶ take the partial derivative with respect to x of both sides of the equation, remembering to use the chain rule and multiply by $\frac{\partial z}{\partial x}$ when differentiating z, and ❷ solve the resulting equation for $\frac{\partial z}{\partial x}$.

Example 29 *Suppose z is defined implicitly as a function of the variables x and y by $x^2 + y^2 - z^2 = 47$. Find $\frac{\partial z}{\partial x}$.*

The equation $x^2 + y^2 - z^2 = 47$ is a quadric surface, in particular a hyperboloid of one sheet. Implicit partial differentiation can help us determine slopes, as well as other properties to be discussed in the remaining sections of this chapter.

Solution   ❶ Take the partial derivative with respect to x of both sides of the equation, treating y as a constant:

$$2x + 0 - 2z \cdot \frac{\partial z}{\partial x} = 0.$$

❷ Solve the resulting equation for $\frac{\partial z}{\partial x}$:

$$2x = 2z\frac{\partial z}{\partial x}$$

$$\frac{2x}{2z} = \frac{\partial z}{\partial x}$$

$$\frac{\partial z}{\partial x} = \frac{x}{z}.$$

∎

Notice how differently the three variables are treated in the solution to example 29. The variable x is treated as the variable we are differentiating with respect to, so that the derivative of the term x^2 is $2x$. The variable y is treated as a constant, so that the derivative of the term y^2 is 0. The variable z is treated as a function of x and y, so that the derivative of z^2 is $2z \cdot \frac{\partial z}{\partial x}$, an application of the chain rule. Remembering these distinctions is key to performing implicit differentiation correctly.

All the usual derivative rules apply, including the product rule.

Example 30 *Use implicit partial differentiation to find $\frac{\partial z}{\partial x}$ given $4x^3 + 2xyz^2 - z^4 = y^2 + 1$.*

Solution   Before differentiating, notice the term $2xyz^2$. While y is treated as a constant, there are two variables present, requiring the use of the product rule. We may use $2xy$ as "first," where $2y$ is a constant and x is the variable, and z^2 as "second."

Ans. to reading exercise 12:
  1

❶ Take the partial derivative with respect to x of both sides of the equation, treating y as a constant:

$$12x^2 + \left(2xy \cdot 2z \frac{\partial z}{\partial x} + 2y \cdot z^2\right) - 4z^3 \frac{\partial z}{\partial x} = 0.$$

❷ Solve the resulting equation for $\frac{\partial z}{\partial x}$:

$$4xyz \frac{\partial z}{\partial x} - 4z^3 \frac{\partial z}{\partial x} = -12x^2 - 2yz^2$$

$$\frac{\partial z}{\partial x}\left(4xyz - 4z^3\right) = -12x^2 - 2yz^2$$

$$\frac{\partial z}{\partial x} = \frac{-12x^2 - 2yz^2}{4xyz - 4z^3}.$$

∎

The quantity inside the parentheses is the result of using the product rule on the term $2xy \cdot z^2$. The form of the product rule used here is first $\cdot$ second$'$ + first$'$ $\cdot$ second. While $\frac{\partial z}{\partial x}$ is present in second$'$, where the derivative is taken, it is not present in second, where the derivative is not taken.

The algebraic steps here are identical to those typically used in implicit differentiation. In line 1 terms involving $\frac{\partial z}{\partial x}$ are gathered on one side of the equation with all other terms on the other side; in line 2 $\frac{\partial z}{\partial x}$ is factored out; in line 3 we divide.

Of course, the procedure for finding $\frac{\partial z}{\partial y}$ is the same, treating x as a constant and y as the variable.

3.5.4   Higher-order partial derivatives

Because $f_x(x, y)$ and $f_y(x, y)$ are functions of two variables, we may take their partial derivatives with respect to x and with respect to y, yielding a total of four *second partial derivatives*. The order of the notations varies. The partial derivatives of f_x are $(f_x)_x$ and $(f_x)_y$, written f_{xx} and f_{xy}, respectively. Similarly, the partial derivatives of f_y with respect to x and y are written f_{yx} and f_{yy}, respectively. With Leibniz notation the order is switched: the derivatives of $\frac{\partial f}{\partial x}$ with respect to x and y are

$$\frac{\partial}{\partial x}\left(\frac{\partial f}{\partial x}\right) = \frac{\partial^2 f}{\partial x^2} \text{ and } \frac{\partial}{\partial y}\left(\frac{\partial f}{\partial x}\right) = \frac{\partial^2 f}{\partial y\, \partial x},$$

and the derivatives of $\frac{\partial f}{\partial y}$ are

$$\frac{\partial}{\partial x}\left(\frac{\partial f}{\partial y}\right) = \frac{\partial^2 f}{\partial x\, \partial y} \text{ and } \frac{\partial}{\partial y}\left(\frac{\partial f}{\partial y}\right) = \frac{\partial^2 f}{\partial y^2}.$$

Notice that

$$f_{xy} = \frac{\partial^2 f}{\partial y\, \partial x}$$

and

$$f_{yx} = \frac{\partial^2 f}{\partial x\, \partial y},$$

which switches the order of x and y for the two notations. Although this can be confusing, thinking about how the second partial derivative notations are derived can help.

Example 31 *Find the second partial derivatives of $f(x, y) = 3x^2y + x^4y^5 + \sin y - e^x$.*

Solution   To find the second partial derivatives, we must begin by finding the first partial derivatives. Treating y as a constant,

$$f_x(x, y) = 3y \cdot 2x + y^5 \cdot 4x^3 + 0 - e^x = 6xy + 4x^3y^5 - e^x,$$

and treating x as a constant,

$$f_y(x, y) = 3x^2 + x^4 \cdot 5y^4 + \cos y - 0 = 3x^2 + 5x^4y^4 + \cos y.$$

Next, we find the partial derivatives of $f_x(x, y) = 6xy + 4x^3y^5 - e^x$. Treating y as a constant,

$$f_{xx}(x, y) = 6y \cdot 1 + 4y^5 \cdot 3x^2 - e^x = 6y + 12x^2y^5 - e^x,$$

and treating x as a constant,

$$f_{xy}(x, y) = 6x \cdot 1 + 4x^3 \cdot 5y^4 - 0 = 6x + 20x^3y^4.$$

Finally, we find the partial derivatives of $f_y(x, y) = 3x^2 + 5x^4y^4 + \cos y$. Treating y as a constant,

$$f_{yx}(x, y) = 6x + 5y^4 \cdot 4x^3 + 0 = 6x + 20x^3y^4,$$

and treating x as a constant,

$$f_{yy} = 0 + 5x^4 \cdot 4y^3 - \sin y = 20x^4y^3 - \sin y.$$

The answer consists of the four functions f_{xx}, f_{xy}, f_{yx}, and f_{yy}. ∎

Reading Exercise 13   Find $f_{yy}(x, y)$ for $f(x, y) = y\tan^{-1}x + 3y^7$.

There are, of course, third partial derivatives, of which there are eight, namely $f_{xxx}, f_{xxy}, f_{xyx}, f_{xyy}, f_{yxx}, f_{yxy}, f_{yyx}$, and f_{yyy}. There are 16 fourth partial derivatives, and so on.

Example 32   *Find* $f_{xxy}(x, y)$ *for* $f(x, y) = 3x^2y + x^4y^5 + \sin y - e^x$.

Solution   This is the same function f from example 31, where we calculated $f_{xx}(x, y) = 6y + 12x^2y^5 - e^x$. We take the partial derivative again, this time with respect to y, treating x as constant:

$$f_{xxy}(x, y) = 6 + 12x^2 \cdot 5y^4 - 0 = 6 + 60x^2y^4.$$ ∎

You may have already noticed in example 31 that $f_{xy}(x, y) = f_{yx}(x, y)$. This is happily nearly always the case for the types of functions we study, making the confusion between the two types of notation much less of a concern.

Theorem 2 CLAIRAUT'S THEOREM *If f is defined on a disk D that contains (a, b), and if f_{xy} and f_{yx} are both continuous throughout D, then $f_{xy}(a, b) = f_{yx}(a, b)$.*

Why is Clairaut's Theorem true? While there are some technical subtleties in the computations below that are beyond the scope of the course, they present the basic idea.

For infinitesimals α and β,

$$f_{xy}(x,y) = \frac{\partial}{\partial y} f_x(x,y) = \frac{\partial}{\partial y}\left(\frac{f(x+\alpha,y)-f(x,y)}{\alpha}\right)$$

$$= \frac{\frac{f(x+\alpha,y+\beta)-f(x,y+\beta)}{\alpha} - \frac{f(x+\alpha,y)-f(x,y)}{\alpha}}{\beta}$$

$$= \frac{f(x+\alpha,y+\beta)-f(x,y+\beta)-f(x+\alpha,y)+f(x,y)}{\alpha\beta}.$$

Line 1 rewrites $f_x(x,y)$ using the definition. Line 2 applies the definition of partial derivative with respect to y to the expression inside the parentheses at the end of line 1, which is where the technical subtleties lie. Line 3 uses algebra to rewrite the expression.

Also,

$$f_{yx}(x,y) = \frac{\partial}{\partial x} f_y(x,y) = \frac{\partial}{\partial x}\left(\frac{f(x,y+\beta)-f(x,y)}{\beta}\right)$$

$$= \frac{\frac{f(x+\alpha,y+\beta)-f(x+\alpha,y)}{\beta} - \frac{f(x,y+\beta)-f(x,y)}{\beta}}{\alpha}$$

$$= \frac{f(x+\alpha,y+\beta)-f(x+\alpha,y)-f(x,y+\beta)+f(x,y)}{\alpha\beta}.$$

Line 1 rewrites $f_y(x,y)$ using the definition. Line 2 applies the definition of partial derivative with respect to x to the expression inside the parentheses at the end of line 1, which is where the technical subtleties lie. Line 3 uses algebra to rewrite the expression.

The two expressions are equal, and assuming the technical details are taken care of, Clairaut's Theorem holds.

Calculating both $f_{xy}(x,y)$ and $f_{yx}(x,y)$ can help avoid errors; if both are continuous functions then they should be equal.

EXERCISES 3.5

1–6. Rapid response: rewrite using the other notation.

1. $f_{xy}(x,y)$
2. $f_{yxz}(x,y,z)$
3. $\dfrac{\partial^2 f}{\partial x\,\partial y}$
4. $\dfrac{\partial^2 f}{\partial s\,\partial t}$
5. $\dfrac{\partial^3 f}{\partial z\,\partial y\,\partial x}$
6. $\dfrac{\partial^3 s}{\partial \theta\,\partial t\,\partial w}$

7–16. Find all (first) partial derivatives of the function.

7. $f(x,y,z) = x^2 y^2 z^2$
8. $f(x,y,z) = x^3 y - z y^2$
9. $g(x,y,z) = 3x \sin z + y e^{2x+3z}$
10. $h(x,y,z) = \ln\left(\frac{xy}{z}\right) - \sqrt{5x - 7z}$

11. $k(r, s, t) = 3r - 4t$

12. $f(a, b, c, d) = ad - bc$

13. $f(x_1, y_1, x_2, y_2) = \frac{y_2 - y_1}{x_2 - x_1}$

14. $g(x, y, z) = (2x + 3y)^z$

15. $f(a, b, c) = \frac{-b + \sqrt{b^2 - 4ac}}{2a}$

Ans. to reading exercise 13:
$$f_{yy}(x, y) = 126y^5$$

16. $f(x_1, y_1, x_2, y_2) = \sqrt{(x_2 - x_1)^2 + (y_2 - y_1)^2}$

17-26. Find all second partial derivatives of the function.

17. $f(x, y) = x^5 + y^7$

18. $f(x, y) = x^2 y^3$

19. $s(r, t) = r^2 + t^3 \sin 4r$

20. $z = \cos(x - 2y)$

21. $f(x, y) = xe^y$

22. $f(y, z) = \tan y - z^3$

23. $f(u, v) = (2uv - v^2)^4$

24. $f(x, y) = x \ln y$

25. $f(x, y, z) = x \sin z + 3y^2 - 7x$

26. $f(x, y, z) = x^2 + y \cosh z$

27-34. Find the slope at the indicated point in (a) the positive x-direction, (b) the negative x-direction, (c) the positive y-direction, and (d) the negative y-direction.

27. $f(x, y) = 3x^2 - 4xy + 1$, at input $(2, 7)$

28. $f(x, y) = x^2 y^3 - \frac{y}{x}$, at $\left(2, 1, \frac{7}{2}\right)$

29. $f(x, y) = \frac{x + 2y}{3y^2 + 7}$, at $\left(1, 0, \frac{1}{7}\right)$

30. $f(x, y) = (2x - 4y)^5$, at input $(2, 1)$

31. $g(x, y) = \cos xy - x \sin y$, at input $\left(3, \frac{\pi}{2}\right)$

32. $f(x, y) = x^2 e^y - y^2 e^x$, at input $(-1, 1)$

33. $f(x, y) = (3x - y^2)\sqrt{x^2 - y}$, at $(2, 1, 5\sqrt{3})$

34. $g(x, y) = \frac{\tan^{-1}(x + y)}{3 - xy}$, at $(0, 0, 0)$

35-44. Find the indicated partial derivative.

The notation $\frac{\partial^3 z}{\partial x^2 \, \partial y}$ means the same as $\frac{\partial^3 z}{\partial x \, \partial x \, \partial y}$.

35. $\frac{\partial^3 z}{\partial x^2 \, \partial y}$ for $z = x^2 \sin y$

36. $f_{xyy}(x, y)$ for $f(x, y) = x^{15} + y^{27} + 7x^2 y^2$

37. $f_{xyzy}(x, y, z)$ for $f(x, y, z) = x^2 + y^2 + z^2 + x^2 y^2 z^2$

38. $f_{xyyx}(1, 2)$ for $f(x, y) = x^3 y^3$

39. $\frac{\partial^3 w}{\partial x \, \partial z \, \partial y}$ for $w = x^3 z^4 + x^2 y$

40. $\frac{\partial^7 z}{\partial x^4 \, \partial y^3}$ for $z = 13x^3 y + 87y^2 - 155x^3 + 18x^2 y^2$

41. $f_{xy}(x, y)$ for $f(x, y) = \frac{\sqrt{x^2 - 7e^{4x}}}{\tan^{-1} x} + \left(\sec\left(y^2 + 5\ln(y^{14} - 3y^5)\right)\right) + y^{1.38}$

42. $\dfrac{\partial^3 w}{\partial y\,\partial x\,\partial z}$ for $w = x^2 y z^3$

43. $f_{xzy}(x, y, z)$ for $f(x, y, z) = (2x + 3y)^z$

44. $f_{xyxxy}(x, y)$ and $f_{40x\,|\,62y}(x, y)$ for $f(x, y) = (\sin 2x)(\cosh 3y)$

45. Given that z is defined implicitly as a function of x and y by
$x^2 y^2 + x^2 z^2 + y^2 z^2 = 5$, find $\dfrac{\partial z}{\partial y}$.

46. Given that z is defined implicitly as a function of x and y by
$z^3 + y\cos z - x^2 y = 1$, find $\dfrac{\partial z}{\partial x}$.

47. Given that z is defined implicitly as a function of x and y by
$x^6 + y^{17} + z^4 - 5xyz + 4e^y = 0$, find $\dfrac{\partial z}{\partial x}$.

48. Given that z is defined implicitly as a function of x and y by
$e^{xy} + e^{yz} = x + 1$, find $\dfrac{\partial z}{\partial x}$.

49. Given that z is defined implicitly as a function of x and y by
$\dfrac{4-\sin z}{3+\cos x} + \ln(y^2 + 1) = y$, find $\dfrac{\partial z}{\partial y}$.

50. Given that z is defined implicitly as a function of x and y by
$x\sqrt{y^2 + z} = z^2 + y$, find $\dfrac{\partial z}{\partial y}$.

51. Given that s is defined implicitly as a function of t and r by
$\dfrac{s^2 - 2r}{t + s} = 1$, find $\dfrac{\partial s}{\partial t}$.

52. Given that s is defined implicitly as a function of t and θ by
$s^2 + s\cos t\theta - \sin t\theta = t^2$, find $\dfrac{\partial s}{\partial \theta}$.

53. Given that z is defined implicitly as a function of x and y by
$2x^2 y^3 - xz^2 + 3y^2 z = 4$, find the slope in the positive x-direction
at $(1, 1, 1)$.

54. Given that z is defined implicitly as a function of x and y by
$x^2 \cos z + y^2 - y - z^2 = x - \pi^2$, find the slope in the negative y-direction at $(1, 2, \pi)$.

55. Many settings in physics give rise to functions that satisfy Laplace's
equation $\dfrac{\partial^2 f}{\partial x^2} + \dfrac{\partial^2 f}{\partial y^2} = 0$. Which of the following functions satisfy
Laplace's equation?

    (a) $f(x, y) = x^2 + y^2$
    (b) $f(x, y) = x^2 - y^2$
    (c) $f(x, y) = e^x \sin y$
    (d) $f(x, y) = e^{2x} \cos 2y$

56. Given the truth of Clairaut's Theorem, the corresponding
theorems for third and higher-order partial derivatives can
be proven with a little algebraic ingenuity. For instance, if
$f_{xy}(x, y) = f_{yx}(x, y)$, then their partial derivatives with respect to x
are equal, that is, $f_{xyx}(x, y) = f_{yxx}(x, y)$. Assuming that the appropriate hypotheses hold, prove the following: (a) $f_{xyx}(x, y) = f_{xxy}(x, y)$;
(b) $f_{xxyx}(x, y) = f_{xxxy}(x, y)$; and (c) $f_{xyxy}(x, y) = f_{yxxy}(x, y)$.

When checking answers to implicit partial derivative questions, remember that $\dfrac{-a}{-b} = \dfrac{a}{b}$. In particular, $\dfrac{c-d}{k-n} = \dfrac{d-c}{n-k}$.

The three-dimensional Laplace's equation is $\dfrac{\partial^2 f}{\partial x^2} + \dfrac{\partial^2 f}{\partial y^2} + \dfrac{\partial^2 f}{\partial z^2} = 0$.

57. The *total degree* of a term $x^a y^b$ is $a + b$. The total degree of a polynomial is the largest degree of any term of the polynomial. Let $f(x, y) = x^5 y^7 + 57y$.

    (a) What is the total degree of f?
    (b) Find $f_x(x, y)$ and $f_y(x, y)$.
    (c) What is the total degree of the partial derivatives of f?
    (d) Find all second partial derivatives of f.
    (e) What is the total degree of the second partial derivatives of f?
    (f) Find $f_{xxxxxx}(x, y)$.
    (g) Must all of the nth partial derivatives have the same total degree?
    (h) Which order of partial derivative n is the first where all of the nth partial derivatives are zero?

58. Let $f(x, y) = xe^y$. Recognizing a pattern in the calculations of exercise 21, quickly find (a) $f_{yyyyyyyyyy}(x, y)$, (b) $f_{yyxyyyy}(x, y)$, and (c) $f_{yyxyyyx}(x, y)$.

3.6 Differentiability and Local Linearity

For a function f of a single variable x, the existence of the derivative $f'(k)$ is synonymous with local linearity at $x = k$. Zoom in far enough and the graph of f looks linear; at an infinitesimal scale, the function and its tangent line are infinitely close. Is the same true for a function of two variables? We begin exploring this question by looking at tangent planes.

3.6.1 Tangent planes

Consider the tangent line to the curve $y = f(x)$ passing through the point $(2, 1)$, with $f'(2) = \frac{3}{2}$, as in Figure 3.67.

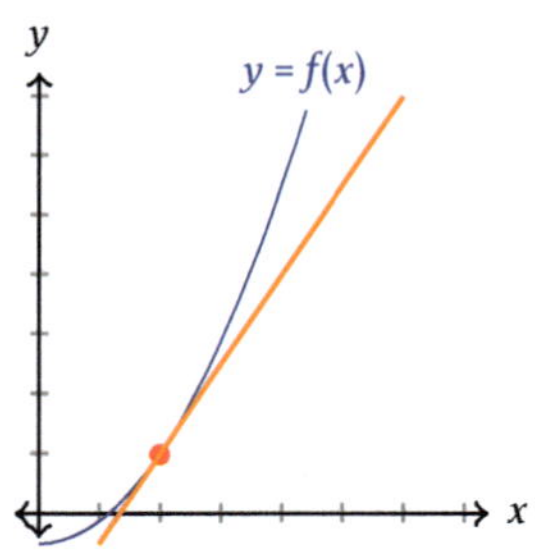

Figure 3.67 *A curve $y = f(x)$ (blue) and its tangent line at the point $(2, 1)$ (orange)*

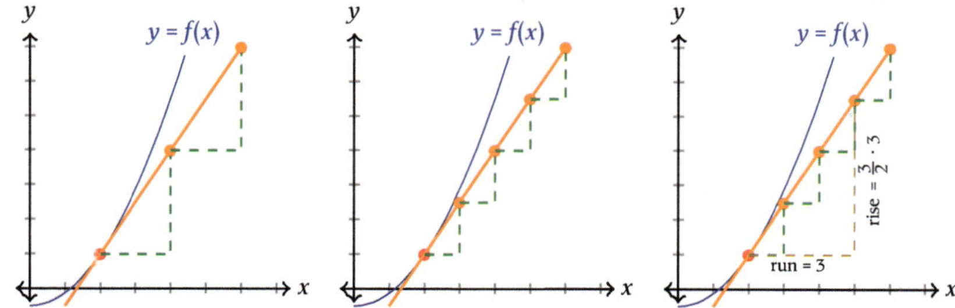

Figure 3.68 *A curve $y = f(x)$ (blue), its tangent line at the point $(2, 1)$ (orange), and using the slope to find additional points on the tangent line (dashed)*

The slope of the tangent line, $f'(2) = \frac{3}{2}$, is $\frac{\text{rise}}{\text{run}}$, so each time that we move two units to the right and three units up we find another point on the tangent line, as in Figure 3.68, left. We can also rewrite the slope $\frac{3}{2}$ as $\frac{3/2}{1}$ and use a run of 1 and rise of $\frac{3}{2}$, as in Figure 3.68, middle. In general, the rise is the slope times the run (Figure 3.68, right). For a generic x-coordinate on the tangent line, the run from the point $(2, 1)$ is $x - 2$, and the rise is $\frac{3}{2}(x - 2)$, so that the y-coordinate is $1 + \frac{3}{2}(x - 2)$. More generically, with the point of tangency $(a, f(a))$ instead of $(2, f(2))$, the equation of the tangent line is $y = f(a) + f'(a)(x - a)$, which you might recognize as the formula for the linearization of f at $x = a$.

Now consider a surface $z = f(x, y)$ passing through the point $(a, b, f(a, b))$. The partial derivatives $f_x(a, b)$ and $f_y(a, b)$ give the slopes of the tangent lines in the x- and y-directions, respectively (Figure 3.69, left). But how do we find the z-coordinate of some other point in the tangent plane (Figure 3.69, right)? The answer is that we determine the rise in both the x- and y-directions. The slopes in the x- and y-directions remain the same throughout the plane, so the rise in the x-direction from the point of tangency is slope · run, or $f_x(a, b)(x - a)$, and then the rise in the y-direction is slope · run, or $f_y(a, b)(y - b)$. Starting at the z-coordinate $f(a, b)$ and following those rises gives the z-coordinate

$$z = f(a, b) + f_x(a, b)(x - a) + f_y(a, b)(y - b).$$

TANGENT PLANE

Let f be a function of the two variables x and y, and let (a, b) be in the domain of f. If $f_x(a, b)$ and $f_y(a, b)$ both exist, we call

$$z = f(a, b) + f_x(a, b)(x - a) + f_y(a, b)(y - b)$$

the *tangent plane* of f at (a, b). We also call the equation the *linearization of f at (a, b)* and the *standard linear approximation of f at (a, b)*.

For the linearization we do not simplify, but rather we leave the equation in the form given here.

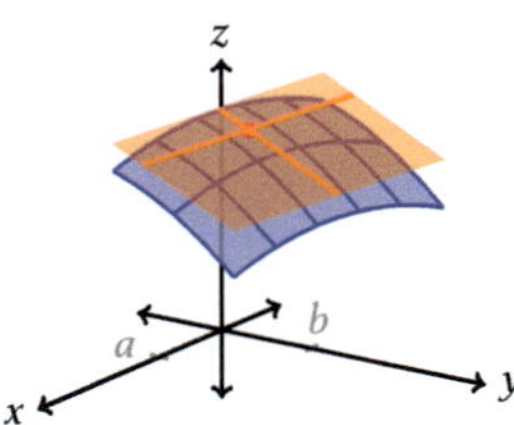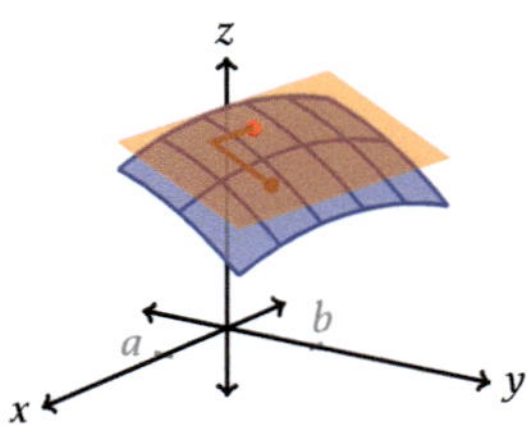

Figure 3.69 *A surface $z = f(x, y)$ (blue), its tangent plane (semi-transparent orange) at the point $(a, b, f(a, b))$ (red) and (left) the tangent lines in the x- and y-directions (orange) and (right) tracing a path to a different point in the tangent plane, first in the x-direction and then in the y-direction (darker orange)*

Example 33 *Find the equation of the tangent plane to the function $f(x, y) = 2x^3 + 3y^2 - 7$ at the input $(1, 2)$.*

Solution   To use the formula for the equation of the tangent plane, we set $(a, b) = (1, 2)$. Next, we calculate the partial derivatives and evaluate them at $(1, 2)$:

$$f_x(x, y) = 6x^2 \qquad f_x(1, 2) = 6$$

$$f_y(x, y) = 6y \qquad f_y(1, 2) = 12$$

We also need $f(1, 2) = 2(1^3) + 3(2^2) - 7 = 7$. Then applying the formula with $a = 1$ and $b = 2$ gives

$$z = 7 + 6(x - 1) + 12(y - 2),$$

which can be simplified to

$$z = 6x + 12y - 23.$$

■

If we are asked instead for the linearization of f at the input $(1, 2)$, we do not simplify and write $L(x, y) = 7 + 6(x - 1) + 12(y - 2)$, with the L signifying the linearization function, just as in the single-variable case.

Reading Exercise 14   Find the equation of the tangent plane to the function $f(x, y) = x^2 + y^2 + 1$ at the input $(1, 2)$.

The tangent plane in Figure 3.70 lies below the surface (except at the point of tangency), which in the single-variable case is associated with concave up.

Multiple views of the function and tangent plane of example 33, produced in Mathematica, appear in Figure 3.70. An edge-on view, as in Figure 3.70, middle, makes the appearance of the tangent plane like that of a tangent line.

Zooming out (Figure 3.71), we notice that a tangent plane can cut across the surface, just as in the single-variable case where a tangent line can cut across a curve away from the point of tangency.

The concept of local linearity also applies, this time to a function resembling its tangent plane as we zoom in. Figure 3.72 shows the surface and its tangent plane

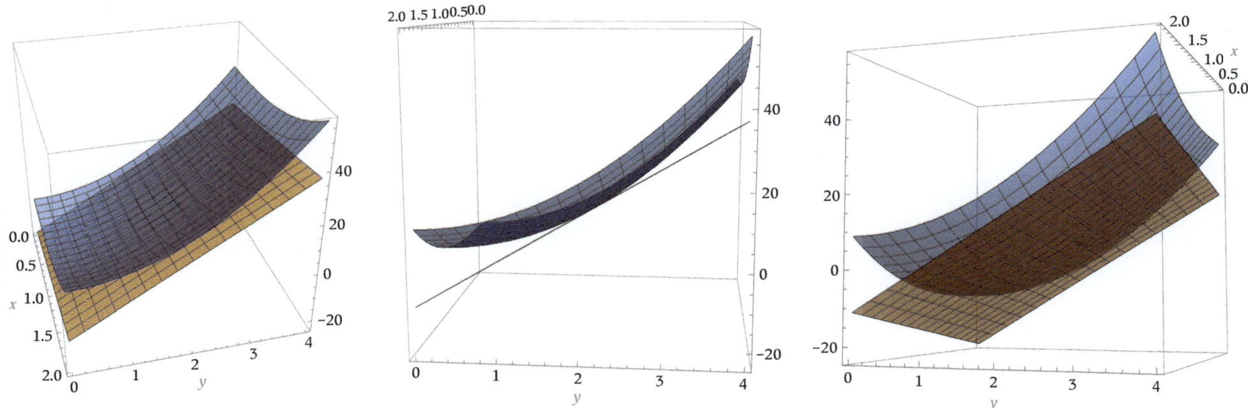

Figure 3.70 *The surface $z = 2x^3 + 3y^2 - 7$ (blue) and its tangent plane at $(1, 2)$ (orange/brown)*

close (left) and closer yet (right) to the point of tangency. When zooming in, the surface seems to flatten; on the right, there is very little difference except that the surface curls up just slightly at the edges and corners. And if we zoom in even further, the surface and tangent plane appear indistinguishable (Figure 3.73); views from above and below make it seem as if there is just one plane, with different colors on top and bottom.

3.6.2 Local linearity: existence of partial derivatives is insufficient

Example 34 *Find the equation of the tangent plane to the function*
$$f(x, y) = \begin{cases} \dfrac{xy}{x^2+y^2}, & (x, y) \neq (0, 0) \\ 0, & (x, y) = (0, 0) \end{cases} \quad \text{at the input } (0, 0).$$

Solution To use the formula for the equation of the tangent plane, we set $(a, b) = (0, 0)$. Because f is piecewise-defined with one "piece" at $(0, 0)$, we calculate the partial derivatives at $(0, 0)$ using definition 4 rather than derivative rules:

$$f_x(0, 0) = \frac{f(0 + \alpha, 0) - f(0, 0)}{\alpha} = \frac{\frac{\alpha \cdot 0}{\alpha^2 + 0^2} - 0}{\alpha} = 0$$

$$f_y(0, 0) = \frac{f(0, 0 + \beta) - f(0, 0)}{\beta} = \frac{\frac{0 \cdot \beta}{0^2 + \beta^2} - 0}{\beta} = 0.$$

Ans. to reading exercise 14:
  $z = 2x + 4y - 4$; alternate answer $z = 6 + 2(x - 1) + 4(y - 2)$

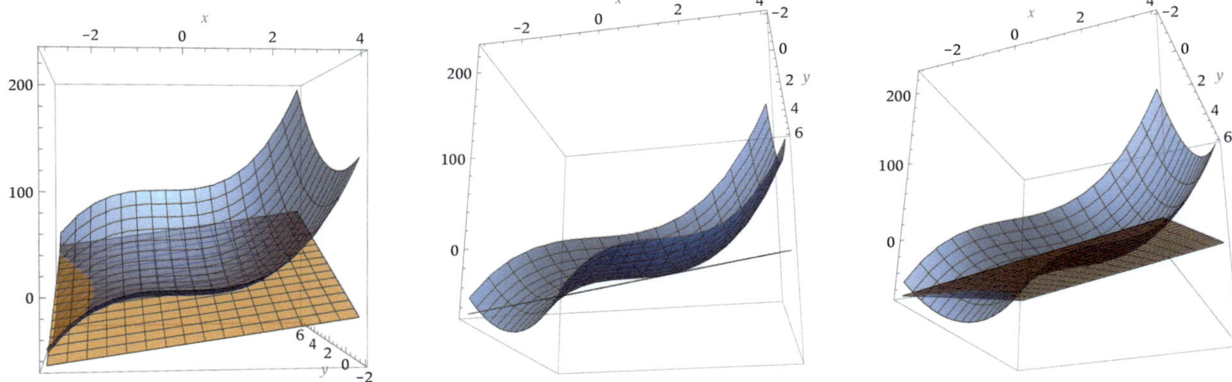

Figure 3.71 *Additional views of the surface $z = 2x^3 + 3y^2 - 7$ (blue) and its tangent plane at $(1, 2)$ (orange/brown)*

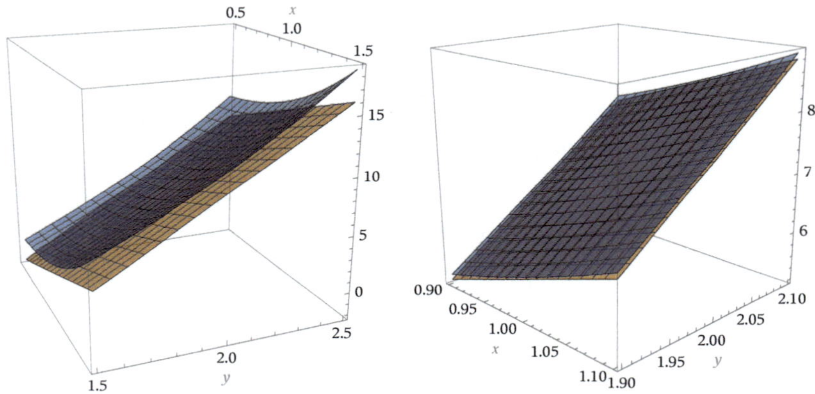

Figure 3.72 *The surface $z = 2x^3 + 3y^2 - 7$ (blue) and its tangent plane at $(1, 2)$ (orange/brown), showing views closer to the point of tangency*

Applying the formula with $f(0, 0) = 0$, $f_x(0, 0) = 0$, $f_y(0, 0) = 0$, $a = 0$, and $b = 0$, the equation of the tangent plane is

$$z = 0 + 0(x - 0) + 0(y - 0) = 0,$$

which is a horizontal plane (the xy-plane). ∎

The graph of the function f from example 2, along with its tangent plane $z = 0$ at $(0, 0)$, is in Figure 3.74.

The tangent plane doesn't appear to be tangent at all! But notice that the tangent plane does appear to match the surface along the input lines $x = 0$ and $y = 0$;

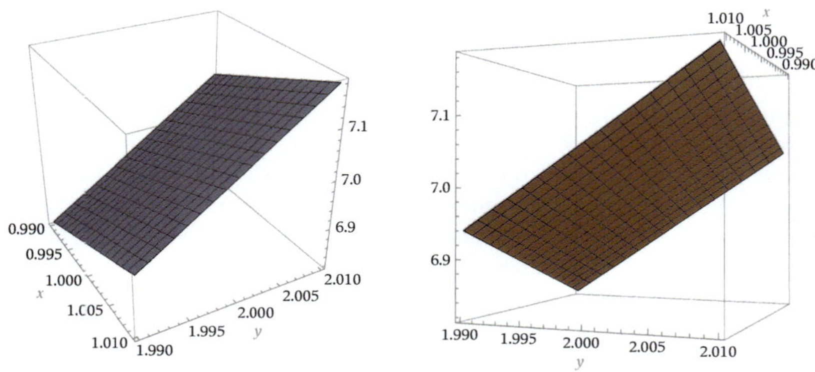

Figure 3.73 *Views from above and below of the surface $z = 2x^3 + 3y^2 - 7$ (blue) and its tangent plane at $(1, 2)$ (orange/brown) very near the point of tangency, appearing virtually indistinguishable*

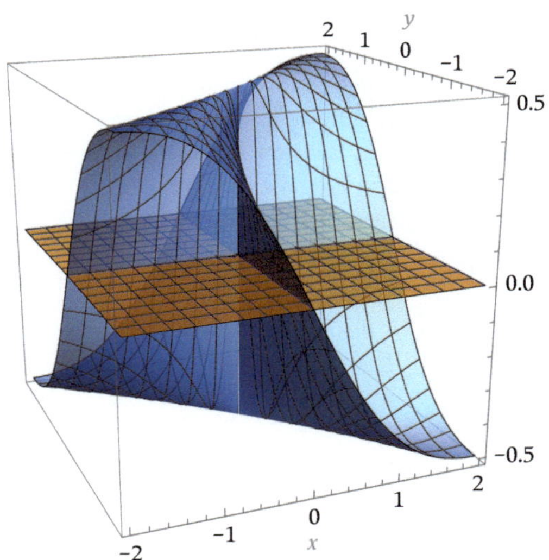

Figure 3.74 *The surface* $f(x, y) = \begin{cases} \dfrac{xy}{x^2 + y^2}, & (x, y) \neq (0, 0) \\ 0, & (x, y) = (0, 0) \end{cases}$ *(blue) and its tangent plane at $(0, 0)$ (orange)*

as designed, the so-called tangent plane has the correct slopes in the x- and y-directions. It just doesn't match the slopes (if they are even defined) in the infinitely many other directions. The existence of the partial derivatives is not enough to guarantee local linearity.

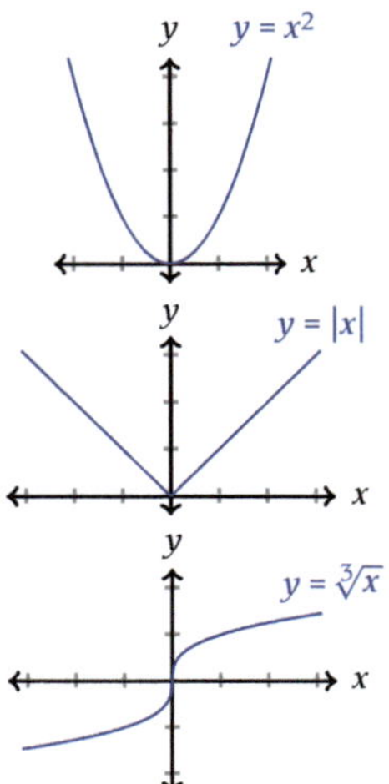

Figure 3.75 *(Top) a function differentiable at $x = 0$, continuous with no corner or vertical tangent; (middle) a function not differentiable at $x = 0$, with a corner; (bottom) a function not differentiable at $x = 0$, with a vertical tangent*

Compare to Figure 3.73, right, in which the surface and tangent plane are virtually indistinguishable and the viewpoint is from below the surface/tangent.

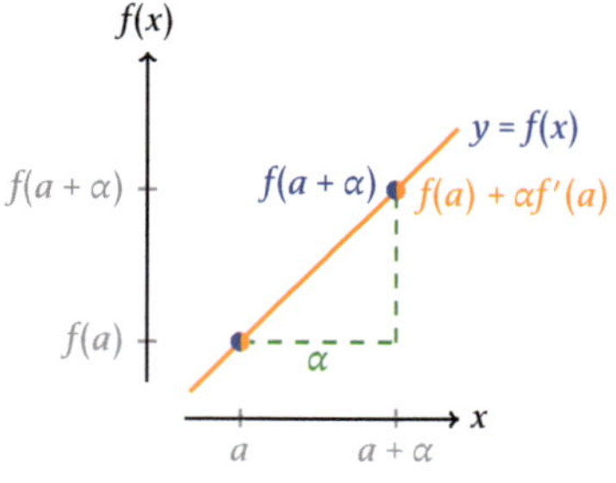

Figure 3.76 *A differentiable function $y = f(x)$ (blue) is indistinguishable from its tangent line (orange) at an infinitesimal scale. If we scoot over an infinitesimal amount α from the point of tangency, the difference (not visible) in the y-coordinates of the function and the tangent line at $x = a + \alpha$ must be on a lower level than α (visible), a fact derived from the definition of derivative*

3.6.3 Differentiability

Geometrically, if a function f of one variable is differentiable at $x = a$, then the function is not only continuous at $x = a$ but also has no corner or vertical tangent at $x = a$ (Figure 3.75). Look at Figure 3.70; at the point of tangency, the function is continuous and there are no corners or vertical tangents. But in Figure 3.74, the features near the point $(0, 0)$ appear similar to both a corner and a vertical tangent, and perhaps also a tear in the fabric; nor does it appear to be continuous (differing limits along various paths). If the existence of f_x and f_y are not enough to guarantee local linearity, how do we define differentiability? How do we avoid disruptive features such as tears, corners, and vertical tangents, and guarantee a smooth surface?

The key to defining differentiability, which we equate with local linearity, is to ensure local linearity (and not just existence of slopes) in every direction, not just the x- and y-directions. Just as a differentiable function of one variable is indistinguishable from its tangent line at an infinitesimal scale (Figure 3.76), differentiability requires the same for a function of two variables and its tangent plane.

To that end, suppose the function f is differentiable at the input (a, b), so that the surface and its tangent plane at (a, b) are indistinguishable at an infinitesimal scale (Figure 3.77). If we scoot over an infinitesimal amount α in the x-direction and β in the y-direction, then in total we have moved over a distance of $\sqrt{\alpha^2 + \beta^2}$ (the length of the vector $\langle \alpha, \beta \rangle$). At this new input $(a + \alpha, b + \beta)$, the z-coordinate of the function is $f(a + \alpha, b + \beta)$. Using the tangent plane formula with $x = a + \alpha$, $y = b + \beta$, the z-coordinate of the tangent plane is

$$z = f(a, b) + f_x(a, b)(a + \alpha - a) + f_y(a, b)(b + \beta - b)$$

$$= f(a, b) + f_x(a, b)\alpha + f_y(a, b)\beta.$$

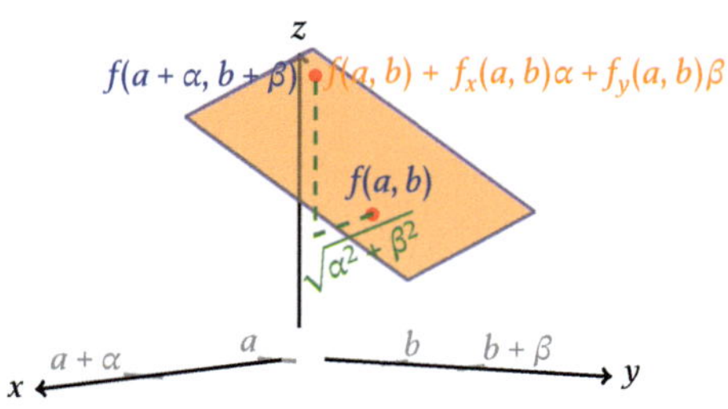

Figure 3.77 *A surface $z = f(x, y)$ (blue but covered), indistinguishable from its tangent plane at the input (a, b) (orange, visible). The view is from below; we are looking up at the surface/tangent plane. A second point and its coordinates on the surface and tangent plane is also given*

Then the difference between the z-coordinates of the function and tangent plane (a quantity which is too small to be seen in Figure 3.77) must be on a lower level than the infinitesimal distance $\sqrt{\alpha^2 + \beta^2}$ that we scooted over (a quantity which can be seen in Figure 3.77). Hence

$$\frac{f(a + \alpha,\ b + \beta) - (f(a,b) + f_x(a,b)\alpha + f_y(a,b)\beta)}{\sqrt{\alpha^2 + \beta^2}} \doteq 0,$$

because the numerator is on a lower level than the denominator and therefore the fraction is infinitesimal.

Definition 6 DIFFERENTIABLE (FUNCTION OF TWO VARIABLES) *A function* $z = f(x, y)$ *of two variables is* differentiable *at* (a, b) *if*

$$\frac{f(a + \alpha,\ b + \beta) - (f(a,b) + f_x(a,b)\alpha + f_y(a,b)\beta)}{\sqrt{\alpha^2 + \beta^2}} \doteq 0$$

for every choice of infinitesimals α *and* β.

When f is differentiable at (a, b) and α and β are infinitesimals, the difference between $f(a + \alpha,\ b + \beta)$ and $f(a, b) + f_x(a, b)\alpha + f_y(a, b)\beta$ must be an infinitesimal on a lower level than $\sqrt{\alpha^2 + \beta^2}$, and therefore on a lower level than either α or β. Thus

$$f(a + \alpha,\ b + \beta) \approx f(a, b) + f_x(a, b)\alpha + f_y(a, b)\beta,$$

as long as the approximation principle is not violated.

LOCAL LINEARITY FORMULA (FUNCTION OF TWO VARIABLES)

If $z = f(x, y)$ is differentiable at (a, b), α and β are infinitesimals, and not all of $f(a, b), f_x(a, b)$, and $f_y(a, b)$ are zero, then

$$f(a + \alpha,\ b + \beta) \approx f(a, b) + f_x(a, b)\alpha + f_y(a, b)\beta.$$

The local linearity formula is useful for proving theorems in the remainder of this chapter.

3.6.4   Using the definition of differentiable

We will soon have a theorem that can greatly simplify our recognition of a differentiable function. But for now, let's return to examples 33 and 34 to demonstrate how the definition of differentiability works.

The numerator consists of the difference in the z-coordinates of the function ($f(a + \alpha,\ b + \beta)$) and the tangent plane ($f(a, b) + f_x(a, b)\alpha + f_y(a, b)\beta$). The denominator is the distance moved from the input (a, b) to the input $(a + \alpha,\ b + \beta)$.

It is implied that if any of $f(a, b), f_x(a, b)$, or $f_y(a, b)$ does not exist, then the function is not differentiable at (a, b).

Example 35 *Use the definition to show that $f(x, y) = 2x^3 + 3y^2 - 7$ is differentiable at $(1, 2)$.*

Solution   Several quantities that are needed here are calculated in example 33, including $f(1, 2) = 7, f_x(1, 2) = 6,$ and $f_y(1, 2) = 12.$ We also need

$$f(1 + \alpha, 2 + \beta) = 2(1 + \alpha)^3 + 3(2 + \beta)^2 - 7$$
$$= 2(1 + 3\alpha + 3\alpha^2 + \alpha^3) + 3(4 + 4\beta + \beta^2) - 7$$
$$= 2 + 6\alpha + 6\alpha^2 + 2\alpha^3 + 12 + 12\beta + 3\beta^2 - 7$$
$$= 7 + 6\alpha + 12\beta + 6\alpha^2 + 2\alpha^3 + 3\beta^2$$

and

$$f(1, 2) + f_x(1, 2)\alpha + f_y(1, 2)\beta = 7 + 6\alpha + 12\beta.$$

Then for any infinitesimals α and β,

Note that in the numerator not only do the real-level terms go away, but the terms on the α and β levels also subtract out, leaving only terms on lower levels. This is indicative of a differentiable function.

$$\frac{f(1 + \alpha, 2 + \beta) - (f(1, 2) + f_x(1, 2)\alpha + f_y(1, 2)\beta)}{\sqrt{\alpha^2 + \beta^2}} = \frac{6\alpha^2 + 2\alpha^3 + 3\beta^2}{\sqrt{\alpha^2 + \beta^2}}.$$

If it is not yet clear that the numerator is on a lower level than the denominator, there are three possibilities for comparing α and β.

❶ Suppose $\alpha \ll \beta$. Continuing the calculation,

The denominator $|\beta|$ is on the β level, whereas the numerator $3\beta^2$ is on the β^2 level. With the numerator on a lower level than the denominator, the fraction is infinitesimal and renders zero.

$$\frac{6\alpha^2 + 2\alpha^3 + 3\beta^2}{\sqrt{\alpha^2 + \beta^2}} \approx \frac{3\beta^2}{\sqrt{\beta^2}} = \frac{3\beta^2}{|\beta|} \doteq 0.$$

❷ Suppose $\alpha \gg \beta$. Continuing the calculation,

$$\frac{6\alpha^2 + 2\alpha^3 + 3\beta^2}{\sqrt{\alpha^2 + \beta^2}} \approx \frac{6\alpha^2}{\sqrt{\alpha^2}} = \frac{6\alpha^2}{|\alpha|} \doteq 0.$$

❸ Suppose $\beta = k\alpha$. Continuing the calculation,

Unlike many limits discussed in Section 3.3, there will never be a need for a "final case" when checking differentiability using the definition, because the denominator is never 0 no matter the value of k.

$$\frac{6\alpha^2 + 2\alpha^3 + 3\beta^2}{\sqrt{\alpha^2 + \beta^2}} = \frac{6\alpha^2 + 2\alpha^3 + 3k^2\alpha^2}{\sqrt{\alpha^2 + k^2\alpha^2}}$$
$$\approx \frac{6\alpha^2 + 3k^2\alpha^2}{\sqrt{\alpha^2(1 + k^2)}} = \frac{3\alpha^2(2 + k^2)}{|\alpha|\sqrt{1 + k^2}} = \frac{3(2 + k^2)}{\sqrt{1 + k^2}} \frac{\alpha^2}{|\alpha|} \doteq 0.$$

Because we render the real result 0 for any infinitesimals α and β, f is differentiable at $(1, 2)$.

∎

Notice that in all three cases ($\alpha \ll \beta, \alpha \gg \beta, \beta = k\alpha$), the quantity $\sqrt{\alpha^2 + \beta^2}$ is either on the α level or the β level (or both), whichever is highest.

Example 36 *Use the definition to show that* $f(x,y) = \begin{cases} \frac{xy}{x^2+y^2}, & (x,y) \neq (0,0) \\ 0, & (x,y) = (0,0) \end{cases}$

is not differentiable at $(0,0)$.

Solution Several quantities that are needed here are calculated in example 34, including $f(0,0) = 0, f_x(0,0) = 0,$ and $f_y(0,0) = 0$. We also need

$$f(0+\alpha,\, 0+\beta) = f(\alpha,\beta) = \frac{\alpha\beta}{\alpha^2 + \beta^2}$$

and

$$f(0,0) + f_x(0,0)\alpha + f_y(0,0)\beta = 0 + 0\alpha + 0\beta = 0.$$

Then for any infinitesimals α and β,

$$\frac{f(0+\alpha,\, 0+\beta) - (f(0,0) + f_x(0,0)\alpha + f_y(0,0)\beta)}{\sqrt{\alpha^2 + \beta^2}} = \frac{\frac{\alpha\beta}{\alpha^2+\beta^2} - 0}{\sqrt{\alpha^2 + \beta^2}}$$

$$= \frac{\alpha\beta}{(\alpha^2 + \beta^2)^{3/2}}.$$

Notice that, in contrast to the same place in example 35, nothing subtracts out in the numerator. It is not clear that the numerator is on a lower level than the denominator.

There are three possibilities for comparing α and β, one of which is $\beta = k\alpha$. Continuing the calculation,

$$\frac{\alpha\beta}{(\alpha^2 + \beta^2)^{3/2}} = \frac{\alpha k\alpha}{(\alpha^2 + k^2\alpha^2)^{3/2}}$$

$$= \frac{k\alpha^2}{(\alpha^2)^{3/2}(1 + k^2)^{3/2}} = \frac{k\alpha^2}{(\sqrt{\alpha^2})^3 (1 + k^2)^{3/2}}$$

$$= \frac{k\alpha^2}{|\alpha|^3 (1 + k^2)^{3/2}}.$$

Line 1 replaces β by $k\alpha$; line 2 factors out α^2 from inside the parentheses and then rewrites the $\frac{3}{2}$ power as the square root cubed; line 3 rewrites $\sqrt{\alpha^2}$ as $|\alpha|$.

Because the numerator is on the α^2 level and the denominator is on the α^3 level, the numerator is on a higher level than the denominator, and the expression is infinite, definitely not rendering 0. The function f is not differentiable at $(0,0)$. ∎

Because α^2 and $|\alpha|^3$ are both positive, it would be possible to continue the calculation as $= \frac{k}{|\alpha|(1+k^2)^{3/2}} = \frac{k}{(1+k^2)^{3/2}}|A|$, which is infinite ($\doteq \infty$ if $k > 0, \doteq -\infty$ if $k < 0$).

3.6.5   A helpful theorem

The following theorem is often very helpful for establishing differentiability without the need for using the definition of differentiability.

Theorem 3 DIFFERENTIABILITY OF NICELY-BEHAVED TWO-VARIABLE FUNCTIONS *If the partial derivatives f_x and f_y exist in a neighborhood of (a, b) and are continuous at (a, b), then f is differentiable at (a, b).*

To demonstrate the easy application of theorem 3, we return again to the function of examples 33 and 35.

Example 37 *Use theorem 3 to show that $f(x, y) = 2x^3 + 3y^2 - 7$ is differentiable at $(1, 2)$.*

Solution First, we calculate the partial derivatives:

$$f_x(x, y) = 6x^2$$

$$f_y(x, y) = 6y.$$

Because f_x and f_y are polynomials, they exist and are continuous everywhere. By theorem 3, f is differentiable not only at $(1, 2)$ but everywhere else, as well. ∎

Because elementary functions of two variables are continuous where defined, as long as (a, b) is in the domain and not at the edge of the domain of f, f_x, and f_y, then f is differentiable at (a, b).

Example 38 *Is $f(x, y) = \sqrt{x^2 + y^2}$ differentiable at $(0, 0)$?*

Solution We begin by calculating the partial derivatives:

$$f_x(x, y) = \frac{1}{2} \left(x^2 + y^2 \right)^{-1/2} \cdot 2x = \frac{x}{\sqrt{x^2 + y^2}},$$

which is undefined at $(0, 0)$ for reason of division by zero. By the definition, f is not differentiable at $(0, 0)$. ∎

Two views of the graph of the function from example 38 are in Figure 3.78. The function is defined and continuous at $(0, 0)$, but it is not differentiable there. Notice the "corner" in the graph, a higher-dimensional version of the single-variable corner in Figure 3.75, middle.

Example 39 *Is $f(x, y) = e^x y^2$ differentiable at $(0, 1)$?*

Solution We begin by calculating the partial derivatives:

$$f_x(x, y) = e^x y^2$$

$$f_y(x, y) = 2e^x y.$$

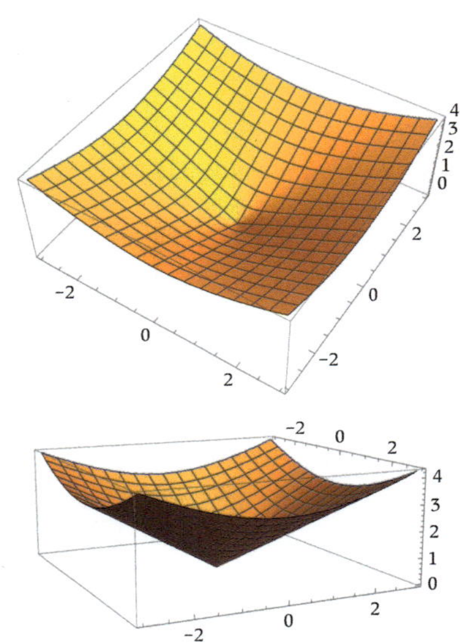

Figure 3.78 *Two views of the function $f(x, y) = \sqrt{x^2 + y^2}$ showing the "corner" at the input $(0, 0)$*

Because f_x and f_y are elementary functions defined everywhere, by theorem 3 f is differentiable at $(0, 1)$, and everywhere else as well. ∎

The graph of the function of example 39, which is differentiable everywhere, has no discontinuities, corners, or vertical tangents (Figure 3.79). Because the function is defined everywhere and the page is finite, the entire graph cannot be shown. The edge of what is shown in the graph should not be interpreted as a corner.

Reading Exercise 15 Is $f(x, y) = \sqrt{xy}$ differentiable at $(0, 0)$?

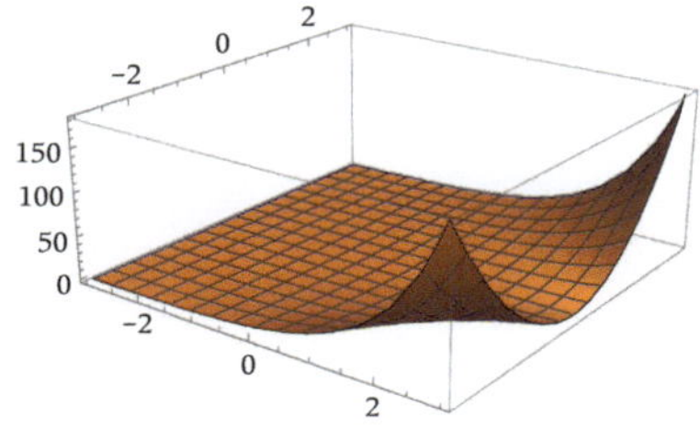

Figure 3.79 *A portion of the graph of the function* $f(x, y) = e^x y^2$ *of example 39*

3.6.6 Linearization and estimation

As mentioned previously, the linearization of a function of two variables is the equation of the tangent plane presented in a particular format.

Example 40 *Find the linearization of* $f(x, y) = y^2 e^{x+1} + 3$ *at the input* $(-1, 3)$.

Solution We need to determine values of the function and both partial derivatives at $(-1, 3)$. To that end, we calculate the partial derivatives:

$$f_x(x, y) = y^2 e^{x+1}$$
$$f_y(x, y) = 2y e^{x+1}.$$

Then the values of the function and partial derivatives are

$$f(-1, 3) = 3^2 e^0 + 3 = 12$$
$$f_x(-1, 3) = 3^2 e^0 = 9$$
$$f_y(-1, 3) = 2(3)e^0 = 6.$$

The linearization is

$$L(x, y) = 12 + 9(x + 1) + 6(y - 3).$$

For the linearization, leave the answer in the format presented in the solution. Do not simplify to $z = 9x + 6y + 3$.

∎

The function f of example 40 and its partial derivatives f_x and f_y are all elementary functions that are defined everywhere, and therefore by theorem 3, f is differentiable everywhere. Local linearity applies and as long as we stick close to the point of tangency, the value of the function is close to the value of the tangent plane (the linearization). Although it is simple to calculate values of f using a calculator, such values can also be estimated by hand using the linearization.

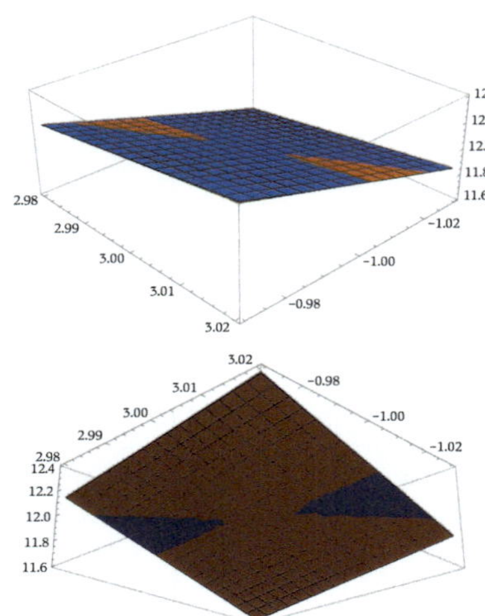

Figure 3.80 *Views of the function* $f(x,y) = y^2 e^{x+1} + 3$ *(blue) and its tangent plane (orange/brown) at* $(-1, 3)$, *(top) from above and (bottom) from below. The view sticks close enough to the point of tangency that the function and its tangent plane are indistinguishable.*

Ans. to reading exercise 15:
    No (neither f_x nor f_y are defined at $(0, 0)$)

The estimated increment is found by following the tangent plane from (a, b) to $(a + dx, b + dy)$.

The first edition was by Granville, 1904.

Example 41 *For* $f(x, y) = y^2 e^{x+1} + 3$, *estimate* $f(-0.97, 3.02)$ *without using a calculator.*

Solution   This is the same function as in example 40, where we calculated the linearization at $(-1, 3)$ to be

$$L(x, y) = 12 + 9(x + 1) + 6(y - 3).$$

Notice that $(-0.97, 3.02)$ is fairly close to $(-1, 3)$, so the linearization should give a good estimate of the value of the function:

$$L(-0.97, 3.02) = 12 + 9(-0.97 + 1) + 6(3.02 - 3)$$
$$= 12 + 9(0.03) + 6(0.02) = 12.39,$$

which is a simple calculation by hand without a calculator.   ∎

The actual value of the quantity estimated in example 41 is $f(-0.97, 3.02) = 3.02^2 e^{-0.97+1} + 3 = 12.398\ldots$, which is close to the estimated value (Figure 3.80). But the further we stray from the point of tangency, the worse the estimates become.

Reading Exercise 16   Find the linearization of $f(x, y) = x \sin y$ at $(4, \pi)$.

Yet another way of describing example 41 is that if the x-coordinate is incremented by 0.03 units (from -1 to -0.97) and the y-coordinate is incremented by 0.02 units (from 3 to 3.02), then the function is incremented by an estimated 0.39 units (from 12 to 12.39). The idea of estimating the total increment is important in scientific and engineering applications where measurement error or manufacturing tolerances are present.

ESTIMATING THE TOTAL INCREMENT

Let f be a function of two variables, x and y. Let dx and dy represent small (real) increments in the x- and y-directions, respectively, from the input (a, b). Then an estimate of the total increment in the value of the function is

$$dz = f_x(a, b)\, dx + f_y(a, b)\, dy.$$

The following example is an exercise in the classic text *Elements of the Differential and Integral Calculus,* by Granville, Smith, and Longley, 1941 revised edition, Ginn and Company.

Example 42 *The specific gravity of a solid is given by the formula $s = \dfrac{P}{w}$, where P is the weight in a vacuum and w is the weight of an equal volume of water. How is the computed specific gravity affected by an error of $\pm\dfrac{1}{10}$ weighing P and $\pm\dfrac{1}{20}$ weighing w, assuming $P = 8$ and $w = 1$ in the experiment, (a) if both errors are positive? (b) if one error is negative? (c) What is approximately the largest percentage error?*

Solution We begin by finding the partial derivatives:

$$\frac{\partial s}{\partial P} = \frac{1}{w}$$

$$\frac{\partial s}{\partial w} = -Pw^{-2} = \frac{-P}{w^2}.$$

These correspond to $f_x(x, y)$ and $f_y(x, y)$ in the notation of the formula.

We then evaluate the partial derivatives at the given input:

$$\frac{\partial s}{\partial P}(8, 1) = \frac{1}{1} = 1$$

$$\frac{\partial s}{\partial w}(8, 1) = \frac{-8}{1^2} = -8.$$

These correspond to $f_x(a, b)$ and $f_y(a, b)$ in the formula.

(a) The given increments are $dP = 0.1$ and $dw = 0.05$. Then the total increment is

$$ds = \frac{\partial s}{\partial P}(8, 1)\, dP + \frac{\partial s}{\partial w}(8, 1)\, dw = 1(0.1) - 8(0.05) = -0.3.$$

Specific gravity is unitless. Granville, Smith, and Longley (GSL) state the answer in terms of absolute error, 0.3.

(b) If $dP = -0.1$ and $dw = 0.5$, then

$$ds = \frac{\partial s}{\partial P}(8, 1)\, dP + \frac{\partial s}{\partial w}(8, 1)\, dw = 1(-0.1) - 8(0.05) = -0.5.$$

Once again, the answer in GSL is given in terms of absolute error, 0.5.

If $dP = 0.1$ and $dw = -0.5$, then $ds = 0.5$.

(c) The largest possible error is given in (b). The largest percent relative error is therefore

$$\frac{\pm 0.5}{8} = \pm 0.0625 = \pm 6.25\%.$$

■

GSL gives the answer as absolute percent relative error, 6.25%.

3.6.7 Proving the theorem

A proof of theorem 3, without fully justifying all of the details, proceeds as follows. Suppose that f is a function of two variables for which f_x and f_y both exist in a neighborhood of (a, b) and are continuous at (a, b). We wish to show that the definition of differentiable is satisfied, that is,

$$\frac{f(a + \alpha,\, b + \beta) - (f(a, b) + f_x(a, b)\alpha + f_y(a, b)\beta)}{\sqrt{\alpha^2 + \beta^2}} \doteq 0$$

Ans. to reading exercise 16:
$$L(x, y) = 0 + 0(x - 4) - 4(y - \pi)$$

for every choice of infinitesimals α and β. To that end, let α and β be arbitrary infinitesimals.

Because $f_x(a, b)$ exists, $\frac{f(a+\alpha,\, b) - f(a,b)}{\alpha} \doteq f_x(a, b)$, hence

$$\frac{f(a + \alpha,\ b) - f(a, b)}{\alpha} = f_x(a, b) + \gamma_1$$

for some infinitesimal γ_1. Rearranging,

$$f(a + \alpha,\ b) = f(a, b) + \alpha f_x(a, b) + \alpha \gamma_1,$$

which is (single-variable) local linearity in the x-direction applied to f at (a, b).

Applying the same reasoning, local linearity in the x-direction applied to f_y at (a, b) should give

$$f_y(a + \alpha,\ b) = f_y(a, b) + \alpha f_{yx}(a, b) + \alpha \gamma_2,$$

for some infinitesimal γ_2.

Assuming local linearity applies to f in the y-direction at $(a + \alpha,\ b)$,

$$f(a + \alpha,\ b + \beta) = f(a + \alpha,\ b) + \beta f_y(a + \alpha,\ b) + \beta \gamma_3,$$

for some infinitesimal γ_3.

Fitting these three equations together,

$$
\begin{aligned}
f(a + \alpha,\ b + \beta) &= f(a + \alpha,\ b) + \beta f_y(a + \alpha,\ b) + \beta \gamma_3 \\
&= f(a, b) + \alpha f_x(a, b) + \alpha \gamma_1 + \beta f_y(a + \alpha,\ b) + \beta \gamma_3 \\
&= f(a, b) + \alpha f_x(a, b) + \alpha \gamma_1 \\
&\quad + \beta \left(f_y(a, b) + \alpha f_{yx}(a, b) + \alpha \gamma_2 \right) + \beta \gamma_3 \\
&= f(a, b) + \alpha f_x(a, b) + \beta f_y(a, b) + \alpha \beta f_{yx}(a, b) \\
&\quad + \alpha \gamma_1 + \beta \gamma_3 + \alpha \beta \gamma_2.
\end{aligned}
$$

Then checking the quantity from the definition of differentiable,

$$\frac{f(a + \alpha,\ b + \beta) - \left(f(a, b) + f_x(a, b)\alpha + f_y(a, b)\beta \right)}{\sqrt{\alpha^2 + \beta^2}}$$

$$= \frac{\alpha \beta f_{yx}(a, b) + \alpha \gamma_1 + \beta \gamma_3 + \alpha \beta \gamma_2}{\sqrt{\alpha^2 + \beta^2}}$$

$$\doteq 0,$$

because the denominator is on the α level or the β level, whichever is higher, but in the numerator each α and β is multiplied by an infinitesimal and therefore is on a lower level. The definition is satisfied, and the function f is differentiable at (a, b).

Here is the first place where the hypotheses are needed and details are omitted.

Here is another place where the hypotheses are needed and details are omitted.

Line 1 repeats the local linearity expression for $f(a + \alpha,\ b + \beta)$; line 2 uses the local linearity expression for $f(a + \alpha,\ b)$; lines 3 and 4 use the local linearity expression for $f_y(a + \alpha,\ b)$; lines 5 and 6 expand and rearrange.

EXERCISES 3.6

1–10. Find the equation of the tangent plane to the function at the given input.

1. $f(x,y) = 5x^2 - y^3 + 3x$, at $(1,3)$
2. $f(x,y) = 3x^3 - 5y^2 + 7$, at $(-1,1)$
3. $f(x,y) = 4x^7y^5 + 4x^3y - 11y$, at $(0,0)$
4. $f(x,y) = 3xy$, at $(9,5)$
5. $f(x,y) = \sqrt{x^2 + y^2}$, at $(3,4)$
6. $f(x,y) = \sin xy$, at $(1,\pi)$
7. $f(x,y) = \tan(2x - y)$, at $(1,2)$
8. $z = e^x - \dfrac{1}{y}$, at $(0,2)$
9. $f(x,y) = 2 + y\ln x$, at $(1,1)$
10. $f(x,y) = x^2y^2\cosh(x - y)$, at $(2,2)$

11–14. (a) Use a CAS to produce a picture showing both the surface and its tangent plane at the given input, in the style of Figure 3.70. (b) Zoom in to show the surface and the tangent plane as very close and virtually indistinguishable, in the styles of Figures 3.72 and 3.73.

11. $f(x,y) = e^x + e^y$, at $(-1,0)$
12. $f(x,y) = x^y$, at $(2,1)$
13. $f(x,y) = \sin x + \sin y$, at $\left(\dfrac{\pi}{2}, \dfrac{\pi}{4}\right)$
14. $f(x,y) = \dfrac{x}{y}$, at $(1,1)$

15–20. (a) Find the linearization of the function at the given input. (b) Estimate the stated value of the function without using a calculator.

15. (a) $f(x,y) = \sin x \cosh y$, at $(\pi,0)$; (b) $f(\pi + 0.01,\ -0.03)$
16. (a) $f(x,y) = x^2y - 3\ln y$, at $(2,1)$; (b) $f(2.05, 1.03)$
17. (a) $f(x,y) = x^2 \ln y$, at $(3,1)$; (b) estimate $2.997^2 \ln 1.08$
18. (a) $f(x,y) = 3\cosh(x^2 + y^2)$, at $(0,0)$;
     (b) $f(0.03277, -0.001749)$
19. (a) $f(x,y) = \sqrt{x + y^2}$, at $(3,1)$; (b) $f(3.03, 1.02)$
20. (a) $f(x,y) = e^x y^2$, at $(0,1)$; (b) estimate $1.01^2 e^{0.03}$

21–28. Determine whether the function is differentiable at the given input.

21. $f(x,y) = 0.1\sqrt{x^2 + y}$, at $(-1,-1)$
22. $f(x,y) = 0.1\sqrt{x^2 + y}$, at $(1,1)$
23. $f(x,y) = \dfrac{x^2}{x^2 + y^2}$, at $(0,0)$
24. $f(x,y) = \sqrt[3]{x + y}$, at $(0,0)$

25. $f(x, y) = \sin(x + y)$, at $(0, \pi)$
26. $f(x, y) = \ln(x + y)$, at $(-1, 1)$
27. $f(x, y) = \ln(x + y)$, at $(1.7, 14.9)$
28. $f(x, y) = 20xye^{-x^2 - y^2}$, at $(3e, 4 \ln 2)$

29–34. Use the definition of differentiable to determine whether the function is differentiable at the given input.

29. $f(x, y) = 5x - 4xy^2$, at $(3, 5)$
30. $f(x, y) = xy + 4x - 3y$, at $(2, 1)$
31. $f(x, y) = \begin{cases} 0, & y \neq 0 \\ x, & y = 0 \end{cases}$, at $(0, 0)$
32. $f(x, y) = \sin(x + y)$, at $(0, 0)$
33. $f(x, y) = \sqrt{x^2 y^2}$, at $(0, 0)$
34. $f(x, y) = \sqrt{|xy|}$, at $(0, 0)$

Exercise 32: Watch out for violations of the approximation principle. You may need to use an approximation formula derived from the Maclaurin series for sine, $\sin \gamma \approx \gamma - \dfrac{\gamma^3}{6}$ for any infinitesimal γ.

Exercises 35–38 are from Granville, Smith, and Longley.

35. If specific gravity is determined by the formula $s = \dfrac{A}{A - W}$, where A is the weight in air and W the weight in water, what is (a) approximately the largest error in s if A can be read within 0.01 lb and W within 0.02 lb, the actual readings being $A = 9$ lb, $W = 5$ lb? (b) the largest relative error?

36. The resistance of a circuit was found by using the formula $C = \dfrac{E}{R}$, where $C = $ current and $E = $ electromotive force. If there is an error of $\dfrac{1}{10}$ ampere in reading C and $\dfrac{1}{20}$ volt in reading E, (a) what is the approximate [largest] error in R if the readings are $C = 15$ amperes and $E = 110$ volts? (b) What is the percentage error?

GSL intends these to be interpreted as absolute errors, that is, the errors could be in the positive or negative direction.

37. The acceleration of a particle down an inclined plane is given by $a = g \sin i$. If g varies by 0.1 ft per second per second, and i, which is measured as $30°$, may be in error $1°$, what is the approximate error in the computed value of a? Take the normal value of g to be 32 ft per second per second.

Recall that the modern notation for ft per second per second is ft/s².

38. The period of a pendulum is $P = 2\pi \sqrt{\dfrac{\ell}{g}}$. (a) What is the greatest approximate error in the period if there is an error of ± 0.1 ft in measuring a 10-foot suspension and g, taken as 32 ft per second per second, may be in error by 0.05 ft per second per second? (b) What is the percentage error?

39. Show that the function $f(x, y) = \begin{cases} \dfrac{xy}{x^2 + y^2}, & (x, y) \neq (0, 0) \\ 0, & (x, y) = (0, 0) \end{cases}$ of examples 34 and 36 does not satisfy the hypotheses of theorem 3 by using the following steps.

Note: because theorem 3 is an "if-then" statement and not an "if and only if" statement, this exercise does not show that f is not differentiable at $(0, 0)$. See example 36 for a proof of nondifferentiability.

(a) Find $f_x(x, y)$ for $(x, y) \neq (0, 0)$.
(b) Find $\displaystyle\lim_{(x,y)\to(0,0)} f_x(x, y)$. What does this mean about the continuity of f_x at $(0, 0)$?
(c) Conclude that the hypotheses of theorem 3 are not met.

40. Let $f(x,y) = \sqrt{x^2 y^2}$.

(a) Investigate $f_x(a,0)$ and $f_y(a,0)$ for $a \neq 0$.
(b) Investigate $f_x(0,b)$ and $f_y(0,b)$ for $b \neq 0$.
(c) Use a CAS to graph the surface and comment on the features corresponding to parts (a) and (b).
(d) Comment on how the result of exercise 33 relates to part (c).

41–46. A function $z = f(x,y)$ has the *opposite corners property* on a rectangular region $ABCD$ in the xy-plane if $f(A) + f(C) = f(B) + f(D)$, that is, if the sum of the function's values on opposite corners is the same for each pair of opposite corners. A plane has the opposite corners property on every rectangular region (the converse is also true). Then if a surface is to be indistinguishable from its tangent plane as in Figure 3.77, the sum of its values at one pair of opposite corners, $f(a,b) + f(a+\alpha, b+\beta)$, must match the sum of its values at the other pair, $f(a+\alpha, b) + f(a, b+\beta)$, or at least differ only by an amount on a lower level than $\sqrt{\alpha^2 + \beta^2}$. This gives rise to an alternate definition of differentiability, that f is differentiable at (a,b) if

$$\frac{f(a+\alpha,\, b) + f(a,\, b+\beta) - f(a+\alpha,\, b+\beta) - f(a,b)}{\sqrt{\alpha^2 + \beta^2}} \doteq 0$$

for every choice of infinitesimals α and β. Use this alternate definition to determine whether the function is differentiable at the given input, and compare to the work required in exercises 29–34.

41. $f(x,y) = 5x - 4xy^2$, at $(3,5)$
42. $f(x,y) = xy + 4x - 3y$, at $(2,1)$
43. $f(x,y) = \begin{cases} 0, & y \neq 0 \\ x, & y = 0 \end{cases}$, at $(0,0)$
44. $f(x,y) = \sin(x+y)$, at $(0,0)$
45. $f(x,y) = \sqrt{x^2 y^2}$, at $(0,0)$
46. $f(x,y) = \sqrt{|xy|}$, at $(0,0)$

I am grateful to my colleague Troy Riggs for alerting me to the opposite corners property of a plane.

This also gives rise to an alternate local linearity formula,

$$f(a+\alpha,\, b+\beta)$$
$$\approx f(a+\alpha,\, b) + f(a,\, b+\beta) - f(a,b).$$

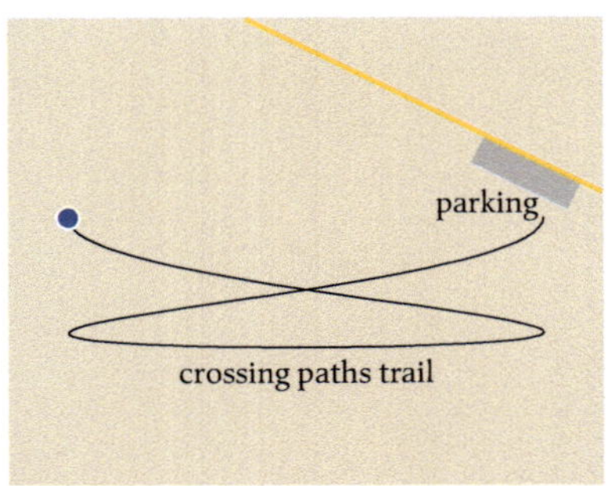

Figure 3.81 *A path taken by a hiker, as plotted on a map produced by an app. The blue dot represents the current position, and the hiker's path up until now is black*

The rate of change of elevation with respect to time t is

$$\frac{\text{change in elevation}}{\text{change in time}} = \frac{dz}{dt},$$

which is a function of t.

3.7　Chain Rule

Suppose that a hiker is walking a trail in the hills, using an app to map the path taken, and the path is shown on a two-dimensional map, perhaps as in Figure 3.81. The resulting path can be represented by a parametrically defined curve, or as a vector-valued function $\mathbf{r}(t) = \langle x(t), y(t) \rangle$, where the variable t represents time and the output $\langle x(t), y(t) \rangle$ represents the location at time t in the xy-plane as plotted on the map. The graph of the function $\mathbf{r}$, without the map's features, is in Figure 3.82, left. Suppose also that the hillside upon which the hiker is traveling, which can be represented as a surface in a three-dimensional graph as in Figure 3.82, right, is given by the function $z = f(x, y)$; the input to f is a map location (a location in the xy-plane), and the output is elevation.

Then the composition $z = f(\mathbf{r}(t))$ (Figure 3.83) represents elevation at time t. The output of $\mathbf{r}$ tells us the location (x, y) on the map, which serves as the input into f, which tells us the hiker's elevation. Question: how fast is the hiker's elevation changing at time t? That is, what is $\frac{dz}{dt}$?

One method for determining $\frac{dz}{dt}$ is to form the composition so that we have z as a function of t, and then take the derivative. This method is illustrated in example 43.

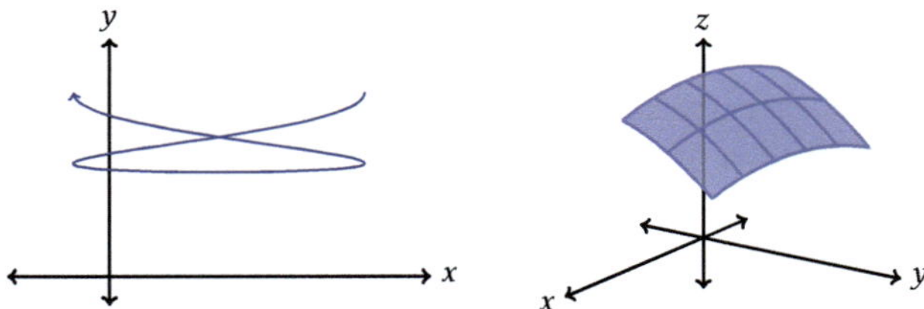

Figure 3.82 *(Left) a parametrically defined curve $\mathbf{r}(t) = \langle x(t), y(t) \rangle$ and (right) a surface $z = f(x, y)$*

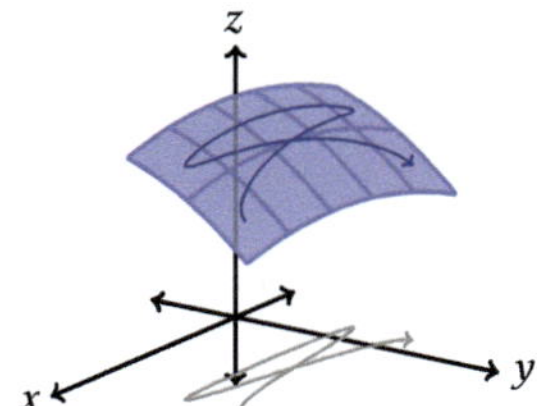

Figure 3.83 *The composition $z = f(\mathbf{r}(t))$ gives the elevation of the hiker at time t, the graph of which is the hiker's path on the surface. The shadow (gray) in the xy-plane gives the map coordinates $\mathbf{r}(t) = (x(t), y(t))$ of the hiker, which serve as the input into the elevation function. The shadow in the xy-plane is the path shown by the app in Figure 3.81*

Example 43 *Let $r(t) = \langle \cos t, \sin t \rangle$ and $z = f(x, y) = x^2 + y$. Find $\frac{dz}{dt}$.*

Solution We begin by expressing z as a function of t. We have $z = x^2 + y, x = \cos t$, and $y = \sin t$. Therefore

$$z = (\cos t)^2 + \sin t.$$

Then

$$\frac{dz}{dt} = 2(\cos t)^1(-\sin t) + \cos t = -2\cos t \sin t + \cos t. \qquad \blacksquare$$

3.7.1 Multivariable chain rule: first case

Notice the use of the chain rule when calculating the solution to example 43. After all, we are calculating the derivative of a composition, and that's what the chain rule is for. As a reminder, in Leibniz notation, the single-variable chain rule states that if z is a function of x and x is a function of t, then $\frac{dz}{dt} = \frac{dz}{dx} \cdot \frac{dx}{dt}$. But we have a problem: when z is a function of two variables, x and y, we do not have a $\frac{dz}{dx}$, but rather a $\frac{\partial z}{\partial x}$, as well as a $\frac{\partial z}{\partial y}$. But by putting the two "parts" together, the chain rule still works.

> Recall that one way to remember the single-variable chain rule is to treat the expression as a fraction of differentials, so that the dx's cancel.

Theorem 4 CHAIN RULE CASE 1 *If $z = f(x, y)$ is a differentiable function of x and y and $x = g(t)$ and $y = h(t)$ are differentiable functions of t, then z is a differentiable function of t and*

$$\frac{dz}{dt} = \frac{\partial z}{\partial x} \cdot \frac{dx}{dt} + \frac{\partial z}{\partial y} \cdot \frac{dy}{dt}.$$

> More precisely, if g and h are differentiable at $t = k$, and f is differentiable at $(g(k), h(k))$, then the composition $z = f(g(t), h(t))$ is differentiable at $t = k$.

Proof. Using the single-variable local linearity formula $g(x + \alpha) \approx g(x) + \alpha g'(x)$ (similarly for h) and the two-variable local linearity formula $f(x + \alpha, y + \beta) \approx f(x, y) + f_x(x, y)\alpha + f_y(x, y)\beta$, and assuming that f, g, and h are differentiable at the appropriate inputs, we have

> Notice that some of the derivatives in the rule are partial derivatives and others are regular derivatives.

$$\frac{dz}{dt} = \frac{d}{dt} f(g(t), h(t))$$

$$= \frac{f(g(t + \alpha), h(t + \alpha)) - f(g(t), h(t))}{\alpha}$$

Line 1 rewrites z as a composition; line 2 uses the definition of derivative; line 3 uses the single-variable local linearity formula on g and h, each with the real number t and infinitesimal α; lines 4 and 5 (split because of length) use the two-variable local linearity formula with real numbers $g(t)$ and $h(t)$ and infinitesimals $\alpha g'(t)$ and $\alpha h'(t)$; line 6 simplifies; and line 7 rewrites using Leibniz notation, with the understanding that the partial derivatives are functions of t.

Details such as accounting for possible violations of the approximation principle are omitted.

$$\approx \frac{f(g(t) + \alpha g'(t),\, h(t) + \alpha h'(t)) - f(g(t),\, h(t))}{\alpha}$$

$$\approx \frac{f(g(t),\, h(t)) + f_x(g(t),\, h(t))\alpha g'(t)}{\alpha}$$

$$+ \frac{f_y(g(t),\, h(t))\alpha h'(t) - f(g(t),\, h(t))}{\alpha}$$

$$= f_x(g(t),\, h(t))g'(t) + f_y(g(t),\, h(t))h'(t)$$

$$= \frac{\partial f}{\partial x} \cdot \frac{dx}{dt} + \frac{\partial f}{\partial y} \cdot \frac{dy}{dt}.$$

Changing notation,

$$\frac{dz}{dt} = \frac{\partial z}{\partial x} \cdot \frac{dx}{dt} + \frac{\partial z}{\partial y} \cdot \frac{dy}{dt}. \qquad \blacksquare$$

Chain rule case 1 can be used to provide an alternate solution to example 43.

Alternate solution   Given $z = x^2 + y$, $x = \cos t$, and $y = \sin t$, the derivatives and partial derivatives are

$$\frac{\partial z}{\partial x} = 2x \qquad\qquad \frac{\partial z}{\partial y} = 1$$

$$\frac{dx}{dt} = -\sin t \qquad\qquad \frac{dy}{dt} = \cos t$$

Using the chain rule case 1,

Line 2 uses the partial derivatives as calculated in terms of the variables x and y; line 3 replaces the variables x and y with their expressions in terms of the variable t.

$$\frac{dz}{dt} = \frac{\partial z}{\partial x} \cdot \frac{dx}{dt} + \frac{\partial z}{\partial y} \cdot \frac{dy}{dt}$$

$$= 2x(-\sin t) + 1(\cos t)$$

$$= 2\cos t(-\sin t) + \cos t,$$

matching the original solution. $\qquad \blacksquare$

Because the task is to find $\frac{dz}{dt}$, the solution should be written in terms of the variable t. The solution should not contain the intermediate variables x and y.

Tree diagrams can be helpful for remembering various cases of the chain rule. The tree diagram for case 1, where z is a function of x and y and each of x and y are functions of t, is in Figure 3.84. The dependent variable z is represented by a node (a dot) on the top row, the intermediate variables x and y are represented by nodes on the middle row, and the independent variable t is represented by a node on the bottom row. Edges from z to each of x and y are marked by partial derivatives, and edges from each of x and y to t are marked by ordinary derivatives. The chain rule is formed by multiplying the derivatives (ordinary and/or partial) along each path

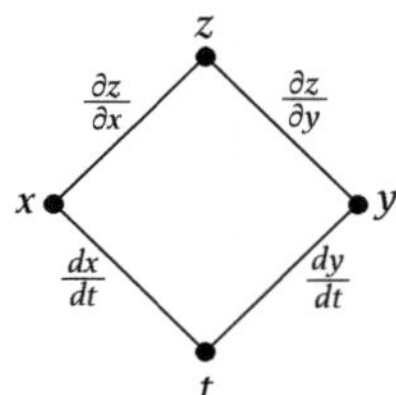

Figure 3.84 *A tree diagram for chain rule case* 1

from the top variable z to the bottom variable t, and adding the results of those paths together. For the left-hand path we multiply $\frac{\partial z}{\partial x}$ and $\frac{dx}{dt}$; for the right-hand path we multiply $\frac{\partial z}{\partial y}$ and $\frac{dy}{dt}$. Adding the results together gives the chain rule case 1,

The principle of the diagram remains the same no matter how many variables are represented in the middle and bottom rows. See example 45 for another case.

$$\frac{dz}{dt} = \frac{\partial z}{\partial x} \cdot \frac{dx}{dt} + \frac{\partial z}{\partial y} \cdot \frac{dy}{dt}.$$

3.7.2   Using the chain rule case 1

Because there are multiple cases of the chain rule, each with different formulas, drawing the tree diagram each time is highly recommended.

Example 44 *Use the chain rule to find* $\frac{dz}{dt}$ *for* $z = 4x^2 - 7xy$, $x = 4\sin t$, $y = \ln t$.

Solution First, we draw the tree diagram and write the chain rule. We wish to find $\frac{dz}{dt}$, so we place a node for z on the top row and a node for t on the bottom row. The intermediate variables x and y are placed in the middle row. Because z is a function of x and y, we draw edges from z to each of x and y. Because x is a function of t, we draw and edge from x to t. Because y is a function of t, we draw an edge from y to t. We label each edge with the appropriate derivative or partial derivative. The result is in Figure 3.85.

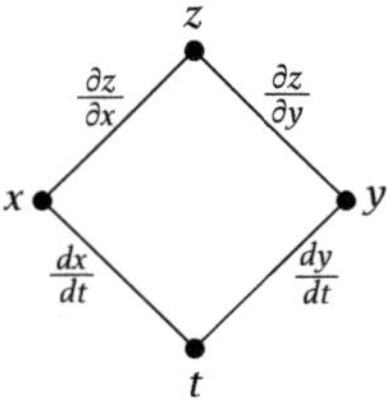

Figure 3.85 A *tree diagram for example 44*

Multiplying the derivatives along each path from z to t and adding the results from the paths, the chain rule is

$$\frac{dz}{dt} = \frac{\partial z}{\partial x} \cdot \frac{dx}{dt} + \frac{\partial z}{\partial y} \cdot \frac{dy}{dt}.$$

Next, we calculate the partial and ordinary derivatives indicated in the chain rule:

$$\frac{\partial z}{\partial x} = 8x - 7y \qquad \frac{\partial z}{\partial y} = -7x$$
$$\frac{dx}{dt} = 4\cos t \qquad \frac{dy}{dt} = \frac{1}{t}.$$

Line 2 uses the partial derivatives as calculated in terms of the variables x and y; line 3 replaces the variables x and y with their expressions in terms of the variable t.

Then

$$\frac{dz}{dt} = \frac{\partial z}{\partial x} \cdot \frac{dx}{dt} + \frac{\partial z}{\partial y} \cdot \frac{dy}{dt}$$

$$= (8x - 7y) \cdot 4 \cos t + (-7x) \cdot \frac{1}{t}$$

$$= (32 \sin t - 7 \ln t) 4 \cos t - \frac{28 \sin t}{t}. \qquad \blacksquare$$

Reading Exercise 17 Use the chain rule to find $\frac{dz}{dt}$ for $z = x^2 - 4y^2$, $x = t^2 + 1$, $y = \cosh t$.

3.7.3 Multivariable chain rule, case 2

Example 45 *Draw a tree diagram and write the chain rule for finding $\frac{\partial z}{\partial r}$ and $\frac{\partial z}{\partial \theta}$, given that z is a function of x and y, x is a function of r and θ, and y is a function of r and θ.*

Solution We wish to find $\frac{\partial z}{\partial r}$ and $\frac{\partial z}{\partial \theta}$, so we place a node for z on the top row and nodes for r and θ on the bottom row. We place nodes for the intermediate variables x and y in the middle row. Because z is a function of x and y, we draw edges from z to both x and y. Because x is a function of r and θ, we draw edges from x to both r and θ. Because y is a function of r and θ, we draw edges from y to both r and θ. We then label the edges with the appropriate partial derivatives, arriving at the tree diagram in Figure 3.86.

To write the chain rule for calculating $\frac{\partial z}{\partial r}$, we identify all the (descending-only) paths from z to r (Figure 3.87), multiply the (partial) derivatives along each path, and add the results of the paths:

$$\frac{\partial z}{\partial r} = \frac{\partial z}{\partial x} \cdot \frac{\partial x}{\partial r} + \frac{\partial z}{\partial y} \cdot \frac{\partial y}{\partial r}.$$

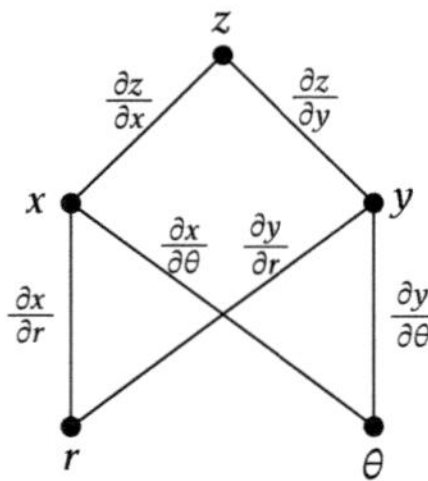

Figure 3.86 *A tree diagram for example 45*

To write the chain rule for calculating $\frac{\partial z}{\partial \theta}$, we identify all the paths from z to θ (Figure 3.88), multiply the derivatives along each path, and add the results of the paths:

$$\frac{\partial z}{\partial \theta} = \frac{\partial z}{\partial x} \cdot \frac{\partial x}{\partial \theta} + \frac{\partial z}{\partial y} \cdot \frac{\partial y}{\partial \theta}.$$

■

The method used in example 45 is sufficient in practice for determining the correct version of the chain rule to use, but it doesn't constitute an actual proof of the formula. A formal proof is presented next.

Theorem 5 CHAIN RULE CASE 2 *If $z = f(x, y)$ is a differentiable function of x and y and $x = g(r, \theta)$ and $y = h(r, \theta)$ are differentiable functions of r and θ, then z is a differentiable function of r and θ,*

$$\frac{\partial z}{\partial r} = \frac{\partial z}{\partial x} \cdot \frac{\partial x}{\partial r} + \frac{\partial z}{\partial y} \cdot \frac{\partial y}{\partial r},$$

and

$$\frac{\partial z}{\partial \theta} = \frac{\partial z}{\partial x} \cdot \frac{\partial x}{\partial \theta} + \frac{\partial z}{\partial y} \cdot \frac{\partial y}{\partial \theta}.$$

Proof. As in the proof of theorem 4, local linearity plays a key role, but with a slight wrinkle. The two-variable local linearity formula is

$$f(a + \alpha, b + \beta) \approx f(a, b) + f_x(a, b)\alpha + f_y(a, b)\beta.$$

But what if only one variable is incremented, and the other is not? Then the other increment is replaced by zero:

$$f(a + \alpha, b + 0) \approx f(a, b) + f_x(a, b)\alpha + f_y(a, b) \cdot 0 = f(a, b) + f_x(a, b)\alpha.$$

Using the two-variable local linearity formula for g and h, and then for f, and assuming that f, g, and h are differentiable at the appropriate inputs, we have

$$\frac{\partial z}{\partial r} = \frac{\partial}{\partial r} f(g(r, \theta), h(r, \theta))$$
$$= \frac{f(g(r + \alpha, \theta), h(r + \alpha, \theta)) - f(g(r, \theta), h(r, \theta))}{\alpha}$$
$$\approx \frac{f(g(r, \theta) + g_r(r, \theta)\alpha, h(r, \theta) + h_r(r, \theta)\alpha) - f(g(r, \theta), h(r, \theta))}{\alpha}$$

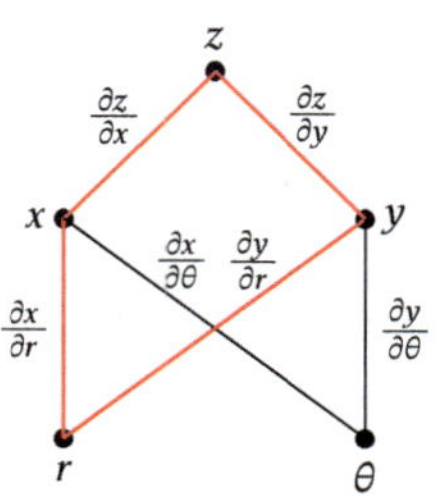

Figure 3.87 *the paths from z to r (red) used to write the chain rule for $\frac{\partial z}{\partial r}$*

More precisely, if both g and h are differentiable at (k, m), and f is differentiable at $(g(k, m), h(k, m))$, then the composition $z = f(g(r, \theta), h(r, \theta))$ is differentiable at (k, m).

Ans. to reading exercise 17:
$\frac{dz}{dt} = 2(t^2 + 1)2t - 8 \cosh t \sinh t = 4t^3 + 4t - 8 \cosh t \sinh t$

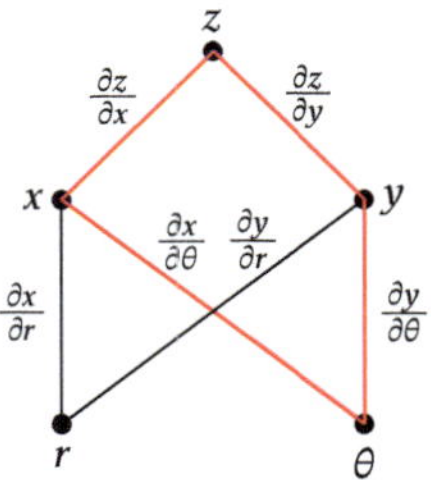

Figure 3.88 *The paths from z to θ (red) used to write the chain rule for $\frac{\partial z}{\partial \theta}$*

Line 1 rewrites z as a composition; line 2 uses the definition of partial derivative, incrementing only the value of r; line 3 uses the two-variable local linearity formula for the function g with real numbers r and θ, infinitesimal increment α, and second increment 0, and the same for the function h; lines 4 and 5 (split because of length) use the two-variable local linearity formula for the function f with real numbers $g(r,\theta)$ and $h(r,\theta)$ and infinitesimals $g_r(r,\theta)\alpha$ and $h_r(r,\theta)\alpha$; line 6 simplifies; line 7 rewrites using Leibniz notation, with the understanding that $\frac{\partial f}{\partial x}$ and $\frac{\partial f}{\partial y}$ are functions of r and θ; and line 8 changes notation from f to z.

$$\approx \frac{f(g(r,\theta),\, h(r,\theta)) + f_x(g(r,\theta),\, h(r,\theta)) \cdot g_r(r,\theta)\alpha}{\alpha}$$

$$+ \frac{f_y(g(r,\theta),\, h(r,\theta)) \cdot h_r(r,\theta)\alpha - f(g(r,\theta),\, h(r,\theta))}{\alpha}$$

$$= f_x(g(r,\theta),\, h(r,\theta)) \cdot g_r(r,\theta) + f_y(g(r,\theta),\, h(r,\theta)) \cdot h_r(r,\theta)$$

$$= \frac{\partial f}{\partial x} \cdot \frac{\partial x}{\partial r} + \frac{\partial f}{\partial y} \cdot \frac{\partial y}{\partial r}$$

$$= \frac{\partial z}{\partial x} \cdot \frac{\partial x}{\partial r} + \frac{\partial z}{\partial y} \cdot \frac{\partial y}{\partial r},$$

as promised. The proof of the formula for $\frac{\partial z}{\partial \theta}$ is similar. ∎

Example 46 *Let $z = x^2 y + y^2 + 1$, $x = r\cos\theta$, and $y = -2r\sin 3\theta$. Use the chain rule to find $\frac{\partial z}{\partial r}$ and $\frac{\partial z}{\partial \theta}$.*

Solution The tree diagram and chain rule formulas for z a function of x and y and x and y functions of r and θ are developed in example 45; see Figure 3.89.

The six partial derivatives written in the diagram are

$$\frac{\partial z}{\partial x} = 2xy \qquad\qquad \frac{\partial z}{\partial y} = x^2 + 2y$$

$$\frac{\partial x}{\partial r} = \cos\theta \qquad\qquad \frac{\partial x}{\partial \theta} = -r\sin\theta$$

$$\frac{\partial y}{\partial r} = -2\sin 3\theta \qquad\qquad \frac{\partial y}{\partial \theta} = -6r\cos 3\theta$$

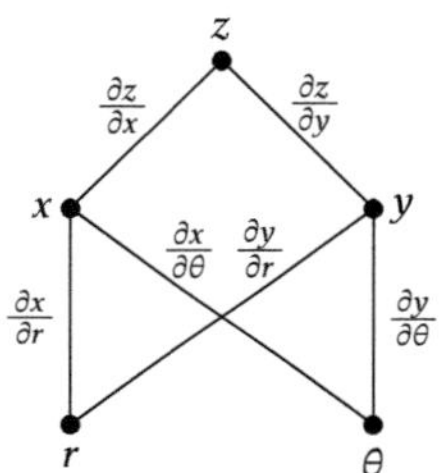

Figure 3.89 *The tree diagram for example 46*

Then

$$\frac{\partial z}{\partial r} = \frac{\partial z}{\partial x} \cdot \frac{\partial x}{\partial r} + \frac{\partial z}{\partial y} \cdot \frac{\partial y}{\partial r}$$

$$= 2xy\cos\theta + (x^2 + 2y)(-2\sin 3\theta)$$

$$= 2r\cos\theta(-2r\sin 3\theta)\cos\theta + (r^2\cos^2\theta - 4r\sin 3\theta)(-2\sin 3\theta)$$

and

$$\frac{\partial z}{\partial \theta} = \frac{\partial z}{\partial x} \cdot \frac{\partial x}{\partial \theta} + \frac{\partial z}{\partial y} \cdot \frac{\partial y}{\partial \theta}$$

$$= 2xy(-r\sin\theta) + (x^2 + 2y)(-6r\cos 3\theta)$$

$$= 2r\cos\theta(-2r\sin 3\theta)(-r\sin\theta) + (r^2\cos^2\theta - 4r\sin 3\theta)(-6r\cos 3\theta).$$

The results can be simplified if desired. ∎

The same results can be obtained by calculating the composition, that is, rewriting z as a function of r and θ, and then calculating the partial derivatives directly.

The direct method is often messier and more error-prone, which is one reason why the chain rule is often preferred.

There are, of course, additional cases of the multivariable chain rule. Some of these cases are explored in the exercises.

Reading Exercise 18   Draw a tree diagram and state the chain rule formula for $\frac{\partial z}{\partial s}$ if z is a function of x and y while x and y are each functions of s and t.

3.7.4   Implicit differentiation revisited

In single-variable calculus, implicit differentiation can be used to find $\frac{dy}{dx}$ when y is defined implicitly as a function of x. For instance, given

$$x^2 y^3 - 3y + 5x - 7 = 0,$$

we begin by differentiating both sides of the equation implicitly with respect to x:

$$x^2 \cdot 3y^2 \cdot \frac{dy}{dx} + 2x \cdot y^3 - 3\frac{dy}{dx} + 5 = 0.$$

The form of the product rule used here is first $\cdot$ second$'$ + first$'$ $\cdot$ second.

Notice that the product rule is used for $x^2 \cdot y^3$, and when taking the derivative of y^3, we use the power rule to get $3y^2$, but because it is "not just plain x" to a power, we need to multiply by the derivative of what's inside, which is $\frac{dy}{dx}$. Then we solve the resulting equation for $\frac{dy}{dx}$:

The expression y' can be used in place of $\frac{dy}{dx}$ as long as the notation is clear in context.

$$3x^2 y^2 \frac{dy}{dx} - 3\frac{dy}{dx} = -2xy^3 - 5$$

$$\frac{dy}{dx}\left(3x^2 y^2 - 3\right) = -2xy^3 - 5$$

$$\frac{dy}{dx} = \frac{-2xy^3 - 5}{3x^2 y^2 - 3}.$$

Line 1 gathers terms involving $\frac{dy}{dx}$ on one side of the equation and terms not involving $\frac{dy}{dx}$ on the other side; line 2 factors out $\frac{dy}{dx}$; line 3 divides to solve for $\frac{dy}{dx}$. These algebraic steps are typical.

It turns out that the chain rule can be used to develop an alternative procedure that is much simpler.

Notice that the expression $x^2 y^3 - 3y + 5x - 7$ is an expression in two variables x and y; let's call that expression $F(x, y)$. Then, the equation defining y implicitly as a function of x has the form $z = F(x, y) = 0$. To apply the multivariable chain rule, we draw the tree diagram to determine the appropriate chain rule formula. We place a node for z on the top row, nodes for x and y in the middle row, and then a node for x in the bottom row as well. Because z is a function of x and y, we draw edges from z to x and y in the middle row. Because y is (implicitly) a function of x, we draw an edge from y to the x in the bottom row. And we also draw an edge from x in the middle row to the x in the bottom row. Next, we label the edges with

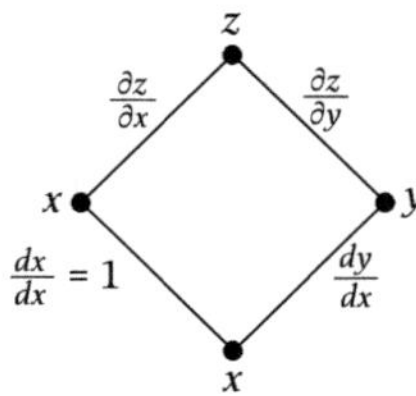

Figure 3.90 *A tree diagram for implicit differentiation*

Notice that certain details, such as a proof of the relevant chain rule formula, have been omitted in the derivation of this formula. A formal proof would need to include such details.

Compare to the single-variable calculus procedure reviewed at the beginning of this discussion. Using partial differentiation (using the implicit differentiation formula) is much more efficient.

derivatives or partial derivatives as appropriate, noting that $\frac{dx}{dx} = 1$. The result is in Figure 3.90.

Multiplying the derivatives along each path from z to the bottom x and adding the path results, the chain rule formula is

$$\frac{dz}{dx} = \frac{\partial z}{\partial x} \cdot 1 + \frac{\partial z}{\partial y} \cdot \frac{dy}{dx}.$$

But notice that because $z = F(x, y) = 0$, the derivative of z is 0, that is, $\frac{dz}{dx} = 0$. Then, we have

$$0 = \frac{\partial z}{\partial x} + \frac{\partial z}{\partial y} \cdot \frac{dy}{dx}.$$

We are interested in $\frac{dy}{dx}$, so let's solve for it:

$$-\frac{\partial z}{\partial x} = \frac{\partial z}{\partial y} \cdot \frac{dy}{dx}$$

$$-\frac{\frac{\partial z}{\partial x}}{\frac{\partial z}{\partial y}} = \frac{dy}{dx}.$$

Rewriting the partial derivatives in terms of F, we have

$$\frac{dy}{dx} = -\frac{F_x}{F_y}.$$

IMPLICIT DIFFERENTIATION FORMULA

If $z = F(x, y)$ is differentiable such that $F(x, y) = 0$ defines y implicitly as a function of x that is also differentiable, then wherever $F_y \neq 0$,

$$\frac{dy}{dx} = -\frac{F_x}{F_y}.$$

Example 47 Find $\frac{dy}{dx}$ given $x^2 y^3 - 3y + 5x - 7 = 0$, assuming that y is a function of x.

Solution Using the implicit differentiation formula with $F(x, y) = x^2 y^3 - 3y + 5x - 7$,

$$\frac{dy}{dx} = -\frac{F_x}{F_y} = -\frac{2xy^3 + 5}{3x^2 y^2 - 3}.$$

■

Notice that to apply the implicit differentiation formula, the equation must be in the form $F(x, y) = 0$. If nonzero algebraic quantities are on both sides of the equation such as $x^2 y^3 + 5x = 3y + 7$, rearrange the equation before determining F; in this case, the rearrangement is $x^2 y^3 + 5x - 3y - 7 = 0$.

Reading Exercise 19 Find $\frac{dy}{dx}$ for $x^2 y - 7y^2 + 2 = 0$.

Implicit partial differentiation formulas can be developed in a similar manner.

IMPLICIT PARTIAL DIFFERENTIATION FORMULAS

If $F(x, y, z) = 0$ defines z implicitly as a function of x and y and certain conditions are met, then wherever $F_z \neq 0$,

$$\frac{\partial z}{\partial x} = -\frac{F_x}{F_z} \quad \text{and} \quad \frac{\partial z}{\partial y} = -\frac{F_y}{F_z}.$$

Ans. to reading exercise 18:

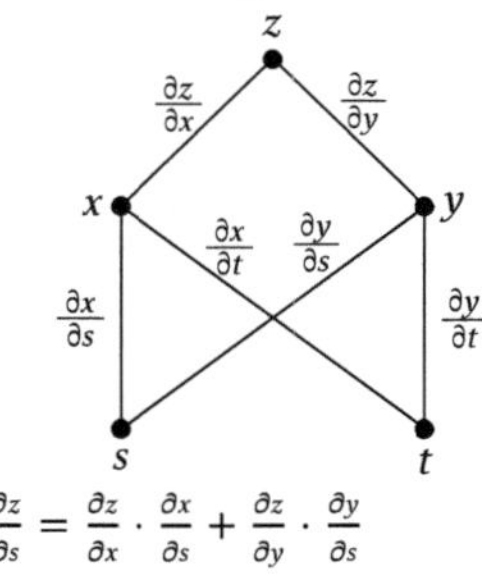

$$\frac{\partial z}{\partial s} = \frac{\partial z}{\partial x} \cdot \frac{\partial x}{\partial s} + \frac{\partial z}{\partial y} \cdot \frac{\partial y}{\partial s}$$

The unnamed conditions mentioned in the statement of the implicit partial differentiation formula are nearly universally met for the types of functions encountered in calculus.

Example 48 *Suppose z is defined implicitly as a function of the variables x and y by $x^2 + y^2 - z^2 = 47$. Find $\frac{\partial z}{\partial x}$.*

Compare to example 29 of Section 3.5.

Solution Using the implicit partial differentiation formula with $F(x, y, z) = x^2 + y^2 - z^2 - 47$,

$$\frac{\partial z}{\partial x} = -\frac{F_x}{F_z} = -\frac{2x}{-2z} = \frac{x}{z}.$$

∎

EXERCISES 3.7

1–6. Find the requested derivative or partial derivative by first finding the composition and then differentiating, as demonstrated in example 43. Do not use the multivariable chain rule.

1. $\frac{dx}{dt}$, where $x = r^2 \sin 5\theta$, $r = t^3 + t$, and $\theta = 3 - 4t$

2. $\frac{dz}{dt}$, where $z = x^2 y - 3x$, $x = t^2$, and $y = \ln t$

3. $\dfrac{\partial x}{\partial t}$, where $x = r^2 \sin 5\theta$, $r = t^3 + s^2$, and $\theta = 3s + 5t$

4. $\dfrac{\partial z}{\partial s}$, where $z = x^2 y + 4x^3 - 7y^2 + 0.047$, $x = r^2 s - s$, and $y = 3rs^2$

5. $\dfrac{\partial z}{\partial s}$, where $z = x^4 - 3y^2 + 4xy$, $x = 3 + 5r + 2s$, and $y = s^2 r^3$

6. $\dfrac{dz}{dt}$, where $z = y^2 \tan^{-1} x$, $x = \cos t$, and $y = e^{t^2}$

7-16. Draw a tree diagram and write the chain rule formula for the requested derivative or partial derivative.

7. $\dfrac{dx}{dt}$, given that x is a function of two variables r and θ, and each of r and θ are functions of t

8. $\dfrac{ds}{dx}$, given that s is a function of two variables u and v, and each of u and v are functions of x

9. $\dfrac{dw}{dt}$, given that w is a function of three variables x, y, and z, and each of x, y, and z are functions of t

10. $\dfrac{dz}{dt}$, given that z is a function of four variables x, y, u, and v, and each of x, y, u, and v are functions of t

11. $\dfrac{\partial s}{\partial x}$, given that s is a function of two variables u and v, and each of u and v are functions of x and y

12. $\dfrac{\partial s}{\partial \theta}$, given that s is a function of three variables u, v, and w, and each of u, v, and w are functions of r and θ

13. $\dfrac{\partial w}{\partial r}$ and $\dfrac{\partial w}{\partial \theta}$, given that w is a function of three variables x, y, and z, and each of x, y, and z are functions of two variables r and θ

14. $\dfrac{\partial z}{\partial r}, \dfrac{\partial z}{\partial s}$, and $\dfrac{\partial z}{\partial t}$, given that z is a function of two variables x and y, and each of x and y are functions of three variables r, s, and t

15. $\dfrac{dz}{dt}$, given that z is a function of two variables x and y, each of x and y are functions of two variables r and θ, and each of r and θ are functions of t

16. $\dfrac{dw}{dt}$, given that w is a function of three variables x, y, and z, each of x and z are functions of two variables r and θ, and each of y, r, and θ are functions of t

17-30. Use the chain rule to find the requested derivative or partial derivative.

17. $\dfrac{dx}{dt}$, where $x = r^2 \sin 5\theta$, $r = t^3 + t$, and $\theta = 3 - 4t$

18. $\dfrac{dz}{dt}$, where $z = x^2 y - 3x$, $x = t^2$, and $y = \ln t$

19. $\dfrac{\partial x}{\partial t}$, where $x = r^2 \sin 5\theta$, $r = t^3 + s^2$, and $\theta = 3s + 5t$

20. $\dfrac{\partial z}{\partial s}$, where $z = x^2 y + 4x^3 - 7y^2 + 0.047$, $x = r^2 s - s$, and $y = 3rs^2$

21. $\dfrac{\partial z}{\partial s}$, where $z = x^4 - 3y^2 + 4xy$, $x = 3 + 5r + 2s$, and $y = s^2 r^3$

22. $\dfrac{dz}{dt}$, where $z = y^2 \tan^{-1} x$, $x = \cos t$, and $y = e^{t^2}$

Ans. to reading exercise 19:
$$\dfrac{dy}{dx} = -\dfrac{2xy}{x^2 - 14y}$$

23. $\dfrac{dw}{dt}$, where $w = x^2 - 3yz$, $x = 4t^3$, $y = \sqrt{5t + 1}$, and $z = 2^t$

24. $\dfrac{dz}{dt}$, where $z = s^2 r \cos\theta$, $s = \dfrac{t}{t+1}$, $r = t^2 + 1$, and $\theta = \pi t$

25. $\dfrac{\partial w}{\partial r}$, where $w = x^2 - 3yz$, $x = r\cos\theta$, $y = r^2$, and $z = r + 4\theta$

26. $\dfrac{\partial z}{\partial s}$, where $z = x^2 y^3$, $x = rs + t$, and $y = st + r$

27. $\dfrac{dx}{dy}$, where $x = 4t^2 + s^3 - st$, $t = \sin y$, and $s = \cos y$

28. $\dfrac{\partial x}{\partial s}$, where $x = 4uv - 3w$, $u = 3t^2$, $v = 5s^2 + \sqrt{t}$, and $w = 14\sin^2(14 - 5t)$

29. $\dfrac{dz}{dt}$, where $z = 3x^2 - 5y^2 + 6y - 18$, $x = 3r\sin\theta$, $y = r + \tan\theta$, $r = t^3 + 6t$, and $\theta = 3\pi t$

30. $\dfrac{dz}{dt}$, where $z = ab - cd$, $a = 3t + 7$, $b = t^2 + 1$, $c = 4\cosh t$, and $d = -\sinh 2t$

31–50. Find $\dfrac{dy}{dx}$, given that y is implicitly defined as a function of x.

Exercises 31–50 are identical to exercises 11–30 of Section 2.6 of the single-variable calculus textbook *Calculus Set Free*.

31. $7x^2 - 5y^3 = y + 53$

32. $7xy = y^3 + 22$

33. $4xy - y^3 = 7x$

34. $2x^2 + 3y^2 = 5x + 9y$

35. $\dfrac{7x}{y} = y^2 + 27$

36. $5x^3 y^2 = 4x - 17y$

37. $4y^3 + 2x^2 + y = 1$

38. $x + y^2 - x^3 = y^4$

39. $y + 1 = 5x - xy^2$

40. $\dfrac{x-1}{y^2} = 4 - x$

41. $(2y - 7)^3 = y - x^5$

42. $x^2 y - 5x = y^3 + \sqrt{7}$

43. $3xy = y^2 - 4x + \sqrt[3]{5}$

44. $\tan y = y + \sec x$

45. $\cos y + x^2 = x$

46. $4 + y^{4/3} = 5x - y^3$

47. $11 + \sqrt{y - 5} = x^2 + y^2$

48. $(6y - 19)^4 + y^2 = 7x$

49. $x\sin y = y\sin x$

50. $(5x - 2)(3y - 8) = y^2$

51. Given that z is defined implicitly as a function of x and y by $x^2 y^2 + x^2 z^2 + y^2 z^2 = 5$, find $\dfrac{\partial z}{\partial y}$.

Exercises 51–58 are repeated from Section 3.5 exercises 45–52.

52. Given that z is defined implicitly as a function of x and y by $z^3 + y\cos z - x^2 y = 1$, find $\dfrac{\partial z}{\partial x}$.

53. Given that z is defined implicitly as a function of x and y by $x^6 + y^{17} + z^4 - 5xyz + 4e^y = 0$, find $\dfrac{\partial z}{\partial x}$.

54. Given that z is defined implicitly as a function of x and y by $e^{xy} + e^{yz} = x + 1$, find $\dfrac{\partial z}{\partial x}$.

55. Given that z is defined implicitly as a function of x and y by $\dfrac{4 - \sin z}{3 + \cos x} + \ln(y^2 + 1) = y$, find $\dfrac{\partial z}{\partial y}$.

56. Given that z is defined implicitly as a function of x and y by $x\sqrt{y^2 + z} = z^2 + y$, find $\dfrac{\partial z}{\partial y}$.

57. Given that s is defined implicitly as a function of t and r by $\frac{s^2 - 2r}{t+s} = 1$, find $\frac{\partial s}{\partial t}$.

58. Given that s is defined implicitly as a function of t and θ by $s^2 + s\cos t\theta - \sin t\theta = t^2$, find $\frac{\partial s}{\partial \theta}$.

59. Find the slope of the tangent line to the curve $xy^3 = x^2 - y$ at the point $(2, 1)$.

60. Find the slope of the tangent line to the curve $y^2 - \sqrt{x} = 4y - 4x$ at the point $(1, 3)$.

61. Find the equation of the tangent line to the curve $4xy - x^2 - y^2 = -13$ at the point $(-1, 2)$.

62. Find the equation of the tangent line to the curve $\sin x \cos y = y - \pi$ at the point $(0, \pi)$.

63. Finish the proof of theorem 5 by proving the formula for $\frac{\partial z}{\partial \theta}$.

64. Suppose that w is a differentiable function of three variables x, y, and z, and that each of x, y, and z are differentiable functions of t. Use the following steps to derive a chain rule formula for determining $\frac{dw}{dt}$.

    (a) Using the two-variable local linearity formula, by analogy write a three-variable local linearity formula.

    (b) Using the calculation in the proof of theorem 4 as a guide, use single-variable local linearity and the result of part (a) to calculate $\frac{dw}{dt}$.

    (c) Compare to exercise 9. Which derivation, that of exercise 9 or this exercise part (b), best constitutes a proof of the formula?

65. Compare the work required for exercises 17–22 to the work required for the corresponding exercises 1–6. Also comment on the form of the answers given by each approach.

3.8   The Gradient and Directional Derivatives

Slopes in the x- and y-directions are discussed in Section 3.5. In this section, we learn about slopes in other directions.

3.8.1   The gradient

The following definition will be helpful in our discussion of slopes.

Definition 7 THE GRADIENT *The gradient of $z = f(x, y)$ is given by*

$$\nabla f = \nabla f(x, y) = \left\langle \frac{\partial f}{\partial x}, \frac{\partial f}{\partial y} \right\rangle = f_x(x, y)\mathbf{i} + f_y(x, y)\mathbf{j}.$$

The notation ∇f is read "del f" or "grad f" or "gradient f."

Example 49 *Find ∇f for $f(x, y) = x^2 - 3xy^2 + y$.*

Solution The partial derivatives of f are

$$\frac{\partial f}{\partial x} = 2x - 3y^2 \quad \text{and} \quad \frac{\partial f}{\partial y} = -6xy + 1.$$

Therefore, the gradient is

$$\nabla f(x, y) = \langle 2x - 3y^2, \, -6xy + 1 \rangle. \qquad \blacksquare$$

The gradient of a function of two variables is itself a function of two variables, and it can be evaluated in the usual manner.

Example 50 *Find $\nabla f(2, 4)$ for $f(x, y) = x \ln y$.*

Solution We first find $\nabla f(x, y)$. The partial derivatives of f are

$$f_x(x, y) = \ln y \quad \text{and} \quad f_y(x, y) = \frac{x}{y}.$$

Therefore, the gradient is

$$\nabla f(x, y) = \left\langle \ln y, \, \frac{x}{y} \right\rangle.$$

Finally, we evaluate the gradient at $(2, 4)$ to find the requested quantity, which is a vector:

$$\nabla f(2, 4) = \left\langle \ln 4, \, \frac{2}{4} \right\rangle = \left\langle \ln 4, \, \frac{1}{2} \right\rangle. \qquad \blacksquare$$

Notice that because the gradient is a vector, it has both direction and magnitude.

Reading Exercise 20 Find ∇f for $f(x, y) = x^2 - x \sin y$.

3.8.2 Directional derivatives

Consider a surface $z = f(x, y)$ and a point on that surface, $(a, b, f(a, b))$, where f is differentiable, as in Figure 3.91. If we are standing at the point and begin to travel in the positive x-direction, the slope we experience, which is the slope of the tangent line through the point in the positive x-direction, is given by $f_x(a, b)$. Similarly, if we are standing at the point and begin instead to travel in the positive y-direction, the slope we experience, the slope of the tangent line in that direction, is given

The nomenclature "del" is short for "delta," and is pronounced the same as the first syllable in "delta." The first syllable of "gradient" sounds like the word "grade," but "grad" is pronounced the same as the first syllable of "graduate."

The phrase "tangent line in the x-direction" does not mean that the tangent line and a vector in the positive x-direction are parallel. A vector in the positive x-direction does not change z-coordinates, whereas the tangent line in the positive x-direction may change z-coordinates.

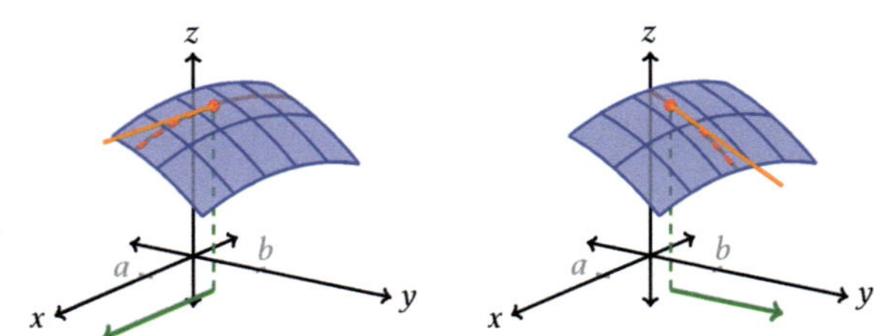

Figure 3.91 *A surface $z = f(x, y)$ (blue), a point $(a, b, f(a, b))$ on the surface (red), and (left) a vector in the positive x-direction (green) along with the tangent line (orange) in the x-direction, and (right) a vector in the positive y-direction (green) along with the tangent line (orange) in the y-direction*

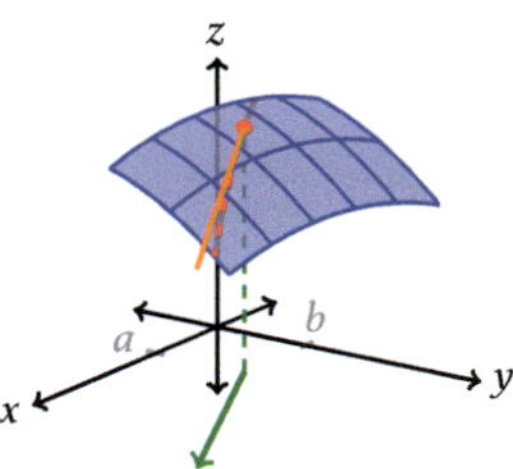

Figure 3.92 *A surface $z = f(x, y)$ (blue), a point $(a, b, f(a, b))$ on the surface (red), the vector $\langle c, d \rangle = \langle 2, 1 \rangle$ (green), and the tangent line (orange) in the direction of the vector*

The vector $\langle c\omega, d\omega \rangle = \omega \langle c, d \rangle$ is a scalar multiple of the vector $\langle c, d \rangle$, so the two vectors have the same direction.

by $f_y(a, b)$. But experience tells us that if we are standing on a hillside, we are not naturally restricted to traveling only north, south, east, or west. We can strike out in any direction we choose. What if we desire to travel in the direction of the vector $\langle 2, 1 \rangle$, as in Figure 3.92? How would we determine the slope of the tangent line in that direction? The principle is the same as always when finding the slope—we move an infinitesimal amount in the chosen direction and calculate the slope using the slope formula.

To that end, suppose that f is differentiable at (a, b) and we wish to move in the direction of the vector $\langle c, d \rangle$. Let ω be a positive infinitesimal. Then, we can move by $\langle c\omega, d\omega \rangle$ from the input (a, b) to the input $(a + c\omega, b + d\omega)$, which is a "run" of length

$$\|\langle c\omega, d\omega \rangle\| = \sqrt{c^2\omega^2 + d^2\omega^2} = \omega\sqrt{c^2 + d^2} = \|\langle c, d \rangle\|\omega.$$

The "rise" is calculated by the difference in z-coordinates, which are values of the function. Then, the slope is

$$\begin{aligned}
\text{slope} &= \frac{f(a + c\omega,\, b + d\omega) - f(a, b)}{\|\langle c, d \rangle\|\omega} \\[1.2em]
&\approx \frac{f(a, b) + c\omega f_x(a, b) + d\omega f_y(a, b) - f(a, b)}{\|\langle c, d \rangle\|\omega} \\[1.2em]
&= \frac{c\omega f_x(a, b) + d\omega f_y(a, b)}{\|\langle c, d \rangle\|\omega} \\[1.2em]
&= \frac{c f_x(a, b) + d f_y(a, b)}{\|\langle c, d \rangle\|} \\[1.2em]
&= \frac{\langle f_x(a, b), f_y(a, b) \rangle \cdot \langle c, d \rangle}{\|\langle c, d \rangle\|} \\[1.2em]
&= \frac{\nabla f(a, b) \cdot \langle c, d \rangle}{\|\langle c, d \rangle\|}.
\end{aligned}$$

Line 1 uses the slope formula (the "rise" is the numerator, which is the difference in z-coordinates at the inputs $(a + c\omega, b + d\omega)$ and (a, b)); line 2 uses the two-variable local linearity formula with real numbers a and b and infinitesimals $c\omega$ and $d\omega$; line 3 simplifies the numerator; line 4 factors out the ω in the numerator and cancels with the ω in the denominator; line 5 rewrites the numerator as a dot product of two vectors; and line 6 rewrites using the definition of gradient.

Even better, if the direction we wish to move is expressed as a unit vector $\mathbf{u} = \langle c, d \rangle$, then the slope is given by

$$\nabla f(a, b) \cdot \mathbf{u}.$$

If you wondered in Section 3.4 why we only looked at derivatives in the x- and y-directions rather than find a more general derivative in any direction, this formula provides the answer. We certainly can look at derivatives in any direction, but they can be calculated using the partial derivatives f_x and f_y, which in turn can be calculated using derivative rules.

Definition 8 DIRECTIONAL DERIVATIVE *Let* $\mathbf{u}$ *be a unit vector and let* $z = f(x, y)$ *be differentiable at* (a, b). *Then the* directional derivative *of* f *at* (a, b) *in direction* $\mathbf{u}$ *is*

$$D_{\mathbf{u}} f(a, b) = \nabla f(a, b) \cdot \mathbf{u}.$$

Example 51 *Find the directional derivative of* $f(x, y) = x^2 y$ *at* $(2, 5)$ *in the direction* $\mathbf{u} = \left\langle \frac{3}{5}, \frac{4}{5} \right\rangle$.

Ans. to reading exercise 20:
$$\nabla f(x, y) = \langle 2x - \sin y, -x \cos y \rangle$$

Solution Notice that $\mathbf{u} = \left\langle \frac{3}{5}, \frac{4}{5} \right\rangle$ is a unit vector, so definition 8 applies. We first calculate the gradient:

If the given direction vector is not a unit vector, normalize it first (divide by its norm).

$$\nabla f(x, y) = \langle f_x(a, b), f_y(a, b) \rangle = \langle 2xy, x^2 \rangle.$$

Next, we evaluate the gradient at the desired input $(2, 5)$:

$$\nabla f(2, 5) = \langle 2(2)(5), 2^2 \rangle = \langle 20, 4 \rangle.$$

Then using the definition of directional derivative, we have

$$D_{\mathbf{u}} f(2, 5) = \nabla f(2, 5) \cdot \mathbf{u} = \langle 20, 4 \rangle \cdot \left\langle \frac{3}{5}, \frac{4}{5} \right\rangle = \frac{60}{5} + \frac{16}{5} = \frac{76}{5} = 15.2.$$

■

One interpretation of the result of example 51 is that the slope of the surface $f(x, y) = x^2 y$ at the input $(2, 5)$ in the direction $\langle 0.6, 0.8 \rangle$ is 15.2.

3.8.3 Directional derivative examples

Example 52 *Find the slope of the surface* $f(x, y) = \ln x - y^2$ *at* $(1, 2)$ *in the direction* $\mathbf{u} = \left\langle \frac{5}{13}, \frac{12}{13} \right\rangle$.

Solution The slope in the direction $\mathbf{u} = \left\langle \frac{5}{13}, \frac{12}{13} \right\rangle$ is a directional derivative, which requires the calculation of the gradient. But first we verify that $\mathbf{u}$ is a unit vector:

$$\left\| \left\langle \frac{5}{13}, \frac{12}{13} \right\rangle \right\| = \sqrt{\frac{25}{169} + \frac{144}{169}} = 1.$$

The gradient is

$$\nabla f(x,y) = \left\langle \frac{1}{x}, -2y \right\rangle.$$

Next, we evaluate at the given point (input):

$$\nabla f(1,2) = \langle 1, -4 \rangle.$$

Finally, we use the directional derivative formula:

$$D_{\mathbf{u}} f(1,2) = \langle 1, -4 \rangle \cdot \left\langle \frac{5}{13}, \frac{12}{13} \right\rangle = \frac{5}{13} - \frac{48}{13} = -\frac{43}{13}. \qquad \blacksquare$$

Reading Exercise 21   Find the slope of the surface $f(x,y) = x^2 - y^3$ at $(1,1)$ in the direction $\left\langle \frac{3}{5}, -\frac{4}{5} \right\rangle$.

Example 53 *Find the rate of change of $z = \frac{y}{x}$ in the direction $\mathbf{u} = \left\langle -\frac{1}{\sqrt{2}}, \frac{1}{\sqrt{2}} \right\rangle$ when $x = 1, y = 2$.*

Solution   The rate of change is the value of the derivative, and the direction $\left\langle -\frac{1}{\sqrt{2}}, \frac{1}{\sqrt{2}} \right\rangle$ is given, so we use a directional derivative. The vector $\mathbf{u}$ is easily verified to be a unit vector.

Rewriting z as $z = yx^{-1}$ can simplify the calculation of $\frac{\partial z}{\partial x}$.

The gradient is

$$\nabla f(x,y) = \left\langle \frac{-y}{x^2}, \frac{1}{x} \right\rangle.$$

The desired input is $(1,2)$, and

$$\nabla f(1,2) = \langle -2, 1 \rangle.$$

Then

$$D_{\mathbf{u}} f(1,2) = \langle -2, 1 \rangle \cdot \left\langle -\frac{1}{\sqrt{2}}, \frac{1}{\sqrt{2}} \right\rangle = \frac{2}{\sqrt{2}} + \frac{1}{\sqrt{2}} = \frac{3}{\sqrt{2}}.$$

The rate of change is $\frac{3}{\sqrt{2}}$. $\qquad \blacksquare$

3.8.4   Gradients and directional derivatives for $f(x,y,z)$

For a function of three variables, it turns out that the gradient and directional derivatives merely need the third coordinate; the same ideas as for functions of two variables apply.

GRADIENT AND DIRECTIONAL DERIVATIVE, THREE-VARIABLE FUNCTIONS

If $w = f(x, y, z)$ is differentiable at (x_0, y_0, z_0) and $\mathbf{u}$ is a unit vector in $\mathbf{R}^3$, then

$$\nabla f(x, y, z) = \left\langle \frac{\partial f}{\partial x}, \frac{\partial f}{\partial y}, \frac{\partial f}{\partial z} \right\rangle$$

$$D_\mathbf{u} f(x_0, y_0, z_0) = \nabla f(x_0, y_0, z_0) \cdot \mathbf{u}.$$

Example 54 *Find the directional derivative of $f(x, y, z) = xe^y - zx^2$ at $(1, 0, 2)$ in the direction $\mathbf{u} = \left\langle \frac{1}{3}, \frac{2}{3}, -\frac{2}{3} \right\rangle$.*

Solution A directional derivative is requested, so we first verify that $\mathbf{u}$ is a unit vector:

$$\left\| \left\langle \tfrac{1}{3}, \tfrac{2}{3}, -\tfrac{2}{3} \right\rangle \right\| = \sqrt{\tfrac{1}{9} + \tfrac{4}{9} + \tfrac{4}{9}} = 1.$$

The gradient is

$$\nabla f(x, y, z) = \left\langle \frac{\partial f}{\partial x}, \frac{\partial f}{\partial y}, \frac{\partial f}{\partial z} \right\rangle = \left\langle e^y - 2xz, \ xe^y, \ -x^2 \right\rangle.$$

Evaluating the gradient at the desired input gives

$$\nabla f(1, 0, 2) = \left\langle e^0 - 2(1)(2), \ 1 \cdot e^0, \ -1^2 \right\rangle = \langle -3, 1, -1 \rangle.$$

Finally,

$$D_\mathbf{u} f(1, 0, 2) = \langle -3, 1, -1 \rangle \cdot \left\langle \tfrac{1}{3}, \tfrac{2}{3}, -\tfrac{2}{3} \right\rangle = -\frac{3}{3} + \frac{2}{3} + \frac{2}{3} = \frac{1}{3}. \qquad \blacksquare$$

Ans. to reading exercise 21:
$$\frac{18}{5}$$

Reading Exercise 22 Find ∇f for $f(x, y, z) = x^2 + yz$.

3.8.5 Fastest change

Suppose $z = f(x, y)$ is differentiable at (a, b). We can calculate the directional derivative, which represents the slope of the surface or the rate of change of the function, in any direction we wish. A natural question is in which direction is the slope the steepest, that is, in which direction is the rate of change the largest?

We wish to find the maximum value of the directional derivative, that is, the largest value of

$$D_\mathbf{u} f(a, b) = \nabla f(a, b) \cdot \mathbf{u},$$

where **u** is a unit vector. The point (a, b) is given, so the value of $\nabla f(a, b)$ is fixed; only **u**, the direction, is changing. The question, then, is which value of **u**, which direction, makes the directional derivative at (a, b) the largest. The key to answering this question is to apply one of the angle between vectors formulas, namely $\|\mathbf{v}\|\|\mathbf{w}\|\cos\theta = \mathbf{v}\cdot\mathbf{w}$, where θ is the angle between **v** and **w**. We have

Line 1 uses the definition of directional derivative; line 2 uses the angle between vectors formula for vectors **u** and $\nabla f(a, b)$; line 3 uses the fact that **u** is a unit vector, that is, $\|\mathbf{u}\| = 1$.

$$D_{\mathbf{u}} f(a, b) = \nabla f(a, b) \cdot \mathbf{u}$$

$$= \|\nabla f(a, b)\| \, \|\mathbf{u}\| \cos\theta$$

$$= \|\nabla f(a, b)\| \cos\theta,$$

where θ is the angle between **u** and $\nabla f(a, b)$. Again, because the point (a, b) is fixed, the value of $\|\nabla f(a, b)\|$ does not change; the quantity is also nonnegative because it's a norm. Then the largest the directional derivative can be must occur when $\cos\theta$ is largest. That, of course, occurs when $\theta = 0°$, where $\cos\theta = 1$. If $\theta = 0°$, then the angle between the direction vector **u** and the gradient vector $\nabla f(a, b)$ is $0°$, meaning that the two vectors have the same direction. Therefore, the direction of the largest rate of change, the largest slope of the surface, is the direction of the gradient! And that maximum value is

$$\|\nabla f(a, b)\| \cos 0° = \|\nabla f(a, b)\|,$$

the magnitude of the gradient.

What about the minimum value of the directional derivative? That must occur when $\cos\theta = -1$, that is, where $\theta = 180°$. In that case, the vectors **u** and $\nabla f(a, b)$ are pointing in opposite directions. The direction of fastest decrease of the function (the direction of the minimum value of the slope) is the direction of $-\nabla f(a, b)$, and the magnitude of the minimum slope is $-\|\nabla f(a, b)\|$.

FASTEST INCREASE AND DECREASE

Let $z = f(x, y)$ be differentiable at (a, b). Then the direction of fastest increase of f at (a, b) is the direction of $\nabla f(a, b)$, and the direction of fastest decrease of f at (a, b) is the direction of $-\nabla f(a, b)$. Furthermore, the magnitude of fastest increase of f at (a, b) is $\|\nabla f(a, b)\|$, and the magnitude of fastest decrease of f at (a, b) is $-\|\nabla f(a, b)\|$.

Example 55 *Find the maximum rate of change of $f(x, y) = 2xy^2$ at $(1, 3)$ and the direction in which it occurs.*

Ans. to reading exercise 22:
$\nabla f(x, y, z) = \langle 2x, z, y \rangle$

Solution   We need the direction and magnitude of the gradient vector at the input $(1, 3)$. The gradient is

$$\nabla f(x, y) = \langle 2y^2, \, 4xy \rangle.$$

Next, we evaluate at the desired input:

$$\nabla f(1,3) = \langle 18, 12 \rangle.$$

Then, the maximum rate of change is

$$\|\nabla f(1,3)\| = \sqrt{18^2 + 12^2} = \sqrt{468} = 6\sqrt{13} \approx 21.63,$$

and the maximum change is in the direction of $\langle 18, 12 \rangle$. If desired, we can write the direction as a unit vector:

$$\frac{1}{6\sqrt{13}}\langle 18, 12 \rangle = \left\langle \frac{18}{6\sqrt{13}}, \frac{12}{6\sqrt{13}} \right\rangle = \left\langle \frac{3}{\sqrt{13}}, \frac{2}{\sqrt{13}} \right\rangle. \qquad \blacksquare$$

3.8.6 Fastest change examples

Example 56 *Find the maximum rate of change of $f(x,y) = \ln(x+y)$ at $(1,0)$.*

Solution Because we want the maximum rate of change, we want the magnitude of the gradient vector at $(1,0)$. The gradient is

$$\nabla f(x,y) = \left\langle \frac{1}{x+y} \cdot 1, \frac{1}{x+y} \cdot 1 \right\rangle = \left\langle \frac{1}{x+y}, \frac{1}{x+y} \right\rangle.$$

Then

$$\nabla f(1,0) = \left\langle \frac{1}{1+0}, \frac{1}{1+0} \right\rangle = \langle 1, 1 \rangle$$

and

$$\|\nabla f(1,0)\| = \sqrt{1^2 + 1^2} = \sqrt{2}.$$

The maximum rate of change of f at $(1,0)$ is $\sqrt{2}$. (The direction of the fastest change was not requested.) $\qquad \blacksquare$

Reading Exercise 23 Find the direction of fastest change of $f(x,y) = x^2 - y^3$ at $(1,1)$.

Example 57 *You are standing at the point $(5,10,-1.5)$ on the surface $z = 3 + 0.02x^2 - 0.05y^2$. In what direction should you begin walking to go downhill most quickly?*

Solution To go downhill most quickly corresponds to the fastest decrease, so we wish to know $-\nabla f(5,10)$. The gradient is

$$\nabla f(x,y) = \langle 0.04x, -0.10y \rangle.$$

Then

$$\nabla f(5, 10) = \langle 0.2, -1 \rangle,$$

and the direction of fastest decrease is the direction of

$$-\langle 0.2, -1 \rangle = \langle -0.2, 1 \rangle.$$

An alternate answer would be the direction of the unit vector $\left\langle \frac{-0.2}{\sqrt{1.04}}, \frac{1}{\sqrt{1.04}} \right\rangle$.

To go downhill most quickly we begin walking in the direction of $\langle -0.2, 1 \rangle$.

∎

3.8.7   Gradient orthogonal to level curves

Along a level curve, the slope is zero; the "elevation" is not changing. So in the direction $\mathbf{u}$ of a level curve, the directional derivative $D_{\mathbf{u}} f$ is zero, that is, $\nabla f \cdot \mathbf{u} = 0$, meaning that the gradient and a vector in the direction of the level curve are orthogonal.

THE GRADIENT AND LEVEL CURVES

Where f is differentiable, the gradient is orthogonal to a level curve.

In a practical sense, this says that if you are walking a flat path on a hillside, then the steepest slope (fastest uphill, quickest downhill) is at a right angle to the path. We can also locate on a topographical map the direction of the gradient vector. We simply need to draw a vector orthogonal to a level curve through the point, in the uphill direction.

Example 58 *Two dots appear in the topographical map (Figure 3.93). Draw a vector in the direction of the gradient at each dot.*

Solution We draw vectors orthogonal to the level curves at each dot, in the direction of higher elevation. The length of the vectors is unspecified, but we have chosen to keep them from crossing the next level curve on the map to avoid confusion. See Figure 3.94.

∎

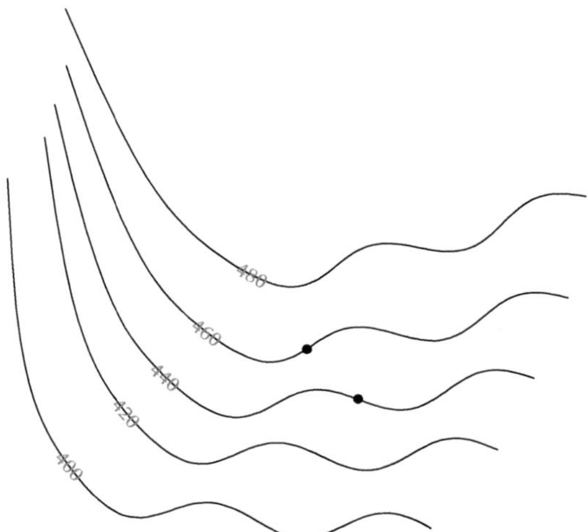

Figure 3.93 *The topographical map for example 58*

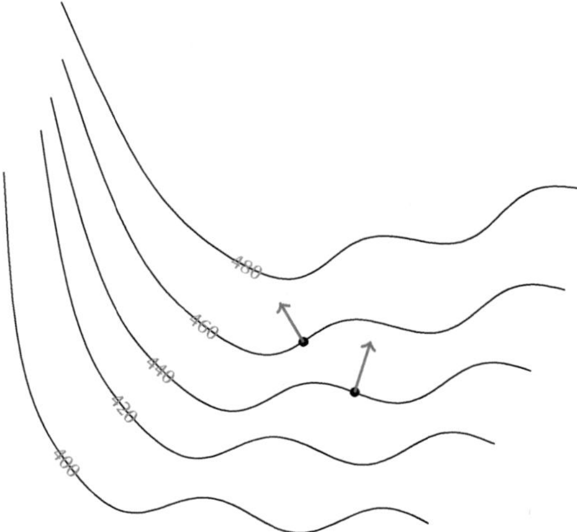

Figure 3.94 *The topographical map for the solution to example 58, including a vector in the direction of the gradient (green) at each dot on the map*

EXERCISES 3.8

1–8. Find ∇f.

1. $f(x, y) = 4x^2 y - 7xy^2$
2. $f(x, y) = x^3 - 5y$
3. $f(x, y) = x^y + \ln x^y$
4. $f(x, y, z) = xyz$
5. $f(x, y, z) = 6x \cosh y + \tan^{-1} z^2$
6. $f(u, v) = u^3 - (v^2 + 8)e^u$
7. $f(r, \theta) = 3r^2 \cos \theta - \tan \theta \sec \theta$
8. $f(x, y) = \dfrac{\sqrt{e^x + e^y}}{e^{xy}}$

9–16. Find the value of the gradient at the given input.

9. $f(x, y) = 3x^2 - 5y^2$, at $(1, 5)$
10. $f(x, y) = \ln(5x)$, at $\left(\frac{1}{5}, 14 \right)$
11. $f(x, y) = 6x - 87y$, at $(-4.7, 3.3)$
12. $f(x, y) = \dfrac{3x + 2y}{5x - 4y}$, at $(1, 1)$
13. $f(x, y, z) = \cos x + \sin(y + z)$, at $(0, \pi, -\pi)$
14. $f(x, y) = \dfrac{y}{x + 1}$, find $\nabla f(3, 7)$
15. $g(x, y) = \sin^{-1} xy$, find $\nabla g\left(1, \frac{1}{2}\right)$
16. $x(u, v) = u^3 v^5$, at $(1, 0)$

17–24. Find the directional derivative of the function at the given point in the specified direction.

17. $f(x, y) = 3 - x^2 y^3$, $(a, b) = (2, -1)$, $\mathbf{u} = \left\langle \frac{3}{5}, \frac{4}{5} \right\rangle$
18. $f(x, y) = \sqrt{2x - 4y}$, $(a, b) = (10, 1)$, $\mathbf{u} = \left\langle -\frac{1}{\sqrt{2}}, \frac{1}{\sqrt{2}} \right\rangle$
19. $D_{\mathbf{u}} f(-3, 1)$, where $f(x, y) = \cos \pi x + \sin \pi y$, $\mathbf{u} = \langle 0, -1 \rangle$
20. $\mathbf{u} = \left\langle \frac{2}{\sqrt{29}}, \frac{5}{\sqrt{29}} \right\rangle$, $f(x, y) = 5x - y^2$, find $D_{\mathbf{u}}(3, 1)$
21. $\mathbf{u} = \left\langle -\frac{3}{7}, \frac{6}{7}, -\frac{2}{7} \right\rangle$, $f(x, y, z) = 3yz - xy^2$, find $D_{\mathbf{u}} f(7, 1, 0)$
22. $f(x, y, z) = \dfrac{x^2}{y - 5z}$, $\mathbf{u} = \left\langle \frac{3}{4}, \frac{1}{2}, \frac{\sqrt{3}}{4} \right\rangle$, find $D_{\mathbf{u}} f(2, 1, -1)$
23. $f(x, y) = (3x^2 y - 1)^2$, at $(2, 1)$, in the direction of $\langle 6, 5 \rangle$
24. $f(x, y) = (x^2 + 7y) \ln x$, at $(2, 0)$, in the direction of $\langle 4, 4 \rangle$

25. Find the rate of change of $f(x, y) = 2x\sqrt{y - 1}$ at $(3, 5)$ in the direction $\mathbf{u} = \left\langle \frac{-5}{13}, \frac{12}{13} \right\rangle$.

26. Find the slope of the surface $f(x, y) = e^{x-y}$ at $(1, 1)$ in the direction
    $$\mathbf{u} = \left\langle \frac{1}{5}, \frac{\sqrt{24}}{5} \right\rangle.$$

27. Find the slope of the surface $f(x, y) = (x - y)^2$ at $(0, 3)$ in the direction of
    (a) $\left\langle \frac{3}{5}, \frac{4}{5} \right\rangle$, (b) $\left\langle -\frac{3}{5}, \frac{4}{5} \right\rangle$, (c) $\left\langle \frac{3}{5}, -\frac{4}{5} \right\rangle$, and (d) $\left\langle -\frac{3}{5}, -\frac{4}{5} \right\rangle$.

28. Find the rate of change of $z = x - 3y^2 + 71$ at $(2, 1)$ in the direction of
    (a) $\left\langle \frac{1}{\sqrt{2}}, \frac{1}{\sqrt{2}} \right\rangle$; (b) $\langle 1, 0 \rangle$; (c) $\langle 0, 1 \rangle$; and (d) $\langle 0, -1 \rangle$.

29. Find the direction of the steepest slope on the surface $f(x, y) = x^2 + y^2$ at $(2, -5)$.

30. Find the direction of the steepest slope on the surface $f(x, y) = \sqrt{x + y^2}$ at $(1, -1)$.

31. Find the greatest rate of change of $f(x, y) = x \sin y$ at $\left(3, \frac{\pi}{2}\right)$.

32. Find the greatest rate of change of $f(x, y) = y \ln(2 - x)$ at $(1, 7)$.

33. Find the direction of fastest decrease of $f(x, y) = \dfrac{2x}{y + 5x}$ at $(1, 0)$.

34. Find the direction of fastest decrease of $f(x, y) = x^3 - 4y^2$ at $(-1, 1)$.

35. Find the maximum rate of change of $f(x, y) = x^3 + 0.5y^2$ at $(2, 1)$ and the direction in which it occurs.

36. Find the minimum rate of change of $f(x, y) = 20xye^{-x^2-y^2}$ at $(1, 0)$ and the direction in which it occurs.

37. Find the minimum rate of change of $f(x, y) = \ln |x^3 + y^2|$ at $(2, -3)$ and the direction in which it occurs.

38. Find the maximum rate of change of $f(x, y) = 0.1\sqrt{x^2 + y}$ at $(0, 3)$ and the direction in which it occurs.

39. You are hiking along ground described by the surface $z = 0.02x^2 - 0.05y^2 + 905$. You are currently standing on the point $(10, 10, 902)$ and decide to move northwest. Will you start moving uphill, downhill, or level?

40. A spring bubbles up on the surface $z = \sqrt{x} - 5xy$ at $(4, 2)$. In what direction will the water begin to flow?

41. At each dot in the topographical map, draw a vector in the direction of the gradient.

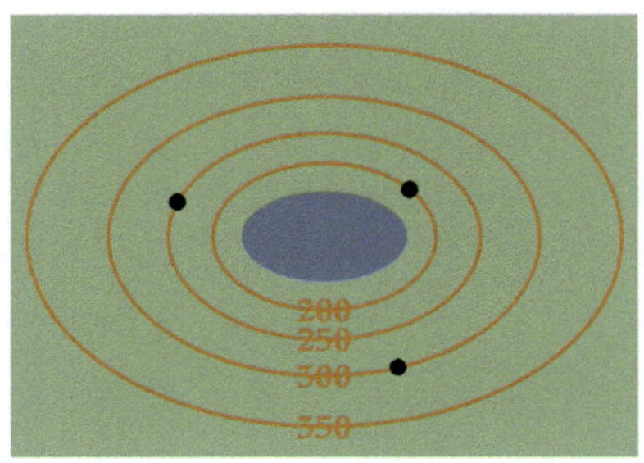

42. At each dot in the topographical map, draw a vector in the direction of the gradient.

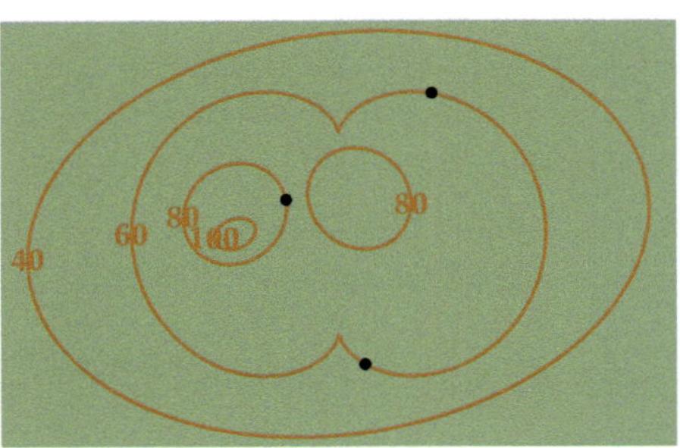

43. (a) Use the calculation presented prior to definition 8 to write a formula for $D_\mathbf{v}f(a, b)$ where $\mathbf{v}$ is not a unit vector.
   (b) Use the result of part (a) to find the directional derivative of $f(x, y) = x^2 y$ at $(2, 5)$ in the direction $\langle 3, 4 \rangle$.
   (c) Compare the calculation in part (b) to the solution to example 51. Is there any real difference between (1) ensuring that the direction vector is a unit vector for the purpose of using definition 8 and (2) using the formula from part (a)?

44. Suppose $z = f(x, y)$ is differentiable at (a, b). Show that the vector $\langle -f_y(a, b), f_x(a, b) \rangle$ points along a level curve.

Exercises 45–50 were suggested by Matthew Dawson of SECIHTI/CIMAT Unidad Mérida.

45–48. The directional derivative $D_\mathbf{u}f(a, b) = \nabla f(a, b) \cdot \mathbf{u}$ is the slope of the surface f at (a, b) in the direction of the unit vector $\mathbf{u}$. Suppose instead that we are traveling along the surface f, that at the point $(a, b, f(a, b))$ our velocity vector is $\mathbf{v}$, and that we wish to know the rate of change in elevation (z-coordinates) with respect to time. If the vector $\mathbf{v}$ happens to be a unit vector, then the rate of change in elevation is identical to the slope of the surface. If not, then the rate of change in elevation is the slope times the speed of travel, which is the norm of the velocity vector, $\|\mathbf{v}\|$. Instead of taking the dot product of the gradient with the unit vector $\dfrac{\mathbf{v}}{\|\mathbf{v}\|}$ to take into account only the direction, multiplying by the speed means that we use $\dfrac{\mathbf{v}}{\|\mathbf{v}\|} \cdot \|\mathbf{v}\| = \mathbf{v}$ instead; in other words, the rate of change of elevation is $\nabla f(a, b) \cdot \mathbf{v}$, the same formula as the directional derivative, but without the stipulation that it be a unit vector. We call this the *total derivative* of f at (a, b) along the vector $\mathbf{v}$ and write it as

$$(f'(a, b))(\mathbf{v}) = \nabla f(a, b) \cdot \mathbf{v}.$$

In this notation for the total derivative, $f'(a, b)$ is the name of the function and $\mathbf{v} = \langle x, y \rangle$ is the input to the function. The total derivative $f'(a, b)$ is therefore a function of two variables (x and y, the components of the vector $\mathbf{v}$). As long as f is differentiable at (a, b), the domain of the total derivative function $f'(a, b)$ is all of $\mathbf{R}^2$, because we can use any velocity vector $\mathbf{v}$ as input. Analogous formulas are valid in three or even more dimensions.

For each surface, point, and vector, calculate the total derivative $(f'(a, b))(\mathbf{v})$.

45. $f(x, y) = 3x^2 + 4y^3, (a, b) = (1, 2), \mathbf{v} = \langle 3, -1 \rangle$
46. $f(x, y) = 3x^2 + 4y^3, (a, b) = (1, 2), \mathbf{v} = \langle 6, -2 \rangle$
47. $f(x, y) = 3x^2 + 4y^3, (a, b) = (4, 0), \mathbf{v} = \langle 3, -1 \rangle$
48. $f(x, y) = x + \ln xy, (a, b) = (1, 2), \mathbf{v} = \langle 3, -1 \rangle$

49. Use properties of the dot product to show that

$$(f'(a, b))(\mathbf{v} + \mathbf{w}) = (f'(a, b))(\mathbf{v}) + (f'(a, b))(\mathbf{w}).$$

50. Use properties of the dot product to show that for any constant c,

$$(f'(a, b))(c\mathbf{v}) = c\left((f'(a, b))(\mathbf{v})\right).$$

In exercise 48, the point $(a, b) = (1, 2)$ is in the domain of f. The vector $\mathbf{v} = \langle 3, -1 \rangle$ is in the domain of $f'(a, b)$. The fact that the point $(3, -1)$ is not in the domain of f is irrelevant.

Together, exercises 49 and 50 show that the total derivative of f at (a, b) is linear.

3.9   Extreme Values and Saddle Points

You may recall the procedures for finding local extrema (maxima, minima) for functions of one variable. Critical numbers are found first, and then either the first derivative test or the second derivative test may be used. Adding a dimension complicates matters somewhat, but the ideas and procedures are similar.

The words "extrema," "maxima," and "minima" are plural; singular forms are "extremum," "maximum," and "minimum."

3.9.1   Local extrema

Many theorems and definitions of single-variable calculus utilize an open interval around a number, such as in Figure 3.95. For a function of two variables x and y, we instead use an *open disk* containing the point (Figure 3.96).

Figure 3.95 *An open interval (green) containing the number a*

Definition 9 OPEN DISK *An* open disk *containing the point (a, b) is a set of points of the form*

$$\{(x, y) \mid (x - a)^2 + (y - b)^2 < r\},$$

where $r > 0$.

A more general concept can be defined, but doing so is not necessary for our purposes.

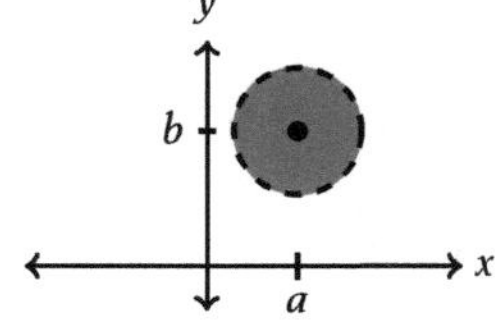

In Figure 3.96, imagine taking a horizontal slice of the disk through the point. The result should resemble Figure 3.95.

Figure 3.96 *An open disk (green) containing the point (a, b)*

The "top of the hill" need not be a single point where all around it is lower elevation. The point can be on a ridge that is level, or even on a mesa. This is the reason for using $\leq$ and $\geq$ instead of $<$ and $>$ in definition 10.

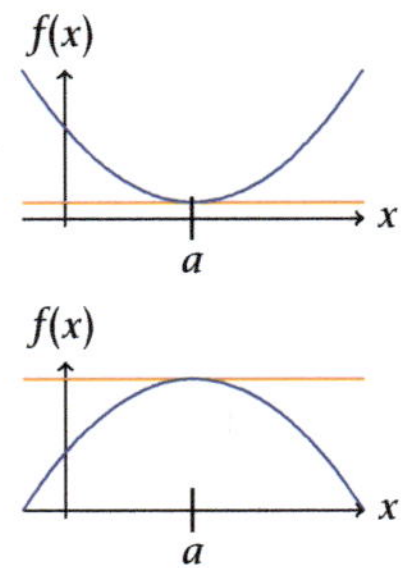

Figure 3.97 *Local extrema where the derivative is zero, indicating a horizontal tangent line*

This is the two-variable version of the single-variable Fermat's theorem (not to be confused with Fermat's last theorem), and the proof is similar.

A critical point for a function of two variables is analogous to a critical number for a function of one variable.

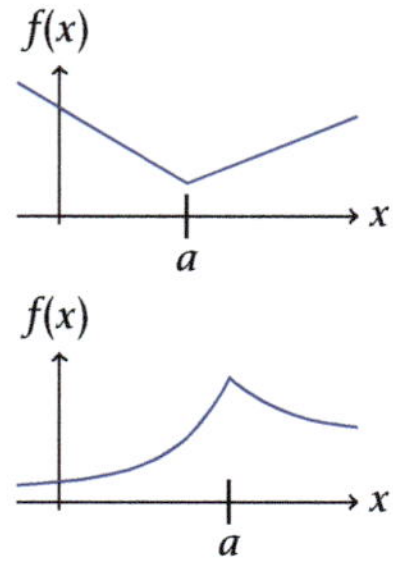

Figure 3.98 *Local extrema where the derivative is undefined and a corner is present*

Imagine yourself standing on the top of a hill. You might see a taller hill in the distance, but you can draw a circle around your location on the map (as long as the circle is not too wide) and know that you are at the highest elevation anywhere inside that circle; you are at a local maximum. Conversely, if you are not at the top of a hill, then no matter how small a circle you draw around yourself, there are higher-elevation points inside that circle. This forms the basis for a definition of local extrema.

Definition 10 LOCAL EXTREMA *A function $z = f(x, y)$ has a* local maximum *at (a, b) if $f(a, b) \geq f(x, y)$ for every point (x, y) in an open disk containing (a, b). A function $z = f(x, y)$ has a* local minimum *at (a, b) if $f(a, b) \leq f(x, y)$ for every point (x, y) in an open disk containing (a, b).*

For a function of one variable, local extrema can only occur where the derivative is zero (Figure 3.97) or undefined (Figure 3.98). There is only one derivative to check, only two sides of the critical number to look.

By contrast, with a function of two variables, there are infinitely many directions to look, infinitely many directional derivatives to check (Figure 3.99). But if the function $z = f(x, y)$ is differentiable at (a, b), the tangent plane must be horizontal at a maximum (or minimum) point; otherwise, there is a direction where one could go uphill to a higher point (or downhill to a lower point). Hence, all the directional derivatives must be zero, forcing the gradient to be zero. The only other possibility is if the function is not differentiable.

Theorem 6 POSSIBLE LOCATIONS OF LOCAL EXTREMA *If $z = f(x, y)$ has a local maximum or local minimum at (a, b) and both $f_x(a, b)$ and $f_y(a, b)$ exist then $f_x(a, b) = f_y(a, b) = 0$, i.e., $\nabla f(a, b) = 0$.*

Definition 11 CRITICAL POINTS *A function of two variables $z = f(x, y)$ has a* critical point *at (a, b) if (a, b) is in the domain of f and either $\nabla f(a, b) = 0$ or $\nabla f(a, b)$ does not exist.*

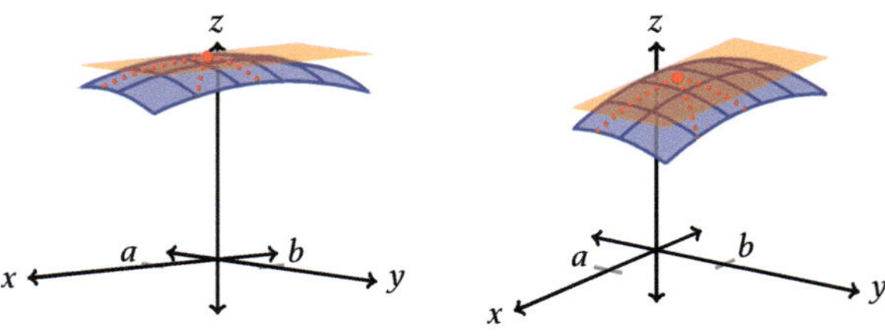

Figure 3.99 *Two views of a local maximum point (red) and a horizontal tangent plane (orange, parallel to the xy-plane) at that point, along with three of the infinitely many directions one can look along the surface from the point (red dotted)*

Rephrasing theorem 6, local extrema may only occur at critical points. The theorem does not guarantee that there is a local extremum at each critical point; rather, it says that critical points are the only places to look.

Finding critical points often requires solving equations with two variables. Any algebraic technique you have learned for solving simultaneous equations with two variables may be used.

Example 59 *Find all critical points of* $f(x, y) = x^3 - xy - x + y + 13$.

Solution We are seeking the points at which the gradient is zero or undefined. The gradient is

$$\nabla f(x, y) = \langle 3x^2 - y - 1, \, -x + 1 \rangle,$$

which is always defined. Setting $\nabla f(x, y) = \mathbf{0}$, we have the simultaneous equations

$$3x^2 - y - 1 = 0 \quad \text{and} \quad -x + 1 = 0.$$

The easiest to solve is $-x + 1 = 0$, which has solution $x = 1$. Substituting $x = 1$ into the other equation, we have

$$3(1^2) - y - 1 = 0$$
$$2 - y = 0$$
$$y = 2.$$

We conclude that there is one critical point, $(1, 2)$. ∎

There can, of course, be more than one critical point.

Example 60 *Find the critical points of* $f(x, y) = x^2 y - 4y$.

Solution The gradient is

$$\langle 2xy, \, x^2 - 4 \rangle,$$

which is defined everywhere. Setting the gradient equal to zero, our equations to solve are

$$2xy = 0 \quad \text{and} \quad x^2 - 4 = 0.$$

Again one of the equations has only one variable, so we solve it first; $x^2 - 4 = 0$ has two solutions, $x = 2$ and $x = -2$. Taking them one at a time, if $x = 2$ we substitute into the other equation to get $2(2)y = 0$, which has solution $y = 0$. Therefore, $(2, 0)$

See example 61 for an example where both equations contain both variables.

There can even be infinitely many critical points, as demonstrated in example 63.

If it is not the case that the gradient is defined for all values of the variables x and y, then we must look for critical points for which f is defined but ∇f is not. As with functions of one variable, root functions are examples of when this can apply.

is a critical point. Using $x = -2$, we have $-4y = 0$ and therefore $y = 0$, and $(-2, 0)$ is a critical point. ∎

Reading Exercise 24 Find the critical points of $f(x, y) = x^3 - 3x + 4y^2$.

3.9.2 Second derivative test

Once we have obtained the critical points, we need to test each critical point to determine extrema.

If f' changes from positive to negative at a critical number a, then there is a local maximum at $x = a$; if f' changes from negative to positive, then there is a local minimum at $x = a$.

For functions of one variable we often use the first derivative test, which consists of using critical numbers as endpoints of intervals, determining the sign of the derivative on each interval, and looking for changes in sign. This works because there are only two directions to look from the critical number a, to the left and to the right (consult Figures 3.97 and 3.98). But for a function of two variables, there are infinitely many different directions to look (as in Figure 3.99, where three of the infinitely many directions are indicated in red), making the equivalent of the first derivative test impractical.

The second derivative test for a function of one variable consists of determining the concavity at the critical point. If f has a critical point at $x = a$ and $f''(a) > 0$, then the function is concave up at $x = a$ and the function has a local minimum at $x = a$, as illustrated in Figure 3.97, top. If instead $f''(a) < 0$, then the function is concave down at the critical point and the function has a local maximum at $x = a$, as illustrated in Figure 3.97, bottom. There is one other possibility, $f''(a) = 0$, for which the test is inconclusive.

Some versions of the second derivative test use even higher-order derivatives to enable a conclusion to be drawn under some circumstances.

For a function of two variables there is not just one second derivative, but rather four, complicating matters a bit. We will discuss the proof of the following theorem at the end of this section.

If the hypotheses of theorem 7 are met, then so are the hypotheses of Clairaut's theorem, meaning that $f_{xy} = f_{yx}$. Therefore, a more symmetric way of writing (and remembering) the formula for D is

$$D = f_{xx}(a, b)f_{yy}(a, b) - f_{xy}(a, b)f_{yx}(a, b).$$

Theorem 7 SECOND DERIVATIVE TEST *If $z = f(x, y)$ and its first and second-order partial derivatives are all continuous on an open disk containing (a, b), $\nabla f(a, b) = 0$, and*

$$D = f_{xx}(a, b)f_{yy}(a, b) - \left(f_{xy}(a, b)\right)^2,$$

then

(1) if $D > 0$ and $f_{xx}(a, b) < 0$, then f has a local maximum at (a, b);
(2) if $D > 0$ and $f_{xx}(a, b) > 0$, then f has a local minimum at (a, b); and
(3) if $D < 0$, then f has a saddle point at (a, b).
If $D = 0$, the test is inconclusive.

By the same concavity, we mean the direction (sign) of the concavity, either concave up or concave down. The magnitude of the concavity may vary from one slice to another.

It turns out that a positive value of D indicates that all slices of the surface through (a, b) have the same concavity. Then, if $f_{xx}(a, b) < 0$ (case (1)), all the slices through (a, b) are concave down and the function has a local maximum at (a, b).

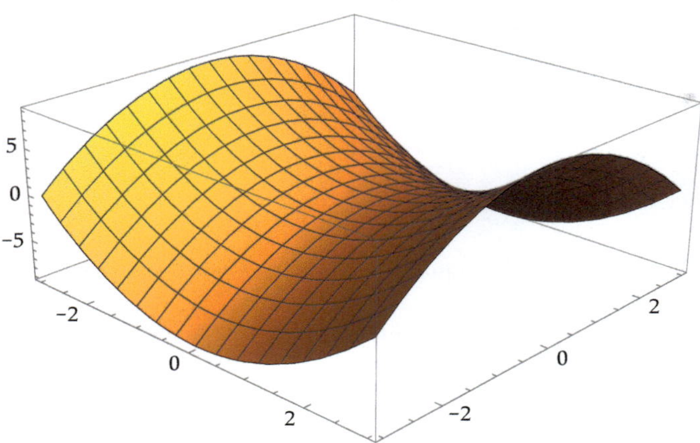

Figure 3.100 *The graph of the hyperbolic paraboloid $z = x^2 - y^2$, which has a saddle point at $(0, 0)$. Some slices through $(0, 0)$ are concave up, whereas other slices through $(0, 0)$ are concave down*

Similarly, if $D > 0$ and $f_{xx}(a, b) > 0$ (case (2)), then all the slices through (a, b) are concave up and the function has a local minimum at (a, b).

    A negative value of D indicates that some slices through (a, b) are concave up while others are concave down, similar to the shape of a hyperbolic paraboloid. In that case, we say there is a *saddle point* at (a, b), because it can resemble a saddle; see Figure 3.100, which is repeated from Section 3.1. For instance, if $f_{xx}(a, b) > 0$ and $f_{yy}(a, b) < 0$ so that slices in the x- and y-directions have opposite concavity, then a sign analysis of D is $D = (+ \cdot -) - ()^2$, which must be negative.

Because all the slices have the same concavity, f_{yy} could be used in place of f_{xx}.

A saddle point is not an extremum. It is neither a maximum nor a minimum, even though it appears to be a minimum on some slices and a maximum on other slices. The zenith of a mountain pass often resembles a saddle point, because the road reaches its highest elevation at the saddle point but the mountain is higher on either side of the saddle point.

3.9.3 Second derivative test examples

To determine local extrema and saddle points for a function of two variables, the steps are ❶ find critical points; ❷ evaluate D at each critical point; and ❸ use the second derivative test to classify each critical point, if possible.

Ans. to reading exercise 24:
    $(1, 0)$ and $(-1, 0)$

Example 61 *Find the local extrema and saddle points, if any, of $f(x, y) = 2x^2 - xy + 4y^2$.*

Solution ❶ Find critical points. The gradient is

$$\nabla f(x, y) = \langle 4x - y, \, -x + 8y \rangle.$$

Setting the gradient equal to zero, we arrive at the simultaneous equations

$$4x - y = 0 \quad \text{and} \quad -x + 8y = 0.$$

Using the addition method, we multiply (both sides of) the first equation by 8, add the equations, and solve:

$$\begin{aligned} 32x - 8y &= 0 \\ -x + 8y &= 0 \\ \hline 31x \qquad\;\; &= 0 \\ x &= 0. \end{aligned}$$

Replacing x with 0 in the first equation yields $4(0) - y = 0$, hence $y = 0$. There is one critical point, $(0, 0)$.

❷ Evaluate D at each critical point. We have

$$\begin{aligned} f_{xx}(x, y) &= 4 & f_{xx}(0, 0) &= 4 \\ f_{xy}(x, y) &= -1 & f_{xy}(0, 0) &= -1 \\ f_{yy}(x, y) &= 8 & f_{yy}(0, 0) &= 8 \end{aligned} \;.$$

For the critical point $(0, 0)$,

$$D = f_{xx} f_{yy} - (f_{xy})^2 = 4(8) - (-1)^2 = 31.$$

❸ Use the second derivative test to classify each critical point, if possible. For the critical point $(0, 0)$, $D > 0$, so all the slices have the same concavity. Because $f_{xx}(0, 0) > 0$, the slices are all concave up. Therefore, there is a local minimum at $(0, 0)$, by case (2) of the second derivative test. ∎

The graph of the function f from example 61 is in Figure 3.101. The local minimum at $(0, 0)$ is seen in the graph, and it is apparent that a slice in any direction containing that local minimum point is concave up.

Example 62 *Find the local extrema and saddle points, if any, of* $f(x, y) = 3x - x^3 - \frac{1}{2}y^2.$

Solution ❶ Find critical points. The gradient is

$$\nabla f = \langle 3 - 3x^2, -y \rangle.$$

Setting each component of the gradient equal to zero, we have

$$3 - 3x^2 = 0 \quad \text{and} \quad -y = 0.$$

Therefore, $y = 0$, and solving the first equation yields $3 = 3x^2$, $1 = x^2$, and $x = \pm 1$. The two critical points are $(1, 0)$ and $(-1, 0)$.

❷ Evaluate D at each critical point. We have

$$\begin{aligned} f_{xx}(x, y) &= -6x & f_{xx}(1, 0) &= -6 & f_{xx}(-1, 0) &= 6 \\ f_{xy}(x, y) &= 0 & f_{xy}(1, 0) &= 0 & f_{xy}(-1, 0) &= 0 \\ f_{yy}(x, y) &= -1 & f_{yy}(1, 0) &= -1 & f_{yy}(-1, 0) &= -1 \end{aligned} \;.$$

Here, f_{xx} is written as a shortcut for $f_{xx}(0, 0)$, and the same is true for f_{xy} and f_{yy}. The formula for D is more easily recalled using this shortcut notation.

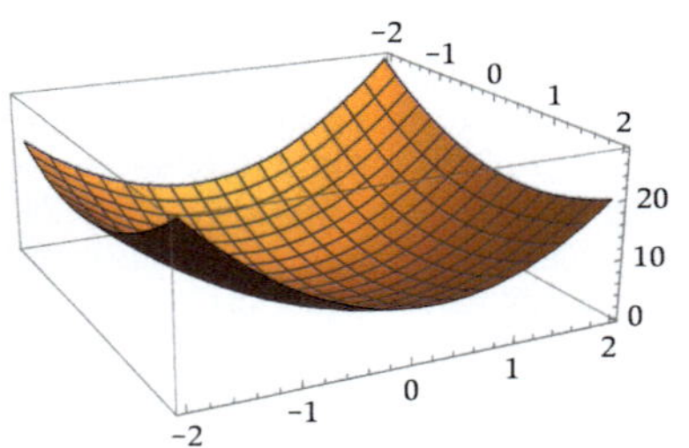

Figure 3.101 *The function from example 61*

For the critical point $(1, 0)$,

$$D = f_{xx}f_{yy} - (f_{xy})^2 = (-6)(-1) - 0^2 = 6,$$

and for the critical point $(-1, 0)$,

$$D = f_{xx}f_{yy} - (f_{xy})^2 = (6)(-1) - 0^2 = -6.$$

❸ Use the second derivative test to classify each critical point, if possible. For the critical point $(1, 0)$, $D > 0$, so all slices have the same concavity. Because $f_{xx}(1, 0) < 0$, all the slices are concave down. There is a local maximum at $(1, 0)$.

For the critical point $(-1, 0)$, $D < 0$, meaning that some slices are concave up while others are concave down. There is a saddle point at $(-1, 0)$. ■

The graph of the function f from example 62 is in Figure 3.102. Both the local maximum and saddle point can be seen on the surface. Slices through the local maximum appear concave down. Some slices through the saddle point appear concave up and other slices appear concave down.

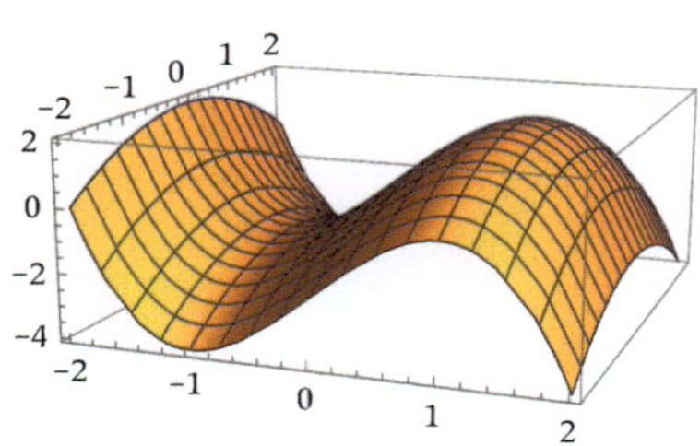

Figure 3.102 *The function from example 62*

"Slices" should be interpreted "locally." Further from the point of interest, the concavity may change, but that is irrelevant.

Reading Exercise 25 Calculate D given $f_{xx}(a, b) = 4$, $f_{xy}(a, b) = 3$, and $f_{yy}(a, b) = 2$. What conclusion may we draw?

Example 63 *Find the local extrema and saddle points, if any, of $f(x, y) = x^2 - 2xy + y^2$.*

Solution **❶** Find critical points. The gradient is

$$\nabla f(x, y) = \langle 2x - 2y, \; -2x + 2y \rangle.$$

Setting the components of the gradient equal to zero gives

$$2x - 2y = 0 \quad \text{and} \quad -2x + 2y = 0,$$

which are the same equation! We have $2x = 2y$ for both equations, or $x = y$, which means there are infinitely many critical points, all of the form (a, a).

❷ Evaluate D at each critical point. We have

$$f_{xx}(x, y) = 2$$
$$f_{xy}(x, y) = -2$$
$$f_{yy}(x, y) = 2$$

For any critical point,

$$D = f_{xx}f_{yy} - (f_{xy})^2 = (2)(2) - (-2)^2 = 0.$$

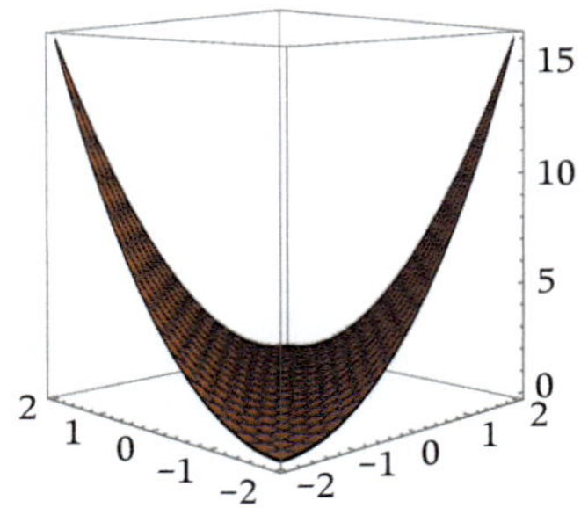

Figure 3.103 *The function from example 63*

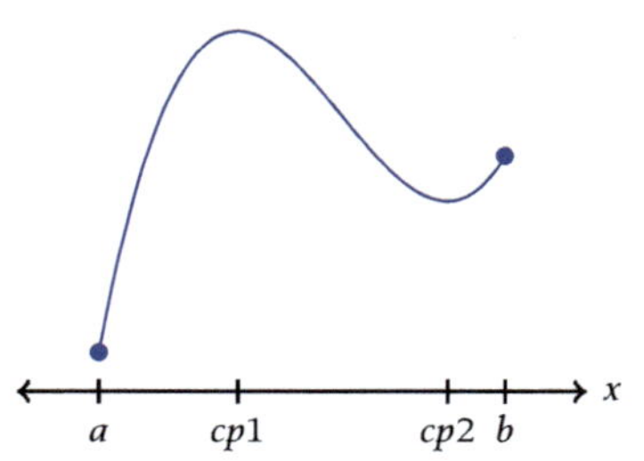

Figure 3.104 *A continuous function on a closed, bounded interval $[a, b]$, with critical numbers $cp1$ and $cp2$*

Ans. to reading exercise 25:

$D = -1; f$ has a saddle point at (a, b). Remark: notice that in this reading exercise, the slices in both the x- and y-directions are concave up, even though there are other directions in which the slices are concave down. Knowing that $f_{xx}(a, b)$ and $f_{yy}(a, b)$ have the same sign is not enough information to classify the critical point.

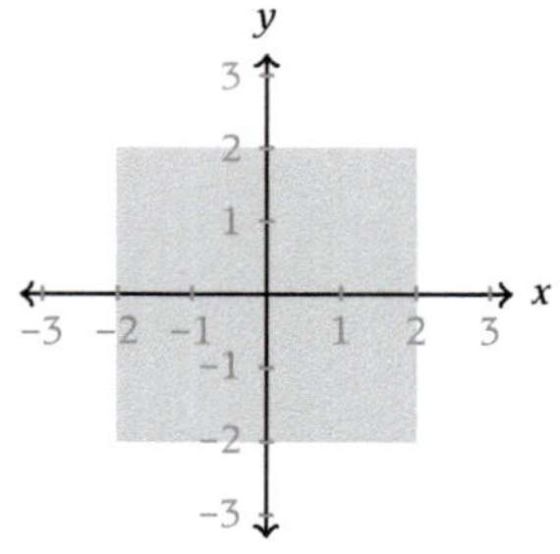

Figure 3.105 *The closed and bounded domain $-2 \le x \le 2, -2 \le y \le 2$*

③ Use the second derivative test to classify each critical point, if possible. Because $D = 0$ for every critical point, the second derivative test is inconclusive. ∎

So what is going on with the graph of $f(x, y) = x^2 - 2xy + y^2$? Notice that the expression can be factored as $(x - y)^2$, so that $f(x, y) = (x - y)^2 \ge 0$. Whenever $x = y, f(x, y) = (x - x)^2 = 0$, which is the minimum value. At each critical point, the function has a local (and absolute) minimum. The graph is in Figure 3.103. Collectively, the local minima form what can be described as a "trough." The surface is a parabolic cylinder.

3.9.4 Absolute extrema

If $y = f(x)$ is a continuous function defined on (or restricted to) a closed and bounded interval $[a, b]$, then f must attain an absolute maximum value and an absolute minimum value (the overall largest and smallest values of the function, like the tallest peak in a mountain range) somewhere in the interval (Figure 3.104). These absolute extrema must occur either at a critical number (absolute max located at $x = cp1$ in Figure 3.104) or on the boundary of the domain, that is, an endpoint of the interval (absolute min located at $x = a$ in Figure 3.104). The "closed interval method" is to determine critical numbers and then evaluate the function at the critical numbers and endpoints of the interval; the largest of these function values must be the absolute maximum and the least must be the absolute minimum.

What changes when we move to a function of two variables, $z = f(x, y)$? Instead of the domain being an interval, now the domain can be a two-dimensional set of input points. Consider, for instance, the graphs in Figures 3.101, 3.102, and 3.103. Each function is graphed on the square domain $-2 \le x \le 2, -2 \le y \le 2$. Suppose that we restrict each function to that domain only. Each is a continuous function throughout that domain (they are all polynomial functions). Each appears to have a lowest point (or points) and a highest point (or points). Sometimes these absolute extrema occur at a critical point, as in the absolute minimum value in Figure 3.101. At other times an absolute extrema occurs along the boundary of the function, which can occur at a corner of the domain (Figure 3.103) or elsewhere along the boundary (Figure 3.102).

The domain of the function need not be rectangular like Figure 3.105; it can be circular, or any other closed and bounded set of inputs. A continuous function on a closed and bounded set of inputs must attain maximum and minimum values, and these values must occur either at critical points or along the boundary. The issue is that the boundary contains infinitely many points, not just two. But for a rectangular boundary, a variation of the closed interval method can still be used! The method shall be illustrated with an example.

Example 64 *Find the absolute maximum and minimum values of* $f(x,y) = x^4 + y^4 - 4xy + 1$ *restricted to the rectangular domain* $-2 \leq x \leq 2$, $-2 \leq y \leq 2$.

Solution We are given a continuous function (a polynomial) on a closed and bounded set of inputs, so the function must reach an absolute maximum and an absolute minimum value somewhere along the boundary or at a critical point. Let's check values along the boundary first.

The boundary consists of four line segments (Figure 3.106). Consider the top segment, which is labeled with its equation $y = 2$. What are the values of the function along this segment? Because we are really only interested in the absolute maximum and minimum values of f, we only need to find the largest and smallest values of the function along this segment, not all infinitely many of them. If $y = 2$, then we have

$$z = f(x, 2) = x^4 + 2^4 - 4x(2) + 1 = x^4 - 8x + 17.$$

But the values of x are also restricted to be between -2 and 2. In other words, we wish to find the absolute maximum and minimum values of $g(x) = x^4 - 8x + 17$, which is a single-variable function, on the interval $[-2, 2]$, which can be found using the closed interval method!

To determine the critical numbers of g, we calculate the derivative:

$$g'(x) = 4x^3 - 8.$$

Setting this equal to zero, we solve to find the critical numbers:

$$4x^3 - 8 = 0$$
$$4x^3 = 8$$
$$x^3 = 2$$
$$x = \sqrt[3]{2},$$

which is in the interval $[-2, 2]$. Next we evaluate g at its endpoints and the critical number:

$$g(-2) = (-2)^4 - 8(-2) + 17 = 49$$
$$g(\sqrt[3]{2}) \approx 9.44$$
$$g(2) = 2^4 - 8(2) + 17 = 17.$$

Therefore, the largest value of f along the top of the boundary is 49, which occurs at $x = -2$, $y = 2$, that is, the input $(-2, 2)$. The smallest is approximately 9.44, which

A closed set of inputs must include all of its boundary points. A domain such as $-2 < x \leq 2$, $-2 \leq y \leq 2$ does not include all of its boundary points (notice the $<$ instead of $\leq$) and is not closed; a function restricted to such a domain might not attain an absolute maximum or minimum value.

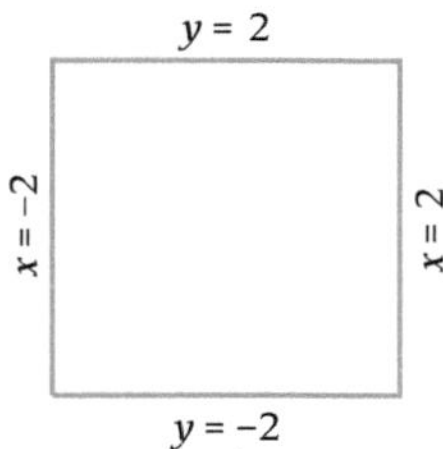

Figure 3.106 *The boundary of the closed and bounded domain* $-2 \leq x \leq 2$, $-2 \leq y \leq 2$

Recall that if a critical number lies outside the interval, it is not considered.

Derived from previous calculations, g is defined as $g(x) = f(x, 2)$. In other words, the values of the function g are values of the function f with $y = 2$.

Recall that along this top segment, $y = 2$.

occurs at $x = \sqrt[3]{2}, y = 2$, that is, the input $(\sqrt[3]{2}, 2)$. These are candidates for the absolute maximum and minimum values of f, but we still need to consider the rest of the boundary and the critical points to see if there are yet larger or smaller values.

Next, we consider the bottom segment, $y = -2$. On this segment,

$$z = f(x, -2) = x^4 + (-2)^4 - 4x(-2) + 1 = x^4 + 8x + 17.$$

Calling this function h, we wish to find the absolute maximum and minimum values of h on the interval $[-2, 2]$. The derivative is

$$h'(x) = 4x^3 + 8,$$

and setting the derivative equal to zero, the solution is $x = -\sqrt[3]{2}$, which is in the interval. Evaluating at the critical number and the endpoints yields

$$h(-2) = (-2)^4 + 8(-2) + 17 = 17$$

$$h(-\sqrt[3]{2}) \approx 9.44$$

$$h(2) = 2^4 + 8(2) + 17 = 49.$$

Once again, the largest of these is 49, which occurs at $x = 2$, $y = -2$, that is, the input $(2, -2)$, and the least is approximately 9.44, which occurs at $x = -\sqrt[3]{2}, y = -2$, the input $(-\sqrt[3]{2}, -2)$.

The right-side segment is $x = 2$, between $y = -2$ and $y = 2$. Then

$$z = f(2, y) = 2^4 + y^4 - 4(2)y + 1 = y^4 - 8y + 17,$$

The fact that the variable is y instead of x only changes the name of the variable.

and we wish to find the absolute extrema of $g(y) = y^4 - 8y + 17$ on $[-2, 2]$. But that is identical to the function g from above, on the same interval! Therefore, the maximum value is 49 at $y = -2, x = 2$, which is the input $(2, -2)$. This input, which is at the bottom right corner of the domain, has already been identified. The minimum value is approximately 9.44 at $y = \sqrt[3]{2}, x = 2$, which is the input $(2, \sqrt[3]{2})$.

Finally, we check the left side of the boundary, which is $x = -2$ between $y = -2$ and $y = 2$. Then

$$z = f(-2, y) = (-2)^4 + y^4 - 4(-2)y + 1 = y^4 + 8y + 17,$$

Because a corner is on two segments, the same point can be the greatest or least value along two segments.

and we wish to find the absolute extrema of $h(y) = y^4 + 8y + 17$ on $[-2, 2]$. This is also a repeat. The maximum value is 49 at $y = 2, x = -2$, which is the input $(-2, 2)$. This has already been identified, and is at the top left corner of the domain. The minimum value is approximately 9.44 and occurs at $y = -\sqrt[3]{2}, x = -2$, which is the input $(-2, -\sqrt[3]{2})$.

To summarize, along the boundary the largest function value is 49, which occurs at $(-2, 2)$ and $(2, -2)$. The least value is approximately 9.44, which occurs at four locations. Again, these are candidates for the absolute maximum and minimum values, but we still need to consider the critical points of f.

To find the critical points, we calculate the gradient:

$$\nabla f(x, y) = \langle 4x^3 - 4y,\ 4y^3 - 4x \rangle.$$

Setting the components of the gradient equal to zero, we have the simultaneous equations

$$4x^3 - 4y = 0 \quad \text{and} \quad 4y^3 - 4x = 0.$$

Using the substitution method, we solve the left-hand equation for the variable y:

$$4x^3 = 4y$$
$$x^3 = y.$$

We then substitute the result into the right-hand equation:

$$4y^3 - 4x = 0$$
$$4(x^3)^3 - 4x = 0$$
$$x^9 - x = 0$$
$$x(x^8 - 1) = 0.$$

Setting each factor equal to zero, we have $x = 0$ and $x^8 = 1$, which has solutions $x = \pm 1$. Finally, we substitute each value of x into the equation $y = x^3$ to get

$$
\begin{array}{lll}
x = 0 & y = x^3 = 0^3 = 0 & \text{point } (0, 0) \\
x = 1 & y = 1^3 = 1 & \text{point } (1, 1) \\
x = -1 & y = (-1)^3 = -1 & \text{point } (-1, -1).
\end{array}
$$

Each of these three points lies in the domain, so we consider all three. The values of the function at these three critical points are

If a critical point does not lie in the domain, it is not considered.

$$f(0, 0) = 1$$
$$f(1, 1) = 1 + 1 - 4 + 1 = -1$$
$$f(-1, -1) = (-1)^4 + (-1)^4 - 4(-1)(-1) + 1 = 1 + 1 - 4 + 1 = -1.$$

None of these match or exceed 49, so the absolute maximum value of f is 49 at $(-2, 2)$ and $(2, -2)$. But the least of the values at the critical points is -1, which is less than the least of the values along the boundary, 9.44. Therefore, the absolute minimum value of f is -1 at $(1, 1)$ and $(-1, -1)$.

Although the solution to example 64 is quite lengthy, for a rectangular domain the work itself is rather straightforward once the explanatory narrative is removed.

■

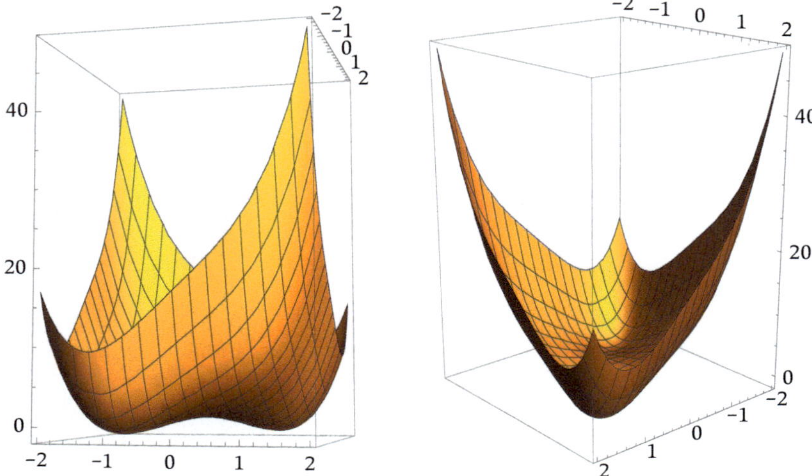

Figure 3.107 *Two views of the function f from example 64*

Two views of the graph of the function f from example 64 are in Figure 3.107. The absolute maximum value of 49 at opposite corners of the domain can be easily seen in both views. The absolute minimum value at two points can be seen in the view on the left. On each edge of the boundary, the function has a least value along that edge, all four of which can be seen in the view on the right; these values are not the absolute minimum of the function on the domain, however, as can be seen in the view on the left.

To find absolute extrema of a continuous function, the closed and bounded domain need not be rectangular. The work (and creativity) required varies depending on the shape of the boundary.

3.9.5 Proof of the second derivative test

We end this section with a proof of the second derivative test, although some details of the proof of part (3) are omitted.

Proof. The idea is to find directional second derivatives at the critical point (a, b) in all directions. If they are all positive, then all slices are concave up and the function must have a local minimum at the critical point. If all slices are concave down, then the function must have a local maximum at the critical point. If some slices are concave up and some are concave down, then the function has a saddle point at the critical point.

We wish to find directional second derivatives of $z = f(x, y)$ at (a, b) in the direction $\mathbf{u} = \langle c, d \rangle$ (assume $\mathbf{u}$ is a unit vector).

For a continuous function of one variable, if the function possesses two local minima, there must be a local maximum between them. Notice that the function of example 64 possesses two local minima, and the bottom profile of the view at the left in Figure 3.107 makes it appear as if there may be a local maximum between them. But closer inspection shows that instead of a local maximum between the two local minima, there is a saddle point. This is a difference from the single-variable case; a function can have more than one local minimum and yet have no local maximum.

It is implied that the function and partial derivatives are evaluated at (x, y).

First-order directional derivatives have the formula

$$D_{\mathbf{u}}f = \nabla f \cdot \mathbf{u} = cf_x + df_y,$$

which is a function of x and y. We may therefore take its derivative in the direction of $\mathbf{u}$,

$$D_{\mathbf{u}}\left(D_{\mathbf{u}}f\right) = D_{\mathbf{u}}^2 f$$

$$= c\left(\frac{\partial}{\partial x}(cf_x + df_y)\right) + d\left(\frac{\partial}{\partial y}(cf_x + df_y)\right)$$

$$= c\left(cf_{xx} + df_{yx}\right) + d\left(cf_{xy} + df_{yy}\right)$$

$$= c^2 f_{xx} + cdf_{yx} + dcf_{xy} + d^2 f_{yy}$$

$$= c^2 f_{xx} + 2cdf_{xy} + d^2 f_{yy},$$

provided that the hypotheses of Clairaut's theorem are satisfied.

Next, we complete the square on the variable c:

$$= f_{xx}\left(c^2 + 2cd\frac{f_{xy}}{f_{xx}}\right) + d^2 f_{yy}$$

$$= f_{xx}\left(c^2 + 2cd\frac{f_{xy}}{f_{xx}} + d^2\frac{(f_{xy})^2}{(f_{xx})^2}\right) + d^2 f_{yy} - d^2\frac{(f_{xy})^2}{(f_{xx})^2}\cdot f_{xx}$$

$$= f_{xx}\left(c + d\frac{f_{xy}}{f_{xx}}\right)^2 + \frac{d^2}{f_{xx}}\left(f_{xx}f_{yy} - (f_{xy})^2\right).$$

Let $D = f_{xx}f_{yy} - (f_{xy})^2$, which is the quantity inside the second-term parentheses.

For case (2), suppose that when evaluated at (a, b), $D > 0$ and $f_{xx} > 0$. Then the second directional derivative $D_{\mathbf{u}}^2 f$ is of the form

$$f_{xx}(\quad)^2 + \frac{1}{f_{xx}}d^2(\text{positive}),$$

which is positive. Because this is true for any direction vector $\mathbf{u}$, the second directional derivative in every direction is positive. Therefore, every slice through the critical point is concave up, meaning that the function has a local minimum at (a, b).

For case (1), suppose that when evaluated at (a, b), $D > 0$ and $f_{xx} < 0$. Then the second directional derivative $D_{\mathbf{u}}^2 f$ is of the form

$$f_{xx}(\quad)^2 + \frac{1}{f_{xx}}d^2(\text{positive}),$$

Line 1 introduces notation for the second-order directional derivative; line 2 takes the derivative of $(cf_x + df_y)$ in the direction $\mathbf{u}$; line 3 uses the partial derivative sum and constant multiple rules inside the outer parentheses; line 4 distributes; and line 5 applies Clairaut's theorem.

The hypotheses of the second derivative test are sufficient to meet the hypotheses of Clairaut's theorem.

Line 1 factors out the coefficient on c^2 from the terms involving c; line 2 completes the square by taking half the coefficient on the c term, which would be $d\frac{f_{xy}}{f_{xx}}$, squares that quantity, and adds it inside the parentheses and subtracts it back, outside the parentheses, times the factor outside the parentheses; line 3 factors the perfect square, and from the remaining terms factors out $\frac{d^2}{f_{xx}}$.

One detail: if d happens to be 0, then c cannot be 0 (since $\mathbf{u}$ is a unit vector), and therefore $f_{xx}\left(c + d\frac{f_{xy}}{f_{xx}}\right)^2 = f_{xx}c^2$ cannot be zero and is positive.

which is negative because f_{xx} is negative. This means that the second directional derivative in every direction is negative, so that every slice through the critical point is concave down, meaning that the function has a local maximum at (a, b).

For case (3), suppose that when evaluated at (a, b), $D < 0$. Through laborious algebraic details that shall be omitted, it can be shown that the expression is positive for some directions $\mathbf{u}$ and negative for others, resulting in a saddle point. ∎

EXERCISES 3.9

1–10. Find all critical points of the function.

1. $f(x, y) = x^3 - 4x^2 - y^2 + 3y + 11$
2. $f(x, y) = x^2 + 4x - y^2$
3. $f(x, y) = x^3 - 4x^2 + 3y + 11$
4. $f(x, y) = x^3 + 7x^2 - 9x + 22y + 866$
5. $f(x, y) = 2x^2 + xy + y^3$
6. $f(x, y) = x^{2/3} - y^2 + 5y$
7. $f(x, y) = 4xy - \tan^{-1} x - 2y$
8. $f(x, y) = x^2 y - 4y + 7$
9. $f(x, y) = x^3 - 12x + \sqrt[3]{y}$
10. $f(x, y) = x^2 + xy^2 + 4y + 12$

11–20. (a) Use the values of the second partial derivatives to calculate D for the critical point (a, b). (b) What conclusion can we draw?

11. $f_{xx}(a, b) = 4, f_{xy}(a, b) = 5, f_{yy}(a, b) = 9$
12. $f_{xx}(a, b) = 7, f_{xy}(a, b) = 5, f_{yy}(a, b) = 3$
13. $f_{xx}(a, b) = 0, f_{xy}(a, b) = 2, f_{yy}(a, b) = 18$
14. $f_{xx}(a, b) = 7, f_{xy}(a, b) = 4, f_{yy}(a, b) = 3$
15. $f_{xx}(a, b) = 4, f_{xy}(a, b) = 6, f_{yy}(a, b) = 9$
16. $f_{xx}(a, b) = -5, f_{xy}(a, b) = 3, f_{yy}(a, b) = -2$
17. $f_{xx}(a, b) = -3, f_{xy}(a, b) = 1, f_{yy}(a, b) = -2$
18. $f_{xx}(a, b) = 2, f_{xy}(a, b) = 4, f_{yy}(a, b) = 8$
19. $f_{xx}(a, b) = 4, f_{xy}(a, b) = 0, f_{yy}(a, b) = -2$
20. $f_{xx}(a, b) = 3, f_{xy}(a, b) = 1, f_{yy}(a, b) = -4$

21–30. Find the local extrema and saddle points, if any, of the given function. (In the event that the second derivative test is inconclusive, say so; it is not necessary to do further work to classify the critical point.)

21. $f(x, y) = x^3 - y^4 - 3x + 32y + 77$
22. $f(x, y) = x^2 + 4xy + y^2$
23. $f(x, y) = x^2 + 4y^2 + 5x - 2y$
24. $f(x, y) = x^3 - 3x^2 + y^2 - 6y$

25. $f(x, y) = x^3 - 12x + y^3 - 27y - 5$

26. $f(x, y) = \dfrac{1}{x^2 + y^2 + 1}$

27. $f(x, y) = \dfrac{2}{3}x^3 - y^4 + xy$

28. $f(x, y) = x^2 y^3 - 8x^2 - 4y^3 + 6y$

29. $f(x, y) = x^3 y^4 - y^4 - 16x^3$

30. $f(x, y) = x^2 \ln y - y^3$

31–38. Find (a) the absolute minimum and absolute maximum values of the given function on the given domain and (b) where those values occur.

31. $f(x, y) = 3x - x^3 - \frac{1}{2}y^2$, on $-2 \le x \le 2, -2 \le y \le 2$

32. $f(x, y) = 2x^2 - xy + 4y^2$, on $-2 \le x \le 2, -2 \le y \le 2$

33. $f(x, y) = xy^2 - \ln x$, on $1 \le x \le 3, -1 \le y \le 1$

34. $f(x, y) = x^3 + 0.5y^2, -3 \le x \le 3, -3 \le y \le 3$ (funky chair)

35. $f(x, y) = 20xye^{-x^2 - y^2}, -3 \le x \le 3, -3 \le y \le 3$ (peaks and pits)

36. $f(x, y) = \sqrt{x^4 - y^4}, 3 \le x \le 5, 0 \le y \le 2$

37. $f(x, y) = x^2 - y^2$, on the circular domain $x^2 + y^2 \le 9$

38. $f(x, y) = x^2 - y^2 + 4y$, where the domain is the region bounded by $y = x^2$ and $y = 4$

39. Using a CAS to perform the calculations, explore the function
$$f(x, y) = \dfrac{x - 2y}{x^2 + y^2 + 3} \text{ as follows.}$$

(a) Graph the function.
(b) Find the critical points of the function.
(c) Calculate D for each critical point. What can we conclude?

40. Using a CAS to perform the calculations, explore the function $f(x, y) = -0.4x^2 + \sin 5y$ as follows.

(a) Graph the function.
(b) Find the critical points of the function.
(c) Calculate D for each critical point. What can we conclude?

41. Using a CAS to perform the calculations, explore the function
$$f(x, y) = \dfrac{x^2 y}{x^4 + y^2} \text{ as follows.}$$

(a) Graph the function.
(b) Find the critical points of the function.
(c) Calculate D for each critical point. What can we conclude?
(d) Compute the value of the function at each critical point. Using the graph of the function as a guide, what appears to be true?

42. Investigate the graph of $f(x, y) = x^3 y^4 - y^4 - 16x^3$ near the critical point $(0, 0)$, for which the second derivative test is inconclusive, using the following steps.

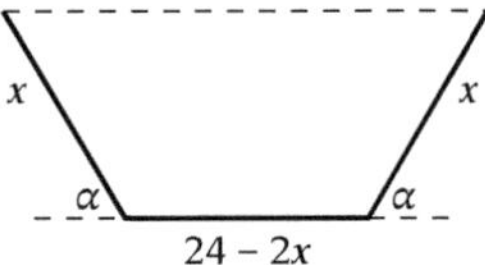

Figure 3.108 *The figure for exercise 43*

Exercise 43 is an example in *Elements of the Differential and Integral Calculus*, by Granville, Smith, and Longley, 1941 revised edition, Ginn and Company.

(a) For the slice $y = 0$, what is the concavity at, to the left of, and to the right of $x = 0$?

(b) For the slice $x = 0$, what is the concavity at, to the left of, and to the right of $y = 0$?

(c) For an arbitrary real number k, form the slice $y = kx$, and call its equation $g(x)$. Find $g''(x)$, $g''(-\omega)$, and $g''(\omega)$ for an arbitrary positive infinitesimal ω. What can be said about the concavity of this slice at, to the left of, and to the right of $x = 0$?

(d) Is there a local maximum, a local minimum, a saddle point, or none of the above at the critical point $(0, 0)$?

(e) Use a CAS to graph the function f and observe the results of parts (a)–(d) on the graph.

43. A long piece of tin 24 inches wide is to be made into a trough by bending up two sides (Figure 3.108). Find the width and inclination of each side if the carrying capacity is a maximum.

3.10 Lagrange Multipliers

We have seen how partial derivatives can simplify finding derivatives of implicitly defined functions. Partial derivatives can also be used to simplify certain optimization problems.

3.10.1 Rethinking an optimization problem

See *Calculus Set Free* Section 3.8.

The following optimization problem is an example in *Calculus Set Free*.

Example 65 *A farmer's child has purchased a piglet. The farmer has given the child 60 feet of fencing left over from another project. Using the side of the barn as one side of a rectangular pig pen, the child wishes to enclose the largest area possible. What dimensions should be used?*

Whenever possible with an applied problem, the first step is to draw a diagram (Figure 3.109). Labeling the sides of the rectangular pig pen ℓ and w (for length and width), the area of the pig pen, which we wish to maximize, is $A = \ell w$. In *Calculus Set Free*, where only single-variable calculus is used, the expression for area must be rewritten in terms of only one variable. This is accomplished by using the other information given, namely that the amount of available fencing is 60 feet. The fencing is used on three sides, and we must have $2\ell + w = 60$. Solving for w (we could also solve for ℓ) gives $w = 60 - 2\ell$, and we may rewrite area as $A = \ell(60 - 2\ell)$, a function of one variable, and proceed from there.

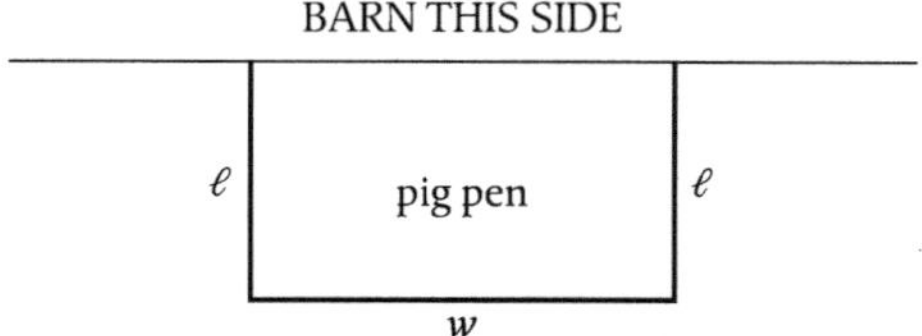

Figure 3.109 *The pig pen of example 65*

The single-variable procedure works just fine, but now that we have multi-variable calculus at our disposal, the single-variable procedure seems somewhat unsatisfying, because the area is inherently a function of two variables:

$$\text{area} = f(\ell, w) = \ell w.$$

We also must take into account the *constraint*, namely that the amount of fencing must be 60 feet:

$$2\ell + w = 60.$$

This equation is called the *constraint equation.*

This type of problem is sometimes called *constrained optimization.*

To set up an inherently two-variable method, consider a second function of two variables representing the constraint. Given any constraint equation, we may move everything to one side of the equation with zero on the other side:

$$2\ell + w - 60 = 0.$$

Then, we can define a constraint function using this expression:

$$g(\ell, w) = 2\ell + w - 60.$$

Our aim, our objective, is to maximize the function f; hence the name "objective function."

The constraint $2\ell + w - 60 = 0$ can then be written as $g(\ell, w) = 0$.

Now consider a simultaneous contour plot for f (which we call the *objective function*) and g, as given in Figure 3.110. Only one contour is shown for the constraint function, namely $g(\ell, w) = 0$, because that particular contour represents our constraint; any solution to the problem must lie on that contour (the green line in Figure 3.110). We want the largest possible area given our constraint, that is, the objective function contour with the largest value that intersects the constraint contour; in Figure 3.110, it is apparent this largest value is 450, where the constraint contour and a level curve (contour 450) are tangent. In fact, at the point where the objective function is maximized, the constraint contour and objective function contour must be tangent, or there would be a direction to move along the constraint contour that would lead to a larger value.

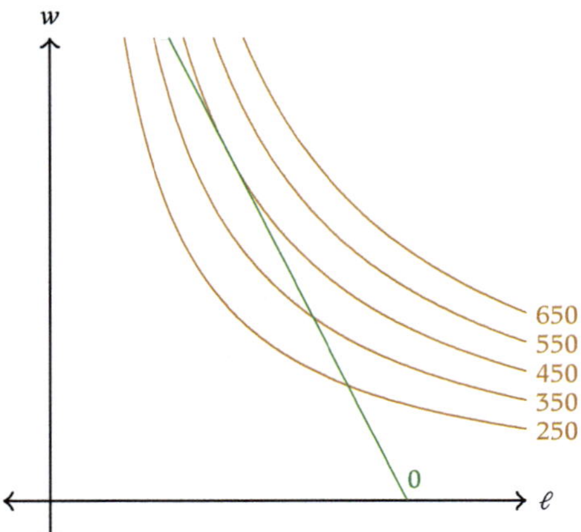

Think of Figure 3.110 as a topographical map, and consider the point where the green trail is at elevation 350. If we move northwest along the trail we are traveling uphill toward higher elevations; this point is not the highest-elevation point along the green trail. Meanwhile, at the point along the green trail where the elevation is 450, where the trail is tangent to a level curve, moving in either direction moves toward lower elevations; this is the highest point along the green trail.

Figure 3.110 *A simultaneous contour plot for the objective function f (brown) and the constraint function g (green)*

Recall that the gradient is orthogonal to a level curve. At the maximizing point the level curves lie in the same direction, and therefore the gradients of f and g must point in the same (or opposite) directions! Therefore, the gradient vector $\nabla f(a, b)$ must be a scalar multiple of the gradient vector $\nabla g(a, b)$, that is, $\nabla f(a, b) = \lambda \nabla g(a, b)$ at the maximizing point (a, b), for some real number λ. So, we are seeking a point (a, b) that is a solution to the equation

$$\nabla f(\ell, w) = \lambda \nabla g(\ell, w),$$

which can be rewritten in terms of partial derivatives as

$$f_\ell(\ell, w) = \lambda \cdot g_\ell(\ell, w)$$
$$f_w(\ell, w) = \lambda \cdot g_w(\ell, w).$$

The lowercase Greek letter λ is pronounced "lambda."

The scalar λ is called a *Lagrange multiplier*. Computing the partial derivatives of the objective function f and the constraint function g gives

$$f_\ell(\ell, w) = w \quad g_\ell(\ell, w) = 2$$
$$f_w(\ell, w) = \ell \quad g_w(\ell, w) = 1.$$

Then, we must have

$$w = \lambda \cdot 2$$
$$\ell = \lambda \cdot 1.$$

Solving both equations for λ gives

$$\lambda = \frac{w}{2}$$

$$\lambda = \ell.$$

Equating the two expressions for lambda gives

$$\frac{w}{2} = \ell.$$

Knowing that the constraint equation must also be satisfied, we substitute $\frac{w}{2}$ for ℓ in the constraint equation:

$$2\ell + w = 60$$

$$2\frac{w}{2} + w = 60$$

$$2w = 60$$

$$w = 30.$$

Using $\ell = \frac{w}{2}$ gives $\ell = \frac{30}{2} = 15$. This solution matches the solution given by the single-variable method in *Calculus Set Free*. Notice that there is no need to find the value of the Lagrange multiplier λ; it is merely a help in solving for the variables of interest.

Once the method of Lagrange multipliers is understood, the actual work to carry out the method is often easier than the work required for the single-variable method.

METHOD OF LAGRANGE MULTIPLIERS

---

❶ Form the objective function $f(x, y)$ and the constraint function $g(x, y)$.
❷ Solve the simultaneous equations

$$f_x(x, y) = \lambda \cdot g_x(x, y)$$

$$f_y(x, y) = \lambda \cdot g_y(x, y)$$

$$g(x, y) = 0.$$

❸ If there is more than one solution, evaluate f at each solution from step **❷**. The largest of these values is the maximum and the smallest is the minimum.

The method of Lagrange multipliers requires that some hypotheses be met. One obvious hypothesis is that the partial derivatives must exist, and another is that a maximum or minimum value (whichever is being sought) must exist. But the continuity inherent in most applied settings guarantees that all the necessary hypotheses are met.

As demonstrated when solving example 65, step **❷** can often be accomplished by the following substeps.

2 *a* Solve each of the equations

$$f_x(x,y) = \lambda \cdot g_x(x,y)$$
$$f_y(x,y) = \lambda \cdot g_y(x,y)$$

for λ.

2 *b* Equate the two expressions for λ from step **2** *a* and solve for one of the variables x or y.

2 *c* Substitute the result of step **2** *b* into the constraint equation and solve for the remaining variable.

2 *d* Use the results of steps **2** *b* and **2** *c* to determine the value of the other variable.

3.10.2 Lagrange multiplier examples

The following example is also from *Calculus Set Free*.

Example 66 *A cylindrical can must have volume 100 cm³. What dimensions should be used to minimize the amount of material used?*

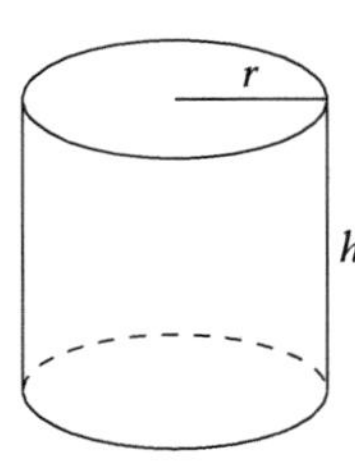

Figure 3.111 *A cylindrical can with radius r and height h*

The quantity $2\pi r^2$ is for the area of the bottom and top of the can; $2\pi rh$ is the lateral (side) surface area.

In practice, some of this writing may be condensed rather easily.

Solution **1** Form the objective function $f(x,y)$ and the constraint function $g(x,y)$. A diagram is often helpful in physical situations; see Figure 3.111.

We wish to minimize the amount of material used, so the objective function represents the amount of material used. The amount of material used is proportional to the surface area of the cylinder. Therefore, the objective function is

$$f(r,h) = 2\pi r^2 + 2\pi rh.$$

The constraint is that the volume of the can must be 100 cm³, or $\pi r^2 h = 100$. The constraint function is

$$g(r,h) = \pi r^2 h - 100.$$

2 Solve the simultaneous equations

$$f_r(r,h) = \lambda \cdot g_r(r,h)$$
$$f_h(r,h) = \lambda \cdot g_h(r,h)$$
$$g(r,h) = 0.$$

The partial derivatives are

$$f_r(r,h) = 4\pi r + 2\pi h \quad g_r(r,h) = 2\pi rh$$
$$f_h(r,h) = 2\pi r \quad\quad\quad g_h(r,h) = \pi r^2.$$

Substituting into the first two of the simultaneous equations,

$$4\pi r + 2\pi h = \lambda \cdot 2\pi r h$$

$$2\pi r = \lambda \cdot \pi r^2.$$

Solving the first for λ gives

$$\lambda = \frac{4\pi r + 2\pi h}{2\pi r h} = \frac{2}{h} + \frac{1}{r}.$$

Solving the second gives

$$\lambda = \frac{2\pi r}{\pi r^2} = \frac{2}{r}.$$

We equate the results and solve for h:

$$\frac{2}{h} + \frac{1}{r} = \frac{2}{r}$$

$$2r + h = 2h$$

$$2r = h.$$

Next we substitute this result into the constraint equation and solve for r:

$$\pi r^2 h = 100$$

$$\pi r^2 (2r) = 100$$

$$r^3 = \frac{100}{2\pi} = \frac{50}{\pi}$$

$$r = \sqrt[3]{\frac{50}{\pi}} \approx 2.515 \text{ cm}.$$

Finally, we substitute $\sqrt[3]{\frac{50}{\pi}}$ for r in the previous equation $h = 2r$,

$$h = 2\sqrt[3]{\frac{50}{\pi}} \approx 5.031 \text{ cm}.$$

There is only one solution (one critical value), which must therefore be the required minimum. The dimensions of the can should be radius $\sqrt[3]{\frac{50}{\pi}} \approx 2.515$ cm and height $2\sqrt[3]{\frac{50}{\pi}} \approx 5.031$ cm. ∎

Sometimes two variables can be used even if using one variable might seem more natural. Once again, the example is taken from *Calculus Set Free*.

Step **2** *a*

Step **2** *b* : line 1 equates the two results for λ, and line 2 multiplies both sides of the equation by the common denominator rh to clear the fractions.

The main geometric conclusion is readily apparent here, that to minimize the amount of material needed, the height and diameter should be the same.

Step **2** *c* : line 1 is the constraint equation; using the previous result, line 2 substitutes $2r$ for h; line 3 simplifies and divides both sides by 2π; and line 4 takes the cube root of both sides.

Step **2** *d*

If the applied setting guarantees the presence of a minimum (or maximum), and the method of Lagrange multipliers yields only one possible point at which that minimum can occur, there is no need for further verification other than checking that calculations are accurate. If additional verification is desired, one way of double-checking that the proposed solution really is a minimum is to use a CAS to graph the function and observe the presence of a minimum. A less-desirable method is to choose a value of r near the critical value, compute the value of h required by the constraint, and evaluate f at both the proposed solution and the new point. The value at the proposed solution should be the smaller value.

Example 67 *A utility company needs to run a pipeline from its power plant on one side of a straight, 900-meter-wide river to a manufacturer on the other side of the river, 3000 meters downstream. It costs $50 per meter to install the pipeline along the ground, but $200 per meter to install the pipeline across (under) the river. At what point should the pipe emerge from the river in order to minimize the cost of the pipeline?*

Solution We begin by drawing a diagram (Figure 3.112).

➊ Form the objective function $f(x, y)$ and the constraint function $g(x, y)$. Labeling variables in the diagram can be accomplished in more than one way. For our purposes, we label the distance downstream from the utility at which the pipeline emerges x, and the distance from that point to the manufacturer y, as in Figure 3.113. (For the one-variable solution method, we label the distance from the point the pipeline emerges to the manufacturer $3000 - x$ instead of y.) The length of pipe under the river is then found by the Pythagorean theorem to be $\sqrt{900^2 + x^2}$. We wish to minimize the cost of the pipeline, so the cost of the pipeline is the objective function:

$$f(x, y) = 200\sqrt{x^2 + 900^2} + 50y.$$

The constraint is $x + y = 3000$, so the constraint function is

$$g(x, y) = x + y - 3000.$$

The constraint is always a relationship between the two variables that is present in the application. In the single-variable procedure, the constraint is used to eliminate one of the variables. If you are having difficulty identifying the constraint, consider looking for the relationship that allows you to eliminate a variable.

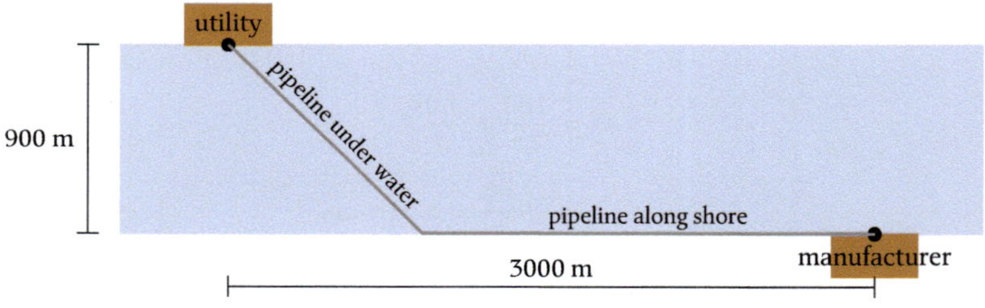

Figure 3.112 *An illustration of example 67*

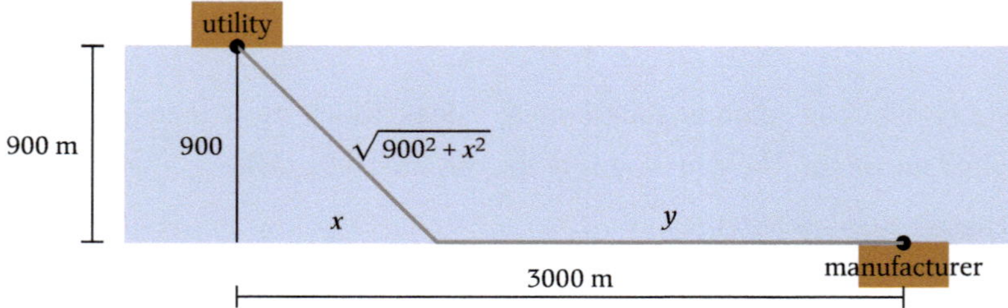

Figure 3.113 *Two variables are placed in the diagram*

The labeling of variables is where choices affect not only the method to use (single-variable or Lagrange multipliers), but also the messiness of the algebra. For instance, we could choose to label the length of the pipeline under water as x, in which case the third side of the triangle would be $\sqrt{x^2 - 900^2}$ (in place of the x in Figure 3.113). This would change both the objective and constraint functions, complicating the solution.

❷ Solve the simultaneous equations

$$f_x(x, y) = \lambda \cdot g_x(x, y)$$
$$f_y(x, y) = \lambda \cdot g_y(x, y)$$
$$g(x, y) = 0.$$

The partial derivatives are

$$f_x(x, y) = \frac{200}{2\sqrt{x^2 + 900^2}} \cdot 2x \quad g_x(x, y) = 1$$
$$f_y(x, y) = 50 \qquad\qquad g_y(x, y) = 1.$$

Substituting into the first of the two simultaneous equations,

$$\frac{200x}{\sqrt{x^2 + 900^2}} = \lambda \cdot 1$$
$$50 = \lambda \cdot 1.$$

The equations are already solved for λ, so we equate the results and solve for x.

$$\frac{200x}{\sqrt{x^2 + 900^2}} = 50$$
$$200x = 50\sqrt{x^2 + 900^2}$$
$$4x = \sqrt{x^2 + 900^2}$$
$$16x^2 = x^2 + 900^2$$
$$15x^2 = 900^2$$
$$x^2 = \frac{900^2}{15}$$
$$x = \pm\frac{900}{\sqrt{15}}$$

The negative solution would mean the pipeline under the river angles away from the manufacturer, which obviously would add cost rather than minimize cost, so we only consider the positive solution $x = \frac{900}{\sqrt{15}} \approx 232.4$ m downstream, where the pipe should emerge from the river to minimize the cost. Although not necessary to answer the question that was asked, we may also solve for y to obtain the distance the pipe should emerge from the manufacturer, $y = 3000 - \frac{900}{\sqrt{15}} \approx 2767.6$ m.

■

It is readily apparent that $\lambda = 50$. Although the value of λ is irrelevant to the solution procedure, it does have an interpretation as a rate of change; λ is the number of additional objective units for each additional constraint unit (at the solution point), although the sign ($+$ or $-$) might be incorrect. In this example, if the constraint is increased by one unit to $x + y = 3001$, that is, if the manufacturer is one meter further downstream, then the total cost of the pipeline is increased by \$50. A quick analysis should help this interpretation make perfect sense for this example.

Line 1 equates the two expressions for λ; line 2 clears the fraction; line 3 divides both sides by 50; line 4 squares both sides; line 5 subtracts x^2 from both sides; line 6 divides both sides by 15; and line 7 takes the square root of both sides.

Notice that leaving the quantity 900^2 alone and not computing its value is convenient in this example.

Before stating the answer, always re-read the question to determine the quantity being requested.

3.10.3 Constrained optimization solutions as critical points

In examples 66 and 67, the solution is not a critical point of the objective function. In example 66, both f_r and f_h evaluated at the solution point are positive, rather than zero; the same is true for f_x and f_y in example 67. This is because we are not maximizing or minimizing the function f overall, but rather under a constraint. As a result, the second derivative test does not apply.

There is, however, a manner in which we can consider the solutions to be critical points. Given an objective function $f(x, y)$ and a constraint function $g(x, y)$, form a function of the three variables x, y, and λ as follows:

$$h(x, y, \lambda) = f(x, y) - \lambda \cdot g(x, y).$$

For the single-variable procedure the single-variable second derivative test applies, and the closed-interval method may also apply. The fact that neither equivalent applies for the two-variable case explains the need for the side note at the end of example 66.

Then $h_x(x, y, \lambda) = f_x(x, y) - \lambda \cdot g_x(x, y)$, $h_y(x, y, \lambda) = f_y(x, y) - \lambda \cdot g_y(x, y)$, and $h_\lambda(x, y, \lambda) = -g(x, y)$. Then a critical point of h occurs when $h_x(x, y, \lambda) = 0$, $h_y(x, y, \lambda) = 0$, and $h_\lambda(x, y, \lambda) = 0$ simultaneously, which is the same as satisfying the simultaneous equations

$$f_x(x, y) = \lambda \cdot g_x(x, y)$$
$$f_y(x, y) = \lambda \cdot g_y(x, y)$$
$$g(x, y) = 0,$$

which is the solution provided by Lagrange multipliers.

3.10.4 Additional notes

While the main point of this section is to solve applied problems, a few notes are in order.

(1) When removed from an applied setting, the hypotheses for the method of Lagrange multipliers might not be met. For instance, try finding extrema of the objective function $f(x, y) = x^2 + y^2$ under the constraint $y = \frac{1}{x}$. It turns out that f has a minimum value but no maximum value under this constraint, and the two points identified by the method of Lagrange multipliers are both locations of minimum values.

(2) The substeps ❷ *a* and ❷ *b* might not reveal all solutions to the simultaneous equations; one scenario where this can occur is dividing by a variable when solving for λ. These substeps are merely a guide that can be helpful for solving the simultaneous equations.

(3) The method of Lagrange multipliers also applies for functions of three variables; we merely need to add $f_z(x, y, z) = \lambda \cdot g_z(x, y, z)$ to the list of simultaneous equations, as well as adjust the other equations for the third variable.

(4) For functions of three variables, a second constraint can be added as well; we would use $\nabla f(x, y, z) = \lambda_1 \nabla g_1(x, y, z) + \lambda_2 \nabla g_2(x, y, z)$.

EXERCISES 3.10

It is implied that Lagrange multipliers should be used to solve each exercise.

1. A rural homeowner has found 400 feet of unused wire fencing behind the tractor shed. The goat needs to be fenced in to keep it from eating the garden. The homeowner decides to fence a rectangular area using one side of the tractor shed for one side of the rectangle, and the wire fencing for the other three sides. What dimensions should be used to maximize the area inside the fence?

2. An enclosure is to be made from 160 feet of straight fencing material. Using the sheer rock face of a cliff as one side of a rectangle, the fencing will be used to form the other three sides of the rectangular enclosure. What dimensions should be used to enclose the largest possible area?

3. A dog owner has 90 feet of fencing to make a rectangular pen. The garage will be used as one side of the pen. What dimensions should be used to fence in the largest possible area?

4. For the situation in exercise 3, if a semicircular fence is made instead, would the enclosed area be larger than the best rectangular pen? By how much?

5. A cylindrical can must have volume 200 in³. What dimensions should be used to minimize the amount of material used?

6. A cylindrical can must have volume 320 in³. What dimensions should be used to minimize the amount of material used?

7. An oil refinery is located on the north bank of a straight river that is 2 km wide. A pipeline is to be constructed from the refinery to storage tanks located on the south bank of the river 6 km east of the refinery. The cost of laying the pipe is \$400 000/km over land to a point P on the north bank and \$800 000/km under the river to the tanks. To minimize the cost of the pipeline, where should P be located?

8. Repeat exercise 7 using a cost of only \$600 000/km under the river.

9. A farmer with a small number of cattle wants to fence a rectangular field with the field divided in half by a fence parallel to one of the sides of the rectangle. The farmer wishes to enclose a total area of 450 000 ft². What dimensions should be used for the rectangle to minimize the amount of fencing used?

10. A rectangular field of area 28 800 m² is to be enclosed and divided into three lots by parallels to one of the sides. What should be the dimensions of the field if the amount of fencing is to be a minimum?

Exercises 1–22 are repeated from Section 3.8 of *Calculus Set Free*, but not necessarily in the same order.

Exercises in this section marked (Love) are from *Differential and Integral Calculus* by Clyde E. Love, The MacMillan Company, 1926.

Exercises in this section marked (GSL) are from *Elements of the Differential and Integral Calculus*, by William Granville, Percey Smith, and William Longley, Ginn and Company, 1941 revised edition.

11. (Love) A Norman window consists of a rectangle surmounted by a semi-circle. If the perimeter of the window is to be 20 feet, what dimensions give the most light?

12. (Love) A person in a rowboat 6 miles from shore desires to reach a point on the shore at a distance of 10 miles from the person's current position. If the individual can walk 4 mph and row 3 mph, where should the person land to reach the destination in the shortest possible time?

13. A college student is staying at a house on a beach that is along a straight shoreline. The student receives a call from a friend who is on an island 6 miles down the shore and $\frac{1}{2}$ mile offshore, challenging the student to run and swim to the island in less than 1 hour. If the student can swim at a pace of 25 min/mi and run at a pace of 8 min/mi, can the student meet the challenge? How far should the student run down the shoreline before beginning to swim toward the island to minimize the time required?

14. An athlete who can swim 6 km/h and run 20 km/h is competing in a race with a starting line on an island's beach 2 km off the straight mainland shore to a finish line 10 km down that shore. (If an athlete crosses the water swimming directly toward the shore, the race involves 2 km of swimming and 10 km of running). How far from the finish line should the athlete emerge from the water to minimize the time required to complete the race?

For exercise 15, assume that the can is cylindrical in shape.

15. (GSL) What should be the diameter of a tin can holding 1 quart (58 in^3) and requiring the least amount of tin if the can is open at the top?

16. A square-based, open-top box needs to contain a volume of 25 ft^3. A metal alloy forms the sides and base; there is no top to the container. What dimensions should be used for the container to minimize the material used?

17. If we produce x mini-widgets each month, then we can sell them for $p = 660 - 0.5x$ dollars each. The cost of making x mini-widgets in a month is given by $C(x) = 35\,000 + 540x - 1.5x^2 + 0.004x^3$. (a) When the correct number of mini-widgets is produced to maximize profit, what is the price of a mini-widget? (b) What is the maximum monthly profit?

18. Repeat exercise 17 for a cost function of $C(x) = 44\,000 + 540x - 2.5x^2 + 0.004x^3$ (higher fixed costs, but greater efficiency of mass production—that is, costs per widget that decrease more rapidly with greater production).

19. (GSL) A steel plant is capable of producing x tons per day of a low-grade steel and y tons per day of a high-grade steel, where $y = \dfrac{40 - 5x}{10 - x}$. If the fixed market price of low-grade steel is half that of high-grade steel, how many tons of low-grade steel should be produced per day for maximum revenue?

20. (GSL) A radio manufacturer finds that it can sell x instruments per week at p dollars each, where $5x = 375 - 3p$. The cost of production is $(500 + 15x + \frac{1}{5}x^2)$ dollars. Show that the maximum profit is obtained when the production is about 30 instruments per week.

21. (GSL) Assuming that the strength of a beam with a rectangular cross section varies as the product of the breadth and the square of the depth, what are the dimensions of the strongest beam that can be sawed out of a round log with a diameter of 1 m?

22. (Love) The stiffness of a rectangular beam is proportional to the product of the breadth and the cube of the depth. Find the shape of the stiffest beam that can be cut from a log of diameter 1 m.

23. A side note in this section claims that λ is a rate of change. Verify this for example 66 using the following steps.

    (a) Rework example 66 using a volume of $(100 + \omega)$ cm, where ω is a positive infinitesimal.

    (b) Call the solution (r, h) in example 66 (a_1, b_1) and the solution in part (a) (a_2, b_2). State why the quantity $\dfrac{f(a_2, b_2) - f(a_1, b_1)}{\omega}$ is the rate of change being sought.

    (c) Compute $\dfrac{f(a_2, b_2) - f(a_1, b_1)}{\omega}$. (Hint: you will need to use a variation of multiplying the numerator and the denominator by the conjugate of the expression in the numerator, for cube roots instead of square roots. The method is based on the algebraic identity $a^3 - b^3 = (a - b)(a^2 + ab + b^2)$ instead of $a^2 - b^2 = (a - b)(a + b)$.)

    (d) Compute the value of λ in example 66.

    (e) Verify that the answers to parts (c) and (d) are identical.

24. Three electical generating plants need to produce a combined 800 megawatts of power. Incremental cost data (the cost of producing one more megawatt hour (MWh) at the given level of output) for each plant is given in the tables below. Notice that as output rises, so does the cost of generating additional power.

Exercise 24, although slightly modified, is due to Michael F. Stemper.

A. S. Queen		Coal River		Gr. Knoll	
MW	**$/MWh**	**MW**	**$/MWh**	**MW**	**$/MWh**
187	27.6814	91	27.4532	490	27.6741
189	27.7108	93	27.5017	492	27.6907
191	27.7401	95	27.5502	494	27.7074
193	27.7695	97	27.5988	496	27.7240
195	27.7988	99	27.6473	498	27.7407
197	27.8282	101	27.6958	500	27.7573
199	27.8576	103	27.7443	502	27.7740
201	27.8869	105	27.7929	504	27.7906

Use the following steps to determine how much each plant should produce in order to minimize the total cost.

(a) Let x be the output level (number of megawatts generated) at A. S. Queen, and let $Q(x)$ be the cost of producing x MWh of power. Then $Q'(x)$ is the incremental (marginal) cost at the output level x. The table gives values of the derivative $\frac{dQ}{dx}$, for instance, $Q'(187) = 27.6814$ and $Q'(201) = 27.8869$. Use technology to plot the data points and observe that they appear to be linear.

(b) Either (i) determine the equation of the line through the first and last data points for A. S. Queen or (ii) use technology and least squares regression to determine an equation of the line. Use the result as the equation for $\frac{dQ}{dx}$.

(c) Let y be the output level at Coal River, and let $CR(y)$ be the cost of producing y MWh of power. Repeat (b) to determine an equation for $\frac{dCR}{dy}$.

(d) Let z be the output level at Gr. Knoll, and let $GK(z)$ be the cost of producing z MWh of power. Repeat (b) to determine an equation for $\frac{dGK}{dz}$.

(e) Determine the constraint equation $g(x, y, z)$. Write the total cost function $C(x, y, z)$ in terms of Q, CR, and GK.

(f) Using the components of the gradient ∇C found in (b), (c), and (d), write the system of four simultaneous equations to solve when using Lagrange multipliers.

(g) Use technology to solve the system of equations from (f). How much power should be generated by each plant to minimize the total cost?

(h) Repeat (f) and (g) supposing that we need to generate 803 MW. Do all three plants see an increase from the 800 MW solution? Which plant sees the greatest increase? Can you explain why using the answers to parts (b), (c), and (d)?

Multiple Integrals

4

4.1 Slices, Volumes, and Introduction to Double Integrals

In Chapter 3, we wound our way through the calculus of functions of two variables, mirroring the development of single-variable calculus from graphs to limits to (partial) derivatives and their applications, such as slopes in various directions, extrema, and optimization. Next on the list is integration! Just as with limits and differentiation, the multivariable setting for integration is richer, and a joy to explore.

4.1.1 Slice areas

Consider a surface $z = f(x, y)$ and a slice of that surface corresponding to $x = a$, as in Figure 4.1. The slice $z = f(a, y)$ is a curve rather than a surface, and the region underneath the slice is akin to the region under a curve. The region is two-dimensional and therefore has area. Other than placement in a three-dimensional coordinate system, this appears identical to the area of a region under a curve, and the area can be found by integration. In fact, because $z = f(a, y)$ for a constant a, the slice is a function of just one variable, y, and the area can be found using single-variable methods.

By "the region underneath the slice" we mean the vertical region between the slice and the xy-plane. For this to make sense, the z-coordinates of the surface must be non-negative. Compare to the single-variable calculus concept of "area under the curve" or "area of the region under a curve," where the function is nonnegative and the region extends from the axis to the curve.

When the surface dips below the xy-plane, we can instead use the concept of net area under the curve to make sense of the result.

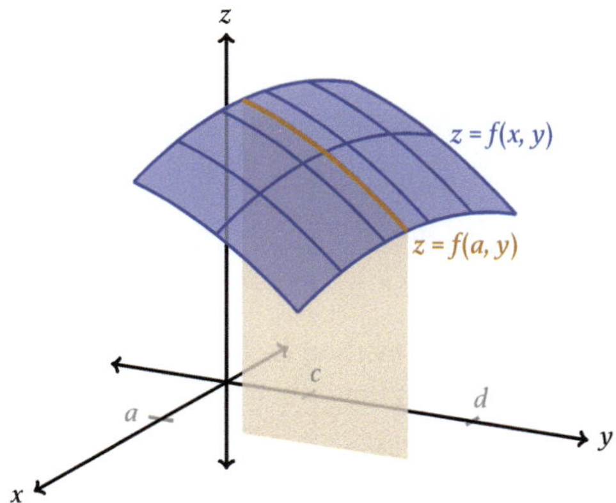

Figure 4.1 *A surface $z = f(x, y)$ (blue), a slice $x = a$ (brown), and the region under the slice (tan)*

Multivariable Calculus Set Free. Charles Bryan Dawson, Oxford University Press.
© C. Bryan Dawson (2026). DOI: 10.1093/oso/9780198984269.003.0005

Example 1 *Find the area under the surface $z = f(x, y) = xy^2$ in the slice $x = 2$ from $y = 0$ to $y = 4$.*

Solution   For the slice $x = 2$, the curve is $z = f(2, y) = 2y^2$. We can therefore find the region under this curve by integrating $2y^2$ from $y = 0$ to $y = 4$:

$$\text{area} = \int_0^4 2y^2 \, dy$$

$$= 2\frac{y^3}{3}\bigg|_0^4$$

$$= 2 \cdot \frac{64}{3} - 0 = \frac{128}{3} \text{ units}^2.$$

$\blacksquare$

Although the surface in Figure 4.1 is not the same surface as in example 1, the figure is still useful for visualization of setting up the integral. We determine the integrand by replacing x with a, which is 2 in this example. The equivalent of the interval of integration is from $y = c$ to $y = d$, which in this example is $y = 0$ to $y = 4$. Because the interval of integration consists of values of y, the variable of integration is y.

Next, we repeat example 1 for the slice $x = \frac{1}{5}$.

Example 2 *Find the area under the surface $z = f(x, y) = xy^2$ in the slice $x = \frac{1}{5}$ from $y = 0$ to $y = 4$.*

Solution   For the slice $x = \frac{1}{5}$, the curve is $z = f(\frac{1}{5}, y) = \frac{1}{5}y^2$. We can therefore find the region under this curve by integrating $\frac{1}{5}y^2$ from $y = 0$ to $y = 4$:

$$\text{area} = \int_0^4 \frac{1}{5}y^2 \, dy$$

$$= \frac{1}{5} \cdot \frac{y^3}{3}\bigg|_0^4$$

$$= \frac{1}{5} \cdot \frac{64}{3} - 0 = \frac{64}{15} \text{ units}^2.$$

$\blacksquare$

It is apparent that for any value of x, the calculation works the same way as examples 1 and 2. Now let's repeat these examples for a generic value of x, considering x as a constant.

Example 3 *Find the area under the surface $z = f(x, y) = xy^2$ in the slice x, where x is an unspecified constant, from $y = 0$ to $y = 4$.*

Solution   Following the solution method for examples 1 and 2, we integrate xy^2, where x is an unspecified constant, from $y = 0$ to $y = 4$:

$$\text{area} = \int_0^4 xy^2 \, dy$$

$$= x\frac{y^3}{3}\bigg|_0^4$$

$$= x \cdot \frac{64}{3} - 0 = \frac{64}{3}x \text{ units}^2.$$

$\blacksquare$

Reading Exercise 1   Find the area under the surface $z = f(x, y) = xy^2$ in the slice
$x = 3$ from $y = 0$ to $y = 4$.

Notice how the antiderivative in example 3 is calculated. The variable x is treated as a constant, so that an antiderivative of xy^2 is $x \cdot \frac{y^3}{3}$, just as an antiderivative of $2y^2$ is $2 \cdot \frac{y^3}{3}$ in example 1. But treating x as a constant means we are performing *partial antidifferentiation*. Do the reverse–find the partial derivative of $x\frac{y^3}{3}$ with respect to y:

$$\frac{\partial}{\partial y}\left(x\frac{y^3}{3}\right) = \frac{x}{3} \cdot 3y^2 = xy^2,$$

therefore a partial antiderivative of xy^2 with respect to y is $x\frac{y^3}{3}$.

PARTIAL ANTIDERIVATIVES

To find a partial antiderivative of $f(x, y)$ with respect to y, treat x as a constant and use antiderivative formulas, rules, and techniques applied to the variable y.
   To find a partial antiderivative of $f(x, y)$ with respect to x, treat y as a constant and use antiderivative formulas, rules, and techniques applied to the variable x.

4.1.2   Slice area examples

The result of example 3, area $= \frac{64}{3}x$ units^2, can be considered a function of the variable x. When $x = 2$, the area is $\frac{64}{3} \cdot 2 = \frac{128}{3}$ units^2, matching the solution to example 1. Similarly, if $x = \frac{1}{5}$, the area is $\frac{64}{3} \cdot \frac{1}{5} = \frac{64}{15}$ units^2, matching the solution to example 2. For whichever slice we want, the solution to example 3 gives us the area under the surface in that slice.

SLICE AREA FUNCTION

Let $z = f(x, y)$ be a surface. Then

$$\int_{y=a}^{y=b} f(x, y)\, dy$$

The formula for the area under the surface for a specific slice can be written similarly; for instance, the area under the surface for the slice $x = 2$ is $\int_{y=a}^{y=b} f(2, y)\, dy$. More generally, the area under the surface for the slice $x = k$ is $\int_{y=a}^{y=b} f(k, y)\, dy$.

is a function of x representing the area under the surface in the slice x, for any value of x (where defined). Similarly,

$$\int_{x=a}^{x=b} f(x, y)\, dx$$

is a function of y representing the area under the surface in the slice y, for any value of y (where defined).

Example 4 *(a) Find the area under the surface $z = x\sin y$ in the slice $x = 2$ from $y = 0$ to $y = \pi$. (b) Find the area under the surface $z = x\sin y$ in the slice $x = 3$ from $y = 0$ to $y = \pi$. (c) Find the area under the surface $z = x\sin y$ for any slice x from $y = 0$ to $y = \pi$.*

Solution

(a) The requested area is the area under the curve $z = f(2, y) = 2\sin y$ from $y = 0$ to $y = \pi$:

Line 1 uses the formula for area under a curve; line 2 finds an antiderivative; line 3 evaluates the antiderivative at π, at 0, and subtracts, and then completes the arithmetic.

$$\text{area} = \int_0^\pi 2\sin y\, dy$$

$$= -2\cos y\,\Big|_0^\pi$$

$$= -2\cos\pi - (-2\cos 0) = (-2)(-1) + 2(1) = 4\,\text{units}^2.$$

(b) The requested area is the area under the curve $z = f(3, y) = 3\sin y$ from $y = 0$ to $y = \pi$:

Line 1 uses the formula for area under a curve; line 2 finds an antiderivative; line 3 evaluates the antiderivative at π, at 0, and subtracts, and then completes the arithmetic.

$$\text{area} = \int_0^\pi 3\sin y\, dy$$

$$= -3\cos y\,\Big|_0^\pi$$

$$= -3\cos\pi - (-3\cos 0) = (-3)(-1) + 3(1) = 6\,\text{units}^2.$$

(c) The requested area is the area under the curve $z = f(x, y) = x\sin y$ from $y = 0$ to $y = \pi$, where x is an unspecified constant:

Line 1 uses the slice area function formula; line 2 finds a partial antiderivative with respect to y, treating x as a constant; line 3 evaluates the antiderivative at $y = \pi$, at $y = 0$, and subtracts, and then simplifies, the final answer being a function of x.

$$\text{area} = \int_0^\pi x\sin y\, dy$$

$$= -x\cos y\,\Big|_0^\pi$$

$$= -x\cos\pi - (-x\cos 0) = (-x)(-1) + x(1) = 2x\,\text{units}^2.$$

An alternate solution is to complete part (c) first, then evaluate the result with $x = 2$ for (a) and $x = 3$ for (b). ∎

Notice in example 4 part (c) that if x is negative, the slice "area" is negative. This corresponds to the idea of "net area under the curve" for a single-variable function.

Example 5 *Find the net area under the surface $z = x^3y^2 - y$ (a) in the slice $y = 1$ from $x = -2$ to $x = 2$, (b) in the slice $y = -2$ from $x = -2$ to $x = 2$, and (c) for any slice y from $x = -2$ to $x = 2$.*

Solution For efficiency, we answer part (c) first. Using the slice area function formula,

$$\text{area} = \int_{-2}^{2} (x^3y^2 - y)\, dx$$

$$= \left(\frac{x^4}{4}y^2 - xy \right)\Big|_{-2}^{2}$$

$$= \frac{16}{4}y^2 - 2y - \left(\frac{16}{4}y^2 - (-2)y \right)$$

$$= -4y \text{ units}^2.$$

Line 1 uses the slice area function formula for a slice y; line 2 finds a partial antiderivative with respect to x, treating y as a constant; line 3 evaluates the antiderivative at $x = 2$, at $x = -2$, and subtracts; line 4 simplifies. The final answer is a function of y.

Then for part (a) the net area under the curve for the slice $y = 1$ is found by evaluating the slice area function at $y = 1$:

$$-4y\Big|_{y=1} = -4(1) = -4 \text{ units}^2.$$

For part (b) we evaluate the slice area function at $y = -2$; the net area under the curve is

$$-4y\Big|_{y=-2} = -4(-2) = 8 \text{ units}^2.$$ ∎

Reading Exercise 2 Find the area under the surface $z = x \cos y$ for any slice x from $y = 0$ to $y = \frac{\pi}{2}$.

4.1.3 Volumes

How might we determine the volume under a surface? In particular, suppose we have a surface $z = f(x, y)$ over a rectangular region $c \leq x \leq d, a \leq y \leq b$ (Figure 4.2) and wish to find the volume under its surface.

Consider again the slice in Figure 4.1. The slice is two-dimensional and possesses area, but not volume. But what if we endowed the slice with a thickness Δx to make it three-dimensional? The result would no longer be a slice, but rather a box

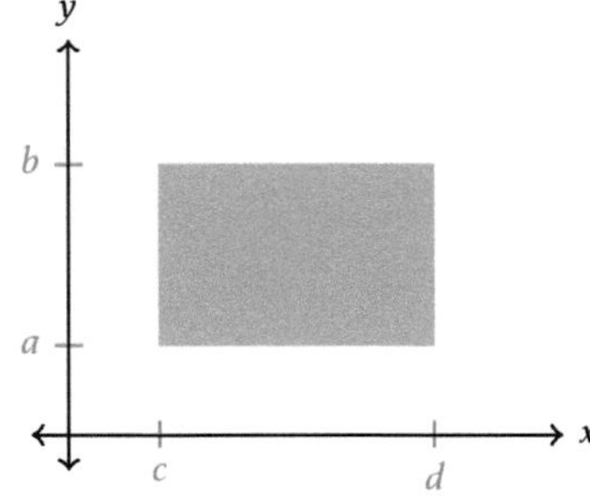

Figure 4.2 *The rectangular region $c \leq x \leq d, a \leq y \leq b$ (gray)*

Although we are assuming in this discussion that f is nonnegative throughout the rectangular region, see the discussion of "net volume" later in this section.

Figure 4.3 *A slice endowed with a thickness* Δx, *forming a box*

with the slice as one face, a parallel and congruent face, and four connecting faces, as in Figure 4.3. The volume of the box would be the area of the slice times the thickness of the box, or

$$\text{box volume} = \text{slice area} \cdot \text{thickness}$$

$$= \left(\int_{y=a}^{y=b} f(x_k, y)\, dy \right) \cdot \Delta x,$$

for the slice x_k.

Return to the volume under the surface $z = f(x, y)$ (Figure 4.4). One idea for determining the volume is to partition the interval $x = c$ to $x = d$ into intervals of width Δx, as done when developing formulas for the area under a curve, arc length, and many other geometric and physical settings. Then the volume under the surface can be approximated by the total volumes of the resulting boxes, each resembling the box in Figure 4.3. With finitely many such boxes as in Figure 4.4, the boxes might miss some of the volume under the surface and might include extra volume not under the surface. But if we use infinitely many equal-width boxes (Figure 4.5), that is, slices endowed with infinitesimal thickness, the differences should be infinitesimal and the result should render the exact volume under the surface. Using Ω slices and right-hand endpoints, the volume of the region under the surface is

$$\text{volume} = \text{total volume of slice boxes}$$

$$= \sum_{k=1}^{\Omega} \left(\int_{y=a}^{y=b} f(x_k, y)\, dy \right) \cdot \Delta x,$$

which is an omega sum! As always, the omega sum can be rewritten as a definite integral by replacing the summation with the integral symbol, dropping

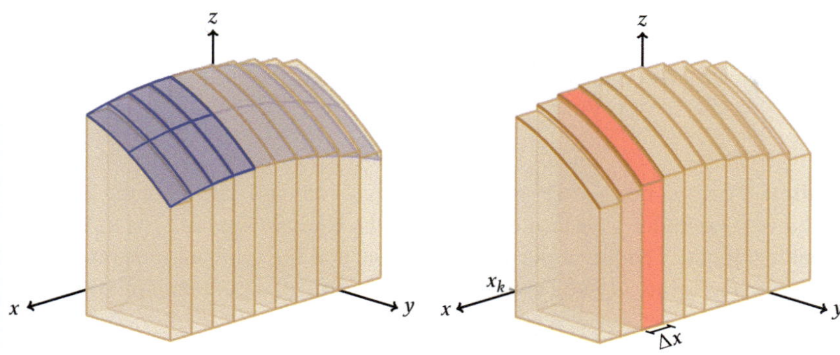

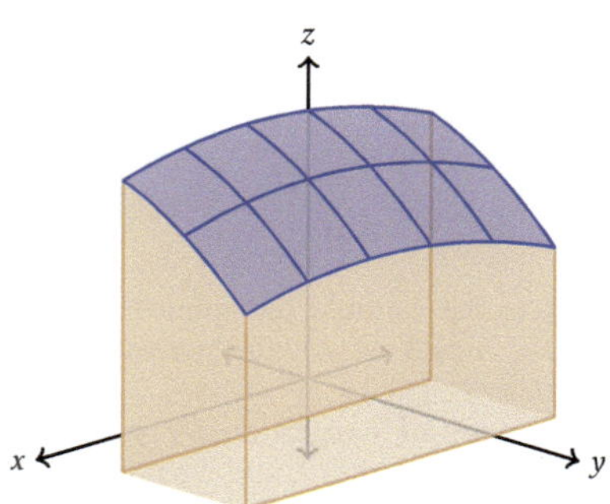

Figure 4.4 *(Left) a surface $z = f(x, y)$ (blue) along with boxes for approximating the volume under the surface (brown). The surface appears darker when it lies above the tops of the boxes, and lighter when it lies below the tops of the boxes. (right) The approximating boxes with the surface removed. The approximating box with heights matching the surface at slice x_k is highlighted; the width of each box is Δx*

Figure 4.5 *A surface $z = f(x, y)$ (blue) along with Ω boxes approximating the region under the surface (brown), where Ω is an infinite hyperreal integer*

the subscripts, replacing Δx with dx, and inserting the limits of integration. The volume is

$$V = \int_{x=c}^{x=d} \left(\int_{y=a}^{y=b} f(x, y)\, dy \right) dx.$$

The *double integral* $\int_{x=c}^{x=d} \left(\int_{y=a}^{y=b} f(x, y)\, dy \right) dx$ is an *iterated integral*, meaning that the integral inside the parentheses is evaluated first. Because the inside integral is a slice area for the slice x, the "inside" integral is evaluated by finding a partial antiderivative, treating x as a constant. The result of the inside integral is a function of x, which can therefore be integrated (the "outside" integral) with respect to x.

$$\underbrace{\int_{x=c}^{x=d} \left(\underbrace{\int_{y=a}^{y=b} f(x, y)\, dy}_{\substack{\text{do first} \\ \text{treat } x \text{ as a constant} \\ \text{eliminates } y}} \right) dx}_{\text{do second}}$$

VOLUME UNDER A SURFACE, RECTANGULAR DOMAIN

Suppose $z = f(x, y)$ is continuous on a rectangular domain $c \le x \le d, a \le y \le b$. Then, the (net) volume under the surface on this rectangular domain is

$$V = \int_{x=c}^{x=d} \left(\int_{y=a}^{y=b} f(x, y)\, dy \right) dx.$$

Ans. to reading exercise 2:
x units^2

4.1.4 Volume examples

Example 6 *Find the volume under the surface $z = xy^2$ over the region $1 \le x \le 3$,*
$0 \le y \le 4$.

Solution The function $z = xy^2$ is continuous (polynomial), therefore the formula
for the volume under a surface on a rectangular domain applies. The volume is
a double integral. The integrand is the function, and the limits of integration are
taken from the values of x and y that define the rectangular domain. The volume is

Line 1 is the formula for volume under a surface over a rectangular domain; line 2 replaces $f(x, y)$ by xy^2, writes 0 and 4 as the limits of integration for y (from $0 \le y \le 4$), and writes 1 and 3 as the limits of integration for x (from $1 \le x \le 3$); line 3 works inside the parentheses; a partial antiderivative of xy^2 with respect to y is found by treating x as a constant; line 4 still works inside the parentheses; the partial antiderivative is evaluated at $y = 4$ and at $y = 0$, the latter being subtracted; after simplification, we see an integral of the variable x; line 5 finds an antiderivative with respect to x; line 6 evaluates at $x = 3$, at $x = 1$, and subtracts, writing units^3 because volume is three-dimensional.

$$
\begin{aligned}
V &= \int_{x=c}^{x=d} \left(\int_{y=a}^{y=b} f(x,y)\, dy \right) dx \\[2mm]
&= \int_{1}^{3} \left(\int_{0}^{4} xy^2\, dy \right) dx \\[2mm]
&= \int_{1}^{3} \left(x \cdot \frac{y^3}{3} \Big|_{0}^{4} \right) dx \\[2mm]
&= \int_{1}^{3} \left(\frac{64}{3}x - 0 \right) dx = \int_{1}^{3} \frac{64}{3} x\, dx \\[2mm]
&= \frac{64}{3} \cdot \frac{x^2}{2} \Big|_{1}^{3} = \frac{32}{3} x^2 \Big|_{1}^{3} \\[2mm]
&= \frac{32}{3}(9) - \frac{32}{3}(1) = \frac{256}{3}\ \text{units}^3 .
\end{aligned}
$$

$\blacksquare$

Although not required, using a computer algebra system (CAS) to graph the
surface and the region below it can be helpful for building visualization skills and
intuition. The graph of the surface and region whose volume is found in example 6
is in Figure 4.6.

Example 7 *Find the net volume under the surface $f(x, y) = \sin x + \cos y$ over the*
rectangular region $0 \le x \le \pi, 0 \le y \le \pi$.

Solution The function is continuous (both sine and cosine are defined everywhere), so the formula for the net volume under a surface on a rectangular domain
applies. The net volume is a double integral. The integrand is the function, and the
limits of integration are taken from the values of x and y that define the rectangular
domain. Then

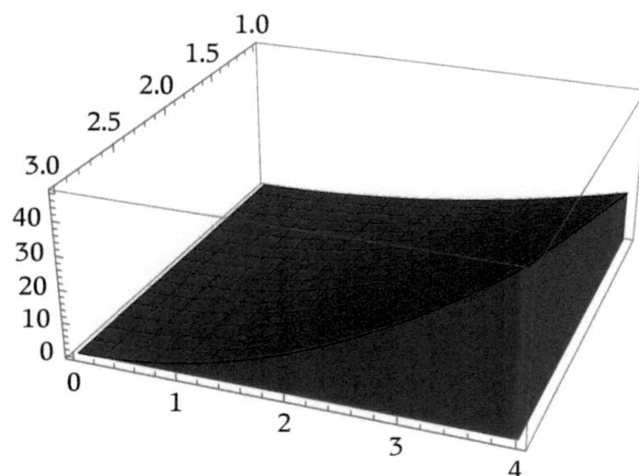

Figure 4.6 *The surface (blue) and region under the surface (brown) of example 6. The volume of the brown region is $\dfrac{256}{3}$ units^3*

$$V = \int_{x=c}^{x=d} \left(\int_{y=a}^{y=b} f(x,y)\, dy \right) dx$$

$$= \int_{0}^{\pi} \left(\int_{0}^{\pi} (\sin x + \cos y)\, dy \right) dx$$

$$= \int_{0}^{\pi} \left((y \sin x + \sin y)\Big|_{0}^{\pi} \right) dx$$

$$= \int_{0}^{\pi} (\pi \sin x + \sin \pi - (0 + \sin 0))\, dx = \int_{0}^{\pi} \pi \sin x\, dx$$

$$= -\pi \cos x \Big|_{0}^{\pi}$$

$$= -\pi \cos \pi - (-\pi \cos 0) = -\pi(-1) + \pi(1) = 2\pi \text{ units}^3.$$

The net volume under the surface is 2π units^3. ∎

Line 1 is the formula for volume under a surface over a rectangular domain; line 2 replaces $f(x,y)$ by $\sin x + \cos y$, writes 0 and π as the limits of integration for y (from $0 \le y \le \pi$), and writes 0 and π as the limits of integration for x (from $0 \le x \le \pi$); line 3 works inside the (outer) parentheses; a partial antiderivative of $\sin x + \cos y$ with respect to y is found by treating x as a constant (if x is a constant, then so is $\sin x$, so its antiderivative is the constant times the variable y); line 4 still works inside the (outer) parentheses; the partial antiderivative is evaluated at $y = \pi$ and at $y = 0$, the latter being subtracted; line 5 finds an antiderivative; line 6 evaluates at $x = \pi$, at $x = 0$, and subtracts.

The graph of the surface of example 7, along with the region between the surface and the xy-plane, is in Figure 4.7. Some of the surface lies above the xy-plane, and some lies below. Where the surface lies above the xy-plane, the region between the surface and the xy-plane contributes positively to the net volume (brown region in Figure 4.7). Where the surface lies below the xy-plane, the region between the surface and the xy-plane contributes negatively to the net volume (red region in Figure 4.7). Because the net volume is positive, more volume lies in the brown region than is subtracted out by the red region.

Think of the surface in Figure 4.7 as the surface of the ground at a construction site that must be flattened so that the entire site is at z-coordinate 0. Some of the excess dirt (brown in Figure 4.7) can be used to fill in the low region (red in Figure 4.7), but there will still be some dirt left over because the net volume is positive. In fact, the amount of leftover dirt that must be removed from the site is the net volume 2π units^3. If the net volume had been negative, we would need to bring in extra dirt in the amount indicated by the net volume.

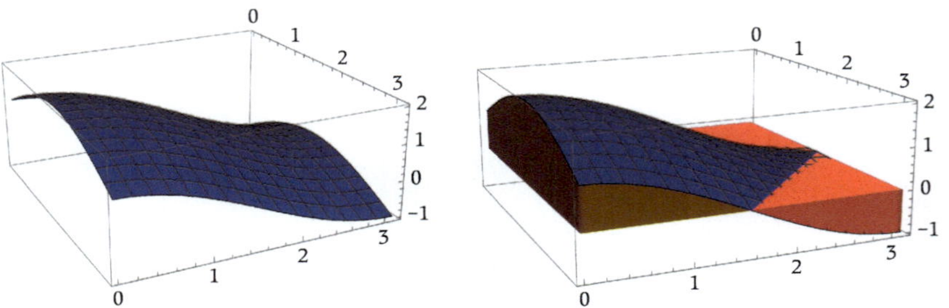

Figure 4.7 *(Left) the surface (blue) of the function of example 7. (right) The brown region lies between the surface and the xy-plane where function is positive; the red region lies between the surface and the xy-plane where the function is negative*

This is the same function as example 6, but on a different region (a different domain).

Reading Exercise 3   Find the volume under the surface $z = xy^2$ over the region $0 \leq x \leq 2, 0 \leq y \leq 3$.

4.1.5   Double integral examples

Double integrals will be more precisely defined in the next section. However, we have already practiced their computation when finding volume, and the computation method does not change when the formal definition is introduced.

The double integral

$$\int_{x=c}^{x=d} \left(\int_{y=a}^{y=b} f(x, y) \, dy \right) dx$$

is often written as

$$\int_{c}^{d} \int_{a}^{b} f(x, y) \, dy \, dx,$$

with the understanding that the innermost integral's limits of integration are values of y and are used first, and the outermost integral's limits of integration are values of x and are used last. When written without parentheses, parentheses around the inside integral are implied.

Example 8 *Evaluate* $\displaystyle\int_{0}^{\frac{\pi}{2}} \left(\int_{0}^{1} (xe^y + \cos x) \, dy \right) dx.$

Solution   We begin by evaluating the integral on the inside. Because the inside integral is with respect to y, we treat x as a constant:

$$\int_0^{\frac{\pi}{2}} \left(\int_0^1 (xe^y + \cos x)\, dy \right) dx$$

$$= \int_0^{\frac{\pi}{2}} \left((xe^y + (\cos x)y) \Big|_0^1 \right) dx$$

$$= \int_0^{\frac{\pi}{2}} \left(xe^1 + (\cos x)\cdot 1 - (xe^0 + (\cos x)\cdot 0) \right) dx$$

$$= \int_0^{\frac{\pi}{2}} (xe + \cos x - x)\, dx = \int_0^{\frac{\pi}{2}} (x(e-1) + \cos x)\, dx.$$

Line 1 is the double integral we seek to evaluate; line 2 works inside the parentheses, finding an antiderivative with respect to y, treating x as a constant (if x is a constant, so is $\cos x$, so its antiderivative is the constant times the variable y); line 3 works inside the parentheses, evaluating at $y = 1$, at $y = 0$, and subtracting; line 4 simplifies.

Next, we evaluate the remaining integral, which is a familiar single-variable integral, with only the variable x, and not the variable y:

$$= \left((e-1)\frac{x^2}{2} + \sin x \right) \Big|_0^{\frac{\pi}{2}}$$

$$= (e-1)\frac{\frac{\pi^2}{4}}{2} + \sin \frac{\pi}{2} - (0 + \sin 0)$$

$$= (e-1)\frac{\pi^2}{8} + 1.$$

The value of the double integral is $(e-1)\frac{\pi^2}{8} + 1$. ∎

No units are attached to the answer because the integral is not placed in a context.

All of our double integrals so far have y as the inside integral's variable of integration, with x as the outside integral's variable of integration. That is not a requirement. Just as variables other than x and y may be used, any variable can serve as the inside integral's variable of integration, and likewise for the outside integral.

Example 9 *Evaluate* $\displaystyle\int_{-1}^{3} \int_{-1}^{1} \frac{y^2}{1+x^2}\, dx\, dy.$

Solution The double integral can be rewritten by making the implied parentheses explicit:

$$\int_{-1}^{3} \left(\int_{-1}^{1} \frac{y^2}{1+x^2}\, dx \right) dy.$$

Although it is not required to insert the implied parentheses, doing so can reduce errors; the parentheses are helpful for keeping track of which variable of integration is being used at each stage of the computation.

We integrate with respect to x first, treating y as a constant, working inside the parentheses:

$$\int_{-1}^{3} \left(\int_{-1}^{1} \frac{y^2}{1+x^2}\, dx \right) dy = \int_{-1}^{3} \left(\int_{-1}^{1} y^2 \cdot \frac{1}{1+x^2}\, dx \right) dy$$

$$= \int_{-1}^{3} \left(y^2 \tan^{-1} x \Big|_{-1}^{1} \right) dy$$

$$= \int_{-1}^{3} \left(y^2 \tan^{-1} 1 - y^2 \tan^{-1}(-1) \right) dy$$

$$= \int_{-1}^{3} \left(y^2 \frac{\pi}{4} - y^2 \left(-\frac{\pi}{4} \right) \right) dy = \int_{-1}^{3} \frac{\pi}{2} y^2\, dy.$$

Line 1 rewrites the fraction to make it clear that the integrand is the constant y^2 times the quantity $\frac{1}{1+x^2}$; line 2 finds a partial antiderivative with respect to x, treating y as a constant; line 3 evaluates at $x = 1$, at $x = -1$, and subtracts; line 4 simplifies.

The result of the inner integral is a function of y, and the variable x is eliminated. We finish by integrating with respect to y:

$$\int_{-1}^{3} \frac{\pi}{2} y^2\, dy = \frac{\pi}{2} \cdot \frac{y^3}{3} \Big|_{-1}^{3}$$

$$= \frac{\pi}{2} \left(\frac{27}{3} \right) - \frac{\pi}{2} \left(-\frac{1}{3} \right) = \frac{14}{3} \pi.$$

Ans. to reading exercise 3: $18\, \text{units}^3$

The value of the double integral is $\frac{14}{3}\pi$. ∎

Any of the techniques of integration that you have learned, such as substitution, may be required to evaluate a double integral.

Example 10 *Evaluate* $\displaystyle\int_{1}^{4} \int_{0}^{2} y(x + y^2)^5\, dy\, dx.$

Solution We first rewrite the double integral using parentheses as

$$\int_{1}^{4} \left(\int_{0}^{2} y(x + y^2)^5\, dy \right) dx.$$

Working inside the parentheses, we need to find an antiderivative with respect to y, treating x as a constant. Because the expression $(x + y^2)^5$ is "not just plain y" to a power, we need substitution. When using a technique of integration to evaluate an inside integral, it is sometimes helpful to write it by itself, and afterward insert the result inside the parentheses of the original integral. To that end, we desire to evaluate $\int_{0}^{2} y(x + y^2)^5\, dy$, and use the substitution

Because we are treating x as a constant, this is similar to recognizing that $(4 + y^2)^5$ is "not just plain y" to a power.

$$u = x + y^2$$

Because x is treated as a constant, the procedure for finding the differential du is the same as partial differentiation, times dy.

$$du = 2y\, dy.$$

When substituting, we need to replace values of y with values of u:

$$y = 2 : \quad u = x + 4$$
$$y = 0 : \quad u = x$$

Completing the substitution and integrating, we have

$$\int_0^2 y(x + y^2)^5 \, dy = \frac{1}{2} \int_0^2 2y(x + y^2)^5 \, dy$$

$$= \frac{1}{2} \int_x^{x+4} u^5 \, du$$

$$= \frac{1}{2} \cdot \frac{u^6}{6} \Big|_x^{x+4}$$

$$= \frac{1}{12}(x + 4)^6 - \frac{1}{12}x^6.$$

Line 1 adjusts for the constant 2 that is needed for du; line 2 completes the substitution, replacing $2y \, dy$ with du, replacing $x + y^2$ with u, and replacing the limits of integration for y with limits of integration for u; line 3 finds an antiderivative; line 4 evaluates at $u = x + 4$, at $u = x$, and subtracts. Notice that this is a definite integral and the answer is a constant (x is treated as a constant); this is not an indefinite integral, so we do not insert $+C$.

Now that we have evaluated the inside integral, we insert the result in the parentheses of the outer integral and proceed with the evaluation.

$$\int_1^4 \left(\frac{1}{12}(x + 4)^6 - \frac{1}{12}x^6 \right) dx = \int_1^4 \frac{1}{12}(x + 4)^6 \, dx - \int_1^4 \frac{1}{12}x^6 \, dx.$$

Splitting into two integrals using the antiderivative difference (or sum) rule is useful when one portion of the expression requires substitution and another part does not, or when two different substitutions are indicated.

The integral on the left is "not just plain x" to a power, so we use substitution; the integral on the right does not require substitution. For the integral on the left, we use $u = x + 4$ and $du = dx$. Continuing, we have

$$= \int_5^8 \frac{1}{12}u^6 \, du - \int_1^4 \frac{1}{12}x^6 \, dx$$

$$= \frac{1}{12} \cdot \frac{u^7}{7} \Big|_5^8 - \frac{1}{12} \cdot \frac{x^7}{7} \Big|_1^4$$

$$= \frac{1}{84}(8^7 - 5^7) - \frac{1}{84}(4^7 - 1^7) = 23\,841.$$

Because the "not just plain x" expression is of the form $x + k$, we may skip substitution and evaluate as if it is "just plain x." However, being in the habit of writing these substitutions can reduce errors overall.

The value of the double integral is $23\,841$.

■

EXERCISES 4.1

1–10. Find the (net) area under the surface for the indicated slice.

    1. surface $z = x^3 - y$, (a) slice $y = 2$, (b) slice $y = -1$, (c) any slice y, all from $x = 0$ to $x = 3$

2. surface $z = x^2 y^3$, (a) slice $x = -1$, (b) slice $x = 3$, (c) any slice x, all from $y = -1$ to $y = 2$

3. surface $z = x^3 - y$, (a) slice $x = 2$, (b) slice $x = -1$, (c) any slice x, all from $y = 0$ to $y = 3$

4. surface $z = x^2 y^3$, (a) slice $y = -1$, (b) slice $y = 2$, (c) any slice y, all from $x = 0$ to $x = 5$

5. surface $z = x^y$, (a) slice $x = 1$, (b) slice $x = 3$, (c) any slice x where $x > 0$ and $x \neq 1$, all for $-1 \leq y \leq 1$

6. surface $z = \dfrac{1}{(\sin x)\sqrt{y}}$, (a) slice $x = \dfrac{\pi}{2}$ for $1 \leq y \leq 4$, (b) slice $y = 1$, for $\dfrac{\pi}{4} \leq x \leq \dfrac{\pi}{2}$

7. surface $z = x^y$, (a) slice $y = -1$, (b) slice $y = 2$, (c) any slice y where $y \neq -1$, all from $x = 1$ to $x = 3$

8. surface $z = \dfrac{x}{\sqrt{1 - y^2}}$, (a) slice $x = 0$, (b) slice $x = 1$, (c) any slice x, from $y = -\dfrac{1}{\sqrt{2}}$ to $y = \dfrac{1}{\sqrt{2}}$

9. surface $z = xy \cos x^2 y$, (a) slice $y = 2$ from $x = 0$ to $x = 2$, (b) slice $x = 3$ from $y = 0$ to $y = \pi$

10. surface $z = \dfrac{x}{x^2 - xy - 2}$, (a) slice $y = 0$ from $x = 3$ to $x = 4$, (b) slice $y = 1$ from $x = 3$ to $x = 4$

11–20. Find the (net) volume under the surface over the given rectangular region.

11. surface $f(x, y) = xy + 4x + 5y$, region $0 \leq x \leq 2, 1 \leq y \leq 4$

12. surface $f(x, y) = 3x^2 y^2 + 4$, region $-1 \leq x \leq 1, 0 \leq y \leq 1$

13. surface $f(x, y) = \sin x \cos y$, region $0 \leq x \leq \pi, 0 \leq y \leq \dfrac{\pi}{2}$

14. surface $f(x, y) = x + y^2 \sqrt{x}$, region $0 \leq x \leq 2, 1 \leq y \leq 4$

15. surface $z = xe^y$, region $[0, 2] \times [\ln 2, \ln 3]$

16. surface $z = y^2 \sinh x$, region $[0, 2] \times [-1, 1]$

17. surface $f(x, y) = x^3 + 0.5y^2$, region $\{(x, y) \mid -3 \leq x \leq 3, -3 \leq y \leq 3\}$ (funky chair)

18. surface $f(x, y) = 2 \sin x \sin y$, region $\{(x, y) \mid -2\pi \leq x \leq 2\pi, -2\pi \leq y \leq 2\pi\}$ (egg crate mattress)

19. surface $z = xy \cosh x^2$, region $1 \leq x \leq 1.02, 0 \leq y \leq 2$

20. surface $z = \sec^2 y + \dfrac{\ln x}{x}$, region $1 \leq x \leq 2, 0 \leq y \leq \dfrac{\pi}{4}$

21–30. Evaluate the double integral.

21. $\displaystyle \int_{-1}^{1} \int_{0}^{5} (x^2 - y^2)\, dy\, dx$

22. $\displaystyle \int_{1}^{2} \int_{0}^{1} 4x^2 y^3\, dy\, dx$

23. $\displaystyle \int_{-2}^{3} \int_{0}^{1} (4x - xy^3)\, dy\, dx$

The notation $[0, 2] \times [\ln 2, \ln 3]$ means the set of points (x, y) for which $0 \leq x \leq 2$ and $\ln 2 \leq y \leq \ln 3$.

24. $\displaystyle\int_0^3 \int_1^2 (xy - 3y^2)\, dy\, dx$

25. $\displaystyle\int_0^1 \int_{-2}^3 (4x - xy^3)\, dx\, dy$

26. $\displaystyle\int_1^2 \int_0^3 (xy - 3y^2)\, dx\, dy$

27. $\displaystyle\int_{-2}^3 \int_0^1 (4x - xy^3)\, dx\, dy$

28. $\displaystyle\int_0^1 \int_0^\pi (e^{2x} + \sin y)\, dy\, dx$

29. $\displaystyle\int_0^{\frac{\pi}{2}} \int_0^{\sqrt{\pi}} x \sin(x^2 - y)\, dx\, dy$

30. $\displaystyle\int_0^1 \int_0^1 xye^{x^2 y}\, dx\, dy$

31. Use a CAS to graph the surface $z = x^y$ along with the slice $x = 3$ for $-1 \le y \le 1$. Repeat for the slices $x = 1$ and $x = \frac{1}{2}$.

32. Use a CAS to graph the surface $z = \dfrac{x}{\sqrt{1-y^2}}$ along with the slice $x = 1$ for $-\frac{1}{\sqrt{2}} \le y \le \frac{1}{\sqrt{2}}$. Repeat for the slices $x = 0$ and $x = -1$.

33. Use a CAS to produce a graph of the type in Figure 4.6 for the setting of (a) exercise 11 and (b) exercise 13.

34. Use a CAS to produce a graph of the type in Figure 4.6 for the setting of (a) exercise 14 and (b) exercise 12.

To complete exercise 31 using Desmos, graph the slice using a restriction. For instance, try $x = 3\{0 < z < x^y\}$, where the braces represent the restriction, to color the portion of the slice $x = 3$ that lies between the xy-plane and the surface.

4.2 Double Integrals, Continued: Definition, Properties, Applications

Although we have discussed how to calculate a double integral and we have given a formula for net volume under a surface, we have not yet stated a formal definition of double integral. The definition of double integral over a rectangular region can be motivated by reconsidering the volume under a surface.

4.2.1 Definition of double integral

Suppose $z = f(x,y)$ is defined on a rectangular region $a \le x \le b,\ c \le y \le d$ (Figure 4.8). We wish to find the volume under the surface over this rectangular region.

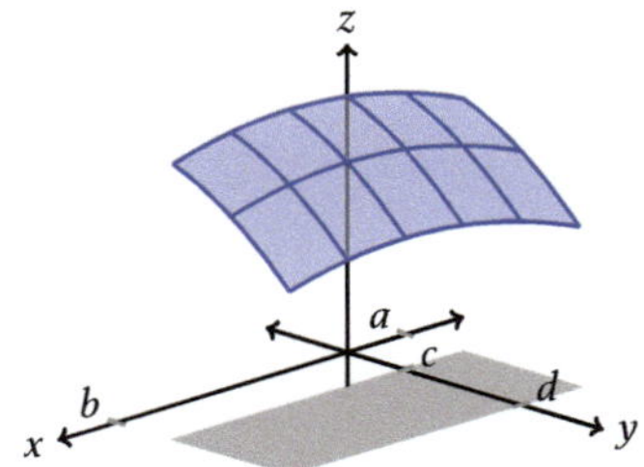

Figure 4.8 *A surface $z = f(x,y)$ (blue) defined on a rectangular region $a \le x \le b,\ c \le y \le d$ (gray), shown in the xy-plane; the rectangular region resembles the shadow of the surface, if the sun is directly overhead and the xy-plane is the ground*

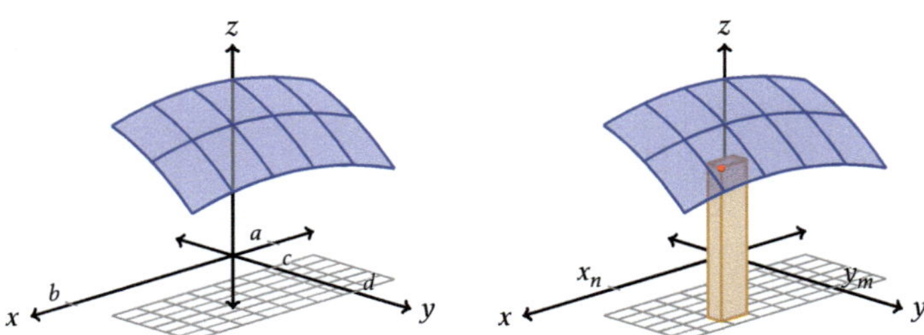

Figure 4.9 *(Left) a surface $z = f(x, y)$ (blue) over a rectangular region partitioned into smaller rectangles; (right) the n-mth rectangular box is added to approximate the volume under the surface over the n-mth rectangle*

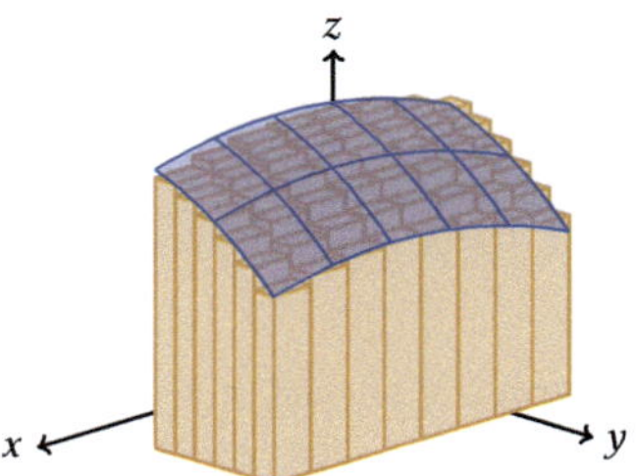

Figure 4.10 *A surface $z = f(x, y)$ (blue) over a rectangular region partitioned into smaller rectangles, along with rectangular boxes approximating the volume under the surface*

As usual, Ω is a positive infinite hyperreal integer.

Following the development of the definite integral as the area under a curve, we partition the interval $a \le x \le b$ into Ω subintervals, using $\Delta x = \frac{b-a}{\Omega} = (b - a)\omega$ and $x_k = a + k\Delta x$ for $k = 0, 1, \ldots, \Omega$. But with an added dimension, we have another interval to partition. Again dividing into Ω subintervals, $\Delta y = \frac{d-c}{\Omega} = (d - c)\omega$ and $y_k = c + k\Delta y$ for $k = 0, 1, \ldots, \Omega$. Then, we have effectively divided the rectangular region into smaller rectangles; see Figure 4.9, left, where the intervals are only partitioned into finitely many pieces for the purpose of visibility. On each of these smaller rectangles, approximate the volume under the surface by a rectangular box (a rectangular parallelepiped) whose height matches the height of the surface at the front right corner of the rectangle, as in Figure 4.9, right, for the *n-m*th approximating box, and Figure 4.10 for all of the approximating boxes. Because only finitely many approximating boxes are shown in Figure 4.10, the total volume of the boxes might not match the volume under the surface. But when infinitely many are used, so that each of the smaller rectangles in Figure 4.9, left, have infinitesimal area, then the difference between the total volume of the boxes and the volume under the surface should also be infinitesimal.

Using the front right corner is analogous to using right-hand endpoints.

The task, then, is to write a formula for the sum of the volumes of the boxes. The volume of a box is length times width times height. The width and length of the boxes are Δx and Δy. The height of the box is the value of the function at the front right corner, $f(x_n, y_m)$. Therefore, the volume of the n-mth box is

$$f(x_n, y_m)\Delta x \Delta y.$$

We may also write the area of each of the smaller rectangles as $\Delta A = \Delta x \Delta y$ and the volume of the n-mth box as $f(x_n, y_m)\Delta A$.

The volume under the surface is found by adding the volumes of all of the boxes. But with two dimensions to sum (the boxes are not all in one row, but in rows and columns), how is the sum accomplished?

Just as with a two-dimensional table of numbers to add, we may add the numbers in each row to get a row total and then add the row totals, or we may add the numbers in each column to get a column total and then add the column totals. Either way we should arrive at the same result. As long as certain conditions hold, the same is true even with sums to infinite hyperreal integers such as Ω. The total volume of the boxes is

Add the numbers in a 4-by-3 table:

			total
2	1	5	8
3	7	4	14
5	5	2	12
1	3	1	5
total 11	16	12	39

Adding the row totals or the column totals gives the same result.

$$\sum_{m=1}^{\Omega} \sum_{n=1}^{\Omega} f(x_n, y_m)\Delta x \Delta y = \sum_{n=1}^{\Omega} \sum_{m=1}^{\Omega} f(x_n, y_m)\Delta y \Delta x.$$

These sums are called *double omega sums.*

As long as the sums are equal and we render the same real result for every such Ω, this serves as the value of the double integral.

Definition 1 DOUBLE INTEGRAL OVER A RECTANGULAR REGION *Let* $f(x, y)$ *be defined throughout the rectangular region* $a \leq x \leq b, c \leq y \leq d$. *If*

$$\sum_{m=1}^{\Omega} \left(\sum_{n=1}^{\Omega} f(x_n, y_m)\Delta x \right) \Delta y = \sum_{n=1}^{\Omega} \left(\sum_{m=1}^{\Omega} f(x_n, y_m)\Delta y \right) \Delta x \doteq L$$

for every positive infinite hyperreal integer Ω *(where* $\Delta x = (b - a)\omega$, $\Delta y = (d - c)\omega$, $x_n = a + n\Delta x$, *and* $y_m = c + m\Delta y$), *then we write*

$$\int_c^d \int_a^b f(x, y)\, dx\, dy = L \ \ or \ \ \int_a^b \int_c^d f(x, y)\, dy\, dx = L$$

The use of the iterated integral notation is justified later in this section. An alternate notation is

$$\iint_R f(x, y)\, dA,$$

and call $\int_c^d \int_a^b f(x, y)\, dx\, dy$ *the double integral of* f *over the region* $a \leq x \leq b$, $c \leq y \leq d$.

When L *is a real number (i.e., not* ∞ *or* $-\infty$), *then we say that* f *is* integrable *on the region.*

where R is the (name of the) rectangular region and the A is for "area."

Theorem 1 INTEGRABILITY OF CONTINUOUS FUNCTIONS *If* $f(x, y)$ *is continuous throughout the rectangular region* $a \leq x \leq b, c \leq y \leq d$, *then* f *is integrable on that region.*

The proof of theorem 1 is difficult and will not be presented here. The theorem can also be generalized to include all bounded functions that are discontinuous only on a finite number of smooth curves.

Calculating double integrals and using double integrals to calculate net volume under a surface has already been practiced and will not change. But a little more work remains to connect this definition to those previous calculations.

4.2.2 Fubini's theorem

Suppose $z = f(x, y)$ is an integrable function and consider the double omega sum

$$\sum_{m=1}^{\Omega} \left(\sum_{n=1}^{\Omega} f(x_n, y_m)\Delta x \right) \Delta y.$$

Consider adding the numbers in the 4-by-3 table in a previous margin note. For each row, we add the numbers in the row and obtain a row total. Inside the parentheses here, we are obtaining the row total for $m = 1$, then for $m = 2$, and so on. Each row total is an omega sum where m is a constant. Therefore, each row total is given by an integral, inside the parentheses. Then the remaining omega sum adds the row totals.

Inside the parentheses, looking only at the summation with variable n, the value of m does not vary; y_m can be considered as a constant in the inside summation.

Then the expression inside the parentheses is an omega sum in the variable x and therefore it can be rewritten as an integral (assuming that the single-variable function $f(x, y_m)$ is integrable). We replace the summation with the integral symbol, drop the subscript n, replace Δx with dx, and insert the limits of integration for the variable x, yielding

Remembering that y_m is a constant, the only variable in $f(x, y_m)$ is x.

$$\sum_{m=1}^{\Omega} \left(\int_a^b f(x, y_m)dx \right) \Delta y.$$

Notice that the subscript m remains, because inside the parentheses y_m is a constant. The integral inside the parentheses therefore treats the value of y (written here as y_m) as a constant, just as we have done previously!

The integral inside the parentheses gives the area under the surface for the slice $y = y_m$.

What remains is also an omega sum. This time, the variable is y. The expression inside the parentheses may not look like the type of expression that we have encountered previously in an omega sum, but that doesn't matter; the expression still contains a y_m and varies according to the value of m. We can therefore rewrite this as an integral also, by replacing the summation with an integral symbol, dropping the subscript m, replacing Δy with dy, and inserting the limits of integration for the variable y, resulting in

$$\int_c^d \left(\int_a^b f(x, y)\, dx \right) dy.$$

Therefore, the double integral of definition 1 truly is the iterated integral practiced in Section 4.1.

For an integrable function the order of the omega sums can be switched (by definition), thereby switching the order of the iterated integral. Indeed, the above analysis becomes

$$\sum_{n=1}^{\Omega}\left(\sum_{m=1}^{\Omega} f(x_n, y_m)\Delta y\right)\Delta x = \sum_{n=1}^{\Omega}\left(\int_c^d f(x_n, y)\,dy\right)\Delta x$$

$$= \int_a^b\left(\int_c^d f(x, y)\,dy\right)dx.$$

For the omega sum inside the parentheses, line 1 replaces the summation with an integral symbol, drops the subscript m, replaces Δy with dy, and inserts limits of integration for the variable y; for the remaining omega sum, line 2 replaces the summation with an integral symbol, drops the subscript n, replaces Δx with dx, and inserts limits of integration for the variable x.

Theorem 2 FUBINI'S THEOREM *If f is a continuous function (or any other integrable function), then the double integral is an iterated integral and can be calculated using either order;*

$$\int_c^d\int_a^b f(x, y)\,dx\,dy = \int_c^d\left(\int_a^b f(x, y)\,dx\right)dy$$

$$= \int_a^b\left(\int_c^d f(x, y)\,dy\right)dx = \int_a^b\int_c^d f(x, y)\,dy\,dx.$$

The conclusion of Fubini's theorem can be illustrated by calculating the value of a double integral of a continuous function using both orders and noting that the result is the same.

Example 11 *Calculate the value of the double integral $\int_3^5\int_0^2(x^2 + 3xy)\,dx\,dy$ by calculating (a) the iterated integral $\int_3^5\left(\int_0^2(x^2 + 3xy)\,dx\right)dy$ and (b) the iterated integral $\int_0^2\left(\int_3^5(x^2 + 3xy)\,dy\right)dx$.*

The rectangular region is $0 \le x \le 2$, $3 \le y \le 5$. Think "inside" and "outside" integrals when interpreting which variable goes with which limits of integration.

Solution Note that the integrand is a polynomial in the variables x and y and is therefore continuous everywhere, so Fubini's theorem applies; the same answer results from either order of integration.

(a) First, we evaluate $\int_3^5\left(\int_0^2(x^2 + 3xy)\,dx\right)dy$:

$$\int_3^5\left(\int_0^2(x^2 + 3xy)\,dx\right)dy = \int_3^5\left(\left(\frac{x^3}{3} + 3y\frac{x^2}{2}\right)\Big|_0^2\right)dy$$

$$= \int_3^5\left(\frac{8}{3} + 6y - (0 + 0)\right)dy$$

$$= \left(\frac{8}{3}y + 3y^2\right)\Big|_3^5 = \frac{40}{3} + 75 - (8 + 27)$$

$$= \frac{160}{3}.$$

Line 1 finds an antiderivative with respect to x for the inside integral, treating y as a constant; line 2 evaluates at $x = 2$, at $x = 0$, and subtracts; and line 3 finds an antiderivative with respect to y and then evaluates at $y = 5$, at $y = 3$, and subtracts.

(b) Next, we evaluate $\displaystyle\int_0^2\left(\int_3^5 (x^2 + 3xy)\,dy\right)dx$:

$$\int_0^2\left(\int_3^5 (x^2 + 3xy)\,dy\right)dx = \int_0^2\left(\left(x^2 y + 3x\frac{y^2}{2}\right)\bigg|_3^5\right)dx$$

$$= \int_0^2\left(5x^2 + \frac{75}{2}x - \left(3x^2 + \frac{27}{2}x\right)\right)dx$$

$$= \int_0^2 (2x^2 + 24x)\,dx$$

$$= \left(\frac{2x^3}{3} + 12x^2\right)\bigg|_0^2$$

$$= \frac{16}{3} + 48 - (0 + 0) = \frac{160}{3}.$$

Line 1 finds an antiderivative with respect to y for the inside integral, treating x as a constant; line 2 evaluates the expression at $y = 5$, at $y = 3$, and subtracts; line 3 simplifies the integrand; line 4 finds an antiderivative with respect to x; and line 5 evaluates at $x = 2$, at $x = 0$, and subtracts.

As expected, the answers to (a) and (b) are identical. ∎

Because either order can be used, we may choose which we prefer. Do you have a preference between the calculation of part (a) or part (b) of example 11? My preference is (a); although part (a) is messier in the arithmetic, part (b) is messier in the algebra. Sometimes there is little advantage to using one order over another. At other times, the order can make the difference between successfully evaluating the integral and having extreme difficulty doing so.

In this case, it is easier to evaluate the integral with 0 as a limit of integration first rather than second.

4.2.3   What is the advantage of Fubini's theorem?

Sometimes techniques of integration are required to evaluate a double integral.

Example 12 *Evaluate* $\displaystyle\int_0^4\left(\int_0^1 y e^{xy}\,dx\right)dy.$

Solution The integrand is continuous everywhere, so Fubini's theorem applies and we may switch the order if we wish. With either order (integrate with respect to x first or integrate with respect to y first), the exponent is either "not just plain x" or "not just plain y," so substitution is indicated either way. Seeing no obvious reason to switch the order, we evaluate the integral as is.

This comment holds true for other techniques of integration as well.

Because substitution is needed for the inside integral, it is convenient to calculate the inside integral separately, then insert the result into the outer integral to proceed. To that end, we calculate

$$\int_0^1 y e^{xy}\,dx,$$

treating y as a constant. Because the exponent is "not just plain x," we use substitution, letting u be the expression in the exponent.

$$\text{Let } u = xy.$$

$$\text{Then } du = \frac{\partial u}{\partial x}\, dx = y\, dx.$$

When we make the substitution to the variable u we must change the limits of integration from values of x to values of u:

$$x = 1: \quad u = y$$

$$x = 0: \quad u = 0.$$

Making the substitution and continuing to evaluate,

$$\int_0^1 ye^{xy}\, dx = \int_0^y e^u\, du$$

$$= e^u\big|_0^y$$

$$= e^y - e^0 = e^y - 1.$$

Next we return to the double integral, using the finished calculation of the inside integral to continue:

$$\int_0^4 \left(\int_0^1 ye^{xy}\, dx \right) dy = \int_0^4 (e^y - 1)\, dy$$

$$= (e^y - y)\big|_0^4$$

$$= e^4 - 4 - (e^0 - 0) = e^4 - 5.$$

$\blacksquare$

Next we switch the order, just as we did for part (b) of example 11.

Example 13 *Evaluate* $\displaystyle \int_0^1 \left(\int_0^4 ye^{xy}\, dy \right) dx.$

Solution The integrand is continuous everywhere, so Fubini's theorem applies and we may switch the order if we wish. With either order (integrate with respect to y first or integrate with respect to x first), the exponent is either "not just plain x" or "not just plain y," so substitution is indicated either way. Seeing no obvious reason to switch the order, we evaluate the integral as is.

Recall that when using substitution, we choose u and then calculate the differential du from the chosen expression for u. Writing the differential in the manner shown here can be helpful to keep track of what is the variable and what is treated as a constant.

Line 1 makes the substitution, replacing the exponent xy with u, replacing $y\, dx$ with du, and replacing the limits of integration for x with corresponding values of u; line 2 finds an antiderivative with respect to the variable u; and line 3 evaluates at $u = y$, at $u = 0$, and subtracts. Note that y is considered to be a constant, not a variable, in this calculation.

Line 1 replaces the inside integral with its calculated value; line 2 finds an antiderivative with respect to y, noting that we are now treating y as a variable; and line 3 evaluates at $y = 4$, at $y = 0$, and subtracts.

Because substitution is needed for the inside integral we calculate the inside integral separately, then insert the result into the outer integral to proceed. To that end, we calculate

$$\int_0^4 ye^{xy}\, dy,$$

treating x as a constant. Because the exponent is "not just plain y," we use substitution, letting u be the expression in the exponent.

$$u = xy$$

$$du = \frac{\partial u}{\partial y}\, dy = x\, dy$$

When we make the substitution to the variable u we must change the limits of integration from values of y to values of u:

$$y = 4 : \quad u = 4x$$

$$y = 0 : \quad u = 0.$$

To this point, the steps and the reasoning of example 13 exactly match those of example 12.

Making the substitution requires replacing $x\, dy$ with du, and we do not see an x to go with the dy in the integrand. Therefore, we must adjust for the constant:

$$\int_0^4 ye^{xy}\, dy = \int_0^4 \frac{x}{x} ye^{xy}\, dy = \frac{1}{x}\int_0^4 xye^{xy}\, dy.$$

Recall that we are treating x as a constant; this is like adjusting the integrand for a missing 4 or a missing 7 needed for the du in a substitution. We therefore multiply and divide the integrand by the constant. Finally, the constant multiple rule may be used to move the constant $\frac{1}{x}$ outside the integral. *Reminder: do not move a variable outside the integral.*

Tentatively making the substitution, we replace the $x\, dy$ with du, we replace the exponent xy with u, and we replace the limits of integration for the variable y with values of u—but there is an issue:

$$\frac{1}{x}\int_0^4 xye^{xy}\, dy = \frac{1}{x}\int_0^{4x} ye^{u}\, du.$$

This partial substitution is done for purposes of illustrating the issue that remains to be dealt with. This step is replaced by the full substitution below.

We haven't yet dealt with the y in the expression, which must be replaced with an expression involving u because it is a variable. One method is to solve the substitution expression for y:

$$u = xy$$

$$\frac{u}{x} = y.$$

We can therefore replace y with $\frac{u}{x}$, recalling that x is treated as a constant. Then the substitution becomes

The substitution is made by replacing $x\, dy$ with du, replacing the exponent xy with u, replacing the remaining y with $\frac{u}{x}$, and replacing the limits of integration for y with values of u. Afterward, the constant multiple rule is used to move the denominator x in the integrand, which is treated as a constant, outside the integral.

$$\frac{1}{x}\int_0^4 xye^{xy}\, dy = \frac{1}{x}\int_0^{4x} \frac{u}{x}e^{u}\, du = \frac{1}{x^2}\int_0^{4x} ue^{u}\, du.$$

Now we need an antiderivative for ue^u, and we recognize a product, indicating the use of parts. (Perhaps now is already a good time to decide to give up and switch the order of integration!) Writing the antiderivative problem separately, we wish to integrate

$$\int ue^u \, du.$$

We assign the polynomial part to f and the exponential part to g', and then calculate f' and g.

$$f(u) = u \qquad\qquad g'(u) = e^u$$

$$f'(u) = 1 \qquad\qquad g(u) = e^u$$

Using the integration by parts formula,

$$\int ue^u \, du = ue^u - \int e^u \, du = ue^u - e^u + C.$$

Returning to the integral we are solving,

$$\frac{1}{x^2} \int_0^{4x} ue^u \, du = \frac{1}{x^2} (ue^u - e^u) \Big|_0^{4x}$$

$$= \frac{1}{x^2} (4xe^{4x} - e^{4x}) - \frac{1}{x^2}(0 - e^0)$$

$$= \frac{1}{x^2} (4xe^{4x} - e^{4x}) + \frac{1}{x^2}.$$

The integration by parts formula being used here is

$$\int f(u)g'(u) \, du = f(u)g(u)$$

$$- \int g(u)f'(u) \, du.$$

Line 1 uses the antiderivative we just calculated; line 2 evaluates at $u = 4x$, at $u = 0$, and subtracts; line 3 simplifies, recalling that $e^0 = 1$.

Having finally evaluated the inside integral, we return to the original double integral:

$$\int_0^1 \left(\int_0^4 ye^{xy} \, dy \right) dx = \int_0^1 \left(\frac{1}{x^2} (4xe^{4x} - e^{4x}) + \frac{1}{x^2} \right) dx.$$

$$= \int_0^1 \left(\frac{4e^{4x}}{x} - \frac{e^{4x}}{x^2} + \frac{1}{x^2} \right) dx.$$

Line 1 replaces the inside integral with its calculated value; line 2 rewrites the integrand.

Now we recognize an additional problem–this integral is improper! The integrand is not defined when $x = 0$. What's more, finding an antiderivative for this integrand uses a technique that you have likely not practiced (see exercise 39), and then evaluating the improper integral requires a series-based approximation to $e^{4\omega}$. It's time to quit being stubborn and use Fubini's theorem! ∎

Alternate solution   Switch the order of integration using Fubini's theorem, which we may do because the integrand is continuous throughout the region of integration:

$$\int_0^1 \left(\int_0^4 y e^{xy} \, dy \right) dx = \int_0^4 \left(\int_0^1 y e^{xy} \, dx \right) dy.$$

Now we proceed as in the solution to example 12. ■

STRATEGY

When a double integral gets overly complicated, consider switching the order of integration using Fubini's theorem.

4.2.4   Properties of double integrals

The definite integral sum rule and constant multiple rule are helpful for manipulating integrals. Their counterparts for double integrals may also be helpful, allowing us manipulate integrals in the manner to which we have become accustomed.

DOUBLE INTEGRAL SUM RULE

If both f and g are functions of two variables x and y and are continuous throughout the region of integration, then

$$\int_c^d \int_a^b \left(f(x,y) + g(x,y) \right) dx \, dy$$

$$= \int_c^d \int_a^b f(x,y) \, dx \, dy + \int_c^d \int_a^b g(x,y) \, dx \, dy.$$

The proof of the double integral sum rule is an exercise which may be completed in the same manner as the proof of the double integral constant multiple rule.

DOUBLE INTEGRAL CONSTANT MULTIPLE RULE

If f is a function of two variables x and y and is continuous throughout the region of integration, then for any real number k,

$$\int_c^d \int_a^b kf(x,y)\,dx\,dy = k \int_c^d \int_a^b f(x,y)\,dx\,dy.$$

Here we use k as the constant instead of c, because c is already being used for a different quantity.

Proof. Because f is continuous throughout the region of integration, Fubini's theorem applies and we may evaluate the double integral by treating it as an iterated integral:

$$\int_c^d \int_a^b kf(x,y)\,dx\,dy = \int_c^d \left(\int_a^b kf(x,y)\,dx \right) dy.$$

Consider the inside integral, where we are treating y as a constant. Then the integrand is a function of one variable x, and the definite integral constant multiple rule applies.

$$\int_c^d \left(\int_a^b kf(x,y)\,dx \right) dy = \int_c^d \left(k \int_a^b f(x,y)\,dx \right) dy.$$

The result of evaluating the integral $\int_a^b f(x,y)\,dx$ will be a function of y. Then the definite integral constant multiple rule applies once again, and we can move k out of the outside integral:

See example 12 for an example of how an inner integral with respect to x results in a function of y.

$$\int_c^d \left(k \int_a^b f(x,y)\,dx \right) dy = k \int_c^d \left(\int_a^b f(x,y)\,dx \right) dy.$$

We may then finish by using Fubini's theorem to equate the iterated integral with a double integral;

$$k \int_c^d \left(\int_a^b f(x,y)\,dx \right) dy = k \int_c^d \int_a^b f(x,y)\,dx\,dy.$$

∎

Other double integral rules are explored in the exercises.

4.2.5 Average value of a function

Suppose we wish to find the average value of a continuous function $z = f(x,y)$ on a rectangular region $a \le x \le b, c \le y \le d$. We could try to estimate the average value by finding several values of the function and averaging those values. This is

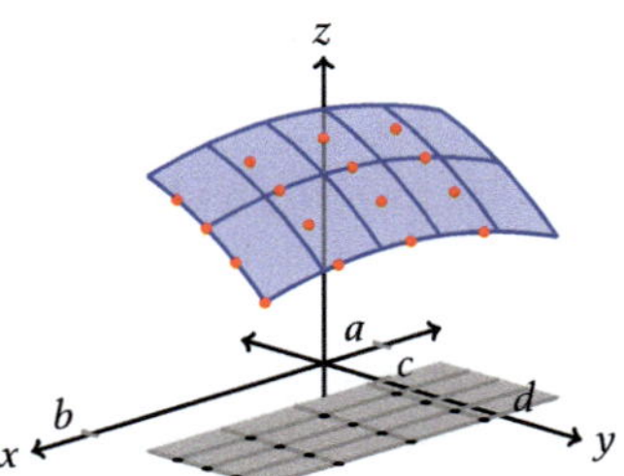

Figure 4.11 *A surface* $z = f(x, y)$ *over a rectangular region, sampled at 16 locations*

pictured in Figure 4.11, where the intervals $a \leq x \leq b$ and $c \leq y \leq d$ are each partitioned into four subintervals, and the value of the function is sampled at the front right corner of each of the 16 smaller rectangles. To average these function values, we add them and divide by 16:

$$\frac{\displaystyle\sum_{n=1}^{4}\sum_{m=1}^{4} f(x_n, y_m)}{4^2}.$$

The interval $a \leq x \leq b$ is partitioned into $a = x_0, x_1, x_2, x_3,$ and $b = x_4$. Similarly, the interval $c \leq y \leq d$ is partitioned into $y_0, ...,$ y_4. Then, the functions values are $f(x_1, y_1),$ $f(x_1, y_2), ..., f(x_4, y_4)$. The numerator adds these function values.

Of course, this estimate likely does not equal the average value we seek, because so many values of the function are not considered. But if we replace 4 with a positive infinite hyperreal integer Ω,

$$\frac{\displaystyle\sum_{n=1}^{\Omega}\sum_{m=1}^{\Omega} f(x_n, y_m)}{\Omega^2},$$

the result effectively samples the function everywhere and should give us the true average value of the function—provided, of course, that we render the same real result for every such Ω.

That sounds much like the definition of double integral. In fact, if we do a little algebraic manipulation, then we can relate the average value to the double integral;

$$\frac{\displaystyle\sum_{n=1}^{\Omega}\sum_{m=1}^{\Omega} f(x_n, y_m)}{\Omega^2} = \frac{\displaystyle\sum_{n=1}^{\Omega}\sum_{m=1}^{\Omega} f(x_n, y_m)(d-c)(b-a)}{\Omega^2(d-c)(b-a)}$$

$$= \frac{\displaystyle\sum_{n=1}^{\Omega}\sum_{m=1}^{\Omega} f(x_n, y_m)(d-c)\omega(b-a)\omega}{(d-c)(b-a)}$$

$$\frac{\displaystyle\sum_{n=1}^{\Omega}\sum_{m=1}^{\Omega} f(x_n, y_m)\Delta y \Delta x}{(d-c)(b-a)}$$

$$=\frac{\displaystyle\int_a^b \int_c^d f(x,y)\,dy\,dx}{(d-c)(b-a)}$$

$$=\frac{\displaystyle\int_a^b \int_c^d f(x,y)\,dy\,dx}{\text{area of region}}.$$

Line 1 begins with the previous formula and multiplies numerator and denominator by the quantity $(d-c)(b-a)$; line 2 uses the reciprocal relationship $\frac{1}{\Omega^2} = \omega^2 = \omega \cdot \omega$, strategically placing the omegas; line 3 replaces $(d-c)\omega$ with Δy and replaces $(b-a)\omega$ with Δx; line 4 rewrites the double omega sum by replacing the summations with integral symbols, dropping the subscripts n and m, replacing Δy with dy and Δx with dx, and inserting the limits of integration; line 5 notes that the denominator is equal to the area of the rectangular region.

As usual, we define the geometric quantity using the formula we have developed.

Definition 2 AVERAGE VALUE OF A FUNCTION OF TWO VARIABLES *Suppose $z = f(x,y)$ is integrable on the region $a \leq x \leq b$, $c \leq y \leq d$. Then the average value of f on the region is*

$$f_{avg} = \frac{1}{(d-c)(b-a)} \int_a^b \int_c^d f(x,y)\,dy\,dx.$$

The order of integration can be switched. Compare this formula to the single-variable formula for the average value of a function.

Note that the value of the double integral is the (net) volume under the surface, so we may think of the average value of the function as the volume under the surface divided by the area of the region.

Think of the space under a surface as being filled with dry sand, contained in a box whose base is the rectangular region. Now shake the box horizontally until the sand is level. The height of the sand should be the average value of the function; the volume is the area of the base times the height of the sand.

Example 14 *Find the average value of $f(x,y) = xy$ over the region $0 \leq x \leq 2$, $0 \leq y \leq 3$.*

Solution   The average value is

$$f_{avg} = \frac{1}{(3-0)(2-0)} \int_0^2 \int_0^3 xy\,dy\,dx$$

$$= \frac{1}{6} \int_0^2 \left(\frac{xy^2}{2}\bigg|_0^3 \right) dx$$

$$= \frac{1}{6} \int_0^2 \left(\frac{9x}{2} - 0 \right) dx$$

$$= \frac{1}{6}\frac{9x^2}{4}\bigg|_0^2$$

$$= \frac{1}{6}(9-0) = \frac{3}{2}.$$

Line 1 applies the average value formula; line 2 finds an antiderivative with respect to y, treating x as a constant; line 3 evaluates at $y = 3$, at $y = 0$, and subtracts; line 4 finds an antiderivative with respect to x; and line 5 evaluates at $x = 2$, at $x = 0$, and subtracts.

∎

4.2.6 Midpoint rule for double integrals

In single-variable calculus, numerical integration is used to estimate the value of a definite integral when we cannot determine an antiderivative. The same is true for multi-variable functions. Several methods are available, including a version of the trapezoid rule. Perhaps the simplest method to complete by hand is the midpoint rule.

The idea is illustrated in Figure 4.12. The interval $a \le x \le b$ is partitioned into p subintervals ($p = 4$ is pictured) and the interval $c \le y \le d$ is partitioned into q subintervals ($q = 4$ is pictured). Using the midpoints of each interval as x- and y-coordinates for points at which to sample the height of the function, there are $p \cdot q$ subrectangles for which we determine a function value. Each function value must be multiplied by the area of the rectangle so as to estimate the volume under the curve over that subrectangle. These volumes are summed to determine the estimate for the (net) volume under the surface. It is as if we are using Figure 4.10, but matching the heights at the center point instead of the front right.

Recall that some functions have nonelementary antiderivatives, that is, their antiderivatives cannot be expressed as a combination or composition of elementary functions. Elementary functions include polynomial, rational, root, trigonometric, inverse trigonometric, exponential, logarithmic, hyperbolic, and inverse hyperbolic functions.

The values of p and q are not required to be equal.

More precisely, $x_n = a + n\Delta x$ for $n = 0, 1, ..., p$, $y_m = c + m\Delta y$ for $m = 0, 1, ..., q$, $\overline{x_n} = \frac{x_{n-1}+x_n}{2}$ for $n = 1, 2, ..., p$, and $\overline{y_m} = \frac{y_{m-1}+y_m}{2}$ for $m = 1, 2, ..., q$. Try not to use these formulas directly.

MIDPOINT RULE FOR DOUBLE INTEGRALS

Let $z = f(x, y)$ be a function of two variables defined throughout the rectangle $a \le x \le b, c \le y \le d$. Then we estimate the value of $\int_c^d \int_a^b f(x, y)\, dx\, dy$ by

$$M_{p,q} = \sum_{m=1}^{q} \sum_{n=1}^{p} f(\overline{x_n}, \overline{y_m})\Delta x\Delta y,$$

where $\Delta x = \frac{b-a}{p}$, $\Delta y = \frac{d-c}{q}$, and $\overline{x_n}$ and $\overline{y_m}$ are midpoints of the subintervals for x and y, respectively.

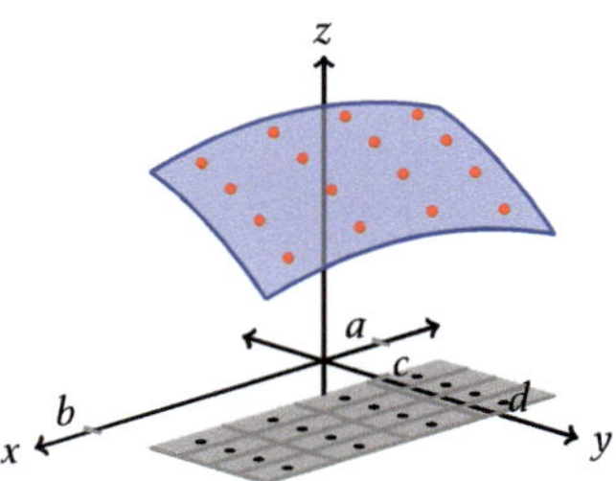

Figure 4.12 *A surface $z = f(x, y)$ over a rectangular region, with midpoints of 16 rectangular subregions in the xy-plane (black dots) and the corresponding values of the function marked on the surface (red dots)*

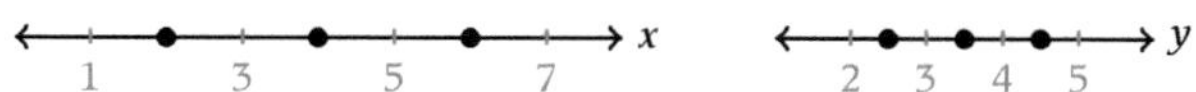

Figure 4.13 *Partitions of the intervals* $1 \le x \le 7$ *and* $2 \le y \le 5$ *into three subintervals each, with the midpoint of each subinterval indicated by a dot*

The formula looks more complicated than it is; just remember to evaluate the function at the center point of each subrectangle and multiply by the area of each subrectangle. An example can help.

Example 15 *(a) Estimate* $\displaystyle\int_1^7 \int_2^5 \sqrt{x^4 + y}\, dy\, dx$ *using the midpoint rule with* $p = 3$, $q = 3$. *(b) Use a CAS to determine the value of the double integral. (c) Determine the error and the percent relative error in using the midpoint rule with* $p = 3, q = 3$.

Solution

The integrand in example 15 is nonelementary.

(a) Before evaluating the function, we must determine the values of $\overline{x_n}$ and $\overline{y_m}$. We have $\Delta x = \frac{7-1}{3} = 2$ and $\Delta y = \frac{5-2}{3} = 1$. Then as illustrated in Figure 4.13, $x_0 = 1, x_1 = 3, x_2 = 5$, and $x_3 = 7$, and the midpoints are $\overline{x_1} = 2, \overline{x_2} = 4$, and $\overline{x_3} = 6$. Also, $y_0 = 2, y_1 = 3, y_2 = 4$, and $y_3 = 5$, and the midpoints are $\overline{y_1} = 2.5$, $\overline{y_2} = 3.5$, and $\overline{y_3} = 4.5$.

We therefore have $p \cdot q = 3 \cdot 3 = 9$ function values to calculate, one corresponding to each of the center points in Figure 4.14, representing each of the combinations of $\overline{x_n}$ with $\overline{y_m}$. For $\overline{x_1} = 2$ the values are

$$f(2, 2.5) = \sqrt{2^4 + 2.5} = 4.3012$$

$$f(2, 3.5) = \sqrt{2^4 + 3.5} = 4.4159$$

$$f(2, 4.5) = \sqrt{2^4 + 4.5} = 4.5277.$$

For $\overline{x_2} = 4$ the values are

$$f(4, 2.5) = \sqrt{4^4 + 2.5} = 16.0779$$

$$f(4, 3.5) = \sqrt{4^4 + 3.5} = 16.1090$$

$$f(4, 4.5) = \sqrt{4^4 + 4.5} = 16.1400.$$

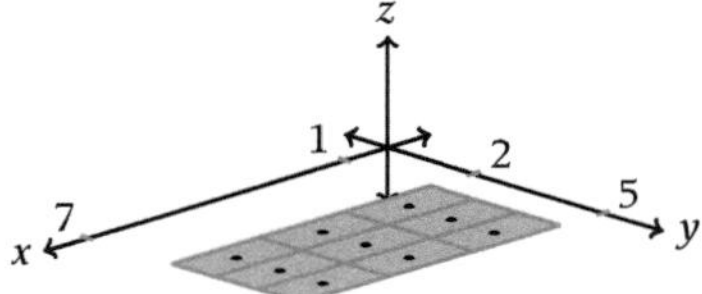

Figure 4.14 *Partitions of the intervals* $1 \le x \le 7$ *and* $2 \le y \le 5$ *into three subintervals each, for a total of 9 subrectangles, with the center point of each subrectangle indicated by a dot*

For $\overline{x_3} = 6$ the values are

$$f(6, 2.5) = \sqrt{6^4 + 2.5} = 36.0347$$

$$f(6, 3.5) = \sqrt{6^4 + 3.5} = 36.0486$$

$$f(6, 4.5) = \sqrt{6^4 + 4.5} = 36.0624.$$

Add these nine values and the sum is 169.7174. Next we multiply by $\Delta x \Delta y = 2 \cdot 1 = 2$, and the estimate is

$$M_{3,3} = 169.7174 \cdot 2 = 339.4348.$$

A numerical value of an integral determined by a CAS (as in part (b)) may be a numerical approximation and not an exact value. It should be accurate to the number of decimal places displayed.

(b) Using a CAS, the value of the integral is 345.957.

(c) The error in using the midpoint rule with $p = 3, q = 3$ is

The same number of decimal places are used for the two estimates. A negative result means that the midpoint rule underestimated the value of the double integral.

$$E_M = M_{3,3} - \text{actual} = 339.435 - 345.957 = -6.522.$$

The percent relative error is given by

$$100 \cdot \frac{\text{estimate} - \text{actual}}{\text{actual}} = 100 \cdot \frac{339.435 - 345.957}{345.957} = -1.885\%.$$

∎

In example 15, the changes in the calculated function values are small when changing from $\overline{y_1}$ to $\overline{y_2}$ to $\overline{y_3}$ for the same $\overline{x_n}$, but the changes are much larger when changing from $\overline{x_1}$ to $\overline{x_2}$ to $\overline{x_3}$ for the same $\overline{y_m}$. If we wish to improve our midpoint rule estimate, it therefore makes sense to increase the value of p to partition x more finely. A spreadsheet can accomplish these calculations rather easily, with values of x_n in rows and y_m in columns, or vice versa. If we use $p = 30, q = 3$, then the estimate is $M_{30,3} = 345.893$, and the error shrinks to $E_M = -0.064$ with percent relative error -0.0185%, two orders of magnitude improvement from using $p = 3$, $q = 3$.

Error bounds for using the midpoint rule for double integrals exist but will not be stated here. The bounds are useful for determining when p and q are large enough to ensure the answer is within the desired error tolerance. Other stopping criteria may also be used by a CAS.

4.2.7   When Fubini's theorem does not apply

When Fubini's theorem applies, either order of integration gives the same result. But if the hypotheses of Fubini's theorem are not met, then the results might differ.

A classic example is $f(x, y) = \frac{x^2 - y^2}{(x^2 + y^2)^2}$ on the rectangular region $0 \leq x \leq 1$, $0 \leq y \leq 1$. At the corner of the domain where $x = 0 = y$, the function is not continuous (division by zero); further investigation shows that f is not bounded. Fubini's theorem does not apply. But it turns out that both orders for the iterated integral can be calculated (see exercise 43):

$$\int_0^1 \left(\int_0^1 \frac{x^2 - y^2}{(x^2 + y^2)^2} \, dy \right) dx = \frac{\pi}{4}$$

$$\int_0^1 \left(\int_0^1 \frac{x^2 - y^2}{(x^2 + y^2)^2} \, dx \right) dy = -\frac{\pi}{4}.$$

So what is the value of the double integral $\int_0^1 \int_0^1 \frac{x^2 - y^2}{(x^2 + y^2)^2} \, dy \, dx$? Is it $\frac{\pi}{4}$ or $-\frac{\pi}{4}$? The answer is it is neither! The function f is not integrable on this region; definition 1 requires that either order of calculating the double omega sums renders the same real result, which is not the case.

Moral of the story: always check for continuity of the integrand (or other appropriate hypotheses) before applying Fubini's theorem!

EXERCISES 4.2

1–10. Calculate the value of the double integral $\int_c^d \int_a^b f(x,y) \, dx \, dy$ by calculating (a) the iterated integral $\int_c^d \left(\int_a^b f(x,y) \, dx \right) dy$ and (b) the iterated integral $\int_a^b \left(\int_c^d f(x,y) \, dy \right) dx$.

1. $\displaystyle \int_0^2 \int_1^3 (x^2 + y^2) \, dx \, dy$

2. $\displaystyle \int_2^3 \int_0^1 (y - x^3) \, dx \, dy$

3. $\displaystyle \int_{-2}^0 \int_1^5 x^2 y \, dx \, dy$

4. $\displaystyle \int_0^2 \int_0^4 \left(\sqrt{x} + e^y \right) dx \, dy$

5. $\displaystyle \int_1^2 \int_0^\pi \left(\frac{1}{y} + \sin x \right) dx \, dy$

6. $\displaystyle \int_1^2 \int_0^3 \frac{x^2}{y^3} \, dx \, dy$

7. $\displaystyle \int_1^3 \int_0^1 \frac{1}{x + y} \, dx \, dy$

8. $\displaystyle \int_\pi^{2\pi} \int_0^1 \sin(3xy) \, dx \, dy$

9. $\displaystyle \int_{\pi}^{2\pi} \int_{0}^{1} \sin(3x + y)\, dx\, dy$

10. $\displaystyle \int_{2}^{4} \int_{0}^{\pi} (y^3 - 5y^2 + 7)\cos x\, dx\, dy$

11-20. Evaluate the double integral.

11. $\displaystyle \int_{0}^{1} \int_{2}^{5} x^3 y^4\, dx\, dy$

12. $\displaystyle \int_{0}^{2} \int_{0}^{1} (x^3 - y^2)\, dx\, dy$

13. $\displaystyle \int_{0}^{1} \int_{2}^{5} x e^y\, dx\, dy$

14. $\displaystyle \int_{0}^{\pi} \int_{1}^{2} x \sin y\, dx\, dy$

15. $\displaystyle \int_{0}^{1} \int_{0}^{\pi} x \cos xy\, dx\, dy$

16. $\displaystyle \int_{\frac{\pi}{4}}^{\frac{\pi}{2}} \int_{0}^{2} x \sin xy\, dx\, dy$

17. $\displaystyle \int_{0}^{1} \int_{0}^{2} y^3 \sqrt{x + y^4}\, dx\, dy$

18. $\displaystyle \int_{4}^{9} \int_{1}^{3} \sqrt{y}\, dx\, dy$

19. $\displaystyle \int_{-1}^{0} \int_{0}^{4} x^2 y \sqrt{(x^3 + 1)y}\, dy\, dx$

20. $\displaystyle \int_{-1}^{1} \int_{0}^{1} x^2 y \cosh x^3 y\, dy\, dx$

21. Find the average value of $f(x, y) = xy^3$ over the region $1 \le x \le 2$, $0 \le y \le 2$.

22. Find the average value of $f(x, y) = x(x^2 + y)$ over the region $1 \le x \le 5$, $0 \le y \le 3$.

23. Find the average value of $g(u, v) = 10uv$ over the region $-3 \le u \le 5$, $\sqrt{3} \le v \le 2\sqrt{3}$.

24. Find the average value of $f(x, y) = xy\sqrt{x^2 + 1}$ over the region $0 \le x \le 3$, $-2 \le y \le 0$.

25. Find the average value of $f(x, y) = \dfrac{x}{y}$ over the region $7 \le x \le 10$, $1 \le y \le e$.

26. Find the average value of $s(r, t) = r^2 - 4t^2$ over the region $4 \le r \le 9$, $2 \le t \le 4$.

27. Find the average value of $h(w, z) = \sin(w + z)$ over the region $0 \le w \le \pi, \pi \le z \le 2\pi$.

28. Find the average value of $f(y, x) = y^2 + 7x$ over the region $2 \le y \le 3, -1 \le x \le 2$.

29–34. (a) Use the midpoint rule to estimate the integral, using the given values of p and q. State the final answer to four significant digits. (b) Use a CAS to estimate the value of the integral to four significant digits. (c) Using the value from the CAS as the actual value, determine the percent relative error in the midpoint rule approximation.

Desmos can be used to complete part (b).

It is good practice to use the same number of decimal places for each computed function value. Using an extra significant digit for intermediate calculations can reduce the accumulation of roundoff error.

29. $\displaystyle\int_2^3 \int_0^1 \frac{\sin x}{x^2 + y}\, dx\, dy$, using $p = 4, q = 4$

30. $\displaystyle\int_1^5 \int_2^4 \sqrt{x^3 + x \ln y}\, dx\, dy$, using $p = 4, q = 4$

31. $\displaystyle\int_{0.1}^{0.4} \int_0^1 e^{x^2 + y^2}\, dx\, dy$, using $p = 5, q = 3$

32. $\displaystyle\int_0^{0.9} \int_0^{0.6} \cos\left(\tfrac{1}{4}x + \tfrac{3}{5}y^3\right) dy\, dx$, using $p = 3, q = 3$

33. $\displaystyle\int_0^1 \int_1^4 \left(\frac{x - y^2}{x^2 + y^2}\right)^{1.6} dx\, dy$, using $p = 3, q = 4$

34. $\displaystyle\int_1^3 \int_0^1 \sqrt{xy^3\, 2^{xy}}\, dx\, dy$, using $p = 4, q = 4$

35. Prove the double integral sum rule.

36. State and prove the double integral difference rule.

37. (a) State a rule for the double integral of a constant. (b) Why does the rule make sense geometrically? (c) Prove the rule.

38. State a double integral equivalent of the additive property of definite integrals for adjacent rectangular regions, where the combined region is also rectangular.

39. Return to the integral of example 13 and act as though your stubbornness–or your curiosity–has not yet been stifled. To evaluate the integral without switching the order of integration, we need to evaluate the improper integral $\displaystyle\int_0^1 \left(\frac{4e^{4x}}{x} - \frac{e^{4x}}{x^2} + \frac{1}{x^2}\right) dx$. Do so using the following steps.

(a) Use the formula
$$\int \frac{f'(x)}{g(x)}\, dx = \frac{f(x)}{g(x)} + \int \frac{f(x)g'(x)}{(g(x))^2}\, dx,$$

which can be derived from the quotient rule for derivatives in the same manner as the integration by parts formula is derived from the product rule, on the integral $\int \dfrac{4e^{4x}}{x}\, dx$. The result should still contain an integral.

(b) Use the result of (a) to evaluate $\displaystyle\int \left(\dfrac{4e^{4x}}{x} - \dfrac{e^{4x}}{x^2}\right) dx$.

(c) The integral we need to evaluate contains three terms, not two as in part (b). Finish finding the antiderivative we need.

(d) Evaluate the improper integral using the antiderivative from (c) and the approximation formula $e^{\alpha} \approx 1 + \alpha$, which can be derived from the Maclaurin series for e^x.

(e) Compare to the solution of example 12.

40. (a) State the double omega sum constant multiple rule.

(b) The derivation of the average value of a function of two variables formula subtly uses the double omega sum constant multiple rule. In which line(s) of the derivation is the rule used?

41. Some double integrals, such as $\displaystyle\int_0^2 \int_0^3 x^2 y^3 \, dy\, dx$, can easily be computed using double omega sums.

(a) Dividing both the x and y intervals into Ω subintervals for a positive infinite hyperreal integer Ω, calculate Δx, Δy, x_n, and y_m.

(b) Calculate the double omega sum $\displaystyle\sum_{n=1}^{\Omega} \left(\sum_{m=1}^{\Omega} f(x_n, y_m)\Delta y \right) \Delta x$, carefully calculating the inside omega sum first and the outside omega sum second.

(c) Calculate the double omega sum $\displaystyle\sum_{m=1}^{\Omega} \left(\sum_{n=1}^{\Omega} f(x_n, y_m)\Delta x \right) \Delta y$, carefully calculating the inside omega sum first and the outside omega sum second.

(d) Calculate the double omega sum $\displaystyle\sum_{n=1}^{\Omega} \sum_{m=1}^{\Omega} f(x_n, y_m)\Delta y \Delta x$ using the sum of powers approximation formulas for both variables in the same step.

(e) Compare (b), (c), and (d).

42. An iterated version of the trapezoid rule can be used to estimate the value of a double integral. Use the following steps to estimate the value of $\displaystyle\int_1^2 \int_4^6 \sin\dfrac{x}{y}\, dx\, dy$ using $p = 4, q = 5$.

(a) Determine Δx, Δy, and the values of x_n and y_m.

(b) Evaluate the integrand at each combination of x_n and y_m, arranging the answers in columns and rows.

(c) Use the trapezoid rule to calculate a value for each row. These values correspond to (estimated) slice areas.

(d) Use the trapezoid rule again on the values from (c). The result is the trapezoid rule estimate $T_{4,5}$.

(e) Using the estimate from a CAS as the actual value of the integral, calculate the percent relative error in the estimate of (d).

Use $x_0, \ldots, x_4$ as column headings and $y_0, \ldots, y_5$ as row headings, or vice versa.

43. Investigate the function $f(x, y) = \dfrac{x^2 - y^2}{(x^2 + y^2)^2}$ on the rectangular region $0 \le x \le 1, 0 \le y \le 1$ as follows.

(a) Investigate the limit $\displaystyle\lim_{(x,y)\to(0,0)} \frac{x^2 - y^2}{(x^2 + y^2)^2}$ and conclude that f is unbounded on the region, which means that Fubini's theorem does not apply.

(b) Calculate $\displaystyle\int \frac{x^2 - y^2}{(x^2 + y^2)^2} \, dy$, treating x as a constant. (Hint: after choosing an appropriate technique of integration, use trig identities to simplify the integrand to use just one trig function. After finding an antiderivative, another trig identity will be required to complete the technique.)

(c) Use the result of (b) to finish evaluating $\displaystyle\int_0^1 \frac{x^2 - y^2}{(x^2 + y^2)^2} \, dy$ for the case $x \ne 0$.

(d) What happens if $x = 0$?

(e) Because the result of (c) is valid only if $x \ne 0$, finish evaluating $\displaystyle\int_0^1 \left(\int_0^1 \frac{x^2 - y^2}{(x^2 + y^2)^2} \, dy \right) dx$ by evaluating

$$\int_\omega^1 (\text{result from (c)}) \, dx.$$

(f) Repeat steps (b)–(e) to calculate $\displaystyle\int_0^1 \left(\int_0^1 \frac{x^2 - y^2}{(x^2 + y^2)^2} \, dx \right) dy$.

The goal of steps (b)–(e) is to calculate the iterated integral

$$\int_0^1 \left(\int_0^1 \frac{x^2 - y^2}{(x^2 + y^2)^2} \, dy \right) dx.$$

4.3   Double Integrals Over General Regions

So far, every double integral we have evaluated has been over a rectangular region. That is about to change.

4.3.1   Double integrals over non-rectangular regions

Suppose we wish to know the volume under the surface $z = f(x, y)$ over the triangular region bounded by $y = x$, $x = 2$, and $y = 0$. We may try the approach of Section 4.1, where we used slice areas and obtained an iterated integral. Figure 4.15 illustrates one slice. As can be visualized in Figure 4.15, these slices do not all have the same length.

We are trying to set up an iterated integral to find the volume under the surface. To craft the inner integral, consider the area for the slice $x = 1$. Look in Figure 4.15, and imagine that the value of a is 1. We integrate $f(1, y)$, where the y-coordinates stretch from $y = 0$ on the left to the line $y = x$ on the right; because $x = 1$, the y-coordinate on that line is 1. The area for the slice $x = 1$ is

$$\int_0^1 f(1, y)\, dy.$$

Reasoning that the triangle (gray in Figure 4.15) is half of a rectangle, it is tempting to think that we can find the volume by taking half the volume over a rectangle, $\frac{1}{2}\int_0^2 \int_0^2 f(x, y)\, dx\, dy$. But if the surface isn't symmetric enough, that is, if it looks different over the two triangular halves of the rectangle, then halving the area over the rectangle gives the wrong answer.

For the slice $x = 2$, which is at the front edge of the region in the viewpoint of Figure 4.15, the y-coordinates stretch from $y = 0$ on the left to $y = 2$ on the right, and the area for the slice $x = 2$ is

$$\int_0^2 f(2, y)\, dy.$$

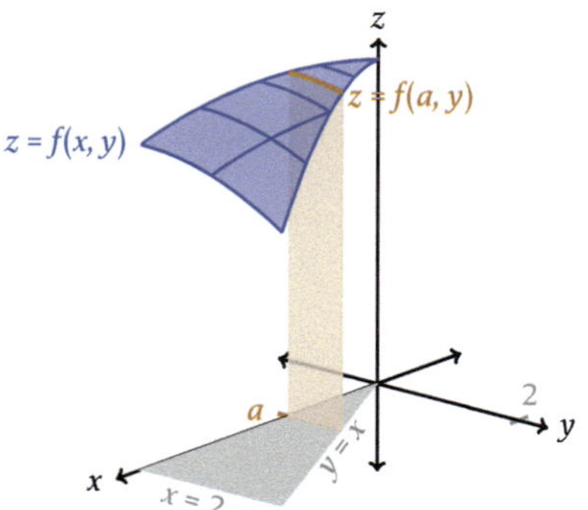

Figure 4.15 *A surface $z = f(x, y)$ (blue) over a triangular region (gray), a slice $x = a$ (brown), and the region under the slice (tan)*

The point is that the upper limit of integration is not the same number for every slice. But not all is lost. Recall that when integrating with respect to y, we treat x as a constant. This means that we can use x when writing the limits of integration! In other words, for every slice (refer again to Figure 4.15), the y-coordinates stretch from $y = 0$ on the left to $y = x$ on the right, making the inner integral

$$\int_0^x f(x, y)\, dy.$$

The inner integral can use a "variable" that is treated like a constant as a limit of integration.

What about the outer integral, with respect to x? The slices in Figure 4.15 stretch from $x = 0$ to $x = 2$. Here, we just need to know where the slices are, to add up the slice areas (technically adding volumes, the slice areas times the thickness Δx as in Section 4.1). So the limits of integration on the outer integral are always specific numbers, representing the locations of the slices, and this will be true no matter the shape of the region we are integrating over. The volume under the surface is

$$\int_0^2 \left(\int_0^x f(x, y)\, dy \right) dx.$$

The outer integral will always use specific numbers for limits of integration.

Let's pause from the theory for a moment. What is different about evaluating a double integral that has a "variable" treated as a constant in a limit of integration? Procedurally, nothing is new.

Example 16 *Evaluate* $\displaystyle \int_0^1 \left(\int_1^{x+1} (2x + y)\, dy \right) dx.$

Solution The double integral is calculated in the usual manner:

$$\int_0^1 \left(\int_1^{x+1} (2x + y)\, dy \right) dx = \int_0^1 \left(\left(2xy + \frac{y^2}{2} \right) \Big|_1^{x+1} \right) dx$$

$$= \int_0^1 \left(2x(x + 1) + \frac{(x + 1)^2}{2} - \left(2x + \frac{1}{2} \right) \right) dx$$

$$= \int_0^1 \left(2x^2 + 2x + \frac{x^2 + 2x + 1}{2} - 2x - \frac{1}{2} \right) dx$$

$$= \int_0^1 \left(\frac{5}{2}x^2 + x \right) dx$$

$$= \left(\frac{5}{2}\frac{x^3}{3} + \frac{x^2}{2} \right) \Big|_0^1$$

$$= \frac{5}{6} + \frac{1}{2} - 0 = \frac{4}{3}.$$

Line 1 finds an antiderivative with respect to y, treating x as a constant; line 2 evaluates at $y = x + 1$, at $y = 1$, and subtracts (still treating x as a constant); lines 3 and 4 simplify the integrand, which is a function of the variable x; line 5 finds an antiderivative with respect to x; line 6 evaluates at $x = 1$, at $x = 0$, and subtracts, then completes the arithmetic.

No context for the integral is given, so we do not include units in the answer. ∎

Reading Exercise 4 Evaluate $\displaystyle \int_0^1 \int_0^x x^2 y\, dy\, dx.$

In Section 4.1, we began calculating volumes using the iterated integral approach, but we did not give the formal definition of double integral until Section 4.2; the connection was made through Fubini's theorem. So what do we do here? The definition given in Section 4.2 was only for the volume over a rectangular region. Does this mean that we must go through the process again for a triangular region, and then again for a circular region, and again and again for every shape we could imagine? What a nightmare that would be! Fortunately, there is a much more efficient approach.

Suppose we wish to find the volume under a surface $z = f(x, y)$ over a bounded region D, such as the oddly-shaped region illustrated in Figure 4.16. The shape does not matter; the region can even consist of two or more unconnected pieces. As long as the region D is bounded, we may draw a rectangle large enough to include all of D. Call this rectangular region R; in Figure 4.17, the rectangle is the region $a \leq x \leq b, c \leq y \leq d$.

If the region is not bounded, that is, if it is infinite in extent, then the double integral would be improper, just as $\int_1^\infty g(x)\, dx$ is an improper integral.

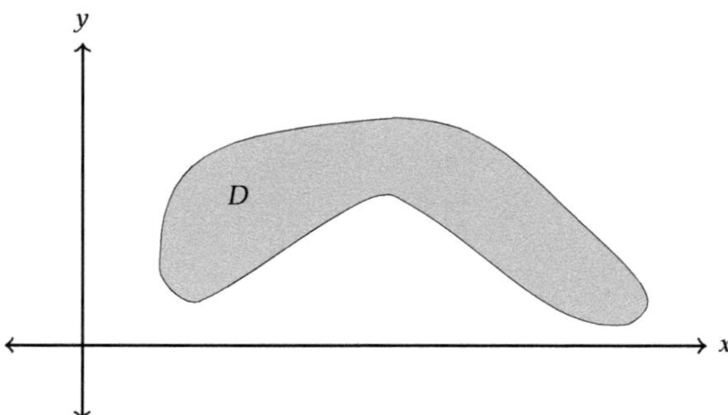

Figure 4.16 *A bounded region D in the xy-plane. Compare to the triangular region in Figure 4.15*

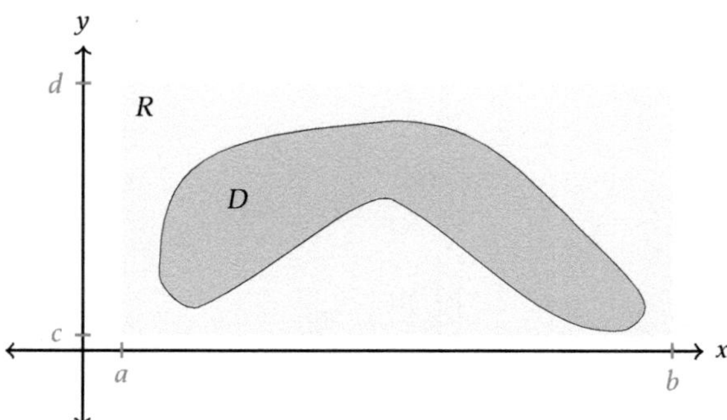

The rectangular region R must contain the entire region D, but it is not required that we use the smallest possible rectangle.

Figure 4.17 *A bounded region D in the xy-plane, completely contained in the rectangular region $a \leq x \leq b, c \leq y \leq d$*

Next, we define a new function H, where the value of H matches the value of f in the region D, but outside the region D the value of H is 0:

$$H(x, y) = \begin{cases} f(x, y), & \text{if } (x, y) \text{ is in } D \\ 0, & \text{otherwise} \end{cases}.$$

An example of the function H for the triangular region of Figure 4.15 is in Figure 4.18. Now consider the volume under the surface $z = H(x, y)$ over the rectangular region R. Where $H(x, y) = 0$, there is no contribution to volume because the "height" is 0. Therefore, the volume under the surface $z = H(x, y)$ over R consists only of the volume under the surface $z = f(x, y)$ over D. Admittedly, this new function H may not be (and usually is not) continuous along the boundary of the region D. But that's okay; definition 1 only requires the function to be defined throughout the rectangular region. This approach allows us to define the double integral of a function f over a general region D.

Definition 3 DOUBLE INTEGRAL OVER A GENERAL REGION *Let $f(x, y)$ be defined throughout a closed region D that is completely contained in the rectangular region $a \le x \le b$, $c \le y \le d$. Let*

$$H(x, y) = \begin{cases} f(x, y), & \text{if } (x, y) \text{ is in } D \\ 0, & \text{otherwise} \end{cases}.$$

Then the double integral of f over the region D is given by

$$\iint_D f(x, y)\, dx\, dy = \int_c^d \int_a^b H(x, y)\, dx\, dy,$$

provided the latter integral exists.

As long as the function f is continuous on the closed region D, Fubini's theorem applies and we may use the iterated integral approach, with either order of integration, to calculate the value of the double integral.

4.3.2 Setting up integrals

Before turning to examples of finding volumes over general regions, we need to discuss how to set up the iterated integrals.

Suppose the region D over which we wish to integrate the function $z = f(x, y)$ is the region between two curves $y = g_1(x)$ and $y = g_2(x)$, between $x = a$ and $x = b$, as pictured in Figure 4.19. If we slice parallel to the yz-plane as in Figure 4.15, then

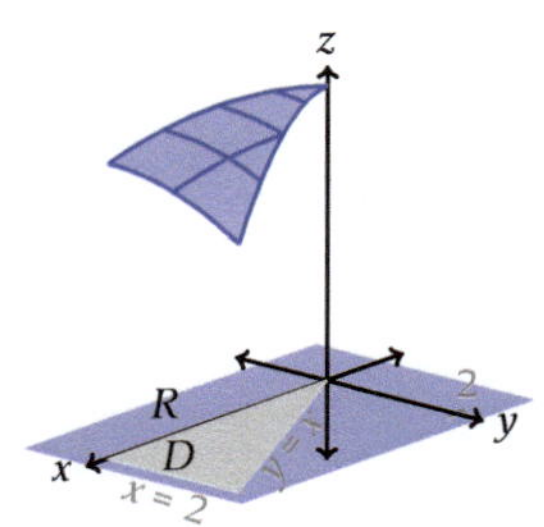

Figure 4.18 *The surface $z = H(x, y)$ (blue) on the rectangular region R. The values of H match the values of the function f from Figure 4.15 on the triangular region D. Outside of the triangular region, the value of H is 0*

A *closed* region D is similar to the idea of a closed interval; the boundary is included in the region.

Ans. to reading exercise 4:
$$\frac{1}{10}$$

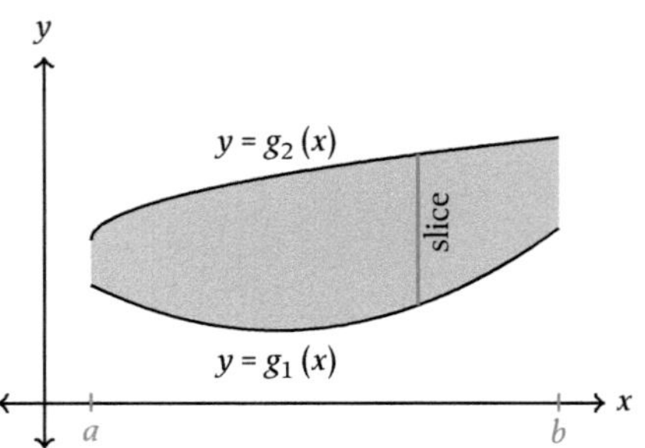

Figure 4.19 *A region (gray) between two curves $y = g_1(x)$ and $y = g_2(x)$, between $x = a$ and $x = b$, along with a slice (darker gray) parallel to the y-axis*

each slice extends from $y = g_1(x)$ (bottom curve, bottom boundary of the region) to $y = g_2(x)$ (top curve, top boundary of the region), and the inner integral is

$$\int_{g_1(x)}^{g_2(x)} f(x, y)\, dy.$$

Here, "top" and "bottom" refer to the diagram of the region in the xy-plane, as in Figure 4.19.

The slices are located from $x = a$ to $x = b$, which indicates the limits of integration for the outer integral. Then the double integral is

$$\int_a^b \left(\int_{g_1(x)}^{g_2(x)} f(x, y)\, dy \right) dx.$$

A succinct way of determining the limits of integration for the order $dy\, dx$ is to determine the y-coordinates, allowing x as part of the formulas, from the bottom curve to the top curve; then determine the extent of the x-coordinates, between two numerical values of x.

DOUBLE INTEGRAL OVER A REGION BETWEEN CURVES, PART 1

This formula further assumes that $g_1(x) \le g_2(x)$ throughout the interval $[a, b]$.

To integrate $z = f(x, y)$ over the region between $y = g_1(x)$ and $y = g_2(x)$ for $a \le x \le b$, use

$$\int_a^b \left(\int_{g_1(x)}^{g_2(x)} f(x, y)\, dy \right) dx.$$

Here, "left" and "right" refer to the diagram of the region in the xy-plane, as in Figure 4.20.

If the region D lies between two curves $x = g_1(y)$ and $x = g_2(y)$ as in Figure 4.20, then each slice extends from $x = g_1(y)$ (left curve, left boundary of the region) to $x = g_2(y)$ (right curve, right boundary of the region), and the inner integral is

$$\int_{g_1(y)}^{g_2(y)} f(x, y)\, dx.$$

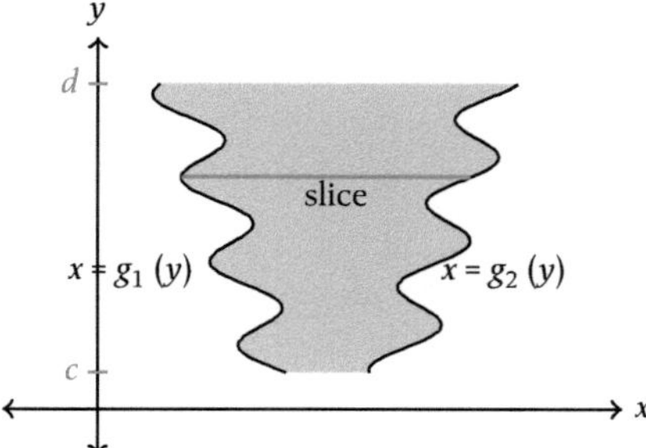

Figure 4.20 *A region (gray) between two curves $x = g_1(y)$ and $x = g_2(y)$, between $y = c$ and $y = d$, along with a slice (darker gray) parallel to the x-axis*

The slices are located between $y = c$ and $y = d$, which indicates the limits of integration for the outer integral. Then, the double integral is

$$\int_c^d \left(\int_{g_1(y)}^{g_2(y)} f(x, y)\, dx \right) dy.$$

To determine the limits of integration for the order $dx\,dy$, determine the x-coordinates, allowing y as part of the formulas, from the left curve to the right curve; then determine the extent of the y-coordinates, between two numerical values of y.

This formula further assumes that $g_1(y) \le g_2(y)$ throughout the interval $[c, d]$.

DOUBLE INTEGRAL OVER A REGION BETWEEN CURVES, PART 2

To integrate $z = f(x, y)$ over the region between $x = g_1(y)$ and $x = g_2(y)$ for $c \le y \le d$, use

$$\int_c^d \left(\int_{g_1(y)}^{g_2(y)} f(x, y)\, dx \right) dy.$$

Example 17 *Set up the double integral (do not evaluate) for determining the volume under the surface $f(x, y) = \sin(x^2 + y^2)$ over the region bounded by $y = \sqrt{x}$, $x = 4$, and $y = 0$, (a) using the order $dy\,dx$ and (b) using the order $dx\,dy$.*

Solution (a) The region over which we need to integrate is in Figure 4.21, including a slice parallel to the y-axis (for the inner integral, with respect to y). The slice is from $y = 0$ to $y = \sqrt{x}$, indicating an inner integral of

$$\int_0^{\sqrt{x}} \sin(x^2 + y^2)\, dy.$$

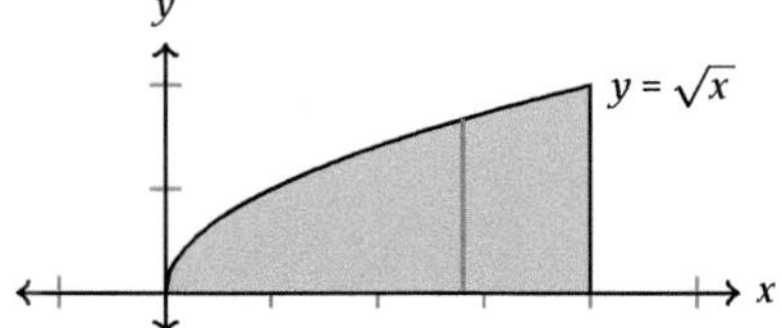

Figure 4.21 *The region (gray) bounded by $y = \sqrt{x}$, $x = 4$, and $y = 0$, along with a slice parallel to the y-axis as a guide for setting up the inner integral with respect to y*

The double integral may be written in this fashion, or parentheses may be included in the style of an iterated integral.

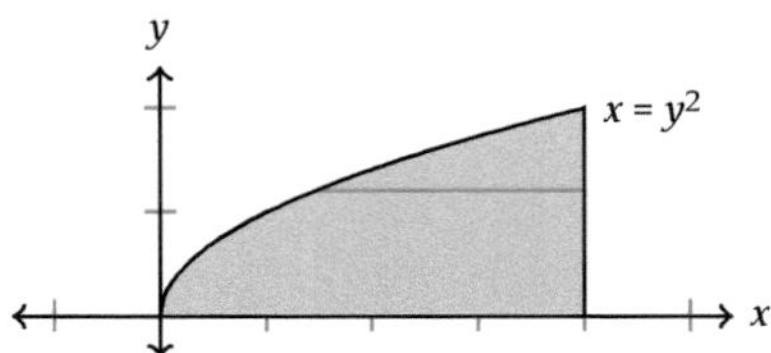

Figure 4.22 *The region (gray) bounded by $y = \sqrt{x}$, $x = 4$, and $y = 0$, along with a slice parallel to the x-axis as a guide for setting up the inner integral with respect to x. Compare to Figure 4.21*

The integrand has a non-elementary antiderivative, making it difficult to evaluate by hand. Using a CAS, the value can be determined as 0.532789, for both part (a) and part (b).

For the outer integral with respect to x, the x-coordinates extend from 0 to 4, and we complete the setup as

$$\int_0^4 \int_0^{\sqrt{x}} \sin(x^2 + y^2)\, dy\, dx.$$

(b) For the order $dx\, dy$, the inner integral is with respect to x, so we draw our slice parallel to the x-axis (Figure 4.22). Then we need to determine the values of x between which the slice lies. On the right, this is clearly $x = 4$. But on the left, we must solve the equation of the curve for the variable x:

$$y = \sqrt{x}$$
$$y^2 = x.$$

Placing this new equation in the diagram (again see Figure 4.22), it is clear that the x-coordinates of the slice are between $x = y^2$ (on the left) and $x = 4$ (on the right). Therefore, the inner integral is

$$\int_{y^2}^4 \sin(x^2 + y^2)\, dx.$$

For the outer integral with respect to y, the y-coordinates extend from $y = 0$ to $y = 2$, and we complete the setup as

$$\int_0^2 \int_{y^2}^4 \sin(x^2 + y^2)\, dx\, dy.$$

∎

Example 17 illustrates the fact that changing from the order $dy\, dx$ to the order $dx\, dy$ for non-rectangular regions is not as simple as swapping the order of the limits of integration; compare to the alternate solution to example 13 of Section 4.2. Changing the order of integration for non-rectangular regions is demonstrated in example 21.

Reading Exercise 5 Set up a double integral (do not evaluate) for finding the volume under the surface $f(x, y) = e^{x^2 y^2}$ over the region bounded by $y = 2x$, $y = 0$, and $x = 3$.

4.3.3 Double integral example: region between curves

Now that we have practiced setting up double integrals over non-rectangular regions (example 17 and reading exercise 5) and evaluating double integrals (example 16 and reading exercise 4), we are ready to put it all together and find volumes under surfaces over these more general regions.

Example 18 *Find the volume under the surface $f(x,y) = ye^x + 10$ over the region bounded by $y = 1 - x^2$ and $y = x^2 - 1$.*

Solution ❶ Graph the region. This task should be familiar, having been used in a few applications before (such as finding the area between two curves). The graphs of the two parabolas enclose a region (Figure 4.23).

❷ Set up a double integral. There are two choices for the order of integration, $dy\,dx$ and $dx\,dy$. Figure 4.23 includes a slice parallel to the y-axis. Notice that the slice is between the same two curves throughout the region, requiring just one integral. However, if we slice parallel to the x-axis (Figure 4.24), the slices are not always between the same two curves and two integrals are required (as with example 20 later in this section). Because we have our choice, we choose to use the order $dy\,dx$.

Referring to Figure 4.23, we see that the region lies between $y = x^2 - 1$ (lower boundary) and $y = 1 - x^2$ (upper boundary). The inner integral is

$$\int_{x^2-1}^{1-x^2} (ye^x + 10)\, dy.$$

To determine the values of x to use, we need to know where the two curves intersect. It may be obvious from the graph of the region, but if not, we may always equate the two functions and solve:

$$x^2 - 1 = 1 - x^2$$
$$2x^2 = 2$$
$$x^2 = 1$$
$$x = \pm 1.$$

The x-coordinates of the region extend from $x = -1$ to $x = 1$, and therefore the double integral is

$$\int_{-1}^{1}\int_{x^2-1}^{1-x^2} (ye^x + 10)\, dy\, dx.$$

❸ Evaluate the double integral. The volume we wish to know is

$$\int_{-1}^{1}\int_{x^2-1}^{1-x^2} (ye^x + 10)\, dy\, dx = \int_{-1}^{1}\left(\left(\frac{y^2}{2}e^x + 10y\right)\Big|_{x^2-1}^{1-x^2}\right) dx$$

$$= \int_{-1}^{1}\left(\frac{(1-x^2)^2}{2}e^x + 10(1-x^2) - \left(\frac{(x^2-1)^2}{2}e^x + 10(x^2-1)\right)\right) dx$$

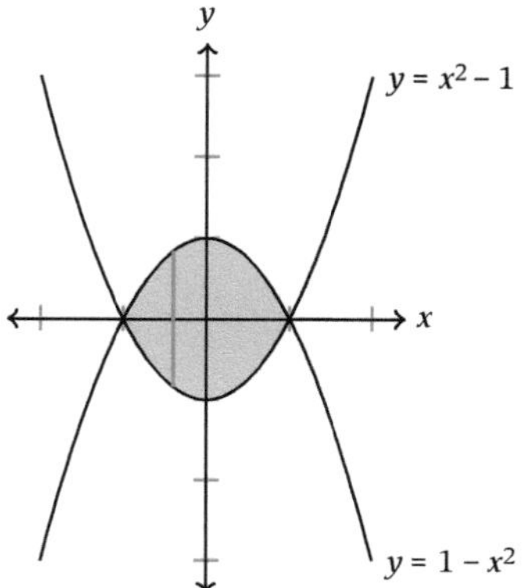

Figure 4.23 *The region (gray) bounded by the parabolas $y = 1 - x^2$ and $y = x^2 - 1$, along with a slice parallel to the y-axis as a guide for setting up the inner integral with respect to y*

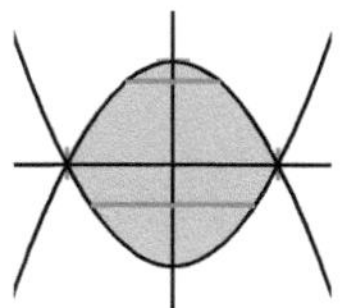

Figure 4.24 *The region (gray) bounded by the parabolas $y = 1 - x^2$ and $y = x^2 - 1$, along with two slices parallel to the x-axis as a guide for setting up the inner integral with respect to x. Two integrals are required for the order $dx\,dy$*

Line 1 finds an antiderivative with respect to y, treating x as a constant; line 2 evaluates at $y = 1 - x^2$, at $y = x^2 - 1$, and subtracts (still treating x as a constant); line 3 expands and line 4 simplifies; line 5 finds an antiderivative with respect to x; line 6 evaluates at $x = 1$, at $x = -1$, and subtracts, then completes the arithmetic and includes units[3] because the context is volume.

$$= \int_{-1}^{1} \left(\frac{1 - 2x^2 + x^4}{2} e^x + 10 - 10x^2 - \frac{x^4 - 2x^2 + 1}{2} e^x - 10x^2 + 10 \right) dx$$

$$= \int_{-1}^{1} (20 - 20x^2)\, dx$$

$$= \left(20x - \frac{20x^3}{3} \right)\Bigg|_{-1}^{1}$$

$$= 20 - \frac{20}{3} - \left(-20 + \frac{20}{3} \right) = \frac{80}{3} \text{ units}^3.$$

$\blacksquare$

Ans. to reading exercise 5:

$$\int_0^3 \int_0^{2x} e^{x^2 y^2}\, dy\, dx; \quad \text{alternate} \quad \text{answer}$$

$$\int_0^6 \int_{y/2}^3 e^{x^2 y^2}\, dx\, dy$$

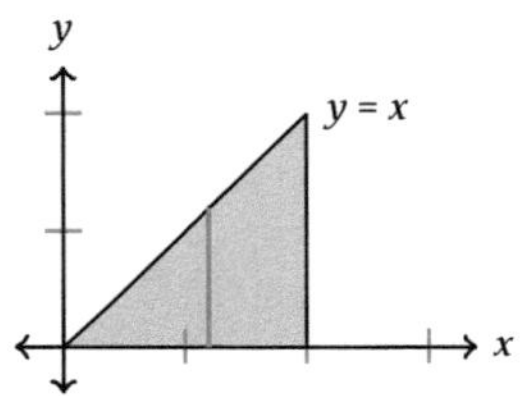

Figure 4.25 *The region (gray) bounded by the lines* $y = x$, $y = 0$, *and* $x = 2$, *along with a slice parallel to the y-axis as a guide for setting up the inner integral with respect to y. Compare to the region of integration in Figure 4.15*

Compare to the discussion prior to example 16.

Line 1 finds an antiderivative with respect to y, treating x as a constant; line 2 evaluates at $y = x$, at $y = 0$, and subtracts (still treating x as a constant); line 3 simplifies; line 4 finds an antiderivative with respect to x; line 5 evaluates at $x = 2$, at $x = 0$, and subtracts, finishes the arithmetic, and includes units for volume.

4.3.4 Double integral example: triangular region

In example 18, there is a clear preference as to which order of integration to use. That is not always the case. We will present two solutions to the next example.

Example 19 *Find the volume under the surface* $f(x, y) = x^2 + y^3$ *over the region bounded by* $y = x$, $y = 0$, *and* $x = 2$.

Solution ❶ Graph the region. The region is graphed in Figure 4.25. Notice that this is the same region of integration used in Figure 4.15, although we are not using the same surface.

❷ Set up a double integral. Figure 4.25 contains a slice parallel to the y-axis. The y-coordinates of the slice are between $y = 0$ and $y = x$, making the inner integral

$$\int_0^x (x^2 + y^3)\, dy.$$

The x-coordinates extend from $x = 0$ to $x = 2$; we complete the setup as

$$\int_0^2 \int_0^x (x^2 + y^3)\, dy\, dx.$$

❸ Evaluate the double integral. The volume is

$$\int_0^2 \int_0^x (x^2 + y^3)\, dy\, dx = \int_0^2 \left(x^2 y + \frac{y^4}{4} \Bigg|_0^x \right) dx$$

$$= \int_0^2 \left(x^2(x) + \frac{x^4}{4} - 0 \right) dx$$

$$= \int_0^2 \left(x^3 + \frac{x^4}{4} \right) dx$$

$$= \left(\frac{x^4}{4} + \frac{x^5}{20} \right)\Big|_0^2$$

$$= \frac{16}{4} + \frac{32}{20} - 0 = \frac{28}{5}\text{units}^3.$$

■

Alternate solution ❶ Graph the region. This time we include a slice parallel to the x-axis; see Figure 4.26.

❷ Set up a double integral. The equation $y = x$ is the same as $x = y$ and is ready for use. Integrating with respect to x first, the slice is between $x = y$ and $x = 2$, and the inner integral is

$$\int_y^2 (x^2 + y^3)\, dx.$$

The y-coordinates extend from $y = 0$ to $y = 2$, and we complete the setup as

$$\int_0^2 \int_y^2 (x^2 + y^3)\, dx\, dy.$$

❸ Evaluate the double integral. The volume is

$$\int_0^2 \int_y^2 (x^2 + y^3)\, dx\, dy = \int_0^2 \left(\left(\frac{x^3}{3} + y^3 x \right)\Big|_y^2 \right) dy$$

$$= \int_0^2 \left(\frac{8}{3} + 2y^3 - \left(\frac{y^3}{3} + y^4 \right) \right) dy$$

$$= \left(\frac{8}{3}y + \frac{2y^4}{4} - \frac{y^4}{12} - \frac{y^5}{5} \right)\Big|_0^2$$

$$= \frac{16}{3} + \frac{1}{2} \cdot 16 - \frac{16}{12} - \frac{32}{5} - 0 = \frac{28}{5} \text{ units}^3,$$

which agrees with the previous solution to the problem.   ■

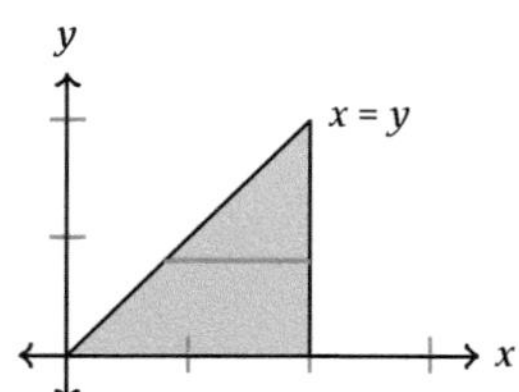

Figure 4.26 *The region (gray) bound-ed by the lines $y = x$, $y = 0$, and $x = 2$, along with a slice parallel to the x-axis as a guide for setting up the inner integral with respect to x. Compare to Figure 4.25*

Think of sliding the slice in Figure 4.26, which is parallel to the x-axis, up and down the region. The slice can go as low as $y = 0$ and as high as $y = 2$.

Line 1 finds an antiderivative with respect to x, treating y as a constant; line 2 evaluates at $x = 2$, at $x = y$, and subtracts (still treating y as a constant); line 3 finds an antiderivative with respect to y; line 4 evaluates at $y = 2$, at $y = 0$, and subtracts, finishes the arithmetic, and includes units for volume.

4.3.5   Double integral example: split region

Sometimes the use of two double integrals cannot be avoided because of the shape of the region.

Example 20 *Find the volume under the surface $z = x^2 + y^2$ over the region bounded by $y = x$, $y = 3x$, and $x + y = 4$.*

Solution ❶ Graph the region. The three lines are graphed in Figure 4.27.

Notice that if the slices are parallel to the *x*-axis instead, we still need two integrals.

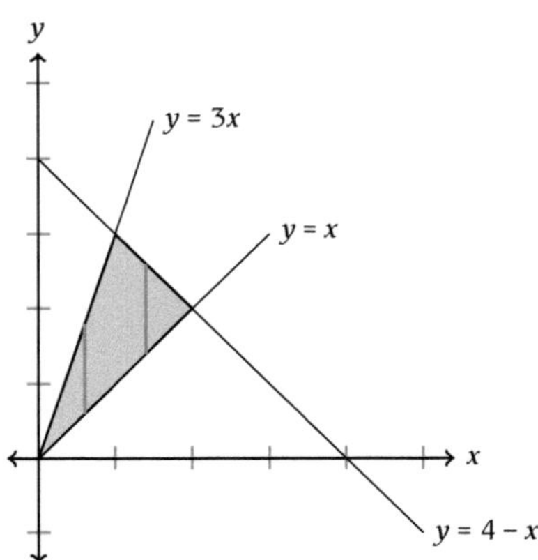

Figure 4.27 *The region (gray) bounded by the lines $y = x$, $y = 3x$, and $x + y = 4$. Two slices parallel to the y-axis are included to illustrate the need for two integrals*

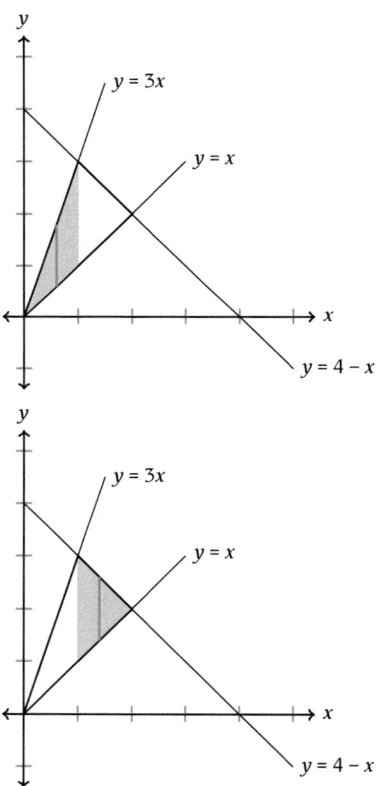

Figure 4.28 *The region of Figure 4.27 split into (top) the region containing slices with top boundary $y = 3x$ and (bottom) the region containing slices with top boundary $y = 4 - x$*

❷ Set up a double integral. For the order $dy\,dx$, the slices (see Figure 4.27) do not all have the same upper boundary. We therefore need to use two integrals, one for each upper boundary. A quick look at the region shows that slicing parallel to the *x*-axis also requires two integrals, and we see no apparent advantage to using the order $dx\,dy$. Continuing with the order $dy\,dx$, we split the region into two pieces, one containing the slices with upper boundary $y = 3x$ and one containing the slices with upper boundary $y = 4 - x$; see Figure 4.28.

It is apparent from Figures 4.27 and 4.28 that we need to know where the lines $y = 3x$ and $y = 4 - x$ intersect, as well as where the lines $y = 4 - x$ and $y = x$ intersect. If not already known from the graphing process, these points of intersection can easily be determined by setting $3x = 4 - x$ and $4 - x = x$, respectively, resulting in $x = 1$ and $x = 2$.

To set up the integral for the first region, we consult the upper portion of Figure 4.28. The slices lie between $y = x$ and $y = 3x$, whereas the *x*-coordinates extend from $x = 0$ to $x = 1$, yielding

$$\int_0^1 \int_x^{3x} (x^2 + y^2)\, dy\, dx.$$

To set up the integral for the second region, we consult the lower portion of Figure 4.28. The slices lie between $y = x$ and $y = 4 - x$, whereas the *x*-coordinates extend from $x = 1$ to $x = 2$, yielding

$$\int_1^2 \int_x^{4-x} (x^2 + y^2)\, dy\, dx.$$

The volume we wish to find is the sum of the volumes over these two regions,

$$\int_0^1 \int_x^{3x} (x^2 + y^2)\, dy\, dx + \int_1^2 \int_x^{4-x} (x^2 + y^2)\, dy\, dx.$$

❸ Evaluate the double integral. The first integral is

$$\int_0^1 \int_x^{3x} (x^2 + y^2)\, dy\, dx = \int_0^1 \left(\left(x^2 y + \frac{y^3}{3} \right)\Big|_x^{3x} \right) dx$$

$$= \int_0^1 \left(x^2(3x) + \frac{27x^3}{3} - \left(x^3 + \frac{x^3}{3} \right) \right) dx$$

$$= \int_0^1 \frac{32x^3}{3}\, dx$$

$$= \frac{32x^4}{12} \Big|_0^1$$

$$= \frac{32}{12} - 0 = \frac{8}{3}.$$

Line 1 finds an antiderivative with respect to y, treating x as a constant; line 2 evaluates at $y = 3x$, at $y = x$, and subtracts (still treating x as a constant); line 3 simplifies; line 4 finds an antiderivative with respect to x; and line 5 evaluates at $x = 1$, at $x = 0$, and subtracts, then finishes the arithmetic.

The second integral is

$$\int_1^2 \int_x^{4-x} (x^2 + y^2)\, dy\, dx = \int_1^2 \left(\left(x^2 y + \frac{y^3}{3} \right)\Big|_x^{4-x} \right) dx$$

$$= \int_1^2 \left(x^2(4 - x) + \frac{(4 - x)^3}{3} - \left(x^3 + \frac{x^3}{3} \right) \right) dx$$

$$= \int_1^2 \left(4x^2 - x^3 + \frac{64 - 48x + 12x^2 - x^3}{3} - \frac{4}{3}x^3 \right) dx$$

$$= \int_1^2 \left(\frac{64}{3} - 16x + 8x^2 - \frac{8}{3}x^3 \right) dx$$

$$= \left(\frac{64}{3}x - 8x^2 + \frac{8}{3}x^3 - \frac{2}{3}x^4 \right)\Big|_1^2$$

$$= \frac{128}{3} - 32 + \frac{64}{3} - \frac{32}{3} - \left(\frac{64}{3} - 8 + \frac{8}{3} - \frac{2}{3} \right) = 6.$$

Line 1 finds an antiderivative with respect to y, treating x as a constant; line 2 evaluates at $y = 4 - x$, at $y = x$, and subtracts (still treating x as a constant); line 3 expands and line 4 simplifies; line 5 finds an antiderivative with respect to x; and line 6 evaluates at $x = 2$, at $x = 1$, and subtracts, then finishes the arithmetic.

The volume we seek is the sum of these integrals,

$$\frac{8}{3} + 6 = \frac{26}{3}\ \text{units}^3.$$

∎

Reading Exercise 6 Set up a double integral (do not evaluate) for finding the volume under the surface $f(x, y) = \cosh(x^2 y^2)$ over the region bounded by $y = x$, $y = 4x$, and $y = 6 - 2x$.

In example 20, we split a region into two pieces, found the volume under the surface over each piece, and then added those two volumes to find the volume under the surface over the combined region. Although this procedure fits our geometric intuition, it is difficult to prove in general, so the proof of the following property will be omitted.

> ### ADDITIVE PROPERTY OF DOUBLE INTEGRALS
>
> Suppose R_1 and R_2 are two closed regions in the plane that do not overlap except along a common boundary, and suppose that $z = f(x, y)$ is continuous on the combined region $R_1 \cup R_2$. Then
>
> $$\iint_{R_1} f(x, y) \, dy \, dx + \iint_{R_2} f(x, y) \, dy \, dx = \iint_{R_1 \cup R_2} f(x, y) \, dy \, dx.$$

The additive property also applies to the order $dx \, dy$.

4.3.6 Switching the order of integration

As demonstrated in Section 4.2, it is sometimes useful to switch the order of integration. If one order of integration is difficult, perhaps the other order is easier. As long as our integrand is continuous throughout the region, Fubini's theorem applies and it is legitimate to change the order; the result will be the same.

If the region is rectangular, switching the order of integration is simple. But if the region is not rectangular, care must be taken to determine the correct limits of integration when switching the order.

Why change the order of integration? Because if x is the variable of integration, the antiderivative of e^{x^2} is non-elementary.

Example 21 *Evaluate* $\displaystyle\int_0^2 \int_{y^2}^4 4ye^{x^2} \, dx \, dy$ *by reversing the order of integration.*

Caution–trying $\displaystyle\int_{y^2}^4 \int_0^2 4ye^{x^2} \, dy \, dx$ does not work! For one thing, the final answer would contain the variable y and would not be numeric.

Solution ❶ Determine the region of integration. This time we are not given curves describing the region, but rather limits of integration for the order $dx \, dy$, and we must use this information to determine the region. Because the inner integral is with respect to x, the slices are parallel to the x-axis. The slices are between $x = y^2$ and $x = 4$. We therefore graph $x = y^2$, which is a parabola oriented horizontally opening to the right, as well as the line $x = 4$. Also note that we only draw the region from $y = 0$ to $y = 2$, so we omit anything below the x-axis. The result is in Figure 4.29.

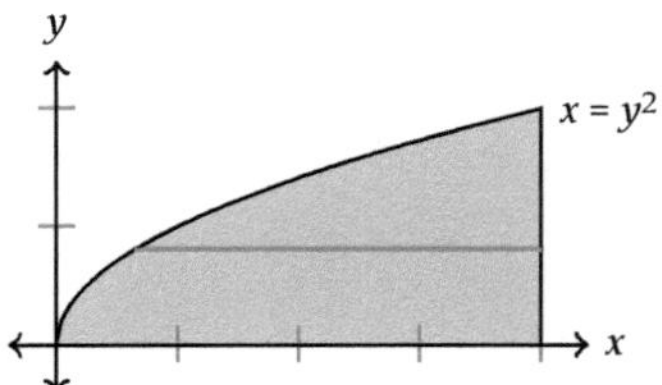

Figure 4.29 *The region (gray) between $x = y^2$ and $x = 4$, from $y = 0$ to $y = 2$. A slice parallel to the x-axis is added to help us visualize the given limits of integration*

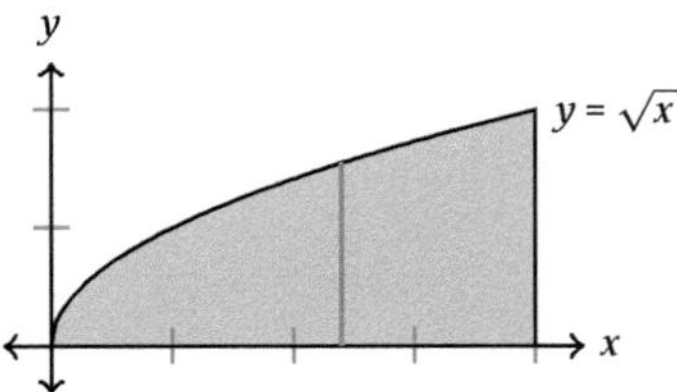

Figure 4.30 *The region (gray) of Figure 4.29, with the curve relabeled and the slice reoriented*

Ans. to reading exercise 6:

$$\int_0^1 \int_x^{4x} \cosh(x^2 y^2)\, dy\, dx \ +$$

$$\int_1^2 \int_x^{6-2x} \cosh(x^2 y^2)\, dy\, dx;\ \text{alternate}$$

answer $\displaystyle\int_0^2 \int_{y/4}^y \cosh(x^2 y^2)\, dx\, dy\ +$

$$\int_2^4 \int_{y/4}^{3-y/2} \cosh(x^2 y^2)\, dx\, dy$$

② Set up a double integral. Because we wish to use the order *dy dx*, we change to a slice parallel to the *y*-axis. We also need to rewrite the equation of the curve in terms of *y*:

$$x = y^2$$

$$\sqrt{x} = y.$$

In line 2, we know to use *y*, as opposed to $-y$ or $\pm y$, because the *y*-coordinates in the region are nonnegative.

These changes are noted in Figure 4.30. Now we see that the inner integral is from $y = 0$ to $y = \sqrt{x}$, and then the *x*-coordinates extend from $x = 0$ to $x = 4$, resulting in the integral

$$\int_0^4 \int_0^{\sqrt{x}} 4y e^{x^2}\, dy\, dx.$$

③ Evaluate the double integral. With this order, we can at least find an antiderivative with respect to *y* for the inner integral:

$$\int_0^4 \int_0^{\sqrt{x}} 4y e^{x^2}\, dy\, dx = \int_0^4 \left(2y^2 e^{x^2} \Big|_0^{\sqrt{x}} \right) dx$$

$$= \int_0^4 \left(2x e^{x^2} - 0 \right) dx.$$

Line 1 finds an antiderivative with respect to *y*, treating *x* as a constant; line 2 evaluates at $y = \sqrt{x}$, at $y = 0$, and subtracts.

Noticing the need for substitution (the exponent is "not just plain x"), we use

$$u = x^2$$

$$du = 2x\,dx,$$

No context for the integral is given, so no units are included.

which will succeed because we now have the x in the integrand, which we did not have before changing the order of integration. We also need to change to values of u; for $x = 4$, $u = 16$ and for $x = 0$, $u = 0$. Making the substitution and continuing,

$$= \int_0^{16} e^u\,du = e^u\big|_0^{16} = e^{16} - 1.$$

■

EXERCISES 4.3

1–8. Set up the double integral(s) for determining the (net) volume under the surface $f(x, y) = x \sin y$ over the specified region, (a) using the order $dy\,dx$ and (b) using the order $dx\,dy$. (c) Use a CAS to evaluate the integral(s) from part (a) or part (b); either choice should yield the same result.

1. region bounded by $y = x + 2$, $x = 0$, and $y = 0$
2. region bounded by $x = 0$, $y = 0$, and $y = 5 - x$
3. region bounded by $y = x$, $y = -2x$, and $y = \frac{1}{2}(3 - x)$
4. region bounded by $y = \sqrt{x}$, $y = 2 - x$, and $y = 0$
5. region bounded by $y = x^2$, $y = 6 - x$, and $x = 0$ (for $x \geq 0$ only)
6. region bounded by $y = x$, $y = -x$, and $y = 4 - 3x$
7. region $\dfrac{x^2}{4} + y^2 \leq 1$
8. region $x^2 + y^2 \leq 4$

9–12. Set up a double integral (do not evaluate) for finding the volume under the surface $f(x, y) = x \sin y$ over the specified region.

9. region bounded by $y = x^2$, $x = 3$, and $y = 0$
10. region bounded by $y = 0$ and $y = 4 - x^2$
11. region bounded by $x = 4 - y^2$ and $x = 1 + 2y^2$
12. region bounded by $y = x^2 + 1$ and $y = 2x^2$

13–18. Evaluate the integral.

13. $\displaystyle \int_{-2}^{0} \int_{0}^{-x} x^2 y^3 \, dy\,dx$

14. $\displaystyle \int_{0}^{1} \int_{0}^{2x} (x^2 + y^2) \, dy\,dx$

15. $\displaystyle\int_{2}^{3}\int_{x}^{5}(x-4y)\,dy\,dx$

16. $\displaystyle\int_{1}^{2}\int_{x}^{3x}(x^{2}y+xy^{2})\,dy\,dx$

17. $\displaystyle\int_{1}^{e}\int_{0}^{3\sqrt{\frac{1}{y^{2}}}}3x^{2}y\,dx\,dy$

18. $\displaystyle\int_{0}^{1}\int_{0}^{y^{2}}\left(\sqrt{x}+y^{2}\right)dx\,dy$

19–24. Choose the result of part (a) or part (b) of the given exercise to evaluate (by hand) to determine the (net) volume.

19. exercise 1
20. exercise 2
21. exercise 5
22. exercise 4
23. exercise 7
24. exercise 8

25. Find the volume under the surface $f(x,y)=x+y^{2}$ over the region bounded by $y=\frac{1}{2}x$, $y=x$, and $x=2$.
26. Find the volume under the surface $f(x,y)=x^{2}y$ over the region bounded by $y=x^{2}$ and $y=4$.
27. Find the volume under the surface $f(x,y)=x+y^{2}$ over the region bounded by $y=\frac{1}{2}x$, $y=x$, and $y=2$.
28. Find the volume under the surface $z=x+3y$ over the region bounded by $y=x$, $y=3x$, and $y=3$.
29. Find the volume under the surface $f(x,y)=xe^{y}$ over the region bounded by $y=x$ and $y=x^{2}$.
30. (Love) Find the volume under the surface $z=\frac{xy}{5}$, whose base is the region in the xy-plane bounded by the curves $y^{2}=5x$, $x+y=10$, and $y=0$.
31. (Love) Find the volume in the first octant bounded by the surfaces $x^{2}+y^{2}=z$ and $x+2y=2$.
32. (Love) Find the volume in the first octant bounded by the surfaces $z=\frac{x^{2}}{2}$, $x^{2}=2y$, and $2x-y=2$.

Exercises in this section marked (Love) are from *Differential and Integral Calculus* by Clyde E. Love, The MacMillan Company, 1926.

Reminder: the first octant is the octant where the x-, y-, and z-coordinates are all positive.

33–42. Evaluate the integral by reversing the order of integration.

33. $\displaystyle\int_{0}^{4}\int_{y}^{4}6ye^{x^{3}}\,dx\,dy$

34. $\displaystyle\int_{0}^{2}\int_{2x}^{4}96x^{2}\sqrt{1+y^{4}}\,dy\,dx$

35. (Love) $\displaystyle\int_0^2 \int_{\frac{1}{2}y}^1 \frac{1}{(1+x^2)^3}\, dx\, dy$

36. $\displaystyle\int_0^4 \int_{\sqrt{y}}^2 \sin x^3\, dx\, dy$

37. (Love) $\displaystyle\int_0^1 \int_y^{\sqrt{y}} \frac{\sin x}{x}\, dx\, dy$

38. (Love) $\displaystyle\int_0^1 \int_{\sqrt{x}}^1 \sqrt{1+y^3}\, dy\, dx$

39. (Love) $\displaystyle\int_{-1}^0 \int_{-x}^1 \frac{x^2}{1+y^4}\, dy\, dx$

40. (Love) $\displaystyle\int_0^4 \int_{\sqrt{y}}^2 \frac{y}{\sqrt{1+x^5}}\, dx\, dy$

41. $\displaystyle\int_0^1 \int_{x^2}^1 \frac{6x^5}{1+y^4}\, dy\, dx$

42. $\displaystyle\int_0^1 \int_0^{\sqrt{1-y}} 2y \cos x\, dx\, dy$

43. The midpoint rule for double integrals can be adapted to non-rectangular regions. Use the following steps to obtain an estimate of the volume under the surface $f(x,y) = \sqrt{x^{1.4} + y^{1.2}}$ over the region bounded by $x = 0$, $y = 0$, and $y = 2 - x$.

    (a) Graph the region.
    (b) Using $p = 4$ and $q = 4$, determine Δx and Δy as if the region is the rectangle $0 \le x \le 2$, $0 \le y \le 2$.
    (c) Determine the values of $\overline{x_n}$ and $\overline{y_m}$ as if the region is rectangular. Plot the points $(\overline{x_n}, \overline{y_m})$ to determine which of these points are located in the region graphed in part (a).
    (d) For each of the points located in the interior of the region, calculate $f(\overline{x_n}, \overline{y_m})\Delta x \Delta y$. Add the results.
    (e) For each of the points located on the boundary of the region, calculate $\frac{1}{2}f(\overline{x_n}, \overline{y_m})\Delta x \Delta y$. Add the results. (Notice that for a point on the boundary, the area of the triangular subregion it represents is $\frac{1}{2}\Delta x \Delta y$, hence the change in the formula.)
    (f) Add the results of parts (d) and (e). This is the midpoint estimate.
    (g) This procedure can be used for other shapes of regions by enclosing the region in a rectangle. As long as p and q are sufficiently large, should this procedure give a reasonable estimate? Use drawings and your intuition to formulate your response.

44. For exercise 24 (exercise 8), plot the function (the integrand) over the region. From what you see, explain the result of exercise 24.
45. Improper double integrals can also be evaluated, in the manner one would expect. Find the volume under the surface $z = xy$ over the region between $y = 0$ and $y = \dfrac{1}{x^3}$, for $x \geq 1$.
46. Find the volume under the surface $z = xy$ over the region between $y = 0$ and $y = \dfrac{1}{x^3}$, for $0 \leq x \leq 1$.

4.4   Double Integrals in Polar Coordinates

The definition of double integral, definition 1 in Section 4.2, relies on the usual rectangular coordinate system. Is there ever an advantage to using a different coordinate system?

4.4.1   A motivating example

Sometimes a relatively simple-looking volume leads to a difficult double integral in the usual rectangular coordinates.

Example 22 *Find the volume under the surface $f(x, y) = 1 - x^2 - y^2$ over the region $x^2 + y^2 \leq 1$.*

Solution   While not part of our solution procedure as practiced in Section 4.3, it can be helpful to visualize the surface (Figure 4.31); think of the paraboloid $z = x^2 + y^2$ reflected across the xy-plane to $z = -x^2 - y^2$ and then shifted one unit in the positive z-direction to $z = 1 - x^2 - y^2$. On the boundary of the region of integration, $x^2 + y^2 = 1$, the value of the function is $1 - (x^2 + y^2) = 1 - 1 = 0$, as pictured. In other words, we wish to find the volume of the inside of a paraboloid bowl (upside-down). This is the volume under a surface over a non-rectangular region. The procedure of examples 18–20 of Section 4.3 applies.

❶ Graph the region. The region is the interior of the unit circle (Figure 4.32). Solving $x^2 + y^2 = 1$ for y yields the equations of the top half and bottom half of the circle, $y = \pm\sqrt{1 - x^2}$.

❷ Set up a double integral. For the order $dy\,dx$, the slices are parallel to the y-axis and the integral is

$$\int_{-1}^{1} \int_{-\sqrt{1-x^2}}^{\sqrt{1-x^2}} (1 - x^2 - y^2)\, dy\, dx.$$

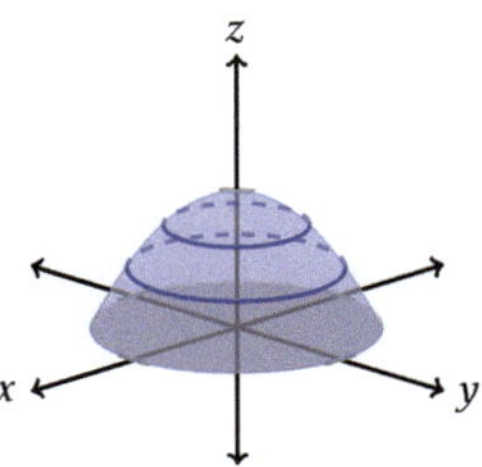

Figure 4.31 *The surface $z = 1 - x^2 - y^2$ (blue) over the region $x^2 + y^2 \leq 1$ (gray)*

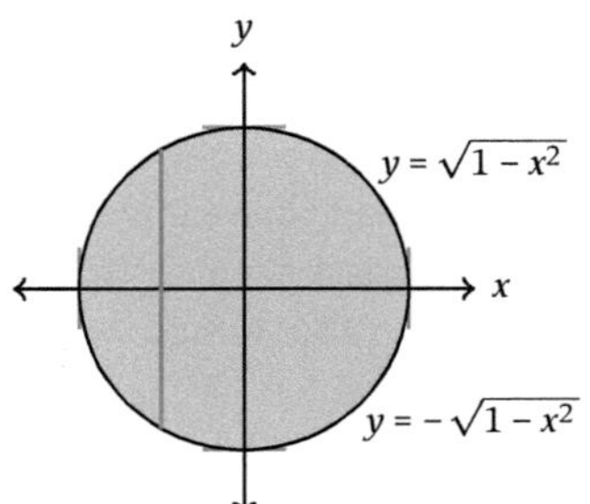

Figure 4.32 *The region $x^2 + y^2 \leq 1$*

3 Evaluate the double integral. Beginning with the inner integral and skipping some details,

$$\int_{-1}^{1} \left(\int_{-\sqrt{1-x^2}}^{\sqrt{1-x^2}} (1 - x^2 - y^2) \, dy \right) dx$$

$$= \int_{-1}^{1} \left((1 - x^2)y - \frac{y^3}{3} \bigg|_{-\sqrt{1-x^2}}^{\sqrt{1-x^2}} \right) dx$$

$$= \cdots = \int_{-1}^{1} \frac{4}{3}(1 - x^2)^{3/2} \, dx.$$

The remaining integral requires trigonometric substitution.

$$x = \sin \theta$$

$$dx = \cos \theta \, d\theta$$

The trig substitution $x = \sin \theta$ is designed to use the interval $-\frac{\pi}{2} \leq \theta \leq \frac{\pi}{2}$, although other intervals could be used instead.

The limits of integration must also be adjusted; if $x = 1$, we have $1 = \sin \theta$, so $\theta = \frac{\pi}{2}$; if $x = -1$, we get $\theta = -\frac{\pi}{2}$. Making the substitution and using the associated trig identity,

$$= \int_{-\frac{\pi}{2}}^{\frac{\pi}{2}} \frac{4}{3}(1 - \sin^2 \theta)^{3/2} \cos \theta \, d\theta$$

$$= \int_{-\frac{\pi}{2}}^{\frac{\pi}{2}} \frac{4}{3}(\cos^2 \theta)^{3/2} \cos \theta \, d\theta$$

$$= \int_{-\frac{\pi}{2}}^{\frac{\pi}{2}} \frac{4}{3} \cos^4 \theta \, d\theta.$$

Now we see an even power of sine, which requires the use of a half-angle identity–squaring because the exponent is 4–and then using the half-angle identity again, eventually arriving at

$$= \cdots = \left(\frac{1}{2}\theta + \frac{1}{3}\sin 2\theta + \frac{1}{24}\sin 4\theta \right)\Bigg|_{-\frac{\pi}{2}}^{\frac{\pi}{2}} = \cdots = \frac{\pi}{2} \text{ units}^3.$$

A simple-looking region. A simple-looking answer. A non-simple calculation. Is there a simpler method? ∎

4.4.2   Double integrals in polar coordinates

Using Cartesian (rectangular) coordinates, the volume under a surface $z = f(x, y)$ over a rectangular region is calculated by summing the volumes of boxes such as those in Figure 4.33. The volume of the box is the area of the base, $\Delta x \Delta y$, times the height of the box, $f(x_n, y_m)$. The rectangular region is of the form $a \le x \le b$, $c \le y \le d$. What changes if instead we use polar coordinates?

Polar coordinates are designed for two dimensions. Leaving the z-axis as is, we replace the rectangular variables x and y with their polar counterparts, r and θ. Consider, then, a surface $z = f(r, \theta)$ over a region $c \le r \le d$, $a \le \theta \le b$, as in Figure 4.34. The base is not rectangular, and when the region is partitioned into smaller regions, these subregions are not rectangular, either. The volume of the "box" is still the area of its base times its height, so our next task is to determine the area of one of these smaller subregions.

Each of the subregions shown in Figure 4.34 is a portion of a ring between two concentric circles, and between two radial lines, as in Figure 4.35. The angle between two radial lines is $\Delta\theta$, the radius of the inner circle is r, and the radius of the

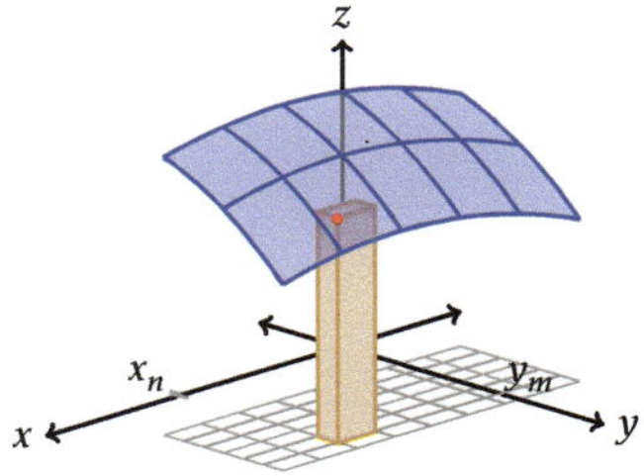

Figure 4.33 *A surface $z = f(x, y)$ (blue) over a rectangular region partitioned into smaller rectangles, along with the n-mth rectangular box (brown). In order to be visible, the box is pictured with non-infinitesimal base area*

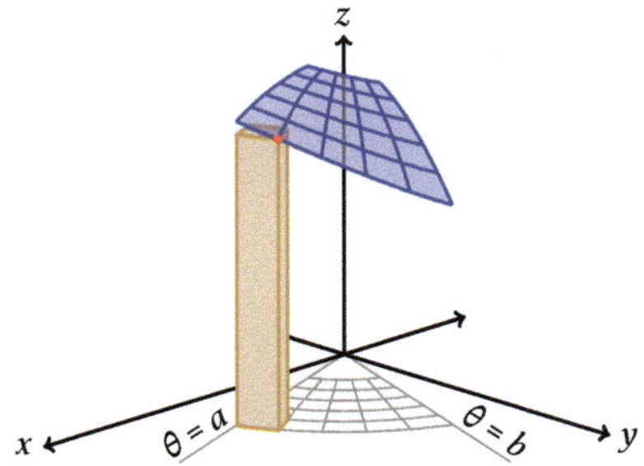

Figure 4.34 *A surface $z = f(r, \theta)$ (blue) over the region $c \le r \le d$, $a \le \theta \le b$, which is partitioned into smaller regions, along with the n-mth box (brown), which has a subtly curved, non-rectangular base. In order to be visible, the box is pictured with non-infinitesimal base area. The x- and y-axes are included for the sake of perspective and comparison*

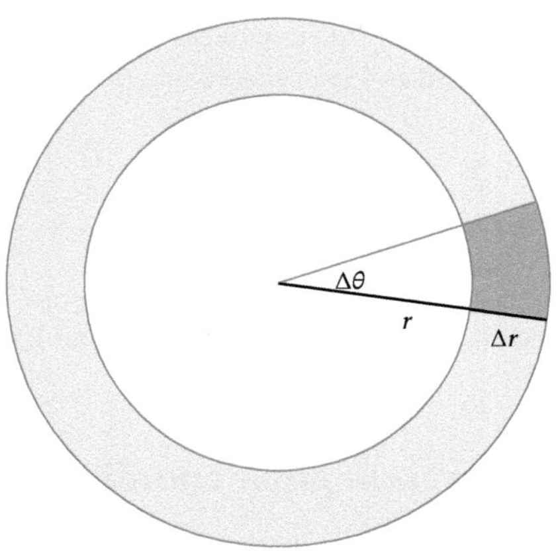

Figure 4.35 *A subregion (darker gray) as a portion of a ring between two concentric circles (lighter gray). The quantities Δθ and Δr are not shown as infinitesimal so that they may be visible*

The darker gray region in Figure 4.35 can be described as a *sector of a washer*, or *washer sector* for short.

outer circle is $r + \Delta r$. The area of the gray ring between the two concentric circles can be found by taking the area inside the larger circle, $\pi(r + \Delta r)^2$, and subtracting the area inside the smaller circle, πr^2:

$$\pi(r + \Delta r)^2 - \pi r^2 = \pi \left(r^2 + 2r\Delta r + (\Delta r)^2\right) - \pi r^2$$

$$= \pi \cdot 2r\Delta r + \pi \left(\Delta r\right)^2$$

$$\approx 2\pi r \Delta r.$$

If we partition the interval $a \le \theta \le b$ into Ω pieces and likewise partition the interval $c \le r \le d$ into Ω pieces, where Ω is a positive infinite hyperreal integer, then both $\Delta\theta$ and Δr are infinitesimal, justifying line 3.

The proportion of the light gray ring that is inside the darker gray subregion is $\frac{\Delta\theta}{2\pi}$, and therefore the area of the subregion is

$$\frac{\Delta\theta}{2\pi} \cdot 2\pi r\Delta r = r \cdot \Delta r\Delta\theta.$$

This formula is the source of the "extra" r in the volume formula below. Whereas the area of the base for using rectangular coordinates (Figure 4.33) is $\Delta x\Delta y$, the area of the base for using polar coordinates is $r \cdot \Delta r\Delta\theta$.

The volume of the box in Figure 4.34 is the height of the box times the area of the base,

$$f(r_n, \theta_m) \cdot r \cdot \Delta r\Delta\theta,$$

and the sum of the volumes of all the boxes is

$$\sum_{n=1}^{\Omega} \sum_{m=1}^{\Omega} f(r_n, \theta_m) \cdot r \cdot \Delta r\Delta\theta = \int_a^b \int_c^d f(r, \theta) \cdot r \, dr \, d\theta.$$

Notice that in Figure 4.35 we are assuming that $r \ge 0$, and this assumption is also used in the formulas that follow.

VOLUME UNDER A SURFACE, POLAR COORDINATES

The volume under the surface $z = f(r, \theta)$ over the region $c \le r \le d, a \le \theta \le b$ is given by

$$V = \int_a^b \int_c^d f(r, \theta) \cdot r \, dr \, d\theta,$$

assuming additionally that $r \ge 0$ (equivalently, that c is nonnegative.).

Don't forget the extra r! Other formulas later in this chapter will also feature a "don't forget!" factor.

Technically, this double integral can be considered to be a double intergral as defined by definition 1, with the function $f(r, \theta) \cdot r$; the particular symbols used to name the variables are immaterial, and the region is described by each variable being between two constants. If $f(r, \theta)$ is continuous on the region, then so is $f(r, \theta) \cdot r$.

Assuming f is continuous, Fubini's theorem still applies with the variables r and θ, and the double integral may be evaluated as an iterated integral using either order of integration.

Example 23 *Find the volume under the surface $f(r, \theta) = 1 + r\theta$ over the region $1 \le r \le 2, 0 \le \theta \le \pi$.*

Solution Using the polar coordinate formula for volume under a surface, the volume is given by the integral

$$V = \int_0^\pi \int_1^2 (1 + r\theta) \cdot r \, dr \, d\theta.$$

Don't forget the extra r!

The integral is then evaluated as usual; the volume is

$$V = \int_0^\pi \int_1^2 (r + r^2\theta) \, dr \, d\theta = \int_0^\pi \left(\left(\frac{r^2}{2} + \frac{r^3}{3}\theta \right)\Big|_1^2 \right) d\theta$$

$$= \int_0^\pi \left(\frac{4}{2} + \frac{8}{3}\theta - \left(\frac{1}{2} + \frac{1}{3}\theta \right) \right) d\theta$$

$$= \int_0^\pi \left(\frac{3}{2} + \frac{7}{3}\theta \right) d\theta$$

$$= \left(\frac{3}{2}\theta + \frac{7}{3}\frac{\theta^2}{2} \right)\Big|_0^\pi$$

$$= \left(\frac{3}{2}\pi + \frac{7}{6}\pi^2 \right) \text{units}^3.$$

Line 1 performs the multiplication by the "don't forget" factor r and then finds an antiderivative with respect to r, treating θ as a constant; line 2 evaluates at $r = 2$, at $r = 1$, and subtracts; line 3 simplifies; line 4 finds an antiderivative with respect to θ; and line 5 evaluates at $\theta = \pi$, at $\theta = 0$, and subtracts.

∎

Reading Exercise 7 Set up the integral for finding the volume under the surface $z = \cos r^2$ over the region $0 \le r \le 3, \frac{\pi}{2} \le \theta \le \pi$. Do not evaluate the integral.

If we desire to change from rectangular coordinates to polar coordinates to seek a simpler integral (or vice versa), the following formulas may be helpful.

See Section 8.4 of *Calculus Set Free.*

POLAR AND RECTANGULAR COORDINATE CONVERSION FORMULAS

$$y = r\sin\theta \qquad x = r\cos\theta$$
$$x^2 + y^2 = r^2 \qquad \tan\theta = \frac{y}{x}$$

We are now ready to return to example 22 and make the switch to polar coordinates, hoping for an easier integral.

Example 24 *Switch to polar coordinates to find the volume under the surface* $f(x,y) = 1 - x^2 - y^2$ *over the region* $x^2 + y^2 \leq 1.$

Solution ❶ Graph the region. Figure 4.36 shows the region, which is the interior of the unit circle. Instead of slices parallel to the y-axis as shown in Figure 4.32, or parallel to the x-axis, the slices are *radial*, extending from the origin to the curve.

The same basic steps as in example 22 may be used, but the details change to use polar coordinates.

It is also possible for the radial segments to go from one curve to another, but always along a line through the origin.

Line 2 makes the conversion to polar coordinates, including rewriting the function as $f(r,\theta)$; line 3 uses a common trig identity to simplify. An alternative is to rewrite f as $f(x,y) = 1 - (x^2 + y^2)$ and use the conversion formula $x^2 + y^2 = r^2$ to obtain $f(r,\theta) = 1 - r^2.$ When appropriate, the use of this conversion formula can eliminate the need for using an identity.

❷ Set up a double integral. If we wish to use polar coordinates, we need to (1) convert the function to polar coordinates and (2) describe the region using polar coordinates.

For (1), we may use $x = r\cos\theta$ and $y = r\sin\theta$ to make the conversion:

$$f(x,y) = 1 - x^2 - y^2$$
$$f(r,\theta) = 1 - (r\cos\theta)^2 - (r\sin\theta)^2 = 1 - r^2\cos^2\theta - r^2\sin^2\theta$$
$$= 1 - r^2(\cos^2\theta + \sin^2\theta) = 1 - r^2.$$

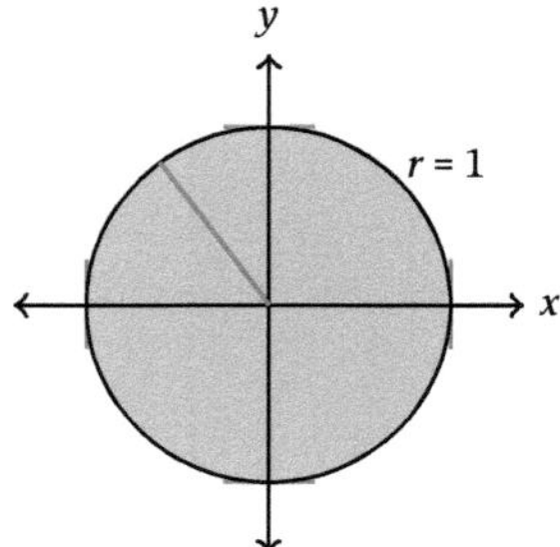

Figure 4.36 *The region* $x^2 + y^2 \leq 1$

For (2), note that $0 \le r \le 1$ along any radial segment (recall that r must be nonnegative, so we cannot be creative and say $-1 \le r \le 0$). For θ, because we are traveling around a complete circle we can use $0 \le \theta \le 2\pi$.

Now we are ready to use the formula for volume under a surface, polar coordinates; the volume is

$$V = \int_0^{2\pi} \int_0^1 (1 - r^2) \cdot r \, dr \, d\theta.$$

We could also use $-\pi \le \theta \le \pi$, $-\frac{\pi}{3} \le \theta \le \frac{5\pi}{3}$, or any other interval of width 2π.

Don't forget the extra r!

❸ Evaluate the double integral. The volume is

$$V = \int_0^{2\pi} \int_0^1 (r - r^3) \, dr \, d\theta$$

$$= \int_0^{2\pi} \left(\left(\frac{r^2}{2} - \frac{r^4}{4} \right) \Big|_0^1 \right) d\theta$$

$$= \int_0^{2\pi} \left(\frac{1}{2} - \frac{1}{4} - 0 \right) d\theta = \int_0^{2\pi} \frac{1}{4} \, d\theta$$

$$= \frac{1}{4} \theta \Big|_0^{2\pi} = \frac{1}{4}(2\pi) - 0 = \frac{\pi}{2} \text{ units}^3.$$

Line 1 performs the multiplication by the "don't forget" factor r; line 2 finds an antiderivative with respect to r, treating θ as a constant; line 3 evaluates at $r = 1$, at $r = 0$, and subtracts, and then simplifies; and line 4 finds an antiderivative with respect to θ, then evaluates at $\theta = 2\pi$, at $\theta = 0$, and subtracts.

No need for trig substitution, no need for half-angle identities–a simpler solution indeed! ∎

Ans. to reading exercise 7:

$$\int_{\frac{\pi}{2}}^{\pi} \int_0^3 (\cos r^2) \, r \, dr \, d\theta =$$

$$\int_{\frac{\pi}{2}}^{\pi} \int_0^3 r \cos r^2 \, dr \, d\theta \text{ (don't forget the}$$

extra r!); alternate answer

$$\int_0^3 \int_{\frac{\pi}{2}}^{\pi} r \cos r^2 \, d\theta \, dr$$

4.4.3 Regions between polar curves

More general polar regions can also be considered, in the same manner as for rectangular coordinates.

DOUBLE INTEGRAL OVER A REGION BETWEEN POLAR CURVES

To integrate $z = f(r, \theta)$ over the region between $r = g_1(\theta)$ and $r = g_2(\theta)$ for $a \le \theta \le b$, use

$$\int_a^b \int_{g_1(\theta)}^{g_2(\theta)} f(r, \theta) \cdot r \, dr \, d\theta.$$

This formula further assumes that $g_1(\theta) \le g_2(\theta)$ throughout the interval $[a, b]$.

Don't forget the extra r!

Example 25 *(a) Set up a double integral for finding the volume under the surface $f(r, \theta) = r^2 - \frac{\sin 2\theta}{r}$ over the region between $r = 2 + \theta$ and $r = 1 + \sin \theta$, for $0 \le \theta \le \frac{\pi}{2}$. (b) Use a CAS to evaluate the double integral.*

For information about graphing polar curves, see Section 8.3 of *Calculus Set Free*. Not all details about the shape of the curves are necessary to set up the integral.

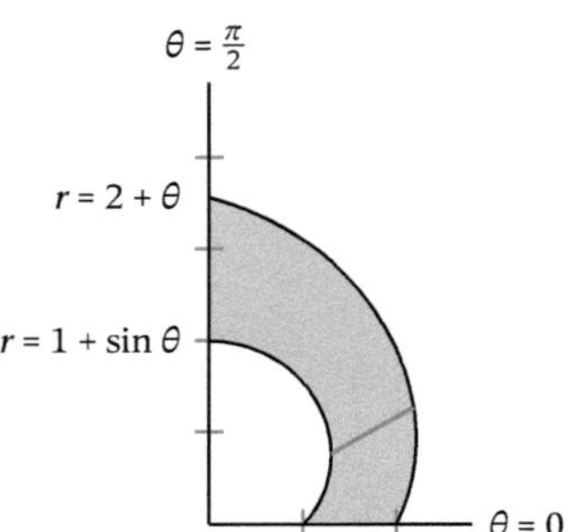

Figure 4.37 *The region between $r = 2 + \theta$ and $r = 1 + \sin\theta$, for $0 \le \theta \le \frac{\pi}{2}$. The radial slice, which is along a radial line from the origin, is drawn only inside the shaded region*

Solution

(a) ❶ Graph the region. The region is graphed in Figure 4.37.

❷ Set up a double integral. The function and the region are in polar form, so we are ready to use the formula for double integral over a region between polar curves;

Line 1: for the inner integral, the inner curve is the lower limit of integration, and the outer curve is the upper limit of integration. Don't forget the extra r! Line 2 simplifies.

$$V = \int_0^{\frac{\pi}{2}} \int_{1+\sin\theta}^{2+\theta} \left(r^2 - \frac{\sin 2\theta}{r}\right) \cdot r\, dr\, d\theta$$

$$= \int_0^{\frac{\pi}{2}} \int_{1+\sin\theta}^{2+\theta} \left(r^3 - \sin 2\theta\right) dr\, d\theta.$$

(b) ❸ Evaluate the double integral. Using a CAS, the volume is

$$\left(\frac{\pi^5}{640} + \frac{\pi^4}{32} + \frac{\pi^3}{4} + \pi^2 + \frac{77\pi}{64} - 2\right) \text{units}^3 = 22.9231 \text{ units}^3.$$

∎

4.4.4 More double integral practice

Example 26 *Evaluate $\displaystyle\int_{-2}^{2} \int_0^{\sqrt{4-x^2}} (x + y)\, dy\, dx$ by changing to polar coordinates.*

The y-coordinates of the function are positive. The bottom half of the circle would have negative coordinates, $y = -\sqrt{4 - x^2}$.

Solution ❶ Graph the region. The region must be inferred from the integral. You may recognize the equation $y = \sqrt{4 - x^2}$ as the top half of a circle; indeed, if we square both sides we have $y^2 = 4 - x^2$, which rearranges to $x^2 + y^2 = 4$, a circle of

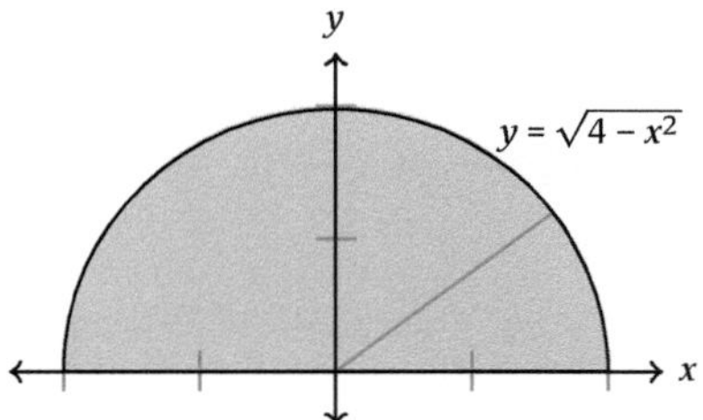

Figure 4.38 *The region between $y = 0$ and $y = \sqrt{4 - x^2}$, extending from $x = -2$ to $x = 2$*

radius 2 centered at the origin. The portion between $x = -2$ and $x = 2$ includes the entire top half of the circle (Figure 4.38).

❷ Set up a double integral. Switching to polar coordinates requires (1) converting the integrand to polar coordinates and (2) describing the region using polar coordinates.

For (1), the integrand is $f(x, y) = x + y$ and using $x = r \cos\theta$ and $y = r \sin\theta$ quite easily yields

$$f(r, \theta) = r\cos\theta + r\sin\theta.$$

For (2), note that $0 \le r \le 2$ along any radial segment, and that the region lies between $\theta = 0$ and $\theta = \pi$. The double integral is rewritten as

$$\int_{-2}^{2} \int_{0}^{\sqrt{4-x^2}} (x + y)\, dy\, dx = \int_{0}^{\pi} \int_{0}^{2} (r\cos\theta + r\sin\theta) \cdot r\, dr\, d\theta$$

$$= \int_{0}^{\pi} \int_{0}^{2} (r^2 \cos\theta + r^2 \sin\theta)\, dr\, d\theta.$$

Line 1: don't forget the extra r! Line 2 simplifies the integrand.

❸ Evaluate the double integral. We have

$$\int_{0}^{\pi} \int_{0}^{2} (r^2 \cos\theta + r^2 \sin\theta)\, dr\, d\theta$$

$$= \int_{0}^{\pi} \left(\left(\frac{r^3}{3}\cos\theta + \frac{r^3}{3}\sin\theta \right)\Big|_{0}^{2} \right) d\theta$$

$$= \int_{0}^{\pi} \left(\frac{8}{3}\cos\theta + \frac{8}{3}\sin\theta - 0 \right) d\theta$$

$$= \frac{8}{3}(\sin\theta - \cos\theta)\Big|_{0}^{\pi}$$

$$= \frac{8}{3}(\sin\pi - \cos\pi) - \frac{8}{3}(\sin 0 - \cos 0) = \frac{8}{3}(1) - \frac{8}{3}(-1) = \frac{16}{3}.$$

Line 2 finds an antiderivative with respect to r, treating θ as a constant; line 3 evaluates at $r = 2$, at $r = 0$, and subtracts; line 4 finds an antiderivative with respect to θ; and line 5 evaluates at $\theta = \pi$, at $\theta = 0$, and subtracts, then completes the arithmetic. No units are included because the integral is not placed in a context.

∎

The original integral of example 26, $\int_{-2}^{2} \int_{0}^{\sqrt{4-x^2}} (x + y)\, dy\, dx$, requires substitution but is otherwise innocuous. It is not always the case that changing to polar coordinates is greatly advantageous, even when the region is simple to describe with polar coordinates.

Reading Exercise 8   Find a polar description of the region used in the integral

$$\int_{0}^{3} \int_{0}^{\sqrt{9-x^2}} f(x, y)\, dy\, dx.$$

4.4.5   Volume of a sphere

One way to derive the formula for the volume of a sphere uses double integrals in polar coordinates.

Consider the top half of a sphere of radius a centered at the origin. The equation of the sphere is $x^2 + y^2 + z^2 = a^2$, which can be rearranged to

$$z^2 = a^2 - x^2 - y^2$$

$$z = \pm\sqrt{a^2 - x^2 - y^2}.$$

The top half of the sphere is $z = \sqrt{a^2 - x^2 - y^2}$. Then the volume of the top hemisphere is the volume under the surface $z = \sqrt{a^2 - x^2 - y^2}$ over the region in Figure 4.39, a circle centered at the origin with radius a.

The polar equation of the top half of the sphere can be derived using the conversion formula $x^2 + y^2 = r^2$:

$$z = \sqrt{a^2 - x^2 - y^2} = \sqrt{a^2 - (x^2 + y^2)}$$

$$f(r, \theta) = \sqrt{a^2 - r^2}.$$

A polar description of the circular region is $0 \le r \le a$, $0 \le \theta \le 2\pi$. Then the volume of the hemisphere is

$$\int_{0}^{2\pi} \int_{0}^{a} \sqrt{a^2 - r^2} \cdot r\, dr\, d\theta = \int_{0}^{2\pi} \int_{0}^{a} r\sqrt{a^2 - r^2}\, dr\, d\theta.$$

Because we use r as a polar variable, we do not use r to also represent a constant, in this case the radius of the sphere.

The equation of a sphere is derived in Section 1.1.

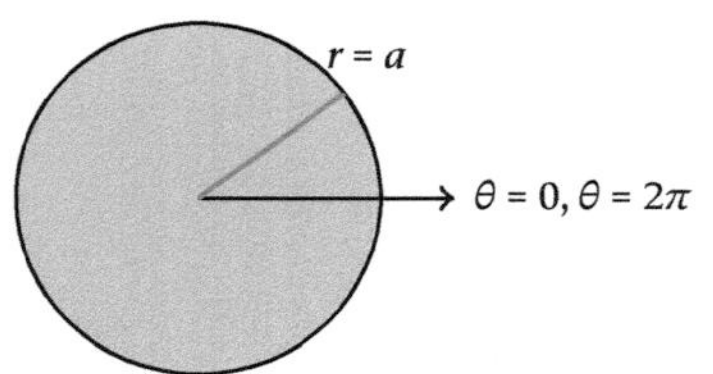

Figure 4.39 *The equatorial cross section of a sphere of radius a, which is a circle of radius a. A radial element for the region is also shown*

Don't forget the extra r! Notice how nicely the "don't forget" factor sets the integral up for substitution.

The inner integral requires the substitution

$$u = a^2 - r^2$$

$$du = -2r\,dr.$$

The new limits of integration are

$$r = a: \quad u = a^2 - a^2 = 0$$

$$r = 0: \quad u = a^2 - 0 = a^2.$$

Making the substitution, the inner integral becomes

$$\int_0^a r\sqrt{a^2 - r^2}\,dr = -\frac{1}{2}\int_0^a -2r\sqrt{a^2 - r^2}\,dr$$

$$= -\frac{1}{2}\int_{a^2}^0 \sqrt{u}\,du$$

$$= -\frac{1}{2}\left.\frac{u^{3/2}}{\frac{3}{2}}\right|_{a^2}^0 = -\frac{1}{3}u^{3/2}\Big|_{a^2}^0$$

$$= 0 - -\frac{1}{3}(a^2)^{3/2} = \frac{1}{3}a^3.$$

Line 1 adjusts for the constant -2 needed for du; line 2 makes the substitution, including changing the limits of integration from values of r to values of u; line 3 finds an antiderivative with respect to u and simplifies; and line 4 evaluates at $u = 0$, at $u = a^2$, and subtracts.

Then the outer integral is

$$\int_0^{2\pi} \frac{1}{3}a^3\,d\theta = \frac{1}{3}a^3\theta\Big|_0^{2\pi} = \frac{1}{3}a^3 \cdot 2\pi - 0 = \frac{2}{3}\pi a^3 \text{ units}^3.$$

That's just a hemisphere, the top half of the sphere; the volume of the entire sphere of radius a is therefore

$$V = \frac{4}{3}\pi a^3 \text{ units}^3,$$

as expected.

Ans. to reading exercise 8:
$0 \le r \le 3, 0 \le \theta \le \frac{\pi}{2}$. Alternate answers include $0 \le r \le 3,\ 2\pi \le \theta \le \frac{5\pi}{2}$. Do not use negative r coordinates to describe the region.

EXERCISES 4.4

1–6. Describe the region using polar coordinates.

1.

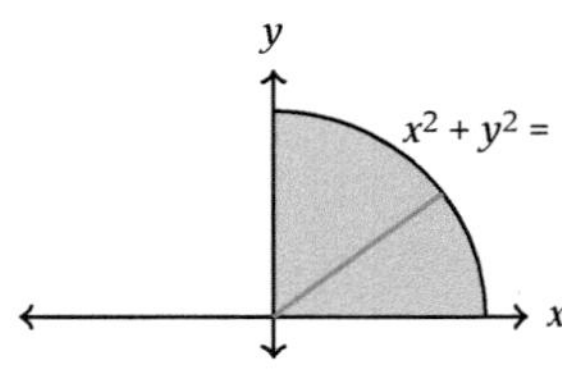

4.

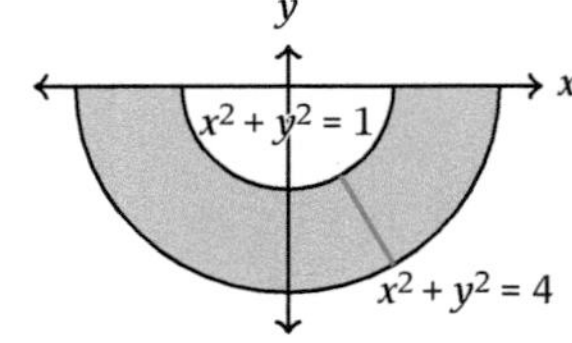

2.

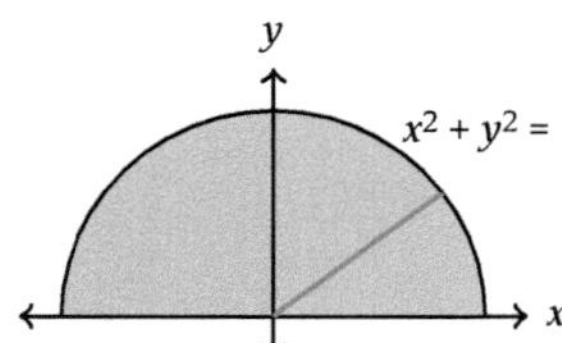

5.

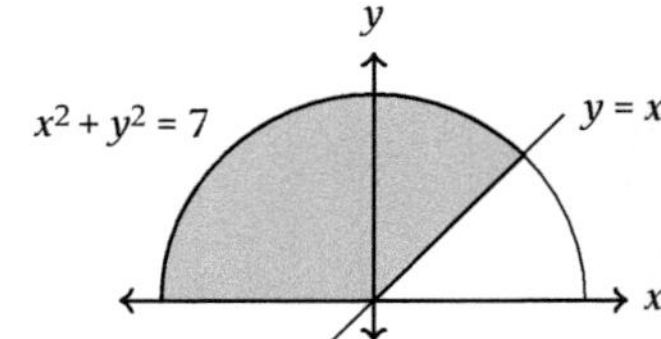

3.

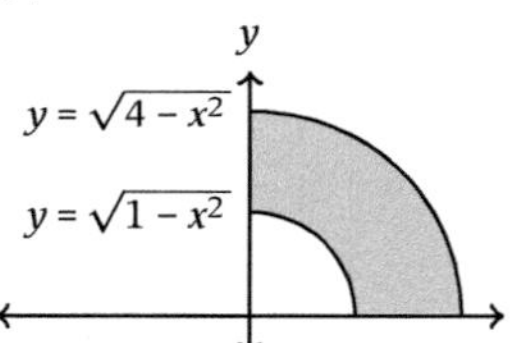

6.

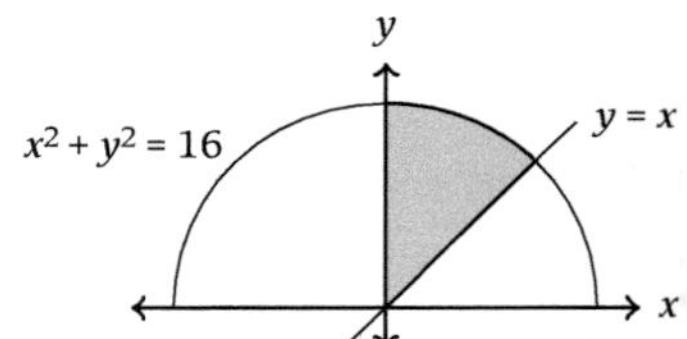

7. (a) Set up a double integral for finding the volume under the surface $f(r, \theta) = r^2 \tan \frac{\theta}{6}$ over the region between $r = \sin \theta$ and $r = 2$, for $0 \le \theta \le \pi$.

(b) Use a CAS to evaluate the double integral.

8. (a) Set up a double integral for finding the volume under the surface $f(r, \theta) = \frac{\theta^2}{r+1}$, between $r = \theta$ and $r = 2$, for $\frac{\pi}{6} \le \theta \le \frac{\pi}{3}$.

(b) Use a CAS to evaluate the double integral.

9. (a) Set up a double integral for finding the volume under the surface $f(r, \theta) = \frac{\sqrt{1+\sin \theta}}{r}$, between $r = 2$ and $r = 3 + \sin \theta$, for $-\frac{\pi}{2} \le \theta \le \frac{\pi}{2}$.

(b) Use a CAS to evaluate the double integral.

10. (a) Set up a double integral for finding the volume under the surface $f(r, \theta) = r + \cos \theta$, between $r = \sec \frac{\theta}{4}$ and $r = 1$, for $0 \le \theta \le \pi$.

(b) Use a CAS to evaluate the double integral.

11. Find the volume under the surface $z = r^2 \sin 2\theta$ over the region $0 \le r \le 3, 0 \le \theta \le \frac{\pi}{2}$.

Although Desmos is not a CAS and cannot compute the exact value of some of these double integrals, it can give calculator-style numerical approximations to the requested values.

12. Find the volume under the surface $z = \frac{1}{r} + \theta^2$ over the region $1 \le r \le 2, \frac{\pi}{2} \le \theta \le \pi$.

13. Find the volume under the surface $f(r,\theta) = 1 + \frac{1}{r}\sec^2\theta$ over the region $1 \le r \le 3, -\frac{\pi}{4} \le \theta \le \frac{\pi}{4}$.

14. Find the volume under the surface $f(r,\theta) = 2r\sin\theta + 4$ over the region $x^2 + y^2 \le 3$.

15. Find the (net) volume under the surface $z = \frac{\sin\theta}{r^2}$ over the region between $x^2 + y^2 = 1$ and $x^2 + y^2 = e^2$.

16. Find the (net) volume under the surface $z = 2\theta\cos r^2$ over the region $2 \le r \le 5, 1 \le \theta \le 2$.

17. (Love) Find the volume inside the first octant bounded by the surfaces $y = x, x = a$, and $xy = az$.

18. Use polar coordinates to find the volume under the paraboloid $z = x^2 + y^2 + 7$ over the region $x^2 + y^2 \le 9$.

19. (Love) Find the volume above the xy-plane, inside the cylinder $x^2 + y^2 = a^2$, and below the paraboloid $x^2 + y^2 = az$.

20. Find the net volume under the surface $f(x,y) = xy$ over the region between the circles $x^2 + y^2 = 1$ and $x^2 + y^2 = 4$.

21–26. Evaluate the integral by changing to polar coordinates.

21. (Love) $\displaystyle\int_0^1 \int_0^{\sqrt{1-x^2}} e^{x^2+y^2}\, dy\, dx$

22. $\displaystyle\int_0^3 \int_{-\sqrt{9-x^2}}^0 (x^2 + y^2)\, dy\, dx$

23. $\displaystyle\int_0^2 \int_{-\sqrt{4-x^2}}^{\sqrt{4-x^2}} (x^2 + y^2)^5\, dy\, dx$

24. (Love) $\displaystyle\int_0^1 \int_0^{\sqrt{1-x^2}} \cos(x^2 + y^2)\, dy\, dx$

25. (Love) $\displaystyle\int_0^5 \int_0^{\sqrt{25-x^2}} x^2\sqrt{x^2 + y^2}\, dy\, dx$

26. $\displaystyle\int_{-3}^0 \int_0^{\sqrt{9-x^2}} \frac{1}{1 + x^2 + y^2}\, dy\, dx$

27. (Love) Evaluate $\displaystyle\int_0^1 \int_y^1 \frac{y}{\sqrt{x^2 + y^2}}\, dx\, dy$ by changing to polar coordinates.

28. (Love) Evaluate $\displaystyle\int_0^1 \int_{2y}^2 y\sqrt{x^2 + y^2}\, dx\, dy$ by changing to polar coordinates.

Exercises in this section marked (Love) are from, or modified from, *Differential and Integral Calculus* by Clyde E. Love, The MacMillan Company, 1926.

Hint: Rewrite $x = 1$ as $r = \frac{1}{\cos\theta}$.

29. (Love) Evaluate $\displaystyle\int_0^{\frac{\pi}{2}}\int_0^{\sec\theta}\frac{r}{1+r^2\sin^2\theta}\,dr\,d\theta$ by changing to Cartesian (rectangular) coordinates.

30. (Love) A round hole of radius b is bored through the center of a sphere of radius a. Find the volume cut out.

31. The area of a region in a plane can be found with a double integral. The volume under the surface $z = 1$ over the region whose area we wish to find is equal to the area of the region times 1; therefore, the double integral yields the correct number. Use this approach to derive the formula for area inside a polar curve $r = g(\theta)$ between $\theta = c$ and $\theta = d$.

    (a) Set up the double integral.
    (b) Evaluate the inner integral.
    (c) Place the result of (b) in the outer integral. This is the formula we seek.
    (d) Compare to the polar area formula in *Calculus Set Free* Section 9.2.

4.5   Physical Applications of Double Integrals

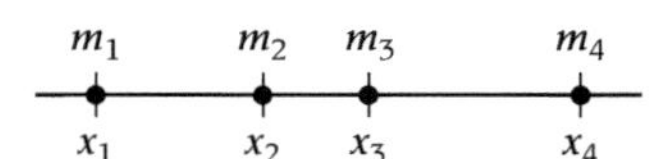

Figure 4.40 *A system of four masses m_1, m_2, m_3, and m_4 placed at coordinates x_1, x_2, x_3, and x_4, respectively*

See Section 9.6 of *Calculus Set Free* for an expanded discussion and derivation of this center of mass formula.

If point masses $m_1, m_2, ..., m_n$ are placed at coordinates $x_1, x_2, ..., x_n$ on a number line as in Figure 4.40 (pictured with $n = 4$ such masses), then the center of mass $\bar{x}$, the balance point on the number line, is given by

$$\bar{x} = \frac{\displaystyle\sum_{k=1}^{n} x_k m_k}{\displaystyle\sum_{k=1}^{n} m_k}.$$

The denominator, $\sum_{k=1}^{n} m_k$, is the total mass, or *system mass*. The numerator is the *system moment about the origin*, which is the magnitude of the tendency for the system to pivot about a fulcrum placed at the origin.

Determining the center of mass of a thin flat plate of constant density is covered in Section 9.6 of *Calculus Set Free*. Next, we consider variable density, utilizing the formula above as a guide.

4.5.1   Center of mass: thin flat plates of variable density

Consider a thin, flat plate covering a region R, such as the one pictured in Figure 4.41. Let the density of the plate at a point (x, y) be given by $\rho(x, y)$, where ρ

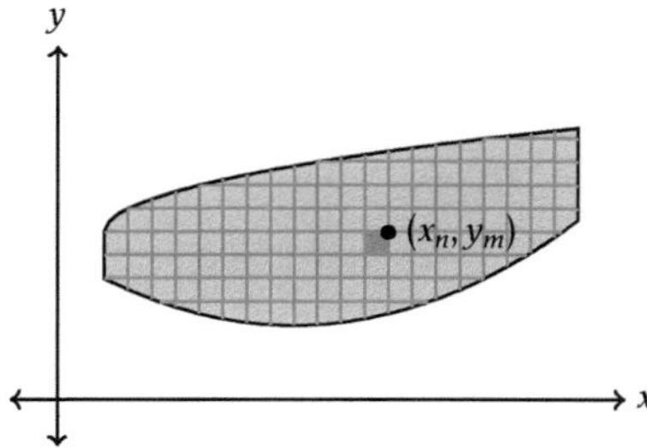

Figure 4.41 *A region R, subdivided into rectangles. The rectangles are shown with non-infinitesimal area so as to be visible. The n-mth rectangle is highlighted, along with its upper-right corner (x_n, y_m)*

is the *density function*, with units of the form $\frac{\text{mass}}{\text{area}}$. If the region is divided into rectangles with infinitesimal side lengths Δx and Δy, then the mass of the plate on the n-mth rectangle should be the density times the area, $\rho(x_n, y_m)\Delta x\Delta y$. Adding the masses on all of the rectangles should give the total mass of the plate. To include all of the rectangles we add across "rows" and "columns," resulting in a double omega sum, which in turn can be rewritten as a double integral:

Technically, following the same pattern as in Section 4.3, we begin with a rectangular region $a \leq x \leq b$, $c \leq y \leq d$ containing the region R, then divide the intervals $[a, b]$ and $[c, d]$ into Ω subintervals each, where Ω is a positive infinite hyperreal integer. The density function ρ has value 0 outside of the region R.

$$\text{total mass} = \sum_{m=1}^{\Omega} \sum_{n=1}^{\Omega} \rho(x_n, y_m)\Delta x\Delta y = \iint_R \rho(x, y)\, dx\, dy.$$

To locate the center of mass $(\bar{x}, \bar{y})$, we need to calculate the numerator for each of $\bar{x}$ and $\bar{y}$. The *system moment about the y-axis* is the magnitude of the tendency for the system to rotate about the y-axis. The larger the x-coordinate, the farther from the y-axis and the larger the contribution to the moment; the larger the mass, the larger the contribution to the moment. As with point masses on a number line, we multiply the x-coordinates by the masses; for the n-mth rectangle this is $x_n \cdot \rho(x_n \cdot y_m)\Delta x\Delta y$. We then add these up to get the moment about the y-axis,

The symbol $\iint_R$ means the double integral over the region R. The R is replaced by appropriate limits of integration determined by the shape of the region, as in Section 4.3.

$$M_y = \sum_{m=1}^{\Omega} \sum_{n=1}^{\Omega} x_n\rho(x_n, y_m)\Delta x\Delta y = \iint_D x\rho(x, y)\, dx\, dy.$$

Then the x-coordinate of the center of mass is given by

Compare to the center of mass formula for a finite system stated at the beginning of this section. For each formula, the denominator is the total mass and the numerator has an extra factor of x.

$$\bar{x} = \frac{M_y}{\text{total mass}} = \frac{\displaystyle\iint_D x\rho(x, y)\, dx\, dy}{\displaystyle\iint_D \rho(x, y)\, dx\, dy}.$$

Rather than memorize these formulas using the moments M_x and M_y, where the x and y are opposite, it is customary to remember that the numerator for $\bar{x}$ has an extra x and the numerator for $\bar{y}$ has an extra y.

In a similar manner, the y-coordinate of the center of mass is given by

$$\bar{y} = \frac{M_x}{\text{total mass}} = \frac{\iint_D y\rho(x,y)\,dx\,dy}{\iint_D \rho(x,y)\,dx\,dy}.$$

CENTER OF MASS, THIN FLAT PLATE OF VARIABLE DENSITY

The center of mass of a thin flat plate with density function $\rho(x,y)$ is

$$(\bar{x},\bar{y}) = \left(\frac{\iint_D x\cdot\rho(x,y)\,dx\,dy}{\iint_D \rho(x,y)\,dx\,dy}, \frac{\iint_D y\cdot\rho(x,y)\,dx\,dy}{\iint_D \rho(x,y)\,dx\,dy} \right).$$

A thin, flat plate is also called a *lamina*.

As long as the density function is continuous (or at least integrable), Fubini's theorem applies and the order of integration can be changed.

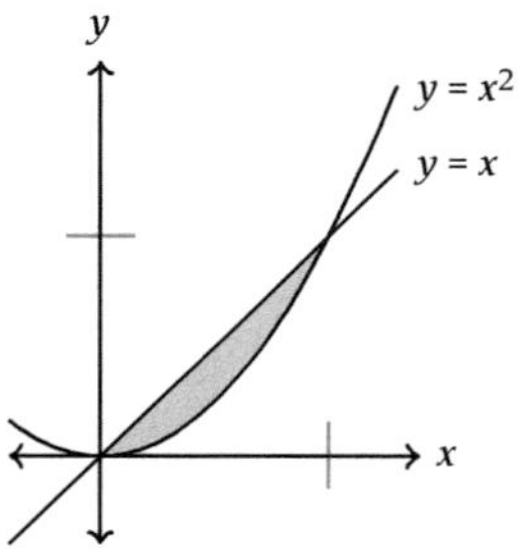

4.5.2 Center of mass example

Example 27 *A thin flat plate can be described as lying on the region bounded by* $y = x$ *and* $y = x^2$ *and having density* $\rho(x,y) = x + y$. *Find its center of mass.*

Solution As with previous exercises involving general regions, it is helpful to graph the region to determine limits of integration (Figure 4.42).

There are three double integrals to calculate: the total mass (the denominator of both $\bar{x}$ and $\bar{y}$), and the system moments about the axes (the numerators of $\bar{x}$ and $\bar{y}$).

The total mass is

$$\iint_D \rho(x,y)\,dx\,dy = \int_0^1 \int_{x^2}^x (x+y)\,dy\,dx$$

$$= \int_0^1 \left(\left(xy + \frac{y^2}{2} \right)\Big|_{x^2}^x \right) dx$$

$$= \int_0^1 \left(x^2 + \frac{x^2}{2} - \left(x^3 + \frac{x^4}{2} \right) \right) dx$$

$$= \int_0^1 \left(\frac{3}{2}x^2 - x^3 - \frac{1}{2}x^4 \right) dx$$

$$= \cdots = \frac{3}{20}.$$

Figure 4.42 *The region containing the thin flat plate of example 27*

Then $\bar{x}$ is

$$\bar{x} = \frac{\displaystyle\iint_D x \cdot \rho(x,y)\,dx\,dy}{\displaystyle\iint_D \rho(x,y)\,dx\,dy} = \frac{\displaystyle\iint_D x \cdot \rho(x,y)\,dx\,dy}{\text{total mass}}$$

$$= \frac{1}{\frac{3}{20}} \int_0^1 \int_{x^2}^x x(x+y)\,dy\,dx$$

$$= \frac{20}{3} \int_0^1 \left(\left(x^2 y + \frac{xy^2}{2} \right)\Big|_{x^2}^x \right) dx$$

$$= \cdots = \frac{20}{3} \int_0^1 \left(-\frac{1}{2}x^5 - x^4 + \frac{3}{2}x^3 \right) dx$$

$$= \cdots = \frac{11}{18}.$$

Having already calculated the total mass, $\bar{x}$ can be determined by calculating the double integral in the numerator and dividing by the total mass (equivalent to multiplying by the reciprocal of the total mass).

Finally, $\bar{y}$ is

$$\bar{y} = \frac{\displaystyle\iint_D y \cdot \rho(x,y)\,dx\,dy}{\displaystyle\iint_D \rho(x,y)\,dx\,dy} = \frac{1}{\frac{3}{20}} \int_0^1 \int_{x^2}^x y(x+y)\,dy\,dx$$

$$= \frac{20}{3} \int_0^1 \int_{x^2}^x (xy + y^2)\,dy\,dx$$

$$= \frac{20}{3} \int_0^1 \left(\left(\frac{xy^2}{2} + \frac{y^3}{3} \right)\Big|_{x^2}^x \right) dx$$

$$= \frac{20}{3} \int_0^1 \left(\frac{x^3}{2} + \frac{x^3}{3} - \left(\frac{x^5}{2} + \frac{x^6}{3} \right) \right) dx$$

$$= \cdots = \frac{65}{126}.$$

Notice the extra y in this integral, as compared to the extra x in the integral for $\bar{x}$.

The center of mass of this plate is $\left(\frac{11}{18}, \frac{65}{126} \right) = (0.611, 0.516)$.

With density $\rho(x,y) = x+y$, the further from the axes the greater the density. The center of mass should therefore be to the right and above the centroid of the region.

$\blacksquare$

Reading Exercise 9 (a) Set up (but do not evaluate) the integral for finding the total mass of a thin flat plate lying on the region bounded by $y = 4$ and $y = x^2$ with density $\rho(x,y) = (x+2)\sqrt{y}$. (b) Set up (but do not evaluate) the integrals for finding $\bar{x}$ and $\bar{y}$ for the same thin flat plate. The total mass is 32.

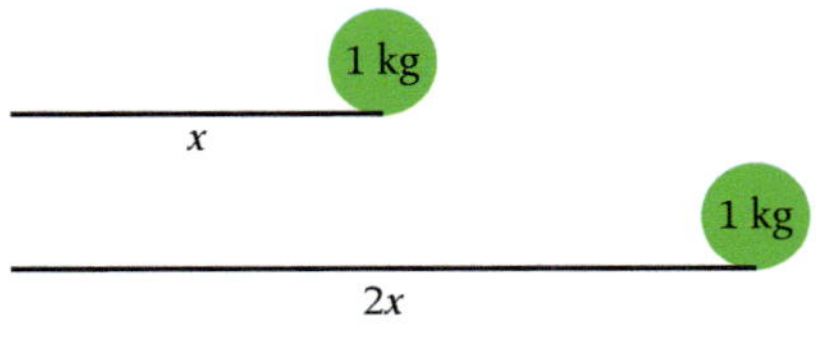

Figure 4.43 1 *kg masses at the ends of rods*

We shall use the word "mass" to refer to the object as well as to the object's mass.

Rotational speed is measured in units such as rotations per minute or radians per second.

4.5.3   **Moment of inertia**

Imagine a 1 kg mass at the end of a rod of length x, and imagine that you are holding the other end of the rod. It takes effort to keep the rod and mass in place. Now imagine another rod with a 1 kg mass, but twice as long (Figure 4.43). It is twice as hard to hold the longer rod; there is twice the leverage working against you from the mass that is twice as far away. Increasing the mass also proportionately increases the leverage working against you. The moment is therefore proportional to both the mass and the distance. In fact, for the system of point masses at the beginning of this section, the system moment, the numerator of the center of mass formula, is given as $\sum_{k=1}^{n} x_k m_k$, the sum of the products of the distances and masses.

A similar situation arises when considering rotational inertia. Imagine two masses placed on massless wheels, one twice as far from the point of rotation as the other, as in Figure 4.44. To rotate the closer mass is not as difficult as rotating the more distant mass; just as with the rods of Figure 4.43, there is greater leverage the further the mass is from the center of rotation. But that's not all. If the rotational speed is the same for the two wheels, the mass further from the center of rotation moves faster than the closer mass, again proportional to the distance from the center; see the red arcs in Figure 4.44. So the inertia of a mass is proportional to its mass m_k times its distance r_k (leverage) times its distance again, or $m_k r_k^2$. Because this quantity uses the distance squared, it is a *second moment*, named for the exponent two.

Now consider a lamina (thin flat plate) covering a region R with density function $\rho(x, y)$. If we subdivide the lamina into rectangles with infinitesimal side length Δx and Δy, as we imagined with Figure 4.41, we can find the *moment of inertia about the origin* of one subrectangle by multiplying its mass $\rho(x_n, y_m)\Delta x\Delta y$ by the square

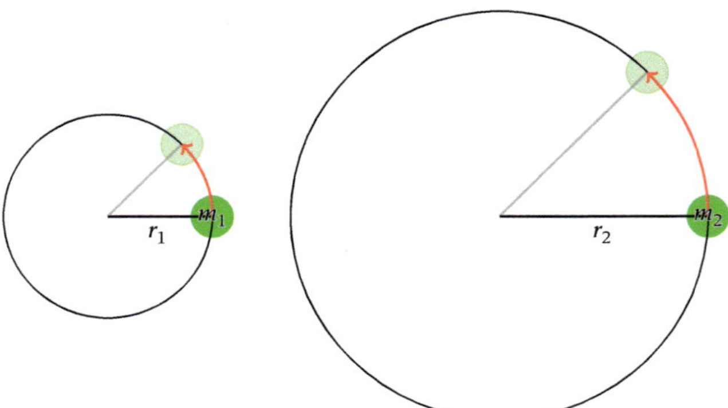

Figure 4.44 *Two masses, m_1 and m_2, rotating from an initial position (darker green) to a later position (lighter green). For each, the center of rotation is the center of the circle*

of its distance from the origin, $\left(\sqrt{x_n^2 + y_m^2}\right)^2 = x_n^2 + y_m^2$, yielding

$$\rho(x_n, y_m)\Delta x \Delta y \left(x_n^2 + y_m^2\right).$$

Sum the moment over all the rectangles and the result is

$$\sum_{n=1}^{\Omega} \sum_{m=1}^{\Omega} \rho(x_n, y_m)\Delta y \Delta x \left(x_n^2 + y_m^2\right) = \iint_R (x^2 + y^2)\rho(x, y)\, dy\, dx.$$

Technically, following the same pattern as in Section 4.3, we begin with a rectangular region $a \leq x \leq b$, $c \leq y \leq d$ containing the lamina's region R, then divide the intervals $[a, b]$ and $[c, d]$ into Ω subintervals each, where Ω is a positive infinite hyperreal integer. The density function ρ has value 0 outside of the region R.

MOMENT OF INERTIA ABOUT THE ORIGIN

The *moment of inertia about the origin* of a lamina on a region R with density function ρ is

$$I_O = \iint_R (x^2 + y^2)\rho(x, y)\, dy\, dx.$$

As always, we may switch to polar coordinates if we so desire. In that case, we must rewrite ρ, rewrite $x^2 + y^2$ as r^2, and remember to include the extra r.

The symbol I_O is derived from the words "inertia" (the I) and "origin" (the O).

Example 28 *Find the moment of inertia about the origin for a thin flat plate on the region bounded by $x = 0$, $y = x$, and $y = \sqrt{4 - x^2}$ with density function $\rho(x, y) = x$.*

Solution We start by graphing the region containing the thin flat plate (Figure 4.45). The region looks easy to describe in polar coordinates, so we make the switch.

To set up the integral using polar coordinates, we use the moment of inertia about the origin formula, with r^2 in place of $x^2 + y^2$ and $\rho(r, \theta) = r\cos\theta$ in place of $\rho(x, y) = x$, and we insert the extra r:

$$I_O = \int_{\frac{\pi}{4}}^{\frac{\pi}{2}} \int_0^2 r^2(r\cos\theta) \cdot r\, dr\, d\theta.$$

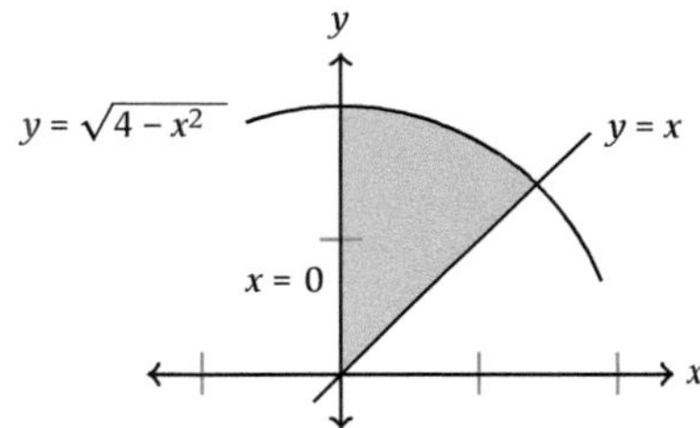

Figure 4.45 *The region for example 28. The curve $y = \sqrt{4 - x^2}$ is part of the circle $x^2 + y^2 = 4$*

"Inserting the extra r" comes from replacing $dx\, dy$ with $r\, dr\, d\theta$.

Next, we evaluate the integral:

$$I_O = \int_{\frac{\pi}{4}}^{\frac{\pi}{2}} \int_0^2 r^4 \cos\theta\, dr\, d\theta = \int_{\frac{\pi}{4}}^{\frac{\pi}{2}} \left(\frac{r^5}{5}\cos\theta \Big|_0^2\right) d\theta$$

$$= \int_{\frac{\pi}{4}}^{\frac{\pi}{2}} \left(\frac{32}{5}\cos\theta - 0\right) d\theta$$

Notice the same workflow we have followed for double integrals over general regions: graph the region, set up the integral, evaluate the integral.

Line 1 simplifies the integrand and then finds an antiderivative with respect to r, treating θ as a constant; line 2 evaluates at $r = 2$, at $r = 0$, and subtracts; line 3 finds an antiderivative with respect to θ; line 4 evaluates at $\theta = \frac{\pi}{2}$, at $\theta = \frac{\pi}{4}$, and subtracts, then simplifies.

$$= \frac{32}{5} \sin \theta \Big|_{\frac{\pi}{4}}^{\frac{\pi}{2}}$$

$$= \frac{32}{5} \sin \frac{\pi}{2} - \frac{32}{5} \sin \frac{\pi}{4} = \frac{32}{5} \left(1 - \frac{\sqrt{2}}{2} \right).$$

∎

The absence of units in the answer to example 28 is due to no units being given; units for I_O are units of mass times units of distance squared.

Example 28 can be solved without switching to polar coordinates. The inner integral with respect to y is not hard, but the outer integral then involves both substitution and trig substitution.

If a region is easily expressible in polar coordinates, it often helps to make the switch.

Reading Exercise 10   Set up (but do not evaluate) the integral for finding the moment of inertia about the origin for a thin flat plate on the region $x^2 + y^2 \le 9$ with density $\rho(x,y) = y$. Use polar coordinates.

4.5.4   Radius of gyration

Ans. to reading exercise 9:

(a) $\displaystyle\int_{-2}^{2} \int_{x^2}^{4} (x + 2)\sqrt{y}\, dy\, dx$;

(b) $\displaystyle\bar{x} = \frac{1}{32} \int_{-2}^{2} \int_{x^2}^{4} x(x + 2)\sqrt{y}\, dy\, dx$,

$\displaystyle\bar{y} = \frac{1}{32} \int_{-2}^{2} \int_{x^2}^{4} y(x + 2)\sqrt{y}\, dy\, dx$

The center of mass can be thought of as the place at which all the mass of the system can be placed to have the same effect as the original system. Placing all the mass at the balance point does not change the balance point, and with the same denominator, the numerator must also be the same, so the system moment is unchanged.

In a similar manner, the *radius of gyration* is the distance from the origin (the center of rotation) where we can place all the mass of the system without changing the moment of inertia about the origin. If m is the total mass of the system, and if we let R be the radius of gyration, then the moment of inertia about the origin for this point-mass system is

As explained earlier, the inertia is the mass times the square of the distance from the center of rotation.

$$I_O = mR^2.$$

Solving for R, we have

$$\frac{I_O}{m} = R^2$$

$$\sqrt{\frac{I_O}{m}} = R.$$

RADIUS OF GYRATION

The radius of gyration R of a lamina about a center of rotation placed at the origin is

$$R = \sqrt{\frac{I_O}{m}},$$

where I_O is the moment of inertia about the origin and m is the total mass of the lamina.

Note that both m and I_O can be found with double integrals, using formulas from this section.

Example 29 *Find the radius of gyration for the lamina of example 28.*

Solution From example 28 we have $I_O = \frac{32}{5}\left(1 - \frac{\sqrt{2}}{2}\right)$. We still need to calculate the total mass m of the lamina, using a formula from earlier in this section, total mass $= m = \iint_R \rho(x, y)\, dx\, dy$. Making the switch to polar coordinates as in example 28, we use $\rho(r, \theta) = r\cos\theta$ and insert the extra r:

Compare the formula for total mass to the formula for the moment of inertia about the origin; the latter has an extra $(x^2 + y^2)$ in the integrand.

$$m = \iint_R \rho(x, y)\, dx\, dy = \iint_R \rho(r, \theta) \cdot r\, dr\, d\theta$$

Compare to the calculation of I_O in example 28.

$$= \int_{\frac{\pi}{4}}^{\frac{\pi}{2}} \int_0^2 r\cos\theta \cdot r\, dr\, d\theta$$

$$= \int_{\frac{\pi}{4}}^{\frac{\pi}{2}} \left(\frac{r^3}{3}\cos\theta \Big|_0^2\right) d\theta$$

$$= \int_{\frac{\pi}{4}}^{\frac{\pi}{2}} \left(\frac{8}{3}\cos\theta - 0\right) d\theta$$

$$= \frac{8}{3}\sin\theta \Big|_{\frac{\pi}{4}}^{\frac{\pi}{2}} = \frac{8}{3}\left(1 - \frac{\sqrt{2}}{2}\right).$$

Then the radius of gyration is

Units for the radius of gyration are units of distance:

$$R = \sqrt{\frac{I_O}{m}} = \sqrt{\frac{\frac{32}{5}\left(1 - \frac{\sqrt{2}}{2}\right)}{\frac{8}{3}\left(1 - \frac{\sqrt{2}}{2}\right)}} = \sqrt{\frac{32}{5} \cdot \frac{3}{8}} = \sqrt{\frac{12}{5}} = 1.549.$$

$$\sqrt{\frac{\text{mass} \cdot \text{distance}^2}{\text{mass}}} = \sqrt{\text{distance}^2}$$

$$= \text{distance}.$$

■ Ans. to reading exercise 10:

$$I_0 = \int_0^{2\pi} \int_0^3 r^2 \cdot r\sin\theta \cdot r\, dr\, d\theta$$

$$= \int_0^{2\pi} \int_0^3 r^4 \sin\theta\, dr\, d\theta$$

EXERCISES 4.5

Even if the CAS can produce an exact value, a decimal approximation is requested. You may need to use a numerical integration command in the CAS. Desmos can also be used.

1–6. (a) Set up a double integral for finding the total mass of a lamina on the given region with the given density function. (b) Use a CAS to determine a decimal approximation to the total mass. (c) Set up double integrals for determining the center of mass $(\bar{x}, \bar{y})$. (d) Use a CAS to determine a decimal approximation to $(\bar{x}, \bar{y})$.

1. region between $y = \sqrt{x}$ and $y = x$, for $1 \le x \le 4$; density $\rho(x, y) = (x^2 + y)\, e^{\sqrt{y}}$

2. region bounded by $y = x^2$, $y = -x^2$, and $x = 3$; density $\rho(x, y) = y^2 \tan^{-1} x^2$

3. region bounded by $x = y^2$ and $x = 4$; density $\rho(x, y) = \sin^2\left(\sqrt{x} + \frac{1}{y+3}\right)$

4. region between $y = x$ and $x = 20 - y^2$, for $0 \le y \le 3$; density $\rho(x, y) = \frac{1}{x + y^4}$

5. region bounded by $y = 0$ and $y = -x^2 + 8x - 12$; density $\rho(x, y) = (x - 4)^2 \sqrt{1 + y^4}$

6. region bounded by $y = x^2$ and $y = x + 2$; density $\rho(x, y) = \sqrt{y}\ln(2 + x)$

7–18. Find (a) the total mass and (b) the center of mass for the lamina (thin flat plate) lying on the region with density function ρ.

7. region $0 \le x \le 2, 0 \le y \le 1; \rho(x, y) = 2xy$

8. region $0 \le x \le 1, 0 \le y \le 3; \rho(x, y) = y + \sqrt{x}$

9. region $-2 \le x \le 2, 0 \le y \le 4; \rho(x, y) = x^2 + y^2$

10. region $1 \le x \le 2, -1 \le y \le 1; \rho(x, y) = x(2 + y)$

11. region bounded by $y = \frac{1}{x^2}, y = 0, x = 1,$ and $x = 4; \rho(x, y) = x^2$

12. region between $y = \sqrt{x}$ and $y = 0$, for $1 \le x \le 4; \rho(x, y) = \frac{1}{\sqrt{x}}$

13. region bounded by $y = x, y = 4,$ and $x = 0; \rho(x, y) = 4 - x$

14. region bounded by $y = 0$ and $y = x^2 - 4; \rho(x, y) = x^2$

15. region $0 \le x \le \pi, 2 \le y \le 5; \rho(x, y) = y + \sin x$

16. region bounded by $x = e, y = 0,$ and $y = \ln x; \rho(x, y) = xe^y$

17. region $x^2 + y^2 \le 4; \rho(x, y) = 3 + y$

18. region bounded by $y = 0$ and $y = \sqrt{25 - x^2}; \rho(x, y) = 4$

19–26. Find (a) the moment of inertia about the origin and (b) the radius of gyration for a lamina lying on the region with density function ρ.

19. region $0 \le x \le 2, 0 \le y \le 1; \rho(x, y) = 2xy$

20. region $0 \le x \le 1, 0 \le y \le 3; \rho(x, y) = y + \sqrt{x}$

21. region enclosed by $y = x, x = 0,$ and $y = -\sqrt{4 - x^2}; \rho(x, y) = 2$

22. region $x^2 + y^2 \le 1; \rho(x, y) = x + 1$

23. region $1 \le r \le 2, 0 \le \theta \le \frac{\pi}{2}; \rho(x, y) = \frac{1}{x^2 + y^2}$

24. region $x^2 + y^2 \le 9$; $\rho(x,y) = 9 - x^2 - y^2$
25. region $x^2 + y^2 \le 4$; $\rho(x,y) = 3 + y$
26. region bounded by $y = 0$ and $y = \sqrt{25 - x^2}$; $\rho(x,y) = 4$

27. (a) For a thin, flat plate of variable density, what effect does doubling the density function have on the location of the center of mass? Re-calculate the center of mass for the plate of example 27 using the density function $\rho(x,y) = 2(x+y)$ to determine your answer.

(b) Use the density function $\rho_{\text{alternate}}(x,y) = k \cdot \rho(x,y)$ and the formula for the center of mass, along with the constant multiple rule for double integrals, to prove that the center of mass does not change.

28. Moments of inertia can be calculated not just about the origin, but also about the y-axis and about the x-axis, using the formulas $I_y = \iint_R x^2 \rho(x,y)\, dy\, dx$ and $I_x = \iint_R y^2 \rho(x,y)\, dy\, dx$, respectively. Find I_y and I_x for the lamina in exercise 19, and note that $I_x + I_y = I_O$.

In particular, this says that the units with which we measure density (or mass) do not effect the location of the center of mass. And although the numbers will look different, the geometric location of the center of mass is similarly not affected by the units we use for distance.

The moments I_y and I_x are also classified as second moments.

4.6 Surface Area

The area of a surface of revolution is discussed in Sections 9.3 and 9.4 of *Calculus Set Free*. With functions of two variables at our disposal, we can now consider the area of a more general surface.

4.6.1 Surface area for $z = f(x,y)$

Consider a surface $z = f(x,y)$, as in Figure 4.46. Perhaps one way to find the area of the surface is to chop up the region into smaller pieces, find the surface area for

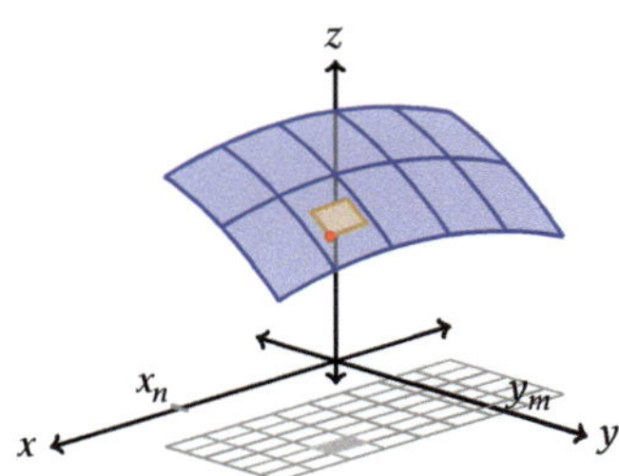

Figure 4.46 *A surface $z = f(x,y)$ (blue) over a rectangular region partitioned into smaller rectangles, along with the n-mth piece of the surface (brown). The subrectangles and the n-mth piece of the surface are shown with non-infinitesimal side lengths so as to be visible*

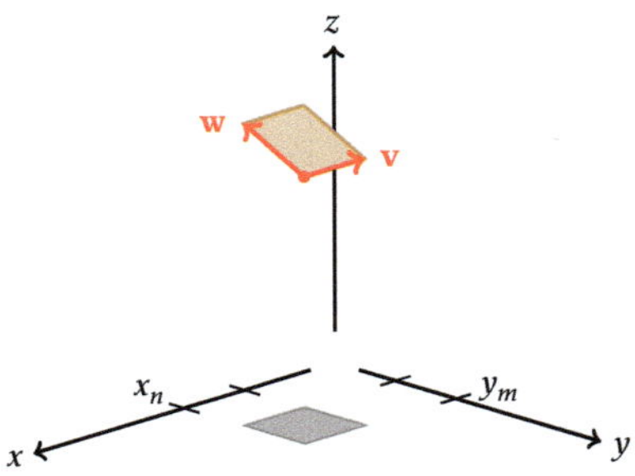

Figure 4.47 *The n-mth piece of the surface, indistinguishable from its tangent plane (brown). Vectors* **v** *and* **w** *(red) are added along adjacent sides of the piece of the tangent plane*

Because we use the front-right corner of each subrectangle as the point of tangency for the tangent plane, we must subtract (rather than add) Δx to stay within the subrectangle.

An exception to the approximation formula is when all of $f(a,b)$, $f_x(a,b)$, and $f_y(a,b)$ are zero. When an exception applies, the details are different but the result is the same.

This is not **w**, for **w** is along an adjacent side.

each piece, and add these surface areas to find the total surface area. As usual, we may partition the x- and y-intervals into Ω pieces each, where Ω is a positive infinite hyperreal integer, and then the subrectangles of the region have infinitesimal side lengths Δx and Δy. As long as the function f is differentiable, local linearity applies, meaning that the surface and its tangent plane are indistinguishable on each infinitesimal-sized piece. The area of the surface above a subrectangle is therefore the area of the tangent plane above the subrectangle (Figure 4.47).

Note, though, that the area of the gray subrectangle in Figure 4.47, which is $\Delta x\,\Delta y$, is not the same as the area of the surface above that subrectangle; the surface may be tilted, in which case its area is larger than $\Delta x\,\Delta y$. And the brown piece in Figure 4.47 may not even be a rectangle, because of the tilt. How, then, do we find its area? The key is to realize that the brown piece is a parallelogram, and we know from Section 1.5 how to find the area of a parallelogram given its side vectors.

Let's begin by calculating the vector **v** shown in Figure 4.47. This vector starts at z-coordinate $f(x_n, y_m)$ and moves along the tangent plane Δx units in the negative x-direction to z-coordinate $f(x_n - \Delta x, y_m)$. Now it is time to use local linearity, which says that where f is differentiable,

$$f(a + \alpha,\, b + \beta) \approx f(a,b) + f_x(a,b)\alpha + f_y(a,b)\beta,$$

where α and β are infinitesimal or zero (see Section 3.6). Applying the formula with $\alpha = -\Delta x$ and $\beta = 0$,

$$f(x_n - \Delta x,\, y_m) \approx f(x_n, y_m) + f_x(x_n, y_m)(-\Delta x) + 0.$$

In other words, **v** moves from the point $(x_n, y_m, f(x_n, y_m))$ to the point $(x_n - \Delta x,\, y_m,\, f(x_n, y_m) - f_x(x_n, y_m)\Delta x)$. Therefore

$$\mathbf{v} = \langle -\Delta x,\, 0,\, -f_x(x_n, y_m)\Delta x \rangle.$$

Next, let's calculate the vector along the opposite side of the brown quadrilateral in Figure 4.47. For this we need

$$f(x_n,\, y_m - \Delta y) \approx f(x_n, y_m) + 0 + f_y(x_n, y_m)(-\Delta y)$$

and

$$f(x_n - \Delta x,\, y_m - \Delta y) \approx f(x_n, y_m) + f_x(x_n, y_m)(-\Delta x) + f_y(x_n, y_m)(-\Delta y).$$

Then, this vector moves from the point $(x_n,\, y_m - \Delta y,\, f(x_n, y_m) - f_y(x_n, y_m)\Delta y)$ to the point $(x_n - \Delta x,\, y_m - \Delta y,\, f(x_n, y_m) - f_x(x_n, y_m)\Delta x - f_y(x_n, y_m)\Delta y)$, and subtraction yields the vector

$$\langle -\Delta x,\, 0,\, -f_x(x_n, y_m)\Delta x \rangle,$$

which is the same as vector **v**. In other words, these opposite sides are parallel! The calculations for the other pair of opposite sides are similar, resulting in

$$\mathbf{w} = \langle 0, -\Delta y, -f_y(x_n, y_m)\Delta y \rangle.$$

The brown piece in Figure 4.47 is a parallelogram. We wish to calculate the area of the parallelogram. First we calculate the cross product,

$$\mathbf{v} \times \mathbf{w} = \begin{vmatrix} \mathbf{i} & \mathbf{j} & \mathbf{k} \\ -\Delta x & 0 & -\Delta x f_x(x_n, y_m) \\ 0 & -\Delta y & -\Delta y f_y(x_n, y_m) \end{vmatrix}$$
$$= \mathbf{i}\,(0 - \Delta x\,\Delta y\,f_x(x_n, y_m)) - \mathbf{j}\,(\Delta x\,\Delta y\,f_y(x_n, y_m) - 0) + \mathbf{k}\,(\Delta x\,\Delta y - 0),$$

and then find its norm,

$$\|\mathbf{v} \times \mathbf{w}\|$$
$$= \sqrt{(\Delta x)^2\,(\Delta y)^2\,(f_x(x_n, y_m))^2 + (\Delta x)^2\,(\Delta y)^2\,(f_y(x_n, y_m))^2 + (\Delta x)^2\,(\Delta y)^2}$$
$$= \Delta x\,\Delta y\sqrt{(f_x(x_n, y_m))^2 + (f_y(x_n, y_m))^2 + 1}.$$

This is the area of one piece of the surface, the *n-m*th piece as illustrated in Figures 4.46 and 4.47. Adding the surface areas of all the pieces gives the double omega sum

$$\sum_{n=1}^{\Omega} \sum_{m=1}^{\Omega} \sqrt{(f_x(x_n, y_m))^2 + (f_y(x_n, y_m))^2 + 1}\,\Delta y\,\Delta x,$$

which can be rewritten as a double integral,

$$\int_a^b \int_c^d \sqrt{(f_x(x, y))^2 + (f_y(x, y))^2 + 1}\ dy\,dx.$$

We can move to a more general region R in the usual manner.

<table>
<tr><td>

SURFACE AREA

The area of a surface $z = f(x, y)$ over a region R, where f is differentiable and f_x and f_y are both continuous on R, is

$$SA = \iint_R \sqrt{(f_x(x, y))^2 + (f_y(x, y))^2 + 1}\ dy\,dx.$$

</td></tr>
</table>

Notice the similarity to the formula for arc length (see *Calculus Set Free*, Section 9.1). The surface area formula is a two-dimensional version of that arc length formula; instead of measuring one dimension (length), we are measuring two dimensions (area). The derivations of the formulas are also similar, each using ideas such as local linearity and calculations such as factoring out Δx.

Requiring f to be differentiable and f_x and f_y to be continuous throughout the region R is one way of ensuring that the integrand in the surface area formula is integrable and Fubini's theorem applies, so that we can evaluate the double integral as an iterated integral using either order of integration. As always, we may switch to polar coordinates when convenient.

4.6.2 Surface area example

Notice the similarity to the wording of problems asking for volume. Carefully reading a problem is an underappreciated skill!

Example 30 *Find the area of the surface $z = 4x + 2y^3$ above the region bounded by*
$$x = 0, y = 1, \text{ and } y = \sqrt[3]{\frac{x}{8}}.$$

Solution As always, we begin by sketching the region (Figure 4.48).

The formula for surface area requires the partial derivatives of $f(x, y) = 4x + 2y^3$, which are

Notice that f_x and f_y are both continuous everywhere, so the hypotheses of the surface area formula are met. The main issue to watch for is if either of f_x or f_y is undefined anywhere in the region R.

$$f_x(x, y) = 4 \text{ and } f_y(x, y) = 6y^2.$$

Then, the surface area formula yields

The points of intersection can be found by solving equations. The curves $y = 1$ and $y = \sqrt[3]{\frac{x}{8}}$ meet where

$$SA = \int_0^8 \int_{\sqrt[3]{\frac{x}{8}}}^1 \sqrt{4^2 + \left(6y^2\right)^2 + 1}\, dy\, dx = \int_0^8 \int_{\sqrt[3]{\frac{x}{8}}}^1 \sqrt{17 + 36y^4}\, dy\, dx.$$

$$1 = \sqrt[3]{\frac{x}{8}}$$

$$1^3 = \frac{x}{8}$$

$$8 = x,$$

The problem is that integrating with respect to y, an antiderivative of $\sqrt{17 + 36y^4}$ is non-elementary. But the expression is a constant with respect to x, so perhaps it will help to reverse the order of integration. To that end, we need to solve $y = \sqrt[3]{\frac{x}{8}}$ for x, which results in $x = 8y^3$. Then the surface area is

and the curves $x = 0$ and $y = \sqrt[3]{\frac{x}{8}}$ meet where

$$y = \sqrt[3]{\frac{0}{8}} = 0.$$

$$SA = \int_0^1 \int_0^{8y^3} \sqrt{17 + 36y^4}\, dx\, dy$$

$$= \int_0^1 \left(\sqrt{17 + 36y^4} \cdot x \right)\Big|_0^{8y^3}\, dy$$

$$= \int_0^1 \left(8y^3 \sqrt{17 + 36y^4} - 0 \right) dy,$$

Line 1 uses the surface area formula; line 2 finds an antiderivative with respect to x, treating y as a constant; and line 3 evaluates at $x = 8y^3$, at $x = 0$, and subtracts.

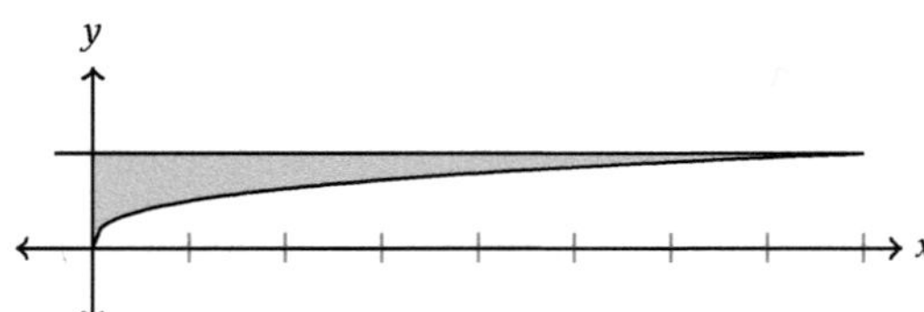

Figure 4.48 *The region for example 30*

which is set up nicely for substitution! Using

$$u = 17 + 36y^4$$

$$du = 144y^3 \, dy,$$

we also calculate limits for u as $17 + 36(1) = 53$ and $17 + 36(0) = 17$. Making the substitution and continuing with the calculation,

$$= \frac{1}{144} \int_0^1 144 \cdot 8y^3 \sqrt{17 + 36y^4} \, dy$$

$$= \frac{8}{144} \int_{17}^{53} \sqrt{u} \, du = \frac{1}{18} \left. \frac{u^{3/2}}{3/2} \right|_{17}^{53}$$

$$= \frac{1}{27} \left(53^{3/2} - 17^{3/2} \right) = 11.69 \, \text{units}^2.$$

Line 1 adjusts for the constant 144; line 2 makes the substitution, then finds an antiderivative with respect to u; and line 3 evaluates at $u = 53$, at $u = 17$, and subtracts.

The surface area is $11.69 \, \text{units}^2$. ∎

Notice that the region and the surface seem set up specially for making the integral calculable by hand. This is not generally the case, as the nature of the formula does not often lend itself to elementary antiderivatives. The use of a CAS to perform numerical approximations to integral values is often required.

Reading Exercise 11   Set up, but do not evaluate, a double integral for finding the area of the surface $z = 4x^3 - 2y^2$ above the region bounded by $y = x, y = 0$, and $x = 2$.

4.6.3   Surface area of a sphere

In Section 4.4, we used double integrals to determine the volume of a sphere. Next, we find the surface area of a sphere.

As with the volume, we begin with the upper half of the sphere $x^2 + y^2 + z^2 = a^2$, which can be rewritten as

Alternately, one can reason that the rate of change of volume as the sphere expands (that is, with respect to the radius) is the surface area, so that $SA = \frac{d}{dr} \left(\frac{4}{3} \pi r^3 \right) = 4\pi r^2$.

$$z = \sqrt{a^2 - x^2 - y^2}.$$

We need the partial derivatives:

$$\frac{\partial z}{\partial x} = \frac{1}{2\sqrt{a^2 - x^2 - y^2}}(-2x) = \frac{-x}{\sqrt{a^2 - x^2 - y^2}}$$

$$\frac{\partial z}{\partial y} = \frac{1}{2\sqrt{a^2 - x^2 - y^2}}(-2y) = \frac{-y}{\sqrt{a^2 - x^2 - y^2}}.$$

Then the integrand in the formula for surface area is

$$\sqrt{\left(\frac{\partial z}{\partial x}\right)^2 + \left(\frac{\partial z}{\partial y}\right)^2 + 1} = \sqrt{\frac{x^2}{a^2 - x^2 - y^2} + \frac{y^2}{a^2 - x^2 - y^2} + 1}$$

$$= \sqrt{\frac{x^2 + y^2 + a^2 - x^2 - y^2}{a^2 - x^2 - y^2}}$$

$$= \sqrt{\frac{a^2}{a^2 - x^2 - y^2}},$$

Line 1 writes the integrand in the surface area formula using Leibniz notation, then applies the formula; line 2 gets a common denominator and performs the addition; and line 3 simplifies.

which if we switch to polar coordinates is rewritten as

$$\sqrt{\frac{a^2}{a^2 - r^2}}.$$

See the discussion of volume of a sphere in Section 4.4 for details.

The polar region is still described as $0 \le r \le a$, $0 \le \theta \le 2\pi$, the same as for finding the volume of the sphere. Making the switch to polar coordinates, the surface area is

Don't forget the extra r!

$$SA = \int_0^{2\pi} \int_0^a \sqrt{\frac{a^2}{a^2 - r^2}} \cdot r \, dr \, d\theta.$$

For the inner integral we use the substitution $u = a^2 - r^2$, $du = -2r \, dr$; the limits of integration change to $a^2 - a^2 = 0$ and $a^2 - 0 = a^2$. Making the substitution,

$$\int_0^a \sqrt{\frac{a^2}{a^2 - r^2}} \cdot r \, dr = -\frac{1}{2} \int_0^a \sqrt{\frac{a^2}{a^2 - r^2}} \cdot -2r \, dr$$

Line 1 writes the inner integral and then adjusts for the constant -2; line 2 makes the substitution, including changing from values of the variable r to values of the variable u, then simplifies; line 3 finds an antiderivative with respect to u, then simplifies; and line 4 evaluates at $u = 0$, at $u = a^2$, and subtracts, then simplifies.

$$= -\frac{1}{2} \int_{a^2}^0 \sqrt{\frac{a^2}{u}} \, du = -\frac{a}{2} \int_{a^2}^0 u^{-1/2} \, du$$

$$= -\frac{a}{2} \frac{u^{1/2}}{\frac{1}{2}} \Bigg|_{a^2}^0 = -a\sqrt{u} \, \Big|_{a^2}^0$$

$$= -a(0) - -a\sqrt{a^2} = a^2.$$

Placing this result in the outer integral,

$$SA = \int_0^{2\pi} a^2 \, d\theta = 2\pi a^2.$$

This is just one hemisphere, so the surface area of an entire sphere of radius a is

$$SA = 4\pi a^2 \text{ units}^2,$$

as expected.

Ans. to reading exercise 11:

$$\int_0^2 \int_0^x \sqrt{144x^4 + 16y^2 + 1} \, dy \, dx;$$

alternate answer

$$\int_0^2 \int_y^2 \sqrt{144x^4 + 16y^2 + 1} \, dx \, dy$$

EXERCISES 4.6

1–10. (a) Set up a double integral for finding the area of the surface on the given region. (b) Use a CAS to determine a decimal approximation to the surface area.

> Even if the CAS can produce an exact value, a decimal approximation is requested. You may need to use a numerical integration command in the CAS. Desmos can also be used.

1. surface $f(x,y) = x^3 + 4x - y^4 + 5y^2$, on the rectangular region $1 \le x \le 14, 3 \le y \le 7$

2. surface $z = \sqrt{x + 4y^3 - 7}$, on the rectangular region $0 \le x \le 5$, $2 \le y \le 8$

3. surface $f(x,y) = x\sin y$, on the region bounded by $y = \sqrt{x}$, $x = 5$, and $y = 0$

4. surface $z = y^2 e^{3x}$, on the region bounded by $y = x$ and $y = x^4$

5. surface $f(x,y) = e^{x^2 + y^2}$, on the region $x^2 + y^2 \le 9$

6. surface $f(x,y) = 3x + y^3$, on the region inside the unit circle

7. surface $f(x,y) = (x^2 + y^2)^{3/2} - \frac{4y}{x}$, on the polar region $-\frac{\pi}{4} \le \theta \le \frac{\pi}{4}, 1 \le r \le 2$

8. surface $z = y^2(x^2 + y^2)^3$ on the polar region $0 \le r \le 5$, $0 \le \theta \le \pi$

9. surface $z = \cosh(x^2 y)\ln(y^3 + 1)$, on the region between $y = \frac{1}{4}x$ and $y = \sqrt{x}$, for $1 \le x \le 2$

10. surface $f(x,y) = \frac{(3x^2 + 4)^2}{y^3 + 7}$, on the region $2 \le x \le 10, 3 \le y \le 6$

11–20. Find the area of the surface on the given region.

11. surface $z = -4 + 2x - 6y$, on the rectangular region $1 \le x \le 3$, $-2 \le y \le 0$

12. surface $z = 3 + 4x + 5y$, over the rectangular region $0 \le x \le 2$, $1 \le y \le 4$

13. surface $f(x,y) = 1$, over the region between $y = x$ and $y = x^3$, for $0 \le x \le 1$

> Notice that exercise 13 is identical to calculating the area of the region.

14. surface $f(x,y) = y - x$, on the region bounded by $y = x$, $y = 4 - x$, and $x = 0$

15. surface $z = \frac{x^2}{2} + \frac{2}{3}y^{3/2}$, on the region between $y = -1$ and $y = -x^2$, for $0 \le x \le 1$

16. surface $z = x^2 - y^2$, where $x^2 + y^2 \le 7$

17. surface $z = \frac{y^2}{2} - \frac{x^2}{2}$, on the region bounded by $y = x, y = 0$, and $x = \sqrt{3 - y^2}$

> Some algebraic tricks necessary for these exercises are essentially the same as the ones employed for some arc length exercises.

18. surface $f(x,y) = \frac{x^4}{2} + \frac{1}{x^2} + \sqrt{15}\,y$, on the rectangular region $1 \le x \le 2, 0 \le y \le 1$

19. surface $f(x,y) = e^x \sin y$, over the region bounded by $x = 0$, $x = \ln 3, y = 0$, and $y = e^{2x}$

Part (a) of this exercise is from *Elements of the Differential and Integral Calculus*, by William Granville, Percey Smith, and William Longley, Ginn and Company, 1941 revised edition.

In reality, the earth is not a sphere; the rotation of the earth causes it to bulge at the equator relative to the poles. So the requested conclusion of part (b) is not quite correct.

There is a sense in which the original double integral is not actually improper. Because the definition of double integral uses "right-front" points to evaluate the function on a subrectangle, the function is never evaluated at $(0, 0)$.

20. surface $z = 4x^{3/2} - 2y^{3/2}$, on the region bounded by $y = 0$, $x = 1$, and $y = -4x$

21. (a) Find the area of that portion of the surface of the sphere $x^2 + y^2 + z^2 = 100$ which lies between the parallel planes $x = -8$ and $x = 6$.

    (b) Notice that the value of the inner integral from part (a) is a constant. Write a formula for the area of the portion of the surface of the sphere between the parallel planes $x = c$ and $x = d$, with $-10 \le c < d \le 10$. Reorienting the two planes, conclude that the surface area of the earth between two planes parallel to the equator depends only on the distance between the two planes.

    (c) Rewrite the answer to part (b) for the sphere $x^2 + y^2 + z^2 = a^2$.

    (d) What is the area of the (lateral) surface of a cylinder of radius a with height h? (This should be a familiar geometric formula and does not need integration.)

    (e) Compare your answers to parts (c) and (d).

22. (a) Find the area of the surface $z = x + \ln y$ over the region bounded by $y = \sqrt{x}$, $x = 0$, and $y = 4$. (Hint: one order of integration is easier than the other; after evaluating the inner integral, use an algebraic manipulation of the remaining integral's integrand to make substitution straightforward.)

    (b) Is there a point in the region where $y = 0$? Are the hypotheses of the surface area formula satisfied? Is the calculation of part (a) justified?

    (c) Using the same integration order as for part (a), try using the "improper integral" strategy. In particular, change the region of integration to start at $y = \omega$ instead of $y = 0$. Does the surface meet the hypotheses of the surface area formula on the adjusted region? If so, complete the calculation of the double integral.

    (d) Examine the calculations of parts (a) and (c). Does the algebraic manipulation seem to make the improperness "disappear"? If so, state where and why. Do you think that the improper integration strategy will always give the same answer as ignoring the improperness when an algebraic manipulation makes the improperness disappear?

4.7   Triple Integrals

For a function of one variable, $y = f(x)$, integration is conceived as giving the area under a curve. The domain is one-dimensional (Figure 4.49, left) and is subdivided into Ω pieces, each of width Δx. Rectangles of width Δx and height $f(x_n)$

(Figure 4.49, center and right) approximate the region under the curve. The area, measured in units^2, is given by the omega sum $\sum\limits_{n=1}^{\Omega} f(x_n)\Delta x$ (sum of the heights times the widths), which is rewritten as the integral $\int_a^b f(x)\,dx$.

Add a dimension: for a function of two variables, $z = f(x, y)$, integration is conceived as giving the volume under a surface. The domain is two-dimensional (Figure 4.50, left) and is subdivided into $\Omega \cdot \Omega$ rectangular pieces, each measuring $\Delta x \cdot \Delta y$. Rectangular boxes of base area $\Delta x\,\Delta y$ and height $f(x_n, y_m)$ (Figure 4.50, center and right) approximate the region under the surface. The volume, measured in units^3, is given by the double omega sum $\sum\limits_{m=1}^{\Omega}\sum\limits_{n=1}^{\Omega} f(x_n, y_m)\,\Delta x\,\Delta y$ (sum of the heights times the areas), which is rewritten as the double integral $\int_c^d \int_a^b f(x, y)\,dx\,dy$; either order can be used under the appropriate conditions.

For this discussion we are assuming that Ω is a positive infinite hyperreal integer, but for the purpose of illustrating the ideas, the figures show subdivision into only finitely many pieces.

Carefully compare this paragraph to the preceding paragraph.

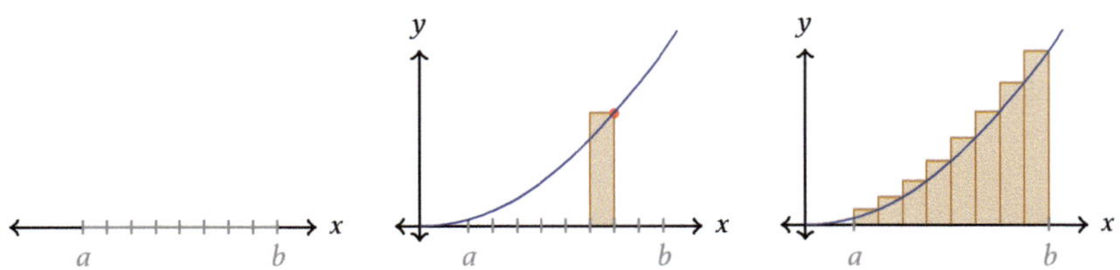

Figure 4.49 *Integration of a function of one variable. The domain of the function (left) is one-dimensional, and the area under the curve (right) is two-dimensional. Subdivision of the interval into finitely many subintervals, instead of infinitely many, is pictured so that the subintervals and rectangles based on those subintervals are visible. The height of each rectangle (center) matches the height of the curve at the right-hand endpoint*

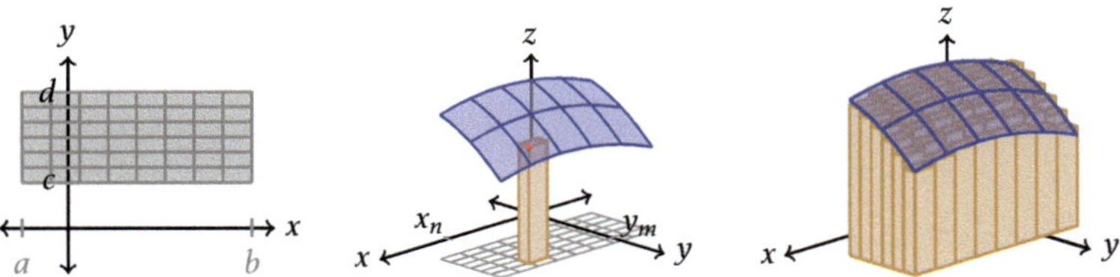

Figure 4.50 *Integration of a function of two variables. The domain of the function (left) is two-dimensional, and the volume under the surface (right) is three-dimensional. Subdivision of the rectangular region into finitely many subrectangles, instead of infinitely many, is pictured so that the subrectangles and rectangular boxes based on those subrectangles are visible. The height of each rectangular box (center) matches the height of the surface at the front-right corner. Compare to Figure 4.49*

Add another dimension: for a function of three variables, $w = f(x, y, z)$, how shall we conceive integration? By adding one dimension to each piece of the puzzle, as we did when moving from one variable to two.

4.7.1   Triple integrals over rectangular boxes

These are similar, but different, uses of the prefix "hyper" than with the word "hyperreal." In this case, the prefix "hyper" is used to indicate an increase in dimension; the alternative would be to come up with a different term for each dimension. That's fine for dimensions we can experience with our senses, making such terms as length, area, and volume easy to understand. But beyond three dimensions, using separate words becomes unwieldy.

For a function of three variables, $w = f(x, y, z)$, integration is conceived as giving the four-dimensional *hypervolume* under a three-dimensional *hypersurface* (a solid). The domain is three-dimensional (Figure 4.51, left) and is subdivided into $\Omega \cdot \Omega \cdot \Omega$ rectangular box pieces, each measuring $\Delta x \cdot \Delta y \cdot \Delta z$. Rectangular *hyperboxes* of base volume $\Delta x\,\Delta y\,\Delta z$ and height $f(x_n, y_m, z_k)$ (no figure this time, sorry) approximate the region under the hypersurface. The hypervolume, measured in units[4], is given by the triple omega sum

$$\sum_{k=1}^{\Omega} \sum_{m=1}^{\Omega} \sum_{n=1}^{\Omega} f(x_n, y_m, z_k)\, \Delta x\,\Delta y\,\Delta z$$

Compare this paragraph to the paragraph describing double integrals.

(sum of the heights times the volumes), which is rewritten as the triple integral

$$\int_g^h \int_c^d \int_a^b f(x, y, z)\, dx\,dy\,dz;$$

any of the six possible orders can be used under the appropriate conditions.

The following definition and theorems are presented somewhat formally and without proof, but the gist of the matter is that triple integrals work the way we would think, using iterated integrals in the same manner as double integrals.

An alternative to the fourth dimension being geometric, as in Figure 4.51 right, is for the fourth dimension to be represented in a non-geometric manner, such as by color. Various such schemes exist but will not be explored here.

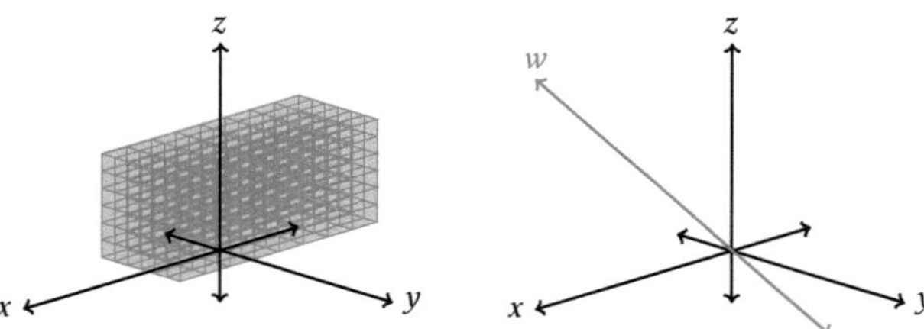

Figure 4.51 *Integration of a function of three variables. The domain of the function (left) is three-dimensional, but the four-dimensional hypervolume under the hypersurface requires four dimensions, which is more than we humans can easily interpret visually (right). Subdivision of the rectangular box region into finitely many subrectangular boxes, instead of infinitely many, is pictured so that the subrectangular boxes are visible. Compare to Figures 4.49 and 4.50.*

Definition 4 TRIPLE INTEGRAL OVER A RECTANGULAR BOX REGION *Let*
$f(x, y, z)$ be defined throughout the rectangular box region $a \le x \le b, c \le y \le d,$
$g \le z \le h$. If

$$\sum_{k=1}^{\Omega} \sum_{m=1}^{\Omega} \sum_{n=1}^{\Omega} f(x_n, y_m, z_k)\Delta x \Delta y \Delta z$$

renders the same real result L for any possible order of interchang-
ing the sums and for every positive infinite hyperreal integer Ω (where
$\Delta x = (b - a)\omega$, $\Delta y = (d - c)\omega$, $\Delta z = (h - g)\omega$, $x_n = a + n\Delta x$, $y_m = c + m\Delta y$
and $z_k = g + k\Delta z$), then we write

$$\int_g^h \int_c^d \int_a^b f(x, y, z)\, dx\, dy\, dz = L,$$

or use any of the other possible orders, and call $\int_g^h \int_c^d \int_a^b f(x, y, z)\, dx\, dy\, dz$ the
triple integral of f over the region $a \le x \le b, c \le y \le d, g \le z \le h$.
    When L is a real number (i.e., not ∞ or $-\infty$), then we say that f is integrable
on the region.

An alternate notation is $\iiint_R f(x, y, z)\, dV$, where R is the region of integration and dV (V for volume) can be $dx\, dy\, dz$ or any of the other orders of integration.

Theorem 3 INTEGRABILITY OF CONTINUOUS FUNCTIONS *If $f(x, y, z)$ is*
continuous throughout the rectangular box region $a \le x \le b, \ c \le y \le d,$
$g \le z \le h$, then f is integrable.

Fubini's theorem also applies to triple integrals.

Theorem 4 FUBINI'S THEOREM FOR TRIPLE INTEGRALS *If f is a continuous*
function (or any other integrable function), then the triple integral is an iterated
integral and can be calculated using any of the six possible orders;

$$\int_g^h \int_c^d \int_a^b f(x, y, z)\, dx\, dy\, dz$$

$$= \int_g^h \left(\int_c^d \left(\int_a^b f(x, y, z)\, dx \right) dy \right) dz = \int_g^h \left(\int_a^b \left(\int_c^d f(x, y, z)\, dy \right) dx \right) dz$$

$$= \int_c^d \left(\int_g^h \left(\int_a^b f(x, y, z)\, dx \right) dz \right) dy = \int_c^d \left(\int_a^b \left(\int_g^h f(x, y, z)\, dz \right) dx \right) dy$$

$$= \int_a^b \left(\int_c^d \left(\int_g^h f(x, y, z)\, dz \right) dy \right) dx = \int_a^b \left(\int_g^h \left(\int_c^d f(x, y, z)\, dy \right) dz \right) dx.$$

4.7.2 Triple integral example

Again, triple integrals work just like double integrals.

Example 31 *Evaluate $\int_0^1 \int_1^2 \int_0^3 (x + y + z)\, dx\, dy\, dz$.*

Solution As with double integrals, we evaluate from the inside out, treating variables as constants when not the variable of integration. Then

Line 1 reinterprets the integral as an iterated integral, where we integrate with respect to x first; line 2 finds an antiderivative with respect to x, treating y and z as constants; line 3 evaluates at $x = 3$, at $x = 0$, and subtracts; line 4 finds an antiderivative with respect to y, treating z as a constant; line 5 evaluates at $y = 2$, at $y = 1$, and subtracts, then simplifies; line 6 finds an antiderivative with respect to z; and line 7 evaluates at $z = 1$, at $z = 0$, and subtracts.

$$\int_0^1 \int_1^2 \left(\int_0^3 (x + y + z)\, dx \right) dy\, dz$$

$$= \int_0^1 \int_1^2 \left(\left(\tfrac{1}{2}x^2 + yx + zx \right)\Big|_0^3 \right) dy\, dz$$

$$= \int_0^1 \left(\int_1^2 \left(\tfrac{9}{2} + 3y + 3z - (0 + 0 + 0) \right) dy \right) dz$$

$$= \int_0^1 \left(\left(\tfrac{9}{2}y + \tfrac{3}{2}y^2 + 3zy \right)\Big|_1^2 \right) dz$$

$$= \int_0^1 \left(9 + 6 + 6z - \left(\tfrac{9}{2} + \tfrac{3}{2} + 3z \right) \right) dz = \int_0^1 (9 + 3z)\, dz$$

$$= \left(9z + \tfrac{3}{2}z^2 \right)\Big|_0^1$$

$$= 9 + \tfrac{3}{2} - 0 = \tfrac{21}{2}.$$

$\blacksquare$

Reading Exercise 12 Evaluate $\int_0^1 \int_0^2 \int_0^3 (1 + y)\, dz\, dy\, dx$.

4.7.3 Triple integrals over general regions

Recall that for double integrals over general regions, the theoretical approach is to enclose the general region D inside a rectangular region R (Figure 4.52), and then define a new function H that matches f for points inside D but takes the value 0 outside D. Integrating this new function H on the rectangular region R gives the value of the double integral on the region D. That's not how we calculate a double integral, but it is how the calculation is justified.

The same approach works for triple integrals. Suppose $w = f(x, y, z)$ is defined throughout the general three-dimensional region D, and that the region is bounded. Enclose the region inside a rectangular box R, as in Figure 4.53. Define the function H by

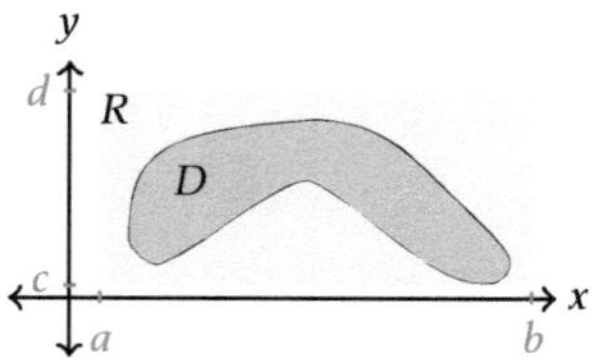

Figure 4.52 *A bounded region D (darker gray) in the xy-plane completely contained in a rectangular region R*

This approach is more of theoretical interest, for justifying the practical procedure. The procedure to use is demonstrated in example 32.

$$H(x,y,z) = \begin{cases} f(x,y,z), & \text{if } (x,y,z) \text{ is in } D \\ 0, & \text{otherwise} \end{cases}.$$

Then the value of the triple integral of H on the rectangular box R is the value of the triple integral of f on the general region D. A formal definition, which we omit, follows the pattern of definition 3.

Setting up a triple integral over a general region can be challenging and often involves the use of more than one diagram. A careful, rather methodical approach is demonstrated next.

Example 32 *Evaluate $\iiint_R 3y\, dV$ where R is the region bounded by $z = 0$ and $z = 9 - x^2 - y^2$.*

Solution   The steps are analogous to those for a double integral, but expanded to account for the additional dimension.

❶ Graph the three-dimensional region. The equation $z = 0$ is the equation of the xy-plane. The equation $z = 9 - x^2 - y^2$ is the equation of a paraboloid opening downward, with vertex at the point $(0, 0, 9)$. The region between these two surfaces is graphed in Figure 4.54.

In particular, $z = 9 - x^2 - y^2$ is an elliptic paraboloid, reflected and shifted from the standard form given in Section 1.7. The vertical cross sections are parabolas, and the horizontal cross sections are ellipses (circles, because $a = b = 1$).

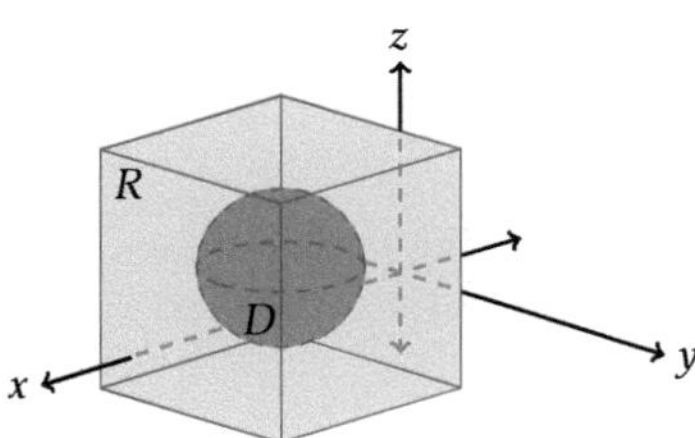

Figure 4.53 *A bounded region D (darker gray) completely contained in a rectangular box R*

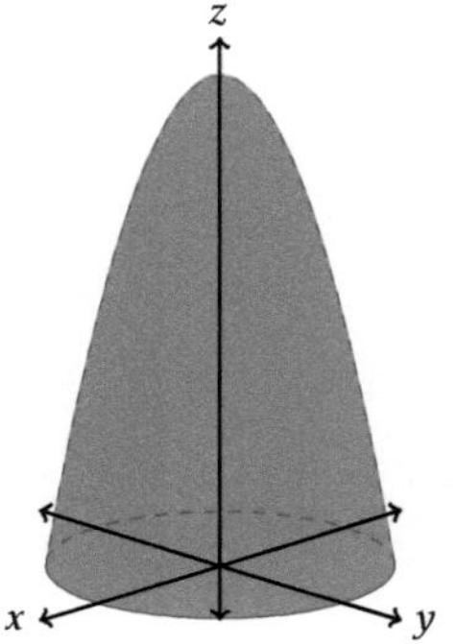

Figure 4.54 *The region R of example 32, viewed from a position in the first octant*

If you wish, try to rearrange the equation $z = 9 - x^2 - y^2$ so that the paraboloid is described as being between $y = \cdots$ and $y = \cdots$. If you are successful, then you can integrate with respect to y first.

Ans. to reading exercise 12:
  12

We could instead solve for x and integrate with respect to x next, then y.

Imagine eliminating the z-dimension and considering only the two-dimensional disk in Figure 4.55, right. Move your drone (your viewpoint) down into the xy-plane, along the negative y-axis, and consider the edge-on view of the disk. What do you see? An object occupying the interval from $x = -3$ to $x = 3$.

❷ Set up the inner integral. The region lies between $z = 0$ and $z = 9 - x^2 - y^2$, so we integrate with respect to z first. The triple integral, with the inner integral set up, is

$$\int \int \int_0^{9-x^2-y^2} 3y \, dz \, d\;\; d\;\;.$$

❸ Graph the extent of the region in the remaining two dimensions. Imagine viewing the three-dimensional region from straight down the z-axis (the axis of the variable we have already used); in your mind's eye, move your drone camera from the first-octant viewing position of Figure 4.54 upward above the mound and look down. The result is shown in Figure 4.55, left. Then rotate the drone until the positive x- and y-axes are seen in their traditional orientation (Figure 4.55, right). The region now appears circular, with its largest extent coming when $z = 0$ (refer again to Figure 4.54). The relationship between x and y that describes the boundary of the circular region is therefore $0 = 9 - x^2 - y^2$, or $x^2 + y^2 = 9$. Solving for y, we have $y = \pm\sqrt{9 - x^2}$. This allows us to set up the middle integral.

❹ Set up the middle integral. The triple integral, adding the setup for the middle integral, becomes

$$\int \int_{-\sqrt{9-x^2}}^{\sqrt{9-x^2}} \int_0^{9-x^2-y^2} 3y \, dz \, dy \, d\;\;.$$

❺ Set up the outer integral. The remaining variable is x, which varies from $x = -3$ to $x = 3$ (see Figures 4.54 and 4.55). The triple integral to evaluate is

$$\int_{-3}^{3} \int_{-\sqrt{9-x^2}}^{\sqrt{9-x^2}} \int_0^{9-x^2-y^2} 3y \, dz \, dy \, dx.$$

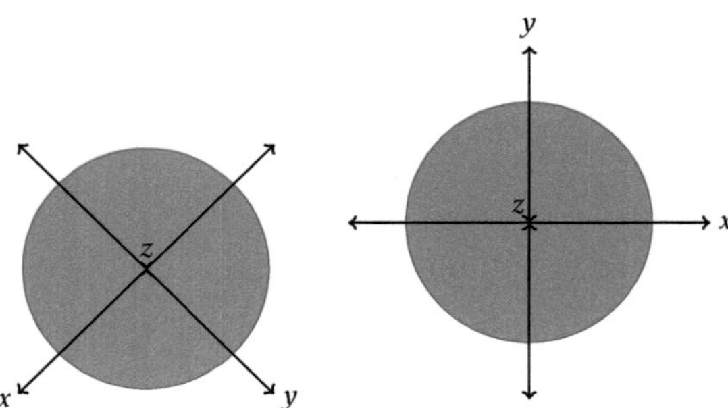

Figure 4.55 *The region of example 32, two views from above*

❻ Evaluate the triple integral. Evaluating from the inside out, we have

$$\int_{-3}^{3}\int_{-\sqrt{9-x^2}}^{\sqrt{9-x^2}}\left(\int_{0}^{9-x^2-y^2} 3y\, dz\right)dy\, dx$$

$$=\int_{-3}^{3}\int_{-\sqrt{9-x^2}}^{\sqrt{9-x^2}}\left(3yz\Big|_{0}^{9-x^2-y^2}\right)dy\, dx$$

$$=\int_{-3}^{3}\int_{-\sqrt{9-x^2}}^{\sqrt{9-x^2}}\left(3y(9-x^2-y^2)-0\right)dy\, dx$$

$$=\int_{-3}^{3}\left(\int_{-\sqrt{9-x^2}}^{\sqrt{9-x^2}}(27y-3x^2y-3y^3)\, dy\right)dx$$

$$=\int_{-3}^{3}\left(\left(\frac{27}{2}y^2-\frac{3}{2}x^2y^2-\frac{3}{4}y^4\right)\Big|_{-\sqrt{9-x^2}}^{\sqrt{9-x^2}}\right)dx$$

$$=\int_{-3}^{3}\left(\frac{27}{2}(9-x^2)-\frac{3}{2}x^2(9-x^2)-\frac{3}{4}(9-x^2)^2\right.$$

$$\left.-\left(\frac{27}{2}(9-x^2)-\frac{3}{2}x^2(9-x^2)-\frac{3}{4}(9-x^2)^2\right)\right)dx$$

$$=\int_{-3}^{3} 0\, dx = 0.$$

All that for nothing!

Line 1 reinterprets the integral as an iterated integral, where we integrate with respect to z first; line 2 finds an antiderivative with respect to z, treating x and y as constants; line 3 evaluates at $z=9-x^2-y^2$, at $z=0$, and subtracts; line 4 simplifies and rewrites as an iterated integral; line 5 finds an antiderivative with respect to y, treating x as a constant; line 6, which is continued in line 7, evaluates at $y=\sqrt{9-x^2}$, at $y=-\sqrt{9-x^2}$, and subtracts, noting that $\left(\sqrt{9-x^2}\right)^2=\left(-\sqrt{9-x^2}\right)^2=9-x^2$; and line 8 simplifies and completes the integration.

Explanation of joke: the word "nothing" is referring to the result of 0.

Reading Exercise 13 Set up a triple integral (do not evaluate) for finding $\iiint_R e^{xyz}\, dV$, where R is the region bounded by $z=0$ and $z=\sqrt{4-x^2-y^2}$.

Example 33 Evaluate $\int_{1}^{4}\int_{0}^{2}\int_{0}^{y} 1\, dz\, dy\, dx$.

Before calculating the solution, let's pause to consider the region of integration (Figure 4.56), which appears to be a box with triangular ends. From the inner integral, we see that the region lies between the planes $z=0$ and $z=y$, seen as the horizontal bottom and slanted top of the box, respectively. From the middle integral, the region lies between $y=0$, which is seen as an edge of the box, and the plane $y=2$, which is seen as a vertical side of the box. From the outer integral, the region lies between the planes $x=1$ and $x=4$, which are seen as the vertical ends of the box.

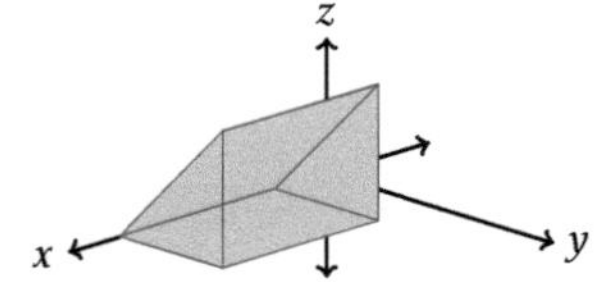

Figure 4.56 *The region of integration for example 33*

If you are setting up this integral and choose to integrate from $z=0$ to $z=y$ first, move your drone view to above the box, and you should see the rectangular region $0\le y\le 2, 1\le x\le 4$.

Solution   Evaluating from inside out, we have

$$\int_1^4 \int_0^2 \left(\int_0^y 1 \, dz \right) dy \, dx$$

$$= \int_1^4 \int_0^2 \left(z\big|_0^y \right) dy \, dx$$

$$= \int_1^4 \left(\int_0^2 (y - 0) \, dy \right) dx$$

$$= \int_1^4 \left(\tfrac{1}{2} y^2 \big|_0^2 \right) dx$$

$$= \int_1^4 (2 - 0) \, dx$$

$$= 2x \big|_1^4 = 8 - 2 = 6.$$

■

Line 1 reinterprets the integral as an iterated integral, integrating with respect to z first; line 2 finds an antiderivative with respect to z, treating x and y as constants; line 3 evaluates at $z = y$, at $z = 0$, and subtracts; line 4 finds an antiderivative with respect to y, treating x as a constant; line 5 evaluates at $y = 2$, at $y = 0$, and subtracts; and line 6 finds an antiderivative with respect to x and finishes the integration.

The order of integration for a general region can be swapped in the same manner as for double integrals.

Example 34 *Set up a triple integral for changing the order of integration of $\int_1^4 \int_0^2 \int_0^y 1 \, dz \, dy \, dx$ to $dy \, dx \, dz$, evaluate the triple integral, and compare to the solution to example 33.*

This is the same triple integral as in example 33.

Solution   ❶ Use the original triple integral to graph the region. See the prior discussion of Figure 4.56.

❷ Set up the inner integral. We wish to integrate with respect to y first. Throughout the region, the y-coordinates extend from the slanted side of the box, which has equation $z = y$, to the vertical rectangular end of the box, which has equation $y = 2$. We therefore integrate from $y = z$ to $y = 2$, and the triple integral with the inner integral set up is

This solution follows the same steps as for example 32.

$$\int \int \int_z^2 1 \, dy \, dx \, dz.$$

It does not matter whether you are looking in the positive y-direction or the negative y-direction.

❸ Graph the extent of the region in the remaining two dimensions. Referring to Figure 4.56, imagine moving the drone camera to view the region looking in the same direction as the y-axis. The region in these dimensions should look like a rectangle (Figure 4.57).

❹ Set up the middle integral. We wish to integrate with respect to x next, so we integrate from $x = 1$ to $x = 4$ as in the original triple integral. The triple integral with the middle integral set up is

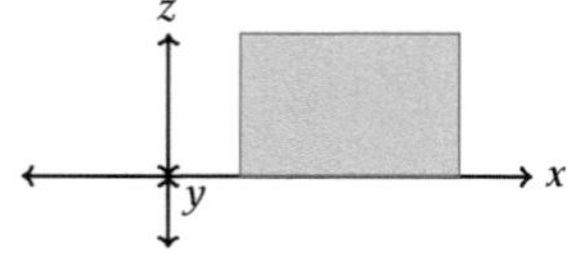

Figure 4.57 *The region of integration for examples 33 and 34, viewed straight down the y-axis*

$$\int \int_1^4 \int_z^2 1 \, dy \, dx \, dz.$$

5 Set up the outer integral. Because we have already integrated with respect to y, we cannot follow the original triple integral and integrate from $z = 0$ to $z = y$. Instead, to determine the upper z-coordinate of the region in Figure 4.57, note that in the original triple integral the upper limit of integration for z is y, and the largest y-coordinate is $y = 2$. Hence, the largest z-coordinate is $z = 2$, which we use as the upper limit of integration. The triple integral is

$$\int_0^2 \int_1^4 \int_z^2 1 \, dy \, dx \, dz.$$

6 Evaluate the triple integral. Evaluating from the inside out, we have

$$\int_0^2 \int_1^4 \left(\int_z^2 1 \, dy \right) dx \, dz$$

$$= \int_0^2 \int_1^4 \left(y|_z^2 \right) dx \, dz$$

$$= \int_0^2 \left(\int_1^4 (2 - z) \, dx \right) dz$$

$$= \int_0^2 \left((2 - z)x|_1^4 \right) dz$$

$$= \int_0^2 ((2 - z)4 - (2 - z)1) \, dz = \int_0^2 (6 - 3z) \, dz$$

$$= \left(6z - \frac{3}{2}z^2 \right)\Big|_0^2 = 12 - 6 - 0 = 6,$$

which matches the solution to example 33, as expected. $\blacksquare$

Line 1 reinterprets the integral as an iterated integral, integrating with respect to y first; line 2 finds an antiderivative with respect to y, treating x and z as constants; line 3 evaluates at $y = 2$, at $y = z$, and subtracts; line 4 finds an antiderivative with respect to x, treating z as a constant (note that $(2 - z)$ is a constant); line 5 evaluates at $x = 4$, at $x = 1$, and subtracts, then simplifies; line 6 finds an antiderivative with respect to z and finishes the integration.

4.7.4 Center of mass of a solid

In Section 4.5, we learned to use double integrals to determine the center of mass of a thin flat plate with variable density. A thin flat plate is treated as a two-dimensional object. But what if we have a three-dimensional object with variable density? The center of mass will then have three coordinates to find, $\bar{x}$, $\bar{y}$, and $\bar{z}$. Instead of reworking the analysis from scratch, we add one dimension to our previous formulas, moving from double integrals to triple integrals.

For a thin flat plate, the density is given by $\rho(x, y)$, with units of the form $\frac{\text{mass}}{\text{area}}$. The total mass m of the plate is given by $m = \iint \rho(x, y) \, dA$, where A is for area and

Ans. to reading exercise 13:

$$\int_{-2}^2 \int_{-\sqrt{4-x^2}}^{\sqrt{4-x^2}} \int_0^{\sqrt{4-x^2-y^2}} e^{xyz} \, dz \, dy \, dx;$$

alternate answer

$$\int_{-2}^2 \int_{-\sqrt{4-y^2}}^{\sqrt{4-y^2}} \int_0^{\sqrt{4-x^2-y^2}} e^{xyz} \, dz \, dx \, dy$$

represents either $dx\,dy$ or $dy\,dx$. For a solid, the density is given by $\rho(x, y, z)$, with units of the form $\frac{\text{mass}}{\text{volume}}$. The total mass m is given by

$$m = \int\int\int \rho(x, y, z)\,dV,$$

where dV represents any of the six possible orders of integration.

For a thin flat plate, the coordinates of the center of mass are given by $\bar{x} = \frac{\int\int x\cdot\rho(x,y)\,dA}{m}$ and $\bar{y} = \frac{\int\int y\cdot\rho(x,y)\,dA}{m}$. For a solid, the coordinates of the center of mass are given by

$$\bar{x} = \frac{\int\int\int x\cdot\rho(x,y,z)\,dV}{m} = \frac{1}{m}\int\int\int x\cdot\rho(x,y,z)\,dV,$$

$$\bar{y} = \frac{\int\int\int y\cdot\rho(x,y,z)\,dV}{m} = \frac{1}{m}\int\int\int y\cdot\rho(x,y,z)\,dV,$$

$$\text{and}\quad \bar{z} = \frac{\int\int\int z\cdot\rho(x,y,z)\,dV}{m} = \frac{1}{m}\int\int\int z\cdot\rho(x,y,z)\,dV.$$

Instead of moments about the y-axis and x-axis, the numerators of $\bar{x}$, $\bar{y}$, and $\bar{z}$ give moments M_{yz} about the yz-plane, M_{xz} about the xz-plane, and M_{xy} about the xy-plane, respectively.

Example 35 *Find the center of mass of a solid with density function $\rho(x, y, z) = 5$ occupying the region $3 \leq x \leq 5$, $1 \leq y \leq x$, $0 \leq z \leq x + y$.*

We cannot use a variable as a limit of integration after we have already integrated that variable. As a result, we cannot integrate with respect to either x or y before integrating with respect to z if we wish to integrate z from $z = 0$ to $z = x + y$. The alternative is to draw our region as a guide for changing the order of integration in the manner of example 34.

Line 1 is the triple integral to evaluate; line 2 is the result of computing the inner integral; line 3 is the result of computing the middle integral; and line 4 is the result of computing the outer integral.

Solution　❶　Calculate the mass of the solid. The formula for mass is $m = \int\int\int \rho(x, y, z)\,dV$. The description of the region appears to already be in a form giving limits of integration for the three variables. But which order of integration should we use? Because we are given $0 \leq z \leq x + y$, we integrate with respect to z first. Next, we integrate with respect to y, because x is its upper limit. Then, we can integrate with respect to x. The mass is

$$\int_3^5 \int_1^x \int_0^{x+y} 5\,dz\,dy\,dx$$

$$= \cdots = \int_3^5 \int_1^x (5x + 5y)\,dy\,dx$$

$$= \cdots = \int_3^5 \left(\frac{15}{2}x^2 - 5x - \frac{5}{2}\right) dx$$

$$= \cdots = 200.$$

② Formulate the integrals for finding $\bar{x}, \bar{y}$, and $\bar{z}$ and evaluate the integrals. Using the formula $\bar{x} = \frac{1}{m} \iiint x \cdot \rho(x, y, z)\, dV$ with the same order of integration used for finding the mass, we have

$$\bar{x} = \frac{1}{200} \int_3^5 \int_1^x \int_0^{x+y} 5x\, dz\, dy\, dx$$

$$= \cdots = \frac{1}{200} \int_3^5 \int_1^x (5x^2 + 5xy)\, dy\, dx$$

$$= \cdots = \frac{1}{200} \int_3^5 \left(\frac{15}{2}x^3 - 5x^2 - \frac{5}{2}x\right) dx$$

$$= \cdots = \frac{251}{60} = 4.1833.$$

Compare to the triple integral calculation for the mass. The difference is that we divide by the mass (the $\frac{1}{200}$ multiplying the triple integral) and there is an extra factor of x in the integrand.

Similarly, the formula for $\bar{y}$ is $\bar{y} = \frac{1}{m} \iiint y \cdot \rho(x, y, z)\, dV$:

$$\bar{y} = \frac{1}{200} \int_3^5 \int_1^x \int_0^{x+y} 5y\, dz\, dy\, dx$$

$$= \cdots = \frac{1}{200} \int_3^5 \int_1^x (5xy + 5y^2)\, dy\, dx$$

$$= \cdots = \frac{1}{200} \int_3^5 \left(\frac{25}{6}x^3 - \frac{5}{2}x - \frac{5}{3}\right) dx$$

$$= \cdots = \frac{163}{60} = 2.7167.$$

Compare to the triple integral calculation for $\bar{x}$. Instead of an extra factor of x, there is an extra factor of y in the integrand. Notice the difference that makes in line 2 compared to the calculations for mass and $\bar{x}$. But this pattern does not continue for calculating $\bar{z}$. Can you think why?

Finally, the formula for $\bar{z}$ uses an extra factor of z in the integrand instead of an extra x or y:

$$\bar{z} = \frac{1}{200} \int_3^5 \int_1^x \int_0^{x+y} 5z\, dz\, dy\, dx$$

$$= \cdots = \frac{1}{200} \int_3^5 \int_1^x \frac{5}{2}(x^2 + 2xy + y^2)\, dy\, dx$$

$$= \cdots = \frac{1}{200} \int_3^5 \left(\frac{35}{6}x^3 - \frac{5}{2}x^2 - \frac{5}{2}x - \frac{5}{6}\right) dx$$

$$= \cdots = \frac{69}{20} = 3.45.$$

Compare to the triple integral calculations for $\bar{x}$ and $\bar{y}$. The pattern does not continue because once we integrate with respect to z, the variable z cannot appear again in our calculations.

③ Write the center of mass as a point in $\mathbf{R}^3$. The center of mass is $(4.1833, 2.7167, 3.45)$.

The calculations for center of mass can quickly become tedious enough that the use of a CAS is desirable. ∎

When a solid has constant density $\rho = C$ as is the case for example 35, then the center of mass is the same as the geometric centroid of the region.

Reading Exercise 14 (a) Set up (but do not evaluate) the integral for finding the total mass of a solid in the region $0 \le x \le 1, 0 \le y \le \sqrt{x}, 0 \le z \le x + y$ with density function $\rho(x, y, z) = x^2 y^4 z^6$. (b) Set up (but do not evaluate) the integrals for finding $\bar{x}, \bar{y},$ and $\bar{z}$ for the same solid. The total mass is $\dfrac{23174383}{113565375}$.

4.7.5 Average value of a function

Another concept that can be expanded by an extra dimension is the average value of a function. Compare the following definition to definition 2 in Section 4.2.

Definition 5 AVERAGE VALUE OF A FUNCTION OF THREE VARIABLES *Suppose $w = f(x, y, z)$ is integrable on the region $a \le x \le b, c \le y \le d, g \le z \le h$. Then the average value of f on the region is*

The order of integration can be changed when Fubini's theorem applies.

$$f_{avg} = \frac{1}{(h - g)(d - c)(b - a)} \int_a^b \int_c^d \int_g^h f(x, y, z)\, dz\, dy\, dx.$$

Example 36 *The temperature of a block of material is given by $T(x, y, z) = (200 - x - y)°K$, for $0 \le x \le 5, 0 \le y \le 3, 0 \le z \le 1$. Find the average temperature of the block.*

°K is degrees Kelvin.

Solution The average temperature of the block is the average value of the function T. Using the formula for the average value of a function of three variables, we have

The region of integration is a rectangular box, so we may use any order of integration that we wish without fear of needing to draw a diagram to determine limits of integration.

$$T_{avg} = \frac{1}{(1 - 0)(3 - 0)(5 - 0)} \int_0^1 \int_0^3 \int_0^5 (200 - x - y)\, dx\, dy\, dz$$

$$= \frac{1}{15} \int_0^1 \int_0^3 \left(\left(200x - \tfrac{1}{2}x^2 - xy \right)\Big|_0^5 \right) dy\, dz$$

$$= \frac{1}{15} \int_0^1 \int_0^3 \left(1000 - \tfrac{25}{2} - 5y \right) dy\, dz$$

$$= \frac{1}{15} \int_0^1 \left(\left((1000 - \tfrac{25}{2})y - \tfrac{5}{2}y^2 \right)\Big|_0^3 \right) dz$$

$$= \cdots = \frac{1}{15} \int_0^1 2940\, dz = \frac{2940z}{15}\Big|_0^1 = 196°K.$$

The average temperature of the block is $196°K$. ∎

EXERCISES 4.7

1–8. Rapid response: what, if anything, is wrong with the following integrals?

1. $\displaystyle\int_1^2\int_3^4\int_5^6 (x+y)\,dx\,dy\,dz$

2. $\displaystyle\int_1^2\int_3^4 (x+y)\,dx\,dy\,dz$

3. $\displaystyle\int_1^2\int_3^4\int_5^6 (x+y)\,dx\,dy$

4. $\displaystyle\int_1^2\int_3^4\int_5^y (x+y)\,dx\,dy\,dz$

5. $\displaystyle\int_1^2\int_3^4\int_5^x (x+y)\,dx\,dy\,dz$

6. $\displaystyle\int_1^2\int_3^4\int_5^6 (x+y)\,dz\,dy\,dz$

7. $\displaystyle\int_1^2\int_3^x\int_x^y (x+y)\,dz\,dy\,dx$

8. $\displaystyle\int_x^y\int_3^x\int_1^2 (x+y)\,dx\,dy\,dz$

9–16. Evaluate the triple integral.

9. $\displaystyle\int_0^{\frac{\pi}{2}}\int_0^{\pi}\int_1^3 x\sin y\cos z\,dx\,dy\,dz$

10. $\displaystyle\int_0^1\int_0^1\int_0^1 yze^x\,dx\,dy\,dz$

11. $\displaystyle\int_1^4\int_1^3\int_0^{\frac{\pi}{4}} \frac{\sqrt{x}}{y}\sec^2 z\,dz\,dy\,dx$

12. $\displaystyle\int_1^2\int_1^2\int_1^e \frac{\ln x}{xy^2z}\,dx\,dy\,dz$

13. $\displaystyle\int_0^1\int_1^3\int_0^z e^{yz}\,dx\,dy\,dz$

14. $\displaystyle\int_0^1\int_0^x\int_0^{x+y} z\,dz\,dy\,dx$

15. $\displaystyle\int_0^4\int_z^2\int_x^z y^3\,dy\,dx\,dz$

16. $\displaystyle\int_0^3\int_0^1\int_0^x \frac{2^z\cosh x}{x}\,dy\,dz\,dx$

17–20. Set up (but do not evaluate) a triple integral for finding $\iiint_R \tan^{-1} xyz\,dV$ for the given region R.

Ans. to reading exercise 14:

(a) $\displaystyle\int_0^1\int_0^{\sqrt{x}}\int_0^{x+y} x^2y^4z^6\,dz\,dy\,dx;$

(b) $\bar{x}=\dfrac{113565375}{23174383}\displaystyle\int_0^1\int_0^{\sqrt{x}}\int_0^{x+y} x^3y^4z^6\,dz\,dy\,dx,$

$\bar{y}=\dfrac{113565375}{23174383}\displaystyle\int_0^1\int_0^{\sqrt{x}}\int_0^{x+y} x^2y^5z^6\,dz\,dy\,dx,$

$\bar{z}=\dfrac{113565375}{23174383}\displaystyle\int_0^1\int_0^{\sqrt{x}}\int_0^{x+y} x^2y^4z^7\,dz\,dy\,dx$

17. region bounded by $z = 0$ and $z = 4 - x^2 - y^2$
18. region bounded by $z = 1$ and $z = 10 - x^2 - \frac{1}{4}y^2$
19. region bounded by $x = 0$ and $x = 9 - y^2 - z^2$
20. region bounded by $y = 4$ and $y = x^2 + z^2$

21–24. (a) Set up and (b) evaluate the triple integral.

21. $\iiint_R x \, dV$, region bounded by $z = 4$, $z = x^2$, $y = 1$, and $y = 3$
22. $\iiint_R z \, dV$, region bounded by $z = 4$, $z = x^2$, $y = 1$, and $y = 3$
23. $\iiint_R (x + 2) \, dV$, region bounded by $z = y$, $z = 2y$, $z = 4$, $x = 0$, and $x = 5$
24. $\iiint_R x^2 \, dV$, region $0 \leq z \leq y$, $x^2 \leq y \leq 9$, $-3 \leq x \leq 3$

25. Evaluate $\iiint_R z \, dV$ where R is the region $x^2 + y^2 + z^2 \leq 9$, $z \geq 0$.
26. Evaluate $\iiint_R e^x \, dV$ where R is the region bounded by $y = x^2$, $y = 4$, $z = 0$, and $z = x + 2$.

27–30. Find the center of mass of the solid occupying the given region with the given density function ρ.

27. $\rho(x, y, z) = x^2 yz$, region $0 \leq x \leq 1$, $0 \leq y \leq 2$, $0 \leq z \leq 3$
28. $\rho(x, y, z) = x + y$, region $0 \leq x \leq 2$, $0 \leq y \leq 1$, $0 \leq z \leq 1$
29. $\rho(x, y, z) = \frac{1}{xyz}$, region $1 \leq x \leq e$, $1 \leq y \leq 5$, $1 \leq z \leq 2$
30. $\rho(x, y, z) = \frac{x}{y}$, region $0 \leq x \leq 1$, $1 \leq y \leq e^2$, $11 \leq z \leq 15$

31–34. Find the average value of the function on the given region.

31. $f(x, y, z) = x^2 + 3yz$, region $0 \leq x \leq 4$, $2 \leq y \leq 5$, $-2 \leq z \leq 0$
32. $f(x, y, z) = x^2 + 3yz$, region $0 \leq x \leq 4$, $2 \leq y \leq 5$, $0 \leq z \leq 2$
33. $f(x, y, z) = x \sin xz + y \sin yz$, region $0 \leq x \leq 1$, $0 \leq y \leq 1$, $0 \leq z \leq \pi$
34. $f(x, y, z) = y^2 e^{2x + 3z}$, region $0 \leq x \leq \ln 2$, $0 \leq y \leq 1$, $0 \leq z \leq \ln 3$

35. Find the average temperature of a block occupying the region $0 \leq x \leq 10$, $0 \leq y \leq 2$, $0 \leq z \leq 2$ with temperature function $T(x, y, z) = 1300 - 10(x - 5)^2 - 10y^2 - 10z^2$, measured in degrees Fahrenheit.

36. Find the average temperature of a block occupying the region $0 \leq x \leq 2$, $0 \leq y \leq 2$, $0 \leq z \leq 3$ with temperature function $T(x, y, z) = x + y + z$, measured in degrees Celsius.

37. Evaluate the improper triple integral $\int_0^\infty \int_0^\infty \int_1^\infty \frac{e^{-y}}{x^2(1+z^2)} \, dx \, dy \, dz$.

38. Evaluate the improper triple integral $\int_0^1 \int_0^1 \int_0^1 \frac{1}{\sqrt{xyz}} \, dx \, dy \, dz$.

39. Rework example 32 using the order of integration $dy \, dz \, dx$ as follows.

(a) Solve the equation $z = 9 - x^2 - y^2$ for y.
(b) Using Figure 4.54 as a visual guide, set up the inner integral.

Exercises 25 and 26 require the use of various techniques of integration.

(c) Imagine moving your drone camera from its first octant position to a point along the positive y-axis to view the region straight down the y-axis. Which axes are seen as we project to the remaining two dimensions? Draw a figure in the style of Figure 4.55, right.

(d) What value of y should we use to capture the largest extent of the two-dimensional region seen in your figure from part (c)? Refer to Figure 4.54 as necessary.

(e) Using the answer to part (d), rewrite the appropriate equation in the form $z = \cdots$.

(f) Set up the middle integral.

(g) Set up the outer integral.

(h) Evaluate the triple integral of part (g) and compare to the answer to example 32.

40. (a) Use a double integral to find the volume under the surface $g(x, y) = y$ over the rectangular region $1 \le x \le 4, 0 \le y \le 2$.

(b) Is the box in Figure 4.56 the same as the region whose volume is sought in part (a)?

(c) Compare your double integral calculation in part (a) to the solution to example 51.

(d) Based on your answer to part (c), write a triple integral to find the volume of the region between $z = 0$ and $z = f(x, y)$ (an arbitrary nonnegative function of two variables) over the domain $a \le x \le b, c \le y \le d$.

(e) Evaluate the inner integral. Does the resulting double integral give the volume being sought in part (d)?

41. (a) Using the result of exercise 40, set up a triple integral to find the volume under the surface $f(x, y) = xy + 4x + 5y$ over the rectangular region $0 \le x \le 2, 1 \le y \le 4$.

(b) Evaluate the triple integral and compare to the answer to exercise 11 of Section 4.1.

42. (a) Using the result of exercise 40, set up a triple integral to find the volume under the surface $f(x, y) = \sin x \cos y$ over the rectangular region $0 \le x \le \pi, 0 \le y \le \frac{\pi}{2}$.

(b) Evaluate the triple integral and compare to the answer to exercise 13 of Section 4.1.

43. The average value of a function $w = f(x, y, z)$ on a non-rectangular-box region can be found by first finding the volume of the region using the formula volume $= \iiint_R 1 \, dV$ and calculating $f_{avg} = \frac{1}{volume} \iiint_R f(x, y, z) \, dV$. Use these formulas to calculate the average value of $f(x, y, z) = \sqrt{x + y + z}$ on the region $0 \le x \le 3, 0 \le y \le x, 0 \le z \le x + y$.

The calculations for exercises 43 and 44 can be completed by hand but are rather messy.

44. Repeat the previous exercise for $f(x, y, z) = x^2 - 3y + \frac{1}{z}$ on the region $1 \le x \le 2, 1 \le y \le z^2, 1 \le z \le x$.

4.8   Triple Integrals in Cylindrical Coordinates

The three-dimensional coordinate system we have employed so far extends the xy-plane by adding a third coordinate, z. Cartesian (rectangular) coordinates are used. But what if we start with the $r\theta$-plane instead? That is, what if we begin with polar coordinates and add a third coordinate, z? The result is *cylindrical coordinates*.

4.8.1   Cylindrical coordinates

For three dimensions using Cartesian coordinates (x, y, z), the plane containing the first two coordinates is the "floor" and the third coordinate gives the elevation above or below the floor. Viewed from the first octant in the usual orientation (Figure 4.58), the x- and y-axes are placed on the page differently than is customary in two dimensions.

For cylindrical coordinates (r, θ, z), we begin the same way–the plane containing the first two coordinates is the "floor" and the third coordinate gives the elevation above or below the floor. The usual two-dimensional representation of polar coordinates (r, θ) places the initial ray along the positive x-axis, with positive values of θ rotating in the direction of the positive y-axis (Figure 4.59). For three dimensions we follow the same positioning, placing the initial ray along the positive x-axis, with positive values of θ rotating in the direction of the positive y-axis (Figure 4.60). Each point in the $r\theta$-plane has infinitely many different coordinates because we can rotate in either the positive or negative direction, and we can add as many rotations as we want as long as we stop at the same terminal ray; that does not change.

Example 37 *Plot the point with cylindrical coordinates* $\left(2, \frac{\pi}{2}, 3\right)$.

Solution   We begin by drawing the positive z-axis and the initial ray (Figure 4.61, left). Next, we rotate $\frac{\pi}{2}$ radians in the positive direction and draw the terminal ray, which in this case happens to be in the same location as the positive y-axis

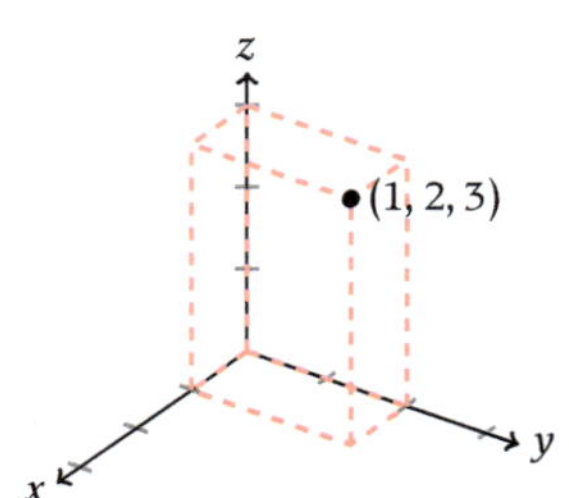

Figure 4.58 *The point with Cartesian coordinates* $(1, 2, 3)$, *viewed from the first octant*

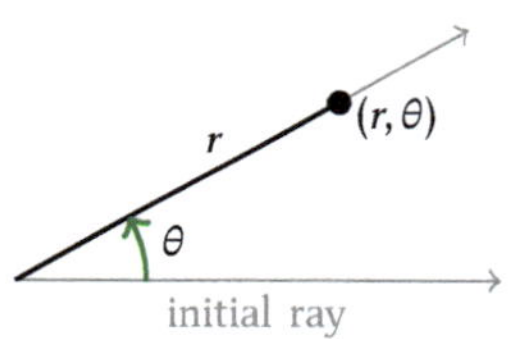

Figure 4.59 *The point with polar coordinates* (r, θ), *with* $r > 0$ *and* $0 < \theta < \frac{\pi}{2}$

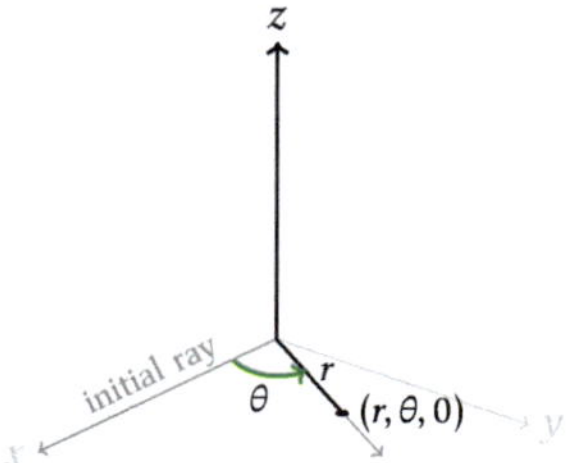

Figure 4.60 *The point with cylindrical coordinates* $(r, \theta, 0)$, *with* $r > 0$ *and* $0 < \theta < \frac{\pi}{2}$, *viewed from the first octant. Compare to Figures 4.58 and 4.59.*

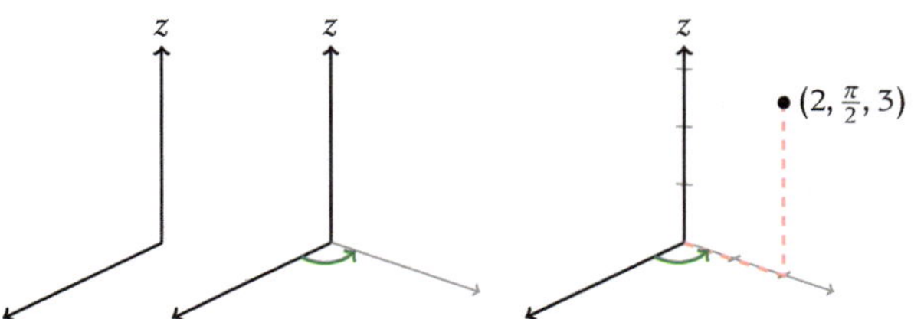

Figure 4.61 *Three stages in drawing the point with cylindrical coordinates* $\left(2, \frac{\pi}{2}, 3\right)$. *The solution to example 37 is the diagram on the right*

(Figure 4.61, center). Finally, we move two units from the origin along the terminal ray and then three units in the z-direction (red dotted segments in Figure 4.61, right) to arrive at the point $\left(2, \frac{\pi}{2}, 3\right)$.

If you wish to mimic the idea of the box in Figure 4.58 to help you locate the point more accurately, draw a ray parallel to the terminal ray emanating from the point $(0, 0, 3)$, shown red dotted in Figure 4.62. ■

Reading Exercise 15 Plot the point with cylindrical coordinates $\left(2, -\frac{\pi}{2}, 3\right)$.

4.8.2   Converting between cylindrical and Cartesian coordinates

To convert coordinates from (x, y, z) to (r, θ, z) or vice versa, we need not change z; it represents the "elevation" above or below the "floor" in either system. This means that we only need to convert between Cartesian coordinates (x, y) and polar coordinates (r, θ), which we already know how to do.

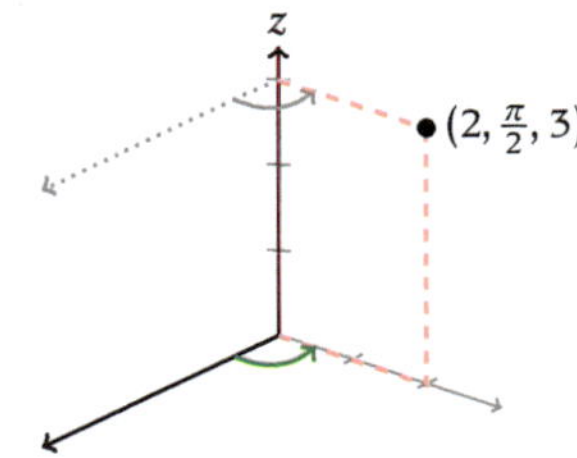

Figure 4.62 *The point with cylindrical coordinates* $\left(2, \frac{\pi}{2}, 3\right)$

CYLINDRICAL AND CARTESIAN COORDINATE CONVERSION FORMULAS

$$y = r\sin\theta \qquad x = r\cos\theta$$
$$x^2 + y^2 = r^2 \qquad \tan\theta = \frac{y}{x}$$

These formulas also appear in Section 4.4 under the name "polar and rectangular coordinate conversion formulas."

Example 38 *Find cylindrical coordinates for a point with Cartesian coordinates* $(1, -1, 4)$.

Solution The task is identical to finding polar coordinates (r, θ) to match the Cartesian coordinates $(1, -1)$. One method begins by plotting the point $(1, -1)$

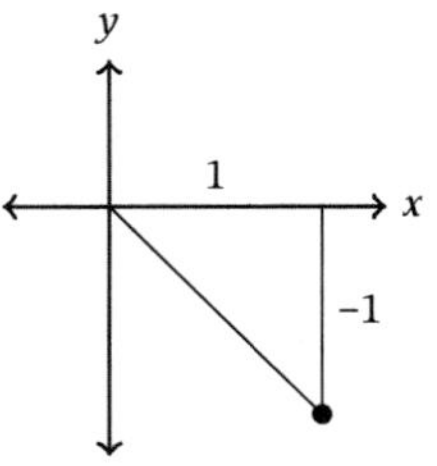

Figure 4.63　*The point with Cartesian coordinates* $(1, -1)$

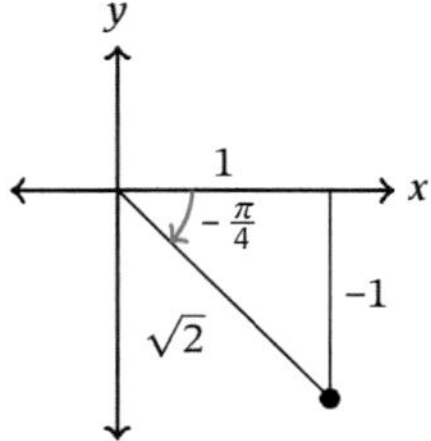

Figure 4.64　*The point with Cartesian coordinates* $(1, -1)$ *has polar coordinates* $\left(\sqrt{2}, -\frac{\pi}{4}\right)$

Because the z-coordinate does not need to be considered when converting from Cartesian coordinates to cylindrical coordinates, two-dimensional diagrams are sufficient. These two-dimensional diagrams are drawn in the usual orientation.

Alternately, we can use $\theta = \tan^{-1}\frac{y}{x} = \tan^{-1}\frac{-1}{1} = -\frac{\pi}{4}$. But remember that the inverse tangent only gives angles in the first or fourth quadrants, so if the point is in the second or third quadrant, an adjustment by π radians or by using a negative value of r is required. See Section 8.4 of *Calculus Set Free*.

Alternately, the conversion formula $x^2 + y^2 = r^2$ can be used, with $x = 1$ and $y = -1$. Which solution to use, the one with $r > 0$ or $r < 0$, depends on which value of θ is chosen.

The z-coordinate does not change and remains the same as given in Cartesian coordinates.

and drawing the associated triangle (Figure 4.63). Next, we use the Pythagorean theorem to calculate the length of the hypotenuse of the right triangle; $r = \sqrt{1^2 + (-1)^2} = \sqrt{2}$. Then, we place $\sqrt{2}$ in the diagram (Figure 4.64) and recognize the angle in the triangle as $\frac{\pi}{4}$. Based on the position of the angle below the x-axis, we use $\theta = -\frac{\pi}{4}$. Cylindrical coordinates for the point are therefore $\left(\sqrt{2}, -\frac{\pi}{4}, 4\right)$. ■

Alternate solutions to example 38 include $\left(\sqrt{2}, \frac{7\pi}{4}, 4\right)$ and $\left(-\sqrt{2}, \frac{3\pi}{4}, 4\right)$. See Figure 4.65.

Reading Exercise 16　Find cylindrical coordinates for a point with Cartesian coordinates $(0, 3, 5)$.

Converting from cylindrical to Cartesian coordinates is more straightforward.

Example 39 *Convert the cylindrical coordinates* $\left(-2, \frac{\pi}{3}, 7\right)$ *to Cartesian coordinates.*

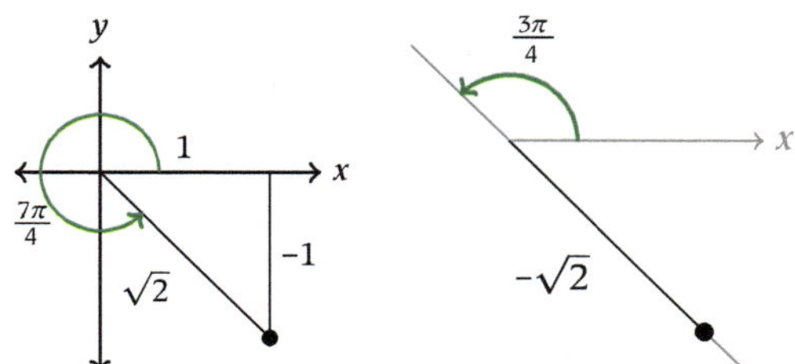

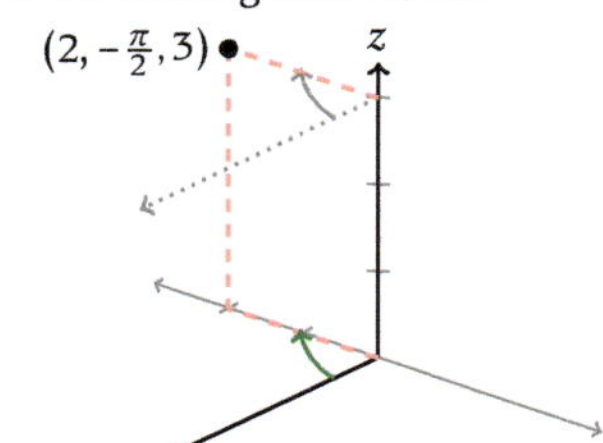

Figure 4.65 *The point with Cartesian coordinates* $(1, -1)$ *(left) represented with polar coordinates* $\left(\sqrt{2}, \frac{7\pi}{4}\right)$ *and (right) represented with polar coordinates* $\left(-\sqrt{2}, \frac{3\pi}{4}\right)$

Solution We use the formulas $x = r\cos\theta$ and $y = r\sin\theta$ to convert (r, θ, z) to (x, y, z):

$$x = r\cos\theta = -2\cos\frac{\pi}{3} = -2\left(\frac{1}{2}\right) = -1$$

$$y = r\sin\theta = -2\sin\frac{\pi}{3} = -2\left(\frac{\sqrt{3}}{2}\right) = -\sqrt{3}.$$

The Cartesian coordinates of the point are $(-1, -\sqrt{3}, 7)$. ■

The z-coordinate does not change and remains the same as given in cylindrical coordinates.

Equations can be rewritten in the same manner.

Example 40 *Rewrite* $f(x, y, z) = 2\sqrt{z} - xy$ *using cylindrical coordinates.*

Solution We replace x by $r\cos\theta$, replace y by $r\sin\theta$, and leave z as is:

$$f(r, \theta, z) = 2\sqrt{z} - r\cos\theta \cdot r\sin\theta$$
$$= 2\sqrt{z} - r^2\cos\theta\sin\theta.$$ ■

4.8.3 Why "cylindrical?"

The reason for the name "cylindrical coordinates" comes from the graph of $r = k$, where k is a constant.

Example 41 *Graph the equation* $r = 3$ *in cylindrical coordinates.*

Solution First let's recall the graph of $r = 3$ in polar coordinates, which is a circle of radius 3 (Figure 4.66).

The z-coordinate is not present in the equation, meaning that whatever z may be, we still have $r = 3$; the slice $z = 2$ has equation $r = 3$, the slice $z = 14.7$ has

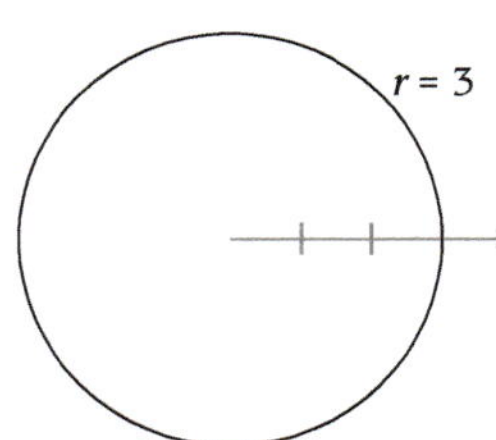

Figure 4.66 *The graph of* $r = 3$ *in polar coordinates*

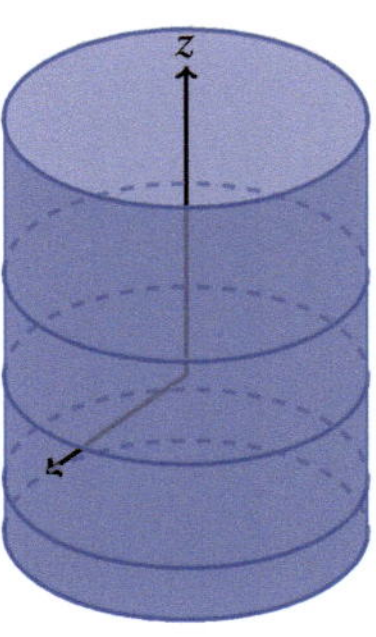

Figure 4.67 *The graph of r = 3 in cylindrical coordinates is a cylinder of radius 3*

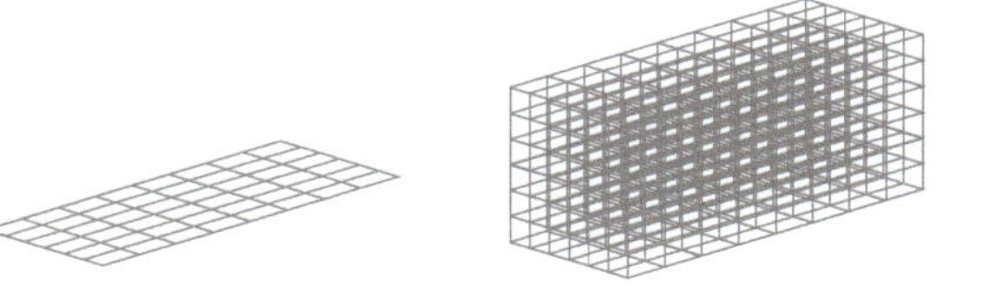

Figure 4.68 *(Left) subdividing a rectangle; (center) adding a dimension to subdivide a rectangular box; (right) one of the rectangular box pieces that are the components of the center rectangular box*

The graph of the equation $r = k$ is a cylinder of radius $|k|$. The central axis of the cylinder is the z-axis.

equation $r = 3$, and so forth. Every horizontal cross-section is $r = 3$. The result is a cylinder (Figure 4.67), but in contrast to the figure the cylinder extends infinitely upward and downward. ∎

4.8.4   Triple integrals in cylindrical coordinates

For a double integral over a rectangular region, we subdivide the rectangle into rectangular pieces of area $\Delta x\,\Delta y$ (Figure 4.68, left). For a triple integral over a rectangular box region, we subdivide into rectangular box pieces (Figure 4.68, center), each of which is a rectangular box (Figure 4.68, right) of base area $\Delta x\,\Delta y$ and height Δz, resulting in a volume of $\Delta x\,\Delta y\,\Delta z$.

Instead of beginning with a double integral using rectangular coordinates, let's use the same process beginning with a double integral using polar coordinates, for which we subdivide into polar regions of area $r\,\Delta r\,\Delta\theta$ (Figure 4.69, left). To use cylindrical coordinates, extend these polar regions in the z-direction (Figure 4.69 center), and each curved-box subregion (Figure 4.69, right) has base area $r\,\Delta r\,\Delta\theta$ and height Δz, resulting in a volume of $r\,\Delta r\,\Delta\theta\,\Delta z$.

Then, instead of the triple integral $\iiint_R f(x, y, z)\,dx\,dy\,dz$, we have the triple integral $\iiint_R f(r, \theta, z) \cdot r\,dz\,dr\,d\theta$.

Ans. to reading exercise 16:
$\left(3, \frac{\pi}{2}, 5\right)$ Hint:

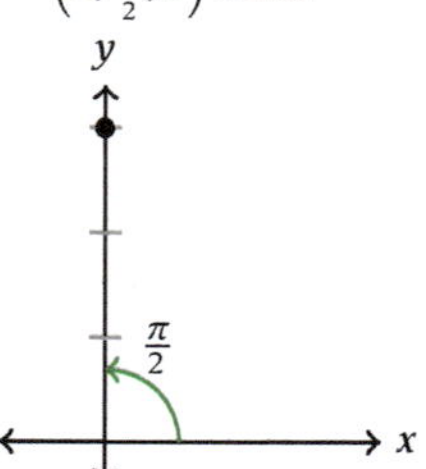

Don't forget the extra r!

It is often convenient to integrate with respect to z first when using cylindrical coordinates.

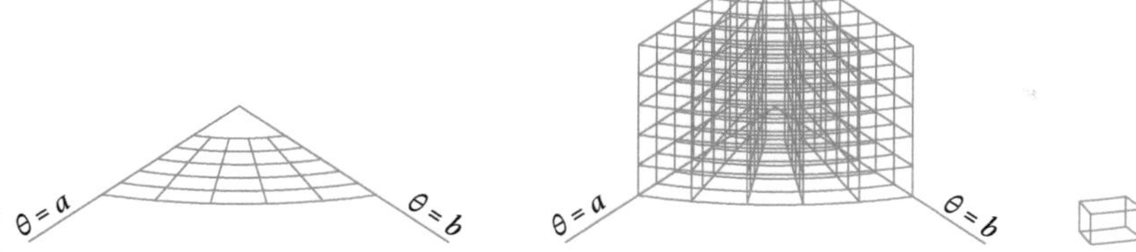

Figure 4.69 *(Left) subdividing a polar region; (center) adding a dimension to subdivide a curved box; (right) one of the curved box pieces that are the components of the center curved box*

TRIPLE INTEGRALS IN CYLINDRICAL COORDINATES

To integrate a function $f(r, \theta, z)$ in cylindrical coordinates over a region $c \leq r \leq d, a \leq \theta \leq b, g \leq z \leq h$, use

$$\int_a^b \int_c^d \int_g^h f(r, \theta, z) \cdot r \, dz \, dr \, d\theta.$$

Don't forget the extra r when using cylindrical coordinates. It can be traced to the "don't forget the extra r" from polar coordinates.

Fubini's theorem applies in the usual circumstances, but it is often convenient to integrate with respect to z first.

4.8.5 Triple integral in cylindrical coordinates example

Example 42 *Find the four-dimensional hypervolume under the three-dimensional hypersurface $f(x, y, z) = 2x + z$ on the region $x^2 + y^2 \leq 4, 0 \leq z \leq 5$.*

Solution This appears to be similar to the task of example 32 of Section 4.7, so we begin by following the same steps.

❶ Graph the three-dimensional region. The equation $x^2 + y^2 = 4$ is a circle centered at the origin with radius 2. Graphing these circular cross-sections for $z = 0$ to $z = 5$ results in a cylinder (Figure 4.70).

❷ Set up the inner integral. The region lies between $z = 0$ and $z = 5$. The triple integral, with the inner integral set up, is

$$\int \int \int_0^5 (2x + z) \, dz \, d \quad d \ .$$

❸ Graph the extent of the region in the remaining two dimensions. Imagine moving the drone camera to view the region from above, straight down the z-axis (Figure 4.71, left). Then rotate to the traditional orientation (Figure 4.71, right). From this viewpoint, the region is circular. Circular regions are easy to describe in

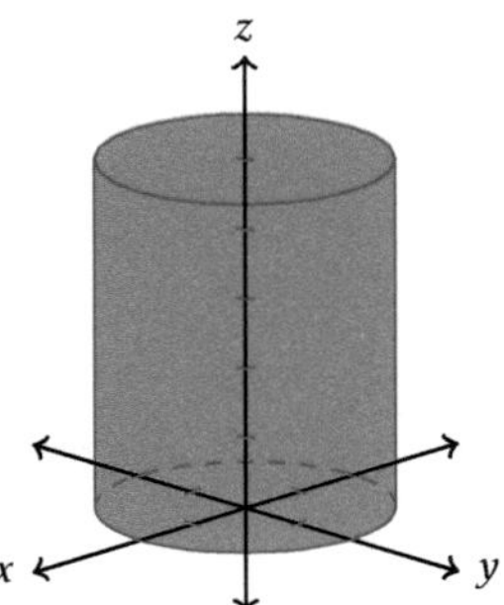

Figure 4.70 *The cylindrical region of integration for example 42*

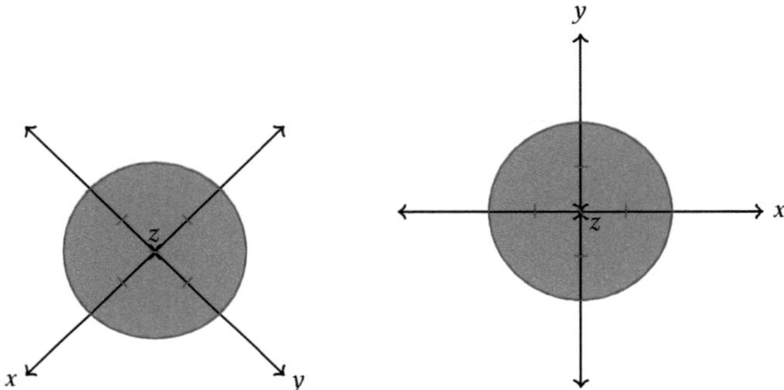

Figure 4.71 *Two views of the cylindrical region of integration for example 42, viewed from above, with the traditional orientation on the right*

polar coordinates, so perhaps we might wish to switch to cylindrical coordinates. The entire region is a cylinder, so perhaps you suspected this was coming from the very beginning!

Let's pause the setup to make the switch to cylindrical coordinates. First, we convert the function from $f(x, y, z)$ (Cartesian coordinates) to $f(r, \theta, z)$ (cylindrical coordinates) in the manner of example 40. To do so, we replace x with $r\cos\theta$ and replace y with $r\sin\theta$, leaving z as is:

$$f(r, \theta, z) = 2r\cos\theta + z.$$

Because the z coordinate does not change, we do not need to change the limits of integration on the inner integral. If the inner integral had been with respect to x or y, we would have needed to start over. And don't forget the extra r!

Next, we rewrite the inner integral using the rewritten function, inserting the extra r required when integrating using cylindrical coordinates:

$$\int \int \int_0^5 (2r\cos\theta + z) \cdot r \, dz \, d \quad d \quad .$$

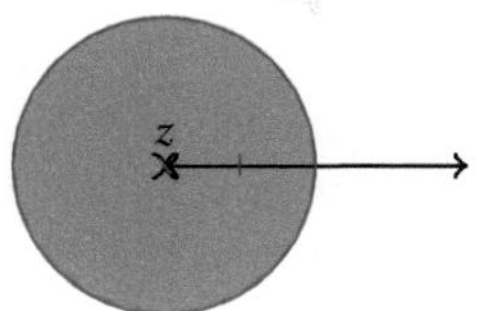

Figure 4.72 *The cylindrical region of integration for example 42, viewed from above, re-marked using cylindrical coordinates*

Now we are ready to resume step ❸. The region of Figure 4.71, using polar coordinates, is described by $0 \le r \le 2, 0 \le \theta \le 2\pi$ (Figure 4.72).

❹ Set up the middle integral. Using $0 \le r \le 2$, we arrive at

$$\int \int_0^2 \int_0^5 (2r\cos\theta + z)r\,dz\,dr\,d \ .$$

Don't forget the extra r!

❺ Set up the outer integral. Using $0 \le \theta \le 2\pi$ (one full rotation), the integral to evaluate is

$$\int_0^{2\pi} \int_0^2 \int_0^5 (2r\cos\theta + z)r\,dz\,dr\,d\theta.$$

Any full rotation may be used, such as $-\pi \le \theta \le \pi$ or $14.3\pi \le \theta \le 16.3\pi$. And don't forget the extra r!

❻ Evaluate the triple integral. Evaluating from the inside out, we have

$$\int_0^{2\pi} \int_0^2 \left(\int_0^5 (2r^2\cos\theta + rz)\,dz \right) dr\,d\theta$$

$$= \int_0^{2\pi} \int_0^2 \left(\left((2r^2\cos\theta)z + r\frac{z^2}{2}\right)\Big|_0^5 \right) dr\,d\theta$$

$$= \int_0^{2\pi} \int_0^2 \left(10r^2\cos\theta + \frac{25}{2}r - 0 \right) dr\,d\theta$$

$$= \int_0^{2\pi} \left(\left(\frac{10}{3}r^3\cos\theta + \frac{25}{4}r^2\right)\Big|_0^2 \right) d\theta$$

$$= \int_0^{2\pi} \left(\frac{80}{3}\cos\theta + 25 - 0 \right) d\theta$$

$$= \left(\frac{80}{3}\sin\theta + 25\theta \right)\Big|_0^{2\pi} = 25(2\pi) - 0 = 50\pi \text{ units}^4.$$

Line 1 simplifies the integrand, multiplying the function's expression by the "extra r"; line 2 finds an antiderivative with respect to z, treating r and θ as constants; line 3 evaluates at $z = 5$, at $z = 0$, and subtracts; line 4 finds an antiderivative with respect to r, treating θ as a constant; line 5 evaluates at $r = 2$, at $r = 0$, and subtracts; and line 6 finds an antiderivative with respect to θ and finishes the integration.

The four-dimensional hypervolume, which uses units^4, is 50π units^4. ∎

If we so desire, we may instead continue step ❸ and beyond with Cartesian coordinates; the answer is the same. So why bother to make the switch to cylindrical coordinates? Because it is often simpler to do so. Cartesian coordinates requires

Any time the horizontal cross sections can be easily described using polar coordinates, switching to cylindrical coordinates is likely to be easier than sticking with Cartesian coordinates.

the evaluation of the triple integral $\int_{-2}^{2} \int_{-\sqrt{4-x^2}}^{\sqrt{4-x^2}} \int_{0}^{5} (2x + z)\, dz\, dy\, dx$, which involves (1) substitution and (2) trig substitution, including use of a half-angle identity, a double-angle identity, and an inverse trig function. The triple integral using cylindrical coordinates is straightforward and utilizes no techniques of integration. Simpler indeed!

4.8.6   Rewriting an integral using cylindrical coordinates

Example 26 of Section 4.4 gives an integral in Cartesian coordinates and requests changing to polar coordinates. The next example is analogous.

Example 43   *Rewrite the triple integral $\int_{-3}^{3} \int_{-\sqrt{9-x^2}}^{\sqrt{9-x^2}} \int_{0}^{9-x^2-y^2} 3y\, dz\, dy\, dx$ using cylindrical coordinates.*

Solution   Because z is the same in both coordinate systems, we may still integrate with respect to z first. But we need to convert the variables x and y to polar.

➊ Rewrite the integrand. We replace x with $r\cos\theta$ and y with $r\sin\theta$, leaving z as is. The function to integrate is

$$f(r,\theta,z) = 3r\sin\theta.$$

Inserting the extra r required for cylindrical coordinates, we begin the new triple integral with

Don't forget the extra r!

$$\int \int \int 3r\sin\theta \cdot r\, dz\, d\,d\,.$$

➋ Rewrite the limits of integration for z using cylindrical coordinates. Because the limits of integration for z include x and y, we must convert to polar coordinates;

Because x and y are treated as constants in the inner integral, the limits of integration for z can include those variables. They must be converted to polar coordinates r and θ when rewriting an integral using cylindrical coordinates.

$$9 - x^2 - y^2 = 9 - (x^2 + y^2) = 9 - r^2.$$

The inner integral can now be completed:

The conversion formula $x^2 + y^2 = r^2$ is used. The conversion formulas $x = r\cos\theta$ and $y = r\sin\theta$ and a Pythagorean identity could also be used to obtain the same result.

$$\int \int \int_{0}^{9-r^2} 3r\sin\theta \cdot r\, dz\, d\,d\,.$$

➌ Graph the region in the xy-plane corresponding to the middle and outer integrals and describe the region using polar coordinates. The equations $y = -\sqrt{9 - x^2}$ and $y = \sqrt{9 - x^2}$ are easily recognizable as the bottom and top halves of a circle of radius 3 centered at the origin (Figure 4.73). The limits of integration for x, which are -3 and 3, match the full circle. In polar coordinates, the region can be described as $0 \leq r \leq 3,\, 0 \leq \theta \leq 2\pi$.

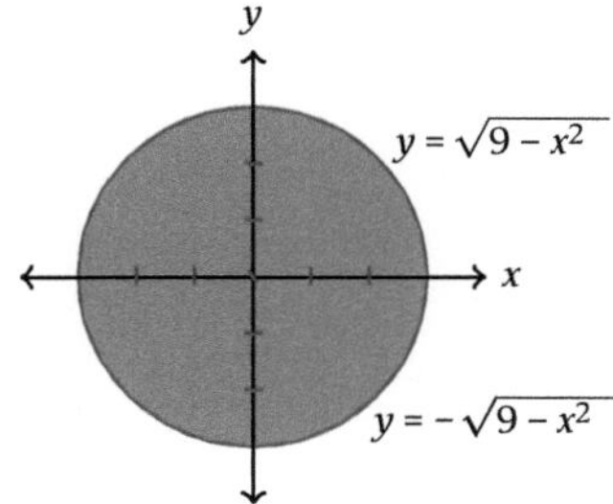

Figure 4.73 *The region in the xy-plane corresponding to the limits of integration for the middle and outer integrals of the original triple integral in Cartesian coordinates*

4 Set up the middle and outer integrals for cylindrical coordinates. Using the results of step **3**, the triple integral is

$$\int_0^{2\pi} \int_0^3 \int_0^{9-r^2} 3r \sin\theta \cdot r \, dz \, dr \, d\theta$$

Don't forget the extra r!

$$= \int_0^{2\pi} \int_0^3 \int_0^{9-r^2} 3r^2 \sin\theta \, dz \, dr \, d\theta.$$

The task of rewriting the triple integral is complete. ∎

Reading Exercise 17 Use cylindrical coordinates to set up a triple integral for integrating $f(x, y, z) = xy$ on the region bounded by $z = 0$ and $z = x^2 + y^2 - 4$. Do not evaluate the integral.

EXERCISES 4.8

1–8. Plot the point with the given cylindrical coordinates.

1. $(2, 0, 3)$
2. $(-3, 0, -1)$
3. $(2, \pi, -3)$
4. $\left(3, \dfrac{\pi}{2}, 2\right)$
5. $\left(1, \dfrac{3\pi}{4}, 0\right)$
6. $\left(-3, -\dfrac{\pi}{2}, 4\right)$
7. $\left(0, -\dfrac{\pi}{2}, 2\right)$
8. $\left(3, -\dfrac{\pi}{4}, 0\right)$

9–16. Find cylindrical coordinates for a point with the given Cartesian coordinates.

9. $(3, 3, 1)$
10. $(0, -3, 41)$
11. $(-2, 0, 55)$
12. $(-2, -2, -2)$
13. $(\sqrt{3}, 1, \pi)$
14. $(1, -\sqrt{3}, 17)$
15. $(3, 4, 5)$
16. (π, π, π)

17-24. A point is given in cylindrical coordinates. Find Cartesian coordinates for the point.

17. $\left(-4, \frac{\pi}{2}, 11.4\right)$
18. $\left(2, \frac{3\pi}{4}, -7\right)$
19. $\left(2, \frac{7\pi}{4}, 18\right)$
20. $\left(-7, -\frac{\pi}{3}, 9\right)$
21. $(\pi, 0, 3)$
22. $(0, \pi, 3)$
23. $\left(5, \frac{11\pi}{3}, 1\right)$
24. $(\pi, 2\pi, 3\pi)$

25-30. Rewrite the function using cylindrical coordinates.

25. $f(x, y, z) = \frac{xz}{y}$
26. $f(x, y, z) = 5x - zy^2$
27. $f(x, y, z) = (x^2 + y^2)^3$
28. $f(x, y, z) = 4e^z - z^2$
29. $f(x, y, z) = 1 + \sin y$
30. $f(x, y, z) = \frac{xy}{x^2 + y^2}$

31. Assuming cylindrical coordinates, match each equation to a description.

(i)   a horizontal plane
(ii) a vertical plane containing the origin
(iii) a vertical plane not containing the origin
(iv) a cylinder of radius 1
(v)   a sphere of radius 1

(a) $\theta = \frac{\pi}{4}$
(b) $r = -1$
(c) $z = 1$
(d) $r \sin \theta = 1$

32. Assuming cylindrical coordinates, match each equation to a description.

(i)   a plane

(ii)   an ellipsoid
(iii) a sphere
(iv) a cone
(v)   a spiral-shaped vertical wall

(a) $z = r$
(b) $z = r^2$
(c) $r = \theta$
(d) $z = 0$

33–36. Set up a triple integral in cylindrical coordinates. Do not evaluate the integral.

33. $\iiint_R (x + y)\, dV$ where R is the region bounded by $x^2 + y^2 = 36$, $z = 0$, and $z = 2$

34. $\iiint_R \ln(z + 1)\, dV$ where R is the region bounded by $z = 0$ and $z = 4 - x^2 - y^2$

35. $\iiint_R \dfrac{xy}{z}\, dV$ where R is the region $1 \le x^2 + y^2 \le 4,\, 1 \le z \le 3$

36. $\iiint_R (x^2 + y^2 + z^2)\, dV$ where R is the region $x^2 + y^2 + z^2 \le 9$

37–42. Find the (net) four-dimensional hypervolume under the three-dimensional hypersurface on the given region.

37. hypersurface $f(x, y, z) = 2z$, region $1 \le x^2 + y^2 \le 4,\, 0 \le z \le 1$

38. hypersurface $f(x, y, z) = 2 + x$, region inside the cylinder $x^2 + y^2 = 4$ between the planes $z = 0$ and $z = 3$

39. hypersurface $f(x, y, z) = z^2 e^{x^2 + y^2}$, region inside the cylinder $x^2 + y^2 = 25$ between the planes $z = 0$ and $z = 3$

40. hypersurface $f(x, y, z) = y + 5$, region bounded by $z = 4$ and $z = x^2 + y^2$

41. hypersurface $f(x, y, z) = 2x$, region in the first octant with $0 \le x^2 + y^2 \le 1$ and $z \le y$

42. hypersurface $f(x, y, z) = z\sqrt{x^2 + y^2 + 1}$, region $-3 \le z \le 4$, $x^2 + y^2 \le 9$

43–46. Rewrite the triple integral using cylindrical coordinates.

43. $\int_{-3}^{3} \int_{0}^{\sqrt{9-x^2}} \int_{0}^{x^2+y^2} xz\sqrt{y}\, dz\, dy\, dx$

44. $\int_{0}^{3} \int_{-\sqrt{9-x^2}}^{\sqrt{9-x^2}} \int_{0}^{x} (x^2 y + z)\, dz\, dy\, dx$

45. $\int_{-3}^{3} \int_{0}^{5} \int_{-\sqrt{9-x^2}}^{\sqrt{9-x^2}} 7\, dy\, dz\, dx$

46. $\int_{0}^{3} \int_{0}^{\sqrt{9-x^2}} \int_{-3y}^{0} xz\sqrt{y}\, dz\, dy\, dx$

47. Find the center of mass of the semi-cylindrical shell occupying the region $0 \le z \le 10,\, 4 \le x^2 + y^2 \le 9,\, y \ge 0$ with density function $\rho(x, y, z) = x^2 + y^2 + z^2$.

48. A koi pond is in the shape of the paraboloid bounded by $z = 0$ and $z = x^2 + y^2 - 9$. It is a cold day, and the pond is warmer at greater depth; in degrees Celsius, the temperature is given by

Ans. to reading exercise 17:

$$\int_{0}^{2\pi} \int_{0}^{2} \int_{r^2-4}^{0} r\cos\theta \cdot r\sin\theta \cdot r\, dz\, dr\, d\theta$$

$$= \int_{0}^{2\pi} \int_{0}^{2} \int_{r^2-4}^{0} r^3 \cos\theta \sin\theta\, dz\, dr\, d\theta$$

Hints: the region lies below the xy-plane; don't forget the extra r!

$T(x, y, z) = 10 - z$. Find the average temperature of the pond as follows.

(a) Find the volume of the pond by determining $\iiint_R 1 \, dV$.

(b) Calculate $T_{\text{avg}} = \dfrac{1}{\text{volume}} \iiint_R T(x, y, z) \, dV$.

When converting to cylindrical coordinates, don't forget the extra r!

4.9   Triple Integrals in Spherical Coordinates

Three dimensions offer quite a lot of flexibility regarding how to describe a location. Cartesian coordinates and cylindrical coordinates are just two of the options. Another natural coordinate system is *spherical coordinates*.

4.9.1   Spherical coordinates

Imagine standing in the middle of the playing field in a sports stadium. Suppose you wish to point to something–perhaps a plane in the sky, a ball on the field, or the scoreboard. How might you point toward the object? First, you turn your body to face the object. Describing the direction your body is turned is a familiar task, the θ of polar and cylindrical coordinates. Next, you raise or lower your arm, pivoting from the shoulder to point toward the object. This can also be described by an angle! For spherical coordinates, we choose to describe the angle of pointing directly upward as 0 and pointing directly downward as π (measured in radians; in degrees it would be $180°$) . That would make pointing "level" (horizontally) $\frac{\pi}{2}$. We shall denote this angle by φ, the lowercase Greek letter "phi." With the shoulder as the origin, the two angles θ and φ describe the direction the arm is pointing. Then, if we wish to describe the location of the plane, ball, or point on the scoreboard, we only lack the distance from the origin, which we will describe using the lowercase Greek letter ρ (pronounced like "row" as in "row your boat"). The quantities ρ, θ, and φ are the first, second, and third coordinates of the spherical coordinate system (Figure 4.74).

There are two common variants of lowercase phi (pronounced like the word "fee"), ϕ and φ. I choose to use φ because it is more easily distinguished from θ when writing by hand.

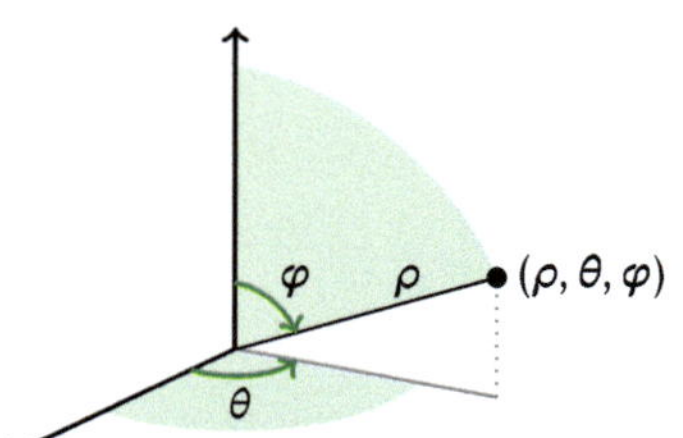

Figure 4.74 *The spherical coordinate system, viewed from the first octant. The segment from the origin to the point, its shadow in the xy-plane, and the rotation of the angle φ all lie in a vertical plane. The shadow and the rotation of the angle θ lie in the (horizontal) xy-plane*

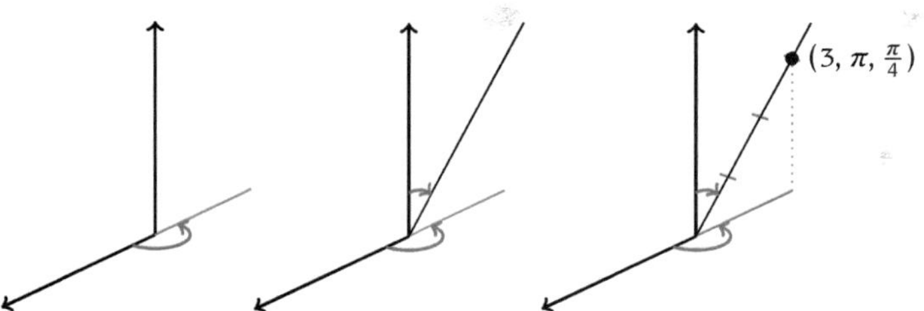

Figure 4.75 *Three stages in drawing the point with spherical coordinates* $\left(3, \pi, \frac{\pi}{4}\right)$. *The solution to example 44 is in the diagram on the right*

Example 44 *Plot the point with spherical coordinates* $\left(3, \pi, \frac{\pi}{4}\right)$.

Solution   We begin by identifying the coordinates; $\rho = 3$, $\theta = \pi$, and $\varphi = \frac{\pi}{4}$. But the order we write the coordinates and the order we process the coordinates when plotting a point are not the same!

First comes the angle θ, which is the same as in cylindrical coordinates. Rotate π radians from the initial ray (Figure 4.75, left). We can think of the terminal side of the angle θ as a shadow, because it lies directly below the ray from the origin to the point (refer to Figure 4.74). The remainder of the diagram is drawn in the vertical plane containing the vertical axis and the shadow.

Next is the angle φ. From the ray pointing vertically upward (the positive z-axis), rotate toward the shadow. In this case, $\varphi = \frac{\pi}{4}$. The result is a ray emanating from the origin that is halfway between vertical and horizontal (halfway between 0 radians and $\frac{\pi}{2}$ radians). See Figure 4.75, center.

Lastly, we use the distance ρ. From the origin along the ray, mark a distance of three units and place the point there (Figure 4.75, right).   ■

In example 44, because the distance from the origin to the point is 3 units, the point lies on a sphere of radius 3 centered at the origin. That is always true of the distance ρ and is one explanation of the name "spherical coordinates."

Example 45 *Plot the point with spherical coordinates* $\left(2, 0, \frac{\pi}{2}\right)$.

Solution   The coordinates are $\rho = 2$, $\theta = 0$, and $\varphi = \frac{\pi}{2}$. Because $\theta = 0$, the shadow is along the initial ray (the positive x-axis). With $\varphi = \frac{\pi}{2}$, the ray emanating from the origin toward the point is horizontal, so it also lies along the positive x-axis. Two units from the origin in that direction is the location of the point (Figure 4.76).   ■

Reading Exercise 18   Plot the point with spherical coordinates $\left(2, \frac{\pi}{2}, \frac{\pi}{4}\right)$.

Example 46 *Plot the point with spherical coordinates* $\left(4, \frac{7\pi}{13}, 0\right)$.

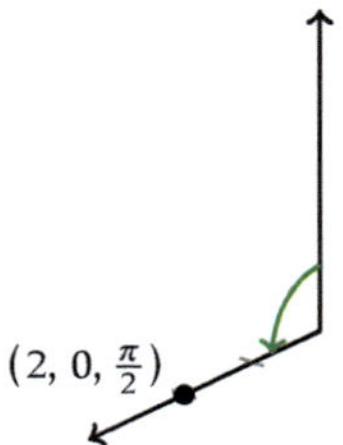

Figure 4.76 *The solution to example 45, the point with spherical coordinates* $\left(2, 0, \frac{\pi}{2}\right)$. *The initial ray, shadow, and ray from the origin to the point are identical. Compare to Figure 4.75.*

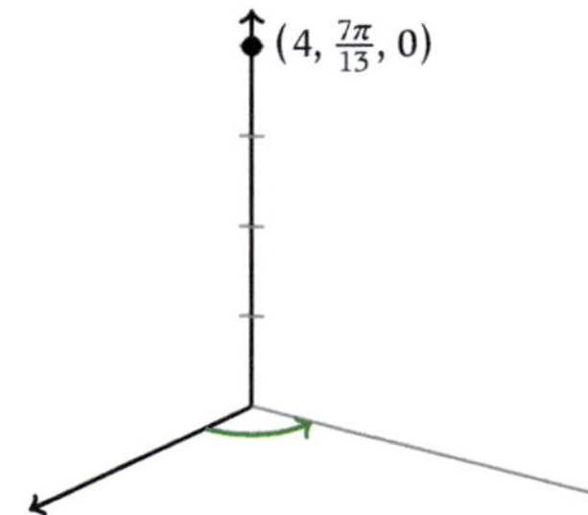

Figure 4.77 *The solution to example 46, the point with spherical coordinates* $\left(4, \frac{7\pi}{13}, 0\right)$

Solution The coordinates are $\rho = 4, \theta = \frac{7\pi}{13}$, and $\varphi = 0$. We rotate $\frac{7\pi}{13}$ radians from the initial ray (to the shadow), rotate 0 radians from the vertical axis toward the shadow to obtain the ray from the origin to the point, and mark four units along that ray to place the point. The result is in Figure 4.77. ∎

Whenever $\varphi = 0$, the value of θ is irrelevant and the point lies on the vertical axis.

Examples 44–46 and Figures 4.74–4.77 all have positive values of ρ. Negative values of ρ are handled in the same manner as negative values of r in polar coordinates, but they are not used in our context because triple integrals in spherical coordinates will require nonnegative values of ρ.

4.9.2 Converting between spherical and Cartesian coordinates

Figure 4.78 shows the location of a point (ρ, θ, φ) with angles and segment lengths marked using both cylindrical and spherical coordinate variables. Note that the triangle with segment lengths r, ρ, and z, as well as the rotation of the angle φ, all

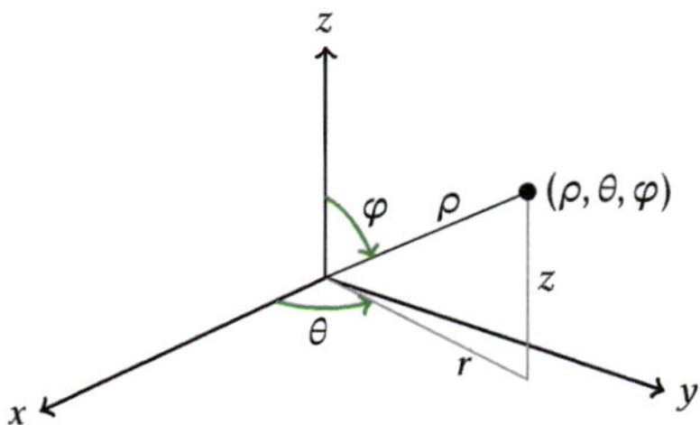

Figure 4.78 *A point $(\rho, \theta, \varphi) = (r, \theta, z)$ in three dimensions*

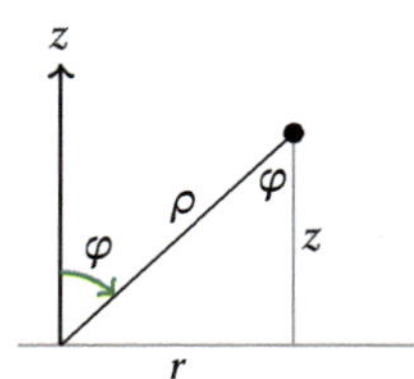

Figure 4.79 *A view of the plane containing the triangle from Figure 4.78*

occur in the same vertical plane, which is redrawn in Figure 4.79, with one addition. Because the vertical axis and the vertical segment of length z are parallel and the segment with length ρ is a transversal, we can use alternate interior angles to conclude that the angle in the triangle at our point has the same measure as the angle φ.

We already know how to convert between Cartesian and cylindrical coordinates, using the polar conversions $x = r \cos \theta$ and $y = r \sin \theta$. Spherical coordinates do not use the quantity r, so our next task is to determine how to rewrite r using spherical coordinate variables. Using the triangle of Figure 4.79,

$$\sin \varphi = \frac{\text{opposite}}{\text{hypotenuse}} = \frac{r}{\rho},$$

which can be rearranged to

$$r = \rho \sin \varphi.$$

Making the substitution of $\rho \sin \varphi$ for r,

$$x = r \cos \theta = \rho \sin \varphi \cos \theta$$
$$y = r \sin \theta = \rho \sin \varphi \sin \theta.$$

Using the triangle again,

$$\cos \varphi = \frac{\text{adjacent}}{\text{hypotenuse}} = \frac{z}{\rho},$$

More precisely, it's the line containing the segment of length z that is parallel to the vertical axis. A similar technical correction applies to the transversal.

It can be shown that these coordinate systems satisfy the axioms of Euclidean geometry, and therefore the converse of the alternate interior angles theorem holds true.

Hint: Which side of the triangle is opposite the angle labeled φ?

Ans. to reading exercise 18:

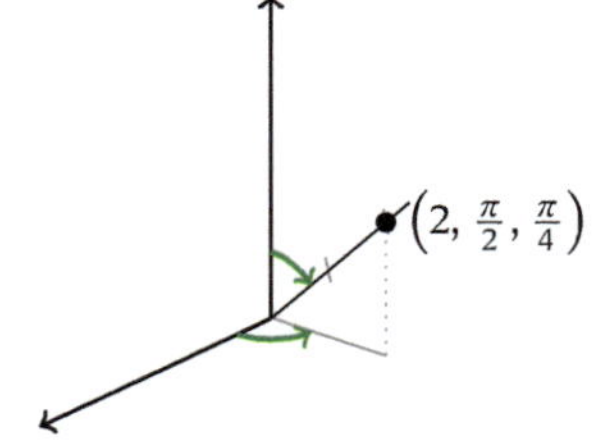

which can be rearranged to

$$z = \rho \cos \varphi.$$

From the Pythagorean theorem applied to the triangle, $r^2 + z^2 = \rho^2$, and applying $r^2 = x^2 + y^2$ yields

$$x^2 + y^2 + z^2 = \rho^2.$$

CARTESIAN–SPHERICAL CONVERSION FORMULAS

$$x = \rho \sin \varphi \cos \theta$$
$$y = \rho \sin \varphi \sin \theta$$
$$z = \rho \cos \varphi \qquad x^2 + y^2 + z^2 = \rho^2$$

Example 47 *Convert the spherical coordinates $\left(3, \pi, \frac{\pi}{4}\right)$ to Cartesian coordinates.*

Solution   We need the x-, y-, and z-coordinates, from $\rho = 3$, $\theta = \pi$, and $\varphi = \frac{\pi}{4}$. Using the conversion formulas,

$$x = 3 \sin \frac{\pi}{4} \cos \pi = 3\left(\frac{\sqrt{2}}{2}\right)(-1) = \frac{-3\sqrt{2}}{2}$$

$$y = 3 \sin \frac{\pi}{4} \sin \pi = 3\left(\frac{\sqrt{2}}{2}\right)(0) = 0$$

$$z = 3 \cos \frac{\pi}{4} = \frac{3\sqrt{2}}{2}.$$

In Cartesian coordinates, the point is $\left(\frac{-3\sqrt{2}}{2}, 0, \frac{3\sqrt{2}}{2}\right)$. ∎

Example 47 illustrates that just as with polar-to-Cartesian coordinates, the conversion to Cartesian is very straightforward. The other direction can involve a little more thought.

Example 48 *Convert the Cartesian coordinates $\left(0, \sqrt{3}, 1\right)$ to spherical coordinates.*

When rewriting in spherical coordinates, there are infinitely many different possible solutions, just as with polar and cylindrical coordinates. We only need to determine one solution, so using a positive value of ρ and ignoring the negative value is sufficient.

Solution   We need the ρ-, θ-, and φ-coordinates, from $x = 0$, $y = \sqrt{3}$, and $z = 1$. Using $x^2 + y^2 + z^2 = \rho^2$,

$$\rho^2 = 0^2 + (\sqrt{3})^2 + 1^2 = 4$$

$$\rho = 2.$$

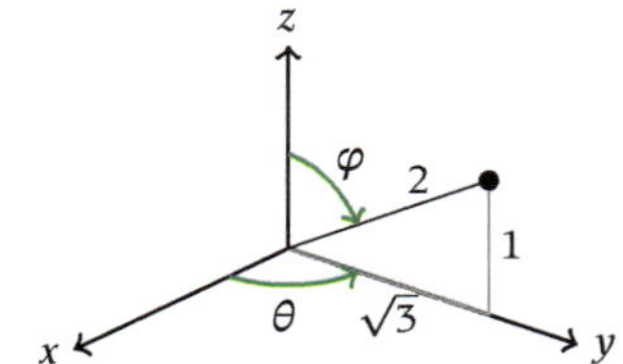

Figure 4.80 *The point of example 48. Compare to Figure 4.78.*

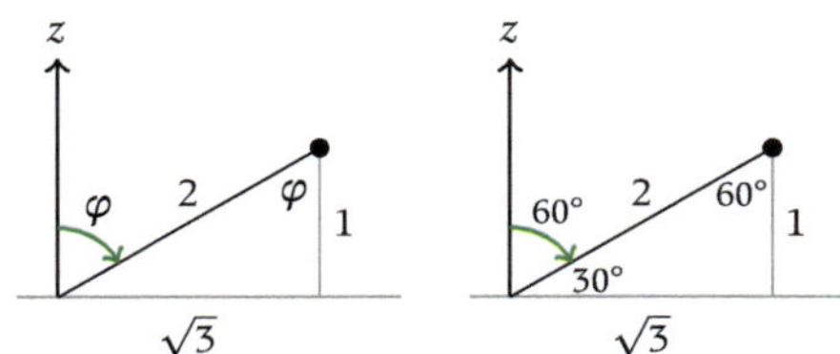

Figure 4.81 *Determining φ. Compare to Figure 4.79.*

Seeing no formulas in our list that give θ and φ directly from x, y, and z, the strategy is to plot the Cartesian point (Figure 4.80) and try to determine θ and φ from the diagram. We add all relevant information, such as the values of x, y, z, ρ, and r. In this case, $x = 0$ and we omit it, and r is the same as y. We also indicate the angles θ and φ. It is clear from Figure 4.80 that the value of θ is $\frac{\pi}{2}$.

We still need to determine φ. Sometimes drawing only the plane containing the triangle can help, in the style of Figure 4.79. The result is Figure 4.81, left. Recognizing 1, $\sqrt{3}$, and 2 as side lengths of a $30°$–$60°$–$90°$ triangle, we determine that $\varphi = 60° = \frac{\pi}{3}$ radians (Figure 4.81, right).

In spherical coordinates, the point is $\left(2, \frac{\pi}{2}, \frac{\pi}{3}\right)$. ∎

Reading Exercise 19 Convert the spherical coordinates $\left(2, \frac{\pi}{2}, \frac{\pi}{4}\right)$ to Cartesian coordinates.

Rewriting a function from Cartesian variables to spherical variables is straightforward.

Example 49 *Rewrite $f(x, y, z) = x^2 + y^2 + z^2 + 2xz + 4$ in spherical coordinates.*

Solution Using the conversion formulas for x, y, and z, and in this case for $x^2 + y^2 + z^2$ as well, we have

$$f(x, y, z) = x^2 + y^2 + z^2 + 2xz + 4$$

$$f(\rho, \theta, \varphi) = \rho^2 + 2 \cdot \rho \sin \varphi \cos \theta \cdot \rho \cos \varphi + 4$$

$$= \rho^2 + 2\rho^2 \sin \varphi \cos \varphi \cos \theta + 4.$$
∎

Recall that $r^2 = x^2 + y^2 = 0^2 + (\sqrt{3})^2 = 3$, so $r = \sqrt{3}$.

When both x and y are nonzero, a diagram similar to Figure 4.81, but for the xy-plane containing x, y, r, and θ, can be helpful.

Another triangle that is helpful to recall is a $45°$–$45°$–$90°$ triangle, which is isosceles. If the leg lengths are equal in a right-angled triangle, the other angles are $\frac{\pi}{4}$ radians each.

Using only the formulas for x, y, and z is sufficient. But when applicable, the formula for $x^2 + y^2 + z^2$ achieves significant simplification rather quickly and is therefore quite convenient.

The next example is even quicker.

Example 50 *Rewrite the equation $x^2 + y^2 + z^2 = 9$ using spherical coordinates.*

Solution Use the conversion formula for $x^2 + y^2 + z^2$:

$$x^2 + y^2 + z^2 = 9$$
$$\rho^2 = 9.$$

∎

The equation in example 50 is the equation of a sphere centered at the origin of radius 3. So is $\rho^2 = 9$, as well as $\rho = 3$. This gives rise to another explanation for the name of this coordinate system. For cylindrical coordinates, $r = k$ is a cylinder; for spherical coordinates, $\rho = k$ is a sphere.

The equation $\rho = -3$ describes the same sphere, but we choose to use nonnegative values of ρ as explained earlier.

Ans. to reading exercise 19:
$$\left(0, \sqrt{2}, \sqrt{2}\right)$$

Reading Exercise 20 Describe the surface $\rho = 4$.

Example 51 *Rewrite the equation $x^2 + y^2 = \dfrac{z^2}{4}$, $z \geq 0$, using spherical coordinates, and simplify.*

Solution Using the conversion formulas for x, y, and z, we have

$$x^2 + y^2 = \frac{1}{4}z^2$$
$$\rho^2 \sin^2 \varphi \cos^2 \theta + \rho^2 \sin^2 \varphi \sin^2 \theta = \frac{1}{4}\rho^2 \cos^2 \varphi$$
$$\sin^2 \varphi \cos^2 \theta + \sin^2 \varphi \sin^2 \theta = \frac{1}{4} \cos^2 \varphi$$
$$\sin^2 \varphi(\cos^2 \theta + \sin^2 \theta) = \frac{1}{4} \cos^2 \varphi$$
$$\sin^2 \varphi = \frac{1}{4} \cos^2 \varphi$$
$$\sin \varphi = \frac{1}{2} \cos \varphi$$
$$\tan \varphi = \frac{1}{2}$$
$$\varphi = \tan^{-1} \frac{1}{2}.$$

∎

Line 2 replaces x, y, and z using their conversion formulas; line 3 divides both sides by ρ^2; line 4 factors out $\sin^2 \varphi$ on the left-hand side; line 5 uses the Pythagorean identity $\cos^2 \theta + \sin^2 \theta = 1$; line 6 takes the square root of both sides, noting that $\sin \varphi \geq 0$ because $0 \leq \varphi \leq \pi$, and $\cos \varphi \geq 0$ because $z \geq 0$, meaning $0 \leq \varphi \leq \frac{\pi}{2}$; line 7 divides both sides by $\cos \varphi$; and line 8 uses the definition of inverse tangent.

The equation in example 51, $x^2 + y^2 = \dfrac{z^2}{4}$, is the equation of a cone (double-napped, vertex at the origin, axis vertical); the restriction $z \geq 0$ yields the top half (one nappe) of the cone. Any equation of the form $\varphi = k$, where $0 < k < \pi$, is the equation of a cone.

An alternate explanation that $\varphi = k$ is a cone can be had from Figure 4.78; fix the angle φ and allow θ to rotate and ρ to vary.

4.9.3   Triple integrals in spherical coordinates

Before conquering spherical coordinates, let's revisit how polar coordinates work. Figure 4.82, which is repeated from Section 4.4, shows a polar subregion, which is curved. But if Δr and $\Delta\theta$ are infinitesimal, how curved is it? At an infinitesimal scale, local linearity takes over; a curve and its tangent line are indistinguishable. Arcs of a circle become straight line segments. From geometry we know that a tangent line to a circle is perpendicular to its radius. The difference in angle between the radii at either end of the subregion is infinitesimal, so they are approximately parallel. The subregion is indistinguishable from a rectangle! With area $r\,\Delta r\,\Delta\theta$ and one of the side lengths Δr, the other side length must be $r\,\Delta\theta$ (Figure 4.83).

Now consider a subregion in spherical coordinates, with increments $\Delta\rho$, $\Delta\theta$, and $\Delta\varphi$ (Figure 4.84). If each increment is infinitesimal, then local linearity takes over and the subregion is indistinguishable from a rectangular box. If we know the side lengths, then we can multiply them to determine the volume. The quantity $\Delta\rho$ is a length, the change in the distance ρ. But $\Delta\theta$ and $\Delta\varphi$ are changes in angles, not lengths.

To determine the lengths of the sides of an infinitesimally-sized spherical subregion, let's begin with the length marked $\Delta\rho$ in Figure 4.84, which is the distance between the "bottom" and "top" of the box. The bottom of the box lies on the sphere of radius $\rho = \rho_n$ centered at the origin (blue in Figure 4.85), and the top of the box

Imagine $\Delta\theta$ shrinking in Figure 4.82. The smaller the angle $\Delta\theta$, the less curved the region.

An alternate argument is that the length of the arc is proportional to the circumference of the circle; the length is

$$2\pi r \cdot \frac{\Delta\theta}{2\pi} = r\Delta\theta.$$

(This is the usual argument for the familiar arc length formula.)

In what follows, recall that for a constant k, $\rho = k$ is a sphere centered at the origin with radius k, $\theta = k$ is a vertical plane containing the vertical axis, and $\varphi = k$ is a cone with vertical axis and vertex at the origin.

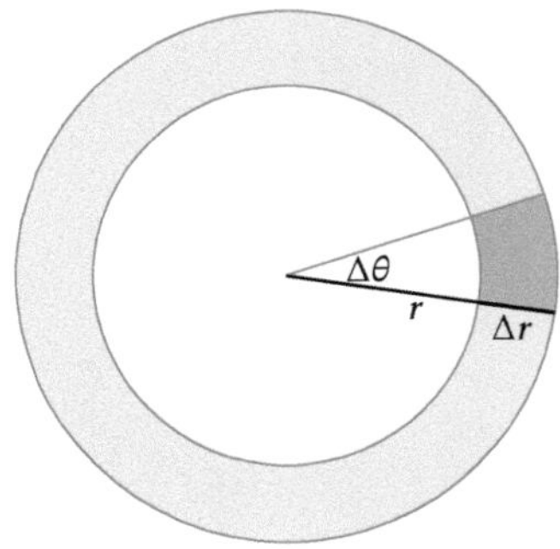

Figure 4.82 *A polar subregion (darker gray), shown with non-infinitesimal Δr and $\Delta\theta$*

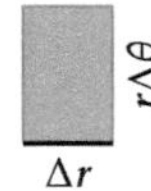

Figure 4.83 *A polar subregion on an infinitesimal scale. Compare to the subregion in Figure 4.82*

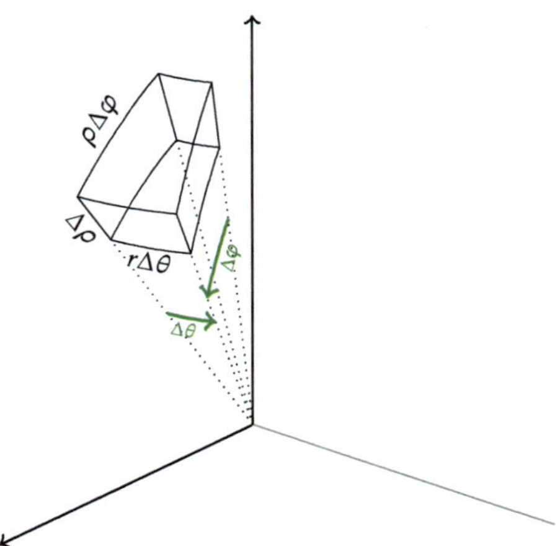

Figure 4.84 *A spherical subregion. Non-infinitesimal $\Delta\rho$, $\Delta\theta$, and $\Delta\varphi$ are shown for the sake of visualization*

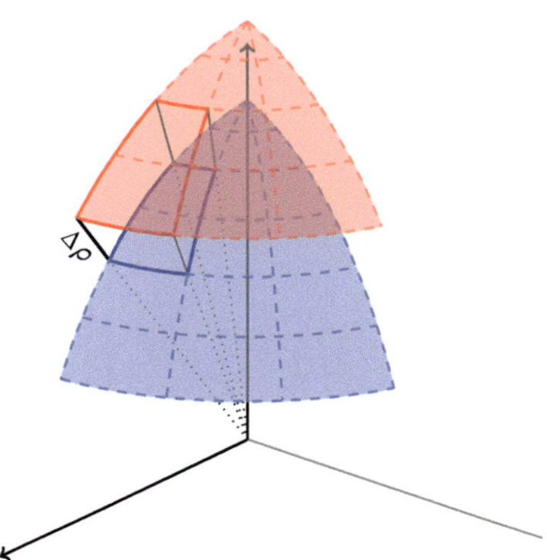

Figure 4.85 *The spherical subregion of Figure 4.84, along with portions of the spheres containing the "bottom" and "top" of the subregion*

Each interval $a \leq \rho \leq b$, $c \leq \theta \leq d$, $g \leq \varphi \leq h$ is subdivided into Ω pieces, with $\Delta\rho = (b-a)\omega$, $\Delta\theta = (d-c)\omega$, $\Delta\varphi = (h-g)\omega$, $\rho_n = a + n\Delta\rho$, $\theta_m = c + m\Delta\theta$, and $\varphi_k = g + k\Delta\varphi$.

lies on the sphere with radius slightly larger, $\Delta\rho$ larger (red in Figure 4.85). The length of a segment between these spheres is $\Delta\rho$.

By contrast, consider the "left" and "right" faces of the subregion. The left side of the box lies in the vertical plane through the origin $\theta = \theta_m$ (blue in Figure 4.86), and the right side of the box lies in a vertical plane where θ is slightly larger, $\Delta\theta$ larger (red in Figure 4.86). The length of a segment (an arc, but indistinguishable from a line segment when the region is infinitesimal) between these planes depends on

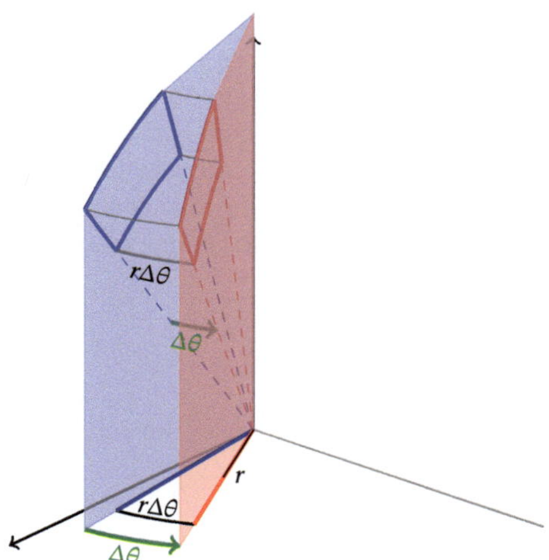

Figure 4.86 *The spherical subregion of Figure 4.84, along with portions of the vertical planes containing the "left" and "right" of the subregion*

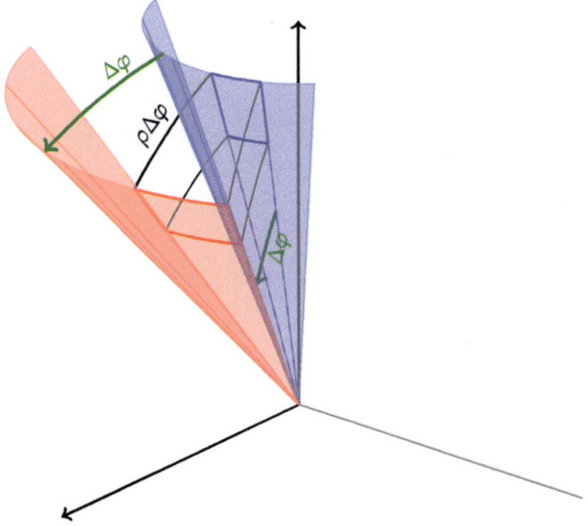

Figure 4.87 *The spherical subregion of Figure 4.84, along with portions of the cones containing the "back" and "front" of the subregion*

the distance from the vertical axis, which is the cylindrical coordinate variable r. As in Figure 4.83, this length is $r\,\Delta\theta$.

Finally, consider the "back" and "front" faces of the subregion. The back side of the box lies in the cone $\varphi = \varphi_k$, whose axis is vertical and whose vertex is at the origin (blue in Figure 4.87). The front side of the box lies in a cone where φ is slightly

Ans. to reading exercise 20:
   a sphere with radius 4 centered at the origin

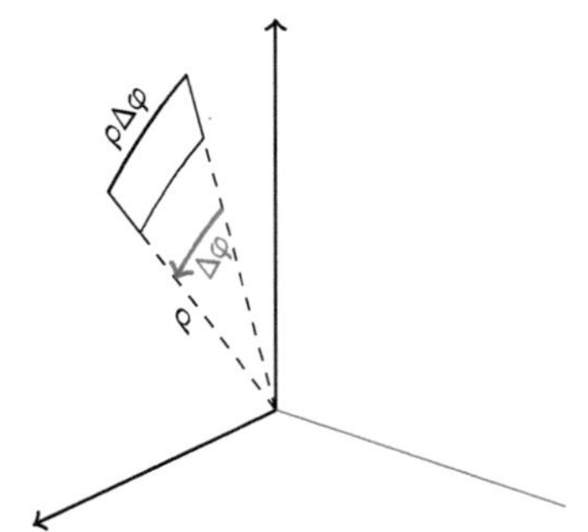

Figure 4.88 *A portion of Figure 4.87, extracted. Compare to Figures 4.82 and 4.83.*

Don't forget the extra $\rho^2 \sin \varphi$ when using spherical coordinates!

larger, $\Delta\varphi$ larger (red in Figure 4.87). The length of a segment (an arc, but indistinguishable from a line segment when the region is infinitesimal) between these cones depends on the distance ρ from the origin; see the relevant part of Figure 4.87 extracted as Figure 4.88. The geometry is identical to that of Figures 4.82 and 4.83, just in a different plane. The length of the segment is therefore $\rho\,\Delta\varphi$.

Referring once again to Figure 4.84, having now determined the lengths of the sides of the subregion, we conclude that the volume of an infinitesimal spherical subregion is

$$\Delta\rho \cdot r\,\Delta\theta \cdot \rho\,\Delta\varphi.$$

Replacing the cylindrical variable r with $\rho \sin \varphi$, the volume using only spherical variables is

$$\Delta\rho \cdot \rho \sin \varphi\,\Delta\theta \cdot \rho\,\Delta\varphi$$
$$= \rho^2 \sin \varphi\,\Delta\rho\,\Delta\theta\,\Delta\varphi.$$

TRIPLE INTEGRALS IN SPHERICAL COORDINATES

To integrate a function $f(\rho, \theta, \varphi)$ in spherical coordinates over a region $a \le \rho \le b, c \le \theta \le d, g \le \varphi \le h$, use

$$\int_g^h \int_c^d \int_a^b f(\rho, \theta, \varphi) \cdot \rho^2 \sin \varphi\, d\rho\, d\theta\, d\varphi.$$

Fubini's theorem applies in the usual circumstances.

4.9.4   Triple integral in spherical coordinates example

Now that we have three coordinate systems to choose from (Cartesian, cylindrical, and spherical), the steps for evaluating a triple integral must include choosing a coordinate system.

Example 52 *Integrate* $f(x, y, z) = \dfrac{2}{x^2 + y^2 + z^2}$ *over the region* $1 \le x^2 + y^2 + z^2 \le 9$, $z \ge 0$.

Solution ❶ Graph the three-dimensional region and choose a coordinate system. We recognize the equations $x^2 + y^2 + z^2 = 1$ and $x^2 + y^2 + z^2 = 9$ as equations of spheres centered at the origin with radii 1 and 3, respectively. The restriction $z \ge 0$ means we just want the top halves of the spheres (Figure 4.89). The region lies between the two hemispheres.

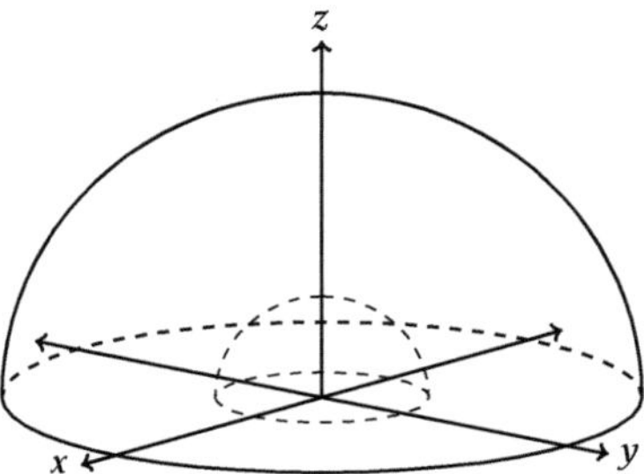

Figure 4.89 *The region of integration for example 52*

Describing the region of integration is very difficult in Cartesian coordinates, because the same surface is not on bottom always, and then the same curve is not always on the left and right, and so forth. Describing the region using cylindrical coordinates is not as difficult but still requires two triple integrals, one where the inner sphere is on bottom and one where the xy-plane is on bottom. But describing the region using spherical coordinates is relatively simple; the region lies between $\rho = 1$ (the inner sphere) and $\rho = 3$ (the outer sphere), where $0 \le \theta \le 2\pi$ and $0 \le \varphi \le \frac{\pi}{2}$ (from straight up to the xy-plane). So we make the switch to spherical coordinates.

If you try cylindrical coordinates, you will encounter great difficulty finding an antiderivative for the middle integral.

② Rewrite the function in the chosen coordinate system. Using $x^2 + y^2 + z^2 = \rho^2$, our function can be rewritten as

$$f(\rho, \theta, \varphi) = \frac{2}{\rho^2}.$$

③ Set up the triple integral in the chosen coordinate system, including the "don't forget" factor as appropriate. Using the rewritten function from **②**, the limits of integration from **①**, and inserting the "don't forget!" factor, the triple integral is

$$\int_0^{\frac{\pi}{2}} \int_0^{2\pi} \int_1^3 \frac{2}{\rho^2} \cdot \rho^2 \sin\varphi \, d\rho \, d\theta \, d\varphi = \int_0^{\frac{\pi}{2}} \int_0^{2\pi} \int_1^3 2\sin\varphi \, d\rho \, d\theta \, d\varphi.$$

④ Evaluate the triple integral. Then

$$\int_0^{\frac{\pi}{2}} \int_0^{2\pi} \left(\int_1^3 2\sin\varphi \, d\rho \right) d\theta \, d\varphi$$

$$= \int_0^{\frac{\pi}{2}} \int_0^{2\pi} \left((2\sin\varphi)\rho\big|_1^3 \right) d\theta \, d\varphi$$

Line 1 rewrites the triple integral as an iterated integral; line 2 finds an antiderivative with respect to ρ, treating θ and φ as constants; line 3 evaluates at $\rho = 3$, at $\rho = 1$, subtracts, and simplifies; line 4 finds an antiderivative with respect to θ, treating φ as a constant; line 5 evaluates at $\theta = 2\pi$, at $\theta = 0$, and subtracts; line 6 finds an antiderivative with respect to φ; line 7 evaluates at $\varphi = \frac{\pi}{2}$, at $\varphi = 0$, subtracts, and finishes the computation.

$$= \int_0^{\frac{\pi}{2}} \left(\int_0^{2\pi} 4 \sin \varphi \, d\theta \right) d\varphi$$

$$= \int_0^{\frac{\pi}{2}} \left((4 \sin \varphi)\theta \big|_0^{2\pi} \right) d\varphi$$

$$= \int_0^{\frac{\pi}{2}} 8\pi \sin \varphi \, d\varphi$$

$$= -8\pi \cos \varphi \big|_0^{\frac{\pi}{2}}$$

$$= -8\pi \cos \frac{\pi}{2} - (-8\pi \cos 0) = 0 - (-8\pi(1)) = 8\pi.$$

Because no context is given, we omit units in the answer. ■

As illustrated by example 52, spherical coordinates are often the easiest choice when the region involves spheres.

Reading Exercise 21　Set up (do not evaluate) a triple integral using spherical coordinates for integrating $f(x, y, z) = x^2 + y^2 + z^2 + 16$ on the region $x^2 + y^2 + z^2 \leq 4$.

EXERCISES 4.9

1–6.　Rapid response: Describe the object whose equation is given.

1. $\rho = 4$
2. $\rho = -3$
3. $\rho = 0$
4. $\theta = \dfrac{\pi}{2}$
5. $\varphi = \dfrac{\pi}{2}$
6. $\varphi = \dfrac{\pi}{6}$

7–18.　Plot the point with the given spherical coordinates.

7. $\left(2, 0, \dfrac{\pi}{3}\right)$
8. $\left(1, \dfrac{\pi}{2}, \dfrac{\pi}{4}\right)$
9. $\left(2, \dfrac{\pi}{4}, \dfrac{\pi}{2}\right)$
10. $(1, \pi, 0.9\pi)$
11. $\left(2.5, -\dfrac{\pi}{2}, \dfrac{3\pi}{4}\right)$
12. $\left(2, \pi, \dfrac{3\pi}{4}\right)$
13. $(3, 3, 0)$
14. $(4, 1, \pi)$
15. $\left(4, \dfrac{\pi}{2}, \dfrac{\pi}{2}\right)$

16. $\left(2, -\dfrac{\pi}{2}, \dfrac{\pi}{2}\right)$

17. $\left(0, \dfrac{\pi}{3}, \dfrac{2\pi}{3}\right)$

18. $\left(27, \dfrac{\pi}{30}, \pi\right)$

19-24. Convert the spherical coordinates to Cartesian coordinates.

19. $\left(5, \dfrac{\pi}{3}, \dfrac{3\pi}{4}\right)$

20. $\left(7, \dfrac{3\pi}{2}, \dfrac{\pi}{4}\right)$

21. $\left(\sqrt{2}, \dfrac{\pi}{4}, \dfrac{\pi}{2}\right)$

22. $(4, 17, \pi)$

23. $\left(3.7, \dfrac{11\pi}{13}, 0\right)$

24. $\left(3, \dfrac{\pi}{3}, \dfrac{2\pi}{3}\right)$

25-30. A point is given in Cartesian coordinates. Determine spherical coordinates for the point.

25. $(0, 6, 6)$

26. $(1, 1, 0)$

27. $(-2, 0, 0)$

28. $(\sqrt{3}, 0, 1)$

29. $(-1, 0, -\sqrt{3})$

30. $(0, 0, 3)$

31-36. Rewrite the function using spherical coordinates. Simplify if appropriate.

31. $f(x, y, z) = 3 - x^2 - y^2 - z^2$

32. $f(x, y, z) = x^2 + y^2 + 2z^2$

33. $f(x, y, z) = x^2 + y^2$

34. $f(x, y, z) = 2x - 5y + 7$

35. $f(x, y, z) = \dfrac{x^2}{4} + \dfrac{y^2}{4} + \dfrac{z^2}{4} - (z + 1)$

36. $f(x, y, z) = \left(\sqrt{x^2 + y^2 + z^2} + 1\right)^3$

37-40. Set up a triple integral in spherical coordinates. Do not evaluate the integral.

37. $\iiint_R (1 - x^2 - y^2 - z^2)\, dV$ where R is the region inside a sphere of radius 16 centered at the origin

38. $\iiint_R \ln(x^2 + y^2 + z^2)\, dV$ where R is the region $1 \le x^2 + y^2 + z^2 \le 4$

39. $\iiint_R (xy - 3z)\, dV$ where R is the region $2x^2 + 2y^2 + 2z^2 \le 8$, $z \le 0$

40. $\iiint_R (x - 2y)\, dV$ where R is the region in the first octant inside the sphere $x^2 + y^2 + z^2 = 9$

41. Integrate $f(x, y, z) = e^{(x^2 + y^2 + z^2)^{3/2}}$ on the region $0 \le \rho \le 1$.

Ans. to reading exercise 21:
$\int_0^\pi \int_0^{2\pi} \int_0^2 (\rho^2 + 16)\rho^2 \sin\varphi \, d\rho \, d\theta \, d\varphi$
Don't forget the extra $\rho^2 \sin\varphi$! Alternate answers include changing the order of integration.

Refer to Section 4.8 exercise 48(b).

Because ρ is used as a spherical coordinate variable, we choose not to use ρ as the name of the density function.

42. Integrate $f(\rho, \theta, \varphi) = \rho \cos\varphi$ over the region $0 \le \rho \le 1$, $0 \le \theta \le \pi$, $\frac{\pi}{4} \le \varphi \le \frac{\pi}{2}$.

43–46. Evaluate the integral using spherical coordinates.

43. $\iiint_R \frac{1}{x^2+y^2+z^2} \, dV$ where R is the region inside the sphere $x^2 + y^2 + z^2 = 4$

44. $\iiint_R \frac{1}{\sqrt{x^2+y^2}} \, dV$ where R is the region $1 \le x^2 + y^2 + z^2 \le 4$

45. $\iiint_R yz \, dV$ where R is the region $x^2 + y^2 + z^2 \le 1$, $z \le 0$

46. $\iiint_R 3 \, dV$ where R is the region $x^2 + y^2 + z^2 \le 6$, $y \ge 0$

47. Find the average value of $f(x, y, z) = 4 + x$ on the hemisphere $x^2 + y^2 + z^2 \le 4$, $x \ge 0$.

48. A spherical paperweight occupying $x^2 + y^2 + z^2 \le 9$ is heavier at the bottom than at the top, with density function $d(x, y, z) = 4 - z$. Find its center of mass.

4.10   Change of Variables in Multiple Integrals

Trigonometric substitution is an example of a change of variables. For instance, to calculate

$$\int \frac{x^2}{\sqrt{9 - x^2}} \, dx,$$

other techniques of integration (parts, partial fractions, and so forth) do not help, so we resort to using the substitution $x = 3\sin\theta$, $dx = 3\cos\theta \, d\theta$, arriving at

$$\int \frac{9\sin^2\theta}{\sqrt{9 - 9\sin^2\theta}} \cdot 3\cos\theta \, d\theta,$$

and continue from there.

What is the form of this change of variables? We begin with

$$\int f(x) \, dx,$$

we change from the variable x to the variable θ by $x = 3\sin\theta$, and the integral becomes

The x is replaced by $3\sin\theta$, so $f(x)$ becomes $f(3\sin\theta)$.

$$\int f(3\sin\theta) \cdot 3\cos\theta \, d\theta.$$

This "don't forget!" factor comes from a derivative, $\frac{dx}{d\theta}$.

This should look somewhat familiar; it's not just that the function is rewritten, but the length element dx is also rewritten, including a "don't forget!" factor of $3\cos\theta$.

The very same form applies to double integrals when switching to polar coordinates. We begin with

$$\iint_R f(x,y)\,dA$$

and make the change of variables $x = r\cos\theta$ and $y = r\sin\theta$ (simultaneously); we rewrite the function in terms of r and θ to get $f(r\cos\theta, r\sin\theta)$, rewrite the region R in terms of r and θ, and include the "don't forget!" factor of r, arriving at

$$\int_a^b \int_c^d f(r\cos\theta, r\sin\theta)\cdot r\,dr\,d\theta.$$

With triple integrals, switching to cylindrical or spherical coordinates has the same form. There are change of variables formulas, which are used to rewrite the function; we rewrite the region in terms of the new variables; and we include a "don't forget" factor.

Can this idea be generalized? Of course! (It would be rather anticlimactic if the answer was "no," wouldn't it?) The proof is presented at the end of this section, but for now we shall describe and demonstrate the method.

Begin with a double integral

$$\iint_R f(x,y)\,dA,$$

where f is continuous throughout the region R. Use the simultaneous change of variables

$$x = g(u,v)$$
$$y = h(u,v),$$

and assume that the first-order partial derivatives of g and h are continuous. We also need the transformation from x and y to u and v to be one-to-one in the interior of the region R. Then ❶ we rewrite the region R in terms of the new variables u and v, ❷ rewrite the function in terms of u and v, and ❸ include the "don't forget!" factor, which turns out to be the absolute value of the *Jacobian*. The rewritten integral becomes

$$\int_a^b \int_c^d f(g(u,v), h(u,v)) \left| \frac{\partial(x,y)}{\partial(u,v)} \right| dv\,du.$$

Let's discuss the three steps individually, beginning with step ❸.

4.10.1   The Jacobian

The Jacobian is a combination of partial derivatives and is denoted $\frac{\partial(x,y)}{\partial(u,v)}$. The notation may be read "the Jacobian of x and y with respect to u and v."

Definition 6 THE JACOBIAN *The Jacobian of the simultaneous change of variables* $x = g(u, v)$ *and* $y = h(u, v)$ *is given by*

$$\frac{\partial(x,y)}{\partial(u,v)} = \begin{vmatrix} \dfrac{\partial x}{\partial u} & \dfrac{\partial x}{\partial v} \\ \dfrac{\partial y}{\partial u} & \dfrac{\partial y}{\partial v} \end{vmatrix} = \frac{\partial x}{\partial u} \cdot \frac{\partial y}{\partial v} - \frac{\partial x}{\partial v} \cdot \frac{\partial y}{\partial u}.$$

Calculating the Jacobian is rather straightforward as long as the format for organizing the partial derivatives in the determinant can be remembered. The rows represent the original variables and the columns represent the new variables. For instance, in the formula in the definition, the first row contains the partial derivatives of x and the second row the partial derivatives of y, and the first column contains the partial derivatives with respect to u whereas the second column contains the partial derivatives with respect to v.

Example 53 *Evaluate the Jacobian for the simultaneous change of variables* $x = 3u - 4v$ *and* $y = u^2 - v^2$.

Solution The partial derivatives are

$$\frac{\partial x}{\partial u} = 3 \qquad \frac{\partial x}{\partial v} = -4$$

$$\frac{\partial y}{\partial u} = 2u \qquad \frac{\partial y}{\partial v} = -2v.$$

Changing the order of the variables swaps rows or columns in the determinant, which can change the sign of the Jacobian. For instance, $\frac{\partial(y,x)}{\partial(u,v)} = 6v - 8u$.

Then the Jacobian is

$$\frac{\partial(x,y)}{\partial(u,v)} = \begin{vmatrix} 3 & -4 \\ 2u & -2v \end{vmatrix} = -6v - (-8u) = 8u - 6v.$$

■

Next, we calculate the Jacobian for the change of variables from Cartesian to polar coordinates.

Example 54 *Evaluate the Jacobian for the simultaneous change of variables* $x = r\cos\theta$ *and* $y = r\sin\theta$.

Solution The partial derivatives are

$$\frac{\partial x}{\partial r} = \cos\theta \quad \frac{\partial x}{\partial \theta} = -r\sin\theta$$

$$\frac{\partial y}{\partial r} = \sin\theta \quad \frac{\partial y}{\partial \theta} = r\cos\theta.$$

Then, the Jacobian is

$$\frac{\partial(x,y)}{\partial(r,\theta)} = \begin{vmatrix} \cos\theta & -r\sin\theta \\ \sin\theta & r\cos\theta \end{vmatrix} = r\cos^2\theta - (-r\sin^2\theta)$$

$$= r\cos^2\theta + r\sin^2\theta = r(\cos^2\theta + \sin^2\theta) = r.$$

■

As expected, the Jacobian for changing from Cartesian to polar coordinates is the "don't forget!" factor of r. Why does this work? See the derivation at the end of this section!

Reading Exercise 22 Evaluate the Jacobian of $x = 4uv$, $y = 5 - u^2$.

4.10.2 Rewriting a region using a change of variables

We illustrate step ❶ with an example.

Example 55 *Rewrite the region bounded by $y = \sqrt{x}$, $y = 8\sqrt{x}$, $y = \frac{1}{x}$, and $y = \frac{4}{x}$, using the transformation $x = \frac{u}{v}$, $y = v$.*

Before presenting the solution, notice that the region (Figure 4.90) is difficult to describe for integrating with respect to x and y. If we integrate with respect to y first, three double integrals are required, as illustrated by the dashed lines in Figure 4.90: one from $y = \frac{1}{x}$ to $y = 8\sqrt{x}$, one from $y = \frac{1}{x}$ to $y = \frac{4}{x}$, and one from $y = \sqrt{x}$ to $y = \frac{4}{x}$. Switching the order and integrating with respect to x first does not help; three double integrals are still required.

Solution As long as the transformation is continuous and one-to-one, we may rewrite the region by rewriting the boundaries of the region. To this end, we convert each boundary equation to the new variables u and v, and simplify as appropriate. We start with $y = \frac{1}{x}$, replacing y by v and x by $\frac{u}{v}$:

Swapping the order of the variables in the Jacobian, which does not affect the actual change of variables, can result in $-r$. But in the integral we shall use the absolute value of the Jacobian, so that the order of variables in the Jacobian does not matter.

The term *transformation* is to be understood as a synonym for "simultaneous change of variables."

Step ❶ is also a good place to check some of the required hypotheses, namely that the transformation is defined throughout the region and one-to-one in the interior of the region. For example 55, because $x = \frac{u}{v}$, we cannot use $v = 0$ as part of the region. Requiring the transformation to be one-to-one means that we must be careful with expressions such as u^2 or $\sin v$ that are not one-to-one unless u and v are appropriately restricted.

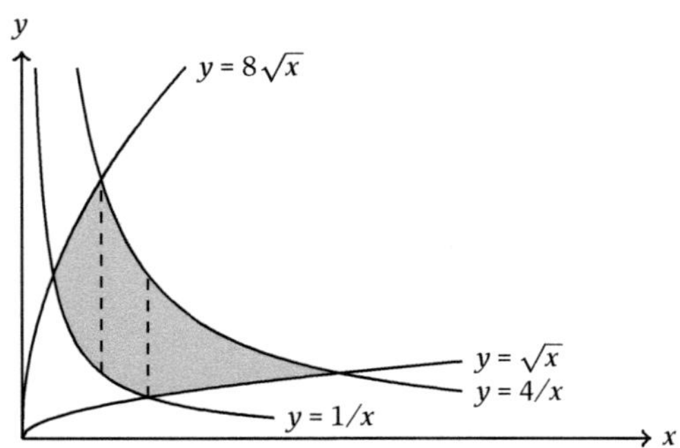

Figure 4.90 *The region of example 55, in terms of the variables x and y*

Line 3 is appropriate only because we already know that we cannot use $v = 0$.

$$y = \frac{1}{x}$$

$$v = \frac{1}{\frac{u}{v}} = \frac{v}{u}$$

$$u = \frac{v}{v} = 1.$$

Similarly, $y = \frac{4}{x}$ becomes $u = 4$. It is not the case that a transformation always yields variable $=$ constant. For the next two we solve for one variable in terms of the other.

Why rewrite as $v = \sqrt[3]{u}$, where v is written as a function of u, instead of stopping at $u = v^3$, where u is written as a function of v? We already have two equations for u, namely $u = 1$ and $u = 4$. Solving the others for v facilitates graphing the new region. In practice, you may need to adjust which variable to solve for based on convenience for graphing or order of integration.

$$y = \sqrt{x}$$

$$v = \sqrt{\frac{u}{v}}$$

$$v^2 = \frac{u}{v}$$

$$v^3 = u$$

$$v = \sqrt[3]{u}.$$

Similarly, $y = 8\sqrt{x}$ yields $v = 4\sqrt[3]{u}$.

Next we graph the region. Which of the new variables should be used for the horizontal axis and which for the vertical axis? Either choice is fine, but because we have v as a function of u for two of our boundary equations, we use v for the vertical axis. Determining the region bounded by the curves $u = 1$, $u = 4$, $v = \sqrt[3]{u}$, and $v = 4\sqrt[3]{u}$ is not a new task (Figure 4.91).

Ans. to reading exercise 22:
$8u^2$

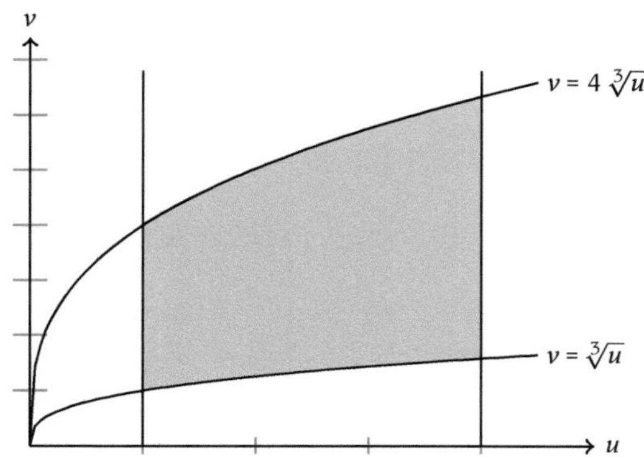

Figure 4.91 *The region of example 55, in terms of the variables u and v*

This region is easier to describe; the double integral has the form

$$\int_1^4 \int_{\sqrt[3]{u}}^{4\sqrt[3]{u}} dv\,du.$$

One double integral versus three. Nice! ∎

Reading Exercise 23   Rewrite the region inside the ellipse $\frac{x^2}{9} + \frac{y^2}{16} \le 1$ using the transformation $x = 3u, y = 4v$.

4.10.3   Rewriting a function using a change of variables

Step ❷ is the same as rewriting a function using polar coordinates, or cylindrical coordinates, or spherical coordinates. We simply apply the change of variables formulas.

Example 56 *Rewrite $f(x, y) = xy - \ln x$ using the simultaneous change of variables $x = e^u, y = 4v^2$.*

Solution   We replace x by e^u, y by $4v^2$, and simplify:

$$f(x, y) = xy - \ln x$$
$$f(e^u, 4v^2) = e^u 4v^2 - \ln e^u$$
$$= 4e^u v^2 - u.$$

Line 2 uses the change of variables formulas; line 3 uses the fact that the natural logarithmic and natural exponential functions are inverses of one another.

∎

Reading Exercise 24   Rewrite $f(x, y) = (x + 2y)e^{x/(y-2)}$ using the transformation $x = uv, y = v + 2$.

4.10.4   Change of variables example

Now that we have practiced each of the steps, we are ready for a complete example.

Example 57 *Evaluate $\iint_R (2x + y)\, dA$, where R is the region bounded by $y = 3x$, $y = \frac{1}{3}x$, and $y = 8 - x$, using the simultaneous change of variables $x = 3u + v$ and $y = u + 3v$.*

We first note that the double integral(s) can be set up and evaluated without too much difficulty using the original variables x and y. The region (Figure 4.92) requires the use of two double integrals, but the resulting integration is not difficult. This relatively simple example is presented for the purpose of illustrating the method.

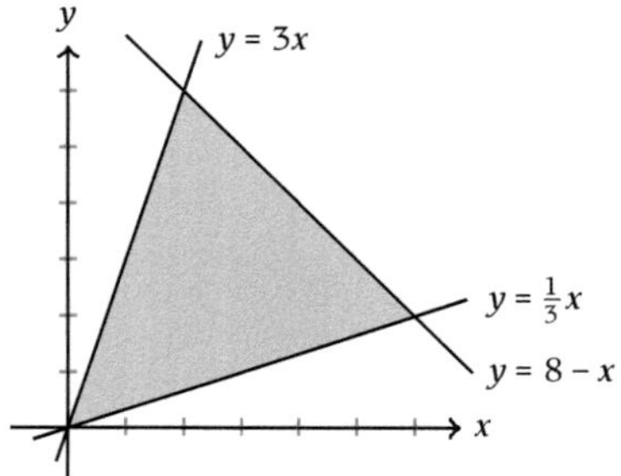

Figure 4.92 *The region of example 57, in terms of the variables x and y*

Solution First, note that the integrand is linear and therefore continuous throughout the region (and everywhere else). The change of variables formulas are also linear and therefore both continuous and one-to-one throughout the region.

❶ Rewrite the region. We rewrite each of the boundary equations using $x = 3u + v$ and $y = u + 3v$. First is $y = 3x$:

$$y = 3x$$
$$u + 3v = 3(3u + v)$$
$$u + 3v = 9u + 3v$$
$$u = 9u$$
$$8u = 0$$
$$u = 0.$$

Next is $y = \frac{1}{3}x$:

$$y = \frac{1}{3}x$$
$$u + 3v = \frac{1}{3}(3u + v)$$
$$u + 3v = u + \frac{1}{3}v$$
$$\frac{8}{3}v = 0$$
$$v = 0$$

Compare this calculation to the one before it. Can you see how this change of variables is designed? Accumulating such tricks is one key to the effective use of the change of variables technique.

Lastly,

$$y = 8 - x$$

$$u + 3v = 8 - (3u + v)$$

$$4u + 4v = 8$$

$$u + v = 2$$

$$v = 2 - u.$$

In line 5 we have the choice of which variable to solve for. We have chosen to solve for v and use v for the vertical axis.

The region in terms of u and v, which is bounded by $u = 0$, $v = 0$, and $v = 2 - u$, is drawn in Figure 4.93.

Drawing the rewritten region is helpful for setting up the double integral.

❷ Convert the function to the new variables. We replace x with $3u + v$ and y with $u + 3v$, then simplify:

$$f(x, y) = 2x + y$$

$$f(3u + v, u + 3v) = 2(3u + v) + (u + 3v) = 6u + 2v + u + 3v$$

$$= 7u + 5v.$$

❸ Compute the Jacobian. The partial derivatives are

$$\frac{\partial x}{\partial u} = 3 \quad \frac{\partial x}{\partial v} = 1$$

$$\frac{\partial y}{\partial u} = 1 \quad \frac{\partial y}{\partial v} = 3.$$

Ans. to reading exercise 23:
$u^2 + v^2 \le 1$ (Note: this transformation has turned an ellipse into a circle, which is easily described using polar coordinates.)

Then the Jacobian is

$$\frac{\partial(x, y)}{\partial(u, v)} = \begin{vmatrix} 3 & 1 \\ 1 & 3 \end{vmatrix} = 9 - 1 = 8.$$

The "don't forget!" factor to use in the double integral is $|8| = 8$.

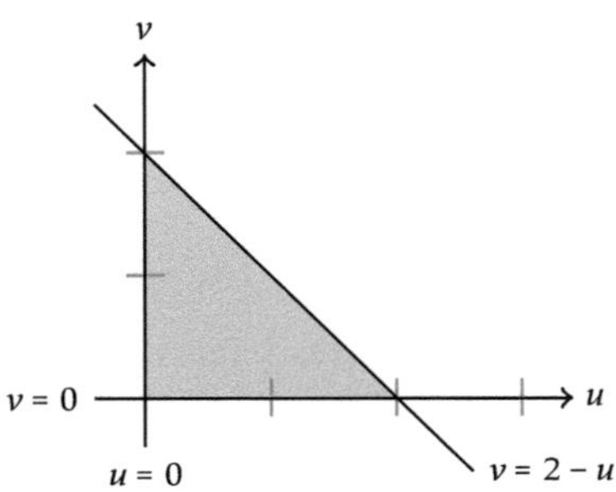

Figure 4.93 *The region of example 57, in terms of the variables u and v*

When the Jacobian is a variable expression, analysis of the values of the variables on the region is necessary to determine whether the Jacobian as calculated in step ❸ is positive (include as is) or negative (negate it).

❹ Set up the double integral. Using Figure 4.93 from step ❶ as a guide, we write the inner integral with respect to v, from $v = 0$ to $v = 2 - u$. Then the outer integral is with respect to u, from $u = 0$ to $u = 2$. For the integrand we use the rewritten function from step ❷ and the "don't forget!" factor from step ❸ . The double integral is

$$\int_0^2 \int_0^{2-u} (7u + 5v) \cdot 8 \, dv \, du = \int_0^2 \int_0^{2-u} (56u + 40v) \, dv \, du.$$

❺ Evaluate the double integral. The value of the integral is

$$\int_0^2 \left(\int_0^{2-u} (56u + 40v) \, dv \right) du$$

$$= \int_0^2 \left((56uv + 20v^2)\big|_0^{2-u} \right) du$$

$$= \int_0^2 \left(56u(2 - u) + 20(2 - u)^2 - 0 \right) du$$

$$= \cdots = \int_0^2 \left(80 + 32u - 36u^2 \right) du$$

$$= (80u + 16u^2 - 12u^3)\big|_0^2$$

$$= 160 + 64 - 96 - 0 = 128.$$

Line 1 rewrites as an iterated integral; line 2 finds an antiderivative with respect to v, treating u as a constant; line 3 evaluates at $v = 2 - u$, at $v = 0$, and subtracts; line 4 simplifies; line 5 finds an antiderivative with respect to u; line 6 evaluates at $u = 2$, at $u = 0$, and subtracts.

$\blacksquare$

4.10.5   Determining a transformation

To find the Jacobian for changing variables from x and y to u and v, the transformation must be in the form $x = g(u, v)$, $y = h(u, v)$ so that we can determine the partial derivatives of x and y with respect to u and v. But when setting up a potentially helpful change of variables, we might instead have u and v as functions of x and y.

Ans. to reading exercise 24:
$f(uv, v + 2) = (uv + 2v + 4)e^u$

Example 58 *For evaluating $\displaystyle\iint_R \frac{4x + 7y}{x - 2y} \, dA$, we decide to try the change of variables $u = 4x + 7y$ and $v = x - 2y$, so that the function can be rewritten as $\frac{u}{v}$. Rewrite this transformation in the form $x = g(u, v)$, $y = h(u, v)$.*

Solution   We need to solve the equations $u = 4x + 7y$ and $v = x - 2y$ for the two variables x and y. That is a familiar task–solving simultaneous equations! In this case, the equations are linear, and any method for solving simultaneous linear equations can be used. These include the addition method, substitution, Cramer's rule, and many more. We'll use the addition method. We start with

$$u = 4x + 7y$$

$$v = x - 2y.$$

Multiply both sides of the latter equation by -4:

$$u = 4x + 7y$$

$$-4v = -4x + 8y.$$

Then add the two equations to obtain

$$u - 4v = 15y$$

and divide both sides by 15:

$$y = \frac{1}{15}u - \frac{4}{15}v.$$

Either of the original equations can be used next, substituting the solution for y:

$$x = 2y + v$$

$$x = 2\left(\frac{1}{15}u - \frac{4}{15}v\right) + v$$

$$= \cdots = \frac{2}{15}u + \frac{7}{15}v.$$

The transformation is $x = \frac{2}{15}u + \frac{7}{15}v, y = \frac{1}{15}u - \frac{4}{15}v.$ ∎

Sometimes a change of variables simplifies the task, and sometimes it does not. To what extent the transformation of example 58 is helpful depends on the details of the region R.

Cleverly predetermining the result of step ❷ is potentially helpful. But don't forget about the factor from step ❸, which might change the difficulty of determining an antiderivative in step ❺. However, anytime u and v are both linear functions of the original variables, the Jacobian will turn out to be a constant.

Line 1 comes from rearranging the equation $v = x - 2y$; line 2 replaces y by $\frac{1}{15}u - \frac{4}{15}v.$

4.10.6 Derivation of the change of variables formula

We now turn our attention to the derivation of the method illustrated in example 57.

Suppose that we wish to evaluate $\iint_R f(x, y)\, dA$, where f is continuous throughout the region R, using the simultaneous change of variables $x = g(u, v)$ and $y = h(u, v)$. Assume that the first-order partial derivatives of g and h exist and are continuous throughout the region R. Then theorem 3 of Chapter 3 applies, and we can use the local linearity formula on both g and h. Furthermore, assume that the transformation is one-to-one in the interior of the region R.

We wish to integrate with respect to the variables u and v. This means that instead of partitioning the region R in the xy-plane into infinitely many subrectangles of infinitesimal area $\Delta x\, \Delta y$, we begin in the uv-plane with infinitely many subrectangles of infinitesimal side lengths Δu and Δv; the *n-m*th subrectangle is pictured in Figure 4.94.

Each point in the *n-m*th rectangle in the *uv*-plane corresponds to a point in the *xy*-plane, using the change of variables equations $x = g(u, v)$ and $y = h(u, v)$. For instance, the corner point (u_n, v_m) of Figure 4.94 corresponds to the point $(g(u_n, v_m), h(u_n, v_m))$ in the *xy*-plane (Figure 4.95). As with changing from Cartesian coordinates to polar coordinates, when an infinitesimal subregion is pictured in the *xy*-plane instead of the *uv*-plane, its sides may be curved—but by how much? Because both g and h are differentiable, the variables x and y move in a fashion indistinguishable from a straight line (at an infinitesimal scale), making the sides of the subregion in Figure 4.95 straight. Labeling adjacent sides as the vectors $\mathbf{w}$ and $\mathbf{z}$, these vectors can be calculated to help determine the area of the quadrilateral. By the local linearity formula,

This is where it is important that the transformation be one-to-one in the interior of the region R.

At an infinitesimal scale, the graph of $x = g(u, v)$ is indistinguishable from its tangent plane (where we picture u and v as horizontal axes and x as the vertical axis; think of a surface above the plane of Figure 4.94). The same is true of $y = h(u, v)$. Because x and y are both linear functions of u and v, they must be linear in relation to one another.

The area of the rectangle in the *uv*-plane is not guaranteed to be the same as the area of the corresponding subregion in the *xy*-plane.

$$g(u_n, v_m + \Delta v) = g(u_n + 0, v_m + \Delta v)$$

$$\approx g(u_n, v_m) + 0 + g_v(u_n, v_m)\Delta v.$$

Likewise, $h(u_n, v_m + \Delta v) \approx h(u_n, v_m) + h_v(u_n, v_m)\Delta v$. Therefore

$$\mathbf{w} = \langle g_v(u_n, v_m)\Delta v, h_v(u_n, v_m)\Delta v \rangle .$$

The *n-m*th subrectangle in Figure 4.94 is the same concept as the *n-m*th subrectangle in Section 4.2 Figure 4.9.

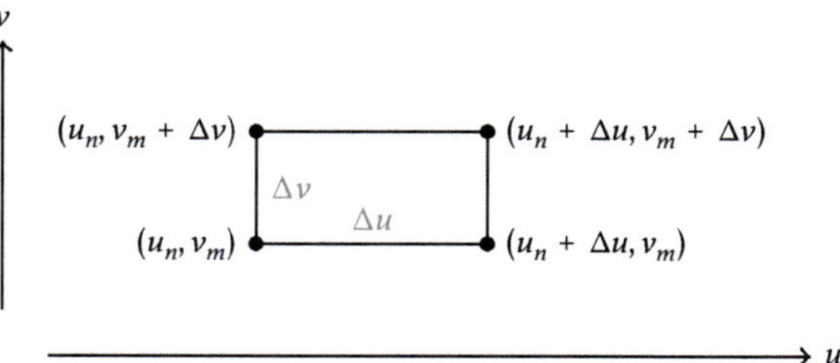

Figure 4.94 *The n-mth subrectangle in the uv-plane, with infinitesimal side lengths Δu and Δv*

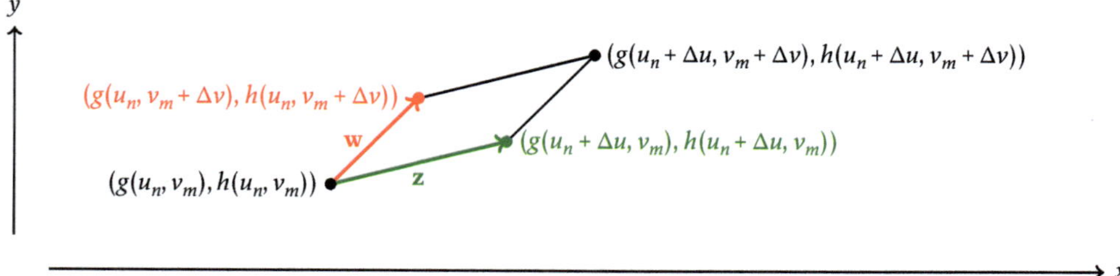

Figure 4.95 *The subregion of Figure 4.94 as it appears in the xy-plane*

For the same reasons,

$$z = \langle g_u(u_n, v_m)\Delta u, h_u(u_n, v_m)\Delta u \rangle.$$

A similar calculation (left as an exercise) shows that the opposite sides are the same vectors $\mathbf{w}$ and $\mathbf{z}$, meaning that the quadrilateral is a parallelogram! In three dimensions, the area of a parallelogram can be calculated as the norm of the cross product of its side vectors. This quadrilateral is in two dimensions, but we simply make the third dimension 0 when we calculate the cross product:

$$\mathbf{w} \times \mathbf{z} = \begin{vmatrix} \mathbf{i} & \mathbf{j} & \mathbf{k} \\ g_v(u_n, v_m)\Delta v & h_v(u_n, v_m)\Delta v & 0 \\ g_u(u_n, v_m)\Delta u & h_u(u_n, v_m)\Delta u & 0 \end{vmatrix}$$

$$= \mathbf{i}(0) - \mathbf{j}(0)$$

$$+ \mathbf{k}(g_v(u_n, v_m)\Delta v \cdot h_u(u_n, v_m)\Delta u - h_v(u_n, v_m)\Delta v \cdot g_u(u_n, v_m)\Delta u)$$

$$= \mathbf{k}\left(g_v(u_n, v_m) \cdot h_u(u_n, v_m) - h_v(u_n, v_m) \cdot g_u(u_n, v_m)\right) \Delta u \, \Delta v.$$

Then the area of the parallelogram is

$$\|\mathbf{w} \times \mathbf{z}\| = \sqrt{\left(\left(g_v(u_n, v_m) \cdot h_u(u_n, v_m) - h_v(u_n, v_m) \cdot g_u(u_n, v_m)\right) \Delta u \, \Delta v\right)^2}$$

$$= \left|\left(g_v(u_n, v_m) \cdot h_u(u_n, v_m) - h_v(u_n, v_m) \cdot g_u(u_n, v_m)\right) \Delta u \, \Delta v\right|$$

$$= \left|\left(\frac{\partial x}{\partial v}\frac{\partial y}{\partial u} - \frac{\partial y}{\partial v}\frac{\partial x}{\partial u}\right)\bigg|_{(u_n, v_m)} \Delta u \, \Delta v\right|$$

$$= \left|\frac{\partial(y, x)}{\partial(u, v)}\bigg|_{(u_n, v_m)}\right| \Delta u \, \Delta v.$$

When rewriting using integral notation (instead of omega sum notation), the area element is $\left|\frac{\partial(x,y)}{\partial(u,v)}\right| dv\,du$, which is the replacement for $dy\,dx$. Expressing the function in terms of the new variables and rewriting the region in terms of the new variables finishes the formula.

Referencing Figure 4.95, the x-component of $\mathbf{w}$ is found by subtracting the x-coordinate of the initial point, $g(u_n, v_m)$, from the x-coordinate of the terminal point, calculated as approximately $g(u_n, v_m) + g_v(u_n, v_m)\Delta v$.

Line 2 is the source of the absolute value in the formula, recalling that $\sqrt{x^2} = |x|$; line 3 rewrites using Leibniz notation; line 4 rewrites using the notation of the Jacobian and makes use of the fact that Δu and Δv are positive quantities.

Because of the absolute value, the order of the variables in the Jacobian may be switched.

CHANGE OF VARIABLES DOUBLE INTEGRAL FORMULA

Suppose that $f(x, y)$ is continuous on the region R. To calculate $\iint_R f(x, y)\, dy\, dx$ using the simultaneous change of variables $x = g(u, v)$ and $y = h(u, v)$, where the first-order partial derivatives of g and h are continuous throughout R and the transformation is one-to-one in the interior of R, use

$$\iint_R f(g(u, v), h(u, v)) \cdot \left|\frac{\partial(x, y)}{\partial(u, v)}\right| dv\, du,$$

where the region R is expressed in terms of the variables u and v.

EXERCISES 4.10

1–8. Evaluate the Jacobian for the simultaneous change of variables.

1. $x = 2u - 5v, y = 3u + 7v$
2. $x = u - v, y = 5u + 8v$
3. $x = u^3 - v, y = v^3 - u$
4. $x = \dfrac{u}{v}, y = v^2$
5. $x = r \cosh t, y = r \sinh t$
6. $x = \ln r, y = rs$
7. $r = e^w, s = e^z$
8. $r = u \sin v, t = u \cos v$

9. Rewrite the region bounded by $y = 5x, y = 2x$, and $y = 6 - x$ using the simultaneous change of variables $x = u + v$ and $y = 2u + 5v$.

10. Rewrite the region bounded by $y = -x, y = -5x$, and $y = x - 12$, using the simultaneous change of variables $x = u - v$ and $y = -5u + v$.

11. Rewrite the region bounded by $y = 4x + 7$, $y = \frac{1}{2}x - 7$, and $y = -x$ using the transformation $x = 2u + v, y = u + 4v$.

12. Rewrite the region bounded by $y = x + 1$, $y = x + 4$, $y = 5x$, and $y = 5x + 6$ using the transformation $x = 5u + v, y = u + 5v$.

13. Rewrite $f(x, y) = y\sqrt{\dfrac{3y}{x+5}}$ using the simultaneous change of variables $x = u - 5$ and $y = uv$.

14. Rewrite $f(x, y) = (x + y)^2 - (x - y)^2$ using the transformation $x = \dfrac{u}{v}$, $y = v$.

15. Rewrite the simultaneous change of variables $u = 2x - 3y, v = 5x + y$ in the form $x = g(u, v), y = h(u, v)$.

16. Rewrite the simultaneous change of variables $u = 2x + 9y, v = x - 7y$ in the form $x = g(u, v), y = h(u, v)$.

17. Evaluate $\iint_R \sqrt{\dfrac{2x+1}{2y-2}}\, dA$ where R is the region between $y = 2x + 2$ and $y = x + \dfrac{3}{2}$ for $2 \le y \le 4$, using the transformation $y = v + 1$, $x = uv - \dfrac{1}{2}$.

18. Evaluate $\iint_R y e^{x/(y-2)}\, dA$ where R is the region bounded by $y = 3$, $y = 4$, $y = x + 2$, and $y = \frac{1}{2}x + 2$, using the transformation $x = uv$, $y = v + 2$.

19. Evaluate $\iint_R e^{(x^2 - y^2)}\, dA$, where R is the region bounded by $y = x$, $y = 2 - x, y = 4 - x$, and $x^2 - y^2 = 4$, using the simultaneous change of variables $x = u + v$ and $y = u - v$.

20. Evaluate $\iint_R \dfrac{2x - y}{y - x}\, dA$ where R is the region bounded by $y = x + 1$, $y = 2x$, and $y = 6 - x$, using the simultaneous change of variables $x = u + v$ and $y = 2u + v$.

Use alphabetical order. For instance, if x and y are given as functions of u and v, calculate $\dfrac{\partial(x,y)}{\partial(u,v)}$.

Because rewriting the region is part of the larger task of integration, your answers to exercises 9–12 should include a sketch of the region using the new variables and the form of the resulting double integral (without the integrand), mimicking example 55.

21. Evaluate $\iint_R xy^2\, dA$ where R is the region bounded by $y = x^2, y = 8x^2$, $y = \dfrac{1}{x}$, and $y = \dfrac{4}{x}$, using the transformation $x = \dfrac{u}{v}, y = v$.

22. Evaluate $\iint_R \dfrac{4x-5y}{2x+y}\, dA$, where R is the region bounded by $4x - 5y = 1$, $4x - 5y = 3$, $2x + y = e$, and $2x + y = e^2$. Use the transformation defined by $u = 4x - 5y$ and $v = 2x + y$.

23. *Log-polar coordinates* use the simultaneous change of variables $x = e^\rho \cos\theta$ and $y = e^\rho \sin\theta$. Calculate the Jacobian $\dfrac{\partial(x,y)}{\partial(\rho,\theta)}$.

24. For a change of variables involving three variables such as changing from Cartesian coordinates to spherical coordinates, the Jacobian is still found using the determinant. Calculate the Jacobian $\dfrac{\partial(x,y,z)}{\partial(\rho,\theta,\varphi)}$ for the simultaneous change of variables $x = \rho \sin\varphi \cos\theta$, $y = \rho \sin\varphi \sin\theta$, and $z = \rho \cos\varphi$.

25. (a) Use the local linearity formula for a function of two variables to approximate the x- and y-coordinates of the upper-right corner of the quadrilateral in Figure 4.95.
    (b) Verify that a vector lying on the side of the quadrilateral opposite vector **z** is identical to **z**.
    (c) Verify that a vector lying on the side of the quadrilateral opposite vector **w** is identical to **w**.
    (d) Conclude that the quadrilateral is a parallelogram.

26. (a) Rewrite the region between $y = \sin x$ and $y = 2\sin x$ for $0 \le x \le \dfrac{\pi}{2}$, using the simultaneous change of variables $x = \sin^{-1} u$ and $y = v$.
    (b) Try repeating part (a) using $0 \le x \le \pi$. What requirement is violated?

27. (a) Following the strategy of reading exercise 23, determine a simultaneous change of variables that transforms the elliptical region $\dfrac{x^2}{25} + \dfrac{y^2}{4} \le 1$ into a circular region.
    (b) Use the transformation of part (a) to help determine the area of the elliptical region R, $\iint_R 1\, dA$.

28. Use a transformation like that of reading exercise 23 to determine a formula for the area of the ellipse $\dfrac{x^2}{a^2} + \dfrac{y^2}{b^2} = 1$.

29. Use the same strategy as the previous exercise to find a formula for the volume of the ellipsoid $\dfrac{x^2}{a^2} + \dfrac{y^2}{b^2} + \dfrac{z^2}{c^2} = 1$.

5

Vector Calculus

For the comparison, note that $\mathbf{r}(t) = \langle a, t \rangle$, meaning that $y = t$. Then $z = f(a, y) = f(a, t) = f(\mathbf{r}(t))$.

5.1   Line Integrals

In Section 4.1, we considered the area under a surface in a slice such as $x = a$. This region is pictured in Figure 5.1, which is nearly identical to Section 4.1 Figure 4.1. But instead of labeling the surface along the slice $z = f(a, y)$, it is labeled $z = f(\mathbf{r}(t))$, and the bottom of the slice is labeled as a vector-valued function, $\mathbf{r}(t) = \langle a, t \rangle$, where $c \leq t \leq d$.

Figure 5.2 is nearly the same, but with a more general vector-valued function $\mathbf{r}$ in the xy-plane, one that is curved instead of straight. The region under the space-

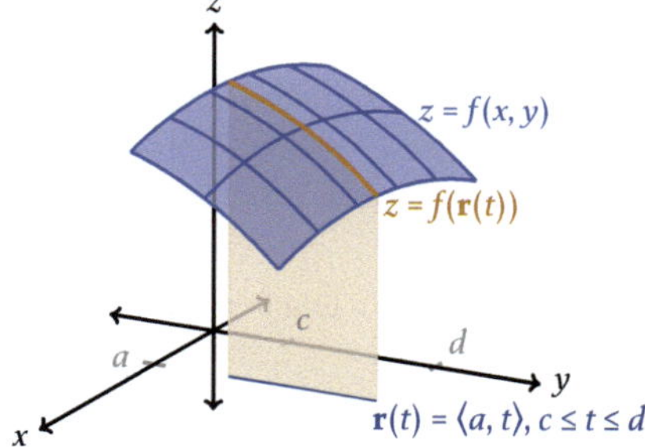

Figure 5.1 *A surface $z = f(x, y)$ (blue), a slice $x = a$ (brown), and the region under the slice (tan). Compare to the labeling of Section 4.1 Figure 4.1*

Observe that $z = f(\mathbf{r}(t))$ is a spacecurve.

The point labeled $t = a$ is the point corresponding to the vector $\mathbf{r}(a)$. Recall that parametrically defined curves are sometimes labeled with "time stamps."

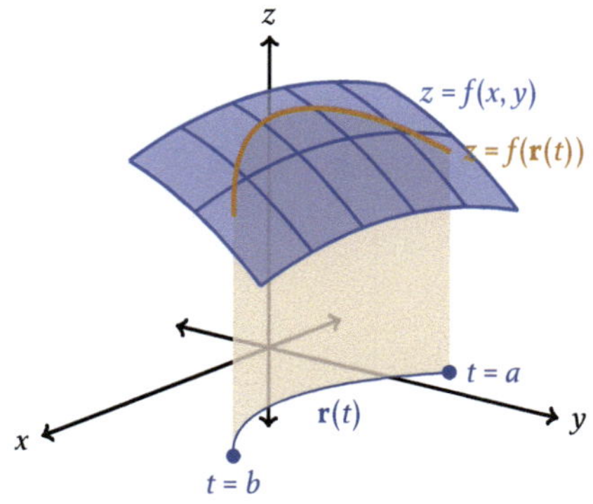

Figure 5.2 *A surface $z = f(x, y)$ (blue), a vector-valued function $\mathbf{r}$ in the xy-plane (blue), and the region under the surface over $\mathbf{r}$*

Multivariable Calculus Set Free. Charles Bryan Dawson, Oxford University Press.
© C. Bryan Dawson (2026). DOI: 10.1093/oso/9780198984269.003.0006

curve $z = f(\mathbf{r}(t))$ is no longer in a plane. How might we find this type of area? If the surface is a horizontal plane such as $z = 3$, as in Figure 5.3, it is apparent that the area should be the length of the curve, $L = \int_a^b \|\mathbf{r}'(t)\|\, dt$, times the height of the region, which is $3 = f(\mathbf{r}(t))$. But what if the height varies as in Figure 5.2? Is the area $\int_a^b f(\mathbf{r}(t))\|\mathbf{r}'(t)\|\, dt$?

See Section 2.3 for the arc length formula.

5.1.1   Line integral formula

Let $z = f(x, y)$ be continuous on a domain D, and suppose a curve C in the xy-plane is described by the vector-valued function $\mathbf{r}(t) = \langle g(t), h(t) \rangle$, $a \le t \le b$, and lies entirely in the domain of f. Suppose also that $\mathbf{r}$ is smooth (that is, both g' and h' are continuous and never simultaneously zero). Subdivide the interval $[a, b]$ into Ω pieces, where Ω is a positive infinite hyperreal integer, and consider the kth approximating rectangle (Figure 5.4). The height of this rectangle is

Think of the tan region in Figure 5.3 as being a rectangular piece of sheet metal that has been curved. The area of the rectangle would be its width times its height, which in Figure 5.3 is the length of the curve $\mathbf{r}(t)$ between $t = a$ and $t = b$ times 3.

The values of t are evenly subdivided. The points $\mathbf{r}(t_k)$ on the curve might not be evenly spaced.

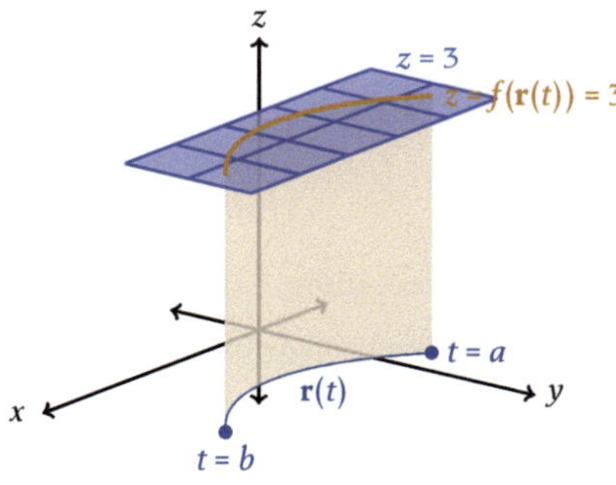

Figure 5.3 *The surface $z = 3$ (blue), a vector-valued function $\mathbf{r}$ in the xy-plane (blue), and the region under the surface over $\mathbf{r}$*

This is yet another example of area = height · width, add the areas of the subdivided regions, and rewrite as an integral.

By local linearity, because Δt is infinitesimal, the portion of the curve C between $t = t_k$ and $t = t_{k-1}$ is straight. By the local linearity formula, $g(t_{k-1}) = g(t_k - \Delta t) \approx g(t_k) - g'(t_k)\Delta t$, so that the difference in x-coordinates is $g'(t_k)\Delta t$. Similarly, the difference in y-coordinates is $h'(t_k)\Delta t$. Therefore, the length of the segment is

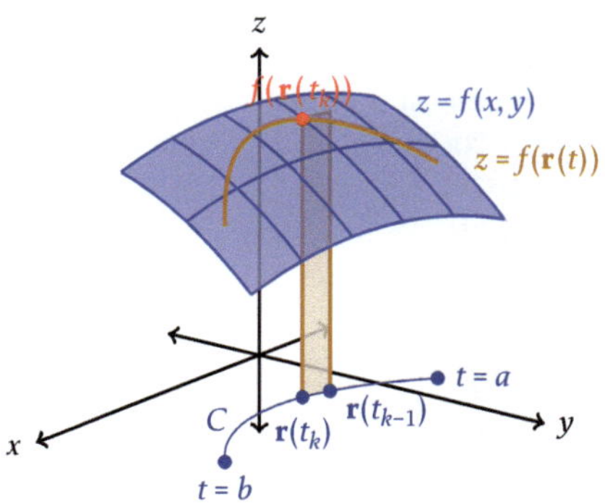

Figure 5.4 *The kth approximating rectangle for the region of Figure 5.2. The width is shown as non-infinitesimal for the purpose of visualization*

$$\|\langle g'(t_k)\Delta t, h'(t_k)\Delta t \rangle\| = \|\mathbf{r}'(t_k)\|\Delta t.$$

$f(\mathbf{r}(t_k)) = f(g(t_k), h(t_k))$, and the width of the rectangle is the arc length element, $\|\mathbf{r}'(t_k)\|\Delta t$, which is the length of the segment between $\mathbf{r}(t_{k-1})$ and $\mathbf{r}(t_k)$. Multiply these quantities to get the area of one rectangle, then sum them all to arrive at a total area of

$$\sum_{k=1}^{\Omega} f(\mathbf{r}(t_k))\ \|\mathbf{r}'(t_k)\|\Delta t.$$

Rewriting as a definite integral, the area is

$$\int_a^b f(\mathbf{r}(t))\|\mathbf{r}'(t)\|\,dt,$$

as expected. We call this a *line integral*, even though "curve integral" or "path integral" would be a better description.

In terms of the component functions $x = g(t)$ and $y = h(t)$, the formula can be rewritten as

$$\int_a^b f(g(t), h(t))\sqrt{\left(g'(t)\right)^2 + \left(h'(t)\right)^2}\,dt$$

$$= \int_a^b f(x(t), y(t))\sqrt{\left(\frac{dx}{dt}\right)^2 + \left(\frac{dy}{dt}\right)^2}\,dt,$$

which are convenient forms for calculations. The latter formula can be shortened by recalling that the arc length differential ds has the formula $\sqrt{\left(\frac{dx}{dt}\right)^2 + \left(\frac{dy}{dt}\right)^2}\,dt$. Writing C for the curve, the formula is often shortened to

$$\int_C f(x, y)\,ds.$$

> The weight of tradition is a type of inertia that is difficult to overcome, even though it has been said that "curve integral" is a better name since long before I was born.

> Line 1 also rewrites the norm using a square root. Line 2 rewrites the component derivatives using Leibniz notation and uses x and y instead of g and h.

> See *Calculus Set Free* Section 9.4.

> The formulas $x = g(t)$ and $y = h(t)$ are also used for this rewrite.

> The formula $\int_a^b f(\mathbf{r}(t))\|\mathbf{r}'(t)\|\,dt$ is called the *vector form* of the line integral. The formula $\int_a^b f(g(t), h(t))\sqrt{\left(x'(t)\right)^2 + \left(y'(t)\right)^2}\,dt$ is called the *component form* of the line integral. The formula $\int_a^b f(x(t), y(t))\sqrt{\left(\frac{dx}{dt}\right)^2 + \left(\frac{dy}{dt}\right)^2}\,dt$ is called the *parametric form* of the line integral. The formula $\int_C f(x, y)\,ds$ is called the *shorthand form* or the *arc length differential form* of the line integral.

LINE INTEGRAL FORMULA

Let $z = f(x, y)$ be continuous on a domain D and let C be a smooth curve $\mathbf{r}(t) = \langle g(t), h(t)\rangle, a \le t \le b$ contained in D. The *line integral* of f along C is given by

$$\int_C f(x, y)\,ds = \int_a^b f(\mathbf{r}(t))\|\mathbf{r}'(t)\|\,dt$$

$$= \int_a^b f(g(t), h(t))\sqrt{\left(g'(t)\right)^2 + \left(h'(t)\right)^2}\,dt.$$

Similarly, if $w = f(x, y, z)$ is continuous on a domain D and C is a smooth spacecurve $\mathbf{r}(t) = \langle g(t), h(t), k(t) \rangle$, $a \le t \le b$, contained in D, then

$$\int_C f(x, y, z)\, ds = \int_a^b f(\mathbf{r}(t)) \|\mathbf{r}'(t)\|\, dt$$

$$= \int_a^b f(g(t), h(t), k(t)) \sqrt{(g'(t))^2 + (h'(t))^2 + (k'(t))^2}\, dt.$$

This illustrates the advantage of the vector form of the line integral; it is the same in any dimension.

5.1.2   Line integral examples

Example 1 *Calculate the line integral of $f(x, y) = x^2 + y$ along the curve $\mathbf{r}(t) = \langle 3t, 1 + 4t \rangle$ from $t = 0$ to $t = 1$.*

Solution   The curve is given in vector form, so the vector form of the line integral is the most convenient to use. To set up the integral, we need $f(\mathbf{r}(t))$ and $\|\mathbf{r}'(t)\|$. First,

$$f(x, y) = x^2 + y$$

$$f(\mathbf{r}(t)) = (3t)^2 + (1 + 4t) = 9t^2 + 4t + 1.$$

Next,

$$\mathbf{r}'(t) = \langle 3, 4 \rangle,$$

and

$$\|\mathbf{r}'(t)\| = \sqrt{3^2 + 4^2} = 5.$$

The quantity $\|\mathbf{r}'(t)\|$ is not always a constant. It can depend on t.

The integral to calculate is

$$\int_C f(x, y)\, ds = \int_a^b f(\mathbf{r}(t)) \|\mathbf{r}'(t)\|\, dt$$

$$= \int_0^1 (9t^2 + 4t + 1) \cdot 5\, dt.$$

Once the line integral has been set up, it is an ordinary integral of the variable t.

Finally,

$$\int_0^1 (9t^2 + 4t + 1) \cdot 5\, dt = 5\left(3t^3 + 2t^2 + t\right)\Big|_0^1 = 5(3 + 2 + 1) - 0 = 30. \quad \blacksquare$$

Reading Exercise 1   Set up the integral (do not evaluate) for calculating the line integral of $f(x, y) = x^2 + y$ along the curve $\mathbf{r}(t) = \langle t, 5t \rangle$ from $t = 0$ to $t = 2$.

Example 2 *Calculate the line integral of $f(x, y) = xy^2$ along the upper-right portion of the unit circle from the point $(1, 0)$ to the point $(0, 1)$.*

We can write the parameterization of the curve in vector form if we wish.

See *Calculus Set Free* Section 8.1 for parameterization formulas for circles and line segments.

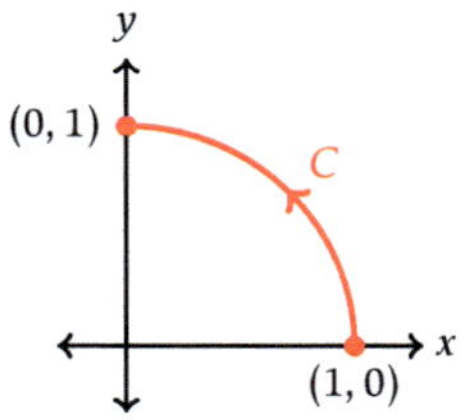

Figure 5.5 *The curve of example 2*

Ans. to reading exercise 1:
$\int_0^2 (t^2 + 5t)\sqrt{26}\, dt$

See *Calculus Set Free* Section 7.3 for integration strategies for trigonometric integrals.

The substitution is carried out in line 1, including replacing limits on the variable t with limits on the variable u.

Solution The vector function **r** is not given; instead the curve is described geometrically (Figure 5.5). We must therefore parameterize the curve ourselves. Recalling that we can use

$$x = \cos t, \quad y = \sin t$$

for a parameterization of the unit circle, all that is left is to set the limits on t, which are $t = 0$ to $t = \dfrac{\pi}{2}$.

Because we now have the curve described parametrically, the parametric form of the line integral is convenient to use. We therefore need to calculate $f(x(t), y(t))$ and $\sqrt{\left(\dfrac{dx}{dt}\right)^2 + \left(\dfrac{dy}{dt}\right)^2}$.

Then

$$f(x, y) = xy^2$$

$$f(x(t), y(t)) = \cos t \sin^2 t.$$

Also,

$$\frac{dx}{dt} = -\sin t, \quad \frac{dy}{dt} = \cos t$$

and

$$\sqrt{\left(\frac{dx}{dt}\right)^2 + \left(\frac{dy}{dt}\right)^2} = \sqrt{\sin^2 t + \cos^2 t} = 1.$$

Therefore, the integral to calculate is

$$\int_C f(x, y)\, ds = \int_a^b f(x(t), y(t)) \sqrt{\left(\frac{dx}{dt}\right)^2 + \left(\frac{dy}{dt}\right)^2}\, dt$$

$$= \int_0^{\frac{\pi}{2}} \cos t \sin^2 t \cdot 1\, dt.$$

Recognizing the need for substitution, we use
$$u = \sin t$$
$$du = \cos t\, dt.$$

The new limits of integration are $u = \sin \dfrac{\pi}{2} = 1$ and $u = \sin 0 = 0$. Then we have

$$= \int_0^1 u^2\, du$$

$$= \frac{1}{3} u^3 \Big|_0^1 = \frac{1}{3}.$$

In the next example we use the same surface but take a different path between the points.

Example 3 *Calculate the line integral of $f(x,y) = xy^2$ along the line segment from the point $(1,0)$ to the point $(0,1)$.*

Solution   This time we use the parameterization

$$x = 1 - t, \quad y = 0 + t$$

with $0 \le t \le 1$. Then

$$f(x,y) = xy^2$$

$$f(x(t), y(t)) = (1 - t)t^2 = t^2 - t^3.$$

The derivatives are

$$\frac{dx}{dt} = -1, \quad \frac{dy}{dt} = 1.$$

Then

$$\sqrt{\left(\frac{dx}{dt}\right)^2 + \left(\frac{dy}{dt}\right)^2} = \sqrt{(-1)^2 + 1^2} = \sqrt{2}.$$

Then the line integral is

$$\int_C f(x,y)\,ds = \int_a^b f(x(t), y(t)) \sqrt{\left(\frac{dx}{dt}\right)^2 + \left(\frac{dy}{dt}\right)^2}\,dt$$

$$= \int_0^1 (t^2 - t^3)\sqrt{2}\,dt$$

$$= \sqrt{2}\left(\tfrac{1}{3}t^3 - \tfrac{1}{4}t^4\right)\Big|_0^1$$

$$= \sqrt{2}\left(\tfrac{1}{3} - \tfrac{1}{4}\right) - 0 = \tfrac{1}{12}\sqrt{2}.$$

■

Notice that the answers to examples 2 and 3 are not the same, even though the surface and the endpoints of the curves are identical. It turns out that the path matters. In this case, the longer path (the curved path of example 2) has a larger line integral than the shorter path (the straight path of example 3). A look at Figures 5.3 and 5.4 might convince you that this is always true for a surface with nonnegative values over the curves. But you would be wrong! A path straight over a mountain-top might give a larger line integral than a curved (longer) path through a valley. Visualizing such a surface and a path is instructive.

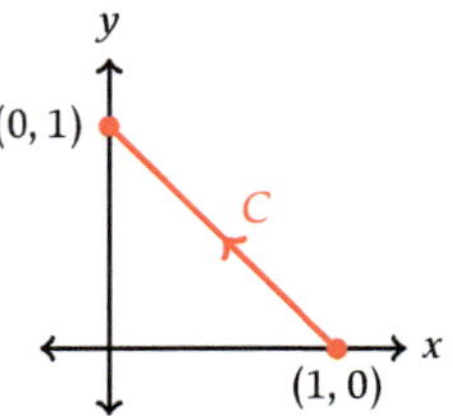

Figure 5.6 *The curve of example 3. Compare to Figure 5.5*

A parameterization of the line segment from the point (h, k) to the point $(h + a, k + b)$ is $x = h + at$, $y = k + bt$, $0 \le t \le 1$. In this case, the initial point is $(h, k) = (1, 0)$, so we use $h = 1$, $k = 0$. The vector between the two points is $\langle a, b \rangle = \langle 0 - 1, 1 - 0 \rangle = \langle -1, 1 \rangle$, so we use $a = -1$, $b = 1$.

This is the reason for the notation $\int_C f(x, y)\,ds$ instead of $\int_{(1,0)}^{(0,1)} f(x, y)\,ds$; the specific curve between the endpoints must be stated.

The path matters. Does the chosen parameterization matter? Look again at Figure 5.2. The area of the region under the spacecurve $f(\mathbf{r}(t))$ should depend on the curve $\mathbf{r}$, but not at what pace the curve is traversed!

Example 4 *Calculate the line integral of $f(x,y) = xy^2$ along the upper-right portion of the unit circle from the point $(1,0)$ to the point $(0,1)$ using the parameterization $x = \cos 2t,\ y = \sin 2t,\ 0 \le t \le \frac{\pi}{4}$.*

If we think of t as time, we are traversing the curve twice as fast as the parameterization of example 2.

Solution This is the same curve C pictured in Figure 5.5, and the surface is also the same as for example 2. What has changed is the parameterization. The function in terms of the variable t is therefore different:

$$f(x,y) = xy^2$$

$$f(x(t), y(t)) = \cos 2t \sin^2 2t.$$

Compare to the calculations of example 2.

The derivatives and arc length element are also different:

$$\frac{dx}{dt} = -2\sin 2t, \qquad \frac{dy}{dt} = 2\cos 2t$$

and

$$\sqrt{\left(\frac{dx}{dt}\right)^2 + \left(\frac{dy}{dt}\right)^2} = \sqrt{4\sin^2 2t + 4\cos^2 2t} = 2.$$

Therefore, the line integral to calculate also appears different,

$$\int_C f(x,y)\,ds = \int_a^b f(x(t), y(t)) \sqrt{\left(\frac{dx}{dt}\right)^2 + \left(\frac{dy}{dt}\right)^2}\,dt$$

$$= \int_0^{\frac{\pi}{4}} \cos 2t \sin^2 2t \cdot 2\,dt.$$

The substitution this time is

$$u = \sin 2t$$

$$du = 2\cos 2t\,dt.$$

The new limits of integration are $u = \sin 2\frac{\pi}{4} = \sin\frac{\pi}{2} = 1$ and $u = \sin 2(0) = 0$, which are the same as for example 2. After substitution we have

$$= \int_0^1 u^2\,du,$$

which is the also same as for example 2! The calculation is identical from there, and value of the line integral is $\frac{1}{3}$. ∎

Theorem 1 INDEPENDENCE OF PARAMETERIZATION *Let $z = f(x, y)$ be continuous on a domain D and let C be a smooth curve contained in D. Then the value of the line integral of f along C is the same for any smooth parameterization of C.*

Although example 4 is not a proof of theorem 1, it does illustrate how it sometimes works.

But what if we reverse the direction of the curve C in Figure 5.5 so that it goes from the point $(0, 1)$ clockwise to the point $(1, 0)$? The curve hasn't changed, we are just traversing it backward. It seems that we can still use $x = \cos t, y = \sin t$ as in example 2, switching the order of the limits of integration to go from $t = \frac{\pi}{2}$ to $t = 0$. We call this curve $-C$. Theorem 1 says we should get the same result, which is $\frac{1}{3}$, but if we calculate

$$\int_{\frac{\pi}{2}}^{0} \cos t \sin^2 t \cdot 1 \, dt = \cdots = -\frac{1}{3},$$

the answer is negated from example 2. What's going on?

It turns out that there is a very subtle error in the calculation that yields $-\frac{1}{3}$. It stems from the way the formula for the line integral is derived. Notice that the derivation assumes that $a < b$, so that Δt is positive. Then $\sqrt{(\Delta t)^2} = \Delta t$. If $b < a$ instead, then $\Delta t < 0$, and the arc length element becomes $\|\mathbf{r}'(t)\|(-\Delta t)$, necessitating the insertion of a negative into the formula, which then causes the above calculation to yield $-(-\frac{1}{3}) = \frac{1}{3}$. The moral of this story is to always set up the parameterization so that $a < b$. The calculations in examples 1–4, as well as all of our subsequent calculations, all use $a < b$.

Reference the discussion under the heading "Line integral formula."

The statement of the line integral formula also requires $a \le b$.

Example 5 *Calculate the line integral of $f(x, y) = xy^2$ along the upper-right portion of the unit circle from the point $(0, 1)$ to the point $(1, 0)$ using the parameterization $x = \cos\left(\frac{\pi}{2} - t\right), y = \sin\left(\frac{\pi}{2} - t\right), 0 \le t \le \frac{\pi}{2}$.*

This parameterization reverses the direction of the curve, but keeps $a < b$. We shall still call the curve $-C$, because of the direction we are traversing it compared to the curve C in examples 2 and 4.

Solution The function in terms of the variable t is

$$f(x, y) = xy^2$$

$$f(x(t), y(t)) = \cos\left(\frac{\pi}{2} - t\right) \sin^2\left(\frac{\pi}{2} - t\right).$$

The derivatives are

$$\frac{dx}{dt} = \left(-\sin\left(\frac{\pi}{2} - t\right)\right)(-1), \quad \frac{dy}{dt} = \left(\cos\left(\frac{\pi}{2} - t\right)\right)(-1)$$

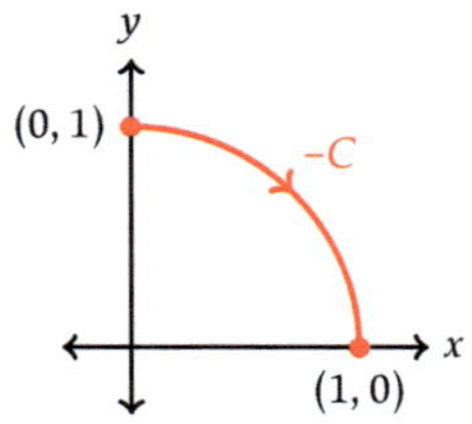

Figure 5.7 *The curve of example 5*

The extra negatives in the derivatives go away when squaring.

As required, the values of t go from the smaller $t = 0$ to the larger $t = \frac{\pi}{2}$, and these are the lower and upper limits of integration, respectively. The reversal of direction of the curve is accomplished instead by using the argument $\frac{\pi}{2} - t$ in the trig functions.

The negative in du, which can be traced to the design of the argument $\frac{\pi}{2} - t$ in the trig functions which is used to reverse the direction of the curve, is the source of the negative in the integrand in line 1. The 1 as the lower limit of integration and the 0 as the upper limit of integration can be traced to the same source. Line 2 shows how these two negate each other, seemingly magically resulting in the same value as example 2.

Notice that the shorthand form of the line integral is convenient for writing and remembering integration facts.

and

$$\sqrt{\left(\frac{dx}{dt}\right)^2 + \left(\frac{dy}{dt}\right)^2} = \sqrt{\sin^2\left(\frac{\pi}{2} - t\right) + \cos^2\left(\frac{\pi}{2} - t\right)} = 1.$$

The line integral to calculate is

$$\int_{-C} f(x, y)\, ds = \int_a^b f(x(t), y(t)) \sqrt{\left(\frac{dx}{dt}\right)^2 + \left(\frac{dy}{dt}\right)^2}\, dt$$

$$= \int_0^{\frac{\pi}{2}} \cos\left(\frac{\pi}{2} - t\right) \sin^2\left(\frac{\pi}{2} - t\right) \cdot 1\, dt.$$

The substitution is

$$u = \sin\left(\frac{\pi}{2} - t\right)$$

$$du = \left(\cos\left(\frac{\pi}{2} - t\right)\right)(-1)\, dt.$$

The new limits of integration are (upper) $u = \sin\left(\frac{\pi}{2} - \frac{\pi}{2}\right) = 0$ and (lower) $u = \sin\left(\frac{\pi}{2} - 0\right) = 1$. After substitution we have

$$\int_1^0 -u^2\, du$$

$$= -\frac{1}{3}u^3 \Big|_1^0 = -\frac{1}{3}(0 - 1) = \frac{1}{3},$$

as promised.   ∎

Example 5 illustrates the following formula.

REVERSING THE DIRECTION OF A CURVE

$$\int_{-C} f(x, y)\, ds = \int_C f(x, y)\, ds.$$

5.1.3   Piecewise smooth curves

A line integral can also be calculated along a curve that is *piecewise smooth*. A piecewise smooth curve consists of a finite number of smooth curves laid end-to-end, as illustrated in Figure 5.8. The line integral along C is simply the sum of the line integrals along the pieces.

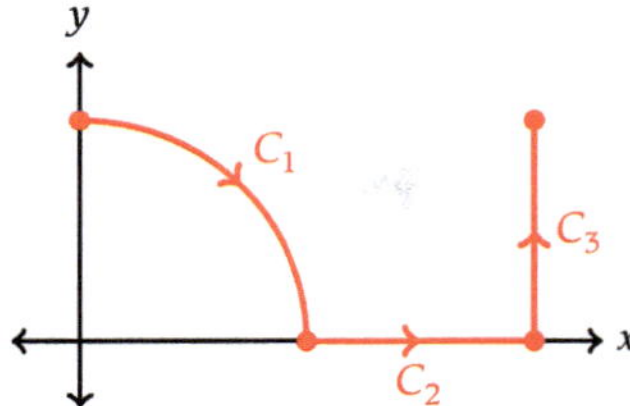

Figure 5.8 *A piecewise smooth curve* $C = C_1 \cup C_2 \cup C_3$

LINE INTEGRAL ALONG A PIECEWISE SMOOTH CURVE

Let $z = f(x, y)$ be continuous on a domain D that contains the curve C. Suppose the curve C is piecewise smooth, consisting of the smooth curves $C_1, C_2, \ldots, C_n$ where the initial point of C_k is the terminal point of $C_{k-1}, k = 2, 3, \ldots, n$. Then

$$\int_C f(x, y)\, ds = \int_{C_1} f(x, y)\, ds + \int_{C_2} f(x, y)\, ds + \cdots + \int_{C_n} f(x, y)\, ds.$$

Example 6 *Calculate the line integral of* $f(x, y, z) = e^{x+y+z}$ *along the curve* $C = C_1 \cup C_2$ *where* C_1 *is the curve* $\mathbf{r}_1(t) = \langle t, t, t \rangle, 0 \le t \le 1$, *and* C_2 *is the curve* $\mathbf{r}_2(t) = \langle 1, 1, t \rangle, 1 \le t \le 2$.

Solution The curve C is graphed in Figure 5.9. Because the curve consists of two pieces, we need to calculate two line integrals—one along C_1 and one along C_2.

The curve C_1 is given in vector form, so we use the vector form of a line integral. We need to calculate $f(\mathbf{r}_1(t))$ and $\|\mathbf{r}_1{}'(t)\|$. First,

$$f(x, y, z) = e^{x+y+z}$$

$$f(\mathbf{r}_1(t)) = e^{t+t+t} = e^{3t}.$$

Next,

$$\mathbf{r}_1{}'(t) = \langle 1, 1, 1 \rangle,$$

and

$$\|\mathbf{r}_1{}'(t)\| = \sqrt{1^2 + 1^2 + 1^2} = \sqrt{3}.$$

Then

$$\int_{C_1} f(x, y, z)\, ds = \int_0^1 e^{3t}\sqrt{3}\, dt$$

$$= \sqrt{3}\,\frac{e^{3t}}{3}\Big|_0^1 = \frac{\sqrt{3}}{3}\left(e^3 - 1\right).$$

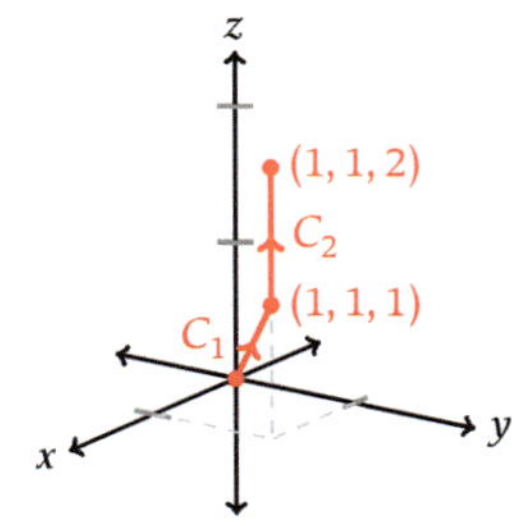

Figure 5.9 *The curve of example 6*

Notice that the procedure for three dimensions is essentially the same as for two dimensions.

When finding the antiderivative of e^{2+t}, the formal use of substitution can be skipped because $\frac{d}{dt}(2+t) = 1$, resulting in $du = dt$.

Similarly, for the curve C_2,

$$f(\mathbf{r}_2(t)) = e^{1+1+t} = e^{2+t}$$

$$\mathbf{r}_2'(t) = \langle 0, 0, 1 \rangle$$

$$\|\mathbf{r}_2'(t)\| = \sqrt{0^2 + 0^2 + 1^2} = 1.$$

Then

$$\int_{C_2} f(x,y,z)\, ds = \int_1^2 e^{2+t} \cdot 1\, dt$$

$$= e^{2+t}\big|_1^2 = e^4 - e^3.$$

The line integral along C is the sum of the line integrals along C_1 and C_2:

$$\int_C f(x,y,z)\, ds = \frac{\sqrt{3}}{3}\left(e^3 - 1\right) + e^4 - e^3.$$

$\blacksquare$

For the remainder of this text, when we use the term *path*, it can be interpreted as a piecewise smooth curve.

Reading Exercise 2   For the curve $C = C_1 \cup C_2$, we have $\int_{C_1} f(x,y)\, ds = 14$ and $\int_{C_2} f(x,y)\, ds = 7$. What is $\int_C f(x,y)\, ds$?

5.1.4   Line integrals, dx and dy

Instead of a line integral with respect to arc length, which uses $ds = \sqrt{\left(\frac{dx}{dt}\right)^2 + \left(\frac{dy}{dt}\right)^2}\, dt$, line integrals can also be computed with respect to x (using $dx = \frac{dx}{dt}\, dt$) and with respect to y (using $dy = \frac{dy}{dt}\, dt$).

The formulas given in the box are in parametric form. The vector form replaces $f(x(t), y(t))$ with $f(\mathbf{r}(t))$.

LINE INTEGRALS WITH RESPECT TO x AND y

Let $z = f(x,y)$ be continuous on a domain D and let C be a smooth curve $r(t) = \langle x(t), y(t) \rangle$, $a \le t \le b$ contained in D. The line integral of f along C with respect to x is given by

$$\int_C f(x,y)\, dx = \int_a^b f(x(t), y(t))\, x'(t)\, dt,$$

and the line integral of f along C with respect to y is given by

$$\int_C f(x,y)\, dy = \int_a^b f(x(t), y(t))\, y'(t)\, dt.$$

Example 7 *Calculate the line integral of $f(x,y) = xy - x^2$ along the curve $\mathbf{r}(t) = \langle t^2, t^3 \rangle$, $0 \le t \le 2$, (a) with respect to x and (b) with respect to y.*

Solution First, we compute the function along the curve:

$$f(x,y) = xy - x^2$$

$$f(\mathbf{r}(t)) = t^2 t^3 - (t^2)^2 = t^5 - t^4.$$

Line 2 replaces x by t^2 and y by t^3, and then simplifies.

(a) We calculate a derivative next:

$$x'(t) = 2t.$$

Using the vector form of the line integral,

$$\int_C f(x,y)\, dx = \int_a^b f(\mathbf{r}(t))x'(t)\, dt$$

Compare this procedure with the procedure used in example 1. The only real difference is the use of $x'(t)$ instead of $\|\mathbf{r}'(t)\|$.

$$= \int_0^2 (t^5 - t^4)\, 2t\, dt$$

$$= \int_0^2 (2t^6 - 2t^5)\, dt$$

$$= \left(\tfrac{2}{7}t^7 - \tfrac{2}{6}t^6 \right)\Big|_0^2$$

$$= \cdots = \frac{320}{21}.$$

(b) Similarly,

$$y'(t) = 3t^2$$

and

$$\int_C f(x,y)\, dy = \int_a^b f(\mathbf{r}(t))y'(t)\, dt$$

$$= \int_0^2 (t^5 - t^4)\, 3t^2\, dt$$

$$= \int_0^2 (3t^7 - 3t^6)\, dt$$

$$= \left(\tfrac{3}{8}t^8 - \tfrac{3}{7}t^7 \right)\Big|_0^2$$

$$= \cdots = \frac{288}{7}.$$

∎

Ans. to reading exercise 2:
21

When reversing the direction of a curve, the line integrals with respect to x and y act differently than with respect to arc length; the value is negated, just as with reversing the limits of integration for an ordinary integral. This is explored further in the exercises.

> ### REVERSING THE DIRECTION OF A CURVE, dx or dy
>
> $$\int_{-C} f(x,y)\,dx = -\int_{C} f(x,y)\,dx$$
>
> $$\int_{-C} f(x,y)\,dy = -\int_{C} f(x,y)\,dy$$

For this reason, we allow $b < a$ when setting up a parameterization of a curve for calculating a line integral with respect to either x or y.

In physical applications, line integrals with respect to x and y can occur together, in the form

$$\int_{C} M(x,y)\,dx + \int_{C} N(x,y)\,dy.$$

It is customary to combine the two integrals and abbreviate the notation as

$$\int_{C} M(x,y)\,dx + N(x,y)\,dy.$$

Example 8 *Calculate $\int_{C} y^2\,dx + x^2\,dy$, where C is the portion of the parabola $y = x^2$ from the point $(1,1)$ to the point $(3,9)$.*

In general, the curve $y = f(x)$ can be parameterized by $x = t$, $y = f(t)$. If $a \le x \le b$, use $a \le t \le b$.

Solution The curve C can be parameterized by $x = t$, $y = t^2$, $1 \le t \le 3$. Then $dx = dt$ and $dy = 2t\,dt$. Substituting for x, y, dx, and dt, the integral is

Line 1 replaces x by t, y by t^2, dx by dt, and dy by $2t\,dt$; line 2 simplifies, including factoring out the dt and placing it at the end of the integral; line 3 finds an antiderivative and line 4 completes the calculation.

$$\int_{C} y^2\,dx + x^2\,dy = \int_{1}^{3} (t^2)^2\,dt + t^2 \cdot 2t\,dt$$

$$= \int_{1}^{3} (t^4 + 2t^3)\,dt$$

$$= \left(\tfrac{1}{5}t^5 + \tfrac{2}{4}t^4\right)\Big|_{1}^{3}$$

$$= \cdots = \frac{442}{5}.$$

The combination of what essentially is two integrals into one integral in line 2 of the calculation illustrates a reason for the notation. ∎

A connection between line integrals with respect to arc length (ds) and line integrals of the type calculated in example 8 will be established in the next section.

5.1.5   Total mass and center of mass

Consider a wire (or a thin rod) lying along a spacecurve C described by $\mathbf{r}(t)$, $a \le t \le b$. The thickness and composition of the wire contribute to the density of the wire, which can be measured in mass per unit length. Let $\rho(x, y, z)$ be the density of the wire at the point (x, y, z). How might we find the total mass of the wire? Because the mass of a small piece is its density times its length, $\rho(\mathbf{r}(t_k)) \cdot \|\mathbf{r}'(t_k)\| \Delta t$ for the kth piece, the development is identical to that of the line integral with respect to arc length. Therefore, the total mass is

Given density $= \frac{\text{mass}}{\text{length}}$, then mass $=$ density $\cdot$ length.

$$\int_C \rho(x, y, z)\, ds.$$

Developing formulas for the center of mass is analogous to the other times we have found a center of mass.

TOTAL MASS AND CENTER OF MASS OF A WIRE

For a wire lying on a spacecurve C with density function $\rho(x, y, z)$,

$$\text{total mass } m = \int_C \rho(x, y, z)\, ds.$$

The moments about the coordinate planes are

$$M_{yz} = \int_C x\rho(x, y, z)\, ds$$

$$M_{xz} = \int_C y\rho(x, y, z)\, ds$$

$$M_{xy} = \int_C z\rho(x, y, z)\, ds.$$

The center of mass is

$$\left(\frac{M_{yz}}{m}, \frac{M_{xz}}{m}, \frac{M_{xy}}{m} \right).$$

Example 9 *A coil spring is more dense at the bottom than at the top. It occupies the curve $x = 4\cos t$, $y = 4\sin t$, $z = 0.3t$, for $0 \le t \le 7\pi$. The density function is $\rho(x, y, z) = 10 - z$. Find the center of mass of the spring by setting up the appropriate integrals and obtaining numerical approximations using a computer algebra system (CAS).*

These integrals can be evaluated by hand, but the work is tedious.

Solution   Following the procedure of previous examples, we calculate $\rho(x(t), y(t), z(t))$:

$$\rho(x, y, z) = 10 - z$$

$$\rho(x(t), y(t), z(t)) = 10 - 0.3t.$$

Also,

$$\frac{dx}{dt} = -4\sin t, \quad \frac{dy}{dt} = 4\cos t, \quad \frac{dz}{dt} = 0.3,$$

and

$$\sqrt{\left(\frac{dx}{dt}\right)^2 + \left(\frac{dy}{dt}\right)^2 + \left(\frac{dz}{dt}\right)^2} = \sqrt{16\sin^2 t + 16\cos^2 t + 0.09} = \sqrt{16.09}.$$

Specific units for mass are not given in the example, so we omit them from the total mass. Mass units are irrelevant for stating the center of mass.

Therefore, the total mass is

$$m = \int_C \rho(x, y, z)\, ds = \int_0^{7\pi} (10 - 0.3t)\sqrt{16.09}\, dt = 591.135.$$

As always, it is often easier to remember to calculate the coordinates of the center of mass directly than it is to remember the correct notation for the moments.

The x-coordinate of the center of mass is

$$\frac{1}{m}\int_C x\rho(x, y, z)\, ds = \frac{1}{591.135}\int_0^{7\pi} (4\cos t)(10 - 0.3t)\sqrt{16.09}\, dt = 0.0163,$$

which is very close to 0. A drawing of the coil might suggest that the x-coordinate of the center of mass should be 0, but the decreasing density as the coil rises means that there is just a little more mass on positive x-coordinates than on negative x-coordinates.

The y-coordinate of the center of mass is

$$\frac{1}{m}\int_C y\rho(x, y, z)\, ds = \frac{1}{591.135}\int_0^{7\pi} (4\sin t)(10 - 0.3t)\sqrt{16.09}\, dt = 0.3638,$$

which strays even further from 0 because there is an extra half rotation that possesses positive y-coordinates (the coil goes from 0 to 7π, three and a half rotations).

The z-coordinate of the center of mass is

$$\frac{1}{m}\int_C z\rho(x, y, z)\, ds = \frac{1}{591.135}\int_0^{7\pi} 0.3t(10 - 0.3t)\sqrt{16.09}\, dt = 2.7574,$$

which is less than half of the highest z-coordinate $0.3 \cdot 7\pi = 6.5973$ because the coil is more dense toward the bottom.

The center of mass is

$$(0.0163, 0.3638, 2.7574).$$

■

EXERCISES 5.1

1–12.　Calculate the line integral of the function along the given curve.

Line integrals are assumed to be with respect to arc length (ds) unless otherwise specified.

1. $f(x, y) = xy$, segment from $(1, 1)$ to $(4, 9)$
2. $f(x, y) = 2x - 3y + 5$, unit circle
3. $f(x, y, z) = z^2 - xy$, segment between $(0, 0, 0)$ and $(4, 4, 7)$
4. $f(x, y) = \frac{1}{x} + y^2$, segment from $(2, 3)$ to $(5, 7)$
5. $f(x, y) = x + y$, left half of the circle $x^2 + y^2 = 9$ from $(0, 3)$ to $(0, -3)$
6. $f(x, y, z) = xyz$, 　curve 　$x = 3 \sin t$, 　$y = 4 \sin t$, 　$z = 5 \cos t$, 　$0 \le t \le \frac{\pi}{2}$
7. $f(x, y) = x\sqrt{y}$, semicircle $x^2 + y^2 = 4$, $y \ge 0$
8. $f(x, y) = \frac{1}{x^2 - 6y + 22}$, segment between $(3, 5)$ and $(2, 4)$
9. $f(x, y) = y$, curve $y = \sqrt{x}$ between $(1, 1)$ and $(4, 2)$
10. $f(x, y, z) = xy - z^2$, 　curve 　$x = 4 \cos t$, 　$y = 4 \sin t$, 　$z = 3t$, 　$0 \le t \le 2\pi$
11. $f(x, y) = x \cos y$, curve $x = 3t, y = 2 + 5t, 0 \le t \le 1$
12. $f(x, y) = xe^y$, curve $\mathbf{r}(t) = \langle t, 4t - 1 \rangle, \frac{1}{4} \le t \le \frac{1}{2}$

13–16.　(a) Set up the line integral of the function along the given curve and (b) evaluate the line integral using a CAS.

13. $f(x, y) = 4x - y^2$, curve $x = t^2, y = 1 - 2t, -1 \le t \le 0$
14. $f(x, y) = \sqrt{x^2 + y}$, curve $x = 5t, y = 3 + t, 0 \le t \le 10$
15. $f(x, y, z) = x^2 - 3xy + z$, 　curve 　$x = 3 \cos t$, 　$y = 3 \sin t$, 　$z = t$, 　$0 \le t \le 10\pi$
16. $f(x, y, z) = x + y + z$, curve $x = t, y = t^2, z = t + 5, -2 \le t \le 7$

17–20.　Calculate (a) $\int_C f(x, y)\, dx$ and (b) $\int_C f(x, y)\, dy$ for the given function and curve.

17. $f(x, y) = 2y - 5x^2$, curve $y = 3x^2$ from $(0, 0)$ to $(2, 12)$
18. $f(x, y) = x^3 y + 2$, curve $y = \frac{1}{x}$ from $\left(\frac{1}{2}, 2\right)$ to $\left(2, \frac{1}{2}\right)$
19. $f(x, y) = xy^3 - x^2 y^2$, segment from $(4, 7)$ to $(0, 0)$
20. $f(x, y) = 4xy^2$, segment from $(1, 1)$ to $(4, 1)$

21–24. Calculate the line integral of the function along the given piecewise curve.

21. $f(x, y) = 4x + 3y$, curve from $(1, 0)$ to $(0, 1)$ counterclockwise along the unit circle, followed by the segment from $(0, 1)$ to $(0, 4)$

22. $f(x, y) = xy$, segment from $(0, 0)$ to $(2\sqrt{2}, 2\sqrt{2})$ followed by the arc of the circle of radius 4 centered at the origin from $(2\sqrt{2}, 2\sqrt{2})$ to $(4, 0)$, clockwise

23. $f(x, y, z) = x^2 y + z^3$, segments from $(0, 0, 0)$ to $(2, 4, 0)$ and then to $(2, 4, 5)$

24. $f(x, y, z) = x^2 + yz$, segments from $(0, 0, 0)$ to $(2, 2, 2)$ and then to $(2, 5, 2)$

25–30. Calculate the given line integral along the given curve.

25. $\int_C x^2 y \, dx + 3x \, dy$, curve $y = x^2$ from $(-1, 1)$ to $(1, 1)$

26. $\int_C y^3 \, dx + \frac{x}{y} \, dy$, curve $y = x^3$ from $(1, 1)$ to $(2, 8)$

27. $\int_C \frac{1}{y} \, dx + \frac{3}{x} \, dy$, curve $x^2 + y^2 = 9$

28. $\int_C x^2 \, dy + y \, dx$, curve $y = \ln x$ from $(1, 0)$ to $(e, 1)$

29. $\int_C xy \, dz + xz \, dy + yz \, dx$, curve $\mathbf{r}(t) = \langle t, t^2, t^3 \rangle$, $1 \le t \le 2$

30. $\int_C y \, dx + 3z \, dy + x^2 \, dz$, curve $\mathbf{r}(t) = \langle 2t, 5t, t^2 \rangle$, $-1 \le t \le 1$

31. A thin rod is placed at $x = 2t$, $y = 1 - t$, $z = 4 + t$, for $0 \le t \le 3$. It has the density function $\rho(x, y, z) = \frac{1}{2}x + z$. Find (a) the total mass of the rod and (b) the center of mass of the rod.

32. A thin rod is placed at $x = t + 1$, $y = 2t$, $z = 5 - 2t$, for $0 \le t \le 1$. It has the density function $\rho(x, y, z) = x + 2y + z$. Find (a) the total mass of the rod and (b) the center of mass of the rod.

33. A wire occupies the curve at $x = 3 \sin t$, $y = 5 \cos t$, $z = 4 \sin t$ for $0 \le t \le \pi$. It has constant density $\rho(x, y, z) = 1$. Find (a) the total mass of the wire and (b) the center of mass of the wire.

34. A wire occupies the curve at $x = \frac{1}{3}t^3$, $y = t^2$, $z = 2t$, for $1 \le t \le 2$. Its density is $\rho(x, y, z) = 3x + 4y - z^2$. Find (a) the total mass of the wire and (b) the center of mass of the wire.

35. (a) Return to the surface and curve of example 2 but compute the line integral with respect to x instead of arc length.
    (b) Return to the surface and curve of example 5 but compute the line integral with respect to x instead of arc length.
    (c) Do the results of part (a) and (b) agree with the claim that $\int_{-C} f(x, y) \, dx = -\int_C f(x, y) \, dx$?

36. Return to example 8. Instead of using the parameterization,

   (a) compute the differential dy from the equation of the curve.
   (b) In the given integral, replace y by x^2 (from the equation of the curve) and dy by the result from part (a), and insert the appropriate limits of integration.
   (c) Compare the result of part (b) to the integral in the solution to example 8. Conclude that parts (a) and (b) constitute an alternate procedure in the case that the curve C is given in the form $y = f(x)$.

5.2   Vector Fields

In Section 2.1, we began exploring functions with one dimension of input and two or three dimensions of output. We called these functions vector-valued, with the idea that the two or three dimensions of output are represented by vectors (Figure 5.10).

In Section 3.1, we began exploring functions with two or three dimensions of input and one dimension of output. For these functions the output is not a vector, but rather a real number (Figure 5.11).

What if we combine these ideas (Figure 5.12)? Why not have two dimensions of input and two dimensions of output, or three dimensions each of input and output? One interpretation of such a function is called a *vector field*. The input is a point in two or three dimensions, and the output is a vector with the same number of dimensions.

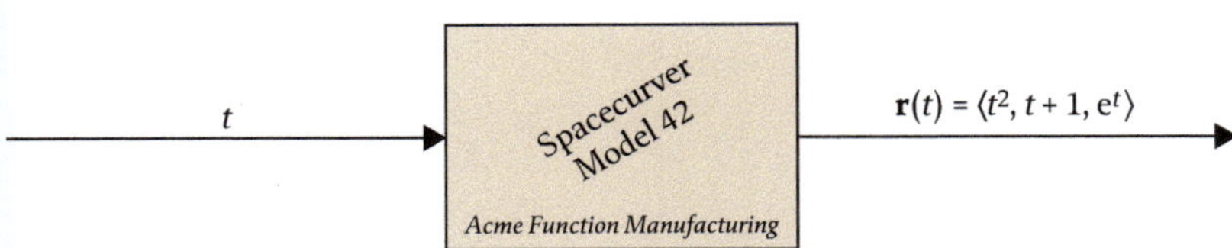

Figure 5.10 *One dimension input, three dimensions output*

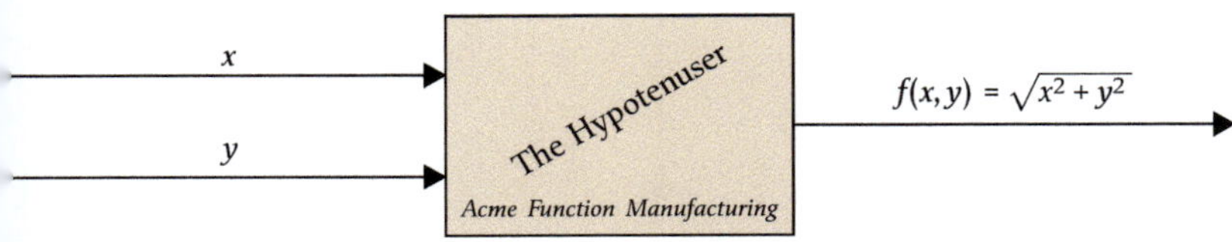

Figure 5.11 *Two dimensions input, one dimension output*

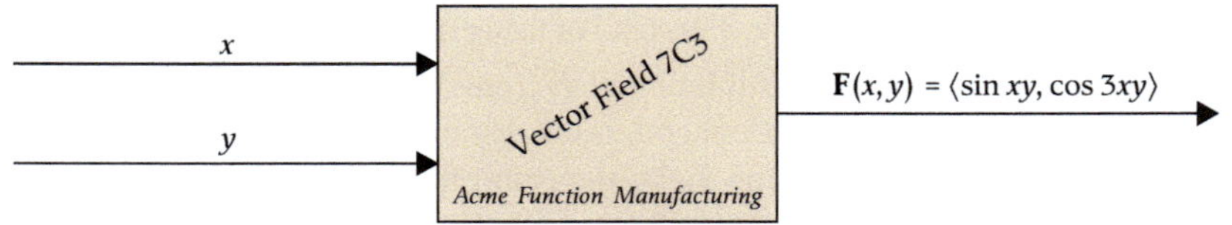

Figure 5.12 *Two dimensions input, two dimensions output*

5.2.1 Vector Fields

In this notation, **F** is a vector-valued function of two variables, and M and N are the *component functions*, each of which is a real-valued function of two variables.

In a two-dimensional vector field, with every point (x, y) in the xy-plane (or a portion of the plane) we associate a two-dimensional vector $\mathbf{F}(x, y) = \langle M(x, y), N(x, y) \rangle$. Two dimensions of input (the points in the plane) and two dimensions of output (the vectors) add up to four dimensions, making it seem difficult to visualize vector fields in the same manner as spacecurves or surfaces. But we know how to draw vectors, and we know how to locate points; an effective strategy for visualization is to draw vectors at their associated points. If the vectors are drawn at enough points and the vector field is continuous, we can infer what is happening between the drawn vectors. An example of a vector field is in Figure 5.13. The vector associated with a point is drawn emanating from the associated point.

A vector field **F** is continuous if each of its component functions is continuous.

Vector fields have many practical uses. For instance, wind velocity at ground level can be represented by a vector field. After all, velocity is a vector, possessing both direction and magnitude. If Figure 5.13 represents wind velocities, we can see strong winds from the southwest in the upper-right portion of the figure, whereas

With location represented by two dimensions and ground-level wind velocity reported in two dimensions, a vector field is a natural manner of representation. Many weather reports use a vector field, or a variation of a vector field, to visualize wind.

The chosen input points in a graph of a vector field are the initial points of the vectors.

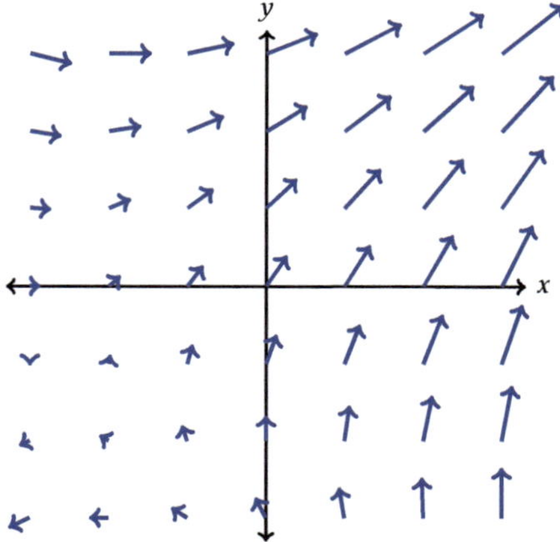

Figure 5.13 *A vector field. The length of an arrow represents the magnitude of the vector*

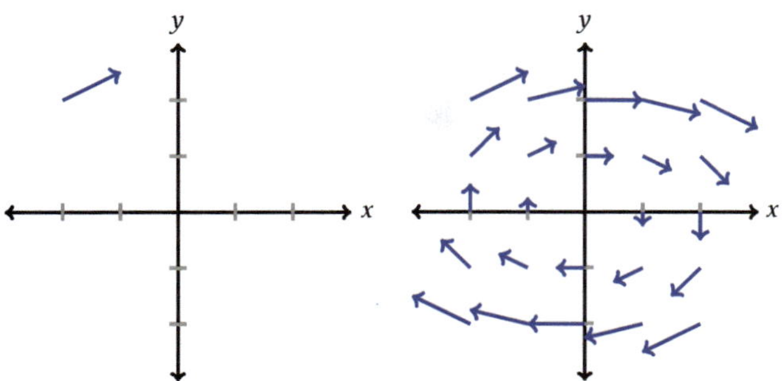

Figure 5.14 *Graphing the vector field of example 10; (left) the first drawn vector; (right) the completed graph*

much calmer winds are in the lower left portion of the figure. A vector field representing velocities is called a *velocity field*. Other types of vector fields include *force fields* (electricity and magnetism, gravity).

Example 10 *Graph the vector field* $\mathbf{F}(x, y) = \left\langle \frac{1}{2}y, -\frac{1}{4}x \right\rangle$.

Solution   First, we must choose the points at which to plot the vectors in the vector field. We choose a grid with $-2 \leq x \leq 2$ and $-2 \leq y \leq 2$ with step size 1; that is, we use $x = -2, -1, \ldots, 2$ and $y = -2, -1, \ldots, 2$, giving $5 \cdot 5 = 25$ total grid points.

For each chosen point, we calculate the vector $\mathbf{F}(x, y) = \left\langle \frac{1}{2}y, -\frac{1}{4}x \right\rangle$. Starting at the upper left corner,

$$\mathbf{F}(-2, 2) = \left\langle \frac{1}{2}(2), -\frac{1}{4}(-2) \right\rangle = \left\langle 1, \frac{1}{2} \right\rangle.$$

Then, we plot the vector $\left\langle 1, \frac{1}{2} \right\rangle$ with initial point $(-2, 2)$, as in Figure 5.14, left.

One down, twenty-four to go! The completed graph is shown in Figure 5.14, right.   ∎

Does the completed vector field look like it is drawn on a rectangular grid? The vectors can give the illusion that it is not, but if you locate the initial points of the left-most vectors in Figure 5.14, right, you will see that they lie on a vertical line, and if you locate the initial points of the vectors at the bottom of the graph, you will see that they lie on a horizontal line. Figure 5.15 shows the grid points to help overcome the illusion.

What should we do if the vectors are long enough that they overlap or interfere with each other? One option is to scale the magnitude of the vectors; that is a common option when using technology. Figure 5.16 shows the vector field of example 10 with the vectors scaled by $\frac{1}{2}$.

Be careful not to interpret the gridlines on the axes as part of the vector arrows in Figure 5.14, right.

"Southwest winds" blow from the southwest toward the northeast.

Winds with low velocity and varying direction are often called "light and variable."

A step size of $\frac{1}{2}$ would mean using $x = -2, -1.5, -1, \ldots, 2$ and $y = -2, -1.5, -1, \ldots, 2$, meaning $9 \cdot 9 = 81$ total grid points.

The rectangular grid used in the solution to example 10 is not the only available choice. A polar grid may be used, or any other set of convenient points.

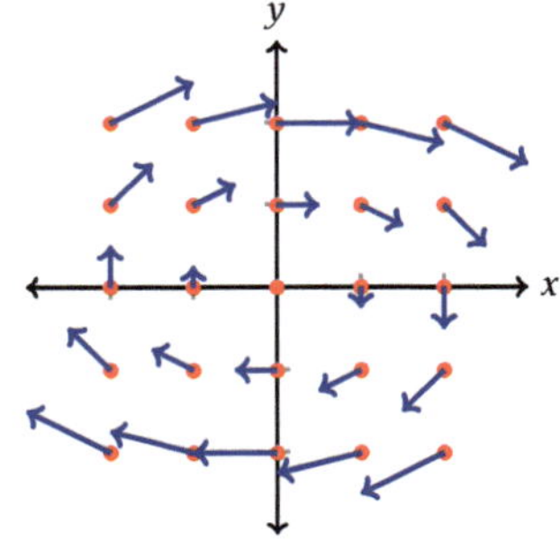

Figure   5.15 *The   vector   field   of example   10,   with   the   grid   points marked (red dots)*

Both CalcPlot3D and Desmos are capable of producing vector fields, although restrictions may apply.

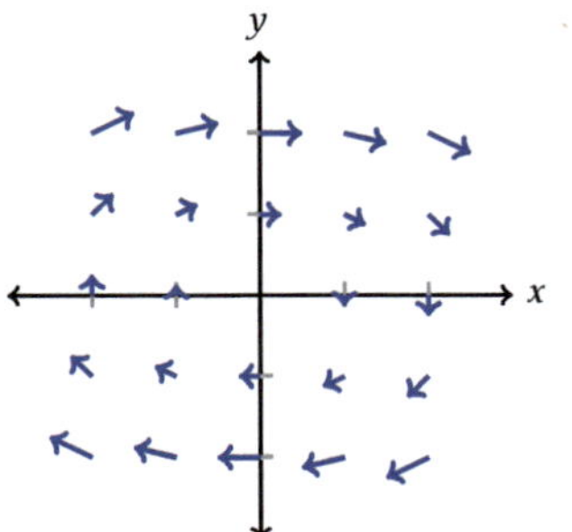

Figure 5.16 *The vector field of example 10 with vectors scaled by $\frac{1}{2}$. Compare to Figure 5.14, right. Notice the more rectangular appearance; the illusion of distortion is less with shorter vectors*

Think of a planet with center at the origin in Figure 5.17. The force of gravity exerted by the planet on an orbiting rock is greater the closer the rock is to the planet, and the direction of the gravitational force is toward the center of the planet.

As demonstrated by example 10, graphing a vector field by hand can be a rather tedious task. While it is worth doing one or two by hand, it is common practice to rely on technology to produce the graphs.

Example 11 *Use technology to graph the vector field* $\mathbf{F}(x, y, z) = \left\langle \frac{-x}{8(x^2+y^2+z^2)^{3/2}}, \frac{-y}{8(x^2+y^2+z^2)^{3/2}}, \frac{-z}{8(x^2+y^2+z^2)^{3/2}} \right\rangle.$

Solution A graph of the vector field produced by Mathematica is in Figure 5.17. Different CAS have different conventions and different options. In this graph, the color denotes the magnitude, as does the size of the arrow. Notice that the vectors all appear to be pointing in the direction of the origin, and when closer to the origin the vectors have larger magnitude. ∎

A vector field of the type in example 11 is called an *inverse square field*. Inverse square fields occur in the study of gravity and electricity. Inverse square fields are one type of force field.

An example–a very important example–of a vector field is the gradient. If $f(x, y) = x^2y - 4y^2$, then

$$\nabla f(x, y) = \langle 2xy, x^2 - 8y \rangle$$

is a vector field; it has two inputs x and y, and the output is a two-dimensional vector. Recall that the gradient at a point is orthogonal to a level curve at that point. So if we superimpose a graph of the vector field ∇f on top of a contour plot for f (Figure 5.18), the vectors in the vector field should appear orthogonal to the level

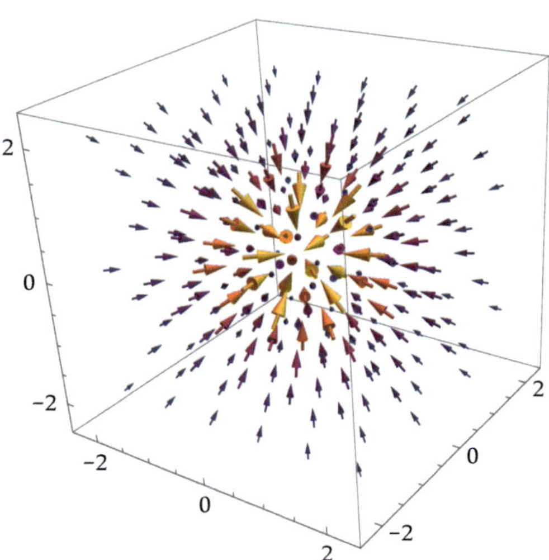

Figure 5.17 *The vector field of example 11, generated using Mathematica*

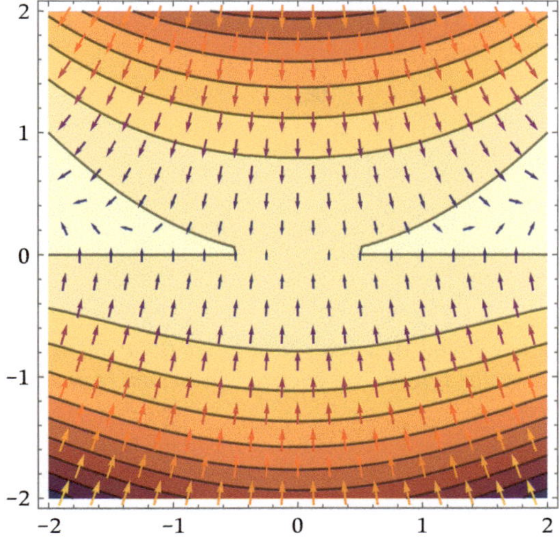

Figure 5.18 *The vector field $\nabla f(x, y)$ and the contour plot of $f(x, y) = x^2 y - 4y^2$*

curves from the contour plot. Be careful how this orthogonality is interpreted–the vector that corresponds to a point is placed with initial point, not terminal point, at the corresponding point, meaning that the direction of the vectors at their initial points, not necessarily at their terminal points where the arrows are, are orthogonal to level curves.

When a vector field comes from a gradient, it is called a *conservative vector field* or a *gradient field*. The vector field of example 11 is conservative; if $f(x, y, z) = \frac{1}{8\sqrt{x^2 + y^2 + z^2}}$, which is $\frac{1}{8d}$ where d is the distance from the point (x, y, z) to the origin, then

The term "conservative" is borrowed from physics.

$$\nabla f(x, y, z) = \mathbf{F}(x, y, z)$$

$$= \left\langle \frac{-x}{8(x^2 + y^2 + z^2)^{3/2}}, \frac{-y}{8(x^2 + y^2 + z^2)^{3/2}}, \frac{-z}{8(x^2 + y^2 + z^2)^{3/2}} \right\rangle.$$

Although the vector field in example 11 is conservative, the vector field in example 10 is not. How to determine if a vector field is conservative is discussed in Section 5.3.

The function f is called the *potential function* for $\mathbf{F}$.

The term "potential" is borrowed from physics.

5.2.2   Line integrals with vector fields: work

Work done by a variable force in one dimension is discussed in *Calculus Set Free* Section 6.5. The basic idea is this: because work = force $\cdot$ distance, we add small bits of force times small bits of distance to get an integral for calculating work. In one dimension, the force function is always in the direction (or opposite direction) of the number line where distance is subdivided. But it can be more

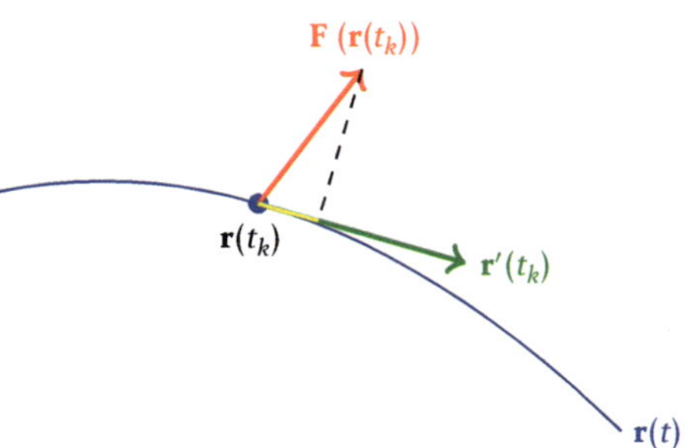

Figure 5.19 *A curve* **r** *(blue) in a force field* **F**. *At the point* **r**(t_k) *the force vector (red) and tangent vector (green) are shown. The yellow segment is the tangential component of force*

In one dimension, everything is parallel. In two or more dimensions, that's not true.

complicated in two or more dimensions. First, we can move along any curve (or spacecurve), not just in one dimension. Second, the force might not be parallel to the curve.

Consider moving a particle along the path $\mathbf{r}(t) = \langle x(t), y(t) \rangle$, $a \le t \le b$, through a force field $\mathbf{F}(x, y)$. Subdivide the interval $a \le t \le b$ (using Ω pieces) and consider the work done from t_{k-1} to t_k. By local linearity, the infinitesimal-length path is straight and work is force times distance on that straight piece. But if the force is not in the direction of the curve, we need to find the component of force in that direction, called the *tangential component of force*; consult Figure 5.19. That's the scalar component of $\mathbf{F}(\mathbf{r}(t_k))$ in the direction of $\mathbf{r}'(t_k)$. Therefore, the force on this piece is

The same discussion applies to three dimensions, that is, $\mathbf{r}(t) = \langle x(t), y(t), z(t) \rangle$ and $\mathbf{F}(x, y, z)$.

Recall the formula $\text{comp}_\mathbf{u} \mathbf{v} = \frac{\mathbf{v} \cdot \mathbf{u}}{\|\mathbf{u}\|}$.

$$\text{comp}_{\mathbf{r}'(t_k)} \mathbf{F}(\mathbf{r}(t_k)) = \frac{\mathbf{F}(\mathbf{r}(t_k)) \cdot \mathbf{r}'(t_k)}{\|\mathbf{r}'(t_k)\|}.$$

The sum is

The distance the particle moves on this piece is its length, which is the usual arc length element $\|\mathbf{r}'(t_k)\|\Delta t$. Therefore, the work performed on this piece, force times distance, is

$$\sum_{k=1}^{\Omega} \mathbf{F}(\mathbf{r}(t_k)) \cdot \mathbf{r}'(t_k)\Delta t.$$

$$\frac{\mathbf{F}(\mathbf{r}(t_k)) \cdot \mathbf{r}'(t_k)}{\|\mathbf{r}'(t_k)\|} \|\mathbf{r}'(t_k)\|\Delta t = \mathbf{F}(\mathbf{r}(t_k)) \cdot \mathbf{r}'(t_k)\Delta t.$$

To rewrite as an integral we replace the summation with the integral symbol, remove the subscript k, replace Δt with dt, and insert the limits of integration.

Add all the pieces and rewrite as an integral, and the work done is

$$\int_a^b \mathbf{F}(\mathbf{r}(t)) \cdot \mathbf{r}'(t)\, dt.$$

> ### WORK ALONG A SMOOTH CURVE IN A FORCE FIELD
>
> Let $\mathbf{F}$ be a continuous force field and let $\mathbf{r}(t), a \leq t \leq b$ be a curve in the domain of $\mathbf{F}$. Then the work done by $\mathbf{F}$ along the curve is
>
> $$W = \int_a^b \mathbf{F}(\mathbf{r}(t)) \cdot \mathbf{r}'(t) \, dt.$$

The definition applies to both two and three dimensions.

The $\cdot$ in the formula is the dot product.

Example 12 *Find the work done by the force* $\mathbf{F}(x, y, z) = \langle xy, z + 1, \sqrt{y} \rangle$ *along the spacecurve* $\mathbf{r}(t) = \langle t, t^2, 3t - 1 \rangle, 0 \leq t \leq 1.$

Solution We first compute the force in terms of the parameter t:

$$\mathbf{F}(\mathbf{r}(t)) = \mathbf{F}(t, t^2, 3t - 1)$$

$$= \left\langle t(t^2), 3t - 1 + 1, \sqrt{t^2} \right\rangle$$

$$= \langle t^3, 3t, t \rangle.$$

Because the work integral is computed with respect to the parameter t, we need to express force in terms of the parameter t.

Line 2 takes $\langle xy, z + 1, \sqrt{y} \rangle$ and replaces x by t, y by t^2, and z by $3t - 1$; line 3 simplifies, noting that t is nonnegative. The result is a vector.

Next, we compute the tangent to the curve:

$$\mathbf{r}'(t) = \langle 1, 2t, 3 \rangle.$$

Then, the integrand of the work integral is the dot product of the results of the previous two computations:

$$\mathbf{F}(\mathbf{r}(t)) \cdot \mathbf{r}'(t) = \langle t^3, 3t, t \rangle \cdot \langle 1, 2t, 3 \rangle$$

$$= t^3 + 6t^2 + 3t.$$

Finally, we compute the integral to find work:

$$W = \int_a^b \mathbf{F}(\mathbf{r}(t)) \cdot \mathbf{r}'(t) \, dt$$

$$= \int_0^1 (t^3 + 6t^2 + 3t) \, dt$$

$$= \left(\frac{1}{4}t^4 + 2t^3 + \frac{3}{2}t^2 \right)\Big|_0^1$$

$$= \frac{1}{4} + 2 + \frac{3}{2} - 0 = \frac{15}{4}.$$

The work performed by $\mathbf{F}$ along $\mathbf{r}$ is $\frac{15}{4}$. $\blacksquare$

We follow the custom of calculus textbooks and omit the determination of units for work.

Writing $\mathbf{F}$ and $\mathbf{r}$ in terms of component functions, that is, $\mathbf{F}(x, y) = M(x, y), N(x, y) \rangle$ and $\mathbf{r}(t) = \langle x(t), y(t) \rangle$, the work formula can be rewritten as

Line 1 is the work formula; line 2 rewrites in terms of the component functions; line 3 computes the dot product; line 4 rewrites as two integrals; line 5 replaces $x'(t)\,dt = \frac{dx}{dt}\,dt$ with dx and $y'(t)\,dt = \frac{dy}{dt}\,dt$ with dy, and also rewrites M and N in terms of x and y to arrive at a traditional form.

$$
\begin{aligned}
W &= \int_a^b \mathbf{F}(\mathbf{r}(t)) \cdot \mathbf{r}'(t)\,dt \\[2mm]
&= \int_a^b \langle M(x(t), y(t)), N(x(t), y(t)) \rangle \cdot \langle x'(t), y'(t) \rangle\,dt \\[2mm]
&= \int_a^b \left(M(x(t), y(t))x'(t) + N(x(t), y(t))y'(t) \right)\,dt \\[2mm]
&= \int_a^b M(x(t), y(t))x'(t)\,dt + \int_a^b N(x(t), y(t))y'(t)\,dt \\[2mm]
&= \int_a^b M(x, y)\,dx + N(x, y)\,dy,
\end{aligned}
$$

which is in the form of line integrals with respect to x and y.

Alternately, noting that for a vector-valued function $\mathbf{r}(t)$ the unit tangent vector is

$$
\mathbf{T}(t) = \frac{\mathbf{r}'(t)}{\|\mathbf{r}'(t)\|},
$$

we can rearrange to find a replacement for $\mathbf{r}'(t)$:

$$
\mathbf{r}'(t) = \mathbf{T}(t)\,\|\mathbf{r}'(t)\|.
$$

Then the work formula can also be rearranged in the form of a line integral with respect to arc length:

Line 1 is the work formula; line 2 replaces $\mathbf{r}'(t)$ with $\mathbf{T}(t)\|\mathbf{r}'(t)\|$ as computed previously; line 3 replaces $\|\mathbf{r}'(t)\|\,dt$, the vector form of the arc length differential, with ds.

$$
\begin{aligned}
W &= \int_a^b \mathbf{F}(\mathbf{r}(t)) \cdot \mathbf{r}'(t)\,dt \\[2mm]
&= \int_a^b \mathbf{F}(\mathbf{r}(t)) \cdot \mathbf{T}(t)\|\mathbf{r}'(t)\|\,dt \\[2mm]
&= \int_a^b \mathbf{F}(\mathbf{r}(t)) \cdot \mathbf{T}(t)\,ds.
\end{aligned}
$$

Line integrals of vector fields along curves have other applications in physics in addition to calculating work, justifying the need for a more general definition.

In fact, if you compare the derivation of the work integral formula with the derivation of the line integral formula, you will find that both integrate over a curve; it is only the function that differs. This motivates the definition of a line integral of a vector field (of any kind, not just a force field) along a curve, connecting the ideas of line integrals with respect to arc length and with respect to the coordinate variables.

Definition 1 LINE INTEGRAL OF A VECTOR FIELD ALONG A SPACE-CURVE *Let* $\mathbf{F}(x, y, z) = \langle M(x, y, z), N(x, y, z), P(x, y, z)\rangle$ *be a continuous vector field and let* $C = \mathbf{r}(t) = \langle x(t), y(t), z(t)\rangle$, $a \leq t \leq b$ *be a smooth spacecurve lying in the domain of* $\mathbf{F}$. *Then the line integral of* $\mathbf{F}$ *along* C *is given by*

Of course, an analogous definition holds for two dimensions.

$$\int_C \mathbf{F} \cdot \mathbf{T}\, ds = \int_a^b \mathbf{F}(\mathbf{r}(t)) \cdot \mathbf{T}(t)\, ds$$

$$= \int_a^b \mathbf{F}(\mathbf{r}(t)) \cdot \mathbf{r}'(t)\, dt$$

$$= \int_a^b M(x, y, z)\, dx + N(x, y, z)\, dy + P(x, y, z)\, dz.$$

Although there are several ways (including some not stated above) to write and remember the work integral, or more generally the line integral of a vector field along a curve, in practice they all rely on expressing the integrand in terms of the parameter t and evaluating an ordinary integral, as demonstrated in example 12.

Example 13 *Compute the line integral of* $\mathbf{F}(x, y, z) = \langle xy, xz, yz\rangle$ *along the curve* $\mathbf{r}(t) = \langle 1, t, t^2\rangle$, $0 \leq t \leq 1$.

Solution First, we compute $\mathbf{F}$ along the curve:

Compare this solution to that of example 12.

$$\mathbf{F}(\mathbf{r}(t)) = \mathbf{F}(1, t, t^2)$$

$$= \langle 1 \cdot t, 1 \cdot t^2, t \cdot t^2\rangle = \langle t, t^2, t^3\rangle.$$

Next, we compute the tangent to the curve:

$$\mathbf{r}'(t) = \langle 0, 1, 2t\rangle.$$

The integrand is

$$\mathbf{F}(\mathbf{r}(t)) \cdot \mathbf{r}'(t) = \langle t, t^2, t^3\rangle \cdot \langle 0, 1, 2t\rangle = 0 + t^2 + 2t^4.$$

The value of the line integral is

$$\int_a^b \mathbf{F}(\mathbf{r}(t)) \cdot \mathbf{r}'(t)\, dt = \int_0^1 (t^2 + 2t^4)\, dt$$

$$= \left(\tfrac{1}{3}t^3 + \tfrac{2}{5}t^5\right)\Big|_0^1$$

$$= \frac{1}{3} + \frac{2}{5} = \frac{11}{15}.$$

∎

Reading Exercise 3   Set up the integral (do not evaluate) for computing the line integral of $\mathbf{F}(x,y) = \langle x + y, 3x^2 \rangle$ along the curve $\mathbf{r}(t) = \langle t, t^2 \rangle$, $0 \le t \le 2$.

The next example requires computation of the vector field.

Example 14   *Let $f(x,y) = 2xy^3$. (a) Compute the line integral of $\nabla f(x,y)$ along C_1, the curve $y = x^2$ from $(0,0)$ to $(1,1)$. (b) Compute the line integral of $\nabla f(x,y)$ along C_2, the line segment from $(0,0)$ to $(1,1)$.*

Solution   First, we compute the vector field $\nabla f(x,y)$:

$$\nabla f(x,y) = \langle 2y^3, 6xy^2 \rangle.$$

(a)   The curve $y = x^2$ from $(0,0)$ to $(1,1)$ may be parameterized by

$$\mathbf{r}(t) = \langle t, t^2 \rangle,$$

with $0 \le t \le 1$. Computing from there using $\nabla f(x,y)$ for $\mathbf{F}(x,y)$ in the formulas,

$$\nabla f(\mathbf{r}(t)) = \nabla f(t, t^2) = \langle 2t^6, 6t^5 \rangle$$
$$\mathbf{r}'(t) = \langle 1, 2t \rangle$$
$$\nabla f(\mathbf{r}(t)) \cdot \mathbf{r}'(t) = \langle 2t^6, 6t^5 \rangle \cdot \langle 1, 2t \rangle = 2t^6 + 12t^6 = 14t^6$$

$$\int_0^1 14t^6 \, dt = 2t^7 \Big|_0^1 = 2.$$

(b)   The segment from $(0,0)$ to $(1,1)$ may be parameterized by

$$\mathbf{r}(t) = \langle t, t \rangle,$$

with $0 \le t \le 1$. Then

$$\nabla f(\mathbf{r}(t)) = \nabla f(t, t) = \langle 2t^3, 6t^3 \rangle$$
$$\mathbf{r}'(t) = \langle 1, 1 \rangle$$
$$\nabla f(\mathbf{r}(t)) \cdot \mathbf{r}'(t) = 2t^3 + 6t^3 = 8t^3$$

$$\int_0^1 8t^3 \, dt = 2t^4 \Big|_0^1 = 2.$$

The two paths give the same result.   ∎

Try any other path from $(0,0)$ to $(1,1)$ for the vector field of example 14 and you will get the same result. But didn't we say that the path matters, with examples 2 and 3 of Section 5.1 taking different paths and getting different results from the line integral? Is there something special about the setup of example 14? Stay tuned–the mystery is solved in the next section.

Example 15 *Let $f(x, y, z) = x^2 y + \ln z$ and let C be the curve $\mathbf{r}(t) = \langle 3t, t^2, t + 1 \rangle$, $0 \le t \le 2$. Compute $\int_C f_x(x, y, z)\, dx + f_y(x, y, z)\, dy + f_z(x, y, z)\, dz$.*

Solution   Let's begin by computing the partial derivatives of f so that we may determine the integral we wish to calculate. The partial derivatives are

$$f_x(x, y, z) = 2xy$$

$$f_y(x, y, z) = x^2$$

$$f_z(x, y, z) = \frac{1}{z}.$$

Compare to the solution to example 14. Other than notation, computing the partial derivatives is the same task as computing the gradient.

Therefore, the line integral we wish to calculate is

$$\int_C 2xy\, dx + x^2\, dy + \frac{1}{z}\, dz,$$

which is a line integral of the type discussed in example 8 of Section 5.1. Using the components of the curve $x = 3t$, $y = t^2$, and $z = t + 1$, the differentials are $dx = 3\, dt$, $dy = 2t\, dt$, and $dz = dt$. Substituting all these quantities into the integral, we have

Compare the remainder of this solution to the solution of Section 5.1 example 8.

$$
\begin{aligned}
&\int_C 2xy\, dx + x^2\, dy + \frac{1}{z}\, dz \\[4pt]
&= \int_0^2 2(3t)(t^2) \cdot 3\, dt + (3t)^2 \cdot 2t\, dt + \frac{1}{t+1}\, dt \\[4pt]
&= \int_0^2 \left(18t^3 + 18t^3 + \frac{1}{t+1} \right) dt \\[4pt]
&= \left(\frac{36t^4}{4} + \ln|t + 1| \right)\Bigg|_0^2 \\[4pt]
&= 9 \cdot 16 + \ln 3 - (0 + \ln 1) = 144 + \ln 3.
\end{aligned}
$$

Notice that we do not try to find antiderivatives with respect to x, y, and z. Line 2 replaces x by $3t$, y by t^2, z by $t + 1$, dx by $3\, dt$, dy by $2t\, dt$, and dz by dt; line 3 simplifies partially, including factoring out the dt; line 4 finds an antiderivative with respect to t and line 5 finishes the computation.

■

Ans. to reading exercise 3:
$\int_0^2 (t + t^2 + 6t^3)\, dt$

EXERCISES 5.2

   1–6. Graph the vector field.

      1. $\mathbf{F}(x, y) = \left\langle 1 - \frac{1}{3}x, \frac{1}{3}y \right\rangle$, $0 \le x \le 3$, $0 \le y \le 3$

      2. $\mathbf{F}(x, y) = \left\langle \frac{1}{4}x^2, \frac{x+y}{4} \right\rangle$, $-2 \le x \le 2$, $-2 \le y \le 2$

It is implied that the vector field should be graphed without the use of technology, as per example 10.

3. $F(x, y) = \langle -y, x \rangle$, $-2 \leq x \leq 2$, $-2 \leq y \leq 2$; scale the vectors by $\frac{1}{2}$

4. $F(x, y) = \langle x + y, x - y \rangle$, $0 \leq x \leq 3$, $0 \leq y \leq 3$; scale the vectors by $\frac{1}{6}$

5. $F(x, y) = \left\langle \frac{5 - y^2}{5}, \frac{1}{2} \sin x \right\rangle$, $-\pi \leq x \leq \pi, -2 \leq y \leq 2$

6. $F(x, y) = \left\langle \frac{\pi}{8}, \cos x \right\rangle$, $0 \leq x \leq \pi, -2 \leq y \leq 2$

7–12. Use technology to graph the vector field.

7. $F(x, y) = \langle x^2 + y^2, x^2 - y^2 \rangle$
8. $F(x, y) = \langle \sqrt{x^2 + y^2 + 1}, 3x - 2y \rangle$
9. $F(x, y) = \langle e^x, e^y \rangle$
10. $F(x, y) = \langle \ln xy, x - y \rangle$, domain $x \geq 1, y \geq 1$
11. $F(x, y, z) = \langle x, y, z \rangle$
12. $F(x, y, z) = \langle 1, 1, \cos x \rangle$

13–18. Find the work done by the force F along the curve r.

13. $F(x, y) = \langle 2x, 3y^2 \rangle$, $r(t) = \langle t^2, t \rangle, 0 \leq t \leq 2$
14. $F(x, y) = \langle x^2 y, xy^2 \rangle$, curve $r(t) = \langle t^3, t^2 \rangle, 0 \leq t \leq 1$
15. $F(x, y) = \left\langle \frac{x}{y}, x^2 \right\rangle$, $r(t) = \langle \sqrt{t}, t \rangle, 1 \leq t \leq 4$
16. $F(x, y) = \left\langle \frac{e^y}{\sqrt{x}}, \frac{1}{1 + \sqrt{x(y+1)^2}} \right\rangle$, curve $r(t) = \langle t^2, t - 1 \rangle, 1 \leq t \leq 2$
17. $F(x, y, z) = \langle xy, y^2, z^2 \rangle$, $r(t) = \langle \cos t, \sin t, t \rangle, 0 \leq t \leq 4\pi$
18. $F(x, y, z) = \langle yz, xz, xy \rangle$, $r(t) = \langle t, t^2, t^3 \rangle, 0 \leq t \leq 1$

19–22. Compute the line integral of F along the curve.

19. $F(x, y, z) = \langle x, y, 1 \rangle$, segment from $(1, 1, 1)$ to $(2, 3, 4)$
20. $F(x, y, z) = \langle y + z, x + z, x + y \rangle$, segment from $(0, 1, 0)$ to $(2, 2, -1)$
21. $F(x, y) = \langle x \cos y, \sin y \rangle$, curve $y = x^2$ from $(0, 0)$ to $(2, 4)$
22. $F(x, y) = \langle 2x + 3, y - 1 \rangle$, top half of unit circle counterclockwise from $(1, 0)$ to $(-1, 0)$

23. Let $f(x, y) = xe^y$. (a) Compute the line integral of $\nabla f(x, y)$ along the segment from $(0, 0)$ to $(2, 4)$. (b) Compute the line integral of $\nabla f(x, y)$ along the segment from $(2, 4)$ to $(0, 0)$. (c) How do the answers to parts (a) and (b) compare?

24. Let $f(x, y) = x^3 y^2 - \ln x$. Compute the line integral of $\nabla f(x, y)$ along the curve $y = x^2$ from $(1, 1)$ to $(3, 9)$.

25. Let $f(x, y, z) = xy - z^2$ and let C be the curve $r(t) = \langle t, t^3, 2 \rangle$, $1 \leq t \leq 3$. Compute $\int_C f_x(x, y, z)\, dx + f_y(x, y, z)\, dy + f_z(x, y, z)\, dz$.

26. Let $f(x, y, z) = x^2$ and let C be the curve $r(t) = \langle \cos t, \sin t, t^{15} e^t \rangle$, $0 \leq t \leq \pi$. Compute $\int_C f_x(x, y, z)\, dx + f_y(x, y, z)\, dy + f_z(x, y, z)\, dz$.

27. Another form of a line integral of a vector field that could have been given in definition 1 is $\int_C M\,dx + N\,dy + P\,dz$. What happens if we reverse the direction of the curve? That is, how does $\int_{-C} M\,dx + N\,dy + P\,dz$ compare with $\int_C M\,dx + N\,dy + P\,dz$?

28. (a) Determine a parameterization $\mathbf{r}(t)$ of the line segment from $(0, 0, 1)$ to $(6, 4, 3)$.

    (b) Repeat example 15 for the curve in part (a). Using this path between the points, is the answer the same as for example 15?

5.3 The Fundamental Theorem of Line Integrals

In Section 5.1, examples 2 and 3 demonstrated that the path matters for a line integral; different paths between the same two points might give different results. But example 14 of Section 5.2 showed that sometimes different paths give the same result. The vector field of that example is conservative, which turns out to be the key to path independence.

5.3.1 Open, connected, simply connected

The properties of a region D turn out to be important when stating theorems for the remainder of this chapter. This is true in single-variable calculus; for instance, the mean value theorem requires the function f to be continuous on the closed interval $[a, b]$ and differentiable on the open interval (a, b). Open intervals do not contain their endpoints, whereas closed intervals do. In two or more dimensions, the same idea holds; an *open region* does not contain its boundary points, whereas a *closed region* does (Figure 5.20). An example of an open region is $x^2 + y^2 < 4$, and an example of a closed region is $x^2 + y^2 \leq 4$. For each point in an open region D, there is an open disk (see Section 3.9) containing the point that lies entirely in D.

See *Calculus Set Free* Section 3.2.

If a region contains some, but not all, of its boundary points, the region is neither closed nor open, similar to the interval $[2, 5)$.

Figure 5.20 *(Left) an open region does not contain its boundary points, as indicated by the dashed boundary; (right) a closed region contains its boundary points, as indicated by the solid boundary*

A *connected region D* has the property that for any two points A and B in D, there exists a smooth curve C from A to B, lying entirely in D. Visually, the difference between a connected region and a disconnected region is easy to distinguish (Figure 5.21), with a connected region consisting of one piece and a disconnected region consisting of two or more pieces.

A *closed curve* (Figure 5.22) is a curve whose initial and terminal endpoints are identical. A *simple closed curve* is a closed curve that does not intersect itself (except at its identical endpoints).

Visually, a connected region is *simply connected* if it has no holes (Figure 5.23). Every simple closed curve contained in a simply connected region can be shrunk to a single point without ever leaving the region, but a simple closed curve surrounding a hole would have to pass through the hole, leaving the region, in order to shrink to a single point (Figure 5.24).

Figure 5.21 *(Left) a connected region; (right) a region that is not connected*

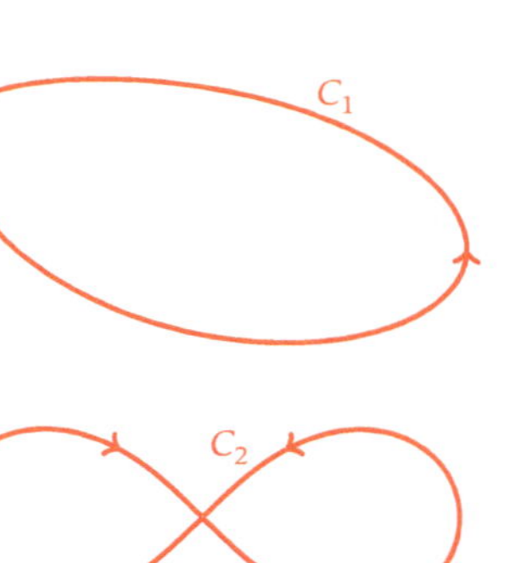

Figure 5.22 *(Top) a simple closed curve C_1; (bottom) a closed curve C_2 that is not simple*

Figure 5.23 *(Left) a simply connected region; (right) a connected, but not simply connected, region*

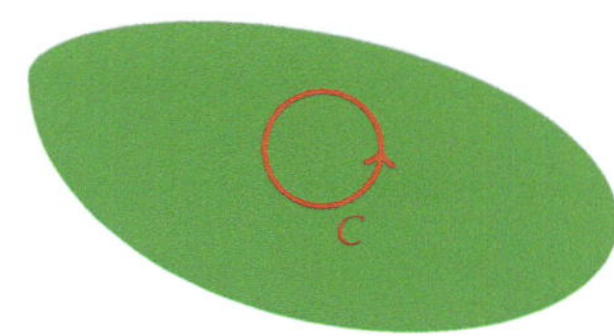
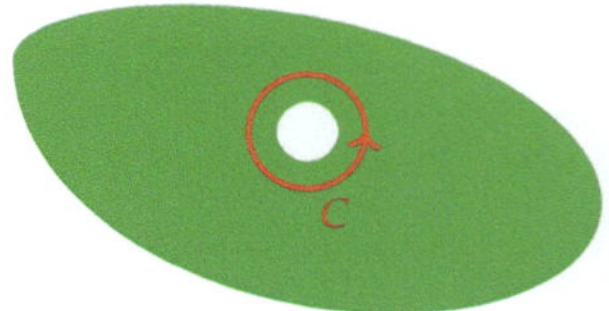

Figure 5.24 *(Left) any simple closed curve can shrink to a point without leaving the region; (right) a simple closed curve that cannot shrink to a point without leaving the region*

5.3.2   Path independence

Suppose **F** is a conservative vector field on an open, connected region D, that is, there is some function f such that $\mathbf{F}(x,y) = \nabla f(x,y) = \langle f_x(x,y), f_y(x,y) \rangle$. Suppose $\mathbf{r}(t) = \langle x(t), y(t) \rangle$, $a \le t \le b$ is a smooth curve from $A = \mathbf{r}(a)$ to $B = \mathbf{r}(b)$ lying entirely within D. Then the composition of **F** and **r**,

$$\mathbf{F}(\mathbf{r}(t)) = \langle f_x(x(t), y(t)), f_y(x(t), y(t)) \rangle,$$

The point A is the initial point of the curve, $(x(a), y(a))$, and the point B is the terminal point of the curve, $(x(b), y(b))$.

is a function of t. To calculate the line integral of **F** along the curve, we calculate an integral with respect to t, namely

$$\int_a^b \mathbf{F}(\mathbf{r}(t)) \cdot \mathbf{r}'(t)\, dt.$$

The composition of f and **r**,

$$f(\mathbf{r}(t)) = f(x(t), y(t)),$$

is also a function of t. So we may ask what is its derivative, $\frac{d}{dt} f(\mathbf{r}(t))$? The multivariable chain rule, case 1, of Section 3.7 applies to this composition:

Line 1 is the multivariable chain rule, case 1, where f is a function of x and y, and each of x and y are functions of t; line 2 rewrites the derivatives and partial derivatives as functions of t; line 3 rewrites as a dot product of two vectors (check by calculating the dot product of line 3 to get the result of line 2); and line 4 uses the previously calculated formula for $\mathbf{F}(\mathbf{r}(t))$.

$$\frac{d}{dt} f(\mathbf{r}(t)) = \frac{\partial f}{\partial x}\frac{dx}{dt} + \frac{\partial f}{\partial y}\frac{dy}{dt}$$

$$= f_x(x(t), y(t))x'(t) + f_y(x(t), y(t))y'(t)$$

$$= \langle f_x(x(t), y(t)), f_y(x(t), y(t)) \rangle \cdot \langle x'(t), y'(t) \rangle$$

$$= \mathbf{F}(\mathbf{r}(t)) \cdot \mathbf{r}'(t),$$

which is the integrand of the line integral! Therefore, $f(\mathbf{r}(t))$ is an antiderivative with respect to t, and the line integral is

$$\int_a^b \mathbf{F}(\mathbf{r}(t)) \cdot \mathbf{r}'(t)\, dt = f(\mathbf{r}(t))\Big|_a^b$$

Line 1 finds an antiderivative with respect to t; line 2 evaluates at $t = b$, at $t = a$, and subtracts.

$$= f(\mathbf{r}(b)) - f(\mathbf{r}(a)) = f(B) - f(A).$$

This means that whatever the path from the point A to the point B, as long as we stay within the region, the answer is the same, $f(B) - f(A)$. The integral is path independent, and we can evaluate the antiderivative at the terminal point, at the initial point, and subtract, just as with the fundamental theorem of calculus.

An analogous theorem applies for three dimensions. The theorem also applies to piecewise smooth curves.

Theorem 2 FUNDAMENTAL THEOREM OF LINE INTEGRALS *Suppose A and B are two points in an open, connected region D. Suppose further that* $F = \langle f_x(x,y), f_y(x,y) \rangle$ *is a conservative vector field whose components are continuous throughout D. Then the line integral of F from A to B is path independent; if* $\mathbf{r}(t)$, $a \le t \le b$, *is any smooth curve from A to B lying entirely within D, then*

$$\int_A^B \nabla f \cdot d\mathbf{r} = \int_a^b F(\mathbf{r}(t)) \cdot \mathbf{r}'(t)\, dt = f(B) - f(A).$$

The notation $d\mathbf{r}$ is short for $\mathbf{r}'(t)\, dt$.

Let's revisit the computation of Section 5.2 example 14, using the fundamental theorem.

Example 16 *Let* $f(x,y) = 2xy^3$. *Compute the line integral of* $\nabla f(x,y)$ *along any path from* $(0,0)$ *to* $(1,1)$.

Compare to the solution of Section 5.2 example 14.

Solution Because f is a polynomial in the variables x and y, its partial derivatives exist and are continuous everywhere, so we may take D to be the entire xy-plane. Then for any path $\mathbf{r}$ from $(0,0)$ to $(1,1)$,

In line 1, the function f serves as an antiderivative for ∇f, with respect to any curve $\mathbf{r}$; line 2 evaluates at $(x,y) = (1,1)$, at $(x,y) = (0,0)$, and subtracts.

$$\int_{(0,0)}^{(1,1)} \nabla f \cdot d\mathbf{r} = 2xy^3 \Big|_{(0,0)}^{(1,1)}$$

$$= 2(1)(1^3) - 2(0)(0^3) = 2.$$

$\blacksquare$

How marvelously simple! But we have an advantage in example 16; we already know that the vector field is conservative, and we already know f. Instead, what if we are given the vector field F directly, without knowledge of f? How can we tell if the field is conservative, and how can we calculate a potential function f?

Clairaut's theorem can help with the first question.

5.3.3   When is a vector field conservative?

For $F(x,y,z) = \langle M(x,y,z), N(x,y,z), P(x,y,z) \rangle$, in addition to $\frac{\partial M}{\partial y} = \frac{\partial N}{\partial x}$, we also need $\frac{\partial M}{\partial z} = \frac{\partial P}{\partial x}$ and $\frac{\partial N}{\partial z} = \frac{\partial P}{\partial y}$.

Theorem 3 COMPONENT TEST FOR CONSERVATIVE FIELDS *Suppose* $F(x,y) = \langle M(x,y), N(x,y) \rangle$ *is a vector field for which M and N have continuous partial derivatives throughout a simply connected region D. Then F is conservative if and only if*

$$\frac{\partial M}{\partial y} = \frac{\partial N}{\partial x}.$$

Proof. Part 1. Suppose that $\mathbf{F}$ is conservative, that is, $\mathbf{F} = \nabla f$ for some function $f(x, y)$ on D. Then $M = f_x$ and $N = f_y$. The hypothesis that M and N have continuous partial derivatives means that f_{xy} and f_{yx} are both continuous. By Clairaut's theorem, $f_{xy} = f_{yx}$, that is, $\dfrac{\partial M}{\partial y} = \dfrac{\partial N}{\partial x}$.

See Section 3.5.

Part 2 of the proof is partially presented in Section 5.9.

Example 17 *Is the vector field* $\mathbf{F}(x, y) = \langle 2y^3, 6xy^2 \rangle$ *conservative?*

Solution Following the component test for conservative fields, we calculate two partial derivatives:

$$\frac{\partial}{\partial y} 2y^3 = 6y^2$$

$$\frac{\partial}{\partial x} 6xy^2 = 6y^2.$$

Line 1 computes the partial derivative of the x-component with respect to y; line 2 computes the partial derivative of the y-component with respect to x.

Because the two partial derivatives are the same, the vector field is conservative. ∎

Notice that the vector field of example 17 is the same vector field as for example 16. The solution to example 17 shows how we could start if we are given the vector field F directly instead of beginning with f.

Reading Exercise 4 Is the vector field $\mathbf{F}(x, y) = \langle 4xy, 3x^2 \rangle$ conservative?

Example 18 *Is the vector field* $\mathbf{F}(x, y, z) = \langle xy, xz, yz \rangle$ *conservative?*

Solution For three dimensions, there are three pairs of partial derivatives to check. First, we choose to check the x- and z-components:

$$\frac{\partial}{\partial z} xy = 0$$

$$\frac{\partial}{\partial x} yz = 0.$$

Line 1 computes the partial derivative of the x-component with respect to z; line 2 computes the partial derivative of the z-component with respect to x.

These two partial derivatives are the same. Before we can conclude that the vector field is conservative, there are two more pairs to check. Next, we check the x-and y-components:

$$\frac{\partial}{\partial y} xy = x$$

$$\frac{\partial}{\partial x} xz = z.$$

Line 1 computes the partial derivative of the x-component with respect to y; line 2 computes the partial derivative of the y-component with respect to x.

Because these two partial derivatives are not the same, the vector field is not conservative. There is no need to check the last pair; if any pair does not match, the vector field is not conservative. ∎

The vector field of example 18 is the same vector field as example 13 of Section 5.2. Because the vector field is not conservative, the fundamental theorem of line integrals does not apply and we cannot calculate the line integral of Section 5.2 example 13 in the simple manner of example 16. The approach used in Section 5.2 is our only option.

In other words, learning the approach of Section 5.2 is necessary.

5.3.4 Calculating potential functions

The second question remains: how can we calculate a potential function f? We explain using the next example.

Example 19 *The vector field* $\mathbf{F}(x, y) = \langle 8xy^2, 8x^2y - 3y^2 \rangle$ *is conservative. Calculate a potential function f.*

Solution We know that the x-component of $\mathbf{F}(x, y)$ is $f_x(x, y)$, the partial derivative of f with respect to x. Going backward, then, we can find an antiderivative of the x-component with respect to x, treating y as a constant:

$$f(x, y) = \int f_x(x, y)\, dx = \int 8xy^2\, dx = 4x^2y^2 + g(y),$$

Note that $\frac{\partial}{\partial x}(4x^2y^2 + g(y)) = 8xy^2 + 0$, which is the x-component of $\mathbf{F}$. Whatever y-stuff is in $g(y)$ doesn't matter for finding f_x. But g can't include anything with x, because that would affect our calculation of f_x.

where $g(y)$, a constant with respect to the variable x, is used in place of a constant C. This gives a tentative version of f, one for which f_x is correct. What is left is to determine the value of $g(y)$.

We also need $f_y(x, y)$ to be the y-component of $\mathbf{F}(x, y)$. To that end, calculate f_y from the tentative f:

The derivative of $g(y)$ with respect to y is $g'(y)$.

$$f_y(x, y) = 8x^2y + g'(y).$$

Now we may set this result equal to the y-component of $\mathbf{F}(x, y)$, and solve for $g'(y)$:

$$8x^2y + g'(y) = 8x^2y - 3y^2$$

$$g'(y) = -3y^2.$$

Finally, we find an antiderivative of g' with respect to y,

$$g(y) = \int g'(y)\, dy = \int -3y^2\, dy = -y^3 + C.$$

Substituting this expression for $g(y)$ in the tentative f gives

With C serving as any constant, this is the collection of all possible potential functions.

$$f(x, y) = 4x^2y^2 - y^3 + C.$$

Then $\nabla f(x, y) = \langle 8xy^2, 8x^2y - 3y^2 \rangle = \mathbf{F}(x, y)$, as desired.

Instead of starting with x, we could start with y. The alternate solution is presented more compactly, with less explanation.

Alternate solution   Integrate the y-component of $\mathbf{F}$ to determine a tentative f:

$$f(x,y) = \int f_y(x,y)\, dy = \int (8x^2y - 3y^2)\, dy = 4x^2y^2 - y^3 + g(x).$$

Because x is treated as a constant, we use $g(x)$ as the constant.

Calculate f_x from the tentative f:

$$f_x(x,y) = 8xy^2 - 0 + g'(x).$$

Set the result equal to the x-component of $\mathbf{F}$ and solve for $g'(x)$:

$$8xy^2 + g'(x) = 8xy^2$$
$$g'(x) = 0.$$

Integrate the result to determine $g(x)$:

$$g(x) = \int g'(x)\, dx = \int 0\, dx = 0 + C.$$

Ans. to reading exercise 4:

no

Substitute the result into f:

$$f(x,y) = 4x^2y^2 - y^3 + C,$$

matching the first solution.   ■

The procedure for three dimensions is analogous.

Example 20   *The vector field* $\mathbf{F}(x,y,z) = \left\langle \ln y, \dfrac{x}{y} - z^2, -2yz \right\rangle$ *is conservative. Calculate a potential function* f.

Solution   Integrate the x-component of $\mathbf{F}$ to determine a tentative f:

Alternately, we could start with the y- or z- component.

$$f(x,y,z) = \int f_x(x,y,z)\, dx = \int \ln y\, dx = x\ln y + g(y,z).$$

Both y and z are treated as constants when integrating with respect to x, so we use $g(y,z)$ as the constant.

Calculate f_y from the tentative f:

We could calculate f_z next instead.

$$f_y(x,y,z) = \frac{x}{y} + g_y(y,z).$$

The derivative of $g(y,z)$ with respect to y is $g_y(y,z)$.

Set the result equal to the y-component of $\mathbf{F}$ to solve for $g_y(y,z)$:

$$\frac{x}{y} + g_y(y,z) = \frac{x}{y} - z^2$$
$$g_y(y,z) = -z^2.$$

Because z is treated as a constant, we use $h(z)$ as the constant.

Integrate to determine $g(y, z)$:

$$g(y, z) = \int g_y(y, z)\, dy = \int -z^2\, dy = -yz^2 + h(z).$$

Rewrite the tentative f using the result:

$$f(x, y, z) = x\ln y - yz^2 + h(z).$$

From the previously calculated $f(x, y, z) = x\ln y + g(y, z)$, we replace $g(y, z)$ with $-yz^2 + h(z)$.

Calculate f_z next:

$$f_z(x, y, z) = 0 - 2yz + h'(z).$$

Set the result equal to the z-component of $\mathbf{F}$ to solve for $h'(z)$:

$$-2yz + h'(z) = -2yz$$
$$h'(z) = 0.$$

Integrate to determine $h(z)$:

$$h(z) = \int h'(z)\, dz = \int 0\, dz = 0 + C.$$

Use the result to write the potential function:

From the previously calculated $f(x, y, z) = x\ln y - yz^2 + h(z)$, we replace $h(z)$ with C.

$$f(x, y, z) = x\ln y - yz^2 + C. \qquad\blacksquare$$

5.3.5   Putting it all together

Given a vector field, we can first test to see if it is conservative. If it is, we can find a potential function and calculate the desired integral using the fundamental theorem.

Compare this solution to the solutions to examples 17, 19, and 16.

Example 21 *Determine the line integral of* $\mathbf{F}(x, y) = \langle 3x, 4y^2 \rangle$ *from* $(1, 1)$ *to* $(4, 4)$, *or state that the field is not conservative and the path matters.*

Solution First, we test to determine if the field is conservative:

What can you conclude about a vector field of the form $\mathbf{F}(x, y) = \langle h(x), k(y) \rangle$?

$$\frac{\partial}{\partial y} 3x = 0$$

$$\frac{\partial}{\partial x} 4y^2 = 0.$$

The partial derivatives match, so the vector field is conservative. Next, we find a potential function:

$$f(x, y) = \int 3x\, dx = \frac{3}{2}x^2 + g(y)$$

$$f_y(x, y) = 0 + g'(y)$$

$$0 + g'(y) = 4y^2$$

$$g(y) = \int 4y^2\, dy = \frac{4}{3}y^3 + C$$

$$f(x, y) = \frac{3}{2}x^2 + \frac{4}{3}y^3 + C.$$

Finally, we calculate the line integral using the fundamental theorem:

$$\int_{(1,1)}^{(4,4)} \mathbf{F} \cdot d\mathbf{r} = \left(\frac{3}{2}x^2 + \frac{4}{3}y^3\right)\Bigg|_{(1,1)}^{(4,4)}$$

$$= \frac{3 \cdot 16}{2} + \frac{4 \cdot 64}{3} - \left(\frac{3}{2} + \frac{4}{3}\right) = \cdots = \frac{213}{2}.$$

■

A slight change in **F** makes for a quite different conclusion.

Example 22 *Determine the line integral of* $\mathbf{F}(x, y) = \langle 4y^2, 3x \rangle$ *from* $(1, 1)$ *to* $(4, 4)$, *or state that the field is not conservative and the path matters.*

Solution First, we test to determine if the field is conservative:

$$\frac{\partial}{\partial y}4y^2 = 8y$$

$$\frac{\partial}{\partial x}3x = 3.$$

The partial derivatives do not match, so the field is not conservative. The path matters.

■

Try the procedure for calculating a potential function for the vector field of example 22. First we integrate the x-component of **F** with respect to x:

$$f(x, y) = \int 4y^2\, dx = 4xy^2 + g(y).$$

Then, calculate f_y from the tentative f:

$$f_y(x, y) = 8xy + g'(y).$$

Line 1 finds a tentative f by integrating the x-component of **F** with respect to x; line 2 calculates f_y from the tentative f; line 3 sets the calculated f_y equal to the y-component of **F**; line 4 integrates the result to find $g(y)$; line 5 substitutes $g(y)$ into the tentative f, giving the potential function.

What can you conclude about a potential function for a vector field of the form $F(x, y) = \langle h(x), k(y) \rangle$?

By theorem 2, any potential function can be used as an antiderivative. Including the $+C$ is not necessary. Line 2 evaluates at $(x, y) = (4, 4)$, at $(x, y) = (1, 1)$, and subtracts.

Compare to example 21; the x- and y-components are switched.

What can you conclude about a vector field of the form $F(x, y) = \langle k(y), h(x) \rangle$?

Because the path matters and we are not given a specific path from $(1, 1)$ to $(4, 4)$, we cannot calculate the value of the line integral.

Set the result equal to the y-component of $\mathbf{F}$ and solve for $g'(y)$:

$$8xy + g'(y) = 3x$$

$$g'(y) = 3x - 8xy.$$

Now we see a problem. The function g' is a function of the variable y only, and cannot contain any reference to the variable x. This is where the procedure breaks down, and no such potential function f exists for which $\nabla f = \mathbf{F}$. Although it is possible to skip the component test for conservative fields and use this analysis to conclude that a vector field is not conservative, it is not recommended; it does not always save effort (compare to the solution to example 22) and is more error-prone.

5.3.6 Technical results

The implication of example 22 is that if a vector field is not conservative, then line integrals are not path independent. Is this always true? Equivalently, if all line integrals are independent of path, must the vector field be conservative?

Start with a continuous vector field $\mathbf{F}(x, y) = \langle M(x, y), N(x, y)\rangle$ on an open, connected region D, and suppose that line integrals of $\mathbf{F}$ on curves lying entirely within D are path independent. Choose a point (x_0, y_0) in D. Then for any point (x, y) in D, we may define

This is where the hypothesis of D being connected is used.

$$f(x, y) = \int_{(x_0, y_0)}^{(x, y)} \mathbf{F} \cdot d\mathbf{r}.$$

Because the path does not matter, the value of $f(x, y)$ depends only on the point (x, y) and not the path.

To show that $\mathbf{F}$ is conservative, we shall show that $\nabla f(x, y) = \mathbf{F}(x, y) = \langle M(x, y), N(x, y)\rangle$.

Let (a, b) be a point in D. Using the definition of partial derivative, for any infinitesimal α,

This is where the hypothesis of D being an open region is used.

$$f_x(a, b) = \frac{f(a + \alpha, b) - f(a, b)}{\alpha}$$

Line 1 is the definition of f_x at the point (a, b); line 2 uses the definition of the function f; line 3 uses additivity of line integrals and path independence (details in exercise 47); line 4 rewrites as an integral with respect to t.

$$= \frac{1}{\alpha}\left(\int_{(x_0, y_0)}^{(a+\alpha, b)} \mathbf{F} \cdot d\mathbf{r} - \int_{(x_0, y_0)}^{(a, b)} \mathbf{F} \cdot d\mathbf{r} \right)$$

$$= \frac{1}{\alpha} \int_{(a, b)}^{(a+\alpha, b)} \mathbf{F} \cdot d\mathbf{r}$$

$$= \frac{1}{\alpha} \int_{(a, b)}^{(a+\alpha, b)} \mathbf{F}(\mathbf{r}(t)) \cdot \mathbf{r}'(t)\, dt,$$

For this path, $t = 0$ corresponds to the point (a, b), and $t = \alpha$ corresponds to the point $(a + \alpha, b)$.

for any path $\mathbf{r}$ from (a, b) to $(a + \alpha, b)$, because line integrals of $\mathbf{F}$ in D are path independent. Therefore, we may use the segment between the points, $\mathbf{r}(t) = \langle a + t, b\rangle$, $0 \le t \le \alpha$. Proceeding as if we are calculating the integral, we have

$$\mathbf{F}(\mathbf{r}(t)) = \mathbf{F}(a + t, b) = \langle M(a + t, b), N(a + t, b) \rangle$$

$$\mathbf{r}'(t) = \langle 1, 0 \rangle$$

$$\mathbf{F}(\mathbf{r}(t)) \cdot \mathbf{r}'(t) = M(a + t, b) \cdot 1 + N(a + t, b) \cdot 0 = M(a + t, b).$$

For line 2, recall that a and b are constants, t is the variable.

Then the integral becomes

$$\frac{1}{\alpha} \int_{(a,b)}^{(a+\alpha,b)} \mathbf{F}(\mathbf{r}(t)) \cdot \mathbf{r}'(t)\, dt = \frac{1}{\alpha} \int_0^{\alpha} M(a + t, b)\, dt.$$

To continue, we need an antiderivative of $M(a + t, b)$, which is a function of the variable t. Suppose $g(t)$ is such an antiderivative. Then continuing the calculation for $f_x(a, b)$,

Here is where the hypothesis of $\mathbf{F}$ being continuous is used.

$$f_x(a, b) = \frac{1}{\alpha} \int_0^{\alpha} M(a + t, b)\, dt$$

$$= \frac{1}{\alpha} g(t) \Big|_0^{\alpha}$$

$$= \frac{1}{\alpha} \left(g(\alpha) - g(0) \right)$$

$$= g'(0)$$

$$= M(a + 0, b) = M(a, b).$$

Line 1 is the progress made so far in calculating $f_x(a, b)$; line 2 finds an antiderivative with respect to t; line 3 evaluates at $t = \alpha$, at $t = 0$. and subtracts; line 4 uses the definition of derivative (the equation is true for every infinitesimal, so it is true for α); for line 5, because $g(t)$ is an antiderivative of $M(a + t, b)$, then $g'(t) = M(a + t, b)$.

Because this is true for any point (a, b) in D, $f_x(x, y) = M(x, y)$ throughout D. A similar argument shows that $f_y(x, y) = N(x, y)$, and we have succeeded in showing that $\nabla f(x, y) = \mathbf{F}(x, y)$, meaning that $\mathbf{F}$ is conservative.

Theorem 4 PATH INDEPENDENCE *Let $\mathbf{F}$ be a continuous vector field on an open, connected region D. Then all line integrals $\int_C \mathbf{F} \cdot d\mathbf{r}$, where C lies in D, are independent of path if and only if $\mathbf{F}$ is conservative.*

This theorem holds in both two and three dimensions. The proof of the "only if" direction has just been completed; the proof of the "if" direction is presented immediately preceding the fundamental theorem.

Suppose that $\mathbf{F}$ is a vector field on a region D and that line integrals of $\mathbf{F}$ are path independent. Consider a closed curve C lying in D, and let A and B be two distinct points of the curve C (Figure 5.25). Let C_1 be the portion of the curve from A to B, and let C_2 follow the curve back to A. Then C_1 and $-C_2$ are both curves from A to B, and by path independence

We also assume that C is piecewise smooth.

$$\int_{C_1} \mathbf{F} \cdot d\mathbf{r} = \int_{-C_2} \mathbf{F} \cdot d\mathbf{r}.$$

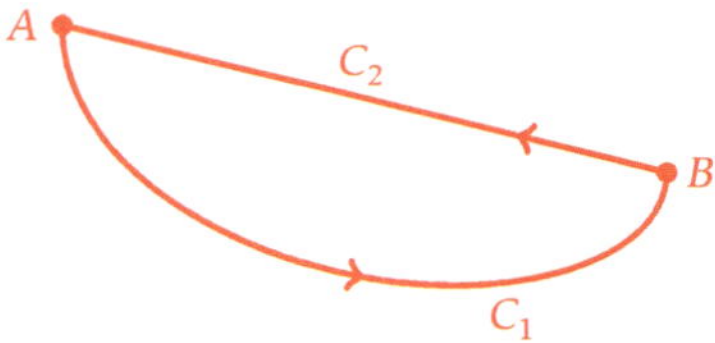

Figure 5.25 *A curve C consisting of C_1 followed by C_2*

The line integral along a closed curve C is sometimes written as $\oint_C \mathbf{F} \cdot d\mathbf{r}$.

Then

$$\int_C \mathbf{F} \cdot d\mathbf{r} = \int_{C_1} \mathbf{F} \cdot d\mathbf{r} + \int_{C_2} \mathbf{F} \cdot d\mathbf{r}$$

$$= \int_{-C_2} \mathbf{F} \cdot d\mathbf{r} + \int_{C_2} \mathbf{F} \cdot d\mathbf{r}$$

$$= -\int_{C_2} \mathbf{F} \cdot d\mathbf{r} + \int_{C_2} \mathbf{F} \cdot d\mathbf{r} = 0.$$

In other words, if all line integrals are path independent, then the integral along any closed curve must be 0.

Conversely, suppose the integral along any closed curve is 0. Let C_1 and C_2 be two distinct curves from A to B (Figure 5.26). Let C be the curve C_1 followed by $-C_2$. Then C is a closed curve, and by the assumption,

$$0 = \int_C \mathbf{F} \cdot d\mathbf{r} = \int_{C_1} \mathbf{F} \cdot d\mathbf{r} + \int_{-C_2} \mathbf{F} \cdot d\mathbf{r} = \int_{C_1} \mathbf{F} \cdot d\mathbf{r} - \int_{C_2} \mathbf{F} \cdot d\mathbf{r}.$$

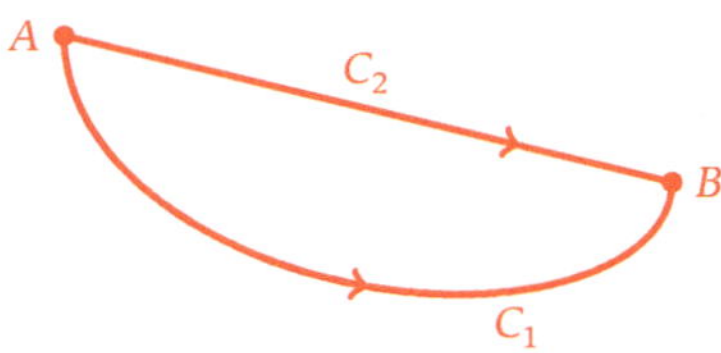

Figure 5.26 *Two curves from A to B*

Therefore

$$\int_{C_1} \mathbf{F} \cdot d\mathbf{r} = \int_{C_2} \mathbf{F} \cdot d\mathbf{r},$$

and because this is true for any two paths between A and B, line integrals are path independent in D.

This fact, combined with the previous theorem, yields the following.

Theorem 5 LINE INTEGRALS OF CLOSED CURVES *Let* $\mathbf{F}$ *be a continuous vector field on an open, connected region* D. *Then the following three statements are equivalent:*

(1) $\mathbf{F}$ *is conservative.*
(2) *Line integrals of* $\mathbf{F}$ *are path independent in* D.
(3) *If* C *is a closed curve in* D, *then* $\int_C \mathbf{F} \cdot d\mathbf{r} = 0$.

EXERCISES 5.3

Exercises 1–4 are repeated (or modified) from Section 5.2 exercises 23–26.

1. Let $f(x, y) = xe^y$. (a) Compute the line integral of $\nabla f(x, y)$ along any path from $(0, 0)$ to $(2, 4)$. (b) Compute the line integral of $\nabla f(x, y)$ along any path from $(2, 4)$ to $(0, 0)$. (c) How do the answers to parts (a) and (b) compare?

2. Let $f(x, y) = x^3 y^2 - \ln x$. Compute the line integral of $\nabla f(x, y)$ along any path from $(1, 1)$ to $(3, 9)$ with $x > 0$.

3. Let $f(x, y, z) = xy - z^2$ and let C be the curve $\mathbf{r}(t) = \langle t, t^3, 2 \rangle$, $1 \leq t \leq 3$. Compute $\int_C f_x(x, y, z)\, dx + f_y(x, y, z)\, dy + f_z(x, y, z)\, dz$.

4. Let $f(x, y, z) = x^2$ and let C be the curve $\mathbf{r}(t) = \langle \cos t, \sin t, t^{15} e^t \rangle$, $0 \leq t \leq \pi$. Compute $\int_C f_x(x, y, z)\, dx + f_y(x, y, z)\, dy + f_z(x, y, z)\, dz$.

5–10. Compute the line integral of ∇f along the path, or say why the fundamental theorem does not apply.

Notice that f is given, not ∇f.

5. $f(x, y) = \sin x + \cos 3y$, any path from $(0, \pi)$ to $(\pi, 0)$

6. $f(x, y) = \dfrac{y}{x + 1}$, any path from $(0, 0)$ to $(2, 2)$ for which $x \geq 0$

7. $f(x, y, z) = \dfrac{x + y}{3z^2}$, any path from $(1, 2, 1)$ to $(5, 7, 2)$

8. $f(x, y, z) = xe^y - z^2 y$, any path from $(0, 0, 1)$ to $(2, 1, -2)$

9. $f(x, y) = \sin x - \tan y$, any path from $(\pi, 0)$ to $(0, \pi)$

10. $f(x, y) = \dfrac{y}{x - 1}$, any path from $(0, 0)$ to $(2, 2)$ for which $x \geq 0$

11–22. Is the vector field conservative?

11. $\mathbf{F}(x, y) = \langle x^3, e^y \rangle$

12. $\mathbf{F}(x, y) = \langle 4xy, 6x^2 y^2 \rangle$

13. $\mathbf{F}(x, y) = \langle 2xy - 3, x^2 + 7 \rangle$

14. $\mathbf{F}(x, y) = \langle \sin x, (y - 2)^3 \rangle$

15. $\mathbf{F}(x, y) = \langle 3 - 4y, xy^2 \rangle$

16. $\mathbf{F}(x, y) = \left\langle \dfrac{3}{4y}, -\left(\dfrac{3x-2}{4}\right) y^{-2} \right\rangle$

17. $\mathbf{F}(s, t) = \langle -(\sin s)(\sin t), (\cos s)(\cos t) \rangle$

18. $\mathbf{F}(s, t) = \langle 2st, 2s^2 \rangle$

19. $\mathbf{F}(x, y, z) = \langle 3x^2 yz, x^3 z - 2yz^3, x^3 y - 3y^2 z^2 \rangle$

20. $\mathbf{F}(x, y, z) = \langle e^{yz}, xze^{yz}, xye^{yz} \rangle$

21. $\mathbf{F}(x, y, z) = \langle 3y^2, 6xy, 5z + 2y \rangle$

22. $\mathbf{F}(x, y, z) = \langle yz, xz, xy + y \rangle$

23–30. Find a potential function for the conservative vector field.

Each of the vector fields in exercises 23–30 is conservative.

23. $\mathbf{F}(x, y) = \langle 3x, y^2 \rangle$

24. $\mathbf{F}(x, y) = \langle e^x, \cosh y \rangle$

25. $\mathbf{F}(x, y) = \langle 2xy - y^2, x^2 - 2xy \rangle$

26. $\mathbf{F}(x, y) = \langle 3y + x, 3x - y \rangle$

27. $\mathbf{F}(x, y, z) = \langle 4y, 4x, 2z \rangle$

28. $\mathbf{F}(x, y, z) = \langle 4x, 4y, 2z \rangle$

29. $\mathbf{F}(x, y, z) = \langle 2xy - 3z, x^2 + 4z^2, -3x + 8yz + 5 \rangle$

30. $\mathbf{F}(x, y, z) = \left\langle 4 \ln y, \dfrac{4x}{y} + 6y - 2z^4, -8yz^3 \right\rangle$

31–40. Determine the line integral of $\mathbf{F}$ from A to B, or state that the field is not conservative and the path matters.

31. $\mathbf{F}(x, y) = \langle 4x, y^2 \rangle$, $A = (1, 0)$, $B = (0, 4)$

32. $\mathbf{F}(x, y) = \langle y^2, 3x^2 \rangle$, $A = (1, -2)$, $B = (2, 3)$

33. $\mathbf{F}(x, y) = \left(\dfrac{-2y}{x^3} - 2\right) \mathbf{i} + \left(\dfrac{1}{x^2} + 3\right) \mathbf{j}$, $A = (4, 5)$, $B = (1, 3)$

34. $\mathbf{F}(x, y) = (\sin y)\mathbf{i} + (x \cos y)\mathbf{j}$, $A = \left(2, \dfrac{\pi}{2}\right)$, $B = \left(1, \dfrac{3\pi}{2}\right)$

35. $F(x, y) = \langle \ln x, x \ln y \rangle$, $A = (1, 1)$, $B = (e, e)$
36. $F(x, y) = \langle 2x - 2y, 2y - 2x \rangle$, $A = (0, 0)$, $B = (4, 4)$
37. $F(x, y, z) = \langle 3yz, 3xz, 3xy + 1 \rangle$, $A = (2, 1, 0)$, $B = (1, 2, 5)$
38. $F(x, y, z) = \left\langle 2xy\sqrt{z}, x^2\sqrt{z}, \frac{x^2 y}{2\sqrt{z}} \right\rangle$, $A = (1, 1, 1)$, $B = (2, 2, 4)$
39. $F(x, y, z) = (2^x \ln 2)\,i + 2y\,j + 2k$, $A = (0, 1, 1)$, $B = (-1, 2, 5)$
40. $F(x, y, z) = (2xy^3)\,i + (3x^2 y^2)\,j + (x + y)k$, $A = (2, -1, 1)$, $B = (1, 4, 2)$

41. Find $\int_C 3x^2\, dx - z^2\, dy - 2yz\, dz$ along any curve C from $(1, 2, 3)$ to $(2, -1, 1)$.
42. Find $\int_C 2x^2 z\, dz + 21y^2\, dy + x^3\, dz$ along any curve C from $(10, 2, 2)$ to $(1, 10, 10)$.
43. Determine the work done by the force $F(x, y, z) = \langle 3z, 1, 3x \rangle$ in moving a particle from $(2.0, 1.7, 0.3)$ to $(0.5, 1.0, 0.9)$.
44. Determine the work done by the force $F(x, y, z) = \langle 2x, 2y + 3, 2z \rangle$ in moving a particle from $(0, 0, 0)$ to $(1, 5, 1)$.
45. Return to example 21. After determining that the vector field $F(x, y) = \langle 3x, 4y^2 \rangle$ is conservative, we could use any curve between the points $(1, 1)$ and $(4, 4)$ to determine the value of the line integral.

    (a) Find a parameterization r of the line segment from $(1, 1)$ to $(4, 4)$.
    (b) Use the methods of Section 5.2 to compute the line integral of F along the curve from part (a).
    (c) Compare the work of parts (a) and (b) to the solution to example 21. Is one method more efficient that the other?
    (d) Compare the integral evaluated in part (b) to the integral evaluated in the solution to example 21.

46. Return to example 15 of Section 5.2.

    (a) Compute the integral using the fundamental theorem.
    (b) Compare the work of part (a) to the solution to example 15 of Section 5.2. Is one method more efficient that the other?
    (c) Compare the work of part (a) to the work of solving exercise 28 of Section 5.2. Is one method more efficient that the other?

47. The proof of theorem 4 contains the calculation

$$\frac{1}{\alpha} \left(\int_{(x_0, y_0)}^{(a+\alpha, b)} F \cdot dr - \int_{(x_0, y_0)}^{(a, b)} F \cdot dr \right) = \frac{1}{\alpha} \int_{(a, b)}^{(a+\alpha, b)} F \cdot dr.$$

Justify the calculation as follows.

    (a) Draw a diagram with the points (x_0, y_0), (a, b), and $(a + \alpha, b)$. (Hint: the two points with y-coordinate b should have the same y-coordinate in your diagram.) Draw a curve C_1 from (x_0, y_0) to $(a + \alpha, b)$ and a curve C_2 from $(a + \alpha, b)$ to (a, b).

(b) Using your diagram as a guide, relabel $\int_{(x_0,y_0)}^{(a,b)} \mathbf{F} \cdot d\mathbf{r}$ as a line integral along C_1, C_2, or some combination thereof.

(c) If you have not already done so, rewrite the line integral from (b) as the sum of two line integrals.

(d) Finish justifying the calculation.

(e) What is the corresponding part of the proof of the fundamental theorem of calculus, part 1? (See *Calculus Set Free* Section 4.5.)

48. (a) Does example 17 require "part 1" or "part 2" of theorem 3? (Refer to the parts of the proof.)

    (b) Repeat part (a) for example 18.

49. We wish to compute $\int_C \nabla f \cdot d\mathbf{r}$, where $f(x,y) = x\sqrt[3]{y}$ and C is any path from $(1,-1)$ to $(2,1)$.

    (a) Compute $\nabla f(x,y)$ and identify the issue with using the fundamental theorem.

    (b) For any positive infinitesimal ω, compute $\int_{C_1} \nabla f \cdot d\mathbf{r}$ for any curve C_1 from $(1,-1)$ to $(2,-\omega)$.

    (c) For any positive infinitesimal ω, compute $\int_{C_2} \nabla f \cdot d\mathbf{r}$ for any curve C_2 from $(2,\omega)$ to $(2,1)$.

    (d) Add the answers from parts (b) and (c). Observe that this appears consistent with the idea of an improper integral of type 2, from $(1,-1)$ to $(2,1)$, through the point $(2,0)$.

    (e) Repeat parts (b)–(d) for a path crossing $y = 0$ at $(k,0)$ instead of $(2,0)$. Does your answer agree with your answer to part (d)?

    (f) Does this improper line integral appear to have the same answer for every path from $(1,-1)$ to $(2,1)$?

50. We wish to compute $\int_C \nabla f \cdot d\mathbf{r}$, where $f(x,y) = \frac{x+3}{y}$ and C is any path from $(1,-1)$ to $(2,1)$.

    (a) Identify the issue with using the fundamental theorem.

    (b) Ignoring the answer to part (a), compute $\int_C \nabla f \cdot d\mathbf{r}$ incorrectly by finding $f(2,1) - f(1,-1)$.

    (c) For any positive infinitesimal ω, compute $\int_{C_1} \nabla f \cdot d\mathbf{r}$ for any curve C_1 from $(1,-1)$ to $(2,-\omega)$.

    (d) Following the same concepts as for single-variable improper integrals, what can you conclude about $\int_C \nabla f \cdot d\mathbf{r}$?

51. We wish to compute $\int_C \nabla f \cdot d\mathbf{r}$, where $f(x,y) = \tan^{-1} x + \frac{3}{y}$.

    (a) Compute the line integral along any path C from $(0,1)$ to $(\Omega,1)$, where Ω is positive infinite.

    (b) Compute the line integral along any path C from $(0,1)$ to $(0,\Omega)$, where Ω is positive infinite.

52. An example of an inverse square field is the electrostatic force between two oppositely charged bodies. *Coulomb's law* states that

The term "inverse square" comes from the denominator (inverse) d^2 (distance squared).

the magnitude of the force is $\|\mathbf{F}\| = \frac{k|q_1 q_2|}{d^2}$, where k is a constant, q_1 and q_2 are the quantities of the charges, and d is the distance between the two bodies.

(a) Suppose A, with charge q_1, is fixed at the origin $(0, 0, 0)$ and B, with charge q_2, is located at a point (x, y, z). Rewrite the magnitude of the force experienced by B in terms of x, y, and z.

(b) Determine a unit vector in the direction of the force on B, assuming the force attracts.

(c) Combine the results of (a) and (b) to determine the force $\mathbf{F}(x, y, z)$ on B.

(d) Determine a potential function f.

(e) Determine the work done by the electrostatic force in moving B from $(2, 3, 3)$ to $(1, 1, 1)$.

5.4   Green's Theorem

In Section 5.3 we learned that for a conservative vector field, the line integral along a closed curve is 0. But what if the vector field isn't conservative?

5.4.1   Green's theorem: turning a line integral into a double integral

Any simple closed curve C encloses a connected region D, as in Figure 5.27. Such a curve can be traversed in two directions, clockwise or counterclockwise. The *positive orientation* is counterclockwise, the same as Olympic track races, where the region D is to the left as the curve is traversed. The clockwise direction, where the region D is to the right as the curve is traversed, has a *negative orientation*.

In an Olympic race, the region inside the track is to the runner's left.

If the closed curve C is not simple, as in Figure 5.28, the orientation can change when the curve crosses itself. And if the region D does not contain its boundary, it isn't connected. These are reasons to use simple closed curves.

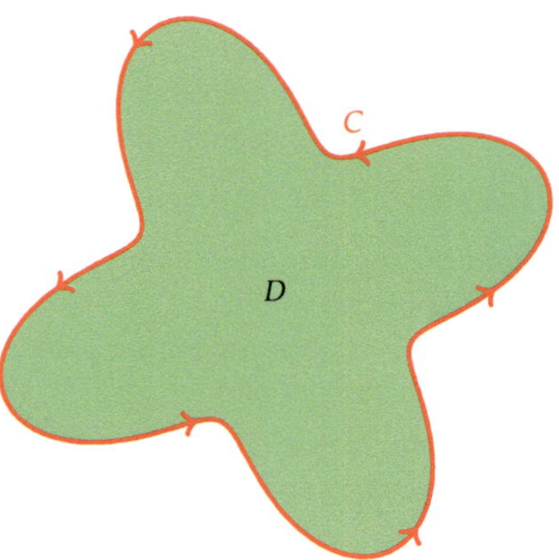

Figure 5.27 *A closed curve C (red) with positive orientation, enclosing a region D (green)*

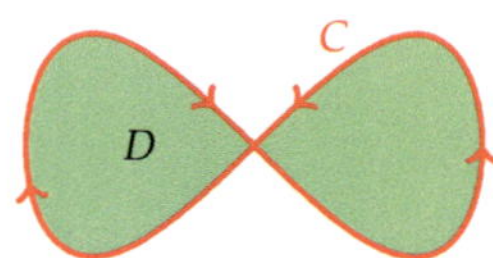

Figure 5.28 *A non-simple closed curve C (red), enclosing a region D (green). The curve appears to have positive orientation on one side of the crossing point, and negative orientation on the other side*

Because the beginning and ending point of the curve doesn't really matter–any point on the curve could be designated as such–it makes sense to have a notation that avoids naming such a point. Therefore

$$\oint_C M(x,y)\,dx + N(x,y)dy$$

Notice the circle in the center of the integration symbol. A circle is an example of a closed curve.

is used to denote the line integral along the closed curve C, traversed exactly once positively (counterclockwise).

The *region enclosed by* C is the closed region bounded by C, including the boundary C.

Reading Exercise 5   Is the curve positively oriented or negatively oriented?

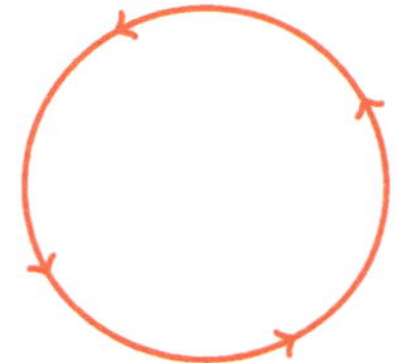

With notation and terminology in hand, we are now ready to state Green's theorem.

Theorem 6 GREEN'S THEOREM *Let C be a positively oriented, piecewise smooth, simple closed curve enclosing the region D. If M and N are functions of x and y that have continuous partial derivatives throughout an open region R containing D, then*

It is traditional to state Green's theorem using M and N instead of $M(x,y)$ and $N(x,y)$.

$$\oint_C M\,dx + N\,dy = \iint_D \left(\frac{\partial N}{\partial x} - \frac{\partial M}{\partial y} \right) dA.$$

Giving a full proof, including for regions shaped like those of Figure 5.27, is left for advanced courses. We shall prove the theorem for any region that can be expressed as both of the form $g_1(x) \le y \le g_2(x)$, $a \le x \le b$, and of the form $h_3(y) \le x \le h_4(y)$, $c \le y \le d$, as in Figure 5.29. This is accomplished in two pieces: the M portions of the formula, $\oint_C M\,dx = \iint_D -\frac{\partial M}{\partial y}\,dA$, and the N portions of the formula, $\oint_C N\,dy = \iint_D \frac{\partial N}{\partial x}\,dA$.

First, using the left side of Figure 5.29, the curve C is comprised of two curves, $C = C_1 \cup C_2$. The curve C_1 can be parameterized as $\mathbf{r}_1(x) = \langle x, g_1(x) \rangle$, from $x = a$ to $x = b$, and the curve C_2 can be parameterized as $\mathbf{r}_2(x) = \langle x, g_2(x) \rangle$, from $x = b$ to $x = a$. Then

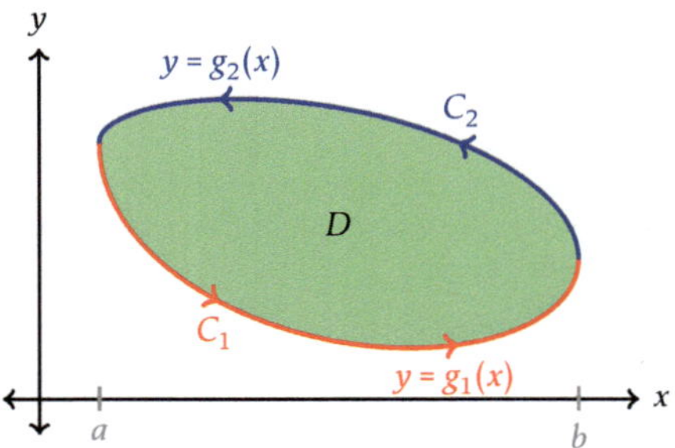

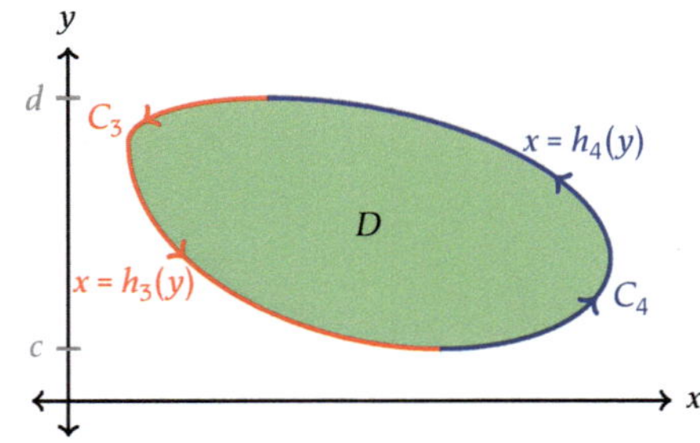

Figure 5.29 *Two views of a region that can be expressed (left) as between the curves $y = g_1(x)$ and $y = g_2(x)$, and (right) as between the curves $x = h_3(y)$ and $x = h_4(y)$*

Line 1 is the line integral of M, along the curves C_1 and C_2; lines 2 and 3 use the parameterizations; line 4 reverses the limits of integration for the second integral.

$$\oint_C M(x,y)\,dx = \int_{C_1} M(x,y)\,dx + \int_{C_2} M(x,y)\,dx$$

$$= \int_{C_1} M(\mathbf{r}_1(x))\,dx + \int_{C_2} M(\mathbf{r}_2(x))\,dx$$

$$= \int_a^b M(x, g_1(x))\,dx + \int_b^a M(x, g_2(x))\,dx$$

$$= \int_a^b M(x, g_1(x))\,dx - \int_a^b M(x, g_2(x))\,dx.$$

We can also compute $\iint_D -\dfrac{\partial M}{\partial y}\,dA$, integrating with respect to y first:

$$\iint_D -\frac{\partial M}{\partial y}\,dA = \int_a^b \int_{g_1(x)}^{g_2(x)} -M_y(x,y)\,dy\,dx$$

Line 1 uses the region D as in the left half of Figure 5.29, changing notation for the partial derivative; line 2 finds an antiderivative with respect to y; line 3 evaluates at $y = g_2(x)$, at $y = g_1(x)$, and subtracts; and line 4 rewrites as two integrals.

$$= \int_a^b \left(-M(x,y)\Big|_{g_1(x)}^{g_2(x)} \right) dx$$

$$= \int_a^b \left(-M(x, g_2(x)) - (-M(x, g_1(x))) \right) dx$$

$$= -\int_a^b M(x, g_2(x))\,dx + \int_a^b M(x, g_1(x))\,dx,$$

which is the same result as $\oint_C M(x,y)\,dx$, as promised.

Halfway there—we simply need to repeat using the right side of Figure 5.29. This time, $C = C_3 \cup C_4$. We may parameterize C_3 as $\mathbf{r}_3(y) = \langle h_3(y), y \rangle$ from $y = d$ to $y = c$ and C_4 as $\mathbf{r}_4(y) = \langle h_4(y), y \rangle$ from $y = c$ to $y = d$. Then

$$
\begin{aligned}
\oint_C N(x,y)\, dy &= \int_{C_3} N(x,y)\, dy + \int_{C_4} N(x,y)\, dy \\
&= \int_{C_3} N(\mathbf{r}_3(y))\, dy + \int_{C_4} N(\mathbf{r}_4(y))\, dy \\
&= \int_d^c N(h_3(y), y)\, dy + \int_c^d N(h_4(y), y)\, dy \\
&= -\int_c^d N(h_3(y), y)\, dy + \int_c^d N(h_4(y), y)\, dy.
\end{aligned}
$$

Finally, we compute $\iint_D \frac{\partial N}{\partial x}\, dA$, integrating with respect to x first:

$$
\begin{aligned}
\iint_D \frac{\partial N}{\partial x}\, dA &= \int_c^d \int_{h_3(y)}^{h_4(y)} N_x(x,y)\, dx\, dy \\
&= \int_c^d \left(N(x,y) \Big|_{h_3(y)}^{h_4(y)} \right) dy \\
&= \int_c^d \left(N(h_4(y), y) - N(h_3(y), y) \right) dy \\
&= \int_c^d N(h_4(y), y)\, dy - \int_c^d N(h_3(y), y)\, dy,
\end{aligned}
$$

which matches the computation for $\oint_C N(x,y)\, dy$.

Putting the two halves of the computation together, we have

$$
\begin{aligned}
\oint_C M(x,y)\, dx + N(x,y)\, dy &= \oint_C M(x,y)\, dx + \oint_C N(x,y)\, dy \\
&= \iint_D -\frac{\partial M}{\partial y}\, dA + \iint_D \frac{\partial N}{\partial x}\, dA \\
&= \iint_D \left(\frac{\partial N}{\partial x} - \frac{\partial M}{\partial y} \right) dA.
\end{aligned}
$$

We shall use Green's theorem for all shapes of the region D.

Compare to the first half of the computation, and note the differences that lead to the absence of the negative sign in the second half.

Line 1 is the line integral of N, along the curves C_3 and C_4; lines 2 and 3 use the parameterizations; line 4 reverses the limits of integration for the first integral.

Line 1 uses the region D as in the right half of Figure 5.29, changing notation for the partial derivative; line 2 finds an antiderivative with respect to x; line 3 evaluates at $x = h_4(y)$, at $x = h_3(y)$, and subtracts; and line 4 rewrites as two integrals.

Ans. to reading exercise 5:
positively oriented

5.4.2   Green's theorem examples

Recall that a vector field $\mathbf{F}(x, y) = \langle M(x, y), N(x, y)\rangle$ is conservative if $\frac{\partial M}{\partial y} = \frac{\partial N}{\partial x}$. In that case, the line integral along a closed curve is 0. In Green's theorem, the integrand of the double integral is $\frac{\partial N}{\partial x} - \frac{\partial M}{\partial y}$, which is zero if the field is conservative! This motivates a strategy for computing the line integral of a vector field along a closed curve.

Example 23 *Use Green's theorem to evaluate the line integral of* $\mathbf{F}(x, y) = \langle ye^x, 2e^x\rangle$ *along the rectangle with corners* $(0, 0)$ *to* $(2, 0)$ *to* $(2, 1)$ *to* $(0, 1)$ *to* $(0, 0)$.

A rectangle is an example of a piecewise smooth simple closed curve.

Solution We first check to see if the vector field is conservative:

$$\frac{\partial M}{\partial y} = \frac{\partial}{\partial y} ye^x = e^x$$

$$\frac{\partial N}{\partial x} = \frac{\partial}{\partial x} 2e^x = 2e^x.$$

If the vector field is conservative, the value of the line integral is 0 by theorem 5, and there is no need to use Green's theorem.

The partial derivatives are unequal and the field is not conservative. The integrand of the double integral is

$$\frac{\partial N}{\partial x} - \frac{\partial M}{\partial y} = 2e^x - e^x = e^x.$$

The wording "$(0, 0)$ to $(2, 0)$" indicates the direction of travel between the two points.

Next, we sketch the closed curve (Figure 5.30), marking the direction of travel, and note that the curve has positive orientation, as in the hypotheses of Green's theorem. We also mark the region D inside the curve. Next, we set up the double integral over the region D. The rectangle is described by $0 \leq x \leq 2, 0 \leq y \leq 1$, and either order of integration is easy to use. We choose

$$\int_0^2 \int_0^1 e^x \, dy \, dx.$$

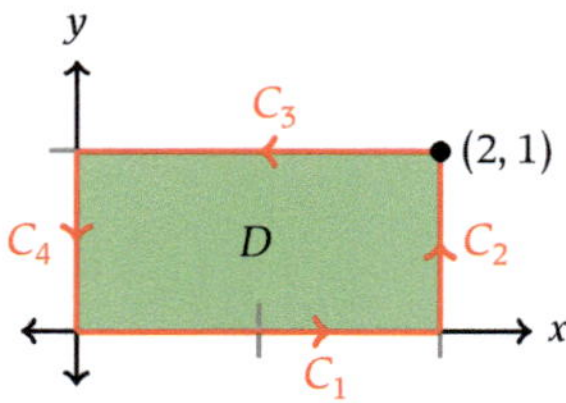

Figure 5.30 *The piecewise smooth, simple closed curve* $C = C_1 \cup C_2 \cup C_3 \cup C_4$ *of example 23, and the region D enclosed by the curve*

Then, we calculate the double integral:

$$\int_0^2 \int_0^1 e^x \, dy \, dx = \int_0^2 \left(e^x y \big|_0^1 \right) dx$$

$$= \int_0^2 (e^x - 0) \, dx$$

$$= e^x \big|_0^2 = e^2 - 1.$$

Line 1 finds an antiderivative with respect to y, treating x as a constant; line 2 evaluates at $y = 1$, at $y = 0$, and subtracts; line 3 finds an antiderivative with respect to x and completes the calculation.

By Green's theorem, the value of the line integral is

$$\oint_C \mathbf{F} \cdot d\mathbf{r} = e^2 - 1.$$

◼

Because the vector field of example 23 is not conservative, we cannot use the fundamental theorem as an alternate solution. We could, however, use the method of Section 5.2, that is, parameterize each of the curves C_1, C_2, C_3, and C_4 of Figure 5.30, calculate the four line integrals, and add the results. Try it and you will experience the advantage of using Green's theorem.

What if the simple closed curve has a negative orientation? We simply negate the result calculated from Green's theorem.

Example 24 *Use Green's theorem to evaluate the line integral of* $\mathbf{F}(x, y) = \langle x^2 y, 3xy \rangle$ *along the triangle from* $(0, 0)$ *to* $(0, 2)$ *to* $(3, 0)$ *to* $(0, 0)$.

Solution We first check to see if the vector field is conservative:

$$\frac{\partial}{\partial y} x^2 y = x^2$$

$$\frac{\partial}{\partial x} 3xy = 3y.$$

Rather than write the check for a conservative field in the same manner as the previous section, we may skip directly to the calculation of the integrand of the Green's theorem double integral. If the integrand is 0, the field is conservative and the value of the line integral is 0. We do so in the remaining examples.

The field is not conservative, so we continue with the Green's theorem strategy. The integrand for the double integral is

$$\frac{\partial N}{\partial x} - \frac{\partial M}{\partial y} = 3y - x^2.$$

Next we sketch the closed curve (Figure 5.31), marking the stated direction of travel, and note that the curve has negative orientation, which will affect the final answer. We also mark the region D inside the curve.

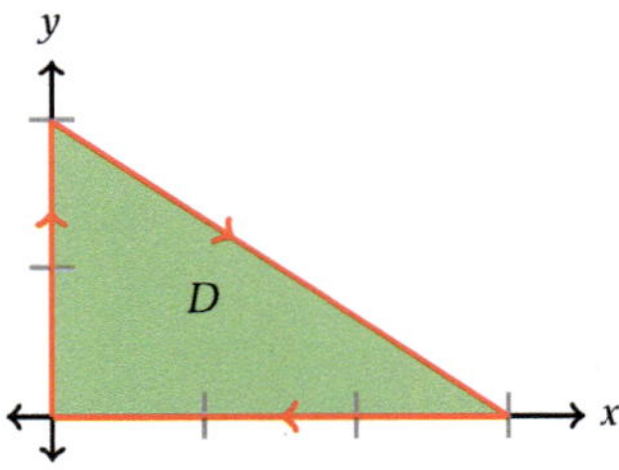

Figure 5.31 *The piecewise smooth, simple closed curve of example 24, and the region D enclosed by the curve*

To set up the double integral, we need to know the equation of the line through the points $(0, 2)$ and $(3, 0)$. The slope of the line is $m = \frac{0-2}{3-0} = -\frac{2}{3}$, and the y-intercept is 2, so the equation is $y = 2 - \frac{2}{3}x$. Integrating with respect to y first we have

Line 1 uses Green's theorem, noting that $-C$ has positive orientation, using the previously calculated integrand, and setting up the limits of integration consulting Figure 5.31; line 2 finds an antiderivative with respect to y, treating x as a constant; line 3 evaluates at $y = 2 - \frac{2}{3}x$, at $y = 0$, subtracts, and simplifies; line 4 finds an antiderivative with respect to x; and line 5 finishes the calculation.

$$\oint_{-C} \mathbf{F} \cdot d\mathbf{r} = \int_0^3 \int_0^{2-\frac{2}{3}x} (3y - x^2) \, dy \, dx$$

$$= \int_0^3 \left(\left(\tfrac{3}{2}y^2 - x^2 y \right) \Big|_0^{2-\frac{2}{3}x} \right) dx$$

$$= \cdots = \int_0^3 \left(6 - 4x - \tfrac{4}{3}x^2 + \tfrac{2}{3}x^3 \right) dx$$

$$= \left(6x - 2x^2 - \tfrac{4}{9}x^3 + \tfrac{1}{6}x^4 \right) \Big|_0^3$$

$$= \cdots = \frac{3}{2}.$$

Using Green's theorem gives the answer for a positively oriented curve. If the curve has negative orientation, then we negate the answer given by Green's theorem.

Adjusting for the negative orientation of the curve,

$$\oint_C \mathbf{F} \cdot d\mathbf{r} = -\frac{3}{2}.$$

Reading Exercise 6 Set up the double integral for using Green's theorem to evaluate the line integral of $\mathbf{F}(x, y) = \langle 3x + 2y, 8x + 5y \rangle$ along the rectangle from $(0, 0)$ to $(5, 0)$ to $(5, 2)$ to $(0, 2)$ to $(0, 0)$.

5.4.3   Areas using Green's theorem in reverse

Let C be a positively oriented, piecewise smooth, simple closed curve enclosing a region D, as in Figure 5.32. These are the hypotheses of Green's theorem, so the theorem applies. If we have a parameterization of the curve C, it would be nice to

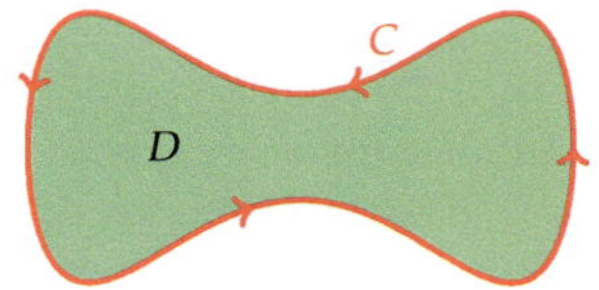

Figure 5.32 *A positively oriented, smooth, simple closed curve*

have a formula for finding the area enclosed by the curve. So far, our only option is to try to eliminate the parameter, write pieces of the curve using Cartesian coordinates, and calculate the area from the Cartesian equations, perhaps using more than one integral. Green's theorem provides another option.

In Green's theorem, let $\mathbf{F}(x, y) = \langle 0, x \rangle$, that is, let $M(x, y) = 0$ and $N(x, y) = x$. Then $\frac{\partial M}{\partial y} = 0$ and $\frac{\partial N}{\partial x} = 1$. By Green's theorem,

$$\oint_C 0\,dx + x\,dy = \iint_D (1 - 0)\,dA,$$

A parameterization of the curve in Figure 5.32 is $x = \cos t$, $y = \left(\frac{1}{5} + \cos^2 t\right) \sin t$, $0 \le t \le 2\pi$. Eliminating the parameter yields the equations $y = \pm\left(\frac{1}{5} + x^2\right)\sqrt{1 - x^2}$, $-1 \le x \le 1$. The integral for determining area requires trig substitution and multiple uses of trig identities.

which is the area of the region D.

Similarly, if $M(x, y) = -y$ and $N(x, y) = 0$, then $\frac{\partial M}{\partial y} = -1$, $\frac{\partial N}{\partial x} = 0$, and

$$\oint_C -y\,dx + 0\,dy = \iint_D (0 - (-1))\,dA,$$

The volume under the surface $z = 1$ over the region D is equal to the area of the region times 1.

This formula turns out to be the simplest to use for the curve of Figure 5.32.

which again is the area of the region D.

Although each of those formulas can be useful, it turns out that an often more useful formula is found by averaging the two formulas:

$$A = \frac{1}{2}\oint_C x\,dy - y\,dx.$$

Add the two line integrals and divide by two to find the average.

AREA INSIDE A PARAMETERIZED SIMPLE CLOSED CURVE

Let C be a positively oriented, piecewise smooth, simple closed curve. Then the area enclosed by the curve is

$$A = \oint_C x\,dy = \oint_C -y\,dx = \frac{1}{2}\oint_C x\,dy - y\,dx.$$

Which of these formulas is easiest to employ depends on the specific curve. Most often, the third formula is easiest.

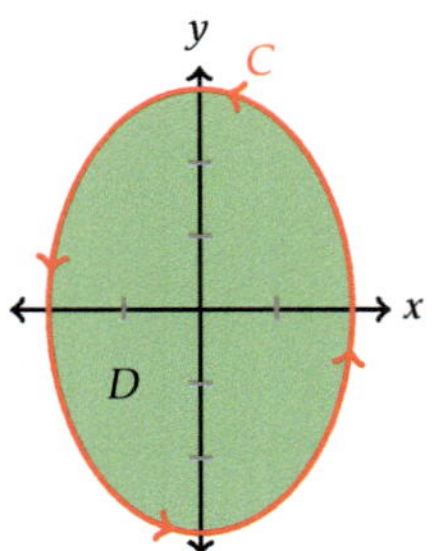

Figure 5.33 *The ellipse* $x = 2\cos t$, $y = 3\sin t$, $0 \le t \le 2\pi$

Line 1 is the third area formula; line 2 uses the parameterization, replacing x by $a\cos t$, dy by $b\cos t\,dt$, y by $b\sin t$, and dx by $a(-\sin t)\,dt$; line 3 simplifies; line 4 uses a familiar pythagorean identity; and line 5 integrates.

Compare this solution to the effort required to solve Section 4.10 exercise 27.

Although the third formula is often advantageous, it is not always the easiest. But it is a good starting point. If the integrand of the third formula does not simplify, choose whichever portion looks easiest, and use the first or second formula accordingly.

Ans. to reading exercise 6:
$\int_0^5 \int_0^2 6\,dy\,dx$ or $\int_0^2 \int_0^5 6\,dx\,dy$

Example 25 *Find the area enclosed by the ellipse $x = a\cos t$, $y = b\sin t$, $0 \le t \le 2\pi$.*

Before presenting the solution, a reminder is in order. The line integrals in the area formulas require the curve to be traversed exactly once, with positive orientation. See *Calculus Set Free* Section 8.1 for a discussion of how to determine values of the parameter for traversing the curve exactly once. Although not necessary for the solution, it can be helpful to graph the curve (Figure 5.33).

Solution The ellipse $x = a\cos t$, $y = b\sin t$, $0 \le t \le 2\pi$ is simple, smooth, positively oriented, and traverses the curve exactly once. To use the third area formula, we need to calculate dy and dx:

$$dx = a(-\sin t)\,dt$$
$$dy = b\cos t\,dt.$$

Then

$$A = \frac{1}{2}\oint_C x\,dy - y\,dx$$
$$= \frac{1}{2}\int_0^{2\pi} a\cos t \cdot b\cos t\,dt - b\sin t \cdot a(-\sin t)\,dt$$
$$= \frac{1}{2}\int_0^{2\pi} (ab\cos^2 t + ab\sin^2 t)\,dt$$
$$= \frac{1}{2}\int_0^{2\pi} ab\,dt$$
$$= \frac{1}{2}abt\Big|_0^{2\pi} = \pi ab.$$

(Notice that when $a = b = r$, the ellipse is a circle with radius r and has area πr^2.)

Example 25 illustrates why the third area formula is often advantageous. If we use the first area formula, we have

$$A = \oint_C x\,dy = \int_0^{2\pi} ab\cos^2 t\,dt,$$

and if we use the second area formula, we have

$$A = \oint_C -y\,dx = \int_0^{2\pi} ab\sin^2 t\,dt.$$

While neither integral is especially difficult, they are less simple than the combination.

EXERCISES 5.4

1–4. Rapid response: does the simple closed curve have positive orientation or negative orientation?

1.

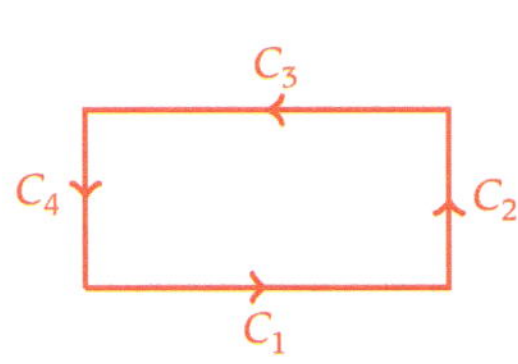

3.

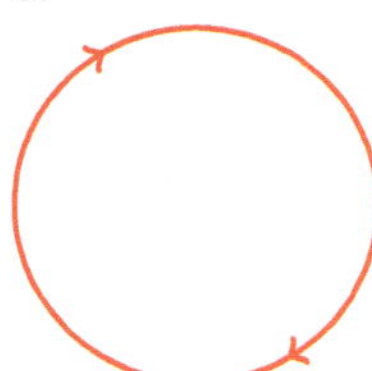

2.

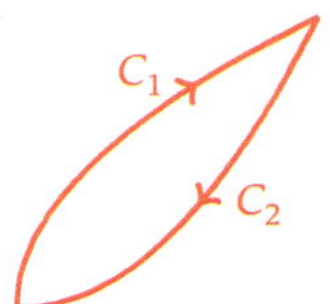

4.

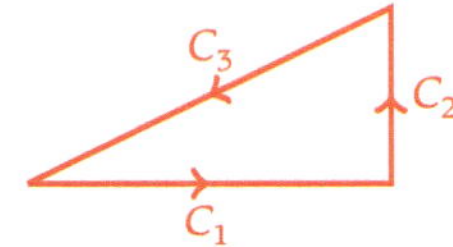

5–14. Use Green's theorem to evaluate the line integral of **F** along the given piecewise smooth simple closed curve.

5. $\mathbf{F}(x,y) = \langle 2x + 4y, 9x + y \rangle$, along the square from $(0,0)$ to $(2,0)$ to $(2,2)$ to $(0,2)$ to $(0,0)$

6. $\mathbf{F}(x,y) = \langle 4y^2, 5x \rangle$ along the rectangle from $(0,0)$ to $(10,0)$ to $(10,1)$ to $(0,1)$ to $(0,0)$

7. $\mathbf{F}(x,y) = \langle x + y, 3y^2 \rangle$, along the parabola $y = x^2$ from $(0,0)$ to $(1,1)$, and then along the line segment from $(1,1)$ back to $(0,0)$

8. $\mathbf{F}(x,y) = \langle x + y, 3y^2 \rangle$, along the square root curve $y = \sqrt{x}$ from $(0,0)$ to $(1,1)$, and then along the line segment from $(1,1)$ back to $(0,0)$

9. $\mathbf{F}(x,y) = \langle y\sqrt{x}, xy^2 \rangle$, along the triangle from $(0,0)$ to $(3,3)$ to $(3,0)$ to $(0,0)$

10. $\mathbf{F}(x,y) = \langle 4x^2 e^x - 5y, 8\left(\cos y^3\right) - 5x \rangle$, along the triangle from $(0,0)$ to $(1,0)$ to $(1,1)$ to $(0,0)$

11. $\mathbf{F}(x,y) = \left\langle 3x^2 \ln y, \dfrac{x^3}{y} + 4\sinh y \right\rangle$, along the triangle from $(1,1)$ to $(2,1)$ to $(2,2)$ to $(1,1)$

12. $\mathbf{F}(x,y) = \langle 5x^3 - 7y, 3x - y^7 \rangle$, along the curve $x = \cos t, y = \sin t$, $0 \le t \le 2\pi$

13. $\mathbf{F}(x,y) = \langle \sin x, \cos x + \cos y \rangle$ along the parabola $y = x^2$ from $(0,0)$ to $(2,4)$, then along the line segment to $(0,4)$ and the line segment back to $(0,0)$

14. $\mathbf{F}(x, y) = \langle e^{xy}, 4 \rangle$, along the rectangle with corners $(1, 1)$, $(4, 1)$, $(4, 7)$, and $(1, 7)$, traversed counterclockwise

15–22. Find the area enclosed by the parameterized curve.

15. $x = 3 \cos 2t$, $y = 5 \sin 2t$, $0 \le t \le \pi$

16. $x = 1 + \cos 3t$, $y = 2 + \sin 3t$, $0 \le t \le \frac{2\pi}{3}$

17. $x = \cos t$, $y = 2 \sin t \cos t$, $-\frac{\pi}{2} \le t \le \frac{\pi}{2}$

18. $x = 4 \cos^3 t$, $y = 4 \sin^3 t$, $0 \le t \le 2\pi$ (astroid)

19. $x = 2 \cos t - \cos 2t$, $y = 2 \sin t - \sin 2t$, $0 \le t \le 2\pi$ (cardioid)

20. $x = 2 \cos t + \cos 2t$, $y = 2 \sin t - \sin 2t$, $0 \le t \le 2\pi$ (tricuspoid)

21. curve $C = C_1 \cup C_2$, where C_1 is $x = t$, $y = -t^2 + 8t - 12$, from $(2, 0)$ to $(6, 0)$, and C_2 follows the x-axis from $(6, 0)$ to $(2, 0)$

22. $x = t^2$, $y = \sin \pi t$, $-1 \le t \le 1$

23. If a smooth, simple closed curve is finite and encloses a bounded region, the area formulas from Green's theorem may work even if the resulting integral is improper. Find the area inside the loop of the folium of Descartes $x = \frac{3t}{1+t^3}$, $y = \frac{3t^2}{1+t^3}$, $t \ge 0$, that is, $t = 0$ to $t = \infty$.

24. (a) Find the area enclosed by the five-star cookie $x = (3 + \sin 5t) \cos t$, $y = (3 + \sin 5t) \sin t$, $0 \le t \le 2\pi$.

(b) Graph the five-star cookie from part (a).

(c) Graph the aster-like flower $x = (3 + \sin 16t) \cos t$, $y = (3 + \sin 16t) \sin t$, $0 \le t \le 2\pi$. Which do you think encloses more area, the 5-star (cookie) or the 16-star (aster)?

(d) Repeat part (a) for the 16-star.

(e) Which encloses more area? Comparing the calculations, can you determine the reason for your answer?

25. Suppose a positively oriented, piecewise smooth, simple closed curve C encloses a region D.

(a) Rewrite the double integrals for the center of mass of a thin flat plate (Section 4.5) using the density function $\rho(x, y) = 1$. The result is a formula for the centroid of the region D.

(b) Let A be the area of the region D. Use Green's theorem to write $\frac{1}{2A} \oint_C x^2 \, dy$ as a double integral.

(c) Use Green's theorem to write $-\frac{1}{2A} \oint_C y^2 \, dx$ as a double integral.

(d) Use the results of (b) and (c) to write a formula for the centroid of the region D.

26. Use the following steps to determine the centroid of the semicircular region D described as inside the unit circle $x^2 + y^2 = 1$ above the x-axis.

    (a) What is the area of this region? (An integral is not necessary.)
    (b) Determine a parameterization for the upper half of the unit circle. Call this curve, which is not closed, C_1.
    (c) Draw the region D, and determine a parameterization for the curve C_2 that borders the region along the x-axis, so that the simple closed curve $C = C_1 \cup C_2$ encloses the region D.
    (d) Use the formula from exercise 25(d) to determine the centroid of the region.
    (e) Does the line integral along the curve C_2 contribute anything to the centroid?

27. Use the following steps to determine the centroid of the region D that is the portion of the region inside the ellipse $\frac{x^2}{36} + \frac{y^2}{9} = 1$ that lies in the first quadrant.

    (a) Determine a parameterization of this portion of the ellipse. Call this curve, which is not closed, C_1.
    (b) Use the answer to example 25 to determine the area A of this region.
    (c) Draw the region D, and determine parameterizations for curves C_2 and C_3 that border the region along the y-axis and x-axis, respectively, so that the simple closed curve $C = C_1 \cup C_2 \cup C_3$ encloses the region D.
    (d) Use the formula from exercise 25(d) to determine the centroid of the region.
    (e) Do the line integrals along the curves C_2 and C_3 contribute anything to the centroid?
    (f) Compare to the work required to complete *Calculus Set Free* Section 9.6 exercise 19.

28. If a non-simple closed curve has one crossing point, the total area enclosed can be found by adding the area enclosed by the portion with positive orientation and the area enclosed by the portion with negative orientation.

    (a) Graph the curve $x = \cos t, y = \cos t \sin t, -\frac{\pi}{2} \le t \le \frac{3\pi}{2}$. Is this a simple closed curve?
    (b) Determine the values of t for which the curve has positive orientation and the values of t for which the curve has negative orientation.
    (c) Find the area enclosed by the portion of the curve with positive orientation.

(d) Find the area enclosed by the portion of the curve with negative orientation.

(e) Determine the total area enclosed by the curve.

(f) The integrals you evaluated in parts (c) and (d) should be the same except for the limits of integration. What happens if you try one integral for the entire curve, evaluating from $t = -\dfrac{\pi}{2}$ to $t = \dfrac{3\pi}{2}$?

5.5   Divergence and Curl

Functions have derivatives.

For a function with one input and one output, such as $g(x) = x^2$, the output of the derivative is a scalar (a number), in this case $g'(x) = 2x$. For a function with three inputs and one output, such as $f(x, y, z) = x^2 + y^2 + z^2$, there are partial derivatives, whose outputs are scalars; in this case, $f_x(x, y, z) = 2x$, $f_y(x, y, z) = 2y$, and $f_z(x, y, z) = 2z$. Another way to look at a "derivative" of such a function is the gradient, which combines the partial derivatives to output a vector, in this case $\nabla f(x, y, z) = \langle 2x, 2y, 2z \rangle$.

Each idea of derivative, whether a scalar or a vector, has an interpretation. For instance, f_x, f_y, and f_z are the slopes in the x-, y-, and z-directions, whereas ∇f gives both the direction and magnitude of fastest increase.

What might a "derivative" look like for a vector-valued function, such as $F(x, y, z) = \langle x^2, y^2, z^2 \rangle$? Should the output be a scalar or a vector? Given $F(x, y, z) = \langle M(x, y, z), N(x, y, z), P(x, y, z) \rangle$, how do we combine the nine possible partial derivatives (each of M, N, and P with respect to each of x, y, and z) to get our scalar or vector? There is more than one answer–in fact, there is more than one answer that has interpretations in the physical world and is therefore worthy of study.

5.5.1   Divergence

One simple way of defining a "derivative" of a vector-valued function $F(x, y, z) = \langle M(x, y, z), N(x, y, z), P(x, y, z) \rangle$ is to add the partial derivatives of each component with respect to its variable,

The divergence is a scalar function of three (or two) variables.

$$\operatorname{div} F(x, y, z) = M_x(x, y, z) + N_y(x, y, z) + P_z(x, y, z).$$

We call this the *divergence* of F. If we write the gradient operator symbolically as

$$\nabla = \left\langle \frac{\partial}{\partial x}, \frac{\partial}{\partial y}, \frac{\partial}{\partial z} \right\rangle,$$

and $F(x, y, z) = \langle M(x, y, z), N(x, y, z), P(x, y, z) \rangle$ as

$$F = \langle M, N, P \rangle,$$

and then manipulate symbolically to find their dot product, we get

$$\nabla \cdot \mathbf{F} = \left\langle \frac{\partial}{\partial x}, \frac{\partial}{\partial y}, \frac{\partial}{\partial z} \right\rangle \cdot \langle M, N, P \rangle = \frac{\partial M}{\partial x} + \frac{\partial N}{\partial y} + \frac{\partial P}{\partial z},$$

which is the same as div F.

DIVERGENCE

The *divergence* of a vector field F is

$$\operatorname{div} \mathbf{F} = \nabla \cdot \mathbf{F} = \frac{\partial M}{\partial x} + \frac{\partial N}{\partial y} + \frac{\partial P}{\partial z},$$

provided the partial derivatives exist.

Although the formula for div F is not difficult to remember, the notation $\nabla \cdot \mathbf{F}$ gives another way to remember the formula, especially in conjunction with a contrasting formula later in this section.

For a function of two variables $\mathbf{F}(x, y) = \langle M(x, y), N(x, y) \rangle$, the divergence is

$$\operatorname{div} \mathbf{F} = \nabla \cdot \mathbf{F} = \frac{\partial M}{\partial x} + \frac{\partial N}{\partial y}.$$

Example 26 *Find the divergence of* $\mathbf{F}(x, y, z) = \langle x^2 + yz, y^2z + 1, 3z + \sqrt{ye^x} \rangle$.

Solution Using the formula and calculating the partial derivatives,

$$\operatorname{div} \mathbf{F} = \frac{\partial M}{\partial x} + \frac{\partial N}{\partial y} + \frac{\partial P}{\partial z}$$

$$= 2x + 2yz + 3. \qquad \blacksquare$$

Notice that the output of divergence is a scalar, not a vector.

The traditional notations div F and $\nabla \cdot \mathbf{F}$ have the advantage that they are compact and they work for both two and three variables. However, they somewhat obscure the fact that the divergence is a scalar function of the type studied in Chapter 3. The solution to example 26 can be written as

$$\operatorname{div} \mathbf{F}(x, y, z) = 2x + 2yz + 3.$$

Reading Exercise 7 Find the divergence of $\mathbf{F}(x, y, z) = \langle 3x, 4y^2, 7 \rangle$.

One of the physical interpretations of divergence has to do with fluid flow. We shall use the next two examples in the explanation.

Example 27 *Find the divergence of* $\mathbf{F}(x, y, z) = \langle -y, x, 0 \rangle$.

Solution Using the formula and calculating the partial derivatives,

$$\operatorname{div} \mathbf{F} = \frac{\partial M}{\partial x} + \frac{\partial N}{\partial y} + \frac{\partial P}{\partial z}$$

$$= 0 + 0 + 0 = 0. \qquad \blacksquare$$

Example 28 *Find the divergence of* $\mathbf{F}(x, y, z) = \langle xy, x, 0 \rangle$.

Solution Using the formula and calculating the partial derivatives,

$$\operatorname{div} \mathbf{F} = \frac{\partial M}{\partial x} + \frac{\partial N}{\partial y} + \frac{\partial P}{\partial z}$$

$$= y + 0 + 0 = y.$$

∎

Suppose that the vector field $\mathbf{F}$ gives velocities of particles in a fluid. For the field $\mathbf{F}(x, y, z) = \langle -y, x, 0 \rangle$ of example 27, the z-coordinate is immaterial; nothing is moving in the z-direction (the z-component of velocity is 0) and the field is identical at every elevation z, so we may picture the flow of the fluid in a plane, the same as if it is $\mathbf{F}(x, y) = \langle -y, x \rangle$. The graph of this vector field is in Figure 5.34. Notice that the fluid seems to be circulating, with larger speeds further from the origin. At any point such as A or B, the arrows heading into the point and out from the point appear to have the same length. The density of the fluid at the point is not changing, because the same mass is flowing into and out of the point. This is what happens when the divergence is 0. We call such a fluid flow *incompressible*.

> If the fluid is moving, we might say that it is "flowing," hence the phrase *fluid flow*.

> In this context, density is mass per unit volume. Divergence measures the (net) rate of change, with respect to time, of density.

> More mass is flowing out of A than is flowing in.

Contrast this to example 28, where the divergence is not zero; the graph of the vector field $\mathbf{F}(x, y) = \langle xy, x \rangle$ is in Figure 5.35. At point A, the arrows heading into the point are shorter than the arrows heading out. This means that the fluid is getting less dense at point A; it is decompressing. Meanwhile, at point B,

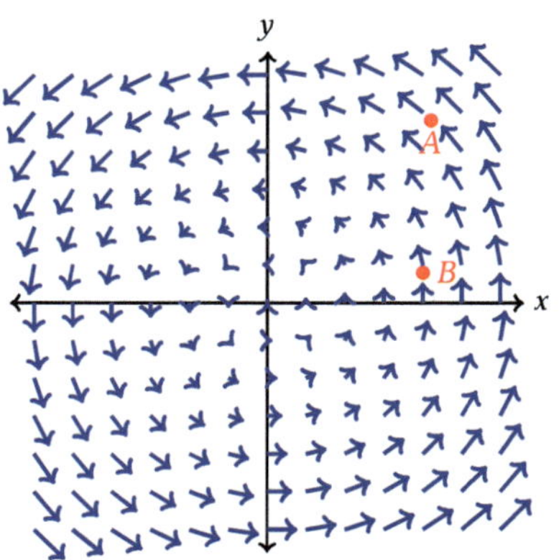

Figure 5.34 *A horizontal slice of the vector field* $\mathbf{F}(x, y, z) = \langle -y, x, 0 \rangle$ *of example 27, which has divergence 0, along with two points A and B*

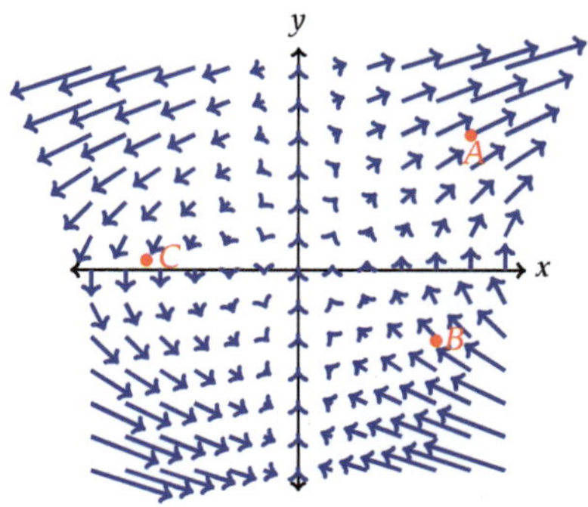

Figure 5.35 *A horizontal slice of the vector field* $\mathbf{F}(x, y, z) = \langle xy, x, 0 \rangle$ *of example 28, which has nonzero divergence, along with three points A, B, and C*

the arrows heading in are longer than those exiting the point, and the fluid is getting denser; it is compressing. But consider the point C, close to the x-axis. The divergence of this vector field is $\operatorname{div}\mathbf{F} = y$, which is zero when $y = 0$. The arrows pointing in and pointing out at C are about the same length, lending credence to the idea that divergence measures the rate of change of density of the fluid.

Key to this interpretation is that the vector field is a velocity field. If the field represents something else, such as a force field, the interpretation is different.

A subtlety of this interpretation of divergence is explored in exercise 43.

Ans. to reading exercise 7:
$$\operatorname{div}\mathbf{F}(x, y, z) = 3 + 8y$$

5.5.2   Derivative rules

The product rule, which has versions for derivatives and partial derivatives, also has a version for divergence, namely

$$\nabla \cdot [f\mathbf{F}] = f[\nabla \cdot \mathbf{F}] + [\nabla f] \cdot \mathbf{F}.$$

Interpreting the symbol ∇ as derivative and the brackets as grouping symbols, we see that "the derivative of the product is equal to the first times the derivative of the second plus the derivative of the first times the second," which is the usual product rule!

To parse the notation for the purpose of proving the rule, we first tackle $f\mathbf{F}$. This is the scalar product of the scalar $f(x, y, z)$ and the vector

$$F(x, y, z) = \langle M(x, y, z), N(x, y, z), P(x, y, z) \rangle, \text{ that is,}$$

$$f(x, y, z)\langle M(x, y, z), N(x, y, z), P(x, y, z) \rangle$$
$$= \langle f(x, y, z)M(x, y, z), f(x, y, z)N(x, y, z), f(x, y, z)P(x, y, z) \rangle,$$

which makes clear the desirability of not writing (x, y, z) all the time; instead, we write

This is the same as the previous calculation, just without the (x, y, z).

$$f\mathbf{F} = f\langle M, N, P \rangle = \langle fM, fN, fP \rangle.$$

Then $\nabla \cdot [f\mathbf{F}]$ is

$$\text{div}\langle fM, fN, fP \rangle = \frac{\partial}{\partial x}(fM) + \frac{\partial}{\partial y}(fN) + \frac{\partial}{\partial z}(fP),$$

which requires the partial derivatives of products! Using the product rule and continuing,

Line 1 uses the product rule for all three partial derivatives; line 2 reorders and groups the terms; line 3 factors f from the first group and writes the second group as the dot product of two vectors; line 4 rewrites $M_x + N_y + P_z$ as the divergence of $\mathbf{F}$ (using $\nabla\cdot$ as divergence) and rewrites the vector $\langle f_x, f_y, f_z \rangle$ as the gradient ∇f, using $\cdot$ as the dot product.

$$= fM_x + f_xM + fN_y + f_yN + fP_z + f_zP$$
$$= (fM_x + fN_y + fP_z) + (f_xM + f_yN + f_zP)$$
$$= f[M_x + N_y + P_z] + \langle f_x, f_y, f_z \rangle \cdot \langle M, N, P \rangle$$
$$= f[\nabla \cdot \mathbf{F}] + [\nabla f] \cdot \mathbf{F},$$

as desired.

DIVERGENCE PRODUCT RULE

If all of the partial derivatives associated with the scalar function f and the vector function $\mathbf{F}$ exist,

$$\nabla \cdot [f\mathbf{F}] = f[\nabla \cdot \mathbf{F}] + [\nabla f] \cdot \mathbf{F}.$$

There are other derivative rules and other identities, some of which are explored in the exercises. Some of the rules are of more theoretical than computational use, including the divergence product rule; direct computation without use of the rule is fairly straightforward.

Example 29 *Let* $f(x,y,z) = xy^2z^3$ *and let* $\mathbf{F}(x,y,z) = \langle x^2, y^2, z^2 \rangle$. *Calculate*
$\nabla \cdot [f\mathbf{F}] = \text{div}[f\mathbf{F}]$.

Solution First, we calculate $f\mathbf{F}$:

$$f(x,y,z)\mathbf{F}(x,y,z) = xy^2z^3\langle x^2, y^2, z^2 \rangle$$
$$= \langle x^3y^2z^3, xy^4z^3, xy^2z^5 \rangle.$$

The scalar product of the scalar xy^2z^3 and the vector $\langle x^2, y^2, z^2 \rangle$ is found by multiplying each component of the vector by the scalar.

Then, the quantity $\nabla \cdot [f\mathbf{F}] = \text{div}[f\mathbf{F}]$ is the divergence of $f\mathbf{F}$:

$$\text{div}\langle x^3y^2z^3, xy^4z^3, xy^2z^5 \rangle = 3x^2y^2z^3 + 4xy^3z^3 + 5xy^2z^4.$$

The divergence of the vector is the sum of the partial derivatives of each component with respect to its component variable.

■

5.5.3   Curl

Curl is another way of defining a "derivative" of a vector-valued function, combining the six partial derivatives that are not used by the divergence. Whereas divergence can be remembered by writing ∇ symbolically as $\nabla = \left\langle \frac{\partial}{\partial x}, \frac{\partial}{\partial y}, \frac{\partial}{\partial z} \right\rangle$ and $\mathbf{F}$ as $\langle M, N, P \rangle$ and taking their dot product, curl can be remembered the same way but with the cross product instead.

CURL

The *curl* of a vector field $\mathbf{F}$ is

$$\text{curl }\mathbf{F} = \nabla \times \mathbf{F} = \begin{vmatrix} \mathbf{i} & \mathbf{j} & \mathbf{k} \\ \frac{\partial}{\partial x} & \frac{\partial}{\partial y} & \frac{\partial}{\partial z} \\ M & N & P \end{vmatrix}$$

$$= \left\langle \frac{\partial P}{\partial y} - \frac{\partial N}{\partial z}, -\left(\frac{\partial P}{\partial x} - \frac{\partial M}{\partial z} \right), \frac{\partial N}{\partial x} - \frac{\partial M}{\partial y} \right\rangle,$$

provided the partial derivatives exist.

The cross product is inherently three-dimensional, but curl for a two-dimensional field can be found by including a third component of zero, in the manner of examples 27 and 28.

Although the divergence formula may be more easily remembered without using the $\nabla \cdot \mathbf{F}$ notation, the curl formula is more easily remembered using $\nabla \times \mathbf{F}$.

Example 30 *Find the curl of* $\mathbf{F}(x,y,z) = \langle e^x + 4yz^2, \sin y + 7x^3z, \sqrt{\ln z} + 5x^2y^4 \rangle$.

Solution Using the formula and calculating the partial derivatives,

$$\text{curl }\mathbf{F} = \nabla \times \mathbf{F} = \begin{vmatrix} \mathbf{i} & \mathbf{j} & \mathbf{k} \\ \frac{\partial}{\partial x} & \frac{\partial}{\partial y} & \frac{\partial}{\partial z} \\ e^x + 4yz^2 & \sin y + 7x^3z & \sqrt{\ln z} + 5x^2y^4 \end{vmatrix}$$

$$= \langle 20x^2y^3 - 7x^3, -(10xy^4 - 8yz), 21x^2z - 4z^2 \rangle.$$

In line 2, the x-component is $\frac{\partial}{\partial y}(\sqrt{\ln z} + 5x^2y^4) - \frac{\partial}{\partial z}(\sin y + 7x^3z)$, the y-component is $-\left(\frac{\partial}{\partial x}(\sqrt{\ln z} + 5x^2y^4) - \frac{\partial}{\partial z}(e^x + 4yz^2) \right)$, and the z-component is $\frac{\partial}{\partial x}(\sin y + 7x^3z) - \frac{\partial}{\partial y}(e^x + 4yz^2)$.

Note that the output of curl is a vector, in contrast to the output of divergence, which is a scalar.

■

Reading Exercise 8 Find the curl of $F(x, y, z) = \langle x + 2y, 3y + 4z, 5z + 6x \rangle$.

According to the component test for conservative fields, a vector field $F = \langle M, N, P \rangle$ is conservative if and only if $\frac{\partial N}{\partial z} = \frac{\partial P}{\partial y}$, $\frac{\partial M}{\partial z} = \frac{\partial P}{\partial x}$, and $\frac{\partial M}{\partial y} = \frac{\partial N}{\partial x}$. Those are the exact conditions for the x-component, y-component, and z-component, respectively, of curl F to be zero!

The hypotheses used here are the same as those of the component test for conservative fields (theorem 3). Because theorem 7 relies on theorem 3, for which a portion of the proof delayed, the proof of theorem 7 is not yet considered complete.

Theorem 7 CURL OF A CONSERVATIVE VECTOR FIELD *Let* $F(x, y, z) = \langle M(x, y, z), N(x, y, z), P(x, y, z) \rangle$ *be a vector field for which M, N, and P have continuous partial derivatives throughout a simply connected region D. Then* F *is conservative if and only if* $curl\, F = 0$.

Theorem 7 gives us another way to write the test for whether a vector field is conservative: compute the curl and see if we get the zero vector. We now reprise example 17, example 18, and reading exercise 4 of Section 5.3.

Compare to example 17 of Section 5.3. Here, the z-component of 0 is inserted to allow the curl to be computed.

Example 31 *Is the vector field* $F(x, y, z) = \langle 2y^3, 6xy^2, 0 \rangle$ *conservative?*

Solution The curl of the vector field is

$$
curl\, F = \nabla \times F = \begin{vmatrix} \mathbf{i} & \mathbf{j} & \mathbf{k} \\ \frac{\partial}{\partial x} & \frac{\partial}{\partial y} & \frac{\partial}{\partial z} \\ 2y^3 & 6xy^2 & 0 \end{vmatrix}
$$

Observe that the z-component of curl contains the partial derivatives computed in Section 5.3 example 17.

$$
= \langle 0 - 0, -(0 - 0), 6y^2 - 6y^2 \rangle
$$

$$
= 0.
$$

Note the use of the zero vector **0**, not the number 0.

Because $curl\, F = 0$, the vector field is conservative.

Example 32 *Is the vector field* $F = \langle xy, xz, yz \rangle$ *conservative?*

Solution The curl of the vector field is

Compare to Section 5.3 example 18. This solution computes more partial derivatives (although the computation can be abandoned once a nonzero component is found), but the cross product makes the computation easier to remember.

$$
curl\, F = \nabla \times F = \begin{vmatrix} \mathbf{i} & \mathbf{j} & \mathbf{k} \\ \frac{\partial}{\partial x} & \frac{\partial}{\partial y} & \frac{\partial}{\partial z} \\ xy & xz & yz \end{vmatrix}
$$

$$
= \langle z - x, -(0 - 0), z - x \rangle
$$

$$
\neq 0.
$$

Because $curl\, F \neq 0$, the vector field is not conservative.

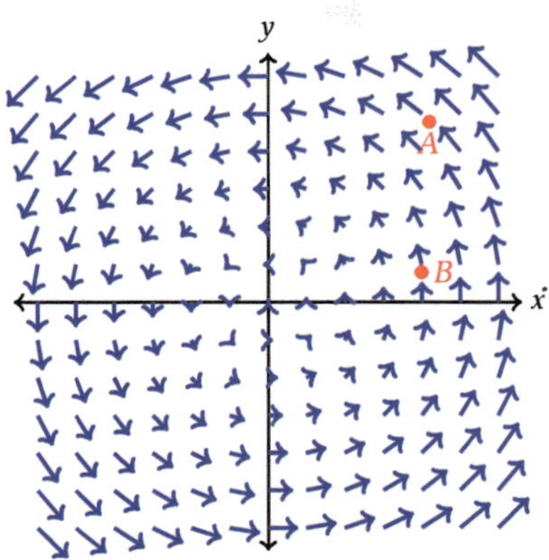

Figure 5.36 *A horizontal slice of the vector field* $\mathbf{F}(x, y, z) = \langle -y, x, 0 \rangle$ *of example 27, which has nonzero curl, along with two points A and B*

Reading Exercise 9 Is the vector field $\mathbf{F}(x, y, z) = \langle 4xy, 3x^2, 0 \rangle$ conservative?

Curl also has an interpretation in fluid flow, although there are more subtleties when inspecting pictures of vector fields than with divergence. The curl of the vector field of example 27, $\mathbf{F}(x, y, z) = \langle -y, x, 0 \rangle$, is $\operatorname{curl} \mathbf{F} = \langle 0, 0, 2 \rangle \neq \mathbf{0}$. For convenience, Figure 5.36 repeats the vector field picture.

Consider a leaf placed on the fluid at each of points A and B. Each leaf will move along a circle centered at the origin, as indicated by the velocity arrows. Notice, however, that for each leaf, the velocities to the right of its position are larger than the velocities to the left, meaning that the leaf will also spin, in a counterclockwise direction, in the same manner as an orbiting planet might spin on its axis. By contrast, for a vector field whose curl is always zero, there is no spinning and the fluid flow is *irrotational*.

Ans. to reading exercise 8:
$$\langle -4, -6, -2 \rangle$$

An alternate description to irrotational is *nonspinning*.

5.5.4 Interaction

Because the output of curl is a vector, we can find its divergence. It turns out to always be 0. The proof is similar in style to the proof of the divergence product rule.

Theorem 8 DIVERGENCE OF CURL IS ZERO *If* $\mathbf{F}(x, y, z) = \langle M(x, y, z), N(x, y, z), P(x, y, z) \rangle$ *is a vector field and each of the component functions has continuous second-order partial derivatives, then*

$$\operatorname{div} \operatorname{curl} \mathbf{F} = 0.$$

Line 1 uses the formula for curl $\mathbf{F}$, changing notation and omitting the (x, y, z); line 2 uses the formula for divergence, taking the sum of the partial derivatives of the x-component with respect to x, the y-component with respect to y, and the z-component with respect to z; line 3 reorders the terms; and line 4 applies Clairaut's theorem, which requires that the second partial derivatives be continuous, three times.

Proof. Using the formula for curl and then for divergence, and applying Clairaut's theorem, we have

$$\text{div}(\text{curl }\mathbf{F}) = \text{div}\,\langle P_y - N_z, -(P_x - M_z), N_x - M_y\rangle$$
$$= (P_{yx} - N_{zx}) - (P_{xy} - M_{zy}) + (N_{xz} - M_{yz})$$
$$= P_{yx} - P_{xy} - N_{zx} + N_{xz} + M_{zy} - M_{yz}$$
$$= 0.$$

$\blacksquare$

Theorem 8 means that if a vector field $\mathbf{G}$ has nonzero divergence, that is, if $\text{div }\mathbf{G} \neq 0$, then $\mathbf{G} \neq \text{curl }\mathbf{F}$ for any other vector field $\mathbf{F}$; for if it was, its divergence would have to be 0.

Example 33 *Show that $\mathbf{G}(x, y, z) = \langle 2x, 3z^2, 5\sqrt{y}\rangle$ is not the curl of any other vector field.*

Solution The divergence of $\mathbf{G}$ is

$$\text{div }\mathbf{G} = \text{div}\langle 2x, 3z^2, 5\sqrt{y}\rangle = 2 + 0 + 0 = 2.$$

Because $\text{div }\mathbf{G} \neq 0$, $\mathbf{G}$ is not the curl of any other vector field.

5.5.5 Laplace's equation

Many settings in physics give rise to functions satisfying Laplace's equation,

$$\frac{\partial^2 f}{\partial x^2} + \frac{\partial^2 f}{\partial y^2} + \frac{\partial^2 f}{\partial z^2} = 0.$$

A shortcut notation can be found using divergence. For a differentiable function $f(x, y, z)$, its gradient is a vector field, $\nabla f(x, y, z) = \langle f_x(x, y, z), f_y(x, y, z), f_z(x, y, z)\rangle$. Then (omitting the (x, y, z)) we may find the divergence of that vector field,

$$\nabla \cdot \nabla f = \text{div }\nabla f = \text{div}\,\langle f_x, f_y, f_z\rangle$$
$$= f_{xx} + f_{yy} + f_{zz}$$
$$= \frac{\partial^2 f}{\partial x^2} + \frac{\partial^2 f}{\partial y^2} + \frac{\partial^2 f}{\partial z^2}.$$

Ans. to reading exercise 9: not conservative, because $\text{curl }\mathbf{F} = \langle 0, 0, 2x\rangle \neq \mathbf{0}$.

Writing $\nabla^2 f$ for $\nabla \cdot \nabla f = \text{div } \nabla f$, Laplace's equation becomes

$$\nabla^2 f = 0.$$

The operator ∇^2 is sometimes called the *Laplacian*, and functions which satisfy Laplace's equation are sometimes called *harmonic*.

The notation $\nabla^2 f$ can be read "the Laplacian of f."

Example 34 *Is the function $f(x, y, z) = x^2 + y^2 - 2z^2$ harmonic?*

Solution   One method is to determine whether the Laplacian of f is zero:

$$\nabla^2 f = \text{div } \nabla f$$
$$= \text{div}\langle 2x, 2y, -4z \rangle$$
$$= 2 + 2 - 4 = 0.$$

Because the Laplacian is zero, f is harmonic.   ∎

5.5.6   A vector form of Green's theorem

You may have already noticed that the integrand of Green's theorem, $\dfrac{\partial N}{\partial x} - \dfrac{\partial M}{\partial y}$, is identical to the z-component of curl. Is it possible to rewrite the statement of Green's theorem using this fact? The problem is that Green's theorem is in two dimensions, whereas curl is in three dimensions. Thankfully, we have already learned how to overcome this problem, having used this tactic to state example 31.

Begin by setting up to use Green's theorem. Let C be a positively oriented, piece-wise smooth, simple closed curve in the xy-plane enclosing the region D. Let M and N be functions of x and y that have continuous partial derivatives throughout an open region R containing D. To move to three dimensions, let

$$\mathbf{F}(x, y, z) = \langle M(x, y), N(x, y), 0 \rangle.$$

Then

$$\text{curl } \mathbf{F} = \text{curl}\langle M(x, y), N(x, y), 0 \rangle$$

$$= \begin{vmatrix} \mathbf{i} & \mathbf{j} & \mathbf{k} \\ \dfrac{\partial}{\partial x} & \dfrac{\partial}{\partial y} & \dfrac{\partial}{\partial z} \\ M(x, y) & N(x, y) & 0 \end{vmatrix}$$

$$= \left\langle 0 - 0, -(0 - 0), \dfrac{\partial N}{\partial x} - \dfrac{\partial M}{\partial y} \right\rangle.$$

Line 2 uses the formula for curl; line 3 computes the partial derivatives, noting that $\frac{\partial}{\partial z} N(x, y) = 0$ and $\frac{\partial}{\partial z} M(x, y) = 0$.

To separate out the z-component, we take the dot product with the unit vector $\mathbf{k}$,

$$(\operatorname{curl}\mathbf{F})\cdot\mathbf{k} = \left\langle 0,0,\frac{\partial N}{\partial x}-\frac{\partial M}{\partial y}\right\rangle\cdot\langle 0,0,1\rangle$$

$$= 0+0+\frac{\partial N}{\partial x}-\frac{\partial M}{\partial y}.$$

VECTOR FORM OF GREEN'S THEOREM

The regions D and R lie within the xy-plane; "open" is interpreted as open in the xy-plane.

The forms of the line integral are from definition 1 (along with the notation $d\mathbf{r}$ for $\mathbf{r}'(t)\,dt$) and are equivalent to $\oint_C M\,dx + N\,dy + 0\,dz$.

Let C be a positively oriented, piecewise smooth, simple closed curve $\mathbf{r}(t)$ in the xy-plane enclosing the region D. Let M and N be functions of x and y that have continuous partial derivatives throughout an open region R containing D. Let $\mathbf{F}(x,y,z) = \langle M(x,y), N(x,y), 0\rangle$. Then

$$\oint_C \mathbf{F}\cdot\mathbf{T}\,ds = \oint_C \mathbf{F}\cdot d\mathbf{r} = \iint_D (\operatorname{curl}\mathbf{F})\cdot\mathbf{k}\,dA.$$

EXERCISES 5.5

1–10. Find div $\mathbf{F}$.

1. $\mathbf{F}(x,y,z) = \left\langle \sin x - y^2\cosh\sqrt{z^5+1}, y^2 + \tan^{-1}x^3\ln z^4, xyz^2 + \cos^2 xy^{-4}\right\rangle$

2. $\mathbf{F}(x,y,z) = \left\langle e^x - 2^{yz/(y^2-z)}, \dfrac{x}{y} + \tanh(x^2 - 4z^5), 3z - (x^4 + 3y)^{11}\right\rangle$

3. $\mathbf{F}(x,y,z) = \left\langle x^2yz, xy^2z, xyz^2\right\rangle$

4. $\mathbf{F}(x,y,z) = \left\langle 3xy, 4y^2, -11yz\right\rangle$

5. $\mathbf{F}(x,y,z) = \left\langle x\sin^2 z, y\cos^2 z, -z\right\rangle$

6. $\mathbf{F}(x,y,z) = \left\langle e^{xy}, e^{xy}, ze^{xy}\right\rangle$

7. $\mathbf{F}(x,y) = \left\langle \left(y^4 + y\right)^3, x^{3x+1}\right\rangle$

8. $\mathbf{F}(x,y) = \left\langle 5xy^6, 15x^2y^5\right\rangle$

9. $\mathbf{F}(x,y) = \left\langle 2\sqrt{2x+y}, -\sqrt{2x+y}\right\rangle$

10. $\mathbf{F}(x,y) = \left\langle 4y^7, 15x^9\right\rangle$

Some of these vector fields are the same as some of the vector fields in exercises 1–10. Take note of when one of div and curl is zero but the other is not.

11–20. Find curl $\mathbf{F}$.

11. $\mathbf{F}(x,y,z) = \left\langle x^2yz, xy^2z, xyz^2\right\rangle$

12. $\mathbf{F}(x,y,z) = \left\langle 3xy, 4y^2, -11yz\right\rangle$

13. $\mathbf{F}(x,y,z) = \left\langle x\sin^2 z, y\cos^2 z, -z\right\rangle$

14. $\mathbf{F}(x,y,z) = \langle -\cos y, x\sin y, \cos z\rangle$

15. $\mathbf{F}(x,y,z) = \left\langle 0, \ln z, \dfrac{y}{z} \right\rangle$
16. $\mathbf{F}(x,y,z) = \langle e^{xy}, e^{xy}, ze^{xy} \rangle$
17. $\mathbf{F}(x,y,z) = \langle y^2 z^3, 2xyz^3, 3xy^2 z^2 \rangle$
18. $\mathbf{F}(x,y,z) = \langle 5xy^6, 15x^2 y^5, 0 \rangle$
19. $\mathbf{F}(x,y,z) = \langle 2\sqrt{2x+y}, -\sqrt{2x+y}, 0 \rangle$
20. $\mathbf{F}(x,y,z) = \langle 4y^7, 15x^9, 0 \rangle$

21–26. Is the vector field **F** conservative?

21. $\mathbf{F}(x,y,z) = \langle 3x^2 yz, x^3 z - 2yz^3, x^3 y - 3y^2 z^2 \rangle$
22. $\mathbf{F}(x,y,z) = \langle e^{yz}, xze^{yz}, xye^{yz} \rangle$
23. $\mathbf{F}(x,y,z) = \langle 3y^2, 6xy, 5z + 2y \rangle$
24. $\mathbf{F}(x,y,z) = \langle yz, xz, xy + y \rangle$
25. $\mathbf{F}(x,y) = \langle 2xy - 3, x^2 + 7 \rangle$
26. $\mathbf{F}(x,y) = \langle x + 4y, 4x - 17y^3 \rangle$

Some of these exercises are identical to exercises in Section 5.3.

27–30. Show that **G** is not the curl of any other vector field.

27. $\mathbf{G}(x,y,z) = \langle 2x, 3y, 5z \rangle$
28. $\mathbf{G}(x,y,z) = \langle x^2 - y^2, y^2 - z^2, z^2 - x^2 \rangle$
29. $\mathbf{G}(x,y,z) = \langle 4y^5, 7z^2, 8xz \rangle$
30. $\mathbf{G}(x,y,z) = \langle \sin x, \cos y, 9 \rangle$

31. For $f(x,y,z) = x^2 + yz$ and $\mathbf{F}(x,y,z) = \langle 3x, 5y, z \rangle$, calculate $\nabla \cdot [f\mathbf{F}]$.

32. For $f(x,y,z) = xyz$ and $\mathbf{F}(x,y,z) = \langle e^x, 3xy, y\sqrt{z} \rangle$, calculate $\nabla \cdot [f\mathbf{F}]$.

33–38. Is the function harmonic?

33. $f(x,y,z) = e^x \sin z$
34. $f(x,y,z) = y^2 - z^2$
35. $f(x,y,z) = 3x^2 + 3y^2 - 3z^2$
36. $f(x,y,z) = x^3 - xy^2 - 2xz^2$
37. $f(x,y,z) = yz + xz + xy$
38. $f(x,y,z) = y^2 z + z^2 x + x^2 y$

39–42. (a) Verify the identity; (b) give an appropriate name for the identity.

39. $\operatorname{div}(\mathbf{F} + \mathbf{G}) = \operatorname{div}\mathbf{F} + \operatorname{div}\mathbf{G}$
40. $\operatorname{curl}(\mathbf{F} + \mathbf{G}) = \operatorname{curl}\mathbf{F} + \operatorname{curl}\mathbf{G}$
41. $\operatorname{div}(\mathbf{F} \times \mathbf{G}) = \mathbf{G} \cdot \operatorname{curl}\mathbf{F} - \mathbf{F} \cdot \operatorname{curl}\mathbf{G}$
42. $\operatorname{curl}[f\mathbf{F}] = f\operatorname{curl}\mathbf{F} + [\nabla f] \times \mathbf{F}$

Use the style of the derivation of the divergence product rule and the proof of theorem 8.

43. The vector field $\mathbf{F}(x, y, z) = \langle y, x, 0 \rangle$ is pictured. Note that $\operatorname{div} \mathbf{F} = 0$.

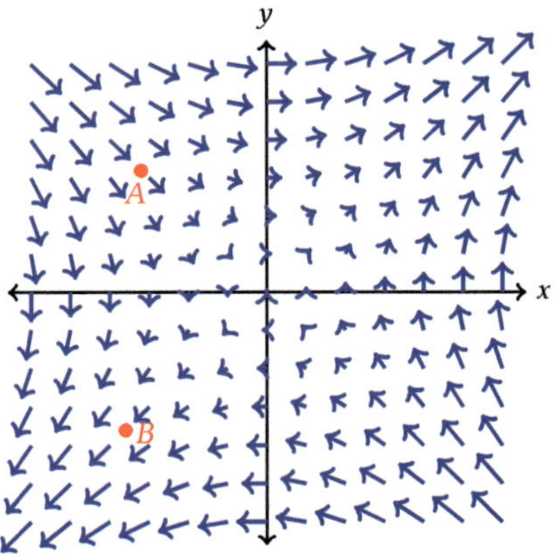

(a) The arrows pointing into A are longer than the arrows pointing out. With $\operatorname{div} \mathbf{F} = 0$, it must be the case that at A the total outflow and total inflow match. Use the diagram to explain how this is the case.

(b) The arrows pointing into B are shorter than the arrows pointing out. Use the diagram to explain how the total outflow and total inflow match.

44. Use the divergence product rule to calculate $\operatorname{div}[f\mathbf{F}]$ for example 29, and compare the effort required to that of the solution to example 29.

5.6 Parametric Surfaces

In many ways, the theme of multivariable calculus is to add a dimension, add a variable, or generalize. Vectors, parametrically defined functions, and functions of more than one variable are key tools in the process. It's time to expand our horizons once again, and it won't be the last time in this chapter, either.

5.6.1 Parametric surfaces

In Section 1.6, parametric equations are helpful for writing equations of lines and planes. A line requires one parameter; the line through the point $(1, 4, 7)$ with direction vector $\langle 2, -5, 7 \rangle$ is described parametrically as $x = 1 + 2t$, $y = 4 - 5t$, $z = 7 + 7t$. The line can also be written in the form $\mathbf{r}(t) = \langle 1 + 2t, 4 - 5t, 7 + 7t \rangle$, a vector-valued function.

A plane, meanwhile, requires two parameters; the plane containing the point $(1, 4, 7)$ with direction vectors $\langle 2, -5, 7\rangle$ and $\langle 1, 6, 3\rangle$ is $x = 1 + 2t + s$, $y = 4 - 5t + 6s$, $z = 7 + 7t + 3s$. Then why not study vector-valued functions of two variables, such as $\mathbf{r}(t, s) = \langle 1 + 2t + s, 4 - 5t + 6s, 7 + 7t + 3s\rangle$? With two variables (parameters), a plane is a type of surface. In general, a two-parameter (two-variable) vector-valued function is a surface, just as a Cartesian function of two variables is a surface.

Recall that describing a plane with two direction vectors requires nonparallel vectors.

Example 35 *Use technology to graph the parametric surface* $\mathbf{r}(u, v) = \langle \cos u, v,$ $\sin u + \sin v\rangle$, $0 \leq u \leq 2\pi$, $\pi \leq v \leq 5\pi$.

Solution The graph produced by Mathematica is in Figure 5.37.

Although the parameter t is traditional for a vector-valued function of one variable, and the parameters t and s are traditional for writing the equation of a plane, the parameters u and v are traditional for vector-valued functions of two variables.

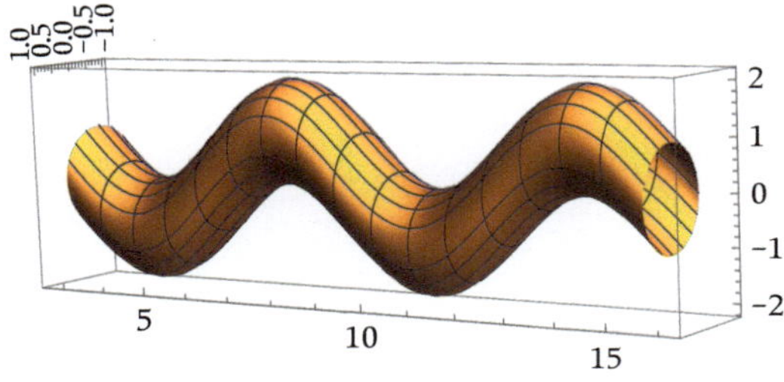

Figure 5.37 *The graph of* $\mathbf{r}(u, v) = \langle \cos u, v, \sin u + \sin v\rangle$, $0 \leq u \leq 2\pi$, $\pi \leq v \leq 5\pi$

CalcPlot3D and Desmos are both capable of producing graphs of parametric surfaces.

While graphing a few very simple parametric surfaces by hand is not beyond the skill set of some readers, use of technology is highly recommended.

■

Why does the graph in Figure 5.37 have the look that it does? The answer can be found by analyzing the *grid curves*. Figure 5.38, left, shows the familiar idea of grid lines in the uv-plane covering the region D, where D is the domain of $\mathbf{r}$. The blue grid line $u = \dfrac{\pi}{6}$ corresponds to the blue spacecurve $\mathbf{r}\left(\dfrac{\pi}{6}, v\right) = \left\langle \dfrac{\sqrt{3}}{2}, v, \dfrac{1}{2} + \sin v\right\rangle$ in Figure 5.38, right, because it has the form $z = \dfrac{1}{2} + \sin y$. In fact, holding u constant will always produce a sine wave in a plane parallel to the yz-plane.

Similarly, the horizontal grid line $v = 3\pi$ corresponds to the spacecurve $\mathbf{r}(u, 3\pi) = \langle \cos u, 3\pi, \sin u\rangle$, which is a circle in the plane $y = 3\pi$ (Figure 5.39). Holding v constant will always produce a circle in a plane parallel to the xz-plane.

In either case, holding one variable constant reduces the number of parameters to one, producing a spacecurve. Such spacecurves are called grid curves, and they are used by many CAS to help produce the graphs of parametric surfaces.

Parametric curves (spacecurves) can cross themselves. So can parametric surfaces.

When u is a constant there is only one variable remaining (v), and the result is a spacecurve.

The components are $x = \dfrac{\sqrt{3}}{2}$, $y = v$, and $z = \dfrac{1}{2} + \sin v$. Substituting y for v in the z-component gives $z = \dfrac{1}{2} + \sin y$.

For a parametric surface $\mathbf{r}(u, v)$, the domain is in the uv-plane, as pictured at left in Figures 5.38 and 5.39. The values of the function are in $\mathbf{R}^3$, as pictured at right in Figures 5.38 and 5.39. There are two input variables (u and v) and three output variables (x, y, and z).

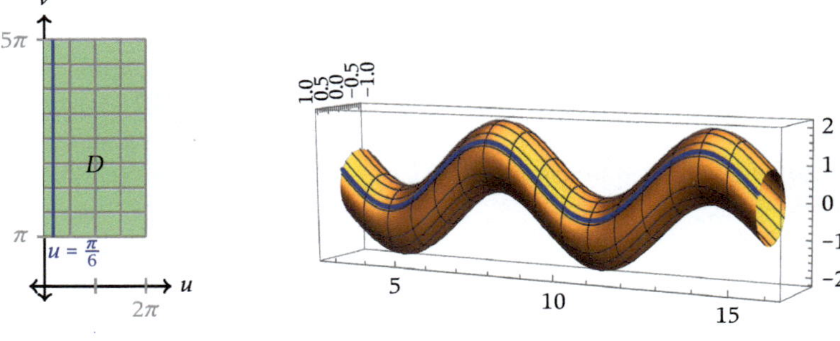

Figure 5.38 *(Left) the rectangular region $0 \le u \le 2\pi$, $\pi \le v \le 5\pi$ and (right) the graph of $\mathbf{r}(u, v) = \langle \cos u, v, \sin u + \sin v \rangle$, $0 \le u \le 2\pi$, $\pi \le v \le 5\pi$*

In Figure 5.39 the surface is shown with lower opacity so that the back side of the blue curve can be seen.

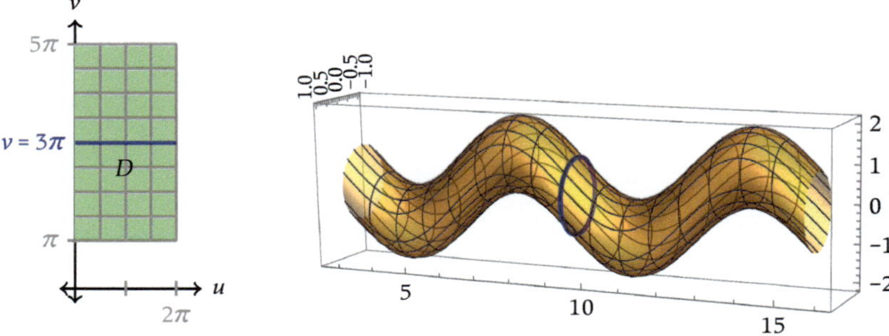

Figure 5.39 *(Left) the rectangular region $0 \le u \le 2\pi$, $\pi \le v \le 5\pi$ and (right) the graph of $\mathbf{r}(u, v) = \langle \cos u, v, \sin u + \sin v \rangle$, $0 \le u \le 2\pi$, $\pi \le v \le 5\pi$*

Example 36 *Use technology to graph the parametric surface $\mathbf{r}(u, v) = \langle \cos u, \sin u + \cos v, \sin v \rangle$, $0 \le u \le 2\pi$, $0 \le v \le \pi$.*

Solution Two views of the graph, each produced by Mathematica, are in Figure 5.40.

Most CAS allow the user to spin the graph to view it from any direction. Doing so can help with perspective, comprehending the shape of the surface, and analyzing its properties.

Just as any curve $y = f(x)$ can be parameterized by $x = t$, $y = f(t)$, any surface $z = f(x, y)$ can be parameterized by $x = u$, $y = v$, $z = f(u, v)$. Therefore, parametric surfaces are a generalization of the surfaces studied in Chapters 3 and 4.

Example 37 *(a) State a parameterization of the surface $z = 0.4x^2 + 0.1y^2$. (b) Use technology to graph the parametric surface.*

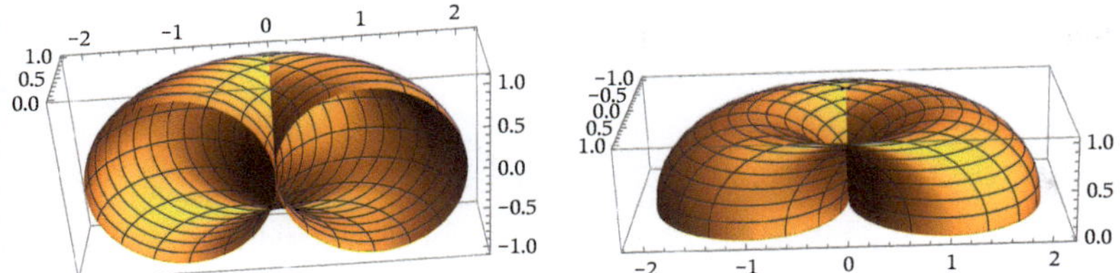

Figure 5.40 *The graph of* $\mathbf{r}(u,v) = \langle \cos u, \sin u + \cos v, \sin v \rangle$, $\quad 0 \le u \le 2\pi$, $0 \le v \le \pi$

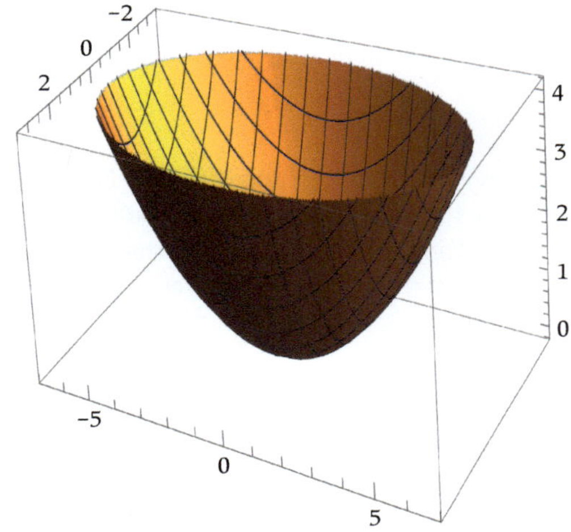

Figure 5.41 *A graph of* $\mathbf{r}(u,v) = \langle u, v, 0.4u^2 + 0.1v^2 \rangle$

Solution

(a) A parameterization of the surface is $x = u$, $y = v$, $z = 0.4u^2 + 0.1v^2$, or in vector notation $\mathbf{r}(u,v) = \langle u, v, 0.4u^2 + 0.1v^2 \rangle$.

(b) A graph produced by Mathematica is in Figure 5.41. ∎

Because there are no domain restrictions on x and y, the parameterization uses $-\infty < u < \infty$ and $-\infty < v < \infty$. Of course, bounded values of u and v must be chosen for viewing purposes. A non-rectangular selection was used to produce Figure 5.41.

Reading Exercise 10 State a parameterization of the surface $z = x^2 y - 3y^4$.

Recall that for spherical coordinates φ is always between 0 and π, and $0 \le \theta \le 2\pi$ gives a full rotation for θ.

You should recognize the equation of example 37 and its graph in Figure 5.41 as an elliptic paraboloid. Every quadric surface has a parameterization, not just the ones that are also graphs of functions of two variables. For instance, a sphere of radius 2 has the equation $\rho = 2$ in spherical coordinates. Then the Cartesian-spherical conversion formulas are $x = 2 \sin \varphi \cos \theta$, $y = 2 \sin \varphi \sin \theta$, and $= 2 \cos \varphi$, which give the three coordinate variables in terms of two parameters and θ. More generally, changing the value of ρ from one variable to the next accomplishes a parameterization of an ellipsoid.

Example 38 *(a) State a parameterization of the ellipsoid $\frac{x^2}{36} + \frac{y^2}{9} + \frac{z^2}{25} = 1$. (b) Use technology to graph the parametric surface.*

Solution

The Cartesian-spherical conversion formulas are used with φ replaced by u, θ replaced by v, and ρ replaced by the appropriate principal semiaxis.

(a)   The principal semi-axis (radius) for x is 6, for y is 3, and for z is 5. Then a parameterization of the ellipsoid is

$$\mathbf{r}(u, v) = \langle 6 \sin u \cos v, 3 \sin u \sin v, 5 \cos u \rangle,$$

for $0 \leq u \leq \pi, 0 \leq v \leq 2\pi$.

(b)   A graph produced by Mathematica is in Figure 5.42.

■

Surfaces of revolution are studied in *Calculus Set Free* Section 9.3. Some quadric surfaces are also surfaces of revolution.

Replacing x by its parameter u, we have $y = f(u) \cos v$ and $z = f(u) \sin v$.

Surfaces of revolution can also be parameterized. For instance, to revolve $y = f(x)$ about the x-axis, the circular cross-sections are parallel to the yz-plane and have radius $f(x)$. Applying the usual parameterization of a circle, the cross-sections are represented by $y = f(x) \cos v$, $z = f(x) \sin v$, $0 \leq v \leq 2\pi$. The variable x gets its own parameter, $x = u$.

Look at the grid curves on the ellipsoid in Figure 5.42. They resemble lines of latitude and longitude on a globe! The grid curve for each constant value of u, which ranges from 0 to π, is a line of latitude, representing a constant value of the variable z. The grid curve for each constant value of v, which ranges from 0 to 2π, is a line of longitude, passing through the z-axis twice (the "north pole" when $u = 0$ and the "south pole" when $u = \pi$). Just as latitude and longitude serve as coordinates locating our position on earth, the values of u and v serve as coordinates locating position on this ellipsoid. This is true for any surface—a parameterization serves as a coordinate system on the surface.

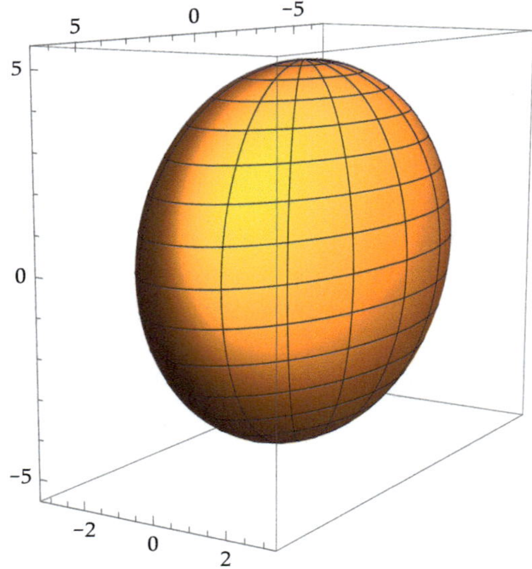

Figure 5.42 *The graph of the ellipsoid $\mathbf{r}(u, v) = \langle 6 \sin u \cos v, 3 \sin u \sin v, 5 \cos u \rangle$*

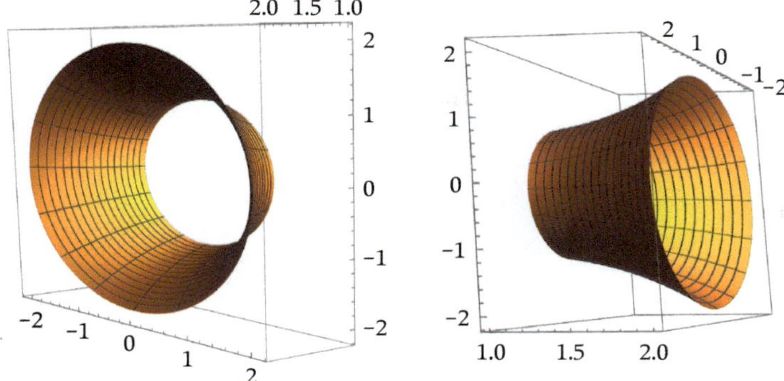

Figure 5.43 *The graph of the surface of revolution* $\mathbf{r}(u, v) = \left\langle u, \left(\frac{1}{8}u^3 + 1\right)\cos v, \left(\frac{1}{8}u^3 + 1\right)\sin v\right\rangle$, *(left) viewed from the first octant and (right) viewed from a perspective often used for visualizing surfaces of revolution*

Example 39 *(a) State a parameterization of the surface of revolution formed by rotating* $y = \frac{1}{8}x^3 + 1$, $1 \le x \le 2$, *about the x-axis. (b) Use technology to graph the parametric surface.*

Solution

(a) A parameterization is

$$\mathbf{r}(u, v) = \left\langle u, \left(\frac{1}{8}u^3 + 1\right)\cos v, \left(\frac{1}{8}u^3 + 1\right)\sin v\right\rangle,$$

An alternate answer swaps $\sin v$ and $\cos v$.

for $1 \le u \le 2$ (the values given for x) and $0 \le v \le 2\pi$ (one complete rotation).

(b) Two views of the graph, each produced by Mathematica, are in Figure 5.43.

■

Reading Exercise 11   State a parameterization of the surface of revolution formed by rotating $y = \sqrt{x}$, $1 \le x \le 4$, about the x-axis.

5.6.2 Tangent planes

Ans. to reading exercise 10:
$$\mathbf{r}(u, v) = \langle u, v, u^2v - 3v^4\rangle; \quad \text{alternate}$$
answers include $\mathbf{r}(u, v) = \langle v, u, v^2u - 3u^4\rangle$

Consider a point $\mathbf{r}(a, b)$ on a parametric surface (Figure 5.44). How might we find an equation of the tangent plane to the surface at that point? A plane can be described by a point and two direction vectors. How might we find two direction vectors? Do we need to somehow find tangent vectors to the surface in the x- and y-directions, as we did for surfaces of the form $z = f(x, y)$? Not necessarily—any two nonparallel direction vectors will do. One strategy is to use grid curves. Grid curves are spacecurves, and we already know how to find tangent vectors to spacecurves. If we find tangent vectors $\mathbf{r}_v(a, b)$ for the grid curve $\mathbf{r}(a, v)$ and $\mathbf{r}_u(a, b)$ for the grid curve

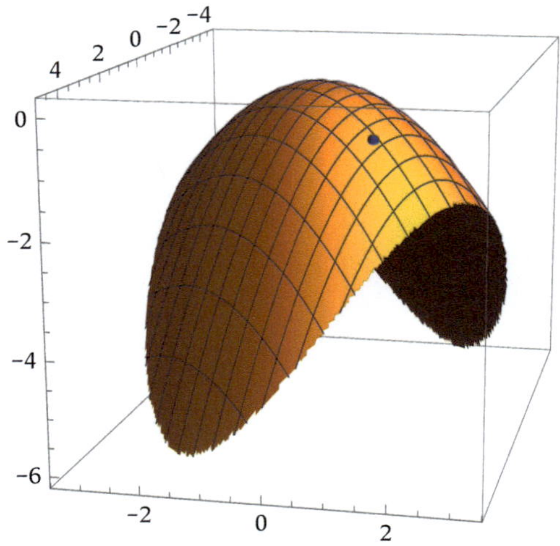

Figure 5.44 *A point $\mathbf{r}(a, b)$ (blue) on a parametric surface*

$\mathbf{r}(u, b)$, as pictured in Figure 5.45, then as long as the two tangent vectors are not parallel (that is, as long as $\mathbf{r_u}(a, b) \times \mathbf{r_v}(a, b) \neq \mathbf{0}$), then we have our two direction vectors and can form the equation of the tangent plane.

Details of the procedure are illustrated by the next example.

Example 40 *Find an equation of the tangent plane to the parametric surface*
$$\mathbf{r}(u, v) = \left\langle u + v, u - 0.6v, -\frac{1}{4}u^2 - \frac{1}{2}v^2 \right\rangle \text{ at the point } \mathbf{r}\left(\frac{3}{2}, \frac{1}{2}\right).$$

Solution   ❶ Determine coordinates of the point of tangency. Evaluating the function gives

$$\mathbf{r}\left(\frac{3}{2}, \frac{1}{2}\right) = \cdots = \langle 2, 1.2, -0.6875 \rangle,$$

which is in the form of a position vector. If desired, we may write the point as $(2, 1.2, -0.6875)$.

❷ Determine $\mathbf{r_u}(a, b)$. Before demonstrating the calculation, an explanation is in order. When $v = b$, the grid curve $\mathbf{r}(u, b) = \langle M(u, b), N(u, b), P(u, b) \rangle$ (bottom in Figure 5.45) has one variable, u; the value of v is the constant b. It's as if we wish to find $\mathbf{r}'(u)$, the tangent vector to the spacecurve. Taking the derivative with respect to the variable u and treating v as a constant is finding a partial derivative; we calculate $\mathbf{r_u}$ by

$$\mathbf{r_u}(u, v) = \langle M_u(u, v), N_u(u, v), P_u(u, v) \rangle.$$

Returning to the calculation, we first calculate $\mathbf{r_u}(u, v)$, taking the derivative of each component with respect to u, treating v as a constant:

$$\mathbf{r_u}(u, v) = \langle 1, 1, -\frac{1}{2}u \rangle.$$

Ans. to reading exercise 11:
$$\mathbf{r}(u, v) = \langle u, \sqrt{u} \cos v, \sqrt{u} \sin v \rangle,$$
$$1 \leq u \leq 4, 0 \leq v \leq 2\pi.$$

This serves as the definition of $\mathbf{r_u}(u, v)$.

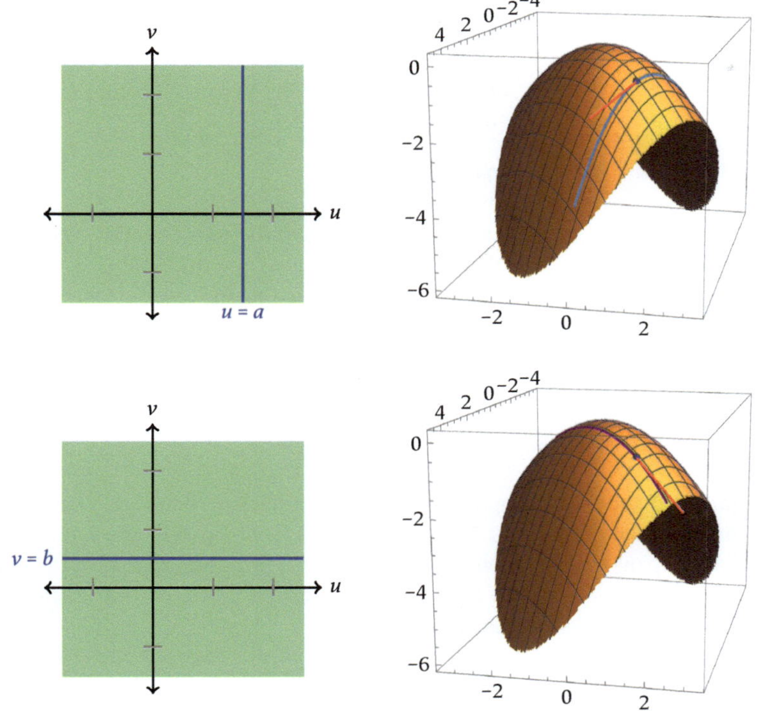

Grid curves for u and v need not be orthogonal, because they do not need to be parallel to the x- and y-axes. Observe the intersections of grid curves (black) in Figure 5.45; they do not meet at right angles.

Figure 5.45 *(Top left) the line $u = a$ in the uv-plane, (top right) its corresponding grid curve $\mathbf{r}(a, v)$ (blue) and its tangent vector $\mathbf{r}_v(a, b)$ (red), (bottom left) the line $v = b$ in the uv-plane, and (bottom right) its corresponding grid curve $\mathbf{r}(u, b)$ (purple) and its tangent vector $\mathbf{r}_u(a, b)$ (red)*

Then one direction vector for the tangent plane is

$$\mathbf{r}_u\left(\tfrac{3}{2}, \tfrac{1}{2}\right) = \left\langle 1, 1, -\tfrac{3}{4}\right\rangle.$$

❸ Calculate $\mathbf{r}_v(a, b)$. Taking the derivative of each component with respect to v and treating u as a constant,

$$\mathbf{r}_v(u, v) = \langle 1, -0.6, -v\rangle.$$

Then the second direction vector for the tangent plane is

$$\mathbf{r}_v\left(\tfrac{3}{2}, \tfrac{1}{2}\right) = \left\langle 1, -0.6, -\tfrac{1}{2}\right\rangle.$$

This step is necessary if we wish to find a Cartesian equation of the plane. This calculation is also used verify that the two vectors are nonparallel.

An alternate solution is to state parametric equations of the tangent plane. In this case, inspection of $\mathbf{r_u}(a, b)$ and $\mathbf{r_v}(a, b)$ can verify that they are nonparallel, rendering step ❹ unnecessary. An equation of the tangent plane is $x = 2 + t + s$, $y = 1.2 + t - 0.6s, z = -0.6875 - \frac{3}{4}t - \frac{1}{2}s.$

The Cartesian equation of a plane is discussed in Section 1.6.

❹ Find $\mathbf{r_u}(a, b) \times \mathbf{r_v}(a, b)$. The cross product, which is a normal vector to the plane, is

$$\mathbf{r_u} \times \mathbf{r_v} = \begin{vmatrix} \mathbf{i} & \mathbf{j} & \mathbf{k} \\ 1 & 1 & -\frac{3}{4} \\ 1 & -0.6 & -\frac{1}{2} \end{vmatrix} = \cdots = \langle -0.95, -0.25, -1.6 \rangle.$$

Because the cross product is not the zero vector, the two direction vectors are nonparallel and the tangent plane exists.

❺ Write the Cartesian equation of the plane. Using the results of steps ❶ and ❹, the equation of the tangent plane is

$$-0.95(x - 2) - 0.25(y - 1.2) - 1.6(z + 0.6875) = 0.$$

Figure 5.45 shows the graph of the parametric surface, point, grid curves, and tangent vectors of example 40. The tangent plane is shown in Figure 5.46.

What happens when the two tangent vectors $\mathbf{r_u}(a, b)$ and $\mathbf{r_v}(a, b)$ are parallel? There could be a corner or a ridge or crease. When $\mathbf{r_u} \times \mathbf{r_v} \neq \mathbf{0}$ throughout the domain of $\mathbf{r}$, we say that $\mathbf{r}$ is *smooth*.

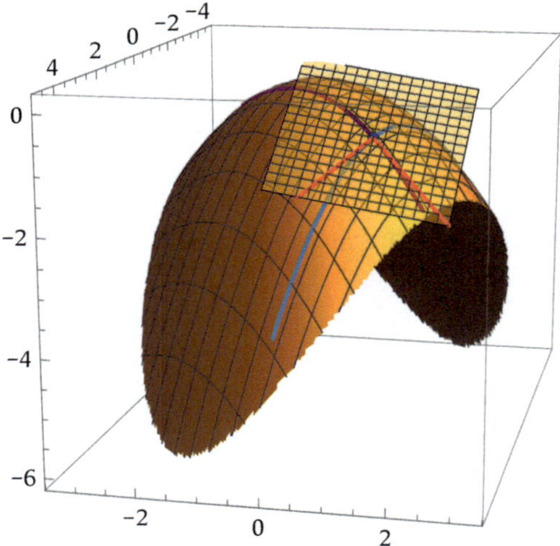

Figure 5.46 *The parametric surface, point, grid curves* $\mathbf{r}\left(\frac{3}{2}, v\right)$ *(blue) and* $\mathbf{r}\left(u, \frac{1}{2}\right)$ *(purple), tangent vectors (red), and tangent plane (transparent yellow) of example 40.*

Example 41 *Verify that the parametric surface* $\mathbf{r}(u, v) = \langle u + v, u - 0.6v,$
$-\frac{1}{4}u^2 - \frac{1}{2}v^2 \rangle$ *is smooth.*

Solution This is the same parametric surface as in example 40, where we calculated $\mathbf{r}_u(u, v) = \langle 1, 1, -\frac{1}{2}u \rangle$ and $\mathbf{r}_v(u, v) = \langle 1, -0.6, -v \rangle$. Their cross product is

$$\mathbf{r}_u \times \mathbf{r}_v = \begin{vmatrix} \mathbf{i} & \mathbf{j} & \mathbf{k} \\ 1 & 1 & -\frac{1}{2}u \\ 1 & -0.6 & -v \end{vmatrix} = \cdots = \langle -v - 0.3u, v - \frac{1}{2}u, -1.6 \rangle.$$

An alternate procedure for example 40 is to skip evaluating $\mathbf{r}_u$ and $\mathbf{r}_v$ at the point of tangency, and instead calculate this cross product and evaluate it at the point of tangency.

Because the z-component of the cross product is -1.6, the cross product is never the zero vector. The surface is smooth. $\blacksquare$

5.6.3 Surface area

Section 4.6 contains a discussion of how to find the area of a surface $z = f(x, y)$. Can we do the same for a parametric surface $\mathbf{r}(u, v)$? Yes, and the same reasoning applies.

Assuming a rectangular domain D, partition the u- and v-intervals into Ω pieces each, where Ω is a positive infinite hyperreal integer. We wish to find the surface area of the portion of the parametric surface corresponding to the n-mth subrectangle (Figure 5.47, left), which we call the n-mth patch of the surface (Figure 5.47, right). Assuming a smooth surface (implying that the component functions of $\mathbf{r}$ are differentiable and local linearity applies), the parametric surface is indistinguishable from its tangent plane, and we just need to know the area of this tangent plane patch. We know that $\mathbf{r}_u(u_n, v_m)$ and $\mathbf{r}_v(u_n, v_m)$ are direction vectors for the tangent plane. By a similar calculation to that of Section 4.6, the tangent

If the domain D is not rectangular, the same strategies that work for double integrals over general regions apply here.

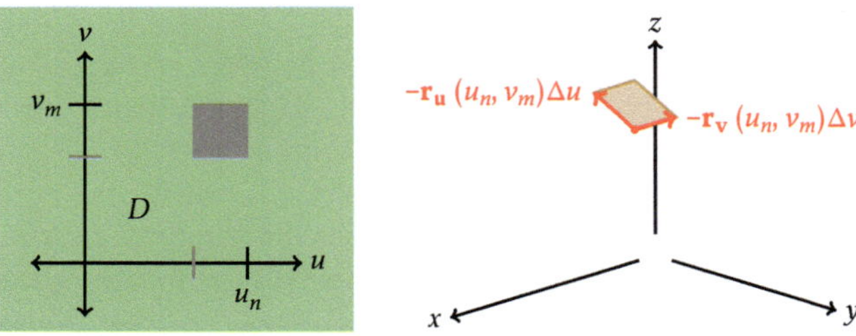

Figure 5.47 *(Left) the n-mth subrectangle in the uv-plane and (right) the corresponding portion of the parametric surface, indistinguishable from its tangent plane (brown). The subrectangle and portion of the surface are shown with noninfinitesimal side lengths so as to be visible. The vectors along the sides are also shown (red). Compare to Section 4.6 Figures 4.46 and 4.47*

See exercise 45 for details.

Recall that a scalar can be factored out of a cross product (property (2) of the cross product) and out of a norm.

plane patch is a parallelogram with side vectors $-\mathbf{r}_\mathbf{u}(u_n, v_m)\Delta u$ and $-\mathbf{r}_\mathbf{v}(u_n, v_m)\Delta v$. Then the area of the patch is the norm of the cross product of the side vectors, $\|\mathbf{r}_\mathbf{u}(u_n, v_m) \times \mathbf{r}_\mathbf{v}(u_n, v_m)\|\,\Delta u\,\Delta v$. To find the area of the entire surface, we add the areas of all the surface patches, obtaining the double omega sum

$$\sum_{n=1}^{\Omega} \sum_{m=1}^{\Omega} \|\mathbf{r}_\mathbf{u}(u_n, v_m) \times \mathbf{r}_\mathbf{v}(u_n, v_m)\|\,\Delta v\,\Delta u.$$

Rewriting as a double integral, the area of the parametric surface is

$$\iint_D \|\mathbf{r}_\mathbf{u}(u, v) \times \mathbf{r}_\mathbf{v}(u, v)\|\,dv\,du.$$

It is implied that the parametric surface is covered exactly once on the domain D.

This traditional presentation of the formula omits the (u, v) and replaces $du\,dv$ or $dv\,du$ with dA. It can also be written as $\iint_D dS$, where $dS = \|\mathbf{r}_u \times \mathbf{r}_v\|\,dA$ is the *surface area element*, analogous to the arc length element ds.

PARAMETRIC SURFACE AREA

Suppose $\mathbf{r}(u, v) = \langle M(u, v), N(u, v), P(u, v)\rangle$ is a smooth parametric surface with domain D. Then the *area of the parametric surface* is

$$SA = \iint_D \|\mathbf{r}_\mathbf{u} \times \mathbf{r}_\mathbf{v}\|\,dA,$$

where $\mathbf{r}_\mathbf{u} = \langle M_u, N_u, P_u\rangle$ and $\mathbf{r}_\mathbf{v} = \langle M_v, N_v, P_v\rangle$.

As is the case with double integrals for finding the area of a surface $z = f(x, y)$, finding the area of a parametric surface often results in an integral that must be approximated numerically.

Example 42 *(a) Set up a double integral for finding the area of the parametric surface $\mathbf{r}(u, v) = \langle \cos u, v, \sin u + \sin v\rangle$, $0 \le u \le 2\pi$, $\pi \le v \le 5\pi$. (b) Use a CAS to determine a decimal approximation to the surface area.*

Before presenting the solution, notice that this is the same parametric surface as example 35 and Figures 5.37–5.39. In Figure 5.39 we see that vertical cross sections are circles of radius 1. With $\pi \le v \le 5\pi$, imagine a circular cylinder of length $\ell = 4\pi$ and radius $r = 1$, which has surface area $2\pi r\ell = 2\pi \cdot 1 \cdot 4\pi = 8\pi^2 \approx 78.96\ \text{units}^2$. Because the parametric surface $\mathbf{r}$ wanders the length of 4π rather than being straight, we suspect that the surface area is larger than $78.96\ \text{units}^2$.

In general, we should check that (1) the parameterization is smooth and (2) the values of u and v cover the surface exactly once. In practice, as long as the component functions are continuous functions, this can be accomplished by examining graphs of the surface produced by a CAS.

Solution

(a) Figures 5.37–5.39 and the accompanying analysis verify that the surface is smooth and covered exactly once. Proceeding according to the formula, we perform the following calculations:

$$\mathbf{r_u}(u, v) = \langle -\sin u, 0, \cos u \rangle$$

$$\mathbf{r_v}(u, v) = \langle 0, 1, \cos v \rangle$$

$$\mathbf{r_u} \times \mathbf{r_v} = \begin{vmatrix} \mathbf{i} & \mathbf{j} & \mathbf{k} \\ -\sin u & 0 & \cos u \\ 0 & 1 & \cos v \end{vmatrix}$$

$$= \langle -\cos u, -(-\sin u \cos v), -\sin u \rangle$$

$$\|\mathbf{r_u} \times \mathbf{r_v}\| = \sqrt{\cos^2 u + \sin^2 u \cos^2 v + \sin^2 u} = \sqrt{1 + \sin^2 u \cos^2 v}.$$

Then the area of the parametric surface can be found by the double integral

$$\int_0^{2\pi} \int_\pi^{5\pi} \sqrt{1 + \sin^2 u \cos^2 v} \, dv \, du.$$

The usual conditions for Fubini's theorem apply to the double integral. An alternate solution reverses the order of integration.

(b) Using a CAS to approximate the value of the integral, the area of the surface is 87.771 units^2.

The CAS that I used was unsuccessful in an attempt to calculate the value exactly. The inner integral is nonelementary in either order.

∎

As suspected, the area of the surface in example 42 is greater than 78.96 units^2.

Example 43 *Find the area of the parametric surface* $\mathbf{r}(u, v) = \langle v^2, uv, v \rangle$, $0 \leq u \leq 1$, $0 \leq v \leq 2$.

Solution A graph can verify that the surface is smooth and covered only once. Proceeding according to the formula,

The fact that each component is strictly increasing for positive values of u and v is evidence that the surface does not repeat.

$$\mathbf{r_u}(u, v) = \langle 0, v, 0 \rangle$$

$$\mathbf{r_v}(u, v) = \langle 2v, u, 1 \rangle$$

$$\mathbf{r_u} \times \mathbf{r_v} = \begin{vmatrix} \mathbf{i} & \mathbf{j} & \mathbf{k} \\ 0 & v & 0 \\ 2v & u & 1 \end{vmatrix} = \langle v, 0, -2v^2 \rangle$$

At the end of line 4, note that $v \geq 0$, so we may use $\sqrt{v^2} = v$. If we had $-2 \leq v \leq 0$ instead, we would use $\sqrt{v^2} = -v$.

$$\|\mathbf{r_u} \times \mathbf{r_v}\| = \sqrt{v^2 + 4v^4} = \sqrt{v^2(1 + 4v^2)} = v\sqrt{1 + 4v^2}.$$

Then the area of the parametric surface is

Because u is not in the integrand, the antiderivative with respect to u is the integrand times u, and we evaluate at $u = 1$, at $u = 0$, and subtract. This is a calculation shorthand worth knowing.

$$\int_0^2 \int_0^1 v\sqrt{1 + 4v^2} \, du \, dv = \int_0^2 (1 - 0)v\sqrt{1 + 4v^2} \, dv.$$

Because u is one of the original variables, it is not a good choice as the substitution variable. Use w instead.

Using the substitution $w = 1 + 4v^2$, $dw = 8v\,dv$, we continue as

$$= \frac{1}{8} \int_0^2 8v\sqrt{1 + 4v^2}\,dv$$

$$= \frac{1}{8} \int_1^{17} \sqrt{w}\,dw$$

$$= \frac{1}{8} \left.\frac{w^{3/2}}{3/2}\right|_1^{17}$$

$$= \frac{1}{12} \left(17^{3/2} - 1\right).$$

The area of the parametric surface is $\frac{1}{12}\left(17^{3/2} - 1\right)$ units^2.

EXERCISES 5.6

1–4. Rapid response: is the parametric surface smooth?

1.

3.

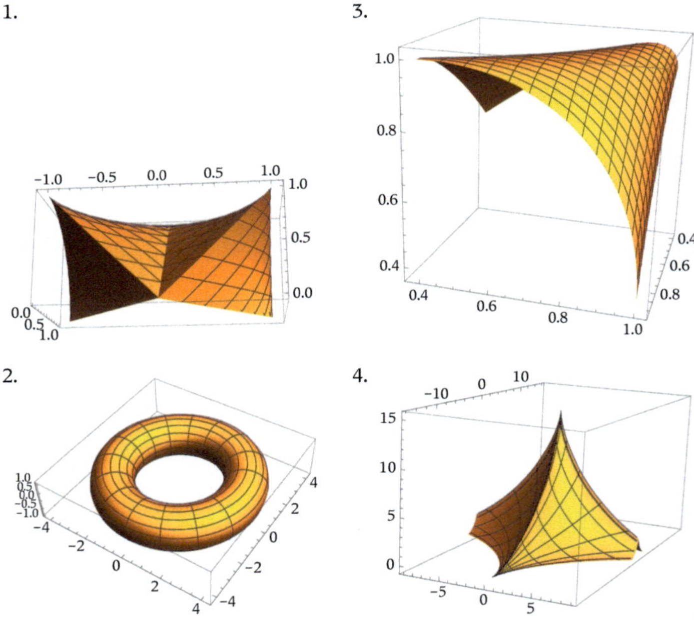

2.

4.

5–10. Use technology to graph the parametric surface.

    5. $\mathbf{r}(u, v) = \langle v \cos u, v \sin u, u \rangle, 0 \le u \le 4\pi, 0 \le v \le 8$

    6. $\mathbf{r}(u, v) = \langle u - \sin u \cosh v, 1 - \cos u \cosh v, 4 \sin(\frac{1}{2}u) \rangle,$
       $-4\pi \le u \le 4\pi, 0 \le v \le 2$ (Catalan's surface)

    7. $\mathbf{r}(u, v) = \langle 4 \cos u \cos v, 4 \sin u \sin v, 4 \sin u \rangle, 0 \le u \le 2\pi,$
       $0 \le v \le 2\pi$

    8. $\mathbf{r}(u, v) = \langle (0.8 + \cos v) \cos u, (0.8 + \cos v) \sin u, \sin v \rangle,$
       $0 \le u \le 2\pi, \pi \le v \le 2\pi$

    9. $\mathbf{r}(u, v) = \langle \sin u, \sin v, \sin(u + v) \rangle, 0 \le u \le 2\pi, 0 \le v \le 2\pi$

    10. $\mathbf{r}(u, v) = \langle uv, uv^2, uv^3 \rangle, -1 \le u \le 1, -1 \le v \le 1$

11–14. (a) State a parameterization of the surface. (b) Use technology to graph the parametric surface.

    11. $z = \sqrt{x^2 - y^2}, 4 \le x \le 7, -3 \le y \le 3$

    12. $f(x, y) = x \ln y - y \ln x, 0.2 \le x \le 5, 0.2 \le y \le 5$

    13. $f(x, y) = 0.7x \sin x + 2 \cos y, 0 \le x \le 3\pi, \pi \le y \le 5\pi$

    14. $z = \sin^2 x - 4\sqrt{y}$

15–18. (a) State a parameterization of the quadric surface. (b) Use technology to graph the parametric surface.

You may need to perform a search to find an appropriate formula.

    15. $\dfrac{y^2}{16} - \dfrac{x^2}{9} = \dfrac{z}{5}$
            17. $\dfrac{z^2}{25} - \dfrac{x^2}{9} - \dfrac{y^2}{16} = 1$

    16. $\dfrac{x^2}{9} + \dfrac{y^2}{16} - \dfrac{z^2}{25} = 1$
          18. $\dfrac{x^2}{9} + \dfrac{z^2}{25} = \dfrac{y}{4}$

19–22. (a) State a parameterization of the surface of revolution formed by rotating the curve about the axis. (b) Use technology to graph the parametric surface.

    19. $y = \sqrt{x}, 0 \le x \le 4$, about x-axis

    20. $y = \ln x, 1 \le x \le 5$, about x-axis

    21. $y = \sqrt{x}, 0 \le x \le 4$, about y-axis

    22. $y = x + 1, 0 \le y \le 4$, about y-axis

23–28. Find an equation of the tangent plane to the surface at the given point.

    23. $\mathbf{r}(u, v) = \langle u^2, v^2, u + v \rangle$, point $\mathbf{r}(1, 2)$

    24. $\mathbf{r}(u, v) = \langle u^2, v^2, u + v \rangle$, point $\mathbf{r}(2, 1)$

    25. $\mathbf{r}(u, v) = \langle \sin u, \cos u, \sin(u + v) \rangle$, point $\mathbf{r}(\pi, 2\pi)$

    26. $\mathbf{r}(u, v) = \langle uv, uv^2, uv^3 \rangle$, point $\mathbf{r}(5, 1)$

    27. $\mathbf{r}(u, v) = \langle \ln uv, \ln v, u^2 \rangle$, point $\mathbf{r}(1, 1)$

    28. $\mathbf{r}(u, v) = \langle \cos 2u, \cos 3v, \sin v \rangle$, point $\mathbf{r}\left(\dfrac{\pi}{4}, \dfrac{\pi}{4}\right)$

29–32. Verify that the parametric surface is smooth.

29. $\mathbf{r}(u, v) = \langle u^2, v^2, u + v \rangle$, $v \geq 1$
30. $\mathbf{r}(u, v) = \langle uv, uv^2, uv^3 \rangle$, $u \geq 1$, $v \geq 1$
31. $\mathbf{r}(u, v) = \langle v \cos u, v \sin u, u \rangle$, $v \geq 1$
32. $\mathbf{r}(u, v) = \langle e^u, e^v, v \rangle$

33–36. (a) Set up a double integral for finding the area of the parametric surface. (b) Use a CAS to determine a decimal approximation to the surface area.

33. $\mathbf{r}(u, v) = \langle u^2, v^2, u + v \rangle$, $1 \leq u \leq 2$, $1 \leq v \leq 3$
34. $\mathbf{r}(u, v) = \langle uv, uv^2, uv^3 \rangle$, $1 \leq u \leq 2$, $1 \leq v \leq 2$
35. $\mathbf{r}(u, v) = \langle u, u^2 + \cos v, u^2 + \sin v \rangle$, $-2 \leq u \leq 2$, $0 \leq v \leq 2\pi$
36. $\mathbf{r}(u, v) = \left\langle \ln \frac{u}{v}, u, 4v \right\rangle$, $1 \leq u \leq 2$, $1 \leq v \leq 2$

37–40. Find the area of the parametric surface.

37. $\mathbf{r}(u, v) = \langle v^2 u, \cos u, \sin u \rangle$, $\pi \leq u \leq 2\pi$, $\pi \leq v \leq 2\pi$
38. $\mathbf{r}(u, v) = \langle \cos v, u + v, \sin v \rangle$, $0 \leq u \leq 2\pi$, $0 \leq v \leq 4\pi$
39. $\mathbf{r}(u, v) = \langle \cos^2 u, \sin^2 u, v\sqrt{2} + u \rangle$, $\frac{\pi}{6} \leq u \leq \frac{\pi}{3}$, $0 \leq v \leq 2$
40. $\mathbf{r}(u, v) = \langle 1 + \cos u, 3 + \sin u, \cos v \rangle$, $0 \leq u \leq 2\pi$, $0 \leq v \leq \pi$

41. (a) Set up a double integral for finding the (parametric) surface area of the ellipsoid of example 38.
   (b) Use a CAS to find a numerical approximation to the surface area.
   (c) There is no "nice" formula for the exact surface area of an ellipsoid; known formulas use nonelementary integrals. But there is an approximate formula,

$$SA = 4\pi \cdot \sqrt[1.6]{\frac{(ab)^{1.6} + (ac)^{1.6} + (bc)^{1.6}}{3}},$$

   where a, b, and c are the principal semi-axes of the ellipsoid. Use this approximation formula to approximate the surface area.
   (d) Assuming the answer in (b) is correct, what is the percent relative error of the approximation formula?

Different exponents are better depending on the shape of the ellipsoid. The exponent 1.6 is optimal for nearly spherical ellipsoids, whereas for very flat ellipsoids (one principal semiaxis much smaller than the others) the exponent $\log_2 3 \approx 1.585$ is better.

42. Section 9.3 of *Calculus Set Free* contains a single-variable method for finding the area of a surface of revolution.
   (a) State a parameterization of the surface of revolution formed by rotating the curve $y = x^3$ between $x = 0$ and $x = 1$ about the x-axis.

(b) Set up a double integral for finding the area of the parametric surface. Have the inner integral be with respect to the variable not present in the integrand.

(c) Evaluate the integral to find the surface area.

(d) Compare the calculation in part (c) to the single-variable method. At what stage of the process does the integral become the same (except for variable name) as the integral for the single-variable method?

43. Use the following steps to show that the surface area formula of Section 4.6 is a special case of the parametric surface area formula.

(a) Suppose $z = f(x, y)$ is a surface over a region D, where f is differentiable and both f_x and f_y are continuous on D (the hypotheses of the surface area formula). For the parametric surface $\mathbf{r}(u, v) = \langle u, v, f(u, v)\rangle$, calculate $\mathbf{r_u}(u, v)$ and $\mathbf{r_v}(u, v)$.

(b) Calculate $\mathbf{r_u} \times \mathbf{r_v}$.

(c) Calculate $\|\mathbf{r_u} \times \mathbf{r_v}\|$.

(d) Are the hypotheses of the parametric surface area formula satisfied?

(e) Write the integral for finding the parametric surface area and compare it to the surface area formula.

44. The parametric surface of examples 35 and 42 and Figures 5.37–5.39 looks like a sine wave tube, making one wonder whether the surface area is the length of the sine wave times 2π, the circumference of a circle of radius 1.

(a) Use the arc length formula of *Calculus Set Free* Section 9.1 to set up an integral for finding the length of the curve $z = \sin y$ for $\pi \le y \le 5\pi$.

(b) Use a CAS to find a decimal approximation to the length in part (a).

The integral in part (a) is nonelementary.

(c) Multiply the result of part (b) by 2π.

(d) Compare the answer to part (c) to the surface area found in example 42(b). Explain why the answer to (c) is larger.

45. The derivation of the parametric surface area formula contains a statement that the tangent plane patch is a parallelogram with side vectors $-\mathbf{r_u}(u_n, v_m)\Delta u$ and $-\mathbf{r_v}(u_n, v_m)\Delta v$. Show this is correct using the following steps.

(a) The corners of the n-mth subrectangle (Figure 5.47, left) are (u_n, v_m), which we shall call point A; $(u_n - \Delta u, v_m)$ (point B); $(u_n, v_m - \Delta v)$ (point C); and $(u_n - \Delta u, v_m - \Delta v)$ (point D). Use the local linearity formula on the component functions M, N, and P to approximate the vector between corners of the surface patch, from $\mathbf{r}(A)$ to $\mathbf{r}(B)$.

This assumes that $\mathbf{r}(u, v) = \langle M(u, v), N(u, v), P(u, v)\rangle$.

These calculations follow the pattern of the corresponding portion of the derivation of the surface area formula.

(b)   Repeat for the vector from $\mathbf{r}(A)$ to $\mathbf{r}(C)$.
(c)   Repeat for the vector from $\mathbf{r}(C)$ to $\mathbf{r}(D)$.
(d)   Repeat for the vector from $\mathbf{r}(B)$ to $\mathbf{r}(D)$.
(e)   Which answers are identical? Does this mean that the surface patch is (approximately) a parallelogram?

5.7   Surface Integrals

Once again, it's time to add a dimension to previous ideas.

5.7.1   Total mass: thin curved plates of variable density

Curved instead of flat.

In Section 4.5, thin flat plates of variable density are discussed. Such a plate is located in a region R in the xy-plane, as in Figure 5.48 (repeated from Section 4.5). A density function $\rho(x, y)$ is given, representing its density at any point in the xy-plane with units $\frac{\text{mass}}{\text{area}}$. Then the mass on any subrectangle is $\rho(x_n, y_m)\Delta x\Delta y$, leading to the double-integral formula for total mass, $\iint_R \rho(x, y)\, dx\, dy$.

This also includes a non-horizontal plane, which is not curved but nevertheless is a surface.

Now suppose that instead of being flat and contained in the xy-plane, the thin plate is curved, lying along a surface $z = f(x, y)$ in three dimensions, as in Figure 5.49 (repeated from Section 4.6). Now the density function, still given with units $\frac{\text{mass}}{\text{area}}$, must be of the form $\rho(x, y, z)$, giving the density at any point (x, y, z) along the surface. The mass of the n-mth piece is now the density times the area of the piece of surface (brown in Figure 5.49). The area of the n-mth piece, as determined in Section 4.6, is $\sqrt{(f_x(x_n, y_m))^2 + (f_y(x_n, y_m))^2 + 1}\, \Delta x\, \Delta y$. The point at which we desire the density is the point on the surface, $(x_n, y_m, f(x_n, y_m))$, and the density is given by $\rho(x_n, y_m, f(x_n, y_m))$. Therefore the mass of the piece is

$$\rho(x_n, y_m, f(x_n, y_m))\sqrt{(f_x(x_n, y_m))^2 + (f_y(x_n, y_m))^2 + 1}\, \Delta x\, \Delta y.$$

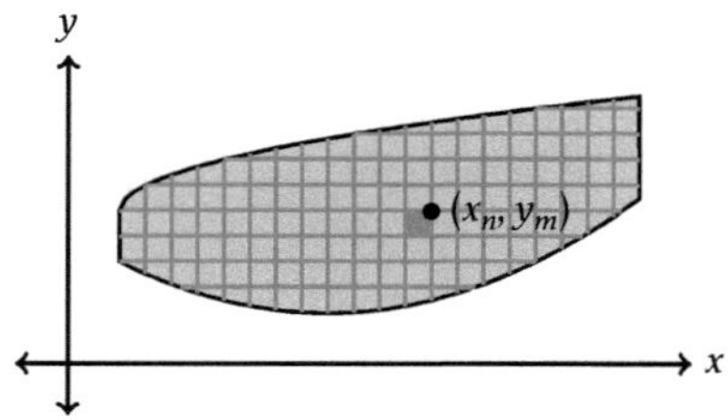

Figure 5.48 *A region R in the xy-plane, subdivided into rectangles. The rectangles are shown with non-infinitesimal area so as to be visible. The n-mth rectangle is highlighted*

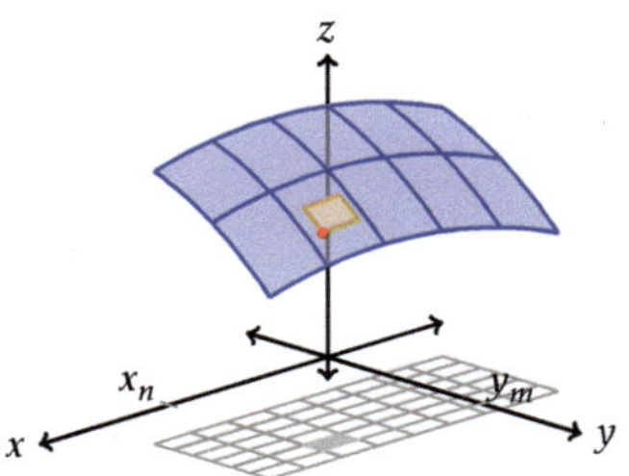

Figure 5.49 *A surface $z = f(x, y)$ (blue) over a rectangular region partitioned into subrectangles, along with the n-mth piece of the surface (brown). The subrectangles and the n-mth piece of the surface are shown with non-infinitesimal side lengths so as to be visible*

Adding up all the masses for each subrectangle of the region R as pictured in Figure 5.49, and writing the double omega sum as a double integral, the total mass is

$$\iint_R \rho(x, y, f(x, y))\sqrt{(f_x(x, y))^2 + (f_y(x, y))^2 + 1}\, dx\, dy.$$

The region in the xy-plane is still divided into subrectangles. Instead of the thin plate lying on the region as in Figure 5.48, it lies above (or below) the region. But the integration is still over the two-dimensional region and results in a double integral.

MASS OF A THIN CURVED PLATE OF VARIABLE DENSITY

The mass of a plate occupying the surface $z = f(x, y)$ over the region R with density function $\rho(x, y, z)$ is

$$m = \iint_R \rho(x, y, f(x, y))\sqrt{(f_x(x, y))^2 + (f_y(x, y))^2 + 1}\, dx\, dy.$$

If the plate lies in the xy-plane as in Figure 5.48, $z = f(x, y) = 0$ and $\sqrt{(f_x(x, y))^2 + (f_y(x, y))^2 + 1} = \sqrt{0 + 0 + 1} = 1$, reducing the integral formula to $\iint_R \rho(x, y, 0)\, dx\, dy$, matching the formula for total mass of a thin flat plate.

Example 44 *A thin plate occupies the surface $z = 3x + 5y$ over the region $0 \le x \le 2$, $0 \le y \le 2$, measured in meters. Its density function is $\rho(x, y, z) = xyz$, measured in $\frac{kg}{m^2}$. Find the mass of the plate.*

Solution The density to include in the integrand is

$$\rho(x, y, z) = \rho(x, y, 3x + 5y) = xy(3x + 5y) = 3x^2y + 5xy^2.$$

Although the plate of example 44 lies along a plane and is therefore flat rather than curved, it is tilted rather than horizontal, and the formula for a curved plate must be used.

The formula for the surface is used, replacing z by $3x + 5y$.

The partial derivatives are $\frac{\partial}{\partial x}(3x + 5y) = 3$ and $\frac{\partial}{\partial y}(3x + 5y) = 5$. Then the mass is

Line 1 uses the formula for the mass of a thin curved plate of variable density; line 2 simplifies and uses the double integral constant multiple rule; line 3 finds an antiderivative with respect to y, treating x as a constant; line 4 evaluates at $y = 2$, at $y = 0$, and subtracts; lines 5 and 6 complete the integration with respect to the variable x.

$$
\begin{aligned}
m &= \int_0^2 \int_0^2 (3x^2y + 5xy^2)\sqrt{3^2 + 5^2 + 1}\,dy\,dx \\
&= \sqrt{35} \int_0^2 \int_0^2 (3x^2y + 5xy^2)\,dy\,dx \\
&= \sqrt{35} \int_0^2 \left(\left(\frac{3x^2y^2}{2} + \frac{5xy^3}{3}\right)\Big|_0^2\right) dx \\
&= \sqrt{35} \int_0^2 \left(6x^2 + \frac{40}{3}x\right) dx \\
&= \sqrt{35} \left(2x^3 + \frac{20}{3}x^2\right)\Big|_0^2 \\
&= \sqrt{35}\left(16 + \frac{80}{3}\right) = \frac{128}{3}\sqrt{35}\,\text{kg}.
\end{aligned}
$$

Units of $3x^2y + 5xy^2$, which is $\rho(x, y, z)$, are $\frac{\text{kg}}{\text{m}^2}$. Units of $\sqrt{35}\,dy\,dx$, which comes from the area of the surface, are m^2. Their product is kg.

∎

5.7.2   Surface integrals

The double integral for finding the mass of a thin curved plate of variable density is an example of a *surface integral*, which is a higher-dimensional version of a line integral. Our first encounter with a line integral, in Section 5.1, begins with a function f of two variables, but instead of a double integral over a two-dimensional region, the line integral integrates f over a curve, producing a single integral. Now add a dimension to all of that: begin with a function f of three variables, but instead of a triple integral over a three-dimensional region, the surface integral integrates f over a surface, producing a double integral.

For the mass of a thin curved plate of variable density, the function of three variables is the density function ρ, which is integrated over a surface, producing a double integral.

The next formula is merely a name change.

SURFACE INTEGRAL, CARTESIAN FORM

The *surface integral* of $\rho(x, y, z)$ along the surface $z = f(x, y)$ over the region R is

$$
\iint_R \rho(x, y, f(x, y))\sqrt{(f_x(x, y))^2 + (f_y(x, y))^2 + 1}\,dx\,dy.
$$

Once we have a double integral, anything we have learned about double integrals is fair game. We can integrate over non-rectangular regions. We can change to polar coordinates.

Example 45 *Compute the surface integral of* $\rho(x,y,z) = \frac{y}{\sqrt{z}}$ *along the surface* $z = x^2 + y^2$ *over the region* $1 \le x^2 + y^2 \le 4, y \ge 0$.

Solution As with any double integral, sketching the region (or seeing it in your mind) is helpful. The equation $x^2 + y^2 = 1$ is a circle of radius 1, and $x^2 + y^2 = 4$ is a circle of radius 2. With the restriction $y \ge 0$, we only want the top half of the region between the circles (Figure 5.50).

It is not usually necessary to visualize the surface over the region, but we'll do it anyway. The surface $z = x^2 + y^2$ is an elliptic (circular in this case) paraboloid. Graphed only over the region of integration, it appears similar to a half-shell stadium (Figure 5.51).

Next, we need to calculate the integrand. Along the surface, the function we are integrating is

$$\rho(x,y,z) = \rho(x,y,x^2+y^2) = \frac{y}{\sqrt{x^2+y^2}}.$$

The partial derivatives of the surface are $\frac{\partial}{\partial x}(x^2+y^2) = 2x$ and $\frac{\partial}{\partial y}(x^2+y^2) = 2y$. Then the area element is

$$\sqrt{(2x)^2 + (2y)^2 + 1} = \sqrt{4x^2 + 4y^2 + 1},$$

and the integrand is

$$\frac{y}{\sqrt{x^2+y^2}}\sqrt{4(x^2+y^2)+1}.$$

The region of integration appears difficult to describe in Cartesian coordinates, but easy to describe in polar coordinates, so we make the switch. The region can be described as $1 \le r \le 2, 0 \le \theta \le \pi$. Replacing x^2+y^2 by r^2 and y by $r\sin\theta$, the integral to evaluate is

$$\int_0^\pi \int_1^2 \frac{r\sin\theta}{\sqrt{r^2}}\sqrt{4r^2+1}\, r\,dr\,d\theta = \int_0^\pi \int_1^2 \sin\theta\sqrt{4r^2+1}\,r\,dr\,d\theta.$$

To evaluate the inner integral we see the need for substitution, and let $u = 4r^2 + 1$; then $du = 8r\,dr$. The inner integral becomes

$$\frac{1}{8}\int_5^{17} \sin\theta\sqrt{u}\,du = \frac{1}{8}\sin\theta\frac{u^{3/2}}{3/2}\Big|_5^{17}$$

$$= \frac{1}{12}\left(17^{3/2} - 5^{3/2}\right)\sin\theta.$$

Compare to the statement of example 1 of Section 5.1. Instead of a line integral, a surface integral; instead of a function of two variables, a function of three variables; instead of a curve, a surface; instead of an interval, a two-dimensional region.

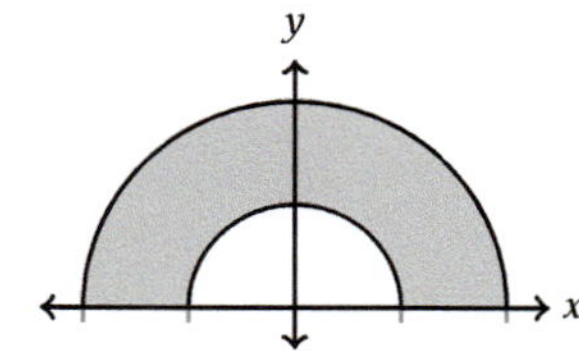

Figure 5.50 *The region of integration for example 45*

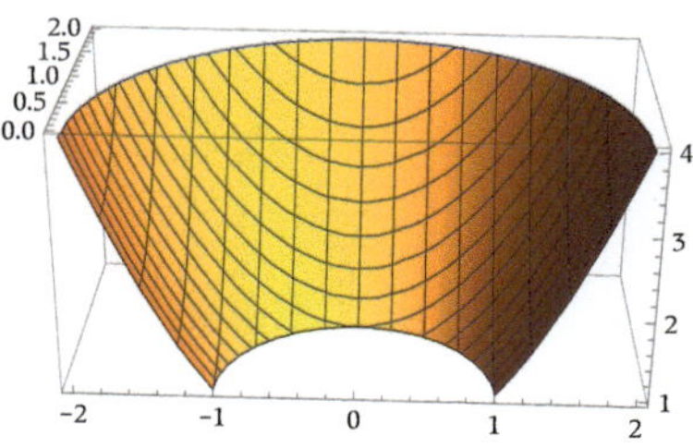

Figure 5.51 *The surface for example 45*

Don't forget the extra r when switching to polar coordinates!

Line 1 starts by completing the substitution, including adjusting for the constant and replacing values of r by values of u; line 1 continues by integrating with respect to u, treating θ as a constant; line 2 evaluates at $u = 17$, at $u = 5$, subtracts, and rearranges.

Continuing with the outer integral,

$$\int_0^\pi \tfrac{1}{12}\left(17^{3/2} - 5^{3/2}\right)\sin\theta\,d\theta = \tfrac{1}{12}\left(17^{3/2} - 5^{3/2}\right)\left(-\cos\theta\right)\Big|_0^\pi$$

$$= \cdots = \tfrac{1}{6}\left(17^{3/2} - 5^{3/2}\right).$$

∎

As involved as example 45 is, it is relatively tame compared to most, which require the use of technology to give a decimal approximation.

5.7.3   Surface integrals: parametric form

Recall that a surface $z = f(x,y)$ can be parameterized, so this approach generalizes the type of surface we may consider.

With line integrals, the curve is usually described parametrically as a vector-valued function of one variable, $\mathbf{r}(t)$. Because surface integrals are one dimension higher, perhaps we could describe the surface parametrically as a vector-valued function of two variables, $\mathbf{r}(u,v)$, allowing us to consider any parametric surface. As we learned in Section 5.6, the surface area element for a parametric surface is $dS = \|\mathbf{r}_\mathbf{u} \times \mathbf{r}_\mathbf{v}\|\,dA$, which replaces the square root in the Cartesian surface integral formula; the parametric form of the surface integral formula is

$$\iint_R \rho(\mathbf{r}(u,v))\|\mathbf{r}_\mathbf{u} \times \mathbf{r}_\mathbf{v}\|\,dA.$$

In the following formula, we make the switch of writing the function as f instead of ρ, and write dS for the surface area element, analogous to ds for the arc length element in a line integral.

Notation note: $\iint_S$ represents the integral over the surface S. If we think of u and v as representing coordinates on the surface, $\iint_S$ means that we are integrating with respect to those coordinate variables along the entire surface S.

SURFACE INTEGRAL, PARAMETRIC FORM

The *surface integral* of a differentiable function $f(x,y,z)$ along a smooth parametric surface $\mathbf{r}(u,v)$ with domain D is

$$\iint_S f(x,y,z)\,dS = \iint_D f(\mathbf{r}(u,v))\|\mathbf{r}_\mathbf{u} \times \mathbf{r}_\mathbf{v}\|\,dA.$$

The surface integral of a scalar function f is independent of the chosen parameterization.

Example 46 *Compute the surface integral of $f(x, y, z) = x^2 z$ along the surface* $\mathbf{r}(u, v) = \langle u, (u^3 + 1)\cos v, (u^3 + 1)\sin v\rangle$, $1 \le u \le 2$, $0 \le v \le 2\pi$.

The surface in example 46 is a surface of revolution. See exercise 25 for a more general version of this example.

Solution   To use the parametric form of the surface integral formula, we need to calculate $f(\mathbf{r}(u, v))$ and $\|\mathbf{r}_u \times \mathbf{r}_v\|$, set up the double integral, and evaluate it.

❶ Calculate $f(\mathbf{r}(u, v))$. We have

$$f(\mathbf{r}(u, v)) = f(u, (u^3 + 1)\cos v, (u^3 + 1)\sin v) = u^2(u^3 + 1)\sin v.$$

This uses the formula for the surface and then the formula for f.

❷ Calculate $\|\mathbf{r}_u \times \mathbf{r}_v\|$. First, we calculate $\mathbf{r}_u$ and $\mathbf{r}_v$:

$$\mathbf{r}_u(u, v) = \langle 1, 3u^2 \cos v, 3u^2 \sin v\rangle$$

$$\mathbf{r}_v(u, v) = \langle 0, (u^3 + 1)(-\sin v), (u^3 + 1)\cos v\rangle.$$

Then

$$\mathbf{r}_u \times \mathbf{r}_v = \begin{vmatrix} \mathbf{i} & \mathbf{j} & \mathbf{k} \\ 1 & 3u^2 \cos v & 3u^2 \sin v \\ 0 & -(u^3 + 1)\sin v & (u^3 + 1)\cos v \end{vmatrix}$$

$$= \langle 3u^2(u^3 + 1)\cos^2 v + 3u^2(u^3 + 1)\sin^2 v, -(u^3 + 1)\cos v,$$

$$- (u^3 + 1)\sin v\rangle$$

$$= \langle 3u^2(u^3 + 1), -(u^3 + 1)\cos v, -(u^3 + 1)\sin v\rangle.$$

Finally, the norm is

$$\|\mathbf{r}_u \times \mathbf{r}_v\| = \sqrt{9u^4(u^3 + 1)^2 + (u^3 + 1)^2 \cos^2 v + (u^3 + 1)^2 \sin^2 v}$$

$$= \sqrt{9u^4(u^3 + 1)^2 + (u^3 + 1)^2}$$

$$= \sqrt{(9u^4 + 1)(u^3 + 1)^2}$$

$$= (u^3 + 1)\sqrt{9u^4 + 1},$$

because $u^3 + 1 \ge 0$ for $1 \le u \le 2$.

❸ Set up the double integral. The integrand is the product of the quantities computed in steps **❶** and **❷**. But which order of integration do we use, u first or v first? It certainly appears that integrating with respect to u is difficult. So we choose to integrate with respect to v first:

If we integrate with respect to u first, the inner integral turns out to be nonelementary.

$$\int_1^2 \int_0^{2\pi} u^2(u^3 + 1)(\sin v) \cdot (u^3 + 1)\sqrt{9u^4 + 1}\, dv\, du.$$

Line 1 simplifies and rearranges; line 2 finds an antiderivative with respect to v, treating u as a constant; line 3 evaluates at $v = 2\pi$, at $v = 0$, and subtracts; line 4 simplifies using $-\cos 2\pi + \cos 0 = -1 + 1 = 0$ and finishes the integration.

If we have $0 \le v \le \pi$ instead, the integrand does not simplify to zero and the remaining integral is nonelementary.

4 Evaluate the integral. Beginning with the inner integral,

$$\int_1^2 \int_0^{2\pi} u^2(u^3 + 1)^2\sqrt{9u^4 + 1}\, \sin v \, dv \, du$$

$$= \int_1^2 \left(u^2(u^3 + 1)^2\sqrt{9u^4 + 1}(-\cos v)\Big|_0^{2\pi} \right) du$$

$$= \int_1^2 \left(u^2(u^3 + 1)^2\sqrt{9u^4 + 1}(-\cos 2\pi - (-\cos 0)) \right) du$$

$$= \int_1^2 0 \, du = 0.$$

The value of the surface integral is 0. ∎

5.7.4   Surface integrals with vector fields: flux

Suppose you are standing in a river, in flowing water, and have a mesh screen in your hands. Holding the mesh screen parallel to the flow of water is easy, as there is no force of water flowing through the screen. (Ignore the force of water on the frame as we are imagining a mesh screen of zero thickness.) Now tilt the screen slightly and it becomes more difficult to hold, because some water is flowing through the screen. Finally, hold the mesh screen perpendicular to the flow of water. This is the most difficult!

Now think of the flowing water as a vector field $\mathbf{F}$, and the mesh screen as a surface. Assuming the flow of water is horizontal, when holding the screen parallel to the flow (horizontal), a normal to the surface is vertical, and the scalar component of the flow of water $\mathbf{F}$ in the direction of the normal is zero, corresponding to no water flowing through the screen (Figure 5.52). With the screen tilted slightly against the flow, nearly horizontal, a normal is nearly vertical and the scalar component of the flow of water in the normal direction is small. When holding the

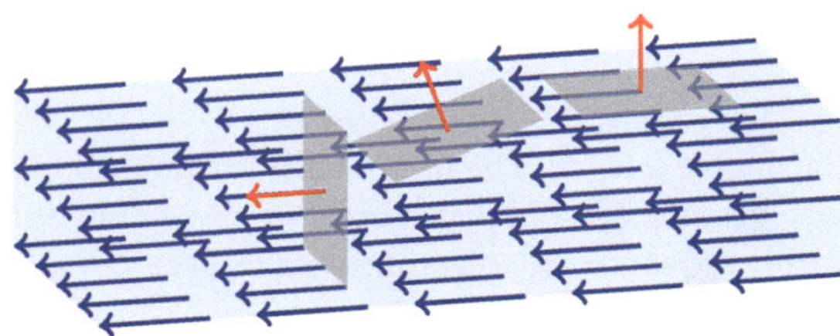

Figure 5.52 *Mesh screen in a river*

screen perpendicular to the flow so that a normal is parallel to the flow, the scalar component of **F** in the direction of the normal is at its greatest, corresponding to the greatest water flow through the screen. We call the amount of water (the mass of the water) flowing through the screen per unit of time the *flux* of the vector field **F** across the surface S.

So how do we calculate the flux? Suppose the surface is given parametrically as $\mathbf{r}(u, v)$. At any point $\mathbf{r}(u_n, v_m)$, the normal to the surface is the vector $\mathbf{r_u}(u_n, v_m) \times \mathbf{r_v}(u_n, v_m)$, and the value of the vector field at that point is $\mathbf{F}(\mathbf{r}(u_n, v_m))$. Then the scalar component of the vector field in the direction of the normal to the surface, omitting the (u_n, v_m), is

$$\operatorname{comp}_{\mathbf{r_u} \times \mathbf{r_v}} \mathbf{F}(\mathbf{r}) = \frac{\mathbf{F}(\mathbf{r}) \cdot (\mathbf{r_u} \times \mathbf{r_v})}{\|\mathbf{r_u} \times \mathbf{r_v}\|}.$$

This is the rate (per unit of time per unit of area) at which water is flowing across the surface at the point $\mathbf{r}(u_n, v_m)$. We know from Section 5.6 that the area of the n-mth piece of the surface is $\|\mathbf{r_u} \times \mathbf{r_v}\| \, \Delta u \, \Delta v$ (once again omitting the (u_n, v_m)). Then the amount of water flowing across the n-mth piece of the surface (per unit of time) is the rate times its area, or

$$\frac{\mathbf{F}(\mathbf{r}) \cdot (\mathbf{r_u} \times \mathbf{r_v})}{\|\mathbf{r_u} \times \mathbf{r_v}\|} \|\mathbf{r_u} \times \mathbf{r_v}\| \, \Delta u \, \Delta v = \mathbf{F}(\mathbf{r}) \cdot (\mathbf{r_u} \times \mathbf{r_v}) \, \Delta u \, \Delta v.$$

Add the amounts across all the pieces of the surface, and we get

$$\iint_D \mathbf{F}(\mathbf{r}) \cdot (\mathbf{r_u} \times \mathbf{r_v}) \, dA,$$

where dA represents $du \, dv$ or $dv \, du$ and D is the domain of the surface.

SURFACE INTEGRAL OF A VECTOR FIELD

The surface integral (flux) of a continuous vector field **F** across a smooth, orientable surface S parameterized by $\mathbf{r}(u, v)$ with domain D is

$$\iint_S \mathbf{F} \cdot d\mathbf{S} = \iint_D \mathbf{F}(\mathbf{r}) \cdot (\mathbf{r_u} \times \mathbf{r_v}) \, dA.$$

The surface integral of a vector field across a surface is also called a *flux integral*.

Electromagnetic flux is another of the myriad applications of surface integrals in physics.

Swapping the order of vectors in a cross product negates the result, pointing the vector in the opposite direction; this can be accomplished here by swapping the parameters u and v. In Figure 5.52, consider the vertically placed screen. The normal is pointing in the direction of the flow of water, and the flux of **F** across S in the direction of the normal is positive. If, instead, the normal points the opposite direction, against the flow of water, the flux of **F** across S in the direction of this normal is negative. For this reason, the surface integral of a vector field across a surface is not independent of parameterization. In practice, the surface must be parameterized so that its normals are pointing in the desired direction. This requires an *orientable* surface, in which a parameterization gives rise to continuously varying normals.

Notation notes: the surface area element $dS = \|\mathbf{r_u} \times \mathbf{r_v}\| \, dA$ is a scalar. The quantity $d\mathbf{S} = (\mathbf{r_u} \times \mathbf{r_v}) \, dA$ is a vector. If we write **n** for the unit normal $\frac{\mathbf{r_u} \times \mathbf{r_v}}{\|\mathbf{r_u} \times \mathbf{r_v}\|}$, then

$$\mathbf{n} \, dS = \frac{\mathbf{r_u} \times \mathbf{r_v}}{\|\mathbf{r_u} \times \mathbf{r_v}\|} \|\mathbf{r_u} \times \mathbf{r_v}\| \, dA = (\mathbf{r_u} \times \mathbf{r_v}) \, dA = d\mathbf{S},$$

and the surface integral may also be written as $\iint_S \mathbf{F} \cdot \mathbf{n} \, dS$. This formula is called the *normal form* of the surface integral of a vector field.

An alternate wording is "Compute the surface integral of **F**... across the surface *S*...."

If **F** represents the flow of water, there is no flow in the *y*- or *z*-directions, but in the *x*-direction only, similar to Figure 5.52 but less consistent. Instead of a flat mesh screen, the surface *S* is a surface of revolution with axis of rotation parallel to the flow.

Example 47 *Calculate the flux of the vector field* $\mathbf{F}(x, y, z) = \langle 3 + 3y + 3z, 0, 0 \rangle$ *across the surface S parameterized by* $\mathbf{r}(u, v) = \langle u, u^2 \cos v, u^2 \sin v \rangle, 0 \le u \le 1, 0 \le v \le \pi.$

Solution To use the surface integral of a vector field formula, we need to calculate $\mathbf{F}(\mathbf{r})$, $\mathbf{r}_u \times \mathbf{r}_v$, and their dot product, set up the double integral, and evaluate it.

1 Calculate $\mathbf{F}(\mathbf{r})$. We have

$$\mathbf{F}(\mathbf{r}(u, v)) = \mathbf{F}(u, u^2 \cos v, u^2 \sin v)$$

$$= \langle 3 + 3u^2 \cos v + 3u^2 \sin v, 0, 0 \rangle$$

$$= \langle 3 + 3u^2(\cos v + \sin v), 0, 0 \rangle.$$

2 Calculate $\mathbf{r}_u \times \mathbf{r}_v$. First we calculate $\mathbf{r}_u$ and $\mathbf{r}_v$:

$$\mathbf{r}_u(u, v) = \langle 1, 2u \cos v, 2u \sin v \rangle$$

$$\mathbf{r}_v(u, v) = \langle 0, -u^2 \sin v, u^2 \cos v \rangle.$$

Then

$$\mathbf{r}_u \times \mathbf{r}_v = \begin{vmatrix} \mathbf{i} & \mathbf{j} & \mathbf{k} \\ 1 & 2u \cos v & 2u \sin v \\ 0 & -u^2 \sin v & u^2 \cos v \end{vmatrix}$$

$$= \langle 2u^3 \cos^2 v + 2u^3 \sin^2 v, -u^2 \cos v, -u^2 \sin v \rangle$$

$$= \langle 2u^3, -u^2 \cos v, -u^2 \sin v \rangle.$$

3 Calculate their dot product. The integrand is

$$\mathbf{F}(\mathbf{r}) \cdot (\mathbf{r}_u \times \mathbf{r}_v) = \langle 3 + 3u^2(\cos v + \sin v), 0, 0 \rangle \cdot \langle 2u^3, -u^2 \cos v, -u^2 \sin v \rangle$$

$$= 6u^3 + 6u^5(\cos v + \sin v) + 0 + 0.$$

4 Set up the double integral. The order of integration appears to not matter much, and we choose the order *dv du*:

$$\int_0^1 \int_0^\pi (6u^3 + 6u^5(\cos v + \sin v)) \, dv \, du.$$

5 Evaluate the double integral. The flux is

$$= \int_0^1 \left(\left(6u^3 v + 6u^5(\sin v - \cos v) \right)\big|_0^\pi \right) du$$

$$= \int_0^1 \left(6\pi u^3 + 6u^5(1) - (0 + 6u^5(-1)) \right) du = \int_0^1 (6\pi u^3 + 12u^5) \, du$$

$$= \left(\tfrac{6}{4}\pi u^4 + 2u^6 \right)\Big|_0^1 = \tfrac{3}{2}\pi + 2.$$

The numerical value of the flux is $\tfrac{3}{2}\pi + 2$. No context is given in the problem statement, so no units are included in the answer.

■

5.7.5 Closed surfaces, piecewise surfaces

If you order a package to be delivered, you expect that the package will arrive in some type of sealed container, such as a box or a padded envelope. The outer surface of such a container is called a *closed surface*. Another example is the outer surface of a piñata, which is a type of closed container.

You are disappointed if the package arrives open, for instance if the top of the box is missing. In that case, the outer surface of the container is not a closed surface.

Spheres and ellipsoids are examples of closed surfaces; their interior could contain something. So is the surface of a cube. On the other hand, a surface of the form $z = f(x, y)$ is not a closed surface (Figure 5.53).

A closed curve has two orientations, positive (counterclockwise) and negative (clockwise). An orientable surface also has two orientations, whether or not it is closed, based on the directions the normals can point. For a closed surface, if all the normals point outward we say that the surface is *positively oriented*; if all the normals point inward, we call it *negatively oriented* (Figure 5.54).

A more geometrical definition is that a closed surface is the boundary of a solid in $\mathbf{R}^3$.

Similar to a piecewise-defined curve, a surface can come in more than one piece. For instance, the easiest parameterization of the closed box of Figure 5.53, left, is in six pieces.

The final example of this section consists of finding the flux across a closed, piecewise-defined surface, using the positive orientation.

 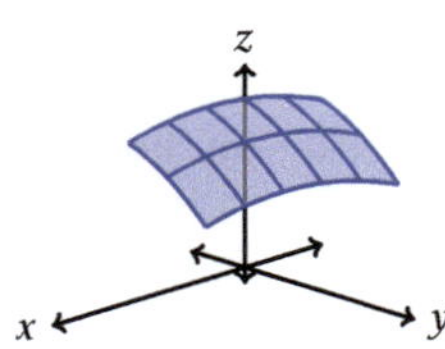

The box in Figure 5.53, left, can contain a cat; the box without a top in Figure 5.53, middle, cannot contain a cat; the surface in Figure 5.53, right, cannot contain anything.

Figure 5.53 *(Left) the exterior of a package is a closed surface; (middle) a box with the top missing is not a closed surface; (right) a surface $z = f(x, y)$ is not a closed surface*

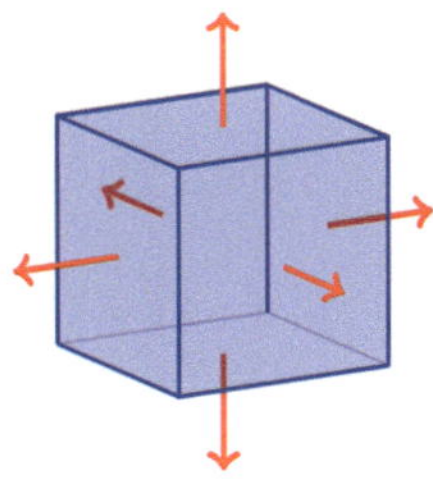
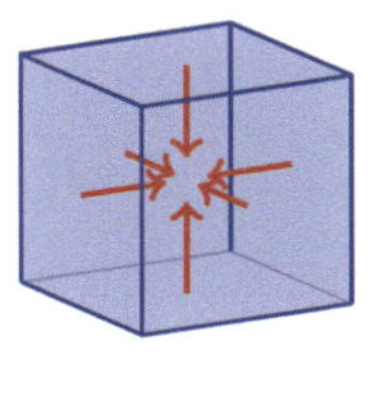

Figure 5.54 *(Left) a closed surface with positive orientation; (right) a closed surface with negative orientation*

The surface S_1 is visible in Figure 5.55, left, and comprises the top half of the piecewise-defined surface S. The surface S_2 is visible in Figure 5.55, right, and comprises the bottom half of the piecewise-defined surface S.

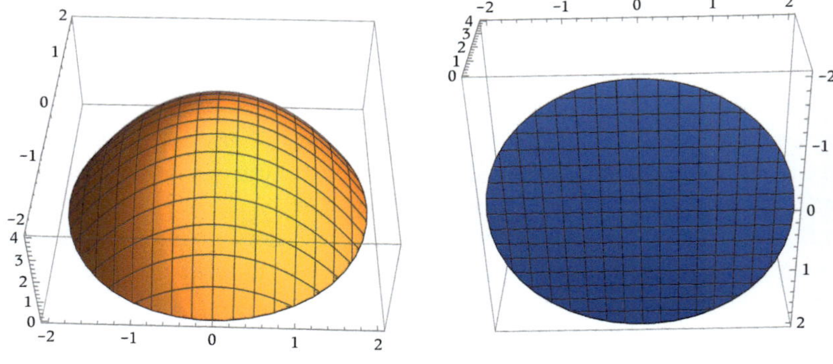

Figure 5.55 *The surface of example 48, (left) viewed from above and (right) viewed from below*

Example 48 *A closed surface S consists of two pieces, the surface S_1 given by $z = f(x, y) = 4 - x^2 - y^2$ where $z \geq 0$, and the surface S_2 consisting of the portion of the xy-plane for which $x^2 + y^2 \leq 4$. Determine the flux of the vector field $F(x, y, z) = \langle x, y, z + 1 \rangle$ across the surface S. Use the positive orientation for S.*

Solution A graph of the surface S is in Figure 5.55. For the positive orientation we need to ensure that the parameterizations we use have normals that point outward. For S_1, the top portion of the surface, that means normals with positive z-component; for S_2, the bottom portion of the surface, that means normals with negative z-component.

To verify that the surface is closed, note that the surface $z = 4 - x^2 - y^2, z \geq$ meets the xy-plane where $z = 0$, that is, where $0 = 4 - x^2 - y^2$, easily rearranged to $x^2 + y^2 = 4$, matching S_2.

The surface S is piecewise-defined, and therefore we must find the flux across each portion of the surface and add the results. We begin with S_1.

Because S_1 is of the form $z = f(x, y)$, a parameterization of the surface is

$$\mathbf{r}(u, v) = \langle u, v, 4 - u^2 - v^2 \rangle,$$

on the region $u^2 + v^2 \leq 4$. Next we calculate the normals:

This region corresponds to $x^2 + y^2 \leq 4$.

$$\mathbf{r}_u(u, v) = \langle 1, 0, -2u \rangle$$

$$\mathbf{r}_v(u, v) = \langle 0, 1, -2v \rangle$$

$$\mathbf{r}_u \times \mathbf{r}_v = \begin{vmatrix} \mathbf{i} & \mathbf{j} & \mathbf{k} \\ 1 & 0 & -2u \\ 0 & 1 & -2v \end{vmatrix} = \langle 2u, 2v, 1 \rangle,$$

which has a positive z-component; therefore the orientation is correct. We also need to compute $\mathbf{F}(\mathbf{r})$:

$$\mathbf{F}(\mathbf{r}(u, v)) = \mathbf{F}(u, v, 4 - u^2 - v^2)$$

$$= \langle u, v, 4 - u^2 - v^2 + 1 \rangle = \langle u, v, 5 - u^2 - v^2 \rangle.$$

As verified in exercise 23, the parameterization $\mathbf{r}(u, v) = \langle u, v, f(u, v) \rangle$ always results in normals with positive z-component, facing upward from the surface. For normals with negative z-component, facing downward from the surface, use $\mathbf{r}(u, v) = \langle v, u, f(u, v) \rangle$ instead.

Finally, we compute the dot product:

$$\mathbf{F}(\mathbf{r}) \cdot (\mathbf{r}_u \times \mathbf{r}_v) = \langle u, v, 5 - u^2 - v^2 \rangle \cdot \langle 2u, 2v, 1 \rangle$$

$$= 2u^2 + 2v^2 + 5 - u^2 - v^2 = 5 + u^2 + v^2.$$

To set up the double integral, notice that the region of integration, $u^2 + v^2 \leq 4$, is easily described in polar coordinates as $0 \leq r \leq 2$, $0 \leq \theta \leq 2\pi$. The integrand is $5 + u^2 + v^2 = 5 + r^2$. Then the integral to evaluate is

With variables u and v, the polar conversion formulas become $u = r\cos\theta$, $v = r\sin\theta$, and $r^2 = u^2 + v^2$.

$$\int_0^{2\pi} \int_0^2 (5 + r^2) r \, dr \, d\theta.$$

Don't forget the extra r when switching to polar coordinates!

The flux across the surface S_1 is

$$\int_0^{2\pi} \int_0^2 (5r + r^3) \, dr \, d\theta = \int_0^{2\pi} \left(\left(\tfrac{5}{2} r^2 + \tfrac{1}{4} r^4 \right) \Big|_0^2 \right) d\theta$$

$$= \int_0^{2\pi} (10 + 4 - 0) \, d\theta$$

$$= 14(2\pi - 0) = 28\pi.$$

Line 3 uses the definite integral of a constant rule; see *Calculus Set Free* Section 4.4. The meaning of a positive flux is that the (net) flux across this surface is with the direction of the normals.

The process must be repeated for S_2. To parameterize the surface $z = 0$, $x^2 + y^2 \leq 4$, we use

$$\mathbf{r}(u, v) = \langle u, v, 0 \rangle,$$

on the region $u^2 + v^2 \le 4$. Next we calculate the normals:

$$\mathbf{r_u}(u, v) = \langle 1, 0, 0 \rangle$$

$$\mathbf{r_v}(u, v) = \langle 0, 1, 0 \rangle$$

$$\mathbf{r_u} \times \mathbf{r_v} = \begin{vmatrix} \mathbf{i} & \mathbf{j} & \mathbf{k} \\ 1 & 0 & 0 \\ 0 & 1 & 0 \end{vmatrix} = \langle 0, 0, 1 \rangle,$$

You should recognize this cross product without the need for calculation as $\mathbf{i} \times \mathbf{j} = \mathbf{k}$.

which has a positive z-component, which is the wrong orientation; for outward-pointing normals, these normals must have a negative z-component (consult Figure 5.55). So we revise our parameterization by swapping u and v:

$$\mathbf{r}(u, v) = \langle v, u, 0 \rangle,$$

on the region $u^2 + v^2 \le 4$. Recalculating the normals yields

$$\mathbf{r_u}(u, v) = \langle 0, 1, 0 \rangle$$

$$\mathbf{r_v}(u, v) = \langle 1, 0, 0 \rangle$$

$$\mathbf{r_u} \times \mathbf{r_v} = \begin{vmatrix} \mathbf{i} & \mathbf{j} & \mathbf{k} \\ 0 & 1 & 0 \\ 1 & 0 & 0 \end{vmatrix} = \langle 0, 0, -1 \rangle,$$

which has a negative z-component; therefore the orientation is correct. We also need to compute $\mathbf{F(r)}$:

One reason to wait to calculate $\mathbf{F(r)}$ is to ensure that we are using the correct parameterization first.

$$\mathbf{F}(\mathbf{r}(u, v)) = \mathbf{F}(v, u, 0) = \langle v, u, 1 \rangle.$$

The dot product is

$$\mathbf{F(r)} \cdot (\mathbf{r_u} \times \mathbf{r_v}) = \langle v, u, 1 \rangle \cdot \langle 0, 0, -1 \rangle = -1.$$

Once again we make the switch to polar coordinates; the flux is

In line 1, don't forget the extra r! The meaning of a negative flux is that the (net) flux across this surface is against the direction of the normals.

$$\int_0^{2\pi} \int_0^2 (-1) r \, dr \, d\theta = \int_0^{2\pi} \left(-\tfrac{1}{2} r^2 \Big|_0^2 \right) d\theta$$

$$= \int_0^{2\pi} -2 \, d\theta = -2(2\pi - 0) = -4\pi.$$

Then the total flux of **F** across the surface S is the sum of the flux across the two pieces of the surface,

$$28\pi + (-4\pi) = 24\pi.$$

∎

Again imagine standing in a river such as Figure 5.52, but this time with a spherical mesh screen, a closed surface. If the flow of the river is constant, the amount of water flowing into the spherical screen matches the amount of water flowing out; the flux across the surface is 0.

What, then, is the meaning of a positive flux across the closed surface in example 48? It means that the total outflow across the surface S is greater than the total inflow. This means that there must be a *source* of **F** within S; in the context of water flow, this could be a natural spring or a faucet. If the flux across a closed surface is negative, then there is more inflow than outflow; we say that there is a *sink* of **F** within S. In the context of water flow, this could be a drain.

In an electrical context, negative and positive flux across a closed surface could be due to charging or discharging a battery.

EXERCISES 5.7

1–4. Rapid response: does the surface appear to be closed?

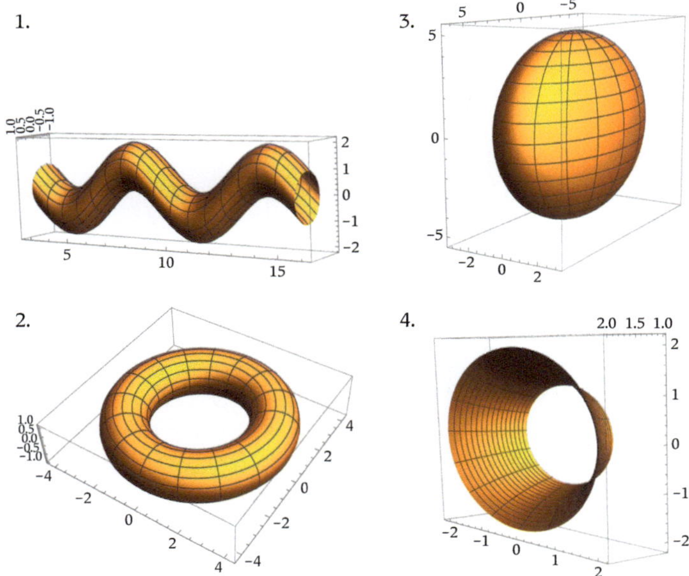

1.

2.

3.

4.

5. (a) Set up the surface integral of $\rho(x, y, z) = xz^2$ along the surface $z = \sqrt{xy}$ over the region $0 \le x \le 1, 0 \le y \le 1$.
   (b) Use a CAS to evaluate the surface integral.

6. (a) Set up the surface integral of $\rho(x, y, z) = 3x \ln z$ along the surface $z = 4 + x^2 - y^2$ over the region $0 \le x \le 2, 0 \le y \le 1$.
   (b) Use a CAS to evaluate the surface integral.

7. A thin plate occupies the surface $z = 2x + 5y$ over the region bounded by $y = \sqrt{x}, y = 0$, and $x = 4$, measured in meters. Its density function is $\rho(x, y, z) = (3x + z)\frac{\text{kg}}{\text{m}^2}$. Find the mass of the plate.

8. A thin plate occupies the surface $z = xy + 1$ over the region bounded by $y = x, y = 0$, and $x = 2$, measured in meters. Its density function is $\rho(x, y, z) = (z - 1)\frac{\text{kg}}{\text{m}^2}$. Find the mass of the plate.

9. A thin plate occupies the surface $z = \sqrt{3}x + y^2$ over the rectangular region $0 \le x \le 1, \sqrt{3} \le y \le \sqrt{8}$, measured in meters. Its density function is $\rho(x, y, z) = 2xy\frac{\text{kg}}{\text{m}^2}$. Find the mass of the plate.

10. A thin plate occupies the surface $z = \sqrt{8}x + 3\cosh y$ over the region $0 \le x \le 1, 0 \le y \le 1$, measured in meters. Its density function is $\rho(x, y, z) = x^2 y\frac{\text{kg}}{\text{m}^2}$. Find the mass of the plate.

<table>
<tr><td style="vertical-align:top; width:30%">These surfaces are the same (except possibly for the domain) as exercises 37–40 of Section 5.6.</td><td style="vertical-align:top">

11–14. Compute the surface integral of f along the surface parameterized by $\mathbf{r}$.

11. $f(x, y, z) = x + y^2 + z^2$, $\mathbf{r}(u, v) = \langle v^2 u, \cos u, \sin u \rangle$, $1 \le u \le 2$, $1 \le v \le 5$

12. $f(x, y, z) = xz + y$, $\mathbf{r}(u, v) = \langle \cos v, u + v, \sin v \rangle$, $0 \le u \le 2\pi$, $0 \le v \le 4\pi$

13. $f(x, y, z) = y^3$, $\mathbf{r}(u, v) = \langle \cos^2 u, \sin^2 u, v\sqrt{2} + u \rangle$, $\frac{\pi}{6} \le u \le \frac{\pi}{3}$, $0 \le v \le 2$

14. $f(x, y, z) = x + y + z$, $\mathbf{r}(u, v) = \langle 1 + \cos u, 3 + \sin u, \cos v \rangle$, $0 \le u \le 2\pi, 0 \le v \le \pi$

</td></tr>
</table>

15–18. Calculate the flux of the vector field $\mathbf{F}$ across the surface S parameterized by $\mathbf{r}$.

15. $\mathbf{F}(x, y, z) = \langle y, z, x^2 \rangle$, $\mathbf{r}(u, v) = \langle u, uv, uv^2 \rangle$, $0 \le u \le 1$, $-1 \le v \le 1$

16. $\mathbf{F}(x, y, z) = \langle x^2 z, x, y \rangle$, $\mathbf{r}(u, v) = \langle \sin u, 3v, v^2 \rangle$, $0 \le u \le \frac{\pi}{2}$, $1 \le v \le 3$

17. $\mathbf{F}(x, y, z) = yz\mathbf{i} + y\mathbf{j} + 2z\mathbf{k}$, $\mathbf{r}(u, v) = (u + v)\mathbf{i} + u^3\mathbf{j} + v^2\mathbf{k}$, $-1 \le u \le 1, 0 \le v \le 2$

18. $\mathbf{F}(x, y, z) = \langle x, xy, xz \rangle$, $\mathbf{r}(u, v) = \langle u, u^2 \cos v, u^2 \sin v \rangle$, $0 \le u \le \sqrt{3}, 0 \le v \le 2\pi$

19. Find $\iint_S \mathbf{F} \cdot d\mathbf{S}$ for the vector field $\mathbf{F}(x, y, z) = \langle x, y, z \rangle$ and the surface S parameterized by $\mathbf{r}(u, v) = \langle u^2 + 1, v^2 + 1, uv \rangle$, $u^2 + v^2 \leq 1$, $v \geq 0$.

20. Find $\iint_S \mathbf{F} \cdot d\mathbf{S}$ for the vector field $\mathbf{F}(x, y, z) = x\mathbf{i} - x\mathbf{j} + z^2\mathbf{k}$ and the surface S parameterized by $\mathbf{r}(u, v) = (\ln u)\mathbf{i} + (\ln v)\mathbf{j} + uv\mathbf{k}$, $2 \leq u \leq 4, 2 \leq v \leq 12$.

21. Evaluate the surface integral of $\mathbf{F}(x, y, z) = (3 + x - y)\mathbf{k}$ across the closed surface $S = S_1 \cup S_2$, oriented positively, where S_1 is the region of the xy-plane $x^2 + y^2 \leq 9$ and S_2 is the portion of the paraboloid $z = x^2 + y^2 - 9$ lying below the xy-plane.

22. Evaluate the surface integral of $\mathbf{F}(x, y, z) = \langle x, y, 7 \rangle$ across the closed surface $S = S_1 \cup S_2$, where S_1 is the surface $z = x^2 + y^2$, $z \leq 4$, and S_2 is the relevant portion of the plane $z = 4$. Use the negative orientation.

23. A surface $z = f(x, y)$ can be parameterized in multiple ways.

    (a) For the parameterization $\mathbf{r}(u, v) = \langle u, v, f(u, v) \rangle$, compute $\mathbf{r}_u \times \mathbf{r}_v$. In which direction (upward or downward) do the normals point?

    (b) For the parameterization $\mathbf{r}(u, v) = \langle v, u, f(u, v) \rangle$, compute $\mathbf{r}_u \times \mathbf{r}_v$. In which direction (upward or downward) do the normals point?

"Upward" and "downward" are to be interpreted as having increasing z-coordinates or decreasing z-coordinates, respectively.

24. The classic example of a nonorientable surface is a *Möbius strip*. Construct a Möbius strip (if you do not know how already, there are a multitude of videos available to tell you how), mark a line around the strip, and mark a point on that line. At the point, use a pencil or similar object to represent a normal vector, and note which direction the normal vector is pointing. Move the pencil continuously along the line around the strip, representing the normal vector. When the pencil first returns to the marked point, is the normal pointing the same direction as at first, or the opposite direction? Conclude that there is no parameterization for which the normals vary continuously, meaning that the surface is not orientable.

25. The surface of revolution S formed by rotating $y = f(x)$, $a \leq x \leq b$, about the x-axis can be parameterized by $\mathbf{r}(u, v) = \langle u, f(u) \cos v, f(u) \sin v \rangle$, $a \leq u \leq b, 0 \leq v \leq 2\pi$.

    (a) Compute $\|\mathbf{r}_u \times \mathbf{r}_v\|$, assuming $f(x) \geq 0$ on $[a, b]$.
    (b) Compute the surface integral of $g(x, y, z) = x^2 z$ along the surface S.
    (c) Compare (a) and (b) to example 46.

26. (a) Compute the surface integral of $g(x, y, z) = x$ along the surface of revolution formed by revolving $y = \sqrt{x}$, $1 \leq x \leq 4$, about the x-axis.

    (b) Compare to the previous exercise. Why isn't the answer to part (a) 0?

27. A thin plate occupies the surface $z = x^2 + y^2, 0 \le x \le 1, 0 \le y \le 1$. Its density function is $\rho(x, y, z) = z$. Find its center of mass using the following steps.

    (a) Set up a double integral for the mass m of the plate.
    (b) Use a CAS to find a decimal approximation to the mass m.
    (c) Set up a double integral for finding $\bar{x}$ in the usual way, by including an extra x in the integrand and dividing by m.
    (d) Use a CAS to find a decimal approximation of $\bar{x}$.
    (e) Repeat parts (c) and (d) for $\bar{y}$.
    (f) Repeat parts (c) and (d) for $\bar{z}$, using $x^2 + y^2$ for z.
    (g) Is the center of mass a point in the thin plate?

The answer to part (g) is key to a high jumper being able to pass over the bar while their center of mass passes under the bar.

28. The normal form $\iint_S \mathbf{F} \cdot \mathbf{n} \, dS$ can sometimes be helpful for calculating a surface integral. Suppose $\mathbf{F}$ is an inverse square field, that is, $\mathbf{F}(x, y, z) = c \left\langle \dfrac{x}{(x^2+y^2+z^2)^{3/2}}, \dfrac{y}{(x^2+y^2+z^2)^{3/2}}, \dfrac{z}{(x^2+y^2+z^2)^{3/2}} \right\rangle$ for some constant c. Let S be the unit sphere, that is, a sphere of radius 1 centered at the origin.

    (a) Simplify $\mathbf{F}(x, y, z)$ for a point (x, y, z) on S.
    (b) What do the outward-facing normals to a sphere look like? Write the unit normal $\mathbf{n}(x, y, z)$ for a point on the sphere.
    (c) Use the results of parts (a) and (b) to compute $\mathbf{F}(x, y, z) \cdot \mathbf{n}(x, y, z)$ for a point (x, y, z) on the sphere.
    (d) Set up the surface integral $\iint_S \mathbf{F} \cdot \mathbf{n} \, dS$ and observe that it is the integral of a constant.
    (e) Compute the surface integral of part (d) by observing that the integral of a constant over a surface is the constant times the area of the surface.

5.8 The Divergence Theorem

Example 48 of the previous section has a rather lengthy solution. It entails two parameterizations, a dozen partial derivatives, two cross products, two dot products, compositions of vector-valued functions, and two double integrals, each converted to polar coordinates, with their answers added. It turns out that there is a more efficient solution.

5.8.1 The divergence theorem: surface integrals across closed surfaces

We shall not formally define a simple closed surface. The idea is analogous to that of a simple closed curve.

Green's theorem is used to simplify the calculation of a line integral along a piecewise smooth simple closed curve with positive orientation. Instead of evaluating the line integral, we evaluate a double integral over the region inside the closed curve. Add a dimension, and the idea is to simplify the calculation of a surface integral along a piecewise smooth simple closed surface with positive orientation, replacing it with a triple integral over the region inside the closed surface.

Theorem 9 THE DIVERGENCE THEOREM *Let S be a piecewise smooth, simple closed surface with positive orientation, let R be the (simple) three-dimensional region inside the surface, and let* **F** *be a vector field whose component functions have continuous partial derivatives on an open region containing R. Then*

$$\iint_S \mathbf{F} \cdot d\mathbf{S} = \iiint_R \operatorname{div} \mathbf{F}\, dV.$$

Other names for the divergence theorem include *Ostrogradsky's theorem* and *Gauss' theorem*.

The divergence theorem formula says that the triple integral of the divergence of **F** on the region R is the surface integral of **F** on the boundary of R.

Although the proof of the divergence theorem for all of the applicable shapes of regions R with boundary S is beyond the scope of this book, justifying a portion of the theorem for the shapes of regions we most commonly use is not too difficult.

Begin with $\mathbf{F}(x, y, z) = \langle M(x, y, z), N(x, y, z), P(x, y, z)\rangle$, a vector field whose component functions M, N, and P all have continuous partial derivatives. The surface integral can then be written as

$$\iint_S \mathbf{F} \cdot d\mathbf{S} = \iint_S \mathbf{F} \cdot \mathbf{n}\, dS = \iint_S (M\mathbf{i} + N\mathbf{j} + P\mathbf{k}) \cdot \mathbf{n}\, dS$$

$$= \iint_S M\mathbf{i} \cdot \mathbf{n}\, dS + \iint_S N\mathbf{j} \cdot \mathbf{n}\, dS + \iint_S P\mathbf{k} \cdot \mathbf{n}\, dS,$$

Line 1 uses the normal form of the surface integral of a vector field (for reasons that will be clear later) and rewrites **F** using its component functions. Line 2 uses a variation of property (2) of the dot product (theorem 4 of Chapter 1) and the double integral sum rule.

and the triple integral can be written as

$$\iiint_R \operatorname{div} \mathbf{F}\, dV = \iiint_R (M_x + N_y + P_z)\, dV$$

$$= \iiint_R M_x\, dV + \iiint_R N_y\, dV + \iiint_R P_z\, dV.$$

Line 2 uses the triple integral sum rule.

If we can prove equality of the three corresponding pieces,

$$\iint_S M\mathbf{i} \cdot \mathbf{n}\, dS = \iiint_R M_x\, dV$$

$$\iint_S N\mathbf{j} \cdot \mathbf{n}\, dS = \iiint_R N_y\, dV$$

$$\iint_S P\mathbf{k} \cdot \mathbf{n}\, dS = \iiint_R P_z\, dV,$$

then the surface integral of **F** on the boundary of R is the triple integral of div **F** on R, as desired. We shall prove the third piece for certain shapes of the region R.

Suppose R is the region $f_1(x, y) \leq z \leq f_2(x, y)$, where f_1 and f_2 have the domain D in common, as pictured in Figure 5.56. Then the boundary of region R is a closed surface S. The surface S consists of the lower surface S_1 with equation $z = f_1(x, y)$, the upper surface S_2 with equation $z = f_2(x, y)$, and one or more vertical surfaces (Figure 5.57). To be positively oriented, the upper surface has normals pointing upward, the lower surface has normals pointing downward, and the vertical surfaces have normals that are horizontal.

As in the previous section, by upward we mean increasing z-coordinates, not necessarily vertical.

We wish to prove $\iint_S P\mathbf{k} \cdot \mathbf{n}\, dS = \iiint_R P_z\, dV$, and we'll start by trying to calculate the surface integral. The surface S consists of multiple pieces, and we start with S_1. The vector field can be written as

This is scalar multiplication, the quantity $P(x, y, z)$ being the scalar.

$$P\mathbf{k} = P(x, y, z)\langle 0, 0, 1 \rangle = \langle 0, 0, P(x, y, z) \rangle.$$

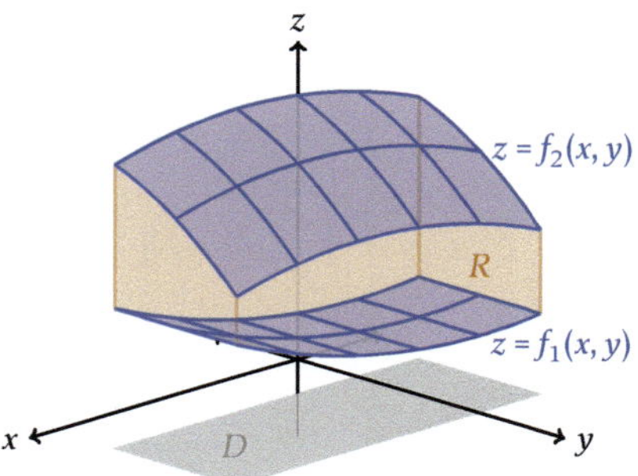

Figure 5.56 *A three-dimensional region R (brown) between two surfaces $z = f_1(x, y)$ and $z = f_2(x, y)$ (blue), having common domain D (gray) in the xy-plane*

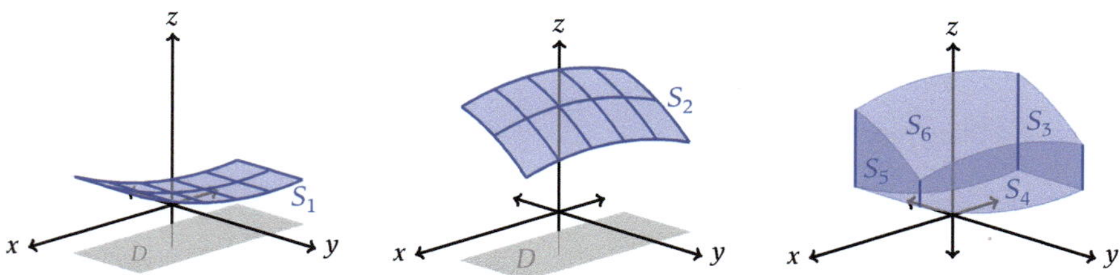

Figure 5.57 *Portions of the piecewise smooth, closed surface S that is the boundary of the region R of Figure 5.56. The lower surface S_1 (left) is $z = f_1(x, y)$ with domain D; the upper surface S_2 (middle) is $z = f_2(x, y)$ with domain D; and the rest of the surfaces (right) are vertical*

The surface $z = f_1(x, y)$ can be parameterized by

$$\mathbf{r}(u, v) = \langle u, v, f_1(u, v)\rangle,$$

and then (see exercise 23 of Section 5.7)

$$\mathbf{r}_u \times \mathbf{r}_v = \langle -(f_1)_u(u, v), -(f_1)_v(u, v), 1\rangle.$$

Because these normals point upward but we need them to point downward, we instead use their negation $-\mathbf{r}_u \times \mathbf{r}_v$. We also need the composition of the vector field with the surface, $P\mathbf{k}(\mathbf{r}(u, v)) = \langle 0, 0, P(u, v, f_1(u, v))\rangle$. Taking the dot product,

$$\begin{aligned}
P\mathbf{k}(\mathbf{r}(u, v)) &\cdot (-\mathbf{r}_u \times \mathbf{r}_v) \\
&= \langle 0, 0, P(u, v, f_1(u, v))\rangle \cdot \langle (f_1)_u(u, v), (f_1)_v(u, v), -1\rangle \\
&= -P(u, v, f_1(u, v)).
\end{aligned}$$

Then the surface integral across S_1 is

$$\iint_{S_1} P\mathbf{k} \cdot \mathbf{n}\, dS = \iint_D -P(u, v, f_1(u, v))\, dA.$$

The calculation for the surface S_2 is very much the same, but without the need to negate the normal because we need the normals to point upward. The surface integral across S_2 is

$$\iint_{S_2} P\mathbf{k} \cdot \mathbf{n}\, dS = \iint_D P(u, v, f_2(u, v))\, dA.$$

What remains are vertical surfaces such as those in Figure 5.57, right, if any. For any of these vertical surfaces, the normals $\mathbf{n}$ are horizontal and therefore orthogonal to $\mathbf{k}$. Therefore, $P\mathbf{k} \cdot \mathbf{n} = 0$, and

$$\iint_{S_3\ldots} P\mathbf{k} \cdot \mathbf{n}\, dS = \iint_{S_3\ldots} 0\, dS = 0.$$

Adding the surface integrals over the various pieces,

$$\begin{aligned}
\iint_S P\mathbf{k} \cdot \mathbf{n}\, dS &= \iint_{S_1} P\mathbf{k} \cdot \mathbf{n}\, dS + \iint_{S_2} P\mathbf{k} \cdot \mathbf{n}\, dS + \iint_{S_3} P\mathbf{k} \cdot \mathbf{n}\, dS + \cdots \\
&= \iint_D (P(u, v, f_2(u, v)) - P(u, v, f_1(u, v)))\, dA.
\end{aligned}$$

If S_1 and S_2 meet, such as at the equator of a sphere, there may not be a vertical surface.

It is this calculation of the surface integrals on the vertical surfaces that is the reason for using the normal form of the surface integral.

In line 1, $dz\,dA$ can be either $dz\,dy\,dx$ or $dz\,dx\,dy$. The limits of integration for the variables x and y are whatever is necessary for covering the region D. Line 2 finds an antiderivative of P_z with respect to z, treating x and y as constants, to get P; line 3 evaluates at $z = f_2(x,y)$, at $z = f_1(x,y)$, and subtracts.

Compare the lengthy calculation of $\iint_S P\mathbf{k}\cdot\mathbf{n}\,dS$ to the much shorter calculation of $\iiint_R P_z\,dV$ to arrive at the same double integral. The potential calculational advantage should be clear!

For setting up the triple integral over the three-dimensional region R, refer to Figure 5.56. Integrating with respect to z first and not caring which order the other variables use, we may write

$$\iiint_R P_z\,dV = \iint_D \int_{f_1(x,y)}^{f_2(x,y)} P_z(x,y,z)\,dz\,dA$$

$$= \iint_D \left(P(x,y,z)\big|_{f_1(x,y)}^{f_2(x,y)} \right) dA$$

$$= \iint_D \left(P(x,y,f_2(x,y)) - P(x,y,f_1(x,y)) \right) dA,$$

which is identical–other than variable names–to $\iint_S P\mathbf{k}\cdot\mathbf{n}\,dS$.

That's as far as we'll go for the proof. If the region R happens to be describable as being between two surfaces in the x-direction and in the y-direction as well, such as a rectangular solid or a spherical solid, then the proofs of the portions of the formula for M and N are the same as for P, and the formula holds for those shapes. The proof for more complicated regions can be found in more advanced texts.

5.8.2   Divergence theorem examples

Applying the divergence theorem to example 48 of Section 5.7 illustrates the efficiency of switching to a triple integral.

Example 49 *A closed surface S consists of two pieces, the surface S_1 given by $z = f(x,y) = 4 - x^2 - y^2$ where $z \geq 0$, and the surface S_2 consisting of the portion of the xy-plane for which $x^2 + y^2 \leq 4$. Determine the flux of the vector field $\mathbf{F}(x,y,z) = \langle x, y, z+1 \rangle$ across the surface S. Use the positive orientation for S.*

Solution   The hypotheses of the divergence theorem are met, and the requested surface integral is equal to the triple integral of $\operatorname{div}\mathbf{F}$ on the three-dimensional region. The divergence is

$$\operatorname{div}\mathbf{F} = 1 + 1 + 1 = 3.$$

The three-dimensional region (see Figure 5.58, repeated from Section 5.7) is between $z = 0$ and $z = 4 - x^2 - y^2$, over a circular region in the xy-plane, setting up perfectly for cylindrical coordinates! For $x^2 + y^2 \leq 4$, we use $0 \leq r \leq 2, 0 \leq \theta \leq 2\pi$. We also rewrite $z = 4 - x^2 - y^2$ as $z = 4 - r^2$. Then the flux is

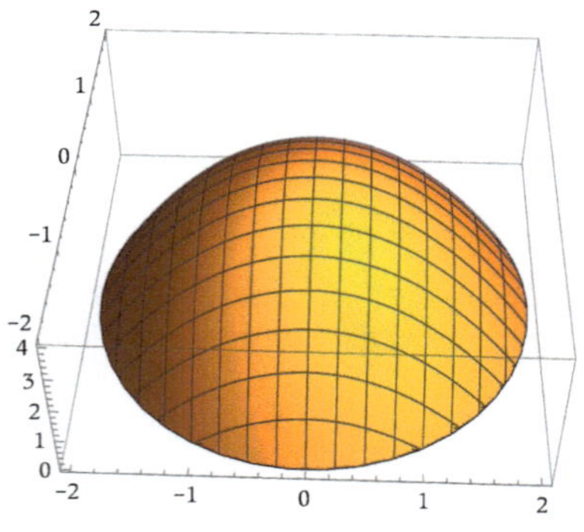

Figure 5.58 *A top view of the surface and region of example 49*

$$\iiint_R \operatorname{div} \mathbf{F}\, dV = \int_0^{2\pi} \int_0^2 \int_0^{4-r^2} 3r\, dz\, dr\, d\theta$$

$$= \int_0^{2\pi} \int_0^2 \left(3rz\big|_0^{4-r^2} \right) dr\, d\theta$$

$$= \int_0^{2\pi} \int_0^2 (3r(4 - r^2) - 0)\, dr\, d\theta = \int_0^{2\pi} \int_0^2 (12r - 3r^3)\, dr\, d\theta$$

$$= \int_0^{2\pi} \left(\left(6r^2 - \tfrac{3}{4}r^4 \right)\Big|_0^2 \right) d\theta$$

$$= \int_0^{2\pi} (24 - 12 - 0)\, d\theta = \int_0^{2\pi} 12\, d\theta = 24\pi.$$

In line 1, don't forget the extra r! Line 2 finds an antiderivative with respect to z, treating r and θ as constants; line 3 evaluates at $z = 4 - r^2$, at $z = 0$, and subtracts, then simplifies; line 4 finds an antiderivative with respect to r, treating θ as a constant; line 5 evaluates at $r = 2$, at $r = 0$, and subtracts, then finishes the integration.

∎

Compare the solution to example 49 to the solution to example 48 of Section 5.7, and the efficiency of this method is clear. But that's nothing compared to the next example. If you worked exercise 21 of Section 5.7, prepare to be amazed. And perhaps annoyed!

This example is identical to exercise 21 of Section 5.7.

Example 50 *Evaluate the surface integral of* $\mathbf{F}(x, y, z) = (3 + x - y)\mathbf{k}$ *across the closed surface* $S = S_1 \cup S_2$, *oriented positively, where* S_1 *is the region of the xy-plane* $x^2 + y^2 \le 9$ *and* S_2 *is the portion of the paraboloid* $z = x^2 + y^2 - 9$ *lying below the xy-plane.*

Solution The hypotheses of the divergence theorem are met, so we choose to evaluate the triple integral of the divergence of $\mathbf{F}$. We have

$$\operatorname{div} \mathbf{F} = \operatorname{div}\langle 0, 0, 3 + x - y \rangle = 0 + 0 + 0 = 0.$$

Then the flux is

When the integrand is 0, there is no need to analyze the region to determine limits of integration; the value of the integral is 0. All we need to know is that the region meets the hypotheses of the divergence theorem.

$$\iiint_R 0\, dV = 0.$$

∎

Example 51 *Evaluate the surface integral of* $\mathbf{F}(x, y, z) = \langle 2x + y^2, 3y - z^2, z^2 + x^2 \rangle$ *across the surface of the cube* $0 \le x \le 1,\ 0 \le y \le 1,\ 0 \le z \le 1$, *oriented positively.*

Solution To calculate the surface integral directly requires six separate surface integrals, one for each face of the cube. Instead, we choose to use the divergence theorem and evaluate one triple integral.

Carefully setting up the parameterizations to ensure the normals face outward, the fluxes across the individual faces are $-\frac{1}{3}, \frac{4}{3}, \frac{1}{3}, \frac{8}{3}, -\frac{1}{3}$, and $\frac{7}{3}$. The vertical faces do not have zero flux because the vector field is not of the form $P\mathbf{k} = \langle 0, 0, P \rangle$.

Line 3 uses the definite integral of a constant rule twice, once for integrating with respect to y and once for integrating with respect to x.

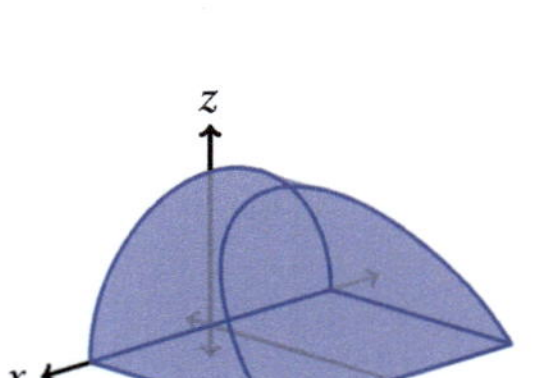

Figure 5.59 *The surface and region of example 52. This shape can be obtained by cutting a semi-circular cylinder at a 45° angle*

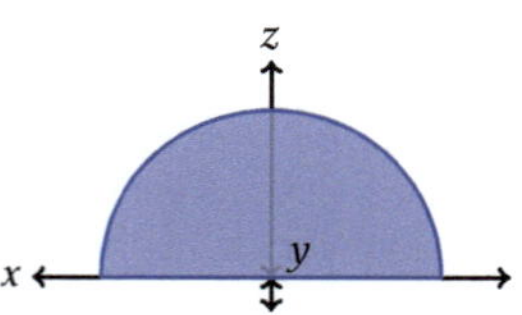

Figure 5.60 *The surface and region of example 52, viewed from straight down the y-axis*

Don't forget the extra r!

The divergence is

$$\nabla \cdot \mathbf{F} = 2 + 3 + 2z = 5 + 2z.$$

Then the flux across the cube is

$$\int_0^1 \int_0^1 \int_0^1 (5 + 2z) \, dz \, dy \, dx = \int_0^1 \int_0^1 \left((5z + z^2)|_0^1 \right) dy \, dz$$

$$= \int_0^1 \int_0^1 (5 + 1 - 0) \, dy \, dx$$

$$= 6(1 - 0)(1 - 0) = 6.$$

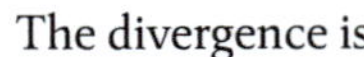

The next example features a four-part closed surface where two of the surface pieces are difficult to parameterize.

Example 52 *Evaluate the surface integral of $\mathbf{F}(x, y, z) = \langle 2x - y^2, z^3, 4y + x^4 \rangle$ across the surface of the solid bounded by the cylinder $z = \sqrt{4 - x^2}$ and the planes $y = 0, z = 0$, and $z = 3 - y$. Use the positive orientation.*

Solution A graph of the surface is in Figure 5.59. The semicircle on the left side of the region is in the plane $y = 0$, the slanted region on the right is in the plane $z = 3 - y$, the bottom is in the plane $z = 0$, and the top shell-like surface is a portion of the cylinder $z = \sqrt{4 - x^2}$. Instead of four surface integrals, two of which are difficult to set up, we use the divergence theorem and evaluate one triple integral.

The divergence is

$$\operatorname{div} \mathbf{F} = 2 + 0 + 0 = 2.$$

Describing the region is a little tricky. Tilt your head to the right and it appears that cylindrical coordinates are in order, but with y the unchanged variable instead of z. The region lies between $y = 0$ and $y = 3 - z$ (rearranged from $z = 3 - y$). Eliminating y by viewing directly down the y-axis (Figure 5.60), we see a semicircle of radius 2 (from $z = \sqrt{4 - x^2}$). This can be described in polar coordinates as $0 \le r \le 2, 0 \le \theta \le \pi$. Because we are rewriting the xz-plane in polar coordinates, the conversion formulas become $x = r \cos \theta, z = r \sin \theta$. We therefore rewrite y as lying between $y = 0$ and $y = 3 - r \sin \theta$. The triple integral to evaluate is

$$\int_0^\pi \int_0^2 \int_0^{3 - r \sin \theta} 2r \, dy \, dr \, d\theta.$$

Then the flux across the surface is

$$= \int_0^\pi \int_0^2 \left(2ry \big|_0^{3-r\sin\theta} \right) dr\, d\theta$$

$$= \int_0^\pi \int_0^2 (6r - 2r^2 \sin\theta - 0)\, dr\, d\theta$$

$$= \int_0^\pi \left(\left(3r^2 - \tfrac{2}{3}r^3 \sin\theta \right) \Big|_0^2 \right) d\theta$$

$$= \int_0^\pi \left(12 - \tfrac{16}{3} \sin\theta - 0 \right) d\theta$$

$$= \left(12\theta + \tfrac{16}{3} \cos\theta \right) \Big|_0^\pi$$

$$= 12\pi + \tfrac{16}{3}(-1) - \left(0 + \tfrac{16}{3}(1) \right) = 12\pi - \tfrac{32}{3}.$$

Line 1 finds an antiderivative with respect to y, treating r and θ as constants; line 2 evaluates at $y = 3 - r\sin\theta$, at $y = 0$, and subtracts; line 3 finds an antiderivative with respect to r, treating θ as a constant; line 4 evaluates at $r = 2$, at $r = 0$, and subtracts; line 5 finds an antiderivative with respect to θ; line 6 evaluates at $\theta = \pi$, at $\theta = 0$, and subtracts.

■

Consider again the solid of example 52. How do we find its volume? By integrating $f(x,y,z) = 1$ over the region. Making the same switch to cylindrical coordinates, the volume is

$$V = \int_0^\pi \int_0^2 \int_0^{3-r\sin\theta} 1r\, dy\, dr\, d\theta = \cdots = 6\pi - \frac{16}{3}.$$

Compare to the triple integral for finding the flux in example 52.

In other words, when the divergence is a constant k (as in examples 49, 50, and 52), the flux is k times the volume of the region. Knowing this can bring additional simplifications.

For example 52, $k = 2$.

Example 53 *Evaluate the surface integral of* $F(x,y,z) = \langle 2x - y^2, z^3, 4y + x^4 \rangle$ *across the sphere centered at* $(0,1,1)$ *with radius* $\tfrac{1}{2}$, *oriented positively.*

Solution The divergence is quickly calculated as

$$\mathrm{div}\, F = 2 + 0 + 0 = 2.$$

The strategy is to use the divergence theorem.

Because the divergence is a constant, we need only multiply the divergence by the volume of the region. The volume of the region inside the sphere is

$$V = \tfrac{4}{3}\pi \left(\tfrac{1}{2} \right)^3 = \tfrac{1}{6}\pi.$$

Then the flux across the sphere is

$$(\mathrm{div}\, F)V = 2 \cdot \tfrac{1}{6}\pi = \tfrac{1}{3}\pi.$$

Because the volume of a sphere is a well-known formula, there is no need to evaluate a triple integral. By contrast, in example 49 there is no well-known formula for the volume of the region, and the use of a triple integral is required. In example 51, the divergence is not constant, so this shortcut does not apply.

■

5.8.3 Solids with holes

Consider a vector field **F** and spheres of radius 1 and 2 with the same center, as in Figure 5.61. The solid R lying between the two spheres can be described as a sphere with a hole inside. It has a boundary surface S consisting of two parts, the outer sphere S_1 and the inner sphere S_2. How might we find the flux of **F** across S?

An outward flow from the solid consists not just of crossing S_1 facing outward, but also of crossing S_2 facing inward. Leaving the solid R and going into the hole inside leaves R just as surely as leaving the solid R and going beyond its outer shell.

The flux across the surface S_1, with outward-facing normals, is $\iint_{S_1} \mathbf{F} \cdot \mathbf{n}\, dS$, which represents leaving the solid going beyond the outer shell. The flux across the surface S_2, with outward-facing normals, is $\iint_{S_2} \mathbf{F} \cdot \mathbf{n}\, dS$. But we want the flux going the other way, leaving the solid R and going into the hole (that is, inside S_2); we need inward-facing normals. The flux across S_2 is therefore $\iint_{S_2} \mathbf{F} \cdot (-\mathbf{n})\, dS = -\iint_{S_2} \mathbf{F} \cdot \mathbf{n}\, dS$. Then the total flux across the surface S, written in the usual form instead of normal form, is

Considering S_2 only, outward-facing normals from S_2 face outward from the region inside the sphere S_2.

$$\iint_{S_1} \mathbf{F} \cdot d\mathbf{S} - \iint_{S_2} \mathbf{F} \cdot d\mathbf{S}.$$

Each of these surface integrals can be computed using the divergence theorem, leading to

$$\iiint_{R_1} \operatorname{div} \mathbf{F}\, dV - \iiint_{R_2} \operatorname{div} \mathbf{F}\, dV,$$

where R_1 is the region inside sphere S_1 and R_2 is the region inside sphere S_2. Take another look at Figure 5.61. Isn't the integral over the region R between the spheres found by taking the integral over the region inside the outer sphere, R_1, and subtracting the part that's not included, the integral over the region inside the inner sphere, R_2?

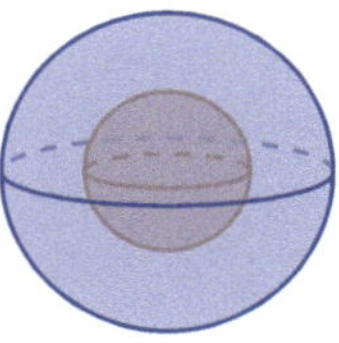

Figure 5.61 *An outer sphere (blue) and an inner sphere (brown)*

SOLIDS BETWEEN SURFACES

Let R be the region between two piecewise smooth, closed, positively oriented surfaces S_1 (the outer surface) and S_2 (the inner surface), where S_2 is contained entirely within the region R_1 enclosed by S_1. Let $F(x, y, z)$ be a vector field whose component functions have continuous partial derivatives on an open region containing R. Then the flux of F across the combined boundary surface S, using outward-facing normals for S_1 and inward-facing normals for S_2, is

$$\iint_S F \cdot dS = \iint_{S_1} F \cdot dS - \iint_{S_2} F \cdot dS$$

$$= \iiint_{R_1} \operatorname{div} F \, dV - \iiint_{R_2} \operatorname{div} F \, dV$$

$$= \iiint_R \operatorname{div} F \, dV.$$

Example 54 *A solid R lies between the surface S_1 of example 52 and the surface S_2 of example 53. Determine the flux of the vector field $F(x, y, z) = \langle 2x - y^2, z^3, 4y + x^4 \rangle$ across the surface of R.*

Solution   A graph of the solid is in Figure 5.62. The vector field is the same as in examples 52 and 53. Using the results of those examples,

$$\iint_{S_1} F \cdot dS = 12\pi - \tfrac{32}{3}$$

and

$$\iint_{S_2} F \cdot dS = \tfrac{1}{3}\pi.$$

Using the formula for solids between surfaces, the flux across the surface S is

$$\iint_S F \cdot dS = \iint_{S_1} F \cdot dS - \iint_{S_2} F \cdot dS$$

$$= 12\pi - \frac{32}{3} - \frac{1}{3}\pi = \frac{35}{3}\pi - \frac{32}{3}.$$

$\blacksquare$

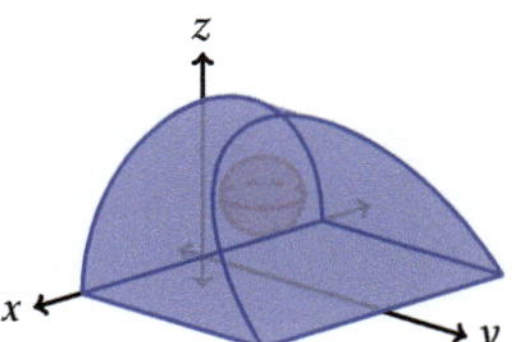

Figure 5.62 *The sphere of example 53 (brown) lies within the solid of example 52*

EXERCISES 5.8

1–12. Evaluate $\iint_S \mathbf{F} \cdot d\mathbf{S}$ for the given vector field $\mathbf{F}$ and surface S.

1. $\mathbf{F}(x, y, z) = \left\langle 3x, z \ln y, \frac{1}{xy} \right\rangle$, surface S of the rectangular solid
$1 \le x \le 4, 1 \le y \le e^2, 0 \le z \le 1$

2. $\mathbf{F}(x, y, z) = \langle x + y + z, x^2 + 3y + z^5, e^{xy} + z \rangle$, surface S of solid
bounded by $x^2 + y^2 = 9, z = 0$, and $z = 4$

3. $\mathbf{F}(x, y, z) = \langle 4x, 5y, 6z \rangle$, surface S of the sphere of radius 2
centered at the origin

4. $\mathbf{F}(x, y, z) = \langle 5y, 6z, 4x \rangle$, surface S of the sphere of radius 2
centered at $(4, 4, 4)$

5. $\mathbf{F}(x, y, z) = \langle \sin^2 y, \sin x \cos z, \cos^2 y \rangle$, surface S of solid bounded by $z = 0, z = x \cosh y, y = 4x, y = 0$, and $x = 3$

6. $\mathbf{F}(x, y, z) = \langle x^2, y^2, z^2 \rangle$, surface S of the rectangular solid
$0 \le x \le 1, 0 \le y \le 2, 0 \le z \le 3$

7. $\mathbf{F}(x, y, z) = \langle 2xy, y^2 + z^2, 4x\sqrt{y} \rangle$, surface S of solid bounded by
$z = 5 + x + y, z = 2 - x - y, y = 0, y = 4, x = 0$, and $x = 2$

8. $\mathbf{F}(x, y, z) = \langle 2e^z, 3y, 4\sqrt{x^3 - 7} \rangle$, surface S of solid bounded by
$z = 9 - x^2 - y^2$ and the xy-plane

9. $\mathbf{F}(x, y, z) = \langle yz, x^2 - 4^z, z^2 \rangle$, surface S of solid bounded by
$z = 4$ and $z = x^2 + y^2$

10. $\mathbf{F}(x, y, z) = \langle x, y^2, z \rangle$, surface S of solid bounded by $y = \sqrt{x}$,
$y = 0, x = 1, z = 0$, and $z = x^2 + y^2$

11. $\mathbf{F}(x, y, z) = \langle 4yz, x + 3z, 2yz \rangle$, surface S of solid bounded by
$x = 0, z = 5 - x, z = 0$, and $z = 1 - y^2$

12. $\mathbf{F}(x, y, z) = \langle 7y, x^2 + z^2, 4z \rangle$, surface S of the solid bounded by
the cylinder $x^2 + z^2 = 9$ and the planes $z = y - 7$ and $y = 0$

13. A solid R consists of the region between the surface of the
cube $-2 \le x \le 2, -2 \le y \le 2, -2 \le z \le 2$ and the surface
of the cube $0 \le x \le 1, 0 \le y \le 1, 0 \le z \le 1$. Find the flux of
$\mathbf{F}(x, y, z) = \langle x^2, y, 3z \rangle$ across the surface of R.

14. A solid R consists of the region between the ellipsoids
$\frac{x^2}{25} + \frac{y^2}{81} + \frac{z^2}{144} = 1$ and $\frac{x^2}{4} + \frac{y^2}{9} + \frac{z^2}{16} = 1$. Find the flux of
$\mathbf{F}(x, y, z) = \langle 3z, 3y, 3x \rangle$ across the surface of R.

15. Assuming appropriate hypotheses, show that $\iint_S \operatorname{curl} \mathbf{F} \cdot d\mathbf{S} = 0$.

16. Suppose S is a surface satisfying the hypotheses of the divergence
theorem. If $\mathbf{c}$ is a constant vector, then $\iint_S \mathbf{c} \cdot d\mathbf{S} = 0$.

17. Suppose $\mathbf{F}$ is an inverse square field, that is, $\mathbf{F}(x, y, z) =$
$c \left\langle \frac{x}{(x^2 + y^2 + z^2)^{3/2}}, \frac{y}{(x^2 + y^2 + z^2)^{3/2}}, \frac{z}{(x^2 + y^2 + z^2)^{3/2}} \right\rangle$ for some constant c.
In exercise 28 of Section 5.7, the flux of $\mathbf{F}$ across the unit sphere
S_1 centered at the origin is calculated to be $4\pi c$. Use the following
steps to show that $\iint_S \mathbf{F} \cdot d\mathbf{S} = 4\pi c$ for any closed surface S

with the origin in its interior, as long as the surface satisfies the hypotheses of the divergence theorem.

(a) Calculate div **F**.
(b) Are the hypotheses of the divergence theorem met for **F** on the region R contained by the closed surface S? Can we continue as we did in example 50?
(c) Suppose the sphere S_1 lies completely within the region enclosed by S. Consider the region Q between the two surfaces. Observe that Q does not contain the origin. Now are the hypotheses met?
(d) Use the solids between surfaces formula to show that $\iint_S \mathbf{F} \cdot d\mathbf{S} = \iint_{S_1} \mathbf{F} \cdot d\mathbf{S} = 4\pi c.$

If not, we can use a smaller sphere that does. The calculation of Section 5.7 exercise 28 can be redone for a sphere of any radius centered at the origin, with the same result.

5.9   Stokes' Theorem

It's time to free Green's theorem from the plane.

5.9.1   Stokes' theorem: a generalization of Green's theorem

The vector form of Green's theorem, presented at the end of Section 5.5, equates the line integral of **F** along a positively oriented, piecewise smooth, simple closed curve C in the xy-plane with the double integral of curl $\mathbf{F} \cdot \mathbf{k}$ in the region D enclosed by the curve:

$$\oint_C \mathbf{F} \cdot \mathbf{T}\, ds = \oint_C \mathbf{F} \cdot d\mathbf{r} = \iint_D (\operatorname{curl} \mathbf{F}) \cdot \mathbf{k}\, dA.$$

Although the closed curve C and the region D are in the xy-plane, the normal to the plane, **k**, is in the z-direction, leading to the three-dimensional diagram in Figure 5.63.

Recall that for the vector form of Green's theorem, we extend the vector field $\mathbf{F}(x, y) = \langle M(x, y), N(x, y) \rangle$ to three dimensions as $\mathbf{F}(x, y, z) = \langle M(x, y), N(x, y), 0 \rangle$ so that curl **F** is defined.

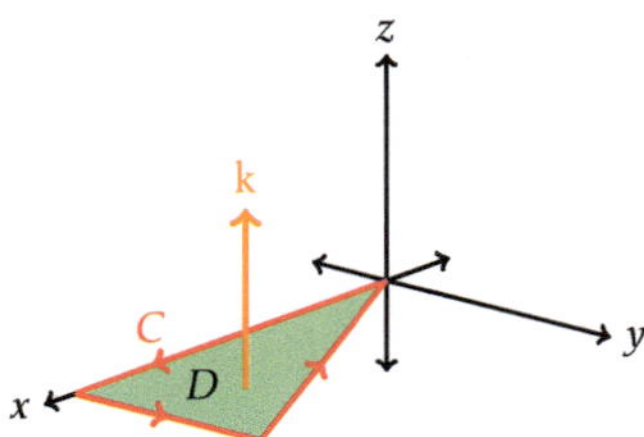

Figure 5.63 *A positively oriented, piecewise smooth, simple closed curve C (red) enclosing a region D (green), all in the xy-plane, along with the normal* **k** *(orange) to the region D. The vector fields* **F** *and curl* **F** *are not pictured, but including one or the other can be imagined if desired*

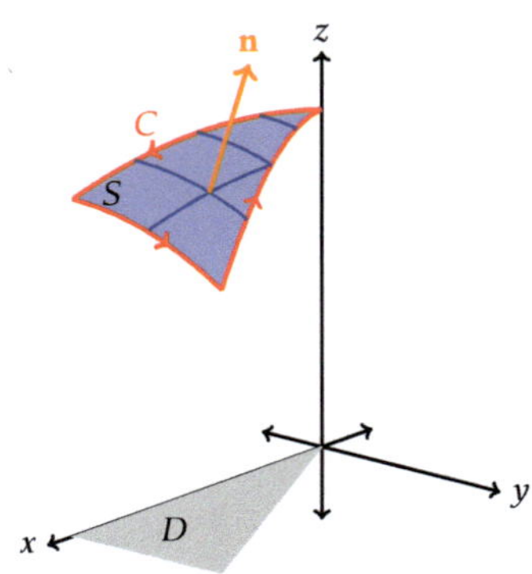

Figure 5.64 *A smooth, orientable surface S (blue) whose boundary is a positively oriented, piecewise smooth, simple closed curve C (red), along with a normal **n** (orange) to the surface S. The vector fields **F** and curl **F** are not pictured, but including one or the other can be imagined if desired. The shadow D (gray) of the surface in the xy-plane is also included, although the surface need not be of the form $z = f(x, y)$. Compare to Figure 5.63*

Now suppose that instead of using a region D lying in the plane, we move to a smooth, orientable surface S, and (with appropriate hypotheses) let C be the boundary of the surface S instead of the boundary of the region D, as in Figure 5.64. Making this adjustment, what might be the analogues to the objects in the vector form of Green's theorem? Instead of the normal **k** to the region D, we must generalize to the normals **n** to the surface S. And instead of the double integral on the region D, we change to the surface integral across the surface S. The formula becomes

The surface S lies in $\mathbf{R}^3$.

Notice that C is a spacecurve.

The term "boundary" can be interpreted in many ways. Here it is interpreted as for a two-dimensional object S (having area but not volume), with one-dimensional boundary C (having length but not area).

$$\int_C \mathbf{F} \cdot \mathbf{T}\, ds = \int_C \mathbf{F} \cdot d\mathbf{r} = \iint_S (\operatorname{curl} \mathbf{F}) \cdot \mathbf{n}\, dS = \iint_S (\operatorname{curl} \mathbf{F}) \cdot d\mathbf{S},$$

This says that the line integral of **F** along the boundary curve C is the same as the surface integral of curl **F** across the surface bounded by C.

which is known as *Stokes' theorem.*

Theorem 10 STOKES' THEOREM *Let S be an oriented, piecewise smooth surface that is bounded by a positively oriented, piecewise smooth, simple closed spacecurve C. Let **F** be a vector field whose component functions have continuous partial derivatives on an open region containing S. Then*

The open region is to be interpreted as open in three dimensions, that is, open in $\mathbf{R}^3$.

$$\int_C \mathbf{F} \cdot d\mathbf{r} = \iint_S (\operatorname{curl} \mathbf{F}) \cdot d\mathbf{S}.$$

The complete proof Stokes' theorem is beyond the level of this text, but a partial proof for a common type of surface S is given after a few examples.

Stokes' theorem can be used in either direction. If the line integral is easier to set up, use the line integral; if the surface integral is easier, use the surface integral. For our first example, we calculate both ways and verify that the result is the same.

Example 55 *Verify Stokes' theorem for the vector field* $F(x, y, z) = \langle xy, y, yz \rangle$ *and the surface S described by* $z = f(x, y) = xy$, $0 \le x \le 1$, $0 \le y \le 1$, *with positive orientation.*

Solution To verify Stokes' theorem means to calculate the line integral $\int_C F \cdot dr$, calculate the surface integral $\iint_S (\text{curl } F) \cdot dS$, and observe that the results are equal. First, we calculate the surface integral of curl F across the surface S. The curl is

Either order may be used.

$$\text{curl } F = \begin{vmatrix} \mathbf{i} & \mathbf{j} & \mathbf{k} \\ \frac{\partial}{\partial x} & \frac{\partial}{\partial y} & \frac{\partial}{\partial z} \\ xy & y & yz \end{vmatrix} = \langle z, 0, -x \rangle.$$

Because the surface S (Figure 5.65) is of the form $z = f(x, y)$ with positive orientation, a parameterization is

$$r(u, v) = \langle u, v, f(u, v) \rangle = \langle u, v, uv \rangle,$$

with $0 \le u \le 1$, $0 \le v \le 1$. Then

$$\text{curl } F(r(u, v)) = \langle uv, 0, -u \rangle.$$

Next, we find the normals:

$$r_u(u, v) = \langle 1, 0, v \rangle$$

$$r_v(u, v) = \langle 0, 1, u \rangle$$

$$r_u \times r_v = \begin{vmatrix} \mathbf{i} & \mathbf{j} & \mathbf{k} \\ 1 & 0 & v \\ 0 & 1 & u \end{vmatrix} = \langle -v, -u, 1 \rangle.$$

The integrand is

$$\text{curl } F(r(u, v)) \cdot (r_u \times r_v) = -uv^2 - u.$$

The surface integral is

$$\iint_S (\text{curl } F) \cdot dS = \int_0^1 \int_0^1 (-uv^2 - u)\, dv\, du$$

$$= \int_0^1 \left(\left(-\frac{1}{3}uv^3 - uv \right)\Big|_0^1 \right) du$$

$$= \int_0^1 \left(-\frac{1}{3}u - u - 0 \right) du = \int_0^1 -\frac{4}{3}u\, du$$

$$= -\frac{2}{3}u^2 \Big|_0^1 = -\frac{2}{3}.$$

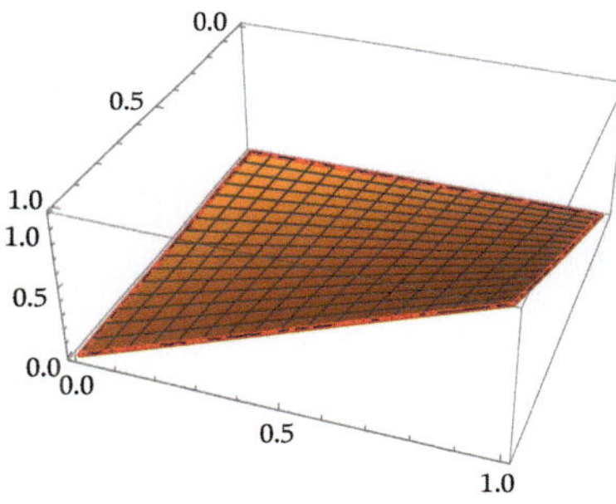

Figure 5.65 *The surface of example 55, with boundary marked in red, viewed from the first octant*

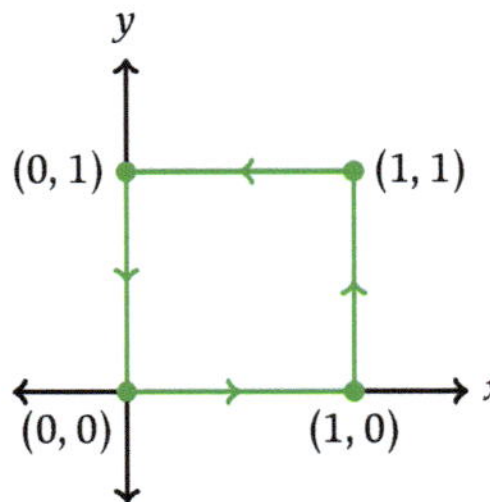

Figure 5.66 *The shadow of the boundary, with positive orientation*

To parameterize the segment from (a, b) to (c, d), use $\mathbf{r}(t) = \langle a + (c - a)t, b + (d - b)t \rangle$, $0 \le t \le 1$ (verify for yourself that $\mathbf{r}(0) = \langle a, b \rangle$ and $\mathbf{r}(1) = \langle c, d \rangle$). For instance, from $(1, 1)$ to $(0, 1)$, we use $\mathbf{r}(t) = \langle 1 + (0 - 1)t, 1 + (1 - 1)t \rangle = \langle 1 - t, 1 \rangle$, which are the x- and y-components of $\mathbf{r}_3(t)$. The z-component is found using the formula for the surface, $f(x, y) = xy$. Because $f(1 - t, 1) = (1 - t)(1) = 1 - t$, $\mathbf{r}_3(t) = \langle 1 - t, 1, 1 - t \rangle$.

The next task is to calculate the line integral of **F** along the closed curve C, whose value should be $-\frac{2}{3}$ to match the surface integral. A quick glance at Figure 5.65 verifies that there are four pieces to the closed curve C. It is sometimes easiest to view the shadow of the curve in the xy-plane, shown in Figure 5.66, in order to help parameterize the various pieces using the correct orientation. The usual parameterization of a line segment can be used for the x- and y-components, and the z-component is calculated from the formula for the surface:

shadow path	surface boundary parameterization ($0 \le t \le 1$)
$(0, 0)$ to $(1, 0)$	$\mathbf{r}_1(t) = \langle t, 0, f(t, 0) \rangle = \langle t, 0, 0 \rangle$
$(1, 0)$ to $(1, 1)$	$\mathbf{r}_2(t) = \langle 1, t, f(1, t) \rangle = \langle 1, t, t \rangle$
$(1, 1)$ to $(0, 1)$	$\mathbf{r}_3(t) = \langle 1 - t, 1, f(1 - t, 1) \rangle = \langle 1 - t, 1, 1 - t \rangle$
$(0, 1)$ to $(0, 0)$	$\mathbf{r}_4(t) = \langle 0, 1 - t, f(0, 1 - t) \rangle = \langle 0, 1 - t, 0 \rangle$

For the first piece, $\mathbf{F}(\mathbf{r}_1(t)) = \mathbf{F}(t, 0, 0) = \langle 0, 0, 0 \rangle$, and the line integral is 0. For the second piece,

$$\mathbf{F}(\mathbf{r}_2(t)) = \mathbf{F}(1, t, t) = \langle t, t, t^2 \rangle$$

$$\mathbf{r}_2{}'(t) = \langle 0, 1, 1 \rangle$$

$$\mathbf{F}(\mathbf{r}_2(t)) \cdot \mathbf{r}_2{}'(t) = \langle t, t, t^2 \rangle \cdot \langle 0, 1, 1 \rangle = t + t^2$$

$$\int_0^1 (t + t^2)\, dt = \left(\tfrac{1}{2}t^2 + \tfrac{1}{3}t^3 \right)\Big|_0^1 = \tfrac{5}{6}.$$

For the third piece,

$$\mathbf{F}(\mathbf{r}_3(t)) = \mathbf{F}(1 - t, 1, 1 - t) = \langle 1 - t, 1, 1 - t \rangle$$

$$\mathbf{r}_3{}'(t) = \langle -1, 0, -1 \rangle$$

$$\mathbf{F}(\mathbf{r}_3(t)) \cdot \mathbf{r}_3{}'(t) = \langle 1 - t, 1, 1 - t \rangle \cdot \langle -1, 0, -1 \rangle$$

$$= -(1 - t) - (1 - t) = 2t - 2$$

$$\int_0^1 (2t - 2)\, dt = \left(t^2 - 2t \right)\Big|_0^1 = -1.$$

For the fourth piece,

$$\mathbf{F}(\mathbf{r}_4(t)) = \mathbf{F}(0, 1 - t, 0) = \langle 0, 1 - t, 0 \rangle$$

$$\mathbf{r}_4{}'(t) = \langle 0, -1, 0 \rangle$$

$$\mathbf{F}(\mathbf{r}_4(t)) \cdot \mathbf{r}_4{}'(t) = \langle 0, 1 - t, 0 \rangle \cdot \langle 0, -1, 0 \rangle = t - 1$$

$$\int_0^1 (t - 1)\, dt = \left(\tfrac{1}{2}t^2 - t \right)\Big|_0^1 = -\tfrac{1}{2}.$$

Then the value of the line integral is

$$\int_C \mathbf{F} \cdot d\mathbf{r} = 0 + \frac{5}{6} - 1 - \frac{1}{2} = -\frac{2}{3},$$

matching the value of the surface integral, as expected from Stokes' theorem.   ∎

As demonstrated in example 55, when the line integral has several pieces and the surface integral does not, it may be easier (less tedious) to calculate the surface integral. The next example illustrates a situation in which the line integral is easiest.

Example 56 *Use Stokes' theorem to determine the value of $\iint_S (\operatorname{curl}\mathbf{F}) \cdot d\mathbf{S}$, where $\mathbf{F}(x,y,z) = \langle ye^z, x^3\sqrt{y}\sin z, x^2y^2z^2\rangle$ and the positively oriented surface is the portion of the paraboloid $z = 4 - x^2 - y^2$ that lies on or above the xy-plane (that is, for $z \geq 0$).*

Solution The implication of the instruction to use Stokes' theorem to calculate the surface integral of $\operatorname{curl}\mathbf{F}$ is that we should calculate the line integral $\int_C \mathbf{F} \cdot d\mathbf{r}$ instead. Key to this process is determining the boundary spacecurve C and a parameterization thereof. A graph of the surface is in Figure 5.67, where it is apparent that the boundary of the surface is where the surface meets the xy-plane, that is, where $0 = 4 - x^2 - y^2$. An equation of the boundary spacecurve is $x^2 + y^2 = 4$, $z = 0$, a circle in the xy-plane centered at the origin of radius 2, which can be parameterized as $\mathbf{r}(t) = \langle 2\cos t, 2\sin t, 0\rangle$, $0 \leq t \leq 2\pi$.

Following the usual order of calculation for a line integral, we have

$$\mathbf{F}(\mathbf{r}(t)) = \langle 2(\sin t)e^0, (2\cos t)^3\sqrt{2\sin t}\sin 0, (2\cos t)^2(2\sin t)^2 0^2\rangle$$

$$= \langle 2\sin t, 0, 0\rangle$$

$$\mathbf{r}'(t) = \langle -2\sin t, 2\cos t, 0\rangle$$

$$\mathbf{F}(\mathbf{r}(t)) \cdot \mathbf{r}'(t) = -4\sin^2 t$$

$$\int_C \mathbf{F} \cdot d\mathbf{r} = \int_0^{2\pi} -4\sin^2 t\, dt$$

$$= \int_0^{2\pi} -4\left(\tfrac{1}{2}(1 - \cos 2t)\right) dt = \int_0^{2\pi} (-2 + 2\cos 2t)\, dt$$

$$= \left(-2t + 2\frac{\sin 2t}{2}\right)\Big|_0^{2\pi} = -4\pi.$$

he value of the line integral, and by Stokes' theorem the value of the requested surface integral, is -4π.   ∎

To find the line integral along the entire closed curve C, we add the line integrals along the four pieces.

Another observation to make in the solution to example 55 is to recall that a parameterization of a surface is a two-variable function $\mathbf{r}(u, v)$, whereas a parameterization of a curve is a one-variable function, $\mathbf{r}(t)$. Consistent use of the traditional parameters (u and v for a surface, t for a curve) can be helpful.

Calculating this surface integral directly is difficult. The curl is $\operatorname{curl}\mathbf{F}(x,y,z) = \langle 2yx^2z^2 - x^3\sqrt{y}\cos z, -2xy^2z^2 + ye^z, 3x^2\sqrt{y}\sin z - e^z\rangle$, a parameterization is $\mathbf{r}(u, v) = \langle u, v, 4 - u^2 - v^2\rangle$, and then $\operatorname{curl}\mathbf{F}(\mathbf{r}(u, v))$ is quite a mess, and we are still a ways from finding the integrand of the double integral, illustrating the advantage of knowing Stokes' theorem.

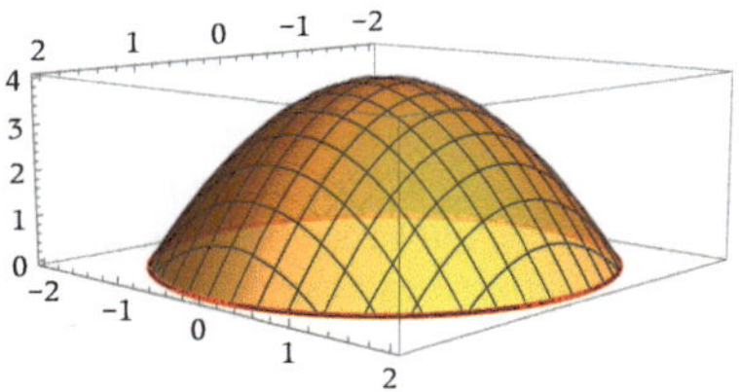

Figure 5.67 *The surface of example 56, with boundary marked in red, viewed from the first octant*

Seeing that the positive orientation of a surface $z = f(x,y)$ corresponds to a positive orientation of the boundary curve is an application of the right-hand rule. Looking at Figure 5.67, let the fingers of your right hand curl in the direction of the boundary curve, with your thumb pointing upward corresponding to a positive orientation of the surface. Viewed from above, your fingers are curling in the counterclockwise direction, corresponding to a positive orientation of the boundary curve.

Another piece of calculation magic stemming from Stokes' theorem is that when finding the surface integral of curl $\mathbf{F}$, we can switch to another, simpler surface that has the exact same boundary curve (and the same orientation). To see this, suppose that S_1 and S_2 are two different surfaces with the same boundary curve C and the same orientation, and suppose that $\mathbf{F}$ is a vector field. Assuming that the hypotheses of Stokes' theorem are met,

The same curve C serves as the boundary of both S_1 and S_2, meaning Stokes' theorem results in the same value.

$$\iint_{S_1} (\text{curl}\,\mathbf{F}) \cdot d\mathbf{S} = \int_C \mathbf{F} \cdot d\mathbf{r},$$

$$\iint_{S_2} (\text{curl}\,\mathbf{F}) \cdot d\mathbf{S} = \int_C \mathbf{F} \cdot d\mathbf{r},$$

and therefore

$$\iint_{S_1} (\text{curl}\,\mathbf{F}) \cdot d\mathbf{S} = \iint_{S_2} (\text{curl}\,\mathbf{F}) \cdot d\mathbf{S}.$$

Example 57 *Calculate $\iint_S (\text{curl}\,\mathbf{F}) \cdot d\mathbf{S}$ for the vector field $\mathbf{F}(x, y, z) = \langle x^3, 3xz^2, xy \rangle$ and the positively oriented surface consisting of the portion of the paraboloid $z = x^2 + y^2$ that lies on or below the plane $z = 9$.*

Solution The paraboloid $z = x^2 + y^2$ opens upward with vertex at the origin. Then the boundary of the surface is where the paraboloid meets the plane $z = 9$, that is, where $x^2 + y^2 = 9$. Instead of parameterizing the surface of the paraboloid, we can parameterize the portion of the plane $z = 9$ for which $x^2 + y^2 \leq 9$, because the boundary is the same:

This is an application of Stokes' theorem, as outlined before this example.

We could use a parameterization of the original surface, $\mathbf{r}(u, v) = \langle u, v, u^2 + v^2 \rangle$, $u^2 + v^2 \leq 9$. The difference between these two parameterizations is the z-component. A constant component usually leads to a simpler computation than a non-constant component, motivating the switch.

$$\mathbf{r}(u, v) = \langle u, v, 9 \rangle, \quad u^2 + v^2 \leq 9.$$

The curl is

Another option is to use Stokes' theorem to switch to the line integral, which eventually entails evaluating the integral $\int_0^{2\pi} (-243 \cos^3 t \sin t + 2187 \cos^2 t)\, dt$, which is not especially difficult but takes more effort than the featured solution.

$$\text{curl}\,\mathbf{F} = \begin{vmatrix} \mathbf{i} & \mathbf{j} & \mathbf{k} \\ \dfrac{\partial}{\partial x} & \dfrac{\partial}{\partial y} & \dfrac{\partial}{\partial z} \\ x^3 & 3xz^2 & xy \end{vmatrix} = \langle x - 6xz, -y, 3z^2 \rangle,$$

and composed with the parameterization the curl is

$$\text{curl}\,\mathbf{F}(\mathbf{r}(u, v)) = \langle u - 6u(9), -v, 3(9^2) \rangle = \langle -53u, -v, 243 \rangle.$$

The computation of $\mathbf{r}_u \times \mathbf{r}_v$ should be immediate, remembering that $\mathbf{i} \times \mathbf{j} = \mathbf{k}$. In fact, even lines 1 and 2 are unnecessary, recalling that $\mathbf{k}$ is orthogonal to a horizontal plane and the surface is positively oriented.

Next we find the normals:

$$\mathbf{r}_u(u, v) = \langle 1, 0, 0 \rangle$$

$$\mathbf{r}_v(u, v) = \langle 0, 1, 0 \rangle$$

$$\mathbf{r}_u \times \mathbf{r}_v = \langle 0, 0, 1 \rangle.$$

The integrand is

$$\operatorname{curl} \mathbf{F}(\mathbf{r}(u, v)) \cdot (\mathbf{r}_u \times \mathbf{r}_v) = 243.$$

The surface integral is

$$\iint_S (\operatorname{curl} \mathbf{F}) \cdot d\mathbf{S} = \iint_S 243 \cdot d\mathbf{S},$$

which is 243 times the area of the surface. The surface is a disk of radius 3, therefore

A disk consists of a circle and its interior.

$$\iint_S (\operatorname{curl} \mathbf{F}) \cdot d\mathbf{S} = 243 \cdot \pi(3^2) = 2187\pi.$$

∎

5.9.2   Partial proof of Stokes' theorem

Examples 55, 56, and 57 all involve a surface S of the form $z = f(x, y)$, with no unusual complications. It is this type of surface for which a proof of Stokes' theorem is computational enough to present here.

Suppose the surface S has the form $z = f(x, y)$ over a simply connected plane region D and that f has continuous second-order partial derivatives. Suppose further that the boundary curve C of the surface S is a positively oriented, piecewise smooth, simple closed spacecurve, and suppose that its shadow in the xy-plane, forming the boundary of D, is a curve for which the hypotheses of Green's theorem are met. Finally, suppose that $\mathbf{F}(x, y, z) = \langle M(x, y, z), N(x, y, z), P(x, y, z) \rangle$ is a vector field for which M, N, and P have continuous second-order partial derivatives on an open region containing S.

These hypotheses are primarily used to ensure that the partial derivatives used in the proof exist, and that Green's theorem and Clairaut's theorem can be used.

A parameterization of the surface S is

$$\mathbf{r}(u, v) = \langle u, v, f(u, v) \rangle.$$

This is the same parameterization used in the examples.

By exercise 43 of Section 5.6,

$$\mathbf{r}_u \times \mathbf{r}_v = \left\langle -\frac{\partial f}{\partial u}, -\frac{\partial f}{\partial v}, 1 \right\rangle.$$

While the switch to variables u and v can be helpful in working examples, it is more convenient for this proof to keep the variable names x and y, that is, to use the parameterization

$$\mathbf{r}(x, y) = \langle x, y, f(x, y) \rangle.$$

We also need the curl,

$$\operatorname{curl} \mathbf{F}(x,y,z) = \left\langle \frac{\partial P}{\partial y} - \frac{\partial N}{\partial z}, \frac{\partial M}{\partial z} - \frac{\partial P}{\partial x}, \frac{\partial N}{\partial x} - \frac{\partial M}{\partial y} \right\rangle.$$

When calculating the partial derivatives to determine curl **F**, the partial derivatives are taken before the composition with the parameterization **r**.

Then the integrand for the surface integral is

$$(\operatorname{curl} \mathbf{F}) \cdot (\mathbf{r}_x \times \mathbf{r}_y)$$

Use of Leibniz notation is convenient for this proof, for the purpose of keeping track of variables and using the multivariable chain rule.

$$= \left\langle \frac{\partial P}{\partial y} - \frac{\partial N}{\partial z}, \frac{\partial M}{\partial z} - \frac{\partial P}{\partial x}, \frac{\partial N}{\partial x} - \frac{\partial M}{\partial y} \right\rangle \cdot \left\langle -\frac{\partial f}{\partial x}, -\frac{\partial f}{\partial y}, 1 \right\rangle$$

$$= -\left(\frac{\partial P}{\partial y} - \frac{\partial N}{\partial z} \right) \frac{\partial f}{\partial x} - \left(\frac{\partial M}{\partial z} - \frac{\partial P}{\partial x} \right) \frac{\partial f}{\partial y} + \frac{\partial N}{\partial x} - \frac{\partial M}{\partial y},$$

and

Because we do not have a specific surface and vector field, we cannot evaluate the surface integral further than this. Instead, we must seek to get the same expression from the line integral.

$$\iint_S \operatorname{curl} \mathbf{F} \cdot d\mathbf{S}$$

$$= \iint_D \left(-\left(\frac{\partial P}{\partial y} - \frac{\partial N}{\partial z} \right) \frac{\partial f}{\partial x} - \left(\frac{\partial M}{\partial z} - \frac{\partial P}{\partial x} \right) \frac{\partial f}{\partial y} + \frac{\partial N}{\partial x} - \frac{\partial M}{\partial y} \right) dA.$$

Next, we need to calculate the line integral and show that it can be manipulated into the same form as the just-completed surface integral. To this end, we need a parameterization of the boundary curve C, which has the form

$$\mathbf{r}_1(t) = \langle x(t), y(t), z(t) \rangle,$$

where $z = f(x,y)$. Then by the chain rule,

Because $z = f(x,y)$, $\frac{\partial z}{\partial x}$ and $\frac{\partial f}{\partial x}$ are two different notations for the same partial derivative.

$$\frac{\partial z}{\partial t} = \frac{\partial z}{\partial x} \frac{dx}{dt} + \frac{\partial z}{\partial y} \frac{dy}{dt} = \frac{\partial f}{\partial x} \frac{dx}{dt} + \frac{\partial f}{\partial y} \frac{dy}{dt}.$$

Therefore

$$\mathbf{r}_1{}'(t) = \left\langle \frac{dx}{dt}, \frac{dy}{dt}, \frac{\partial f}{\partial x} \frac{dx}{dt} + \frac{\partial f}{\partial y} \frac{dy}{dt} \right\rangle.$$

Also,

$$\mathbf{F}(\mathbf{r}_1(t)) = \langle M(x(t), y(t), z(t)), N(x(t), y(t), z(t)), P(x(t), y(t), z(t)) \rangle.$$

Then, using shortcut notation, the integrand for the line integral is

$$\mathbf{F}(\mathbf{r}_1(t)) \cdot \mathbf{r}_1{}'(t) = M\frac{dx}{dt} + N\frac{dy}{dt} + P\left(\frac{\partial f}{\partial x} \frac{dx}{dt} + \frac{\partial f}{\partial y} \frac{dy}{dt} \right)$$

$$= \left(M + P\frac{\partial f}{\partial x} \right) \frac{dx}{dt} + \left(N + P\frac{\partial f}{\partial y} \right) \frac{dy}{dt}.$$

The line integral is therefore

$$\int_C \mathbf{F} \cdot d\mathbf{r} = \oint \left[\left(M + P\frac{\partial f}{\partial x} \right) \frac{dx}{dt} + \left(N + P\frac{\partial f}{\partial y} \right) \frac{dy}{dt} \right] dt$$

$$= \oint \left(M + P\frac{\partial f}{\partial x} \right) dx + \left(N + P\frac{\partial f}{\partial y} \right) dy.$$

This, in turn, is set up perfectly for using Green's theorem! The result is a double integral,

$$\iint_D \left(\frac{\partial}{\partial x} \left(N + P\frac{\partial f}{\partial y} \right) - \frac{\partial}{\partial y} \left(M + P\frac{\partial f}{\partial x} \right) \right) dA.$$

Now comes the tricky part: the partial derivative of N with respect to x is not just $\frac{\partial N}{\partial x}$, as when calculating curl $\mathbf{F}$. In the context of this double integral, N is treated as a function of two variables x and y, not three variables x, y, and z. The partial derivatives in this step are calculated after the composition with $\mathbf{r}_1$, so that the variable z is also a function of x and y (see the tree diagram in Figure 5.68). By the multivariable chain rule,

$$\frac{\partial}{\partial x} N = \frac{\partial N}{\partial x} + \frac{\partial N}{\partial z} \frac{\partial f}{\partial x}.$$

We also need the partial derivative with respect to x of the product $P\frac{\partial f}{\partial y}$, which by the product rule and multivariable chain rule is

$$\frac{\partial}{\partial x} \left(P\frac{\partial f}{\partial y} \right) = P\left(\frac{\partial}{\partial x} \frac{\partial f}{\partial y} \right) + \left(\frac{\partial}{\partial x} P \right) \frac{\partial f}{\partial y}$$

$$= P\frac{\partial^2 f}{\partial x \partial y} + \left(\frac{\partial P}{\partial x} + \frac{\partial P}{\partial z} \frac{\partial f}{\partial x} \right) \frac{\partial f}{\partial y}$$

$$= P\frac{\partial^2 f}{\partial x \partial y} + \frac{\partial P}{\partial x} \frac{\partial f}{\partial y} + \frac{\partial P}{\partial z} \frac{\partial f}{\partial x} \frac{\partial f}{\partial y}.$$

    Applying the same reasoning to $-\frac{\partial}{\partial y}\left(M + P\frac{\partial f}{\partial x} \right)$, the integrand of the double integral becomes

$$\frac{\partial N}{\partial x} + \frac{\partial N}{\partial z} \frac{\partial f}{\partial x} + P\frac{\partial^2 f}{\partial x \partial y} + \frac{\partial P}{\partial x} \frac{\partial f}{\partial y} + \frac{\partial P}{\partial z} \frac{\partial f}{\partial x} \frac{\partial f}{\partial y}$$

$$- \left(\frac{\partial M}{\partial y} + \frac{\partial M}{\partial z} \frac{\partial f}{\partial y} + P\frac{\partial^2 f}{\partial y \partial x} + \frac{\partial P}{\partial y} \frac{\partial f}{\partial x} + \frac{\partial P}{\partial z} \frac{\partial f}{\partial y} \frac{\partial f}{\partial x} \right)$$

$$= \frac{\partial N}{\partial x} + \frac{\partial N}{\partial z} \frac{\partial f}{\partial x} + \frac{\partial P}{\partial x} \frac{\partial f}{\partial y} - \frac{\partial M}{\partial y} - \frac{\partial M}{\partial z} \frac{\partial f}{\partial y} - \frac{\partial P}{\partial y} \frac{\partial f}{\partial x}$$

$$= \left(\frac{\partial N}{\partial z} - \frac{\partial P}{\partial y} \right) \frac{\partial f}{\partial x} + \left(\frac{\partial P}{\partial x} - \frac{\partial M}{\partial z} \right) \frac{\partial f}{\partial y} + \frac{\partial N}{\partial x} - \frac{\partial M}{\partial y},$$

The square brackets in line 1 are grouping symbols (like parentheses); line 2 manipulates the differentials, replacing $\frac{dx}{dt}\, dt$ by dx and $\frac{dy}{dt}\, dt$ by dy (details of the justification omitted).

Now we have a double integral over the region D, which is the same type of integral we are trying to match.

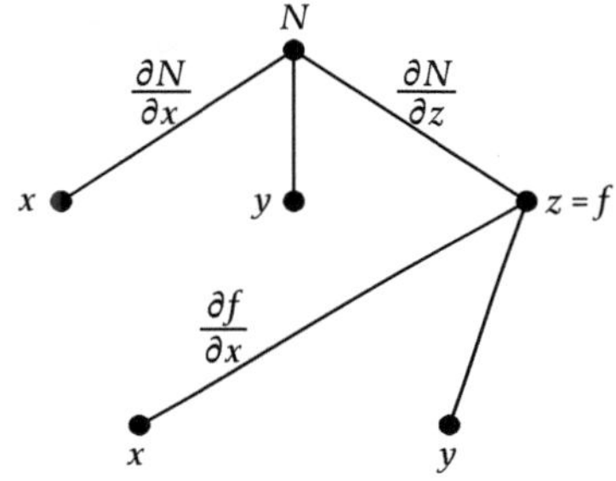

Figure 5.68 *The tree diagram for determining $\frac{\partial}{\partial x} N$, where N is a function of x, y, and z and z is a function of x and y*

Line 1 uses the product rule; line 2 uses the multivariable chain rule, in the same manner as the previous calculation of $\frac{\partial}{\partial x} N$; line 3 simplifies. Notice also that $\frac{\partial}{\partial x} \frac{\partial f}{\partial y}$ is a composition, the partial derivative of $\frac{\partial f}{\partial y}$ with respect to x, whereas the quantity $\frac{\partial f}{\partial x} \frac{\partial f}{\partial y}$ of line 3 is a product.

Lines 1 and 2 apply the previous calculations to the integrand; line 3 simplifies, including a use of Clairaut's theorem; line 4 uses factoring to rearrange.

which matches the integrand of the surface integral, and Stokes' theorem holds for the type of surface used in examples 55, 56, and 57.

5.9.3 Circulation and an interpretation of the curl vector

The hypotheses needed for a mathematical proof, or for mathematical accuracy, are due to the freedom afforded by the abstract generality of mathematics. In the physical sciences, the constraints of the physical laws present in this wonderful universe we inhabit nearly always ensure that the hypotheses are met.

We have usually chosen **F** as the name of the vector field; this time we are using **v** instead.

Consider a positively oriented, piecewise smooth, simple closed curve C in a velocity vector field **v**, where the velocity vector field represents fluid flow. If the unit tangent vector **T** of the curve generally points in the same direction as the fluid flow velocity vectors **v** such as in Figure 5.69, left, then the line integral

$$\int_C \mathbf{v} \cdot d\mathbf{r} = \int_C \mathbf{v} \cdot \mathbf{T}\, ds$$

is positive, because the values of $\mathbf{v} \cdot \mathbf{T}$ are positive. Visually, we see the fluid flowing around the curve in the same direction as the curve. On the other hand, if the fluid flow velocity vectors generally point in the opposite direction of the unit tangent vectors of the curve (Figure 5.69, right), so that the fluid appears to be flowing against the direction of the curve, the values of $\mathbf{v} \cdot \mathbf{T}$ are negative and the line integral is negative. This means that the line integral measures the tendency of the fluid to circulate in the same direction as the curve. We therefore call the value of the line integral the *circulation* of the vector field **v** along (around) the curve C.

So far, this discussion of circulation could have taken place in Section 5.2 when we first encountered line integrals of vector fields. But Stokes' theorem allows us to go further. Consider a surface S that consists of a disk of radius a centered at

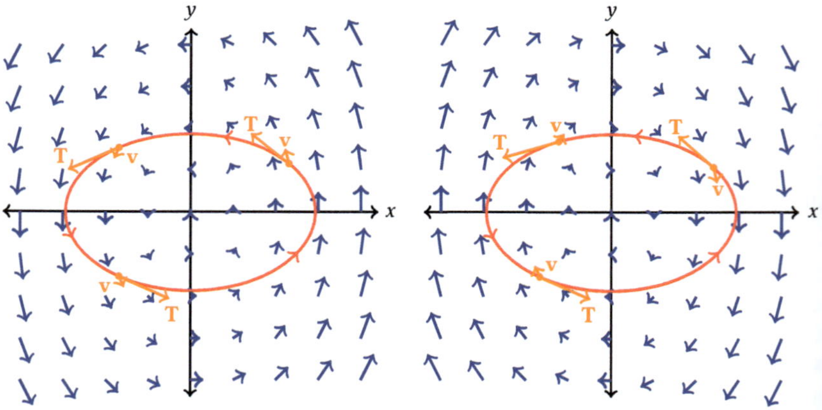

Figure 5.69 *(Left) a closed curve (red) in a vector field with positive circulation; (right) a closed curve (red) in a vector field with negative circulation. Two dimensions are shown for simplicity; the discussion is valid in three dimensions*

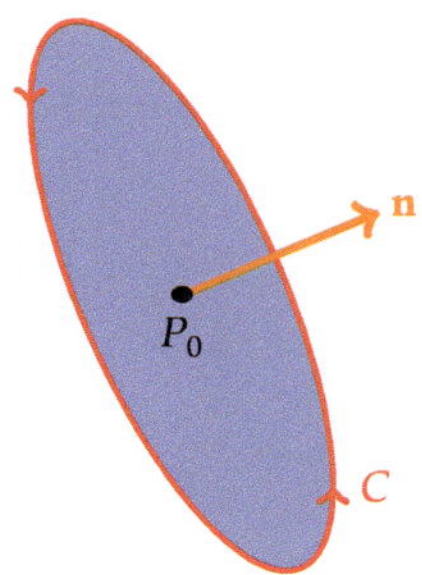

Figure 5.70 *A disk S with boundary curve C, centered at P_0, with radius a and normal* **n**

$P_0 = (x_0, y_0, z_0)$, in some plane containing P_0, as pictured in Figure 5.70. Its boundary curve is a circle C. Then by Stokes' theorem, the circulation of the vector field v around the curve C is

$$\int_C \mathbf{v} \cdot d\mathbf{r} = \iint_S \operatorname{curl} \mathbf{v} \cdot d\mathbf{S} = \iint_S \operatorname{curl} \mathbf{v} \cdot \mathbf{n} \, dS.$$

If the radius $a = \omega$ of the disk is infinitesimal, then the area of the disk is $\pi\omega^2$ and the integrand, being continuous, is approximately the same throughout the disk, the same as at P_0, which is $\operatorname{curl} \mathbf{v}(P_0) \cdot \mathbf{n}(P_0)$. Then

The curve C_ω is a circle centered at P_0 of radius ω.

$$\int_{C_\omega} \mathbf{v} \cdot d\mathbf{r} \approx \iint_S \operatorname{curl} \mathbf{v}(P_0) \cdot \mathbf{n}(P_0) \, dS,$$

which is the constant, $\operatorname{curl} \mathbf{v}(P_0) \cdot \mathbf{n}(P_0)$, times the area of the surface, $\pi\omega^2$:

$$\int_{C_\omega} \mathbf{v} \cdot d\mathbf{r} \approx (\operatorname{curl} \mathbf{v}(P_0) \cdot \mathbf{n}(P_0)) \, \pi\omega^2.$$

Rearranging,

Notice that this equation is true regardless of the plane in which C_ω lies. While the plane does not affect the value of the vector $\operatorname{curl} \mathbf{F}(P_0)$, it does affect the value of the vector $\mathbf{n}(P_0)$.

$$\frac{1}{\pi\omega^2} \int_{C_\omega} \mathbf{v} \cdot d\mathbf{r} \approx \operatorname{curl} \mathbf{v}(P_0) \cdot \mathbf{n}(P_0),$$

for any infinitesimal ω. Look again at Figure 5.70. This says that the value of $\operatorname{curl} \mathbf{v}(P_0) \cdot \mathbf{n}(P_0)$ is the circulation of v around the axis **n**, scaled to the size of the circle. Because $\operatorname{curl} \mathbf{v}(P_0) \cdot \mathbf{n}(P_0)$ is largest when $\operatorname{curl} \mathbf{v}(P_0)$ and $\mathbf{n}(P_0)$ have the same direction, we can conclude that at any point P_0, the vector $\operatorname{curl} \mathbf{v}$ indicates the axis of greatest circulation of **v**.

This is where curl gets its name. The greatest "curl," the fastest "circling" around the point P_0, is around the axis containing $\operatorname{curl} \mathbf{v}(P_0)$.

Example 58 *Find the axis of greatest circulation of the vector field* $F(x, y, z) =$ $\langle xy, xy^2, yz^2 \rangle$ *(a) at the point* $(0, 1, 3)$ *and (b) at the point* $(2, -1, 4)$.

Solution   We first calculate the curl:

$$\text{curl } \mathbf{F} = \begin{vmatrix} \mathbf{i} & \mathbf{j} & \mathbf{k} \\ \frac{\partial}{\partial x} & \frac{\partial}{\partial y} & \frac{\partial}{\partial z} \\ xy & xy^2 & yz^2 \end{vmatrix} = \langle z^2, 0, y^2 - x \rangle.$$

(a) An axis is a line. To represent the line parametrically, we need a position vector, which in this case is $\langle 0, 1, 3 \rangle$, and a direction vector, which is given by the curl:

$$\text{curl } \mathbf{F}(0, 1, 3) = \langle 9, 0, 1 \rangle.$$

Then parametric equations for the axis of greatest circulation of $\mathbf{F}$ at the point $(0, 1, 3)$ are $x = 0 + 9t$, $y = 1 + 0t$, and $z = 3 + t$.

(b) The position vector is $\langle 2, -1, 4 \rangle$, and the direction vector is

$$\text{curl } \mathbf{F}(2, -1, 4) = \langle 16, 0, -1 \rangle.$$

Parametric equations for the axis of greatest circulation of $\mathbf{F}$ at the point $(2, -1, 4)$ are $x = 2 + 16t$, $y = -1$, and $z = 4 - t$.

5.9.4   Penultimate comments

One last loose end. We are now ready to present, without full details, the second half of the proof of theorem 3, the three-dimensional version of which is theorem 7. Let $\mathbf{F}$ be a vector field whose component functions have continuous partial derivatives throughout a simply connected region D, and suppose curl $\mathbf{F} = \mathbf{0}$ throughout D. What remains is to show that $\mathbf{F}$ is conservative. Suppose C is a piecewise smooth simple closed spacecurve in D. Let S be a surface in D whose boundary is the spacecurve C (although easily imagined to always exist, proving so is difficult and requires advanced mathematics). Then by Stokes' theorem,

The proof is still incomplete, pending the proof that such a surface S always exists, and that we can extend to any closed curve C in D.

$$\int_C \mathbf{F} \cdot d\mathbf{r} = \iint_S \text{curl } \mathbf{F} \cdot d\mathbf{S} = \iint_S \mathbf{0} \cdot d\mathbf{S} = 0.$$

According to theorem 5, if $\int_C \mathbf{F} \cdot d\mathbf{r} = 0$ for every closed curve C, then $\mathbf{F}$ is conservative.

This section on Stokes' theorem is a culmination of much of multivariable calculus. Material from every chapter of this text is used in this section. In fact material from a large proportion of the individual sections is used! Mathematics is often said to build on itself, and this section is certainly testament to that idea.

Before closing, I ask your indulgence for one more comparative idea.

5.9.5 The fundamental theorem of calculus: in the right circumstances, integrals are determined by the values of an antiderivative along the boundary

What better way to end our study of calculus than by recapping its fundamental theorem?

In single-variable calculus, part 2 of the fundamental theorem says that as long as f is continuous throughout all of $[a, b]$,

$$\int_a^b f(t)\, dt = H(b) - H(a),$$

where H is an antiderivative of f. In other words, we need only know the value of an antiderivative along the boundary, the endpoints of the interval $[a, b]$ (Figure 5.71). How is that possible? Because the antiderivative carries the information about the function at every point in between! For instance, consider the integral of $f(x) = x^2$ on $[1, 4]$:

$$\int_1^4 x^2\, dx = \left.\tfrac{1}{3}x^3\right|_1^4 = \tfrac{64}{3} - \tfrac{1}{3} = \tfrac{63}{2} = 21.$$

Although we only evaluate the antiderivative $\tfrac{1}{3}x^3$ at the boundary inputs $x = 4$ and $x = 1$, the antiderivative $H(x) = \tfrac{1}{3}x^3$ contains information about the function $f(x) = x^2$ throughout the interval $[1, 4]$. We aren't truly ignoring the values of the function on all the "inside" inputs between the endpoints, we're just disguising where that information lies, taking advantage of the cumulative "summing" effect of the integral that produces the antiderivative. The antiderivative rules are where the real calculational magic occurs.

In Chapter 4, this magic continued with double and triple integrals, even on general regions. The limits of integration describe the boundary (including the shape) of the region, and we evaluate antiderivatives, "partial" antiderivatives in this case, one at a time, at those limits of integration.

In Section 5.3, the fundamental theorem of line integrals states that under the right circumstances,

$$\int_A^B \nabla f \cdot d\mathbf{r} = f(B) - f(A).$$

Figure 5.71 *The interval $[a, b]$ on the one-dimensional number line. With only one dimension, the points between a and b are interior, or in-between, points; the boundary of the interval consists of the two points a and b*

Even the term "limits of integration" describes the idea that we only need evaluate the antiderivative along the boundary of the region. Perhaps, then, a better term would be "bounds of integration."

The gradient ∇f is a type of derivative of f, therefore f is a type of antiderivative of ∇f.

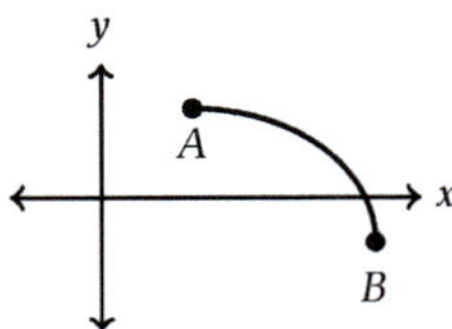

Figure 5.72 *The curve lies in two dimensions, but the curve itself is one-dimensional, meaning that its boundary consists of the two points A and B. Compare to Figure 5.71*

Because $\frac{\partial N}{\partial x}$ and $\frac{\partial M}{\partial y}$ are partial derivatives of N and M, respectively, then N and M are partial antiderivatives of $\frac{\partial N}{\partial x}$ and $\frac{\partial M}{\partial y}$.

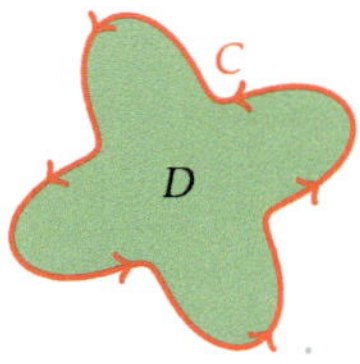

Figure 5.73 *A two-dimensional region D (green) bounded by a curve C (red)*

Compare the difference from Figure 5.63 to Figure 5.64 to the difference from Figure 5.71 to Figure 5.72.

Considering f as a type of antiderivative of ∇f, once again we only need to evaluate the antiderivative at the boundary, the two endpoints, of the curve (Figure 5.72) or spacecurve. As demonstrated in examples in Section 5.3, the meat of the calculation of such a line integral is the calculation of a potential function, an antiderivative of the vector-valued function.

Green's theorem is yet another example. In the original form presented in Section 5.4,

$$\oint_C M\,dx + N\,dy = \iint_D \left(\frac{\partial N}{\partial x} - \frac{\partial M}{\partial y} \right) dA,$$

the expression $M\,dx + N\,dy$ is a type of antiderivative of $\left(\frac{\partial N}{\partial x} - \frac{\partial M}{\partial y} \right)$, so the double integral on the region D is the line integral of an antiderivative along the boundary curve C (Figure 5.73).

Similarly, because curl $\mathbf{F}$ is a type of derivative of the vector field $\mathbf{F}$, we may consider $\mathbf{F}$ to be a type of antiderivative of curl $\mathbf{F}$, and in the vector form of Green's theorem,

$$\oint_C \mathbf{F} \cdot d\mathbf{r} = \iint_D (\text{curl}\,\mathbf{F}) \cdot \mathbf{k}\,dA,$$

we still have that the double integral on the region D is the line integral of an antiderivative along the boundary curve C (refer again to Figure 5.63).

Stokes' theorem continues the theme; in the formula

$$\int_C \mathbf{F} \cdot d\mathbf{r} = \iint_S (\text{curl}\,\mathbf{F}) \cdot d\mathbf{S},$$

the surface integral across the surface S is the line integral of an antiderivative along the boundary curve C (refer again to Figure 5.64).

Even the divergence theorem follows the pattern. The divergence div $\mathbf{F}$ is another type of derivative of a vector field $\mathbf{F}$, and in the formula

$$\iint_S \mathbf{F} \cdot d\mathbf{S} = \iiint_R \text{div}\,\mathbf{F}\,dV,$$

the triple integral on the three-dimensional region R is the surface integral of an antiderivative across the boundary surface S.

There are common themes among the hypotheses required for each formula as well. All require something of the antiderivative function, such as continuity to be true throughout the entire interval or region. And all help, at least in certain circumstances, to make the calculation of an integral easier.

Stokes' theorem, Green's theorem, the divergence theorem, the fundamental theorem of line integrals, and the calculation of double and triple integrals are a versions of the same phenomenon, the fundamental theorem of calculus, applied in higher dimensions to multivariable functions.

I sincerely hope that you have enjoyed this journey through calculus and higher-dimensional thinking.

EXERCISES 5.9

1–16. Rapid response: is the expression an integral, double integral, triple integral, line integral, surface integral, or something else?

1. $\iint_D f(x,y)\,dA$

2. $\int_C f(x,y)\,ds$

3. $\iint_S f(x,y,z)\,dS$

4. $\iint_D \left(\dfrac{\partial N}{\partial x} - \dfrac{\partial M}{\partial y}\right) dA$

5. $\iiint_R f(x,y,z)\,dV$

6. $\iint_D (\operatorname{curl}\mathbf{F}) \cdot \mathbf{k}\,dA$

7. $\iint_S (\operatorname{curl}\mathbf{F}) \cdot d\mathbf{S}$

8. $\int_C \mathbf{F} \cdot d\mathbf{r}$

9. $\int_C M\,dx + N\,dy$

10. $\iint_D \mathbf{F}(\mathbf{r}) \cdot (\mathbf{r}_u \times \mathbf{r}_v)\,dA$

11. $\int_a^b f(x)\,dx$

12. $\int_A^B \nabla f \cdot d\mathbf{r}$

13. $\iint_S \mathbf{F} \cdot d\mathbf{S}$

14. $\int_C \mathbf{F} \cdot \mathbf{T}\,ds$

15. $\iiiint_R f(w,x,y,z)\,dH$

16. $\iiint_R \operatorname{div}\mathbf{F}\,dV$

17–24. Find the axis of greatest circulation of the vector field $\mathbf{F}$ at the given point.

17. $\mathbf{F}(x,y,z) = \langle 4xy, 3z, -5y \rangle$, point $(3,2,1)$

18. $\mathbf{F}(x,y,z) = \langle xyz, x^2yz, x^3yz \rangle$, point $(2,1,\frac{1}{2})$

19. $\mathbf{F}(x,y,z) = \langle \ln yz, \frac{1}{x}, 3x \rangle$, point $(2,4,-1)$

20. $\mathbf{F}(x,y,z) = \langle z\sin xy, z\cos xy, yz^5 \rangle$, point $(\pi, \frac{1}{3}, 2)$

21. $\mathbf{F}(x,y,z) = \langle x\sqrt{y}, y\sqrt{z}, z\sqrt{x} \rangle$, point $(4,9,25)$

22. $\mathbf{F}(x,y,z) = \langle e^{xyz}, 2e^{xyz}, 3e^{xyz} \rangle$, point $(5,8,0)$

23. $\mathbf{F}(x,y,z) = \langle 1+y, 2-z, 7+x \rangle$, point $(-1,2,-7)$

24. $\mathbf{F}(x,y,z) = \langle -y^2, x^3, z \rangle$, point $(-4,5,\sqrt{13})$

25–30. Verify Stokes' theorem for the vector field $\mathbf{F}$ and the surface S.

Use positive orientation for surfaces and curves in the exercises for this section.

25. $\mathbf{F}(x,y,z) = \langle z, x+y, 4 \rangle$, S is described by $z = 3y$, $x^2 + y^2 \leq 1$

26. $\mathbf{F}(x,y,z) = \langle 2xz, -4z, 3y \rangle$, S is the plane $z = 5$ over the region $0 \leq x \leq 1, 0 \leq y \leq 1$

27. $\mathbf{F}(x,y,z) = \langle 3x^2z, 4-y^2, -y \rangle$, S is the portion of the surface $z = y - x^2$ over the region $0 \leq x \leq 1, 0 \leq y \leq 1$

28. $\mathbf{F}(x,y,z) = \langle y, z, x \rangle$, S is described by $z = x + y$, $x^2 + y^2 \leq 4$

Some exercises require the use of various techniques of integration.

29. $F(x, y, z) = \langle xz, x^2 z, -y^2 \rangle$ and the surface S described by $z = 3 - x, x^2 + y^2 \leq 9$

30. $F(x, y, z) = \langle xy, x + y, z^2 \rangle$, S is the portion of the surface $z = x^2 y$ inside the cylinder $x^2 + y^2 = 1$

31–34. Use Stokes' theorem to calculate $\int_C F \cdot dr$, where C is the boundary of the surface S.

31. $F(x, y, z) = \langle 2 - y, 3 + z, x - 2 \rangle$, S is described by $z = 4 - y$, $x^2 + y^2 \leq 7$

32. $F(x, y, z) = \langle y + 2z, z + 3x, x + 4y \rangle$, S is described by $z = x^2 y$, $0 \leq x \leq 1, 1 \leq y \leq 3$

33. $F(x, y, z) = \langle x^2 + y, y^3 - 2z, z^2 + 5x \rangle$, S is the portion of $z = 4x - y^2$ for which $-1 \leq x \leq 1$ and $-1 \leq y \leq 1$

34. $F(x, y, z) = \langle e^x, 3 \cosh y, 4 \sinh z \rangle$, S is the portion of the hyperbolic paraboloid $z = x^2 - 3y^2$ lying inside the cylinder $x^2 + y^2 \leq 25$

35–40. Calculate $\iint_S (\text{curl } F) \cdot dS$.

35. $F(x, y, z) = \langle -2yz, 5y^3 - z, 2^x y \rangle$, S is the portion of $z = -3(x^2 + y^2)$ lying on or above the plane $z = -12$

36. $F(x, y, z) = \langle y^2, 3z, 4x \rangle$, S is the top and sides of the cube with corners $(0, 0, 0)$, $(1, 0, 0)$, $(0, 1, 0)$, $(1, 1, 0)$, $(0, 0, 1)$, $(1, 0, 1)$, $(0, 1, 1)$, and $(1, 1, 1)$

37. $F(x, y, z) = \langle -yz \cos^2 z, xz \sin^2 z, y^2 \rangle$, S is the portion of the paraboloid $z = 10 - x^2 - y^2$ lying on or above the plane $z = 1$

38. $F(x, y, z) = \langle y^2, x^3 y^4 (z - 1)^5, zx^4 y^4 \rangle$, S is the portion of the paraboloid $z = x^2 + y^2$ lying inside the cylinder $x^2 + y^2 = 1$

39. $F(x, y, z) = \langle -yx^2, xy^2, xyz^2 \rangle$, S is the cone with vertex $(0, 0, 10)$ and base $x^2 + y^2 = 25$ in the xy-plane

40. $F(x, y, z) = \langle x^3 z^3, \frac{x}{z+3}, \sqrt{1 + y^2} \rangle$, S is the hemisphere $x^2 + y^2 + z^2 = 9, z \geq 0$

This exercise uses calculations similar to those from the discussion of an interpretation of the curl vector.

41. Suppose $F(x, y, z)$ is a vector field (perhaps representing fluid flow) and that all necessary hypotheses are met.

    (a) Consider a closed surface S_ω that consists of a sphere of infinitesimal radius ω centered at $P_0 = (x_0, y_0, z_0)$. Let R_ω represent the region inside the sphere. Use the divergence theorem to calculate the flux across the surface S_ω.

    (b) Use the result of (a) to solve for $\text{div } F(P_0)$ in terms of the flux.

    (c) Use the result of (b) to determine an interpretation of $\text{div } F(P_0)$.

    (d) Using the terms "source" and "sink," write an interpretation of the sign of the divergence at P_0.

Epilogue

When I was in graduate school, a friend from church asked me what I had to do for a dissertation in mathematics. When I told him that I had to come up with new mathematics, to prove new theorems, he thought I was joking. "It's all known," he said. Nothing I tried could convince him that I was being serious.

Mathematics is a living, breathing subject! New mathematics is being discovered every day. And not just at the frontiers of knowledge, in the mainstream research requiring years of background study. Even in a subject as old and well-understood as calculus, fresh new ideas, new theorems, and clever applications are still being discovered. The impetus for writing this book lies in this truth.

Infinitesimals have been studied for many centuries, yet our understanding is still evolving. The approximation relation in the hyperreal numbers uses much the same reasoning as Euler and others who developed calculus, but the precise definition of approximation and exploration of its behind-the-scenes subtleties was not possible until relatively recently. The subject of algebra, as taught in high schools, was developed across many cultures for hundreds of years before finally being distilled into its present simple, compact form. Various concepts of trigonometry, as applied to astronomy and navigation, were studied worldwide for millennia before the current understanding took shape. Why should we expect calculus to be different?

No one calculus textbook contains all of calculus. There are far too many applications to physics, engineering, economics, and many other subjects for one person to learn, even in a lifetime of study. At some point, each of us must move on to other things.

So, what lies beyond calculus? That's not even a well-formed question, because the line between what is and is not called calculus is blurred. Many branches of calculus have become named subjects in their own right. *Differential equations* is the study of equations involving derivatives, such as $y''(x) + 3e^x y'(x) + 5y(x) = 4$. *Numerical methods* looks at how to compute numerical approximations of quantities that defy exact solution, for example, the midpoint rule for double integrals in Section 4.2. *Complex analysis* studies calculus in the realm of complex numbers. *Mathematical modeling* practices taking a real-life situation and answering questions using mathematics, including calculus, to at least attempt to improve knowledge and make somewhat-informed decisions. And the list goes on! There are many non-calculus mathematical subjects as well.

Infinitesimals continue to make appearances. I will close with two such examples.

The *Dirac delta function*, which is useful in engineering, is often studied in differential equations. The idea is that a force with large magnitude is applied for

Multivariable Calculus Set Free. Charles Bryan Dawson, Oxford University Press.
C. Bryan Dawson (2026). DOI: 10.1093/oso/9780198984269.003.0007

a very short period of time, as in a hammer striking an object. Let ω be a specific infinitesimal and let t_0 be the moment of impact. Then such a force can be modeled by

$$\delta(t - t_0) = \begin{cases} 0, & 0 \le t \le t_0 - \omega \\ \frac{1}{2}\Omega, & t_0 - \omega \le t \le t_0 + \omega \\ 0, & t \ge t_0 + \omega. \end{cases}$$

Then, for any real-number value of t,

$$\delta(t - t_0) \doteq \begin{cases} \infty, & t = t_0 \\ 0, & t \ne t_0 \end{cases}$$

$$\int_0^\infty \delta(t - t_0)\, dt = 1.$$

Because the integral's value is 1, the function is called a *unit impulse*.

The final example comes from the subject of *analysis*, which is a higher-level study of calculus theory. A useful concept is uniform continuity: a function $y = f(x)$ is *uniformly continuous* on an interval I if, for every hyperreal number x in I and for any infinitesimal α, the quantity $f(x + \alpha) - f(x)$ is always infinitesimal. For example, consider $f(x) = x^2$. The quantity $f(x + \alpha) - f(x) = (x + \alpha)^2 - x^2 = x^2 + 2x\alpha + \alpha^2 - x^2 = 2x\alpha + \alpha^2$. If I is the interval $[3, 7]$, then any number x in the interval I is on the real level, the quantity $2x\alpha + \alpha^2$ is infinitesimal, and the function is uniformly continuous on the interval. But if I is the interval $(2, \infty)$, then x can be infinitely large; if $x = \Omega$ and $\alpha = \omega$, the quantity $2x\alpha + \alpha^2 = 2\Omega\omega + \omega^2 = 2 + \omega^2$, which is not infinitesimal, and the function is not uniformly continuous on the interval. Uniform continuity has powerful consequences.

There is no end to mathematics that can be explored and researched. In fact, it has been proved that even in a subject as simple as arithmetic, no finite set of axioms can describe all true theorems. There will always, even with an infinite amount of time at our disposal, be more mathematics to learn! The fun never ends.

Answers to Odd-Numbered Exercises

Introduction

Section 0.1

1 infinitesimal

3 infinitesimal

5 neither

7 neither

9 infinite

11 infinite

13 $14 - 8\omega$

15 $-\frac{1}{4}\omega + \frac{\omega^2}{4}$

17 $9\Omega + 55$

19 $16 - 8\omega + \omega^2$

21 $8 + 12\omega + 6\omega^2 + \omega^3$

23 $3\Omega - 1$

25 $4\Omega - 4$

27 Ω^4

29 Ω^2

31 $\Omega^2 - 2\sqrt{2} + 2\omega^2$

33 $2894756382394578556\sqrt{\omega}$

35 ω

37 $12\Omega^3$

39 $\omega^{3/2}$ level

41 Ω^9 level

43 Ω level

45 real level

47 true

49 true

51 true

53 infinitesimal (because $\ln 3^\omega = \omega \ln 3$)

55 45ω cm (the units are required)

57 4ω cm (the units are required)

Section 0.2

1 -0.00293%

3 -4.63%

5 $4 + \omega$

7 $3\Omega - 5$

9 $\sqrt{\omega} - \omega$

11 14

13 9Ω

15 5

17 $\frac{4}{5}$

19 $2\omega^{3/2}$

21 ω

23 10

25 $-\frac{3}{2}$

27 4ω

29 $\Omega^{3/2}$

31 $-\frac{1}{28}\sqrt{\Omega}$

33 -1

35 $\frac{1}{4}\Omega$

37 $2\sqrt{\Omega}$

39

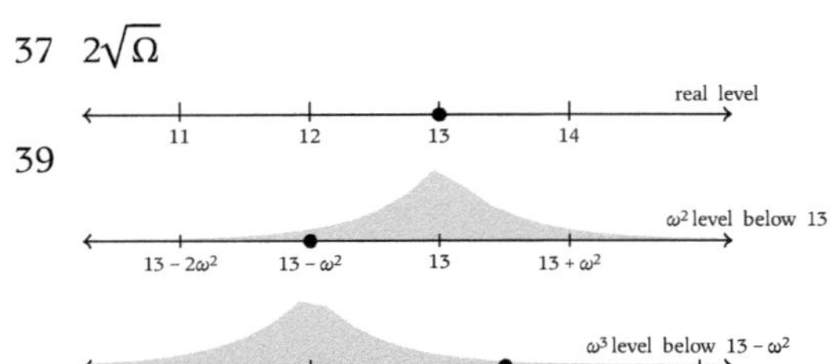

41 (a) $\approx \ln \Omega$ (by theorem 6);

(b) theorems do not apply;

(c) theorems do not apply (alternate answer $e^7 e^\Omega$);

(d) $\approx \sqrt{\Omega}$ (by theorem 5)

43 (a) $\approx \cosh 4$ (by theorem 7);

(b) $\approx \tan^{-1} 4$ (by theorem 7);

(c) theorems do not apply ($\tan \frac{\pi}{2}$ is undefined by reason of division by zero);

(d) theorems do not apply (if we say $\approx \cos \frac{\pi}{2} = 0$ we are violating the approximation principle)

45 (a) $\cos(\omega^2 + 3\omega^4) \approx \cos \omega^2 \approx \cos 0 = 1$;

(b) this theorem does not apply because $\sin 0 = 0$

(c) $e^{\omega^2 + 3\omega^4} \approx e^{\omega^2} \approx e^0 = 1$

Section 0.3

1 0

3 0

5 0

7 ∞

9 A, stuck

11 $\frac{1}{5}$

13 0

15 undefined

17 undefined

19 $-\infty$

21 1

23 $\ln 5$

25 $\ln \frac{1}{2}$ (equivalent answer: $- \ln 2$)

27 $\ln 4$

29 ∞

31 $-\infty$

33 0

35 $\frac{1}{e}$

37 ∞

39 2

41 $f(3 + \omega) = 0, f(3 - \omega) = 0$

43 $f(\Omega) = \infty, f(-\Omega) = -\infty$

45 $f(\Omega) = 1, f(-\Omega) = -1$

47 2

Section 0.4

1 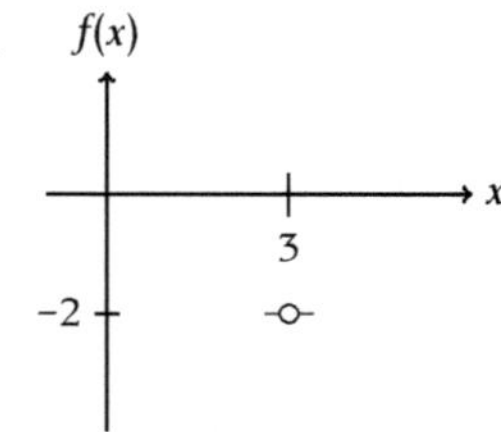

3 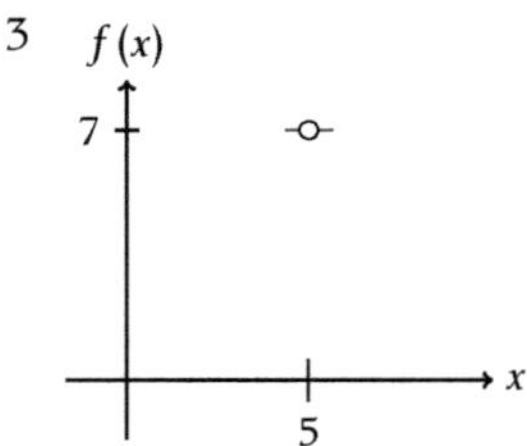

5   $f(x)$

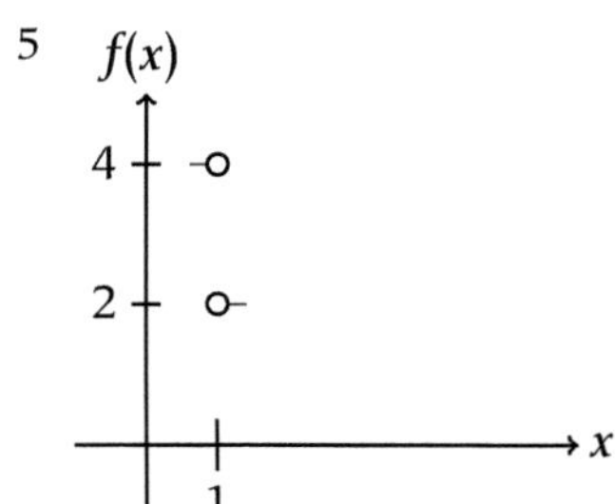

7   $f(x)$

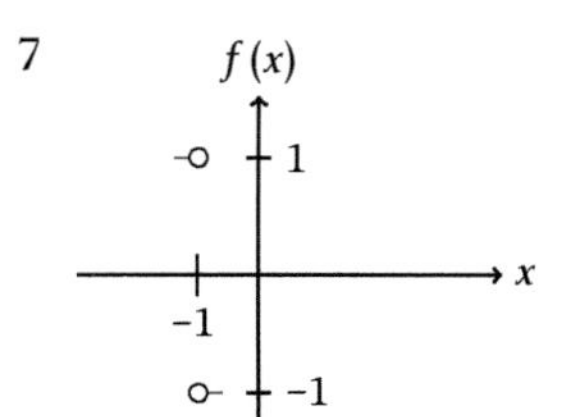

9   $f(x)$

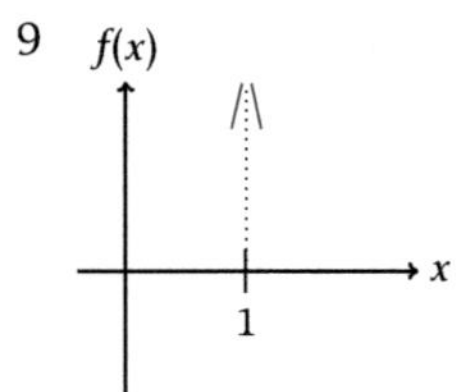

11   $f(x)$

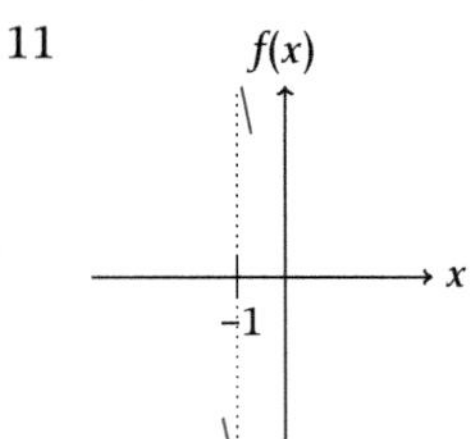

13   $f(x)$

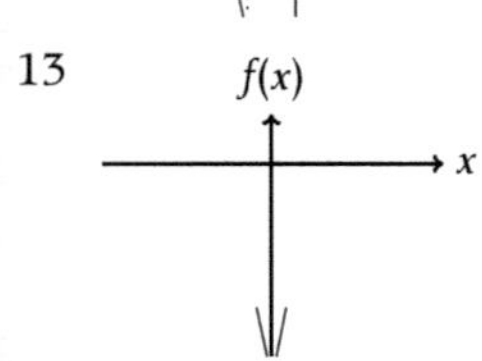

15   $y = \frac{1}{4}$ (on the right)

17   $y = 14$ (on the left)

19   no horizontal asymptote

21   no horizontal asymptote; the limit is not as $x \to \infty$
     or as $x \to -\infty$

23   $\frac{5}{2}$

25   $-\infty$

27   ∞

29   4

31   $-\infty$

33   $\frac{5}{7}$

35   $\frac{4}{9}$

37   $\frac{-3}{2}$

39   $\frac{1}{6}$

41   -1

43   8

45   -1

47   12

49   0

51   1

53   $-\frac{1}{4}$

55   $\frac{4}{7}$

57   0

59   0

61   $\frac{1}{3}$

63   $-\infty$

65   ∞

67   ∞

69   ∞

71   $\lim_{x \to 0} \frac{4}{x}$ DNE, $\lim_{x \to 0^+} \frac{4}{x} = \infty$, $\lim_{x \to 0^-} \frac{4}{x} = -\infty$

73   $\lim_{x \to 3} \frac{x^2-9}{|x-3|}$ DNE, $\lim_{x \to 3^+} \frac{x^2-9}{|x-3|} = 6$,
     $\lim_{x \to 3^-} \frac{x^2-9}{|x-3|} = -6$

75   $y = 4$ (on both sides)

77   $y = 3$ (on the right) and $y = -3$ (on the left)

79   0

81   0

83   ∞

85   0

Section 0.5

1   $f'(x) = 2x$

3   $f'(x) = 3$

5   $f'(x) = 2x + 3$

7   $g'(x) = \dfrac{(4x-9)2x-(x^2-7)4}{(4x+9)^2} = \dfrac{4x^2+18x+28}{(4x+9)^2}$

9   $f'(t) = 5t^4 - 21t^2 + 4$

11   $\dfrac{dy}{dx} = \dfrac{(x^3+5x)(3)-(3x-9)(3x^2+5)}{(x^3+5x)^2} = \dfrac{-6x^3+27x^2+45}{(x^3+5x)^2}$

13   $y' = 5 + 7x^{-2} = 5 + \dfrac{7}{x^2}$

15   $\dfrac{df}{dx} = \dfrac{5}{2}x^{3/2}$

17   $f'(x) = \dfrac{(5x^3+3x)4-(1+4x)(15x^2+3)}{(5x^3+3x)^2} = \dfrac{-40x^3-15x^2-3}{(5x^3+3x)^2}$

19   $y' = \dfrac{(22t-1)(2t^2-\frac{1}{4})-(\frac{2}{3}t^3-\frac{1}{4}t)22}{(22t-1)^2} = \dfrac{\frac{88}{3}t^3-2t^2+\frac{1}{4}}{(22t-1)^2}$

21   $y' = x^2 \sec x \tan x + 2x \sec x$

23   $f'(x) = \dfrac{(\sec x)(12x^2)-(4x^3-7)\sec x \tan x}{\sec^2 x} = \dfrac{12x^2-(4x^3-7)\tan x}{\sec x}$

25   $g'(y) = \dfrac{(\csc y)6-6y(-\csc y \cot y)}{\csc^2 y} = \dfrac{6+6y\cot y}{\csc y}$

27   $\dfrac{dz}{dx} = \dfrac{(x^3+9x)(x\sec x \tan x+\sec x)-(x\sec x)(3x^2+9)}{(x^3+9x)^2}$
$\qquad = \dfrac{-2x^3\sec x+(x^4+9x^2)\sec x \tan x}{(x^3+9x)^2}$

29   $g'(t) = -(t^2 - 4t)\csc^2 t + (2t - 4)\cot t$
     (Hint: $g(t) = (t^2 - 4t)\cot t$)

31   $y' = -4\sin(4x - 3)$

33   $f'(x) = 28(7x + 5)^3$

35   $g'(x) = \dfrac{1}{2}(\tan 5x)^{-\frac{1}{2}} \cdot \sec^2 5x \cdot 5 = \dfrac{5\sec^2 5x}{2\sqrt{\tan 5x}}$

37   $y' = \dfrac{(x^3-5)\cdot3(2x+1)^2\cdot2-(2x+1)^3(3x^2)}{(x^3-5)^2} = \dfrac{6(x^3-5)(2x+1)^2-3x^2(2x+1)^3}{(x^3-5)^2}$

39   $f'(t) = t^3(-\sin(6t + 1)) \cdot 6 + (\cos(6t + 1))3t^2$
$\qquad = -6t^3 \sin(6t + 1) + 3t^2 \cos(6t + 1)$

41   $g'(x) = \sec((x^2 + 5)^3) \tan((x^2 + 5)^3) \cdot 3(x^2 + 5)^2 \cdot 2x$
$\qquad = 6x(x^2 + 5)^2 \sec((x^2 + 5)^3) \tan((x^2 + 5)^3)$

43   $y' = 4(\tan(x^2 + 1))^3 \cdot (\sec^2(x^2 + 1)) \cdot 2x$
$\qquad = 8x \sec^2(x^2 + 1) \tan^3(x^2 + 1)$

45   $f'(x) = 7(5 + \cos x^2)^6(-\sin x^2)(2x)$
$\qquad = -14x(5 + \cos x^2)^6 \sin x^2$

47   $y' = -\sin\tan\theta \cdot \sec^2 \theta = -(\sec^2 \theta)(\sin\tan\theta)$

49   $\dfrac{dx}{dt} = t^2 \cdot \sec(5t - 8)^4 \tan(5t - 8)^4 \cdot 4(5t - 8)^3 \cdot 5$
$\qquad + \sec(5t - 8)^4 \cdot 2t = 20t^2(5t - 8)^3$
$\qquad \sec(5t - 8)^4 \tan(5t - 8)^4 + 2t\sec(5t - 8)^4$
$\qquad = 2t(\sec(5t - 8)^4)(10t(5t - 8)^3 \tan(5t - 8)^4 + 1)$

51   $y' = \dfrac{\ln x - x \cdot \frac{1}{x}}{(\ln x)^2} = \dfrac{\ln x - 1}{(\ln x)^2}$ (Note: $\dfrac{\ln x - 1}{2\ln x}$ is incorrect)

53   $f'(x) = \dfrac{2}{x-1} - \dfrac{3}{3x+7};$
     alternate answer $\dfrac{1}{\left(\frac{(x-1)^2}{3x+7}\right)} \cdot \dfrac{(3x-7)\cdot2(x-1)-(x-1)^2\cdot3}{(3x+7)^2}$ (the
     first answer is obtained by expanding the expression
     before differentiating, whereas the alternate answer
     is for not using the laws of logarithms)

55   $y' = x^2 \cdot \dfrac{1}{5x^2} \cdot 10x + (\ln 5x^2)(2x) = 2x + 2x\ln 5x^2$

57   $f'(x) = 2x$ (Note: e^4 is a constant)

59   $y' = \dfrac{(4x+3)e^x-e^x\cdot4}{(4x+3)^2} = \dfrac{(4x-1)e^x}{(4x+3)^2}$

61   $g'(x) = 20e^{4x+3}$

63   $y'' = -4$

65   $\dfrac{d^2y}{dx^2} = \dfrac{3}{2}x$

67   $f^{(4)}(t) = 840t^3$

69   $y'' = (-\csc x)(-\csc^2 x) + (\cot x)(-(-\csc x \cot x))$
$\qquad = \csc^3 x + \csc x \cot^2 x$

71   $f^{(4)}(x) = \sin x$

73   $f^{(4000)}(x) = \sin x$

75   $-25\sin 5x$

77   $y = -\dfrac{5}{16}x - \dfrac{1}{4}$

79   $y = 10x + 29$

81   $y = -x + \dfrac{\pi}{2}$

83   $y = 2x - 2$

85   $y' = 3^{4x+1}4\ln 3$

87   $g'(x) = x^{7x+1}\left(\dfrac{7x+1}{x} + 7\ln x\right)$

89   $y' = (\sin x)^{2x}\left(\dfrac{2x\cos x}{\sin x} + 2\ln\sin x\right)$

91   $y' = t^2 \sinh t + 2t\cosh t$

Section 0.6

1 $\frac{1}{2} + \frac{1}{3} + \frac{1}{4} + \frac{1}{5}$

3 $(-2)^2 + (-1)^2 + 0^2 + 1^2$

5 30

7 0

9 $\displaystyle\sum_{k=5}^{125} k^4$

11 $\displaystyle\sum_{k=1}^{14} \sqrt{k\omega}$

13 $\displaystyle\sum_{k=5}^{99} (4 + k\omega)^3$

Section 0.7

1 yes

3 no; the integrand is not continuous at $x = 0$, which is in the interval $[-5, 8]$

5 yes

7 no; the integrand is not continuous at several places in the interval, including $x = \pm\frac{\pi}{2}$

9 yes; although the integrand is not continuous at $x = 3$, that number is not in the interval $[-4, -1]$.

11 yes

13 no (integration by parts is needed instead)

15 no

17 yes (technically no, if you are willing to find the eighth power of $4x - 7$)

19 yes

21 $u = 5x - 7$

23 $u = 5x - 7$

25 $u = 5x - 7$

27 $u = x^2 + 4$

29 $u = \sin x$

31 $u = \ln x$

33 $\int (x-4)(x+1)\, dx = \int (x^2 - 3x - 4)\, dx$
$= \frac{x^3}{3} - \frac{3x^2}{2} - 4x + C$

15 10

17 $\frac{19}{3}$

19 $\frac{1}{10}$

21 -39

23 24

25 $\int_3^{19} (x^5 - 4x + 1)\, dx$ (All portions of the answer, including the dx, are required.)

27 $\int_{\pi/2}^{3\pi/2} (\cos(x + \pi) + \sin(4x))\, dx$

29 $\int_0^1 4x\sqrt{1 - x^4}\, dx$

31 $\int_a^b ((f(x))^2 + (g(x))^2)\, dx$

35 $\int \frac{x^2 + 5x}{x}\, dx = \int (x + 5)\, dx = \frac{x^2}{2} + 5x + C$ (comment: because the original integrand is undefined at $x = 0$, the interval over which this antiderivative is valid cannot contain $x = 0$.)

37 $\int (x-1)^2\, dx = \int (x^2 - 2x + 1)\, dx = \frac{x^3}{3} - x^2 + x + C$

39 $\int \sqrt{(3x+1)^4}\, dx = \int (3x+1)^2\, dx$
$= \int (9x^2 + 6x + 1)\, dx = 3x^3 + 3x^2 + x + C$

41 $\int \tan x \cos x\, dx = \int \frac{\sin x}{\cos x}\cos x\, dx$
$= \int \sin x\, dx = -\cos x + C$

43 $\int \sec x \frac{\sin x}{\cos x}\, dx = \int \sec x \tan x\, dx = \sec x + C$

45 $\int \frac{x^2 + 3}{\sqrt{x}}\, dx = \int \left(\frac{x^2}{\sqrt{x}} + \frac{3}{\sqrt{x}}\right) dx$
$= \int (x^{3/2} + 3x^{-1/2})\, dx = \frac{2}{5}x^{5/2} + 6x^{1/2} + C$

47 succeed

49 fail

51 succeed

53 fail

55 succeed

57 $\frac{x^4}{4} - \frac{4x^3}{3} + \frac{7x^2}{2} - x + C$

59 $\frac{1}{2}\sin(2x + 5) + C$

61 $x^2 + 5\sin x + C$ (this integral does not require substitution)

63 $2\ln|x+1| + C$

65 $-6\cot x + C$

67 $-\dfrac{1}{6}\cot(6z - 15) + C$

69 $\dfrac{4}{5}e^{5x} + C$

71 $\int x^3(4x - 9)\,dx = \int(4x^4 - 9x^3)\,dx = \dfrac{4x^5}{5} - \dfrac{9x^4}{4} + C$

73 $\dfrac{-3}{4x} + C$

75 $\ln|x^2 + 4| + C = \ln(x^2 + 4) + C$

77 $\dfrac{x^3}{3} - \dfrac{7x^2}{2} + 4\ln|x| + C$ (rewrite integrand as $x^2 - 7x + \dfrac{4}{x}$)

79 $\int(5x - 2)^2\,dx = \int(25x^2 - 20x + 4)\,dx = \dfrac{25x^3}{3} - 10x^2 + 4x + C$

81 $x\sin x + \cos x + C$

83 $-\dfrac{1}{3}\cos x^3 + C$

85 $\dfrac{1}{7}\sec 7x + C$

87 $\int \dfrac{x^3 - 7x}{4x}\,dx = \int\left(\dfrac{1}{4}x^2 - \dfrac{7}{4}\right)dx = \dfrac{1}{12}x^3 - \dfrac{7}{4}x + C$

89 $e^{\sqrt{x}} + C$ (substitution, $u = \sqrt{x}$)

91 not possible with available rules, formulas, and techniques

93 $\int\left(\dfrac{1}{\sin^2 x} + \dfrac{1}{x^2}\right)dx = \int(\csc^2 x + x^{-2})\,dx = -\cot x - \dfrac{1}{x} + C$

95 $\dfrac{x^2}{2}\ln x - \dfrac{1}{4}x^2 + C$

97 2

99 DNE (improper, each side of the discontinuity is infinite)

101 0 (upper and lower limits of integration are identical)

103 $\dfrac{255}{8}$ (substitution, $u = x^2 + 1$. If you answered $\dfrac{1}{8}$, you forgot to switch the limits of integration to values of u.)

105 2

107 $\dfrac{1}{4}(e^3 - e^{-1}) = 4.92941$

109 $\dfrac{1}{3}$

111 $-\dfrac{3}{16}$ (substitution, $u = \sec x$. If you encountered division by zero, you forgot to switch the limits of integration to values of u.)

113 13 (rewrite as $\int_2^5 \dfrac{1}{3}x^2\,dx$)

115 $\dfrac{7}{6}$

117 $\dfrac{3}{10}5^{5/3} = \dfrac{3}{2}5^{2/3} = 4.38603$ (substitution, $u = x^2 - 4$. If you answered 0.919635, you forgot to switch the limits of integration to values of u.)

119 -2π (integration by parts)

121 $\dfrac{1}{4}$ (improper; replace ∞ by an arbitrary positive infinite hyperreal Ω and follow the EAR procedure)

123 $\dfrac{1}{3}(\ln 2)^3$ (if your answer is $\dfrac{7}{3}$ you forgot to change to values of u; if your answer is $2\ln 2 = \ln 4$ then you incorrectly changed $(\ln x)^2$ to $2\ln x = \ln x^2$; if your answer is 2 you made both mistakes)

125 The available antiderivative formulas, rules, and techniques are insufficient to determine an antiderivative.

127 $\dfrac{1}{2}$ (rewrite as $\int_{-1}^0 \dfrac{1}{(x+2)^2}\,dx$, substitution, $u = x + 2$. If you encountered division by zero, you forgot to switch the limits of integration to values of u.)

129 $\dfrac{162}{5} = 32.4$

131 6 (Note that $\sqrt{x^2} = |x|$, and because the interval of integration contains only negative values of x, $|x| = -x$ on this interval. Use $\int_{-4}^{-2} -x\,dx$.)

133 2 units^2

135 $\dfrac{670}{3}\text{ units}^2 = 223.33\text{ units}^2$

Chapter 1

Section 1.1

1

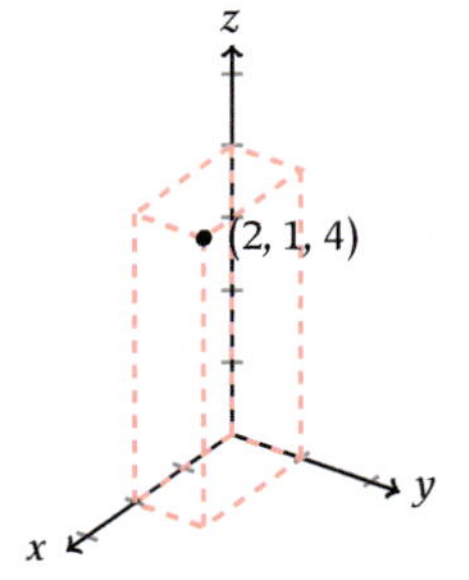

3

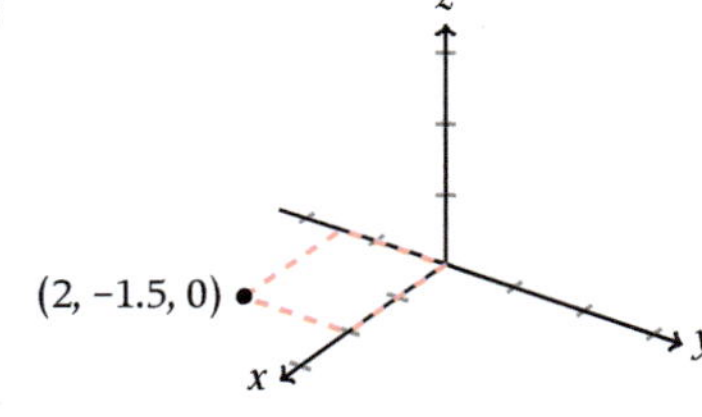

5

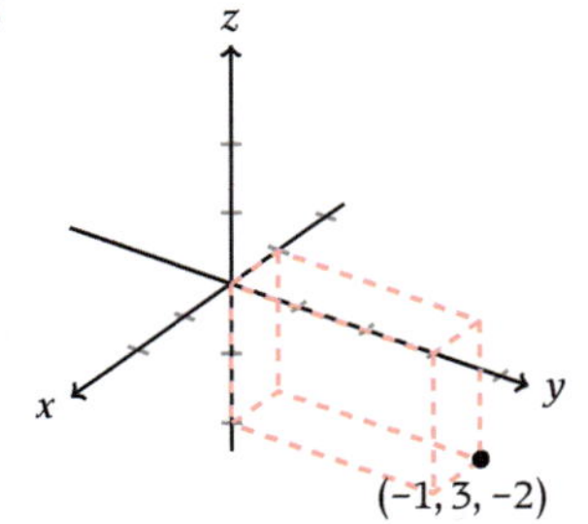

7

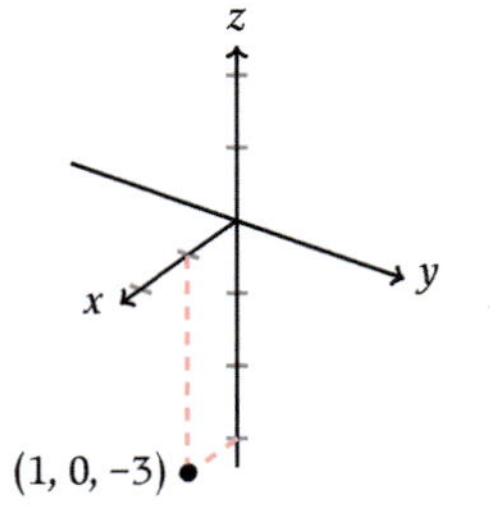

9

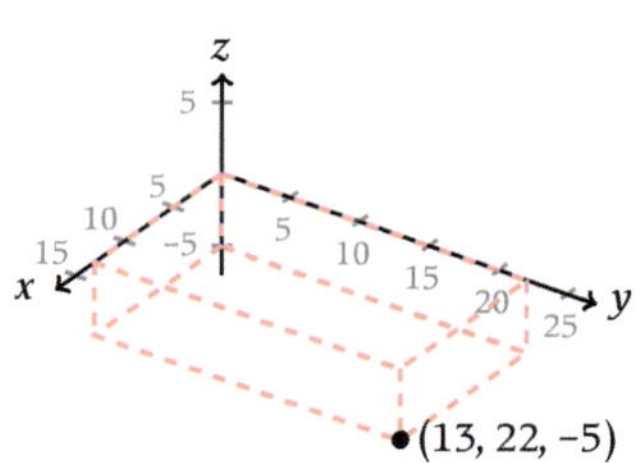

11

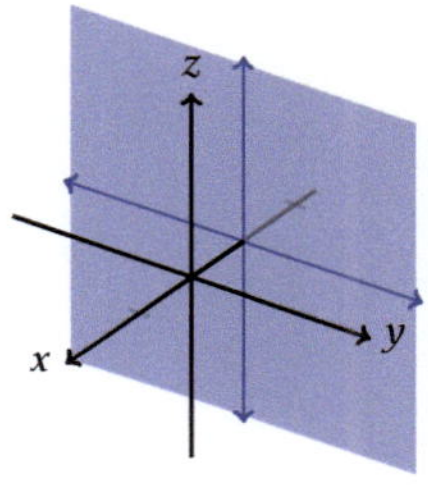

13

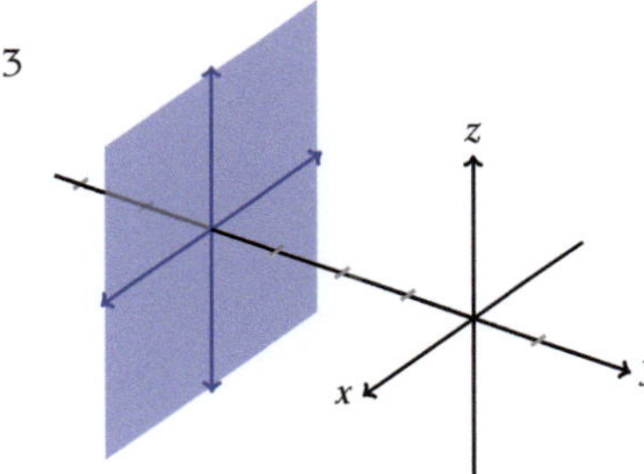

15

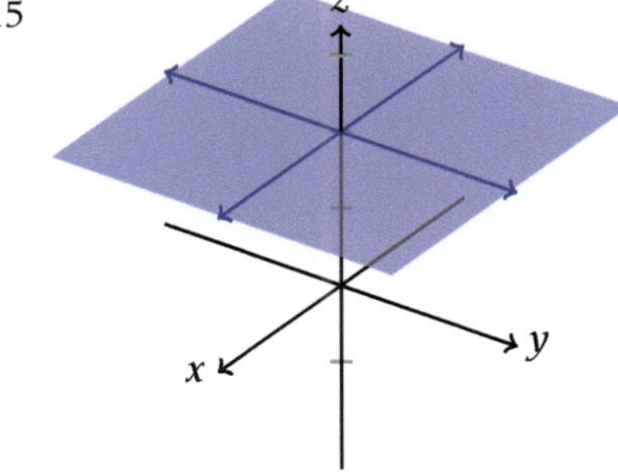

17

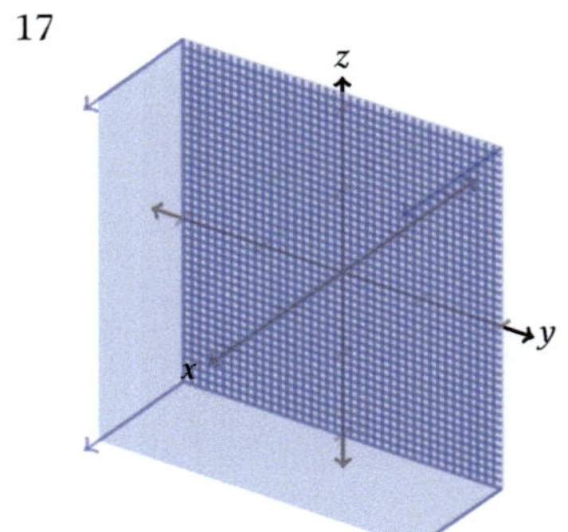

19

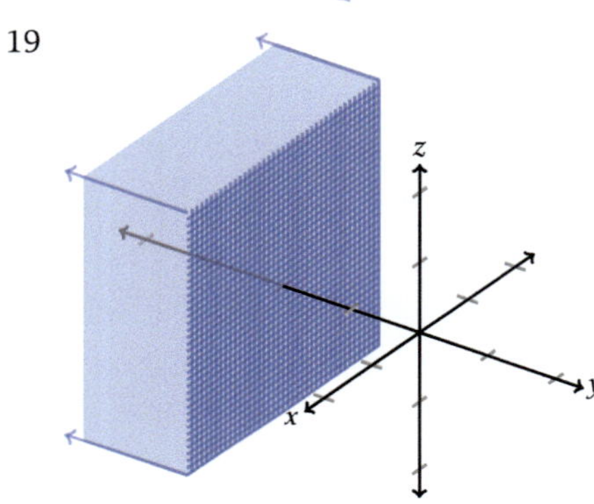

21 $\sqrt{41}$ or 6.4031

23 11

25 9

27 $\sqrt{407.4}$ or 20.184

29 $\left(\frac{3}{2}, 2, -2\right)$

31 $(x - 2)^2 + (y - 9)^2 + (z + 3)^2 = 121$

33 $(x - 1)^2 + (y + 2)^2 + (z - 4)^2 = 14$

35 $(x - 2)^2 + (y - 2)^2 + (z - 2)^2 = 14$

37 $(x - 2.04)^2 + (y - 1.67)^2 + (z - 4)^2 = 16$

39 $(x - 2)^2 + (y - 2)^2 + (z - 9)^2 = 4$

41 center $(3, -2, 11)$, radius 6

43 center $(2, -3, 4)$, radius 5

45 center $(0, 1, -2)$, radius $\sqrt{17}$ or 4.1231

47 center $\left(-\frac{3}{2}, 0, \frac{11}{2}\right)$, radius $\sqrt{\frac{73}{2}} = \frac{\sqrt{146}}{2} \approx 6.0415$

51

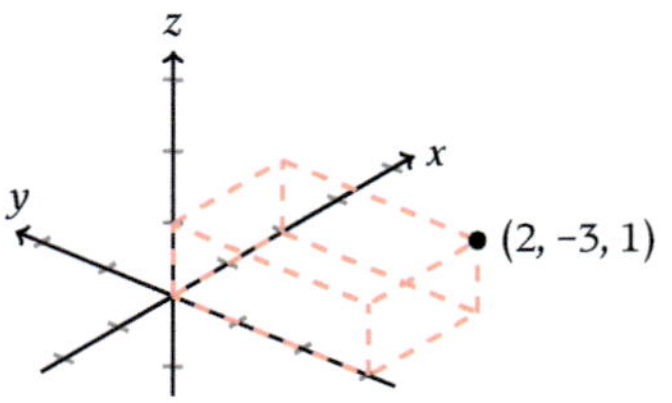

53 shape: cylinder, central axis is the z-axis, radius
infinite height

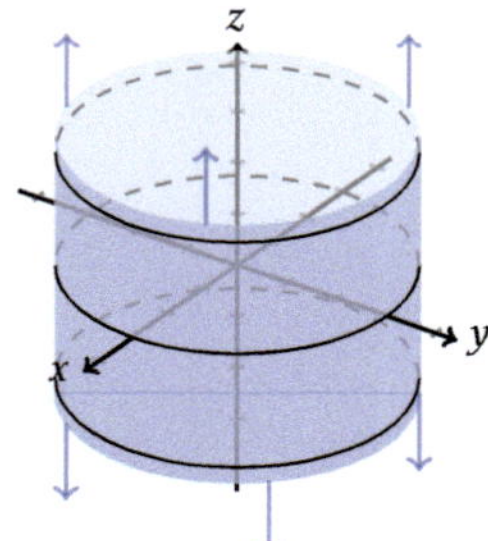

55 equation of x-axis: $y = 0, z = 0$

Section 1.2

1 yes

3 yes

5 no (no direction)

7 no (which endpoint is the initial point and which is
the terminal point is not specified, so the direction
is not given)

9 no (no direction given)

11 yes (the zero vector does not need to specify a direction)

13 (a)

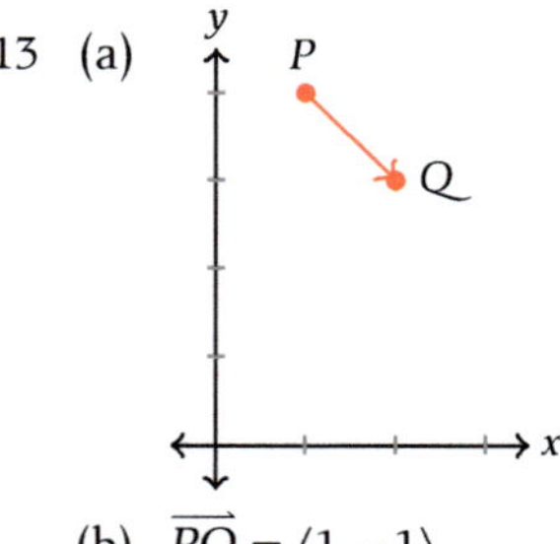

(b) $\overrightarrow{PQ} = \langle 1, -1 \rangle$

15 (a)

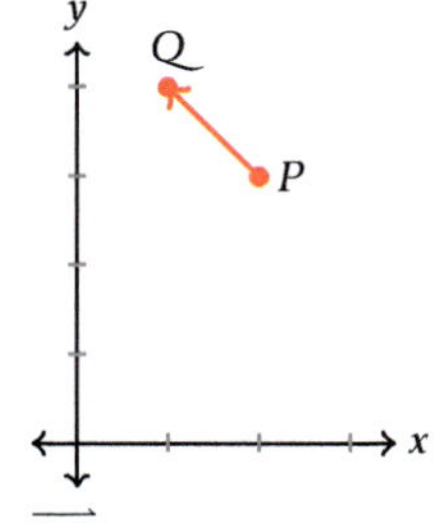

(b) $\overrightarrow{PQ} = \langle -1, 1 \rangle$

17 (a)

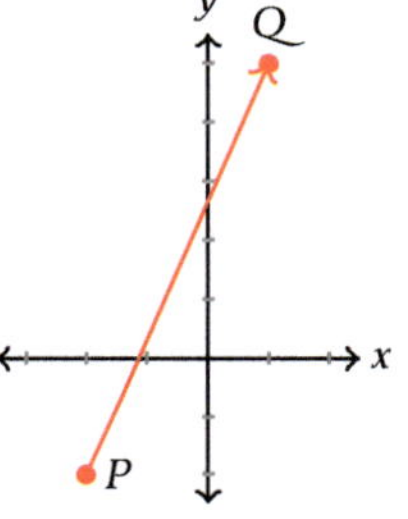

(b) $\overrightarrow{PQ} = \langle 3, 7 \rangle$

19 (a)

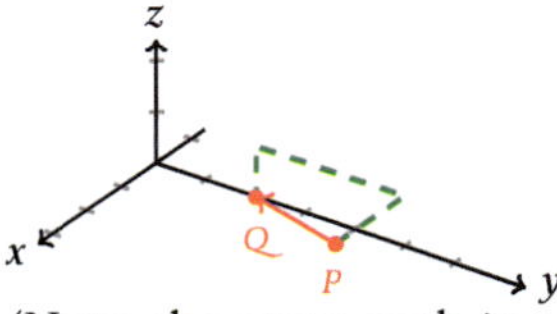

(Note: the green path is not required for the answer; it is provided to help with interpretation as in Figure 1.28.)

(b) $\overrightarrow{PQ} = \langle -2, -3, -1 \rangle$

21 (a)

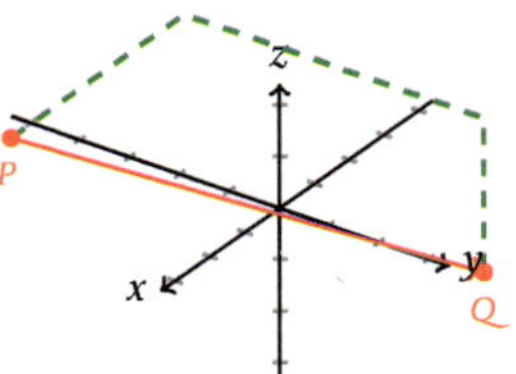

(Note: the green path is not required for the answer; it is provided to help with interpretation as in Figure 1.28.)

(b) $\overrightarrow{PQ} = \langle -5, 6, -3 \rangle$

23 $\|\mathbf{u}\| = 13$

25 $\|\mathbf{w}\| = \sqrt{89} \approx 9.434$

27 $\|\mathbf{b}\| = 5$

29 $\|\mathbf{a}\| = 9$

31 $\|\mathbf{w}\| = 9$

33 $\sqrt{10} \approx 3.162$

35 Answers will vary; one possibility is

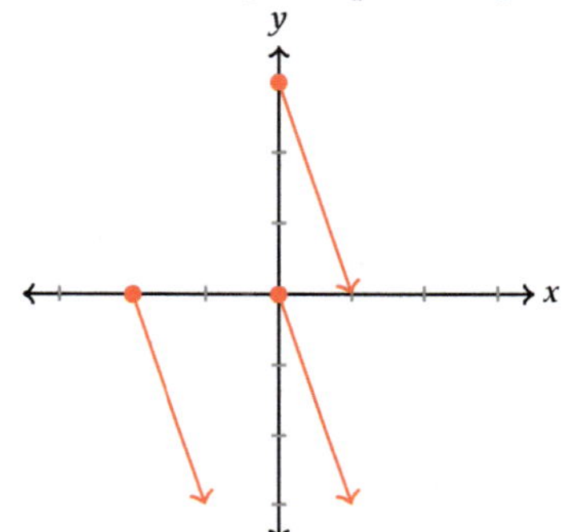

37 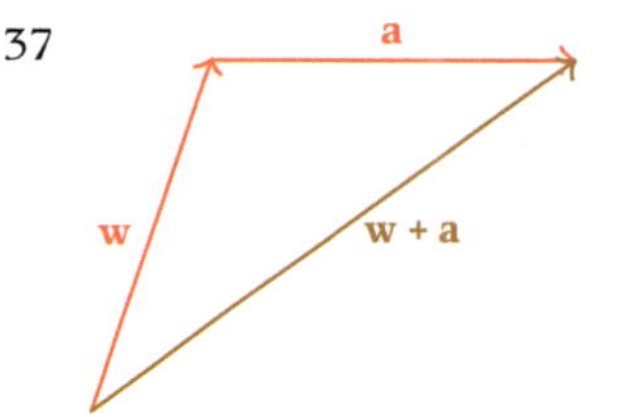

39 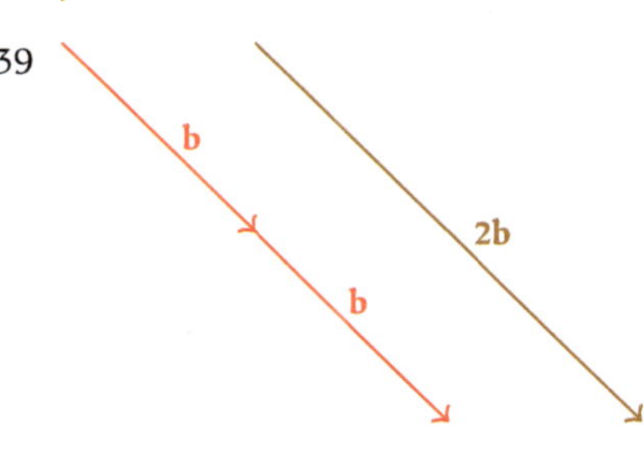

41 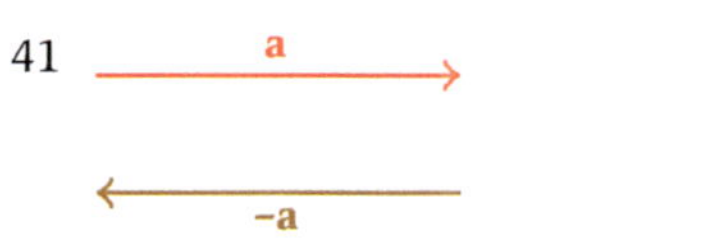

43 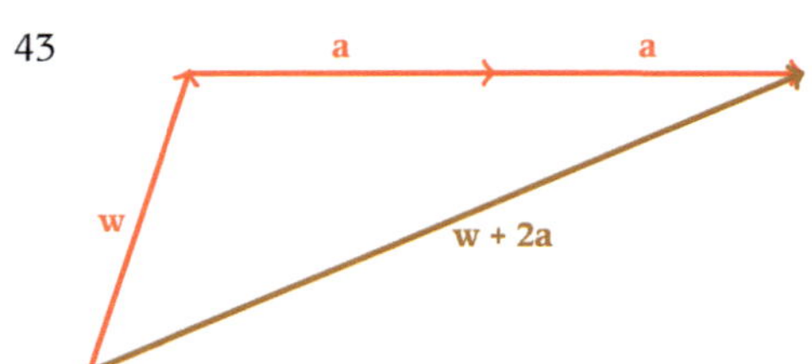

45 0b •

47 **v**

49 $\overrightarrow{AC}$

51 **0**

Section 1.3

1 $\mathbf{u} - \mathbf{v} = \langle 3, 3 \rangle$

3 $\mathbf{v} - \mathbf{u} = \langle -3, -3 \rangle$

5 $3\mathbf{u} - \mathbf{j} = \langle 3, 20 \rangle$

7 $2\mathbf{v} + 3\mathbf{u} - \mathbf{w} = \langle -9, 22 \rangle$ (Hint: the order is switched)

9 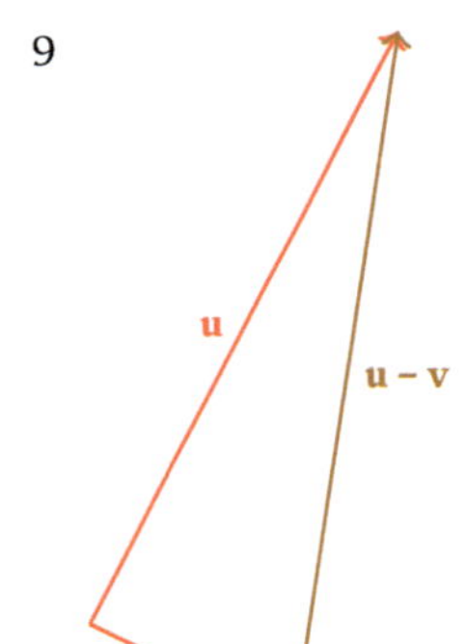

11

53 $\mathbf{u} + \mathbf{w} = \langle 10, -1, 16 \rangle$

55 $-\mathbf{v} = \langle -7, 1.4, 5 \rangle$

57 $\frac{1}{2}\mathbf{w} = \langle 4, 0, 6 \rangle$

59 $2\mathbf{u} + \mathbf{v} = \langle 11, -3.4, 3 \rangle$

61 $\|\mathbf{u}\| = \sqrt{21} \approx 4.583$

63 $\frac{1}{\|\mathbf{u}\|}\mathbf{u} = \left\langle \frac{2}{\sqrt{21}}, \frac{-1}{\sqrt{21}}, \frac{4}{\sqrt{21}} \right\rangle$

65 $\|2\mathbf{u}\| = \sqrt{84} = 2\sqrt{21}$ (compare to the answer to exercise 61)

67 $-3\|\mathbf{u}\| = -3\sqrt{21}$ (compare to the answers to exercises 61 and 65)

13 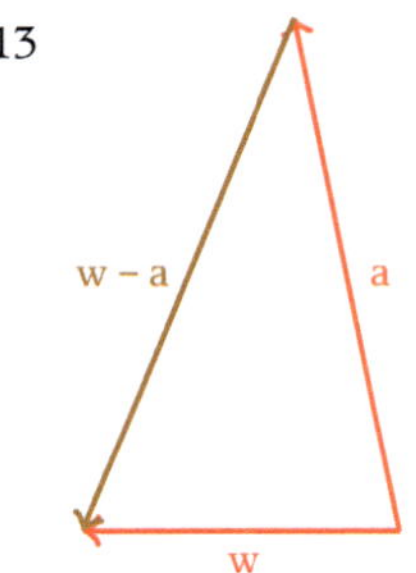

15 $3\mathbf{i} + 9\mathbf{j} - 4\mathbf{k}$

17 $19\mathbf{k}$

19 $2\mathbf{i} - 5\mathbf{j} + 3\mathbf{k}$

21 $\|10\mathbf{k}\| = 10$

23 $\|4\mathbf{i} - 6\mathbf{j}\| = \sqrt{52} = 2\sqrt{13} \approx 7.211$

25 $\| -\mathbf{i} + 5\mathbf{j} + \mathbf{k}\| = \sqrt{27} = 3\sqrt{3} \approx 5.196$

27 yes

29 yes

31 no

33 $\left\langle \frac{3}{\sqrt{35}}, \frac{-1}{\sqrt{35}}, \frac{5}{\sqrt{35}} \right\rangle$

35 $-\mathbf{j}$

37 $\left\langle \frac{-4}{14}, \frac{6}{14}, \frac{-12}{14} \right\rangle = \left\langle \frac{-2}{7}, \frac{3}{7}, \frac{-6}{7} \right\rangle$

39 $\left\langle \frac{3}{\sqrt{58}}, \frac{7}{\sqrt{58}} \right\rangle$

41 $\left\langle \frac{-10}{\sqrt{584}}, \frac{22}{\sqrt{584}} \right\rangle$

43 $5\mathbf{i} + 10\mathbf{j} - 3\mathbf{k}$

45 $7\mathbf{i} + 3\mathbf{j}$ (Hint: the order isn't the same)

47 associative property of vector addition

49 additive inverses exist for vector addition

51 distributive property (real number addition distributes over scalar multiplication)

53 Let $\mathbf{u} = \langle u_1, u_2 \rangle$. Then the left-hand side is

$$(c + d)\mathbf{u} = (c + d)\langle u_1, u_2 \rangle$$
$$= \langle (c + d)u_1, (c + d)u_2 \rangle$$
$$= \langle cu_1 + du_1, cu_2 + du_2 \rangle,$$

and the right-hand side is

$$cu + du = c\langle u_1, u_2 \rangle + d\langle u_1, u_2 \rangle$$
$$= \langle cu_1, cu_2 \rangle + \langle du_1, du_2 \rangle$$
$$= \langle cu_1 + du_1, cu_2 + du_2 \rangle.$$

Because the expressions are identical, $(c + d)\mathbf{u} = c\mathbf{u} + d\mathbf{u}$.

55 Let $\mathbf{u} = \langle u_1, u_2 \rangle$. Then

$$\mathbf{u} + \mathbf{0} = \langle u_1, u_2 \rangle + \langle 0, 0 \rangle$$
$$= \langle u_1 + 0, u_2 + 0 \rangle$$
$$= \langle u_1, u_2 \rangle$$
$$= \mathbf{u}.$$

Therefore $\mathbf{u} + \mathbf{0} = \mathbf{u}$.

57 no (Hint: try, for instance, $\mathbf{u} = \langle 5, 0 \rangle$ and $\mathbf{w} = \langle 0, 2 \rangle$. Calculate $\mathbf{u} - \mathbf{w}$ and then calculate $\|\mathbf{u} - \mathbf{w}\|$.)

Section 1.4

1 scalar

3 vector

5 vector

7 scalar

9 scalar

11 vector

13 the number of components is not the same

15 this is the dot product of a scalar and a vector, which is not defined (the dot product requires two vectors)

17 division by a vector is not defined

19 32

21 17

23 2

25 1

27 0

29 (a) 2.820, (b) 161.57°

31 (a) 0.368, (b) 21.06°

33 (a) 0.711, (b) 40.75°

35 (a) $\frac{\pi}{2} \approx 1.571$, (b) 90°

37 (a) 1.284, (b) 73.57°

39 yes

41 no

43 yes

45 yes

47 (a) $\text{comp}_{\mathbf{u}}\mathbf{v} = 12$, (b) $\text{proj}_{\mathbf{u}}\mathbf{v} = \langle 12, 0 \rangle = 12\mathbf{i}$

49 (a) $\text{comp}_{\mathbf{u}}\mathbf{v} = \frac{55}{\sqrt{50}} = \frac{55}{5\sqrt{2}} = \frac{11}{\sqrt{2}}$,
(b) $\text{proj}_{\mathbf{u}}\mathbf{v} = \langle 5.5, 0, 5.5 \rangle$

51 (a) $\text{comp}_{\mathbf{u}}\mathbf{v} = \frac{33}{\sqrt{18}} = \frac{33}{3\sqrt{2}} = \frac{11}{\sqrt{2}}$,
(b) $\text{proj}_{\mathbf{u}}\mathbf{v} = \langle 5.5, 0, 5.5 \rangle$ (Compare to exercise 49. The vector $\mathbf{v}$ is identical, and the vector $\mathbf{u}$ has the same direction. Therefore, the projection is the same. See Figures 1.49 and 1.50.)

53 (a) $\text{comp}_{\mathbf{u}}\mathbf{v} = \frac{-3}{\sqrt{362}}$, (b) $\text{proj}_{\mathbf{u}}\mathbf{v} = \left\langle \frac{-3}{362}, \frac{-57}{362} \right\rangle$

55 (a) $\text{comp}_{\mathbf{u}}\mathbf{v} = 0$, (b) $\text{proj}_{\mathbf{u}}\mathbf{v} = \mathbf{0}$ (the zero vector; projecting the zero vector doesn't make it longer!)

57 $\alpha = 50.19°$, $\beta = 39.81°$, $\gamma = 90°$

59 $\alpha = 120.47°$, $\beta = 99.73°$, $\gamma = 147.69°$

61 $\alpha = 54.74°, \beta = 54.74°, \gamma = 54.74°$

63   Let $\mathbf{u} = \langle u_1, u_2 \rangle$ and $\mathbf{v} = \langle v_1, v_2 \rangle$. Then the left-hand side is

$$\mathbf{u} \cdot \mathbf{v} = \langle u_1, u_2 \rangle \cdot \langle v_1, v_2 \rangle$$
$$= u_1 v_1 + u_2 v_2$$

and the right-hand side is

$$\mathbf{v} \cdot \mathbf{u} = \langle v_1, v_2 \rangle \cdot \langle u_1, u_2 \rangle$$
$$= v_1 u_1 + v_2 u_2$$
$$= u_1 v_1 + u_2 v_2.$$

Because the expressions are identical, $\mathbf{u} \cdot \mathbf{v} = \mathbf{v} \cdot \mathbf{u}$.

65   Let $\mathbf{u} = \langle u_1, u_2 \rangle$, $\mathbf{v} = \langle v_1, v_2 \rangle$ and $\mathbf{w} = \langle w_1, w_2 \rangle$. Then the left-hand side is

$$(\mathbf{v} - \mathbf{w}) \cdot \mathbf{u} = (\langle v_1, v_2 \rangle - \langle w_1, w_2 \rangle) \cdot \langle u_1, u_2 \rangle$$
$$= \langle v_1 - w_1, v_2 - w_2 \rangle \cdot \langle u_1, u_2 \rangle$$
$$= (v_1 - w_1)u_1 + (v_2 - w_2)u_2$$
$$= v_1 u_1 - w_1 u_1 + v_2 u_2 - w_2 u_2$$

and the right-hand side is

$$\mathbf{v} \cdot \mathbf{u} - \mathbf{w} \cdot \mathbf{u} = \langle v_1, v_2 \rangle \cdot \langle u_1, u_2 \rangle - \langle w_1, w_2 \rangle \cdot \langle u_1, u_2 \rangle$$
$$= v_1 u_1 + v_2 u_2 - (w_1 u_1 + w_2 u_2)$$
$$= v_1 u_1 + v_2 u_2 - w_1 u_1 - w_2 u_2$$
$$= v_1 u_1 - w_1 u_1 + v_2 u_2 - w_2 u_2.$$

Because the expressions are identical, $(\mathbf{v} - \mathbf{w}) \cdot \mathbf{u} = \mathbf{v} \cdot \mathbf{u} - \mathbf{w} \cdot \mathbf{u}.$

67   (a)   Let $\mathbf{a} = \langle a_1, a_2, a_3, a_4 \rangle$ and $\mathbf{b} = \langle b_1, b_2, b_3, b_4 \rangle$. The dot product of $\mathbf{a}$ and $\mathbf{b}$ is

$$\mathbf{a} \cdot \mathbf{b} = a_1 b_1 + a_2 b_2 + a_3 b_3 + a_4 b_4.$$

(b) 23

69   2.67% (proportion: $\dfrac{\text{length of shadow}}{\text{length of w}} \approx 0.0267$)

Section 1.5

1 -5

3 -37

5 $\langle 29, 11, -17 \rangle$

7 $\langle 290, 110, -170 \rangle$ (this is 10 times (scalar mutliplication) the answer to exercise 5; see property of the cross product (2))

9 $\langle -1, -1, -1 \rangle$

11 $\langle 0, 0, 0 \rangle$ (a vector cross product with itself is the zero vector $\mathbf{0}$; see property of the cross product (5))

13 $\langle -32, -20, -11 \rangle$

15 -15

17 -324 (if your answer is 324, you may have switched the order of $\mathbf{u}$, $\mathbf{v}$, and $\mathbf{w}$)

19 $\langle -25, 34, 22 \rangle$ or any scalar multiple thereof (including $\langle 25, -34, -22 \rangle$)

21 $\langle 0, 36, -8 \rangle$ or any scalar multiple thereof (including $\langle 0, -36, 8 \rangle$)

23 $\langle 0, 0, 17 \rangle$ or any scalar multiple thereof, including k which could be noticed immediately without calculation (the instructions cannot be accomplished in two dimensions, so it must be the case that three dimensions is implied)

25   No, because their cross product is not the zero vector   ($\langle 8, 11, 15 \rangle \times \langle -4, -5, -7 \rangle = \langle -2, -4, 4 \rangle$;   once you have calculated one nonzero component, the answer to the question is "no" and it is not necessary to finish calculating the other components).

27   Yes, because their cross product is the zero vector (Alternately, because $\langle 6, 10.5, -3 \rangle = 1.5 \langle 4, 7, -2 \rangle$ so they have the same direction.)

29   Yes, the three vectors are coplanar (the scalar triple product is 0). (Alternately, because $\frac{1}{2} \langle 2, 8, 8 \rangle + \frac{1}{3} \langle 3, -12, 9 \rangle = \langle 2, 0, 7 \rangle$)

31   No, the three vectors are not coplanar (their scalar triple product is $-0.27 \neq 0$).

33 $\sqrt{1273}$ units$^2 \approx 35.68$ units^2

35 2 units^2 (this one can be accomplished by recognizing that the parallelogram is a rectangle with side lengths 1 and 2)

37 $\frac{1}{2}\sqrt{1273}$ units$^2 \approx 17.84$ units^2

39 $\frac{1}{2}\sqrt{3860}$ units$^2 = \sqrt{965}$ units$^2 \approx 31.06$ units^2

41 1 unit^3

43 18 units^3

45 347.7 ft·lb

47 296.7 N·m

49 Let $\mathbf{u} = \langle u_1, u_2, u_3 \rangle, \mathbf{v} = \langle v_1, v_2, v_3 \rangle,$ and $\mathbf{w} = \langle w_1, w_2, w_3 \rangle$. Then the left-hand side is

$$\langle u_1, u_2, u_3 \rangle \cdot (\langle v_1, v_2, v_3 \rangle \times \langle w_1, w_2, w_3 \rangle)$$
$$= \langle u_1, u_2, u_3 \rangle \cdot \langle v_2 w_3 - v_3 w_2,$$
$$v_3 w_1 - v_1 w_3, v_1 w_2 - v_2 w_1 \rangle$$
$$= u_1(v_2 w_3 - v_3 w_2) + u_2(v_3 w_1 - v_1 w_3)$$
$$+ u_3(v_1 w_2 - v_2 w_1)$$
$$= u_1 v_2 w_3 - u_1 v_3 w_2 + u_2 v_3 w_1 - u_2 v_1 w_3$$
$$+ u_3 v_1 w_2 - u_3 v_2 w_1,$$

and the right-hand side is

$$(\langle u_1, u_2, u_3 \rangle \times \langle v_1, v_2, v_3 \rangle) \cdot \langle w_1, w_2, w_3 \rangle$$
$$= \langle u_2 v_3 - u_3 v_2, u_3 v_1 - u_1 v_3, u_1 v_2 - u_2 v_1 \rangle$$
$$\cdot \langle w_1, w_2, w_3 \rangle$$
$$= (u_2 v_3 - u_3 v_2)w_1 + (u_3 v_1 - u_1 v_3)w_2$$
$$+ (u_1 v_2 - u_2 v_1)w_3$$
$$= u_2 v_3 w_1 - u_3 v_2 w_1 + u_3 v_1 w_2 - u_1 v_3 w_2$$
$$+ u_1 v_2 w_3 - u_2 v_1 w_3$$
$$= u_1 v_2 w_3 - u_1 v_3 w_2 + u_2 v_3 w_1 - u_2 v_1 w_3$$
$$+ u_3 v_1 w_2 - u_3 v_2 w_1.$$

Because the two expressions are identical, $\mathbf{u} \cdot (\mathbf{v} \times \mathbf{w}) = (\mathbf{u} \times \mathbf{v}) \cdot \mathbf{w}$.

51 Answers will vary; nearly any choice of vectors will do. One possibility is $\mathbf{a} = \langle 1, 0, 0 \rangle, \mathbf{b} = \langle 0, 1, 0 \rangle,$ and $\mathbf{c} = \langle 2, 3, 4 \rangle$. (Then $(\mathbf{a} \times \mathbf{b}) \times \mathbf{c} = \langle -3, 2, 0 \rangle$ and $\mathbf{a} \times (\mathbf{b} \times \mathbf{c}) = \langle 0, 2, 0 \rangle$.)

53 Let $\mathbf{u} = \langle u_1, u_2, u_3 \rangle,$ $\mathbf{v} = \langle v_1, v_2, v_3 \rangle,$ and $\mathbf{w} = \langle w_1, w_2, w_3 \rangle$. Then the left-hand side is

$$\mathbf{u} \cdot (\mathbf{v} \times \mathbf{w})$$
$$= \langle u_1, u_2, u_3 \rangle \cdot (\langle v_1, v_2, v_3 \rangle \times \langle w_1, w_2, w_3 \rangle)$$
$$= \langle u_1, u_2, u_3 \rangle$$
$$\cdot \langle v_2 w_3 - v_3 w_2, v_3 w_1 - v_1 w_3, v_1 w_2 - v_2 w_1 \rangle$$
$$= u_1(v_2 w_3 - v_3 w_2) + u_2(v_3 w_1 - v_1 w_3)$$
$$+ u_3(v_1 w_2 - v_2 w_1),$$

and the right-hand side is

$$\begin{vmatrix} u_1 & u_2 & u_3 \\ v_1 & v_2 & v_3 \\ w_1 & w_2 & w_3 \end{vmatrix}$$
$$= u_1 \begin{vmatrix} v_2 & v_3 \\ w_2 & w_3 \end{vmatrix} - u_2 \begin{vmatrix} v_1 & v_3 \\ w_1 & w_3 \end{vmatrix} + u_3 \begin{vmatrix} v_1 & v_2 \\ w_1 & w_3 \end{vmatrix}$$
$$= u_1(v_2 w_3 - v_3 w_2) - u_2(v_1 w_3 - v_3 w_1)$$
$$+ u_3(v_1 w_2 - v_2 w_1)$$
$$= u_1(v_2 w_3 - v_3 w_2) + u_2(-v_1 w_3 + v_3 w_1)$$
$$+ u_3(v_1 w_2 - v_2 w_1).$$

Because the two expressions are equal,
$$\mathbf{u} \cdot (\mathbf{v} \times \mathbf{w}) = \begin{vmatrix} u_1 & u_2 & u_3 \\ v_1 & v_2 & v_3 \\ w_1 & w_2 & w_3 \end{vmatrix}.$$

55 They must all be orthogonal to each other and have the same magnitude. (Note: to say that one vector must be the cross product of the other two is insufficient, for (1) the other two might not be orthogonal and (2) the magnitude of the cross product might not be correct; for instance, $2\mathbf{i} \times 2\mathbf{j} = 4\mathbf{k}$, and it would not be a cube.)

57 wrenches twice as long

59 $|v_1 w_2 - v_2 w_1|$

Section 1.6

Note: answers to exercises 1–10 will vary depending on point chosen, direction vector chosen, and sometimes other factors as well. In any case, the direction vector should be parallel to the one given in these answers.

1 (a) $\langle 2, 4, 7 \rangle + t\langle -3, 0, 9 \rangle$; (b) $x = 2 - 3t$, $y = 4$, $z = 7 + 9t$

3 (a) $\langle 0, 0, 0 \rangle + t\langle 1, 3, 7 \rangle$; (b) $x = t, y = 3t, z = 7t$

5 (a) $\langle 8, 0, -3 \rangle + t\langle -10, 5, 8 \rangle$; (b) $x = 8 - 10t, y = 5t, z = -3 + 8t$

7 (a) $\langle 9, 1, -7 \rangle + t\langle -6, -4, -2 \rangle$; (b) $x = 9 - 6t$, $y = 1 - 4t, z = -7 - 2t$

9 (a) $\langle 3, 2, 1 \rangle + t\langle -19, 9, -1 \rangle$; (b) $x = 3 - 19t$, $y = 2 + 9t, z = 1 - t$

Note: answers to exercises 11–18 will vary depending on the point chosen for the position vector and the points chosen to calculate direction vectors, and sometimes other factors as well.

11 (a) $\langle -2, 4, -1 \rangle + t\langle 2, 5, 10 \rangle + s\langle 5, 1, 12 \rangle$;
(b) $50(x - 0) + 26(y - 9) - 23(z - 9) = 0$

13 (a) $2(x - 3) + 9y - 6z = 0$; (b) $x = 3 - 3t - 3s$, $y = \frac{2}{3}s, z = -t$ (Note: if your answer differs, see if the points $(0, 0, -1)$ and $(0, \frac{2}{3}, 0)$ can result from your equations.)

15 $x = -2t + 2s, y = 1 + 3t + s, z = 5 + 4t - 3s$

17 possible answers include parametric: $x = 1 - t + s$, $y = 2t + 7s$, $z = 5 - 3t - 3s$; vector: $\langle 2, 7, 2 \rangle + t\langle -1, 2, -3 \rangle + s\langle 1, 7, -3 \rangle$; and Cartesian $15(x - 1) - 6y - 9(z - 5) = 0$

19 $\frac{\sqrt{5210}}{\sqrt{99}}$ units ≈ 7.254 units (Hint: you should get $\overrightarrow{QP} \times \mathbf{w} = \langle -45, -28, 49 \rangle$)

21 $\sqrt{68}$ units ≈ 8.246 units (Hint: $\overrightarrow{QP} \times \mathbf{w} = \langle 8, -2, 0 \rangle$)

23 $\sqrt{13}$ units ≈ 3.606 units (Hint: $\overrightarrow{QP} \times \mathbf{w} = \langle -3, 2, 0 \rangle$)

25 $\left(\frac{23}{7}, \frac{30}{7}, \frac{47}{7} \right)$

27 $\left(0, 0, \frac{596}{7} \right)$

29 yes, the planes are parallel (The point $(8, 1, -5)$ is in the first plane but not the second plane, so the planes are not identical.)

31 no, the planes are not parallel

33 parallel (The point $(0, 5, 6)$ is in the first line but not the second line, so the lines are not identical.)

35 skew

37 intersecting (at the point $(5, 2, 3)$)

39 (a) 0.8768 radians; (b) 50.24°

41 (a) 1.134 radians; (b) 64.98° (if your answers are 115.02° and 2.008 radians, your angle is obtuse. Change one of your normal vectors to the opposite direction.)

43 use different points on the line such as $(3, -6, 0)$ or $(4, -9, -2)$ to get $\mathbf{v} + t\mathbf{w} = \langle 3 + t, -6 - 3t, 0 - 2t \rangle$ or $\mathbf{v} + t\mathbf{w} = \langle 4 + t, -9 - 3t, -2 - 2t \rangle$, or take scalar multiple of the direction vector such as $\langle 2, -6, -4 \rangle$ to get $\mathbf{v} + t\mathbf{w} = \langle 1 + 2t, -6t, 4 - 4t \rangle$.

45 yes; one possible answer is $\mathbf{v} + t\mathbf{w} + s\mathbf{u} = \langle 1 - t + 8s, 2 - 4t - 2s, 3 - 4t, 4 + t + 2s \rangle$; answers vary depending on which points are chosen to calculate direction vectors, etc.

47 Note: answers will vary based on which point, direction vector is chosen in (a), which affects later answers as well. But in (d), your direction vector should be parallel to the direction vector shown.
(a) $x = 1 + t$, $y = 2 - t$, $z = 2 - 2t$; (b) $(0, 3, 4)$
(c) $x = 1 + 3s$, $y = 2$, $z = 2 + s$, point $\left(\frac{22}{13}, 2, \frac{29}{13} \right)$
(d) $\left\langle \frac{22}{13}t, 3 - t, 4 - \frac{23}{13}t \right\rangle$

49 highlights of derivation: choose Q in plane, such as $\left(0, 0, -\frac{k}{c} \right)$. Then $\overrightarrow{QP} = \left\langle x_0, y_0, z_0 + \frac{k}{c} \right\rangle$. Finally,

$$d = \left| \text{comp}_{\mathbf{n}} \overrightarrow{QP} \right|$$
$$= \left| \frac{\overrightarrow{QP} \cdot \mathbf{n}}{\|\mathbf{n}\|} \right|$$
$$= \left| \frac{ax_0 + by_0 + c(z_0 + \frac{k}{c})}{\sqrt{a^2 + b^2 + c^2}} \right|$$
$$= \frac{|ax_0 + by_0 + cz_0 + k|}{\sqrt{a^2 + b^2 + c^2}}.$$

51 $\frac{4}{\sqrt{134}}$ units ≈ 0.3455 units

53 $\frac{19}{\sqrt{78}}$ units ≈ 2.151 units

55 $\dfrac{7}{\sqrt{56}} = \dfrac{\sqrt{14}}{4}$ units ≈ 0.9354 units (Hint: the distance between two parallel planes is the same at all points, so pick a point in one plane and find its distance to the other plane.)

Section 1.7

1 elliptic paraboloid

3 hyperboloid of two sheets

5 elliptic cone

7 elliptic cone

9 hyperboloid of two sheets

11 not applicable; there are no solutions to this equation

13 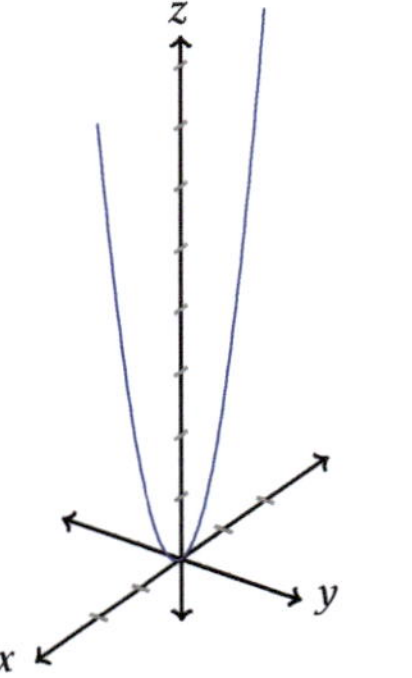

15 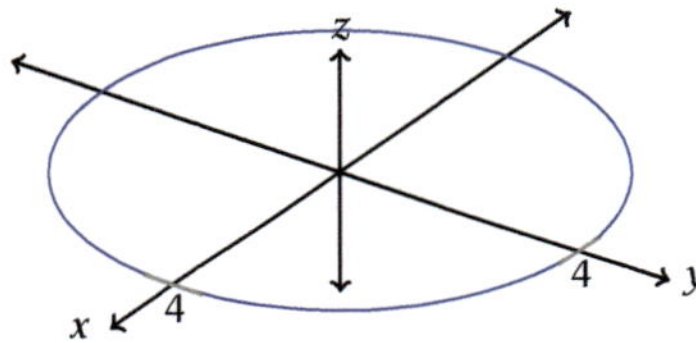

17 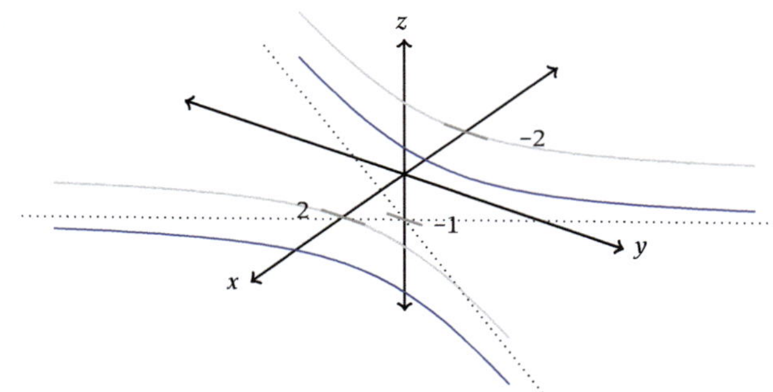

19 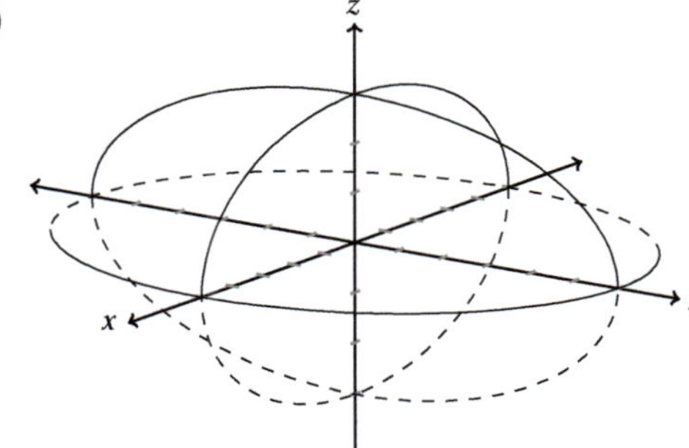

21 (a) ellipsoid;

(b) 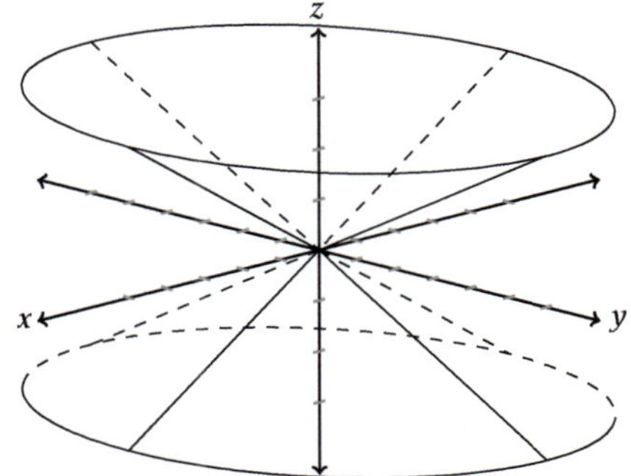

23 (a) elliptic cone;

(b)

25 (a) hyperbolic paraboloid;

(b)

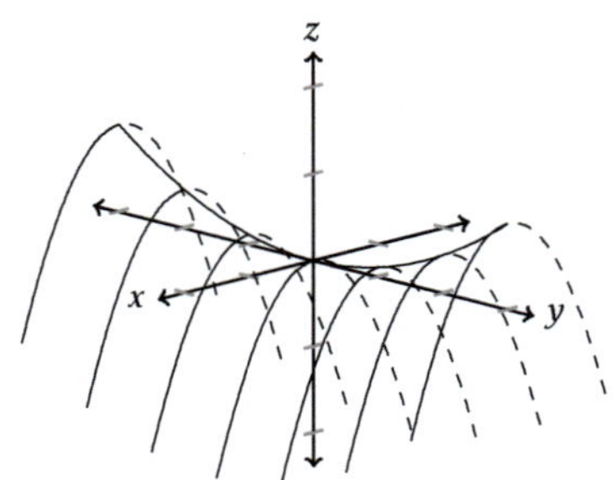

27 (a) elliptic paraboloid (opening in the positive *x*-direction;

(b)

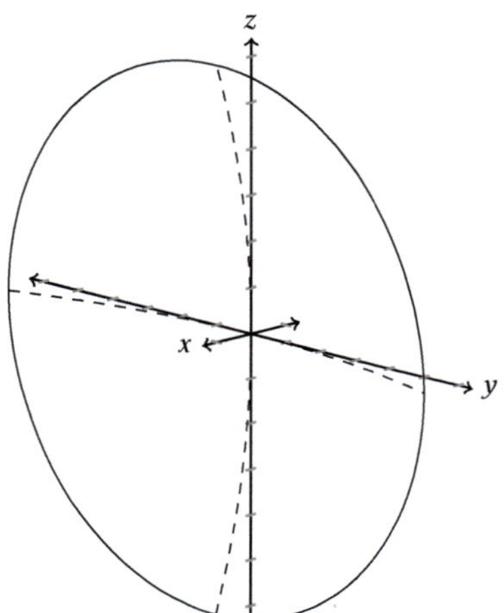

29 (a) hyperboloid of two sheets (opening in the *y*-direction);

(b)

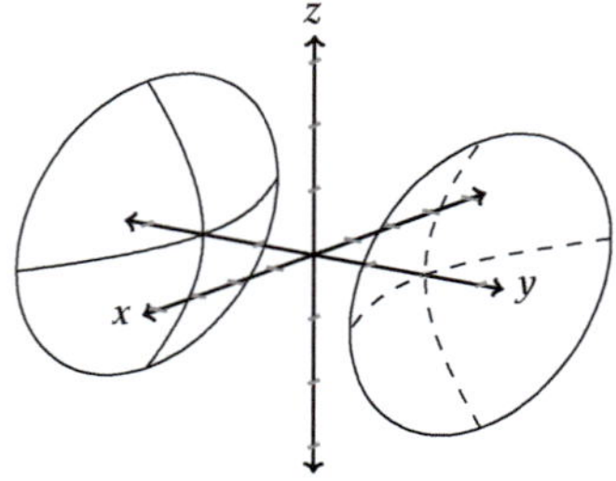

31 (a) hyperboloid of one sheet;

(b)

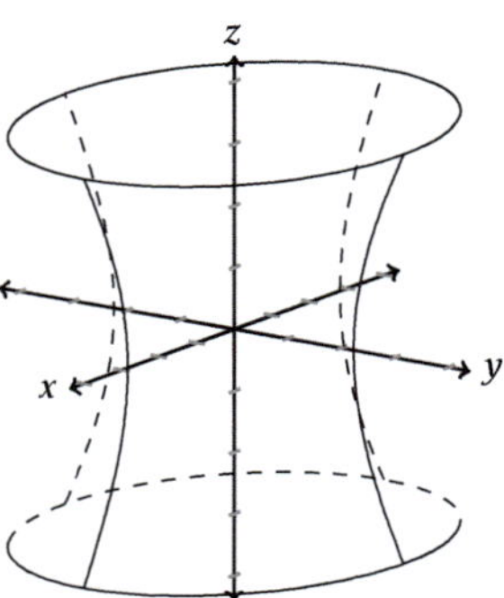

33 (a) ellipsoid (sphere) centered at $(-2, 1, -2)$;

(b) version 1

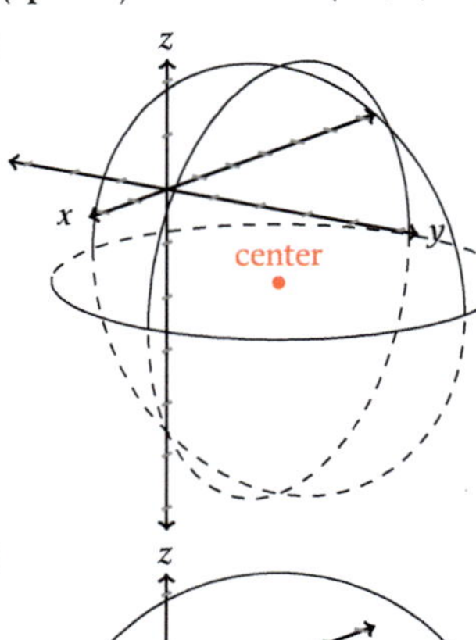

version 2

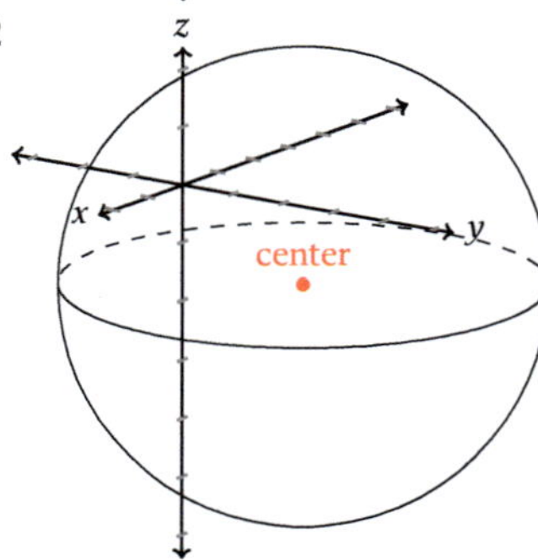

35 (a)

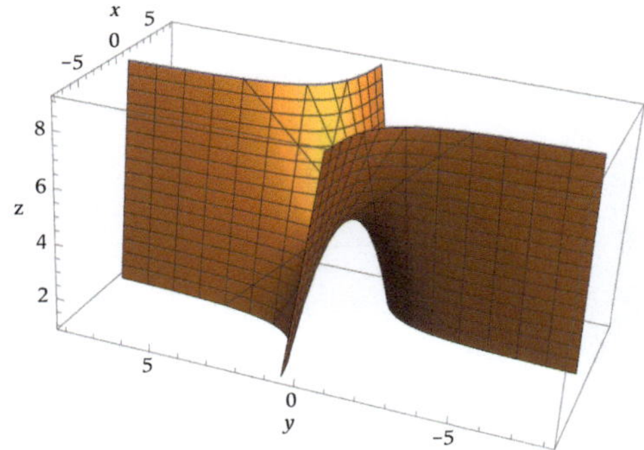

(b) hyperbolic paraboloid

37 (a)

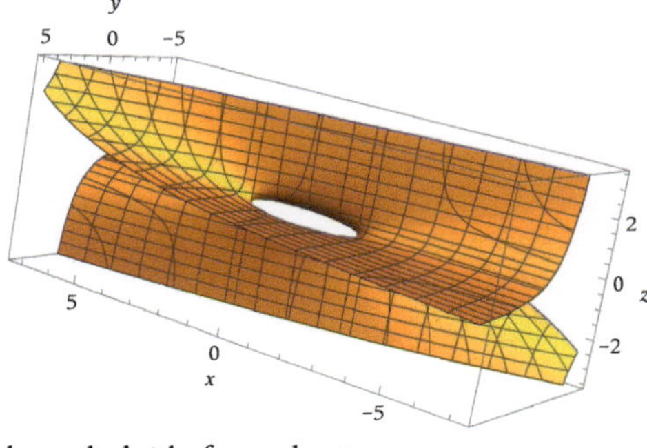

(b) hyperboloid of one sheet

39 (a)

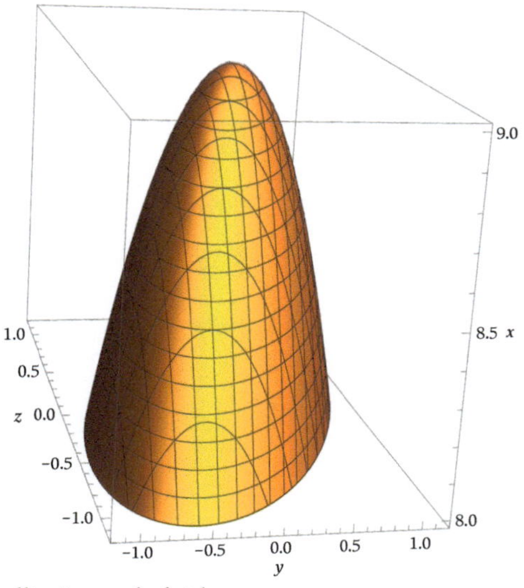

(b) elliptic paraboloid

41 (a)

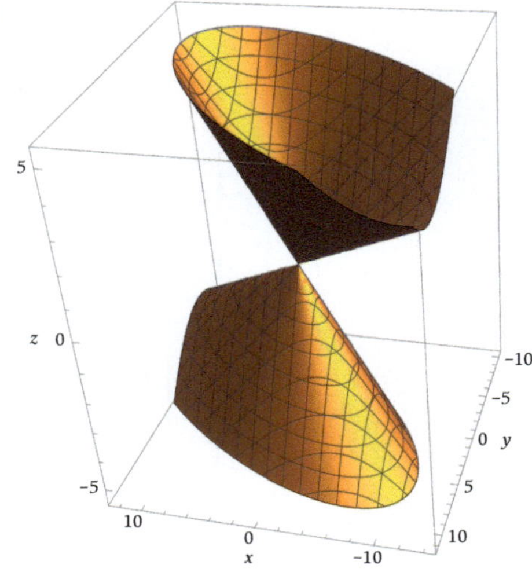

(b) elliptic cone

43 (a) version 1

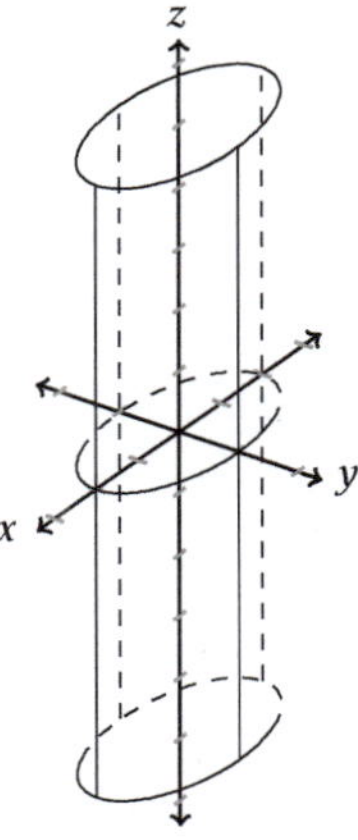

version 2

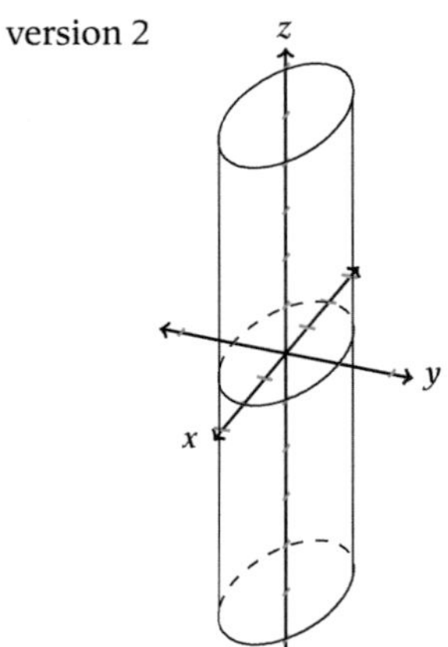

(b) yes

45 (a)

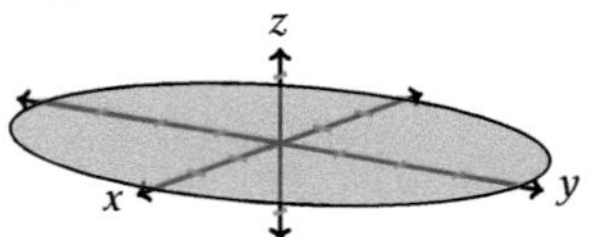

(b) no

47 center $(-1, 2, -4)$

49 yes; hyperbolic paraboloid

51 The cross section $y = k$ has the equation $\frac{x^2}{a^2} + \frac{k^2}{b^2} - \frac{z^2}{c^2} = 1$, which can be rearranged to
$$\frac{x^2}{a^2} - \frac{z^2}{c^2} = 1 - \frac{k^2}{b^2}.$$

This is a quadratic equation in two variables (x and z) with squared terms for both variables with different signs, so it is a hyperbola. If $k > b$, then $\frac{k^2}{b^2} > 1$ and $1 - \frac{k^2}{b^2} < 0$, and the coefficient of z^2 is also negative, so the hyperbola opens in the z-direction. If $k < b$, then $\frac{k^2}{b^2} < 1$ and $1 - \frac{k^2}{b^2} > 0$, and the coefficient of x^2 is also positive, so the hyperbola opens in the x-direction.

53 All the coefficients on the terms with variables are positive, so swapping the roles of the variables doesn't change anything.

55 Try $\frac{x^2}{9} + \frac{y^2}{16} + \frac{z^2}{\omega^2} = 1$. It is the ellipse $z = 0$, $\frac{x^2}{9} + \frac{y^2}{16} = 1$ and its interior.

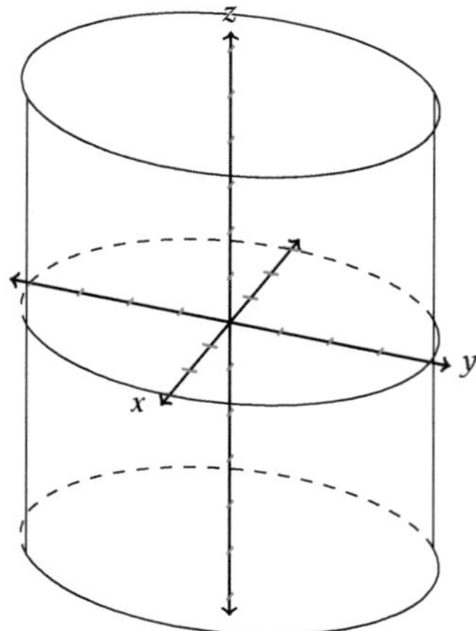

Why? Any value of z would have to satisfy $|z| < \omega$, so $z \doteq 0$. If $z = \sqrt{k}\omega$ with $0 \leq k \leq 1$, then $\frac{x^2}{9} + \frac{y^2}{16} + \frac{k\omega^2}{\omega^2} = 1$, which can be rearranged to
$$\frac{x^2}{9(1-k)} + \frac{y^2}{16(1-k)} = 1,$$

so any interior point can be achieved.

57 Try $\frac{x^2}{9} + \frac{y^2}{16} - \frac{z^2}{\Omega^2} = 1$. For any finite k, the slice $z = k$ has equation
$$\frac{x^2}{9} + \frac{y^2}{16} = 1 + \frac{k^2}{\Omega^2}.$$

Because k is finite (a real number), $\frac{k^2}{\Omega^2}$ is infinitesimal and $1 + \frac{k^2}{\Omega^2} \approx 1$. Thus, we have (approximately) the same ellipse $\frac{x^2}{9} + \frac{y^2}{16} = 1$ in every horizontal cross section.

It is approximately the cylinder $\frac{x^2}{9} + \frac{y^2}{16} = 1$.

59 (a) $y = x$, a line with slope 1 (in the plane $z = 0$)
(b) traces $z = \pm 1$: $y = x + 1$, a line with slope

traces $z = \pm 2$: $y = x + 4$, a line with slope 1; (c) yes; yes; (d)

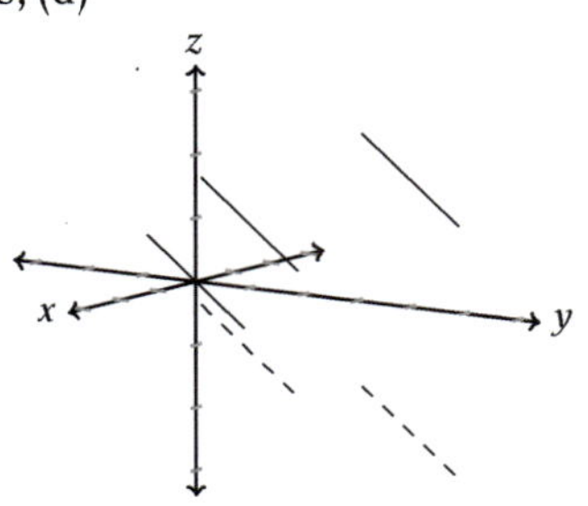

Chapter 2

Section 2.1

1 no

3 yes

5 $(0, 1]$

7 $(-\infty, -\frac{1}{3}) \cup (-\frac{1}{3}, 3]$

9 $\mathbf{R}$ (same as $(-\infty, \infty)$)

11 $\langle -\infty, 0, 6 \rangle$

13 $\langle 1, 0, \infty \rangle$

15 $\langle \frac{11}{3}, \infty, \infty \rangle$

17 (a) two asymptotes: (1) in z-direction, equation $x = 0$, $y = 1$ (from $t \to 0$); (2) in x-direction, equation $y = 0$, $z = 0$ (from $t \to -\infty$);

(b) none

19 (a) two asymptotes: (1) in x-direction, equation $y = 3$, $z = 0$ (from $t \to 0$); (2) in y-direction, equation $x = 0$, $z = \frac{1}{2}$ (from $t \to -1$);

(b) stall point $(1, 2, 1)$

21 (a) one asymptote: in x-direction, equation $y = \frac{1}{17}$, $z = \sin 4$ (from $t \to 4$);

(b) none (recall that stall points are only as $t \to \pm\infty$; also $\sin \Omega$ can be any value between -1 and 1)

61 An elliptic paraboloid of height h, semimajor axis a, and semiminor axis b has parametric equations $x = a\sqrt{u}\cos v$, $y = b\sqrt{u}\sin v$, $z = u$, for $0 \le v < 2\pi$ and $0 \le u \le h$. Using $a = 4$, $b = 5$, and choosing $h = 5$ yields the graph included here.

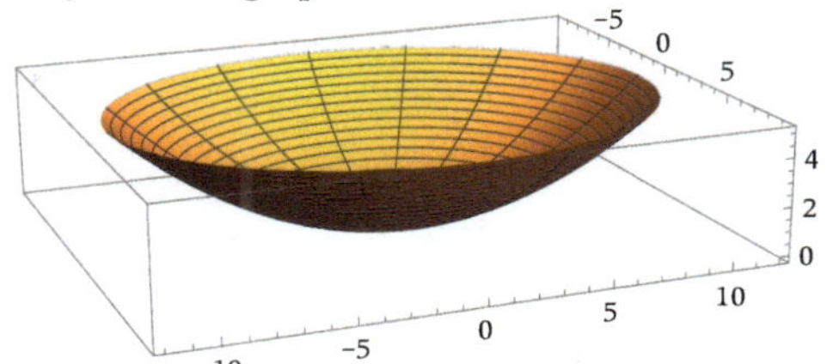

23 (a) two asymptotes: (1) in y-direction, equation $x = 0$, $z = 1$ (from $t \to 0$); (2) in x-direction, equation $y = 0$, $z = 0$ (from $t \to \infty$); notice domain is $t > 0$ so we can't do $t \to -\infty$;

(b) none

25 (a) two asymptotes: (1) in x-direction, equation $y = -1$, $z = 109$ (from $t \to 2^+$); (2) in x-direction, equation $y = 1$, $z = 65$ (from $t \to -2^-$); notice that $\cos((2+\Omega)\frac{\pi}{4})$ can be any number between -1 and 1;

(b) none

27 (a) (Hint: plot points for $t = 0, t = \frac{\pi}{6}, ..., t = 2\pi$)

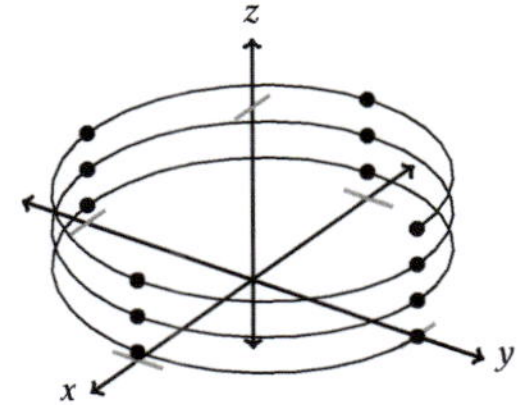

(b)

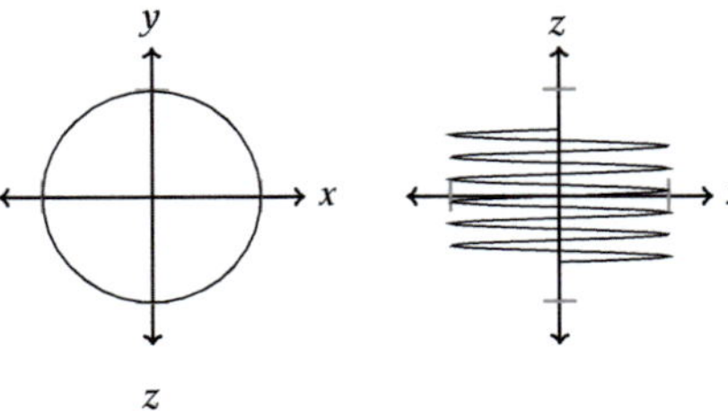

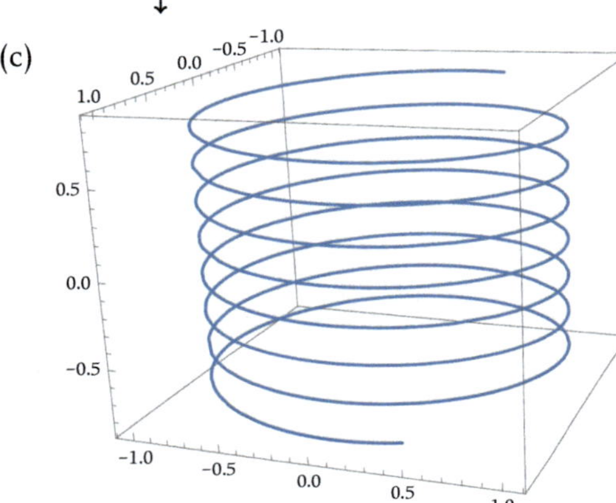

(c)

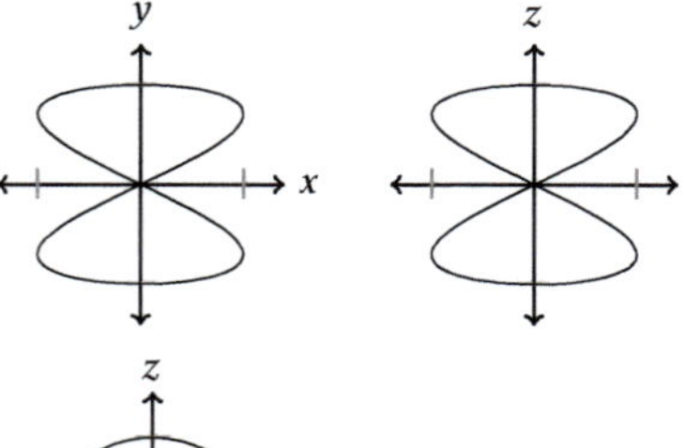

(d) an infinite Slinky (answers may vary); a name
of this curve is a *helix.*

29 (a)

(b)

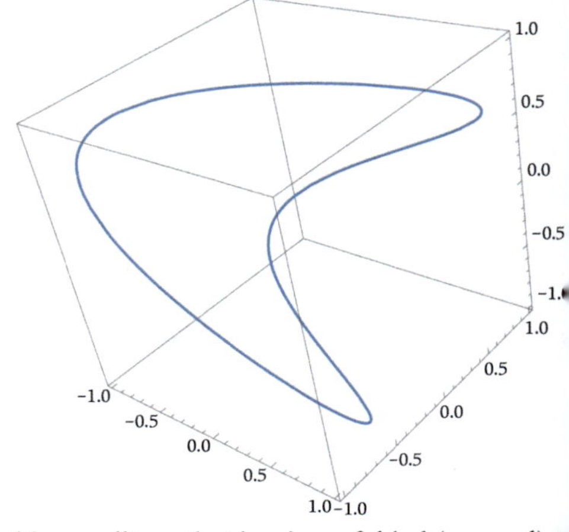

(c) like an ellipse that has been folded (warped) i
the middle, along the minor axis (answers ma
vary)

31 (a)

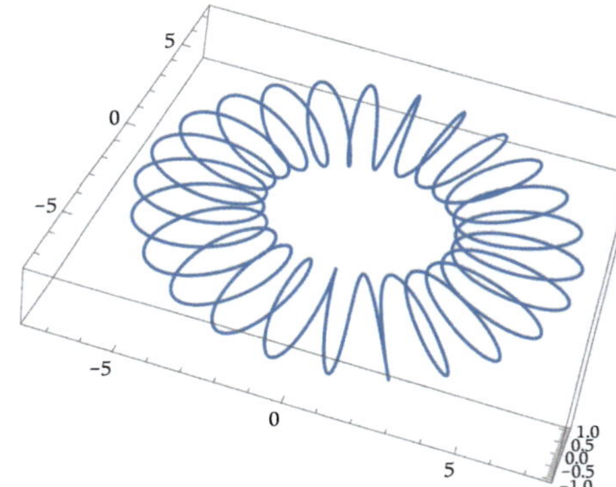

(b) like a Slinky whose ends have been put togeth
er (answers may vary); a name of this curve
a *toroidal spiral.*

33 (a)

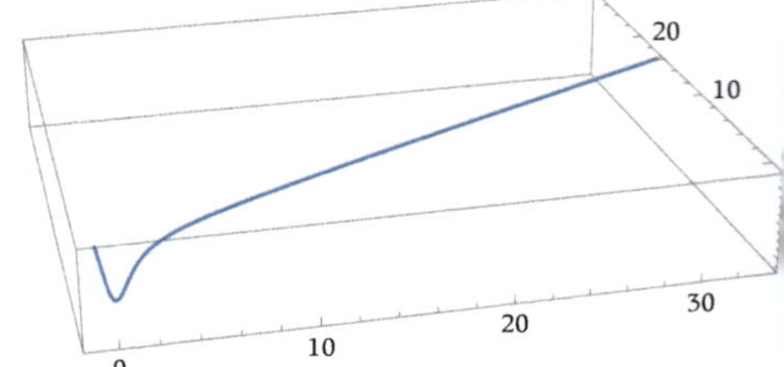

(b) like a carving tool, dental hygienist's tool,
hollowing tool for a lathe, but without the ha
dle (answers may vary)

35 yes; $\langle 0,0,0\rangle + t\langle 9,2,0\rangle$ (Note: any scalar multiple of the direction vector will work)

37 yes; $\langle 0,0,5\rangle + t\langle 1,\frac{1}{2},0\rangle$ (Hint: use definition of cosh; any scalar multiple of the direction vector will work)

39 no (Neither $\langle 0,3,0\rangle + \Omega\langle\Omega,0,1\rangle$ nor $\langle 0,3,0\rangle + \Omega^2\langle 1,0,\frac{1}{\Omega}\rangle$ work because one component of the direction vector is not a real number. But we could do $\langle 0,3,0\rangle + \langle t^2,0,t\rangle$ with $t = \Omega$, which is the equation of the parabola $x = z^2$ in the plane $y = 0$, which could be called a parabolic asymptote.)

41 (a) $\langle 0,12,95\rangle + 5A^3\langle 1,0,0\rangle$;

  (b) $\langle 1,0,0\rangle$; yes

Section 2.2

1 $(-\infty, -\pi) \cup (-\pi, 2)$

3 $[-3,-2] \cup [2,3]$

5 $\mathbf{r}'(t) = \left(\sec(4t+1)\tan(4t+1)\cdot 4, \ \dfrac{(t^2+6)(3)-(3t-7)(2t)}{(t^2+6)^2}, \ 2^{1-t}(\ln 2)(-1)\right) = \left(4\sec(4t+1)\tan(4t+1), \ \dfrac{-3t^2+14t+18}{(t^2+6)^2}, \ 2^{1-t}(-\ln 2)\right)$

7 $\mathbf{r}'(t) = \mathbf{i}(3\cosh 3t) + \mathbf{j}\left(\dfrac{1}{\sqrt{1-t^2}}\right) - \mathbf{k}\left(\dfrac{3}{3t-1}\right)$

9 $\left(3,4,-\frac{1}{4}\right)$

11 $\langle -1,0,3\rangle$

13 $\langle 6,15,125\rangle + t\langle 1,3,75\rangle$

15 $\langle 1,1,0\rangle + t\langle 2,5,7\rangle$

17 $x = 0, y = -1 + 3t, z = \frac{1}{3} + \left(\frac{1}{3}\ln 3\right)t$

19 $\sqrt{229.25}$ m/s ≈ 15.141 m/s

21 (a) $\mathbf{T}(t) = \left(\dfrac{2t+3}{\sqrt{4t^2+12t+61}}, \ \dfrac{4}{\sqrt{4t^2+12t+61}}, \ \dfrac{6}{\sqrt{4t^2+12t+61}}\right)$;

  (b) $\mathbf{T}(0) = \left(\dfrac{3}{\sqrt{61}}, \dfrac{4}{\sqrt{61}}, \dfrac{6}{\sqrt{61}}\right)$

23 (a) $\mathbf{T}(t) = \left(\dfrac{2\sqrt{t}}{|t+2|}, \dfrac{t}{|t+2|}, \dfrac{2}{|t+2|}\right)$

  (Hint: $\sqrt{4t + t^2 + 4} = \sqrt{(t+2)^2}$);

  (b) $\mathbf{T}(0) = \langle 0,0,1\rangle$

25 $\left(\dfrac{-2}{\sqrt{29}}, \dfrac{4}{\sqrt{29}}, \dfrac{3}{\sqrt{29}}\right)$

27 $\dfrac{2}{t} + 5t \cdot 2^t \ln 2 + 5 \cdot 2^t + 10t^2 \sec t\tan t + 20t \sec t$

  $= \dfrac{2}{t} + 2^t(5t\ln 2 + 5) + \sec t(10t^2\tan t + 20t)$

29 $\mathbf{i}\left(5\sec t + 5t\sec t\tan t - (20t\cdot 2^t + 10t^2 \cdot 2^t\ln 2)\right) - \mathbf{j}\left(2\sec t\tan t - (20t\ln t + 10t)\right) + \mathbf{k}\left(2^{t+1}\ln 2 - (5\ln t + 5)\right)$

31 $\left(-\cos t, \ \dfrac{-\cos 2t}{2}, \ \dfrac{\sin 3t}{3}\right) + \mathbf{C}$

33 $\left(\tan^{-1} t, \ \sin^{-1} t, \ \frac{1}{2}\ln(1 + t^2)\right) + \mathbf{C}$

35 $\langle 42, 42, 24\rangle$

37 $\left(\dfrac{2^{5/2}}{\frac{5}{2}}, e^2 + 1, 2\right) = \left(\frac{8}{5}\sqrt{2}, e^2 + 1, 2\right)$

39 $\mathbf{r}(t) = \left(\dfrac{3t^2}{2} + t - \frac{1}{2}, \ 2t^3 + \frac{5}{2}t^2 + \frac{1}{2}, \ \frac{2}{3}t^{3/2} + \frac{31}{3}\right)$

41 $\mathbf{r}(t) = \langle \sin t, \ -\cos t, \ \ln|\sec t|\rangle + \langle 1,0,4\rangle$

43 $\mathbf{r}(t) = \langle 2t + 14, \ t + 72, \ -16t^2 + 100t + 3\rangle$

45 (a) $\dfrac{d}{dt}(\mathbf{u}(t)\cdot\mathbf{v}(t))$

  $= \dfrac{d}{dt}(u_1(t)v_1(t) + u_2(t)v_2(t) + u_3(t)v_3(t))$

  $= u_1'(t)v_1(t) + u_1(t)v_1'(t)$

  $\quad + u_2'(t)v_2(t) + u_2(t)v_2'(t)$

  $\quad + u_3'(t)v_3(t) + u_3(t)v_3'(t).$

  (b) $\mathbf{u}'(t)\cdot\mathbf{v}(t) + \mathbf{u}(t)\cdot\mathbf{v}'(t)$

  $= u_1'(t)v_1(t) + u_2'(t)v_2(t) + u_3'(t)v_3(t)$

  $\quad + u_1(t)v_1'(t) + u_2(t)v_2'(t) + u_3(t)v_3'(t).$

  (c) The results in parts (a) and (b) are equal, therefore $\dfrac{d}{dt}(\mathbf{u}(t)\cdot\mathbf{v}(t)) = \mathbf{u}'(t)\cdot\mathbf{v}(t) + \mathbf{u}(t)\cdot\mathbf{v}'(t).$

47 Sum rule: $\dfrac{d}{dt}(\mathbf{u}(t) + \mathbf{v}(t)) = \mathbf{u}'(t) + \mathbf{v}'(t)$

  Proof:

  $\dfrac{d}{dt}(\mathbf{u}(t) + \mathbf{v}(t))$

  $= \dfrac{d}{dt}(\langle u_1(t) + v_1(t), \ u_2(t) + v_2(t), \ u_3(t) + v_3(t)\rangle)$

$$= \langle u_1'(t) + v_1'(t),\ u_2'(t) + v_2'(t),\ u_3'(t) + v_3'(t)\rangle$$
$$= \langle u_1'(t),\ u_2'(t),\ u_3'(t)\rangle + \langle v_1'(t),\ v_2'(t),\ v_3'(t)\rangle$$
$$= \mathbf{u}'(t) + \mathbf{v}'(t).$$

49
$$\frac{d}{dt}\,(f(t)\mathbf{u}(t))$$
$$= \frac{d}{dt}\,(f(t)\langle u_1(t),\ u_2(t),\ u_3(t)\rangle)$$
$$= \frac{d}{dt}\,(\langle f(t)u_1(t),\ f(t)u_2(t),\ f(t)u_3(t)\rangle)$$
$$= \langle f'(t)u_1(t) + f(t)u_1'(t),$$
$$f'(t)u_2(t) + f(t)u_2'(t),$$
$$f'(t)u_3(t) + f(t)u_3'(t)\rangle$$

$$= \langle f'(t)u_1(t),\ f'(t)u_2(t),\ f'(t)u_3(t)\rangle$$
$$+ \langle f(t)u_1'(t),\ f(t)u_2'(t),\ f(t)u_3'(t)\rangle$$
$$= f'(t)\langle u_1(t),\ u_2(t),\ u_3(t)\rangle$$
$$+ f(t)\langle u_1'(t),\ u_2'(t),\ u_3'(t)\rangle$$
$$= f'(t)\mathbf{u}(t) + f(t)\mathbf{u}'(t).$$

51 Definition. Let $\mathbf{r} = \langle f(t), g(t), h(t)\rangle$ be a vector-valued function and let k be a real number in the domain of $\mathbf{r}$. If $\frac{\mathbf{r}(k+\alpha)-\mathbf{r}(k)}{\alpha}$ is defined and each component of $\mathbf{r}$ renders the same real result a_x, b_y, or c_z respectively, for every infinitesimal α, then we write $\mathbf{r}'(k) = \langle a_x, b_y, c_z\rangle$ and call $\mathbf{r}'(k)$ the derivative of $\mathbf{r}$ at k. When each of a_x, b_y, and c_z are real numbers (not ∞ or $-\infty$), then we say that $\mathbf{r}$ is differentiable at k.

Section 2.3

1 $7\sqrt{9.01} = 21.012$ units

3 $4\sqrt{56} = 8\sqrt{14} = 29.933$ units

5 $\frac{5}{3}$ units

7 (a) $\displaystyle\int_0^\pi \sqrt{4\cos^2 2t + 1}\,dt$;

   (b) 5.2704 units

9 (a) $\displaystyle\int_{-3}^4 \sqrt{e^{2t} + \sinh^2 t + \frac{4}{(1+t^2)^2}}\,dt$;

   (b) 70.958 units

11 (a) $\sqrt{3} + \frac{1}{2}\ln(2+\sqrt{3})$ units;

   (b) 2.3905 units;

   (c) $S_{10} = 2.3905$ units

13 $\kappa(t) = \dfrac{9}{9.01} = 0.99889$

15 $\kappa(4) = 0$

17 $\kappa(\sqrt{3}) = \dfrac{1}{8}$

19 $\kappa(1) = \dfrac{\sqrt{\frac{3}{2}}}{8} = 0.15309$

21 (a) $\mathbf{T}(t) = \left\langle 0,\ \frac{1}{\sqrt{1+t^2}},\ \frac{t}{\sqrt{1+t^2}}\right\rangle$;

   (b) $\mathbf{N}(t) = \left\langle 0,\ \frac{-t}{\sqrt{1+t^2}},\ \frac{1}{\sqrt{1+t^2}}\right\rangle$; (c) $\mathbf{B}(t) = \langle 1, 0, 0\rangle$;

   (d) $x = 1, y = 1 + \frac{1}{\sqrt{2}}t - \frac{1}{\sqrt{2}}s, z = \frac{1}{2} + \frac{1}{\sqrt{2}}t + \frac{1}{\sqrt{2}}s$;

   (e) $x = 1 + s, y = 1 - \frac{1}{\sqrt{2}}t, z = \frac{1}{2} + \frac{1}{\sqrt{2}}t$

23 (a) $\mathbf{T}(t) = \left\langle \frac{\cos t+\sin t}{\sqrt{2}},\ 0,\ \frac{\cos t-\sin t}{\sqrt{2}}\right\rangle$;

   (b) $\mathbf{N}(t) = \left\langle \frac{\cos t-\sin t}{\sqrt{2}},\ 0,\ \frac{-\cos t-\sin t}{\sqrt{2}}\right\rangle$;

   (c) $\mathbf{B}(t) = \langle 0, 1, 0\rangle$;

   (d) $x = \frac{1}{\sqrt{2}}t + \frac{1}{\sqrt{2}}s, y = 3, z = 1 + \frac{1}{\sqrt{2}}t - \frac{1}{\sqrt{2}}s$;

   (e) $x = \frac{1}{\sqrt{2}}t, y = 3 + s, z = 1 - \frac{1}{\sqrt{2}}t$

25 (a) 0 (see exercise 16, for instance); (b) $\frac{1}{\Omega} = \omega \doteq 0$
(c) A circle with infinite radius will look completely flat, like a line, when viewed on a real-number scale

27 (a) $\mathbf{r}'(t) = \langle 1, f'(t)\rangle$; (b) $\|\mathbf{r}(t)\| = \sqrt{1+(f'(t))^2}$;

   (c) $\dfrac{f'(t)}{\sqrt{1+(f'(t))^2}}$;

   (d) $\dfrac{\sqrt{1+(f'(t))^2}\,f''(t) - f'(t)\,\frac{1}{2\sqrt{1+(f'(t))^2}}\,2f'(t)f''(t)}{1+(f'(t))^2}$;

   (e) $\dfrac{f''(t)}{(1+(f'(t))^2)^{3/2}}$; (f) numerator positive, denominator positive because $1 + (f'(t))^2$ must be positive, for y-component of $\mathbf{N}(t)$ we divide this positive quantity by another positive quantity

29 (a) $\kappa(t) = \dfrac{\sqrt{729.81}}{(\sqrt{9.01})^3} = 0.99889$; (b) the alternate formula is more work; the definition is easier this time

Section 2.4

1 (a) velocity $\mathbf{v}(t) = \langle t - 1, t^2 - 1, -\pi \sin \pi t \rangle$;

$$\text{speed } \|\mathbf{v}(t)\| = \sqrt{(t-1)^2 + (t^2-1)^2 + \pi^2 \sin^2 \pi t}$$
$$= \sqrt{t^4 - t^2 - 2t + 2 + \pi^2 \sin^2 \pi t};$$
$$\mathbf{a}(t) = \langle 1, 2t, -\pi^2 \cos \pi t \rangle$$

(b) $\mathbf{v}(0) = \langle -1, -1, 0 \rangle; \|\mathbf{v}(0)\| = \sqrt{2}; \mathbf{a}(0) = \langle 1, 0, -\pi^2 \rangle$;

(c) yes; at point $\left(\frac{5}{2}, -\frac{2}{3}, -1 \right)$; $\mathbf{a}(1) = \langle 1, 2, \pi^2 \rangle$

3 (a) $\mathbf{r}(1) = \langle 3, 1, 0 \rangle; \mathbf{v}(1) = \langle 3, -1, 1 \rangle; \|\mathbf{v}(1)\| = \sqrt{11}$; $\mathbf{a}(1) = \langle 0, 2, -1 \rangle$;

(b) no, the particle never stops;

(c) $0 < t < 1$

5 $\text{comp}_{\mathbf{T}}\mathbf{a} = 0, \text{comp}_{\mathbf{N}}\mathbf{a} = 9$

7 $\text{comp}_{\mathbf{T}}\mathbf{a} = \frac{56t}{\sqrt{56t^2}} = 2\sqrt{14}\frac{t}{|t|}$ (the answer $\sqrt{56} = 2\sqrt{14}$ is incorrect because if $t < 0$ then $|t| = -t$), $\text{comp}_{\mathbf{N}}\mathbf{a} = 0$ (except undefined when $t = 0$)

9 (a) $\text{comp}_{\mathbf{T}}\mathbf{a} = \frac{t}{\sqrt{1+t^2}}, \text{comp}_{\mathbf{N}}\mathbf{a} = \frac{1}{\sqrt{1+t^2}}$;

(b) at $t = 1, \text{comp}_{\mathbf{T}}\mathbf{a} = \frac{1}{\sqrt{2}}, \text{comp}_{\mathbf{N}}\mathbf{a} = \frac{1}{\sqrt{2}}$

11 (a) $\text{comp}_{\mathbf{T}}\mathbf{a} = \frac{3}{4\sqrt{t}} - \frac{1}{4t^{3/2}}$,

$$\text{comp}_{\mathbf{N}}\mathbf{a} = \frac{\sqrt{\frac{3}{32}t^{-3} + \frac{27}{32}t^{-1} + \frac{9}{16}t^{-2}}}{\frac{3}{2}\sqrt{t} + \frac{1}{2\sqrt{t}}} = \frac{\sqrt{\frac{3}{2}}}{2|t|};$$

(b) at $t = 4, \text{comp}_{\mathbf{T}}\mathbf{a} = \frac{11}{32}$,

$$\text{comp}_{\mathbf{N}}\mathbf{a} = \frac{\sqrt{\frac{507}{2048}}}{\frac{13}{4}} = \frac{\sqrt{\frac{3}{2}}}{8} \approx 0.1531$$

13 (Note: no info is given about launch position, so we assume it is launched from the origin)

(a) 236.13 m;

(b) 21.485 m;

(c) 51.125 m (surprisingly far!)

15 (a) 2444.3 ft;

(b) yes, it will easily clear the tree;

(c) 2511.3 ft (Hint: launch point $(0, 45)$), or 67 ft further!

17 $\alpha = 83.36°, \alpha = 6.65°$ (Hint: solve x-component of $\mathbf{r}(t)$ for $\cos \alpha$, use triangle and Pythagorean theorem to determine $\sin \alpha$, substitute into y-component formula, in resulting equation isolate square root and square both sides, result is fourth-degree polynomial that is quadratic in form, use quadratic formula to find t^2, take square root, use x-component to solve for α using inverse cosine)

19 (a) $\mathbf{r}(2.1757) = \langle -906.3, -422.6, 815.9 \rangle$;

(b) 7588.8 m

21 (a) $(678.05, 678.05, 0)$ (leaving out the z-coordinate is acceptable);

(b) 52.444 m

23 (a) $\|\mathbf{v}_0\| = 39.11$ ft/s;

(b) $\alpha = 25.96°$ or $\alpha = 58.36°$ (see hint for exercise 17);

(c) for $\alpha = 24.96°$, lands 58.99 ft downrange; for $\alpha = 26.96°$, 61.05 ft; for $\alpha = 57.36°$, 61.09 ft; for $\alpha = 59.36°$, 58.85 ft; yes; the smaller launch angle, 25.96°, is better

25 $\mathbf{v}_0 = \langle x_0 \|\mathbf{v}_0\| \cos \alpha, y_0 \|\mathbf{v}_0\| \cos \alpha, \|\mathbf{v}_0\| \sin \alpha \rangle$

27 (a) -2.56 ft;

(b) looking below the landing spot;

(c) 125 ft

Chapter 3

Section 3.1

1 8

3 $\ln 2 + \ln \frac{1}{2} = 0$

5 (a) $-\frac{2}{7}$;

(b) slope formula function

7 (a) $\sqrt{53} \approx 7.2801$;

   (b) distance formula function

9 (a) $3x \neq y$;

   (b) 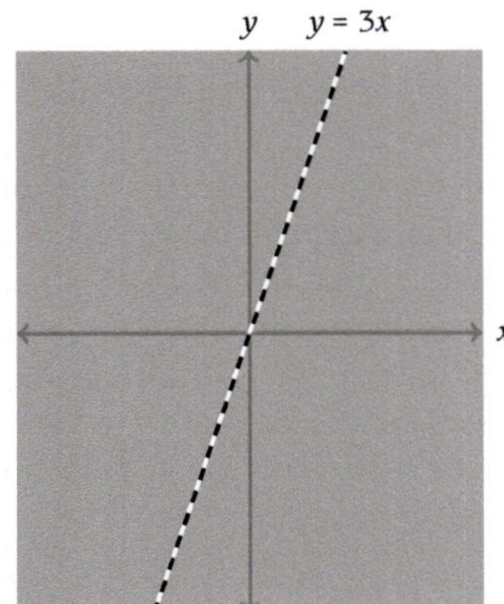

11 (a) $x^2 + y^2 \geq 4$;

   (b) 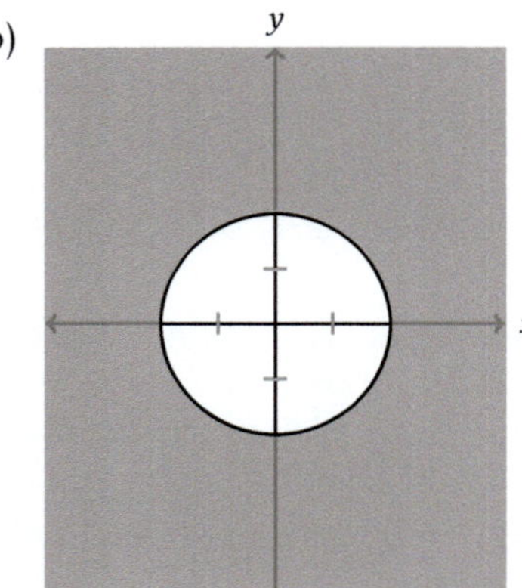

13 (a) $\mathbf{R}^2$;

   (b) NA

15 (a) $x \geq -3$ and $y \neq -3$

   (b) 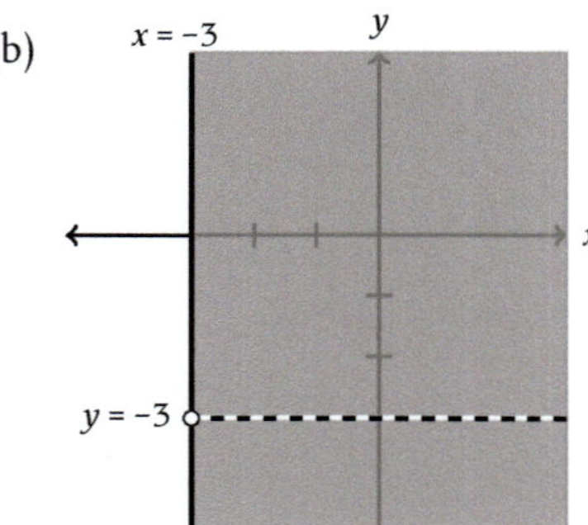

17 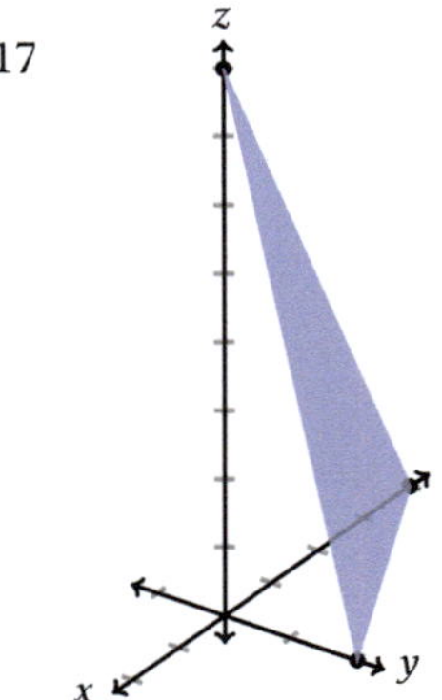

19 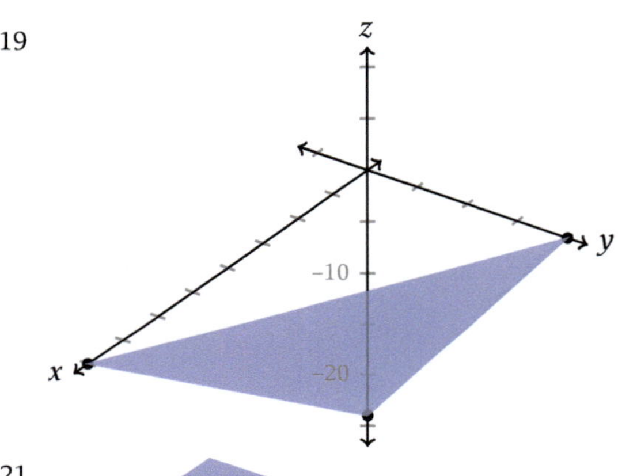

21 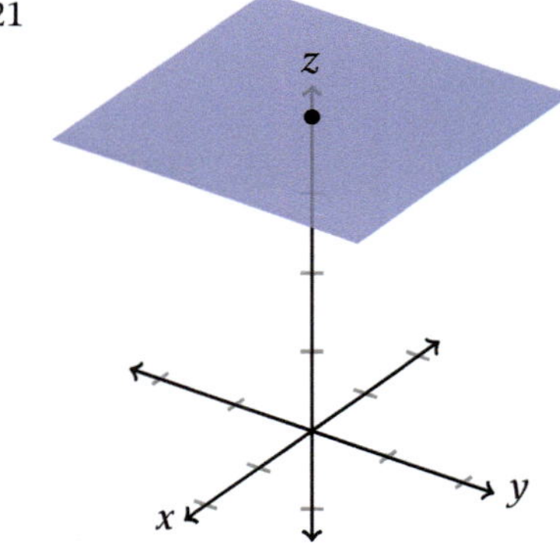

23 (a) (iii);

   (b) (v);

   (c) (iv);

   (d) (vi);

   (e) (ii);

   (f) (i)

25 a, c

27 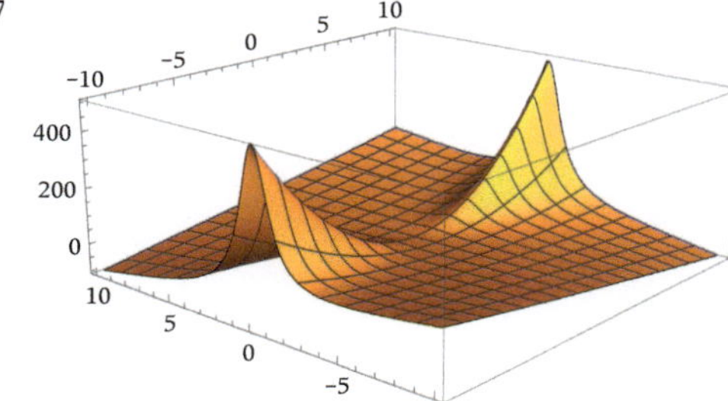

29 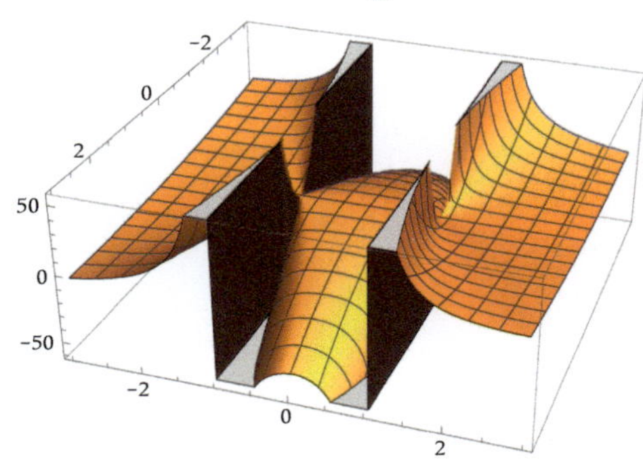

31 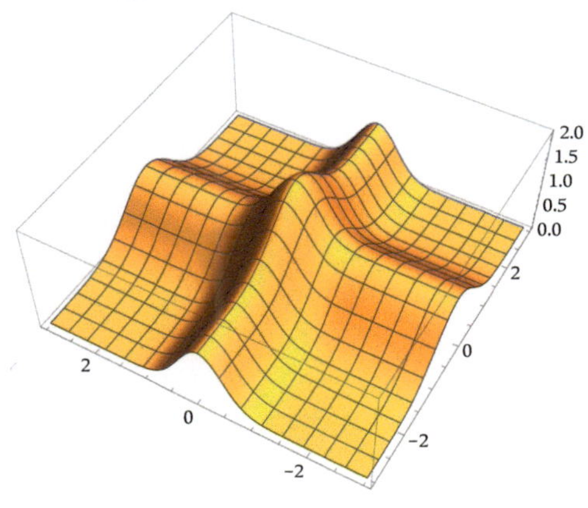

Section 3.2

1. (a) $x^2 - y^2 = 1$;

   (b) $y^2 - x^2 = 1$;

   (c) $y = \pm x$

3. (a) $x = 0$, line (vertical);

   (b) $y = \ln x$, logarithmic curve;

   (c) $y = \ln \frac{x}{2}$, logarithmic curve, same as part (b) stretched horizontally by a factor of 2;

   (d) $y = \ln(-x)$, logarithmic curve, same as part (b) reflected across the y-axis

5. $\dfrac{x^2}{4} + \dfrac{y^2}{4} - \dfrac{z^2}{4} = 1$, hyperboloid of one sheet; $\dfrac{z^2}{9} - \dfrac{x^2}{9} - \dfrac{y^2}{9} = 1$, hyperboloid of two sheets

7. 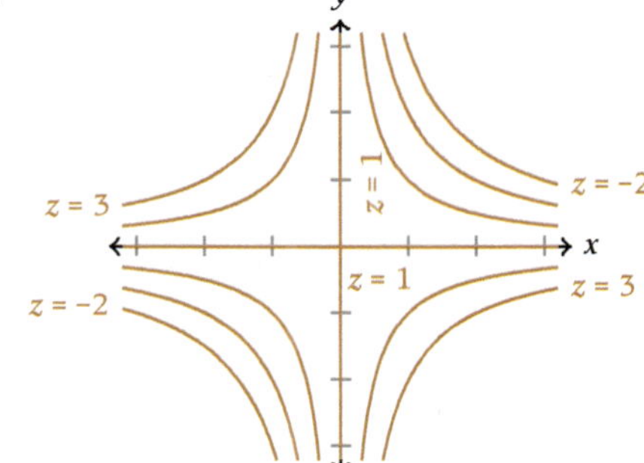

9. 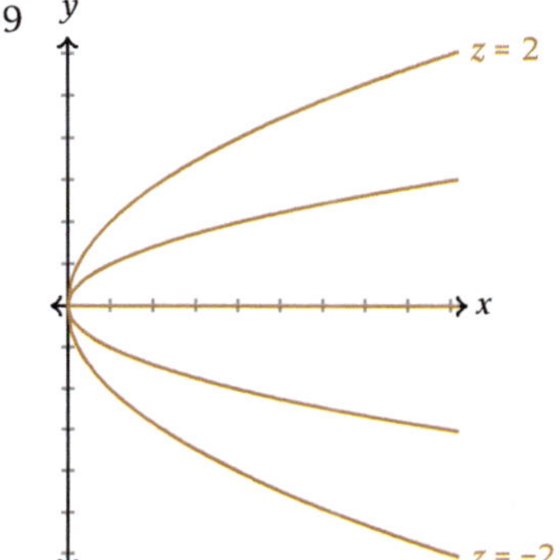

11

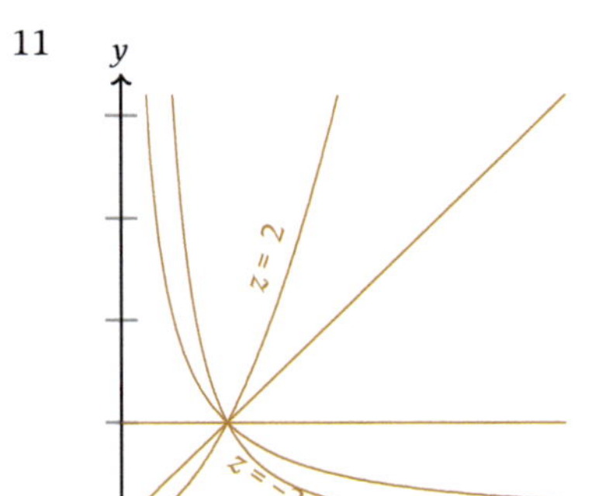

13 (a) (v);
   (b) (i);
   (c) (ii);
   (d) (iv);
   (e) (vi);
   (f) (iii)

15

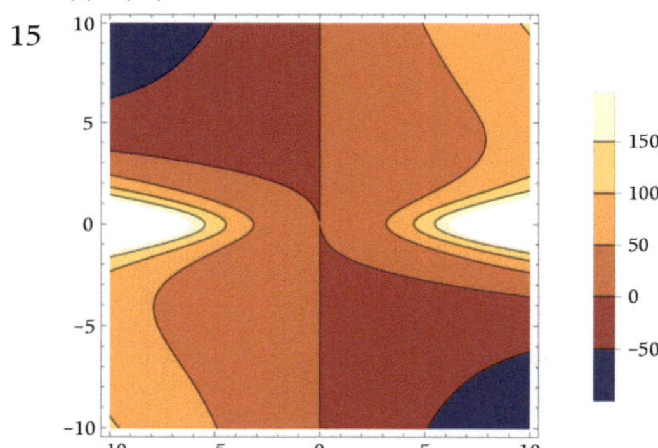

17

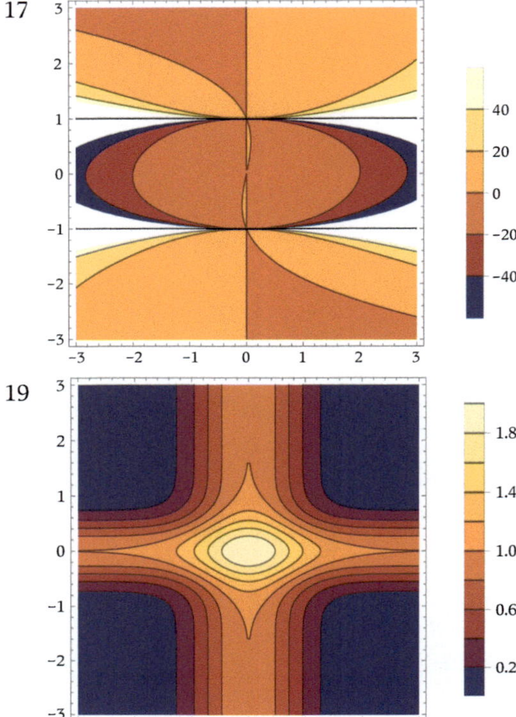

Section 3.3

1 no

3 no

5 DNE $\left(\alpha \ll \beta, \frac{1}{2}; \alpha \gg \beta, \text{infinite}; \alpha = k\beta, \frac{4k+1}{2}\right)$

7 $-\frac{3}{4}$

9 2 (evaluating using continuity works)

11 ∞

13 DNE $\left(\alpha \ll \beta, \frac{3}{4}; \alpha \gg \beta, 1; \alpha = k\beta, \frac{-2k+3}{-2k+4}\right)$

15 0

17 $-\infty$

19 $\frac{\pi}{2}$

21 $f(3,5) = \sqrt{9-5} = 2;$ $\displaystyle\lim_{(x,y)\to(3,5)} \sqrt{x^2 - y}$

$\sqrt{(3+\alpha)^2 - (5+\beta)} = \sqrt{9 + 6\alpha + \alpha^2 - 5 - \beta}$
$\sqrt{4} \quad = \quad 2;$ because $f(3,5) = \displaystyle\lim_{(x,y)\to(3,5)} f(x,y),$ f

continuous at the input $(3,5)$

23 $f(1,-1) = 1 - 3 + 1 + 1 = 0;$ $\displaystyle\lim_{(x,y)\to(1,-1)} (x + 3y + x^2$

$1) = 1 + \alpha + 3(-1+\beta) + (1+\alpha)^2 + 1 = 3\alpha + \alpha^2 + 3\beta \doteq$

because $f(1,-1) = \lim\limits_{(x,y)\to(1,-1)} f(x,y)$, f is continuous at the input $(1,-1)$

25 $f(2,4) = \sin(2(2)-4) = \sin 0 = 0$; $\lim\limits_{(x,y)\to(2,4)} \sin(2x-y) = \sin(2(2+\alpha)-(4+\beta)) = \sin(2\alpha-\beta) \approx 2\alpha-\beta \doteq 0$; because $f(2,4) = \lim\limits_{(x,y)\to(2,4)} f(x,y)$, f is continuous at the input $(2,4)$

27 $f(5,-2) = \dfrac{5+2}{-2-5} = -1$; $\lim\limits_{(x,y)\to(5,-2)} \dfrac{x+2}{y-5} = \dfrac{5+\alpha+2}{-2+\beta-5} \approx \dfrac{7}{-7} = -1$; because $f(5,-2) = \lim\limits_{(x,y)\to(5,-2)} f(x,y)$, f is continuous at the input $(5,-2)$

29 $x \geq 0, y \neq 1$

31 $x \neq 0, y \neq 0$; alternate answer $\{(a,b) \in \mathbf{R}^2 \mid a \neq 0, b \neq 0\}$

33 $x > 0, y > 0$ and $x < 0, y < 0$ (quadrants I and III); alternate answer $\{(a,b) \in \mathbf{R}^2 \mid a > 0, b > 0\} \cup \{(a,b) \in \mathbf{R}^2 \mid a < 0, b < 0\}$

35 ∞

37 DNE $(f(A,B) \doteq 0; f(A,k) \approx e^{-3k^2}; f(k,B) \approx e^{-k^2})$

39 DNE (could be either ∞ or $-\infty$; it's also possible to get real numbers, using the "last case" $A^3 = -B^4 + c$ for a real number c, which is on a lower level than B^4)

41 1

43 $\lim\limits_{(x,y)\to(0,b)} \dfrac{\sin x}{(1+y^2)\sqrt{x^2}} \approx \dfrac{\alpha}{(1+b^2)|\alpha|}$; if $\alpha > 0$, $\dfrac{1}{1+b^2}$; if $\alpha < 0$, $\dfrac{-1}{1+b^2}$; a jump discontinuity at every point of the form $(0,b)$

45 $\lim\limits_{(x,y)\to(a,-a)} \dfrac{x^2-y^2}{x+y} = \cdots = 2a$, which is a real number, so the discontinuity is removable at every point on the line $y = -x$ (Note: if $a = 0$, the approximation principle is violated. But then the expression is $\dfrac{\alpha^2-\beta^2}{\alpha+\beta}$, and the cases $\alpha \ll \beta$, $\alpha \gg \beta$, $\alpha = k\beta$, and the "last case" $\alpha = -\beta + \gamma$ where $\gamma \ll \beta$, all render 0, as required.)

47 $\lim\limits_{(x,y)\to(a,e^a)} \dfrac{\sin(y-e^x)}{y-e^x} = 1$, so the discontinuity is removable for every point on the

curve $y = e^x$. (Notes: $\lim\limits_{(x,y)\to(a,e^a)} \dfrac{\sin(y-e^x)}{y-e^x} = \dfrac{\sin(e^a+\beta-e^{a+\alpha})}{e^a+\beta-e^{a+\alpha}} = \dfrac{\sin(e^a+\beta-(e^a+\gamma))}{e^a+\beta-(e^a+\gamma)} = \dfrac{\sin(\beta-\gamma)}{\beta-\gamma} \approx \dfrac{\beta-\gamma}{\beta-\gamma} = 1$. Because the same quantity is inside the sine function and in the denominator, either $\beta - \gamma$ is infinitesimal and the calculation is as stated or $\beta - \gamma = 0$ and is not in the domain, so that the exception in the definition of limit applies.)

49 $f(a,b) = a^2 + b^2$; $\lim\limits_{(x,y)\to(a,b)} (x^2+y^2) = (a+\alpha)^2+(b+\beta)^2 = a^2+2a\alpha+\alpha^2+b^2+2b\beta+\beta^2 \approx a^2+b^2$; because $f(a,b) = \lim\limits_{(x,y)\to(a,b)} f(x,y)$, f is continuous at the input (a,b). (Note: if $a^2 + b^2 = 0$, then the expression is $\alpha^2 + \beta^2 \doteq 0 = a^2 + b^2$.)

51 $f(a,b) = ca + db + k$; $\lim\limits_{(x,y)\to(a,b)} (cx + dy + k) = c(a+\alpha)+d(b+\beta)+k = ca+c\alpha+db+d\beta+k \approx ca+db+k$; because $f(a,b) = \lim\limits_{(x,y)\to(a,b)} f(x,y)$, f is continuous at the input (a,b).

53 maximum $\dfrac{1}{2}$ (at $x = 1$), minimum $-\dfrac{1}{2}$ (at $x = -1$); Figure 3.51 shows a slit at the input $(0,0)$ with nearby z-coordinates ranging from $-\dfrac{1}{2}$ to $\dfrac{1}{2}$.

55 $\dfrac{\alpha^2-\beta^2}{\alpha+\beta} = \dfrac{(-\beta+\gamma)^2-\beta^2}{-\beta+\gamma+\beta} = \cdots = -2\beta + \gamma \doteq 0$

57 Let $f(x,y,z)$ be a function of three variables and let a, b, and c be real numbers. If $f(a+\alpha, b+\beta, c+\gamma)$ renders the same real result L for every choice of infinitesimals α, β, and γ for which $f(a+\alpha, b+\beta, c+\gamma)$ is defined, then we write
$$\lim\limits_{(x,y,z)\to(a,b,c)} f(x,y,z) = L.$$

If L is a real number (i.e., not ∞ or $-\infty$), then we say that the limit exists.

59 (a) 0;

(b) $k = -1, \beta = -\sin\alpha + \gamma$, with $\gamma \ll \alpha$;

(c) if $\gamma \gg \alpha^2$, 1; if $\gamma \ll \alpha^2$, infinite; the limit does not exist.

61 (a) $\dfrac{1}{c}$;

(b) $\frac{1}{\gamma}$, infinite;

(c) the limit does not exist;

(d) the answer to (a) shows that we can travel along various level curves near the parabolic asymptote infinitely far out; the answer to (b) shows that we can climb the surface near the parabolic asymptote infinitely high as we travel infinitely far out

63 DNE (Because the function is piecewise-defined, the quantities $f(a+\alpha, b)$ and $f(a, b+\beta)$ must be checked.)

65 (a) 1 (Hint: use $r^2 = x^2 + y^2$; notice that after converting to polar the limit is single-variable);

(b) 1

67 (a) 0 (Hint: use $x = r\cos\theta$, $y = r\sin\theta$, simplify cancel; then evaluate using continuity, noting that the numerator is 0 and the denominator is non-zero);

(b) 0

Section 3.4

1 derivative

3 derivative

5 partial derivative

7 y

9 u

11 z

13 4

15 $4\sqrt{3x^2+4x-7}\left(e^{(x^5)}\sin(x^3-6)\right)$

17 $\frac{4}{\sqrt{3x^2+4x-7}\left(e^{(x^5)}\sin(x^3-6)\right)}$

19 $f_x(x,y) = 2x - \cos x; f_y(x,y) = 3y^2 + 2^y \ln 2$

21 $f_x(x,y) = -14x\sin(7x^2); f_y(x,y) = \frac{1}{2\sqrt{y^3+y}}(3y^2+1)$

23 $f_x(x,y) = \frac{(3x-1)2x-(x^2+2)3}{(3x-1)^2} = \frac{3x^2-2x-6}{(3x-1)^2}; f_y(x,y) = y\cosh y + \sinh y$

25 $f_x(x,y) = 2x + 3y + 2; f_y(x,y) = 3x - 14y - 3$

27 $f_x(x,y) = 2xy - 7y^3 + 4\sin y; f_y(x,y) = x^2 - 21xy^2 + 4x\cos y$

29 $f_x(x,y) = y^3\cos(xy)$; $f_y(x,y) = xy^2\cos(xy) + 2y\sin(xy)$

31 $f_x(x,y) = \frac{\frac{x}{x+y}-\ln(x+y)}{x^2}; f_y(x,y) = \frac{1}{x^2+xy}$

33 $\frac{\partial z}{\partial x} = 2x\tan y^2; \frac{\partial z}{\partial y} = 2x^2 y\sec^2 y^2$

35 $\frac{\partial z}{\partial x} = \frac{2xy^4}{1+x^4y^8}; \frac{\partial z}{\partial y} = \frac{4x^2y^3}{1+x^4y^8}$

37 $\frac{\partial z}{\partial t} = k(\sin t)^{k-1}\cos t; \frac{\partial z}{\partial k} = (\sin t)^k \ln(\sin t)$

39 $f_u(u,v) = \frac{u}{\sqrt{u^2+v^2}}; f_v(u,v) = \frac{v}{\sqrt{u^2+v^2}}$

41 $\frac{\partial z}{\partial x} = (y^2+1)\left((x+3y)2x + (x^2-7)\right); \frac{\partial z}{\partial y} = (x^2 - 7)\left((x+3y)2y + (y^2+1)3\right)$

43 $g_x(x,y) = \frac{(y^3+2y+3x)(y^2+2)-(xy^2+2x)3}{(y^3+2y+3x)^2}$; $g_y(x,y) = \frac{(y^3+2y+3x)(2xy)-(xy^2+2x)(3y^2+2)}{(y^3+2y+3x)^2}$ (Hint: rewrite to eliminate compound fraction before finding partia derivatives)

45 $\frac{\partial s}{\partial u} = \sec(ut-1)\sec^2(u^2+t)\cdot 2u + \tan(u^2+t)\sec(ut-1)\tan(ut-1)\cdot t; \frac{\partial s}{\partial t} = \sec(ut-1)\sec^2(u^2+t) + \tan(u^2 + t)\sec(ut-1)\tan(ut-1)\cdot u$

47 $\frac{\partial z}{\partial x} = \frac{(x+\alpha)^2+3(x+\alpha)-y^2+5y-(x^2+3x-y^2+5y)}{\alpha}$ $\frac{x^2+2x\alpha+\alpha^2+3x+3\alpha-y^2+5y-x^2-3x+y^2-5y}{\alpha} = \frac{2x\alpha+\alpha^2+3\alpha}{\alpha}$ $\frac{2x\alpha+3\alpha}{\alpha} = \frac{\alpha(2x+3)}{\alpha} = 2x+3$

49 $f_x(x,y) = \frac{\frac{x+\alpha}{y}-\frac{x}{y}}{\alpha} = \frac{\frac{x+\alpha-x}{y}}{\alpha} = \frac{\frac{\alpha}{y}}{\alpha} = \frac{1}{y}$

51 $f_y(x,y) = \frac{x^2+\sqrt{y+\beta}-(x^2+\sqrt{y})}{\beta} = \frac{\sqrt{y+\beta}-\sqrt{y}}{\beta}\cdot\frac{\sqrt{y+\beta}+\sqrt{y}}{\sqrt{y+\beta}+\sqrt{y}}$ $\frac{y+\beta-y}{\beta(\sqrt{y+\beta}+\sqrt{y})} = \frac{1}{\sqrt{y+\beta}+\sqrt{y}} \approx \frac{1}{2\sqrt{y}}$

53 $f_y(x,y) = \frac{\sin(x(y+\beta))-\sin(xy)}{\beta} = \frac{\sin(xy+\beta x)-\sin(xy)}{\beta}$ $\frac{\sin(xy)\cos(\beta x)+\cos(xy)\sin(\beta x)-\sin(xy)}{\beta}$ $\frac{\sin(xy)\cdot 1+\cos(xy)\cdot\beta x-\sin(xy)}{\beta} = x\cos(xy)$

Section 3.5

1 $\dfrac{\partial^2 f}{\partial y \partial x}$

3 $f_{yx}(x, y)$

5 $f_{xyz}(x, y, z)$

7 $f_x(x, y, z) = 2xy^2 z^2$;
$f_y(x, y, z) = 2x^2 yz^2; f_z(x, y, z) = 2x^2 y^2 z$

9 $g_x(x, y, z) = 3 \sin z + 2ye^{2x+3z}$; $g_y(x, y, z) = e^{2x+3z}$;
$g_z(x, y, z) = 3x \cos z + 3ye^{2x+3z}$

11 $k_r(r, s, t) = 3; k_s(r, s, t) = 0; k_t(r, s, t) = -4$

13 $f_{x_1}(x_1, y_1, x_2, y_2) = \dfrac{y_2 - y_1}{(x_2 - x_1)^2}$;
$f_{y_1}(x_1, y_1, x_2, y_2) = \dfrac{-1}{x_2 - x_1}$;
$f_{x_2}(x_1, y_1, x_2, y_2) = \dfrac{y_1 - y_2}{(x_2 - x_1)^2}$;
$f_{y_2}(x_1, y_1, x_2, y_2) = \dfrac{1}{x_2 - x_1}$

15 $f_a(a, b, c) = \dfrac{b}{2a^2} + \dfrac{\frac{-4ac}{\sqrt{b^2-4ac}} - 2\sqrt{b^2-4ac}}{4a^2} = \dfrac{b}{2a^2} + \dfrac{2ac - b^2}{2a^2 \sqrt{b^2-4ac}}$;
$f_b(a, b, c) = -\dfrac{1}{2a} + \dfrac{b}{2a\sqrt{b^2-4ac}}; f_c(a, b, c) = \dfrac{-1}{\sqrt{b^2-4ac}}$

17 $f_{xx}(x, y) = 20x^3; f_{xy}(x, y) = 0; f_{yx}(x, y) = 0; f_{yy}(x, y) = 42y^5$

19 $s_{rr}(r, t) = 2 - 16t^3 \sin 4r$; $s_{rt}(r, t) = 12t^2 \cos 4r$;
$s_{tr}(r, t) = 12t^2 \cos 4r; s_{tt}(r, t) = 6t \sin 4r$

21 $f_{xx}(x, y) = 0; f_{xy}(x, y) = e^y; f_{yx}(x, y) = e^y; f_{yy}(x, y) = xe^y$

23 $f_{uu}(u, v) = 48v^2 (2uv - v^2)^2; f_{uv}(u, v) = 48 (uv - v^2)(2uv - v^2)^2 + 8 (2uv - v^2)^3; f_{vu}(u, v) = 48 (uv - v^2)(2uv - v^2)^2 + 8 (2uv - v^2)^3; f_{vv}(u, v) = 48(u - v)^2 (2uv - v^2)^2 - 8 (2uv - v^2)^3$

25 $f_{xx}(x, y, z) = 0$; $f_{xy}(x, y, z) = 0$; $f_{xz}(x, y, z) = \cos z$;
$f_{yx}(x, y, z) = 0$; $f_{yy}(x, y, z) = 6$; $f_{yz}(x, y, z) = 0$;
$f_{zx}(x, y, z) = \cos z; f_{zy}(x, y, z) = 0, f_{zz}(x, y, z) = -x \sin z$

27 (a) -16;
(b) 16;
(c) -8;
(d) 8

29 (a) $\frac{1}{7}$;
(b) $-\frac{1}{7}$;
(c) $\frac{2}{7}$;
(d) $-\frac{2}{7}$

31 (a) $\frac{\pi}{2} - 1$;
(b) $1 - \frac{\pi}{2}$;
(c) 3; (d) -3

33 (a) $\dfrac{10}{\sqrt{3}} + 3\sqrt{3} = \dfrac{19}{\sqrt{3}}$;
(b) $-\dfrac{10}{\sqrt{3}} - 3\sqrt{3} = -\dfrac{19}{\sqrt{3}}$;
(c) $\dfrac{-5}{2\sqrt{3}} - 2\sqrt{3} = \dfrac{-17}{2\sqrt{3}}$;
(d) $\dfrac{5}{2\sqrt{3}} + 2\sqrt{3} = \dfrac{17}{2\sqrt{3}}$

35 $\dfrac{\partial^3 z}{\partial x^2 \partial y} = 2 \cos y$

37 $f_{xyzy}(x, y, z) = 8xz$

39 $\dfrac{\partial^2 w}{\partial z \partial y} = 0$

41 $f_{xy}(x, y) = 0$

43 $f_{xzy}(x, y, z) = \dfrac{6z(2x+3y)^{z-1}}{2x+3y} + 6z(z - 1)(2x + 3y)^{z-2} \ln(2x + 3y) + 6(z - 1)(2x + 3y)^{z-2} = 6(2x + 3y)^{z-2} \left(-1 + 2z + z(z - 1) \ln(2x + 3y)\right)$

45 $\dfrac{\partial z}{\partial y} = \dfrac{-x^2 y - yz^2}{x^2 z + y^2 z}$

47 $\dfrac{\partial z}{\partial x} = \dfrac{5yz - 6x^5}{4z^3 - 5xy}$

49 $\dfrac{\partial z}{\partial y} = \left(1 - \dfrac{2y}{y^2 + 1}\right) \dfrac{3 + \cos x}{-\cos z}$

51 $\dfrac{\partial s}{\partial t} = \dfrac{-2r + s^2}{2st + s^2 + 2r}$

53 $\dfrac{\partial z}{\partial x}\Big|_{(1,1,1)} = -3$

55 (a) does not satisfy Laplace's equation;
(b) satisfies Laplace's equation;
(c) satisfies Laplace's equation;
(d) satisfies Laplace's equation

57 (a) 12;
(b) $f_x(x, y) = 5x^4 y^7, f_y(x, y) = 7x^5 y^6 + 57$;
(c) 11;
(d) $f_{xx}(x, y) = 20x^3 y^7, f_{xy}(x, y) = 35x^4 y^6,$
$f_{yx}(x, y) = 35x^4 y^6, f_{yy}(x, y) = 42x^5 y^5$;
(e) 10;
(f) $f_{xxxxxx}(x, y) = 0$;
(g) No!;
(h) 13

Section 3.6

1. $z = -19 + 13(x - 1) - 27(y - 3) = 13x - 27y + 49$

3. $z = -11y$

5. $z = 5 + \frac{3}{5}(x - 3) + \frac{4}{5}(y - 4) = \frac{3}{5}x + \frac{4}{5}y$

7. $z = 0 + 2(x - 1) - 1(y - 2) = 2x - y$

9. $z = 2 + 1(x - 1) + 0(y - 1) = x + 1$

11. (a)

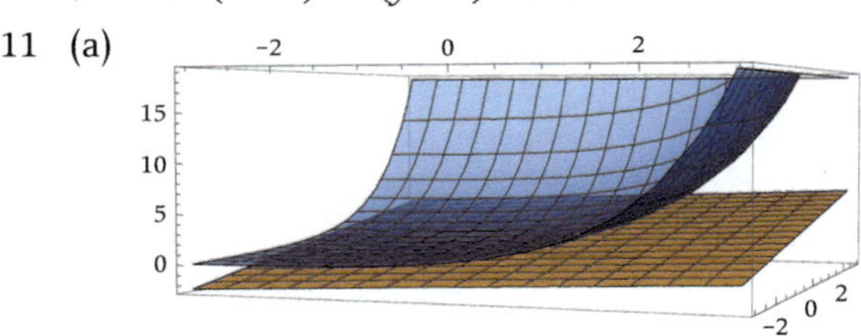

(b)

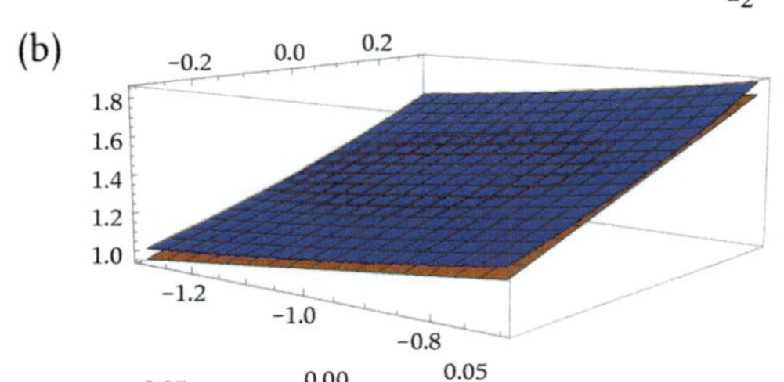

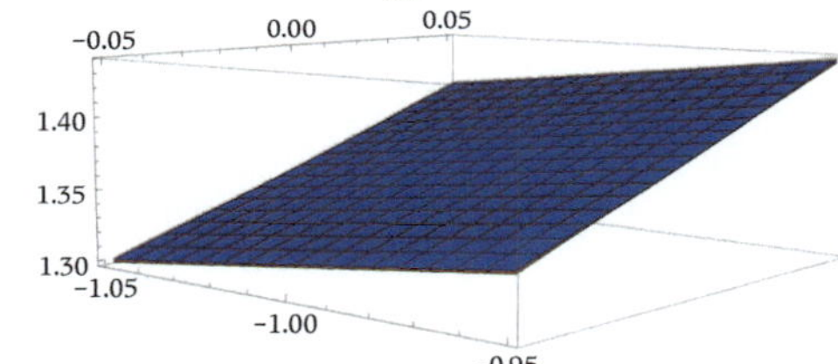

13. (a)

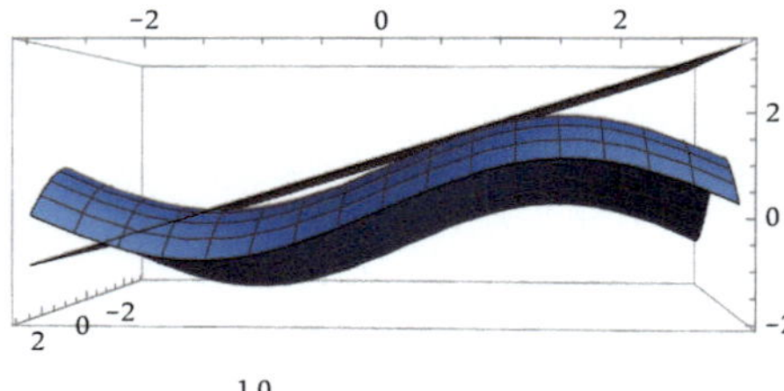

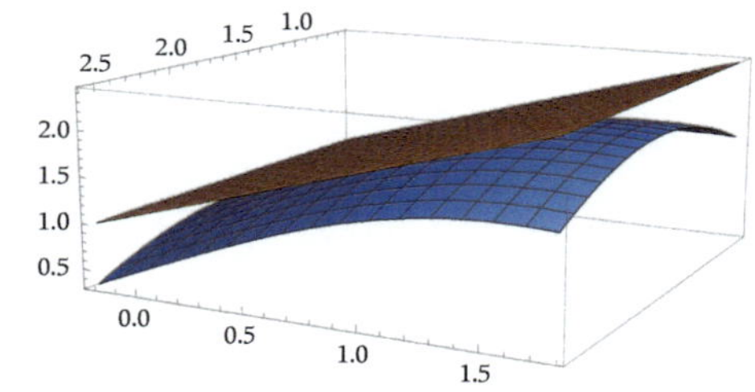

(b)

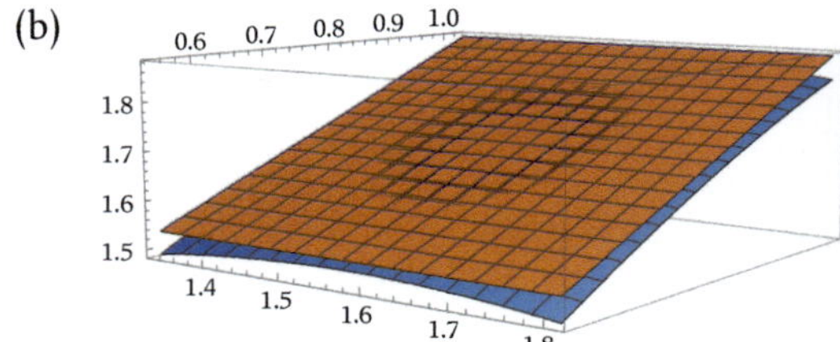

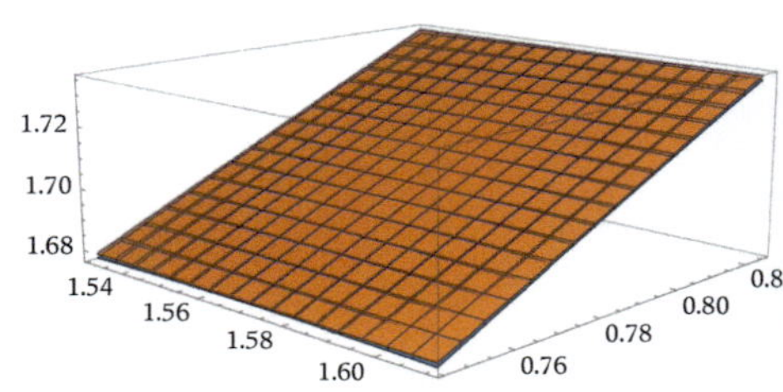

15. (a) $L(x, y) = 0 - 1(x - \pi) + 0(y - 0)$;

    (b) -0.01

17. (a) $L(x, y) = 0 + 0(x - 3) + 9(y - 1)$;

    (b) 0.72

19. (a) $L(x, y) = 2 + \frac{1}{4}(x - 3) + \frac{1}{2}(y - 1)$;

    (b) 2.0175

21. not differentiable (f_x and f_y are not defined a
$(-1, -1)$)

23. not differentiable (f is not defined at $(0, 0)$)

25. differentiable (by theorem 3)

27. differentaible (by theorem 3)

29. differentiable, $\dfrac{f(3+\alpha, 5+\beta) - (f(3,5) + f_x(3,5)\alpha + f_y(3,5)\beta)}{\sqrt{\alpha^2 + \beta^2}}$

$\dfrac{-12\beta^2 - 40\alpha\beta - 4\alpha\beta^2}{\sqrt{\alpha^2 + \beta^2}} \doteq 0$

31. not differentiable $\Big(\dfrac{f(\alpha,\beta) - (f(0,0) + f_x(0,0)\alpha + f_y(0,0)\beta)}{\sqrt{\alpha^2 + \beta^2}}$

$\dfrac{-\alpha}{\sqrt{\alpha^2 + \beta^2}}$, with case $\beta = k\alpha$ results vary, not always (
Note: this function has a slope in every radial dire
tion from $(0, 0)$, so existence of slopes in all (radia
directions is not enough to guarantee differentiabi
ity (local linearity).

33. differentiable $\Big(\dfrac{f(\alpha,\beta) - (f(0,0) + f_x(0,0)\alpha + f_y(0,0)\beta)}{\sqrt{\alpha^2 + \beta^2}}$

$\dfrac{\sqrt{\alpha^2 \beta^2}}{\sqrt{\alpha^2 + \beta^2}}$, which always renders 0; Hint: calcula

$f_x(0,0)$ and $f_y(0,0)$ using the definition of partial derivative); see also exercise 40

35 (a) $\dfrac{0.23}{16} = 0.014375$;

(b) $\dfrac{0.23}{36} = 0.00639$

37 0.5337 ft/s^2 (reminder: convert degrees to radians)

39 (a) $f_x(x,y) = \dfrac{y^3 - x^2 y}{(x^2 + y^2)^2}$;

(b) DNE (infinite), f_x is not continuous at $(0,0)$;

(c) The hypotheses are not met because f_x is not continuous at $(0,0)$

41 differentiable, $\dfrac{f(3+\alpha,5)+f(3,5+\beta)-f(3+\alpha,5+\beta)-f(3,5)}{\sqrt{\alpha^2+\beta^2}} = \dfrac{40\alpha\beta+4\alpha\beta^2}{\sqrt{\alpha^2+\beta^2}} \doteq 0$; this method is slightly messier algebraically

43 not differentiable, $\dfrac{f(\alpha,0)+f(0,\beta)-f(\alpha,\beta)-f(0,0)}{\sqrt{\alpha^2+\beta^2}} = \dfrac{\alpha}{\sqrt{\alpha^2+\beta^2}}$, and case $\alpha \gg \beta$ does not always render 0; this method is much simpler

45 differentiable, $\dfrac{f(\alpha,0)+f(0,\beta)-f(\alpha,\beta)-f(0,0)}{\sqrt{\alpha^2+\beta^2}} = \dfrac{-\sqrt{\alpha^2\beta^2}}{\sqrt{\alpha^2+\beta^2}}$, which always renders 0; this method is simpler

Section 3.7

1 $\dfrac{dx}{dt} = (t^3 + t)^2 (\cos(5(3 - 4t)))(-20) + 2(t^3 + t)(3t^2 + 1)\sin(5(3 - 4t))$

3 $\dfrac{\partial x}{\partial t} = (t^3 + s^2)^2 (\cos(5(3s + 5t)))(25) + 2(t^3 + s^2)3t^2 \sin(5(3s + 5t))$

5 $\dfrac{\partial z}{\partial s} = 4(3 + 5r + 2s)^3 \cdot 2 - 6(s^2 r^3)2sr^3 + 4(3 + 5r + 2s)2sr^3 + 4(2)s^2 r^3$; with simplifcation beforehand, $\dfrac{\partial z}{\partial s} = 4(3 + 5r + 2s)^3 \cdot 2 - 12s^3 r^6 + 24sr^3 + 40sr^4 + 24s^2 r^3$

7

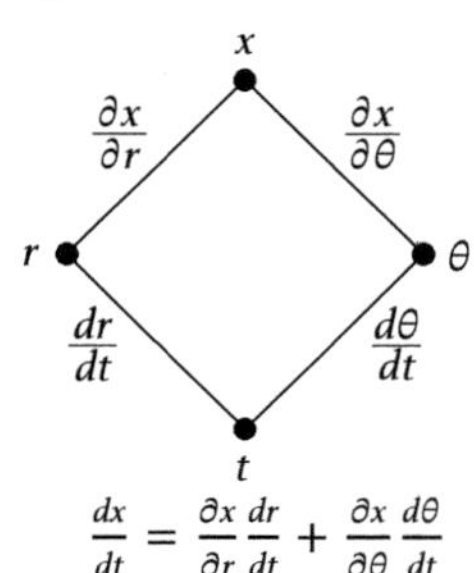

$$\dfrac{dx}{dt} = \dfrac{\partial x}{\partial r}\dfrac{dr}{dt} + \dfrac{\partial x}{\partial \theta}\dfrac{d\theta}{dt}$$

9

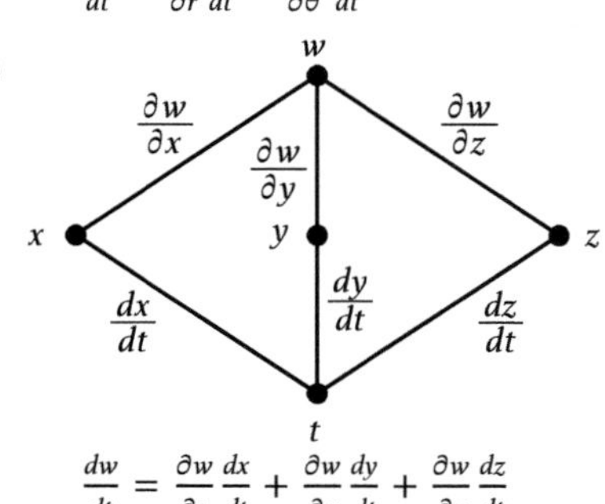

$$\dfrac{dw}{dt} = \dfrac{\partial w}{\partial x}\dfrac{dx}{dt} + \dfrac{\partial w}{\partial y}\dfrac{dy}{dt} + \dfrac{\partial w}{\partial z}\dfrac{dz}{dt}$$

11

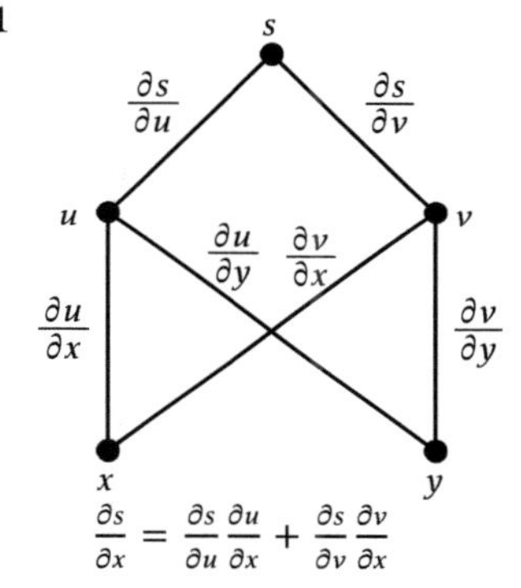

$$\dfrac{\partial s}{\partial x} = \dfrac{\partial s}{\partial u}\dfrac{\partial u}{\partial x} + \dfrac{\partial s}{\partial v}\dfrac{\partial v}{\partial x}$$

13

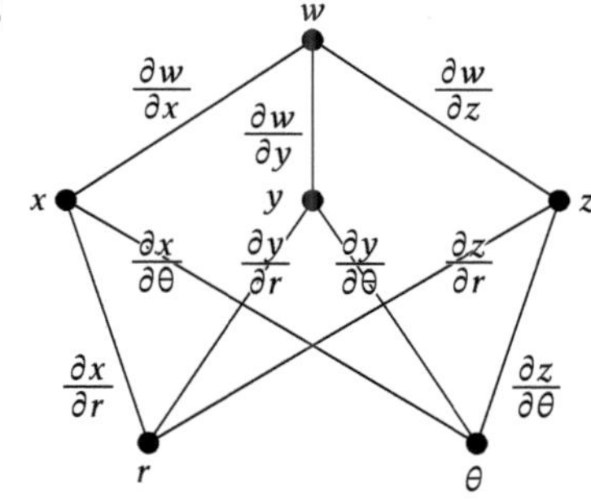

$$\dfrac{\partial w}{\partial r} = \dfrac{\partial w}{\partial x}\dfrac{\partial x}{\partial r} + \dfrac{\partial w}{\partial y}\dfrac{\partial y}{\partial r} + \dfrac{\partial w}{\partial z}\dfrac{\partial z}{\partial r}$$
$$\dfrac{\partial w}{\partial \theta} = \dfrac{\partial w}{\partial x}\dfrac{\partial x}{\partial \theta} + \dfrac{\partial w}{\partial y}\dfrac{\partial y}{\partial \theta} + \dfrac{\partial w}{\partial z}\dfrac{\partial z}{\partial \theta}$$

15

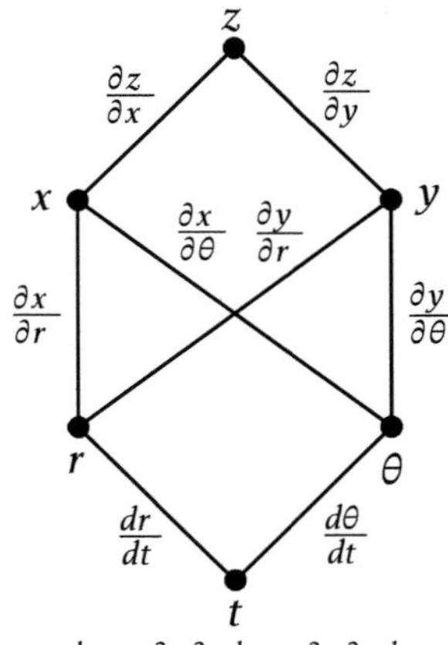

$$\frac{dz}{dt} = \frac{\partial z}{\partial x}\frac{\partial x}{\partial r}\frac{dr}{dt} + \frac{\partial z}{\partial y}\frac{\partial y}{\partial r}\frac{dr}{dt} + \frac{\partial z}{\partial x}\frac{\partial x}{\partial \theta}\frac{d\theta}{dt} + \frac{\partial z}{\partial y}\frac{\partial y}{\partial \theta}\frac{d\theta}{dt}$$

17 $\dfrac{dx}{dt} = 2(t^3 + t)\left(\sin(5(3 - 4t))\right)(3t^2 + 1) - 20(t^3 + t)^2\cos(5(3 - 4t))$

19 $\dfrac{\partial x}{\partial t} = 2(t^3 + s^2)\left(\sin(5(3s + 5t))\right)3t^2 + 25(t^3 + s^2)^2\cos(5(3s + 5t))$

21 $\dfrac{\partial z}{\partial s} = 8(3 + 5r + 2s)^3 + 8s^2r^3 + (-6s^2r^3 + 4(3 + 5r + 2s))2sr^3$

23 $\dfrac{dw}{dt} = 96t^5 - \dfrac{15}{2}\dfrac{2^t}{\sqrt{5t+1}} - 3\sqrt{5t + 1}\cdot 2^t\ln 2$

25 $\dfrac{\partial w}{\partial r} = 2r\cos^2\theta - 6r(r + 4\theta) - 3r^2$

27 $\dfrac{dx}{dy} = (3\cos^2 y - \sin y)(-\sin y) + (8\sin y - \cos y)\cos y$

29 $\dfrac{dz}{dt} = (54(t^3 + 6t)\sin^2 3\pi t + 6 - 10(t^3 + 6t) - 10\tan 3\pi t)(3t^2 + 6) + 162\pi(t^3 + 6t)^2\sin 3\pi t\cos 3\pi t + (6 - 10(t^3 + 6t) - 10\tan 3\pi t)3\pi\sec^2 3\pi t$

31 $\dfrac{14x}{15y^2 + 1}$

33 $-\dfrac{4y - 7}{4x - 3y^2}$

35 $\dfrac{\frac{7}{y}}{2y + \frac{7x}{y^2}} = \dfrac{7y}{2y^3 + 7x}$

37 $\dfrac{-4x}{12y^2 + 1}$

39 $\dfrac{5 - y^2}{1 + 2xy}$

41 $\dfrac{-5x^4}{3(2y - 7)^2(2) - 1}$

43 $-\dfrac{3y + 4}{3x - 2y}$

45 $-\dfrac{2x - 1}{-\sin y} = \dfrac{2x - 1}{\sin y}$

47 $\dfrac{-2x}{2y - \frac{1}{2\sqrt{y-5}}}$

49 $-\dfrac{\sin y - y\cos x}{x\cos y - \sin x}$

51 $-\dfrac{2x^2y + 2yz^2}{2x^2z + 2y^2z} = -\dfrac{x^2y + yz^2}{x^2z + y^2z}$

53 $-\dfrac{6x^5 - 5yz}{4z^3 - 5xy}$

55 $-\dfrac{\frac{1}{y^2+1}\cdot 2y - 1}{\frac{-\cos z}{3 + \cos x}}$

57 $-\dfrac{-\frac{(s^2 - 2r)\cdot 1}{(t+s)^2}}{\frac{(t+s)2s - (s^2 - 2r)\cdot 1}{(t+s)^2}} = \dfrac{s^2 - 2r}{2st + s^2 + 2r}$

59 $\dfrac{3}{7}$

61 $y = \dfrac{5}{4}x + \dfrac{13}{4}$

63 $\dfrac{\partial z}{\partial \theta} = \dfrac{\partial}{\partial \theta}f(g(r, \theta), h(r, \theta))$

$$= \dfrac{f(g(r, \theta + \beta), h(r, \theta + \beta)) - f(g(r, \theta), h(r, \theta))}{\beta}$$

$$\approx \dfrac{f(g(r, \theta) + g_\theta(r, \theta)\beta, h(r, \theta) + h_\theta(r, \theta)\beta) - f(g(r, \theta), h(r, \theta))}{\beta}$$

$$\approx \dfrac{\begin{array}{c}f(g(r, \theta), h(r, \theta))\\ + f_x(g(r, \theta), h(r, \theta))\cdot g_\theta(r, \theta)\beta\end{array}}{\beta}$$

$$+ \dfrac{\begin{array}{c}f_y(g(r, \theta), h(r, \theta))\cdot h_\theta(r, \theta)\beta\\ -f(g(r, \theta), h(r, \theta))\end{array}}{\beta}$$

$$= f_x(g(r, \theta), h(r, \theta))\cdot g_\theta(r, \theta)$$
$$+ f_y(g(r, \theta), h(r, \theta))\cdot h_\theta(r, \theta)$$

$$= \dfrac{\partial f}{\partial x}\cdot\dfrac{\partial x}{\partial \theta} + \dfrac{\partial f}{\partial y}\cdot\dfrac{\partial y}{\partial \theta}$$

$$= \dfrac{\partial z}{\partial x}\cdot\dfrac{\partial x}{\partial \theta} + \dfrac{\partial z}{\partial y}\cdot\dfrac{\partial y}{\partial \theta}$$

65 The work is comparable, but if the tree diagram [is] not produced, perhaps the chain rule is a little m[ore] efficient. The form of the answers is similar, only t[he] order of terms and/or factors is rearranged.

Section 3.8

1 $\nabla f(x, y) = \langle 8xy - 7y^2,\ 4x^2 - 14xy \rangle$

3 $\nabla f(x, y) = \left\langle yx^{y-1} + \frac{y}{x},\ x^y \ln x + \ln x \right\rangle$

5 $\nabla f(x, y, z) = \left\langle 6 \cosh y,\ 6x \sinh y,\ \frac{2z}{1+z^4} \right\rangle$

7 $\nabla f(r, \theta) = \left\langle 6r \cos \theta,\ -3r^2 \sin \theta - \sec \theta \tan^2 \theta - \sec^3 \theta \right\rangle$

9 $\nabla f(1, 5) = \langle 6, -50 \rangle$

11 $\nabla f(-4.7, 3.3) = \langle 6, -87 \rangle$

13 $\nabla f(0, \pi, -\pi) = \langle 0, 1, 1 \rangle$

15 $\nabla g\left(1, \frac{1}{2}\right) = \left\langle \frac{1}{2\sqrt{\frac{3}{4}}},\ \frac{1}{\sqrt{\frac{3}{4}}} \right\rangle = \left\langle \frac{1}{\sqrt{3}},\ \frac{2}{\sqrt{3}} \right\rangle$

17 $D_{\mathbf{u}}f(2, -1) = -\frac{36}{5} = -7.2$

19 $D_{\mathbf{u}}f(-3, 1) = \pi$

21 $D_{\mathbf{u}}f(7, 1, 0) = \frac{-87}{7}$

23 $D_{\mathbf{u}}f(2, 1) = \frac{2904}{\sqrt{61}}$ (Hint: use $\mathbf{u} = \left\langle \frac{6}{\sqrt{61}},\ \frac{5}{\sqrt{61}} \right\rangle$)

25 $\frac{-2}{13}$

27 (a) $\frac{6}{5}$;

   (b) $\frac{42}{5}$;

   (c) $\frac{-42}{5}$;

   (d) $\frac{-6}{5}$

29 the direction of $\langle 4, -10 \rangle$; alternate answer $\left\langle \frac{4}{\sqrt{116}},\ \frac{-10}{\sqrt{116}} \right\rangle$

31 1

33 the direction of $\left\langle 0,\ \frac{2}{25} \right\rangle$; alternate answers include the direction of $\langle 0, 1 \rangle$ or the positive y-direction

35 maximum rate of change $\sqrt{145}$, in the direction of $\langle 12, 1 \rangle$

37 minimum rate of change $-\sqrt{\frac{180}{289}}$, in the direction of $\left\langle -\frac{12}{17},\ \frac{6}{17} \right\rangle$

39 downhill

41 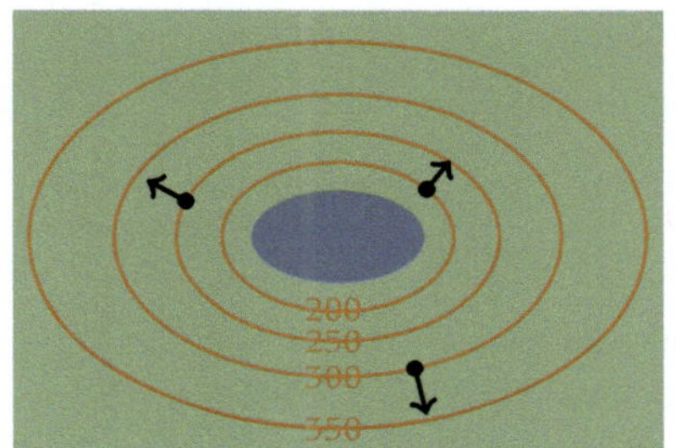

43 (a) $D_{\mathbf{v}}f(a, b) = \frac{\nabla f(a,b) \cdot \mathbf{v}}{\|\mathbf{v}\|}$; (b) $D_{\mathbf{v}}f(a, b) = \frac{\langle 20, 4 \rangle \cdot \langle 3, 4 \rangle}{\sqrt{3^2 + 4^2}} = \frac{76}{5}$; (c) no real difference

45 -30

47 72

49 Using property (2) of the dot product, $(f'(a, b))(\mathbf{v} + \mathbf{w}) = \nabla f(a, b) \cdot (\mathbf{v} + \mathbf{w}) = \nabla f(a, b) \cdot \mathbf{v} + \nabla f(a, b) \cdot \mathbf{w} = (f'(a, b))(\mathbf{v}) + (f'(a, b))(\mathbf{w})$.

Section 3.9

1 $\left(0, \frac{3}{2}\right)$ and $\left(\frac{8}{3}, \frac{3}{2}\right)$

3 none (no critical points)

5 $(0, 0)$ and $\left(-\frac{1}{48}, \frac{1}{12}\right)$

7 $\left(\frac{1}{2}, \frac{1}{5}\right)$

9 $(a, 0)$ for any real number a

11 (a) 11;

   (b) local minimum at (a, b)

13 (a) -4;

   (b) saddle point at (a, b)

15 (a) 0;

   (b) no conclusion can be drawn (the second derivative test is inconclusive)

17 (a) 5;

   (b) local maximum at (a, b)

19 (a) -8;

   (b) saddle point at (a, b)

21 saddle point at $(1, 2)$, local maximum at $(-1, 2)$

23 f has a local minimum at $\left(-\frac{5}{2}, \frac{1}{4}\right)$

25 local minimum at $(2, 3)$, local maximum at $(-2, -3)$, saddle points at $(2, -3)$ and $(-2, 3)$

27 saddle point at $(0, 0)$, local maximum at $(-\frac{1}{2}, -\frac{1}{2})$

29 inconclusive at $(0, 0)$, saddle points at $(1, 2)$ and $(1, -2)$ (see exercise 42 for further exploration of the critical point $(0, 0)$)

31 absolute maximum 2 at $(1, 0)$ and at $(-2, 0)$, absolute minimum -4 at $(-1, 2)$, $(-1, -2)$, $(2, 2)$, and $(2, -2)$

33 absolute minimum $-\ln 3$ at $(3, 0)$, absolute maximum $3 - \ln 3$ at $(3, -1)$ and at $(3, 1)$

35 absolute maximum $\frac{10}{e}$ at $\left(\frac{1}{\sqrt{2}}, \frac{1}{\sqrt{2}}\right)$ and at $\left(-\frac{1}{\sqrt{2}}, -\frac{1}{\sqrt{2}}\right)$, absolute minimum $-\frac{10}{e}$ at $\left(-\frac{1}{\sqrt{2}}, \frac{1}{\sqrt{2}}\right)$ and at $\left(\frac{1}{\sqrt{2}}, -\frac{1}{\sqrt{2}}\right)$

37 absolute minimum -9 at $(0, 3)$ and at $(0, -3)$, absolute maximum 9 at $(-3, 0)$ and at $(3, 0)$

39 (a)

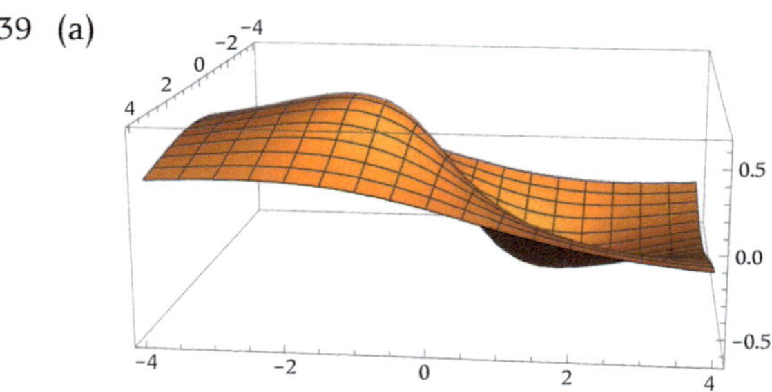

(b) $\left(\sqrt{\frac{3}{5}}, -2\sqrt{\frac{3}{5}}\right)$ and $\left(-\sqrt{\frac{3}{5}}, 2\sqrt{\frac{3}{5}}\right)$

(c) $D = \frac{5}{108}$ for both critical points; f has a local maximum at $\left(\sqrt{\frac{3}{5}}, -2\sqrt{\frac{3}{5}}\right)$ and a local minimum at $\left(-\sqrt{\frac{3}{5}}, 2\sqrt{\frac{3}{5}}\right)$

41 (a)

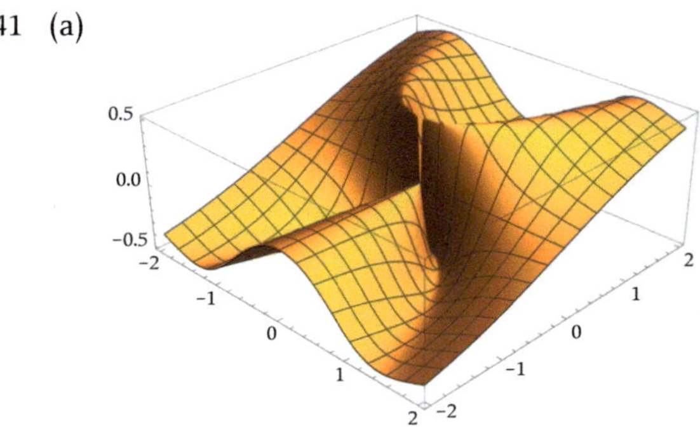

(b) all points of the form (a, a^2), $(a, -a^2)$, and $(0, a)$ except $(0, 0)$ (from $y = \pm x^2$ and from $x = 0$, but $(0, 0)$ is not in the domain of the function)

(c) for each critical point $D = 0$, inconclusive

(d) $f(a, a^2) = \frac{1}{2}$, $f(a, -a^2) = -\frac{1}{2}$; it appears that f has a local (and absolute) maximum value of $\frac{1}{2}$ at (a, a^2) for each real number a except $a = 0$ and that f has a local (and absolute) minimum value of $-\frac{1}{2}$ at $(a, -a^2)$ for each real number a except $a = 0$. It also appears that f has a local maximum value of 0 at $(0, a)$ for each $a < 0$ and a local minimum value of 0 at $(0, a)$ for each $a > 0$. (Notes: (1) the possibility of maximum and minimum values when $D = 0$, as well as the possibility explored in exercise 42, are enough to prove that the second derivative test is inconclusive when $D = 0$. (2) Along a "ridge" of extrema, D will always be 0 where defined (see also example 63), because along the direction of the ridge the function is flat rather than concave up or down.)

43 width $x = 8$ in, inclination $\alpha = 60°$ (Hint: express $\dfrac{\partial A}{\partial x}$ and $\dfrac{\partial A}{\partial \alpha}$ in terms of x and $\cos\alpha$, and solve the two simultaneous equations for the variables x and $\cos\alpha$.)

Section 3.10

1 100 ft by 200 ft (the 200 ft side is parallel to the tractor shed)

3 22.5 ft by 45 ft (the 45 ft side is parallel to the garage)

5 radius $\sqrt[3]{\dfrac{100}{\pi}}$, height $2\sqrt[3]{\dfrac{100}{\pi}}$

7 $6 - \dfrac{2}{\sqrt{3}} \approx 4.845$ km east of the refinery

9 two sides and divider $100\sqrt{30} \approx 547.7$ ft, other two sides $150\sqrt{30} \approx 821.6$ ft

11 width $\dfrac{20}{2+\frac{\pi}{2}} \approx 5.6$ ft, height $\dfrac{10}{2+\frac{\pi}{2}} \approx 2.8$ ft

13 yes (barely–the minimum time is 59.84 minutes); the student should run $6 - \dfrac{4}{\sqrt{561}} = 5.8311$ miles along the shoreline

15 diameter $2\sqrt[3]{\dfrac{58}{\pi}} \approx 5.286$ in

17 (a) \$553.25;

    (b) \$$-$2725 (perhaps we should stop making mini-widgets!)

19 $10 - \sqrt{20} \approx 5.528$ tons of low-grade steel

21 breadth $\dfrac{1}{\sqrt{3}} \approx 0.5774$ m, depth $\sqrt{\dfrac{2}{3}} \approx 0.8165$ m

23 (a) $r = \sqrt[3]{\dfrac{100+\omega}{2\pi}}, h = 2\sqrt[3]{\dfrac{100+\omega}{2\pi}}$

    (b) $(a_1, b_1) = \left(\sqrt[3]{\dfrac{100}{2\pi}}, 2\sqrt[3]{\dfrac{100}{2\pi}}\right)$, $(a_2, b_2) = \left(\sqrt[3]{\dfrac{100+\omega}{2\pi}}, 2\sqrt[3]{\dfrac{100+\omega}{2\pi}}\right)$. The numerator of the quantity is the change in f (the objective function) if the volume (constraint unit) is increased by ω; dividing by ω makes the quantity the rate of change.

    (c) $\dfrac{200}{\frac{100^{4/3}}{(2\pi)^{4/3}}} = 2\sqrt[3]{\dfrac{2\pi}{100}}$ (Note: $f(a_1, b_1) = 6\pi\left(\dfrac{100}{2\pi}\right)^{2/3}, f(a_2, b_2) = 6\pi\left(\dfrac{100+\omega}{2\pi}\right)^{2/3}$)

    (d) $\lambda = \dfrac{2}{\sqrt[3]{\frac{100}{2\pi}}} = 2\sqrt[3]{\dfrac{2\pi}{100}}$

    (e) The answers to c and d are equal.

Chapter 4

Section 4.1

1 (a) $\dfrac{57}{4}$ units^2;

    (b) $\dfrac{93}{4}$ units^2;

    (c) $\left(\dfrac{81}{4} - 3y\right)$ units^2

3 (a) $\dfrac{39}{2}$ units^2;

    (b) $-\dfrac{15}{2}$ units^2;

    (c) $\left(3x^3 - \dfrac{9}{2}\right)$ units^2 (Note: a negative net area means that more area is below the axis than above it.)

5 (a) 2 units^2;

    (b) $\dfrac{8}{3\ln 3}$ units^2;

    (c) $\left(\dfrac{x}{\ln x} - \dfrac{\frac{1}{x}}{\ln x}\right)$ units^2

7 (a) $\ln 3$ units^2;

    (b) $\dfrac{26}{3}$ units^2;

    (c) $\dfrac{3^{y+1}-1}{y+1}$ units^2

9 (a) $\dfrac{1}{2}\sin 8$ units^2;

    (b) $-\dfrac{2}{27}$ units^2

11 114 units^3

13 2 units^3

15 2 units^3

17 54 units^3

19 $(\sinh 1.0404 - \sinh 1)$ units$^3 \approx 0.06332$ units^3

21 -80

23 $\dfrac{75}{8}$

25 $\dfrac{75}{8}$

27 $\dfrac{15}{8}$

29 1

31

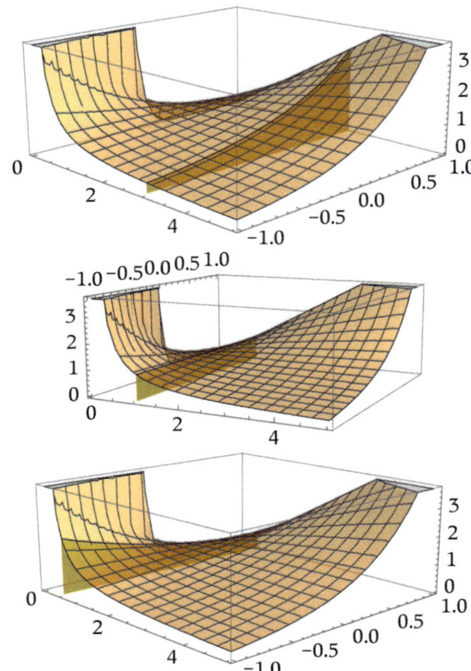

33 (a)

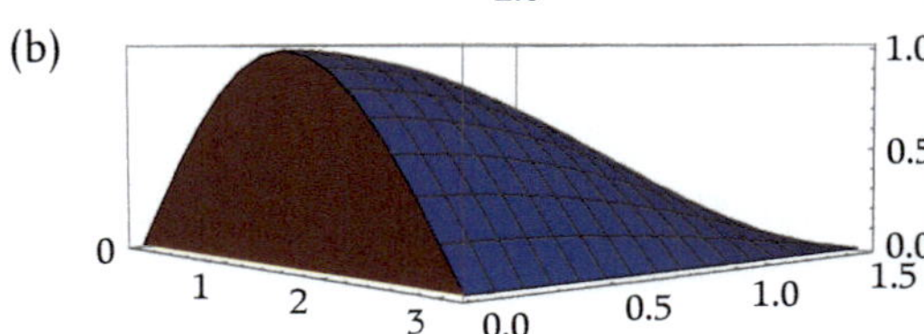

(b)

Section 4.2

1   $\dfrac{68}{3}$

3   $-\dfrac{248}{3}$

5   $\pi \ln 2 + 2 = 4.1776$

7   $4\ln 4 - 3\ln 3 - 2\ln 2 = \ln \dfrac{64}{27} = 0.86305$

9   $\dfrac{1}{3}\sin(\pi + 3) - \dfrac{1}{3}\sin 3 = -\dfrac{2}{3}\sin 3 = -0.09408$

11   $\dfrac{609}{20} = 30.45$

13   $\dfrac{21}{2}(e - 1) = 18.042$

15   2

17   $\dfrac{1}{15}(9\sqrt{3} - 4\sqrt{2} - 1) = 0.5954$

19   $\dfrac{128}{45}$

21   3 (if your answer is $\dfrac{15}{4}$, you switched the limits on x and y)

23   $\dfrac{45}{\sqrt{3}} = 15\sqrt{3} = 25.981$

25   $\dfrac{51}{6(e-1)} = 4.9468$ (if your answer is $\dfrac{1}{6}(e + 1)\ln \dfrac{10}{7} =$ 0.2210, you switched the limits on x and y)

27   0

29   (a) $M_{4,4} = 0.1579$;

    (b)   0.1570;

    (c)   0.573%

31   (a) $M_{5,3} = 0.4677$;

    (b)   0.4711;

    (c)   −0.722%

33   (a) $M_{3,4} = 0.6027$;

    (b)   0.6162;

    (c)   −2.19%

35   Because f and g are continuous throughout the region of integration, Fubini's theorem applies and we may evaluate the double integral by treating as an iterated integral, and then we may use the

single-variable definite integral sum rule:

$$\int_c^d \int_a^b (f(x,y) + g(x,y))\, dx\, dy$$

$$= \int_c^d \left(\int_a^b (f(x,y) + g(x,y))\, dx \right) dy$$

(by Fubini's theorem)

$$= \int_c^d \left(\int_a^b f(x,y)\, dx + \int_a^b g(x,y)\, dx \right) dy$$

(by the definite integral sum rule, inside integral)

$$= \int_c^d \left(\int_a^b f(x,y)\, dx \right) dy + \int_c^d \left(\int_a^b g(x,y)\, dx \right) dy$$

(by the definite integral sum rule, outside integral)

$$= \int_c^d \int_a^b f(x,y)\, dx\, dy + \int_c^d \int_a^b g(x,y)\, dx\, dy$$

(by Fubini's theorem).

37 (a) For any real number k, $\displaystyle \int_c^d \int_a^b k\, dx\, dy = k(b-a)(d-c)$.

(b) The integral represents the volume of a rectangular box, with width $(b-a)$, length $(d-c)$, and height k.

(c) $$\int_c^d \int_a^b k\, dx\, dy = \int_c^d \left(\int_a^b k\, dx \right) dy$$

(constant functions are continuous, so

Fubini's theorem applies)

$$= \int_c^d k(b-a)\, dy$$

Section 4.3

1 (a) $\displaystyle \int_{-2}^0 \int_0^{x+2} x \sin y\, dy\, dx$;

(b) $\displaystyle \int_0^2 \int_{y-2}^0 x \sin y\, dx\, dy$;

(c) $-2\cos^2 1 = -0.5839$

(by the definite integral of a constant rule)

$$= k(b-a)(d-c)$$

(by the definite integral of a constant rule).

39 (a) $\displaystyle \int \frac{4e^{4x}}{x}\, dx = \frac{e^{4x}}{x} + \int \frac{e^{4x}}{x^2}\, dx$;

(b) $\displaystyle \int \left(\frac{4e^{4x}}{x} - \frac{e^{4x}}{x^2} \right) dx = \frac{e^{4x}}{x} + C$;

(c) $\displaystyle \int \left(\frac{4e^{4x}}{x} - \frac{e^{4x}}{x^2} + \frac{1}{x^2} \right) dx = \frac{e^{4x}}{x} - \frac{1}{x} + C$;

(d) $e^4 - 5$;

(e) the answers are the same.

41 (a) $\Delta x = 2\omega,\ \Delta y = 3\omega,\ x_n = 2\omega n,\ y_m = 3\omega m$;

(b) 54;

(c) 54;

(d) 54;

(e) the answers are identical

43 (a) Any of the following cases suffice: case $\alpha \ll \beta$, the expression renders $-\infty$; case $\beta \ll \alpha$, the expression renders ∞; case $\beta = k\alpha$, the expression simplifies to $\frac{1-k^2}{1+k^2}A^2$, which is infinite except when $k \approx \pm 1$; the function is unbounded and Fubini's theorem does not apply (notes: it is not necessary to use all three of those cases, one is sufficient to draw the desired conclusion).

(b) $\frac{y}{x^2+y^2} + C$;

(c) $\frac{1}{x^2+1}$;

(d) $\doteq -\infty$;

(e) $\frac{\pi}{4}$;

(f) $\frac{-x}{x^2+y^2} + C,\ \frac{-1}{1+y^2},\ \infty,\ -\frac{\pi}{4}$

3 (a) $\displaystyle \int_{-1}^0 \int_{-2x}^{\frac{1}{2}(3-x)} x \sin y\, dy\, dx + \int_0^1 \int_x^{\frac{1}{2}(3-x)} x \sin y\, dy\, dx$;

(b) $\displaystyle \int_0^1 \int_{-\frac{y}{2}}^{y} x \sin y\, dx\, dy + \int_1^2 \int_{-\frac{y}{2}}^{3-2y} x \sin y\, dx\, dy$;

(c) -0.04310

5 (a) $\displaystyle\int_0^2 \int_{x^2}^{6-x} x \sin y \, dy \, dx;$

   (b) $\displaystyle\int_0^4 \int_0^{\sqrt{y}} x \sin y \, dx \, dy + \int_4^6 \int_0^{6-y} x \sin y \, dx \, dy;$

   (c) -0.2782

7 (a) $\displaystyle\int_{-2}^2 \int_{-\sqrt{1-\frac{1}{4}x^2}}^{\sqrt{1-\frac{1}{4}x^2}} x \sin y \, dy \, dx;$

   (b) $\displaystyle\int_{-1}^1 \int_{-\sqrt{4-4y^2}}^{\sqrt{4-4y^2}} x \sin y \, dx \, dy;$

   (c) 0

9 $\displaystyle\int_0^3 \int_0^{x^2} x \sin y \, dy \, dx;$

   alternate answer $\displaystyle\int_0^9 \int_{\sqrt{y}}^3 x \sin y \, dx \, dy$

11 $\displaystyle\int_{-1}^1 \int_{1+2y^2}^{4-y^2} x \sin y \, dx \, dy$

13 $\dfrac{32}{7}$

15 $\dfrac{-187}{6}$

17 1

19 $(-1 - \cos 2)\ \text{units}^3$ ($dy \, dx$ is easiest; parts)

21 $\left(\dfrac{5}{2}\sin 4 - \cos 4 + \cos 6\right)\ \text{units}^3 = -0.2782\ \text{units}^3$
    ($dy \, dx$ is easiest; substitution, parts)

23 $0\ \text{units}^3$ ($dx \, dy$ is easiest)

25 $\dfrac{5}{2}\ \text{units}^3$

27 $8\ \text{units}^4$

29 $\left(\dfrac{3}{2} - \dfrac{1}{2}e\right)\ \text{units}^3 = 0.1409\ \text{units}^3$

31 $\dfrac{5}{6}$

Section 4.4

1 $0 \le r \le 3, 0 \le \theta \le \dfrac{\pi}{2}$

3 $1 \le r \le 2, 0 \le \theta \le \dfrac{\pi}{2}$

5 $0 \le r \le \sqrt{7}, \dfrac{\pi}{4} \le \theta \le \pi$

7 (a) $\displaystyle\int_0^\pi \int_{\sin\theta}^2 r^2 \tan\dfrac{\theta}{6} \cdot r \, dr \, d\theta;$

33 $e^{64} - 1$

35 $\dfrac{3}{8}$

37 $1 - \sin 1 = 0.1585$

39 $\dfrac{\ln 2}{12} = 0.0578$

41 $\dfrac{1}{4}\ln 2 = 0.1733$

43 (a) 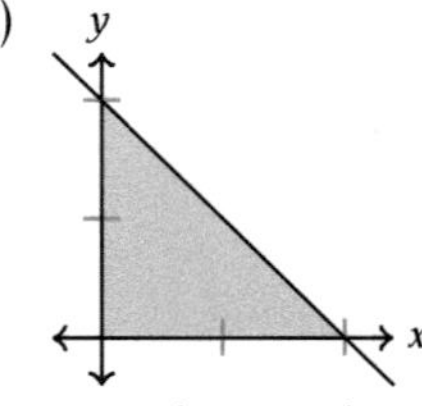

   (b) $\Delta x = \dfrac{1}{2}, \Delta y = \dfrac{1}{2}$

   (c)

$x_1 = \dfrac{1}{4}, x_2 = \dfrac{3}{4}, x_3 = \dfrac{5}{4}, x_4 = \dfrac{7}{4}, y_1 = \dfrac{1}{4}, y_2 = \dfrac{3}{4}$
$y_3 = \dfrac{5}{4}, y_4 = \dfrac{7}{4};$

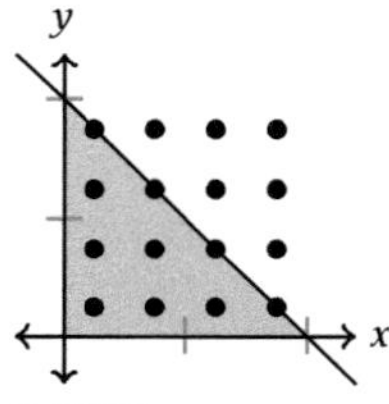

   (d) 1.5128;

   (e) 0.7297;

   (f) 2.2425;

   (g) yes

45 $\dfrac{1}{8}$

   (b) $\left(\dfrac{-23473}{295680} + 4\ln\dfrac{64}{27}\right)\ \text{units}^3 = 3.3728\ \text{units}^3$

9 (a) $\displaystyle\int_{-\frac{\pi}{2}}^{\frac{\pi}{2}} \int_2^{3+\sin\theta} \sqrt{1+\sin\theta} \, dr \, d\theta;$

   (b) $\dfrac{8\sqrt{2}}{3}\ \text{units}^3 = 3.77124\ \text{units}^3$

11 $\frac{81}{4}$ units^3

13 $(2\pi + 4)$ units^3

15 0 units^3 (the volume above the *xy*-plane beneath the surface, and the volume below the *xy*-plane above the surface, are identical, leading to a net volume under the surface of 0.)

17 $\frac{1}{8}a^3$ units^3

19 $\frac{\pi}{2}a^3$ units^3

21 $\frac{\pi}{4}(e - 1) = 1.3495$

23 $\frac{1024}{3}\pi$

25 $\frac{625}{4}\pi$

Section 4.5

1 (a) $\int_1^4 \int_{\sqrt{x}}^x (x^2 + y)\, e^{\sqrt{y}}\, dy\, dx$;

(b) 170.44;

(c) $\bar{x} = \frac{1}{170.44} \int_1^4 \int_{\sqrt{x}}^x x\,(x^2 + y)\, e^{\sqrt{y}}\, dy\, dx$,
$\bar{y} = \frac{1}{170.44} \int_1^4 \int_{\sqrt{x}}^x y\,(x^2 + y)\, e^{\sqrt{y}}\, dy\, dx$;

(d) $(3.3528, 2.6647)$

3 (a) $\int_{-2}^2 \int_{y^2}^4 \sin^2\left(\sqrt{x} + \frac{1}{y+3}\right) dx\, dy$;

(b) 8.26911;

(c) $\bar{x} = \frac{1}{8.26911} \int_{-2}^2 \int_{y^2}^4 x \sin^2\left(\sqrt{x} + \frac{1}{y+3}\right) dx\, dy$,
$\bar{y} = \frac{1}{8.26911} \int_{-2}^2 \int_{y^2}^4 y \sin^2\left(\sqrt{x} + \frac{1}{y+3}\right) dx\, dy$;

(d) $(2.2498, 0.10477)$

5 (a) $\int_2^6 \int_0^{-x^2+8x-12} (x - 4)^2 \sqrt{1 + y^4}\, dy\, dx$;

(b) 21.5173;

(c) $\bar{x} = \frac{1}{21.5173} \int_2^6 \int_0^{-x^2+8x-12} x(x - 4)^2 \sqrt{1 + y^4}\, dy\, dx$,
$\bar{y} = \frac{1}{21.5173} \int_2^6 \int_0^{-x^2+8x-12} y(x - 4)^2 \sqrt{1 + y^4}\, dy\, dx$;

(d) $(4, 1.87554)$

7 (a) 2;

(b) $\left(\frac{4}{3}, \frac{2}{3}\right)$

9 (a) $\frac{320}{3}$;

(b) $\left(0, \frac{14}{5}\right)$

27 $\frac{\sqrt{2}}{2} - \frac{1}{2}$

29 $\frac{\pi}{2}$ (Hints: Don't forget to eliminate a factor of *r*. The region is infinite, but rectangular; evaluate this improper integral in the order *dy dx*.)

31 (a) $\int_c^d \int_0^{g(\theta)} 1 \cdot r\, dr\, d\theta$;

(b) $\frac{(g(\theta))^2}{2}$;

(c) $\int_c^d \frac{1}{2}(g(\theta))^2\, d\theta$;

(d) it is the same, except for the name of the function

11 (a) 3;

(b) $\left(\frac{5}{2}, \frac{1}{8}\right)$

13 (a) $\frac{64}{3}$;

(b) $\left(1, \frac{5}{2}\right)$

15 (a) $\frac{21}{2}\pi + 6$;

(b) $\left(\frac{\pi}{2}, \frac{39\pi+21}{6+\frac{21}{2}\pi}\right) = (1.5708, 3.6813)$

17 (a) 12π;

(b) $\left(0, \frac{1}{3}\right)$ (Hint: switch to polar for some, but not all, of these double integrals)

19 (a) $I_O = 5$;

(b) $R = \sqrt{\frac{5}{2}}$

21 (a) $I_O = 2\pi$;

(b) $R = \sqrt{2}$

23 (a) $I_O = \frac{3}{4}\pi$;

(b) $R = \sqrt{\frac{3}{2\ln 2}} = 1.4711$

25 (a) $I_O = 24\pi$;

(b) $R = \sqrt{2}$

27 (a) center of mass $\left(\frac{11}{18}, \frac{65}{126}\right)$; there is no effect on the center of mass.

(b)

$$(\bar{x},\bar{y})$$

$$= \left(\frac{\iint_D x \cdot k\rho(x,y)\,dx\,dy}{\iint_D k\rho(x,y)\,dx\,dy}, \frac{\iint_D y \cdot k\rho(x,y)\,dx\,dy}{\iint_D k\rho(x,y)\,dx\,dy} \right)$$

$$= \left(\frac{k\iint_D x \cdot \rho(x,y)\,dx\,dy}{k\iint_D \rho(x,y)\,dx\,dy}, \frac{k\iint_D y \cdot \rho(x,y)\,dx\,dy}{k\iint_D \rho(x,y)\,dx\,dy} \right)$$

$$= \left(\frac{\iint_D x \cdot \rho(x,y)\,dx\,dy}{\iint_D \rho(x,y)\,dx\,dy}, \frac{\iint_D y \cdot \rho(x,y)\,dx\,dy}{\iint_D \rho(x,y)\,dx\,dy} \right).$$

Section 4.6

1 (a) $\int_1^{14} \int_3^7 \sqrt{(3x^2+4)^2 + (-4y^3+10y)^2 + 1}\,dy\,dx$;

   (b) $31\,725$ units^2

3 (a) $\int_0^5 \int_0^{\sqrt{x}} \sqrt{\sin^2 y + x^2 \cos^2 y + 1}\,dy\,dx$;

   (b) 16.947 units^2

5 (a) $\int_{-3}^{3} \int_{-\sqrt{9-x^2}}^{\sqrt{9-x^2}} \sqrt{\left(2xe^{x^2+y^2}\right)^2 + \left(2ye^{x^2+y^2}\right)^2 + 1}\,dy\,dx$;

   alternate answer

$$\int_0^{2\pi} \int_0^3 \sqrt{\left(2re^{r^2}\cos\theta\right)^2 + \left(2re^{r^2}\sin\theta\right)^2 + 1} \cdot r\,dr\,d\theta;$$

   (b) $143\,664$ units^2

7 (a) $\int_{-\frac{\pi}{4}}^{\frac{\pi}{4}} \int_1^2 \sqrt{\left(3r^2\cos\theta + \frac{4}{r}\tan\theta\sec\theta\right)^2 + \left(3r^2\sin\theta - \frac{4}{r}\sec\theta\right)^2 + 1} \cdot r\,dr\,d\theta$;

   (b) 20.052 units^2

9 (a) $\int_1^2 \int_{\frac{x}{4}}^{\sqrt{x}} \sqrt{4x^2y^2(\ln(y^3+1))^2\sinh^2(x^2y) + \left(\frac{3y^2}{y^3+1}\cosh(x^2y) + x^2\ln(y^3+1)\sinh(x^2y)\right)^2 + 1}\,dy\,dx$;

   (b) 37.501 units^2

11 $4\sqrt{41}$ units^2 = 25.612 units^2

13 $\frac{1}{4}$ units^2

15 $\frac{1}{2}$ units^2

17 $\frac{7\pi}{12}$ units^2

19 $\frac{1}{3}\left(10\sqrt{10} - 2\sqrt{2}\right)$ units^2 = 9.598 units^2

21 (a) 280π units^2;

   (b) $20\pi(d-c)$ units^2;

   (c) $2\pi a(d-c)$ units^2;

   (d) $2\pi a h$ units^2;

   (e) the two formulas are the same, interpreting the distance between planes as the distance between the planes containing the top and bottom of the cylinder! (Hints for part (a): (1) don't forget to include both the upper and lower halves of the sphere; (2) don't switch to polar; (3) try the trig substitution $y = \sqrt{100 - x^2}\sin\theta$ (not the same θ as polar; it might look unappealing but the integrand simplifies to a constant.)

Section 4.7

Note: alternate answers to these exercises are not always stated.

1 nothing is wrong

3 For the outer integral there is no matching d(variable).

5 The upper limit of integration on the inner integral is the same as the variable being integrated.

7 nothing is wrong

9 8

11 $\frac{14}{3}\ln 3 = 5.1269$

13 $\frac{1}{3}e^3 - e + \frac{2}{3} = 4.6436$

15 $-\frac{608}{15} = -40.5333$

17 $\int_{-2}^{2} \int_{-\sqrt{4-x^2}}^{\sqrt{4-x^2}} \int_0^{4-x^2-y^2} \tan^{-1} xyz\,dz\,dy\,dx$;

   alternate answer

$$\int_{-2}^{2} \int_{-\sqrt{4-y^2}}^{\sqrt{4-y^2}} \int_0^{4-x^2-y^2} \tan^{-1} xyz\,dz\,dx\,dy$$

19 $\int_{-3}^{3} \int_{-\sqrt{9-y^2}}^{\sqrt{9-y^2}} \int_0^{9-y^2-z^2} \tan^{-1} xyz\,dx\,dz\,dy$;

   alternate answer

$$\int_{-3}^{3} \int_{-\sqrt{9-z^2}}^{\sqrt{9-z^2}} \int_0^{9-y^2-z^2} \tan^{-1} xyz\,dx\,dy\,dz$$

21 (a) $\int_1^3 \int_{-2}^2 \int_{x^2}^4 x\, dz\, dx\, dy$;

  (b) 0

23 (a) $\int_0^5 \int_0^4 \int_{\frac{z}{2}}^z (x+2)\, dy\, dz\, dx$;

  (b) 90; alternate answers to part (a) include $\int_0^4 \int_{\frac{1}{2}z}^z \int_0^5 (x+2)\, dx\, dy\, dz$ and the sum of two triple integrals $\int_0^2 \int_y^{2y} \int_0^5 (x+2)\, dx\, dz\, dy + \int_2^4 \int_y^4 \int_0^5 (x+2)\, dx\, dz\, dy$

25 $\frac{81}{4}\pi$

27 $\left(\frac{3}{4}, \frac{4}{3}, 2\right)$

29 $\left(e-1, \frac{4}{\ln 5}, \frac{1}{\ln 2}\right) = (1.718, 2.485, 1.443)$

31 $-\frac{31}{6}$

33 $\frac{2}{\pi}$

35 1190°F

37 $\frac{\pi}{2}$

39 (a) $y = \pm\sqrt{9 - x^2 - z}$;

  (b) $\int \int \int_{-\sqrt{9-x^2-z}}^{\sqrt{9-x^2-z}} 3y\, dy\, d\, d$;

(c)

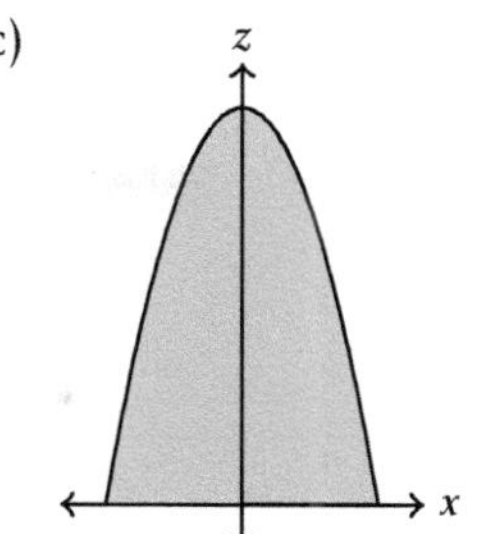

  (d) $y = 0$;

  (e) $z = 9 - x^2$;

  (f) $\int \int_0^{9-x^2} \int_{-\sqrt{9-x^2-z}}^{\sqrt{9-x^2-z}} 3y\, dy\, dz\, d$;

  (g) $\int_{-3}^3 \int_0^{9-x^2} \int_{-\sqrt{9-x^2-z}}^{\sqrt{9-x^2-z}} 3y\, dy\, dz\, dx$;

  (h) 0 (the inner integral evaluates to 0)

41 (a) $\int_0^2 \int_1^4 \int_0^{xy+4x+5y} 1\, dz\, dy\, dx$;

  (b) 114 units^3; after evaluating the inner integral, the solution is identical to that of exercise 11 of Section 4.1.

43 $\frac{16}{2835}\left(\frac{1}{8}(12^{7/2}) - \frac{3}{4}(6^{7/2}) + 3^{7/2}\right)$
$= \frac{16}{2835}\left(459\sqrt{3} - 162\sqrt{6}\right) = 2.2473$ (the volume of the region is $\frac{27}{2}$)

Section 4.8

Note: alternate answers to these exercises are not always stated.

1

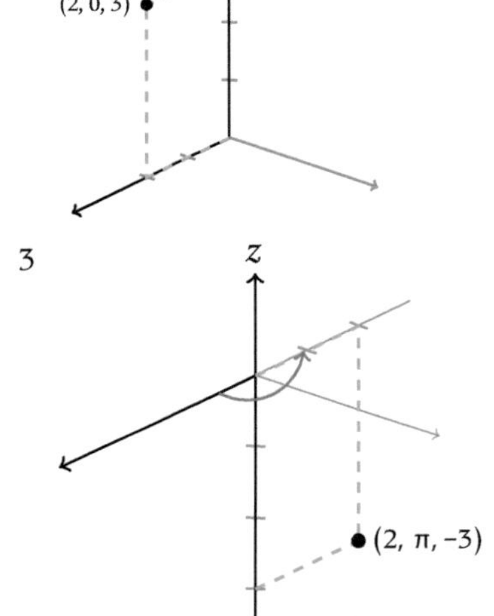

3

5

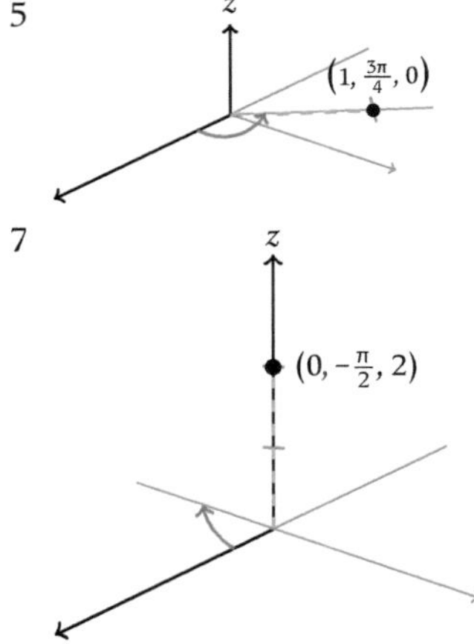

7

9 $\left(3\sqrt{2}, \frac{\pi}{4}, 1\right)$; alternate answers include $\left(-3\sqrt{2}, \frac{5\pi}{4}, 1\right)$

11 $(2, \pi, 55)$; alternate answers include $(2, -\pi, 55)$ and $(-2, 0, 55)$

13 $\left(2, \frac{\pi}{6}, \pi\right)$; alternate answers include $\left(2, \frac{13\pi}{6}, \pi\right)$

15 $\left(5, \tan^{-1}\frac{4}{3}, 5\right)$

17 $(0, -4, 11.4)$

19 $\left(\sqrt{2}, -\sqrt{2}, 18\right)$

21 $(\pi, 0, 3)$

23 $\left(\frac{5}{2}, -\frac{5\sqrt{3}}{2}, 1\right)$

25 $f(r, \theta, z) = \dfrac{r\cos\theta \cdot z}{r\sin\theta} = z\cot\theta$

27 $f(r, \theta, z) = r^6$

29 $f(r, \theta, z) = 1 + \sin(r\sin\theta)$

31 (a) (ii);

    (b) (iv);

    (c) (i);

    (d) (iii)

33 $\int_0^{2\pi}\int_0^6\int_0^2 (r\cos\theta + r\sin\theta)\cdot r\,dz\,dr\,d\theta$
$= \int_0^{2\pi}\int_0^6\int_0^2 r^2(\cos\theta + \sin\theta)\,dz\,dr\,d\theta$
(Don't forget the extra r!)

35 $\int_0^{2\pi}\int_1^2\int_1^3 \dfrac{r^3\cos\theta\sin\theta}{z}\,dz\,dr\,d\theta$

Section 4.9

1 a sphere of radius 4 centered at the origin

3 the origin (one point)

5 the xy-plane

7

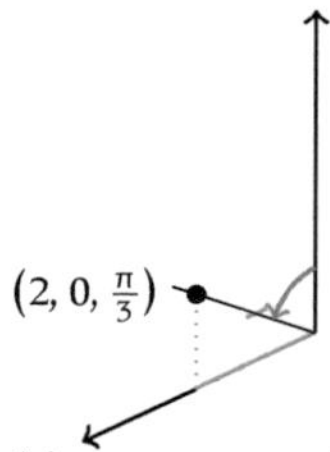

$\left(2, 0, \frac{\pi}{3}\right)$

(the point is in the xz-plane)

9

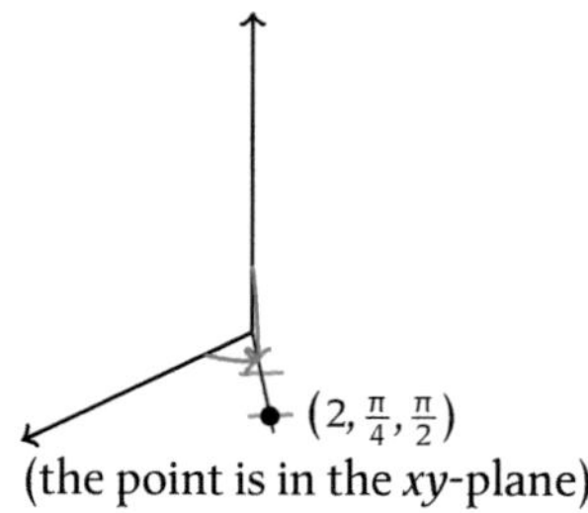

$\left(2, \frac{\pi}{4}, \frac{\pi}{2}\right)$

(the point is in the xy-plane)

(Don't forget the extra r!)

37 3π units^4 (If your answer is 2π units^4, you might have forgotten the extra r)

39 $9\pi\left(e^{25} - 1\right)$ units^4 (Using Cartesian coordinates results in a nonelementary integral; cylindrical coordinates are easiest. If you tried cylindrical coordinates and still obtained a nonelementary integral, you may have forgotten the extra r.)

41 $\frac{1}{4}$ units^4 (Although not terribly difficult in cylindrical coordinates, it is simpler this time not to make the switch, sticking with Cartesian coordinates. If your answer is $\frac{1}{3}$ units^4, you may have forgotten the extra r.)

43 $\int_0^{\pi}\int_0^3\int_0^{r^2} r^{5/2}z\cos\theta\sqrt{\sin\theta}\,dz\,dr\,d\theta$ (Don't forget the extra r!)

45 $\int_0^{2\pi}\int_0^3\int_0^5 7r\,dz\,dr\,d\theta$ (Don't forget the extra r!)

47 $\left(0, \dfrac{91192}{17925\pi}, \dfrac{1695}{239}\right) = (0, 1.6194, 7.0921)$ (Note: the total mass is $m = \dfrac{5975\pi}{6}$)

11

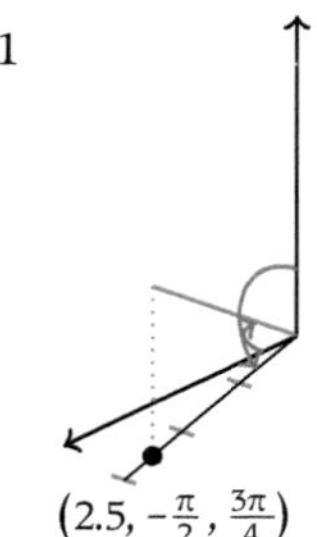

$\left(2.5, -\frac{\pi}{2}, \frac{3\pi}{4}\right)$

(the point is below the xy-plane and is in the yz-plane)

13

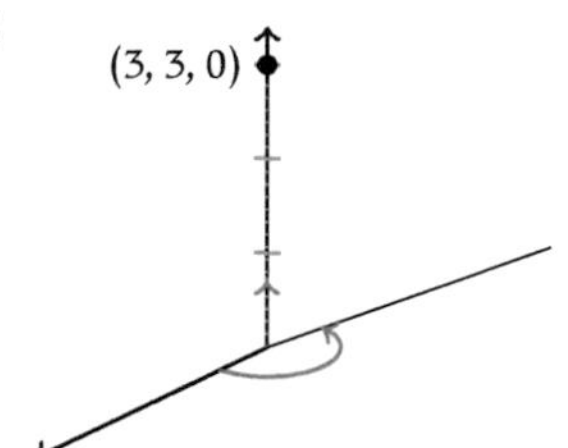

$(3, 3, 0)$

(the point is on the positive z-axis)

15 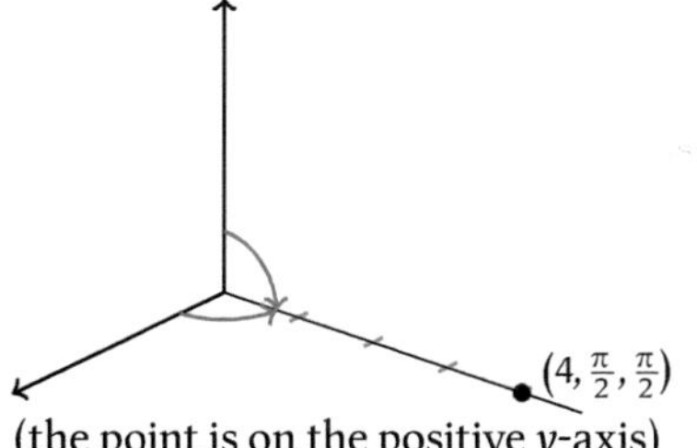

(the point is on the positive *y*-axis)

17 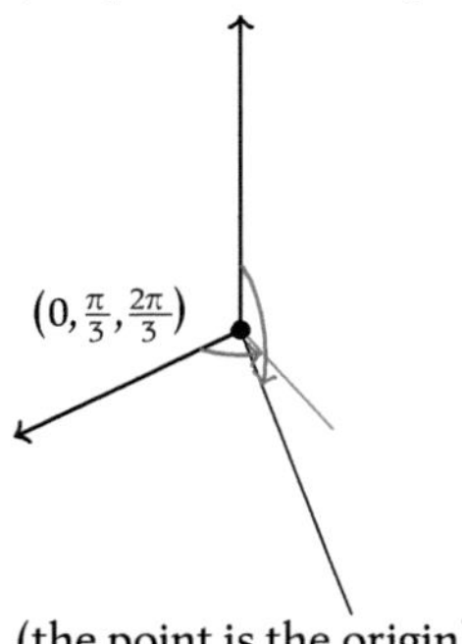

$\left(0, \frac{\pi}{3}, \frac{2\pi}{3}\right)$

(the point is the origin)

19 $\left(\dfrac{5}{2\sqrt{2}}, \dfrac{5\sqrt{3}}{2\sqrt{2}}, -\dfrac{5}{\sqrt{2}}\right) = \left(\dfrac{5\sqrt{2}}{4}, \dfrac{5\sqrt{6}}{4}, -\dfrac{5\sqrt{2}}{2}\right)$

Section 4.10

1 29

3 $9u^2v^2 - 1$

5 r (Hint: use a hyperbolic identity to simplify)

7 e^{w+z}

9 $\int_0^2 \int_0^{1-\frac{1}{2}u} dv\, du$

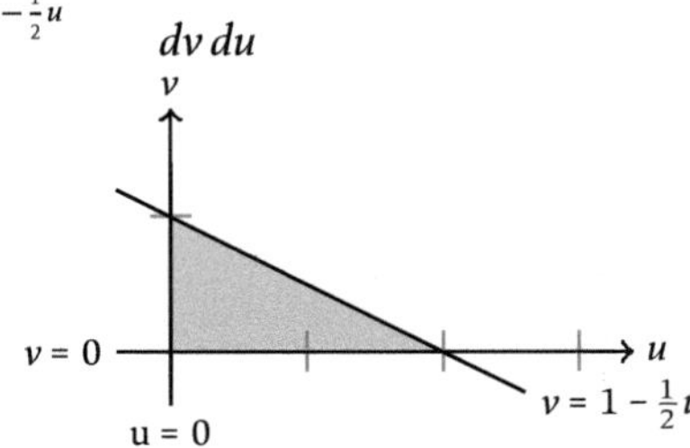

21 $(1, 1, 0)$

23 $(0, 0, 3.7)$

25 $\left(6\sqrt{2}, \dfrac{\pi}{2}, \dfrac{\pi}{4}\right)$

27 $\left(2, \pi, \dfrac{\pi}{2}\right)$

29 $\left(2, \pi, \dfrac{5\pi}{6}\right)$

31 $f(\rho, \theta, \varphi) = 3 - \rho^2$

33 $f(\rho, \theta, \varphi) = \rho^2 \sin^2 \varphi$

35 $f(\rho, \theta, \varphi) = \dfrac{1}{4}\rho^2 - \rho \cos \varphi - 1$

37 $\int_0^\pi \int_0^{2\pi} \int_0^{16}(1 - \rho^2)\rho^2 \sin \varphi\, d\rho\, d\theta\, d\varphi$ (the order of integration may be changed)

39 $\int_{\frac{\pi}{2}}^\pi \int_0^{2\pi} \int_0^2 \left(\rho^2 \sin^2 \varphi \cos \theta \sin \theta - 3\rho \cos \varphi\right) \cdot \rho^2 \sin \varphi\, d\rho\, d\theta\, d\varphi$ (the order of integration may be changed)

41 $\dfrac{4}{3}\pi(e - 1)$

43 8π

45 0

47 $\dfrac{19}{4} = 4.75$

alternate answer $\int_0^1 \int_0^{2-2v} du\, dv$

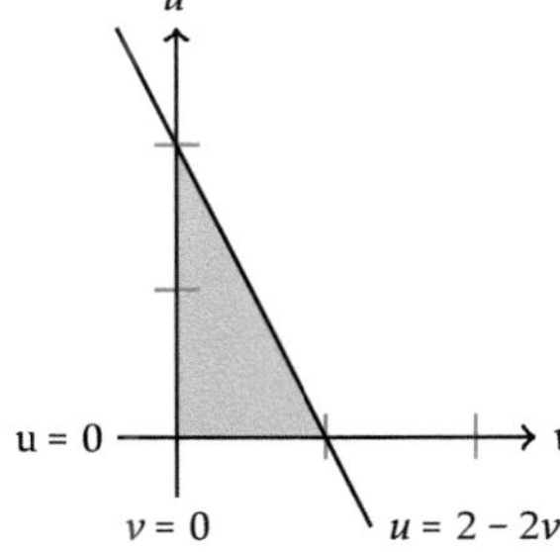

11 $\int_{-1}^{\frac{10}{3}} \int_{-2}^{-\frac{3}{5}u} dv\,du$

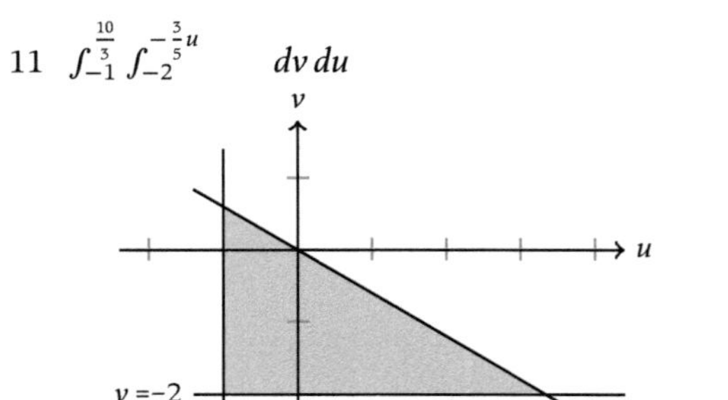

alternate answer $\int_{-2}^{\frac{3}{5}} \int_{-1}^{-\frac{5}{3}v} du\,dv$

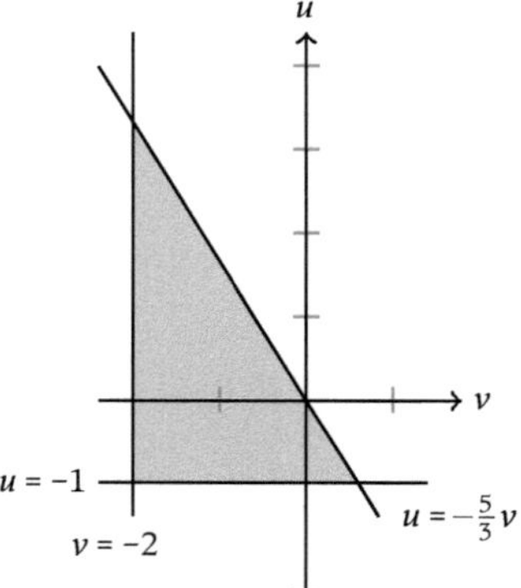

13 $\sqrt{3}uv^{3/2}$

Chapter 5

Section 5.1

1 $\frac{29}{2}\sqrt{73}$

3 99

5 -18

7 0

9 $\frac{1}{12}\left(17^{3/2} - 5^{3/2}\right)$

11 $\frac{3\sqrt{34}}{5}\sin 7 + \frac{3\sqrt{34}}{25}\cos 7 - \frac{3\sqrt{34}}{25}\cos 2$

13 (a) $\int_{-1}^{0}(4t - 1)2\sqrt{t^2 + 1}\,dt;$

(b) $\frac{1}{3}\left(8 - 19\sqrt{2} - 3\ln(1 + \sqrt{2})\right) = -7.1714$

15 $x = \frac{1}{17}u + \frac{3}{17}v,\, y = -\frac{5}{17}u + \frac{2}{17}v$

17 $\frac{8}{3}\left(1 - \frac{1}{2\sqrt{2}}\right) = 1.724$

19 $\frac{1}{2}(e^4 - 1)\ln 2 = 18.576$

21 $\frac{3}{8}\left(4^{8/3} - 1\right) = 14.744$

23 $\frac{\partial(x,y)}{\partial(\rho,\theta)} = e^{2\rho}$

25 (a) x-coordinate $g(u_n + \Delta u, v_m + \Delta v) \approx g(u_n, v_m) \cdot g_u(u_n, v_m)\Delta u + g_v(u_n, v_m)\Delta v,$
y-coordinate $h(u_n + \Delta u, v_m + \Delta v) \approx h(u_n, v_m) \cdot h_u(u_n, v_m)\Delta u + h_v(u_n, v_m)\Delta v;$

(b) vector $\langle g_u(u_n, v_m)\Delta u, h_u(u_n, v_m)\Delta u\rangle = \mathbf{z};$

(c) vector $\langle g_v(u_n, v_m)\Delta v, h_v(u_n, v_m)\Delta v\rangle = \mathbf{w};$

(d) opposite sides are parallel, so the quadrilateral is a parallelogram (alternate answer: opposite sides are congruent, so the quadrilateral is parallelogram)

27 (a) $x = 5u, y = 2v;$

(b) 10π units^2 (Hint: the Jacobian is 10, switch to polar coordinates after using the change of variables)

29 $\frac{4\pi}{3}abc$ units^3 (Hints: use $z = c\sqrt{1 - \frac{x^2}{a^2} - \frac{y^2}{b^2}}$ and double the result; switch to polar coordinates after using the change of variables)

15 (a) $\int_0^{10\pi}\left(9\cos^2 t - 27\cos t\sin t + t\right)\sqrt{10}\,dt;$

(b) $5\sqrt{10}\pi(9 + 10\pi) = 2007.6$

17 (a) $\frac{8}{3};$

(b) 24

19 (a) $-\frac{2352}{5} = -470.4;$

(b) $-\frac{4116}{5} = -832.2$

21 $\frac{59}{2};$ alternate answer $7 + \frac{45}{2}$

23 $8\sqrt{5} + \frac{945}{4}$

25 $\dfrac{22}{5}$

27 4π

29 63

31 (a) $m = 21\sqrt{6}$ (mass units are not identified in the exercise);

    (b) $\left(\dfrac{24}{7}, -\dfrac{5}{7}, \dfrac{40}{7}\right)$

33 (a) $m = 5\pi$ (mass units are not identified in the exercise);

    (b) $\left(\dfrac{6}{\pi}, 0, \dfrac{8}{\pi}\right)$

35 (a) $-\dfrac{1}{4}$;

    (b) $\dfrac{1}{4}$; (c) yes

Section 5.2

1 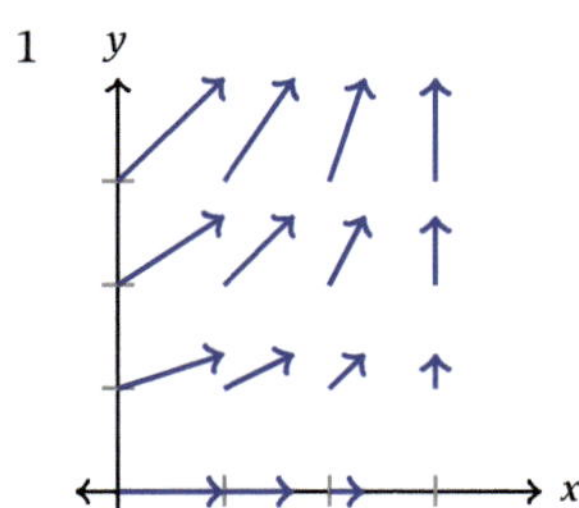

3 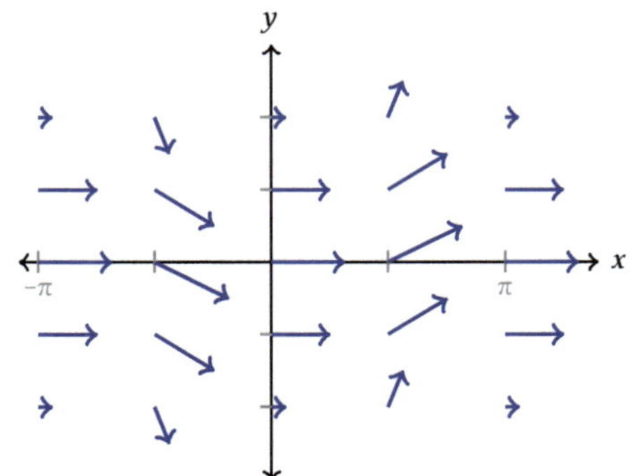

5

7 Answers will vary depending on the technology used, the options chosen, and the domain chosen;

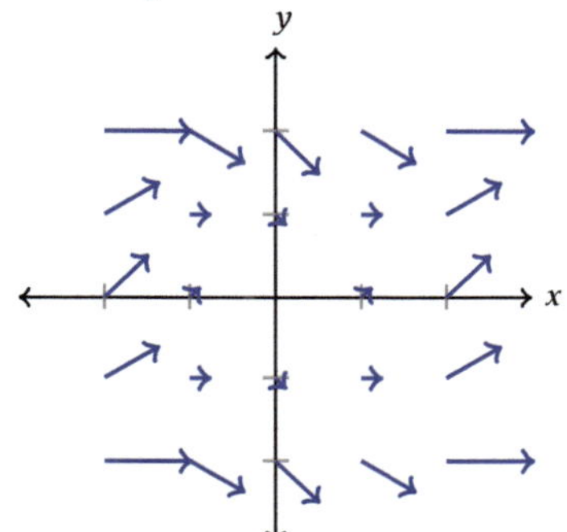

9 Answers will vary depending on the technology used, the options chosen, and the domain chosen;

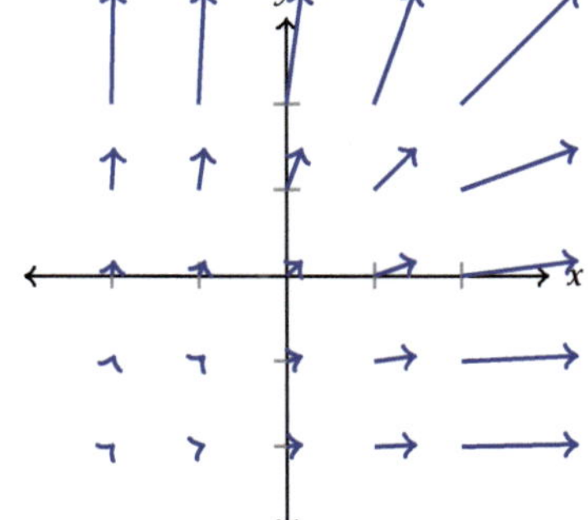

11 Answers will vary depending on the technology used, the options chosen, and the domain chosen;

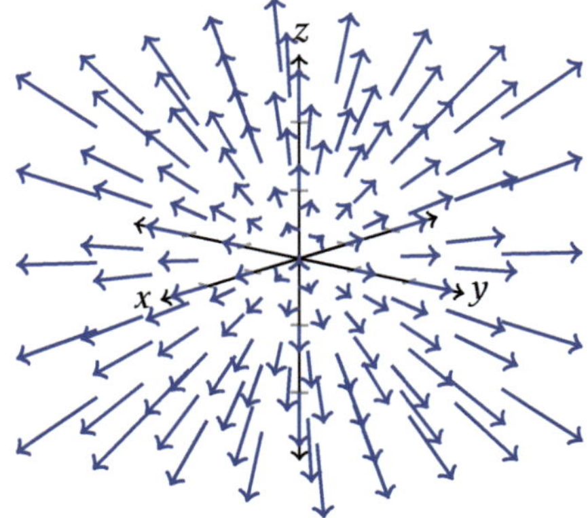

Section 5.3

1 (a) $2e^4$;

   (b) $-2e^4$,

   (c) they are negatives of one another

3 80

5 2

7 0

9 The fundamental theorem does not apply because f (and ∇f) is not defined when $y = \frac{\pi}{2}$, which must be crossed on a path from $(\pi, 0)$ to $(0, \pi)$.

11 yes, conservative

13 yes, conservative

15 no, not conservative

17 yes, conservative

19 yes, conservative

21 no, not conservative

23 $f(x, y) = \frac{3}{2}x^2 + \frac{1}{3}y^3 + C$

25 $f(x, y) = x^2 y - xy^2 + C$

27 $f(x, y, z) = 4xy + z^2 + C$

29 $f(x, y, z) = x^2 y - 3xz + 4yz^2 + 5z + C$

31 $\frac{58}{3}$ (conservative, $f(x, y) = 2x^2 + \frac{1}{3}y^3 + C$)

33 $\frac{43}{16}$ (conservative, $f(x, y) = \frac{y}{x^2} - 2x + 3y + C$)

35 not conservative, the path matters

13 24

15 $\frac{15}{2} + \frac{1}{2}\ln 4 = \frac{15}{2} + \ln 2$

17 $\frac{64}{3}\pi^3$

19 $\frac{17}{2}$

21 $1 + \frac{1}{2}\sin 4 - \cos 4$

23 (a) $2e^4$;

   (b) $-2e^4$;

   (c) they are negatives of one another

25 80

27 $\int_{-C} M\,dx + N\,dy + P\,dz = -\int_C M\,dx + N\,dy + P\,dz$

37 35 (conservative, $f(x, y, z) = 3xyz + z + C$)

39 $\frac{21}{2}$ (conservative, $f(x, y, z) = 2^x + y^2 + 2z$)

41 26 (Comment: this is the same type of question a 31–40, written in a different form. The expression $3x^2\,dx - z^2\,dy - 2yz\,dz$ is called a *total differentia* and when a total differential represents a conserva tive field the total differential is called *exact*.)

43 -1.15 (conservative, $f(x, y, z) = 3xz + y + C$)

45 (a) $\mathbf{r}(t) = \langle t, t \rangle$, $1 \le t \le 4$ (alternate solutio $\mathbf{r}(t) = \langle 1 + 3t, 1 + 3t \rangle, 0 \le t \le 1$);

   (b) $\frac{213}{2}$;

   (c) no, the two methods are comparable (if yo used the alternate solution to (a), you migh conclude that this approach is longer messier);

   (d) both x and y are replaced by t, otherwise the are the same (if you used the alternate solutio to (a), you might have a different conclusion)

47 (a)

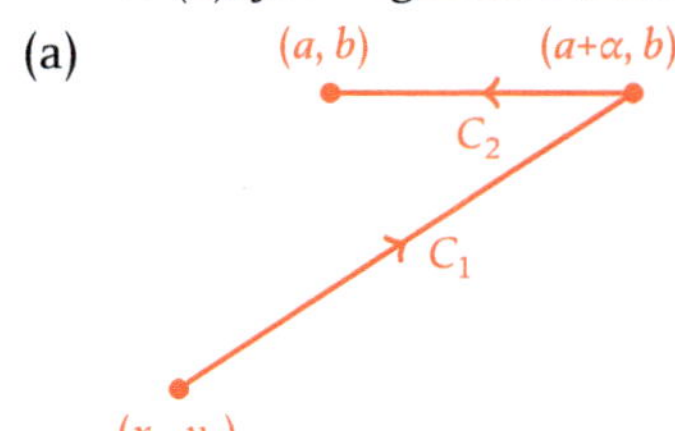

(b) $\int_{(x_0,y_0)}^{(a,b)} \mathbf{F} \cdot d\mathbf{r} = \int_{C_1 \cup C_2} \mathbf{F} \cdot d\mathbf{r}$;

(c) $= \int_{C_1} \mathbf{F} \cdot d\mathbf{r} + \int_{C_2} \mathbf{F} \cdot d\mathbf{r}$;

(d) $\frac{1}{\alpha} \left(\int_{(x_0,y_0)}^{(a+\alpha,b)} \mathbf{F} \cdot d\mathbf{r} - \int_{(x_0,y_0)}^{(a,b)} \mathbf{F} \cdot d\mathbf{r} \right)$
$= \frac{1}{\alpha} \left(\int_{C_1} \mathbf{F} \cdot d\mathbf{r} - \left(\int_{C_1} \mathbf{F} \cdot d\mathbf{r} + \int_{C_2} \mathbf{F} \cdot d\mathbf{r} \right) \right)$
$= \frac{1}{\alpha} \left(-\int_{C_2} \mathbf{F} \cdot d\mathbf{r} \right) = \frac{1}{\alpha} \left(\int_{-C_2} \mathbf{F} \cdot d\mathbf{r} \right)$
$= \frac{1}{\alpha} \int_{(a,b)}^{(a+\alpha,b)} \mathbf{F} \cdot d\mathbf{r}$;

(e) $\frac{1}{\alpha} \left(\int_a^{x+\alpha} f(t)\, dt - \int_a^x f(t)\, dt \right) = \frac{1}{\alpha} \int_x^{x+\alpha} f(t)\, dt$

49 (a) $\nabla f(x,y) = \left\langle \sqrt[3]{y}, \frac{x}{3y^{2/3}} \right\rangle$, the expression is undefined for $y = 0$, which must be crossed between
$(1,-1)$ and $(2,1)$;

(b) ≈ 1;

(c) ≈ 2;

(d) 3;

(e) $\int_{(1,-1)}^{(k,-\omega)} \nabla f \cdot d\mathbf{r} \approx 1$, $\int_{(k,\omega)}^{(2,1)} \nabla f \cdot d\mathbf{r} \approx 2$, 3, yes it
agrees; (f) yes

51 (a) $\frac{\pi}{2}$;

(b) -3

Section 5.4

1 positive

3 negative

5 20

7 $-\frac{1}{6}$

9 $-\frac{81}{12} + \frac{2}{5} \cdot 3^{5/2} = -\frac{27}{4} + \frac{18}{5}\sqrt{3}$

11 0 (the field is conservative)

13 $4\sin 2 + 2\cos 2 - 6$

15 15π units^2

17 $\frac{4}{3}$ units^2

19 6π units^2

21 $\frac{32}{3}$ units^2

23 $\frac{3}{2}$ units^2

25 (a) $(\bar{x},\bar{y}) = \left(\frac{\iint_D x\, dx\, dy}{\iint_D 1\, dx\, dy}, \frac{\iint_D y\, dx\, dy}{\iint_D 1\, dx\, dy} \right)$;

(b) $\frac{1}{A} \iint_D x\, dA = \frac{\iint_D x\, dA}{\iint_D 1\, dA}$;

(c) $\frac{1}{A} \iint_D y\, dA = \frac{\iint_D y\, dA}{\iint_D 1\, dA}$;

(d) $(\bar{x},\bar{y}) = \left(\frac{1}{2A} \oint_C x^2\, dy, -\frac{1}{2A} \oint_C y^2\, dx \right)$

27 (a) $x = 6\cos t, y = 3\sin t, 0 \le t \le \frac{\pi}{2}$;

(b) $A = \frac{1}{4} \cdot 3 \cdot 6 \cdot \pi = \frac{9}{2}\pi$;

(c)
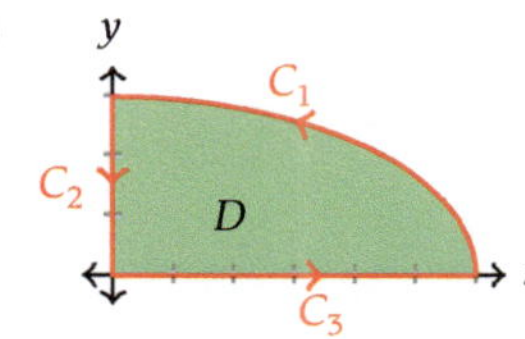

$C_2: x = 0, y = 3 - t, 0 \le t \le 3$, $C_3: x = t, y = 0$,
$0 \le t \le 6$;

(d) centroid $\left(\frac{8}{\pi}, \frac{4}{\pi} \right)$;

(e) no

Section 5.5

1 $\operatorname{div} \mathbf{F} = \cos x + 2y + 2xyz$

3 $\operatorname{div} \mathbf{F} = 6xyz$

5 $\operatorname{div} \mathbf{F} = 0$

7 $\operatorname{div} \mathbf{F} = 0$

9 $\operatorname{div} \mathbf{F} = \frac{3}{2\sqrt{2x+y}}$

11 $\operatorname{curl} \mathbf{F} = \langle xz^2 - xy^2, x^2y - yz^2, y^2z - x^2z \rangle$

13 $\operatorname{curl} \mathbf{F} = \langle 2y\cos z \sin z, 2x\sin z \cos z, 0 \rangle$
$= \langle y\sin 2z, x\sin 2z, 0 \rangle$

15 $\operatorname{curl} \mathbf{F} = 0$

17 $\operatorname{curl} \mathbf{F} = 0$

19 $\operatorname{curl} \mathbf{F} = \left\langle 0, 0, \frac{-2}{\sqrt{2x+y}} \right\rangle$

21 yes, conservative ($\operatorname{curl} \mathbf{F} = 0$)

23 no, not conservative ($\operatorname{curl} \mathbf{F} \ne 0$)

25 yes, conservative ($\operatorname{curl} \mathbf{F} = 0$)

27 $\operatorname{div} \mathbf{G} = 10 \ne 0$; because $\operatorname{div} \mathbf{G} \ne 0$, $\mathbf{G}$ is not the curl
of any other vector field.

29 $\operatorname{div}\mathbf{G} = 8x \neq 0$; because $\operatorname{div}\mathbf{G} \neq 0$, $\mathbf{G}$ is not the curl of any other vector field.

31 $\nabla \cdot [f\mathbf{F}] = 9x^2 + 3yz + 5x^2 + 10yz + x^2 + 2yz = 15x^2 + 15yz$

33   yes, harmonic ($\nabla^2 f = 0$)

35   no, not harmonic ($\nabla^2 f = 6 \neq 0$)

37   yes, harmonic ($\nabla^2 f = 0$)

39   (a) $\operatorname{div}(\mathbf{F} + \mathbf{G}) = \operatorname{div}\langle M + Q, N + R, P + S\rangle = M_x + Q_x + N_y + R_y + P_z + S_z = (M_x + N_y + P_z) + (Q_x + R_y + S_z) = \operatorname{div}\mathbf{F} + \operatorname{div}\mathbf{G}$;

   (b)   divergence sum rule

41   (a) $\operatorname{div}(\mathbf{F}\times\mathbf{G}) = \operatorname{div}\langle NS - RP, -(MS - PQ), MR - NQ\rangle = NS_x + SN_x - (RP_x + PR_x) - (MS_y + SM_y - (PQ_y + QP_y)) + MR_z + RM_z - (NQ_z + QN_z);$ $\mathbf{G}\cdot\operatorname{curl}\mathbf{F} - \mathbf{F}\cdot\operatorname{curl}\mathbf{G} = \langle Q, R, S\rangle\cdot\langle P_y - N_z, -(P_x - M_z), N_x - M_y\rangle - \langle M, N, P\rangle \cdot \langle S_y - R_z, -(S_x - Q_z), R_x - Q_y\rangle = QP_y - QN_z - RP_x + RM_z + SN_x - SM_y - (MS_y - MR_z + NQ_z - NS_x + PR_x - PQ_y)$ because the two expressions are equal, $\operatorname{div}(\mathbf{F}\times\mathbf{G}) = \mathbf{G}\cdot\operatorname{curl}\mathbf{F} - \mathbf{F}\cdot\operatorname{curl}\mathbf{G}$;

   (b)   divergence cross product rule

43   (a)   The outflow includes some to the sides, and not just straight out;

   (b)   the inflow includes some from the sides, and not just straight in

Section 5.6

1   no (ridges and creases are present)

3   yes

5   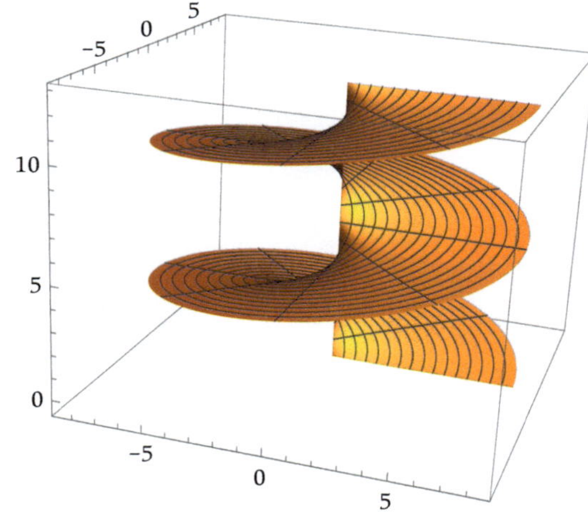

7   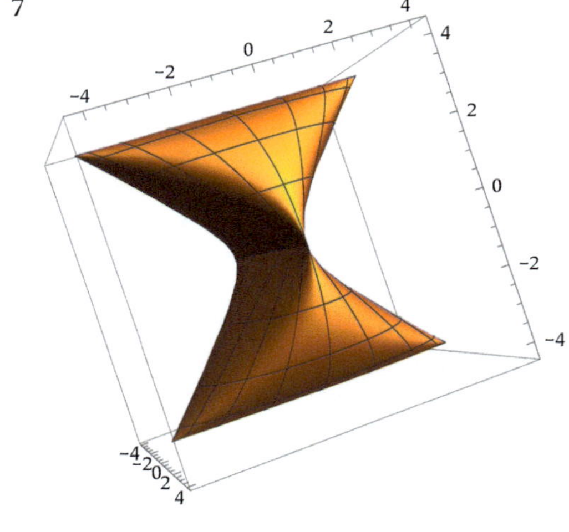

9

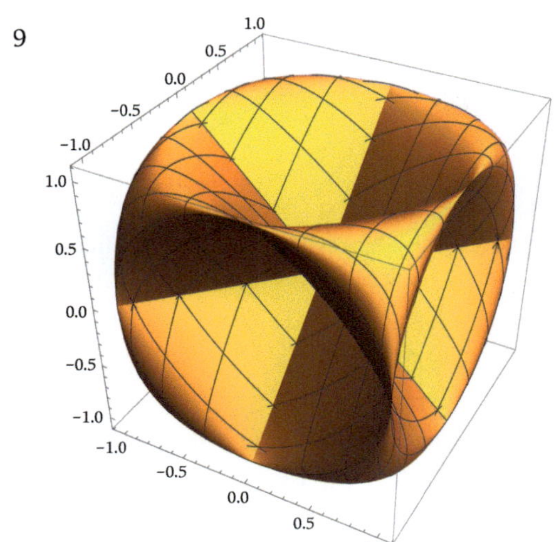

11 (a) $x = u, y = v,$
$z = \sqrt{u^2 - v^2}, 4 \leq u \leq 7, -3 \leq v \leq 3;$

(b)

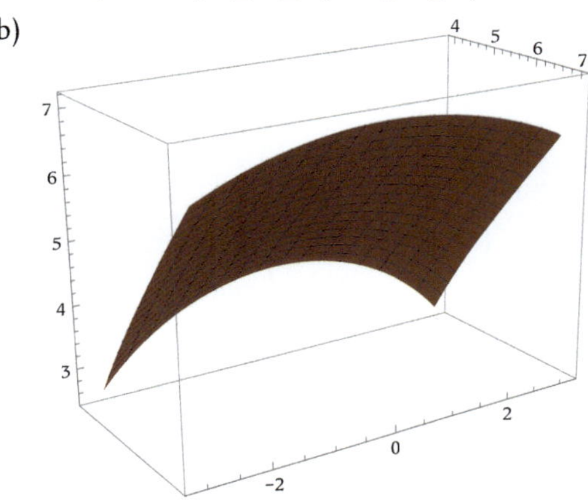

13 (a) $x = u, \qquad y = v, \qquad z = 0.7u\sin u + 2\cos v,$
$0 \leq u \leq 3\pi, \pi \leq v \leq 5\pi;$

(b)

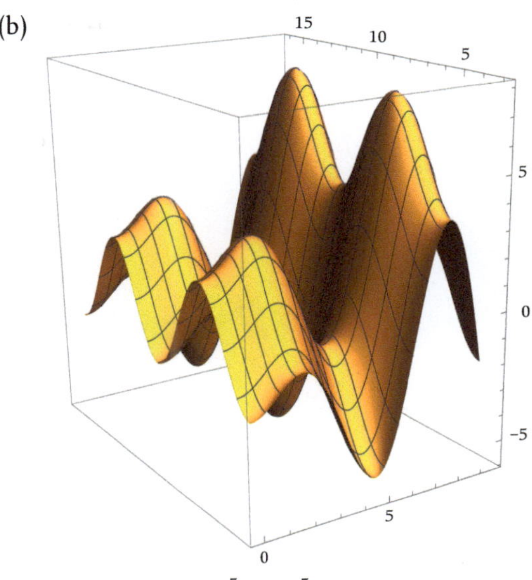

15 (a) $x = u, y = v, z = \dfrac{5}{16}v^2 - \dfrac{5}{9}u^2;$

(b)

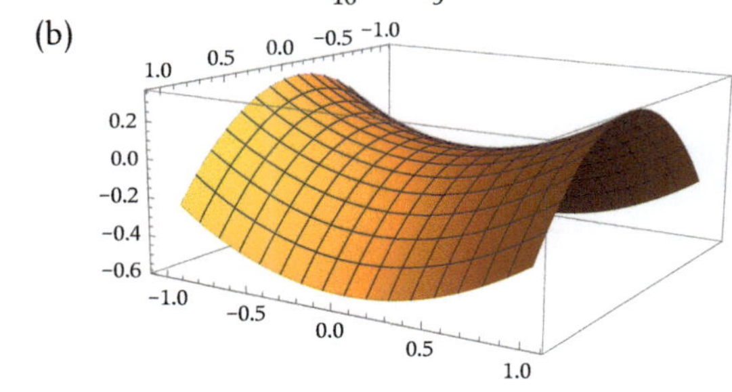

17 (a) $x = 3\sinh v\cos u, y = 4\sinh v\sin u, z = 5\cosh v,$
$v \geq 0, 0 \leq u \leq 2\pi$ (one sheet), $x = 3\sinh v\cos u,$
$y = 4\sinh v\sin u,\quad z = -5\cosh v,\quad v \geq 0,$
$0 \leq u \leq 2\pi$ (second sheet)

(b)
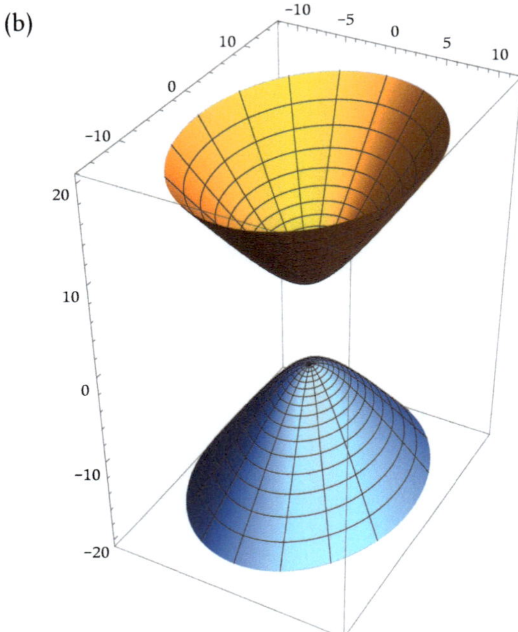

19 (a) $x = u,\quad y = \sqrt{u}\cos v,\quad z = \sqrt{u}\sin v,\quad 0 \leq u \leq 4,$
$0 \leq v \leq 2\pi;$

(b)
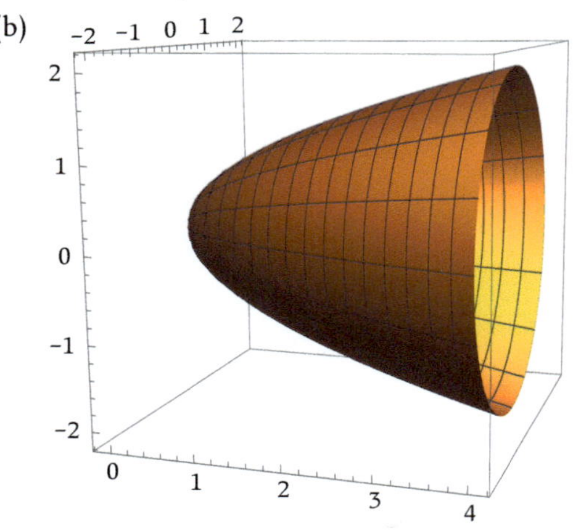

21 (a) $x = u^2\cos v,\quad y = u,\quad z = u^2\sin v,\quad 0 \leq u \leq 2,$
$0 \leq v \leq 2\pi;$

(b)
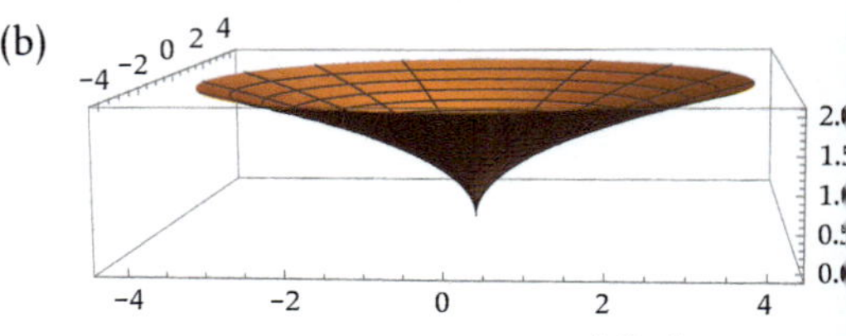

23 $-4x - 2y + 8z - 12 = 0;$ unsimplified   answe
$-4(x - 1) - 2(y - 4) + 8(z - 3) = 0$

25 $y = -1;$ unsimplified answer $0(x - 0) - 1(y + 1)$
$0(z - 0) = 0$

27 $-2x + 2y + z - 1 = 0$

29 $\mathbf{r}_u \times \mathbf{r}_v = \langle -2v, -2u, 4uv \rangle \neq \mathbf{0}$ because $v \geq 1$

31 $\mathbf{r}_u \times \mathbf{r}_v = \langle -\sin u, \cos u, -v \rangle \neq \mathbf{0}$ because $v \geq$
alternate answer because $\sin u, \cos u$ cannot simu
taneously be zero

33 (a) $\int_1^3 \int_1^2 2\sqrt{v^2 + u^2 + 4u^2v^2}\, du\, dv;$

(b) 26.1066 units^2

35 (a) $\int_{-2}^2 \int_0^{2\pi} \sqrt{1 + 4u^2 + 8u^2\cos v\sin v}\, dv\, du;$

(b) 55.283 units^2

37 $\frac{9}{2}\pi^4$ units^2

39 2 units^2

41 (a) $\dfrac{\int_0^{2\pi} \int_0^\pi \sqrt{225\sin^4 u\cos^2 v + 900\sin^4 u\sin^2 v}}{324\cos^2 u\sin^2 u}\, du\, dv;$

(b) 271.086 units^2; (c) 271.233 units^2; (d) 0.0542

43 (a) $\mathbf{r}_u(u, v) = \langle 1, 0, f_u(u, v)\rangle, \mathbf{r}_v(u, v) = \langle 0, 1, f_v(u,$

(b) $\langle -f_u(u, v), -f_v(u, v), 1\rangle;$

(c) $\sqrt{(f_u(u, v))^2 + (f_v(u, v))^2 + 1};$

(d) yes;

(e) $\iint_D \sqrt{(f_u(u, v))^2 + (f_v(u, v))^2 + 1}\, du\, dv;$ it is t
same formula other than variable names

45 (a) $\overrightarrow{\mathbf{r}(A)\mathbf{r}(B)} = -\Delta u\mathbf{r}_u(u_n, v_m);$

(b) $\overrightarrow{\mathbf{r}(A)\mathbf{r}(C)} = -\Delta v\mathbf{r}_v(u_n, v_m);$ (c) $\overrightarrow{\mathbf{r}(C)\mathbf{r}(D)}$
$-\Delta u\mathbf{r}_u(u_n, v_m);$

(d) $\overrightarrow{\mathbf{r}(B)\mathbf{r}(D)} = -\Delta v\mathbf{r}_v(u_n, v_m);$

(e) answers (a) and (c) are the same vect
answers (b) and (d) are the same vector; yes
is (approximately) a parallelogram

Section 5.7

1 no, not closed

3 yes, closed

5 (a) $\int_0^1 \int_0^1 x^2 y \sqrt{1 + \frac{y}{4x} + \frac{x}{4y}}\, dx\, dy$; (b) 0.2114

7 $84\sqrt{30}$ kg

9 $\frac{38}{3}$ kg

11 764

13 $\frac{5}{16}$

15 $\frac{1}{2}$

17 $-\frac{32}{3}$

19 $-\frac{1}{2}\pi$

21 0 (Note: flux across S_1 is 27π, flux across S_2 is -27π)

23 (a) $\langle -f_u(u,v), -f_v(u,v), 1\rangle$, upward;

    (b) $\langle f_v(u,v), f_u(u,v), -1\rangle$, downward

25 (a) $f(u)\sqrt{(f'(u))^2 + 1}$;

    (b) 0 (Note: $g(\mathbf{r}(u,v)) = u^2 f(u)\sin v$);

    (c) This is the same, using $f(x) = x^3 + 1$, $a = 1$, $b = 2$.

27 (a) $\int_0^1 \int_0^1 (x^2 + y^2)\sqrt{4x^2 + 4y^2 + 1}\, dy\, dx$;

    (b) 1.42856;

    (c) $\frac{1}{1.42856}\int_0^1 \int_0^1 x(x^2 + y^2)\sqrt{4x^2 + 4y^2 + 1}\, dy\, dx$;

    (d) $\bar{x} = 0.6545$;

    (e) $\frac{1}{1.42856}\int_0^1 \int_0^1 y(x^2 + y^2)\sqrt{4x^2 + 4y^2 + 1}\, dy\, dx$, $\bar{y} = 0.6545$;

    (f) $\frac{1}{1.42856}\int_0^1 \int_0^1 (x^2 + y^2)^2\sqrt{4x^2 + 4y^2 + 1}\, dy\, dx$, $\bar{z} = 1.0026$;

    (g) no (Hint: calculate z at $(\bar{x}, \bar{y})$ and compare to $\bar{z}$.) (Note: the center of mass is above the plate.)

Section 5.8

1 $9e^2 - 6$

3 160π

5 0

7 $\frac{1984}{3}$

9 $\frac{128}{3}\pi$

11 0 (Note that div $\mathbf{F} = 2y \neq 0$, but the surface integral is still 0)

13 251

15 $\iint_S \text{curl}\,\mathbf{F}\cdot d\mathbf{S} = \iiint_R \text{div curl}\,\mathbf{F}\cdot d\mathbf{S} = \iiint_R 0\, d\mathbf{S} = 0$

17 (a) 0;

    (b) no, the component functions are not defined at the origin; no;

    (c) yes;

    (d) $\iint_{S \cup S_1} \mathbf{F}\cdot d\mathbf{S} = \iint_S \mathbf{F}\cdot d\mathbf{S} - \iint_{S_1} \mathbf{F}\cdot d\mathbf{S}$ $= \iiint_Q \text{div}\,\mathbf{F}\, dV = \iiint 0\, dV = 0$. Therefore $\iint_S \mathbf{F}\cdot d\mathbf{S} = \iint_{S_1} \mathbf{F}\cdot d\mathbf{S} = 4\pi c$.

Section 5.9

1 double integral

3 surface integral

5 triple integral

7 surface integral

9 line integral

11 integral

13 surface integral

15 something else (quadruple integral)

17 $x = 3 - 8t, y = 2, z = 1 - 12t$

19 $x = 2, y = 4 - 4t, z = -1 - \frac{1}{2}t$

21 $x = 4 - \frac{9}{10}t, y = 9 - \frac{25}{4}t, z = 25 - \frac{2}{3}t$

23 $x = -1 + t, y = 2 - t, z = -7 - t$

25 -2π $(\mathbf{r}(t) = \langle \cos t, \sin t, 3\sin t\rangle$; the line integral is $\int_0^{2\pi}(-3\sin^2 t + \cos^2 t + \cos t \sin t + 12\cos t)\, dt$; curl $\mathbf{F} = \langle 0, 1, 1\rangle$; $\mathbf{r}(u,v) = \langle u, v, 3v\rangle$, the surface integral is $\iint_D -2\, dA$)

27 -2 (the line integral is $-\frac{3}{5} + \frac{19}{6} - \frac{7}{5} - \frac{19}{6}$, in the order of example 55; curl $\mathbf{F} = \langle -1, 3x^2, 0 \rangle$; the surface integral is $\int_0^1 \int_0^1 (-2u - 3u^2)\, dv\, du$)

29 $-\frac{243}{4}\pi$ (the surface integral is $\int_{-3}^3 \int_{-\sqrt{9-u^2}}^{\sqrt{9-u^2}} (-2v + 6u - 3u^2)\, dv\, du$, which requires a nontrivial trig substitution, or switch to polar coordinates for a simpler integral; the line integral is $\int_0^{2\pi} (-27\cos t \sin t + 27\cos^3 t \sin t + 81\cos^3 t - 27\sin^3 t - 81\cos^4 t)\, dt$)

31 0

33 -36

35 -96π (switching to the surface $z = -12$ for $x^2 + y^2 \le 4$, which has the same boundary curve, is easiest)

37 9π (switching to the surface $z = 1$ for $x^2 + y^2 \le 9$, which has the same boundary curve, is easiest)

39 $\frac{625}{2}\pi$ (switching to the line integral is easiest)

41 (a) $\iint_{S_\omega} \mathbf{F} \cdot d\mathbf{S} = (\operatorname{div}\mathbf{F}(P_0))\frac{4}{3}\pi\omega^3$;

    (b) $\operatorname{div}\mathbf{F}(P_0) = \frac{3}{4\pi\omega^3} \iint_{S_\omega} \mathbf{F} \cdot d\mathbf{S}$;

    (c) The divergence of $\mathbf{F}$ at P_0 is the (net outward) flux at P_0, scaled to the size of the sphere. That is, it's proportional to the net outward flux at P_0 per unit volume.

    (d) If $\operatorname{div}\mathbf{F}(P_0) > 0$, then P_0 is a source; if $\operatorname{div}\mathbf{F}(P_0) < 0$, then P_0 is a sink.

Index